作者简介

　　张福明，博士，全国工程勘察设计大师，教授级高级工程师，享受国务院政府特殊津贴。长期从事冶金材料工程研究设计、关键技术装备研发创新工作，致力于冶金工程基础理论、现代钢铁厂工程设计与工程技术创新研究。历任北京首钢设计院炼铁设计室主任，副总工程师、副院长；北京首钢国际工程技术有限公司董事、副总经理、党委委员，董事长、党委书记；现任首钢集团有限公司副总工程师。作为国家重大工程——首钢京唐钢铁项目工程总设计师，主持设计了我国第一个沿海靠港千万吨级现代化钢铁厂工程和新一代可循环钢铁制造流程；主持并具体参加我国首座5500m^3超大型高炉、550m^2大型烧结机、504m^2带式焙烧机球团生产线工程设计；带领设计科研团队在工程创新、科技研发、技术攻关等领域形成百余项工程科技创新成果。主持或具体参加12项国家课题研究，获得国家和省部级科技奖40余项，其中国家科技进步奖1项，省部级一等奖18项、二等奖15项；授权专利103项（其中发明专利42项）；发表学术论文280余篇，其中45篇被SCI、EI收录，出版《现代高炉长寿技术》等学术著作10部；获得全国优秀工程设计金奖1项、银奖3项。获得光华工程科技青年奖、魏寿昆青年冶金奖、全国杰出工程师奖和中国冶金青年科技奖；入选国家百千万人才工程；获得国家有突出贡献中青年专家、北京学者、首都科技创新领军人物、北京市有突出贡献的科学技术管理人才、北京市劳动模范等荣誉称号。

> ## 北京学者指导老师

作者与学术导师殷瑞钰院士合影

作者与学术导师张寿荣院士合影

作者与学术导师徐德龙院士合影

作者及夫人周宏女士与学术导师干勇院士合影　　作者与学术导师谢建新院士合影

作者与学术导师
毛新平院士合影

典型工程设计

首钢京唐 550m² 烧结工程全景图

首钢京唐 504m² 带式焙烧机球团生产线全景图

作者主持设计的首钢京唐 3 座 5500m³ 超大型高炉全景

首钢京唐400万吨/年带式焙烧机球团工程

作者主持研发设计的超大型高风温顶燃式热风炉

首钢京唐2250mm热轧生产线

作者主持设计的首钢迁钢工程

作者主持设计的首钢迁钢1号高炉

作者主持设计的首钢京唐工程全貌拂晓

作者主持设计的首钢 2 号高炉（2002 年）

作者组织设计的首钢迁钢 3 号 4000m³ 高炉

作者主持设计的太钢 3 号 1800m³ 高炉

作者主持设计的湘钢 4 号 1800m³ 高炉

学术交流

作者在德国蒂森 2 号高炉（5513m^3）现场考察

作者和首钢京唐前总经理王毅先生在日本新日铁技术本部交流考察

2019 年 9 月作者在美国安塞洛米塔尔印第安纳厂考察

作者在第六届国际炼铁科技大会做学术报告

作者在2014年美国钢铁技术年会（AIST2014）做大会学术报告

作者在第十二届中国钢铁年会上做学术报告

作者和团队成员曹朝真博士在美国钢铁年会现场合影

作者在德国 TKS 8.43m 焦炉考察

2015 年工程方法论研究钢铁冶金工程设计方法课题启动会合影

"作者团队开展首钢京唐工程设计与建设"

作者在查阅资料

作者作为工程总设计师在首钢京唐钢铁项目可行性研究论证会上进行项目总体报告

作者在计算核定关键设计参数

作者在审定首钢京唐钢铁厂工艺设计图

作者与主持设计的首钢京唐钢铁厂模型

作者在汇报首钢京唐顶层设计

作者在首钢京唐工程现场与施工单位领导沟通交流

作者在首钢京唐现场与团队成员研究工作

作者在首钢京唐工程现场处理工程难题

作者及团队主要成员与首钢领导在首钢京唐 1 号高炉开炉庆典现场合影

作者及团队主要成员

荣誉证书

全国工程勘察设计大师证书

2015 年当选北京学者

国家百千万人才工程、有突出贡献中青年专家证书

2009 年国务院政府特殊津贴

2012年光华工程科技奖（第九届青年奖）证书

2016年全国杰出工程师奖证书

2017年魏寿昆青年冶金奖证书

张福明冶金工程设计
与技术创新文集

张福明 著

北 京
冶金工业出版社
2022

内 容 提 要

本书是作者团队从事冶金工程设计研究、技术创新的理论研究、技术研发和实践研究的总结，收录了作者从事冶金工程设计工作 30 年来发表的 100 余篇学术论文，总体上可分为冶金工程设计研究和钢铁冶金工程技术创新两部分，包括现代钢铁冶金工程设计理论、方法和实践，高炉炼铁工程设计，高风温技术，冶金煤气干法除尘技术，烧结球团，冶金技术装备，洁净钢生产体系构建等跨学科、多专业的内容。

本书可供冶金工程设计领域的设计研究、科研开发、技术应用、生产管理和教学研究的相关人员阅读参考。

图书在版编目（CIP）数据

张福明冶金工程设计与技术创新文集/张福明著. —北京：冶金工业出版社，2022.2
ISBN 978-7-5024-8969-4

Ⅰ.①张… Ⅱ.①张… Ⅲ.①冶金工业—设计—文集 Ⅳ.①TF-53

中国版本图书馆 CIP 数据核字（2021）第 237890 号

张福明冶金工程设计与技术创新文集

出版发行	冶金工业出版社	电　　话	(010)64027926
地　　址	北京市东城区嵩祝院北巷 39 号	邮　　编	100009
网　　址	www.mip1953.com	电子信箱	service@mip1953.com

责任编辑　刘小峰　曾　媛　美术编辑　彭子赫　版式设计　禹　蕊
责任校对　李　娜　责任印制　李玉山
北京捷迅佳彩印刷有限公司印刷
2022 年 2 月第 1 版，2022 年 2 月第 1 次印刷
787mm×1092mm 1/16；65.5 印张；8 彩页；1618 千字；1032 页
定价 300.00 元

投稿电话　(010)64027932　投稿信箱　tougao@cnmip.com.cn
营销中心电话　(010)64044283
冶金工业出版社天猫旗舰店　yjgycbs.tmall.com

（本书如有印装质量问题，本社营销中心负责退换）

序

　　张福明是我国冶金工程设计界的后起之秀。

　　我认识张福明是在21世纪初首钢搬迁到曹妃甸建设新一代钢铁基地之时。当时他作为首钢设计院的总工程师，向我介绍了新基地的总体设想方案，颇具前瞻性、创新性，给人以深刻的印象。继而，他主持设计了新一代可循环钢铁厂及其相关技术装备的研发和设计，形成了具有自主知识产权的一系列技术和装备，建立整体性的新一代钢铁制造流程及装备，促进了我国冶金工程设计、装备制造的科技进步，且具有示范性，引起了全国乃至全世界瞩目。

　　张福明除了主持曹妃甸京唐钢铁公司的总体性、结构性设计以外，还直接主持设计了 $5500m^3$ 高炉为代表的大型冶金技术装备，研发了高风温顶燃式热风炉、大型高炉无料钟炉顶装备、高炉煤气全干法除尘装置和全干式炉顶煤气余压发电装置、铁水罐多功能化（"一罐到底"）"界面"技术等重大技术创新。

　　在长期实践的基础上，张福明十分重视理论学习和研究，经常和我的团队融合在一起，研究、探索冶金流程工程学的理论和冶金流程集成理论与方法，而且善于理论联系实际，并自觉地运用到工程设计过程之中，特别是在全流程尺度上，注意构建"三个功能"的钢厂，取得了不少新成果。由此当选为第八届全国工程设计大师，荣获第九届光华工程科技青年奖，第四届魏寿昆冶金青年奖，直至第二届全国杰出工程师奖。

作为一代后起之秀，头角渐露。

本书是张福明及其团队主要成员，围绕钢铁冶金工程关键共性技术问题，开展的一系列工程技术研究和设计方法创新工作的总结。书中收录的代表性论文，时间跨度长达20余年，体现了我国钢铁工业从追赶到创新的发展进程，以及我国设计工程师的工程理念、工程思维的转变、演化和发展，特别是现代工程设计理论与方法的形成和发展。希望本书的出版，能为从事钢铁冶金工程设计研究、科研开发、技术创新和教学研究的科技人员提供参考与引导。

中国工程院院士 殷瑞钰

2021年6月8日

前　言

作者参加工作30年来，长期在一线从事工程设计、科学研究、技术研发和科技管理工作，虽然工作岗位时有变化，但始终能够坚守初心，勤奋治学，辛勤耕耘，矢志不渝。时光荏苒，物换星移，弹指之间，卅年已过，步入知天命之年，已是双鬓微霜，不知不觉间一大半的职业生涯已经悄然而过，即将步入人生新的阶段。

20世纪90年代初，作者进入首钢设计院（现北京首钢国际工程技术有限公司）参加炼铁工程设计工作。时值首钢大发展时期，首钢1~4号高炉相继进行了现代化新技术大修改造，高炉总容积从原来的4139m^3扩大到9934m^3。1990年5月，2号高炉率先大修改造建成投产，高炉有效容积由原来的1327m^3扩大到1726m^3；1992年5月，4号高炉大修改造建成投产，高炉有效容积由原来的1200m^3扩大到2100m^3；1993年6月，3号高炉扩容大修改造建成投产，高炉有效容积由原来的1036m^3扩大到2536m^3；1994年8月，1号高炉扩容大修改造建成投产，高炉有效容积由原来的576m^3扩大到2536m^3，当时为了扩大生铁产能，保留了原来的3号高炉（1036m^3），并将其更名为5号高炉。20世纪90年代，首钢炼铁系统的结构调整和技术升级，创新并形成了30余项炼铁新工艺、新技术和新设备，奠定了首钢快速发展的工艺技术装备基础；在此期间首钢还改造了第二炼钢厂、新建了第三炼钢厂、增加了轧钢生产线，形成了年产粗钢800万吨的生产能力。1994年，首钢粗钢产量达到824万吨，位列全国第一。每年一座大型高炉的大修改造工程设计，设计任务十分繁重而且时间紧迫，每天都要加班加点工作。

20世纪90年代初，举国上下都在改革开放的热潮中奋勇争先，几乎所有的冶金设计院工作任务都非常饱满。所幸的是作者参加工作以后，正好赶上了这个全国钢铁行业和首钢大发展的历史机遇，在老前辈们的具体指导下，很快熟悉了冶金工程设计工作，参加高炉本体系统工程设计，开启了作者的"工程设计人生"。回顾起来，令作者深有感悟的是，作为一名新的从业者，能够加入到一个具有创新灵魂、拼搏奋进、团结协作的团队，对一个人的成长是至关重要的。作者深切感恩、感念、感谢在那个时期传授知识和方法的老领导、老前辈和同事们！他们当中有作者的授业恩师徐和谊、黄晋、刘兰菊、李钦、曾继奋、陈明时等前辈师长，也有早于作者入职的同辈学长江云祺、苏维、倪苹、王建涛、姚轶等诸多同事，他们对作者事业上的学习和长进，都曾给予过很多支持和帮助。除此之外，就是一个时代的重大历史发展机遇，对于一个人的成长而言，也是极其重要的，甚至从某种程度上讲是可望而不可即的。这一点感悟，在21世纪初，首钢战略性搬迁结构调整、曹妃甸首钢京唐钢铁厂的设计建设过程中，再一次得到了验证。时至今日，忆往昔峥嵘岁月稠，抚今追昔，作者时常怀念过往的美好时光，衷心感谢中国改革开放迅猛发展的伟大时代，感恩那些陪伴作者成长的前辈师长和同事们。

20世纪90年代的十年间，是作者学术生涯的起始阶段和奠定基础的时期。这一时期，作者有幸参加了首钢几座高炉的现代化新技术扩容大修改造，参加了高炉高效长寿技术的研究与开发，具体参加了热压小块炭砖和陶瓷杯炉缸内衬结构的设计、高炉软水密闭循环冷却系统设计以及新型冷却壁的研发，积累了设计、科研、生产、应用的理论和实践经验。20世纪90年代末期，作者担任项目技术负责人，承担了首钢喷煤系统技术改造的设计工作，为了学习先进企业的成功经验，先后赴鞍钢、宝钢、武钢、攀钢、马钢等企业考察交流，结识了许多炼铁界知名的专家学者，他们给予的支持和帮助让作者至今记忆犹新。这些前辈师长中，有我国著名冶金学家、教育家，东北大学的杜鹤桂教授，先生当时身体健朗、精神矍铄，潜心研究高炉喷煤技术而且造诣极高，是享誉国际的知名学者和炼铁技术专家。90年代中期，在一次全国炼铁学术会议上，作者邀请杜鹤桂教授来首钢讲学，先生欣然允诺。一次先生专程来到北京，在首钢设计院十四楼学术报告厅为首钢炼铁技术人员，做了一场关于高炉喷煤技术的精彩讲座，使参会者眼界大开，都感到收获巨大、受益匪浅。这件事在杜鹤桂教授95岁高龄时出版的《铸剑扶犁七十年——我的炼铁生涯》中也有专门的回忆。当时对作者在学术上给予过帮助或产生重要影响的前辈师长有：武钢总工程师张寿荣院士；冶金部科技司总工程师徐矩良教授；北京钢铁设计研究总院（现中冶京诚工程有限技术公司）吴启常大师、唐文权教授、王泽愍教授；钢铁研究总院叶才彦教授、张士敏教授；北京科技大学刘述临教授、杨天钧教授、孔令坛教授、王筱留教授、李士琦教授、顾飞教授、高征铠教授、苍大强教授等；东北大学杜鹤桂教授、王文忠教授、蔡九菊教授等；重庆钢铁设计研究院（现中冶赛迪工程技术股份有限公司）项钟庸大师、巩伍权教授、陈茂熙教授；武汉钢铁设计研究院（现中冶南方工程有限技术公司）银汉教授；首钢刘云彩教授、李文秀教授、魏升明教授、张伯鹏教授等；以及鞍钢汤清华教授、武钢于仲杰教授、宝钢李维国教授、济钢温燕明教授、太钢杨子柱教授等一大批工作在钢铁企业的技术专家。

2000年年初，作者开始担任首钢设计院炼铁设计室主任。当年，首钢2号高炉（1726m³）开始进行新一轮的大修技术改造工程设计，在原有技术的基础上，对高炉炉型进行了优化调整，高炉有效容积调整为

1780m³，采用了高风温内燃式热风炉、集中制粉长距离直接喷煤、国产化铜冷却壁和新型软水密闭循环冷却技术等多项先进技术，高炉投产后取得了很好的应用实绩，技术指标一直处于国内同级别高炉的先进行列。后来，以首钢2号高炉为参照，又相继承担了湘钢、太钢、新钢、重钢、首秦等多座同级别高炉的设计工作。这期间通过工程设计创新和研究攻关，在高炉长寿、高风温、喷煤、高炉煤气干法除尘、高炉无料钟炉顶设备等领域取得了许多创新性的工程设计和研究成果。

进入21世纪以来，为成功举办北京2008年奥运会，改善首都环境质量，落实北京城市总体发展规划，国家决定首钢进行战略性搬迁和结构调整。2002年以后，首钢开始先行探索实践，在河北迁安依托首钢既有矿山，建设以板带材为主导产品的钢铁制造基地（迁钢）；在秦皇岛依托首钢既有的秦皇岛中板厂，建设以生产高品质中厚板为主导产品的钢铁制造基地（首秦）。通过异地新建钢铁厂，统筹规划、总体设计、流程创新，提高工艺技术装备总体水平，实现钢铁制造流程和工业技术装备的嬗变；与此同时彻底改变长期以来首钢以棒线材为主的产品结构，实现由长材产品向板材产品的转型升级。

2002年以后，作者担任北京首钢设计院副总工程师，并担任首钢迁钢工程总设计师和项目负责人，专业领域由原来的高炉炼铁，拓展到烧结、球团、焦化、炼钢、连铸、轧钢，以及电力、热力、燃气、给排水、自动化、总图运输、采暖通风、技术经济等十多个专业领域，有很多知识都需要学习、领会和实践。那时作者正值而立之年，年富力强，基本上所有的业余时间都用于读书、看图、学习、计算、研究，但凡有看不懂的、搞不清楚的就虚心向专业人士请教、学习、交流。几年下来，认真钻研了几十本专业著作，研读了数百篇的学术论文，还学习研阅了上百套相关专业的设计图纸，记录并整理了几十本读书笔记，对相关专业的知识理解更加清晰、透彻，学术基础也更加坚实。

当时正值首钢搬迁发展的重大历史机遇期，除了全面负责迁钢工程设计和技术创新工作以外，还分工负责首秦工程铁前系统的工程设计工作。2004年，首秦、迁钢两大工程相继建成投产，初步完成了首钢搬迁结构调整"练兵场"和"试验基地"的工程建设。在这两个新建钢铁厂的总体设计过程中，通过学习、考察、调研和交流，形成了许多新的感悟、认识和理念，掌握了许多新知识、新理论和新方法，了解了很多新技术、新工艺和新装备。在原有高炉炼铁工程设计的视野中，有了许多新的认知和发现，知识领域也随之拓展，对钢铁冶金制造流程的认识和研究也相应跨越了一个新的平台。

2004年年末，作者开始担任曹妃甸首钢京唐钢铁厂工程总设计师，全面主持钢铁厂总体设计工作，从编制项目建设方案、可行性研究报告，到参与工程重大技术方案决策，曾多次向徐匡迪、殷瑞钰、吴溪淳、翁宇庆、干勇、张寿荣、陆钟武、王国栋、石启荣、李世俊等老一辈院士专家汇报首钢京唐工程建设方案、可行性研究报告和工程初步设计。历经三年时间主持完成各工序初步设计和施工图设计，再到常驻曹妃甸工程现场四年有余，组织协调、指挥处理工程设计与施工建设过程中的各类问题，统筹管理各设计单位的现场施工配合工作。四载寒暑，五度春秋，作者亲眼目睹了从一片荒滩，到沿海临港千万吨级现代化超大型钢铁基地的建成过程，具体参与并见证了引领我国钢铁工业发展的示范基地的设计建设。这是作者人生最精彩和宝贵的十年，从而立之年步入不惑之秋，陪伴作者成长的是中国第一个临海靠港千万吨钢铁基地的顺利建成投产、中国第一座$5500m^3$超大型高炉拔地而起、中国第一台$550m^2$

烧结机建成投产、中国第一个采用铁水"全三脱"预处理工艺的洁净钢生产平台的建成投产……

在作者主持曹妃甸首钢京唐钢铁厂工程总体设计期间，与国内外多家钢铁工程技术公司联合合作，建立了良好的合作关系和深厚的工作友谊。首钢京唐钢铁厂项目作为国家"十一五"期间重大工程，得到了高度重视和普遍关注。中国工程院徐匡迪院士、殷瑞钰院士、干勇院士、张寿荣院士、王国栋院士、翁宇庆院士等一大批院士大师、专家学者，都参与了首钢京唐钢铁厂工程的方案论证和项目决策，对钢铁厂工程设计和建设，给予了很多具体的支持和帮助，提出了非常宝贵的意见和建议，应当说首钢京唐钢铁厂的设计建设，凝聚了我国几代钢铁工作者的学识、智慧和经验。

还要特别感谢的是我国工程科技界和工程勘察设计的许多院士、大师对作者的大力支持和热情帮助！2016年，由吴启常、项钟庸、韩国瑞、李龙珍、潘国友五位全国工程勘察设计大师提名举荐，作者顺利当选第八批全国工程勘察设计大师。2015年作者当选第二批北京学者以后，聘请了殷瑞钰、干勇、王国栋、徐德龙、谢建新、毛新平六位院士担任作者的指导老师，他们指导作者在新的学术领域不断开拓、不断进步。

近年来，作者在殷瑞钰院士、张寿荣院士的直接带领下，参加了钢铁冶金工程方法论、知识论等工程哲学课题研究，对钢铁冶金工程设计有了更加深刻的认识和领悟，结合作者多年从事钢铁冶金工程设计研究与技术创新实践，撰写了一些关于钢铁冶金工程设计理论和方法论方面的论文和报告。两位院士都是国际上具有重要影响力的著名冶金学家和战略科学家，他们对作者的学术研究和事业成长，都给予了极大的支持和无私的帮助，恩师殷瑞钰院士在百忙之中亲自为本书作序并几经修改，特别关心本书的出版发行，提携晚辈可谓不遗余力！作者在此谨向两位恩师致以崇高的敬意和诚挚的感谢！

针对高炉长寿技术这一专题，作者团队历时五年时间进行资料整理、汇总、提炼，于2012年出版了"十一五"国家重点图书《现代高炉长寿技术》。这本书也让作者深刻体会到著作的重要性。该书的出版，不仅大大提升了团队总结知识的能力，而且令高炉长寿理念在炼铁科技工作者中深入人心。

本书是作者从事冶金工程设计和技术研究30年来有代表性的论文合集，涉及烧结、球团、炼铁、炼钢、轧钢、能源、环保，以及工程哲学、工程方法论和工程知识论等多个学科领域，汇集了作者在重要期刊和会议上发表的主要学术论文，包括现代钢铁冶金工程设计理论、方法和实践，高炉炼铁工程设计，高风温技术，冶金煤气干法除尘技术，烧结球团，冶金技术装备，洁净钢生产体系构建等跨学科、多专业的内容，基本涵盖了现代钢铁冶金工程主要的专业技术领域，是作者从事冶金工程设计研究、技术创新的理论研究、技术研发和实践研究的总结。

本书面向从事钢铁冶金工程规划、设计和研究工作的设计人员；高等院校从事钢铁冶金工程教学研究和人才培养的教师以及在校学生；科研单位从事钢铁冶金工程研究、技术研发的研究人员；钢铁企业从事生产管理、技术管理和技术研发的工程技术人员等。

作者编著本书，前后历经大约五年时间，都是利用繁忙工作之外的业余时间完成的。周末和节假日的大部分时光在书桌或电脑度过，弹指击键，很多时候也是废寝忘食。文集中收录学术论文都是经过精挑细选、反复推敲确定的。这些入选文章除了作者独著的论文以外，还有一部分是作者和部分团队成员合作发表的论文。这些论文的合作者都是设计科研团队的重要成员，是作者从事冶金工程设计研究和技术创新工作的重要合作者，对很多工程设计和技术创新发挥过重要的作用，也是作者学术研究活动的重要参与者和见证者。他们包括：北京科技大学程树森教授，首钢国际工程公司张建、毛庆武、颉建新、李欣、曹朝真、李林、孟祥龙、刘兰菊、苏维、姚轼、钱世崇、张德国、银光宇、梅丛华、胡祖瑞、王渠生等。这些论文的合作者中，有的是作者的老师，有的是作者同事，也有的是作者的学生，对他们的学术贡献，作者在此一并表示诚挚的感谢！

在本书成稿过程中，李林博士帮助作者对论文进行了汇集整理，做了许多细致周到的工作；首钢国际工程公司颉建新部长给予了工作上的大力支持和帮助；孟祥龙高工和曹朝真博士为本书付出了许多辛劳；首钢技术研究院刘清梅博士帮助作者收集并整理了部分资料和数据。作者对他们给予的支持和帮助，表示衷心的感谢！

与此同时，作者还要感谢首钢集团、首钢国际工程公司、首钢京唐公司、首钢迁钢公司等单位的有关领导、专家和工程技术人员！特别要感谢的是王天义、王毅两位老领导，他们曾先后担任首钢京唐公司总经理，对作者团队的许多科技创新工作都给予了大力支持和帮助！是他们的大力支持、指导和帮助，才使得很多创新技术和装备得以成功应用。

本书的出版发行，还要感谢由北京市人才工作局、北京市人力资源和社会保障局，为作者提供了北京学者培养计划基金的支持，作者对此表示衷心的感谢！

2021 年 8 月

目 录

低碳绿色与智能高效冶金工程设计综合论述 1

首钢新 3 号高炉设计 3
高炉炼铁技术的结构优化 9
迁钢 1 号高炉采用的新技术 14
首钢京唐 5500m³ 高炉采用的新技术 22
当代高炉炼铁技术若干问题的认识 33
钢铁厂流程结构优化与高炉大型化 41
21 世纪初巨型高炉的技术特征 57
高风温低燃料比高炉冶炼工艺技术的发展前景 69
面向未来的低碳绿色高炉炼铁技术发展方向 80
首钢绿色低碳炼铁技术的发展与展望 89
中国高炉炼铁技术装备发展成就与展望 99
大型带式焙烧机球团技术装备设计与应用 110
我国 5000m³ 级高炉技术进步及运行实绩 123
中国钢铁产业发展与展望 137
智能化钢铁制造流程信息物理系统的设计研究 145
钢铁冶金从技艺走向工程科学的演化进程研究 157
Development Orientation of Low Carbon and Greenization BF Ironmaking Technology 169
New Technological Progress of Ironmaking in Shougang 183
Design,Construction and Application of 5,500m³ Blast Furnace at Shougang Jingtang 199

长寿高效高炉设计研究 219

我国大型高炉长寿技术发展现状 221
新型优质耐火材料在首钢高炉上的应用 227
大型长寿高炉设计的探讨 233
延长大型高炉炉缸寿命的认识与方法 240
热压炭砖—陶瓷杯技术在首钢 1 号高炉上的应用 249
首钢 1 号高炉陶瓷杯砌筑质量管理 254

首钢 2 号高炉长寿技术设计 ……………………………………………………… 258
首钢高炉炉缸内衬设计与实践 …………………………………………………… 265
首钢高炉高效长寿技术设计与应用实践 ………………………………………… 272
铜冷却壁的设计研究与应用 ……………………………………………………… 281
高炉炉缸炉底温度场控制技术 …………………………………………………… 288
基于自组织理论的高炉长寿技术 ………………………………………………… 296
涟钢 2800m^3 高炉无料钟炉顶齿轮箱水冷系统的设计 ………………………… 306
Design and Operation Control for Long Campaign Life of Blast Furnaces ……… 313
Design and Application of Blast Furnace Longevous Copper Cooling Stave …… 325
Research and Practice on Inner Profile Design of Modern Blast Furnace ………… 333
Research and Application on Temperature Distribution Control Technology of Blast
　Furnace Hearth and Bottom ……………………………………………………… 346

高风温工艺与高效长寿热风炉设计研究 …………………………………… 361

我国大型顶燃式热风炉技术进步 ………………………………………………… 363
高炉高风温技术发展与创新 ……………………………………………………… 370
首钢京唐 5500m^3 高炉 BSK 顶燃式热风炉设计研究 …………………………… 378
高风温内燃式热风炉的设计理论和应用实践 …………………………………… 388
当代高炉高风温技术理念与工程实践 …………………………………………… 395
大型高炉热风炉技术的比较分析 ………………………………………………… 407
高炉热风炉高温预热工艺设计与应用 …………………………………………… 417
首钢京唐钢铁厂 5500m^3 高炉热风炉系统长寿型两级双预热系统设计 ……… 423
高效长寿顶燃式热风炉燃烧技术研究 …………………………………………… 429
Dome Combustion Hot Blast Stove for Huge Blast Furnace ……………………… 436
Research on High Efficiency Energy Conversion Technology for Modern Hot Blast
　Stove ……………………………………………………………………………… 448
Study on Heat Transfer Process of Dome Combustion Hot Blast Stove ………… 466
Research and Application on Waste Heat Recycling and Preheating Technology of
　Ironmaking Hot Blast Stove in China …………………………………………… 478

高炉富氧喷煤与煤气干法除尘设计研究 …………………………………… 493

浅谈首钢高炉喷煤技术的发展方向 ……………………………………………… 495
关于高炉富氧喷煤技术问题的探讨 ……………………………………………… 502
大型高炉紧凑型长距离喷煤技术 ………………………………………………… 511
引进技术在迁钢 3 号高炉喷煤系统中的应用 …………………………………… 519

首钢高炉喷煤技术发展与创新 526
大型高炉煤气干式布袋除尘技术研究 534
大型转炉煤气干法除尘技术研究与应用 543
冶金煤气干法除尘技术创新与成就 557
首钢京唐1号5500m³高炉煤气干法除尘自动化控制系统的创新设计与实现 566
Research and Application on Blast Furnace Gas Dry Cleaning Technology of Huge Blast Furnace 573
Technical Development of Low Carbon and Greenization Blast Furnace Ironmaking 587
Pulverized Coal Injection Technology in Large-Sized Blast Furnace of Shougang 594

非高炉炼铁工程设计研究 607

气基竖炉直接还原技术的发展现状与展望 609
我国首座HIsmelt工业装置的设计优化与技术进展 624
转底炉直接还原技术的进展 631
焦炉煤气二氧化碳重整热力学规律研究 641
利用焦炉煤气生产直接还原铁关键技术分析 648
低碳绿色炼铁技术的发展前景与展望 655
Application on HIsmelt Smelting Reduction Process in China 670

现代钢铁冶金工程设计与应用研究 683

冶金工程设计的发展现状及展望 685
新一代钢铁厂循环经济发展模式的构建 697
钢铁厂工程设计的创新与实践 706
首钢京唐炼钢厂新一代工艺流程与应用实践 718
首钢京唐2250mm热轧生产线采用的先进技术 729
大型带式焙烧机球团技术创新与应用 736
首钢京唐钢铁公司绿色低碳钢铁生产流程解析 745
钢铁流程固废资源化利用逆向供应链体系探讨 758
首钢京唐烧结厂降低生产工序能耗的实践 768
首钢冶金石灰技术创新与工程实践 777
现代钢铁冶金工程设计方法研究 786
钢铁冶金工程知识研究与展望 796
协同设计管理在三维工厂设计中的研究与应用 814
钢铁制造流程智能制造与智能设计的研究 819
冶金流程工程学的典型应用 827

Innovation and Application on Pelletizing Technology of Large Travelling Grate Induration Machine ……………………………………………………………………………… 840

Research and Optimization of BF-BOF Interface Technology …………………… 857

钢铁冶金工程技术创新 ……………………………………………………………… 863

现代大型高炉关键技术的研究与创新 …………………………………………… 865
现代高炉高风温关键技术问题的认识与研究 …………………………………… 873
首钢京唐BSK顶燃式热风炉的燃烧技术 ………………………………………… 886
顶燃式热风炉高温低氧燃烧技术 ………………………………………………… 896
热风炉送风期格子砖温度分布计算 ……………………………………………… 906
顶燃式与内燃式热风炉燃烧过程物理量均匀性的定量比较 …………………… 918
首钢京唐 $5500m^3$ 高炉BSK顶燃式热风炉燃烧器分项冷态测试研究 ………… 925
高效长寿热风炉格子砖传热研究 ………………………………………………… 931
霍戈文内燃式热风炉传输现象的研究 …………………………………………… 938
热风炉格子砖活面积优化设计 …………………………………………………… 948
首钢高炉出铁场设备的开发与应用 ……………………………………………… 957
CFD Study on Flow Characteristics of BOF Gas in Evaporating Cooler …………… 965
Green Iron and Steel Manufacturing Process of Shougang Jingtang Plant ………… 972
Innovation and Application on High Blast Temperature Technology of Shougang Blast Furnace …………………………………………………………………… 984
Construction and Practice on Energy Flow Network of New Generation Recyclable Iron and Steel Manufacturing Process ………………………………………… 1003
Analysis of Pulverized Coal and Natural Gas Injection in $5500m^3$ Blast Furnace in Shougang Jingtang ……………………………………………………………… 1013
Development and Application on Large Annular Shaft Kiln at Shougang ………… 1027

低碳绿色与智能高效冶金工程设计综合论述

首钢新 3 号高炉设计

摘　要：首钢新 3 号高炉（$V_u=2536m^3$，$H_u/D=1.985$）属超矮胖型高炉，适应超高冶炼强度操作。设计采用皮带上料、新型无料钟炉顶、软水密闭循环冷却、环形出铁场、大型顶燃热风炉、煤气干法布袋除尘、新型计算机控制系统和监测技术等先进技术，投产近一年来，生产状况良好，月平均利用系数达 $2.641t/(m^3\cdot d)$。

关键词：高炉设计；无钟炉顶；软水冷却；顶燃热风炉；布袋除尘

1　引言

新 3 号高炉设计是在 2 号、4 号高炉大修技术改造投产后进行的。设计要求该炉技术装备水平高于 2、4 号高炉，并以精料为基础实现超强化操作。为此，提高了原燃料准备与供应系统的能力，采用了超矮胖炉型。为实现强化冶炼高炉长寿目标（一代产铁 $8000t/m^3$，炉龄 3～10 年），选用了新型优质耐火材料和软水密闭循环冷却等技术。

新 3 号高炉与国内几座高炉设计指标见表 1。

表 1　首钢新 3 号高炉与国内几座高炉主要设计指标

项目	首钢新 3 号高炉	首钢 2 号高炉	宝钢 2 号高炉	武钢新 3 号高炉	马钢 $2500m^3$ 高炉
有效容积/m^3	2536	1726	1063	3200	2545
利用系数/$t\cdot(m^3\cdot d)^{-1}$	2.8 最大 3.0	2.8 最大 3.0	2.36	2.0	1.96
焦比/$kg\cdot t^{-1}$	400	450	430	440	430
煤比/$kg\cdot t^{-1}$	160	100	72	100	90
富氧率/%	4.0	4.0	≤3		2～3
渣铁比/$kg\cdot t^{-1}$	400	400	330	470	427
烧结率/%	85	85	82	80～90	90
热风温度/℃	1050～1100	1200	≥1250	1200	1150～1300
炉顶压力/MPa	0.25	0.25	≥0.2	0.2，最大 0.25	0.2942
日产铁量/t	7100，最大 7600	4800，最大 5100	9600	6400，最大 8000	5000
年产量/$\times10^4 t$	249，最大 266	169，最大 181	325	224 最大 250	175
入炉品位/%	58	58	约 57	53.5	约 55

本文作者：张福明。原刊于《炼铁》，1994，13（5）：36-39。

2 设计特点及采用的新技术

2.1 高炉平面布置

限于总图布置条件,新 3 号高炉设计中尽可能减少高炉各系统的占地面积,进行合理布置,平面布置如图 1 所示。

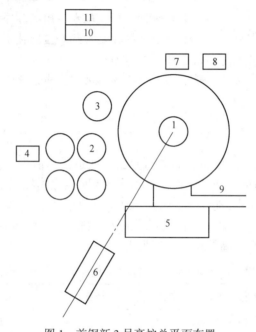

图 1 首钢新 3 号高炉总平面布置
1—高炉;2—热风炉;3—重力除尘器;4—布袋除尘器;5—高炉主控室;
6—料仓;7—喷煤塔;8—鼓风机站;9—公路;10—新 3 号高炉水渣池;11—新 1 号高炉水渣池

2.2 供料系统

(1) 工艺流程。设 6 个烧结矿仓,6 个焦炭仓,3 个球团矿仓,3 个杂矿仓。考虑总图布置较紧,并减少原燃料在输送过程中的机械粉碎,设计中取消了中间称量罐。原燃料在仓下筛分后,筛上物(>5mm 烧结矿,>25mm 焦炭)进入仓下称量罐,称量后直接排放到主胶带上送至高炉炉顶料罐内。主胶带宽 $B=1600$mm,带速 2.0m/s,水平投影长 519m,倾角 8°26′48″,与料仓布置在同一条中心线上。控制系统对仓下设备和运输胶带进行联锁控制,并对原燃料实行重量和水分补偿。

(2) 增设减重法称量设施。为保证每批原燃料在仓下称量的准确性,烧结矿和焦炭称量罐下的给料机具有调速给料功能,可保证给料的精度,调速给料量占每个称量罐装满量(20m^3)的 5%~10%,经过称量罐排放到主胶带上的原燃料数量均能通过计算机的 CRT 显示出来。而且每批料之间可通过计算机自学习系统进行校正,即本批料排放的料量可对上一批进行重量补偿。

(3) 焦丁回收。将仓下过筛后的碎焦经过二次筛分，分出焦丁和焦末，用胶带机将10~25mm的焦丁送到主胶带上方的焦丁称量罐内，将焦丁均匀地布在烧结矿料面上。

(4) 新型振动筛。取消中间称量罐后，仓下设备的负荷加大，设计中仓下设备的能力是按高炉利用系数 $3.0t/(m^3 \cdot d)$（日产量7600t）并留20%的富余能力来考虑的。为此开发研制了新型惯性双质体共振筛，烧结矿筛处理量400t/h，实际筛分效率75%~80%；焦炭筛处理量250t/h，实际筛分效率75%~80%。

新3号高炉投产以来实践证明：上料系统的能力能满足高炉强化冶炼的要求。对于并罐式无料钟炉顶，取消中间罐，采用无中继站式上料工艺是完全可行的。

2.3 新型无料钟炉顶

新3号高炉采用了首钢自行研制的 SGWZ-1000-1 型并罐式无料钟炉顶设备。该设备完善了多环布料配套和程控技术，可指令各个位置准确布料。炉顶装置的冷却、密封系统以及料罐电子称量装置也作了改进，可实现料流比例调节阀的恒料流控制。实践表明，该炉顶设备结构简单，易于控制，投资省，作业率高，能满足高炉超强化冶炼时的布料要求。为克服并罐无料钟设备料流偏析的缺点，生产中采用了矿、焦定期倒罐，调整布料溜槽正反转和料流调节阀的开度等技术措施。无料钟炉顶主要技术性能与规格参数如下：中心喉管直径700mm；布料溜槽长度3500mm；溜槽旋转速度10r/min；溜槽倾动速度0.278r/min（可调速）；布料方式为定点、扇形、单环、多环（螺旋）。

2.4 超矮胖炉型

首钢30多年的高炉生产实践证明：矮胖型高炉可以提高冶炼强度，促进高炉顺行。新3号高炉有效容积为2536m^3，炉缸直径11.56m，30个风口，3个铁口，无渣口。高径比 $H_u/D=1.985$，是目前国内同级别高炉中最为矮胖的一座。国内同级高炉炉型主要尺寸见表2。

表2 首钢新3号高炉与国内几座高炉的内型尺寸

项目	首钢新3号	马钢新1号	鞍钢11号	武钢4号
V_u/m^3	2536	2545	2580	2516
d/mm	11560	11100	11050	10800
D/mm	13000	12000	12200	11900
H_u/mm	25800	29400	29620	30000
h_0/mm	2200	1608	1206	700
h_1/mm	4200	4300	3700	3700
风口数/个	30	30	30	24
铁口数/个	3	3	2	2
H_u/D	1.985	2.45	2.428	2.521

2.5 炉体长寿技术

(1) 选用新型优质耐火材料。根据对首钢高炉炉衬破损特征的分析，为了延长炉体寿命，新 3 号高炉继 2、4 号高炉采用了综合炉底，在"蒜头状"侵蚀区采用了美国 UCAR 公司的 NMA 热压小块炭砖，将死铁层加深至 2.2m，并设计了"简易陶瓷杯"结构；风口、铁口均采用莫来石-碳化硅质的组合砖砌筑；在炉身下部、炉腰、炉腹区域采用 NMD 热压小块炭砖。

（2）软水密闭循环冷却系统。炉身部位（第 11~15 段冷却壁）采用了该系统，其特点是：软水的供、回水均为双路系统，事故状态下，单路供水可满足供水总量的 70%；冷却壁的凸台管、前排管、后排管均由单独的供水环管供水；软水泵站内设 2 个膨胀罐和 1 个 N_2 稳压罐，用来控制整个系统的操作压力；软水泵站内设有柴油泵事故供水系统以保证软水系统安全运行；为防止管道中形成"气塞"，在供回水环管及总管上都设有排气阀，炉顶平台设有 2 个卧式脱气罐；设有多层热电偶检测冷却壁温度，软水密闭循环系统的检测和运行实现微机控制。新 3 号高炉着重对软水系统检漏做了较大改进，采用了新型四通球阀专利产品和均速管流量计检测系统，可以通过微机在线监测冷却壁水管的进出水量、水温、水速，从而判断水管是否泄漏或损坏，并进行及时处理。

软水系统循环水量 2100m^3/h，总循环压力 0.60~0.70MPa，进水温度 55℃，出水温度 57~59℃，软水系统平均补水量为 6m^3/h，最大补水量为 21m^3/h，水质为一级软水。

2.6 圆形出铁场

新 3 号高炉圆形出铁场直径 77.9m，共有 18 根"Γ"形柱，出铁场平台上方设置一个宽 6m 的风口环桥，与风口平台在同一平面上，汽车通过公路引桥直接开到风口环桥。

出铁场内设置了两台新开发的多功能环形吊车，起重量 30/5t，跨度 20.6m，提升高度 21/23m，环形吊车可在 360° 范围内运行。炉前采用了新型 SGXP-400 矮式液压泥炮和 SGK-1 全液压开口机。泥炮结构紧凑，操作简单。

为延长渣铁沟寿命，设计采用了浇注料。主沟使用寿命可达 30 天左右，通铁量 10 万吨，铁沟通铁量 20 万吨，与捣打料相比，提高沟衬寿命 10 倍左右。炉前采用水力冲渣，并备有干渣罐，水渣采用过滤池法过滤。

2.7 大型顶燃热风炉

新 3 号高炉采用 4 座顶燃式热风炉，在燃烧单一高炉煤气的条件下，设计风温 1050~1100℃，每座热风炉设 3 个大功率短焰燃烧器，燃烧能力 32000m^3/h；拱顶用硅砖，蓄热室上部用低蠕变高铝砖；高温区各孔口采用了组合砖。为回收换炉时炉内 40%~50% 废风，同时减少换炉风压波动，缩短换炉时间，设计了用一座热风炉的废风给另一座热风炉充压的新工艺。设置整体式热管换热器，回收烟气余热，预热助燃空气至 100~150℃，为延长热风炉高温阀门寿命，均采用软水冷却。热风炉主要技术参数如下：热风炉全高 48350mm；热风炉钢壳内径 8900mm；高炉有效容积加热面积为 95.58m^2/m^3；助燃风机 2 台（一用一备），风量 160000m^3/h，风压 10kPa，电机功率 800kW。

2.8 大型鼓风机

由瑞士苏尔寿公司引进 AV100-19 静叶片可调式轴流鼓风机。主要性能：风量 7000m^3/min；工况出口压力 0.45MPa（绝压）；最高出口温度 313℃；主电机功率 36140kW。

2.9 煤气干法布袋除尘

采用干法布袋除尘新工艺（BDC），并配置湿式塔-文清洗系统作为高炉开炉初期运行时的补充措施。系统设有8个布袋室。每个布袋室内装有耐热尼龙（NOMEX）针刺毡内滤式布袋（$\phi 306mm \times 10m$）78条。该系统最大处理煤气量为530000m^3/h，最大过滤负荷88.35$m^3/(m^2 \cdot h)$。为严格控制进入布袋除尘器的煤气温度不超过185℃，在高炉炉顶设置了喷水降温装置。当顶温高于250℃时，开始向炉内喷水冷却，当炉顶温度降至200℃以下时停止喷水，另外，重力除尘器内也设置了高压冷却水雾化喷嘴，当布袋除尘器后净煤气温度高于180℃时，该喷水装置启动，低于160℃时停止喷水。布袋除尘器的反吹-过滤工艺是将布袋过滤后净煤气引出一部分，经反吹风机加压至8.0~10kPa后送至布袋出口管使煤气反向流动进行反吹。反吹-过滤操作正常情况由自动程序控制。

按正常情况，湿法系统只作为补充措施和BDC故障、检修时运行，应以BDC工艺为主，但目前由于首钢生产紧张，布袋除尘尚未投入使用。

2.10 喷煤工艺

新3号高炉设计煤比150~200kg/t，喷煤量48~64t/h，设计采用国内先进的三罐三系列多管路喷吹煤粉新工艺。为解决煤粉倒罐时连续计量问题，在中间罐和喷吹罐之间设计了4根刚性拉杆，用以克服由于浮力产生的称量误差。煤粉采用集中制备方式，系统按喷吹无烟煤设计。

2.11 高炉自动化控制

为保证高炉生产的连续、稳定，进一步实现生产各主要环节的自动化、高效化，设计了新型计算机控制系统和监测技术。高炉计算机网络由一级网络MODBUS，PLUS对等通讯网，二级网络VAX-4400-200型计算机，三级网络炼铁厂王安机生产管理网络组成。3号高炉的上位机通过同步通信与公司现有的王安机网络相通，传输生产运行状况。

在煤气自动快速分析、炉身温度及压力检测、炉喉十字测温、炉缸炉底测温、多料种自动上料、自动多环布料等方面共设置了2000多个"开关量"控制点。通过DCS（分布式控制系统）完成生产各个环节的逻辑联锁控制，实行生产过程的全自动控制，全部监视操作都集中在主控室的彩色CRT屏幕上。

各生产主要环节设置了1200多个监测点，在主控室可以观察到高炉、热风炉、上料、炉顶、喷煤等系统的各种生产运行状态，各种参数及各参数的趋势曲线。计算机通信网络相互联锁、沟通，覆盖高炉生产的各个环节，高速交换各种控制信息。

首钢开发的专利技术——人工智能高炉冶炼专家系统，能对生产各过程参数进行数学分析、建立数学模型，进行计算、预测并指导整个冶炼过程，准确率达90%以上。

3 投产后生产概况

新3号高炉1993年6月2日投产，近一年来，生产稳步上升，最高日平均利用系数达2.641$t/(m^3 \cdot d)$，最高日产量达7200t，近期主要技术经济指标见表3。

表3 首钢新3号高炉主要生产指标

时间	利用系数 /t·(m³·d)⁻¹	焦比 /kg·t⁻¹	焦丁 /kg·t⁻¹	煤比 /kg·t⁻¹	风温 /℃	顶温 /℃	顶压 /MPa	入炉品位 /%	一级品率 /%	[Si] /%	[S] /%	η_{CO} /%	休风率 /%
1993年4季度	2.547	522.7	9	40.0	1001	218	0.193	58.06	91.23	0.301	0.021	38.99	1.26
1994年2月	2.545	509.1	20	26.5	953	195	0.188	57.40	86.38	0.416	0.021	41.17	3.10
1994年3月	2.641	462.1	23	54.9	1015	202	0.189	56.70	97.23	0.333	0.018	41.75	0.40
1994年4月	2.625	447.8	27	57.2	1021	210	0.188	57.92	94.47	0.336	0.019	42.54	1.68

注：烧结矿率77%~80%；球团矿率17%~24%。

实践表明，设计中采用的新技术基本上发挥了应有的功能，为高炉正常生产提供了强有力的技术保证，为首钢提前实现年产1000万吨钢的目标奠定了物质基础。

新3号高炉设计中的不足之处是总平面布置比较拥挤，炉前5条铁路线均为尽头线，给炉前铁水的运输带来了一定的困难，在新建的高炉设计应给予足够的重视。

Design of new No. 3 BF of Shoudu Iron & Steel Co.

Zhang Fuming

Abstract: The new No. 3 BF of Shoudu Iron & Steel Co. is designed with ultra-low shaft, effective working volume of 2536m³ and $H_u/D = 1.985$. It can be adapted to ultra-high smelting intensity operation. This blast furnace is equipped with charging belt conveyor, new-type bell less top, closed loop soft water cooling system, circular casthouse, large sized top combustion hot stove, bag house dry dust collector, new-type computer control system and monitoring technology, etc. It was blown-in on June 2, 1993 and has been operating smoothly. A monthly average utilization coefficient was reached 2.641t/(m³·d).

Keywords: blast furnace design; bell less top; soft water cooling; top combustion hot stove; bag house dust collector

高炉炼铁技术的结构优化

摘　要：本文介绍了首钢高炉生产现状，分析了高炉炼铁技术的发展前景，提出了高炉炼铁技术结构优化的重点，即大力发展高炉喷煤技术，提高精料水平和风温，延长高炉寿命，降低工序能耗。通过技术改造提高高炉整体水平，全面实现高炉炼铁技术的结构优化。

关键词：高炉；炼铁；结构优化

1　引言

20世纪90年代以来，首钢高炉相继进行了现代化新技术大修改造，高炉总容积由4139m^3扩大到9934m^3，年生产能力达到750万吨，高炉总体技术装备达到国内先进水平。

21世纪的现代高炉炼铁技术，将以降低生产成本和减少污染为目标，以低焦比、高喷煤量、低渣量、高利用系数、长寿为技术特征。从首钢的生产经营现状和长远发展规划分析，首钢高炉生产的规模和格局将不会有大的变革，5座高炉同时生产的状况将持续到21世纪。提高高炉工艺流程的竞争力和经济效益的关键是进行结构调整和优化，这是一项长期的系统工程，关系到高炉的生存和发展。

2　炼铁技术结构的优化

近年来，首钢高炉生产坚持以优质、高产、低耗、长寿的方针，取得较好的实绩。高炉主要技术参数见表1，1997年主要技术经济指标见表2，近年主要生产指标的变化情况如图1所示。

首钢高炉系统的结构优化，一方面将通过高炉现代化新技术大修改造来实现，另一方面则是在现有高炉装备的基础上，以高炉喷煤为突破口，通过技术改造和技术进步，逐步实现炼铁系统的全面结构优化。

表1　首钢高炉主要技术参数和概况

炉号	1	2	3	4	5
有效容积/m^3	2536	1726	2536	2100	1036
炉缸直径/m	11.56	9.6	11.56	10.4	7.4

本文作者：张福明。原刊于《首钢科技》，1998（12）：11-13。

续表1

炉号	1	2	3	4	5
设计日产铁量/t	5800	4000	5800	4800	2400
风口个数	30	26	30	28	15
铁口个数	3	2	3	2	1
上料系统	胶带机	胶带机	胶带机	胶带机	料车
无料钟炉顶	SG并罐式	SG并罐式	SG并罐式	SG并罐式	SG并罐式
热风炉（顶燃式）	（4座）	（4座）	（4座）	（4座）	3座内燃式
喷煤（多管路）	（3罐3系列）	（2罐2系列）	（3罐3系列）	（2罐2系列）	1系列总管加分配器
煤气净化系统	干法布袋除尘	干法布袋除尘	干法布袋除尘	干法布袋除尘	湿法除尘
	湿法塔~文系统	湿法塔~文系统	湿法塔~文系统	湿法塔~文系统	
炉体冷却系统	冷却壁	冷却壁	冷却壁	冷却壁	冷却壁
炉底炉缸风口	开路工业水	开路工业水	开路工业水	开路工业水	开路工业水
炉腹至炉身上部	软水密闭循环	软水密闭循环	开路工业水和软水密闭循环	软水密闭循环	开路工业水
投产时间	1994年8月	1991年5月	1993年6月	1992年5月	1995年5月

表2　1997年首钢高炉主要技术经济指标

炉号	1	2	3	4	5	全厂平均
平均日产铁量/t	5383.54	3342.53	5307.01	4265.81	2148.92	20447.81
利用系数/t·(m^3·d)$^{-1}$	2.124	1.938	2.093	2.032	2.074	2.059
入炉焦比/kg·t^{-1}	408.5	468.1	418.4	448.2	467	435.2
煤比/kg·t^{-1}	104	93.2	100.8	104.4	90.9	100.1
燃料比/kg·t^{-1}	522.3	563.7	530.5	552.6	562.9	541.7
风温/℃	1013	1005	1021	1010	961	1002
入炉矿品位/%	57.87	57.06	57.3	56.75	56.88	57.29
炉顶压力/MPa	0.183	0.155	0.185	0.180	0.123	0.165
生铁合格率/%	99.98	99.98	99.97	100	100	99.98
生铁一级品率/%	94.26	94.83	97.04	94.48	96.07	95.31
休风率/%	2.46	2.81	3.82	3.60	1.63	2.87
渣铁比/kg·t^{-1}	348	374	375	405	394	376

2.1　大力发展高炉喷煤技术

首钢1号高炉（576m^3）早在1966年就创造了喷煤的世界纪录，月平均煤比达到

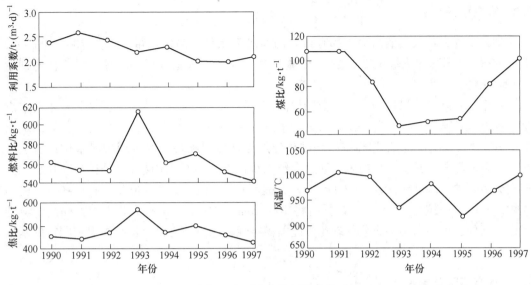

图1 主要生产指标的变化

279kg/t。20世纪90年代以来,由于生产规模的扩大,生产设施不配套,加上制粉能力不足,以及喷煤系统技术装备落后等原因,目前煤比停留在100~130kg/t,与国际先进水平相比尚有较大差距。实践表明,煤比提高10kg/t,可以节约焦炭8kg/t(见图2),降低生铁成本约2元/t,节能降耗效益显著。

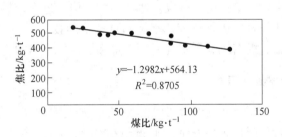

图2 首钢高炉焦比和煤比关系

笔者认为,发展首钢高炉喷煤技术应分层次进行。第一层次:保持现状,使煤比稳定在100kg/t以上,入炉焦比降到420kg/t以下,风温达到1000~1100℃,鼓风不用富氧;第二层次:通过喷煤系统技术改造,将多管路系统改为总管加分配器系统,提高制粉能力,改善高炉操作,使煤比达到150kg/t,入炉焦比降到400~350kg/t,风温达到1100~1150℃,富氧率达到1%~2%;第三层次:制粉系统技术改造,具备喷吹混煤(或烟煤)的条件,六制粉车间改为直接喷吹工艺,改善入炉原燃料条件,优化高炉操作技术,强化煤粉在风口前的燃烧,提高煤粉燃烧率和置换比,使煤比达到200kg/t,入炉焦比降到350kg/t,风温达到1150~1200℃,富氧率达到2%~4%(或采用氧煤枪)。应该指出,由于5座高炉存在装备能力上的差异,因此第三层次可在基础条件较好的高炉上优先实现。

2.2 提高精料水平

国外实现大喷煤量的高炉，都有很好的原燃料条件，高炉渣量一般为 200~300kg/t，瑞典 SSAB 公司高炉渣量仅 150kg/t。低渣量操作为高炉稳定顺行、大量喷煤、长寿创造了有利条件。

首钢随着高炉喷煤量和利用系数的提高，需发展高炉精料技术。发展精料技术的重点是，提高入炉矿品位、改善烧结矿质量；积极采用球团（小球）烧结新技术，降低烧结矿中 SiO_2 和 FeO 含量；采用高碱度烧结矿配加酸性球团和块矿的合理炉料结构；加强原料入炉前的筛分，减少入炉粉末。通过上料系统改造，尽早采用小粒度烧结矿回收工艺，实现分级入炉。降低焦炭灰分（$A<10\%$），提高焦炭强度（$M_{40}>80\%$，$M_{10}<8\%$），提高入炉焦炭的平均粒度（>40mm）。同时将 10~25mm 的焦丁与矿石混装入炉，提高炉料透气性。此外，提高原燃料成分及理化性能的稳定性，减少波动是保证高炉稳定顺行的基本条件。

2.3 提高热风温度

1200~1300℃ 风温是 21 世纪现代化高炉的标准。目前首钢高炉风温一直徘徊在 1000℃ 左右，月平均风温尚未突破 1100℃，这已成为提高喷煤量、改善高炉生产指标的主要障碍。笔者认为，提高风温也应分为三个层次。第一层次：采用现有的热风炉烟气余热预热助燃空气技术，改进热风炉操作，保证风温稳定在 1000~1100℃；第二层次：通过对现有煤气干法布袋除尘工艺的技术改造，使其尽早投产，改善煤气质量，提高煤气热值，使风温达到 1100~1150℃；第三层次：通过高炉大修改进完善热风炉结构形式，采用助燃空气、煤气双预热技术，优化热风炉燃烧和自动控制技术，采用交错并联送风操作，实现 1150~1200℃ 风温。提高风温是实现大喷煤量和炼铁技术结构优化的必要条件。

2.4 发展高炉长寿技术延长寿命

21 世纪现代化大型高炉的寿命应达到 10~12 年，最终实现高炉寿命 15 年。提高高炉技术装备水平的重点就是延长高炉寿命。20 世纪 90 年代以来，首钢通过新技术大修改造和一系列高炉长寿技术的实施，高炉寿命大幅度延长。今后一个时期，重点是要抓好现役高炉的操作管理，加强炉体监测与维护，采用炉衬喷补、压浆造衬等新技术实现高炉长寿。特别是 1 号、3 号、4 号高炉，应采取优化操作、强化管理、加强监测维护等综合措施，使高炉寿命达到 8~10 年（无中修）。此外，在新一周期的高炉大修改造中，要积极采用高炉长寿技术，优化炉体长寿设计，对炉型、炉衬耐火材料、冷却器、冷却系统及炉体自动化检测的设计，要充分结合生产实践，认真加以完善和改进。同时在设备制造和施工建设中，提高质量意识，加强质量管理，为高炉寿命达到 10~12 年（无中修）奠定良好的基础。

2.5 降低工序能耗，加强环境保护

炼铁工序能耗在吨钢能耗中占 50% 以上，因而钢铁企业降低消耗的主要潜力在炼铁系统。要改善高炉各项生产指标，尤其要降低燃料消耗，减少入炉焦比，同时加强能源的

回收利用。

1997年，首钢炼铁工序能耗为476kg/t，比日本高约80kg/t。今后应努力提高煤气利用率，减少高炉煤气放散；利用热风炉烟气余热，预热助燃空气和煤气，提高风温；提高炉顶压力，采用炉顶余压发电技术（TRT）；改进和完善煤气干法布袋除尘工艺；降低吨铁风耗。

应进一步提高环保意识，真正做到环保和生产建设协调发展。要加强炼铁系统污水、污泥的治理；加强原料、上料和出铁场等系统的粉尘净化；降低噪声污染；减少污染源，加大治理力度，改善高炉环境。

3 结语

面向21世纪的首钢高炉炼铁技术，将以工艺结构优化为核心，不断提高高炉技术装备水平和竞争力。节能降耗、降低生产成本、全面改善高炉技术经济指标，将是21世纪首钢高炉炼铁技术的发展目标。通过提高喷煤量、精料水平、风温和延长高炉寿命等一系列技术进步，全面实现首钢炼铁系统的结构优化和两个根本性转变，推动炼铁系统的可持续发展。

迁钢 1 号高炉采用的新技术

摘　要：迁钢 1 号高炉设计中以"长寿、高效、低耗、清洁"作为设计原则，在精料、长寿、高风温、喷煤、清洁生产等方面，积极采用当今国内外高炉炼铁先进技术，如焦丁回收技术、炉料分布与控制技术、高效长寿综合技术、高风温技术、首钢并罐无料钟炉顶及炉前现代化设备等，使高炉整体技术装备达到国内外同级别高炉的先进水平。

关键词：高炉；设计；新技术；应用

1　引言

2003 年 5 月，首钢总公司按照产品结构调整总体实施规划，部署实施首钢搬迁转移 400 万吨钢生产能力的方案——建设首钢迁钢工程，包括炼铁、炼钢、热轧及配套公辅设施。

首钢迁钢炼铁工程分成两期建成，一期工程建设一座 2650m^3 高炉（1 号高炉），二期工程再建一座 2650m^3 高炉（2 号高炉），最终形成一期、二期年产生铁合计 445 万吨生产规模。

迁钢 1 号高炉设计以"长寿、高效、低耗、清洁"作为设计思想和指导方针，采用了多项国内外先进技术和工艺，如：焦丁回收系统；水冷并罐式无料钟炉顶设备；软水密闭循环冷却系统；3 段国产铜冷却壁；UCAR 热压炭砖和 SAVOIE 大型风口组合砖；首钢设计研制的矮式液压泥炮及液压开口机；3 座改进型内燃式热风炉，并采用分离式热管换热器进行预热助燃空气和高炉煤气，在掺烧极少量焦炉煤气的条件下，使风温达到 1250℃；中速磨制粉、总管+分配器长距离直接喷吹工艺；采用螺旋法水渣处理工艺及长寿渣沟；煤气清洗采用串联文氏管湿法煤气清洗工艺，并采用压差发电技术；电动大型静叶可调轴流鼓风机。为提高高炉自动化控制水平，实现高效化生产，设计完善的高炉温度、压力、流量的检测，并预留人工智能专家冶炼系统接口。为实现清洁化生产，降低环境污染，对高炉上料、炉前等系统优化了除尘系统设计。

2　设计基本原则及设计指导思想

迁钢 1 号高炉设计中采用国内外先进、可靠、实用的新技术、新工艺、新设备及新材料，以我国和首钢高炉的设计与生产实践为基础，使新技术应用后的高炉整体技术装备具有国内领先水平。在满足工艺流程短捷、顺畅、合理的情况下，使总图布置紧凑合理，占地面积尽可能减小。在尽量节约投资的条件下，引进部分国外先进、国内目前尚不能生产

本文作者：毛庆武，张福明，张建，黄晋，姚轼。原刊于《炼铁》，2006，25（5）：5-9。

的关键部位的耐火材料和自动化控制系统和设备，使高炉寿命在不中修的条件下，达到一代炉龄15年以上。迁钢1号高炉设计以"长寿、高效、低耗、清洁"作为指导思想和方针，积极采用长寿、精料、高风温、大喷煤、适量富氧等先进技术和工艺，实现高炉长寿化、高效化、现代化、自动化、清洁化。

3 主要设计指标

高炉有效容积2650m^3，年平利用系数2.365t/(m^3·d)，燃料比495kg/t，焦比335kg/t，煤比160kg/t，综合焦比463kg/t，综合入炉矿品位≥59%，熟料率≥85%，热风温度1250℃，炉顶压力0.2~0.25MPa，高炉寿命一代炉龄无中修达到15年。

4 采用的新技术

4.1 精料技术

本系统采用传统原料场和高炉料仓合并建设的联合料仓、无中继站胶带上料工艺，料仓为双列布置，烧结矿直接入称量罐的工艺布置形式。烧结矿、球团矿、块矿、焦炭在仓下分散筛分，分散称量。杂矿仓下只设称量斗，分散称量。称量后的所有物料均通过N1-2及N1-1主胶带机送往炉顶装料设备。烧结矿、焦炭采用24台高效振动筛，强化仓下炉料的筛分，提高处理能力和筛分效率，使小于5mm的入炉烧结矿控制在5%以内。增加了焦丁回收装置，回收10~25mm的焦丁，与矿石混装入炉，提高高炉透气性，降低焦比。

4.2 炉料分布控制技术

采用首钢自行开发研制的水冷气封并罐式无料钟炉顶设备，布料流槽的悬挂装置采用了新型的锁紧装置，彻底杜绝了溜槽脱落的发生，避免了因溜槽脱落而发生的高炉休风的现象，提高了高炉作业率。并设料流调节阀，在自动控制下实现环行（多环）和螺旋布料的功能，在控制室人工控制下完成环形、点状和扇形布料。可以根据炉况变化，及时调整布料制度，抑制边缘煤气流的过分发展，保护炉衬和冷却器。采用多环布料技术可以提高高炉煤气利用率，降低焦比，延长高炉寿命。传动齿轮箱采用新型水冷结构，冷却水量提高到10t/h以上，使氮气消耗量降低到约500Nm^3/h，提高了冷却效率，延长设备使用寿命，改善煤气质量，提高煤气发热值。

4.3 高炉长寿技术

（1）在总结国内外同类容积高炉内型尺寸的基础上，根据迁安矿山地区的原燃料条件和操作条件，以适应高炉强化生产的要求，设计合理的矮胖炉型。设计中对高炉炉型进行了优化，加深了死铁层深度，以减轻铁水环流对炉缸内衬的冲刷侵蚀；适当加大了炉缸高度和炉缸直径，以满足高炉大喷煤操作和高效化生产的要求；降低了炉腹角、炉身角和高径比，使炉腹煤气顺畅上升，改善料柱透气性，稳定炉料和煤气流的合理分布，抑制高温煤气流对炉腹至炉身下部的热冲击，减轻炉料对内衬和冷却器的机械磨损。国内几座

2500m³ 级高炉内型尺寸比较见表1。

表1 国内几座2500m³级高炉内型尺寸比较

项目	单位	迁钢1号高炉	首钢1、3号高炉	鞍钢7号高炉	宝钢不锈钢2号高炉	唐钢3号高炉	武钢4号高炉	本钢5号高炉
有效容积 V_u	m³	2650	2536	2580	2500	2560	2516	2600
炉缸直径 d	mm	11500	11560	11500	11100	11000	11200	11000
炉腰直径 D	mm	12700	13000	13000	12200	12200	12200	12000
炉喉直径 d_1	mm	8100	8200	8200	8200	8300	8200	8300
死铁层高度 h_0	mm	2100	2200	2004	2300	2200	2004	1603
炉缸高度 h_1	mm	4200	4200	4100	4100	4600	4500	4300
炉腹高度 h_2	mm	3400	3400	3600	3600	3400	3400	3600
炉腰高度 h_3	mm	2400	2900	2000	2000	1800	1900	2000
炉身高度 h_4	mm	16600	13500	17500	17400	17500	17400	17000
炉喉高度 h_5	mm	2200	1800	2300	2000	2000	2300	2000
有效高度 H_u	mm	28800	25800	29500	29100	29300	29500	28900
炉腹角 α		79°59′31″	78°02′36″	78°13′54″	81°18′49″	79°59′31″	81°38′02″	75°57′49″
炉身角 β		82°06′42″	79°55′09″	82°11′27″	83°26′34″	83°38′30″	83°26′34″	82°17′42″
风口数	个	30	30	30	30	30	28	28
铁口数	个	3	3	3	3	3	2	3
渣口数	个	无	无	无	无	无	无	无
风口间距	mm	1204	1211	1204	1162	1152	1257	1234
H_u/D		2.268	1.985	2.269	2.385	2.402	2.418	2.258
V_u/A		25.78	24.16	27.04	26.46	27.12	26.32	28.08
d_1^2/d^2		0.496	0.503	0.508	0.546	0.570	0.536	0.555
D^2/d^2		1.22	1.265	1.278	1.208	1.230	1.187	1.354

（2）根据首钢多年的设计和生产实践，在炉缸、炉底交界处至铁口中心线以上，引进UCAR公司的热压小块炭砖，适当减薄炉缸内衬厚度，提高冷却系统的能力。在炉底采用国产优质的莫来石质陶瓷垫，炉底满铺国产大型微孔炭砖和高导热大块半石墨质高炉炭砖，并采用软水冷却；风口采用SAVOIE大型组合砖。

（3）高炉炉腹以上冷却壁采用软水密闭循环冷却系统，以延长冷却器的使用寿命。

（4）在炉腹、炉腰、炉身下部采用3段铜冷却壁，材质为TU_2轧制铜板，冷却通道钻孔成型，铜冷却壁厚度125mm，铜冷却壁沟槽内镶填SiC捣料，以提高冷却效率，这是一种新型无过热长寿冷却壁[1]。

（5）在炉身中上部采用高效单排管冷却壁，冷却壁本体厚度250mm，材质为球墨铸铁QT400-20。冷却壁沟槽内镶填SiC捣料，以提高冷却壁的挂渣性能。

（6）在炉身上部至炉喉钢砖下沿，增加1段"C"形球墨铸铁水冷壁，水冷壁直接与炉料接触，取消了耐火材料内衬。

（7）炉腹、炉腰、炉身下部区域采用 Si_3N_4-SiC 砖和高密度黏土砖组合砌筑，砖衬总厚度 400mm；炉身中上部采用高密度黏土砖。

（8）采用最新开发设计的送风装置，以适应 1250℃ 高风温的要求。加强了送风组件的密封，对送风支管结构进行了改进和优化。

（9）采用新型十字测温装置及炉顶高温摄像仪，在线监测炉内煤气流的分布和温度变化，配合多环布料技术，使高炉操作稳定顺行，提高煤气利用率，延长高炉寿命。炉体系统设计完善的高炉温度、压力、流量的检测，设置煤气取样自动分析装置，以加强高炉各系统的监视，为操作人员提供准确可靠的参数和信息，并预留人工智能专家冶炼系统接口及界面。

4.4 提高炉前机械化水平

（1）采用圆形出铁场，其最大外径为 77.9m，铁口标高为 10.2m，渣铁沟内衬采用浇注料，主沟采用贮铁式结构。出铁场设有公路引桥，出铁场平坦化布置，便于炉前机械操作及运输。出铁场内设 2 台 30t/5t 环行起重机，L_K=20.6m，轨面标高为 21.95m，用于出铁场内的日常生产操作及检修时使用。

（2）采用首钢设计院开发研制的矮式液压泥炮，采用新型炮嘴组合机构，进一步提高炮嘴寿命。

（3）采用首钢设计院开发研制的新一代多功能全液压开口机。

4.5 热风炉高风温技术

采用 3 座达涅利·康利斯（DCE）公司的改进型内燃式热风炉，一列式布置。利用热风炉烟气余热预热助燃空气和高炉煤气，同时掺加极少量焦炉煤气，使风温达到 1250℃ 以上[2]，为提高喷煤量降低焦比创造条件。热风炉主要阀门采用软水密闭循环冷却，以提高冷却强度，延长阀门寿命，节约能源。

首钢采用的高风温内燃式热风炉主要技术性能参数见表 2。

表 2 首钢高风温热风炉主要技术性能

项　　　目	首钢 2 号高炉	迁钢 1 号高炉
高炉容积/m³	1780	2650
热风炉数量/座	3	3
热风炉操作方式	两烧一送	两烧一送
送风时间/min	45	45
燃烧时间/min	80	75
换炉时间/min	10	15
设计风温/℃	1250	1250
拱顶温度/℃	1420	1400
加热风量/Nm³·min⁻¹	4200	5500
炉壳内径/m	9.2	10.2

续表2

项　　目	首钢2号高炉	迁钢1号高炉
热风炉总高度/m	41.59	41.66
蓄热室断面积/m²	35.8	44
燃烧室断面积/m²	9.7	10.9
每座热风炉总蓄热面积/m²	48879	64839
单位高炉容积的加热面积/m²·m⁻³	82.38	73.4
格子砖高度/m	32.4	31.3
燃烧器长度/m	3.4	4.4

4.6 紧凑型长距离制粉喷煤技术

迁钢1号高炉的喷煤工艺采用了紧凑型长距离制粉喷煤技术，采用直接喷吹工艺，将制粉和喷吹合建在一个厂房内。新的喷煤工艺综合了国内外高炉喷煤的先进技术，具有如下优点：

（1）采用直接喷煤工艺，简化了喷煤流程，喷吹烟煤时更为安全。

（2）采用封闭式干燥炉，减少了系统的漏风率，降低了系统的氧含量，在喷吹烟煤时更为安全。

（3）采用中速磨煤机制粉，降低了制粉的运行费用，从而减少了煤粉的生产成本。

（4）采用高效低压脉冲煤粉收集器一级收粉工艺，既简化了流程，提高了煤粉收集效率，而且使排尘浓度大大降低，废气出口浓度不超过 $30mg/Nm^3$，减少了环境污染。

（5）贮煤罐与喷煤罐之间设置了压力平衡式波纹补偿器，提高了连续喷煤过程中的计量精度，实现了喷煤全过程的连续计量。

（6）采用自动可调煤粉给料机和高精度煤粉分配器，以流化喷吹为前提，实现时间过程的均匀喷吹，消除了脉动煤流。

（7）自动化控制水平较高，实现了喷煤倒罐自动控制和调节。

4.7 螺旋法水渣处理工艺及长寿渣沟

（1）螺旋法水渣工艺为机械脱水工艺的一种方法。由于螺旋法水渣工艺关键设备只有一台螺旋机，所以其维护检修工作量小，需要检修较多的是两个轴承，设计时考虑了方便的检修措施。采用了在水渣贮水池上加设小平流池的工艺，设置抓渣吊车，将沉淀下来的细渣进行清除，降低了冲渣水中的细渣含量，减轻其对管道的磨损和冲渣喷嘴的堵塞现象，同时降低了贮水池中沉淀物的堆积速度，为系统正常运转创造了必要的条件。螺旋法水渣工艺较传统的渣池节省占地面积，能耗低，运行费用低；工艺流程简单，布置较灵活。

（2）为了提高水渣沟衬板的使用寿命，减少检修维护量，在设计中采用新型的复合衬板代替普通的耐磨铸铁衬板。新型复合衬板是在普通Q235-A钢板的表面采用等离子喷焊工艺喷焊Ni60+WC工作层，钢板厚度为25mm，耐磨层厚度为8mm。新型复合衬板硬度极高（硬度可以达到HRC70-80），使用寿命可以达到3年，是普通耐磨铸铁衬板使用寿命的3倍以上。

4.8 湿式煤气除尘及压差发电（TRT）技术

煤气净化采用湿式双文煤气清洗系统；考虑干法除尘工艺先进性，是国内外高炉煤气净化的发展方向，在总图布置上预留干法布袋除尘设施占地。高炉煤气清洗设施采用湿式双文除尘并加精脱水工艺，系统由一级文氏管、一文脱水器、二级文氏管、二文脱水器、减压阀组、灰泥捕集器及给排水管道等组成。

炉顶压差发电（TRT）设施是冶金行业重要的节能和环保设施，它可以提供高炉鼓风站所需电能的 1/3，同时减少了由于减压阀组所引起的噪声，减少了对大气的污染，并提高了能源的综合利用率。

4.9 节水技术

新建联合泵站，设常压水供水系统、高压水供水系统、软水密闭循环系统、高炉鼓风机净循环系统、水冲渣浊循环系统、煤气洗涤水浊循环系统及高炉安全供水系统等。高炉采用软水密闭循环冷却，热风炉高温阀门采用软水密闭循环冷却；煤气清洗和水力冲渣的水，循环使用。通过以上节水措施，可以实现炼铁生产过程用水"零"排放，水循环利用率为 97.38%，吨铁耗用新水不超过 $0.71m^3/t$。

4.10 大型电动轴流鼓风机及交变频启动控制技术

高炉鼓风机站内设置 1 台 AV100-19 全静叶可调电动轴流式压缩机及其配套辅机，并预留 2 台鼓风机的位置。鼓风机设计流量 $7000Nm^3/min$、风压 0.43MPa，完全能够满足定风量、定风压操作的要求。鼓风机采用交变频启动控制技术，具有效率高、操作迅速、运行简便、结构紧凑、调节性能好的特点。

4.11 大型高炉自动化控制技术

高炉自动化实现电气、仪表和控制三电一体化，设计完善的高炉温度、压力、流量的检测，设置基础自动化和过程自动化两级自动化控制。基础自动化主要采用 QUANTUM 可编程逻辑控制器及工业微机来完成高炉冶炼过程的数据采集以及各种控制和操作等。过程自动化主要完成高炉冶炼过程的监控、数据处理、生产管理及生产报表的打印等功能。取消了常规仪表、操作台和模拟屏，并预留人工智能专家系统的接口和界面。

4.12 清洁生产技术

设计了供料、料仓、炉前等系统的除尘装置；为减小二次扬尘，重力除尘器卸灰采用加湿卸灰机；在所有风机的进风口和放散阀处，均设置了消音器，降低噪声污染；在铁口区域侧吸的基础上增设顶吸装置，有效地解决开、堵铁口时的烟尘外溢问题。上料及炉前系统除尘技术的应用，实现了高炉清洁化生产，改善了劳动条件，有利于环保。

5 生产实践

迁钢 1 号高炉工程于 2003 年 12 月完成施工图设计，2004 年 10 月 8 日竣工投产。1

号高炉开炉顺利，运行良好、生产稳定顺行，经过1年多的生产运行，迁钢1号高炉取得了良好的实绩。首钢3座2500m³级高炉主要技术经济指标对比见表3。

表3 首钢3座2500m³级高炉主要技术经济指标对比

项目	迁钢1号设计指标	迁钢1号 2005年	首钢1号 2005年	首钢3号 2005年	迁钢1号 2006年1月	迁钢1号 2006年2月	迁钢1号 2006年3月	迁钢1号 2006年4月
利用系数/t·(m³·d)⁻¹	2.16~2.5	2.3	2.29	2.312	2.45	2.47	2.53	2.46
入炉矿品位/%	59	59.02	59.35	59.27				
熟料率/%	90	90.3	87.79	87.69	89.6	93.79	92.5	91.55
入炉焦比/kg·t⁻¹	335	375	359	362.9	340	336	333	314
煤比/kg·t⁻¹	160	111.35	114	119.3	135	138	133	144
焦丁/kg·t⁻¹		34.8	22.6	22.4	27	23	22	22
综合焦比/kg·t⁻¹	463	499.8	484	492.7	472	467	459	449
燃料比/kg·t⁻¹	490	521.16	496	505	502	497	488	480
综合冶强/t·(m³·d)⁻¹	0.984~1.156	1.16	1.14	1.158	1.156	1.153	1.161	1.105
富氧率/%	0	0.2	0.23	0.31	1.53	1.2	1.56	1.6
风温/℃	1200~1250	1141	1119	1111	1173	1198	1200	1211
风压/MPa	0.37	0.359	0.331	0.327	0.36	0.36	0.364	0.364
顶压/MPa	0.2	0.193	0.197	0.197	0.195	0.195	0.198	0.197
顶温/℃	150~200	226	198	207				
渣量/kg·t⁻¹	290	306.4	310	324				
煤气利用率/%	50	48.12	42.1	43.09				
[Si]/%	0.35	0.35	0.49	0.5			0.65	0.49
[S]/%	0.025	0.025	0.024	0.024			0.02	0.032
休风率/%		1.1	2.851	1.716	2	0.99	0.07	2.42

6 结语

高炉精料及焦丁回收技术、炉料分布与控制技术、高炉高效长寿综合技术、高效铜冷却壁技术、软水密闭循环冷却技术、热风炉余热回收及高风温长寿技术、紧凑型长距离制粉喷煤技术、大型高炉风机交变频启动控制技术、炉顶压差发电（TRT）技术、首钢先进的并罐无钟炉顶装料设备及炉前现代化设备等综合技术在迁钢1号高炉上应用，提高了1号高炉整体技术装备水平。生产实践表明，迁钢1号高炉的设计是合理的，技术水平已达到国内领先水平。

参考文献

[1] 张福明，姚轼，等．铜冷却壁的设计研究与应用［C］．铜冷却壁技术研讨会论文集，2003：39-42.
[2] 吴启常，张建梁，苍大强．我国热风炉的现状及提高风温的对策［J］．炼铁，2002（5）：1-4.

New Technology for Qiansteel No. 1 Blast Furnace

Mao Qingwu   Zhang Fuming   Zhang Jian   Huang Jin   Yao Shi

Abstract: Following the principle of "long campaign, high efficiency, low consumption and cleanness", some advanced technologies of ironmaking at home and aboard are adopted such as coke nut recovery, distribution and control technology of burden, long campaign and high efficiency technology, high temperature blasting air, bell-less top with parallel hopper, modern equipment at cast house.

Keywords: BF; design; new technology; application

首钢京唐5500m³高炉采用的新技术

摘　要：本文介绍了首钢京唐钢铁厂5500m³高炉设计特点和采用的先进技术。高炉设计中采用合理炉料结构、炉料分级入炉技术；自主设计开发了并罐式无料钟炉顶设备和炉料分布控制技术；采用纯水密闭循环冷却、铜冷却壁、薄壁内衬等高炉综合长寿技术，高炉设计寿命25年；采用高风温顶燃式热风炉和助燃空气高温预热技术，设计风温1300℃；采用平坦化出铁场和铁水直接运输工艺、环保型螺旋法渣处理工艺；设计开发了特大型高炉煤气全干法布袋除尘技术；采用并罐式浓相喷煤技术；设计了完备的自动化检测和控制系统以及环境除尘系统，使高炉生产实现了"高效、低耗、长寿、清洁"的目标。

关键词：高炉；炼铁；无料钟炉顶；顶燃式热风炉；煤气干法布袋除尘

1　引言

首钢京唐钢铁厂项目是中国钢铁工业结构调整，提升钢铁企业整体技术装备水平的重大项目。钢铁厂年生产能力为970万吨，建设2座5500m³高炉，生产规模为年产898.15万吨生铁。这是我国首次建设5000m³以上的特大型高炉，设计中分析研究了世界上5000m³以上的特大型高炉的设计、技术装备特点，全面实施自主创新，自主设计开发了无料钟炉顶、顶燃式热风炉、煤气全干法布袋除尘、螺旋法渣处理工艺等一系列具有创新的先进技术和工艺装备。

2　设计理念与主要技术指标

2.1　设计理念

首钢京唐钢铁厂的设计、建设按照循环经济理念，以工艺现代化、装备大型化、流程紧凑化、生产集约化、资源和能源循环化、经济效益最佳化为目标，积极采用当今国际一流的先进工艺技术装备，使我国第一座5500m³高炉成为具有21世纪国际先进水平的特大型高炉[1]。

高炉设计中以"高效、低耗、优质、长寿、清洁"为设计理念，采用先进实用、成熟可靠、节能环保、高效长寿的工艺技术装备，全面研究分析日本、欧洲5000m³级特大型高炉[2~4]以及我国近年投产的4000m³级大型高炉设计和生产经验[5~7]，结合首钢京唐

基金项目：国家"十一五"科技支撑计划项目。

本文作者：张福明，钱世崇，张建，毛庆武，苏维。原刊于《钢铁》，2011, 46 (2)：12-17。

钢铁厂自主创新的总体要求，自主设计、自主研制、集成优化了一系列具有国际先进水平的工艺技术和装备[8]。

优化高炉炼铁工艺流程，总图布置紧凑合理，工艺流程短捷顺畅，充分考虑各个单元工序的系统性和整体性，使生产运行达到协调统一。采用完善的自动化检测和控制系统，实现高炉生产的全自动化控制。

2.2 主要技术指标

设计中综合分析了当今世界先进的大型高炉生产技术指标，并结合高炉的原燃料条件和技术装备水平，确定了 5500m³ 特大型高炉先进的设计指标，高炉设计主要技术经济指标见表1。

表1 高炉主要技术经济指标

项 目	设计指标
高炉有效容积/m³	5500
利用系数/t·(m³·d)⁻¹	2.3
焦比/kg·t⁻¹	290
煤比/kg·t⁻¹	200
熟料率/%	90
入炉矿品位/%	≥61
富氧率/%	3.5
送风温度/℃	1300
炉顶压力/MPa	0.28
渣量/kg·t⁻¹	250
一代炉龄/a	25

3 高炉精料和炉料分布控制技术

3.1 高炉精料技术

采用合理炉料结构，烧结矿率为70%，球团矿率为20%，块矿率为10%。综合入炉矿品位为61%，熟料率为90%。烧结矿品位≥58.7%，碱度（CaO/SiO_2）≥1.8，转鼓强度（$TI_{+6.35mm}$）≥78%；球团矿品位为66%，抗压强度≥2500N/球；焦炭抗碎强度（M_{40}）≥89%，耐磨强度（M_{10}）≤6.0%，焦炭反应后强度（CSR）≥68%，热反应性（CRI）≤23%。

高炉的原料、燃料分别由烧结厂、球团厂、焦化厂和原料场通过胶带机运送到高炉料仓。两座高炉共用座联合料仓，焦仓与矿仓并列布置。烧结矿、球团矿、块矿和焦炭在仓下进行筛分，采用分散筛分、分散称量、无集中称量站、胶带机直接上料工艺。该工艺上料能力大，缩短了物料的转运流程，减少了物料的破碎和入炉粉末，为改善高炉透气性、促进高炉稳定顺行创造了有利条件。

采用烧结矿、焦炭分级入炉技术，烧结矿、焦炭各按不同粒度分级入炉，回收 10~25mm 的焦丁和 3~8mm 的小粒度烧结矿。烧结矿分级入炉技术，可以合理调整入炉原料粒度、控制炉内不同粒度原料的分布，从而提高煤气利用率和炉料的透气性，有利于高炉操作和控制煤气流合理分布，实现高炉顺行、长寿。烧结矿按照粒度级别分为两级，大粒度烧结矿粒度为 20~50mm，中粒度烧结矿粒度为 8~20mm，回收 3~8mm 的小粒度烧结矿，提高原料利用率。焦炭按照粒度级别分为 25~60mm 和 60mm 以上级，回收 10~25mm 的焦丁与矿石混装入炉。

3.2 无料钟炉顶设备和炉料分布控制技术

无料钟炉顶装料设备是现代高炉的关键装备。5500m^3 的特大型现代化高炉炉顶装料设备，不但要满足装料能力以外，而且还要满足高炉操作对炉料分布精准控制的要求，实现炉料分级入炉和中心加焦。

并罐式无料钟炉顶设备可以在一个装料周期内将 5 种不同粒度、不同种类的炉料装入高炉，实现炉料分级入炉和中心加焦。首钢并罐式无料钟设备优点是成熟应用的可靠技术，操作维护经验丰富。针对具有国际先进水平的 5500m^3 特大型高炉国内首创自主研制的无料钟炉顶装备，满足了高效、精准、可靠、长寿、降低维护量的生产要求，完全具有自主知识产权，实现重大关键装备技术国产自主化，大幅度降低设备投资。表 2 是无料钟炉顶设备主要性能，图 1 是首钢并罐式无料钟炉顶设备的三维仿真设计图。

表 2　无料钟炉顶设备主要性能

项　目	数　值
料罐有效容积/m^3	2×80
料罐设计压力/MPa	0.30
上密封阀直径/mm	1100
下密封阀直径/mm	1100
料流调节阀直径/mm	1000
料流调节阀排料能力/$m^3 \cdot s^{-1}$	0.7
中心喉管直径/mm	730
料流调节阀开闭精度/(°)	控制精度±0.3；计量精度 0.1
布料溜槽长度/mm	4500
布料溜槽转数/$r \cdot min^{-1}$	8
布料溜槽倾动范围/(°)	2~53
布料溜槽倾动精度/(°)	控制精度±0.3；计量精度 0.1

图1 无料钟炉顶设备三维仿真设计

4 高炉高效长寿综合技术

延长高炉寿命是现代高炉的技术发展趋势。高炉设计寿命25年,一代炉龄产铁量达到20t/m³以上。设计中运用现代高炉长寿技术理论,采用国际先进的高炉高效长寿综合技术,确定合理的高炉炉型,采用无过热冷却器和纯水密闭循环冷却技术,优化炉体内衬结构,设置完善的高炉自动化检测与控制系统,以实现高炉生产的稳定顺行、长寿高效。

4.1 优化高炉炉型

研究特大型长寿高炉内型的发展趋势,根据高炉冶炼进程的传热、传质、动量传输和化学反应原理,结合高炉的原燃料条件,以提高炉料透气性、改善煤气能量利用、促进高炉稳定顺行为目的进行设计优化,高炉有效容积5500m³,炉缸直径15.5m,炉腰直径17.0m,死铁层深度3.2m,炉缸高度5.4m,有效高度32.8m,炉腹角79°22′49″,炉身角81°2′36″,高径比1.93,设4个铁口,42个风口。

4.2 采用"无过热、低应力"炉体长寿结构

炉缸炉底内衬采用炭砖-陶瓷垫组合结构,炉底采用高导热石墨砖、微孔炭砖、超微孔炭砖和陶瓷垫组合结构,炉缸壁和炉缸、炉底交界处采用热压炭砖,风口、铁口采用组合砖结构。为实现高炉寿命25年以上,在炉缸"象脚状"侵蚀区和铁口区采用铜冷却壁。高炉炉体采用全冷却壁结构,炉腹、炉腰、炉身下部采用4段高效铜冷却壁,炉身中上部采用7段镶砖铸铁冷却壁,炉喉钢砖下部设1段"C"形水冷壁。炉腹至炉身为冷却

壁与砖衬一体化薄壁结构，冷却壁镶砖热面直接喷涂耐火材料。高炉本体炉底、冷却壁、风口全部采用纯水密闭循环冷却系统，图2是高炉本体结构。

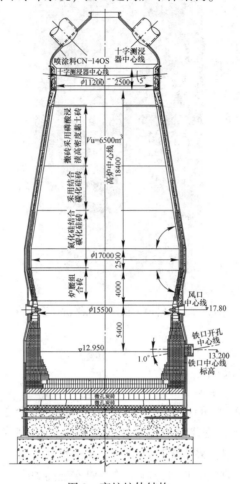

图2 高炉炉体结构

4.3 采用完善的高炉自动化检测与控制系统

炉缸、炉底设置12层548个热电偶监测炉底部位的温度分布，推断炉缸、炉底的侵蚀状况，指导炉缸、炉底维护操作。在风口区以上设置16层460个热电偶，用以监测冷却壁壁体工作温度，监测炉衬和冷却壁的侵蚀状况，计算炉体热负荷，推断操作炉型。设4层炉体静压力检测装置，用以推断软熔带位置，指导高炉布料操作。设置炉喉煤气十字测温装置用以在线监测炉身上部的煤气分布，为优化高炉布料提供可靠的信息。高炉冷却水系统设置进出口压力、流量、温度检测和记录，进行区域热负荷计算，同时加强对冷却环路的监控，推断风口及冷却设备的破损状况及泄漏监测。设置炉顶布料红外摄像、炉顶压力控制、炉顶煤气自动分析以及风口监测等自动化监测装置。

高炉操作采用布料计算模型、炉缸仿真模型、炉缸侵蚀模型、炉身仿真模型、出铁控制模型等数学模型和首钢自行开发的高炉冶炼专家系统，为优化高炉操作、促进高炉稳定顺行、延长高炉寿命创造了有利条件。

5 平坦化出铁场和炉渣处理工艺

特大型高炉的出铁场要满足高炉连续出铁、及时处理大量渣铁的生产要求，同时要便于出铁操作和设备维护，提高设备操作、检修的机械化水平，改善出铁场环境，实现清洁化生产。

5.1 出铁场结构和主要设备

设置2个对称布置的平坦化矩形出铁场，每个出铁场设2个铁口，每个铁口设有各自独立的泥炮、开口机、移盖机和摆动流槽等设备，泥炮和开口机均布置在风口平台之下，位于铁口同侧，铁口另一侧设置移盖机。出铁场一侧与公路引桥相连，汽车可直接通往出铁场。出铁场设2层平台，渣沟、铁沟设置在下层平台上，渣沟、铁沟的活动盖板与上层平台连为一体，实现了出铁场工作平台平坦化，检修设备在工作平台上作业。风口平台为架空式钢结构形式，可以实现设备维护和检修机械化。

风口平台设置装卸风口机械，可实现机械化更换风口设备；出铁场设有可远程操作的液压开口机、液压泥炮、移盖机、摆动流槽、吊车等。

5.2 铁水运输方式

为了优化高炉—转炉之间的界面技术，实现铁水运输的短流程，降低铁水热量损失和铁水倒罐的环境污染，设计开发了铁水直接运输"一罐到底"工艺技术。每个出铁场下方设有4条铁路线，采用炼钢铁水罐直接将铁水运送到炼钢厂。铁水运输采用300t铁水罐车取代了传统的鱼雷罐，优化了铁水运输流程，取消铁水倒罐作业，降低了铁水温降，减少了环境污染，加快了生产节奏。

5.3 出铁场环境保护

铁口一侧设置除尘吸风口，铁口顶部设一个顶部除尘罩，防止开、堵铁口时的二次烟尘污染环境；在渣铁分离器及渣沟、铁沟上方设置活动式平板沟盖，在出铁场双层平台之间设置除尘管道，捕集在出铁过程中产生的烟尘；每个出铁场内的2个铁水摆动流槽的两侧和铁水罐上方设置吸风口。出铁场除尘设置3个系统，每个出铁场设1个布袋除尘系统，每个系统的处理风量为$700km^3/h$，4个铁口设置1套二次烟尘布袋除尘系统，处理风量为$1\times10^6 m^3/h$，出铁场烟尘经除尘净化后，排放浓度小于$20mg/m^3$。

5.4 环保型螺旋法渣处理工艺

采用环保型螺旋法炉渣处理技术，回收冲渣蒸汽，冲渣水循环使用，减少SO_2、H_2S排放量和水量消耗。炉渣全部水淬粒化，干渣坑仅作为渣处理设备故障时备用，粒化后的水渣经过超细磨生产水泥掺和料。

高炉熔渣经过渣沟进入冲渣喷嘴，从冲渣喷嘴喷出的高速水流使熔渣水淬冷却，形成颗粒状水渣，渣水混合物经水渣沟通过冷凝塔输送到水渣池，渣水混合物在水渣池中经过螺旋机分离出水渣，水渣经胶带机输送到堆渣场。细颗粒渣则采用滚筒过滤器进行分离，冲渣水在水渣池中经过过滤器过滤后循环使用。

冲渣过程中产生的大量蒸汽和水渣池中产生的蒸汽排入冷凝塔内，冷凝塔中设 2 层喷淋装置，喷淋水和蒸汽冷凝水返回集水槽循环使用。

6 高风温顶燃式热风炉

高风温是现代高炉炼铁的重要技术特征，提高风温可以有效地降低燃料消耗，提高高炉能源利用效率。设计中对改造型内燃式、外燃式、顶燃式 3 种结构形式的热风炉技术进行了研究分析，在首钢顶燃式热风炉技术和卡鲁金式顶燃式热风炉技术的基础上，综合 2 种技术的优势，设计开发了 BSK（Beijing Shougang Kalugin）型顶燃式热风炉技术，将顶燃式热风炉技术首次应用在 5000m^3 级特大型高炉上。

BSK 顶燃式热风炉主要技术特征是：（1）高风温顶燃式热风炉的陶瓷燃烧器设置在热风炉拱顶部位，具有较广泛的工况适应性，可以满足煤气和助燃空气多工况条件的运行，而且燃烧功率大、燃烧效率高、使用寿命较长。陶瓷燃烧器采用了特殊的旋流燃烧技术，保证了空气和煤气的充分混合和燃烧，提高了理论燃烧温度和拱顶温度；（2）利用拱顶空间作为燃烧室，取消了独立的燃烧室结构，加强了炉体结构的热稳定性，燃烧器设在拱顶，高温烟气在旋流状态下分布均匀，有效地提高了高温烟气在蓄热室格子砖表面的均匀性和传热效率；（3）蓄热室采用高效格子砖，适当缩小格子砖孔径，提高格子砖的加热面积，改善了热风炉传热过程。

高炉配置 4 座 BSK 型顶燃式高风温长寿热风炉，设计风温 1300℃，最高拱顶温度 1420℃，热风炉高温区采用硅砖，设计寿命 25 年以上。热风炉燃料为单一高炉煤气，采用烟气余热回收装置预热煤气和助燃空气，配置 2 座小型顶燃式热风炉单独预热助燃空气，使助燃空气温度达到 520℃ 以上，热风炉高温阀门采用纯水密闭循环冷却，热风炉系统燃烧、送风、换炉实现自动控制。BSK 顶燃式热风炉主要技术性能见表 3。

表 3 BSK 顶燃式热风炉主要技术性能

项　目	数　值
热风炉座数/座	4
热风炉高度/m	50.10
热风炉直径/m	12.50
一座热风炉加热面积/m^2	95885
单位体积格子砖加热面积/$m^2 \cdot m^{-3}$	48.0
格子砖孔直径/mm	30
格子砖高度/m	21.48
蓄热室截面积/m^2	93.21
热风温度/℃	1300
拱顶温度/℃	1420（格子砖顶部最高温度为 1450）
烟气温度/℃	最大 450，正常平均 368
助燃空气预热温度/℃	520~600

续表 3

项　　目	数　　值
煤气预热温度/℃	约 215
冷风风量/$m^3 \cdot min^{-1}$	9300
冷风压力/MPa	0.54
单位鼓风蓄热面积/$m^2 \cdot (m^3 \cdot min)^{-1}$	41.24
单位高炉容积蓄热面积/$m^2 \cdot m^{-3}$	69.73

7　高炉煤气干法布袋除尘技术

高炉煤气干法布袋除尘技术是 21 世纪高炉实现节能减排、清洁生产的重要技术创新，可以显著降低炼铁生产过程的新水消耗、减少环境污染，已成为现代高炉炼铁技术的发展方向。

采用首钢自主设计开发的高炉煤气全干法低压脉冲布袋除尘技术，完全取消了备用的煤气湿式除尘系统。研究开发了煤气温度控制、煤气含尘量在线监测、除尘灰浓相气力输送、管道系统防腐、数字化控制等核心技术，使我国在大型高炉煤气全干法布袋除尘技术达到国际先进水平[9]。

高炉煤气全干法脉冲布袋除尘系统采用 15 个直径为 6.2m 的除尘箱体，箱体为双列布置方式，2 列箱体中间设置荒煤气和净煤气管道，煤气管道按等流速原理设计，使进入各箱体的煤气量分配均匀，整个系统工艺布置紧凑、流程短捷顺畅、设备检修维护便利。

采用低滤速设计理念，确保系统运行安全可靠。每个箱体设滤袋 409 条，滤袋规格 $\phi 160mm \times 7000mm$，单箱过滤面积 $1439m^2$，总过滤面积 $21586m^2$。设计中加大了滤袋的直径和长度，高径比降低，滤袋结构尺寸更加合理；扩大了箱体直径，使除尘单元的处理能力提高，减少了箱体数量、建设投资和占地面积。表 4 是高炉煤气布袋除尘工艺的主要技术参数。

表 4　高炉煤气干法布袋除尘工艺技术参数

项　　目	数　　值
高炉煤气流量/$km^3 \cdot h^{-1}$	760（最大 870）
炉顶压力/MPa	0.28（最大 0.30）
操作温度/℃	100~200
除尘器数量/个	15
箱体直径/m	6.2
滤袋规格/mm	$\phi 160 \times 7000$
单箱过滤面积/m^2	1439
总过滤面积/m^2	21586
标况过滤负荷/$m^3 \cdot m^{-2} \cdot min^{-1}$	35.2（最大 40.3）
工况过滤负荷/$m^3 \cdot m^{-2} \cdot min^{-1}$	14.0（最大 16.0）
净煤气含尘量/$mg \cdot m^{-3}$	≤5

8 高炉喷煤技术

高炉设计煤比为200kg/t（最大设计能力为250kg/t），正常喷煤量为116t/h，喷吹煤种为烟煤或混煤。2座高炉的喷煤和制粉设施集中合建一起，采用中速磨制粉、袋式煤粉收集器、直接喷吹工艺。中速磨能力为2×75t/h，高炉喷煤系统采用3罐并列式、总管分配器的浓相直接喷吹技术。

每座高炉采用3个喷煤罐交替喷煤，设2根喷煤总管和2个煤粉分配器。采用浓相输送技术，固气比大于40kg/kg，喷煤管道内煤粉流速为2~4m/s。在每根总管上设置煤粉流量计和调节阀，调节和计量精度小于4%。

煤粉仓有效容积1200m³，煤粉仓下设置3个喷煤罐；喷煤罐有效容积90m³，正常喷吹周期为30min。喷煤罐下部设流化罐，采用流态化喷煤技术，每个流化罐设2根喷煤支管，6根喷煤支管汇总为2根喷煤总管，煤粉经喷煤总管将煤粉输送到高炉的2个煤粉分配器中，再经42根喷煤支管喷入风口。喷煤罐的充压、煤粉的输送、系统的流态化及防爆吹扫全部采用氮气。每根喷煤支管设有喷吹状态监测装置，喷煤支管堵塞后自动切断吹扫。

9 高炉投产后的生产实践

首钢京唐1号高炉于2009年5月21日送风投产，经过10个月的生产实践，高炉生产稳定顺行，各项生产技术指标不断提升，高炉最高日产量达到14290t/d，月平均利用系数达到2.37t/(m³·d)，燃料比480kg/t，焦比269kg/t，煤比175kg/t，风温1300℃，达到了国际5000m³级高炉生产的先进水平。表5是首钢京唐1号高炉投产后的主要生产技术指标。

表5 首钢京唐1号高炉主要生产技术指标

时间	平均日产量 /t·d^{-1}	利用系数 /t·m^{-3}·d^{-1}	焦比 /kg·t^{-1}	煤比 /kg·t^{-1}	燃料比 /kg·t^{-1}	风温 /℃	工序能耗 /kgce·t^{-1}
2009年5月	4840	0.88	551	83	634	914	799
2009年6月	7425	1.35	503	62	565	998	538
2009年7月	8525	1.55	483	49	532	1063	461
2009年8月	11000	2.01	372	94	481	1166	409
2009年9月	11660	2.12	354	101	483	1212	419
2009年10月	12210	2.22	340	117	488	1262	414
2009年11月	12500	2.27	299	145	484	1276	406
2009年12月	12694	2.31	288	149	479	1281	393
2010年1月	12657	2.30	307	137	482	1259	388
2010年2月	12847	2.34	287	161	482	1277	375
2010年3月	13035	2.37	269	175	480	1300	373

10 结语

（1）首钢京唐5500m³高炉设计是按照循环经济理念和动态精准设计体系设计的新一代特大型高炉，设计中集成采用了当今国际炼铁技术领域的60多项先进技术，其中高炉精料和炉料分布控制技术、高炉综合长寿技术、顶燃式热风炉高风温长寿技术、出铁场平坦化和环境清洁化技术、环保型螺旋法渣处理工艺、高炉煤气全干法除尘技术、数字化自动检测与控制等方面具有创新性，在高炉生产中发挥了重要的支撑作用，使高炉整体技术装备达到国际领先水平。

（2）合理的炉料结构、炉料分级入炉技术和炉料分布控制技术为高炉稳定顺行提供了保障，自主设计制造的无料钟炉顶设备和自主开发的炉料分布控制技术首次应用于5000m³级特大型高炉。

（3）合理炉型、纯水密闭循环冷却、铜冷却壁、薄壁内衬结构等高炉长寿综合技术的采用，为高炉达到25年寿命奠定了基础。

（4）采用助燃空气高温预热技术和高风温顶燃式热风炉技术，在使用单一高炉煤气燃烧的条件下，高炉月平均风温达到了1300℃以上。

（5）高炉煤气干法除尘、环保型渣处理、环境保护技术和完善的自动化控制系统，使高炉操作实现了清洁化和自动化。

参考文献

[1] Zhang F M, Qian S C, Zhang, et al. Design of 5500m³ Blast Furnace at Shougang Jingtang [J]. Journal of Iron and Steel Research International, 2009, 16 (Supplement2): 1029.

[2] Stephenson E D, Parratt J E, Bowler K, et al. Towards the Year 2000—the Second Campaign of Redcar Blast Furnace [C] // 2nd International Congress on the Science and Technology of Ironmaking and 57th Ironmaking Conference Proceedings. Toronto: Ironmaking Division of the Iron and Steel Society, 1998: 625.

[3] Evans J L. Design and Construction of Lining of No. 4 Blast Furnace Port Talbot Works [J]. Ironmaking and Steelmaking, 1994, 21 (2): 101.

[4] Macauley D. Current Blast Furnace Design Philosophies [J]. Steel Times International, 1995, 19 (2): 20.

[5] 陆熔，李维国，陶荣尧. 宝钢炼铁20年的回顾及展望 [J]. 炼铁, 2005, 24（增刊）: 2.

[6] 高成云. 马钢4000m³级高炉的主要技术特点及装备水平 [J]. 炼铁, 2007, 26 (4): 1.

[7] 吴建洲，徐少兵. 宝钢4号高炉设计采用的新技术及其特点 [J]. 炼铁, 2006, 25 (2): 1.

[8] 张福明. 现代大型高炉关键技术的研究与创新 [J]. 钢铁, 2009, 44 (4): 1.

[9] Zhang F M. Study on Dry Type Bag Filter Cleaning Technology of BF Gas at Large Blast Furnace [J]. Journal of Iron and Steel Research International, 2009, 16 (Supplement2): 608.

New Technologies of 5500m³ Blast Furnace at Shougang Jingtang

Zhang Fuming　Qian Shichong　Zhang Jian　Mao Qingwu　Su Wei

Abstract: The technical features and advanced technologies of the 5500m³ blast furnace in Shougang Jingtang Iron and Steel Co., Ltd. were introduced. Reasonable BF burden design and classified burden charging technologies were applied. Bell-less top equipment was self designed and developed, and burden distribution control technology was applied. High-efficiency long campaign life technology of blast furnace including closed loop soft water cooling system, copper stave, thin wall lining etc. was applied, and the design campaign life of the BF was 25 years. The high blast temperature dome combustion hot blast stove and high temperature preheating of combustion air technologies were applied, and the design blast temperature was 1300℃. Flat casthouse, direct transportation of hot metal, environment-friendly screw type slag treatment technologies were applied. BF gas dry bag filter dedusting technology in extra large BF was designed and developed. The high density pulverized coal injection technology with parallel hoppers was applied. The complete automation control system and environmental cleaning system were applied. With all the new technologies mentioned above, the BF operation has achieved the target of high efficiency, low consumption, long campaign and cleanness.

Keywords: blast furnace; ironmaking; bell-less top; dome combustion hot blast stove; BF gas dry bag filter dedusting

当代高炉炼铁技术若干问题的认识

摘　要：结合21世纪以来高炉炼铁生产现状，认为科学合理确定钢铁厂产能，进而再据此确定高炉的产能、数量和容积，采用合理炉料结构，改善焦炭质量，促进低品质资源的高效利用；提高风温和热风炉的长寿高效，是当代高炉炼铁技术要解决的问题。
关键词：高炉；炉料结构；精料；高风温

1　引言

高炉炼铁工艺经历近200年的创新发展与不断完善，已成为21世纪炼铁工艺的主流技术，高炉炼铁工艺受到自然资源短缺、能源供给不足以及环境保护等方面的制约，面临着较大的发展问题。高炉炼铁如何实现可持续发展，在高效低耗、节能减排、循环经济、低碳冶金、清洁环保等方面取得显著突破，是当今炼铁工作者普遍关注的热点问题。

2　高炉大型化与高炉合理容积的确定

2.1　国内外高炉大型化的发展现状

高炉大型化和长寿化是当今世界高炉炼铁技术的主要发展趋势，高炉大型化的技术优势主要体现在高效集约、节能减排、低耗环保等方面。大型高炉具有劳动生产效率高、能源消耗低、生产成本低、污染物排放少、环境治理效果好等诸多综合优势，大型高炉更有利于实现高炉低碳冶炼和循环经济，更有利于实现优质铁水生产和高效能源转换。因此，大型高炉可以有效地提高炼铁工业的综合技术装备水平，大型化是高炉炼铁实现"高效、优质、低耗、长寿、清洁"的必由之路。

随着装备制造、计算机信息化和新材料产业等相关产业的快速发展，世界范围内的高炉大型化进程加快。20世纪70年代，以日本为代表的工业发达国家，相继建成了一批容积5000m^3以上的特大型高炉，引领了国际高炉炼铁大型化发展的潮流。由于大型高炉具有单位投资省、生产效率高和运行成本低等技术经济优势，因此高炉大型化已成为当今国际炼铁技术发展最显著的技术特征。

高炉大型化使单座高炉的产量大幅度提高，提高单产量则提高了劳动生产率，从而提高了高炉炼铁的竞争力。近20年来，日本运行高炉的数量由1990年的65座减少到28座，高炉数量降低了56.9%，高炉平均容积由1558m^3提高到4157m^3，增长幅度达到166.8%，平均单炉产量达到350万吨/年[1]。欧洲运行高炉的数量由1990年的92座减少

本文作者：张福明。原刊于《炼铁》，2012, 31 (5): 1-5。

到 2008 年的 58 座，高炉数量降低了 37%，高炉平均工作容积由 1630m³ 提高到 2063m³，增长幅度为 26.6%，平均单炉产量由 104 万吨/年增加到 154 万吨/年，增长幅度为 48%[2]。日本和欧洲高炉数量及高炉平均容积的演变，基本代表了工业发达国家近 20 年高炉大型化的发展进程和现状。

20 世纪 90 年代，我国钢铁工业发展迅猛，钢铁产量持续增长，在高效连铸、高炉喷煤等关键共性技术取得重大突破的同时，我国钢铁厂整体流程结构优化，高炉大型化的发展进程也随之加快。进入 21 世纪以来，我国高炉大型化进程加快，重点钢铁企业的高炉构成发生了巨大变化。2000 年 2000m³ 以上的高炉仅有 18 座，到 2010 年 2000m³ 以上的高炉已发展到 109 座，其中 2000~3000m³ 高炉为 74 座，3000m³ 以上高炉为 35 座[3]。宝钢、太钢、马钢、本钢、鞍钢鲅鱼圈、首钢迁钢等一批新建的 4000m³ 级高炉相继投产，首钢京唐 2 座 5500m³ 高炉也分别于 2009 年 5 月和 2010 年 6 月建成投产，沙钢 5800m³ 高炉于 2009 年 10 月建成投产。这些特大型高炉的建成投产，标志着我国高炉大型化已经步入国际先进行列。

2.2 钢铁厂流程结构优化与高炉大型化

钢铁厂生产能力的选择要适应社会发展和市场需求，根据区域和钢铁市场结构需求，结合企业实际情况，因地制宜、科学决策，合理确定钢铁厂的产品定位和生产规模。应根据钢铁厂整体流程结构的合理性、高效性、经济性考虑顶层设计，综合考虑轧机组成并评估钢铁厂合理产能，进而再据此确定与之相对应的高炉产能、数量和容积，同时必须兼顾企业投资取向和企业发展的远景目标。国内外高炉炼铁技术进步与高炉大型化发展密不可分，在钢铁厂流程结构优化条件下的高炉大型化，是当代高炉炼铁技术的发展方向。考虑到物质流、能量流和信息流网络结构的优化，一个钢铁厂配置 2~3 座高炉是适宜的选择。应当以钢铁厂整体流程结构优化为前提，科学合理配置高炉，同时不应片面强调高炉越大越好，盲目攀比第一。

对于当代钢铁厂，高炉的功能不仅是通过还原反应过程而获得优质的炼钢生铁，而且伴随着大量的能量转换和信息的输入与输出。因此，应当运用冶金流程工程学理论[4]，从更高、更宽的视野进行分析，也就是从铁素物质流、能量流和信息流等方面，综合评价钢铁厂的流程优化和高炉大型化。高炉大型化是钢铁厂流程结构优化的重要内涵，关系到整个钢铁厂的物质流、能量流和信息流动态运行的优化。在整个钢铁厂生产流程的尺度上，综合考虑钢铁厂物质流、能量流和信息流网络结构优化的高炉大型化，是当代钢铁厂流程结构优化的必然抉择。

对于高炉产能、数量以及高炉容积的确定，必须因地制宜，不能千篇一律，应统筹考虑钢铁厂流程结构的合理性和高效性，科学确定钢铁厂的产品定位和生产能力。对于已有的钢铁厂，高炉数量和高炉容积的选择，还要结合具体条件，综合考虑原燃料条件、物流运输、炼钢、连铸以及轧钢的设备能力和产品结构等各方面因素。

笔者认为，推动实施我国高炉大型化，要实现以钢铁厂整体流程结构优化为前提的大型化，既不提倡一个钢铁厂有过多数量的高炉，也不主张高炉越大越好，应当按照钢铁厂流程优化的原则择优确定高炉数量、产能和容积。值得指出的是，高炉产能和容积的确定绝不能不顾钢铁厂流程结构的合理性，而盲目追求所谓的高炉大型化；同时更不能因循守旧、仅顾当前利益而建造数量过多的小型高炉。总而言之，高炉大型化不是简单的大型化，是钢铁厂功能、结构、效率协同优化条件下的大型化，不能不顾原燃料条件、生产操作和管理水平，为追求高炉大型化而大型化。实践表明，盲目追求高炉大型化后果适得其反[5]。

3 精料技术

3.1 资源和能源条件

精料是当代高炉生产实现"高效、低耗、优质、长寿"的基础，是高炉炼铁工艺中最重要的支撑技术，也是实施高炉生产"减量化"的重要措施。我国钢产量的快速增长导致进口矿石量大幅攀升，由2000年的69.90Mt增长到2010年的618.64Mt，增长了近8倍；进口铁矿石生产的生铁量由2001年的39%增加到2010年的62%。由此可见，进口铁矿石在我国炼铁工业发展进程中发挥了重要作用。与此同时，我国炼铁工业对进口铁矿石的依赖性日益增加，在国内优质矿石资源短缺、进口铁矿石价格攀升的情况下，使炼铁生产成本大幅度提高。

尽管我国煤炭资源丰富，但经济可开采储量不足，2003年公布的我国煤炭经济可开采储量为1450亿吨，优质炼焦煤资源相对短缺，可开采储量为662亿吨，且分布极不平衡。近年来，不少钢铁企业从国外进口优质主焦煤，主焦煤进口比率正在逐年递增。铁矿石和优质炼焦煤资源的短缺，制约了我国钢铁工业的持续发展。进口矿石和主焦煤价格的大幅攀升，使炼铁制造成本加大。在资源、能源条件制约日益严重的情况下，高炉炼铁应积极应对当前形势，通过实施"减量化"精料技术，提高资源与能源的利用效率，降低高炉炼铁过程含铁物质和能量的耗散。

3.2 采用合理炉料结构

采用合理炉料结构对当代高炉生产作用重大，是保障高炉生产稳定顺行的关键要素。多年以来，我国高炉炉料结构形成了以高碱度烧结矿为主，适量配加酸性球团和少量块矿的模式。当前，国际上主要产钢国在注重改善入炉矿石冶金性能的同时，提高综合入炉矿品位和成分稳定性，结合矿石资源条件和造块生产工艺，以实现资源减量化和最佳化利用为目标，确定经济合理的炉料结构。在优质矿石价格日益攀升的条件下，当前国内外先进高炉炉料结构的一个显著变化趋向，是降低了烧结矿的使用比率，而不同程度地增加了球团矿和块矿使用比率。进入21世纪以来，日本和韩国的钢铁厂开始调整高炉炉料结构，将烧结矿比率降低到66%~80%，增加了块矿使用比率。日本为了降低原料成本，开发并应用了MEBIOS（嵌入式铁矿石烧结）等一系列低品质矿石利用技术。近年来，欧洲高炉的炉料结构也发生了显著变化，1990年烧结矿比率约为80%，块矿比率低于10%；2008年，欧洲主要产钢国的高炉平均炉料结构为烧结矿66.2%，球团矿23.4%，块矿10.4%，烧结矿比率大幅度下降，球团矿使用比率显著提高。荷兰艾莫伊登厂6、7号高炉的炉料结构为烧结矿44%+球团矿52%+块矿4%；瑞典SSAB公司的高炉采用100%球团矿；2011年末，芬兰罗德洛基厂两座高炉的炉料结构，由烧结矿70%~75%、球团矿25%~

30%改变为采用100%球团矿;欧洲还有部分高炉的块矿使用量已达到20%左右。亚洲部分大型高炉炉料结构见表1,2008年欧洲部分国家高炉炉料结构见表2。

表1 亚洲部分大型高炉炉料结构

高 炉	高炉容积/m³	入炉品位/%	烧结矿/%	球团矿/%	块矿/%
首钢京唐1号	5500	58.88	69.2	21.2	9.6
首钢京唐2号	5500	59.39	65.3	24.4	10.3
沙钢华盛1号	5800	59.20	65.1	20.7	14.2
宝钢3号	4350	60.28	63.4	21.6	16.0
宝钢4号	4747	59.91	65.7	18.3	16.0
马钢2号	4000	58.46	73.8	21.4	4.8
首钢迁钢3号	4080	58.92	70.1	23.2	6.7
鞍钢鲅鱼圈1号	4038	59.38	73.7	14.8	11.5
太钢5号	4747	59.22	75.0	19.0	6.0
本钢新1号	4747	59.91	72.2	23.5	4.3
日本大分2号	5775	—	81.0	4.0	15.0
日本君津4号	5555	58.00	80.0	10.0	10.0
日本福山5号	5500	—	66.0~71.0	10.0~5.0	24.0
韩国唐津1号	5270	60.00	80.0	5.0	15.0
韩国光阳2号	4350	—	78.0	0	22.0
韩国光阳4号	5500	60.00	80.0	12.0	8.0
韩国光阳5号	3950	—	82.0	5.0	17.0

表2 2008年欧洲部分国家高炉炉料结构

国 家	烧结矿/%	球团矿/%	块矿/%
奥地利	43.9	32.2	23.9
比利时	88.5	7.3	4.2
芬兰	60.2	38.3	1.5
法国	84.8	9.0	6.2
德国	60.7	25.4	13.9
意大利	55.2	33.8	11.0
荷兰	43.8	51.8	4.4
西班牙	75.7	16.7	7.6
英国	70.7	19.1	10.2
欧盟平均	66.2	23.4	10.4

毋庸置疑,经济合理的炉料结构是保障当代高炉炼铁技术持续发展的关键要素。在当今条件下,炉料结构的确定要兼顾资源可获取性、技术可行性和经济性,通过择优比较,探索适宜企业条件的炉料结构。值得指出,开发低品质矿的利用技术,并不是简单地降低入炉矿品位,而是采用新技术实现低品质资源的合理利用。另外还应提高球团矿使用比

率,球团矿作为优质原料应当进一步扩大使用量。在保障高炉生产稳定顺行的前提下,适度增加块矿比率也可以使炉料成本下降,但这需要统筹考虑块矿比率增加后,对高炉燃料比和辅助原料消耗的影响。企业应建立基于运筹学数学规划的炉料结构优化模型,通过数学模型择优确定合理炉料结构。

3.3 提高焦炭质量

焦炭是高炉赖以生存的重要燃料,其在高炉内的骨架作用仍是其他燃料所无法替代。生产实践表明,随着高炉容积的扩大,高炉料柱中的矿焦比增加,焦炭在高炉内停留时间延长,焦炭在高炉内的负荷增加,而且所受到的破损作用几率更大,因此焦炭在软融带和滴落带的骨架作用更为突出,高炉炼铁对焦炭质量的要求更加提高。特别是高炉大型化以后,对焦炭质量提出更高的要求,焦炭不但具有较高的机械强度($M_{40} \geq 80\%$,$M_{10} \leq 8\%$),热反应性(CRI)和反应后强度(CSR)也成为衡量焦炭质量的重要依据。实践表明,在高炉低燃料比、大喷煤操作条件下,焦炭反应后强度$CSR \geq 66\%$,热反应性$CRI \leq 25\%$,平均粒度$\geq 45mm$,这是高炉大喷煤操作的重要保障条件。

4 高风温技术

4.1 高风温的意义和作用

高炉冶炼所需要的热量,一部分是燃料在炉缸燃烧所释放的燃烧热,另一部分是高温热风所带入的物理热。热风带入的热量越多,所需要的燃料燃烧热就越少。由此可见,提高风温可以显著降低燃料消耗和生产成本。除此之外,提高风温还有助于提高风口前理论燃烧温度,使风口回旋区具有较高的温度,炉缸热量充沛,有利于提高煤粉燃烧率、加大喷煤量,还可以进一步降低焦比。因此,高风温是高炉实现大喷煤操作的关键技术,是高炉降低焦比、提高喷煤量、降低生产成本的重要技术途径,是高炉炼铁发展史上极其重要的技术进步。

高风温技术是一项综合技术,要在整个钢铁厂能量流网络的尺度上进行研究。高风温对于优化钢铁厂能源网络结构、降低生产成本和能源消耗、实现低品质能源的高效利用、减少CO_2排放等都具有重大的现实意义。当前,国内外高风温技术发展水平并不平衡,2010年我国重点钢铁企业的高炉平均风温为1160℃,先进高炉的风温已达到1250~1300℃,技术水平差距很大,进一步提高风温是21世纪高炉炼铁技术的热点研究课题。

4.2 获得高风温的技术途径

多年以来,我国高炉平均风温始终徘徊在1000~1080℃,2001~2010年的10年间,重点钢铁企业高炉平均风温由1081℃提高到1160℃,风温仅提高了79℃,可谓步履维艰,高炉风温是我国高炉和国外先进水平差距最大的技术指标。制约风温提高有许多因素,如何突破这些制约条件,达到1200℃以上的高风温乃是当今我国高炉炼铁的主要技术发展目标。

在当前条件下,提高风温应着力解决以下几个关键问题:

（1）研究应用燃烧高炉煤气获得高风温技术。随着高炉炼铁技术进步，高炉燃料比降低、煤气利用率提高，高炉煤气热值不足 3000kJ/m³。由于钢铁厂高热值的焦炉煤气和转炉煤气供应不足，热风炉在燃烧单一低热值高炉煤气条件下，热风炉理论燃烧温度和拱顶温度不高，很难实现 1200℃ 高风温。因此采用煤气、助燃空气高效双预热技术，不但可以回收热风炉烟气余热，减少热量耗散，还可以有效提高热风炉拱顶温度。在众多的预热技术中，要统筹考虑能量转换效率、技术可靠性、设备使用寿命等因素，择优选用适用可靠的双预热技术。首钢京唐 5500m³ 高炉采用了煤气预热和助燃空气高温预热组合技术，利用热管换热器回收烟气余热预热高炉煤气，设置 2 座小型热风炉预热助燃空气，使煤气预热温度达到 200℃，助燃空气预热温度达到 520~600℃，在燃烧单一高炉煤气条件下，热风炉拱顶温度达到 1420℃，月平均风温达到 1300℃[6]。这项利用低热值高炉煤气实现 1300℃ 高风温技术值得推广。

（2）热风炉结构形式的选择。高炉热风炉是典型的蓄热式加热炉，其工作原理不同于其他的冶金炉窑，是当代钢铁厂燃烧功率最大、能量消耗最高、热交换量最大的单体热工装置。尽管现有的 3 种结构热风炉均有实现 1250℃ 高风温的实绩，但在燃烧工况适应性、气体流动及分布均匀性、热量利用有效性等方面仍存在差异[7]。综合考虑热风炉高效长寿和适应性，当代高炉应采用顶燃式或外燃式热风炉。值得指出的是，近年来顶燃式热风炉取得了重大技术进步，已推广应用于首钢京唐 5500m³ 特大型高炉，并取得了突出的应用效果，月平均风温达到 1300℃。目前，国内开发并应用了各种结构的顶燃式热风炉，成为引领热风炉高风温技术创新的研究热点。应当看到，我国热风炉结构多样化将维持较长时期，对于新建或大修改造的高炉应优先采用顶燃式热风炉技术。

（3）采用高效格子砖。实践表明，缩小热风炉拱顶温度与风温的差值可以显著提高风温，其主要技术措施是强化蓄热室格子砖与气体之间的热交换。在保持格子砖活面积或格子砖重量不变的情况下，通过适当缩小格子砖孔径，可增加格子砖加热面积、提高换热系数而增加热交换量。在热风炉燃烧期，高温烟气可以将更多的热量传递给格子砖，使得烟气温度更低；在热风炉送风期，同样有利于鼓风与格子砖的热交换，使得热风温度更高，热风的温降也更为平缓，在风温保持较高的状态下更加稳定。对于格子砖砖型的选择需要综合考虑择优确定，并不是格子砖孔数越多、孔径越小就越有利，要综合考虑蓄热室热效率、蓄热室有效利用率和格子砖使用寿命等各种因素的影响。

（4）优化热风炉操作。优化热风炉燃烧、送风操作，缩小拱顶温度与风温的差值，适当提高烟气温度，是提高风温的主要操作措施。当前，热风炉燃烧操作应进一步优化，合理设定煤气与助燃空气配比，降低空气过剩系数，保证煤气的燃烧完全。热风炉烟气中残余 CO 和 O_2 含量应尽可能降低，燃烧初期应采取强化燃烧模式，使拱顶温度迅速达到设定数值；燃烧过程中合理调节煤气和助燃空气，在较低的空燃比下使烟气温度达到设定目标。根据拱顶温度、烟气温度、烟气含氧量等参数在线调节煤气和助燃空气，优化热风炉燃烧操作是当前需要引起重视的问题。优化热风炉燃烧不但可以获得高风温，还可以降低能源消耗，提高能源转换效率，减少热量耗散。热风炉送风操作也需要改进，4 座热风炉应采用"两烧两送"交叉并联的工作模式，这样可以提高风温约 25℃。同时应尽量减少冷风混风量，使高风温得到充分利用。3 座热风炉采用"两烧一送"的工作模式，可以适当缩短送风周期、增加换炉次数，同样应减少冷风混风量。改善热风炉操作一方面应开发应用简便实用的热风炉操作数学模型，采用数字化、信息化手段提高热风炉操作水平；另一方面应在交叉并联送风技术的基础上，研究开发热风炉"热交叉并联"送风技术，进一步减少冷风混风量，降低风温波动，使高炉获得更加稳定的高风温。

4.3 高温热风的输送

风温提高以后，高温热风的稳定输送成为制约环节。近年来，不少热风炉的热风支管、热风总管和热风环管出现局部过热、管壳发红、管道窜风，甚至管道烧穿事故，热风管道内衬经常出现破损，极大地限制了高风温技术的发展。因此，优化热风管道系统结构，采用"无过热—低应力"设计体系[8]，合理设置管道波纹补偿器和拉杆，妥善处理管道膨胀以降低管道系统应力。热风管道采用组合砖结构，消除热风管道的局部过热和管道窜风。

热风炉各孔口在多种工况的恶劣条件下工作，成为制约热风炉长寿和提高风温的薄弱环节。热风炉各孔口耐火材料要承受高温、高压作用的同时，还要承受气流收缩、扩张、转向运动所产生的冲击和振动作用。热风出口应采用独立的环形组合砖构成，组合砖之间采用双凹凸榫槽结构进行加强，以减轻上部大墙砖衬对组合砖产生的压应力。热风炉的热风出口、热风管道上的三岔口等关键部位均应采用组合砖结构。热风管道内衬的工作层，应采用抗蠕变、体积密度较低、高温稳定性优良的耐火材料；绝热层应采用体积密度低、绝热性能优良的耐火材料，还应注重合理设计耐火材料膨胀缝及其密封结构。

4.4 延长热风炉寿命

研究表明，当热风炉拱顶温度达到1420℃以上时，燃烧产物中的NO_x生成量急剧升高。燃烧产物中的水蒸气在温度降低到露点以下时冷凝成液态水，NO_x与冷凝水结合形成酸性腐蚀性介质，对热风炉炉壳造成晶间应力腐蚀，降低了热风炉使用寿命，成为制约风温进一步提高的主要限制因素。因此，热风炉一般将拱顶温度控制在1420℃以下，旨在降低NO_x生成量，从而有效抑制炉壳晶间应力腐蚀。应采取有效的晶间应力预防措施，延长热风炉使用寿命。通过合理控制热风炉拱顶温度，抑制热风炉燃烧及送风过程NO_x的大量生成；采取高温区炉壳涂刷防酸涂料和喷涂耐酸喷涂料等防护措施，加强对热风炉炉壳的隔离保护；采用细晶粒炉壳钢板，采用热处理措施消除或降低炉壳制造过程的焊接应力；对于热风炉高温区炉壳采取保温等综合防护措施，预防热风炉炉壳晶间应力腐蚀的发生，以延长热风炉寿命。

5 结语

（1）钢铁厂流程结构优化条件下的高炉大型化，是当代高炉炼铁技术发展方向。应根据钢铁厂整体流程结构的合理性、高效性和经济性，综合考虑顶层设计，科学合理确定钢铁厂产能，进而再据此确定高炉的产能、数量和容积。

（2）精料是当代高炉炼铁的重要基础。在当今资源短缺、能源日益匮乏且价格攀升的状况下，必须因地制宜制定相应的精料战略，采用合理炉料结构，改善焦炭质量，促进低品质资源的高效利用是当前应当着力解决的问题。

（3）提高风温、增大喷煤量、降低燃料消耗是当代高炉炼铁实现可持续发展的重要途径，采用燃烧低热值高炉煤气实现高风温技术具有广阔发展前景。采用高效长寿顶燃式热风炉和高效格子砖，优化热风炉操作，采用"无过热—低应力"热风管道设计体系，采取预防热风炉晶间应力腐蚀措施，实现高风温热风炉高效长寿的目标。

参考文献

[1] Miwa T. Development of Iron-making Technologies in Japan [J]. Journal of Iron and Steel Research International, 2009, 16 (supplement 2): 14-19.

[2] Bodo H, Peters M, P Schmoele. Iron Making in Western Europe [C]// The Chinese Society for Metals, Steel Institute VDEh, 3rd CSM-VDEh Metallurgical Seminar Proceedings. Beijing, 2011: 33-50.

[3] 张寿荣. 进入21世纪后中国炼铁工业的发展及存在的问题 [J]. 炼铁, 2012, 31 (1): 1-6.

[4] 殷瑞钰. 冶金流程工程学（第2版）[M]. 北京：冶金工业出版社，2009.

[5] 刘云彩. 炉料条件与巨型高炉建设 [C]// 中国金属学会, 2005 中国钢铁年会论文集, 2005: 304-308.

[6] 张福明, 钱世崇, 张建, 等. 首钢京唐5500m^3高炉采用的新技术 [J]. 钢铁, 2011, 46 (2): 12-17.

[7] 钱世崇, 张福明, 李欣, 等. 大型高炉热风炉技术的比较分析 [J]. 钢铁, 2011, 46 (10): 1-6.

[8] 张福明, 梅丛华, 银光宇, 等. 首钢京唐5500m^3高炉BSK顶燃式热风炉设计研究 [J]. 中国冶金, 2012, 22 (3): 27-32.

Discussion on Several Technical Issues of Modern BF Ironmaking

Zhang Fuming

Abstract: By analyzing the actual conditions of blast furnace ironmaking production since the beginning of 21 century, the author puts forwards his opinion that main issues to be concerned in modern blast furnace ironmaking are scientific and rational determination of enterprise's production capacity and determination of blast furnace production capacity, number and volume; additionally, optimization of burden structure, improvement of coke quality, effective utilization of low grade resource, increase of blasting air temperature and application of long campaign and highly effective hot stove are also absolutely necessary.

Keywords: blast furnace; burden structure; beneficiated burden; high blast temperature

钢铁厂流程结构优化与高炉大型化

摘　要：高炉是钢厂生产流程中物质、能量最为密集的工艺装置，对钢厂的物质流网络、能量流网络的构建与合理化运行有着重大影响。高炉的功能不仅是通过还原反应过程获得优质的铁水，而且伴随大量的能量转换和信息的输入/输出过程，应当在整个钢铁生产流程结构优化的前提下，综合思考高炉的合理座数、合理容积、合理位置。通过分析国际高炉的发展趋势和首钢京唐钢厂 $5576m^3$ 高炉和迁安钢厂 $4080m^3$ 高炉的比较，建设 $2×5576m^3$ 高炉和 $3×4080m^3$ 高炉可以得到相近的产量，但前者在节省投资、能源节约和信息控制等方面具有明显优势。由此可以看出，为了优化钢厂生产流程，提高市场竞争力，高炉大型化是一种明显的趋势。但是，并不是追求单座高炉越大越好，更不应盲目追求"最大"。应该在产品结构、物质流结构、能量流结构优化和动态运行优化前提下实施高炉大型化。一般的趋势是一个高效益、低成本生产的钢厂应以 2~3 座高炉为宜，并由此得到高炉大型化的合理容积、合理座数及其合理位置。这种发展趋势不仅适合于生产薄板的大型联合企业，而且也适合于生产建筑用棒/线材的中、小型钢厂。

关键词：钢铁厂；生产流程；结构优化；高炉大型化

1 引言

近年来，我国首钢京唐钢铁厂基于新一代可循环钢铁工艺流程，构建了动态有序、连续紧凑、高效协同的生产运行体系，实现了新一代钢铁厂优质产品制造、高效能源转换、消纳废弃物并实现资源化的三大功能[1]。

对于现代钢铁厂，高炉的功能不仅可以获得优质的炼钢生铁，而且伴随着大量的能量转换和信息的输入/输出[2]。因此，应该运用冶金流程工程学理论，在整个钢铁厂生产流程的层次上进行分析，也就是从贯通全厂的铁素物质流、能量流、信息流等方面，评价钢铁厂的流程结构优化和与之相关高炉大型化。要从高炉的四个基本功能出发（即氧化矿物还原和渗碳器、液态金属发生器和连续供应器、能量转换器和冶金质量调控器）[3]，全面思考高炉与钢铁厂流程结构优化的关系。炼铁工序是钢铁厂生产成本和能源利用效率控制的关键环节，高炉炼铁技术要实现高效、低耗、优质、长寿、清洁等综合目标。高效不是简单的生产强化，更要重视其经济效益、环境效益和社会效益；长寿不是简单地延长高炉使用寿命，还要重视其技术的先进性和可持续发展的生存能力[4]。面对当前国内外激烈的市场竞争环境，今后我国钢铁企业流程结构如何优化、钢铁厂如何选择合理工艺流程的问题，摆在了钢铁工作者的面前。本文针对钢铁厂高炉炼铁这个主要关键工序，运用冶金流程工程学理论，对科学合理地确定高炉产能、座数和容积进行了全面系统的分析论

本文作者：张福明，钱世崇，殷瑞钰。原刊于《钢铁》，2012，47（7）：1-9。

证，研究了在同样生产规模条件下，不同高炉配置技术方案的经济性与合理性，希望能为我国钢铁企业流程结构优化和高炉大型化提供参考。

2 高炉炼铁的发展趋势

近20年来，国内外高炉炼铁技术在大型、高效、长寿、低耗、环保方面取得长足技术进步，特别是高炉长寿、高风温、富氧喷煤、煤气干法除尘—TRT等单项技术成就突出。高炉炼铁技术进步主要体现在能耗和效率两个方面，这些技术进步与高炉大型化发展密不可分，在钢铁厂整体流程结构优化前提下的高炉大型化是当前国内外高炉炼铁技术的发展趋势。

考虑到物质流、能量流和信息流网络结构的优化，一个钢铁厂配置2~3座高炉是适宜的选择。应该以钢铁厂整体流程结构优化为前提，考虑高炉产能、座数、合理容积以及合理的位置、合理的平面布置等，同时不应不顾产品市场等因素片面强调某一高炉越大越好，盲目比大比小、盲目追求"最大"。高炉大型化是钢铁生产流程结构优化的重要内涵，关系到钢铁厂的物质流、能量流和信息流动态运行的优化。

2.1 国外高炉大型化发展现状

2.1.1 日本高炉大型化发展现状

日本在役高炉数量由1990年的65座下降到28座，下降幅度为56.9%，高炉的平均有效容积由1558m³上升到4157m³，上升幅度为166.8%。与此同时，日本高炉燃料比已经普遍降低到500kg/t以下，煤比达到120kg/t以上，焦比降低到380kg/t以下。图1是近20年来日本高炉座数和容积的变化[5]，图2是近20年来日本高炉燃料比的变化[6]，图1和图2基本代表了日本高炉炼铁技术的发展状况。目前日本单个钢铁厂一般配置2~3座高炉，也有一些钢铁厂只有1座高炉生产，2010年新日铁大分厂生产能力为963.4万吨/年，仅有2座5775m³高炉运行。

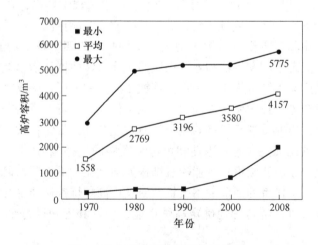

图1 日本高炉座数及容积的变化

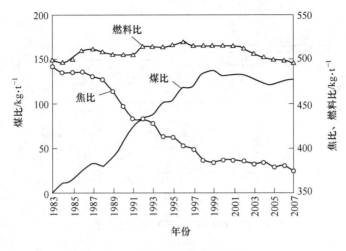

图 2　日本高炉燃料比的变化

2.1.2　欧洲高炉大型化发展现状

欧洲在役高炉数量由 1990 年的 92 座减少到 58 座,下降幅度为 37%。高炉平均工作容积由 1690m³（有效容积约为 2150m³）上升到 2063m³（有效容积约为 2480m³），上升幅度为 22%。欧洲高炉燃料比已经降低到 496kg/t,焦比降低到 351.8kg/t,煤比达到 123.9kg/t 以上,喷吹重油和天然气为 20.3kg/t。图 3 是近 20 年来欧洲高炉座数和容积的变化,图 4 是近 20 年来欧洲高炉燃料比的变化,图 3 和图 4 基本代表了欧洲高炉炼铁技术的发展状况[7]。欧洲单个钢铁厂的高炉数量基本是 2~3 座高炉,如德国蒂森克虏伯（TKS）的施维尔根厂年产量约为 780 万吨,目前仅有 2 座高炉运行（1×4407m³ + 1×5513m³）。

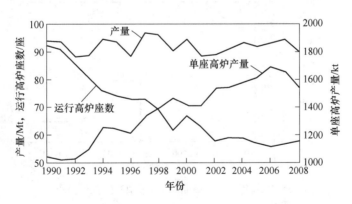

图 3　欧洲 15 国高炉座数及产能的变化

2.2　我国高炉大型化发展现状

日本和欧洲的高炉大型化发展开始于 20 世纪 80 年代,快速发展期为 1990 年以后。在钢铁厂整体流程结构优化的前提下高炉数量减少、高炉容积扩大,单座高炉的产量提

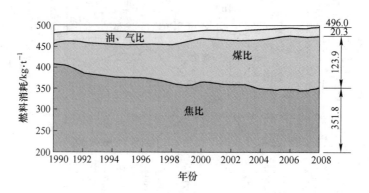

图 4 欧洲 15 国高炉燃料比变化

高。20 世纪 90 年代，我国钢铁工业发展迅猛，钢铁产量持续增长，在高效连铸、高炉喷煤、高炉长寿、连续轧制等关键共性技术取得重大突破的同时，我国钢铁厂整体流程结构优化、高炉大型化的发展进程也随之加快。

1985 年 9 月，宝钢 1 号高炉（4063m³）建成投产，成为我国高炉大型化发展进程的重要里程碑。然而真正大面积推进钢铁厂整体流程结构优化的高炉大型化，应该是在新世纪。据不完全统计，至 2010 年，我国在役和正在建设的 1080m³ 以上高炉数量约为 227 座，高炉总容积约为 429420m³。其中 1080~1780m³ 高炉为 119 座，2000~2500m³ 高炉为 27 座，2500~4080m³ 高炉为 61 座，4063m³ 以上高炉为 20 座。图 5 是 2010 年我国 1000m³ 以上高炉结构分布。

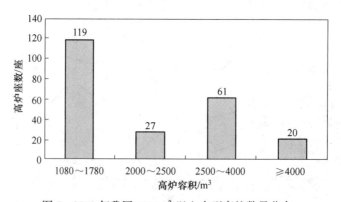

图 5 2010 年我国 1000m³ 以上大型高炉数量分布

实践证实，我国高炉大型化带动了高炉炼铁技术进步。目前，我国重点钢铁企业高炉燃料比已降低到 520kg/t 以下，焦比降低到 370kg/t，煤比达到 150kg/t 以上，炼铁工序能耗降低到 410kgce/t 以下。图 6 是进入 21 世纪以来我国重点钢铁企业高炉燃料消耗和风温的变化情况，由图 6 中可以看出，从 2005 年开始，随着 1080m³ 以上大型高炉数量的增加，高炉燃料比和入炉焦比显著降低。

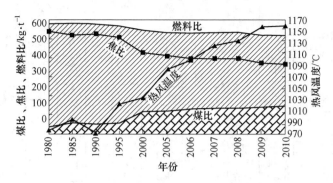

图 6　中国重点钢铁企业高炉燃料比的变化

3　钢铁厂流程结构优化前提下的高炉大型化

钢铁厂生产能力的选择要适应社会发展和市场需求，应根据区域市场需求和产品结构需求的变化，因地制宜，由相关区域的市场容量进行钢铁厂产品的定位和生产规模的优化选择。要根据钢铁厂整体流程结构的合理性、高效性、经济性考虑顶层设计，继而综合考虑轧机组成并评估合理产能，再对与之相应的高炉座数和容积做出初步选择，同时必须兼顾企业投资取向和企业发展的远景目标。

3.1　确定高炉座数与高炉容积选择的准则

高炉座数和容积，除了铁水产能需求以外，还必须考虑高炉的座数、位置对于钢铁厂的物质流网络、能量流网络、信息流网络以及与之相应的动态运行程序有着十分明显的关联性。我们可以通过首钢迁安钢厂和京唐钢厂的实例，对物质流网络、能量流网络进行简要分析（见图7~图10）。由图7~图8可见，高炉座数、容积及其在总平面图中的位置，对钢厂的物质流动态运行的结构和程序有着决定性影响，同时对能量流的结构、转换效率和运行程序也存在着决定性的影响。在相同的产品和产量规模下，高炉大型化、座数和位置合理化有利于企业结构优化和提高市场竞争力。

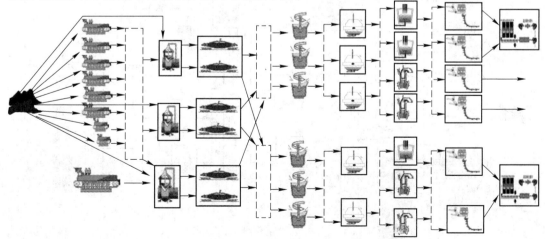

图 7　首钢迁钢物质流网络

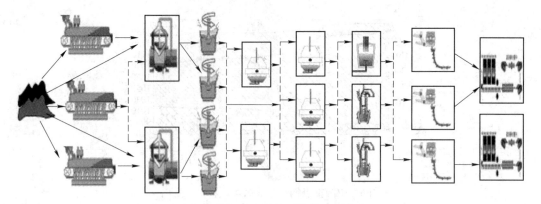

图 8　首钢京唐钢铁厂物质流网络

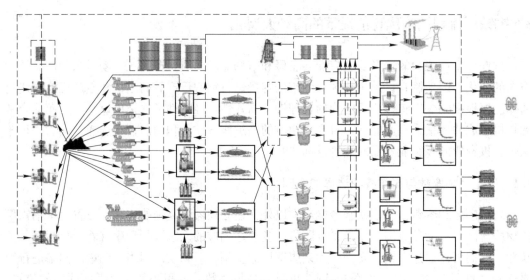

图 9　首钢迁钢能量流网络

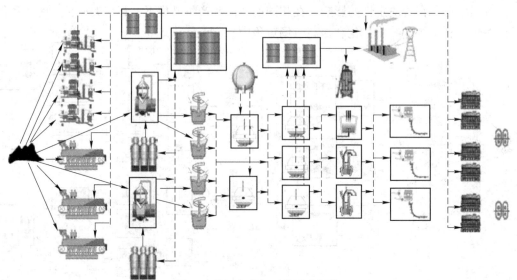

图 10　首钢京唐钢厂能量流网络

笔者认为,推动实施我国高炉大型化,应该是以钢铁厂整体流程结构优化为前提下的大型化,既不提倡一个钢铁厂有过多数量的高炉,也不主张盲目追求高炉越大越好,应当按照钢铁厂流程优化的原则择优确定高炉的座数和产能,而且还应关注其合理位置。值得指出的是,高炉产能和容积的确定绝不能不顾钢铁厂流程结构的合理性,而盲目追求所谓的高炉大型化,同时更不能因循守旧建造数量过多的小高炉。还应避免在评比、设计的过程中片面地"比大比小、凑零凑整"和盲目攀比"第一"。

3.2 高炉生产能力与生产效率

根据目前国内外炼铁技术的发展现状,不同级别的高炉生产能力均有较大幅度提高。按照当前不同级别高炉的技术装备水平、操作水平和原燃料条件,其生产能力已基本确定在一个合理范围内。图11是30年来全国重点钢铁企业高炉平均利用系数变化情况。由图11可见,近30年以来,我国重点钢铁企业高炉平均利用系数提高了$1.03t/(m^3 \cdot d)$。

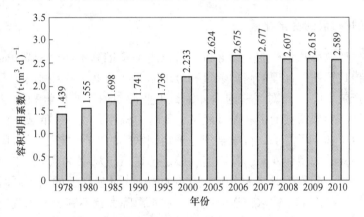

图11 中国重点钢铁企业高炉平均利用系数变化

衡量高炉生产效率,过去一般采用两个技术指标进行评价,即"冶炼强度"和"容积利用系数"。由于炉缸反应是高炉炼铁中十分重要的冶金过程,因此采用高炉炉缸截面积利用系数来衡量高炉生产效率,更具科学性[8],炉缸面积利用系数体现了高炉冶炼的本质特征。

不同级别高炉的容积利用系数和炉缸面积利用系数不同。例如$1260m^3$高炉容积利用系数为$2.5\sim2.7t/(m^3 \cdot d)$,炉缸面积利用系数为$61.16\sim66.05t/(m^2 \cdot d)$;$2500m^3$高炉容积利用系数为$2.4\sim2.6t/(m^3 \cdot d)$,炉缸面积利用系数为$60.93\sim66.01t/(m^2 \cdot d)$;$3200m^3$高炉容积利用系数为$2.3\sim2.5t/(m^3 \cdot d)$,炉缸面积利用系数为$60.98\sim66.28t/(m^2 \cdot d)$;$4080m^3$高炉容积利用系数为$2.2\sim2.4t/(m^3 \cdot d)$,炉缸面积利用系数为$61.51\sim67.10t/(m^2 \cdot d)$;$5500m^3$高炉容积利用系数为$2.1\sim2.3t/(m^3 \cdot d)$,炉缸面积利用系数为$62.04\sim67.95t/(m^2 \cdot d)$。在同等冶炼条件下,小型高炉与大型高炉的容积利用系数不可进行简单的类比。图12为2010年我国不同级别高炉年平均容积利用系数和年平均炉缸面积利用系数。由图12中可以看出,随着高炉容积增加,容积利用系数和炉缸面积利用系数呈现不同的变化趋势。

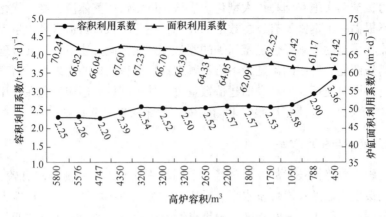

图 12 2010 年我国典型高炉的利用系数

3.3 钢铁厂高炉数量的优化配置

根据高炉生产效率和生产能力的分析，可以进一步推算不同容积高炉的期望年产量，进而确立不同模式钢铁厂合理的高炉数量和合理的高炉容积，表 1 列出了不同级别高炉的生产能力和生产效率[9]。针对当前国内外钢铁工业面临的严峻形势和挑战，保障钢铁工业的可持续发展，实现低碳冶金和循环经济，要着力构建高效率、低成本的洁净钢生产体系。根据产品定位和市场需求，科学合理地确定钢铁厂的生产规模，具有国际影响力和市场竞争力生产薄板的大型钢铁厂，其产能一般定位在 800 万~900 万吨/年，对于钢铁生产流程结构优化而言，配置 2~3 座高炉应是优化的选择。4 座以上的高炉同时运行，会引起物流分散、输送路径拥塞、铁水输送时间长、铁水温降大且不利于铁水脱硫预处理等；同时也将导致能量流网络分散复杂、运行紊乱，将使高炉煤气等二次能源的利用效率降低。高炉容积和座数将直接影响到平面布置的简捷顺畅程度，这就是钢铁企业物质流网络、能量流网络以及信息流网络的优化问题。高炉大型化有利于减少高炉座数，有利于流程简捷顺畅，有利于提高能源效率、节能减排，有利于信息化控制。

表 1 不同级别高炉的生产能力和期望年产量

高炉容积/m^3	1260	1800	2500	3200	4080	4350	5000	5500
容积利用系数/$t \cdot (m^3 \cdot d)^{-1}$	2.5~2.7	2.4~2.6	2.4~2.6	2.3~2.5	2.2~2.4	2.2~2.4	2.1~2.3	2.1~2.3
面积利用系数/$t \cdot (m^2 \cdot d)^{-1}$	61.16~66.05	60.98~66.06	60.93~66.01	60.98~66.28	61.51~67.10	61.32~66.89	61.90~67.79	62.04~67.95
年作业天数/d	350	350	350	355	355	355	355	355
期望年产量/万吨	110~119	151~163	210~227	261~284	312~340	339~370	372~408	410~449

4 不同容积高炉的工艺技术装备比较

高炉大型化的技术优势主要体现在高效集约、节能减排、低耗环保和信息化控制等方面，更具体地讲：有利于实现现代高炉优质铁水生产、高效能源转换和消纳废弃物并实现资源化的三大功能。与此同时，高炉大型化有力地促进着炼铁技术装备的发展，推动大型冶金装备和耐火材料技术的开发，促进信息化技术、高炉长寿、精料、无料钟炉顶、炉前设备、高风温、富氧喷煤、煤气干法除尘—TRT、除尘环保等多项技术的协同发展。

因此，对于不同生产规模和不同产品结构的钢铁厂，高炉生产能力、数量和容积的选择确定具有多种技术方案。对于200万~300万吨/年建筑用棒/线材厂或中/厚板厂、400万~600万吨/年的板材厂（薄板或薄板+中板）、800万~900万吨/年的大型薄板厂等不同产品、不同生产规模的钢铁厂，其高炉数量和容积的确定，必须在整个钢铁厂的层次上综合考虑，以实现物质流、能量流与信息流各自在合理的流程网络上协同高效运行为目标，实现钢铁厂整个生产流程结构优化前提下的高炉大型化。

对于一个生产能力为900万吨/年的钢铁厂，可以配置3座$4080m^3$高炉或配置2座$5576m^3$高炉，为此，针对这两种技术方案进行了对比分析研究。重点对$4080m^3$高炉和$5576m^3$高炉技术经济指标、原燃料适应性、能源及动力消耗、电力装机容量、工程投资、生产管理及运行成本、总图占地、节能环保等多方面进行了分析比较。

4.1 大型高炉主要技术经济指标

表2列出了我国新建的3座大型高炉主要技术经济指标。由表2可以看出，$4000m^3$级高炉与$5500m^3$级高炉的技术指标仍有一定差距，特别是高炉燃料比和炼铁工序能耗两项关键指标，$5500m^3$级高炉分别低10~15kg/t和18kgce/t。

表2 3座大型高炉主要技术经济指标比较

项 目	$4350m^3$高炉		$4080m^3$高炉		$5576m^3$高炉	
	设计值	2009年实际	设计值	2010年7月实际	设计值	2010年3月实际
年产量/万吨	325.0		340.8		449.1	
容积利用系数/t·$(m^3·d)^{-1}$	2.1	2.2	2.4	2.387	2.3	2.37
炉缸面积利用系数/t·$(m^2·d)^{-1}$	57.71	60.46	68.44	68.07	67.07	69.12
入炉焦比/kg·t^{-1}	320	307	305	331.2	290	269
煤比/kg·t^{-1}	200	197	190	167.8	200	175
燃料比/kg·t^{-1}	520	504	495	499	490	480
风温/℃	>1250	1242	1280	1280	1300	1300
炉顶压力/MPa	0.25	0.25	0.25	0.25	0.28	0.28
富氧率/%	≤3	3	3.5~5	4.17	3.5~5	4.0
熟料率/%	95	—	90	90	90	90
工序能耗/kgce·t^{-1}	422	391.5	—	—	404	373
单座高炉占地面积/万平方米	28.200	—			31.925	—

4.2 $4000m^3$以上高炉对原燃料的适应性

通过对日本各主要钢铁厂的考察调研得知，日本高炉大型化和$5000m^3$以上高炉的建造主要是通过高炉扩容大修改造实现的,许多高炉是由原来的 4000m³ 扩大到 5000m³ 以上,虽然高炉在大修期间进行了扩容,但焦化、烧结等工序并未进行全面技术改造,原燃料条件也并未发生根本性改变。日本大型高炉生产实践表明,适应 4000m³ 高炉的原燃料条件基本可以满足 5000m³ 高炉的生产要求。

图 13 是 4350m³ 高炉与 5576m³ 高炉内型和有效高度的比较。2 座高炉容积相差 1226m³,主要是高炉炉缸容积扩大而有效高度仅相差 1.2m,料柱高度相差不大,因此与 4000m³ 级高炉相比,5000m³ 级高炉对焦炭机械强度(M_{40},M_{10})、反应性指数 CRI 及反应后强度 CSR 并无显著苛刻要求。

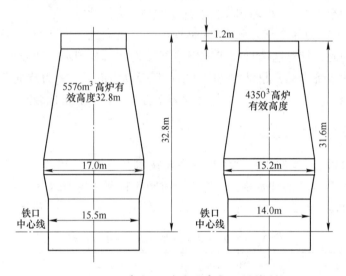

图 13　4350m³ 高炉和 5576m³ 高炉内型比较

根据高炉炼铁工艺设计规范(GB 50427—2008)对高炉原燃料条件要求[10],表 3 和表 4 分别列出 4000m³ 级和 5000m³ 级高炉对原燃料的质量要求。从表 3 和表 4 中可以看出,4000m³ 级和 5000m³ 级高炉的原燃料设计条件差别不是太大。

表 3　4000m³ 高炉和 5000m³ 高炉原料质量要求

	项　目	4000m³ 高炉	5000m³ 高炉
	入炉品位/%	≥59	≥60
	熟料率/%	≥85	≥85
烧结矿	品位波动/%	≤±0.5	≤±0.5
	碱度波动/%	≤±0.08	≤±0.08
	FeO 含量/%	≤8.0	≤8.0
	FeO 含量波动/%	≤±1.0	≤±1.0
	转鼓指数(+6.3mm)/%	≥78	≥78

续表3

项 目		4000m³ 高炉	5000m³ 高炉
球团矿	含铁量/%	≥64	≥64
	转鼓指数（+6.3mm）/%	≥92	≥92
	耐磨指数（-0.5mm）/%	≤4	≤4
	常温耐压强度/N·球$^{-1}$	≥2500	≥2500
	低温还原粉化率（+3.15mm）/%	≥89	≥89
	膨胀率/%	≤15	≤15
块矿	含铁量/%	≥64	≥64
	热爆裂性/%	<1	<1
	铁分波动/%	≤±0.5	≤±0.5

表4 4000m³ 高炉和 5000m³ 高炉焦炭质量要求

项 目	4000m³ 高炉	5000m³ 高炉
抗碎强度 M_{40}/%	≥85	≥86
耐磨强度 M_{10}/%	≤6.5	≤6.0
反应后强度 CSR/%	≥65	≥66
反应性指数 CRI/%	≤25	≤25
焦炭灰分/%	≤12	≤12
焦炭含硫/%	≤0.6	≤0.6
焦炭粒度范围/mm	75~25	75~30
大于上限/%	≤10	≤10
小于下限/%	≤8	≤8

4.3 工艺装备比较

表5列出了2座5576m³高炉和3座4080m³高炉工艺技术装备的比较。通过比较可以看出，2座5576m³高炉比3座4080m³高炉设备数量和质量明显减少，设备数量减少15%，设备总质量减少14820t，设备总质量降低22.6%，上料主胶带机长度缩短118.8m。设备数量的大幅度降低使设备投资、运行维护、备品备件消耗等均相应降低，而且动力消耗、岗位定员、污染物排放等都也相应降低，生产运行成本将大幅度降低。

表5 3×4080m³ 高炉和 2×5576m³ 高炉工艺装备比较

项 目	3×4080m³ 高炉	2×5576m³ 高炉
上料系统工艺设备/套	3	2
上料主胶带机长度/m	1014.0	895.2
炉顶装料设备/套	3	2
高炉本体设备/套	3	2
炉前系统机械设备/套	3	2

续表5

项 目	3×4080m³ 高炉	2×5576m³ 高炉
除尘系统设备/套	9	6
热风炉系统工艺设备/套	12	8
水渣处理工艺设备/套	12	8
煤气除尘系统设备/套	39	30
余压发电设备（TRT）/套	3	2
鼓风机/套	4	3
高炉控制系统/套	3	2
高炉煤气柜/套	3	2
设备总重量/t	65433	50613

4.4 能源动力消耗与节能环保

表6列出了4000m³级高炉与5576m³高炉在能源动力消耗、节能环保及装机容量的比较，由表6可见两者之间存在一定差距。5576m³高炉在水消耗、原燃料消耗、电机装机容量、电力消耗、清洁环保等方面具有显著优势，在节能、环保方面表现突出，主要技术经济指标、信息化水平及清洁化生产等方面均达到国际先进水平，并产生了显著的经济效益、社会效益和环境效益。

表6 4000m³以上大型高炉能源动力消耗及节能环保的比较

项 目	清洁生产一级标准	4350m³ 高炉	4080m³ 高炉	5576m³ 高炉
工艺技术装备				
高炉煤气除尘	全干法	干/湿法并用	全干法	全干法
TRT装机功率/MW		24.3	30.0	36.5
平均热风温度/℃	≥1240	1242	1280	1300
各系统除尘设施	均配备有齐全的环境除尘系统，除尘系统同步运行率达到100%			
资源能源利用指标				
工序能耗/kgce·t^{-1}	≤385	391.5		373
入炉焦比/kg·t^{-1}	≤280	307	331.2	269
入炉焦丁/kg·t^{-1}				36
喷煤比/kg·t^{-1}	≥200	197	167.8	175
燃料比/kg·t^{-1}	≤490	504	499	480
入炉铁矿品位/%	≥59.80	59.3		60.1
新水消耗/m³·t^{-1}	≤1.0	0.9	0.6	0.49
水重复利用率/%	≥98	98	98	98
高炉煤气放散率/%	0	≤5	≤5	≤1
产品指标				
生铁合格率/%	100	100	100	100

续表6

项目	清洁生产一级标准	4350m³ 高炉	4080m³ 高炉	5576m³ 高炉
污染物排放控制指标				
烟粉尘排放量/kg·t⁻¹	≤0.10			≤0.10
SO_2产生量/kg·t⁻¹	≤0.02			≤0.02
废水排放量/m³·t⁻¹	0	0	0	0
渣铁比/kg·t⁻¹	≤280	280	299.6	250~270
废物回收利用指标				
高炉仓下焦丁回收装置		均设有焦丁、矿丁回收装置		
高炉渣回收利用率/%	100	100	100	100
高炉煤气除尘粉尘回收利用率/%	100	100	100	100
单座高炉电机装机容量/kW			26205	26476

4.5 工程投资及总图占地比较

首钢京唐5576m³高炉和首钢迁钢4080m³高炉同属首钢集团，并且由同一家设计单位完成工程设计，在同一时期内建成投产，具有较强的可比性。针对首钢京唐建造2座5576m³高炉或建造3座4080m³高炉技术方案，进行了工程投资和总图占地的对比研究。结果表明，在同口径条件下建造2座5576m³高炉，同建造3座4080m³高炉相比，降低工程投资6.55亿元，降低工程投资12%~14%。同时由于装备大型化、流程紧凑集约，高炉区域占地面积减少15.81万平方米，同比降低20%~24%。节省占地不仅可以节约土地资源，还使物料运输和能源介质输送距离大幅度减低，运行成本降低、生产效率提高。两种技术方案的工程投资与占地比较见表7和表8。

表7 5576m³高炉和4080m³高炉工程投资

项目	迁钢3号高炉	首钢京唐1号高炉
高炉容积/m³	4080	5576
工程投资/亿元	17.9	22.0
单位高炉容积投资/万元·m⁻³	43.87	40.00
单座高炉工程占地面积/万平方米	26.500	31.925

表8 2×5576m³高炉和3×4080m³高炉总图占地与工程投资

项目	2×5576m³ 高炉	3×4080m³ 高炉
占地长度/m	1243	1506
占地宽度/m	695	626
占地面积/万平方米	63.69	79.50
吨铁占地面积/m²·t⁻¹	0.071	0.088
同口径工程投资/亿元	44.0	50.6

5 结论

（1）对于现代钢铁厂，高炉的功能不仅是通过还原反应过程而获得优质的炼钢铁水，而且伴随着大量的能量转换和信息的输入/输出。因此，应从更高、更宽的视野进行分析，也就是应该从铁素物质流、能量流、信息流等方面，综合评价钢铁厂的流程结构优化和与之相关的高炉大型化。

（2）近20年来，国内外高炉炼铁技术在大型、低耗、高效、长寿、环保等方面取得长足进步，高炉长寿、高风温、富氧喷煤、煤气干法除尘-TRT等单体技术成就突出。国内外高炉炼铁技术进步与高炉大型化发展密不可分，在钢铁厂整体流程结构优化下的高炉大型化是当今高炉炼铁技术的重要发展趋势。考虑到物质流、能量流和信息流网络结构的优化，一个钢铁厂配置2~3座高炉应是优化的选择，应该以钢铁厂整体流程结构优化为前提，考虑高炉大型化的合理配置，不应片面强调高炉越大越好，片面比大比小，甚至盲目攀比"最大"；同时必须重视高炉座数及其位置的合理化。

（3）研究分析和生产实践证实，$4000m^3$级高炉和$5000m^3$级高炉的原燃料条件差别不大。相比之下，$5000m^3$级高炉技术指标则更具有优势，特别是高炉燃料比和工序能耗等关键指标。

（4）对于生产规模为800万~900万吨/年的现代大型薄板厂，在钢铁厂流程结构优化的前提下，综合考虑铁素物质流、能量流、信息流的流程网络优化，配置2座$5000m^3$级高炉优于配置3座$4000m^3$级高炉，在节能减排、清洁环保、节省投资、经济效益等方面具有优势。

（5）通过技术方案对比研究，在同等生产规模的条件下，建造2座$5576m^3$高炉比建造3座$4080m^3$高炉，其技术装备数量和质量大幅度减少。设备数量减少15%左右，设备总质量减少14820t，设备总质量降低22.6%，设备数量的大幅度降低使设备投资、运行维护、备品备件消耗等均相应降低，而且动力消耗、岗位定员、污染物排放等都也相应降低，生产运行成本随之相应降低。

（6）在同等生产规模和产品的条件下，建造2座$5576m^3$高炉比建造3座$4080m^3$高炉，可以降低工程投资6.55亿元，即相当于降低工程投资12%~14%。同时由于装备大型化、流程紧凑集约，高炉占地面积减少15.81万平方米，同比降低20%~24%。与此同时，还使高炉运行的动力消耗、物料运输和能源介质输送距离大幅度减低，运行成本降低、生产效率提高。

（7）这种发展趋势对于生产建筑用棒材/线材产品的中、小钢铁厂也同样具有参考价值。

致谢

衷心感谢在本文成文过程中，北京首钢国际工程技术有限公司有关专家给予的大力支持，衷心感谢张春霞等专家的宝贵意见。

参考文献

[1] 殷瑞钰. 论钢厂制造过程中能量流行为和能量流网络的构建 [J]. 钢铁, 2010, 45 (4): 1.

[2] 张春霞,殷瑞钰,秦松,等.循环经济中的中国钢厂[J].钢铁,2011,46(7):1.

[3] 殷瑞钰.冶金流程工程学(第2版)[M].北京:冶金工业出版社,2009:276-277.

[4] 钱世崇,张福明,张建,等.首钢京唐5500m³特大型高炉高效长寿综合技术的设计研究[C]// 2010年全国炼铁学术年会论文集.北京:中国金属学会,2010.

[5] Miwa T. Development of Iron-Making Technologies in Japan [J]. Journal of Iron and Steel Research International, 2009, 16 (S2): 14-19.

[6] Ariyamal T, Uedal S. Current Technology and Future Aspect on CO_2 Mitigation in Japanese Steel Industry [J]. Journal of Iron and Steel Research International, 2009, 16 (S2): 55-62.

[7] Michael P, Bodo L H. Iron Making in Western Europe [J]. Journal of Iron and Steel Research International, 2009, 16 (S2): 20-26.

[8] 唐文权.高炉利用系数探讨[J].冶金管理,2005(8):52.

[9] 钱世崇,张福明,李欣,等.大型高炉热风炉技术的比较分析[J].钢铁,2011,46(10):1.

[10] 中国冶金建设协会.高炉炼铁工艺设计规范(GB50427-2008)[S].北京:中国计划出版社,2008.

Blast Furnace Enlargement and Optimization of Manufacture Process Structure of Steel Plant

Zhang Fuming Qian Shichong Yin Ruiyu

Abstract: The blast furnace is the most intensive process unit for energy and matter in steel plant. It has a significant influence for material flow network construction, energy flow network construction and operation on the rationalization in steel plant. For a modern steel plant, the function of the blast furnace not only produces the high quality hot metal by reduction reaction, but also accompanies energy conversion and a lot of input and output of information. So the reasonable production capacity, reasonable volume and reasonable layout of blast furnace should be evaluated comprehensively under the optimization of manufacture process structure in steel plant. Based on the analysis of international development tendency of the blast furnace, compares the main technical economic specifications of Shougang Jingtang's 5576m³ BF to Qiangang's 4080m³ BF, it can gain similar hot metal output both 2×5576m³ BF and 3×4080m³ BF, but 2×5576m³ BF has obvious advantages in investment reducing, energy saving and information control. Therefore, in order to optimize the steel manufacture process, promote the competitiveness in market, enlargement of blast furnace is a significant tendency. However, that is not better to accelerate enlargement of single blast furnace, moreover to achieve "the biggest". Enlargement of blast furnace should be based on optimization of products structure, material flow structure, energy flow structure and the dynamic operation. The general trend is appropriate that furnish 2 to 3 blast furnaces for a high efficiency and low cost steel plant, and thus according to the top designing, the reasonable volume, rational sets, as well as reasonable layout of blast furnaces will be achieved. This tendency not only suitable for large-scale enterprise for flat products production, but also suitable for medium and small-scale steel plant for bar or wire rods production.

Keywords: steel plants; manufacture process; structural optimization; enlargement of blast furnaces

21 世纪初巨型高炉的技术特征

摘　要：对近10年国内外建成投产的巨型高炉技术装备进行了阐述，研究分析了当代巨型高炉的技术特征。认为当代巨型高炉在工艺技术装备方面取得了长足进步，在高炉大型化、精料、无料钟炉顶设备与炉料分布控制、高炉长寿、高风温、富氧喷煤、煤气干法除尘等方面成就突出。

关键词：巨型高炉；无料钟炉顶；工艺技术

1　引言

进入21世纪以来，国内外有10余座新建或大修改造的5000m^3以上巨型高炉相继建成投产。这批巨型高炉，主要集中在中国、日本和韩国亚洲国家。巨型高炉的设计建造，是当代高炉炼铁技术进步的综合成就，体现了当代高炉在工艺技术、装备制造和工程设计、施工建造等领域的总体水平。

2　巨型高炉发展现状

21世纪初，日本新日铁君津4号高炉进行了扩容大修改造，高炉容积由原来的5151m^3扩大到5555m^3，于2003年5月建成投产；随后，大分2号高炉容积由原来的5245m^3扩大到5775m^3，于2004年5月建成投产，成为当时世界上容积最大的高炉；2009年8月，大分1号高炉容积也由原来的4884m^3扩大到5775m^3，形成了一个钢铁厂同时拥有2座巨型高炉的生产格局；2007年4月，新日铁公司名古屋1号容积由大修前的4650m^3扩大到5443m^3。住友鹿岛厂于2004年9月移地新建的新1号高炉建成投产，容积为5370m^3，这座高炉是日本近10年来唯一新建的巨型高炉；JFE仓敷4号高炉大修改造，容积由4826m^3扩容到5005m^3，于2002年1月投产；京滨2号高炉容积由4052m^3扩大到5000m^3，于2004年3月建成投产；2005年4月，福山5号高炉扩容大修改造，高炉容积由大修前的4664m^3扩大到5500m^3；2006年5月，福山4号高炉扩容大修改造，高炉容积由大修前的4288m^3扩大到5000m^3。2007年5月，神户加古川2号高炉容积也由3850m^3扩容到5400m^3。这10余座巨型高炉，分别归属于日本的4家钢铁公司。由此可见，21世纪初日本在高炉数量不增加的情况下，通过高炉扩容大修改造，提高高炉技术装备水平，实现高炉巨型化和工艺技术现代化。

2005~2010年，首钢京唐钢铁厂工程作为国家"十一五"重大工程，按照新一代可循环钢铁厂工艺流程和装备大型化的总体要求，设计建造了2座5500m^3高炉，这是我国

本文作者：张福明。原刊于《炼铁》，2012，31（2）：1-8。

第一次建造容积5000m³以上的巨型高炉，2座高炉分别于2009年5月和2010年6月建成投产[1]。我国沙钢5800m³高炉于2009年10月建成投产，是迄今为止世界上容积最大的在役高炉[2]。

与此同时，韩国浦项公司光阳4号高炉扩容大修改造，容积由大修前的3800m³扩大到5500m³，于2009年7月建成投产；浦项4号高炉扩容大修改造，容积也由大修前的3800m³扩大到5600m³，于2010年10月建成投产。现代唐津2座新建的5270m³高炉相继于2010年7月和2010年11月建成投产；浦项目前正在设计建造6000m³的巨型高炉也将于近期建成投产。

近10年来国内外建成投产的5000m³以上巨型高炉见表1。

表1　21世纪国内外新建或大修改造的巨型高炉

高炉	高炉容积/m³		投产时间	炉顶装料设备	炉体冷却	热风炉
	大修前	大修后				
中国首钢京唐1号	—	5500	2009-05-21	无料钟	冷却壁	4座BSK顶燃式
中国首钢京唐2号	—	5500	2010-06-26	无料钟	冷却壁	4座BSK顶燃式
中国沙钢	—	5800	2009-10-21	无料钟	冷却壁	3座DME外燃式
韩国现代唐津1号	—	5250	2010-07-21	无料钟	板壁结合	3座DME外燃式
韩国现代唐津2号	—	5250	2010-11-23	无料钟	板壁结合	3座DME外燃式
日本住友鹿岛新1号	3680	5370	2004-09-29	无料钟	冷却壁	3座KOPPERS外燃式
韩国POSCO光阳4号	3800	5500	2009-07-21	无料钟	冷却壁	4座NSC外燃式
韩国POSCO浦项4号	3800	5600	2010-10-08	无料钟	冷却壁	4座NSC外燃式
日本新日铁君津4号	5151	5555	2003-05-08	无料钟	冷却壁	4座NSC外燃式
日本新日铁大分2号	5245	5775	2004-05-15	料钟	冷却壁	4座NSC外燃式
日本新日铁大分1号	4884	5775	2009-08-02		冷却壁	4座NSC外燃式
日本新日铁名古屋1号	4650	5443	2007-04			
日本住友鹿岛3号	5050	5370	2007-05	无料钟	冷却壁	3座KOPPERS外燃式
日本神户加古川2号	3850	5400	2007-05-24			
日本JFE福山4号	4288	5000	2006-05-05	无料钟		
日本JFE福山5号	4664	5500	2005-04-03	无料钟		
日本JFE仓敷4号	4826	5005	2002-01-08	无料钟	板壁结合	
日本JFE京滨2号	4052	5000	2004-03-24	无料钟	冷却壁	

3　先进的设计理念与技术指标

3.1　先进的设计理念

这批巨型高炉的工程设计，总结了5000m³级高炉问世以来的技术发展和生产实践，在已有技术的基础上,不单纯追求工艺技术装备大型化,在技术创新和工艺现代化等方面都具有显著的时代技术特征。当代巨型高炉的功能,也由单一的生铁制造功能演变为优质生铁制造、高效能源转换和消纳废弃物实现资源化的三大功能。高炉合理容积和生产能力的选择确定,不再是盲目追求大型化,而是在整个钢铁联合企业的尺度空间内,结合资源条件和生产规模等综合要素,以建构钢铁企业的物质流、能源流和信息流高效协同运行为优化目标,实现整个钢铁企业功能、结构、效率协同优化的大型化。首钢京唐钢铁厂、韩国现代制铁唐津厂都是本世纪新建的沿海型钢铁厂,均设计建造了2座容积相同的巨型高炉,形成了千万吨级的现代钢铁联合企业。日本新日铁公司大分厂利用高炉大修改造的机遇,将2座高炉容积均扩容到5775m^3,JFE公司福山厂也将2座4000m^3级的高炉扩容为5000m^3级的巨型高炉。

新世纪巨型高炉的工艺技术装备更加注重技术创新,高炉炼铁关键共性技术已形成重大突破,集成创新了许多新世纪的炼铁先进技术,在高炉高效长寿、优质低耗、节能减排、循环经济、低碳清洁等诸多方面取得了长足进步。

3.2 先进的技术指标

在21世纪初巨型高炉的设计中,结合原燃料条件、技术装备、操作技术以及20世纪同级别高炉的生产实践,确定了先进合理的技术经济指标。国内外部分巨型高炉的主要技术经济指标见表2。

表2 国内外部分巨型高炉的主要技术经济指标

高 炉	中国首钢京唐1号	中国沙钢5800m^3	日本新日铁大分2号	日本新日铁君津4号	日本住友鹿岛1号	日本JFE福山5号	韩国现代唐津1号	韩国浦项光阳4号
高炉有效容积/m^3	5500	5800	5775	5555	5370	5550	5250	5500
炉缸直径/m	15.5	15.3	15.5	15.2	15	15.6	14.85	15.6
日产铁量/$t \cdot d^{-1}$	12650	12876	13500	12900	11425	12650	11600	13750
利用系数/$t \cdot (m^3 \cdot d)^{-1}$	2.3	2.22	2.34	2.32	2.13	2.28	2.2	2.5
熟料率/%	90	85	85	90	76		95	85
入炉矿品位/%	61	59	58	58			60	60
燃料比/$kg \cdot t^{-1}$	490	490	475	482		490	490	490
焦比/$kg \cdot t^{-1}$	290	290	355	332		355	310	290
煤比/$kg \cdot t^{-1}$	180	200	120	150		120	180	200
风温/℃	1300	1250	1200	1200	1300	1250	1230	1250
富氧率/%	5.5	4	3.5		3		5.6	10
入炉风量/$Nm^3 \cdot min^{-1}$	9000	8300	8550		7800	8660	7250	7000
热风压力/MPa	0.55	0.52	0.45		0.49	0.42	0.423	0.43
炉顶压力/MPa	0.28	0.25		0.29	0.294	0.275	0.235	0.275
渣量/$kg \cdot t^{-1}$	250	300	290	300	297		290	290
年产铁量/万吨	450	450	480	424	405.6	449	400	490
一代炉龄/a	≥25	20	23.5	25	≥25	≥20	≥20	≥20
单位炉容产铁量/$t \cdot m^{-3}$	≥20000	15000	16500	16500	≥18600	≥16180	≥15700	≥20000

4 精料技术

精料是当代巨型高炉实现高效、低耗、优质、长寿的基础，是高炉炼铁工艺中最主要的支撑技术之一，也是实现高炉生产"减量化"的重要措施。提高精料水平、优化高炉炉料结构是当代巨型高炉必须具备的基础条件。

4.1 采用合理炉料结构

多年以来，亚洲国家高炉炉料结构形成了以高碱度烧结矿为主，适量配加酸性球团和少量块矿的模式。新世纪建造的10多座巨型高炉基本遵循了原有的炉料结构模式，在注重改善入炉矿石冶金性能的同时，提高综合入炉矿品位和熟料率，结合矿石资源条件和造块生产工艺，以实现资源减量化和最佳化利用为目标，确定了经济合理的炉料结构。在优质矿石价格日益攀升的条件下，巨型高炉炉料结构的一个变化趋势是适度增加块矿使用比例。日本和韩国的钢铁企业开始调整高炉炉料结构，在烧结矿比例为70%~80%的条件下，适当增加了块矿的使用比例，同时开发并采用低品质矿石利用技术，以降低原料成本[3]。新世纪部分巨型高炉的炉料结构见表3。

表3 国内外部分巨型高炉的炉料结构　　　　　　　（%）

高　炉	烧结矿	球团矿	块矿
中国首钢京唐1号	70	20	10
中国沙钢5800m³	77	18	5
日本新日铁大分2号	81	4	15
日本新日铁君津4号	80	10	10
日本JFE福山5号	66~71	5~10	24
韩国现代唐津1号	80	5	15
韩国浦项光阳4号	80	12	8

4.2 提高炉料冶金性能

巨型高炉对炉料冶金性能的要求更为严格，巨型高炉精料技术在原燃料资源日益稀缺的条件下，更加注重提高炉料冶金性能的稳定性。巨型高炉精料技术的主要特点是：(1) 入炉矿综合品位达到58%以上，熟料率达到80%~85%。面对当今国际优质铁矿石资源日益匮乏、矿石价格不断攀升的市场环境，巨型高炉并不追求过高的入炉矿品位和熟料率，而更加注重低品质矿石资源的优化利用。日本新日铁、韩国浦项等企业都在开发低品质矿石的综合利用技术，旨在降低高炉原料成本，进一步提高巨型高炉的竞争力。(2) 提高焦炭机械强度（$M_{40} \geq 89\%$，$M_{10} \leq 6.0\%$），特别是焦炭的热态特性指标（$CSR \geq 68\%$，$CRI \leq 25\%$），是保证巨型高炉生产稳定顺行的关键条件。(3) 稳定入炉原燃料成分，减少炉料成分波动，提高炉料成分和理化性能的稳定性，适度提高熟料率和矿石入炉品位，改善炉料冶金性能，采取炉料整粒措施，均取得了显著成效。特别是高炉巨型化以后，对焦炭质量提出更高的要求，焦炭的高温冶金性能、反应性和反应后强度成为衡量焦炭质量的重要指标,提高入炉焦炭平均粒度也成为巨型高炉改善料柱透气性、提高喷煤量的重要技术措施。国内外部分巨型高炉的原燃料技术性能见表4。

表4 国内外部分巨型高炉原燃料技术指标

项 目	中国首钢京唐1号	中国沙钢5800m³	日本新日铁君津4号	韩国现代唐津1号	韩国浦项光阳4号
烧结矿品位/%	58.7	56.94	58	58	57
烧结矿转鼓强度/%	78	79.34	75.3	92.3	93.5
烧结矿碱度 R	1.8	2.11		1.77	1.73
焦炭灰分/%	11.5	12.30	11.2	11.8	11.02
焦炭强度 M_{40}/%	89	89.02	86	89	
焦炭强度 M_{10}/%	6.0	5.83			
焦炭反应后强度 CSR/%	68	68.31		68	68.32
焦炭反应 CRI/%	23	23.43		22~23	
焦炭平均粒度/mm	45~50	51.28		47~50	

4.3 采用炉料分级入炉技术

采用炉料分级入炉技术是现代巨型高炉的重要技术特征。烧结矿和焦炭按照设定的粒度等级分级装入高炉,能实现精准控制炉料分布,从而改善高炉透气性,提高煤气利用率,促进高炉顺行,降低燃料消耗。回收3~5mm的小粒度烧结矿,不仅可以提高资源利用率,还可以用来抑制高炉边缘煤气流过分发展,降低炉墙热负荷,有利于延长高炉寿命。回收10~25mm的焦丁与矿石混装入炉可以改善高炉透气性,降低燃料消耗,稳定高炉操作。首钢京唐、韩国光阳、韩国唐津、日本鹿岛等厂的巨型高炉都采用了炉料分级入炉技术。

首钢京唐2座5500m³高炉采用烧结矿和焦炭分级入炉技术。烧结矿和焦炭按照不同粒度等级分级入炉,回收10~25mm的焦丁和3~8mm的小粒度烧结矿。烧结矿的粒度级别分为两级,大粒度烧结矿粒度为20~50mm,中粒度为8~20mm。回收3~8mm的小粒度烧结矿,可提高原料利用率。焦炭的粒度级别分为两级,即25~60mm和60mm以上。回收10~25mm的焦丁与矿石混装入炉。

5 无料钟炉顶设备与炉料分布控制技术

炉顶装料设备不仅要满足高炉的装料能力,还要满足高炉操作对炉料分布精准控制的要求,如炉料分级入炉以及中心加焦等,为高炉稳定顺行、高效低耗和长寿提供技术保障。21世纪初建成投产的巨型高炉除日本大分2号高炉继续沿用了钟式炉顶设备以外,其余高炉均采用了无料钟炉顶设备。其中,日本君津4号高炉采用串罐式无料钟炉顶,料罐容积为100m³;鹿岛新1号高炉采用在并罐式无料钟炉顶基础上改进的IHI-PW型3罐式无料钟炉顶[4],2个大料罐容积为90m³,小料罐容积为10m³,小料罐主要用于高炉中心加焦和小粒度烧结矿的单独装料。首钢京唐、沙钢、韩国光阳、韩国唐津等厂的巨型高炉均采用并罐式无料钟炉顶设备。

巨型高炉生产效率高，单座高炉的日产量达12000t/d以上，炉料装入量可达25000t/d，无料钟炉顶设备的装料能力要满足高炉高效化生产的要求。巨型高炉炉喉直径达到11m以上，要实现高炉圆周方向和半径方向的炉料均匀分布，抑制炉料形状和粒度偏析，从而获得合理的炉料分布和煤气分布，这对于无料钟炉顶设备提出了更高的技术要求。实践表明，采用炉料分级入炉技术可以减小炉料粒度偏析，改善炉料和煤气分布，使无料钟炉顶的功能得到充分发挥。由于较小粒度的烧结矿具有较大的比表面积和较好的还原性，因此将其分布在靠近炉墙的区域，可以使煤气化学能和热能得到充分利用，抑制边缘煤气流过分发展，保护炉衬和冷却器。采用烧结矿分级入炉技术，在同一装料周期内烧结矿粒度分布偏差缩小，高炉径向煤气流稳定，烧结矿层空隙率提高，块状带透气性得到改善。

计算研究表明，串罐式无钟炉顶在正常上料周期内无法实现多种炉料分级入炉的，一旦出现故障则需要停炉检修，影响高炉正常生产操作。而并罐式无料钟炉顶设备，能在一个装料周期内将5种不同粒度、不同种类的炉料装入高炉，可以实现炉料分级入炉以及中心加焦。

日本为改善高炉操作，提高炉料分布控制的精度，开发了离散元法（DME）炉料分布控制模型，用于控制布料过程的炉料行为。韩国浦项和光阳的巨型高炉重视炉料分布控制研究，在高炉煤比达到200kg/t、矿焦比达到5.5~6.0的条件下，采用烧结矿分级布料和中心加焦技术，优化布料矩阵，控制煤气流合理分布，取得突出的成效。

6 高炉长寿技术

巨型高炉设计寿命均为20~25年，一代炉龄单位容积产铁量将达到15000~20000t/m^3。在总结20世纪高炉长寿技术实践的基础上，当代巨型高炉设计集成了国际先进的高炉高效长寿综合技术，高炉内型合理，采用无过热冷却体系及纯水（软水）密闭循环冷却技术，优化炉缸炉底内衬结构，设置完善的高炉自动化监测及控制系统，以实现高炉生产的稳定顺行、高效低耗及长寿。

6.1 优化高炉炉型

根据高炉冶炼过程传热、传质、动量传输和物理化学反应原理，结合原燃料条件、操作条件和炉体结构，以提高炉料透气性、改善煤气能量利用、促进高炉稳定顺行为目的，在炉型设计方面进行了优化。其主要技术特征是：（1）随着高炉容积的扩大，有效高度变化趋缓，高径比进一步降低。高炉有效高度维持在30~33m，高径比均降低到2.0以下，日本住友鹿岛新1号高炉高径比为1.78，成为目前世界上最矮胖的巨型高炉。（2）为抑制炉缸铁水环流和炉缸炉底"象脚状"异常侵蚀，使高炉寿命达到20年以上，适度增加了死铁层深度。死铁层深度大多达到3m以上，约为炉缸直径的20%~25%。（3）炉缸直径和炉缸高度适度加大。大部分巨型高炉的炉缸直径达到15m以上，炉缸高度达到5m以上，高炉炉缸容积约为高炉有效容积的15%~20%。（4）炉腰直径加大，炉身角和炉腹角缩小。巨型高炉采用强化冷却的薄壁内衬技术，注重设计炉型与操作炉型的趋同，改变了传统的厚壁内衬高炉炉型设计理念，不再依靠侵蚀内衬而形成合理的高炉操作炉型。炉型设计以促进高炉稳定顺行、提高料柱透气性、改善煤气分布、提高煤气利用率、抑制边缘煤气流发展为要素。在高炉有效高度变化不大的情况下，通过扩大炉腰直径，降低高径比。巨型高炉炉腰直径均达到17m以上，炉腹角减小到74°~79°之间。降低炉腹角有利于炉腹煤气的顺畅排升，而且有利于在炉腹区域形成稳定的保护性渣皮，保护炉腹区域冷却器、延长炉腹区域的寿命。巨型高炉炉身角降低到80°~81°之间，在炉腹至炉身下部区域采用高效铜冷却壁的条件下，保护性渣皮可以比较稳固地在铜冷却壁热面形成。即使渣皮脱落也能快速形成新的渣皮，形成自保护的"永久内衬"，从而保护铜冷却壁长期稳定工作。因此，在采用高效铜冷却壁的条件下，适当降低炉身角，有利于炉料的顺利下降和高炉顺行。国内外部分巨型高炉的炉型参数见表5。

表5 国内外部分巨型高炉炉型参数

项 目	中国首钢京唐1号	中国沙钢5800m³	日本新日铁大分2号	日本住友鹿岛1号	日本新日铁君津4号	韩国浦项光阳4号	韩国现代唐津1号
有效容积 V_u/m³	5500	5800	5775	5370	5555	5500	5250
炉缸直径 d/mm	15500	15300	15600	15000	15200	15600	14850
炉腰直径 D/mm	17000	17500	17200	17300	17300	17200	17000
炉喉直径 d_1/mm	11200	11000/11500	11100	11200	10900	11100	11100
死铁层深度 h_0/mm	3100	3200		4500		2953	3700
炉缸高度 h_1/mm	5400	6000	5350	5156	5252	5400	4900
炉腹高度 h_2/mm	4000	4000	4000	4544	4000	4000	4800
炉腰高度 h_3/mm	2500	2400	2500	1800	3000	2600	2500
炉身高度 h_4/mm	18400	18600	18400	17300	18550	17500	17700
炉喉高度 h_5/mm	2500	2200	2625	2000	2200	2000	2500
有效高度 H_u/mm	32800	33200	32875	30800	33002	31500	32400
炉腹角 α	79°22′49″	74°37′26″	78°21′24″	75°47′52″	75°17′30″	78°41′24″	77°22′35″
炉身角 β	81°2′36″	80°50′16″	80°35′17″	80°0′5″	80°12′45″	80°6′48″	80°32′16″
高径比 H_u/D	1.93	1.897	1.91	1.78	1.91	1.83	1.91
风口数/个	42	40	42	40	42	42	42
铁口数/个	4	3	5	4	4	4	4

6.2 采用炉体长寿结构

为了使一代炉役高炉寿命达到20~25年，巨型高炉采取了合理的炉体长寿结构和先进的冷却技术。纯水（软水）密闭循环冷却系统在巨型高炉上得到普遍应用，而且在传统的高炉冷却技术基础上又有所创新。高炉炉底炉缸、炉体冷却器、风口、热风阀等均采用纯水（软水）冷却，工业水仅作为备用或作为冷媒水。值得关注的是，日本大修改造的巨型高炉将原来炉缸喷水冷却结构大部分已改为冷却壁冷却结构，全部采用纯水密闭循环冷却系统。

21世纪初，巨型高炉炉体结构形成了以冷却壁为主的模式。日本君津、大分、鹿岛等厂的巨型高炉，我国首钢京唐、沙钢的巨型高炉，以及韩国现代、浦项的巨型高炉都采用了全冷却壁的薄壁炉体结构。其中，最显著的技术特征是在炉腹至炉身中下部区域大量采用铜冷却壁，依靠高效铜冷却壁及合理配置的冷却系统，形成基于渣皮保护的"永久内衬"以延长高炉寿命。首钢京唐2座5500m³高炉炉体采用全冷却壁结构，炉体共采用18段冷却壁，炉缸采用6段光面冷却壁。其中，第2、3段为铜冷却壁，炉腹、炉腰和炉身下部采用4段高效铜冷却壁，铜冷却壁总高度为10.4m；炉身中上部采用7段镶砖铸铁冷却壁；炉喉钢砖下部设1段"C"形水冷壁。炉腹至炉身采用冷却壁与砖衬一体化的薄壁结构，在铜冷却壁和铸铁冷却壁镶砖热面直接喷涂不定型耐火材料。沙钢5800m³高炉在炉腹、炉腰和炉身下部采用3段钻孔铜冷却壁，总高度为12.7m。炉身下部3段铜冷却壁高度为6.9m，占炉身高度的37.1%。日本君津4号高炉炉缸铁口下部、炉腹至炉身中部共采用11段556块铸铜冷却壁；大分2号高炉在炉腹至炉身中部采用铸铜冷却壁，铸铜冷却壁厚度为100mm，高炉扩容实际上是通过采取减薄冷却壁和砖衬厚度实现的。日本鹿岛1号高炉在炉腹至炉身下部采用4段铜冷却壁，总高度为11.6m，铜冷却壁厚度为145mm。韩国唐津1、2号高炉炉体采用铜冷却壁、铸铁冷却壁和铜冷却板结合的炉体结构，炉缸采用5段铸铁冷却壁；风口区以上至炉腹下部采用铜冷却板；炉腹上部、炉腰和炉身下部采用7段铜冷却壁；炉身中上部至炉喉采用6段镶砖铸铁冷却壁。

值得关注的是日本大型高炉扩容大修改造中，为缩短高炉大修工期，开发了高炉炉体大型模块法的快速大修技术。其特点是炉体采用薄壁结构，炉体模块是炉壳、冷却壁和耐火材料的复合结构，在高炉大修施工时，将多组炉体模块整体安装焊接为一体，大幅度缩短了高炉大修工期。近年来，这项技术在日本新日铁、JFE的高炉大修中得到普遍应用，日本福山5号高炉扩容大修工期仅为58天，仓敷4号高炉大修工期为70天[5]。

6.3 采用合理炉缸炉底内衬结构

21世纪初，巨型高炉的炉缸炉底内衬结构充分吸收了20世纪高炉长寿的经验，通过确定合理的死铁层深度，采用合理的炉缸炉底内衬结构，应用抗铁水渗透性能优异的高导热炭砖，配置合理的冷却系统，抑制炉缸炉底象脚状异常侵蚀，使一代炉役高炉寿命达到20~25年。日本新日铁公司通过对高炉炉缸炉底内衬侵蚀机理的研究，总结已有高炉长寿的经验，君津4号、大分2号等高炉炉缸炉底采用了导热性和抗铁水渗透性优异的含TiC的超微孔大块炭砖，炉底满铺大块炭砖采用立砌结构，其上设置2层陶瓷质的综合炉底；为强化炉缸侧壁冷却，抑制炭砖异常侵蚀，铁口区以下炉缸第3、4段冷却壁采用铜冷却壁。

日本鹿岛1号高炉炉缸炉底采用炭砖-陶瓷杯结构，炉底总厚度为3165mm。炉底采用3层高导热大块炭砖，其上为3层陶瓷垫，炉底满铺炭砖至铁口组合砖以下，炉缸侧壁采用高导热超微孔大块炭砖和陶瓷杯结构，以抑制炉缸炉底交界处的异常侵蚀；炉缸侧壁铁口至风口区域采用铸铁冷却壁，铁口区以下的炉缸侧壁采用喷水冷却。

首钢京唐1、2号高炉炉缸炉底内衬采用炭砖-陶瓷垫组合结构，炉底总厚度为3400mm。炉底满铺炭砖采用1层高导热石墨砖、2层微孔炭砖和1层超微孔炭砖，炉底满铺炭砖之上为2层陶瓷垫，炉缸侧壁和炉缸炉底交界处采用热压小块炭砖，靠近冷却壁采用导热性优异的NMD砖，炉缸侧壁内侧采用抗铁水渗透性能优异的NMA砖，风口、铁口采用组合砖结构。为实现高炉寿命25年以上，在炉缸炉底象脚状异常侵蚀区对应的炉缸第2、3段冷却壁和铁口区采用铜冷却壁。

沙钢5800m³高炉采用炭砖—陶瓷杯复合内衬结构。炉底总厚度为3400mm，炉底最下层平砌300mm厚的国产高导热石墨砖，第2层为厚度500mm的国产高导热超微孔炭砖，其上为2层厚度均为700mm的德国SGL超微孔炭砖9RD-N，炭砖上为厚度1200mm的法国SAVOIE陶瓷垫RL70MLC，炉缸侧壁由SGL超微孔炭砖9RD-N和SAVOIE陶瓷壁CORANIT AL构成。

7 高风温技术

高风温是高炉降低焦比、提高喷煤量、提高产量的重要技术途径，成为高炉炼铁发展史上极其重要的技术进步。高风温是一项综合技术，要在高炉的尺度空间内进行研究，进一步提高风温是21世纪高炉炼铁技术的热点研究课题。巨型高炉在提高风温方面进行了诸多技术创新与实践，取得了显著的技术成效。巨型高炉的设计风温一般为1250~1300℃，提高风温是新世纪巨型高炉的一个显著技术特征。

21世纪以前，5000m³以上的巨型高炉全部采用外燃式热风炉，最具代表性的外燃式热风炉是新日铁式和地得式。21世纪新建设的巨型高炉仍是以外燃式热风炉为主，韩国唐津厂2座5250m³高炉和沙钢5800m³高炉均采用了改进的地得式（DME）外燃热风炉，每座高炉配置3座DME外燃式热风炉；日本新日铁公司巨型高炉则仍采用新日铁式热风炉，君津、大分等厂的巨型高炉均配置4座新日铁式外燃热风炉；住友公司鹿岛厂5370m³高炉采用了3座KOPPERS外燃式热风炉。首钢京唐5500m³高炉开发并应用了BSK顶燃式热风炉，这是世界上首次将顶燃式热风炉应用在5000m³以上的巨型高炉，实现了巨型高炉热风炉结构形式的技术突破。首钢京唐5500m³高炉配置4座BSK顶燃式热风炉，回收热风炉烟气余热预热助燃空气和煤气。同时，设置2座小型顶燃式热风炉预热助燃空气，将助燃空气预热到520~650℃，煤气预热到200~220℃，在此条件下热风炉拱顶温度可以达到1420℃，送风温度达到1300℃。

巨型高炉在热风炉结构形式呈现多样化发展的同时，其主要技术特征体现在：（1）为使风温达到1250~1300℃，利用热风炉烟气余热预热煤气和助燃空气，将煤气和助燃空气温度提高到180~200℃，从而提高热风炉理论燃烧温度100~150℃，在采用焦炉煤气或转炉煤气富化的条件下，可以使热风炉拱顶温度达到1420℃以上，可以获得1250℃以上的高风温。（2）燃烧单一高炉煤气使热风炉拱顶温度达到1420℃。随着高炉操作水平的提高，煤气利用率已提高到50%以上，燃料比降低到450~480kg/t，使高炉煤气热值降低到3000kJ/m³以下。以首钢京唐5500m³高炉为代表，在利用热风炉烟气余热预热煤气和助燃空气的同时，开发应用了两座小型热风炉对助燃空气进行高温预热工艺，使BSK顶燃式热风炉拱顶温度达到1400~1420℃，高炉生产实践中月平均送风温度稳定达到了1300℃。（3）通过优化热风炉燃烧过程及气流运动的研究，提高气流分布的均匀性，采用高效格子砖，提高热风炉换热能力，缩小拱顶温度与风温的差值。通过开发应用小孔径的高效格子砖，增加格子砖的加热面积，加强格子砖与气流之间的热交换。（4）优化热风管道系统结构，采用无过热、低应力设计体系，合理设置管道波纹补偿器和拉杆，妥善处理管道膨胀以降低管系应力，热风管道采用组合砖结构，消除热风管道的局部过热和窜风。(5) 采取有效的炉壳晶间应力预防措施，延长热风炉使用寿命。通过合理控制热风炉拱顶温度，抑制热风炉燃烧过程 NO_x 的大量生成；采取高温区炉壳涂刷防酸涂料和喷涂耐酸喷涂料等防护措施，加强对热风炉炉壳的隔离保护；采用细晶粒炉壳钢板，采用热处理措施消除或降低炉壳制造过程的焊接应力；对于热风炉高温区炉壳采取保温等综合防护措施，预防热风炉炉壳晶间应力腐蚀的发生，以延长热风炉寿命。

8 煤气干法除尘技术

传统的高炉煤气净化普遍采用湿法除尘工艺，高炉煤气经过重力除尘器进行重力除尘以后，再经过一级文氏管、二级文氏管或比肖夫环缝除尘器进行湿法除尘，净化后的煤气再经过 TRT 或调压阀组后回收使用。20 世纪 80 年代，以日本为代表开发成功了高炉煤气干法布袋除尘工艺，采用布袋除尘器对煤气进行净化，利用净化后的煤气对布袋进行反吹清灰。该工艺在日本君津、大分等厂的巨型高炉上应用，取得了较好的应用效果。但仍需设置湿法除尘系统作为干法除尘系统的备用设施，在高炉休风、复风或干法除尘系统故障状态下，仍要采用备用的湿法除尘系统。煤气干法除尘作业率一般可以达到 95%~97%。日本高炉煤气采用干法除尘以后，煤气含尘量可以降低到 $5mg/m^3$ 以下，煤气温度达到 100~140℃，与煤气湿法除尘工艺相比，TRT 发电量可以提高 30% 以上。

首钢京唐 2 座 5500m^3 高炉均采用了煤气全干法布袋除尘工艺，完全取消了备用的湿法除尘系统。采用自主开发研制的低压脉冲清灰布袋除尘技术，利用低压氮气作为布袋清灰介质，采用低压脉冲反吹清灰工艺，可以实现布袋除尘系统的在线清灰，提高了布袋除尘系统的运行效率[6]。首钢京唐 1 号高炉投产 2 年以来，煤气干法除尘系统运行安全稳定，煤气含尘量年平均为 3.66mg/m^3，煤气温度为 145℃，TRT 发电量达到 52.1kW·h/t，比湿法除尘系统 TRT 发电量提高约 45%，布袋使用寿命达到 18 个月[7]。

9 富氧喷煤技术

高炉喷煤是 20 世纪 60 年代以后，高炉炼铁技术的一项重大创新和技术发展。随着稀缺的炼焦煤资源日益匮乏，高炉喷吹非炼焦煤以替代昂贵的焦炭，使当代高炉在减少资源和能源消耗、合理利用资源和能源、降低生产成本和减少环境污染等方面取得重大成效，高炉大量喷煤已成为当代高炉最重要的技术特征之一。当前高炉喷煤量已经达到 200~250kg/t，煤粉成为当代高炉生产工艺中重要的燃料和还原剂。高炉喷煤技术的成功应用，使高炉炼铁工艺又焕发出勃勃生机，这也是目前高炉炼铁工艺仍占主导地位的一个重要因素。高炉喷煤的极限始终是炼铁工作者研究重点的课题，当代大型高炉的喷煤量要达到 180~200kg/t 以上，在大量喷煤操作的条件下，高炉燃料比降低到 450~480kg/t，并且要实现低燃料比条件下喷吹率达到 50% 以上。新世纪巨型高炉喷煤系统一般采用并罐式—总管—分配器的直接喷煤工艺，首钢京唐 2 座高炉采用了浓相喷煤技术，采用氮气作为输送载气，固气比达到 40kg/kg，同时采用了氧煤喷枪，高炉富氧一部分通过鼓风兑入，另外一部分通过氧煤喷枪进入高炉，以改善风口回旋区供氧强度和传质过程，提高煤粉燃烧率。

富氧鼓风是当代高炉强化冶炼的有效技术措施之一,可以充分发挥高风温、喷煤降焦的综合作用。高炉鼓风富氧不仅可以提高产量,还可以提高理论燃烧温度,提高喷煤量和煤粉燃烧率,有效降低炉腹煤气量,改善煤气能量利用和高炉透气性,而且对于提高炉顶煤气热值也会带来益处。理论研究和生产实践表明,富氧对提高煤粉燃烧率和喷煤量的作用十分显著,成为高炉大喷煤操作的重要支撑技术。

新世纪巨型高炉全部采用了高炉富氧喷煤综合技术,设计了完善的制粉喷煤设施,喷煤制粉系统设计能力一般为 200~250kg/t,在生产实践中,高炉煤比可以达到 180kg/t 以上。另一个值得关注的现象是韩国浦项公司浦项厂和光阳厂的巨型高炉,鼓风富氧率达到了 10%,高炉入炉风量一般为 7000~7300m^3/min,在高富氧操作的条件下,高炉煤比达到 200kg/t,入炉焦比 286kg/t,燃料比 487kg/t,高炉炉况稳定顺行,光阳 4 号高炉 2011年 2 月平均利用系数达到 2.79t/(m^3·d),高炉平均日产量达到 15326t/d,最高日产量达到 15613t/d,创造了当今巨型高炉单炉产量的纪录。我国沙钢 5800m^3 高炉投产以后,不断提高鼓风富氧率,2011 年 4 月,鼓风富氧率已达到 9.83%。在保持较高的风温的条件下,通过提高喷煤量和鼓风湿度,维持合理的风口前理论燃烧温度,取得了显著的技术效果[8]。

10 结语

21 世纪初,以日本、中国和韩国为代表的亚洲国家,相继建成投产了一批 5000m^3 以上的巨型高炉,在高炉大型化的同时,采用了高炉精料与炉料分布控制、高炉长寿、高风温、煤气干法除尘、富氧喷煤等一系列具有 21 世纪先进水平的高炉炼铁综合技术,在许多技术领域取得重大突破,具有显著的技术特征。

(1) 高炉大型化是炼铁技术的主流发展趋势,实现整个钢铁企业功能、结构、效率协同优化的大型化是巨型高炉设计建造的基本理念。巨型高炉在生产效率、节能减排、延长寿命、减少污染等方面均具有不可比拟的技术优势。

(2) 巨型高炉在注重改善原燃料条件的同时,以高炉生产稳定顺行为核心,以提高生产效率、降低燃料消耗为目标,以炉料分布控制、提高风温、富氧喷煤等为技术措施,不追求单一技术指标的先进性,更加注重综合技术经济指标和技术水平的整体提升。

(3) 巨型高炉在延长高炉寿命、提高风温、煤气干法除尘、富氧喷煤等领域取得重大技术创新成果,使高炉综合技术装备水平不断提高,引领了 21 世纪高炉炼铁技术进步。

致谢

衷心感谢颉建新、梅丛华、周颂明、毛庆武、钱世崇、苏维、张建、侯健等专家的审阅和宝贵意见。

参考文献

[1] 张福明,钱世崇,张建,等. 首钢京唐 5500m^3 高炉采用的新技术 [J]. 钢铁, 2011, 46 (2): 13-17.

[2] 项明武,周强,张灵,等. 沙钢5800m³高炉工艺技术特点 [J]. 炼铁, 2010, 29 (2): 1-6.
[3] Takashi Miwa. Developmentof Iron-making Technologies in Japan [J]. Journal of Iron and Steel Research International, 2009, 16 (supplement 2): 14-19.
[4] 琴浦. 新第1高炉の設備建設工事の概要 [J]. 石川島播磨技報, 2005, 45 (3): 3-5.
[5] 藤田昌男,德田慶一郎,小島啓孝. 大型高炉の超短期改修技術 [J]. JFE技報, 2006, 11 (4): 1-8.
[6] 张福明. 大型高炉煤气干法除尘技术 [J]. 炼铁, 2011, 30 (1): 1-5.
[7] Zhang Fuming, Mao Qingwu, Zhang Qifu, et al. Research and Application on Dry Type Bag Filter Cleaning Technology of BF Gas for 5500m³ Blast Furnace at Shougang Jingtang [C]// Chinese Society for Metals & Steel Institute VDEh. 3rd CSM-VDEh Metallurgical Seminar Proceedings. Beijing: Chinese Society for Metals, 2011: 99-108.
[8] 刘琦,邱晖. 沙钢5800m³高炉投产20个月实绩及技术思路讨论 [J]. 炼铁, 2011, 30 (6): 15-19.

Technical Features of Super Large Blast Furnace in Earlier 21st Century

Zhang Fuming

Abstract: The paper expounds the technologies and equipment of super large blast furnaces put into operation in the past decade and analyzes the technical characteristics of current super large blast furnaces. It is concluded that the process technologies and equipment of current super large blast furnace achieve great progresses, especially in blast furnace enlargement, beneficiated burden material, bell-less top equipment, burden distribution control, long campaign, high blasting temperature, oxygen enriched pulverized coal injection, dry dedusting of BFG, etc.

Keywords: super large sized blast furnace; bell-less top; process technology

高风温低燃料比高炉冶炼工艺技术的发展前景

摘　要：分析了当前高炉炼铁技术面临的形势和挑战，提出了当代高炉炼铁技术的发展目标。阐述了高风温、富氧喷煤对高炉炼铁的意义和作用。重点分析和论述了实现高风温和高富氧大喷煤的关键技术。提高风温、提高富氧率、增加喷煤量是降低燃料消耗、节约生产成本和实现可持续发展的重要保障。在高风温、低燃料比冶炼条件下，当代高炉炼铁技术具有广阔的发展前景。
关键词：高炉；炼铁；高风温；富氧喷煤；低燃料比

1 引言

当代高炉炼铁技术经过近两个世纪的发展，已经日臻完善，完成了从技艺向工程科学的嬗变。第二次工业革命以后，随着工业化进程加快，20世纪中后期是高炉炼铁技术快速发展时期，氧气顶吹转炉的成功问世，使炼钢生产效率大幅度提高，对高炉炼铁技术发展起到了强有力的引领作用，带动了高炉炼铁技术现代化和工艺装备大型化。20世纪70年代的两次全球石油危机，又极大地促进了高炉喷煤技术的飞跃发展，以廉价易得的非炼焦煤代替焦炭、重油和天然气等资源供给不足的高价值燃料，使现代高炉在优质炼焦煤资源日趋短缺的状况下依然能够保持勃勃生机。回顾现代高炉炼铁技术的发展历程可以看出，在资源制约、能源危机、相关技术领域和相关产业快速发展的情况下，高炉炼铁工艺都会产生重大的技术变革和技术创新。

2 当代高炉炼铁技术展望

进入21世纪以来，高炉炼铁工艺再次受到自然资源短缺、能源供给不足以及环境保护等方面的制约，面临着较大的发展问题。面对当前严峻的形势和挑战，21世纪高炉炼铁工艺要实现可持续发展，必须在高效低耗、节能减排、循环经济、低碳冶金、清洁环保等方面取得显著突破，需要进一步提高风温、降低燃料比，以提高高炉炼铁技术的生命力和竞争力。

高炉炼铁是钢铁厂生产成本和经济效益控制的关键工艺单元，对整个钢铁厂物质流、能量流和信息流流程网络的高效动态运行具有决定性作用。当代高炉炼铁要实现"高效、低耗、优质、长寿、清洁"的总体发展要求，高效不是简单的提高产量和强化冶炼，更要注重其经济效益、环境效益和社会效益；长寿也不是简单的延长高炉寿命，还要重视其技术的先进性和可持续发展的生存能力。面对当前国内外激烈的市场竞争环境，在资源短

本文作者：张福明。原刊于《中国冶金》，2013，23（2）：1-7。

缺、能源供给不足、环境制约的条件下，为保障高炉炼铁工艺的可持续发展，实现低碳冶炼和循环经济，要着力构建高效率、低消耗、低成本、低排放的高炉炼铁生产体系。

近年来，日本、欧洲、中国的高炉燃料比都呈显著的下降趋势。2008年欧洲15国高炉平均燃料比已降低到496kg/t，焦比降低到351.8kg/t，煤比达到123.9kg/t以上，喷吹重油天然气为20.3kg/t[1]。日本28座高炉平均燃料比也达到500kg/t以下，煤比达到120kg/t以上，焦比降低到380kg/t以下[2]。2010年我国重点钢铁企业高炉平均燃料比为518kg/t，煤比达到149kg/t以上，焦比降低到369kg/t以下，燃料比仍高于日本和欧洲的平均水平。图1~图3分别是欧洲、日本和中国高炉平均燃料比、入炉焦比和煤比的变化。

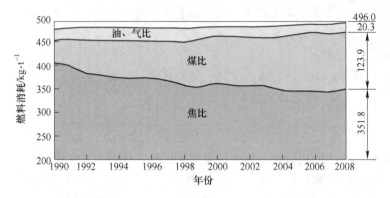

图1 欧洲15国高炉燃料比的变化

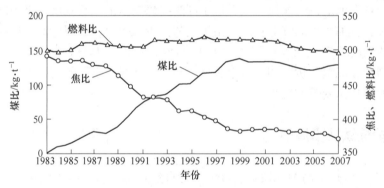

图2 日本高炉燃料比的变化

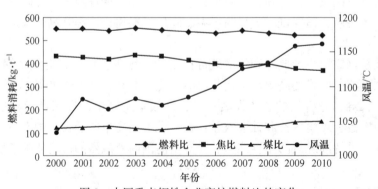

图3 中国重点钢铁企业高炉燃料比的变化

表1列出了中国部分先进大型高炉的主要技术经济指标[3],表2列出了国外先进高炉的主要技术经济指标。

表1 中国部分先进高炉主要技术经济指标（2011年1~6月）

高 炉	京唐1号（2010-03）	迁钢3号	宝钢3号	宝钢4号	武钢8号	沙钢5800m³	太钢5号	马钢1号	鞍钢鲅鱼圈1号
高炉有效容积/m³	5500	4000	4350	4747	3800	5800	4350	4000	4038
平均风温/℃	1300	1280	1236	1254	1192	1230	1243	1225	1225
富氧率/%	3.81	4.42	3.94	1.16	6.47	9.47	4.97	3.63	2.93
燃料比/kg·t⁻¹	480	503		488	532	502	502	514	523
入炉焦比/kg·t⁻¹	269	291	288	292	322	293	305	302	320
焦丁/kg·t⁻¹	36	36	18	26	33	48	10	36	52
煤比/kg·t⁻¹	175	176	185	170	177	161	187	140	151
利用系数/t·(m³·d)⁻¹	2.37	2.39	2.50	2.02	2.69	2.21	2.50	2.22	2.11
吨铁风耗/m³·t⁻¹	917	1205	974	1035	1082	855	—	1065	—

表2 国外部分先进高炉主要技术经济指标

高 炉	日本新日铁大分2号	日本新日铁君津4号	日本住友鹿岛1号	日本JFE福山5号	韩国现代唐津1号	韩国浦项光阳4号	德国施委尔根2号	荷兰艾莫伊登7号
高炉有效容积/m³	5775	5555	5370	5550	5250	5500	5513	4450
炉缸直径/m	15.5	15.2	15	15.6	14.85	15.6	14.9	13.8
日产铁量/t·d⁻¹	13500	12900	11425	12650	11600	13750	10194	10000
利用系数/t·(m³·d)⁻¹	2.34	2.32	2.13	2.28	2.2	2.5	1.85	2.25
燃料比/kg·t⁻¹	475	482	—	490	490	490	497	510
焦比/kg·t⁻¹	355	332	—	355	310	290	345	280
煤比/kg·t⁻¹	120	150	—	120	180	200	152	230
风温/℃	1200	1200	1250	1250	1230	1250	1119	1260
富氧率/%	3.5	—	3	—	5.6	10	3.7	10
入炉风量/Nm³·min⁻¹	8550	—	7800	8660	7250	7000	6952	6400
热风压力/MPa	0.45	—	0.49	0.42	0.423	0.43	0.466	—
炉顶压力/MPa	—	0.29	0.294	0.275	0.235	0.275	0.274	0.23

未来高炉炼铁的发展目标是：

（1）燃料比不超过500kg/t，先进高炉燃料比应不超过480kg/t；入炉焦比不超过300kg/t，先进高炉焦比不超过280kg/t；煤比不小于180kg/t，先进高炉煤比应达到200~250kg/t，喷煤率达到45%~50%。

（2）高炉有效容积利用系数达到2.0~2.3t/(m³·d)，原燃料条件好、技术装备水平高的大型高炉应达到或超过2.5t/(m³·d)。

（3）高炉一代炉役寿命不少于15年，高炉一代炉役单位容积产铁量应达到10000~

15000t/m³;技术装备水平高、原燃料条件好的大型高炉,一代炉役寿命要力争达到20年以上,高炉单位容积产铁量达到15000t/m³以上;热风炉寿命要大于或等于一代高炉寿命。

(4) 热风温度达到1200~1250℃,大型高炉风温应达到1250~1300℃。

(5) 高炉富氧率达到3%~5%,先进高炉富氧率应达到5%~10%。

3 高风温技术

高风温是现代高炉炼铁的主要技术特征之一。提高风温是当前钢铁行业发展循环经济、实现低碳冶金、节能减排和可持续发展的关键共性技术,对于提高高炉综合技术水平、减少CO_2排放、引领行业技术进步具有极其重要的意义。

3.1 高风温的意义和作用

高炉冶炼所需要的热量,一部分是燃料在炉缸燃烧所释放的燃烧热,另一部分是高温热风所带入的物理热。热风带入高炉的热量越多,所需要的燃料燃烧热就越少,即燃料消耗就越低。实践证实,在风温1000~1250℃的范围内,提高风温100℃可以降低焦比约10~15kg/t,由此可见,提高风温可以显著降低燃料消耗和生产成本。除此之外,提高风温还有助于提高风口前理论燃烧温度,使风口回旋区具有较高的温度,炉缸热量充沛,有利于提高煤粉燃烧率、增加喷煤量,还可以进一步降低焦比。因此,高风温是高炉实现大喷煤操作的关键技术,是高炉降低焦比、提高喷煤量、降低生产成本的重要技术途径,是高炉炼铁发展史上极其重要的技术进步。

高风温技术是一项综合技术,涉及到整个钢铁厂物质流、能量流流程网络的动态运行和结构优化,应当在整个钢铁厂流程网络的尺度上进行研究。高风温对于优化钢铁厂能源网络结构、降低生产成本和能源消耗、实现低品质能源的高效利用、减少CO_2排放等都具有重大的现实意义和深远的历史意义。进入21世纪以来,一系列高风温技术相继开发并应用在大型高炉上,高风温技术创新与应用实践取得了显著的技术成效。现代大型高炉的设计风温一般为1200~1300℃,提高风温已成为当前高炉炼铁技术发展的一个显著趋势。

当前,国内外高风温技术发展水平并不平衡,2010年我国重点钢铁企业的高炉平均风温为1160℃,先进高炉的风温已达到1250~1300℃,技术水平差距很大,进一步提高风温仍然是21世纪高炉炼铁技术的热点研究课题。

3.2 获得高风温的关键技术

多年以来,我国高炉平均风温始终徘徊在1000~1080℃,2001~2010年的10年间,重点钢铁企业高炉平均风温由1081℃提高到1160℃,风温仅提高了79℃,可谓步履维艰,高炉风温是我国高炉和国外先进水平差距最大的技术指标。制约风温提高有许多因素,如何突破这些制约条件,达到1200℃以上的高风温乃是当今我国高炉炼铁的主要技术发展目标之一。在当前条件下,提高风温应着力解决利用低热值高炉煤气获得1250℃以上高风温的关键技术难题,通过技术创新实现热风炉高效率、低成本、低排放、长寿命的综合技术目标。

（1）研究开发并应用燃烧高炉煤气获得高风温技术。高炉煤气是高炉冶炼过程中产生的二次能源，热风炉燃烧大约消耗高炉煤气发生量的40%~50%。随着高炉炼铁技术进步，大型高炉燃料比已降低到520kg/t以下，高炉煤气利用率提高到45%以上，煤气低发热值不足3000kJ/m³。由于钢铁厂高热值的焦炉煤气和转炉煤气主要用于炼钢和轧钢等工序，高热值煤气供给不足，绝大部分热风炉只能燃烧低热值高炉煤气，在没有高热值煤气富化的条件下，导致热风炉理论燃烧温度和拱顶温度不高，进而难以实现1200℃高风温，这是当前制约我国高炉提高风温最重要的原因。面对能源供给短缺的现状，采用煤气、助燃空气高效双预热技术，不但可以回收热风炉烟气余热，减少热量耗散，还可以有效提高热风炉拱顶温度。在众多的预热技术中，要统筹考虑能量转换效率、技术可靠性以及设备使用寿命等因素，择优选用适宜可靠的煤气、助燃空气双预热技术。首钢京唐5500m³高炉采用了煤气预热和助燃空气高温预热组合技术，利用分离式热管换热器回收烟气余热预热高炉煤气，设置2座小型热风炉预热助燃空气，使煤气预热温度达到200℃，助燃空气预热温度达到520~600℃，在燃烧单一高炉煤气条件下，热风炉拱顶温度可以达到1420℃，高炉月平均风温达到1300℃[4]。这项利用低热值高炉煤气实现1300℃高风温技术值得推广。

（2）选择合理的热风炉结构形式。高炉热风炉是典型的蓄热式加热炉，其工作原理不同于其他的冶金炉窑，是现代钢铁厂燃烧功率最大、能量消耗最高、热交换量最大的单体热工装置。尽管现有的内燃式、外燃式和顶燃式3种结构热风炉均有实现1250℃以上高风温的实绩，但不同结构的热风炉在燃烧工况适应性、气体流动及分布均匀性、能量利用有效性等方面仍存在差异[5]。综合考虑热风炉高效长寿和工况适应性，现代高炉采用顶燃式或外燃式热风炉是适宜的选择。值得指出的是，近年来顶燃式热风炉取得了重大技术进步，已推广应用于首钢京唐5500m³巨型高炉，并取得了显著的应用效果，月平均风温已稳定保持在1250~1300℃。目前，我国已开发并应用了各种结构的顶燃式热风炉，成为引领热风炉高风温技术创新的研究热点。同时应当看到，我国热风炉结构多样化将维持较长时期，3种结构的热风炉仍然会长期存在，但对于新建或大修改造的高炉，优先采用顶燃式热风炉不失为一种明智之举。

（3）采用高效格子砖。实践证实，缩小热风炉拱顶温度与风温的差值可以显著提高风温，其主要技术措施是强化蓄热室格子砖与气体之间的热交换。在保持格子砖活面积或格子砖重量不变的条件下，适当缩小格子砖孔径，可以增加格子砖加热面积、提高换热系数而增加热交换量。在热风炉燃烧期，高温烟气可以将更多的热量传递给格子砖，热量交换更加充分，使得烟气温度更低；在热风炉送风期，同样有利于鼓风与格子砖的热交换，使得热风温度更高，热风温降也更为平缓，在风温保持较高的状态下更加稳定。对于格子砖砖型的选择需要综合考虑择优确定，并不是格子砖孔数越多、孔径越小就越有利，要综合考虑蓄热室热效率、蓄热室有效利用率和格子砖使用寿命等各种因素的影响。因此采用高效格子砖、提高热风炉热效率是提高风温、降低热风炉燃烧煤气量、减少CO_2排放的关键技术之一。

（4）优化热风炉操作。优化热风炉燃烧、送风操作，缩小拱顶温度与风温的差值，适当提高烟气温度，是提高风温的主要操作措施。当前，热风炉燃烧操作应进一步优化，合理设定煤气与助燃空气配比，降低空气过剩系数，保证煤气的完全燃烧。热风炉烟气中残余CO含量和O_2含量应尽可能降低，燃烧初期应采取强化燃烧模式，使拱顶温度迅速达到设定数值；燃烧过程中合理调节煤气和助燃空气，在较低的空燃比下使烟气温度达到设定目标。根据拱顶温度、烟气温度、烟气O_2含量等参数在线调节煤气和助燃空气，优化热风炉燃烧操作是当前需要引起重视的问题。不少钢铁企业由于低热值高炉煤气资源相对丰富，并不注重节约热风炉燃烧所消耗的高炉煤气，不仅增加了煤气消耗量，造成能源浪费、生产成本升高，也无助于提高风温，而且还增加了CO_2排放量。优化热风炉燃烧不但可以获得高风温，还可以降低能源消耗，提高能源转换效率，减少热量耗散，降低CO_2排放。热风炉送风操作也需要改进，4座热风炉应采用"两烧两送"交叉并联的工作模式，这样可以提高风温约25℃，同时应尽量减少冷风混风量，使高风温得到充分利用。3座热风炉采用"两烧一送"的工作模式，可以适当缩短送风周期、增加换炉次数，同样应减少冷风混风量。改善热风炉操作一方面应开发应用适宜的热风炉操作数学模型，采用信息化、数字化手段提高热风炉操作水平；另一方面应在交叉并联送风技术的基础上，研究开发热风炉"热交叉并联"送风技术，进一步减少冷风混风量，降低风温波动，使高炉获得更加稳定的高风温。

3.3 高温热风的稳定输送

风温提高以后，高温热风的稳定输送成为制约环节。近年来，不少热风炉的热风支管、热风总管和热风环管出现局部过热、管壳发红、管道窜风、甚至管道烧塌事故[6]，热风管道内衬经常出现破损，极大地限制了高风温技术的发展。因此，优化热风管道系统结构，采用"无过热—低应力"设计体系[7]，合理设置管道波纹补偿器和拉杆，妥善处理管道膨胀以降低管道系统应力，热风管道采用组合砖结构，消除热风管道的局部过热和管道窜风。

热风炉各孔口在多种工况的恶劣条件下工作，也是制约热风炉长寿和提高风温的薄弱环节。热风炉各孔口耐火材料要承受高温、高压的作用，还要承受气流收缩、扩张、转向运动所产生的冲击和振动作用。热风出口应采用独立的环形组合砖构成，组合砖之间采用双凹凸榫槽结构进行加强，以减轻上部大墙砖衬对组合砖所产生的压应力。热风炉的热风出口、热风管道上的三岔口等关键部位均应采用组合砖结构。热风管道内衬的工作层，应采用抗蠕变、体积密度较低、高温稳定性优良的耐火材料；绝热层应采用体积密度低、绝热性能优良的耐火材料，还应注重合理设计耐火材料膨胀缝及其密封结构。

值得指出的是，热风管道耐火材料内衬在高温高压热风的流动冲刷作用下，容易出现局部过热、窜风、甚至内衬脱落现象。对于已运行的热风管道应采用表面温度监测系统，可以在线监控热风管道关键部位的管壳温度，并可以进行数据处理和存储，实现信息化动态管理；同时为了监控热风管道受热膨胀而产生的变形情况，设置激光位移监测仪可以在线监测热风管道的膨胀位移。通过数字化在线监控装置，可以提高热风炉管道工作的可靠性，保障高温热风的稳定输送。该项技术已在首钢迁钢大型高炉上得到了成功应用[8]。

3.4 延长热风炉寿命

研究表明，当热风炉拱顶温度达到1420℃以上时，燃烧产物中的NO_x生成量急剧升高，燃烧产物中的水蒸气在温度降低到露点以下时冷凝成液态水，NO_x与冷凝水结合形成酸性腐蚀性介质，对热风炉高温区炉壳造成晶间应力腐蚀，降低了热风炉使用寿命，成为制约风温进一步提高的主要限制因素。因此，热风炉一般将拱顶温度控制在1420℃以下，旨在降低NO_x生成量从而有效抑制炉壳晶间应力腐蚀。应采取有效的晶间应力预防措施，延长热风炉使用寿命。通过合理控制热风炉拱顶温度，抑制热风炉燃烧及送风过程NO_x的大量生成；采取高温区炉壳涂刷防酸涂料和喷涂耐酸喷涂料等防护措施，加强对热风炉炉壳的隔离保护；采用细晶粒炉壳钢板，采用热处理措施消除或降低炉壳制造过程的焊接应力；对于热风炉高温区炉壳采取保温等综合防护措施，预防热风炉炉壳晶间应力腐蚀的发生，以延长热风炉寿命，达到与高炉寿命同步。

4 高富氧大喷煤技术

4.1 高炉喷煤

高炉喷煤是20世纪60年代高炉炼铁技术的一项重大技术创新。1973年第一次石油危机发生以后，高炉喷煤技术得到世界范围的广泛重视，1978年第二次石油危机的爆发，促进了高炉喷煤技术的迅猛发展，20世纪80年代以后，形成了世界范围内高炉喷煤技术的发展高潮。近年来，随着稀缺的炼焦煤资源日益匮乏，高炉喷吹非炼焦煤用以替代昂贵的焦炭，使当代高炉在减少资源和能源消耗、合理利用资源和能源、降低生产成本和减少环境污染等方面取得重大成效，高炉喷煤已成为当代高炉最显著的技术特征。

当前国际先进高炉的喷煤量已经达到200~250kg/t，喷煤率达到40%~45%，煤粉已成为当代高炉炼铁工艺的重要燃料和还原剂。高炉喷煤技术的成功应用，使高炉炼铁工艺对炼焦煤的依赖性下降，高炉燃料结构中约有40%的煤粉替代了资源短缺且价格高昂的焦炭，使高炉炼铁工艺又焕发出新的生机，这也是目前高炉炼铁工艺仍占主导地位的一个重要因素。高炉喷煤的极限始终是炼铁工作者重点研究的课题，当代大型高炉的喷煤量应努力达到200kg/t以上，而且应当实现低燃料比条件下，使喷煤率达到50%。

4.2 鼓风富氧

理论研究和实践证实，鼓风富氧对提高煤粉燃烧率和喷煤量的作用十分显著，已成为高炉大喷煤操作的重要支撑技术。富氧鼓风是当代高炉强化冶炼的有效技术措施之一，可以充分发挥高风温、喷煤降焦的综合作用。高炉鼓风富氧不仅能够提高产量，还可以提高风口前理论燃烧温度，提高煤粉燃烧率和喷煤量，有效降低炉腹煤气量，改善煤气能量利用和高炉透气性，而且对于提高炉顶煤气热值也会带来益处。当前高炉富氧率一般达到3%~5%，部分大型高炉富氧率达到10%左右。随着高炉炼铁技术进步，国内外对高炉鼓风富氧的认识也逐渐发生变化。在当前的冶炼条件下，高炉鼓风富氧的主要目标已不再是追求提高产量、强化冶炼，而更重要的意义在于：

（1）提高煤粉燃烧率，减少未燃煤粉量。研究表明，在风温达到1000℃以上时，富氧对于促进煤粉燃烧的作用比提高风温更为显著。这主要是由于煤粉燃烧动力学特性所决定，在高温条件下，氧的扩散与传质是煤粉燃烧的限制性环节，因此增加鼓风中的氧浓度有利于提高煤粉燃烧率，降低未燃煤粉量，提高煤粉利用率，进而可以有效增加喷煤量。

(2) 提高风口前理论燃烧温度,保持炉缸具有充足的热量。理论燃烧温度对于高炉冶炼进程意义重大,大型高炉应将理论燃烧温度维持在2050~2350℃。理论燃烧温度是高炉炉缸热量储备的重要标志,也是高炉操作的重要参数。高炉喷煤以后,由于水分蒸发、挥发分中碳氢化合物分解吸收大量热量,造成理论燃烧温度下降、炉缸热量不足,因此需要提高风温给予热量补偿。鼓风富氧的实质则是降低了风口回旋区燃烧所产生的炉缸煤气量,进而提高了理论燃烧温度。尽管提高风温和鼓风富氧都可以提高理论燃烧温度,但两者的作用机理却不尽相同。理论计算表明,富氧率提高1%,理论燃烧温度可以提高45~50℃。

(3) 降低炉腹煤气量,改善高炉透气性。鼓风富氧降低了鼓风中N_2含量,使风口前碳素燃烧所消耗的鼓风量下降,使炉缸煤气生成量下降,进而降低了炉腹煤气量。生产实践表现在鼓风富氧以后,入炉风量减少、煤气量降低,富氧率提高1%,吨铁煤气量降低约4%。在高炉冶炼进程中,风口回旋区碳素燃烧所形成的炉缸煤气,和炉缸内直接还原所产生的煤气在炉腹区域汇合形成了炉腹煤气。炉腹煤气是穿透软融带上升的煤气,对高炉冶炼过程的热量传输、质量传输、动量传输和还原反应都有极其重要的意义。鼓风富氧有效地降低了炉腹煤气量,使煤气流速降低,在高炉大喷煤条件下,对于改善料柱透气性意义重大。由此可见,合理的炉腹煤气量是当代高炉大喷煤条件下,保障高炉稳定顺行的根本要素。

(4) 提高煤气热值,减少CO_2排放。鼓风富氧使N_2含量降低,炉顶煤气量减少且N_2含量降低,在一定程度上煤气中的CO体积浓度则相应增加,有助于提高煤气热值,为高炉能量的高效循环利用创造了条件,有利于提高热风炉拱顶温度,进而提高风温。而且还可以降低热风炉燃烧的煤气消耗,降低了CO_2的排放。

4.3 高富氧大喷煤技术

生产实践证实,高炉炼铁系统的能耗占钢铁企业总能耗的70%以上,生产成本约占钢材制造成本的50%以上,降低高炉炼铁燃料消耗和生产成本在当前更具有重要意义。高炉喷煤是降低燃料消耗和生产成本的最重要的技术措施,不但可以显著降低焦比,减轻炼焦过程的环境污染,还可以缓解炼焦煤资源短缺的状况。

鼓风富氧是当代高炉氧-煤强化冶炼的重要技术,是提高生产效率、降低燃料消耗、提高喷煤量、促进高炉稳定顺行的重要支撑条件。高炉富氧率要达到3%~5%以上,有条件的钢铁企业应提高富氧率使之达到5%以上,甚至达到8%~10%。荷兰康力斯艾莫伊登厂、韩国浦项光阳厂以及中国沙钢5800m^3大型高炉,富氧率都达到了10%,取得了良好的技术成效和显著的综合效益。

在当前条件下,推进高富氧大喷煤技术,值得关注以下问题:

(1) 提高精料水平,改善原料条件,这是高炉大喷煤操作的重要基础。在提高入炉矿石冶金性能的同时,采用合理炉料结构,适当增加球团矿比率,采用炉料分级入炉技术都可以获得很好的效果。特别值得指出的是焦炭质量尤为重要,喷煤量达到200kg/t以上,必须提高焦炭质量。焦炭常温抗碎强度M_{40}≥85%,抗磨强度M_{10}≤6%,焦炭反应后强度CSR≥67%,热反应性CRI≤25%,焦炭灰分≤12.5%,焦炭平均粒度≥50mm,这是高炉大喷煤操作的重要保障条件。

（2）提高富氧率，降低吨铁风耗。要转变以"大风量、高冶炼强度"获取高产的传统观念，不应片面追求"大风高产"，要以降低燃料比、提高煤气利用率、保持煤气流合理分布为前提，提高鼓风质量，风温≥1250℃，富氧率≥5%，鼓风湿分≤15g/m³，构建当代高炉炼铁"精料、精风"的技术理念。吨铁风耗降低不但使高炉炉腹煤气量显著降低，有利于高炉稳定顺行，还可以降低鼓风动力消耗，节约生产成本。关于氧气的加入方式，可以采取鼓风兑入和氧煤喷枪局部供氧相结合的方式，以改善风口回旋区供氧强度和传质过程，这样可以获得更高的煤粉燃烧率，从而取得更为理想的应用效果。

（3）优化高炉操作，提高置换比和煤粉利用率。高炉大喷煤条件下，高炉操作难度增加。一方面，由于喷煤量提高，炉腹煤气量增加，风口回旋区扩大，理论燃烧温度下降，炉缸热量需要采取提高风温或富氧鼓风的措施给予补偿；另一方面，会造成高炉料柱压差增高，高炉透气性变差。因此，必须采取合理的高炉操作制度，上下部调剂相适应，以保证高炉大喷煤条件下的高炉稳定顺行。炉料分布控制技术是高炉上部调剂的核心，其技术实质是保证高炉煤气流的合理分布，提高煤气利用率，从而降低燃料比。合理的送风制度的内涵应是保持高炉具有适宜的理论燃烧温度、合理的风速和鼓风动能，使炉缸煤气初始分布合理，具有足够的中心气流，以改善炉缸透气性。特别是高炉大型化以后，炉缸直径和炉缸断面增大，煤气流难于穿透炉缸中心，容易造成边缘气流过分发展，导致燃料比增加、炉墙热负荷提高，不但不利于降低燃料消耗，还会对高炉寿命产生影响。

（4）提高置换比和煤粉利用率。高炉喷煤的首要目的是降低焦比、节约焦炭，随着喷煤量的提高置换比呈下降趋势，表现在煤比提高而焦比降低幅度不大，造成燃料比反而增加。这种现象在高炉大喷煤操作时经常出现，特别是煤比达到200kg/t以上时，这种趋势更为明显。因此必须采取综合措施，提高置换比和煤粉利用率。高风温、富氧可以有效提高煤粉燃烧率达到70%以上；喷吹挥发分20%~25%的混煤既可以保持较高的燃烧率，还可以提高喷煤量。保持较高的置换比，不但有利于降低燃料比，还有利于高炉稳定顺行，同时也有利于高炉接受更多的煤粉。毋庸置疑，在低燃料比冶炼条件下，努力提高喷煤量乃是当今高炉炼铁技术的核心发展目标。

（5）降低制氧成本，保障氧气稳定供应。尽管高炉富氧鼓风的作用早已取得共识，但长期以来我国高炉鼓风富氧技术一直停滞不前，大多数钢铁企业高炉富氧一般是利用炼钢的富裕氧气，不但鼓风富氧率较低，而且氧气供应也得不到保证。目前，高炉鼓风富氧面临最大的问题是氧气的稳定供应，究其原因主要是鼓风富氧量大、制氧成本高所致。因此，需要结合高炉富氧的特点，建设专用的高炉制氧机，保障高炉富氧所需氧气的稳定供应[9]。目前有高纯深冷法、低纯深冷法、变压吸附法（PSA）、膜渗透法等多种制氧方式可供选择，各种制氧方式的生产规模、氧气纯度、工程投资以及制氧电耗也不尽相同，大型钢铁厂选择低纯深冷法可以满足高炉用氧品质和产量的需求，而且设备投资和运行成本也低于高纯深冷法。首钢京唐2座5500m³高炉配置了75000m³/h制氧机，可以满足高炉富氧率5.5%的供氧要求。

5 结语

（1）21世纪初，当代高炉炼铁技术面临着资源短缺、能源供给不足和生态环境的制约，实现高炉炼铁工艺的可持续发展，必须采用高风温、高富氧大喷煤技术，降低燃料消耗和生产成本，这是当代高炉炼铁技术发展的必由之路。

（2）高风温是综合技术，是降低燃料比、提高喷煤量的重要技术保障。在当前条件下，利用低热值高炉煤气实现1250℃以上高风温，是实现低品质能源高效利用和高效能源转换最优化的技术措施。要系统解决高风温的获得、高温热风的稳定输送和高效利用等关键技术问题，采用高效长寿热风炉及高效格子砖，优化热风炉操作，保障高温热风的稳定输送，延长热风炉寿命，使高温热风得到高效利用。

（3）高富氧大喷煤是降低燃料消耗的重要支撑技术。提高富氧率对当代高炉炼铁意义重大，应给予高度重视。在现有条件下必须进一步提高喷煤量，使煤比达到200kg/t以上，燃料比低于500kg/t以下，这是当代高炉炼铁工艺保持生命力和竞争力的根本所在。

（4）到21世纪中期，当代高炉采用高风温、高富氧大喷煤、精料、长寿、环境保护等综合技术，进一步降低燃料比，具有广阔的发展前景。

参考文献

[1] Takashi Miwa. Development of Iron-making Technologies in Japan [J]. Journal of Iron and Steel Research International, 2009, 16 (supplement 2): 14-19.
[2] Hans Bodo, Michael Peters, Peter Schmoele. Iron Making in Western Europe [C]// The Chinese Society for Metals, Steel Institute VDEh, 3rd CSM-VDEh Metallurgical Seminar Proceedings. Beijing, 2011: 33-50.
[3] 刘琦. 国内特大型高炉生产技术点评 [J]. 冶金管理, 2011 (12): 45-51.
[4] 张福明, 钱世崇, 张建, 等. 首钢京唐5500m^{3}高炉采用的新技术 [J]. 钢铁, 2011, 46 (2): 12-17.
[5] 钱世崇, 张福明, 李欣, 等. 大型高炉热风炉技术的比较分析 [J]. 钢铁, 2011, 46 (10): 1-6.
[6] 许秋慧, 常建华, 龚瑞娟, 等. 唐钢3200m^{3}高炉热风炉管道耐火材料塌落的处理 [J]. 炼铁, 2010, 29 (6): 23-25.
[7] 张福明, 梅丛华, 银光宇, 等. 首钢京唐5500m^{3}高炉BSK顶燃式热风炉设计研究 [J]. 中国冶金, 2012, 22 (3): 27-32.
[8] 马金芳, 万雷, 贾国利, 等. 迁钢2号高炉高风温技术实践 [J]. 钢铁, 2011, 46 (6): 26-31.
[9] 沙永志. 高富氧大喷煤技术分析 [J]. 炼铁, 2006, 25 (6): 19-22.

Developing Prospects on High Temperature and Low Fuel Ratio Technologies for Blast Furnace Ironmaking

Zhang Fuming

Abstract: The present status and challenges of blast furnace ironmaking technology are analyzed, and the developing objectives of the contemporary BF ironmaking process are mentioned. The significance and function of high blast temperature and oxygen enriched-pulverized coal injection for BF ironmaking are described. The key technologies of high blast temperature and oxygen enriched-PCI are discussed and evaluated emphatically. Promoting blast temperature and improving oxygen enrichment ratio, as well as enhancing PCI rate which are important guarantee for saving fuel consumption, reducing operation cost, and realizing sustainable development. There are broad developing prospects of contemporary BF process under the condition of high blast temperature and low fuel ratio in ironmaking technologies.

Keywords: blast furnace; ironmaking; high blast temperature; oxygen enrichment-PCI; low fuel ratio

面向未来的低碳绿色高炉炼铁技术发展方向

摘　要：通过对高炉炼铁的功能解析，以及流程集成与结构优化，论述了面向未来的低碳绿色高炉炼铁技术发展方向，指出高炉精料、炉料合理分布控制、高效长寿、高风温、富氧喷煤、煤气干法除尘等关键共性技术，仍是未来高炉低碳绿色发展的重要技术基础、技术支撑和技术保障。认为现代高炉炼铁要构建动态精准、高效运行的生产体系，以高效率、低成本、低排放为目标，提高生产效率、降低能源消耗、减少环境污染，实现高炉炼铁"高效、低耗、长寿、低碳、绿色"的协同发展目标，着力提高综合技术装备水平，大力发展循环经济，实现低碳绿色制造。

关键词：高炉；炼铁；低碳冶金

1　引言

　　进入新世纪以来，高炉炼铁工艺再次受到了自然资源短缺、能源供给不足和生态环境保护等方面的制约，可持续发展面临着较大的问题[1]。在市场低迷、需求疲软的经济形势下，面对当前日益严峻的竞争和挑战[2]，未来高炉炼铁工艺必须在"高效低耗、节能减排、清洁环保、循环经济、低碳冶炼、智能集约、绿色发展"等方面取得显著突破，以低碳绿色发展为主导，进一步优化工艺流程、提高风温、降低燃料比，以提高高炉炼铁工艺的生命力和竞争力[3]。高炉是钢铁制造流程中的关键工序，是铁素物质流转换的核心关键单元，是钢铁制造过程能源转换的核心单元和能量流网络的中枢环节[4]。目前，高炉炼铁生产成本约占钢铁制造总成本的60%~70%，炼铁工序能源消耗约占钢铁综合能耗的60%~70%。由此可见，对于高炉—转炉制造流程的钢铁联合企业，高炉炼铁在钢铁制造流程中，其作用至关重要，其地位举足轻重。

2　功能解析

　　第一次工业革命以后，高炉炼铁工艺历经近200年的发展演进，在大型化、高效化、长寿化、集约化、智能化等方面取得了重大技术进步，已经发展成为当代铁氧化物还原效率最高的铁冶金工艺，高炉炼铁已嬗变成为工程科学，日臻完善。高炉的主要功能也随其技术发展的历程在不断地演化[5]，高炉炼铁功能的解析如图1所示。

　　（1）还原器和渗碳器。高炉是以焦炭作为主要燃料和还原剂，将铁氧化物还原成液态生铁的工艺装置，属于典型的竖炉式反应器和多相复杂巨系统逆向工业反应装置。在下降炉料与上升煤气流的相向运动进程中，实现高炉冶金过程的"三传一反"。高炉高温还

本文作者：张福明。原刊于《炼铁》，2016, 35 (1)：1-6。

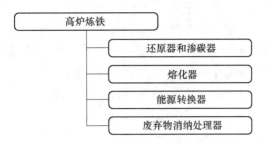

图 1　高炉炼铁功能解析

原过程得以顺利实现的前提,是高炉中焦炭料柱的存在(即骨架作用)和其不可替代性,这就决定了在高炉还原过程中,总是伴随有不同程度的生铁渗碳。

(2)熔化器和质量调控器。高炉最重要的功能就是为转炉提供优质的液态生铁,这是转炉炼钢工艺赖以存在的基础。高炉将固态铁氧化物矿物中的铁素源还原而获得液态生铁,因此其具有熔化器的功能。而且,高炉连续不断还原、生产铁水的方式决定了高炉是一个连续的铁水供应器。与此同时,高炉炼铁工艺过程对铁水成分、质量具有重要的调控功能,可以通过高炉操作稳定控制铁水的温度、成分及其偏差,特别是控制铁水中[S]、[Si]的含量保持在合理的范围内。

(3)能源转化器。高炉生产过程中,伴随着巨大的能源消耗和能量转换过程,其特点是将焦炭、煤粉、天然气等化石质能源转化为产出物(铁水、熔渣和高炉煤气)的物理显热和化学能,因此高炉具有能源转换器的功能。随着高炉炼铁节能技术的进步,TRT和其他余热、余能回收利用技术的实施,使得高炉作为能量转换器的功能愈显突出。高炉工序是目前钢铁制造流程中最大的能源转换工艺单元,也是钢铁厂能量流网络构建的中枢环节。

(4)消纳固体废弃物。高炉炼铁工艺除了具有冶金功能之外,还具有消纳固体废弃物、实现资源化的功能,包括与高炉匹配的焦化、烧结和球团工序,高炉炼铁系统是钢铁厂实施循环经济的重要环节。例如,利用焦化工艺和高炉喷吹处理废塑料;利用烧结或球团工艺将钢铁厂粉尘造块,并回收二次资源,高炉喷吹除尘灰;利用焦炉热解,处理城市生活垃圾等。

3　流程集成与结构优

(1)动态精准设计。新一代钢铁厂工程设计的理论必须建立在符合其动态运行过程物理本质的基础上,特别是生产流程的动态有序、协同连续、稳定高效运行中的运行动力学理论基础上[6]。

现代高炉动态精准设计应以先进的概念研究和顶层设计为指导[7],运用动态甘特图(Gantt Chart),考虑高效协同的界面技术实现动态有序、协同连续/准连续的物质流设计和高效转换、及时回收利用、节能减排的能量流设计,从更高的层次上体现钢铁制造流程的"三大功能",即实现钢铁产品制造、高效能源转换和消纳废弃物并实现资源化。

现代高炉动态精准设计要从高炉生产动态协同运行的总体目标出发,对先进的技术单元进行综合研判、权衡取舍、选择集成，再进行动态整合，把相关的单元技术通过在流程网络化整合和程序化协同，使"物质流""能量流""信息流"的流动、流变过程，在规定的时-空边界内实现动态有序化运行，形成一个动态有序、协同连续、紧凑集约的工程整体集成效应，达到多目标优化的目的。动态精准设计的内涵如图2所示。

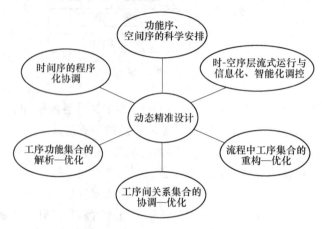

图2 动态精准设计的内涵

（2）流程网络。流程网络的构建是现代高炉设计的一个重要内容。必须要摒弃传统高炉设计"粗放、孤立、静止"的模式和惯例，构建基于动态精准运行的流程网络，实现高炉炼铁的功能优化、结构优化和效率优化。实现钢铁厂流程结构优化条件下的高炉大型化，构建钢铁厂物质流、能量流、信息流流程网络系统优化的动态精准运行体系。现代高炉设计要着力建构基于串联-并联结合、简便-顺畅-高效的物质流网络，基于输入-输出、高效转换、耗散最小的动态能量流网络以及简捷-高效、协同-调控的信息流网络。

（3）界面技术的优化。在现代高炉工程设计中，要将烧结、球团、焦化、高炉以及炼钢等工序的"界面技术"合理化，不仅有利于实现生产过程的组织协调，还有利于实现生产过程、调度过程的信息化，并使单元装置、设备的生产速率和生产效率更高，同时可以实现单位产能的投资最省。主要界面技术优化包括：简捷化的物质流、能量流通路（体现在高炉平面布置图）；工序、装置之间互动关系的缓冲-稳定-协同（体现在工序单元的动态运行甘特图）；制造流程中网络节点优化和节点群优化以及连接器形式优化（体现在装备数量、装置能力和位置合理化、运输方式、运输距离、输送规则优化等）；物质流效率、速率优化；能量流效率优化和节能减排；物质流-能量流-信息流的协同优化等。典型的界面技术包括：以多功能铁水罐"一罐到底"铁水直接运输为代表的炼铁-炼钢界面技术；以集约高效、连续紧凑的无中间料仓原燃料直接运输为代表的烧结-高炉、焦化-高炉界面技术等。

4 关键共性技术

4.1 发展理念

现代高炉炼铁要构建动态精准、高效运行的生产体系，以高效率、低成本、低排放为目标，提高生产效率、降低能源消耗、减少环境污染，实现现代高炉炼铁"高效、低耗、长寿、低碳、绿色"的协同发展目标，提高综合技术装备水平，发展循环经济、实现低碳绿色制造是新世纪高炉炼铁技术的主要技术特征。合理控制高炉运行，建立科学合理的生产指标评价体系，实现高炉炼铁多目标协同优化；要构建高效率、低成本、低消耗、低排放的高炉炼铁技术体系。21世纪大型高炉的工艺技术装备要更加注重技术创新和低碳绿色发展，高炉炼铁关键共性技术要形成重大突破，协同优化、集成创新高炉炼铁先进技术，要在高炉高效长寿、优质低耗、节能减排、循环经济、清洁环保、低碳绿色等诸多领域取得重大技术进步。

4.2 流程集成创新

钢铁厂生产能力的选择必须适应社会发展和市场需求，根据区域和钢铁市场结构需求，结合钢铁企业自身实际，实事求是、因地制宜、科学决策，从而合理地确定钢铁厂的产品定位和生产规模。要遵循钢铁厂整体流程结构的合理性、高效性、经济性综合考量确定钢铁厂产能，进而再据此确定高炉的产能、数量和容积。

考虑到物质流、能量流和信息流网络结构的优化，一个钢铁厂最优化适宜的选择是配置2-3座高炉，必须以钢铁厂整体流程结构优化为基础，科学合理配置高炉，钢铁厂流程结构优化条件下的高炉大型化是未来高炉的主要发展方向[8]。

4.3 优化总图设计

现代高炉炼铁的发展创新和持续演进，完成了由技艺向工程科学的嬗变。以流程工程学的视野和层次来分析，高炉总平面图不仅体现的是平面工艺布置，还体现着空间关系与时间关系，究其实质则是物质流和能量流的运行路径和轨迹。物质和能量在设定的几何空间内的运行，应当实现"路径最短、阻损最低、效率最高、耗散最小"的"层流式运动"。由此可见，高炉总图优化设计的重要意义。

钢铁制造流程的合理性，表现在高效率、低成本、低排放，使铁素物质流、能量流、信息流协同高效动态运行，必须摒弃传统的静态设计理念，实现结构、效率、功能协同优化的高炉大型化。高炉总图设计的原则如图3所示。

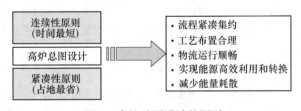

图3 高炉总图设计的原则

4.4 高炉精料技术

精料是现代高炉炼铁实现高效、低耗、优质、长寿的基础，也是实现高炉生产减量化和耗散最小化的重要措施。

（1）采用合理炉料结构。当代高炉炼铁必须要适应资源、能源的可供给、可获取条件，遵循循环经济理念，以减量化为技术特征，努力实现资源、能源的最佳化利用。采用经济合理的炉料结构，是实现高炉炼铁高效率、低成本、可持续发展的必由之路。高炉炉料结构的优化要遵循资源最优化、技术最优化、经济最优化的原则，注重改善炉料综合冶金性能，开发研究低品质矿高效利用技术，实现资源减量化、利用最佳化和环境友好化。

由于资源禀赋和技术传承，世界不同地区的高炉炉料结构不尽相同。亚洲高炉炉料结构大多为高碱度烧结矿为主，配加酸性球团矿和块矿；欧美高炉炉料结构则以较高比率的球团矿为主，部分高炉甚至全部采用球团矿。由于生态环境保护和资源获取等诸多原因，欧洲部分钢铁厂已关闭了烧结厂，高炉全部转而采用球团矿。我国高炉炉料结构的优化，还需要在未来不断探索、不断创新，不可能依然维持传统的炉料结构。钢铁企业应运用运筹学理论，通过建立数学模型优化确定符合企业具体情况的合理炉料结构，以达到技术可行、经济合理、资源节约。值得指出的是，由于我国铁矿石资源的特点和生态环境的制约，我国未来应当大力发展球团工艺，先进的大型高炉应逐渐提高球团矿入炉比率，部分大型高炉还应探索以球团矿为主的炉料结构优化。

（2）改善炉料冶金性能。在铁矿石价格持续波动、市场变化难于预测的形势下，对炉料综合冶金性能的要求却不容忽视，而且应更加注重提高炉料冶金性能的稳定性。未来大型高炉精料技术的主要技术内涵是：1）提高入炉矿品位和熟料率。入炉矿综合品位达到58%以上，熟料率80%~85%，采用经济合理炉料结构，降低原料成本。2）提高焦炭质量。要注重提高用于大型高炉焦炭的机械强度、热强度和平均粒度，M_{40}、M_{10}、CSR、CRI 等指标要满足大型高炉冶炼要求，特别是要保持焦炭平均粒度不小于 45mm 甚至更高。3）提高炉料成分和理化性能的稳定性，减少成分和性能的波动，保证高炉生产稳定顺行。4）有效控制炉料中有害元素的含量。由于资源的可获取性和经济性等原因，炉料中锌、钾、钠等有害元素的含量必须给予高度重视和有效控制。5）开展低品质矿的高效化利用研究，实现资源利用的减量化和最优化。

（3）炉料分布与控制技术。现代大型高炉生产效率高，炉料装入量大。一座 2500m³ 的高炉，每天由炉顶装入的矿石和焦炭就达 10000t/d 以上。炉顶设备不仅要满足高炉装料能力的要求，还要满足高炉操作对炉料分布精准控制的要求。炉料的合理分布与精准控制，是实现高炉煤气流合理分布、煤气化学能和物理显热高效利用的基础，是保障高炉生产稳定顺行、提高煤气利用率、降低燃料消耗的重要技术途径，是实现高炉生产顺稳和长寿的基础，炉料分布控制技术是现代高炉操作中不可或缺的重要调控手段。

4.5 高炉长寿技术

高炉长寿的实质是"在高炉一代炉役期间，保持合理的高炉操作内型"[9]。现代大型高炉设计寿命均为 15 年以上，特大型高炉 20~25 年，高炉一代炉役单位容积产铁量将达到 15000~20000t/m³，实现高炉长寿需要在高炉整个生命周期内采取有效措施，积极采用高炉长寿技术，维护合理的操作内型，形成基于"自组织、自维护、自修复"体系的"动态永久性"炉衬[10]。

（1）优化高炉操作内型。合理的高炉内型是实现高炉炼铁多目标优化的基础和前提。高炉内型在一代炉役期间是不断变化的，初始的设计内型是不断变化的操作内型的基础。因此，设计内型的合理与否，直接决定了高炉操作内型的变化进程及结果。毋庸置疑，高炉操作内型不仅影响着高炉寿命的长短，也对高炉生产稳定顺行、高效低耗具有至关重要作用。随着原燃料条件变化和高炉操作技术进步，现代高炉内型也将持续优化演变。现代高炉高径比渐趋降低，呈"合理矮胖化"发展。高炉有效高度与有效容积的关系并非线性耦合关系，统计表明，对于有效容积 2000m³ 以上的大型高炉，随着高炉容积的增加，其有效高度变化并不显著，特大型高炉高径比（H_u/D）均降低到 2.0 以下。为有效抑制炉缸铁水环流及其破坏作用，应适当增加死铁层深度。为适应富氧大喷煤强化冶炼要求，应适当增加炉缸直径和炉缸高度，改善风口回旋区工作条件和炉缸冶金的效能。为了优化高炉顺行条件，促进炉料下降和煤气流的顺利排升，可以适度增加炉腰直径，适当降低炉身角和炉腹角。

（2）采用长寿炉体结构。21 世纪大型高炉采用了合理的炉体长寿结构和先进的冷却技术，纯水（软水）密闭循环冷却系统在高炉上得到普遍应用，高炉炉底炉缸、炉体冷却器、风口、热风阀等均采用纯水（软水）冷却。现代高炉显著的结构特征是在炉腹至炉身中下部区域大规模采用铜冷却壁，依靠高效铜冷却壁以及配置合理的冷却系统，在风口以上区域形成基于渣皮自保护的"动态永久性内衬"，以延长高炉寿命。

（3）采用合理的炉缸炉底内衬结构。通过设计合理的死铁层深度，采用合理的炉缸炉底内衬结构设计，选用抗铁水渗透、抗铁水熔蚀性能优异的高导热炭砖，同时配置合理的高效冷却结构和冷却系统，有效抑制炉缸炉底"象脚状"异常侵蚀，使高炉寿命达到 15~20 年甚至更长。需要指出，长寿炉缸炉底设计的思考路径与设计方法应当是：1) 优化炉缸内型设计，确定合理的炉缸直径、炉缸高度、死铁层深度、风口个数、铁口个数。建立数学模型，模拟炉缸工作和渣铁排放过程的流动现象及其流场分布，对于精准设计具有重要参考意义。2) 确定炉缸炉底合理的内衬设计结构。无论采用何种结构设计，都应当重视炉缸侧壁厚度、炉底厚度、铁口深度、铁口角度等重要结构参数的选择。对于炭砖-陶瓷杯、陶瓷垫复合结构的内衬设计，应通过传热学仿真模拟，研究分析不同炉役阶段的温度场分布，根据仿真计算结果，优化炉缸炉底温度场分布，合理设计炭砖和陶瓷杯、陶瓷垫的匹配结构。3) 合理确定耐火材料的技术参数。理论研究和生产实践表明，炭砖的使用寿命是决定高炉寿命的关键要素。因此必须高度重视炉缸炉底炭砖物性参数的选择和指标设计，这是多目标的综合考量和权衡集成，不能片面强调某个单一指标的先进。必须兼顾抗铁水渗透、抗铁水熔蚀、抗碱侵蚀以及导热性，同时炭砖气孔率、孔径及其分布也是必须注重的技术指标。4) 重视高效合理冷却体系的配置。对于炉缸炉底而言，冷却是在整个炉役生命周期内都不容忽视的重要环节，特别是炉役后期，冷却系统的功能就更为重要。在炉缸炉底内衬体系确定以后，应通过传热学仿真优化设计，必须建立与炉缸炉底内衬传热体系合理匹配的冷却体系，以保障耐火材料内衬安全稳定工作。

4.6 高风温技术

高风温是高炉降低焦比、提高喷煤量、提高能源转换效率的重要技术途径。目前大型高炉的设计风温一般为 1250~1300℃，提高风温是 21 世纪高炉炼铁的重要技术特征之一。

20 世纪末期，5000m³ 以上的特大型高炉均采用外燃式热风炉，最具代表性的是新日铁式和地得式外燃式热风炉。首钢京唐 5500m³ 高炉开发并应用了顶燃式热风炉，这是世界上首次将顶燃式热风炉应用于 5000m³ 以上的特大型高炉，实现了特大型高炉热风炉结构形式的技术突破[11]。

现代大型高炉热风炉结构形式呈现多样化发展的同时,其主要技术发展特征是[9]:(1)利用空气、煤气低温双预热或富化煤气,使风温达到1250℃甚至更高。(2)采用预热炉预热助燃空气,在燃烧纯高炉煤气的条件下实现1250~1300℃风温。(3)通过优化燃烧过程、气流运动规律研究以及蓄热、传热机理研究,提高气流分布的均匀性;采用高效格子砖,增加传热面积,强化传热过程,缩小热风炉拱顶温度与风温的差值。(4)优化热风管道系统结构,采用无过热-低应力设计体系,合理设置热风管道波纹补偿器和拉杆结构,有效处理管道膨胀以降低管系应力,热风管道采用组合砖结构,消除热风管道的局部过热和窜风。(5)采取有效的炉壳晶间应力预防措施,延长热风炉使用寿命。(6)优化热风炉操作,合理设定热风炉工作周期,提高热风炉换热效率。(7)优化燃烧过程,降低燃料消耗,有效减少NO_x和CO_2的排放,实现节能减排与低碳环保。

4.7 煤气干法除尘与TRT技术

高炉煤气干法除尘技术是21世纪高炉实现高效低耗、节能减排、清洁生产的重要技术创新。高炉煤气干法除尘技术,提高了煤气净化度、煤气温度和热值,不但显著降低炼铁工序的新水消耗和能源消耗,还可以提高二次能源的利用效率、降低环境污染。高炉煤气干法除尘-TRT耦合技术,是高炉炼铁实现节能减排、低碳冶炼和高效能源转换的重要关键技术,已成为现代炼铁工业实施循环经济、实现低碳绿色发展的重要技术发展方向。

目前我国自主创新的高炉煤气干法布袋除尘技术,在设计研究、技术创新、工程集成及生产应用等方面取得重大突破性进展,在5500m^3特大型高炉上已成功应用6年以上,系统运行安全稳定,同湿法除尘技术相比,TRT发电量提高35%~40%。研究开发并形成了脉冲喷吹清灰技术、煤气温度控制技术、煤气含尘量在线监测、除尘灰浓相气力输送、管道系统防腐等关键核心技术,使大型高炉煤气全干法布袋除尘技术日臻完善[12]。

4.8 富氧喷煤技术

20世纪70~80年代,随着两次石油危机的相继爆发,使高炉喷煤技术得到迅猛发展,成为20世纪冶金工程技术进步的核心关键技术。直至目前,高炉喷煤仍是现代高炉重要的节能减排和低碳绿色技术,是现代高炉炼铁必须坚持发展的重点技术。

富氧鼓风是现代高炉提高生产效率的有效技术措施之一,提高富氧率对现代高炉炼铁意义重大,应给予高度重视。高炉富氧-喷煤-高风温集成耦合技术,可以有效改善炉缸风口回旋区工作,提高煤粉燃烧率和喷煤量,有效降低炉腹煤气量,有利于改善高炉透气性,促进高炉稳定顺行,提高煤气利用率,从而有效降低燃料消耗和CO_2排放。

高炉富氧-喷煤-高风温是降低燃料消耗和CO_2排放的重要支撑技术。在当前条件下还应当进一步提高喷煤量,使煤比达到200kg/t甚至更高,燃料比降低到500kg/t以下,这将是现代高炉炼铁工艺保持生命力和竞争力的根本所在。

毫无疑义,21世纪现代大型高炉必须大力推进并创新应用富氧-喷煤-高风温综合技术,这将成为未来高炉低碳绿色发展的重要技术路线。以实现高炉低碳绿色发展、降低高炉燃料消耗为技术目标,以提高风温、提高富氧率、提高喷煤量为技术途径,以精料技术、长寿技术和优化操作为技术保障,高炉富氧-喷煤-高风温综合技术必将具有广阔发展前景[13]。高炉富氧喷煤技术的作用如图4所示。

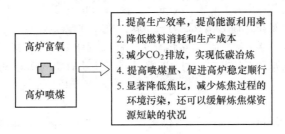

图4 高炉富氧喷煤技术的作用

5 结语

（1）进入21世纪以来，现代高炉炼铁技术面临严峻的挑战，必须采用建立低碳绿色发展理念，选择可持续发展道路，这是现代高炉炼铁技术未来发展的必由之路。现代高炉炼铁要构建动态精准、高效运行的生产体系，以高效率、低成本、低排放为目标，提高生产效率、降低能源消耗、减少环境污染，实现高炉炼铁"高效、低耗、长寿、低碳、绿色"的协同发展目标，着力提高综合技术装备水平，大力发展循环经济、实现低碳绿色制造。

（2）基于冶金流程工程学理论，高炉设计要着力建构物质流、能量流和信息流网络的优化结构，建立动态有序、协同连续、集约高效的流程网络和运行程序。构建现代高炉动态精准设计体系，注重要素、功能、结构和效率的协同优化，实现高炉"三大功能"，大力实施炼铁工业的循环经济、低碳冶炼和绿色发展。

（3）高炉精料、炉料合理分布控制、高效长寿、高风温、富氧喷煤、煤气干法除尘等关键共性技术，仍是未来高炉低碳绿色发展的重要技术基础、技术支撑和技术保障。面向未来，要始终坚持继承创新、集成创新和再创新，在已有技术的基础上，不断完善、持续改进，追求综合技术的集成效应和协同创新。

（4）通过实施循环经济、低碳冶金和绿色发展，现代高炉创新应用精料、长寿、高风温、富氧喷煤、节能减排、低碳冶炼、环境保护等综合技术，进一步降低资源和能源消耗，坚持高效、低成本、低排放和绿色发展理念，促进高炉炼铁的可持续发展，未来高炉炼铁技术仍将具有广阔发展前景。

致谢

衷心感谢殷瑞钰院士、张寿荣院士、干勇院士等老一辈院士专家的悉心指导和谆谆教诲，感谢杨天钧教授、王筱留教授、王维兴教授、吴启常大师、项钟庸大师等前辈师长的大力支持，感谢颉建新、毛庆武、钱世崇、程树森、张建良、沙永志等专家学者对作者学术观点的形成和建立给予的启发和帮助。

参考文献

[1] 张寿荣. 进入21世纪后中国炼铁工业的发展及存在的问题 [J]. 炼铁, 2012, 31 (1)：1-6.

［2］张寿荣，毕学工. 中国炼铁的过去、现在与展望［J］. 炼铁，2015，34（5）：1-6.
［3］张福明. 当代高炉炼铁技术若干问题的认识［J］. 炼铁，2012，31（5）：1-6.
［4］张福明. 21世纪初巨型高炉的技术特征［J］. 炼铁，2012，31（2）：1-8.
［5］殷瑞钰. 冶金流程工程学（第二版）［M］. 北京：冶金工业出版社，2009：274-277.
［6］殷瑞钰. 冶金流程集成理论与方法［M］. 北京：冶金工业出版社，2013：95-103.
［7］殷瑞钰，汪应洛，李伯聪，等. 工程哲学（第二版）［M］. 北京：高等教育出版社，2013：173-184.
［8］张福明，钱世崇，殷瑞钰. 钢铁厂流程结构优化与高炉大型化［J］. 钢铁，2012，47（7）：1-9，19.
［9］张寿荣，于仲洁. 武钢高炉长寿技术［M］. 北京：冶金工业出版社，2010：227-231.
［10］张福明，程树森. 现代高炉长寿技术［M］. 北京：冶金工业出版社，2012：579-581.
［11］张福明. 高炉高风温技术发展与创新［J］. 炼铁，2013，32（6）：1-5.
［12］张福明. 高风温低燃料比高炉冶炼工艺技术的发展前景［J］. 中国冶金，2013，23（2）：1-7.
［13］张福明. 大型高炉煤气干式布袋除尘技术研究［J］. 炼铁，2011，30（1）：1-5.

BF Ironmaking Technological Development Trend Characterized by Green and Low Carbon Emmission

Zhang Fuming

Abstract: By analyzing BF ironmaking functions, process flow integration and structure optimization, the paper expounds the development trend of blast furnace ironmaking technology in green and low carbon emmision and points out that the relevent key technologies are still important technical basis, technical support and technical assurance of green and low carbon emmission of blast furnace in future. These key technologies includes beneficiated burden material of blast furnace, rational distribution control of furnace burden, high efficiency and long campaign, high blast temperature, oxygen enriched PCI, and dry type gas dedusting, etc. The author also indicates it is necessary for the modern blast furnace ironmaking to establish dynamically precise and highly efficient production system in order to increase production efficiency, reduce energy consumption and lessen environment pollution for coordinative development target of high efficiency, low consumption and less emission. Additionally, improvement of comprehensive technological & equipment level and promotion of circular economy and green manufacture are also essential for the modern ironmaking plant.

Keywords: blast furnace; ironmaking; low carbon metallurgy

首钢绿色低碳炼铁技术的发展与展望

摘　要：进入新世纪以来，首钢进行了搬迁调整和结构优化。按照新一代可循环绿色低碳钢铁制造工程理念，在河北唐山地区相继建成了首钢股份和首钢京唐两个现代化钢铁制造基地。阐述了首钢炼铁技术绿色低碳发展理念，提出了发展途径和目标，总结了近年来首钢炼铁技术进步和创新实绩。重点论述了烧结料面喷吹蒸气技术、复合球团制备技术、高比率球团矿高炉冶炼技术以及冶金烟气综合治理与深度净化技术的研究和应用效果。面向未来，结合首钢炼铁技术的发展现状以及绿色低碳发展目标，对首钢炼铁技术的发展趋势进行了探讨和展望。

关键词：炼铁；高炉；球团；烧结；绿色低碳；环境保护

1　引言

进入新世纪以来，为了改善北京环境质量，成功举办 2008 年北京奥运会，落实北京城市总体发展规划，首钢北京地区钢铁生产设施相继停产、实施搬迁，进行产业结构调整和技术装备升级优化，在河北省唐山市迁安和曹妃甸地区，相继建设了两个现代化大型钢铁基地。特别是首钢京唐钢铁厂的设计建设，是基于循环经济和绿色低碳制造理念，构建了新一代可循环钢铁制造流程，以实现现代钢铁制造的先进钢铁产品制造、高效能源转换以及废弃物的消纳—处理和资源化再利用的"三大功能"。

首钢股份和首钢京唐两个现代化钢铁基地的生产能力分别为 800 万吨/年和 1400 万吨/年，产品结构全部为板带材，设计并配置了先进的烧结、球团、焦化、炼铁、炼钢、连铸、热轧、冷轧等单元工序。钢铁厂设计遵循循环经济理念，大力实施绿色、低碳、循环的发展战略，以构建新一代可循环钢铁制造流程和实现现代钢铁制造的"三大功能"为目标，以功能优化、结构优化、效率优化为关注点，着力开展单元工序功能的解析优化、工序之间的集成优化和流程结构的重构优化[1]。以资源、能源的减量化、再循环和再回收为基本原则，在整个钢铁制造流程中，着力建构单元工序内部的物质和能源的"小循环"，不同单元工序之间的物质和能源的"中循环"，以及钢铁企业与社会之间的物质和能源的"大循环"。以构建全流程物质流、能量流和信息流网络及其动态有序、协同连续高效运行为出发点，以期实现钢铁流程的多目标优化。

炼铁系统是钢铁制造流程中相态变化最复杂的工艺过程，炼铁系统的能源消耗约占整个钢铁制造流程的 70%，污染物排放约占整个钢铁制造流程的 70% 以上，同时炼铁系统的工序成本约占整个钢铁制造成本的 60%~70%，由此可见炼铁系统在整个钢铁制造流程中的具有突出的重要地位。因此，要实现绿色低碳钢铁制造，必须以炼铁系统为重点，大

本文作者：张福明。原刊于《钢铁》，2020，55（8）：11-18。

力开展清洁生产、节能减排和循环经济，通过优化炉料结构，提高铁矿资源的利用效率；不断降低能源消耗并加强能源回收利用，通过节能实现低碳和低排放；有效降低高炉燃料比、渣量、CO_2 和污染物排放，实现资源的最佳化利用和环境污染的源头消减。持续开发研究新技术、新工艺，对传统工艺流程及装备进行技术改造、改进和创新，降低工艺过程的污染物生成和排放，实现环境污染的过程控制。积极采用先进工艺和技术，加强烧结、球团、焦化等工序工艺过程烟气或尾气的净化和综合治理，加强烟气的"脱硫—脱硝—除尘"协同治理和深度净化，最大限度减少污染物的排放和对生态环境的影响。

2 主要技术进步与创新

2.1 多功能复合球团的开发与应用

2000 年以来，随着我国钢铁工业的快速发展，进口铁矿石用量不断增加，由于进口铁矿石中 Al_2O_3 质量分数普遍较高，导致高炉炉渣黏度增加、流动性恶化[2,3]。研究表明，加入 MgO 可以有效改善高 Al_2O_3 炉渣的冶金性能，因而需要在烧结矿中配加白云石等含镁熔剂以改善炉渣冶金性能。但是烧结过程中加入过量含镁熔剂以后，造成烧结矿转鼓强度和粒度下降等问题，为高炉生产带来不利影响[4]。因此，研究 MgO 的加入量及其合理的配加方式，提高炉渣综合冶金性能，成为优化高炉渣系的重要课题[5,6]。

长期以来，首钢京唐生产球团一直使用秘鲁磁铁矿粉，由于铁矿粉中 SiO_2 偏低，且 Na_2O 和 K_2O 碱金属质量分数较高，如果不配加高 SiO_2 铁矿粉，球团矿的还原膨胀率异常偏高，无法满足特大型高炉生产要求。首钢京唐带式焙烧机球团生产线投产以后，曾长期采用配加高 SiO_2 铁矿粉的措施，将球团矿中的 SiO_2 质量分数控制在 3.5% 以上，从而有效控制球团矿的还原膨胀率，但其后果则是降低了球团矿的品位和炉料的综合质量。随着高炉高效化生产，以及铁矿石资源的劣质化和钾、钠、锌等有害元素的不断蓄积增加，为防止高炉炉缸炉底内衬出现异常侵蚀，需要在炉料中加入含钛物料进行护炉操作，高炉长期配加含钛炉料进行护炉已成为延长高炉寿命的重要技术措施。为了避免大量使用低品质的钛铁矿或含钛冷固结球团，研究开发在球团矿中添加 TiO_2 的工艺及其方法势在必行。

首钢京唐以 400 万吨/年带式焙烧机球团生产线为依托，充分发挥大型带式焙烧机的技术优势，生产高品质、高性能的球团矿[7]。在含镁熔剂和含钛资源的选择、热工制度的优化控制等方面进行了大量的研究[8]，系统研究了国内外多种含钛铁矿粉的 TiO_2 质量分数及其成球性能的关系。理论与试验研究相结合，在实验室试验和工业试验基础上，在首钢京唐 504 m^2 带式焙烧机上实现了低硅-含钛-含镁多功能复合球团矿的稳定生产[9]。

2014 年 4 月，首钢京唐开始生产含钛球团矿，球团矿 TiO_2 平均质量分数由 0.09% 提高到 0.70%。2015 年，球团矿 TiO_2 质量分数进一步提高到 1.1%，MgO 质量分数为 1.7%，SiO_2 质量分数由 2.9% 降低到 2.8%，球团矿还原膨胀率等冶金性能保持稳定。首钢京唐复合球团矿的技术指标见表 1，首钢京唐配加复合球团后的炉料结构见表 2，图 1 所示为首钢京唐高炉采用复合球团矿前后高炉燃料比和渣量的变化。

表1 首钢京唐球团矿技术指标

项目	主要成分（质量分数）/%				抗压强度/N·球$^{-1}$	还原膨胀率/%	还原度/%
	TFe	SiO$_2$	MgO	TiO$_2$			
普通球团	65.87	3.5	0.51	0.06	2976	18.8	66.5
低硅含镁含钛球团	65.12	2.8	1.72	1.20	2687	19.1	72.3

表2 首钢京唐高炉配加含钛含镁球前后的炉料结构与效果

炉料结构（质量分数）/%					炉渣 MgO 质量分数/%	铁水 Ti 质量分数/%
烧结矿	普通球团矿	含镁含钛球团矿	块矿	钛矿		
63	26	—	9.5	1.5	7.87	0.052
65	—	29	6.0	—	7.92	0.093

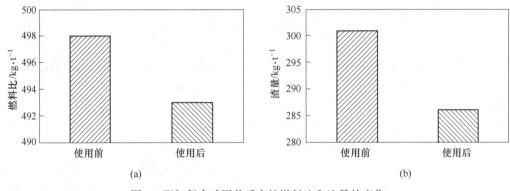

图1 配加复合球团前后高炉燃料比和渣量的变化
(a) 燃料比；(b) 渣量

2.2 烧结料面喷吹蒸汽技术

烧结过程中以配加在混合料中的固体碳作为燃料，通过负压抽风进行燃烧和铁矿粉的烧结。由于燃烧过程不充分，烟气中含有部分未燃尽的 CO 和烧结过程产生的二噁英、NO$_x$，因此研究应用烧结料面喷吹蒸汽技术是实现烧结过程污染物减排的一项重要过程治理技术。通过在烧结料面喷吹蒸汽可以有效改善烧结过程的供氧机制，提高烧结料层燃料燃烧效率和燃尽程度，减少燃烧反应过程对氧的依赖，降低废气中 CO 体积分数[10]。由于固体碳燃烧生成 CO$_2$ 和 CO 的放热量不同，从而有助于降低烧结固体燃料消耗，同时破坏二噁英生成条件，减少烧结过程 CO、NO$_x$ 等污染物排放。上述应用效果已在首钢京唐 550m^2 烧结机上得到验证。

2014年以来，通过建立烧结料面喷吹蒸汽的理论模型，进行了烧结料面喷吹蒸汽与降低固体燃料消耗、污染物排放、提高烧结矿质量的机理研究。针对首钢京唐 550m^2 烧结机工业化应用，研究了烧结料面喷吹蒸汽工艺和生产制度。对烧结料面喷吹蒸汽条件下，大型烧结机提质增效的关键技术进行了研究，实现了烧结过程二噁英和 CO 的协同减排[11]。试验研究和生产实践表明，通过在烧结料面喷吹蒸汽，使水蒸气参与烧结过程的燃烧反应，从而有效降低烧结烟气中残余的 CO；同时改善了烧结料层的传热机制，更有利于将烧结料层上部的热量传递和蓄积到烧结高温区，进而改善烧结料层温度场分布，有效抑制 NO_x 和二噁英的生成与排放，图 2 所示为烧结料面喷吹蒸汽的工艺功能解析。

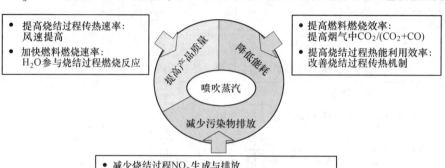

图 2　烧结料面喷吹蒸汽的工艺功能解析

2015 年 5 月，在首钢京唐 1 号 550m^2 烧结机上，进行了喷吹蒸汽工业试验。试验期间，烧结负压降低了约 0.5kPa，主抽风机的废气温度提高了 5~7℃，喷吹蒸汽前后烧结矿的成分基本稳定。喷吹蒸汽后，烧结矿转鼓指数略有提高，返矿率降低了约 0.3%，烧结固体燃料消耗降低了 1.64kg/t，烧结矿的粒度分布有所改善，其中 5~10mm 比例降低了 0.8%。

实践表明，烧结过程料面喷吹蒸汽有助于改善烧结矿质量和降低固体燃耗。总体来看，烧结料面喷吹蒸汽以后，烧结速度和料层透气性得到改善，在降低固体燃耗的同时，烧结矿质量有所改善。烧结烟气中二噁英和 CO 体积分数显著降低，污染物减排效果显著（见表 3）。CO 质量浓度减排率为 25%，二噁英质量浓度减排率为 50%。图 3 所示为首钢京唐 550m^2 烧结机料面喷吹蒸汽实况照片。

表 3　烧结料面喷吹蒸汽前后烧结烟气中二噁英和 CO 的变化

项目	CO 质量浓度/g·m^{-3}	二噁英质量浓度/ng-TEQ·m^{-3}
试验前	5~23	0.033~0.350
试验后	3~18	0.013~0.037

图 3　首钢京唐 550m^2 烧结机料面喷吹蒸汽实况

2.3 高比率球团矿冶炼技术

同烧结矿相比,球团矿具有品位高、成本低、工序能耗低、污染物排放少等综合优势,因此提高炉料中球团矿比率有利于改善高炉操作、优化技术指标、降低环境污染。近年来,首钢结合自有资源条件,充分利用自产秘鲁球团矿粉资源,开展了 $5500m^3$ 特大型高炉高比率球团矿冶炼技术研究。

2015 年 12 月,在首钢京唐 1 号高炉进行高比率球团矿冶炼工业试验,球团矿比率从初始的 33%,按照 41%、46%、51% 三个阶段逐步提高球团矿比率,探索了特大型高炉使用 50% 以上大比率球团矿冶炼的可行性和适应性。工业性试验前高炉生产稳定顺行,矿石批重为 172t/批,焦炭批重为 30.9t/批(干基),焦炭负荷为 5.57t/t,焦丁为 2.1t/批,高炉风量达到 $8400m^3/min$ 以上,铁水温度为 1500℃,风温为 1250℃,喷煤量为 105~110t/h。试验期间球团矿中 MgO 质量分数控制在 1.60%(之前为 1.73%),TiO_2 质量分数控制在 1.15%(之前为 1.27%)。首钢京唐 1 号高炉高比率球团冶炼工业试验效果见表 4。

表 4 首钢京唐 1 号高炉高比率球团冶炼工业试验效果

日期	日产量/$t \cdot d^{-1}$	入炉焦比/$kg \cdot t^{-1}$	煤比/$kg \cdot t^{-1}$	焦丁比/$kg \cdot t^{-1}$	燃料比/$kg \cdot t^{-1}$	铁水一级品率/%
19~21 日平均	12935	283	191	19	493	79
22~24 日平均	12875	287	193	19	499	79
比较	-59	4	2	0	6	0

首钢京唐 1 号高炉高比率球团矿冶炼工业试验期间,高炉透气性指数变化稳定,风量与压差关系平稳,高炉运行状况良好,主要操作指标基本稳定,未出现大幅度炉况波动。试验期间平均风量为 $8419m^3/min$,与试验前的风量(前 3 天平均值)$8435m^3/min$ 基本持平,炉顶煤气温度、压力稳定。实践证实,高比率球团矿冶炼试验取得初步成功,为首钢京唐 3 座特大型高炉全面实施高比率球团矿冶炼进行了探索、积累了经验,奠定了生产实践基础。

2019 年 4 月,首钢京唐 3 号高炉($5500m^3$)建成投产,形成了 3 座 $5500m^3$ 高炉同时生产的工艺流程和生产格局。为了实现 50% 以上高比率球团矿冶炼的目标,开发并生产了碱度为 1.2 以上的碱性球团(见表 5)[12],以适应特大型高炉生产的需要,首钢京唐 3 座特大型高炉炉料中,球团矿入炉平均比率将达到 50% 以上,大幅度提高了资源利用效率和高炉效能,有效降低能耗和污染物排放。

表 5 普通球团和碱性球团的技术指标

项目	化学成分(质量分数)/%						抗压强度/$N \cdot 球^{-1}$	碱度(CaO/SiO_2)	还原度/%
	TFe	FeO	SiO_2	Al_2O_3	CaO	MgO			
普通球	65.15	0.48	3.29	0.78	0.57	0.73	2650	0.17	73.2
碱性球团矿	64.48	0.40	2.02	0.51	2.50	0.80	2828	1.24	83.1

2.4 冶金烟气综合治理与深度净化

钢铁工业的发展必然要受到资源、能源可持续供给和生态环境保护的制约。近10年来，随着生态文明建设的持续推进和各级政府环保标准的不断提高，钢铁行业烟气深度净化与污染物减限值要求排日益严格[13~15]（见表6）。由于烧结、球团工艺过程中烟气排放量大、烟气中 SO_2 和 NO_x 的体积分数高、烟气成分波动大，对烧结、球团烟气净化工艺提出了更高要求。河北唐山地区紧邻京津，钢铁产能集聚，冶金烟气排放总量巨大，大气治理形势严峻。首钢股份、首钢京唐均处于河北唐山地区，烧结、球团、焦化的烟气必须经过深度处理达标排放（见表7），与此同时唐山市地方政府环保部门还规定新建项目必须达到颗粒物低于 $5mg/m^3$、SO_2 低于 $20mg/m^3$、NO_x 低于 $30mg/m^3$ 的污染物排放要求。

表6 钢铁行业烟气排放标准　　　　　　　　　　　　　　　（mg/m^3）

项　目	烟气排放指标		
	国家标准	原河北省地方标准	河北省超低排放标准
颗粒物	50	40	≤10
SO_2	200	180	≤35
NO_x	300	300	≤50

表7 首钢烧结、球团烟气脱硫-脱硝工艺配置及运行情况

项　目	脱硫-脱硝工艺装置	投产时间	污染物排放参数/$mg·m^{-3}$		
			粉尘	SO_2	NO_x
首钢京唐1号烧结机（$550m^2$）	半干法+SCR	2019年1月	1.2	10~20	10~30
首钢京唐2号烧结机（$550m^2$）	半干法+SCR	2019年1月	1.3	10~20	10~30
首钢京唐1号球团（带式焙烧机$504m^2$）	活性焦法	2016年10月	14~23	25~95	170~200
首钢股份烧结一车间（$4×99m^2$）	活性焦法	2019年1月	15	50	160
首钢烧结二车间（$360m^2$）	密相干塔法+SCR	2018年11月	2	10~30	20
首钢股份1号球团 （链算机-回转窑150万吨/年）	密相干塔法+SCR	2018年10月	1.5	10~30	20
首钢股份2号球团 （链算机-回转窑200万吨/年）	活性焦法	2018年12月	3	1	30

冶金烟气的综合治理，首先应当考虑源头消减，有效抑制烟气中 SO_2、NO_x、二噁英的生成，加强工艺过程优化和有效控制，降低污染物的生成和聚集，采取有效措施加强烟气末端深度净化处理，采用先进的脱硫、脱硝、除尘工艺技术和装备，首先要降低烟气排放总量，有效减少污染物排放总量和排放浓度。

在满足 SO_2 减排总量和达标排放的条件下，还要充分考量脱硫副产品的处理和再利用问题，减少二次污染，实现循环经济和废弃物资源的高值转化。

目前，应用比较广泛的烟气脱硫-脱硝工艺主要有干法活性焦/炭工艺和半干法循环流化床脱硫+选择催化还原（SCR）脱硝工艺（见表8）。干法活性焦脱硫-脱硝工艺以物理化学吸附和催化反应原理为基础，可以实现脱硫、脱硝、脱二噁英、脱重金属、脱卤化

氢以及除尘等多功能集成的烟气深度净化和污染物协同脱除，具有清洁环保、协同治理、资源回收、节水显著、运行稳定等优点。适用于烧结和球团烟气处理量大、烟气成分和温度波动大的工况特点，回收烟气中的 SO_2 制备浓硫酸，废水、废渣和废气很少，二次污染较低，是适用于烧结和球团烟气深度净化的主流工艺。半干法循环流化床脱硫工艺基于循环流化床原理，通过对脱硫剂的多次再循环，延长脱硫剂与烟气的接触时间，有效提高了脱硫剂的利用率和脱硫效率。烟气循环流化床脱硫工艺使脱硫塔内达到一种激烈的湍流状态，从而加强了脱硫剂对 SO_2 化学吸收。该工艺采用消石灰加水形成的石灰浆作为脱硫剂，其主要技术优势是工艺简单，无需烟气冷却和加热，设备故障率较低，占地面积小、空间布局紧凑，工程投资和运行费用相对较低，基本无废水排放。其主要缺点是脱硫过程产生的固体废弃物，难于消纳处理实现资源化，还可能造成二次污染。在首钢股份和首钢京唐的烧结、球团烟气脱硫-脱硝工艺选择过程中，根据生产装置特点、原料条件以及已有的技术基础，分别采用了干法活性焦、半干法循环流化床和半干法密相干塔等不同的烟气脱硫工艺，与喷氨法烟气选择性脱硝工艺（SCR）相匹配，以期实现冶金烟气的综合治理和深度净化。

表 8 烟气脱硫-脱硝工艺技术比较

工艺	优 点	缺 点
干法活性焦工艺	（1）可以实现多种污染物（如 SO_2、NO_x、二噁英和重金属等）协同治理和减排。 （2）活性焦法不产生固体废弃物，从根上消除了固体废弃物的二次处理问题。 （3）可实现污染物资源化利用，有利于清洁生产，SO_2 脱除后制备硫酸，实现资源回收利用。 （4）基于干法净化工艺，节水效果显著。 （5）脱硫-脱硝为"一体化"集成工艺，有利于工艺过程总体控制	（1）烟气中 SO_2 质量浓度超高条件下，容易发生局部过热，甚至着火。 （2）烟气中 SO_2 质量浓度高于 $1300mg/m^3$ 时，难以实现超低排放，活性焦消耗比较高，导致运行成本升高。 （3）单一活性焦工艺脱硝效率不高，复杂工况烟气条件下对活性焦催化、吸附能力的影响较大，喷氨后烟囱出现"拖尾"现象，脱硝效率难以保证。 （4）活性焦资源供不应求，质量和采购价格波动较大，难以保障稳定的供应和性能指标。 （5）在活性焦吸附-解析和活性焦输送过程中产出超细碳粉，缺乏有效脱除措施，二次颗粒物的产生难以达到超低排放。 （6）活性焦法解析制酸过程中产生少量废水难于处理；钢铁厂内部硫酸再利用存在问题
半干法脱硫（循环流化床法）+ SCR 脱硝工艺	（1）脱硫-脱硝采用组合式集成工艺，单元工艺相对成熟。 （2）通过增加脱硫剂、脱硝剂、催化剂数量，提高反应温度等措施，可以实现脱硫-脱硝超低排放	（1）SCR 使用后催化剂属于危险废弃物，需要特殊的处理工艺进行无害化处理。 （2）中温 SCR 脱硝需要对烟气进行加热升温，从而增加能耗。 （3）脱硫过程产生的脱硫灰为固体废弃物，需要二次资源化处置，固体废弃物二次处理存在问题。 （4）半干法脱硫+SCR 脱硝工艺对烟气中其他污染物（如二噁英、重金属等）的脱除效果尚待验证

2010 年底，首钢京唐 2 台 $550m^2$ 烧结机烟气脱硫系统建成运行，该项目采取循环流化床工艺，系统脱硫效率达到了 90%，脱硫后烟气中 SO_2 可以降低到 $30\sim50mg/m^3$。2018

年7月对烧结烟气脱硫系统进行升级改造,增加了脱硝和除尘等功能。2018年12月技术改造完成投产运行,运行实绩可以达到SO_2低于$20mg/m^3$、NO_x低于$30mg/m^3$、粉尘低于$5mg/m^3$超净排放地区标准要求(见表9)。

表9 首钢京唐550m^2烧结烟气脱硫-脱硝工艺技术参数

项 目	烟气脱硫-脱硝之前	烟气脱硫-脱硝之后
烟气处理量(标准状况)/$m^3 \cdot h^{-1}$	$2 \times 163 \times 10^4$	$2 \times 163 \times 10^4$
烟气温度/℃	145±15	90~100
SO_2质量浓度/$mg \cdot m^{-3}$	400~600	≤20
NO_x质量浓度/$mg \cdot m^{-3}$	150~450	≤30
粉尘质量浓度/$mg \cdot m^{-3}$	80~100	≤5

2016年12月,首钢京唐504m^2带式焙烧机球团生产线脱硫-脱硝系统建成运行[16],采用干法活性焦脱硫-脱硝工艺,经过近一年的改进完善,烟气处理后出口粉尘质量浓度为14~23mg/m^3,SO_2质量浓度为25~95mg/m^3,NO_x质量浓度为170~200mg/m^3,基本达到了原设计水平和原河北省地方排放标准。2018年10月,政府部门颁布超低排放标准要求粉尘、SO_2、NO_x的小时均值排放标准分别低于$10mg/m^3$、$35mg/m^3$、$50mg/m^3$,且按烟气中氧气体积分数16%折算进行考核。为达到最新环保标准要求,目前首钢京唐正在开展对活性焦脱硫-脱硝工艺进行升级技术改造,增加循环流化床脱硫和SCR喷氨脱硝工艺,以满足唐山地区污染物限值排放标准要求。

3 展望

进入21世纪以来,首钢率先开展战略性搬迁调整和产业升级,基于新一代可循环钢铁制造流程绿色低碳发展理念,自主设计建造了新世纪现代化钢铁厂,采用了国际先进的工艺技术装备,在大型化、现代化、智能化、绿色化等方面具有引领性和创新性。面向未来,以高炉为中心的炼铁系统协同优化和动态有序、协同连续、精准高效运行,是首钢炼铁技术的发展重点。必须加强特大型高炉以稳定顺行为基础的工程运行理念,建立系统性、全局性的工程思维模式,不片面追求所谓的"低成本"和个别技术指标的"领先",摒弃不讲客观、不论条件的盲目攀比,遵循钢铁冶金制造流程的基本规律,科学认识高炉冶炼过程的动态运行规律,不断总结提升,加强知识管理,做好卓越炼铁工程师的培养,造就基础扎实、经验丰富、视野开阔领军人才,形成具有特色的现代化大型炼铁生产、管理的工程思维和工程理念。面向未来经济下行压力加大、市场需求减弱、生态环保严格的宏观形势,未来一个时期,首钢炼铁技术必须遵循绿色低碳的发展理念,主要发展路径可以概括为:

(1)进一步加强精料技术研究,探索并构建以球团矿为主的新型炉料结构,降低整个炼铁流程的碳素消耗和污染物排放,从而在现有基础上进一步降低CO_2排放。

(2)继续推进高风温、富氧喷煤等关键技术的持续创新,进一步降低高炉燃料消耗。

(3)加强特大型高炉操作规律的研究,建立动态有序、协同连续、精准高效的现代高炉操作理念，以高炉生产长期稳定顺行为基础，不追求个别指标先进，不断改善、优化、提升特大型高炉的操作水平。

（4）进一步加强高炉运行过程的炉体维护，采取有效技术措施延长高炉寿命。

（5）构建料场、烧结、球团、焦化、高炉炼铁系统一体化集成智能管控平台，着重解决不同工序的界面技术优化，实现全流程的智能化动态管控。

（6）发挥既有资源优势，开展绿色低碳炼铁新技术的深度研究探索。依托首钢自有超低硅铁矿资源，开展超低渣量和燃料比炉料制备和冶炼技术研究。研究生产超低硅球团矿、超低硅烧结矿，使高炉渣量降低到 200kg/t 左右，燃料比降低到 470kg/t，进而在现有基础上可以实现 CO_2 减排 15%~20%。

4 结论

（1）进入 21 世纪以来，首钢通过搬迁结构调整、工艺技术装备升级，在技术装备大型化、现代化取得了长足进步，基于新一代可循环钢铁制造流程运行取得初步成效。

（2）以绿色低碳为发展理念，在特大型高炉稳定运行、复合球团生产、烧结料面喷吹蒸汽、高炉高比率球团冶炼、冶金烟气综合治理与深度净化等方面取得良好效果。

（3）面向未来，应进一步树立建立以特大型高炉稳定顺行为核心的动态运行工程理念，加强精料、高风温、富氧喷煤、高炉长寿和全流程智能化研究，依托既有资源优势，开发新型炉料结构，开展绿色低碳炼铁技术的探索和研究。

致谢

感谢曹朝真博士、赵志星博士为本文所提供的资料和帮助；感谢首钢京唐、首钢股份、首钢技术研究院、首钢工程技术公司等炼铁同仁们的积极探索和技术创新工作。

参考文献

[1] Zhang F M, Zhao M G, Cao C Z. New technological progress of ironmaking in shougang [C]//Proceedings of the Iron and Steel Technology Conference 2017. Nashville：AISTech，2017：439.

[2] 沈峰满，姜鑫，高强健，等. 高炉炉渣适宜镁铝比的理论基础 [J]. 炼铁，2019，38（2）：17.

[3] 姜鑫，沈峰满，韩宏松，等. 高炉炉渣适宜镁铝比分段管控的分析与应用 [J]. 钢铁，2019，54（10）：12-16.

[4] 程峥明，裴元东，刘伯洋，等. 首钢京唐低硅低镁烧结技术的研究与应用 [J]. 中国冶金，2017，21（11）：41.

[5] 王冬青，马泽军，陈辉，等. 首钢 A 高炉炉渣降低 MgO 的热力学分析 [J]. 中国冶金，2018，28（10）：5-9.

[6] 任立军. 京唐 5500m³ 高炉降低（MgO）生产实践 [J]. 炼铁，2018，37（1）：20.

[7] Zhang F M, Wang Q S, Han Z G. Innovation and application on pelletizing technology of large traveling grate induration machine [C]//Proceedings of the Iron and Steel Technology Conference 2015. Cleaveland：AISTech，2015：402.

[8] 青格勒，吴铿，屈俊杰，等. 不同含镁添加剂对球团矿工艺参数及质量的影响 [J]. 钢铁，2013，48（7）：17.

[9] 青格勒, 王朝东, 侯恩俭, 等. 低硅含镁球团矿抗压强度及冶金性能 [J]. 钢铁研究学报, 2014, 26 (4): 7.

[10] 裴元东, 欧书海, 马怀营, 等. 烧结料面喷吹蒸汽对烧结矿质量和 CO 排放影响研究 [J]. 烧结球团, 2018, 43 (1): 35.

[11] 裴元东, 史凤奎, 吴胜利, 等. 烧结料面喷吹蒸气提高燃料燃烧效率研究 [J]. 烧结球团, 2016, 41 (6): 16.

[12] 田筠清, 青格勒, 刘长江, 等. 使用石灰石生产低硅碱性球团矿试验 [J]. 中国冶金, 2018, 28 (4): 13.

[13] 环境保护部, 国家质量监督检验检疫总局. GB 28662—2012 钢铁烧结、球团工业大气污染物排放标准 [S]. 北京: 中国环境科学出版社, 2012.

[14] 河北省环境保护厅, 河北省质量技术监督局. DB 13/2169—2015 钢铁工业大气污染物排放标准 [S]. 石家庄: 河北省环境保护厅, 2015.

[15] 河北省环境保护厅, 河北省质量技术监督局. DB 13/2169—2018 钢铁工业大气污染物超低排放标准 [S]. 石家庄: 河北省环境保护厅, 2018.

[16] 王代军. 首钢京唐球团烟气治理研究与应用 [C] // 中国环境科学学会学术年会论文集 (2017). 厦门: 中国环境科学学会, 2017: 1329.

Development and Prospect of Green and Low Carbon Ironmaking Technologies in Shougang

Zhang Fuming

Abstract: Since the beginning of the new century, Shougang has carried out relocation adjustment and structural optimization. According to the engineering philosophy of new generation recyclable green and low carbon of iron and steel manufacturing process, two modern iron and steel manufacturing enterprises, Shougang Stock Company and Shougang Jingtang Company, have been built in Tangshan region, Hebei province. The green and low carbon developing the concept of ironmaking technology in Shougang was expounded, the developing approaches and objectives, and summarizes the progress and innovation achievements of the ironmaking technology of Shougang in recent years was proposed. The research and application effects of steam spraying on sintering bed surface technology, composite pelletizing preparation technology, high ratio pellet burden blast furnace operating technology, comprehensive treatment, and deep purification technology of metallurgical flue gas are discussed in detail. Facing the future, combined with the development status of Shougang ironmaking technology as well as the green and low carbon development goal, the development trend of Shougang ironmaking technology is discussed and prospected.

Keywords: ironmaking; blast furnace; sintering; pelletizing; green and low carbon; environmental protection

中国高炉炼铁技术装备发展成就与展望

摘　要：近40年来，我国钢铁工业取得了巨大进步，钢铁产量连续多年居世界第一。我国高炉炼铁技术装备在大型化、现代化、高效化、长寿化等方面发展成就显著。2000年以来，一批5000m^3以上特大型高炉、500m^2以上大型烧结机、7.63m超大容积焦炉和年产400万吨/年以上大型球团生产线相继建成投产，一系列自主研发、集成创新的炼铁关键技术在生产实践中取得重大应用成效。在技术装备大型化的同时，高炉富氧喷煤、无料钟炉顶、煤气干法除尘、顶燃式热风炉及高风温、高效低耗烧结技术、大型清洁炼焦技术等先进技术及其装备研发与应用成效显著，有力推动了炼铁技术装备进步。到21世纪中叶，我国钢铁工业格局和流程结构将发生重大变革，减量化、绿色化、智能化、高效化将是未来一个时期炼铁技术装备的主要发展趋势。

关键词：炼铁；技术装备；高炉；烧结；球团；焦化；绿色化

1　引言

近40年来，我国钢铁工业发展迅猛，钢铁产量逐年增长[1~3]。2018年我国生铁产量达到77105万吨，占世界生铁总产量的62.23%[4]，中国和世界高炉生铁产量变化如图1所示（数据来源于国家统计局网站和世界钢铁协会网站）。在钢铁产量增长的同时，我国冶金技术装备也取得长足进步，在炼铁、烧结、球团、焦化技术装备大型化、现代化、高效化、长寿化等方面成效卓著[5]，经过40多年的自主创新、集成创新和引进消化、再创新，高炉炼铁技术装备发展成就举世瞩目、成就斐然。

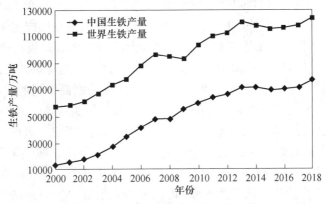

图1　中国和世界高炉生铁产量变化

本文作者：张福明。原刊于《钢铁》，2019，54（11）：1-8。

20世纪80年代，宝钢工程引进日本技术和装备，建设了4063m³高炉、450m²烧结机、6m焦炉等代表20世纪80年代初国际先进水平的大型炼铁技术装备。通过对宝钢工程单元工艺、技术、设备和装置的消化、吸收、移植和再创新，我国炼铁技术装备在已有技术的基础上，不断取得新的技术装备创新成果。

20世纪90年代，我国炼铁技术装备自主集成、自主创新能力显著提升，自主设计建造了一批2000~3000m³高炉、一批260~450m²烧结机、一批年产100万~250万吨/年链箅机-回转窑球团生产线和一批6.0m顶装焦炉，自主集成国际先进技术设计建造了3200和4350m³高炉。

2000年以来，我国炼铁技术装备自主创新能力持续增强，与发达国家"并驾齐驱"，有的甚至领跑世界。近10年，我国自主设计、建造了5000m³以上特大型高炉、500m²以上大型烧结机，自主集成创新了500m²以上带式焙烧机球团生产线和65孔7.0m以上大型焦炉以及260t/h干熄焦装置。这些具有国际先进水平的大型炼铁技术装备，支撑了我国钢铁工业的迅猛发展，提升了我国钢铁工业的综合技术实力。

2 发展现状

2.1 高炉

据不完全统计，2017年我国约有917座高炉，高炉炼铁总能力约为92700万吨/年[6]。我国各级别高炉的数量和产能分布见表1[6]。

表1 我国高炉级别的分布和产能

项目	≥3000m³	2000~3000m³	1200~2000m³	450~1200m³	≤450m³	合计
高炉数量/座	41	76	135	452	213	917
高炉产能/万吨·年⁻¹	12660	15172	17227	36433	11240	92732

1985年9月，通过引进日本技术设计建造的宝钢1号高炉（4063m³）建成投产，这是我国首座4000m³以上的高炉。1990年，武钢通过技术引进和集成创新，采用多项当时国际先进技术的武钢新3号高炉（3200m³）建成投产[7]，随后建成投产的还有宝钢2号高炉（4063m³）、3号高炉（4350m³）等几座大型高炉。在此期间，首钢、鞍钢、马钢、包钢、本钢、唐钢等一大批自主设计建造的2000~2500m³级高炉相继建成投产，至20世纪90年代末期，我国大型钢铁企业的高炉技术装备水平普遍提高。2000年以后，我国高炉技术装备大型化取得飞跃发展，鞍钢、武钢、首钢迁钢、天钢、唐钢、邯钢、梅钢等一大批2500~3200m³级高炉相继建成投产；宝钢、太钢、马钢、鞍钢鲅鱼圈、首钢迁钢、本钢、安钢、包钢等一批4000m³级大型高炉也相继建成投产；特别是自主设计建造的首钢京唐1号5500m³高炉是我国首座5000m³以上特大型高炉，于2009年5月建成投产，其后沙钢5800m³高炉于2009年10月建成投产，京唐2号高炉（5500m³）于2010年6月建成投产。宝钢湛江2座5050m³高炉分别于2015年和2016年建成投产，山钢日照2座5100m³高炉也即将建成投产。

近20年来，我国高炉大型化进程发展迅猛，高炉容积不断扩大，高炉技术装备水平不断提升。我国在4000m³以上高炉的设计、建造和运行等方面，已经完全掌握了核心关键技术，全面实现了自主设计和技术装备国产化。

2017年，我国有23座4000m³以上高炉运行，其中有6座是5000m³以上的特大型高炉[8]。我国高炉大型化和技术装备现代化与淘汰落后、技术升级密切相关。近年来，我国加大产业结构调整和转型升级力度，通过供给侧结构性改革，淘汰落后产能和技术装备，产业结构调整和升级取得显著效果。目前，我国高炉平均容积达到1047m³，高炉产能集中在1000~2000m³级的高炉，这一级别的高炉产能约占生铁总量的35.8%，而在2011年和2014年我国高炉平均容积分别为580m³和770m³[9]。

在高炉技术装备大型化、现代化、高效化的同时，我国高炉炼铁技术创新也取得重大进步[10]。自主设计研发了高炉无料钟炉顶设备及其炉料分布控制技术，在首钢京唐3座5500m³高炉和湛江2座5050m³高炉上得到成功应用；自主创新了特大型高炉煤气全干法布袋除尘技术和炉顶煤气余压发电技术（TRT），净煤气含尘量降低到5mg/m³以下，TRT发电量达到45kW·h/t以上，比传统的煤气湿法除尘工艺提高煤气余压发电量35%以上；采用顶燃式热风炉和高风温综合技术，在燃烧低热值高炉煤气的条件下，高炉风温达到1250~1300℃[11]；通过采用合理高炉炉型、纯水（软水）密闭循环冷却、铜冷却壁、新型优质耐火材料、自动化监测等综合技术，可以延长高炉寿命，我国先进大型高炉寿命已达到15年以上；积极推广采用高炉富氧喷煤技术，先进高炉煤比已达到180~200kg/t甚至更高。大型高炉鼓风机、煤气余压发电装置、无料钟炉顶设备、高效长寿铜冷却壁、高温长寿热风阀等高炉主要技术装备均已实现全面国产化。

近年来，我国高炉技术经济指标不断改善[12~14]，高炉炼铁工序能耗从407.62kgce/t降低到390.8kgce/t，热风温度由1127提高到1142℃，劳动生产率提高到7200t/(人·a)以上。近4年主要钢铁企业高炉技术经济指标见表2。

表2 近4年主要钢铁企业高炉技术经济指标

指 标	2018年	2017年	2016年	2015年
高炉利用系数/t·(m³·d)⁻¹	2.58	2.51	2.48	2.46
燃料比/kg·t⁻¹	536	544	543	526
入炉焦比/kg·t⁻¹	372	363	363	358
煤比/kg·t⁻¹	139	148	140	142
热风温度/℃	1140	1142	1139	1134
入炉矿品位/%	57.4	57.3	57.0	57.0
熟料率/%	89.9	89.1	84.9	89.3
休风率/%	1.86	1.93	2.57	2.27
劳动生产率/t·(人·a)⁻¹	7263.9	7227.2	6998.1	6093.8
炼铁工序能耗/kgce·t⁻¹	392.1	390.8	390.6	387.3

注：统计的钢铁企业为中国钢铁工业协会会员单位。

2.2 烧结

由于资源禀赋和技术传承，我国形成了以高碱度烧结矿配加酸性球团和块矿的高炉炉料结构,铁矿粉造块主要以烧结工艺为主。1985年,宝钢通过引进国外先进技术,建成了我国第一台450m²现代化大型烧结机,带动了我国烧结工艺技术的发展。2000年以后,我国烧结技术进入了前所未有的高速发展时期,一大批180~660m²大型烧结机相继建成投产,2010年投产的太钢660m²烧结机,是当时世界上最大的烧结机。

近20年来,通过对烧结机密封、给料、布料等装置的持续改进,在提高产量、改善质量、降低能耗等方面取得显著成效。近年来,我国烧结机大型化进程加快,2018年我国360m²以上烧结机已经超过100台,主要钢铁企业拥有280多台烧结机,烧结机平均面积已经达到240m²以上[14]。当前我国大型烧结机的设计、制造和运行均已经达到国际先进水平[15],为我国烧结生产技术进步奠定了坚实基础。

随着烧结机装备大型化,烧结新工艺、新技术、新设备的研发应用也取得了显著成效。烧结新型点火、偏析布料、超厚料层烧结、烧结机密封、烧结环冷机余热回收利用、烟气脱硫-脱硝等技术被广泛采用。烧结生产自动化、智能化控制水平不断提高;以针状复合铁酸钙(SFCA)为主的高碱度烧结矿得到普遍应用;烧结料层平均厚度达到688mm,最高可达900mm以上。由于技术装备和生产水平的提高,烧结技术经济指标不断改善。2018年我国主要钢铁企业烧结工序能耗降至48.6kgce/t,烧结矿质量性能稳定提高。近3年主要钢铁企业烧结技术经济指标见表3。

表3 近3年主要钢铁企业烧结技术经济指标

项 目	2018年	2017年	2016年
烧结矿含铁品位/%	55.65	55.79	55.65
烧结矿转鼓指数/%	78.29	77.96	76.69
烧结固体燃料消耗/kg·t⁻¹	53.16	52.87	53.08
烧结利用系数/t·(m²·d)⁻¹	1.28	1.28	1.26
烧结日历作业率/%	93.42	93.58	91.74
烧结工序能耗/kgce·t⁻¹	48.60	48.50	48.53

注:统计的钢铁企业为中国钢铁工业协会会员单位。

2.3 球团

长期以来,我国球团技术装备和产量在世界上均处于相对落后水平。2000年以来,我国球团技术取得了迅猛发展,2017年球团矿产能已增加到17000万吨以上。武钢鄂州、宝钢湛江相继建成了年产500万吨链算机-回转窑球团生产线,引领我国链算机-回转窑装备大型化发展;带式焙烧机原料适应性强、生产规模大、工艺成熟、技术可靠[16]。目前,我国生产运行的带式焙烧机共6台,分别是鞍钢321.6m²带式焙烧机、包钢624m²和162m²带式焙烧机、首钢京唐3台504m²带式焙烧机,我国带式焙烧机球团生产能力已达到2000万吨/年以上。目前,我国链算机-回转窑球团生产线约有100条,链算机-回转窑球团生产能力约占全国球团总产能的一半以上。

同烧结工艺相比,球团生产工艺具有产品品位高、质量高、能耗低、污染小等技术优势。近年来我国加快了球团技术的发展,采用铁矿粉高压滚磨、润磨预处理技术,赤铁矿、镜铁矿生产球团技术,降低膨润土技术、链算机-回转窑防止结圈技术、复合球团造块技术，混合原料复合球团制备与焙烧技术等取得成功应用，首钢京唐采用大型带式焙烧机生产含镁球团、含钛球团取得显著成效，镁质球团、碱性球团、自熔性球团的制备与应用获得成功，为我国高炉实现大比例球团冶炼创造了有利条件。

2.4 焦化

2000年以来，我国焦化总体技术装备水平不断提高，焦炭产量持续增长。2016年，我国独立焦化厂为468家，焦炭产能为43600万吨；钢铁企业联合焦化厂为134家，焦炭产能为25100万吨[17]。2018年，我国焦炭产量达到43820万吨。

我国焦炉大型化及干熄焦（CDQ）技术的推广应用，提高了焦炭质量，降低了炼焦能耗，减少了CO_2和NO_x的排放，目前我国炼焦技术已经进入国际先进行列。我国干熄焦技术发展迅猛，已建成75套干熄焦装置，集成创新的260t/h大型高温高压干熄焦装置达到国际领先。

2000年以来，我国焦炉大型化、炼焦工艺多样化、炼焦自动化、节能环保、拓展炼焦煤资源、降低炼焦成本、焦炉煤气综合利用等方面均取得显著进步。在焦炉装备大型化方面，在4.3m顶装焦炉的基础上，设计建造了6m、7m、7.63m各种系列的大型顶装焦炉；捣固焦技术发展迅速，从4.3m捣固焦炉发展到5.5m、6.25m捣固焦炉。在节能环保方面，积极研发和推广应用干熄焦技术、煤调湿技术、焦炉烟道气余热回收技术、焦化污水深度处理回用技术、焦炉烟道气脱硫-脱硝技术等，特别是在干熄焦技术领域取得显著应用成效，大型干熄焦发电量达到100kW·h/t以上，节能减排效益显著。在拓展炼焦煤资源炼焦技术方面，开发并采用了捣固炼焦工艺多配加低价弱黏结性煤，使焦煤配比降低了9%，肥煤配比降低了5%，气煤、1/3焦煤和瘦煤的比例相应提高3.5%~6.5%，有效降低了炼焦成本。

目前，我国已经具备了7m以上顶装焦炉以及6.2m以上捣固焦炉的自主设计和建造能力。积极推广应用干熄焦技术，可回收80%的红焦显热，不仅降低炼焦能耗50~60kgce/t，还有效提高了焦炭质量。随着焦化污水深度处理及回用技术、干熄焦技术的广泛应用，我国大中型钢铁企业的炼焦生产基本全部实现了焦化污水的净化处理和"近零排放"，焦化生产的工序能耗不断降低。2018年，我国主要钢铁企业焦化工序能耗平均为102.53kgce/t，比2007年降低近40kgce/t。近3年主要钢铁企业焦化技术经济指标见表4。

表4 近3年主要钢铁企业焦化技术经济指标

项目	2018年	2017年	2016年
M_{40}/%	87.64	87.54	87.31
M_{10}/%	6.01	6.06	6.13
灰分/%	12.63	12.54	12.46
硫分/%	0.82	0.78	0.78
冶金焦率/%	90.70	89.30	90.16
洗精煤消耗/kg·t^{-1}	1.37	1.38	1.37
工序能耗/kgce·t^{-1}	102.53	99.67	97.16

注：统计的钢铁企业为中国钢铁工业协会会员单位。

3 主要技术创新成就

3.1 富氧喷煤技术

我国从 20 世纪 60 年代开始在首钢、鞍钢高炉上喷煤,是当时世界上开发应用喷煤技术较早的国家之一。1990 年以后,高炉喷煤技术列入国家科技计划,1999 年宝钢 1 号高炉创造了月平均喷煤量达到 250kg/t 以上的新纪录。目前,全国大中型高炉基本全部采用高炉喷煤技术,取得了显著的经济和环境效益。2017 年,全国高炉喷煤平均达到 143.16kg/t,入炉焦比为 363.83kg/t,宝钢、湛江、首钢京唐、太钢、武钢等部分大型高炉入炉焦比和煤比达到世界先进水平。高炉鼓风富氧率稳步提高,普遍达到 3%~5%,先进大型高炉鼓风富氧率达到了 5%~10%,高炉鼓风量降低到 1000m^3/t 甚至更低,为提高生产效率、改善高炉透气性创造了良好条件。在发展富氧喷煤的同时,还加强煤粉制备、输送、喷吹系统的安全设计和流程优化,首钢京唐 1 号高炉 2013 年 8 月率先使用氧煤枪技术,2 号高炉于 2014 年 6 月开始使用氧煤枪,高炉采用氧煤枪后,高炉顺行状况得到较好改善,风口前煤粉燃烧率和煤粉利用率得到提高。

3.2 高效长寿技术

在我国,高炉长寿化也取得突出进步,长期困扰高炉高效、安全、稳定、顺行、长寿的技术难题得到有效解决。近年来,我国在大型高炉设计体系、核心装备、工艺理论、智能控制等关键技术方面取得了重大进步。我国高炉在炉型设计、耐火材料选用及其结构设计、铜冷却壁、高效冷却方式、自动化监控系统以及高炉维护措施等方面,形成了高炉高效长寿综合技术体系,高炉长寿取得了显著成效,宝钢、武钢、首钢等企业的高炉寿命已达到 15 年以上,其中宝钢 3 号高炉寿命创下近 19 年我国高炉长寿新纪录,进入世界先进行列。应当指出,我国高炉长寿技术发展很不均衡,高炉平均寿命仅为 5~10 年,个别高炉寿命更短,与国外先进高炉相比还存在较大差距。近些年,我国高炉长寿再次面临新的挑战,炉缸侧壁温度异常升高、铜冷却壁过早破损等技术问题仍需要认真研究解决。

3.3 高风温技术

高风温技术是高炉降低焦比、提高喷煤量、提高能源转换效率的重要技术途径。目前,大型高炉的设计风温一般为 1250~1300℃,提高风温是 21 世纪高炉炼铁的重要技术特征之一[18]。目前我国已完全掌握在单烧低热值高炉煤气(热值约为 3000kJ/m^3)条件下风温达到 1250~1300℃的高风温的集成技术,其核心技术是通过预热煤气和助燃空气,使热风炉拱顶温度达到(1380±20)℃,在送风期将热风温度稳定地维持在 1250~1300℃,热风炉拱顶温度与风温之差缩小到 80~100℃。设计研发并应用了多种结构形式高风温热风炉,尤其是首钢京唐 5500m^3 高炉在世界上首次应用了顶燃式热风炉,实现了顶燃式热风炉的大型化;研发多种形式的煤气和空气双预热技术,使燃烧单一高炉煤气的热风炉拱顶温度达到 1400℃;研发应用小孔径高效格子砖,有效提高了热风炉传热效率;优化热风管道结构,合理设置管道波纹补偿器及拉杆;采取有效措施防止炉壳发生晶间应力腐蚀，使热风炉达到长寿、高效。应当看到，我国高风温技术发展并不平衡，先进与落后并存，部分先进大型高炉年平均风温可以保持在 1250℃ 左右，但还有一批高炉年平均风温仍徘徊在 1000~1100℃ 区间，使我国高炉平均风温水平与国际先进指标相比差距较大。

3.4 高炉煤气干法除尘技术

目前我国自主研发的高炉煤气干式布袋除尘技术在设计研究、技术创新、工程集成及生产应用等方面取得突破性进展[19]，自主设计研发了大型高炉煤气全干式脉冲喷吹布袋除尘技术，完全取消高炉煤气湿式除尘备用系统。高炉煤气布袋除尘技术发展迅猛，近期我国新建或大修改造的高炉几乎全部采用煤气布袋除尘工艺，取得了显著的应用效果。首钢京唐 3 座 5500m^3 高炉以及宝钢、湛江、包钢等企业的大型高炉均采用了煤气干法除尘技术。通过自主创新和技术集成，有效解决了高炉煤气干法除尘的关键技术难题，取得了许多技术创新成果。如设计研发了煤气调温装置和喷碱脱氯装置，提高单箱体过滤能力，合理选择滤料，采用双向电磁脉冲喷吹技术，采用压力可调式正压气力输送装置，阀门和管道内壁喷涂耐磨涂层，波纹补偿器采用抗酸抗氯侵蚀的不锈钢材质等。

3.5 高效低耗烧结技术

2000 年以来，我国烧结工艺技术取得重大进展，烧结矿质量和性能显著提高。研发应用了低温烧结、超厚料层（1000mm 料层厚度）烧结、添加适宜 MgO 烧结矿制备技术、热风烧结、复合烧结技术、烧结烟气循环等，均取得较好效果。首钢研发应用了烧结料面喷吹蒸气技术，改善了烧结过程的传热机制，有效降低了烧结过程的固体燃料消耗和污染物的排放；梅钢开发应用了烧结料面喷吹焦炉煤气技术，使燃料消耗有效降低；烧结智能化控制技术取得较好应用；烧结过程多种污染物的协同控制和治理取得明显成效，活性焦、循环流化床烟气脱硫以及 SCR 脱硝技术在中国烧结机上得到推广应用，使烧结烟气中的 SO_2、NO_x、粉尘和二噁英等得到有效脱除。

3.6 球团技术

近年来我国在球团技术领域发展成就突出，建成了一批带式焙烧机和链箅机-回转窑球团生产线，球团技术装备大型化、现代化取得长足进步。我国拥有国际先进的 500 万吨/年带式焙烧机和链箅机—回转窑球团生产线，自主研发的链箅机—回转窑和自主集成的带式焙烧机工程设计、装备制造、生产运行等均取得显著成效。通过技术攻关，有效解决了造球、布料、焙烧、回转窑结圈等技术难题，球团品位、强度等指标达到国际先进水平。积极研究高铁低硅酸性球团、含钛含镁球团、赤铁矿氧化球团，大型带式烧结机生产赤铁矿球团取得显著进展，碱性球团、自熔性球团、高镁低硅复合球团的生产取得成功；球团烟气脱硫-脱硝技术取得明显应用成效。

3.7 大型高效焦化技术

为降低配煤成本，节约优质炼焦煤资源，扩大炼焦煤源开发，我国结合国情开展了一系列研究工作，拓展弱黏结煤在炼焦中的应用。基于煤调试技术和炼焦煤水分调控技术，武钢 7.63m 大型焦炉贫瘦煤配加量达 14%~16%，少用优质瘦煤达 10% 以上，低变质弱黏结性煤配加量达20%以上。马钢、攀钢、沙钢等焦化企业结合大容积焦炉用煤的特殊性，开展炼焦用煤细化分类使用技术及煤岩学配煤研究，实现了优化配煤、提高焦炭质量、降低生产成本。通过配加型煤炼焦技术，将弱黏性煤配加黏结剂后压制成型煤，与其余散状煤混合装入炼焦炉内炼焦。宝钢焦化工序设置了配加型煤设施，型煤配加量达到15%～30%，焦炭质量可以满足5000 m^3 级高炉的要求。同时，我国干熄焦技术应用推广成效突出，总体技术达到国际先进水平；焦炉烟道气余热回收、烟道气脱硫-脱硝技术研发应用取得良好效果。

4 发展趋势与展望

4.1 技术装备对产业发展的推动作用

近40年来，我国高炉、烧结、球团、焦化等工序技术装备的大型化、现代化、高效化，有力支撑了我国钢铁工业的快速发展，迅速提高了我国炼铁系统整体技术水平，为实现炼铁工业高效、低耗、优质、长寿、安全、环保做出贡献，使我国炼铁工业绿色低碳发展取得重要技术进步。与此同时，技术装备水平的提升，对于我国解决技术发展不平衡问题、淘汰落后、流程优化、产业升级、技术进步都具有重要的现实意义和深远的历史意义[20]，特别是近年来我国加大淘汰过剩和落后钢铁产能，钢铁工业供给侧结构性改革，取得了举世瞩目的成就。

面向未来，高炉炼铁的减量化发展仍是一项十分艰巨的任务。近年来，我国加大钢铁工业供给侧结构性改革力度，淘汰、关停了一大批技术装备落后、能耗高、污染大的小高炉、小烧结、小焦炉，通过淘汰落后、削减过剩产能，我国高炉数量明显减小、高炉平均容积逐渐增加，未来1200 m^3 以下的小高炉、100 m^2 以下的烧结机、竖炉球团以及落后的无回收小焦炉等都将淘汰。

4.2 总体发展趋势

到21世纪中叶，我国经济发展将从速度效率型向质量效益型转变，经济社会将保持平稳发展，持续提高发展质量，解决经济过快增长所带来的各种不平衡与不充分。在未来经济发展环境下，钢铁作为首选的功能性结构材料，仍将继续发挥不可替代的作用[21]。

随着我国经济发展的速度减缓和增长方式转变，市场拉动将会降低，加上资源和能源可获取性的制约，以及生态环境治理的限制，直到21世纪中叶，我国钢铁产业发展必须坚持减量化[22]。钢铁产业"去产能"将在未来很长一个时期内成为产业发展的主旋律，必须摒弃过去那种不讲条件、不顾市场、背离产业发展规律的粗放型盲目扩张的发展理念，钢铁项目"一哄而上""盲目攀比"而造成的钢铁产能过度扩张，导致钢铁产能严重过剩，产业发展举步维艰，为经济社会和产业发展带了沉重负担和巨大代价，教训惨痛、代价沉重，必须给予高度的关注，进行深刻的反思。

在钢铁产业减量化、集约化发展的背景下，必须坚持产业发展的科学性，自觉遵循产业发展的客观规律，以工程哲学的视野审视产业未来发展的方向、目标、路径和格局。在优质铁矿石资源和炼焦煤资源不断匮乏的未来，生态环境的制约也会日趋严苛，钢铁产业在未来一个时期内所面临的发展形势是：市场需求拉力减弱，资源、能源供给力不足，生态环境制约力增强。钢铁企业在这样一个发展环境下，如果不坚持绿色发展，建设资源-能源节约型、环境友好型企业，将难以在愈加激烈的市场竞争中生存，更是勿谈发展[23]。

今后，智能化将带动经济社会发展，成为整个经济社会发展的主流，技术装备智能化也将是钢铁产业的重要发展方向之一。对于炼铁工序而言，是大通量的物质流、能量流和信息流输入/输出的过程，基于人工智能的高炉操作专家系统、智能化烧结技术和智能化炼焦技术，将是未来炼铁技术装备发展一个重要方向。

4.3 产业发展格局

至21世纪中叶，我国钢铁产业结构和格局将发生重大变化。主要体现在钢铁产量将在一定时期内保持平稳，不会再出现高速增长，这是我国经济转型发展和经济社会的市场需求所决定的，随着投资拉动效应的减缓，钢铁产量将出现下降趋势，市场供求将保持一种动态平衡。随着产业结构调整、生态环境治理和企业转型升级，将会有一批钢铁企业实现转型发展，转向新的产业领域，寻求新的经济增长点。未来10年间，我国钢铁技术装备水平将进一步提升，高炉炼铁技术装备大型化、现代化将成为主要特征，在生产总量严格控制、总体产能削减的前提下，一大批装备先进、绿色高效的大型高炉、大型烧结和新型焦炉将结合现有炼铁装备的技术改造和结构调整，进行装备升级改造。不仅是单体技术装备的大型化、现代化，而且高炉、烧结机、焦炉的数量将不断减少，总体生产规模和产能将逐步降低，达到合理的规模区间。

未来10年，我国钢铁产业和钢铁制造流程结构也将发生重大变化。随着社会废钢资源的不断积累，一定规模的电炉炼钢流程将会出现，不依赖高比例高炉铁水的电炉钢产量将逐渐增加。而且在可预见的未来，具有一定生产规模的熔融还原、直接还原等不依赖焦炭的非高炉炼铁装备将实现工业化生产。高炉炼铁的产量和比例都将出现下降，高炉炼铁减量化、绿色化、智能化将成为未来炼铁技术发展的主要趋势和方向。

5 结语

（1）近40年来，我国钢铁工业取得了举世瞩目的技术成就，钢铁产量以及主要炼铁技术装备达到或接近国际先进水平，为我国经济社会发展做出了重大贡献。

（2）炼铁技术装备大型化是21世纪以来最显著的产业发展特征，以5000m^3以上特大型高炉的自主设计建造为代表，我国已经在特大型高炉、烧结、焦化等工序完全掌握了工程设计、装备制造、工程建造、运行操作等一系列技术，关键技术及生产指标达到国际先进水平。

（3）高炉喷煤、高风温等一系列关键共性技术，对于炼铁技术装备创新和产业技术进步，起到了至关重要的推动作用，对技术装备大型化和现代化起到了重要的支撑作用。

（4）面向未来，我国钢铁产业将发生重大变革，炼铁技术装备的发展应遵循产业发展战略、不断创新，实现炼铁技术装备的大型化、现代化、绿色化、智能化发展。

参考文献

[1] 张寿荣. 中国炼铁企业的前景 [J]. 中国冶金, 2016, 26 (10): 7.

[2] http://http://data.stats.gov.cn/easyquery.htm?cn=C01.

[3] https://www.worldsteel.org/zh/dam/jcr:e5a8eda5-4b46-4892-856b-00908b5ab492/SSY_2018.pdf.

[4] 王维兴. 2018年我国炼铁生产技术述评 [J]. 世界金属导报, 2019-03-20 (B02).

[5] 王维兴. 2017年我国炼铁技术发展评述 [C]//2018年全国炼铁生产技术暨炼铁学术年会文集. 杭州: 中国金属学会炼铁分会, 2018: 20.

[6] 张寿荣, 于仲洁. 中国炼铁技术60年的发展 [J]. 钢铁, 2014, 49 (7): 8.

[7] 杨天钧, 张建良, 刘征建, 等. 持续改进原燃料质量提高精细化操作水平努力实现绿色高效炼铁生产 [J]. 炼铁, 2018, 37 (3): 1.

[8] 杨天钧, 张建良, 刘征建, 等. 近年来炼铁生产的回顾及新时期持续发展的路径 [J]. 炼铁, 2017, 36 (4): 1.

[9] 张寿荣, 毕学工. 中国炼铁的过去、现在与展望 [J]. 炼铁, 2015, 34 (5): 1.

[10] 张福明. 高炉高风温技术发展与创新 [J]. 炼铁, 2013, 32 (6): 1.

[11] 王维兴. 2016年我国炼铁技术经济指标述评 [N]. 世界金属导报, 2017-02-28 (B02).

[12] 杨天钧, 张建良, 刘征建, 等. 化解产能脱困发展技术创新实现炼铁工业的转型升级 [J]. 炼铁, 2016, 35 (3): 1.

[13] 沙永志, 滕飞, 曹军. 我国炼铁工艺技术的进步 [J]. 炼铁, 2012, 31 (1): 7.

[14] 许满兴, 张玉兰. 新世纪我国烧结生产技术现状与发展趋势 [J]. 烧结球团, 2017, 42 (2): 25.

[15] 张福明, 颉建新. 冶金工程设计的发展现状及展望 [J]. 钢铁, 2014, 49 (7): 41.

[16] 王海风, 裴元东, 张春霞, 等. 中国钢铁工业烧结/球团工序绿色发展工程科技战略及对策 [J]. 钢铁, 2016, 51 (1): 1.

[17] 李超, 郑文华. 我国焦化行业近况、展望及应对 [J]. 河北冶金, 2018, 265 (1): 1.

[18] 张福明. 高风温低燃料比高炉冶炼工艺技术的发展前景 [J]. 中国冶金, 2013, 23 (2): 1.

[19] 张福明. 21世纪初巨型高炉的技术特征 [J]. 炼铁, 2012, 31 (2): 1.

[20] 张福明, 钱世崇, 殷瑞钰. 钢铁厂流程结构优化与高炉大型化 [J]. 钢铁, 2012, 47 (7): 1.

[21] 殷瑞钰. 冶金流程工程学 (第2版) [M]. 北京: 冶金工业出版社, 2009.

[22] 殷瑞钰. 以绿色发展为转型升级的主要方向 [N]. 中国冶金报, 2013-10-31 (1).

[23] 张福明. 面向未来的低碳绿色高炉炼铁技术发展方向 [J]. 炼铁, 2016, 35 (1): 1.

Prospect and Development Achievements of Blast Furnace Ironmaking Technologies and Equipments in China

Zhang Fuming

Abstract: In the past 40 years, the Chinese iron and steel industry has made great progress, and the crude steel production output has ranked first in the world for many years. Chinese blast furnace ironmaking technological equipments have achieved outstanding achievements in large-scale, modernization, high-efficiency, long service life and so on. Since 2000, a batch of super large blast furnaces above $5000m^3$, large sintering machines above $500m^2$, large volume coke ovens with battery height 7.63m and large pelletizing production lines with an annual output of 4 million t/a have been built and put into commission in succession, and a series of the key technologies of integrated innovation and independent research have been achieved great achievements in production practice. At the same time of the technical equipments large sizing lots of advanced technologies and equipments research and application have achieved remarkable results, such as blast oxygen enrichment and pulverized coal injection, bell-less top equipment, BF gas dry dust removal, dome combustion hot blast stove and high blast temperature technology, high efficiency and low consumption sintering technology, large scale clean coking technology, and so on. These achievements promote the progress of ironmaking technologies and equipments. Up to the middle of this century, great changes will take place in the situation and steel manufacturing process structure of China's iron and steel industry. Reduction, greenization, intelligentization and high efficiency will be the main development tendency of ironmaking technology and equipment in the coming period.

Keywords: ironmaking; technical equipment; blast furnace; sintering; pelletizing; coking; greenization

大型带式焙烧机球团技术装备设计与应用

摘　要：球团矿是高炉炼铁的优质炉料，发展球团技术是实现低碳绿色炼铁的重要途径之一。首钢京唐为实现 $5500m^3$ 高炉的稳定顺行和低碳冶炼，大力提高球团矿的入炉比率、冶金性能和产品质量，确定了以球团矿为主的炉料结构，二期工程球团矿入炉比率设计为 55%~70%，相继设计建造了 3 条 400 万吨/年带式焙烧机球团生产线。本文主要探讨了现代高炉炉料结构的演化发展趋势，提出了低碳绿色球团生产的发展理念，研究分析了带式焙烧机球团生产工艺的特点和适用性，并在阐述对带式焙烧机球团工艺流程设计优化的同时，重点介绍了首钢京唐带式焙烧机球团工程设计中应用的先进技术和关键装备，如球团工程能量高效转换、高效燃烧-传热系统、节能减排和智能化控制等方面的设计创新，以及往复式布料器等主要设备的设计开发和应用效果。以期为类似工程提供借鉴与参考。

关键词：球团；带式焙烧机；大型；装备设计；先进技术

1　引言

进入 21 世纪以来，我国钢铁工业发展迅猛，冶金工程技术装备大型化、现代化进程加快，工程应用效果显著[1]。目前，我国约有 25 座 $4000m^3$ 以上特大型高炉建成投产，相应地建设投产了一批大型烧结、球团生产线，为特大型高炉生产稳定顺行奠定了精料基础。近 10 年间，结合铁矿石资源条件和精料技术发展趋势，我国设计建造了一批单体生产能力达到 400 万吨/年以上的大型球团生产装备[2~6]，成为新世纪以来我国高炉炼铁技术发展进程中的一个亮点。

2009 年，我国自主设计建造的首钢京唐钢铁厂建成投产。该工程自主设计建造了我国第一座 $5500m^3$ 高炉及其配套的 $504m^2$ 带式焙烧机球团生产线，着力建构高效低耗、低碳绿色、节能减排的新一代高炉炼铁技术体系[7]，努力提高高炉球团矿入炉比率和入炉矿品位，以满足 $5500m^3$ 高炉高效低耗生产对炉料的要求[8]。本文主要探讨了现代高炉炉料结构的演化发展趋势，提出了低碳绿色球团生产的发展理念，研究分析了带式焙烧机球团生产工艺的特点和适用性，阐述了带式焙烧机球团工艺流程设计优化，重点介绍了首钢京唐带式焙烧机球团工程设计中，创新应用的先进技术和关键装备，包括球团工程能量高效转换技术、高效燃烧-传热系统、节能减排和智能化控制等方面的设计创新，以及往复式布料器等主要设备的设计开发和应用效果。

本文作者：张福明，张卫华，青格勒，王渠生，韩志国。原刊于《烧结球团》，2021, 46（1）：66-75。

2 工程设计研究

2.1 工艺流程选择

长期以来，由于资源条件和技术传承等原因，我国高炉炉料结构主要以高碱度烧结矿配加酸性球团和少量矿块为主，针状（片状）铁酸钙为基的高碱度烧结矿一直是我国高炉的主要原料。2000年以后，随着我国钢铁产量的不断增长和产能扩大，铁矿石对外依存度也逐年上升，近年来铁矿石进口比率已经达到80%，甚至更高。烧结矿由于原料价格相对略低，一直以来在高炉炉料中占主要地位。近年来由于受到优质铁矿粉资源减少、生态环境和资源可获取性的影响，高炉炉料中增加球团矿的用量已经成为一个新的技术发展趋势。在国际铁矿石价格高位起伏波动的全球市场背景下，如何结合我国铁矿资源条件，高效利用高品位细磨精矿粉资源，对于铁矿粉造块工艺而言已经成为一个具有时代性的工程命题。

球团矿作为高炉的主要炉料之一，其生产工艺是铁矿粉造块的一种主要工艺。同烧结工艺相比，该工艺的主要技术优势在于：（1）球团矿具有综合冶金性能优良、粒度均匀、冷态强度高等特点；（2）球团生产工序能耗低（约为烧结工艺的1/2）、系统漏风率低、热风循环利用效率高，生产过程烟气排放少；（3）球团矿含铁品位高，相应可降低高炉渣量，从而有效降低高炉焦比和燃料比；（4）现场实测表明，球团工艺污染物及粉尘排放少，环境清洁，全流程除尘灰总量约为烧结工艺的10%。

目前，世界上用于单机大型化球团生产工艺主要为链箅机—回转窑和带式焙烧机两种。两种工艺技术比较见表1。

表1 带式焙烧机和链箅机—回转窑球团工艺的技术比较

项 目	带式焙烧机	链箅机—回转窑	技 术 特 点
单机能力	最大产量为925万吨/年	最大产量为600万吨/年	带式焙烧机的适度产能规模为300万吨/年以上，回转窑的适度产能规模为60万~300万吨/年
主机数量/台	1	3	带式焙烧机由于流程紧凑集约、热量耗散低、余热循环利用充分，热能利用效率高；工程占地少，工序能耗低，粉尘排放少
原料适应性	强	一般	带式焙烧机适用于磁铁矿、赤铁矿等多种原料条件。回转窑主要以磁铁矿为主，赤铁矿焙烧工艺难度大，设备适用性、可靠性和耐久性不如带式焙烧机
燃料适应性	气体或液体燃料	固体、气体或液体燃料	带式焙烧机工艺一般采用较高热值的气体燃料（如焦炉煤气、天然气、合成煤气等），需要有充沛的高热值气体燃料的供应保障。回转窑一般采用煤粉作为燃料，具有较好的燃料适用性
球团焙烧状态	静止	滚动	带式焙烧机工艺球团静止焙烧，球团破碎和粉末较少、球团成品率高；回转窑工艺球团滚动焙烧，且球团需要转运，易产生粉料和造成回转窑结圈，球团破碎和粉末较多

续表1

项 目	带式焙烧机	链箅机—回转窑	技 术 特 点
生产运行操作	便捷	较复杂	带式焙烧机自动化控制程度高，可实现操作精准调控，操作运行稳定，无结圈、黏结等故障，设备检修维护工作量小；回转窑工艺流程长，热能利用效率低，热工响应速度慢，工艺调整复杂，设备检修维护工作量较大
工程建造投资	较高	较低	回转窑工艺装备技术成熟，主机全部实现国产化，工程建造投资相对较低。带式焙烧机工艺装备要求较高，自动化控制水平高，工程建造投资相对较高

通过对国内外高炉炉料结构发展趋势的研究分析，为奠定超大型高炉生产稳定顺行的精料基础，首钢京唐一期工程2座5500m^3高炉，其炉料结构确定为65%烧结矿、25%球团矿和10%块矿，球团工序生产规模确定为400万吨/年；二期工程投产后，3座高炉炉料结构确定为55%球团矿、40%烧结矿和5%块矿，球团矿总生产规模确定为3×400万吨/年。

在球团工程顶层设计中，基于已有技术的研究成果和生产实践，结合首钢京唐钢铁厂的工程理念和流程结构优化，决策设计建造年产400万吨、504m^2大型带式焙烧机球团生产线。

2.2 工艺流程优化

以冶金流程工程学理论和动态精准设计方法为指导[9]，球团工程顶层设计，是在流程、功能、结构、要素、效率协同优化的前提下，以构建球团工序合理的铁素物质流静态流程网络为核心，通过"物质流、能量流和信息流"流程网络的高效耦合和协同运行，进而实现现代大型球团厂"流程网络和运行程序"的设计优化[10]。通过优化球团生产工艺流程，实现整体工艺布局的高效顺畅、界面紧凑。球团厂全部取消物料运送转运站，球团工程单位占地面积仅为0.016$m^2/(t \cdot a)$。

工程顶层设计中，通过工艺流程优化、工序功能优化、构建铁素物质流和能量流网络，实现了单元工序或装置匹配合理、协同有序。采用动态甘特图（Gantt Chart）等设计方法和物料输运的动态仿真研究，对球团生产的原料准备、配混、造球、焙烧等工序之间的匹配衔接进行协同优化，使主体设备和装置的能力以小时产量为基准，研究了不同设备数量和能力的多种技术方案，不片面追求个别设备的大型化，而是以设备能力耦合匹配、协同连续运行为设计核心，着力优化物流路径和物料运行轨迹，减少物料输送过程的物质和能量耗散，大幅度缩短了物料运距、减少了物料转运和折返。联合设置功能相同或相近的建构筑物，降低非功能区的占地面积，使单元工序之间匹配实现动态有序、协同高效[11]。图1为首钢京唐1号和2号504m^2带式焙烧机球团工程工艺平面布置图。

2.3 技术创新

通过多方案研究对比和权衡比选，决策采用带式焙烧机球团生产工艺，带式焙烧机面积504m^2（宽度为4m、长度为126m），产能为400万吨/年。设计中的主要技术创新体

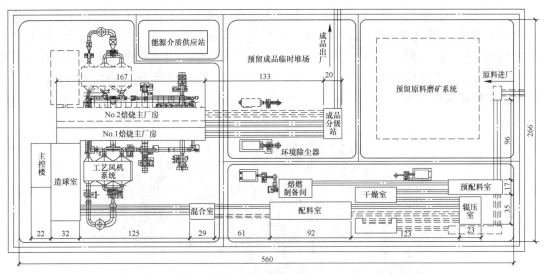

图 1　首钢京唐 504m² 带式焙烧机球团工艺平面布置图

现在如下几个方面。

2.3.1　提高能源利用效率

带式焙烧机球团生产工艺是在带式焙烧机上完成球团的干燥、预热、焙烧、均热和冷却等不同的工艺过程，其典型的热风循环工艺如图 2 所示。由图 2 可见：在带式焙烧机的鼓风干燥段，利用从冷却段抽取的热风对生球进行干燥；在抽风干燥段，则是通过回热风机抽取均热段和焙烧段风箱的热风对生球继续进行干燥；在预热段和焙烧段，利用冷却段的热风和燃料燃烧对球团进行预热和焙烧；在均热段利用冷却段的热风对球团进行均热，不需采用燃料燃烧供热；在冷却一段和冷却二段利用冷却风机鼓入的冷风对球团进行冷却。

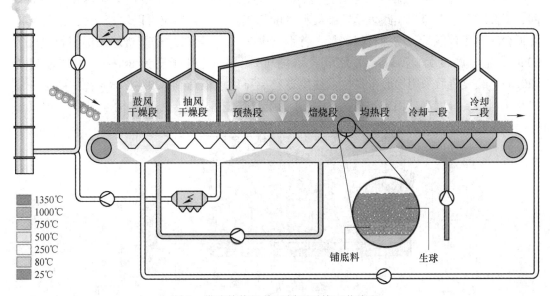

图 2　带式焙烧机热风循环系统工艺流程

工程设计中，采用 CFD 流场数值仿真计算对带式焙烧机热风循环系统进行热风流场模拟（见图3），通过优化设计结构和流场参数使热风循环系统运行实现最优化，以充分回收利用焙烧系统高温烟气的物理显热，从而最大限度地利用工艺余热降低球团热耗。实践表明，带式焙烧机热风梯级循环利用工艺显著地提高了能量的利用效率，大幅度降低了能源消耗和热量耗散[12,13]。

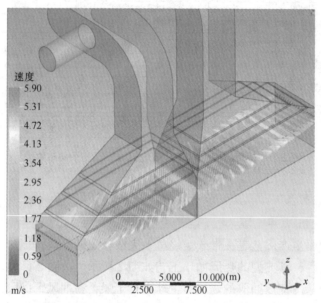

图3 带式焙烧机热风循环系统流场仿真模拟

2.3.2 研发厚料层焙烧技术

厚料层焙烧是提高带式焙烧机能量利用效率的有效技术措施。该技术可以降低铺底料球团的占比，降低带式焙烧机运行速度，延长球团干燥、预热、焙烧、均热和冷却的时间，提高球团矿质量。实践表明：铺底料厚度每减少10mm相应可以增产2%，而且可以降低台车箅条消耗量、增加台时产量。首钢京唐2号、3号带式焙烧机球团铺底料厚度为80mm，生球厚度为370mm；1号带式焙烧机球团铺底料厚度为80mm，生球厚度为320mm，经改造后采用梯形布料技术使生球料层厚度可以达到320~470mm，从而进一步提高了料层的厚度和球团矿质量[14]。图4为带式焙烧机厚料层断面结构。

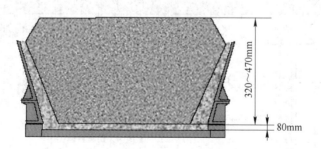

图4 带式焙烧机厚料层断面结构

2.3.3 研发生球分级布料技术

带式焙烧机生球分级布料可以显著改善料层透气性，提高粉料的筛分效率与单位面积焙烧机的风量，缩短反应进程，从而实现高效生产和能量高效利用的目的。在造球区段进行球团分级筛除大球和小球（见图5），焙烧机尾部设置双层辊筛，上层筛缝为12.5mm，下层筛缝为8mm，在料层断面高度方向形成由不同粒级球团所组成的料层分级结构（见图6）。实践表明，采用球团分级布料技术可以降低燃料消耗约10%，提高焙烧机生产能力约18%，同时可以改善球团矿理化性能。

图5 生球分级布料设备三维仿真

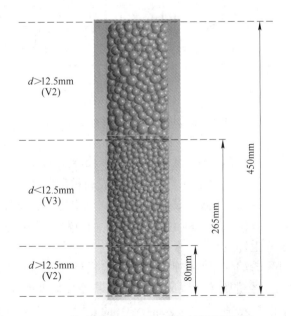

图6 带式焙烧机分级布料的料层断面结构

2.3.4 合理利用钢铁厂二次能源

国外带式焙烧机通常以天然气或重油为燃料，我国的天然气和重油资源并不充沛，钢铁厂结合自身能源结构，可以采用焦炉煤气作为带式焙烧机燃料。首钢京唐1号带式焙烧机配置了32组燃烧器，2号、3号带式焙烧机配置了28组燃烧器，从而可通过精准控制燃烧器的供热能力来满足不同原料的焙烧工艺要求[15,16]。因此利用焦炉煤气替代高热值天然气资源，生产高品质球团，可使钢铁制造过程的煤气实现高效利用和高值转化。

带式焙烧机焦炉煤气燃烧器的主要技术特点是：具有自动点火功能，可以实现燃烧器在线启停；设置了火焰监测装置，实现熄火自动识别、自动吹扫，保证了生产的安全；采用焦炉煤气为燃料的带式焙烧机燃烧器，可以显著降低CO_2的排放，且其空气过剩系数可以降低到15%以下，从而有效地降低NO_x的生成和排放。设计研发了先进的调节阀组和控制系统，可以实现单个或单组燃烧器温度的智能化精准调控[17]。图7为带式焙烧机燃烧器布置结构图。

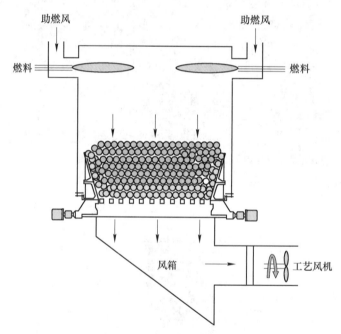

图 7 带式焙烧机燃烧器布置结构

2.3.5 采用半干法循环流化床脱硫+SCR脱硝净化处理工艺

球团生产工艺全程封闭，有效减少了粉尘的生成和扩散。球团的干燥、预热、焙烧、均热、冷却等全部工艺过程在设置密闭罩的带式焙烧机上完成，球团不经过倒运、翻转、装卸等工艺过程，大幅度减少了粉尘的生成、逸散，而且将带式焙烧机整体布置在一个封闭厂房内，有效控制了粉尘外逸。

为实现球团生产过程的烟气超低排放，带式焙烧机烟气采用半干法循环流化床脱硫+SCR脱硝净化处理工艺，经过脱硫、脱硝净化后的带式焙烧机烟气，粉尘颗粒物、SO_x、NO_x的排放浓度分别降低到$2mg/m^3$、$8mg/m^3$、$31mg/m^3$以下（见表2）。

表 2　首钢京唐带式焙烧机球团生产污染物排放指标

吨球团矿烟气排放量/m³·t⁻¹	颗粒物/mg·m⁻³	SO_x/mg·m⁻³	NO_x/mg·m⁻³	吨球团矿污染物排放总量/g·t⁻¹
1720	2	8	31	92.6

3　技术装备研发与创新

3.1　往复式球团布料技术

为适应带式焙烧机球团均匀布料的要求，自主设计研发出往复式布料器+宽胶带+分级布料辊筛一体化联合布料工艺，有效提高了球团的布料精度，实现了动态精准布料[18]。输送带运转驱动采用变频调速，布料器头部伸缩采用移动布料小车，由液压缸驱动；移动小车速度与输送带带速相匹配，实现单向下料，保证料面平整；布料过程生球破损小，有助于提高球团成品率，降低筛分指数。图 8 为往复式布料器三维设计图，图 9 为往复式布料器的实物照片。

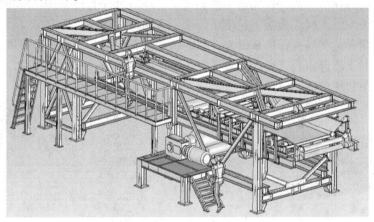

图 8　往复式布料器三维设计图

图 9　往复式布料器实物照片

3.2 造球系统大型化

保证生球质量是球团生产稳定顺行的基础。为了改善造球质量，设计开发了圆盘直径为 7.5m 的新型造球盘，单机造球能力显著提升，提高了生产效能、满足了高效化生产的要求，同时改善了造球效果，提高了生球强度。通过造球工序流程结构的优化，使工艺流程衔接紧凑、运行高效，减小了造球室的占地面积和工程投资。设计了新型固定刮刀和喷水调控机构，有效提高了生球粒径的均匀性、生球强度和质量。

3.3 台车设备设计优化

带式焙烧机的台车承载着球团进行焙烧，在运行过程中承受着球团重力荷载、高温烟气热冲击、化学腐蚀等一系列破坏作用。工程设计中为研究台车在热-力交变条件下的复杂热应力和应变情况，建立了热-力耦合数学模型和数值仿真计算程序，对台车运行工况下的温度场和应力场进行仿真研究。设定台车工况的初始条件和边界条件，通过三维仿真设计和热应力场有限元数值计算研究结果，优化了台车本体和算条的设计结构（见图10）。实践表明，结合数值仿真计算结果对设备设计进行可靠性分析和优化改进是动态精准设计中最重要的环节。依此择优选用性能优良、经济合理的台车材料，使台车本体的化学成分、铸态组织、机械性能等满足使用要求。应用传热过程数学模型对带式焙烧机运行状态的热风流场和温度场进行仿真研究，根据仿真研究结果，优化了台车算条设计结构和形状，进而有利于改善料层透气性、降低气流阻损，使流场分布更为合理，有效地降低了台车主梁和算条的热负荷，减少了关键部件的热应力破损（见图11）。

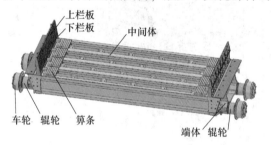

图10 台车三维仿真结构设计

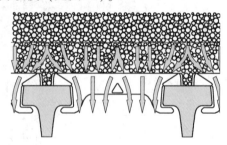

图11 台车纵断面流场分布

4 生产应用

首钢京唐 1 号带式焙烧机于 2010 年 8 月建成投产，2 号和 3 号带式焙烧机分别于 2019 年 5 月和 2019 年 6 月相继建成投产。经过 10 年的生产运行实践，该工艺在不同原料条件和操作条件下，形成了一系列带式焙烧机高效低耗、低碳绿色生产运行关键技术[19]，实现了原料配料、造球、布料、焙烧、冷却等多工序动态有序、协同连续高效运行的目的。

4.1 球团矿质量

生产中，通过优化布料工艺，使生球强度得到有效提高。在球团干燥和预热段，通过精确调控带式焙烧机区段温度分布,有效降低了生球的爆裂。为充分发挥低硅铁矿粉的资源优势,尽最大可能降低造球过程添加大量膨润土,同时将钠基膨润土改为钙基膨润土,从而提高球团矿品位、降低球团的还原膨胀率。2019年,生产自熔性球团矿时,膨润土加入量降到4.5kg/t,球团矿品位提高至66%,抗压强度达到3260N/P。国内外具有代表性的球团矿理化性能见表3。

表3 国内外具有代表性的球团矿理化性能

项目	首钢京唐	巴西Vale	加拿大IOC	美国米诺卡	日本神户	宝钢湛江
$w(TFe)/\%$	66.00	65.90	63.0	62.29	61.90	64.50
$w(SiO_2)/\%$	2.01	2.37	4.00	4.20	2.90	2.80
$w(CaO)/\%$	2.38	2.60	3.60	4.50	3.79	2.40
碱度	1.15	1.10	0.90	1.07	1.31	0.85
抗压强度/N·球$^{-1}$	3260	2800	2200	2300	2500	2800
还原膨胀率/%	16.5	19.8	17.0	12.5	—	19.0

4.2 碱性球团生产

为适应首钢京唐3座5500m³高炉高比率球团冶炼的要求,研发出大型带式焙烧机生产碱性球团矿的成套装备和工艺技术,攻克了生球爆裂、箅条堵塞等工艺和设备难题,实现了大型带式焙烧机高效稳定低耗运行。

建立了带式焙烧机数值模拟系统,优化确定了利用消石灰生产低硅碱性球团矿适宜的干燥预热和焙烧温度场及流场,在工序能耗基本不变的条件下,碱性球团抗压强度达到了3200N/P以上。通过优化带式焙烧机热工参数,解决了生球爆裂问题;采取优化布风板和箅条结构及高频电源等措施,解决了台车和箅条堵塞以及电除尘器故障等技术问题,提高了带式焙烧机的稳定运行。在带式焙烧机上生产低硅碱性球团矿,高炉球团矿入炉比率从28%提高到了55%以上,主要技术经济指标达到预期。碱性球团矿含铁品位达到66%,$w(SiO_2)$降低到2.0%,碱度为1.1~1.2,抗压强度达到3200N/P以上,还原膨胀率为16.5%。单机碱性球团矿日产量达到12750t以上,最高达到12900t,超过了设计指标的6%,工序能耗达到19.8kgce/t。表4为首钢京唐球团矿生产的主要技术经济指标。

表4 首钢京唐球团矿生产主要技术经济指标

球团矿品种	时间	日产量/t·d^{-1}	$w(TFe)$/%	$w(SiO_2)$/%	膨润土量/kg·t^{-1}	平均碱度	抗压强度/N·球$^{-1}$	还原膨胀率/%	工序能耗/kgce·t^{-1}
酸性球团矿（基准期）	2018年全年	12430	65.58	3.30	18.0	0.11	3059	21.6	19.5
高铁低硅碱性球团矿	2019年4~12月	12810	65.95	2.12	4.5	1.17	3238	16.8	19.7
	2020年1~3月	12780	66.02	2.01	4.5	1.18	3265	16.5	19.8

4.3 生产智能化运行

带式焙烧机球团生产工艺流程紧凑、设备集中、高效集约，为物理系统与信息系统（CPS）耦合实现智能化控制创造了有利条件。球团生产智能化控制系统，主要包括：（1）工艺参数检测和过程控制系统。工艺自动化检测使生产全过程的信息实现在线检测、监测、收集、存储，基础控制系统对单元设备或装置进行数字化调控，设备或装置运行全部实现自动化。（2）工序模块控制功能。研发了工艺数学模型，开发生球布料、焙烧温度、热风循环系统的模型控制，使球团生产过程实现优化，年作业率达到98%以上。（3）工艺过程控制功能。采用分布式控制系统和制造执行系统，对造球、布料等单元工序采用可视化实时监控，使带式焙烧机专家控制系统的总体智能化提升，实现了带式焙烧机生产高效低耗、稳定顺行。（4）设备运行全生命周期管理功能。对主体工艺设备建立设备管理模型，进行设备信息和运维动态监控和管理，减少了异常停机操作，有效提高了设备的作业率。

4.4 运行实绩

首钢京唐球团生产的含铁原料，以首钢秘鲁高品位磁铁矿粉为主。经过10余年的生产实践证实，带式焙烧机球团生产工艺技术先进、设备可靠，特别是2020年全球疫情期间，在秘鲁铁矿粉数月"断供"的条件下，带式焙烧机完全可以适应磁铁矿、赤铁矿、褐铁矿甚至是烧结返矿等多种原料条件，为超大型高炉生产高品质、高性能的球团矿提供了保障。高炉采用高比率球团矿冶炼取得显著技术经济实绩，主要技术指标见表5。

表5 高炉高比率球团冶炼主要技术经济指标对比

主要指标		炉料结构/%				高炉主要技术经济指标		
		烧结矿	普通球团矿	低硅球团矿	块矿	入炉品位/%	渣量/kg·t^{-1}	燃料比/kg·t^{-1}
基准期	2018年全年	67	28	—	5	59.5	283	502
试验期	2019年4~12月	43	—	55	2	61.4	218	483
	2020年1~3月	43	—	55	2	61.9	215	475
国际先进	美国印度安纳港厂7号高炉（4838m³）	约39	—	约49	约12	—	260	490
	荷兰艾默伊登厂7号高炉（4500m³）	约40	约60	—	—	—	250	510

5 结论

（1）本文通过探讨现代高炉炉料结构的演化发展趋势与分析，带式焙烧机球团生产工艺的特点和适用性表明：采用带式焙烧机球团生产工艺和装备，大批量稳定生产高品质、高性能球团技术发展前景广阔。

（2）首钢京唐带式焙烧机球团生产线通过工程设计优化、设备研发与创新，在生产时间中取了良好的运行效果，可实现球团-炼铁系统的低碳绿色化运行，此工艺适应性强，也可用来生产碱性球团矿。

（3）采用先进所谓设计理念、理论和方法，实现球团工程的动态精准设计，构建先进的信息物理系统，是我国未来球团工程技术的发展方向。

参考文献

[1] 张福明. 我国高炉炼铁技术装备发展成就与展望 [J]. 钢铁，2019，54（11）：1-8.

[2] 余海钊，廖继勇，范晓慧. 带式焙烧机球团技术的应用及研究进展 [J]. 烧结球团，2020，45（4）：47-54，70.

[3] 智谦，韩志国，易毅辉，等. 包钢624m² 大型带式球团焙烧机设计创新与应用 [J]. 中国冶金，2017，27（4）：61-66.

[4] 季文东，贾西明，赵俊峰，等. 大型带式焙烧机高产低耗技术的发展与应用 [J]. 矿业工程，2019，17（6）：30-33.

[5] 利敏，王纪英，李祥. 我国带式焙烧机技术发展研究与实践 [C]//第八届中国钢铁年会论文集. 北京：冶金工业出版社，2011：8.

[6] 夏雷阁，刘文旺，黄文斌. 大型带式焙烧机在首钢京唐球团的应用 [C]//2011年度全国烧结球团技术交流年会论文集. 烧结球团编辑部，2011：116-120.

[7] 殷瑞钰. 冶金流程工程学（第2版）[M]. 北京：冶金工业出版社，2009：152-158.

[8] Zhang F M，Qian S C. Design，Construction and Application of 5500m³ Blast Furnace at Shougang Jingtang [C]// AISTech 2014 Proceedings of the Iron & Steel Technology Conference. Indianapolis，Ind.，USA：AIST，2014：753-765.

[9] 殷瑞钰. 冶金流程集成理论与方法 [M]. 北京：冶金工业出版社，2013：160-168.

[10] 张福明，王渠生，韩志国，等. 大型带式焙烧机球团技术创新与应用 [C]//第十届中国钢铁年会暨第六届宝钢学术年会论文集. 北京：冶金工业出版社，2015：7.

[11] Zhang F M，Wang Q S，Han Z G. Innovation and Application on Pelletizing Technology of Large Traveling Grate Induration Machine [C]// AISTech 2015 Proceedings of the Iron & Steel Technology Conference. Cleveland，Ohio，USA：AIST，2015：402-412.

[12] 李国玮，夏雷阁，青格勒，等. 京唐带式焙烧机原料方案及热工制度研究 [J]. 烧结球团，2011，36（2）：20-24.

[13] 张卫华. 带式焙烧机球团新技术研究与应用 [N]. 世界金属导报，2019-10-21（B13）.

[14] 夏雷阁，苏步新，李新宇，等. 焙烧温度与球团矿强度和还原性的关系 [J]. 烧结球团，2014，39（1）：21-24.

[15] 夏雷阁，苏步新，李新宇，等. 首钢504m² 带式焙烧机热工制度的试验研究 [J]. 矿冶工程，2014，34（3）：69-75.

[16] 解海波. 504m² 带式焙烧机设计理论模型的分析论证 [J]. 中国冶金，2015，25（2）：21-28.

[17] 韩志国. 大型带式焙烧机球团技术的研究与应用 [N]. 世界金属导报，2013-08-27（B16）.

[18] Zhang F M，Zhao M G，Cao C Z，et al. New Technological Progress of Ironmaking in Shougang [C]// AISTech 2017 Proceedings of the Iron & Steel Technology Conference. Nashville，TN，USA：AIST，2017：439-453.

[19] 张福明. 首钢绿色低碳炼铁技术的发展与展望 [J]. 钢铁，2020，55（8）：11-18.

Design and Application of Pelletizing Technologies and Eqiupments in Large-scale Straight Grate

Zhang Fuming Zhang Weihua Qing Gele Wang Qusheng Han Zhiguo

Abstract: With pellet as a high quality burden for blast furnace ironmaking, developing pelletizing technology has become one of the important approaches to reach the targets of low carbon and green ironmaking. In order to achieve the stable production-smooth and low carbon smelting in the 5500m^3 blast furnace, Shougang Jingtang has made great efforts to improve the charging ratio in burden, metallurgical performance and quality of pellet. As a result, high pellet ratio burden for blast furnace was determined, with a designed 55%-70% pellet burden in phase II which constructed three straight grate pelletizing lines with 4Mt/a capacity each consequently. This article discusses mainly the evolution and development of the modern blast furnace burden and proposes the developing concept of low carbon green pelletizing production. The characteristics and applicability of straight grate pelletizing production process are studied and analyzed. With the introduction of the optimized design of straight grate pelletizing process flow, the article emphasises on the innovations in advanced technologies and key equipments during the design of Shougang Jingtang pelletizing lines, mainly related to the design innovations in aspects as energy efficient conversion of pelletizing process, high efficiency combustion-heat transfer system, energy saving and emission reduction, as well as intelligent control; and the design, development and application effects on main equipments as reciprocationg belt conveyor. This paper is expected to provide necessary reference to the stakeholders involving in such a simlar project.

Keywords: pelletizing technology; straight grate; large-scale; eqiupment design; advanced technology

我国 5000m³ 级高炉技术进步及运行实绩

摘　要：对我国 5000m³ 级高炉技术进步的主要特点及生产运行实绩进行了总结。认为我国 5000m³ 级高炉在设计理念及技术指标、高炉设计内型探索与优化、高炉长寿技术、高风温及顶燃式热风炉技术、自主创新特大型高炉装备与技术等方面，取得了长足进步。生产实践表明，由于注重精料技术研究及应用，积极采用富氧喷煤、高风温、高顶压等高效冶炼技术，实施清洁生产及低碳冶炼，大多数 5000m³ 级高炉的煤气利用率达到 50% 以上，燃料比降低到 500kg/t 以下。

关键词：特大型高炉；炼铁；技术进步；长寿；低碳；绿色化

1　引言

2000 年以来，我国钢铁工业发展迅猛，自主集成创新能力不断提升，在钢铁产量逐年增长的情况下，工艺技术装备大型化、现代化、高效化发展成就举世瞩目[1]。我国钢铁工业经历了高速发展、产能过剩、结构调整、供给侧改革等几个重要的历史阶段，在以生态文明为永续发展理念的指导下，高炉炼铁产业结构调整和优化取得了显著成就，在流程结构优化的前提下，高炉向大型化、现代化、高效化发展，结构调整成效显著，特别是以 5000m³ 级高炉的设计、建造、运行为代表，我国高炉炼铁技术已跃进到国际先进行列。

2　主要发展概况

进入 21 世纪以来，我国高炉炼铁技术取得了跨越式发展。生铁产量已稳居世界首位并逐年攀升，近年来生铁产量约占全球生铁总产量的 50% 甚至更高。高炉大型化、现代化、高效化技术装备升级成就突出[2]。20 年来，我国高炉数量由增到减，高炉平均容积不断提高。2009 年 5 月首钢京唐 1 号高炉建成投产[3]，标志着我国已经完全掌握了 5000m³ 以上巨型高炉设计、建造和生产运行技术，中国已成为继日本、前苏联和德国之后，世界上第四个拥有 5000m³ 级高炉的国家。随后沙钢 5800m³ 高炉、京唐 2 号 5500m³ 高炉、宝钢湛江 2 座 5050m³ 高炉、山钢日照 2 座 5100m³ 高炉、京唐 3 号 5500m³ 高炉相继建成投产。这些巨型高炉是我国钢铁冶金工程科技的集成创新和综合国力的集中体现，我国已成为 5000m³ 高炉数量仅次于日本世界排名第二的国家。面向未来，我国高炉炼铁技术进步和结构优化将进一步围绕高炉大型化、现代化、高效化、长寿化、清洁化、低碳化开展，坚持低碳绿色、智能高效的高质量发展理念[4]，到本世纪中叶成为世界领先的

本文作者：张福明。原刊于《炼铁》，2021，40（1）：1-8。

钢铁大国和钢铁强国。图1显示了中国钢铁产量的变化过程，图2为中国主要钢铁企业高炉炼铁主要技术经济指标的变化。

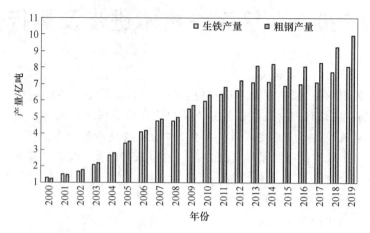

图1　中国2000年以来钢铁产量的变化过程

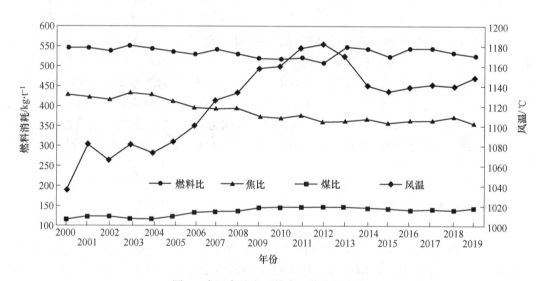

图2　中国高炉主要技术经济指标变化

近20年来，在我国钢铁产量快速增长的同时，通过工艺技术创新和装备技术进步，高炉主要技术经济指标不断得到改善和优化[5]。高炉燃料比、焦比和煤比分别从2000年的547kg/t、429kg/t、118kg/t降低到2019年的528kg/t、356kg/t、145kg/t，燃料比、焦比相应下降了19kg/t、73kg/t，煤比提高了27kg/t；风温从2000年的1034℃提高到1147℃，提高了113℃；高炉利用系数、实物劳动生产率等指标不断优化，炼铁工序能耗、污染物排放等指标持续改进。新世纪以来，我国高炉技术装备大型化成就斐然，据不完全统计，到2019年末2000m^3以上的大型高炉已经达到90余座，其中5000m^3以上巨型高炉8座，4000m^3级特大型高炉18座，3000m^3级高炉18座，2000m^3级高炉48座，近期还将有一批搬迁新建和扩容大修的大型高炉建成投产。

众所周知，高炉装备大型化和技术现代化，对于推动我国高炉炼铁产业创新和整体技术装备的提升，具有重大的现实意义和深远的历史意义，对于解决钢铁产业多层次、不均衡、不充分的发展问题，发挥了无可替代的作用。毫无疑义，5000m³级高炉的自主设计建造和生产运行，已成为我国钢铁工业迈向国际一流最具代表意义的产业创新和技术进步。中国已建成投产的5000m³以上巨型高炉主要技术装备见表1。

表 1 中国5000m³以上巨型高炉主要技术装备

高炉	高炉容积/m³	投产时间	炉顶装料设备	炉体冷却结构	热风炉结构	煤气净化工艺	铁水运输方式
首钢京唐1号	5500	2009-05-21	SG-V型并罐式无料钟	全冷却壁（炉缸象脚侵蚀区和铁口区为铜冷却壁，炉腹至炉身下部4段轧制钻孔铜冷却壁）	4座顶燃式热风炉+2座助燃空气预热炉	全干法布袋除尘	准轨铁路运输"一罐到底"（300t）
首钢京唐2号	5500	2010-06-26	SG-V型并罐式无料钟	全冷却壁（炉缸象脚侵蚀区和铁口区为铜冷却壁，炉腹至炉身下部4段轧制钻孔铜冷却壁）	4座顶燃式热风炉+2座助燃空气预热炉	全干法布袋除尘	准轨铁路运输"一罐到底"（300t）
首钢京唐3号	5500	2019-04-28	SG-V型并罐式无料钟	全冷却壁（炉缸象脚侵蚀区和铁口区为铜冷却壁，炉腹至炉身下部4段轧制钻孔铜冷却壁）	4座顶燃式热风炉+2座助燃空气预热炉	全干法布袋除尘	准轨铁路运输"一罐到底"（300t/200t）
宝钢湛江1号	5050	2015-09-25	BCQS串罐式无料钟	全铸铁冷却壁	4座顶燃式热风炉+2座助燃空气预热炉	全干法布袋除尘	380t鱼雷罐铁路运输
宝钢湛江2号	5050	2016-07-15	BCQS串罐式无料钟	全铸铁冷却壁	4座顶燃式热风炉+2座助燃空气预热炉	全干法布袋除尘	380t鱼雷罐铁路运输
山钢日照1号	5100	2017-12-19	并罐式无料钟	全冷却壁（炉缸铁口区局部铸铜冷却壁；炉腹至炉身下部为轧制钻孔铜冷却壁）	4座顶燃式热风炉+2座助燃空气预热炉	全干法布袋除尘	PBC-380型无轨汽车运输"一罐到底"（210t）
山钢日照2号	5100	2019-04-11	并罐式无料钟	全冷却壁（炉缸铁口区局部铸铜冷却壁；炉腹至炉身下部为轧制钻孔铜冷却壁）	4座顶燃式热风炉+2座助燃空气预热炉	全干法布袋除尘	PBC-380型无轨汽车运输"一罐到底"（210t）
沙钢5800m³高炉	5800	2009-10-21	PW型并罐式无料钟	全冷却壁（炉缸象脚侵蚀区1段和铁口区局部为铸铜冷却壁，炉腹至炉身下部5段轧制钻孔铜冷却壁）	3座PW-DME外燃式热风炉	PW三锥环缝型湿法除尘	准轨铁路运输"一罐到底"（180t）

3 技术进步的主要特点

3.1 先进的设计理念及技术指标

（1）工程设计理念。世界上第一座5000m³巨型高炉于20世纪70年代初在日本建成投产，随后前苏联和德国钢铁工业发达国家也相继建造了5000m³级巨型高炉。从一定意义上讲，建造巨型高炉是一个国家冶金工程科技发展水平的标志和综合国力的体现[6]，由于巨型高炉原燃料条件高、技术装备水平高、操作管理要求高、工程建造投资高，直到20世纪末期的20多年中，世界上巨型高炉并未得到推广。在此期间新建的韩国浦项钢铁厂、光阳钢铁厂和我国宝钢，都没有建造5000m³级巨型高炉。21世纪初，日本开始进行新一轮的钢铁工业结构调整和产业创新，大幅度削减高炉数量，增加单座高炉的生产效能，一批4000m³的高炉经过扩容大修改造，将容积扩大到5000m³以上。截至2008年，世界上运行的5000m³高炉达到13座，其中日本拥有10座。

21世纪初，随着我国钢铁工业的迅猛发展和综合技术实力的不断提升，我国钢铁工作者开始谋划建造5000m³级巨型高炉的构想。在我国钢铁产业协调发展、沿海布局和战略搬迁过程中，自主创新设计建造5000m³巨型高炉的概念研究和顶层设计，在首钢曹妃甸、宝钢湛江、山钢日照等钢铁基地的规划方案中已经显现出来。在循环经济、低碳绿色发展理念的指导下，遵循冶金流程工程学理论和集成方法，以实现现代钢铁厂产品制造、能源转换和消纳处理废弃物功能拓展为目标，在钢铁制造全流程结构优化的前提下，积极推动高炉大型化和现代化，实现了我国5000m³巨型高炉的跨越式发展[7]。

（2）技术指标。新世纪我国巨型高炉的设计，以高效、低碳、优质、长寿、清洁为协同发展目标，主要技术指标参照国际同级别先进高炉生产实践，注重资源能源节约、生产高效低耗、节能减排和低碳绿色，积极采用先进工艺技术和装备，以期实现多目标的协同优化和集成创新[8~10]。表2列出了中国巨型高炉设计的主要技术经济指标。

表2 中国5000m³巨型高炉主要技术经济指标（设计值）

项 目	京唐1、2号5500m³高炉	沙钢5800m³高炉	湛江5050m³高炉	日照5100m³高炉	京唐3号5500m³高炉
高炉有效容积/m³	5500	5800	5050	5100	5500
炉缸直径/m	15.5	15.3	14.6	14.6	15.3
日产铁量/t·d⁻¹	12650	12876	11514	11571	12650
利用系数/t·(m³·d)⁻¹	2.3	2.22	2.28	2.27	2.3
熟料率/%	90	85	85	90	90
入炉矿品位/%	61	59	58	58.8	61
燃料比/kg·t⁻¹	490	490	498	490	490
焦比/kg·t⁻¹	290	290	278	310	290
煤比/kg·t⁻¹	180	200	220	180	180
风温/℃	1300	1250	1300	1260	1300

续表 2

项 目	京唐 1、2 号 5500m³ 高炉	沙钢 5800m³ 高炉	湛江 5050m³ 高炉	日照 5100m³ 高炉	京唐 3 号 5500m³ 高炉
富氧率/%	5.5	4.0	3.5	5.0	7.5
入炉风量（标态）/m³·min⁻¹	9300	8300	8000	8000	9300
热风压力/MPa	0.55	0.52	0.66	0.50	0.55
炉顶压力/MPa	0.28	0.25	0.3	0.28	0.28
年产铁量/万吨·年⁻¹	450	450	411.5	405	450

3.2 高炉设计内型的探索与优化

在高炉-转炉钢铁制造流程中，高炉具有重要的关键作用，是铁前系统生产的核心；是钢铁制造全流程中物质和能源转换最重要的工序[11]。高炉内型的物理本质就是高炉作为冶金反应器的内部形状和结构参数，也是高炉炼铁过程"三传一反"的进程空间和流程结构，高炉内型的设计实质上就是构建高炉炼铁耗散结构的优化过程。

在吸收国外 5000m³ 高炉内型设计经验的基础上，结合长期以来我国高炉炼铁实践以及对高炉设计内型、操作内型的认识和研究，我国 5000m³ 高炉在 10 年的发展历程中，呈现以下技术特征：

（1）优化高炉内型。结合原燃料条件和技术装备条件，为实现高效、低耗、长寿、稳定、顺行等多目标集成优化，经过系统综合、权衡、集成、确定高炉合理的设计内型，特别是与高炉的冷却结构、炉缸和炉底内衬结构相结合，以实现预期合理稳定的操作内型为关注点，注重高炉设计内型各部位的参量关系和协调耦合，在传统高炉内型的设计理念和方法的基础上，通过渐进式的探索和实践，对巨型高炉内型设计进行优化和创新。表 3 为 5000m³ 高炉设计内型的主要参数。

表 3 中国 5000m³ 巨型高炉设计内型主要参数

钢铁厂	首钢京唐			宝钢湛江		山钢日照		沙钢
高炉炉号	1	2	3	1	2	1	2	4
有效容积/m³	5500	5500	5500	5050	5050	5100	5100	5800
炉缸直径/m	15.5	15.5	15.3	14.5	14.5	14.6	14.6	15.3
炉腰直径/m	17.0	17.0	17.5	16.2	16.2	16.8	16.8	17.5
炉喉直径/m	11.2	11.2	11.2	10.7	10.7	11.0	11.0	11.2
死铁层深度/m	3.1	3.1	3.2	3.6	3.6	3.6	3.6	3.2
高径比	1.930	1.930	1.874	2.025	2.025	1.951	1.951	1.897
炉腹角/(°)	79.38	79.38	74.62	79.53	79.53	74.62	74.62	74.62
炉身角/(°)	81.04	81.04	79.98	81.41	81.41	80.85	80.85	80.39
V_u/A	29.16	29.16	29.93	30.60	30.60	30.48	30.48	31.56
风口数量/个	42	42	40	40	40	40	40	40
铁口数量/个	4	4	4	4	4	4	4	3

（2）加深死铁层深度。为了有效抑制炉缸铁水环流的破坏作用，减缓炉缸炉底象脚状异常侵蚀和铁口区两侧下部的凹陷型炉缸侧壁异常侵蚀，在高炉设计时总结经验教训的基础上，死铁层深度从 3.1m 延长到 3.6m，近期投产的高炉已达到炉缸直径的 24.7%[12]，图 3、图 4 为高炉死铁层深度的变化情况。

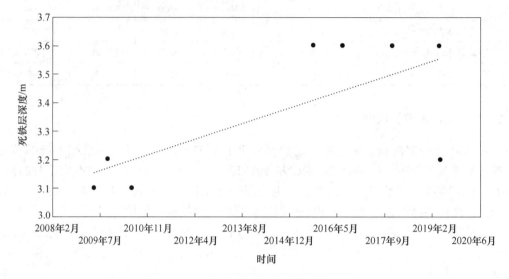

图 3　5000m³ 巨型高炉死铁层深度的变化

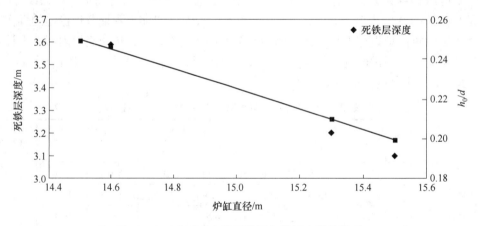

图 4　5000m³ 巨型高炉死铁层与炉缸直径的关系

（3）缩小炉腹角和炉身角。2000 年以后，为有效延长高炉炉腹至炉身下部区域的寿命，高炉铜冷却壁得到了普及应用，薄壁高炉炉体结构成为近 20 年来的主流技术模式。回顾高炉铜冷却壁的发展历程，铜冷却壁应用效果并未完全达到预期效果，有的新建高炉继续全部采用铸铁冷却壁[13]。在总结铜冷却壁过早破损机理和教训的基础上，认识到炉腹角偏大不利于炉腹煤气的顺畅排升，也不利于炉腹区域渣皮的黏结和稳定，吸收国外采用铜冷却壁薄壁高炉的炉型设计特点，大幅度缩小炉腹角至 74.62°，这是我国 5000m³ 级高炉炉型参数发展的一个显著变化。与此同时，炉身角也渐趋缩小趋势，从 81° 缩小到

80°。宝钢湛江 2 座巨型高炉根据既有经验，为了有利于炉况稳定顺行，采用了双炉身角的三段式炉身结构，这对于炉身煤气上升与下降炉料的耦合运动，是一种积极而富有成效的设计优化。应当指出，在采用大比率球团矿冶炼时，由于球团矿的还原膨胀率较高（一般控制≤20%），因此采用双炉身角的炉型设计是合理的。

(4) 控制合理的 H_u/D 和 V_u/A。同常规高炉相比，巨型高炉属于典型的矮胖型高炉。高炉容积的扩大，主要依靠径向尺寸的扩大，而高向尺寸变化不大。近 10 年间，巨型高炉的高径比（H_u/D）渐进变化，这主要取决于炉腰直径和有效高度的微小差异。从高炉内型设计理念上，已经不再追求所谓的"矮胖化"，而更加注重高炉生产的稳定顺行、低燃料比和高煤气利用率。与此同时，高炉内型设计更加关注高炉有效容积与炉缸截面积的比例（V_u/A），以及各部位截面积的合理比例关系，使其更加适应高炉冶炼过程中物质流和能量流的动态运行和变化规律。

3.3 高炉长寿技术

(1) 高炉炉缸炉底复合结构。基于对高炉炉缸内衬破损机理的研究解析，建立了高炉炉缸炉底温度场合理控制的长寿技术理念和方法。高炉炉缸炉底采用高导热、高抗铁水渗透性的优质炭砖，包括热压小块炭砖和微孔大块炭砖，炉底采用微孔/超微孔大块炭砖+陶瓷垫的复合结构；炉缸侧壁有的采用全炭砖结构，有的采用微孔炭砖+陶瓷壁结构[14,15]。高炉炉缸侧壁全部采用冷却壁冷却，部分高炉在铁口区、炉缸炉底交界处或"象脚状"异常侵蚀区，还采用了铜冷却壁。

巨型高炉铁口区的设计结构和材质进行了有效的优化，如采用铜冷却壁+铁口组合砖结构，以适应巨型高炉高出铁强度的要求。此外，结合高炉出铁过程的数字仿真模拟，对铁口布置的夹角、铁口角度和铁口长度等炉缸参数也进行了许多研究工作，对于延长铁口区的炉缸侧壁寿命具有积极效果。值得关注的是，巨型高炉炉缸炉底使用寿命尚未达到预期[16]，部分高炉还出现了炉缸铁口区下部侧壁局部过热和异常破损，这是未来需要重点研究和探索的关键技术难题之一。

(2) 炉腹至炉身下部的冷却结构。2000 年以后，铜冷却壁开始在高炉上得到推广应用。铜冷却壁起源于欧洲，陆续发展到中国、日本、美国、韩国等钢铁大国。我国巨型高炉设计建造时，结合已有技术的实践经验和国外技术的发展趋势，在京唐、沙钢、日照等巨型高炉炉腹至炉身下部高热负荷区域，采用了国产铜板轧制钻孔的铜冷却壁；湛江高炉采用 6 根水管强化型镶砖铁素体球墨铸铁冷却壁。近 10 年来，国内外高炉采用铜冷却壁的应用效果绝大多数不及预期，不少 1000~3000m³ 高炉甚至出现过早破损，3~5 年铜冷却壁大量损坏、被迫更换。关于炉腹至炉身下部高热负荷区域的铜冷却壁的应用问题，仍是今后一个时期需要深入研究、亟待攻克的关键技术难题[17]。

3.4 高风温及顶燃式热风炉技术

(1) 全高炉煤气实现高风温技术。20 世纪 90 年代，我国绝大多数高炉热风炉高热值燃料匮乏、预热技术落后，使得我国的热风炉拱顶温度普遍较低，制约了高炉风温的提高。21 世纪以来，由于顶燃式热风炉技术推广普及和高效双预热技术的广泛应用，我国高炉热风炉拱顶温度不断提高。目前我国已经完全解决了全量燃烧低热值高炉煤气达到

1400℃以上拱顶温度的技术难题，部分配置有助燃空气高温预热的热风炉，拱顶温度完全有能力超过1400℃甚至更高，NO_x的大量生成增加了晶间应力腐蚀的风险，造成生态环境破坏，因此未来热风炉要有节制地提高热风炉拱顶温度，将其控制在1420℃以下，降低拱顶温度和风温的差值（<100℃），进而实现在合理拱顶温度条件下提高风温。

我国巨型高炉热风炉几乎全部采用低热值高炉煤气，回收热风炉烟气余热预热煤气和助燃空气，设置2座前置高温预热炉用于助燃空气的二次预热，可以将煤气温度提高到200℃、助燃空气温度提高到520℃，拱顶温度可以达到1420℃甚至更高，彻底解决了燃烧单一低热值高炉煤气实现（1280±20）℃高风温的技术难题，实现了低品质煤气的高效化利用和钢铁厂能源高效转换[18]。

（2）顶燃式热风炉技术的创新应用。众所周知，顶燃式热风炉是热风炉技术的发展方向。早在20世纪70年代末，首钢率先在国际上将自主设计研发的顶燃式热风炉在首钢2号高炉（1327m^3）上实现工业化应用，取得了良好的应用效果，但由于受到当时综合技术条件限制，这项技术在近20年内没有在大型高炉上推广普及。新世纪以后，随着俄罗斯卡鲁金顶燃式热风炉技术的引进和不同级别高炉的应用，特别是首钢京唐5500m^3高炉采用顶燃式热风炉以后，我国高炉热风炉结构发生巨大变化，新建或改扩建的高炉基本全部采用顶燃式热风炉技术。目前我国运行的8座巨型高炉，有7座采用了顶燃式热风炉（见表4），成为我国特大型高炉高风温技术的一个重要特征。

表4 我国5000m^3高炉热风炉主要技术装备及参数

钢铁厂	宝钢湛江		沙钢	山钢日照		首钢京唐		
高炉炉号	1	2	4	1	2	1	2	3
高炉容积/m^3	5050	5050	5800	5100	5100	5500	5500	5500
炉缸直径/m	14.5	14.5	15.3	14.6	14.6	15.5	15.5	15.3
设计风量/$m^3·min^{-1}$	8000	8000	8300（最大9500）	8000	8000	9300	9300	9300
送风压力/MPa	0.55	0.55	0.52	0.50	0.50	0.54	0.54	0.54
热风炉结构	顶燃式	顶燃式	PW-DME外燃式	顶燃式	顶燃式	顶燃式	顶燃式	顶燃式
热风炉座数/座	4	4	3	4	4	4	4	4
预热工艺	热管换热器+预热炉	热管换热器+预热炉	整体式换热器	板式换热器+预热炉	板式换热器+预热炉	热管换热器+预热炉	热管换热器+预热炉	板式换热器+预热炉
热风炉燃料	全高炉煤气	全高炉煤气	高炉煤气+转炉煤气	全高炉煤气	全高炉煤气	全高炉煤气	全高炉煤气	全高炉煤气
设计风温/℃	1300	1300	1250（最高1310）	1260（最高1300）	1260（最高1300）	1300	1300	1300
年平均实际风温/℃	1251	1240	1217	1246	1233	1234	1201	1245
投产时间	2015-09	2016-07	2009-10	2017-12	2019-04	2009-05	2010-06	2019-04

由于顶燃式热风炉是将环形陶瓷燃烧器置于拱顶，拱顶是顶燃式热风炉的燃烧空间，为充分利用拱顶空间，使煤气燃烧充分，并使高温烟气在蓄热室格子砖内均匀分布，采用

了旋流扩散燃烧技术。为实现烟气的均匀分布，设计合理的环形陶瓷燃烧器预混室与锥形拱顶的几何结构，通过烟气流在拱顶空间内的收缩、扩张、旋流、回流而实现煤气的充分燃烧和高温烟气的均匀分布。创新采用小孔径（20~25mm）高效格子砖，加热面积可达 48~64m²/m³，有效提高了热风炉传热能力和效率。经过一系列技术研发和工程创新，建立了顶燃式热风炉节能和耐火材料配置的国家技术标准，攻克了热风炉燃烧、传热、气体运动、热风输送、热风利用等关键技术难题，使风温稳定达到1250℃以上。

顶燃式热风炉与煤气预热和助燃空气高温预热工艺耦合匹配，是我国在5000m³巨型高炉上实现的重大技术创新，对产业技术进步产生了重要的推动作用。经过10余年的应用实践，顶燃式热风炉高风温技术满足了巨型高炉生产要求。首钢京唐生产实践表明，在1250℃条件下，顶燃式热风炉煤气消耗与外燃式、内燃式相比可降低10%~15%，热风炉系统热效率可达83.8%，每座高炉节约生产成本可达4700万元/年。

3.5 自主创新特大型高炉装备与技术

（1）高炉无料钟炉顶设备。无料钟炉顶设备是巨型高炉的关键设备之一，是高炉炉料分布精准控制的核心装备。我国5000m³巨型高炉无料钟炉顶设备积极采用自主创新研制的国产化装备，取得了令人满意的应用实绩[19]。在大直径炉喉断面上，可以实现"平台+漏斗"、多环往复、中心加焦、炉料分级入炉以及多模式精准布料，料流调节阀、溜槽 α 角和 β 角调节精度等主要参数都已达到或超过国际先进水平[20]。

通过自主设计研制，无料钟炉顶设备的运行可靠性、稳定性和设备制造精度、使用寿命等都已完全满足巨型高炉精准布料的要求，特别是在中心流下料、克服布料过程中炉料偏聚/偏析、控制炉喉断面不同矿焦比、大料批装料提高焦炭负荷、优化布料矩阵、合理调控煤气分布等方面取得了显著进步，为巨型高炉稳定顺行和实现先进技术经济指标，奠定了坚实的装备基础。表5为我国5000m³级高炉典型的无料钟炉顶设备主要技术性能参数。

表5 我国5000m³级高炉无料钟炉顶设备主要技术性能参数

项 目	京唐5500m³高炉	湛江5050m³高炉	沙钢5800m³高炉
无料钟炉顶设备型式	SG-V 并罐式	BCQS 串罐式	PW 并罐式
溜槽倾动（α角）定位精度/(°)	控制精度±0.3（计量精度0.1）	±0.1	
溜槽旋转（β角）定位精度/(°)	±0.3	±0.3	
溜槽倾动速度/(°)·s⁻¹	1~1.6	0~3（能力7）	
溜槽旋转速度/r·min⁻¹	8	8	
上料闸直径/mm	DN1100	DN1400	
上料闸排料能力/m³·s⁻¹	1.0	2.2（O）/1.5（C）	
上密封阀直径/mm	DN1100	DN1600	DN1100
下密封阀直径/mm	DN1100	DN900	DN1100
料流调节阀直径/mm	DN1000	DN750	DN1000
料流调节阀排料能力/m³·s⁻¹	0.7	1.0（O）/0.7（C）	

续表 5

项 目	京唐 5500m³ 高炉	湛江 5050m³ 高炉	沙钢 5800m³ 高炉
中心喉管直径/mm	DN730		DN750
料流调节阀开度（γ角）定位精度/(°)	控制精度±0.3（计量精度 0.1）	±0.1	
布料溜槽长度/mm	4500	4500	4500
布料器密封氮气消耗/$Nm^3 \cdot h^{-1}$	1500	800	
布料器冷却水量/$t \cdot h^{-1}$	8~15	15	

(2) 高炉煤气干法除尘技术。为实现节水节能、提高余压发电效率和煤气热值，其中 7 座巨型高炉采用我国自主研发的高炉煤气干法布袋除尘技术，完全取消了备用的湿法除尘装置。以首钢京唐为例，采用高炉煤气干法除尘，净煤气含尘量长期稳定在 $2 \sim 4mg/m^3$，炉顶煤气余压发电量达到 $45 \sim 50kW \cdot h/t$，比湿法除尘工艺提高 35%~40%，煤气温度提高 80~100℃，煤气热值提高约 7%~10%。自主设计研制出关键系统和核心装备，实现集成创新，主要包括[21]：

1) 开发出基于高效热管换热原理的大功率煤气温度智能化调控技术、系统和方法，攻克了煤气温度变化影响布袋寿命的关键技术难题，布袋寿命达到 3 年以上；

2) 针对氯化物对 TRT 设备和煤气管道的腐蚀破坏，研究发现了高炉煤气中氯化物的生成原因、生成规律及其有效抑制措施，发明了高炉煤气喷碱脱氯装置和管道系统综合防腐方法，全面解决了高炉煤气采用干法除尘技术以后，TRT 叶片、波纹补偿器和煤气管道的异常腐蚀破损的关键技术难题；

3) 针对高过滤负荷条件下超长布袋（>7m）脉冲清灰的技术难题，开发研制出高效超长布袋双排式在线脉冲清灰装置；

4) 研究了高炉煤气全干法除尘与全干式 TRT 高效耦合运行的难题，攻克了 36.5MW 超大功率 TRT 发电机组的稳定运行、智能控制等难题，形成高精度炉顶压力稳定控制技术，实现了清洁能源高效转换。

(3) 大型鼓风机、TRT 等设备全面实现国产化。宝钢湛江 3 号高炉鼓风机采用了自主设计研发的大型鼓风机，主要技术性能和指标达到国际同类装备的一流水平。自主研制的轴流反动式透平机组（TRT），配套水冷无刷励磁同步发电机，额定装机容量为 33MW，在湛江 3 座巨型高炉上均得到成功应用，主要指标达到国际先进水平[22]。首钢京唐 3 座 5500m³ 高炉采用首钢自行设计研发的 SG-V 型无料钟炉顶设备及炉料分布控制技术，已稳定应用 11 年以上；宝钢湛江采用自主研制的 BCQS 型无料钟炉顶设备和布料器液压复合控制技术应用于 2 座 5050m³ 高炉，其布料精度、溜槽使用寿命、冷却水消耗、N_2 消耗等主要技术指标达到国外同类设备的先进水平，完全取代了引进设备。通过巨型高炉的自主设计、技术创新和装备研发，我国已经完全掌握了巨型高炉关键技术装备的设计、制造、运行、维护全面技术，彻底解决了过去"卡脖子"的关键装备难题，并且无料钟炉顶设备、大型鼓风机、TRT 和煤气干法除尘技术装备输出海外，为我国钢铁工业赢得国际声誉。

(4) 铁水直接运输工艺。遵循冶金流程工程学原理，为实现炼铁-炼钢界面技术优化，我国巨型高炉设计过程中，对铁水运输工艺进行了深入研究。大幅度缩短了高炉-转炉的铁水输运距离，首钢京唐1号、2号高炉至炼钢的铁水运输距离为900m，湛江2座高炉的铁水运输距离也是900m。除湛江高炉仍采用鱼雷罐运输铁水之外，其余6座高炉全部采用多功能铁水罐直接运输工艺，其中京唐3座高炉分别采用准轨铁路300t和200t铁水罐运输；日照2座高炉采用PBC-380型超大型汽车运输铁水，铁水罐容量为210t[23]；沙钢5800m³高炉设置3个铁口，采用180t铁水罐运输铁水[24]。实践证实，铁水直接运输工艺成功的，可以提高铁水温度50~100℃，减少了铁水输运的工序环节、热量耗散和环境污染，有利于铁水脱硫预处理，对钢铁制造全流程界面技术优化具有重要作用。

4 生产运行实绩

从首钢京唐1号高炉建成投产至今，我国5000m³高炉已经历了10余年的发展历程。在生产实践中研究探索巨型高炉生产运行规律，结合原燃料条件和操作条件，充分发挥先进技术和大型装备的优势，不断总结高炉操作管理的经验教训，高炉生产和技术指标总体上达到国际先进水平（见表6），有的甚至领跑世界。

表6 我国5000m³以上巨型高炉主要技术经济指标

项 目	首钢京唐			宝钢湛江		山钢日照		沙钢
	1号	2号	3号	1号	2号	1号	2号	5800m³
利用系数/t·(m³·d)⁻¹	2.40	2.29	2.31	2.23	2.36	2.04	2.03	2.21
焦比/kg·t⁻¹	268	281	277	298	291	320	321	308
煤比/kg·t⁻¹	202	191	184	168	167	140	140	161
焦丁比/kg·t⁻¹	25	24	37	27	27	31	31	45
燃料比/kg·t⁻¹	495	496	498	493	485	491	492	514
风温/℃	1247	1228	1230	1244	1248	1248	1232	1187
入炉矿品位/%	61.8	61.2	61.1	59.0	59.1	58.8	58.9	58.8
球团矿比率/%	54.2	47.6	47.8	4.5	5.3	19.6	16.9	12.9

（1）加强精料技术研究和应用。在国际铁矿石市场跌宕起伏的变化过程中，始终坚守高炉精料方针，提高入炉矿综合品位和熟料率，因地制宜采用合理的炉料结构，控制高炉渣量。首钢京唐结合自身资源特点[25]，大幅度提高高品位球团矿入炉比率，开发出自熔性球团、含钛含镁复合球团，球团矿入炉比率达到56%，综合入炉品位达到62%，渣量降低到215kg/t，燃料比降低到475kg/t[26]。宝钢湛江、山钢日照等高炉以高碱度烧结矿为主，配加酸性球团和块矿，在提高精料水平的前提下，获得了优异的经济技术指标。

（2）创新高炉送风新模式。为改善高炉下降炉料和上升煤气流在相向运动过程的"三传一反"，改善高炉透气性、保障高炉稳定顺行，积极采用高富氧、高风温、高顶压、大喷煤、恒湿（脱湿）鼓风等高效冶炼技术和节能减排技术，摆脱了"吹大风求高产"的传统操作思维和生产模式，有效降低吨铁耗风和炉腹煤气量，不少高炉富氧率已经达到5%以上，首钢京唐5500m³高炉、沙钢5800m³高炉的富氧率分别达到7.5%、10%，高炉鼓风量已经降低到1000m³/t以下，甚至达到870m³/t。5000m³级高炉入炉风量基本维持在7800~8500m³/min，保持合理适宜的风速和鼓风动能，风口前理论燃烧温度控制在2150~2350℃合理区间，为保证高炉下部热量充沛、提高炉缸活跃性、改善炉缸死焦柱透气性和透液性奠定了冶炼基础。

（3）探索高炉炉料分布控制技术。近年来，我国注重炉料分布精准控制技术的探索与研究，首钢京唐高炉根据大比率球团冶炼的条件，建立了适合于高比例球团冶炼的精准布料制度，采用中心加焦+水平料面的布料模式，最小矿角向中心收缩以提高单位面积中心煤气通量，确保中心煤气强度，改善中心煤气稳定性；宝钢湛江高炉总结大型高炉长期生产运行经验，不使用中心加焦技术，采用"平台+漏斗"的布料模式，高炉生产稳定顺行，煤气利用率长期保持在50%以上，燃料比降低到495kg/t的先进水平。总体来看，精准控制炉料分布以适应煤气流动态运行规律，大多数高炉的煤气利用率达到50%以上，高炉燃料比降低到500kg/t以下，工序能耗和实物劳动生产率等经济指标达到领先水平。

（4）清洁生产实现超低排放。在构建生态文明的工程理念指导下，大力实施清洁生产、循环经济和低碳冶炼，加强铁前系统料场、烧结、球团、焦化等工序全面实施超低排放治理，固体颗粒物、污水、SO_x、NO_x、VOC等污染物排放达到国际一流水平。宝钢湛江、首钢京唐等企业对料场进行了全封闭改造，固体颗粒物实现超低排放；铁前系统的烧结、球团和焦炉烟气进行全量脱硫、脱硝和除尘超净化处理；建设转底炉固体废弃物处理装置，有效降低了入炉原料中的ZnO含量，实现固体废弃物的资源化高值转化和综合利用。

5 结语

进入新世纪以来，我国炼铁工业通过搬迁结构调整、工艺技术装备升级，在装备大型化、现代化取得了长足进步，自主设计建造的5000m³级高炉取得显著的工程应用实绩。我国在特大型自主设计建造、大型炼铁装备技术自主研发、高炉建造和高炉稳定运行等方面，经过10余年的探索已经取得全面成功，在高炉精料技术、炉料分布控制技术、高风温技术、富氧喷煤及低燃料比操作、节能减排技术等达到国际先进水平。

面向未来至本世纪中叶，在铜冷却壁应用与维护、高炉炉缸长寿、高炉运行智能化精准控制、进一步降低低碳排放等方面还有许多需要持续攻关解决的技术难题。

参考文献

[1] 张寿荣, 毕学工. 中国炼铁的过去、现在与展望 [J]. 炼铁, 2015, 34（5）: 1-6.
[2] 张福明. 我国高炉炼铁技术装备发展成就与展望 [J]. 钢铁, 2019, 54（11）: 1-8.
[3] 张福明. 面向未来的低碳绿色高炉炼铁技术发展方向 [J]. 炼铁, 2016, 35（1）: 1-6.
[4] 张福明, 钱世崇, 张建, 等. 首钢京唐5500m³高炉采用的新技术 [J]. 钢铁, 2011, 46（2）: 12-17.
[5] 李维国. 我国特大型高炉操作和管理改进的思路 [J]. 炼铁, 2017, 36（5）: 1-7.
[6] 张福明. 21世纪初巨型高炉的技术特征 [J]. 炼铁, 2012, 31（2）: 1-8.
[7] 张福明, 钱世崇, 殷瑞钰. 钢铁厂流程结构优化与高炉大型化 [J]. 钢铁, 2012, 47（7）: 1-9, 19.

[8] 项明武,周强,张灵,等.沙钢5800m³高炉工艺技术特点[J].炼铁,2010,29(2):1-6.
[9] 廖建锋,文辉正,邹忠平,等.宝钢湛江5050m³高炉工艺技术特点[J].炼铁,2017,36(3):19-23.
[10] 于国华,陈诚,王冰,等.领先技术支撑特大型高炉绿色智能化运行[N].世界金属导报/2018年9月11日第B01版.
[11] 张福明.当代高炉炼铁技术若干问题的认识[J].炼铁,2012,31(5):1-6.
[12] 敖爱国,廖建锋.湛江1高炉开炉及生产实践[J].钢铁技术,2018(3):3-6.
[13] 周琦,贾海宁,苏威,等.宝钢湛江钢铁高炉长寿技术设计与应用[C]//第十二届中国钢铁年会论文集.北京:中国金属学会,2019:1-5.
[14] 刘行波.沙钢5800m³高炉炉体工艺技术特点[J].炼铁,2008,28(4):1-5.
[15] 敖爱国,梁利生,贾海宁.宝钢湛江钢铁高炉系统耐火材料的配置与应用[J].耐火材料,2018,52(1):35-39.
[16] 汤清华.我国新建和大修高炉中存在的共性问题[J].炼铁,2019,38(1):1-5.
[17] 张福明.延长大型高炉炉缸寿命的认识与方法[J].炼铁,2019,38(6):13-18.
[18] 刘华平,况维良,刘红军,等.5100m³高炉热风炉的设计[J].工业炉,2019,41(1):42-44.
[19] 郑军,刘波,邹达基,等.宝钢湛钢高炉无料钟炉顶系统设计特点[J].钢铁技术,2018(1):7-11.
[20] 刘波.宝钢湛江大型高炉关键设备和技术开发[J].冶金设备,2019(2):14-18.
[21] 张福明.大型高炉煤气干式布袋除尘技术研究[J].炼铁,2011,30(1):1-5.
[22] 钱卫强,苏威,马作仿,等.国产干式TRT在5000m³高炉上的应用[J].冶金动力,2018(7):36-38.
[23] 殷树椿.日照精品基地"一罐到底"铁水运输的优化配置[J].山东冶金,2018,40(6):11-13.
[24] 方音,王卫东,周强,等.沙钢5800m³高炉自主集成创新技术的应用[J].炼铁,2015,34(1):9-13.
[25] 陈艳波,张贺顺,郭艳永,等.首钢京唐高炉炼铁技术的进步[J].炼铁,2016,35(3):42-45.
[26] 张福明.首钢低碳绿色炼铁技术进步与展望[J].钢铁,2020,55(8):11-18.

China's Technical Progress in 5000m³ BF and Operational Performance

Zhang Fuming

Abstract: The main characteristics of China's technical progress in 5000m³ BF and its practical application are summarized. Regarding 5000m³ BF, China has achieved great progress in design concept, technical indicators, exploration and optimization of designed BF internal profile, long campaign technology, high temperature blasting, top-combustion hot stove technology, and super large sized BF equipment and technology. The actual production and operation performance shows that the gas utilization rate of most of the 5000m³ blast furnaces hits over 50% and the fuel rate drops to below 500kg/t through the research and application of beneficiated burden technology, adoption of high-efficiency smelting technologies such as oxygen enriched coal injection, high temperature blasting and high top pressure technologies, and implementation of clean production and low-carbon smelting.

Keywords: super large sized blast furnace; ironmaking; technical progress; long campaign; low carbon; greenization

中国钢铁产业发展与展望

摘　要：钢铁材料在经济社会发展中具有重要的作用和地位，钢铁产业是国民经济重要的基础产业。回顾了中国钢铁产业的发展历程与各时期产业发展特征，论述了关键共性技术和技术进步对钢铁产业发展的推动作用和重要意义。研究指出，面向未来中国钢铁产业必须坚持减量化、绿色化、智能化的可持续发展道路，遵循产业发展的客观规律，严格控制产能规模的盲目扩张和产量增长。钢铁产业的可持续发展要实现钢铁制造的功能拓展，加强产业科学布局和企业重组整合，促进流程结构优化调整，构建钢铁厂物理信息系统，实现高质量转型发展。提出并探讨了到本世纪中叶，钢铁产业的发展方向与路径。

关键词：钢铁产业；工程演化；发展历程；科技进步；关键共性技术

1　钢铁材料的地位与作用

　　第一次工业革命以后，贝塞麦转炉炼钢工艺的发明（1856年），使得钢铁制造从手工作坊真正走向工业化生产。在经历了经济蓬勃发展、经济危机、世界大战和石油危机等重大事件以后，随着氧气转炉、连续铸钢、大型高炉、连续轧制及信息技术的开发应用，钢铁冶金科学理论、工艺技术和工程应用协同发展、相互融合，大力推动了产业创新和进步。近200年来，世界钢铁产量随着全球经济发展不断增长。经过20世纪的发展演进，钢铁冶金已由技艺跨入工程科学的层次[1]。

　　迄今为止，钢铁依然是世界上重要的基础材料、结构材料和功能材料[2]。2019年，世界钢产量已达到18亿吨以上，钢铁及其相关产业已成为经济社会可持续发展的重要基础产业。由于钢铁制造工艺成熟、规模效益显著，在产品竞争性、可靠性和适用性等方面，钢铁材料均具有不可比拟的综合优势。此外钢铁材料还具有优异的加工成型性和循环利用性，是全球经济发展和社会文明进步不可或缺的绿色材料、必选材料或首选材料。预测到21世纪中期，钢铁材料在工业化和信息化的发展进程中仍将发挥着重要的基础支撑作用。

2　中国钢铁产业的发展历程与现状

　　新中国成立以来，经过70多年的发展，目前已成为世界第二大经济体，成就举世瞩目。由于中国是一个农业大国，建国以前工业基础十分薄弱，目前经济社会发展仍处于工业化中后期阶段。因此，在一定时期内，中国钢铁产业仍要以满足国内市场内部需求为主，尤其是要满足市场对高品质、高性能钢材的需求[3]。

本文作者：张福明，李林，刘清梅。原刊于《冶金设备》，2021（1）：1-6, 29。

钢铁产业的迅速发展是中国20世纪重大工程技术成就之一[4]。进入21世纪以来，在中国经济高速发展的带动下，中国钢铁产量迅猛增长，2019年中国钢产量已达到9.96亿吨，占全球钢产量的53.3%。2008年国际金融危机以来，由于钢铁产业的无序化粗放型发展，造成近年来钢铁产能出现严重过剩，钢铁企业经营困难，盈利能力在波动中急剧下滑。在市场竞争日益激烈、资源和能源供给受限以及生态环境保护等多重制约下，中国钢铁产业面临着从粗放型规模扩张到集约型高质量发展的转型升级，同时还面临着大幅度削减过剩产能、推进供给侧结构性改革和生态环境治理等多方面的严峻挑战和考验[5]。图1是新中国成立以来钢铁产量的发展变化历程。

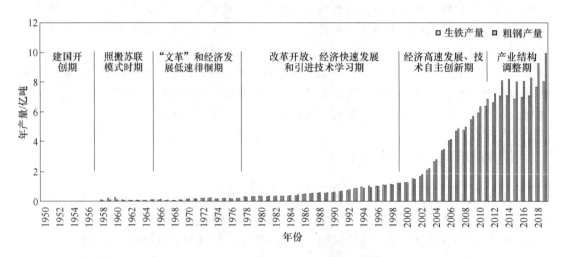

图1 中国钢铁产量的发展变化历程

2.1 新中国钢铁产业创建期

解放前的旧中国是个积贫积弱的农业国家，钢铁工业十分落后，粗钢产量历史最高纪录是1943年的92.3万吨[6]。1949年中国钢产量仅为15.8万吨、铁产量只有25万吨。新中国成立以后，钢铁产业进入起步发展时期，通过学习原苏联的钢铁工业模式和技术，扩建了鞍钢，新建了武钢、包钢等钢铁联合企业，1957年全国钢产量已达到535万吨，处于世界第9位。

1958年举国上下全民"大炼钢铁"，掀起了"大跃进"的发展热潮。回顾历史可以深刻地认识到，这种人为的"跃进式发展"违背了钢铁产业发展的客观规律，为此付出了沉重的代价，教训弥足珍贵，以历史唯物主义的观点反思当时的"大炼钢铁"是不成功的。

20世纪60年代，中国开始自主研发一些冶金技术和装备。在当时经济困难的条件下，自主设计并建设了转炉炼钢厂，自主研制出连铸机，自主开发并成功应用了高炉喷煤技术等。1965年全国钢产量已达到1223万吨，比1957年增长了一倍以上。从1966年开始，在"文化大革命"期间，中国钢铁产业经历了10年艰难的徘徊期。在此期间攀钢工程的建设、武钢热轧-冷轧薄板工程开工建设，对于中国钢铁产业布局和产品质量、品种发展，都具有深远的意义。

从建国以后到改革开放前的近 30 年间，中国钢铁产业发展演进特点是：从建国以后全面引入原苏联的钢铁产业模式，到不顾现实条件和客观规律，盲目追求钢铁产量的"大炼钢铁"，再到"文革"期间一些自主创新技术的研发应用，总体来看，钢铁产业发展始终处于艰难前行、曲折发展的状态之中。工艺技术装备进步出现停顿、徘徊甚至倒退；工程方法基本都是沿用、套用或者照搬原苏联的体系，没有形成完整的钢铁产业方法体系。

2.2 学习引进技术期

1978 年改革开放以后，使中国钢铁产业焕发出发展活力和勃勃生机。在此期间，通过开放合作、技术引进、消化吸收和应用推广，有效提高了中国钢铁技术装备水平，使中国钢铁产业进入了前所未有的发展阶段。以宝钢工程建设为契机，通过引进、消化和吸收宝钢工程的先进技术和装备，对促进中国钢铁产业技术进步具有重要的推动作用，从此中国钢铁产业发展进入到学习国外先进技术的新阶段。20 世纪 80~90 年代，一大批钢铁企业相继进行了改建、扩建和技术升级改造，积极采用先进工艺装备，使中国钢铁产业水平得到迅速提升。

1981 年武钢 1700mm 带钢轧机工程建成投产并通过国家验收，使中国在热轧-冷轧薄板生产综合工艺技术装备方面达到国际先进水平。通过消化吸收宝钢工程建设中引进的先进工艺和技术装备，使钢铁产业整体技术水平得到提高。与此同时，一批连铸、连轧、炉外精炼等技术装备的引进，也促进了中国钢铁工业技术装备水平的总体提高。20 世纪 80 年代以后，中国冶金工作者开始谋划中国钢铁工业科技进步的发展战略，在结合国情总结规律的基础上，积极推进连铸、高炉喷煤等关键共性技术的应用，有效推动了钢铁产业结构调整和技术升级。

1996 年中国钢产量突破 1 亿吨、位居世界首位，成为名副其实的钢铁大国，2000 年中国钢产量再创新高达到 1.29 亿吨。从改革开放以后的 20 年间，中国钢铁产业发生了巨大变化：钢铁产业发展迅猛，钢铁产量逐年攀升；但由于中国经济的高速增长，钢材产品依然供不应求，高端产品还需要大量进口以满足市场需求。

这个时期，中国钢铁产业演进历程和特点是：通过改革开放，引进、消化、吸收和应用国外先进技术，以及宝钢工程的设计建设，带动了中国钢铁产业整体技术水平的提高。工艺技术装备进步体现在引进、消化、吸收、应用国外先进技术，在局部技术领域通过自主集成创新形成了一批共性关键技术，通过集成创新、技术进步推动钢铁产业发展效应显著。工程方法是引入了当时日本和欧美的工程设计理念和方法，开始注重单元工序的理论计算和优化设计，初步形成了单元工序优化设计的技术方法体系。

2.3 自主创新期

进入 21 世纪以后，伴随着中国经济的高速发展，中国钢铁产业发展迅猛，钢铁产量高速增长。中国工业化进程加快、经济发展后发优势突出，拉动了国内市场对钢铁产品的巨大需求。在市场需求和资金驱动双重作用下，钢铁产业的投资规模不断扩大，钢铁产能则是连年增长。这一时期，以中国自主创新设计建设的首钢京唐钢铁厂顺利建成投产为标志，中国钢铁产业开始进入自主创新发展的新阶段。

2006年中国钢产量达到4.2亿吨,已成为钢材净出口国。2007年至今,中国钢铁产业再次发生了重大变化,钢铁产能由于快速增长出现了产能过剩。2008年以后,由于国际金融危机和内需拉动不足的影响,产能过剩问题开始出现。直至目前,在全球经济遭受疫情影响的条件下,国际铁矿石价格依然坚挺、钢铁产品销售价格再度出现低迷、市场需求不旺、市场竞争激烈的错综复杂局面,钢铁企业经营艰难,经济效益显著下降,这种产业发展状态预计还将在一定的时期内维持。

进入2000年以后,中国钢铁产业演进历程和特点是:在引进消化国外先进技术的基础上,结合中国钢铁产业自身的特点,通过自主集成、系统优化,创新并形成了一系列具有国际先进水平的工艺、技术和装备,包括烧结、球团、焦化、炼铁、炼钢、连铸、轧钢全流程系统的关键共性技术的研发与应用。在此期间,殷瑞钰院士等冶金学家基于对现代钢铁制造流程的深刻认识和长期研究,创建冶金流程工程学和冶金流程工程集成理论与方法,提出并构建了新一代可循环钢铁制造流程[7],自主创新、系统集成是这个时期中国钢铁产业最主要的发展特征。除了钢铁工艺技术装备进步,通过自主集成创新实现装备科学合理大型化之外;还创建了冶金流程工程学,形成了一整套钢铁冶金工程科学理念、理论和方法,特别是构建了现代钢铁冶金工程设计理念、理论和方法,形成了冶金工程概念设计、顶层设计和动态精准设计体系[8]。同时应当看到,由于经济粗放型发展和规模化扩张,市场需求和投资拉动,使中国钢铁工业出现了产能过剩、产量过剩的局面,从工程哲学的视角分析,产业发展应遵循其固有的客观规律,综合考虑经济增长、技术进步、资源-能源-环境以及市场-价格-质量-生态等诸多要素,在产业发展层次上必须要建立起工程哲学思维模式,遵循产业发展科学规律,否则将受到客发展规律的制约甚至是惩罚。

3 科技进步对钢铁工业发展的作用与意义

3.1 技术创新推动了钢铁工业的发展

20世纪90年代是中国钢铁产业快速发展的崛起时期,通过结构调整、技术升级和装备改造,在全国范围内推广应用钢铁关键共性技术,为中国钢铁产业快速发展奠定了坚实基础[9]。在此期间,通过研究钢铁工业发达国家的发展轨迹和路径,深入探索中国钢铁产业发展的深层次问题,制订了可实施的战略发展规划。回顾历史可以发现,钢铁产业的整体进步,必须重视科技进步对产业发展的影响和意义,避免无序扩张和简单重复建设。在科技进步发展战略的确定和实施中,必须从注重单项技术或局部机理的研究中摆脱出来,关注关键共性技术的协同突破和推广应用,必须重视战略投资的方向和目标的选择,将其转变为成为钢铁产业整体优化、协同创新的资金驱动力,从而实现产业结构调整、企业转型升级,决不能再让战略投资成为落后工艺装备简单复制、低端产能重复建设的推手。纵观中国钢铁产业创新发展的演化进程,其实就是钢铁技术进步的创新历程[10]。

2010年,在冶金流程工程学理论指导下,首钢京唐钢铁厂全面建成并顺利投产,创新构建了新一代可循环钢铁制造工艺流程,对中国钢铁产业创新树立了新的典范[11,12]。

3.2 关键共性技术对钢铁工业的推动

钢铁制造流程是由多工序构成的复杂系统耗散结构体系,在这个复杂制造流程中,具有核心关键作用、技术关联度大、对整体流程结构影响大,而且在钢铁行业中具有普适性、共性的技术称之为关键共性技术。例如连续铸钢和高炉喷煤就是涉及多工序/装置的关键共性技术,并不是某一个单元技术就能够解决的系统性技术,需要多工序、多专业系统协同才能解决。关键共性技术的选择、集成和应用,需要以整个钢铁制造流程的层次深入研究、理性分析和权衡取舍,确定关键共性技术对钢铁产业不同发展时期的战略主导地位。应当指出,关键共性技术还必须与钢铁企业实际紧密结合,将其有序、有机地嵌入到钢铁制造流程中,促进钢铁生产流程整体结构优化和重构优化。未来钢铁制造流程的结构优化、智能化和绿色化将成为钢铁产业创新和企业转型发展的重要命题。

到本世纪中期,中国钢铁产业发展应积极适应市场发展变化,以产业结构调整和流程结构优化为主题,以增强市场竞争力和实现可持续发展作为总体目标,实现钢铁产业的绿色化、低碳化和智能化高质量发展。钢铁产业发展与科技进步具有密切的关联,科技进步是工程演化的重要推动力;而反过来,工程系统的多目标需求对于技术发明、开发和应用具有拉动作用和限制作用。产业发展和工程演化与科技进步密切相关,科技进步和技术发明是工程创新重要的推动力。实践正是只有技术发明和技术创新有效地嵌入到工程系统中,而且发挥其应有的经济价值和工程效应,才能导致工程创新的发生和协同效应的显现。

4 钢铁产业的未来发展

4.1 钢铁产业与经济、社会的协同发展

钢铁产业的清洁生产、节能减排、循环经济和低碳绿色发展,重点是做好资源能源的节约和高效利用,通过钢铁制造过程的"减量化、再回收和再利用",实现现代钢铁厂的功能拓展。有效降低钢铁生产过程 CO_2 和污染物排放,最大限度降低生态环境污染,并能与其他行业乃至社会实现生态链接,进而实现多产业链接、与经济社会循环发展的模式[13]。

钢铁制造过程,利用来自自然界的铁矿石、煤炭、水、空气等天然的资源和能源,经过一系列复杂的物理-化学变化和生产过程,制造出满足经济社会和人们生活所需的钢铁产品,钢铁制造过程所产生的煤气、蒸气、电力、冶金渣等副产品成为其他产业的资源或能源。与此同时,钢铁制造过程,还必须消纳、处理大宗人们生活或其他产业所产生的废弃物,如废钢、废轮胎、城市污水等,经过无害化、资源化处理或高值化循环利用,使其成为新的再生资源或能源。面向未来,钢铁工业必须建构与自然生态环境、经济社会和相关产业之间和谐友好、相互依存、持续发展的新格局(见图2)。

4.2 发展方向与目标

面向未来直至21世纪中叶,我国经济发展将从速度效率型向质量效益型转变,经济社会将保持平稳发展,持续提高发展质量,解决经济过快增长所带来的各种不平衡与不充分。在未来经济发展环境下,钢铁作为首选的功能性结构材料,仍将在未来的经济社会发展过程中,继续发挥不可替代的重要作用。在中国工业化、城市化的高质量发展进程中,钢铁仍将会对社会发展、经济增长、人们生活等具有重要的不可或缺的作用[14]。

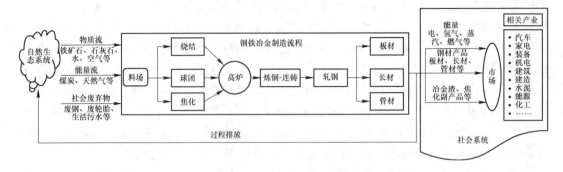

图 2 现代钢铁冶金工程与自然和社会的协同关系

随着中国经济发展的速度减缓和增长方式转变，市场需求拉动将会降低，加之资源和能源可获取性的制约，以及生态环境治理的限制，到 21 世纪中叶，中国钢铁工业发展必须坚持减量化[15]。钢铁工业供给侧结构性改革和"去产能"将在未来很长一个时期内成为产业发展的主旋律、主基调，不容丝毫动摇；必须摒弃那种不讲条件、不顾市场、背离产业发展规律的粗放型盲目扩张的产业发展理念，钢铁项目"一哄而上""盲目攀比""过度增长"而造成的钢铁产能无序扩张，导致钢铁产能严重过剩、产量过度增长，市场竞争混乱无序，甚至出现"劣币驱除良币"现象，最终致使整个产业发展举步维艰，钢铁企业长期濒临在亏损经营的边缘，为经济社会和产业发展带了沉重负担和巨大阻碍，教训惨痛、代价沉重，必须给予高度的重视，进行深刻的反思。

在钢铁产业减量化、高质量发展的形势下，以工程哲学的视野审视产业未来发展的方向、目标、路径和格局。在优质铁矿石资源和炼焦煤资源不断匮乏的未来，生态环境的制约也会日趋严苛，特别是低碳绿色发展将成为全球性的主题，未来面临的发展形势是：市场需求拉力减弱，资源、能源供给力不足，生态环境制约力增强，市场竞争筛选力突显。钢铁企业在这样一个充满不确定性和错综复杂的发展环境下，必须顺应时代潮流，坚持低碳化、绿色化、智能化发展理念，建设资源-能源节约、环境友好的企业发展模式，在愈加激烈的市场竞争中谋求生存发展[16]。

面向未来，智能化将带动经济社会发展，成为整个经济社会发展的主流，技术装备智能化也将是钢铁工业的重要发展方向之一。钢铁制造流程必须构建动态有序、协同连续的信息-物理系统（CPS），实现钢铁制造过程物质、能量和信息的高效协同运行；研发采用基于人工智能系统的高炉操作专家系统、智能化烧结生产控制系统和智能化炼焦技术；建构以连铸、热轧、冷轧等连续化生产线的"数字孪生"系统；构建炼铁-炼钢等界面智能化调控技术，从而实现钢铁生产全流程的动态-有序、协同-连续运行，这将是中国钢铁产业发展的一个重要方向。

4.3 发展路径与前景

至 21 世纪中叶，我国钢铁产业结构和格局将发生重大变化，通过企业重组和战略整合，将形成若干具有国际一流水平的超大型钢铁企业集团，以此进一步优化钢铁产业的结构体系、提升产业集中度，加快淘汰落后产能。通过产业结构调整和重组优化，钢铁工业发展不平衡、不充分的问题，如"先进与落后并存""劣币驱除良币"以及"小、散、乱"的产业不平衡发展格局和长期存在的矛盾将有望逐渐得到解决。主要体现在钢铁产量将在一定时期内保持平稳，不会再出现高速增长，这是中国经济转型发展和经济社会的市场需求所决定的，随着投资拉动效应的减缓，钢铁产量将出现起伏和下降趋势，市场供求将保持一种动态平衡。随着产业结构调整、生态环境治理和企业转型升级，将会有一批钢铁企业实现转型发展，转向新的产业领域，寻求新的经济增长点。未来10年间，我国钢铁技术装备水平将进一步提升，钢铁冶金技术装备大型化、现代化、高效化将成为主要特征，在生产总量严格控制、总体产能削减的前提下，一大批装备先进绿色高效的大型高炉、高效率-低成本洁净钢炼钢连铸生产工艺、智能化高效轧钢生产线将结合现有冶金装备的技术改造和结构调整，进行装备升级改造。其结果不仅是单体技术装备的大型化、现代化，而且高炉、烧结机、焦炉、转炉、轧机的数量将不断减少，生产工艺流程更趋合理，总体生产规模和产能将逐步降低，达到合理的规模区间。

可以预见到21世纪中叶，我国钢铁产业和钢铁制造流程结构也将发生重大变化，随着社会废钢资源的不断蓄积，一定规模的电炉炼钢流程将会出现，不再依赖高比例高炉铁水的电炉钢产量将逐渐增加。与此同时，在可预见的未来，具有一定生产规模的熔融还原、直接还原等不再依赖焦炭资源的非高炉炼铁装备将逐渐实现工业化生产。高炉炼铁的产量和比例都将出现下降，高炉数量减少、高炉平均容积增加，钢铁产业规模减量化、制造绿色化、运行智能化将成为未来发展的主要趋势和方向。

5 结语

（1）钢铁材料是重要的结构材料和功能材料，钢铁产业是经济社会发展的重要基础产业。钢铁制造流程是多工序协同的复杂流程，需要在全产业和全流程的视野中认识和研究钢铁冶金工程。钢铁产业是体现一个国家工业化综合实力水平的代表性产业之一。

（2）面向未来钢铁产业可持续发展，必须以工程哲学的视野，遵循产业发展的客观规律，以减量化、集约化、高效化、绿色化、低碳化多目标优化为关注点，实现钢铁产业和钢铁企业的健康可持续发展。

（3）关键共性技术的研究和应用是实现钢铁产业可持续发展的重要驱动力。钢铁产业未来发展，不仅体现在产能的变化，还要依靠技术进步和科技创新，通过钢铁制造全流程的解析优化、集成优化和重构优化，将关键共性技术与钢铁制造流程有效融合，以全流程功能、结构和效率优化为重点，大力推动钢铁制造流程整体创新和钢铁产业协同进步。

（4）面向未来，中国钢铁产业应以低碳化、绿色化、智能化发展作为产业转型升级的主要方向，重视产业结构调整和企业结构升级。加强钢铁冶金工程的概念设计研究和顶层设计，科学合理规划产业布局和产业发展方向，从战略上解决钢铁产业和企业发展所面临的时代命题，持续提升钢铁产业健康可持续发展核心实力。

参考文献

[1] 徐匡迪. 20世纪——钢铁冶金从技艺走向工程科学 [J]. 上海金属，2002（1）：1-10.
[2] 殷瑞钰. 冶金流程工程学（第2版）[M]. 北京：冶金工业出版社，2009：1-10.
[3] 张寿荣. 论21世纪中国钢铁工业结构调整 [J]. 冶金丛刊，2000（1）：39-44.

[4] 中国工程院编,常平主编.20世纪我国重大工程技术成就[M].广州:暨南大学出版社,2002:117-126.
[5] 张寿荣,于仲洁.中国炼铁技术60年的发展[J].钢铁,2014,49(7):8-14.
[6] 张寿荣,毕学工.中国炼铁的过去、现在与展望[J].炼铁,2015,34(5):1-6.
[7] 张福明,颉建新.冶金工程设计的发展现状及展望[J].钢铁,2014,49(7):41-48.
[8] 殷瑞钰,张寿荣,张福明,等.现代钢铁冶金工程设计方法研究[J].工程研究-跨学科视野中的工程,2016,8(5):502-510.
[9] 张寿荣.钢铁工业与技术创新[J].中国冶金,2005,15(5)1-5.
[10] 徐匡迪.中国钢铁工业的发展和技术创新[J].钢铁,2008,43(2):1-13.
[11] 张福明,钱世崇,殷瑞钰.钢铁厂流程结构优化与高炉大型化[J].钢铁,2012,47(7)1-9.
[12] 张福明,崔幸超,张德国,等.首钢京唐炼钢厂新一代工艺流程与应用实践[J].炼钢,2012,28(2):1-6.
[13] 殷瑞钰.冶金流程工程学(第2版)[M].北京:冶金工业出版社,2009:1-10.
[14] 殷瑞钰.冶金流程集成理论与方法[M].北京:冶金工业出版社,2013:14-16.
[15] 殷瑞钰.以绿色发展为转型升级的主要方向[N].中国冶金报,2013-10-31(1).
[16] 张福明.我国高炉炼铁技术装备发展成就与展望[J].钢铁,2019,54(11):1-8.

Prospect and Development of China's Steel Industry Innovation

Zhang Fuming Li Lin Liu Qingmei

Abstract: Iron and steel materials play an important role and position in economic and social development, and the steel industry is an important basic industry of national economy. The development course of China's steel industry and the characteristics of industrial development in various periods are reviewed, and the role and significance of key common technologies and technological progress in promoting the development of steel industry are discussed. The research points out that facing the future, China's steel industry must adhere to the approach of sustainable development of reduction, green and intelligent, follow the objective law of industrial development, and strictly control the blind expansion and output growth of production capacity scale. The sustainable development of steel industry should realize the function expansion of steel manufacturing, strengthen the scientific layout of industry and the integration of enterprise reorganization, promote the optimization and adjustment of manufacturing process structure, construct cyber physical system of steel plant, and realize the high quality transformation and development. The development direction and path of China's steel industry by the middle of this century are put forward and discussed in this paper.

Keywords: steel industry; engineering evolution; development history; scientific and technical progress; key-common technology

智能化钢铁制造流程信息物理系统的设计研究

摘　要：现代钢铁制造流程是集成烧结、球团、焦化、炼铁、炼钢和轧钢等多工序过程的耗散结构体系。钢铁制造流程是在物质流、能量流和信息流协同运行的条件下，完成一系列复杂的冶金过程和转变。探讨了对钢铁制造过程物理本质以及运行特征的认识和研究，提出了现代钢铁制造流程物理系统的设计理念和方法。阐述了流程工程动态精准设计体系在钢铁冶金工程设计中的应用，讨论了智能化钢铁制造流程的系统层次和构建理念。着重论述了钢铁制造流程信息物理系统的内涵和架构，提出了静态物理系统与信息系统集成、耦合、协同的设计理念和方法，论证了现代钢铁制造流程信息物理系统的智能化设计思路、程序和应用实践。

关键词：钢铁；制造流程；智能化；信息物理系统；设计；耗散结构

1　引言

现代钢铁冶金工艺流程是由烧结、球团、焦化、炼铁、炼钢、轧钢等多个单元工序和单元装置构成的，是通过多工序非线性耦合构成的复杂制造系统。钢铁制造流程中各单元工序或单元装置之间，具有系统层次性和复杂多样性，上游工序的输出即为下游工序的输入，上下游工序之间紧密衔接、相互匹配。现代钢铁制造流程以动态-有序、协同-连续和耗散结构优化为特征，通过智能化推动钢铁产业高质量转型发展，实现钢铁企业高效率-低成本生产运行，不断提高生产效率、产品质量、经济效益和市场竞争力。

2　钢铁制造流程的物理本质和运行特征

2.1　物理本质

现代高炉-转炉钢铁生产过程中，烧结、球团、焦化、高炉、转炉、连铸、轧钢等单元工序基本都是独立运行，工序之间通过胶带机、铁水罐、钢水罐、辊道、吊车等运输（储存或缓冲）设备/装置连接起来，进而形成完整的钢铁制造流程，经过这个复杂的多工序耦合运行的制造流程（单元操作+单元过程），铁矿石经过一系列复杂的冶金过程最终被生产成钢材产品。

现代钢铁冶金工艺过程的实质是一类开放的、远离平衡的、不可逆的、由不同结构-功能的相关单元工序过程经过非线性相互作用、嵌套构建而成的流程系统，是一个多相多态复杂巨系统。在这个流程系统中，铁素物质流和能量流耦合作用、协同运行，按照设定

本文作者：张福明。原刊于《钢铁》，2021，56（6）：1-9。

的运行程序,沿着所设计的复杂流程网络结构运行。与此同时,在这个复杂流程中,还要实现结构、功能、效率的集成优化和运行效果的多目标优化。

由此可见,动态变化是钢铁制造流程运行的核心[1]。进一步研究可以发现,对于高炉-转炉工艺流程,其流程的物理本质是:铁素物质流在碳素能量流的作用和驱动下、在信息流的调控下,沿着设定的流程网络(即设计的工艺流程),按照设定的运行程序(即控制系统、控制模型等)进行动态连续的运行过程,在这个运行过程中实现成本、质量、效率、效益、环境等多目标集成优化[2]。所谓多目标集群包括提高产品质量和性能,降低生产消耗和运行成本,实现生产高效顺行,提高能源转换效率和利用率,降低资源和能源消耗、CO_2及污染物排放,实现钢铁制造过程的循环经济和低碳绿色等。

2.2 运行特征

现代钢铁生产工艺过程是多个单元工序和单元装置相互衔接、前后匹配所构成的一个协同运行的集成系统和制造过程。典型的现代钢铁制造流程如图1所示。

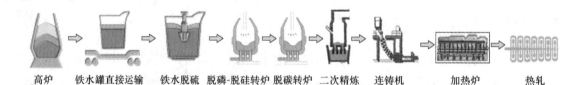

图1 典型的现代钢铁制造流程

整体性和复杂性是现代钢铁制造流程的重要特征[3]。钢铁制造流程的整体性是系统论的具体体现,表现在钢铁制造流程的整体和工序之间的相互关系方面。即当流程整体处于有序状态时,流程整体的功能和效率大于工序之和;但是当流程整体处于无序状态时,流程整体的功能和效率则小于工序之和;与此同时,工序的紊乱和无序也将对钢铁制造流程的整体产生影响效应。现代钢铁制造流程的复杂多样性和层次结构性则是其复杂性的具体体现。

2.3 特征要素

殷瑞钰院士指出"流""流程网络"和"运行程序"是钢铁制造流程动态运行的特征要素[4]。所谓流是现代钢铁制造流程运动态变化中最关键的要素;而所谓流程网络乃是钢铁制造流程的运行路径、结构和时空边界,即流程设计中的钢铁制造工序流程结构以及流程界面;运行程序则是指钢铁制造流程的信息化原则、策略和调控程序等。在钢铁制造流程中,运行程序主要是指流程运行中各类信息的采集、存储、传输、处理和网络化集成,以及网络化的调控操作和运行管理,还涵盖过程检测、程序控制、制造执行、能源管控和企业资源计划等。现代钢铁制造过程运行程序还体现着信息输入"他组织"的特征。

3 静态物理系统的设计

3.1 功能优化

现代钢铁制造流程是基于钢铁制造过程物质、能量和信息的动态耦合、协同运行和系统集成而构建起来的全新型钢铁制造系统[5]。现代钢铁厂的功能已经拓展为钢铁产品制造、能源转换和废弃物消纳处理与再资源化三大功能。

现代钢铁冶金流程设计必须摒弃传统的机械还原论理念和方法,树立动态运行的思维理念,所设计的钢铁制造流程物理系统(工程物理实体)应充分体现钢铁制造要素优化、动态有序、协同连续、运行高效的结构特征,通过物质-能量-信息的流程网络以及相应的动态运行程序的协同运行和耦合匹配,使其构建成为具备高效率-低成本-高质量的钢铁产品制造功能、高效能源转换功能以及资源化处理废弃物三大功能的工程系统。现代钢铁制造流程功能优化如图2所示。

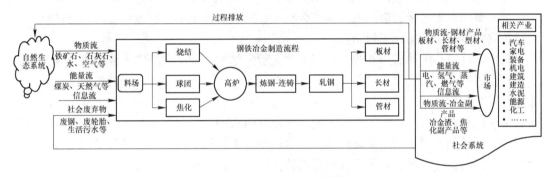

图2 现代钢铁制造流程功能优化

3.2 设计理念

现代钢铁制造流程以现代钢铁冶金工程设计理论和方法指导概念设计和顶层设计,并以优化的顶层设计统筹工序、设备、装置等工艺要素的合理选择和动态集成。顶层设计包括单元工序和装置等要素的优化选择、总体流程结构的形成和建构、流程功能的优化和合理建构。系统优化体现在工序功能集合解析-优化、工序之间关系集合协调-优化、流程工序集合的重构-优化,以及流程动态运行效率、能源转换效率等方面的策划和考量等内涵。

现代钢铁制造流程使钢铁冶金工程从孤立的局部性研究走向开放的动态系统研究,从间歇、等待和随机组合运行的流程走向动态、协同、非线性耦合的动态精准、耗散优化的流程。

钢铁冶金工程演化和现代钢铁制造流程构建,本质上是工程思维模式的转变和创新,是从机械"还原论"思维模式所暴露出的缺失中摆脱出来,探索到了系统性整体集成优化的新思路和新理念。从工程哲学视角分析,在钢铁制造流程设计中,不仅要研究"孤立"和"局部"的"最佳",更要解决流程整体动态运行过程的系统最优化;不能用机械论拆分-还原的方法来解决相关的、异质功能的而又往往是不易同步运行工序/装置的组合集成的复杂问题。钢铁制造流程设计要注重研究钢铁冶金耗散结构动态运行过程的工程科学问题，要厘清工艺表象和物理本质之间的表里关系、因果关系、非线性相互作用和动态耦合关系，并探索出其内在规律。

因此离散型装备制造业所通行的采用信息网络链接实现网络智能化的理念，在钢铁冶金复杂的流程制造系统中，不是简单照搬照抄就能够实现钢铁制造全流程智能化，需要一切从实际出发，结合钢铁制造流程的本构特征和运行特点，必须要设计并构建出结构合理、运行高效、动态有序、协同连续的钢铁制造流程静态物理结构（static physical systems）。

3.3 设计方法

钢铁制造流程设计是基于钢铁冶金基础科学，应用钢铁冶金技术科学和工艺技术、装备技术以及相关专业技术等进行有效的工程集成的过程，是在科学-技术-工程三个层次上进行综合、权衡、集成和建构的过程。流程设计是现代钢铁厂工程设计最重要的关键环节，是承载工程理念的载体。从方法论上来看，流程设计是钢铁冶金工程的要素、结构、功能、效率的权衡选择、协调取舍和综合集成，从而构建出结构、效率、功能协同优化具有可持续竞争能力的实体工程结构。

流程设计核心关注点是要协调处理整体流程的结构、效率和功能集成优化以及动态运行过程中的目标群的优化，使各单元工序和装置之间实现动态-有序和协同-连续，合理构建出物质、能量和信息的流程网络，制定动态-精准的运行程序，协调处理单元工序、单元装备和信息控制系统之间的耦合匹配。因此，钢铁冶金工程设计是跨学科、跨领域的学科分支，需要集成钢铁冶金基础科学、技术科学和工程科学的研究成果和发展趋势进行设计应用和创新，并将其凝聚成工程设计新的认识、理念、理论和方法。

4 钢铁制造流程智能化设计

4.1 设计思维

现代钢铁冶金工程设计是基于系统论、协同学、运筹学、耗散结构理论和冶金流程工程学理论，从传统静态-粗放的设计模式发展到动态-精准的设计体系，基于钢铁制造流程动态运行过程物理本质的深刻认识，以冶金流程工程学为基础所建立的设计理念和设计方法[6]。

钢铁制造流程智能化设计应以概念研究为核心，摈弃传统的比拟放大、参照设计的思维模式和设计模式，充分运用系统论、运筹学、耗散结构理论和动态仿真模拟等先进理论和方法，以全流程高效协同运行为宗旨，着重研究物质、能量和信息的流程网络架构设计和运行程序设计，为了解决全流程的连续化、高效化运行问题，深入研究工序之间和工序内部的界面技术优化问题，以期在钢铁冶金全流程中实现耗散结构优化，并且着力实现现代钢铁厂的三大功能。动态精准设计是构建现代钢铁信息物理系统、实现全流程智能化的基础和关键。

钢铁制造流程智能化设计是现代钢铁制造流程的重要基础，流程设计的先进性、科学性和合理性直接决定钢铁厂的工程投资、运行成本、生产效率和产品质量。实践证实没有卓越的流程设计和工程设计，钢铁企业在未来激烈的市场中就会缺乏可持续竞争力。设计面向未来，设计引领未来，设计创造未来。现代钢铁制造流程智能化的发展框架如图3所示。

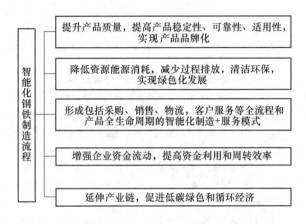

图3 现代钢铁制造流程智能化的发展框架

4.2 设计体系

现代钢铁冶金工程动态精准设计方法属于典型的智能化设计方法[7]，是以构建钢铁制造全流程高效协同、动态精准运行为目标，形成动态-有序-连续运行系统的实体（硬件）-虚体（软件）集成，构建出钢铁冶金全流程具有高度自组织功能，进而实现全流程系统自感知、自适应、自决策、自执行的智能化运行的设计方法。

钢铁冶金工程动态精准设计基于现代科学理论和先进的信息技术，从本质上属于智能化设计方法体系。动态精准设计本身也是智能化的，是必须依靠计算机、大数据工作站以网络化协同设计等先进技术装备、工具、方法和手段才能得以实现的。钢铁流程运行参数的选择、流程结构的确定、流程网络和运行程序的设计，已经不再是传统意义上的经验设计、半理论-半经验设计和静态比拟设计；从简单堆砌、比拟放大跃迁到仿真模拟研究、数字化设计、动态精准设计和智能化设计，进而发展到智能化虚拟现实钢铁冶金制造过程。钢铁冶金工程智能化设计流程与内容见表1。

表1 钢铁冶金工程智能化设计流程与内容

设计流程的层次	设计研究的内容
概念设计	（1）建立现代钢铁冶金动态运行的思维理念。 （2）构建钢铁制造流程的耗散结构和耗散优化过程，制定钢铁冶金制造过程的运行准则。 （3）选择并确定钢铁冶金工艺流程，如高炉+转炉的工艺流程、非高炉炼铁工艺流程或电炉冶金工艺流程等

续表1

设计流程的层次	设计研究的内容
顶层设计	（1）钢铁制造流程产品制造、能换转换和废弃物资源化处理三大功能的设计与优化。 （2）物质流、能量流和信息流流程网络和静态结构的构建。 （3）流程要素优化研究： 1) 设计产能、产品大纲、轧钢产线的工艺设备配置及确定； 2) 炼钢产能、连铸机与热轧机的工艺匹配研究，铁水预处理-炼钢-精炼-连铸工艺的选择和装备配置； 3) 炼铁产能、高炉数量、容积等主要参数的确定； 4) 铁前系统主要单元工序的工艺装备配置和耗散结构研究，如料场、烧结、球团、焦化等工序的产能、工艺产线数量、装备配置、界面连接技术等； 5) 主要工艺界面技术研究，如炼铁-炼钢铁水运输界面、连铸-热轧连铸坯输送界面等； 6) 能量流网络、能源转换装置和系统节能设计研究，如干熄焦（CDQ）、炉顶煤气余压发电（TRT）、燃气-蒸气联合发电（CCPP）工艺设计，烧结、热风炉、加热炉余热回收利用，高炉和转炉冶金煤气干法除尘设计等； 7) 构建现代钢铁厂与上下游企业之间的循环经济产业链，与社会和相关产业形成融合发展格局。 （4）流程结构的构建与优化研究，设计科学合理的流程结构，使钢铁生产高效顺行，形成上下游工序——对应的"层流"运行模式。 （5）现代钢铁厂功能拓展和效率优化。以实现三大功能为优化目标，合理设计布局钢铁厂总平面图、工艺布置图、设备安装图和综合管线图，做到流程集约高效、布局紧凑合理、时间-空间协调有序
动态精准设计	（1）在概念设计和顶层设计的基础上，建立科学合理的时间-空间协调关系，节约工程占地、减少工序和装置冗余、降低过程耗散，确立以时间精准控制的钢铁生产过程动态管控理念。 （2）通过钢铁生产过程动态运行的仿真研究进一步加强流程网络的设计与优化，实现钢铁生产过程耗散的最小化。 （3）利用工序/装置功能集合的解析-优化，最终确定工序功能定位和装置能力选择。 （4）利用工序之间协调-优化和动态运行评价，研究解决工序之间协同、缓冲、连接、互补等问题，实现工序之间的紧凑衔接和耦合匹配，设计研究出集约高效的界面技术。 （5）加强工程集成创新，将先进的工艺、技术和装备合理集成，形成系统的整体协同优势，局部单体技术先进并不能代表全流程的先进，必须合理地嵌入到钢铁制造流程中，才能充分发挥作用，注重利用流程中工序集合的重构优化，构建流程网络和运行程序。 （6）通过计算机仿真研究、动态甘特图编制等，进行钢铁厂产量、设备作业率的核算和校验，对整个钢铁生产流程进行动态运行效果的评估和调整

由表1可以看出，现代钢铁冶金工程设计包含了概念设计、顶层设计和动态精准设计三个层次，每个层次设计研究的内容不尽相同，动态精准设计是钢铁冶金工程智能化设计的基础，也是钢铁制造流程信息物理系统构建的基础。通过表1还可以归纳出钢铁冶金工程智能化设计的思维逻辑、设计体系和设计过程的要素包括：

(1) 建立科学合理的时间-空间的协同关系;
(2) 注重流程网络的构建与优化;
(3) 注重单元工序和单元装置之间的耦合匹配以及界面技术的设计研究;
(4) 突出顶层设计中的集成创新,追求多目标协同优化;
(5) 提高钢铁制造全流程动态精准运行的整体性、系统性、高效性、精准性和协调性。

4.3 智能化信息系统的构建

现代钢铁制造流程智能化在信息化基础架构上,通过人工智能可以实现精准管理与控制[8]。钢铁制造流程信息系统的设计主要应用冶金流程工程学理论和现代钢铁冶金工程动态精准设计方法,首先建立流程动态有序运行的理论架构和物理实体模型及仿真计算模型,针对各单元工序关键工艺参数的确定、主体设备装置能力的确定、工序之间的界面连接方式等进行研究分析,运用现代设计方法进行动态的仿真研究、精准计算和优化配置,将粗放型、经验型或半经验-半理论性的传统设计方法转变为基于智能化研究的精准设计方法,构建现代钢铁冶金工程动态精准、高效协调的制造流程体系[9]。

钢铁制造流程智能化是以自动化、数字化、信息化、网络化为基础[10],具有系统性和结构层次性,并非单纯依靠 ERP 就可以解决钢铁制造复杂流程的智能化问题。对于单元工序、设备或装置,基础的检测和控制(PLC/DCS)是必须的,比如工况运行参数的检测、计量、监测,设备的启-停以及设备之间的联锁控制等都是实现智能化的基础,这也是实现智能化"自感知"的基础[11,12]。单元工序过程控制(PCS)是实现具体工序或系统的模型控制,例如高炉供料-装料-布料的模型控制、转炉冶炼过程的加料-吹氧-测温-定碳-出钢操作控制等。工厂级的制造执行系统(MES)则是控制料场、烧结、球团、焦化、炼铁、炼钢、轧钢等工厂级制造流程的运行,如炼钢厂的制造执行系统就包括铁水预处理-转炉冶炼-二次精炼-连铸等各个主要生产工序环节的综合调控;而对于钢铁企业制造管控中心(总调度室或生产制造部)而言,制造执行系统(MES)则主要调控的是全流程的界面衔接与工序匹配,例如炼铁向炼钢的铁水供应、连铸向热轧供应连铸坯料等。另外,作为全厂性的能源管控中心,能源介质的供应、调配和管控则是通过能源管控系统(EMS)进行的,这与生产制造系统(MES)属于同一层次的控制系统。而对于钢铁企业的供应、销售、财务、资金、订单、合同管理需要通过企业资源计划系统(ERP)进行集成化管理,这是钢铁企业经营管理层面的流程,是直接服务于企业经营决策、战略、管理、计划、运营、财务和资金等范畴的高层级信息化管控,属于钢铁企业经营管理的控制范畴。

图 4 所示为钢铁企业智能化体系的结构与层次;图 5 所示为钢铁生产过程智能化信息管控系统的层级关系和设计框架。

4.4 物理系统与信息系统的耦合

基于对钢铁制造流程物理本质的认识和研究,可以看出钢铁制造流程设计的主要目的,不是简单地设计若干个相互独立的工序单元,然后再进行叠加和拼凑从而形成一个组合的流程结构,而是要设计出满足钢铁制造流程动态运行要求及其特性的优化的流程

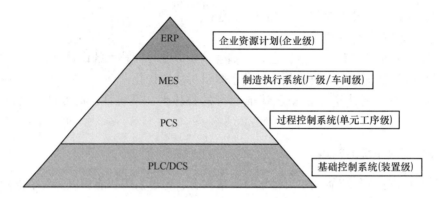

图 4 钢铁企业智能化体系的结构与层次

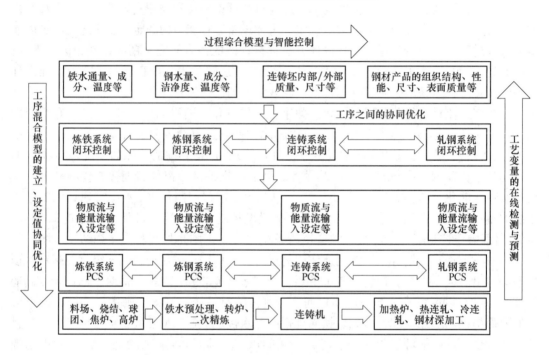

图 5 钢铁生产过程智能化信息系统的设计框架

体系。

信息物理系统（CPS，cyber physical systems）将信息系统和物理实体集成为统一整体，将数据处理、数据通讯及动态控制集成一体，是具有自学习、自生长能力的新一代智能化生态系统[13]。CPS 是实现钢铁制造流程系统的信息感知、动态调控和信息集成，可以使系统更加可靠、高效、精准，可以做到实时协同、动态耦合运行，在流程制造业领域具有重要而广泛的应用前景。

根据钢铁冶金制造流程的技术特征，现代钢铁冶金工厂智能制造的涵义，是以企业生产经营全过程和企业发展全局的智能化、绿色化以及产品质量的品牌化为发展目标，设计研究出的生产经营全过程的信息物理系统。其关键技术是生产工艺/装置技术优化、工艺/装置之间的"界面"技术优化和制造全过程的整合-协同优化，以此为基础嵌入数字信息技术，以物质-能量-信息协同、物质流网络-能量流网络-信息流网络交互融合为切入口，从而构成体现智能特色的信息物理系统——CPS。

钢铁制造流程信息系统与物理系统的耦合建构主要包括以下几个方面：

（1）钢铁冶金工程设计不仅是单元工序、设备和装置的设计，要在整个钢铁制造流程的层次上研究流程设计，不仅要重视物质流和能量流的网络结构设计，也要重视信息流网络的结构设计。新一代钢铁制造流程不同于传统的流程，是面向智能化的流程工程，必须重视信息流的输入、调控和耦合运行，建构与静态物理流程结构具有相互映射关系的"数字孪生"系统，形成"三网协同""三流耦合"交互融合的流程整体和完整的信息-物理系统（CPS），如图6所示。

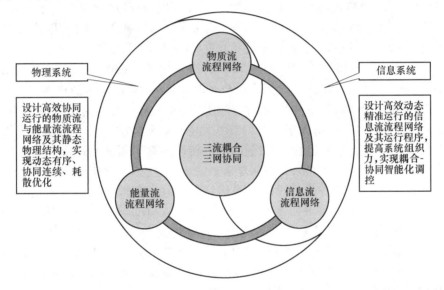

图6 现代钢铁制造流程的信息物理系统

（2）现代钢铁制造流程属于复杂巨系统，在时-空关系、质-能转换、自组织性和他组织性等方面关系错综复杂，复杂性、层次性、多样性、多目标性特征鲜明。基于耗散结构优化理论，钢铁制造流程设计必须要使系统具有高度的自组织性和自组织力[14]，上下游工序之间匹配合理、协同运行。

（3）现代钢铁冶金工程设计是在限定或约束的条件下，结合市场需求和资源能源供给条件、生态环境承载条件等，合理确定钢铁厂生产规模、产品结构、工艺流程和装备配置等关键要素，同时还要进行投资、效益、环境等多目标群进行综合、权衡、协调、决策等集成优化的过程。

（4）现代钢铁冶金工程设计既要实现炼铁、炼钢等单元工序的优化，还要通过系统集成优化和流程重构优化，从而实现钢铁生产全过程的系统优化。

（5）现代钢铁冶金流程智能化设计有别于传统的工程设计，是由钢铁厂系统全局出发自上而下开展的。在实现钢铁生产全过程动态精准、协同高效运行的工程理念统领下，注重概念设计研究和顶层设计深化，进而提出对单元工序和装置进行精准设计的要求，而不是各自独立设计单元工序，再汇总拼凑出流程系统。

（6）现代钢铁冶金工程设计创新必须顺应时代发展要求，以循环经济、低碳绿色、智能高效作为工程设计的发展理念，要拓展现代钢铁厂的功能[15]。同时，不仅重视流程物理实体系统的建构，还必须重视信息虚拟系统的构建，做到"虚拟与现实结合"。

采用现代钢铁冶金工程动态精准设计体系的首钢京唐钢铁厂设计规模为年产粗钢870万～920万吨/年[16]。该工程于2007年3月12日开工建设，2009年5月21日1号高炉点火送风，2010年6月26日全面竣工、顺利投产。首钢京唐一期工程设计构建了以2座高炉、1个炼钢厂、2条热轧生产线和3条冷轧生产线为物理框架的新一代可循环钢铁制造工艺流程。自主设计建造了2座5500m^3高炉、2台550m^2烧结机、1台504m^2带式焙烧机球团生产线、（2+3）座300t转炉高效率-低成本洁净钢生产平台、2250mm和1780mm宽带钢热轧生产线、2230mm和1700mm以及1420mm镀锡板冷轧生产线、5万吨/天海水淡化等一批具有代表性的先进工艺技术装备，形成了近百项具有自主知识产权的工程技术创新成果，对中国现代钢铁厂的设计建设具有重要的引领和示范作用。

首钢京唐工程投产10年来，5500m^3高炉日产量达到12650t/d以上，利用系数达到2.35t/($m^3 \cdot d$)；高炉球团矿入炉比例达到55%以上，入炉矿品位达到60%以上，高炉渣量降低到215kg/t；燃料比低于495kg/t，焦比为287kg/t，煤比为180kg/t，风温达到1250℃以上；高炉煤气利用率大于50%，TRT发电量达到45kW·h/t以上；炼铁工序能耗低于385kgce/t。自主集成、首次在巨型高炉-转炉界面采用铁水运输"一罐到底"技术，铁水温降100℃以下，节能环保效果显著。铁水经过脱硫、脱硅和脱磷预处理，转炉冶炼周期为24～30min，转炉全炉役碳氧积达到0.0016，超低碳钢冶炼出钢温度降低至1648℃；洁净钢生产过程中[S]、[P]、[N]、[H]、[O]等有害元素的质量分数可降低到0.0045%以下；连铸系统采用结晶器液位自动控制、结晶器电磁制动和动态轻压下和动态二次冷却控制技术，低碳钢连铸最高拉速达到2.5m/min；热轧工序连铸坯热装率大于75%；2230mm和1700mm冷轧生产线的冷热转换比达到55%、涂镀比达到51%以上。采用动态精准设计体系，实现了钢铁厂工程设计的集约化和精准化，钢铁厂吨钢占地仅为0.9m^2，炼铁至炼钢的铁水运输距离仅为900m；充分利用钢铁冶金过程的余热、余能，自发电率达到96%以上，吨钢电耗降低到615kW·h，全流程吨钢能源消耗为556kgce；采用海水淡化技术和综合节水技术，吨钢新水消耗仅为2.18t；钢铁制造过程流程紧凑、动态有序、协同连续，从高炉出铁到热轧成材的制造周期仅为345min。经过十年多的生产实践，在钢铁制造流程智能化、绿色化发展进程中达到了预期的运行效果。图7所示为首钢京唐钢铁制造流程智能化管控架构。

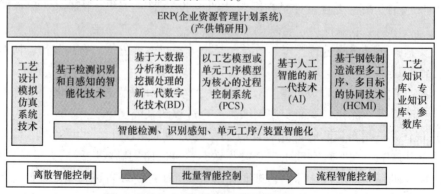

图7　首钢京唐钢铁制造流程智能化管控架构

5 结语

（1）现代钢铁冶金过程是开放的耗散结构体系，由多个单元工序和装置组合而构成的复杂系统，需要认真研究其物理本质和运行规律，不能照搬照抄离散型装备制造业的智能化模式。

（2）现代钢铁制造流程智能化设计的核心是构建信息物理系统（CPS）。基于对钢铁制造流程物理本质和运行特征的深入研究，构建流程结构合理的静态物理系统是基础和前提。

（3）现代钢铁制造流程必须重视物质流和能量流的网络结构设计，也要重视信息流网络的结构设计，提高系统的自组织性，在信息流输入和他组织力的调控下实现动态有序、协同连续和耗散结构优化。

（4）智能化钢铁制造流程设计，对于现代钢铁制造流程而言是至关重要的，卓越的流程设计是企业参与市场竞争的核心要素，是工程创新和技术创新的始端。只有智能化的工程设计才能建构出智能化的工艺流程。

致谢

衷心感谢殷瑞钰院士、张寿荣院士的指导和支持，感谢颉建新、刘清梅、李欣、任绍峰等专家的建议和帮助。

参考文献

[1] 殷瑞钰. 冶金流程工程学（第2版）[M]. 北京：冶金工业出版社，2009.

[2] 殷瑞钰. 冶金流程集成理论与方法 [M]. 北京：冶金工业出版社，2013.

[3] 殷瑞钰，张寿荣，张福明，等. 现代钢铁冶金工程设计方法研究 [J]. 工程研究-跨学科视野中的工程，2016，8（5）：502.

[4] 殷瑞钰. 冶金流程工程学 [M]. 北京：冶金工业出版社，2004.

[5] 殷瑞钰. 过程工程与制造流程 [J]. 钢铁，2014，49（7）：15.

[6] 殷瑞钰，张福明，张寿荣，等. 钢铁冶金工程知识研究与展望 [J]. 工程研究-跨学科视野中的工程，2019，11（5）：438.

[7] 颉建新，张福明. 钢铁制造流程智能制造与智能设计 [J]. 中国冶金，2019，29（2）：1.

[8] 刘玠. 人工智能推动冶金工业变革 [J]. 钢铁，2020，55（6）：1.

[9] 殷瑞钰. 关于智能化钢厂的讨论——从物理系统一侧出发讨论钢厂智能化 [J]. 钢铁，2017，52（6）：1.

[10] 孙彦广. 钢铁工业数字化、网络化、智能化制造技术发展路线图 [J]. 冶金管理，2015（9）：4.

[11] 曾加庆. 关于钢铁流程智能化提升的思考 [J]. 冶金自动化，2019，43（1）：13.

[12] 蔡自兴. 人工智能在冶金自动化中的应用 [J]. 冶金自动化，2015，39（1）：1.

[13] 姚林，王军升. 钢铁流程工业智能制造的目标与实现 [J]. 中国冶金，2020，30（7）：1.

[14] 沈小峰. 混沌初开：自组织理论的哲学探索 [M]. 北京：北京师范大学出版社，2008.

[15] 张福明. 我国高炉炼铁技术装备发展成就与展望 [J]. 钢铁，2019，54（11）：1.

[16] 张福明，颉建新. 冶金工程设计的发展现状及展望 [J]. 钢铁，2014，49（7）：41.

Research and Design on Cyber Physics System of Intelligent Iron and Steel Manufacturing Process

Zhang Fuming

Abstract: The modern iron and steel manufacturing process is a dissipative structure system that integrates multiple production processes of sintering, pelletizing, coking, ironmaking, steelmaking, and rolling, etc. The iron and steel manufacturing process is to achieve a series of complex metallurgical processes and transformations under the condition of coordinated operation of substance flow, energy flow, and information flow. The understanding and research on the physical essence and operation characteristics of iron and steel manufacturing process are discussed; the engineering design philosophy and method of modern steel manufacturing process physical system are put forward. The application of process engineering dynamic precision design system in the iron and steel metallurgical engineering design is described; the systematic level and construction concept of intelligent iron and steel manufacturing process are discussed. The connotation and structure of the cyber-physical system for the iron and steel manufacturing process are emphasized. The design concept and method of the integration, coupling, and cooperation between the static physical system and the cyber system are proposed; the thinking, program, and application practice of intelligent design for the cyber-physical system of modern iron and steel manufacturing process.

Keywords: iron and steel; manufacturing process; intelligentization; cyber physical system; design; dissipative structure

钢铁冶金从技艺走向工程科学的演化进程研究

摘 要：以工程哲学的视野，研究了钢铁冶金的起源、发展和演化进程。阐述了钢铁材料的功能与地位，论证了钢铁对的人类社会进步和文明发展的重要作用。研究了古代钢铁冶金技艺的发展、传承和创新，讨论了 17 世纪以后，文艺复兴以及近代科学的肇始对钢铁冶金技术和科学的影响。论述了近代科学对钢铁冶金从技艺走向工程科学的重大和深远的影响，对以冶金物理化学为代表的钢铁冶金基础科学、创建进行了详细论述。阐述了以冶金反应工程为代表的钢铁冶金技术科学对现代钢铁冶金工程的重大意义；讨论了宏观钢铁冶金工程科学、冶金流程工程学对钢铁产业发展的重要意义。提出了钢铁冶金基础科学、技术科学和工程科学协同发展、相互支撑的学科体系，提出了未来钢铁工业可持续健康发展的目标和方向。

关键词：钢铁冶金；工程演化；物理化学；冶金反应工程；冶金流程工程学；工业革命

1 钢铁材料的功能与地位

长期以来，钢铁由于资源储备丰富、制造成本相对低廉、材料综合性能优越、易于加工而且便于循环利用，成为了当前世界上重要的结构材料，也是世界上消费量最大的功能材料（例如电工钢、不锈钢等）[1]。毋庸置疑，钢铁是伴随着人类文明演化发展的重要材料和重要标志，甚至可以说，人类文明的发展史与钢铁息息相关，钢铁是人类生存发展不可或缺甚至是无可替代的重要基础材料、结构材料和功能材料。

考古学认为铁器时代是人类发展进程中一个极为重要的时代，是继石器时代、青铜器时代之后的一个新时代，铁器时代以人类能够进行铁的冶炼和铁器的制造为标志。人类从石器时代演变到青铜时代和铁器时代，其重要的标志就是人类生产和生活的工具从石器变成了青铜器和铁器。众所周知，无论是旧石器时代、中石器时代还是新石器时代，人类先祖们开始使用三角形、梯形或不规则四边形的尖锐、锋利且容易持握的天然石器作为工具，所用的石材全部来自自然界，仅是对石器进行简单的挑选、打磨和加工而已，谈不上对石器的"深加工"和精雕细刻。究其原因，一方面是人类当时还找不到比石器更加坚硬锋利的材料和工具，另一方面是当时人类还处于从蒙昧向智人的过渡时期，人们对自然的认知能力还极其有限。直到进入新石器晚期，人类才开始使用陶器，人类社会才开始进行农业和畜牧生产，人类农耕文明时代从公元前 8000 年一直延续到 17 世纪[2]。

本文作者：张福明。原刊于《工程研究-跨学科视野中的工程》，2020，12（6）：527-537。

2 古代钢铁冶金技术的发展

我国大约在春秋战国时期，创造了举世闻名的青铜文化，在冶金领域制造和生产出许多绝世经纶、庄重精美的国之重器，如20世纪考古发现的司母戊鼎（1939年出土）、曾侯乙青铜编钟（1978年出土），充分证实了我国古代先民高超的制造技艺和水平[3]。在当时，除了制造精美无比的青铜器制品，还探索了青铜冶炼、铸造的一般规律，进行了初步的理论性总结。我国著名的古代科技名著《考工记》中"金有六齐""改煎金、锡则不耗，不耗然后权之"的论述，阐述了青铜冶金的成分调剂，利用标准量器准确计量，是商周以来积累的青铜合金中铜、锡、铅的成分配比的系统总结和归纳，在世界上属于首次阐述；"铸金之状"总结了冶炼和铸造青铜时的几种情状，以及对"火候"的描述，通过观察火焰颜色变化规律，根据焰色变化规律掌握火候，这可以认为是现代高温冶金工艺中，通过观察冶金过程火焰和温度来判定冶金进程的"滥觞"。

铁器时代是在青铜器时代之后的人类重要文明时代。铁的冶炼和铁器的制造，在今天看来，仍是人类社会发展过程中一个了不起的成就。考古研究认为，人类开始铁的冶炼距今大约3500年以前，而铁器的大量使用大约在距今1500~2000年以前[4]。古代冶铁的工艺和现代大不相同，基本是在青铜冶炼和铸造的基础上发展起来的。囿于当时的工艺和技术水平，人们将铁矿石和木炭放入陶罐中加热，加热的温度只能达到800~1000℃，得到的其实是一种块状海绵铁，然后再经过反复锻打的方法脱除海绵铁中的碳和杂质，这种方法称为"块铁法"，现在从冶金学的学术角度来看，这种方法属于铁矿石直接还原工艺。海绵铁经过反复锻打脱碳后被称为"熟铁"，实际上就是含碳量低的铁，性能更加良好，也就是现代的钢。从含碳量高的生铁，变成含碳量低的钢，要经过反复的加热、锻打和脱碳，工艺十分复杂。古代钢铁冶金的历史，也是不断发展演进的，从"块铁法""炒钢法""百炼钢"到"灌钢法"等许多工艺方法不断发展变革[5]。大约在2000年前，利用竖炉生产生铁的工艺开始问世，这是古代高炉炼铁的起源。中国古代高炉的鼓风设备被称之为"橐"，实际上是一种用皮革制成的皮囊，也就是鼓风用的风箱，利用畜力或者水车对橐进行反复压缩，将空气鼓风送入高炉。

我国北宋著名科学家、政治家沈括（1031~1095）撰写的《梦溪笔谈》[6]，是一部涉及古代中国自然科学、工艺技术及社会历史现象的综合性笔记体著作，在我国乃至世界科技史上具有崇高地位，被英国科技史学家李约瑟评价为"中国科学史上的里程碑"。《梦溪笔谈》是沈括晚年的总结性著作，是他一生学问最精华部分的结晶，包罗万象、广博精深，涉及的门类非常广泛。书中就记载了"灌钢"的炼钢方法，"取精铁锻之百余火，每锻称之，一锻一轻，至累锻而斤两不减则纯钢也，虽百炼不耗矣"。由此可见，早在900多年前的宋代，我国就已经完全掌握了通过反复加热、锻打而脱除碳和其他杂质的"百炼钢"工艺。

我国明代著名科学家宋应星（1587~约1666）在《天工开物》中写道，"凡铁分生、熟，出炉未炒则生，既炒则熟。生、熟相合，炼成则钢"，从工艺制造方法上给出了铁和钢的区别，描述了钢的制造方法，直到今天来看，这些论述都有很高的学术研究价值。"取出加锤，再炼再锤，不一而足。俗名团钢，亦曰灌钢者是也"[7]。《天工开物》中详细记载了古代钢铁冶炼和加工的方法，是记载我国科技发展的世界性科技名著，特别是其中的生产工艺图尤为可贵。在钢铁冶炼工艺的记载中，有很多是我国先民的发明创造，如灌钢法、以煤炼铁、直接将生铁炒制成熟铁、采用大型活塞式风箱鼓风炼铁等，都是我国先民的发明创造和对人类冶金技术的贡献，在人类文明的发展历程中，闪烁着熠熠光辉。

3 近代钢铁冶金技术的突破

3.1 文艺复兴与第一次工业革命

纵观人类科学技术发展史，可以更加清晰地梳理出人类科技进步的历程和现代科技文明起源及其发展的脉络。公元5世纪后期到公元15世纪中期，历史上被称为"黑暗的中世纪"（从公元476年西罗马帝国的灭亡开始，直到公元1453年东罗马帝国的灭亡终止），欧洲经历了长达近千年的黑暗时期，这个时期欧洲没有统一的封建集中政权，封建割据带来频繁的战争，再加上天主教对人们思想的禁锢，造成了科技和生产力发展停滞，人们生活在毫无希望的痛苦之中。中世纪对自然现象缺乏兴趣，漠视个人主张，其根源在于一种超自然的观点、一种向往来世的思想占据支配地位[9]。直到14~16世纪，欧洲开始了文艺复兴运动[10]，使欧洲进入到近现代时期，欧洲文艺复兴运动的代表性人物包括意大利著名诗人但丁，意大利著名人文主义者弗兰齐斯科·彼特拉克，意大利著名诗人、文学家乔万尼·薄伽丘（《十日谈》的作者）；文艺复兴运动最负盛名的意大利著名画家达·芬奇、拉斐尔·桑西，意大利著名雕塑家米开朗基罗·博那罗蒂，以及文艺复兴时期英国著名的文学巨匠莎士比亚。所谓"文艺复兴运动"是由于当时还没有成熟的文化体系取代天主教宗教文化，人们借助复兴古代希腊、罗马文化的形式来表达他们的文化主张。因此，文艺复兴并不是单纯的文化文艺的古典复兴，而是一场尊崇古典、弘扬人本主义的新文化运动，其根本目的是摆脱天主教的神权统治，摈弃宗教禁欲主义对人们思想禁锢和束缚的思想和文化的解放运动，实际上是资产阶级反封建的新文化运动。文艺复兴是欧洲近代三大思想解放运动（文艺复兴、宗教改革与启蒙运动）之一，在人类文明进程中具有极其重要的地位。

直到18世纪末至19世纪初的第一次工业革命以后，高炉开始使用焦炭、机械动力和热风炼铁，开启了近代高炉冶炼的新时代[8]。第一次工业革命是人类科技发展史上的一次巨大革命，开创了以机器代替手工劳动的时代。这不仅是一次技术革命，更是一场深刻的社会变革。以蒸汽机为代表的机械动力的出现，使社会生产力发生了巨大的变化，从而使人类社会步入了工业文明、电气文明和信息文明的新时代。

3.2 近代科学的肇始

17世纪被世界史学界誉为近代科学肇始的时代[11]。欧洲文艺复兴运动后期，近代科学研究开始起源。最具代表性的科学家是意大利天文学家、物理学家和工程师伽利略·伽利雷（Galileo Galilei，1564~1642）。伽利略对17世纪的自然科学的发展起到了重大作用[12]，改变了人类对物质运动和宇宙的认识，被后人誉为"现代物理学之父""科学方法之父""现代科学之父"，爱因斯坦曾经评价："伽利略的发现和他所用的科学推理方法,是人类思想史上最伟大的成就之一,而且标志着物理学的真正开端!"伽利略开创了以实验事实为根据并具有严密逻辑体系的近代科学体系,伽利略的科学发现不仅在物理学史上而且在整个科学史上都占有极其重要的地位。

17 世纪中后期,荷兰物理学家、天文学家、数学家克里斯蒂安·惠更斯(Christiaan Huygens,1629~1695),是继伽利略之后一位重要的物理学先驱,是科学发展史上著名的物理学家之一。他对力学的发展和光学的研究都有杰出的贡献,在数学和天文学方面也有卓越的成就,是近代自然科学的一位重要开拓者。他建立向心力定律,提出动量守恒原理,并改进了计时器。

对人类自然近代科学起源具有重要贡献的代表性人物是英国著名科学家艾萨克·牛顿(Isaac Newton,1643~1727),牛顿出生于伽利略去世以后的第二年。牛顿在系统地总结了伽利略、惠更斯等人的科学研究工作以后(史称"牛顿综合"),1687 年发表论文《自然定律》,提出了万有引力定律和牛顿运动三大定律,提出了光的色散原理、发明了反射式望远镜,牛顿还与莱布尼茨共同发明了微积分,牛顿的科学发现、发明和论述奠定了物理世界的科学观点,他在力学、光学、数学等多学科领域的开创性研究,奠定了自然科学研究的基础,并成为现代技术科学、工程科学的基础,大力并推动了现代科学革命。

1687 年牛顿发表巨作《自然哲学的数学原理》[13],开辟了大科学时代。牛顿被公认为是最具影响的科学家,被誉为"科学的天才"和"物理学之父",他是经典力学基础的牛顿运动定律的建立者。美国著名科学哲学史专家伯纳德·科恩认为,牛顿是一位在历史上独领风骚、卓越非凡的人物,因为他在不同的领域都做出了重要和突出的贡献[14]。他发现的运动三大定律和万有引力定律,为近代物理学和力学奠定了基础,牛顿在科学上的巨大成就连同他的朴素唯物主义哲学观点和一套初具规模的物理学方法论体系,给物理学乃至整个自然科学的发展,特别是对于 18 世纪的工业革命、社会经济变革及机械唯物论思潮的发展带来了巨大和深远的影响。

18 世纪中叶至 19 世纪中叶,自然科学基础理论发展第一次工业革命,有力推动了人类科技进步,也推动人类社会从封建君主专制发展到资本主义共和制[15]。

3.3 近代钢铁冶金技术的突破

钢铁冶金的技术演化历史,既是人类文明进步的历史,还是技术、工程、科学三元论最有力的实证。第一次工业革命以前,直到 1825 年之前,人们并不清楚钢铁冶金的科学理论,也不完全懂得钢铁冶金过程的氧化-还原反应和热量、质量、动量传输和化学反应("三传一反")的传输理论。钢铁冶金术长期以来就是一门技艺和技巧,很多技术的变革和工艺的改进都是工匠们奇思妙想或从反复失败的教训中总结得出,谈不上科学理论依据和理论基础,也更不是在科学理论指导下的实践,完全是一种经验和技艺的总结、提炼、学习、传承和发扬,属于典型的实践先于理论、技术先于科学、工程先于科学。回顾古代冶金技术的发展演化进程,是无数的工匠们经过口传心授、不断摸索和反复实践,冶金技术的知识、经验和技艺的传承基本停留在"默会知识"和"隐性知识"的范畴内,类似于现在非物质文化遗产的传承,高明的工匠在总结前人经验的基础上,敢于探索创新就会产生新的技艺和方法;一旦这些技艺和经验失传,也就很难再复原。因此,千百年间,古代冶金技术和技艺基本没有取得突破性的跃迁和颠覆性的革命,像陶瓷、纺织等大多数生产加工技艺一样,世代相传,延绵不绝。直到 19 世纪中叶,随着物理化学学科的发展,冶金过程热力学、动力学等知识体系的形成与发展,才逐渐形成了现代冶金学及其造块、炼铁、炼钢、连铸、金属压力加工等学科分支,完成了钢铁冶金从技艺到科学的嬗变[16]。

1856 年,英国著名发明家、冶金学家和工程师贝塞麦(Bessemer Sir Hery,1813~1898)发明了转炉炼钢工艺他发明了将熔化的生铁装入转炉内,吹入高压空气便可通过氧化反应脱除生铁所含的硅、锰、磷、碳等元素,将生铁冶炼成钢的工艺方法。这是世界首创大批量炼钢的方法,生产效率大幅度提高。贝塞麦转炉炼钢工艺后来经过马希特(R. Mushet)父子的改进,在钢水中加镜铁(含锰量较低的锰铁)用来脱硫,使转炉炼钢工艺得到进一步完善。贝塞麦转炉炼钢工艺的诞生标志着早期工业革命的"铁时代"向"钢时代"的演变,在冶金技术发展史上具有重要的划时代的意义。后来欧洲、美洲都引进了这一先进炼钢工艺,世界从此进入了现代钢铁时代。直到 20 世纪中叶,顶吹纯氧的转炉炼钢工艺在世界范围内推广普及,极大地推动了炼钢技术的进步。

4 现代钢铁冶金科学、技术和工程的发展

4.1 钢铁是工业革命的基础产业

现代钢铁冶金是人类社会进步不可或缺的重要支撑。18 世纪中下叶的第一次工业革命[17],以英国为代表,开始了将热能通过蒸汽机转化为机械能的机器化工业生产,随后以机器生产为标志的工业革命浪潮席卷欧洲、北美乃至全球。19 世纪以来,随着电力技术的发明和应用,以电力能源技术、无线电通讯技术为核心的第二次工业革命迅速爆发,电力作为全新的二次能源被广泛地应用到工业生产中,内燃机得到了普遍应用,人类的交通工具从马车、火车、轮船,跃进到汽车、飞机时代。20 世纪中叶,电子信息技术的发展突飞猛进,电子计算机技术、集成电路以及信息网络技术的相继问世,使人类社会步入了崭新的时代。毋庸置疑,工业化颠覆了人类传统的生产和生活方式,极大推动了人类文明的快速发展。钢铁材料在第一次和第二次工业革命中都扮演重要的角色,是支撑科技进步、产业革命和经济社会发展的重要结构材料、功能性材料或首选材料,甚至是不可替代的。图 1 是 1900~2019 年世界粗钢产量变化的情况。

4.2 现代冶金科学的创立

第一次革命以后,钢铁冶金科学、技术和工程都取得了飞跃发展,在不同的尺度和层次上,形成了原子、分子为研究对象微观尺度的基础冶金科学,工序、装置中观尺度的冶金工艺技术,以及工厂、钢铁企业宏观尺度的冶金工程,学科发展和学科知识体系日臻完善[18]。

1840~1847 年,德国医生迈尔(1814~1878)、英国物理学家焦耳(1818~1889)和德国物理学家赫尔姆赫兹(1821~1894),相继在卡诺热力学研究的基础上,提出并阐述了热力学第一定律——能量守恒定律。1850 年,德国物理学家克劳修斯(Ruddof Lulius Emanuel Clausius,1822~1888)发表重要论文,首次提出热力学第二定律的概念,即热量

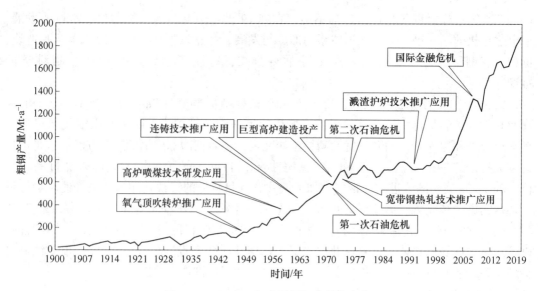

图 1　1900~2019 年世界粗钢产量的变化

不可能从低温的物体转移到高温的物体，要实现这一过程就必须做功；随后克劳修斯又引入熵的概念，因此热力学定律又被后人称之为熵增定律。1851 年，英国物理学家、发明家开尔文（Lord Kelvin，1824~1907）提出了另一种形式的热力学第二定律，即不可能从单一热源吸收热量使其完全做功，而不发生其他变化。与此同时，卡尔文对热力学第一定律和第二定律的公式化做出了重要贡献，是热力学的主要奠基者之一。钢铁冶金热力学的理论基础是建立在经典热力学三大定律——能量守恒（热力学第一定律）、反应进行的可能性及最大限度（热力学第二定律）、绝对零度不能达到（热力学第三定律）的基础上，主要体现是热力学第二定律在钢铁冶金过程中的应用[19]。

1925 年，英国法拉第学会（Faraday Society）在伦敦召开的"炼钢过程物理化学"讨论会，对当时乃至以后的冶金学和钢铁冶金工业的发展，产生了重要和深远的影响。1932 年，德国物理学家申克（R. Schenck，1900~?）出版了学术专著《钢铁冶金过程物理化学导论》，首次提出了钢铁冶金物理化学的概念，奠定了钢铁冶金物理化学的学科基础，新的学科由此诞生。经过 20 世纪近 70 年的发展，钢铁冶金已经初步形成了现代钢铁冶金科学、技术和工程三元体系。冶金学从 20 世纪初开始跨入了现代科学的发展进程。然而，冶金学又是有别于物理学、化学、生物学、地学、天文学、数学等以研究自然物理现象为主要目标的基础科学的。从根本上看，冶金学是属于研究人工物世界的工程科学、技术科学范畴，是重在研究发展现实生产力的工程知识、技术知识和工程科学知识。其知识的来源是多元化、多层次、集成、综合性的，不仅是只来源于基础科学。冶金学的知识重在对各类要素、各类知识集成起来，并能转化为现实的、直接的生产力，发明、集成、综合、转化是其特征。

4.3　现代冶金技术科学的形成

20 世纪是钢铁冶金发展成为科学，特别是成为技术科学的关键时期。20 世纪 20 年代，化学热力学理论应用于钢铁冶金领域中，逐渐发展为冶金过程物理化学；其中化学反应热力学主要解决冶金化学反应过程中分子与分子之间反应问题,包括化学反应进行方向上的可能性,反应平衡的极限,不同元素之间反应的选择性、排序性,反应平衡时的焓变化等。化学反应热力学主要是对冶金过程中化学变化本质的认识和规律的揭示,对冶金学而言,具有划时代的科学意义。上述属于基础科学性质的理论研究,主要是从原子或分子尺度的微观层次上进行一系列的研究,通过合理的简化、假设、典型化,采用"还原论"的研究方法,将发生在原子/分子间的冶金反应过程假设成为"孤立系统"进行研究,再将研究结果"还原"到实际冶金反应过程中。显然,这些基础科学的研究对解释各类冶金-材料方面的现象、理解过程的本质,具有非常重要的理论价值,使冶金-材料的生产过程从手艺、技艺、经验逐步进入科学[20]。

当前,冶金过程化学反应热力学理论已经得到了较为系统的研究,其理论体系集中地表现在:(1)冶金化学反应过程中氧化-还原的基本规律;(2)凝固过程中晶体形核的热力学;(3)金属固态相变热力学;(4)金属再结晶过程热力学;(5)冶金化学反应过程动力学。这些以分子尺度进行研究的微观动力学理论,并不研究反应分子如何达到反应区,也不研究反应产物如何离开反应区。这种化学反应动力学是以反应体系均相分布为前提的,研究纯化学反应的微观机理、步骤和速率等问题,实际上是分子层次上的微观动力学。

在钢铁冶金过程中,多数的冶金化学反应是在相界面进行的多相反应或非均相反应,因此,研究其反应速率和机理时,必然要解析反应物到达反应区以及反应产物离开反应区的物质传输过程,这也是实际的冶金化学反应过程的一部分。对某一反应过程而言,这属于"宏观"动力学。而这种"宏观"是相对于原子/分子尺度的微观而言的,实际上仍然是分子层次的动力学问题,只不过是包括了对反应区附近的传质过程的综合分析。因而,这类关于冶金过程化学反应动力学的研究及其知识,仍然属于冶金基础科学的研究范畴。

钢铁冶金的基础科学主要解决原子、分子尺度上的问题;技术科学主要解决工序、装置、场域尺度上的问题;工程科学主要解决制造流程整体尺度/层次以及流程中工序、装置之间关系的衔接、匹配、整合、协同、优化问题。物理学、化学、金属学、物理化学、传输理论等基础科学和学科知识是钢铁冶金学的理论基础。冶金学科的发展使得人们研究冶金问题的思路和方法越来越开阔,基础学科理论知识奠定了冶金学科的基础。

冶金反应工程学是20世纪70年代兴起的一门新的学科,最初始于20世纪中叶,50年代以后,冶金学家分别在不同的条件下将化学反应工程学的观点、原理和方法应用与冶金过程。以日本的鞭岩和森山昭为代表,合著出版的《冶金反应工程学》开创了化学反应工程学在冶金过程的应用[21]。冶金反应工程学主要研究某些典型冶金反应器的工艺特性及其功能改进,如钢水连铸的中间罐冶金、钢包冶金等装置。这些研究主要是应用数学模型化的方法,首先建立物理模型,进而建立数学模型,给定初始条件和边界条件,选用或开发适用的计算软件(如 CFD、FEM 模型软件)借助于编程和计算机数值计算,解析研究某一类冶金反应器及其系统操作过程的现象、特性、和规律,从而通过对比研究和参数优化,得到优选的解决方案、工艺参数或结构尺寸;与此同时对新工艺、新装置、新设备的开发过程中可以预测其功能的特性,或者进行数值化的动态运行(仿真模拟),进而指导生产操作工艺的改进以及装置的优化设计。而且,对于已运行的冶金装置的革新、改造以及操作优化,冶金反应工程学都能够提供较完备的解决方案和相应的技术信息的支持。

4.4 钢铁冶金工程科学的创建与发展

现代钢铁冶金工程的发展经历了漫长的演变、集成、完善、变革和创新的过程。其中理论体系的形成、建立和发展，技术的发明、开发、应用和革命，生产工艺流程的组合、集成、演变和完善，在第一次工业革命以后大约200年的历史进程中，不断交替出现和相互促进，理论的形成、发展和不断完善，是指导技术发明和技术创新，以及工程集成和工程创新的重要动力。

与此同时，回顾钢铁冶金的发展演进历程不难看出，由于钢铁冶金包括矿物开采与加工、高温冶金过程、凝固-成形过程，金属塑性变形过程与材料性能控制过程，具有工序繁多、功能各异、过程复杂、流程结构多样等特点，因而钢铁冶金基础理论的形成、建立和发展，是多领域、多学科的理论研究和相互交叉发展的过程，呈现出一种典型的解析-组合-再解析-再组合的不断发展和不断完善的过程。

20世纪是钢铁冶金基础科学的完善与深化、技术的集成与进步、工程的演化与创新以及钢铁产业的快速发展的重要时期[22]。转炉炼钢技术的普遍应用和连铸技术的推广普及，有利推动了钢铁工业的技术进步和产业发展。全球钢铁产量随着"颠覆性"冶金新技术的问世、应用和普及，快速攀升并不断增长；钢铁冶金装备大型化和现代化，也推动了钢铁冶金工业的迅猛发展。从近两百年的现代冶金工程技术演化发展的历史中，可以推断出，经济社会的发展和市场的需求是钢铁冶金产业发展的"拉动力"，而技术进步则是钢铁冶金产业发展的"推动力"。图2解析了钢铁工业可持续发展的"五力"模型。未来钢铁工业的科学、健康和可持续发展，必须在这几种力量的平衡协调中寻求发展——走钢铁制造流程绿色化和智能化的新型工业化道路。

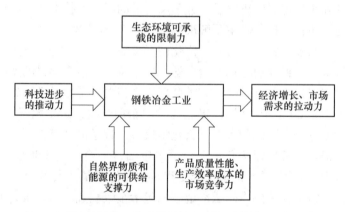

图2 钢铁工业可持续发展的"五力"模型

从20世纪90年代开始至21世纪初，我国冶金学家开始思考钢铁冶金产业层次的宏观科学问题，以钢铁联合企业为关注点，以全局性和系统性的视角，考察钢铁制造流程的运行规律及其特征。2004年，我国著名冶金学家殷瑞钰院士发表了开创钢铁冶金工程科学的学术专著《冶金流程工程学》[23]，基于耗散结构理论对钢铁制造流程的物理本质进行了深刻研究，提出了现代钢铁制造流程应具有产品制造、能源高效转换和消纳-处理废弃物并实现资源化的"三大功能"。经过21世纪初20年的发展，钢铁冶金工程科学体系已经初步形成,在冶金流程工程学理论的指导下,我国第一个自主设计建造的沿海靠港千万吨级新一代可循环钢铁厂——首钢京唐钢铁厂于 2010 年全面建成、顺利投产[24~26]。

经过近百年的探索、研究发展进程,现代钢铁冶金学(冶金科学与工程)也已形成三个层次的知识集成体系,即原子/分子层次上的微观基础冶金学,工序/装置层次上的专业工艺冶金学和全流程/过程群层次上的动态宏观冶金学(见图 3)。可见,随着不同层次科学问题研究的深入,学者们的研究目标、研究领域不断扩宽,认识问题的视野发生了层次性跃迁,并进而嵌套集成为一个新的知识结构,即不囿于经典热力学孤立系统观念,跨入探索冶金企业全流程的过程群的集成优化、结构优化,研究的对象发生了变层次、变轨的跃迁。同时,扩大了研发领域,既引导企业全流程中过程和过程群的自组织结构以及他组织调控过程中,共同形成的耗散结构和耗散过程优化的研究,又引导新的工程设计、工程运行的理论和方法。当代冶金学发展的战略目标也跟着时代的发展,发生了战略性的变化;当代冶金学的战略目标除了制造新一代产品以外,已经聚焦于冶金工厂的绿色化(绿色、低碳、循环发展)和智能化(智能化设计、智能化制造、智能化服务、智能化管理等)。

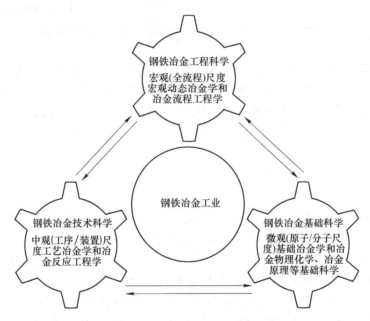

图 3 钢铁冶金科学-科学-工程的集成知识体系

从钢铁冶金的发展演进历程,可以看出钢铁冶金工业现在面临的挑战是多方面的,要解决这些复杂环境下的复杂命题,就必须从战略层面上来思考钢铁厂的要素-结构-功能-效率问题,实质上这是全厂性的生产流程层面上的问题,就必须从生产流程的结构优化及其相关的工程设计等根源着手。进而还可以清晰地认识到,这样的系统性、全局性、复杂性问题,不是依靠技术科学层次上的单元技术革新和技术攻关所能解决的,而是需要以工程哲学的视野,在全产业、全过程和工业生态链等工程科学层次上解决。

钢铁冶金制造过程,利用来自于自然界的铁矿石、煤炭、水、空气等天然的资源和能源,经过一系列复杂的物理-化学变化和生产加工过程,制造出满足经济社会和人们生活所需的钢铁产品，而且钢铁制造过程所产生的煤气、蒸气、电力、冶金渣等副产品也成为其他产业的资源或能源。与此同时，钢铁制造过程，还可以消纳、处理大宗人们生活或其他产业所产生的废弃物，如废钢、废轮胎、城市污水等，经过无害化、资源化处理或高值化循环利用，使其成为新的再生资源或再生能源。面向未来，钢铁工业必须建构与自然生态环境、经济社会和相关产业之间和谐友好、相互依存、持续发展的新格局（见图4）。

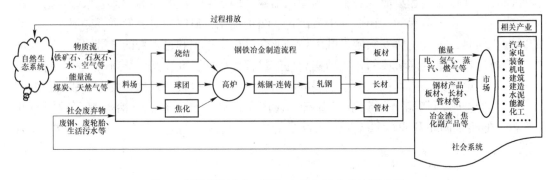

图 4 现代钢铁冶金工程与自然和社会的协同关系

21世纪初国际金融危机以来，使全球经济出现新一轮的衰退，经济发展动力不足、增速缓慢，与此同时，全球政治、经济、金融、贸易形势错综复杂，预见和不可预见的各类风险不断叠加，不确定性和涌现性以及由此带来的突发性，造成全球钢铁产业面临着新的发展挑战和机遇[27]。

5 结语

钢铁无论在过去、现在，乃至未来，都是人类文明发展和经济社会进步不可或缺的基础材料，具有重要的功能和地位。古代钢铁冶金技术主要依靠技艺的学习、传承和传播，囿于经济发展和当时的技术条件，制铁匠人的技艺、技能和技术世代相传，鲜有突破。14~16世纪的欧洲文艺复兴开启了近代思想解放运动，随后哲学、艺术、文学、技术、科学协同发展，17世纪则是近代科学肇始的时代。人类从宗教和神学统治的思想禁锢中解放出来，对自然界的认知从模糊的主观认知走向科学探索和理性研究。钢铁冶金在第一次工业革命以后，在工艺技术上得到飞跃发展，贝塞麦转炉炼钢技术的发明，极大推动了钢铁冶金技术进步。

到20世纪初，钢铁冶金物理化学学科的创立，形成了钢铁冶金工程的理论基础；20世纪70年代，冶金反应工程学的创建，标志着钢铁冶金技术科学走向成熟；21世纪初，冶金流程工程学的创建，则标志着钢铁冶金工程科学的形成，从而形成了钢铁冶金基础科学、技术科学和工程科学的三个学科层次。

面向未来，到21世纪中叶，钢铁工业必须坚持绿色化、智能化的可持续发展道路，在资源、能源、生态环境、技术和市场等多种要素的耦合作用下，实现与经济社会的和谐相融和可持续健康发展。

参考文献

[1] 殷瑞钰. 冶金流程工程学（第2版）[M]. 北京：冶金工业出版社，2009.
[2] 殷瑞钰，李伯聪，汪应洛，等. 工程演化论 [M]. 北京：高等教育出版社，2011：152-167.
[3] 闻人军，译注. 考工记译注 [M]. 上海：上海古籍出版社，2008：41-59.
[4] 杨宽. 中国古代冶铁技术发展史 [M]. 上海：上海人民出版社，2014：1-17.
[5] 姜茂发，车传仁. 中华铁冶志 [M]. 沈阳：东北大学出版社，2005：49-84.
[6] 沈括. 梦溪笔谈 [M]. 施适，校点. 上海：上海古籍出版社，2015：16.
[7] 宋应星. 天工开物译注 [M]. 潘吉星，译注. 上海：上海古籍出版社，2013：111-133.
[8] 吴国胜. 科学的历程（第2版）[M]. 北京大学出版社，北京：2002：362-364.
[9] 王鸿生. 科学技术史 [M]. 北京：中国人民大学出版社，2011：184-186.
[10] 亚·沃尔夫. 十六、十七世纪科学、技术和哲学史 [M]. 周昌忠，等译. 北京：商务印书馆，2016：5-11.
[11] 亚·沃尔夫. 十八世纪科学、技术和哲学史 [M]. 周昌忠，等译. 北京：商务印书馆，2016：811-826.
[12] 伯纳德·科恩. 科学中的革命 [M]. 鲁旭东，赵培杰，译. 北京：商务印书馆，2017.
[13] 艾萨克·牛顿. 自然哲学的数学原理 [M]. 任海洋，译. 重庆：重庆出版社，2015：1-24.
[14] I·伯纳德·科恩. 科学中的革命 [M]. 鲁旭东，赵培杰，译. 北京：商务印书馆，2017：241-261.
[15] 蔡子亮，杨钢，白政民. 现代科学技术与社会发展 [M]. 郑州：郑州大学出版社，2006：36-41.
[16] 徐匡迪. 20世纪——钢铁冶金从技艺走向工程科学 [J]. 上海金属，2002（1）：1-10.
[17] 扬·卢腾·范赞登. 通往工业革命的漫长道路 [M]. 隋福民，译. 杭州：浙江大学出版社，2016：1.
[18] 殷瑞钰，张福明，张寿荣，等. 钢铁冶金工程知识研究与展望 [J]. 工程研究-跨学科视野中的工程，2019，11（5）：438-453.
[19] 沈峰满. 冶金物理化学 [M]. 北京：高等教育出版社，2017：1-18.
[20] 殷瑞钰. 冶金流程集成理论与方法 [M]. 北京：冶金工业出版社，2013：14-16.
[21] 鞭岩，森山昭. 冶金反应工程学 [M]. 北京：科学出版社，1981.
[22] 徐匡迪. 中国钢铁工业的发展和技术创新 [J]. 钢铁，2008，43（2）：1-13.
[23] 殷瑞钰. 冶金流程工程学 [M]. 北京：冶金工业出版社，2004.
[24] 张福明，颉建新. 冶金工程设计的发展现状及展望 [J]. 钢铁，2014，49（7）：41-48.
[25] 张福明，崔幸超，张德国，等. 首钢京唐炼钢厂新一代工艺流程与应用实践 [J]. 炼钢，2012，28（2）：1-6.
[26] 殷瑞钰，张寿荣，张福明，等. 现代钢铁冶金工程设计方法研究 [J]. 工程研究-跨学科视野中的工程，2016，8（5）：502-510.
[27] 张福明. 我国高炉炼铁技术装备发展成就与展望 [J]. 钢铁，2019，54（11）：1-8.

Research on Iron and Steel Metallurgy Evolution Process from Skill to Engineering Science

Zhang Fuming

Abstract: The origin, development, and evolution process of iron and steel metallurgical technology were studied under the view of engineering philosophy. Both the function and status of iron and steel materials were explained, and the significant roles of iron and steel in human society progress and civilization development were demonstrated. Research on ancient metallurgical skills from development and inheritance to innovation was carried out. Further, the influence of Renaissance and modern science initiation on iron and steel metallurgy technology and science since the 17th century was discussed. Modern science has had an impressive and profound impact on iron and steel metallurgy in terms of the skill change to engineering science, which was pursued comprehensively. It also focuses on the foundation of basic iron and steel metallurgical courses and metallurgical physical chemistry. The remarkable and far-reaching influence of modern science on iron and steel metallurgy as a skill to engineering science was discussed. The significance of iron and steel metallurgical technical science represented by metallurgical reaction engineering to modern iron and steel metallurgical engineering was expounded. The discipline system of the basic, technical, and engineering science of iron and steel metallurgy was put forward, in addition to the goal and direction of sustainable and healthy development of the industry in the future.

Keywords: iron and steel metallurgy; engineering evolution; physical chemistry; metallur-gical reaction engineering; metallurgical process engineering; industrial revolution

Development Orientation of Low Carbon and Greenization BF Ironmaking Technology

Zhang Fuming   Cao Chaozhen   Li Xin

Abstract: Development of modern BF ironmaking has experienced for nearly 200 years. Since entering the 21st century, the BF ironmaking technology has developed rapidly, and remarkable technological progress has been achieved. At present, in the face of the changes in the conditions of raw material and fuel, the restraint of ecological environment, the decline of economic situation, huge challenge and threat have been brought to the BF ironmaking technology. The BF ironmaking technology development philosophy in the future should be low carbon and greenization, high efficiency and low energy consumption, intellectualization and integration, as well as to achieve "three functions": hot metal production, energy conversation and waste disposition. The new generation design and optimization of ironmaking process based on the core of the BF, inorder to realize the dynamic-orderliness and continuation-coordination in the whole ironmaking process, will be the key issue of the ironmaking technical innovation in the future. Based on the theory of metallurgical process engineering and the coordinated optimization of BF ironmaking process, the new breakthrough will be achieved in the future. This paper discusses the development orientation of BF ironmaking technology under the concept of circular economy, expounds the technical developemnt route of the future BF, and points out the key common technical innovation of the development of low carbon and green development.

Keywords: blast furnace ironmaking; low carbon metallurgy; green development; engineering design

1 Introduction

Since stepping on 21st century, BF iron making process is faced with many major sustainable development issues on restriction again by aspects of natural resources shortage, undersupply of energy resources, ecological environment protection, etc. [1]. Under the economic situation of market downturn and weak demand, and in the face of increasingly severe competition and challenge[2], it is a must to have remarkable breakthroughs in terms of high efficiency and low consumption, energy conservation and emission reduction, circular economy, clean and environment protection, low carbon smelting, intelligent intensivism, green development, etc.,

---

本文作者：张福明，曹朝真，李欣。原刊于 Proceedings of the Iron & Steel Technology Conference, AISTech, 2016: 155-163。

taking low carbon and green development is regarded as the leading factor to have further optimization of process flow, increase of blast temperature and reduction of fuel ratio in order to improve vitality and competitiveness of BF iron making technology[3].

Blast furnace is a key process of iron and steel manufacture, the core and key element of ferrous substance flow conversion, the core unit of energy conversion in the iron and steel manufacture process and the central link of energy flow network[4]. At present, the cost of BF ironmaking production accounts for approx. 60%-70% of the total of iron and steel manufacture, and the energy consumption of the ironmaking process is approx. 60%-70% of the integrative energy consumption of iron and steel industry. Thus, for integrated iron and steel works with blast furnace-converter manufacture flow, BF ironmaking is extremely important in the iron and steel manufacture process, and in the position of decisive role.

1.1 Function Analysis

After the first industrial revolution, the BF ironmaking process, experienced nearly 200 years of development and evolution, has achieved significant technical progress in respect of enlargement, high efficiency, long life, intensification, intelligence, etc. It has been developed into the ironmaking process with the highest iron oxide reduction efficiency. BF ironmaking has been transmuted into engineering science, which is becoming perfection gradually. Blast furnace main function is evolving all the time in pace with its course of technical development. Refer to Fig. 1 for function analysis of BF ironmaking[5].

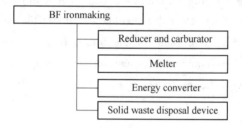

Fig. 1 Function analysis of BF ironmaking

1.2 Reducer and Carburator

Blast furnace, taking coke as the main fuel and reductant, is the process unit which is used for reduction iron oxide into liquid pig iron. It belongs to typical shaft type reactor and reverse industrial reaction device with huge multiphase complex system. During the opposite movement process of burden lowering and gas flow rising, "three-transfer and one-reactor" can be achieved during blast furnace smelting process. Premise for smooth achievement of blast furnace high temperature reduction process is to have coke column (namely skeleton effect) in blast furnace, which is irreplaceable. For this reason, different degree of carburizing process of pig iron always exists during blast furnace reduction process.

1.3 Melter and Quality Governor

The most important function of blast furnace is to provide high quality liquid pig iron to converter, and it is the foundation of converter steelmaking process. Blast furnace reduces ferrous substance in solid iron oxide mineral into liquid pig iron, so that it has function of melter. And continuous reduction and production of hot metal by blast furnace shows blast furnace is a consecutive hot metal supplier. Meanwhile, BF ironmaking process has significant regulating and control function to hot metal composition/quality. And temperature, composition and deviation of hot metal can be stabilized by means of blast furnace operation, especially for control of [S] and [Si] in hot metal in a reasonable range.

1.4 Energy Converter

During blast furnace production, huge energy consumption and energy conversion are executed with characteristics of physical sensible heat and chemical energy of blast furnace products (hot metal, slag and BFG) from conversion of fossil mass energy such as coke, pulverized coal, natural gas and so on, so that blast furnace has function of energy converter. With progress of energy saving of BF ironmaking, as well as implementation of technologies of TRT, recovery of other waste heat and energy, it is more prominent for blast furnace as a energy converter. Blast furnace process is maximum energy conversion process unit in iron and steel manufacture flow at present, and also the central link for establishment of energy flow network in iron and steel plant.

1.5 Solid Waste Disposal

Besides metallurgical functions, BF ironmaking process can have functions of solid waste disposal and changing waste into resources, including coking, sintering and pelletizing which are matched for blast furnace. BF ironmaking system is an important link in achievement of recycle economy in iron and steel plant. For instance, coking process and BF PCI are used to treat waste plastics, sintering or pelletizing process are considered to briquette dust from the iron and steel plant for recovery of secondary resources, collected dust is injected into blast furnace, and coke oven is used to pyrolyze and handle municipal solid waste, etc..

2 Flow Integration and Structure Optimization

2.1 Dynamic Precision Design

Design theory for new generation iron and steel plant must be established on the basis of conformance of physical essential of its dynamic operation process, in particular, on the basis of movable kinetic theory in dynamic-orderliness, continuation-coordination, stable-efficient operation in production flow[6].

Dynamic precision design of modern blast furnace should be directed by advanced concept study and Top-Down Design[7]. Dynamic Gantt Chart is applied, and high efficient collaborative interface technology is considered to achieve material flow design and high efficient conversion with characteristics of dynamic-orderliness, continuation-coordination/quasi-continuous way. Energy flow of timely recovery and energy conservation and emission reduction is designed to show the "three-function" of iron and steel manufacture flow from a higher level, namely, achievement of iron and steel product manufacture, high efficient energy conversion and solid waste disposal, as well as achievement of recycling.

Dynamic precision design of modern blast furnace should be set out from the overall objective of dynamic collaborative operation of blast furnace production. The advanced technical unit is carried out with comprehensive judgements, trade-off, selection and integration, and then dynamic Integration is followed. And the relative unit technology is integrated by means of the flow network and cooperated via programming to make flowing/rheology process of "material flow" "energy flow" and "information flow" achieve dynamic and orderly operation in the specified time and space boundary, and forming a integrated effect with characteristics of dynamic-orderliness, continuation-coordination and impact-intensive process so as to achieve the goal of multi-objective optimization. Refer to Fig. 2 for analysis of connotation of dynamic precision design.

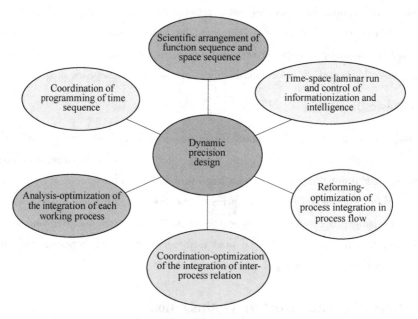

Fig. 2 Connotation of dynamic precision design

2.2 Flow Network

Establishment of flow network is an important item for engineering design of modern BF. Traditional BF design with "extensive, isolated and static" mode and convention must be abandoned, and the flow network is established based on dynamic precision design in order to have function optimization, structure optimization and efficiency optimization of BF ironmaking. BF large-sizing is achieved under optimization condition of the flow structure of iron and steel plant, and the dynamic precision operation system with optimization of flow network system of material flow, energy flow and information flow in iron and steel plant is established. Modern BF design should put forth effort to establish the material flow with series-parallel combination and simpleness-smoothness-high efficiency, the dynamic energy flow network based on input-output, high efficient conversion and minimum dissipation, as well as the information flow with characteristics of coordination-regulating and control.

2.3 Optimization of Interface Technology

In engineering design of modern BF, the "interface technology" with working procedures of sintering/pelletizing, coking, BF, steelmaking, etc., should be rationalized. This is not only conducive to achievement of organization and coordination of production progress, but also conducive to have informationization for achievement of production process and scheduling process, and it can make the production speed and production efficiency of the unit device/equipment higher, while the investment of unit capacity is the most economical. Technology optimization of main interfaces includes: simple material flow, energy flow path (reflected in BF layout); buffer-stabilization-collaboration of interaction between working procedure/device (reflected in Gantt Chart of dynamic operation of working procedure unit); network node optimization in manufacture flow, node cluster optimization and connector type optimization (reflected in equipment quantity, device capacity and position rationalization, transport mode, transport distance, transport regulation optimization, etc.); efficiency optimization of energy flow and energy conservation and emission reduction; collaboration optimization of material flow-energy flow-information flow, etc. Typical interface technology includes interface technology of ironmaking-steelmaking represented by the direct transport mode of hot metal based on multi-function hot metal ladle "One ladle for the whole process" concept; interface technology of sintering-BF and coking-BF represented by the direct transport mode of intensiveness and high efficiency, continuation and compact layout without intermediate hopper for raw material and fuel, etc.

3 Key Common Technologies

3.1 Development Concept

Modern BF ironmaking should establish production system of dynamic precision and high efficient operation. High efficiency, low cost and less emission are considered as the goal to improve production efficiency, decrease energy consumption and reduce environmental pollution so as to achieve the goal of collaborative development of modern BF ironmaking with characteristics of high efficiency, low consumption, long service life, low carbon and greenization. Main technical characteristics of new century BF ironmaking technology is improvement of comprehensive technical equipment level, development of circular economy and achievement of low carbon and green manufacture. BF operation is controlled rationally, and the scientific and reasonable index assessment system for is established to achieve multi-objective collaboration optimization of BF ironmaking; BF ironmaking technology system with characteristics of high efficiency, low cost, low consumption and less emission should be built.

Process technology equipment of large scale BF in 21 century pays more attention to technical innovation and low-carbon green development. Key common technologies for BF ironmaking should be advanced technologies for new BF ironmaking with major breakthrough, collaborative optimization and intensiveness innovation. Great technical progress should be obtained in many fields such as BF high efficiency and long service life, high quality and low consumption, energy conservation and emission reduction, circular economy, clean and environment protection, low carbon and greenization, and so on.

3.2 Integrated Innovation of Flow

Selection of production capacity of iron and steel plant must adapt to social development and market demand. In accordance with structure demand in regional and iron & steel market, combined with the iron and steel enterprise itself, the product positioning and production scale of the iron and steel plant should be determined reasonably with philosophy of seeking the truth from the facts, tailor measures to suit local conditions, as well as scientific decision-making. Reasonability, high efficiency and economy of the whole flow structure of iron and steel plant should be followed and considered to make decision of production capacity of the iron and steel plant, and then based on this to determine BF production capacity, quantity and volume.

With consideration of network structure optimization of material flow, energy flow and information flow, the optimal and most suitable selection of one iron and steel plant is to configure 2-3 blast furnaces, which must be based on structure optimization of the whole flow in iron and steel plant to arrange blast furnace in scientific and reasonable way. BF large sizing under flow structure optimization of iron and steel plant is the main development direction for blast furnace in the future[8].

3.3 Optimization of General Layout Design

Development innovation and continuous evolution of modern BF ironmaking has completed transmutation from technology to engineering science. Analysis is carried out based on view and gradation of process engineering. BF general layout does not only show plane process arrangement, but also relationship between space and time, and its essence is running path and track of the material flow and energy flow. Substance and energy run in the setting geometric space, and "laminar movement" with "minimum path, minimum resistance loss, maximum efficiency and minimum dissipation" should be achieved. Thus, the important significance of optimization design BF general layout can be seen.

Reasonability of iron and steel manufacture flow is reflected in high efficiency, low cost and less emission, making ferrous substance material flow, energy flow and information flow operate in collaborative, high efficient and dynamic way. Traditional static design philosophy must be abandoned to achieve BF large-sizing with characteristics of collaborative optimization of structure, efficiency and function. Refer to Fig. 3 for principle of BF general layout design.

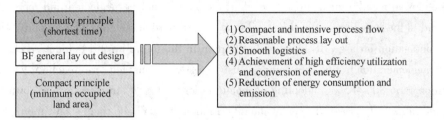

Fig. 3   Principle of BF general layout design

3.4   BF Beneficiated Burden Technology

The beneficiated burden is the foundation for modern BF iron making to realize high efficiency, low consumption, prime quality and long campaign and also an important measure to realize the reduction of BF production and minimization of consumption and emission.

(1) Adoption of rational burden structure. Modern BF iron making must adapt to the availability of resources and energy resources, follow the concept of circulating economy to strive to realize the best utilization of resources and energy resources with technical feature of reduction. Adoption of economic and rational burden structure is the only way to realize the high efficiency, low cost and sustainable developing BF iron making. The optimization of BF burden structure must follow the principle of optimization of resources, technology and economy, pay attention to the improvement of comprehensive metallurgical properties of the burden, develop and research the high efficient utilization of low grade ore to realize reduction of resources, optimization of utilization and environment friendly.

The BF burden structures in different regions of the world are not the same due to the natural property of the resources and the technical inheritance. Most of the Asia BF burden are high alkalinity sinter added with acid pellet and lump ore; and most of Europe and America BF burden are high percent of pellet, some BF even adopted total pellet. Some steel plants in Europe shut down the sintering plant for many reasons like the protection of ecological environment and availability of resources, all BFs turned to pellet. The optimization of Chinese BF burden structure still needs continuous probe and innovation in the future, it is impossible to none the less maintain the traditional burden structure. Steel enterprise shall use operations research theory to optimize and determine the rational burden structure which comply with actual situation of the enterprise by establishing mathematical model to reach the target of feasible technology, reasonable economy and resource conservation. It is worthy of pointing out that confined by the characteristics of our iron ore resources and ecological environment, our country shall in the future energetically develop pelletizing technology, advanced large BF shall gradually increase the pellet charging ratio, some large BF shall also explore the optimization of burden structure with pellet as the main material.

(2) Improve the metallurgical property of the burden. Under the situation that the iron ore price is continuously fluctuating and the market variation is hard to predict, the requirement on the comprehensive metallurgical property of the burden shall not be neglected and more attention shall be paid to the improvement of the stability of the BF burden metallurgical property. The main technical connotation of large BF beneficiated burden in the future are: 1) Increase charged ore grade and agglomeration ratio. The comprehensive grade of charged ore reach 58% and up, agglomeration ratio 80%-85%, adopting economic rational burden structure to low down the cost of raw material. 2) Improve the coke quality. Emphasis shall be made on improvement of mechanical strength, hot strength and average size of coke for large BF, the indices such as M_{40}, M_{10}, CSR, CRI shall meet the requirement of large BF smelting, especially the coke average size shall be kept at $\geqslant 45mm$ or even higher. 3) Improve the stability of burden composition and physical and chemical property, reduce the fluctuation of composition and property to ensure the stable and smooth BF operation. 4) Effectively control the content of harmful element in the burden. Due to the availability and economy of resources, the content of harmful elements such as zinc, potassium and sodium in the burden should be given high attention and effective control. 5) Unfold the research of high efficient utilization of low grade ore to realize the reduction and optimization of utilization of resources.

(3) Burden distribution and control technology. Modern large BF has a high production efficiency and large burden charging volume. For one $2500m^3$ BF, the quantity of ore and coke charged through the top every day are 10000t/d and more. Furnace top equipment shall not only satisfy the requirement for BF charging but also the requirement for precision distribution by BF operation. Rational distribution and precision control of the burden is the foundation to realize the BF gas flow rational distribution and high efficiency utilization of gas chemical energy and physical sensible heat, is the important technical measure to ensure the stable and smooth BF production, increase gas utilization rate and low down fuel consumption and is the foundation to realize stable and smooth BF production and long campaign life, burden distribution control technology is the indispensable important control means in modern BF operation.

3.5 Long-Campaign Life of the BF

The essence of BF long campaign life is "keep rational BF operation inner profile in BF campaign"[9]. The designed campaign life of modern large BF is above 15 years, super-large BF is 20-25 years, yield per volume of one campaign will reach $15000-20000t/m^3$, to realize BF long campaign life needs adoption of effective measures during the whole BF campaign, actively adopt BF long campaign life technology, maintain rational operation inner profile and form the "dynamic permanent" lining based on the system of "self-organization, self-maintenance and self-repair"[10].

(1) Optimize BF inner profile. Rational BF inner profile is the foundation and precondition to

realize multi-target optimization of BF ironmaking. BF inner profile is changing constantly during the campaign, the initial designed inner profile is the foundation for the changing operation inner profile. So, the rationalness of BF designed inner profile will directly determine the progress and result of changing of BF operation inner profile. There is no doubt that the BF operation inner profile will not only affect the BF campaign but also have key function on BF stable and smooth operation, high efficiency and low consumption. As the raw material condition varies and BF operation technology progresses, the modern BF inner profile will also keep optimizing and evolving. Modern BF height to diameter ratio tends to lower down, in "rational low shaft BF" development. The relation between BF effective height and effective volume is not a linear coupling relation, statistics show that for large BF with effective volume above 2000m^3, as the BF volume increases, the effective height did not change much, all the super-large BF height-diameter ratio (H_u/D) had lowered down to below 2.0. In order to effectively suppress hearth peripheral flow of hot metal and its destructive effect, the depth of salamander shall be properly increased. In order to suit the requirement of intensified smelting with oxygen enrichment and large volume of pulverized coal injection, the hearth diameter and height shall be adequately increased to improve working condition of tuyere raceway zone and effectiveness of hearth metallurgy. In order to optimize BF smooth operation condition and promote the free burden movement and smooth ascending of gas flow, the diameter of belly could be adequately increased and the shaft angle and bosh angle could be adequately decreased.

(2) Adoption of long campaign life furnace body structure. 21 century large BF adopts rational furnace body long life structure and advanced cooling technology, pure water (demineralised water) closed circulating cooling system are extensively applied in BF, all BF bottom hearth, furnace body cooler, tuyere and hot blast valve adopted pure water (demineralised water) cooling. The remarkable structure feature of modern BF is that copper stave are adopted from bosh to middle and lower area of shaft in large scale, forming the slag skull self protection based "dynamic permanent lining" in the area above the tuyere to prolong BF campaign life by high efficiency copper stave and rationally configured cooling system.

(3) Adoption of rational hearth bottom lining structure. By designing of rational depth of salamander, adoption of rational hearth bottom lining structure design, selection of anti-hot metal seepage and anti-hot metal erosion high thermal conductivity carbon brick and providing at the same time the rational high efficiency cooling structure and cooling system, abnormal "elephant-foot-shaped" erosion at hearth bottom is effectively suppressed to make the BF campaign life up to 15-20 years and even longer. It is necessary to point out that consideration route and design method for hearth bottom design should be: 1) optimize hearth inner profile design, determine rational hearth diameter, hearth height, depth of salamander, number of tuyere and number of taphole. Establish mathematical model, simulate the hearth work and flowing phenomenon and its flow field distribution for slag discharging process, those are significant reference to precision design. 2) Determine rational lining design structure for hearth bottom. Importances shall be attached to the selection of key structure parameters such as the thickness of hearth side wall,

thickness of bottom, depth of taphole and taphole angle regardless of whatever structure design adopted. For the design of the lining with carbon brick-ceramic cup/ceramic pad composite structure, the temperature field distribution in different campaign stage shall be studied and analysed through heat transfer simulation, based on the result of simulation calculation optimize hearth bottom temperature field distribution and rationally design the matching structure of carbon brick-ceramic cup/ceramic pad. 3) Rationally determine the technical parameters of refractory. Theoretical research and production practice show that the service life of carbon brick is a key element to determine BF life. Therefore, high attention must be paid to the selection of hearth bottom carbon brick physical parameters and indices design, this is a comprehensive consideration and integration of multi-target and balance, advancement of any one index shall not be emphasised one-sided. Must consider at the same time the anti-hot metal seepage, anti-hot metal erosion, anti-alkali erosion and heat conductivity, and also the carbon brick porosity, pore diameter and distribution are the technical indices which must be emphasised. 4) Pay attention to the configuration of high efficiency rational cooling system. For hearth bottom, cooling is the important link which could not be neglected during the whole campaign, especially in late campaign stage the function of the cooling system is more important. After the lining system of hearth bottom is determined, must establish the cooling system that rationally match the hearth bottom lining heat transfer system by heat transfer simulation optimization design to ensure safe and stable operation of refractory lining.

3.6 High Blast Temperature Technology

High blast temperature technology is an important technical approach for BF to low down coke ratio, increase coal injection volume and improve energy conversion efficiency. At present the designed blast temperature of large BF is generally 1250-1300℃, to increase blast temperature is one of the important technical features of the 21 century BF ironmaking.

In late 20 century, super-large BF above 5000m^3 always adopted external combustion hot blast stove, NSC and Didier external combustion hot blast stove are the most representative. Shougang Jingtang 5500m^3 BF developed and used dome combustion hot blast stove, it is the first in the world to use dome combustion hot blast stove in super-large BF above 5000m^3, making a breakthrough in super-large BF hot blast stove structure type[11].

The structure type of modern large BF hot blast stove take on diversified development, the main features of its technical development are: (1) Increase blast temperature to 1250℃ and higher by making use of air gas low temperature double pre-heating or enriched gas. (2) Adopt preheating furnace preheating combustion air, realize 1250-1300℃ blast temperature under condition of burning BFG only. (3) Increase the uniformity of gas flow distribution by optimization of combustion process, study of gas flow movement rule and the study of regeneration/heat transfer; use high efficiency checker brick, increase heat transfer area, intensify heat transfer process and shorten the different value between hot blast stove dome temperature and blast temperature. (4) Optimize hot blast pipeline system structure, use non-over-heat-low stress design system, rationally configure hot blast pipeline bellows expansion joint and tie rod structure, deal effectively with pipeline expansion to reduce pipe system stress, hot blast pipeline use composite brick structure, eliminate local over-heat and air leakage of hot blast pipeline. (5) Take effective measures to prevent the intergranular stress in stove shell to prolong the service life of the hot blast stove. (6) Optimize hot blast stove operation, rationally set up hot blast stove working cycle, increase hot blast stove heat exchange efficiency. (7) Optimize combustion process, reduce fuel consumption. Effectively reduce NO_x and CO_2 emission to realize energy saving and emission control and low carbon environment protection.

3.7 Gas Dry Dedusting and TRT Technology

BF gas dry dedusting technology is the important technical innovation for 21 century BF to realize high efficiency, low consumption, energy saving and emission control and clean production. The BF gas dry dedusting technology increases the purity of gas, gas temperature and calorific value, not only remarkably reduces the fresh water consumption and energy consumption in iron making procedure but also increases the utilization rate of secondary energy and reduce environment pollution. BF gas dry dedusting-TRT coupling technology is the important key technology for BF ironmaking to realize energy saving and emission control, low carbon smelting and high efficiency energy conversion, becoming the important technical developing direction for modern ironmaking industry to perform cycling economy and realize low carbon green development.

At present the BF gas dry bag dedusting technology independently innovated by China have achieved great breaking progress in term of design research, technical innovation, engineering integration and production application, it has been used in super-large 5500m^3 BF for 6 years more, the system operates safely and stably, electric energy production of TRT is increased by 35%-40%, compared with wet dedusting technology. The key core technologies such as impulse injection dust removal technology, gas temperature control technology, on-line detection of gas dust content, dense phase conveying of collected dust and anti-corrosion for pipeline system have been researched and developed to make the large BF gas fully dry bag dedusting technology become better and better day by day[12].

3.8 Oxygen-Enriched PCI Technology

In 70-80s in 20th century, the outburst in succession of two oil crisis helped BF PCI technology dramatically developed and become the key core technology of metallurgical engineering technology progress in 20th century. Up to now, the BF PCI has been still the important energy saving and emission control and low carbon green technology for modern BF and is the major technology that must be kept on development for modern BF ironmaking.

Oxy-enriched blasting is one of the effective technical measures for modern BF to increase production efficiency. To increase oxygen enrichment percentage is significant to modern BF iron making and should deserve high attention. BF oxygen-enrichment-PCI-high blast temperature integrated coupling technology can effectively improve the work of hearth tuyere raceway zone, increase pulverized coal burning rate and coal injection rate, effectively reduce bosh gas volume, improve BF permeability, promote BF stable and smooth operation and increase gas utilization rate so as to effectively reduce fuel consumption and CO_2 emission.

BF oxygen-enrichment-PCI-high blast temperature is the important supporting technology to reduce fuel consumption and CO_2 emission. Under current condition, coal injection rate shall be further increased to make the coal ratio up to 200kg/t and even higher, fuel ratio reduced to below 500kg/t, which is the foundation on which modern BF will keep vitality and competitiveness.

Doubtlessly 21th century modern large BF must strive to push and use oxygen-enrichment-PCI-high blast temperature technology in new ways, which will become the important technical route for BF low carbon green development in future. Taking the BF low carbon green development and reduction of BF fuel consumption as the technical target, taking the increment of blast temperature, oxygen-enrichment ratio and coal injection rate as the technical route and taking the beneficiated burden technology, long campaign life technology and optimized operation as the technical support, the oxygen-enrichment-PCI-high blast temperature technology is bound to have a broad development prospect[13]. Fig. 4 analyzes the function of BF oxygen-enriched PCI technology.

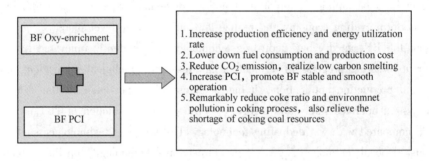

Fig. 4　Function of BF oxygen-enriched PCI

4　Conclusions

(1) Since the 21st century, modern BF ironmaking technology has been facing severe challenge, must adopt the low carbon green development concept and choose the sustainable development route, this is the only way for modern BF ironmaking development. Modern BF ironmaking should establish production system of dynamic precision and high efficient operation. High efficiency, low cost and less emission are considered as the goal to improve production efficiency, decrease energy consumption and reduce environmental pollution so as to achieve the goal of collaborative development of BF ironmaking with characteristics of high efficiency, low consumption, long service life, low carbon and greenization, try hard to improve comprehensive technical equipment level, strive to develop circular economy and achieve low

carbon and green manufacture.

(2) Based on the metallurgical process flow engineering theory, the BF design shall put emphasis on the construction of optimized structure of material flow, energy flow and information flow network and establishment of dynamic orderly, collaboratively continuous and intensively high efficient process flow network and operation program. Build up modern BF dynamic precision design system, pay attention to collaborative optimization of element, function, structure and efficiency, realize BF "three functions", strive to carry out cycling economy, low carbon smelting and green development of ironmaking industry.

(3) The key common technologies of BF beneficiated burden, control of rational burden distribution, high efficiency long campaign, high blast temperature, oxygen enriched PCI and gas dry dedusting are still the important technical foundation, support and guarantee for BF low carbon green development in future. Facing the future must always persist on inheritance innovation, integration innovation and re-innovation, on the basis of existing technologies, continuously improve, modify and pursue integration effect and collaborative innovation of comprehensive technologies.

(4) By execution of comprehensive technologies of cycling economy, low carbon metallurgy and green development, innovated application of beneficiated burden, long campaign, high blast temperature, oxygen enriched PCI, energy saving and emission control, low carbon smelting and environment protection, further lower down resources and energy consumption, insist in high efficiency, low cost, low emission and green development concept, promote the sustainable development of BF ironmaking, the future BF ironmaking still has a broad developing prospect.

Acknowledgements

Thanks for the support and help by the experts of Yin Ruiyu academician, Zhang Shourong academician and Gan Yong academician, thanks for the energetic support by "Beijing Scholar" Program.

References

[1] Zhang S R. The development of China ironmaing industry and the existing problems after entering into 21th century [J]. Ironmaking, 2012, 31 (1): 1-6.

[2] Zhang S R, Bi X G. The past, present and prospect of China ironmaking [J]. ironmaking, 2015, 34 (5): 1-6.

[3] Zhang F M. Understandings on some iron making technologies by modern blast furnace [J]. ironmaking, 2012, 31 (5): 1-6.

[4] Zhang F M. Technical features of super-large blast furnace in early 21th century [J]. ironmaking, 2012, 31 (2): 1-8.

[5] Yin R Y. Metallurgical Process Flow Engineering (Second edition) [M]. Beijing: Metallurgical Industry Press, 2009: 274-277.

[6] Yin R Y. Theory and Method of Metallurgical Process Flow Integration [M]. Beijing: Metallurgical Industry

Press, 2013: 95-103.
[7] Yin R Y, Wang Y L, Li B C, et al. Engineering Philosophy (Second Edition) [M]. Beijing: Higher Education Press, 2013: 173-184.
[8] Zhang F M, Qian S C, Yin R Y. Optimization of steel plant process flow structure and upsizing of blast furnace [J]. Iron & Steel, 2012, 47 (7): 1-9, 19.
[9] Zhang S R, Yu Z J. Long Campaign Life Technology of Wugang Blast Furnace [M]. Beijing: Metallurgical Industry Press, 2010: 227-231.
[10] Zhang F M, Chen S S. Long Campaign Life Technology of Modern Blast Furnace [M]. Beijing: Metallurgical Industry Press, 2012: 579-581.
[11] Zhang F M. The development and innovation of BF high blast temperature technology [J]. Ironmaking, 2013, 32 (6): 1-5.
[12] Zhang F M. The development prospects of high blast temperature low fuel ratio of blast furnace smelting technology [J]. China metallurgy, 2013, 23 (2): 1-7.
[13] Zhang F M. Study on large blast furnace gas dry bag dedusting technology [J]. Ironmaking, 2011, 30 (1): 1-5.

New Technological Progress of Ironmaking in Shougang

Zhang Fuming   Zhao Minge   Cao Chaozhen   Guo Yanyong

Keywords: ironmaking; blast furnace; sintering; pelletizing; coking; energy saving; emission control

1 Introduction

Since entering the new century, in order to ease the non-capital functions, hold the Beijing Olympic Games successfully, implement the Beijing's city development planning, improve the quality of environment in Beijing, Shougang's Beijing region iron and steel production facilities have been shut down in succession, the relocation and adjustment of industrial structure in Qinhuangdao, Qian'an and Caofeidian region of Hebei province. Three new modern iron and steel integrated plants have been built, therefore, Shougang's steel product structure and process structure have been optimized, and the improvement of the steel industry has been realized. Especially the design and construction of Shougang Jingtang steel plant, a new generation of circulating steel manufacturing process based on the concept of green manufacturing to achieve advanced steel products, high efficiency energy conversion, disposal and recycling of waste, to realize the "three functions" for the construction of the project philosophy. It is the first project of new generation circulating iron and steel plant near by the coastal port in China. The blast furnace technologies and equipments of Shougang's three new iron and steel base are shown in Table 1.

Table 1 Main specifications of technology and equipment of Shougang's new blast furnaces

Plant	Shouqin (Qinhuangdao)		Qiangang (Qian'an)			Jingtang (Caofeidian)	
BF number	1	2	1	2	3	1	2
Effective volume/m³	1200	1800	2650	2650	4000	5500	5500
Diameter of hearth/m	8.1	9.7	11.5	11.5	13.5	15.5	15.5
Design daily output/t·d⁻¹	2800	4200	6000	6000	9200	12650	12650
BF proper structure	Thin-lining+ full cooling stave	Thin-lining + full cooling stave	Thin-lining+ full cooling stave	Thin-lining+ full cooling stave	Thin-lining+ full cooling stave	Thin-lining+ full cooling stave	Thin-lining + full cooling stave

本文作者：张福明，赵民革，曹朝真，郭艳勇。原刊于 Proceedings of the Iron & Steel Technology Conference, AISTech, 2017: 439-453。

Continued Table 1

Plant	Shouqin (Qinhuangdao)		Qiangang (Qian'an)			Jingtang (Caofeidian)	
BF cooling system	Full soften water closed cooling circle	Full soften water closed cooling circle	Soften water closed cooling circle and opening cooling loop (for hearth and bottom)	Soften water closed cooling circle and opening cooling loop (for hearth and bottom)	Full soften water closed cooling circle	Full demineralized water closed cooling circle	Full demineralized water closed cooling circle
Bell-less Top	SG-2	SG-2	SG-3	SG-3	SG-4	SG-4	SG-4
BF gas dedusting system	Bag filter	Bag filter	Wet Ventrirui	Bag filter	Bag filter	Bag filter	Bag filter
PCI system	Single pipe-distributor	Single pipe-distributor	Single pipe-distributor	Single pipe-distributor	Single pipe-distributor	High density single pipe-distributor	High density single pipe-distributor
Hot blast stove	Dome combustion type	Dome combustion type	Internal combustion type	Internal combustion type	Internal combustion type	Dome combustion type	Dome combustion type
Quantity of HBS	3	3	3	3	4	4	4
Preheating process of gas and combustion air	Hot pipe heat exchanger+preheating stove	Hot pipe heat exchanger+preheating stove	Hot pipe heat exchanger+preheating stove	Hot pipe heat exchanger+preheating stove	Hot pipe heat exchanger+preheating stove	Hot pipe heat exchanger+preheating stove	Hot pipe heat exchanger+preheating stove
Fuel of HBS	Full BF gas	Full BF gas	Full BF gas	Full BF gas	Full BF gas	Full BF gas	Full BF gas
Design hot blast temperature, degree Celsius	1250	1250	1250	1250	1250-1300	1300	1300
Blew in, y-m	2004-6	2005-1	2004-10	2005-1	2010-1	2009-5	2010-6

2 Sintering

2.1 New Sintering Technology Research and Development

2.1.1 Large sintering machine

Based on independent design and research, the first 500m² large sintering machine is constructed in China. It is the largest sintering machine in China at that time, the production capacity is 5.7 million tons per year. Compared with the conventional advanced 360m² sintering machine, it can reduce the consumption of air volume 5%-10% per ton sinter, reduce energy consumption and dust emission, decrease the fuel consumption approximate 14% per ton sinter; as well as, also reduce the project investment more than 10% per ton sinter.

In order to achieve large scale sintering machine development successfully, the pallet car widening technology is researched and applied to implement sintering machine availability, the pallet car width is increased to 5.5m by the pallet widening technology, compared with the traditional structure of the pallet car, it can reduce the amount of air leakage of sintering pallet side wall zone significantly, improve the sintering condition at the edge of the pallet car side wall, and the productivity of sintering machine can be raised, as well as the quantity of return sinter is reduced, the energy consumption is reduced. By applying this technology, the sintering area is increased 10%, capacity is increased approximate 10%, and the air volume is decreased about 9% per ton sinter, thus the power consumption and comprehensive energy consumption are reduced remarkably.

2.1.2 Circular cooling facility

The large circular cooling facility is configured to meet the large sintering machine production stable and smooth. The large circular cooling facility effective cooling area is $580m^2$. During the design and development for circular cooling facility, widening technology is applied, the width of pallet car is 3.9m, the height of pallet side plate is 1.55m, blast area is $520m^2$, the actual cooling area reaches $580m^2$, capacity of circular cooling facility increases 5%-8.2%, the cooling air volume consumption is reduced 7% per ton sinter. The compact heavy wheel drive technology is designed and developed to achieve stable operation under high speed ratio conditions. The sinter segregation distributing device is researched and adopted, to promote the sinter ore cooling uniformity, improve the quality and to reduce pollution emission. The waste heat recovery system of the circular cooling facility is configured to produce steam for desalination of desalted water preparation.

2.1.3 Key equipment development

The new components and parts of the large sintering machine pallet car are researched and developed to meet the deep bed layer operation, promote the bed layer thickness reach 800mm. The elastic slide plate sealing and flexible sealing device for large sintering machine technology are developed and adopted. The sealing structure design is reasonable with good sealing performance. The air leakage ratio of sintering machine decreased from general 40%-60% to 25%-35%, reach the domestic advanced technical merit.

The sintering intelligent closed-loop control system is researched and developed independently. According to the sintering production process, the mathematical model and intelligent algorithms are established to achieve the intelligent closed-loop control of sintering production process.

2.2 Technical Innovation of Sintering Production

2.2.1 Optimizing the ore blending

In order to reduce the sintering cost of raw materials, dispose the solid waste of steel plant, improve sinter quality by optimizing ore blending. Under the condition of the raw materials cost reducing, increase the amount of solid waste of steel plant, such as collected dust, steel slag and others, the solid waste ratio reached 7%-10%, return ore ratio reached 13%-16%.

While a large amount of solid waste of steel works was added in the blending ore, an important problem is reduce the harmful elements (ZnO, Na_2O, K_2O, etc.) circulation and accumulation. In order to reducing the harmful elements cycling and concentrating, the technological system is developed based on the element analysis combine with the blending optimization, so the contradiction between the harmful elements control and the solid waste disposal is solved effectively.

2.2.2 Improving the sinter quality

For improving the sinter quality, improving the sinter metallurgical performances, improving the slag comprehensive performance, the natural magnesium sintering technology is developed and applied, combine with the development of magnesia pellet, the magnesium in the sinter and pellet rational distribution, promote the performance and quality of the sinter and pellet improvement. The development of low silica sinter, the sinter SiO_2 content was controlled in 4.8%-5.0%, the total ferrous content reach 57.6%, the burden comprehensive grade reach 59.89%, thus reduced the blast furnace coke rate 1.44kg/tHM, decrease the hot metal cost approximate 15 million Yuan RMB per year. The main technical parameters of sinter is shown in Table 2.

Table 2　Main technical index of sinter

Year	Total Fe/%	FeO/%	SiO_2/%	TI/%	RI/%	$RDI_{+3.15}$/%	CaO/SiO_2
2014	56.67	8.53	5.31	83.19	84.50	76.52	2.02
2015	57.65	8.77	4.99	82.61	83.11	74.19	2.02
2016	57.35	8.56	4.81	82.69	84.39	59.96	2.06

2.2.3 Energy saving and emission reducing

To realize green development of iron and steel manufacturing, and to effectively reduce the sintering process of SO_x, NO_x, and dust emissions, the sintering desulfurizing system technological improvement is investigated and researched. The outlet flue gas SO_2 content of desulfurizing tower controlled below 100mg/m³, was superior to the national standard of 180mg/m³. The dust emission concentration of sintering machine feed end part was 15.9mg/m³ in 2015, SO_2 emission concentration was 49.9mg/m³, NO_x emission concentration was 245.6mg/m³. The dust emission concentration was 16.1mg/m³, SO_2 emission concentration was 65.7mg/m³, NO_x emission concentration of 196.2mg/m³ in 2016.

2.3 Development Direction of Sintering Technology

2.3.1 Reducing the air leakage on sintering

To improve the sealing performance of the sintering machine to further reduce the air leakage ratio of sintering machine. Sintering machine sealing performance varies with the production process, the heat deformation, wear, aging, and fatigue of the equipment components are the main reasons to cause the air leakage ratio raise. Development of rational structure, advanced materials, reliable performance, durable new type sealing structure etc. are the developing direction in the future. Meanwhile, to strengthen the equipment management and maintenance, replace the wear, deformation, damage equipment parts timely is also important segment.

2.3.2 Reducing return sinter ratio

To improve the sintering process control, improve the sinter quality and metallurgical performance continually. Investigate and study the difference of grain size and metallurgical properties of sinter at pallet car vertical section, improve the uniformity and stability of sinter grain size and physical-chemical properties on whole sintering section, decrease the ratio of return sinter, improve the sinter production yield and average particle size, reduce the small size sinter. The hot air circulation sintering technology will be researched and applied to reduce the heat dissipation on the surface of the pallet; investigate the oxygen enrichment sintering technology to improve the surface temperature of sintering bed layer, reduce the grain size distribution non-uniformity caused by temperature difference diversification.

2.3.3 Improving sinter quality stability

To optimize the blending ore composition, control the reasonable cost of raw materials, improve the sintering process operational parameters control, and research the microstructure and mineral lithofacies of sinter, improve the comprehensive metallurgical properties of sinter, to meet the huge blast furnace high efficiency low consumption operational requirement.

2.3.4 Green sintering

To modify the electrostatic precipitators, improve the ability of dust catcher and the dust removal efficiency, reduce dust emission. Develop and research the section desulfurization process, the high sulfur high temperature exhaust gas recycling, in order to reduce the emission of dust and SO_2. Study explores steel green development oriented "desulfurization-denitration-dioxin removal-dust removal-heavy metals removal" in the integration of sintering flue gas emission efficient control technology.

Table 3 shows the operational parameters of Shougang Jingtang's sintering plant in 2016.

Table 3 Operational parameters of Jingtang's sintering plant in 2016

Items	Data
Area of sintering machine/m^2	2×500
Productivity/$t \cdot (m^2 \cdot h)^{-1}$	1.17
Thickness of bed layer/mm	812
Coke fine consumption/$kg \cdot t^{-1}$	56.67
Coke oven gas consumption/$MJ \cdot t^{-1}$	41.56
Power generation/$kW \cdot h \cdot t^{-1}$	47.85
Steam recovery/$kg \cdot t^{-1}$	−32.08
Total energy consumption/$MJ \cdot t^{-1}$	1440

3 Pelletizing

3.1 Engineering Innovation

The production capacity of Shougang Jingtang pelletizing plant is 4.00Mt/a. Technology of large travelling grate induration machine with an effective induration area of $504m^2$ is adopted. Process flow of pelletizing production fully absorbs advanced technologies at local and abroad, pursuing easy and high efficient flow, compact and reasonable process[1]. Main process flow includes such procedures as pre-proportioning, drying, roller pressing, preparation of flux and fuel, proportioning, balling, induration, and product classification. The pelletizing process flow sheet is shown in Fig. 1.

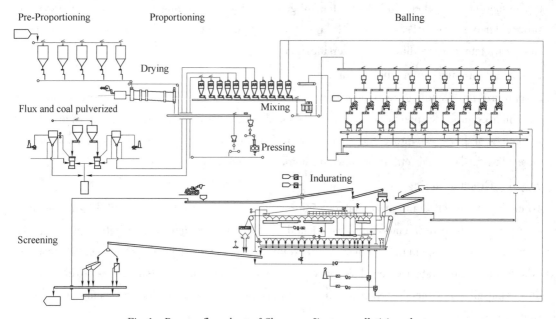

Fig. 1 Process flow sheet of Shougang Jingtang pelletizing plant

The integrated control management technology system of multiple working procedures are established based on material proportioning research, balling, distribution and indurating control, etc.

(1) The perfect automatic test and control system are designed to make information of the overall process of the entire technology monitored online.

(2) The optimal mathematical model is built, as well as the close-loop control for material distribution, close-loop control for temperature, close-loop control for wind system, balance control of hearth layer, and gas safety system are developed to have optimal operation of pelletizing production.

(3) Remote control and unattended operation for equipment are foreseen, so that labour productivity is improved greatly.

3.2 Application

Since Shougang Jingtang pelletizing plant was put into production in August, 2010, the pelletizing production process with large travelling grate induration machine has been studied and explored. The key technologies of stable production, high quality and high performance for pelletizing process have been developed under condition of different materials.

Optimization of production process is very important to ensure stable and smooth production. Innovations in the following aspects are mainly carried out:

(1) Centralized control and regulation by computer is applied in production process. The industrial TV monitor and expert control system management are used in major process systems. Such as fan system, material thickness control system, hearth layer circulating system, burner system, and so on, with advantages of high automation level, stable production and low gas consumption.

(2) A series of optimization and innovations were carried out after it was put into production, for instance, feeding mode from malarial bin, feeding belt conveyor for balling disc, bottom lining of the balling disc, etc. And a crusher for green ball is provided. So that balling performance of the returns is improved and product pellet yield is further increased.

(3) The software for equipment management cycle is developed to have scheduled maintenance, change and service for wearing parts. Maintenance schedule can be adjusted in time as per the service experience. Off-line maintenance should be conducted to the qualified equipment as far as possible, so as to solve any problem before it occurs for guarantee of efficient operability.

(4) Job-site manning is adjusted according to production experience for raising labour productivity in order to achieve 98% or above of the annual availability.

After 6 years' stable operation of the plant since it was put into operation in August, 2010, it has completely reached the designed level: yearly availability is above 8%, pellet total ferrous content is close to 66%, compression strength of pellets gets to 3000N/pellet, energy consumption in the process is 17.11kgce/t and screening index stands on 0.3%. Main technical indices reached to the international advanced level with remarkable environmental benefit and social benefit.

This project is integrated with advanced process and technologies at domestic and abroad with basis of independent design and integration. On basis of more than 10 years' development and research by Shougang, the international advanced process technologies and equipments are integrated, the process flow and key unit technology are optimized, so as to ensure reasonable process flow, complete process configuration, compact plant layout and high efficient production operation. Raw material preparation stage and precise regulation of indurating heat is especially emphasized in design and development process, so as to make it suitable for different kinds of raw material and create advantages to produce multi-variety and high-quality pellets. It has been proved through practices of production that this technology can produce acid and alkali pellets suitable for BFs in a high-efficiency, low energy-consumption, low-pollution way and solve technical difficulties thoroughly from process aspect, such as relatively low heat-resistant temperature of travelling grate and easy ringing of rotary kiln.

After put into operation, 4.00Mt/a pelletizing plant of Shougang Jingtang steel plant mainly treats high-sulfur magnetite from Peru with relatively high alkali metal content. Due to reliable technical performance, good adaptability of the process, flexible control and regulation of large travelling grate machine, it has fitted with many kinds of raw materials, thus stable and fluent production running and advanced technical index are realized. Many kinds of raw materials are adopted to produce high-quality pellets stably. Needs for concentrate burden for high efficiency and low consumption production of two 5500m^3 huge BFs are fully satisfied. Those important indexes as energy consumption of pellet production, pellet quality, and equipment availability all reach advanced international level. The main technical index of pellet is referred in Table 4.

Table 4 Main technical index of pellet

Year	Total Fe/%	FeO/%	SiO_2/%	CaO/%	Al_2O_3/%	MgO/%	CaO/SiO_2	Compression strength/N	Energyconsumption /MJ·t^{-1}
2014	65.00	0.47	2.93	0.76	0.62	1.72	0.26	2699	588
2015	65.06	0.50	2.84	0.71	0.57	1.73	0.25	2582	592
2016	64.58	0.68	2.83	0.57	0.69	1.65	0.20	2645	578

4 Coking

Four batteries of 70 chambers 7.63m large coke oven and two sets of 260t/h Coke Dry Quenching (CDQ) facility are configured to meet the requirements of large blast furnace on coke quality. The coking system annual output of 4.2 million t/a, the main technical characteristics are as follows:

(1) 7.63m coke oven adopts advanced COKEMASTER system introduced from Uhde company automation control technology to realize the coke oven automatic heating and automatic control.

(2) The single chamber pressure control technology (PROVEN) is applied for coke oven. According to the actual coke oven crude gas ratio, adjust each carbonization chamber pressure accurately, to achieve the single chamber pressure automatic adjusting, reduce the dust emissions on coal charging.

(3) The sectional coke oven heating technology is used to achieve a uniform heating of coke oven height direction, it can also reduce NO_x emissions.

(4) The two sets of 260t/h CDQ facility, equipped with 2 sets of 151t/h high temperature (540 dgree Celsius) and high pressure (9.5MPa) waste heat boiler, recover the hot coke heat produce steam, equipped with two sets of 30MW power generators, annual power output of 400 million kW·h/a.

(5) The vacuum potash coke oven gas desulfurization and preparation technology of sulphoacid are applied, after desulfurization H_2S content in gas can reach below 200mg/m^3.

This engineering put into production has been stable operation more than 6 years, the technical indexes reach the advanced level. The 260t/h super high temperature and high pressure CDQ compare with the common two sets of 140t/h CDQ device, reduce engineering investment 20%-25%, reduce operating costs about 14%. This technology can promote the energy recycle efficiency increased by 7%-15%, increase the energy recovery of 49.9kg standard coal per ton coke. The advanced coking technology improves the quality of coke and meet the requirements of the stability of the 5500m³ huge blast furnace production, reduce the blast furnace coke rate of about 2.5%. Effectively reduce the coke quenching process of the emission of pollutants, to improve the ecological environment quality, realize circulating economy and effective energy saving and emission reduction, created the outstanding economic and environmental benefits. The main technical index of coking is shown in Table 5.

Table 5 Main technical index of coke

Items	2014	2015	2016
A_d/%	11.86	11.81	11.85
V_{af}/%	1.24	1.23	1.24
S/%	0.74	0.73	0.74
M_{40}/%	90.92	90.23	91.10
M_{10}/%	5.67	5.71	5.55
CRI/%	19.80	20.11	19.45
CSR/%	72.52	72.36	72.54
25-40mm/%	27.68	23.17	11.84
<25mm/%	1.93	4.71	3.29
Average grain size/mm	57.81	54.83	55.99

5 Blast Furnace

5.1 Engineering Innovation

In the designing, the design philosophy of "high efficiency, low consumption, high quality, long campaign life and clean" is followed, advanced, practicable, mature, reliable, energy saving, environmental friendly, high efficiency and long campaign life technologies, equipments and materials are adopted. The design and operation experience of huge BFs above 5000m³ in Japan and Europe and large BFs over 4000m³ in China which were completed in recent years are researched and studied[2,3]. The advanced technology and equipment are adopted according to general planning of Shougang Jingtang iron and steel plant.

The BF iron making process flow is optimized, general layout arrangement is compact and reasonable, process flow is compact and smooth. The systematicness and completeness for each process sequence is considered thoroughly to achieve harmony of the operation. Complete automation control system and artificial intelligence (AI) expert system are developed and adopted to achieve the full automation of BF operation.

The whole BF ironmaking system of the project include: stock house and feeding system, BLT charging system, BF proper and cooling system, cast house and slag granulation, hot blast stoves, pulverized coal injection system, BF gas dedusting system and top gas pressure recovery turbine system (TRT), blower station and thermal dynamic system, environmental dedusting system, combined pump house, water supply and discharge system, oxygen and nitrogen supply system, power supply and distribution system, general layout and transportation system, and so on.

5.1.1 Beneficiated material technology and stock house

Beneficiated material is the basis of BF operation. BF burden composition optimization and beneficiated material improvement are the necessities for huge BF. Reasonable burden composition, 61% Fe content in burden, 90% agglomeration ratio are considered in the design. The burden composition includes 65% sinter, 25% pellet ore and 10% lump ore. Improving the mechanical strength of coke ($M_{40} \geqslant 89\%$, $M_{10} \leqslant 6.0\%$), especially the coke strength after reaction ($CSR \geqslant 68\%$) and coke thermal reaction performance ($CRI \leqslant 23\%$) are the key conditions to ensure stable operation of huge BF.

The raw material and fuel are transported by belt conveyors from sintering plant, pelletizing plant, coking plant and raw material yard to BF stock house. One combined stock house is adopted for two BFs. Coke bins and raw material bins are arranged in parallel. Sinter, pellet, lump ore and coke are screened under the bins. Individual screening and weighing system is adopted. There is no central weighing station and the material is transported directly by belt conveyor. In this process, the material transferring route is shortened, material crushing and fines are reduced which creates beneficial conditions for improving BF permeability and smooth operation of BF.

5.1.2 Bell-less top and burden distribution control technology

Bell-less top charging device is the key equipment for contemporary BF. The top charging equipment for 5500m³ huge BF shall meet not only the charging capacity, but also the requirements of burden distribution control in order to achieve burden classification, charging and central coke charging.

The bell-less top with parallel hoppers can charge five different sizes and kinds of material into BF within one charging cycle to achieve the burden classification charging and central coke charging. The advantages of bell-less top with parallel hoppers includes matured technology, sufficient charging capacity, simple equipment structure and operation, small equipment maintenance work and low operational cost. Bell-less top with parallel hoppers is the reliable technology self-developed and grasped by Shougang with rich operation and maintenance experience. The application of Shougang type bell-less top with parallel hoppers can achieve complete localization of equipment and reduce significantly the equipment investment.

5.1.3 BF Proper and long campaign life technology

The design campaign life of BF is 25 years, and the hot metal output during one campaign life reaches more than 20,000t/m^3 is desired. In the design, advanced BF long campaign life concept is developed and introduced, a mount of internationally advanced BF high efficiency and long campaign life integrated technology are applied, rational BF inner profile is adopted, non-overheat cooler and purified water closed loop circulating cooling technology are adopted, BF refractory lining is optimized, complete BF automation measuring and control system are provided in order to achieve stable operation and long campaign life of BF.

Reasonable BF inner profile is designed according to the investigation and analysis of large BFs. Based upon heat transfer, mass transfer, momentum transfer and chemical reaction principle during BF operation, combining the raw material and fuel conditions of BF, and in order to improve burden permeability and gas energy utilization and achieve BF stable operation, the design is optimized. The effective volume of BF is 5,500m^3, hearth diameter is 15.5m, belly diameter is 17.0m, well death depth is 3.2m, hearth height is 5.4m, effective height is 32.8m, effective height to belly diameter ratio is 1.93. Four tap holes and 42 tuyeres are configured.

BF long campaign life technology with high quality refractory and high efficiency non-overheat cooling system is adopted. The BF hearth and bottom lining is provided with combined structure of carbon brick and ceramic pad. The BF bottom uses combined structure of high conductivity graphite block, micropore carbon block, super micropore carbon block and ceramic pad. At the hearth side wall and interface of hearth and bottom, hot pressed carbon brick is used. At the position of tuyeres and tap holes, combined brick structure is applied. In order to achieve BF campaign life over 25 years, copper stave is used at hearth "elephant foot" wear zone and tap hole position. Full cooling stave structure is used for the BF proper. 4 rows of high efficiency copper cooling staves are used for BF bosh, belly and lower stack. 7 rows of brick cast iron cooling staves are used for the middle and upper stack. 1 row of C shape cooling stave is used below the throat armor. From BF bosh to stack, there is stave and brick lining integrated thin lining structure. At the hot surface of lining, gunning material is used. At BF bottom cooling pipes, cooling staves and tuyeres, de-mineralized water closed loop circulating cooling system is applied. Fig. 2 shows the structure of BF proper.

5.1.4 Dome combustion hot blast stove

High blast temperature is an important technical characteristic of contemporary BF iron making technology. High blast temperature can reduce effectively fuel consumption and improve BF energy utilization efficiency[4]. On the basis of Shougang type dome combustion hot blast stove (DCHBS) technology and Russian Kalugin type DCHBS technology, a new type DCHBS technology is developed by combining the advantages of these two technologies, and the DCHBS is applied in first time in 5,000m^3 grade huge BFs.

The main technical features of DCHBS are as follows: Ceramic burner for DCHBS is arranged at dome position, which has wide applicability. It can be operated under several operating

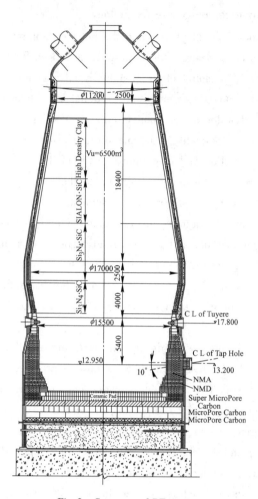

Fig. 2　Structure of BF proper

conditions including gas and combustion air. It has big combustion capacity, high combustion efficiency and long service life. The ceramic burners are provided with special circulating flow technology to ensure the complete mixing and combustion of air and gas and improve the flame temperature and dome temperature. Dome space is used as combustion chamber, and separate combustion chamber structure is canceled, which can improve the thermal stability of hot stove structure. Burner is configured at dome space; high temperature fume is evenly distributed under circulating flow condition, which can improve effectively the evenness and thermal conductivity of high temperature fume on checker bricks surfaces of checker chamber. The checker chamber is provided with high efficiency checker bricks with reduced channel diameter and increased heating area.

　　The BF is provided with 4 sets of dome combustion hot blast stoves with high blast temperature and long service life. The design blast temperature is 1,300 degree Celsius, and the maximum dome temperature is controlled under 1,420 degree Celsius, the smaller temperature difference between dome temperature and hot blast temperature is reached. The high temperature zone of hot blast stove is provided with silica bricks whose design service life is over 25 years. The fuel for hot blast stove is BF gas only. Fume waste heat reclaim unit is applied to preheat gas and combustion air. Two sets of small dome combustion hot stove are configured to preheat combustion air in order to achieve combustion air temperature over 520 degree Celsius. The high temperature valve for hot blast stove is cooled by pure water closed loop circulation cooling system. The combustion, blasting and stove changing of hot stove system is controlled automatically. Fig. 3 shows the preheating process flow of gas and combustion air for hot blast stoves of Jingtang's BF.

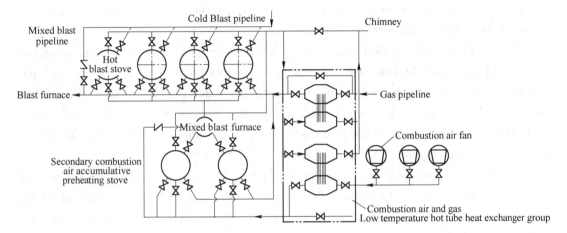

Fig. 3　Preheating process flow sheet of gas and combustion air for hot blast stoves

5.1.5　BF gas dry bag filter dedusting and top gas pressure recovery turbine

BF gas dry type bag filter dedusting technology is an important technical innovation for energy saving, emission reducing and clean manufacture in 21st century. It can reduce significantly the fresh water consumption during iron making process and reduce environmental pollution. It has become the development direction of contemporary BF iron making technology. The self-developed BF gas dry type low pressure pulse bag filter technology is adopted, and standby gas wet type dedusting system is cancelled[5].

Several key technologies including gas temperature control, gas dust content online supervision, collected dust pneumatic transportation, corrosion proof of duct and digital control etc. are developed, which makes the huge BF gas dry bag filter technology to internationally advanced level.

This system uses 15 dedusting vessels with diameter of 6.2m. The vessels are arranged in two lines in parallel. Between these two lines of vessels, crude gas and purified gas ducts are arranged. The gas pipe is designed based upon equal velocity principle to ensure the even distribution of gas volume to each vessel. This system is compact arranged, short and smooth flow with convenient maintenance.

5.2　Operational Application

Shougang Jingtang's No.1 BF blew-in on May 21 of 2009, and No.2 BF blew-in on June

21. After commissioning, BF operated stably and smoothly, the main operational performances promoted constantly. The highest daily production reached 14,350tHM/d, monthly effective volume productivity reached 2.37tHM/(m^3 · d), fuel ratio reduced to 480kg/tHM, coke rate lower to 269kg/tHM, PCI rate reached 175kg/tHM, hot blast temperature reached 1,300 degree Celsius, the international advanced performance of 5,000m^3 grade BF production has been achieved. The main technological achievements are as follows:

(1) Jingtang's ironmaking technological innovation and application in the design and production, in such aspects as high efficiency, low consumption, energy saving, environmental protection have made remarkable achievements.

(2) The BF gas dry bag filter dedusting technology has reached significantly achievement. It is the world's first full dry bag filter dedusting process of large blast furnace, compare with the wet dust removal process, purified gas dustiness long-term stability in 2-4mg/m^3, reduces 80% compare with wet dedusting process, TRT (Top gas pressure Recovery Turbine) power generation reaches to 54kW · h/tHM increase by 45% compare with conventional process, saving the fresh water consumption of 3.8 million t/a.

(3) The high blast temperature dome combustion hot blast stove technology is researched and applied successfully at large BF in the world. Under the condition of full BF gas with fuel for hot blast stove, the monthly 1,300 degree Celsius high blast temperature has achieved. The low quality BF gas high efficiency energy conversion is realized.

(4) The large blast furnace bell-less top charging equipment and burden distribution control technology are developed and applied independently, realize the water cooling-nitrogen sealing rotatory gearing box and distributing chute etc. all kinds of key technology breakthrough, promote the technical progress of metallurgical equipment in China. In the blast furnace operating, establish the "platform-funnel type" burden distribution shape system. On the basis of material balance and heat balance in the blast furnace, to construct the "platform-funnel type" charging system, optimize the gas distribution at upper stack zone, improves the gas utilization efficiency to realize the higher efficient utilization of the gas heat and chemical energy significantly.

(5) The heavy weight of batch ore burden technology is researched and adopted, and the ratio of ore/coke reach to 5.8t/t. The heavy ore batch to control the burden distribution, adjust the gas flow distribution, to achieve the rational gas utilization. The ore batch capacity is increased from 137 tons to 180 tons, greatly improves the BF working condition and stability.

(6) The coal-oxygen lance technique is researched and developed. The oxygen enrichment technology is promoted, and the oxygen concentration of mixture gas in the coal-oxygen lance outsleeve reaches 60%, increases the ratio of oxygen enriched blast by an average of 0.8%, to realize the economic benefit of 16.47 million RMB Yuan per year.

(7) The shortened tuyere blast technology is applied. On the basis of comprehensive analysis and the simulating calculation, the length of the tuyere is shortened reasonably to optimize the gas distribution in the blast furnace, to control the blast furnace pressure difference effectively, to improve the hearth working condition and quality of hot metal. The K_2O+Na_2O content is limited

less than 3.475kg/tHM in the burden, ZnO average content is controlled around 214g/tHM in the burden.

(8) The hazard control technology of harmful element concentration is adopted. Under the circumstances of the raw materials and fuel quality decline, through control the harmful alkali metals (ZnO, K_2O, Na_2O, etc.) content in the burden, to ensure the blast furnace stable and smooth operation, thus prolong the blast furnace lining service life.

Table 6 shows the main operational parameters of Shougang Jingtang's two huge BFs in 2014 to 2015.

Table 6 Main operational parameters of Jingtang's BFs in 2014 to 2016

Items	Year		
	2014	2015	2016
Daily production/tHM · d^{-1}	24585	24612	24292
Productivity/tHM · (m^3 · d)$^{-1}$	2.235	2.238	2.208
Blast temperature/degree Celsius	1200	1229	1233
Blast volume/Nm^3 · min^{-1}	8303	8337	8294
Oxygen enrichment ratio/%	5.10	5.12	5.29
Total Fe in burden/%	59.23	59.97	59.56
Ore/coke ratio/t · t^{-1}	5.28	5.40	5.56
Coke rate/kg · tHM^{-1}	312.50	300.59	292.77
PCI rate/kg · tHM^{-1}	153.89	173.88	191.04
Coke nut/kg · tHM^{-1}	32.04	24.51	20.80
Fuel ratio/kg · tHM^{-1}	498.45	498.87	504.61
CO utilization ratio/%	48.84	48.24	48.0

6 Conclusions

In recent year, Shougang has built 3 modern iron and steel integrated plant in Hebei province. A lots of advanced technologies and equipments have been developed and applied, more than 60 advanced technologies of current international iron making industry are adopted and integrated in the design. The general technological equipment achieves internationally advanced level.

Shougang Jingtang 5,500m^3 BF is a new generation huge BF following circulating economy principle and dynamic accurate design system. Reasonable burden composition and burden distribution control technology ensure the BF stable operation. The self-designed and manufactured bell-less top is applied in huge BF for the first time. BF long campaign life technologies including suitable BF profile, pure water closed loop circulating cooling technology, copper stave, thin wall lining structure lay foundation for the 25 years of campaign life. High blast temperature technologies including combustion air high temperature pre-heating technology and dome combustion hot stove ensure the blast temperature 1,300 degree Celsius. BF gas dry type dedusting technology, slag treatment technology, environmental protection technology and complete automation control system can achieve the clean and automatic BF operation.

Advanced sintering and pelletizing process are designed and adopted to meet the huge BF production requirements, good performance burden and fuel are the base of BF operation. At present, under the situation which raw material and the fuel condition worsen, in order to maintain the BFs operation to be suitable stably and smoothly, reduces the production cost, the lump ore charging quantity is increased, the quality and total ferrous content in burden are lowered. Nevertheless, the good operational performance and better technical parameters of Shougang Jingtang two 5,500m^3 BFs have been still obtained and achieved.

References

[1] Zhang F M, Wang Q S, Han Z G, et al. Innovation and application on pelletizing technology of large traveling grate induration machine [C]. AISTech 2015-Proceedings of the Iron and Steel Technology Conference. 2015: 402-412.

[2] Zhang F M, Qian S C. Design, construction and application of 5500m^3 blast furnace at Shougang Jingtang [C]. AISTech 2014-Proceedings of the Iron and Steel Technology Conference. 2014: 753-765.

[3] Zhang F M, Qian S C, Zhang J, et al. Design of 5500m^3 bast furnace at Shougang Jingtang [J]. Journal of Iron and Steel Research Internationa, 2009, 16 (S2): 1029-1033.

[4] Zhang F M. Dome combustion hot stove for huge blast furnace [J]. Journal of Iron and Steel Research International. 2012, 19 (9): 1-7.

[5] Zhang F M. Study on dry type bag filter cleaning technology of BF gas at large blast furnace [J]. Journal of Iron and Steel Research International, 2009, 16 (S2): 608-612.

Design, Construction and Application of 5,500m³ Blast Furnace at Shougang Jingtang

Zhang Fuming　Qian Shichong

Abstract: Shougang Jingtang iron and steel plant engineering is an important project of China iron and steel industry structure adjustment and improvement of general technology and equipment level. Its annual output is 9.7Mt/a, including two sets of 5,500m³ blast furnace with annual hot metal output of 8.98Mt/a. A number of innovative advanced technologies including bell-less top, dome combustion hot blast stove, blast furnace gas dry bag filter cleaning process and environmental protection screw slag granulation facility etc. are developed. In this paper description is given to the technical features and advanced technologies of the extra large sized blast furnace in Shougang Jingtang. Also introduction is made to blast furnace sinter charging by different fraction sizes, bell-less top equipment and control technology of burden distribution, high efficiency and long campaign life technology of blast furnace, high blast temperature with long campaign life technology of dome combustion hot blast stove, environment-friendly slag treatment technology, oxygen-enrichment pulverized coal injection technology and blast furnace gas dry type bag filter cleaning technology that are applied to the 5,500m³ blast furnace at Shougang Jingtang.

Keywords: blast furnace; bell-less top; dome combustion hot stove; bag filter dedusting

1 Introduction

Shougang Jingtang iron and steel plant engineering is an important project of China iron and steel industry structure adjustment and improvement of general technology and furnishment level. Its rude steel annual output is 9.7 million tons, including 2 sets of 5,500m³ blast furnace (BF) with total annual hot metal (HM) production of 8.98 million tons. They are the first two huge BFs above 5,000m³ in China.[1] In the engineering, the design and technical specifications of huge BFs above 5,000m³ around the world are investigated and studied. Self innovation is implemented completely. A great number of innovative advanced technologies including bell-less top, dome combustion hot blast stove, BF gas dry bag filter dedusting system and new type screw slag granulation facility etc. are developed and applied. The project construction began in March 2007, the No.1 BF blew in on May 2009 and the No.2 BF blew in on June 2010 in

---

本文作者：张福明，钱世崇。原刊于 Proceedings of the Iron & Steel Technology Conference, AISTech, 2014: 753-765。

succession. After commissioning, the major technological parameters and specifications have been achieved advanced technological merit.

2 Project Overview

2.1 Design Philosophy

In the designing, the design philosophy of "high efficiency, low consumption, high quality, long campaign life and clean" is followed, advanced, practicable, mature, reliable, energy saving, environmental friendly, high efficiency and long campaign life technologies, equipments and materials are adopted. The design and operation experience of huge BFs above 5,000m^3 in Japan and Europe and large BFs over 4,000m^3 in China which were completed in recent years are researched and studied.[2,3] The advanced technology and equipment are adopted according to general planning of Shougang Jingtang iron and steel plant.

The BF iron making process flow is optimized, general layout arrangement is compact and reasonable, process flow is compact and smooth. The systematicness and completeness for each process sequence is considered thoroughly to achieve harmony of the operation. Complete automation control system and artificial intelligence (AI) expert system are developed and adopted to achieve the full automation of BF operation.

The whole BF ironmaking system of the project include: stock house and feeding system, BLT charging system, BF proper and cooling system, cast house and slag granulation, hot blast stoves, pulverized coal injection system, gas dedusting system and top gas pressure recycle turbine system (TRT), blower station and thermal dynamic system, environmental dedusting system, combined pump house, water supply and discharge system, oxygen and nitrogen supply system, power supply and distribution system, general layout and transportation system, and so on.

2.2 Advanced Technological Parameters

The operational parameters of internationally advanced large BFs are analyzed and evaluated. Considering BF raw material and fuel conditions and technological furnishment level, the advanced design parameters of 5,500m^3 huge BF are determined. Table 1 shows main technological specifications of the BF.

Table 1 Main technological specifications of the BF

Items	Data
Effective volume	5,500m^3
Working volume	4,670m^3
Diameter of hearth	15.5m

Continued Table 1

Items	Data
Daily output	12,650tHM/d
Annual production	4.49Mt/a
Annual productivity (effective volume/working volume)	2.3/2.7tHM/(m^3·d)
Fuel ratio	490kg/tHM
Coke rate	290kg/tHM
PCI rate	180kg/tHM
Coke nut	20kg/tHM
Agglomeration ratio	90%
Total Fe in burden	≥61%
Composition of burden (sinter, pellet, lump ore)	65%, 25%, 10%
Blast volume	9,300Nm^3/min (max.)
Oxygen enrichment ratio	3.5% (max. 5.5%)
Hot blast temperature	1,300℃ (max.)
Blast pressure	0.55MPa
Top pressure	0.28MPa
Slag volume	250kg/tHM
Campaign life	≥25years

3 Design Highlights

3.1 Layout of Plant

The analysis and study of function of each process and facility are carried out during the project feasibility study. According to the investigative result, the general layout of two blast furnaces is considered completely in the design. The same process and system of two BFs are integrated and combined as possible through analytical study of the function of each process and system. Therefore, the process is integrated and optimized, and the process optimization and general layout reasonable arrangement are achieved.

The characteristics of whole process are provided with dynamic and orderly, continuous compact and efficient coordination. The perfect automatic supervision and control system are configured, production process applying computer for centralized control and adjustment, implementation of blast furnace efficiency, fully automated operation. The stock house, general control room, blower house, PCI facility and the other auxiliary systems of two BFs are integrated and combined together. The completely occupied area is 589,000m², west to east length is 950m, north to south width is 620m, the distance of two BF centre is 320m, and the compact and rational general layout is simplified and implemented. Fig. 1 shows the 3D CAD general layout of two huge BFs.

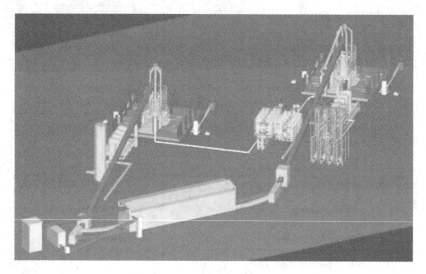

Fig. 1　General layout of two BFs

3.2　Beneficiated Material Technology and Stock House

Beneficiated material is the basis of BF operation. BF burden composition optimization and beneficiated material improvement are the necessities for huge BF. Reasonable burden composition, 61% Fe content in burden, 90% agglomeration ratio are considered in the design. The burden composition includes 65% sinter, 25% pellet ore and 10% lump ore. Improving the mechanical strength of coke ($M_{40} \geqslant 89\%$, $M_{10} \leqslant 6.0\%$), especially the coke strength after reaction ($CSR \geqslant 68\%$) and coke thermal reaction performance ($CRI \leqslant 23\%$) are the key conditions to ensure stable operation of huge BF.

The raw material and fuel are transported by belt conveyors from sintering plant, pelletizing plant, coking plant and raw material yard to BF stock house. One combined stock house is adopted for two BFs. Coke bins and raw material bins are arranged in parallel. Sinter, pellet, lump ore and coke are screened under the bins. Individual screening and weighing system is adopted. There is no central weighing station and the material is transported directly by belt conveyor. In this process, the material transferring route is shortened, material crushing and fines are reduced which creates beneficial conditions for improving BF permeability and smooth operation of BF.

Classification technology is applied for sinter and coke. The sinter and coke are classified and charged into BF with different sizes. Coke nut of 10-25mm and small size sinter of 2-8mm are reclaimed. The classification and charging technology of sinter can regulate charging material size, control the distribution of different sizes in BF, improve gas utilization and burden permeability, benefit the BF operation and control gas distribution, ensure BF smooth operation and long campaign life. The sinter can be classified into two grades, i. e. 20-50mm big size and 8-20mm medium size. The small size sinter of 3-8mm is reclaimed to improve raw material utilization. The coke is classified into two grades, i. e. 25-60mm, and >60mm. The coke nut of 10-25mm is reclaimed and charged into BF together with burden.

3.3 Bell-less Top and Burden Distribution Control Technology

Bell-less top charging device is the key equipment for contemporary BF. The top charging equipment for 5,500m^3 huge BF shall meet not only the charging capacity, but also the requirements of burden distribution control in order to achieve burden classification, charging and central coke charging.

The bell-less top with parallel hoppers can charge five different sizes and kinds of material into BF within one charging cycle to achieve the burden classification charging and central coke charging. The advantages of bell-less top with parallel hoppers includes matured technology, sufficient charging capacity, simple equipment structure and operation, small equipment maintenance work and low operational cost. Bell-less top with parallel hoppers is the reliable technology self-developed and grasped by Shougang with rich operation and maintenance experience. The application of Shougang type bell-less top with parallel hoppers can achieve complete localization of equipment and reduce significantly the equipment investment. Table 2 shows the main technical performances of bell-less top. Fig. 2 shows the 3D CAD simulation of bell-less top.

Table 2 Main technical performances of bell-less top

Items	Data
Effective volume of hopper	2×80m^3
Design pressure of hopper	0.30MPa
Upper sealing valve diameter	1,100mm
Lower sealing valve diameter	1,100mm
Material flow regulating gate valve diameter	1,000mm
Discharging capacity of material flow regulating gate valve	0.7m^3/s
Central throat tube diameter	760mm
α, γ angle accuracy	control accuracy ±0.3°
	metering accuracy 0.1°
Length of distribution chute	4,500mm

Fig. 2 3D CAD simulation of bell-less top

3.4 BF Proper and Cooling System

The design campaign life of BF is 25 years, and the hot metal output during one campaign life reaches more than 20,000t/m³ is desired. In the design, advanced BF long campaign life concept is developed and introduced, a mount of internationally advanced BF high efficiency and long campaign life integrated technology are applied, rational BF inner profile is adopted, non-overheat cooler and pure water closed loop circulating cooling technology are adopted, BF refractory lining is optimized, complete BF automation measuring and control system are provided in order to achieve stable operation and long campaign life of BF.

Reasonable BF inner profile is designed according to the investigation and analysis of large BFs. Based upon heat transfer, mass transfer, momentum transfer and chemical reaction principle during BF operation, combining the raw material and fuel conditions of BF, and in order to improve burden permeability and gas energy utilization and achieve BF stable operation, the design is optimized. The effective volume of BF is 5,500m³, hearth diameter is 15.5m, belly diameter is 17.0m, well death depth is 3.2m, hearth height is 5.4m, effective height is 32.8m, effective height to belly diameter ratio is 1.93. Four tap holes and 42 tuyeres are configured.

BF long campaign life technology with high quality refractory and high efficiency non-overheat cooling system is adopted. The BF hearth and bottom lining is provided with combined structure of carbon brick and ceramic pad. The BF bottom uses combined structure of high conductivity graphite block, micropore carbon block, super micropore carbon block and ceramic pad. At the hearth side wall and interface of hearth and bottom, hot pressed carbon brick is used. At the position of tuyeres and tap holes, combined brick structure is applied. In order to achieve BF campaign life over 25 years, copper stave is used at hearth "elephant foot" wear zone and tap hole position. Full cooling stave structure is used for the BF proper. 4 rows of high efficiency copper cooling staves are used for BF bosh, belly and lower stack. 7 rows of brick cast iron cooling staves are used for the middle and upper stack. 1 row of C shape cooling stave is used below the throat armor. From BF bosh to stack, there is stave and brick lining integrated thin lining structure. At the hot surface of lining, gunning material is used. At BF bottom cooling pipes, cooling staves and tuyeres, de-mineralized water closed loop circulating cooling system is applied. Fig. 3 shows the structure of BF proper.

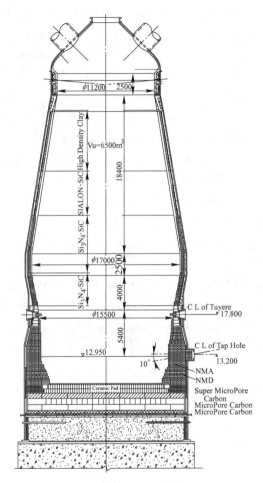

Fig. 3　Structure of BF proper

Pure water closed loop circulation cooling system is adopted in BF proper, the pure water preparation after sea water desalination and demineralized. BF pure water closed circulating cooling system is mainly divided into A and B two systems, system A mainly provides to the cooling stave for bottom, hearth, bosh, belly, and shaft; system B mainly includes the tuyere's big jacket, medium jacket and tuyere nozzle main body. The whole cooling systems according to the different pressure of cooling system is divided into two independent closed loop circulation cooling system. The backwater of each branched system is transported into the closed cooling tower correspondingly, after cooling tower the water through the corresponding pump system to recycle. The cooling system design parameters are as follows:

(1) The pure water volume of closed loop circulation cooling system A for BF proper is 5,900m^3/h, the water supply pressure is 1.1MPa. The water supply temperature is lower than 45℃, the backwater temperature is lower than 55℃, the pressure loss of piping system is 0.3MPa. The normal volume of additional pure water is 6m^3/h, the emergency accident volume of additional pure water is 59m^3/h.

(2) The pure water volume of closed loop circulation cooling system B for tuyeres is 2,900m^3/h, the water supply pressure is 1.7MPa. The water supply temperature is lower than 45℃, the backwater temperature is lower than 50℃, the pressure loss of piping system is 0.65MPa. The normal volume of additional pure water is 3m^3/h, the emergency accident volume of additional pure water is 29m^3/h.

Complete BF proper automatic measuring and control system is applied. 12 layers of thermocouples are provided at BF hearth and bottom to supervise the temperature distribution at BF bottom, conclude the wear conditions of BF hearth and bottom and instruct the maintenance work for BF hearth and bottom. 16 rows of thermocouples are provided above tuyere breast to supervise the working temperature of cooling stave wall, detect the wear conditions of BF lining and cooling stave and calculate the thermal load of BF proper. 4 rows of BF proper static pressure measuring unit is provided to conclude the cohesive or melting layer position and instruct BF distribution operation. BF throat gas crossing temperature measuring unit is provided to detect the gas temperature distribution of BF body upper part. The BF cooling water system is provided with inlet & outlet pressure, flow and temperature measuring and recording to calculate the thermal load, strengthen the supervision of cooling circuit, and conclude the damage of tuyere and cooling equipment and leakage supervision. BF top pressure control, automatic top gas analysis and tuyere detection is provided.

Mathematic models including burden distribution calculation model, BF hearth simulation model, hearth corrosion model, proper simulation model, tapping control model etc. are adopted for BF operation.

3.5 Cast House

Two symmetrically arranged flat rectangular cast houses are configured. Two tap holes are arranged in each cast house. Each tap hole is provided with independent clay gun, driller, cover removing device and tilting chute etc. Clay gun and driller are arranged under tuyere platform at same side of tap hole. At another side of tap hole, cover removing device is configured. Under each cast house, there are 4 ladle car railways. Fig. 4 shows the layout of cast house.

The hot metal is transported directly to the steelmaking plant by ladle instead torpedo car. There are two platforms in the cast house. Slag runner and iron runner is arranged at lower floor. The flexible cover of slag runner and iron runner is connected with upper floor in order to achieve flat

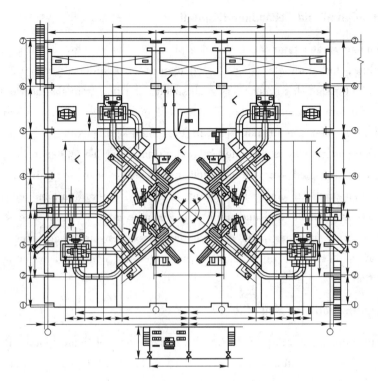

Fig. 4 Layout of cast house

operating platform of cast house. The maintenance equipment can work on the operating platform. The tuyere platform is overhead type steel structure which can achieve mechanization of equipment maintenance.

In order to optimize the interface technology between BF and BOF shop, achieve short hot metal transportation, reduce heat loss and environmental pollution during hot metal ladle changing, hot metal direct transportation technology is developed. 300t hot metal ladle is used to replace conventional torpedo car, which can optimize hot metal transportation, delete the ladle changing work, reduce hot metal temperature drop and environmental pollution and expedite the production cycle.

At one side of tap hole, a dedusting suction hole is provided. One top hood is arranged over the tap hole to prevent the secondary dust pollution during tap hole opening and blocking. Movable flat cover is arranged over the slag iron separator, slag runner and iron runner. Dedusting duct is arranged between two platforms of cast house to collect the dust during tapping. Suction holes are provided at both sides of two tilting chutes and over the hot metal ladle in each cast house. There are three dedusting systems in the cast houses. One set of bag filter dedusting system is provided for each cast house, the treatment capacity of each system is 700,000m^3/h. One set of secondary dust bag filter dedusting system is provided for 4 tap holes, the treatment capacity of dedusting system is 1,000,000m^3/h.

3.6 Slag Granulation and Secondary Disposal

The slag is completely water granulated. Dry slag pit is used as standby slag treatment unit. After super fine grinding, the granulated slag is used to produce addition of cement. Melted BF slag flows to slag nozzle through slag runner. The high speed water flow from slag nozzle makes the melt slag cool down and forms granulated slag. The mixture of slag and water goes through tunnel to condensing tower to water granulation pond. In the pond, the mixture is separated by screw separator. The fine particle slag is separated by rotary drum filter. The water in the pond is filtered and circulated. The granulated slag is transported by belt conveyor to slag pulverizing facility. Two sets of pulverized slag mill are configured with total capacity 120 million tons per year. Super fine of granulating slag is manufactured and applied to concrete additive agent. The steam generated during granulation and water slag pond is discharged to condensing tower. In condensing tower, there are two layers of spraying units. The spraying water and condensed water returns to water tank for circulation.

3.7 Dome Combustion Hot Blast Stove

High blast temperature is an important technical characteristic of contemporary BF iron making technology. High blast temperature can reduce effectively fuel consumption and improve BF energy utilization efficiency.[4] Upon the basis of Shougang type dome combustion hot blast stove (DCHBS) technology and Russian Kalugin type DCHBS technology, BSK (Beijing Shougang Kalugin) type DCHBS technology is developed by combining the advantages of these two technologies, and the DCHBS is applied in first time in 5,000m^3 grade huge BFs.

The main technical features of DCHBS are as follows: Ceramic burner for DCHBS is arranged at dome position, which has wide applicability. It can be operated under several operating conditions including gas and combustion air. It has big combustion capacity, high combustion efficiency and long service life. The ceramic burners are provided with special circulating flow technology to ensure the complete mixing and combustion of air and gas and improve the flame temperature and dome temperature. Dome space is used as combustion chamber, and separate combustion chamber structure is canceled, which can improve the thermal stability of hot stove structure. Burner is configured at dome space; high temperature fume is evenly distributed under circulating flow condition, which can improve effectively the evenness and thermal conductivity of high temperature fume on checker bricks surfaces of checker chamber. The checker chamber is provided with high efficiency checker bricks with reduced channel diameter and increased heating area.

The BF is provided with 4 sets of dome combustion hot blast stoves with high blast temperature and long service life. The design blast temperature is 1,300℃, and the maximum dome temperature is controlled under 1,420℃, the smaller temperature difference between dome temperature and hot blast temperature is reached. The high temperature zone of hot blast stove is provided with silica bricks whose design service life is over 25 years. The fuel for hot blast stove is

BF gas only. Fume waste heat reclaim unit is applied to preheat gas and combustion air. Two sets of small dome combustion hot stove are configured to preheat combustion air in order to achieve combustion air temperature over 520℃. The high temperature valve for hot blast stove is cooled by pure water closed loop circulation cooling system. The combustion, blasting and stove changing of hot stove system is controlled automatically.

Fig. 5 shows the preheating process flow of gas and combustion air for hot blast stoves of Jingtang's BFs. Table 3 shows the major technical parameters of dome combustion hot blast stove, and Table 4 shows the main technical parameters of heat pipe exchanger. Fig. 6 shows the photograph of dome combustion hot blast stoves of No. 1 BF.

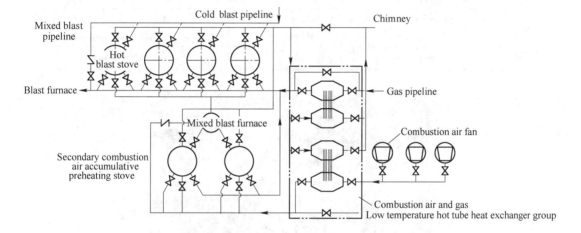

Fig. 5 Preheating process flow of gas and combustion air for hot blast stoves

Table 3 Major technical parameters of dome combustion hot blast stove

Items	Data
Hot stove number	4 sets
Hot stove height	50.1m
Hot stove diameter	12.5m
Checker brick heating surface	95,885m^2
Checker brick heating surface per m^3	48.0m^2
Chanel diameter of checker brick	30mm
Height of checker brick	21.48m
Cross-sectional area of checker chamber	93.21m^2
Hot blast temperature	1,300℃
Dome temperature	1,420℃
Fume temperature	Max. 450℃, Average 368℃
Combustion air preheating temperature	520-600℃
Gas preheating temperature	Approximate 215℃
Cold blast volume	9,300Nm^3/min
Cold blast pressure	0.54MPa

Continued Table 3

Items	Data
Specific blast regenerative surface	41.24m²/(m³·min)
Specific BF volume regenerative surface	67.73m²/m³

Table 4 Main technical parameters of heat pipe exchanger

Items	Fume	Gas	Combustion air
Gasflow volume/Nm³·h⁻¹	700,000	505,349	319,000
Inlet temperature/℃	350	45	20
Outlet temperature/℃	165	215	200
Pressure loss/Pa	≤500	≤550	≤600
Maximum steam temperature inside the heat pipe/℃		260	248
Heat exchange area/m²		6,811+14,224 (high temperature side)	
Heat recovery/kW		21,461+35,373=56,833	

Fig. 6 The dome combustion hot blast stoves of No. 1 BF

3.8 BF Gas Dry Bag Filter Dedusting and Top Pressure Recovery Turbine

BF gas dry type bag filter dedusting technology is an important technical innovation for energy saving, emission reducing and clean manufacture in 21st century. It can reduce significantly the fresh water consumption during iron making process and reduce environmental pollution. It has become the development direction of contemporary BF iron making technology. The self-developed BF gas dry type low pressure pulse bag filter technology is adopted, and standby gas wet type dedusting system is cancelled.[5]

Several key technologies including gas temperature control, gas dust content online supervision, collected dust pneumatic transportation, corrosion proof of duct and digital control etc. are developed, which makes the huge BF gas dry bag filter technology to internationally advanced level.

This system uses 15 dedusting vessels with diameter of 6.2m. The vessels are arranged in two lines in parallel. Between these two lines of vessels, crude gas and purified gas ducts are arranged. The gas pipe is designed based upon equal velocity principle to ensure the even distribution of gas volume to each vessel. This system is compact arranged, short and smooth flow with convenient maintenance.

Low filtering velocity principle is adopted to ensure the safe and stable system operation. Each vessel has 409 bag filters. The bag filter size is $\phi 160 \times 7000$mm. Filtering area for each box is 1,439m^2 and total filtering area is 21,586m^2. In the design, the diameter and length of bag filters are increased, and the bag structure and dimensions are more reasonable. The vessel diameter is increased to improve the treatment capacity of dedusting unit, reduce vessel quantity, investment and occupied area. Table 5 shows the main technical performances of BF gas bag filter dedusting technology. Fig. 7 shows the 3D CAD simulation of BF gas dry bag filter dedusting system. Fig. 8 shows the photograph of the BF gas dry bag filter dedusting system of No. 1 BF.

Table 5 Technical performances of dry bag filter dedusting for BF gas

Items	Data
BF gas flowvolume	760,000 (Max. 870,000) m^3/h
Top pressure	280 (Max 300) kPa
Operating temperature	100-220℃
Vessel number	15sets
Vessel diameter	6.2m
Bag sizes (diameter×length)	160×7,000mm
Total filtering surface	21,586m^2
Filteringsurface per vessel	1,439m^2
Filtering velocity under standard conditions	0.59m/min
Filtering velocity under operating conditions	0.23m/min
Dust content in cleaned gas	≤5mg/m^3

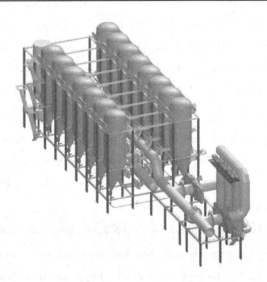

Fig. 7 3D CAD simulation of BF gas bag filter dedusting system

Fig. 8 BF gas dry bag filter dedusting system of No. 1 BF

After bag filter dedusting system, BF gas top pressure recovery turbine (TRT) system is configured, the top gas pressure is recovered and transformed to electrical power. The BF top pressure regulating valve block as TRT alternate device with a parallel arrangement, as an aid to adjust the BF top pressure. A new generation of dry type TRT is introduced to match the BF gas dry dedusting system, and the BF gas heat energy and kinetic energy can be recovered and utilized fully to realize energy conversion with high efficiency. The dry type fully static vane adjustable axial-flow turbine is adopted in TRT system with the generator capacity is 36.5MW, the power generating output of 45kW·h/tHM is designed.

3.9 Pulverized Coal Injection System

The design pulverized coal injection (PCI) rate of the BF is 220kg/tHM (max. design capacity is 250kg/tHM) and normal PCI capacity is 116t/h. The coal injection and pulverization facilities are constructed together for two BFs. Medium speed mill is used to pulverization. Bag filter catcher and direct injection process is adopted. The capacity of medium speed mill is 2×75t/h. The PCI system uses high density phase direct injection technology with three hoppers in parallel and main pipe-distributor process.[6]

Each BF has three injection hoppers working in sequence, two injection main pipes and two distributors. High density phase pneumatic transportation technology is adopted with density is above 40kg/kg. The velocity of coal flow in injection pipe is within 2-4m/s. In each main pipe coal flow meter and regulating valve is provided.

The effective volume of pulverized coal bin is 1,200m^3. Below the pulverized coal bin, there are three coal discharging hoppers. The effective volume of injection vessel is 90m^3 with normal injection cycle of 30min. Under the injection hopper, there are three fluidizing vessels. Fluidizing injection PCI technology and high density phase pneumatic conveying process are adopted for

BFs. Two injection branch pipes are provided for each fluidizing vessel. The pulverized coal is injected through injection main pipe to two distributors, and through 42 injection branch pipes to each tuyere. Nitrogen is used for the pressurizing of injection hopper, transportation of pulverized coal, fluidization and anti-explosion. Injection supervision units are provided for each injection branch pipe. In case of blocking, the branch pipe will be shut off and injecting automatically.

4 Engineering Construction

The construction of Shougang Jingtang's No. 1 BF started from March 12 of 2007, on October 18 of 2008, BF heated up, the whole project took a period with 19 months. The total weight of steel structure is approximately 80,000 tons, the total of 25 bands furnace shell, and the total height is nearly 126m. The BF foundation structure has been completed on April 2, 2007, the project took a period of 84.5 hours, and poured concrete volume is 10,500m^3. On October 2007, the BF proper main structure engineering has been finished, the hot blast circle pipe has been installed, the hot blast stove and cyclone structure have been accomplished, the structure frame of cast house was erected, started to assemble the cooling stave, the hot blast stove proper shell has been erected, started to assemble the grid.

On September 5 of 2007, hot blast stove shell installation was completed, in March of 2008, furnace brick laying was started, on August 15, furnace brick laying work was completed. On January 23 of 2008, the BF cooling staves were installed completely. Since March 2008, the BF hearth and bottom lining laying work were started, which lasted 100 days was completed on June 23. On May 27 of 2008, "five-way ball" connection joint was assembled, and on June 6 of 2008, down comer of BF has been successfully installed. On July 13 of 2008, main belt conveyor corridor installation was competed. The BF bleeding valve was completed on August 20 of 2008. No. 1 and No. 3 hot blast stoves baking work was started on September 8, 2008, No. 2 and No. 4 hot blast stoves heating up work was started on September 9. The blower installation and debugging were completed on September 12 of 2008, and the blower motor commissioning was started. The BF gas dedusting system debugging was finished on October 16 of 2008.

The No. 1 BF heating up work started on October 18 of 2008, that proved the BF engineering has fully completed, as well as that have been provided with the BF commissioning conditions. Foundation construction of No. 1 BF began on March 12 of 2007, the engineering take a period of 19 months to complete construction. Due to the market and business reasons, No. 1 BF blew-in on May 21 of 2009; No. 2 BF blew-in on June 21 of 2010.

Fig. 9 shows the photograph of No. 1 BF during project erection. Fig. 10 shows the photograph of No. 1 BF has been built. Fig. 11 shows the photograph of No. 1 BF after blew-in. Fig. 12 shows the photograph of No. 1 and No. 2 BFs.

Fig. 9　Photograph of No. 1 BF during project erection

Fig. 10　Photograph of No. 1 BF has been built

Fig. 11　Photograph of No. 1 BF after commissioning

Fig. 12 Photograph of No. 1 and No. 2 BFs

5 Operation and Appliction

Shougang Jingtang's No. 1 BF blew-in on May 21 of 2009, after commissioning, BF operated stably and smoothly, the main operational technical performances promoted constantly. The highest daily production reached 14,350tHM/d, monthly effective volume productivity reached 2.37tHM/($m^3 \cdot d$), fuel ratio reduced to 480kg/tHM, coke rate lower to 269kg/tHM, PCR reached 175kg/tHM, hot blast temperature reached 1,300℃, the international advanced performance of 5,000m^3 grade BF production has been achieved.

In order to promote the BF operation performance, and to accept high blast temperature and worse raw material, a series of techniques for BF operation for satisfactory of high temperature production are studied and developed, mainly including:

(1) Application of concentrate burden technology has been adopted. Such as improvement of sinter quality, reduction of chargeable powder, improvement of acceptance ability of high blast temperature by BF, optimization of burden composition, improvement of integrative grade and decrease of slag volume.

(2) Improvement of high top pressure and control of the bosh gas volume index of BF. Jingtang's No. 1 and No. 2 BF top pressure are controlled at nearly 270kPa.

(3) Control of reasonable theoretic flame temperature. After the blast temperature reaches to 1,250℃, measures like improvement of ore-coke burden ratio, PCI rate and adjustment of blast humidity, etc. are taken with theoretic combustion temperature controlled within 2100±50℃.

(4) Strengthening burden distribution control and realization of reasonable distribution of gas when the blast temperature reaches to 1,250℃ and over, measures of increase of ore batch, improvement of ore-coke burden ratio, etc. are taken to promote central gas flow and stabilize gas flow at edges, with gas utilization ratio reaches to 50%-52%.

(5) Improvement of blast temperature stability and reduction of blast temperature fluctuation

BF operation can stabilize ore-coke charging ratio based on fixation of blast temperature and control of theoretic flame temperature to control theoretic flame temperature fluctuation within 50℃.

The optimal operation mode under high temperature condition has been investigated and explored from long term production, practice and research. The blast temperature is kept within 1,250±50℃ in long term stability status. Fuel ratio is decreased and BF high blast temperature has been achieved significant application effectiveness.

Table 6 shows the main operational parameters of Jiangtang's No. 1 BF after commissioning. Table 7 shows the main operational parameters of Jingtang's two huge BFs in 2012 and 2013.

Table 6　Main operational parameters of Jingtang's No. 1 BF after commissioning

Date	Production /tHM·d^{-1}	Productivity /tHM·(m^3·d)$^{-1}$	Coke rate /kg·(tHM)$^{-1}$	PCR /kg·(tHM)$^{-1}$	Fuel ratio /kg·(tHM)$^{-1}$	Hot blast Temp./℃
2009-5	4,840	0.88	551	83	634	914
2009-6	7,425	1.35	503	62	565	998
2009-7	8,525	1.55	483	49	532	1,063
2009-8	11,000	2.01	372	94	481	1,166
2009-9	11,660	2.12	354	101	483	1,212
2009-10	12,210	2.22	340	117	488	1,262
2009-11	12,500	2.27	299	145	484	1,276
2009-12	12,694	2.31	288	149	479	1,281
2010-1	12,657	2.30	307	137	482	1,259
2010-2	12,847	2.34	287	161	482	1,277
2010-3	13,035	2.37	269	175	480	1,300

Table 7　Main operational parameters of Jingtang's BFs in 2012 and 2013

Items	No. 1 BF		No. 2 BF	
Year	2012	2013	2012	2013
Daily production/HM·d^{-1}	11,995	12,552	12,313	12,104
Productivity/tHM·(m^3·d)$^{-1}$	2.21	2.26	2.24	2.20
Blast temperature/℃	1,223	1,235	1,234	1,214
Blast volume/Nm^3·min^{-1}	7,934	8,032	8,268	7,987
Blast pressure/kPa	433	448	447	445
Top pressure/kPa	228	254	260	254
Oxygen enrichment ratio/%	3.75	5.46	3.53	4.86
Total Fe in burden/%	58.94	59.2	58.95	58.7

Continued Table 7

Items	No. 1 BF		No. 2 BF	
Year	2012	2013	2012	2013
Coke rate/kg·(tHM)$^{-1}$	309	307.9	304	316.8
CO utilization ratio/%	50.82	49.3	50.59	48.3
PCR/kg·(tHM)$^{-1}$	148	154.0	157	152.8
Coke nut/kg·(tHM)$^{-1}$	28	32.4	29	30.8
Fuel ratio/kg·(tHM)$^{-1}$	485	494.3	490	500.4

6 Conclusions

Shougang Jingtang 5,500m^3 BF is a new generation huge BF following circulating economy principle and dynamic accurate design system. More than 60 advanced technologies of current international iron making industry are adopted and integrated in the design. The general technological equipment achieves internationally advanced level. Reasonable burden composition and burden distribution control technology ensure the BF stable operation. The self-designed and manufactured bell-less top is applied in huge BF for the first time.

BF long campaign life technologies including suitable BF profile, pure water closed loop circulating cooling technology, copper stave, thin wall lining structure lay foundation for the 25 years of campaign life. High blast temperature technologies including combustion air high temperature pre-heating technology and dome combustion hot stove ensure the blast temperature 1,300℃. BF gas dry type dedusting technology, slag treatment technology, environmental protection technology and complete automation control system can achieve the clean and automatic BF operation.

After blew-in, No.1 BF operated stably and smoothly, the main operation technical performances promoted constantly. The main technical-ecnomic specifications have been achieved the design goal since commissioning 6 months. After No. 2 BF commissioning, good operational performance and condition have been achieved and maintained continuously.

At present, under the situation which raw material and the fuel condition worsen, in order to maintain the BFs operation to be suitable stably and smoothly, reduces the production cost, the lump ore charging quantity is increased, the quality and total ferrous content in burden are lowered. Nevertheless, the good operational performance and better technical parameters of Shougang Jingtang two 5,500m^3 BFs have been still obtained and achieved.

Acknowledgements

The authors would like to thank our colleagues of Mr. Zhang Jian, Mr. Mao Qingwu, Mr. Su Wei, Mr. Mei Conghua, etc. who participate the technical study works on this issue. Thanks for the help from our colleagues of Mr. Meng Xianglong, Dr. Cao Chaozhen, Dr. Li lin, and Dr. Guo Yanyong.

References

[1] Zhang F M, Qian S C, Zhang J. Design of 5500m^3 blast furnace at Shougang Jingtang [J]. Journal of Iron and Steel Research International, 2009, 16: 1029-1033.

[2] Evans J L. Design and construction of lining of no. 4 blast furnace port talbot works [J]. Ironmaking and Steelmaking, 1994, 21 (2): 101-108.

[3] Macauley D. Current blast furnace design philosophies [J]. Steel Times International, 1995, 19 (2) .20-26.

[4] Zhang F M, Research and innovation of the key technologies at modern large blast furnace [J]. Iron and steel, 2009, 44 (4): 1-5.

[5] Zhang F M. Study on dry type bag filter cleaning technology of BF gas at large blast furnace [J]. Journal of Iron and Steel Research International, 2009, 16: 608-612.

[6] Zhang F M, Qian S C, Zhang J. New technologies of 5500m^3 blast furnace at Shougang Jingtang [J]. Iron and steel, 2011, 46 (2): 13-17.

长寿高效高炉设计研究

我国大型高炉长寿技术发展现状

摘　要：论述了我国大型高炉长寿技术的发展现状。在大型高炉设计中，通过优化炉型、采用合理炉缸内衬结构、铜冷却壁、软水密闭循环冷却系统、薄壁内衬等技术为高炉长寿创造条件。通过自动化检测与控制、炉体维护等技术使高炉寿命达到15年以上。对高炉炉缸内衬结构、铜冷却壁、软水密闭循环冷却系统、薄壁内衬的应用进行了评述，对我国大型高炉长寿技术的发展提出了建议。

关键词：高炉；长寿；设计；铜冷却壁

1　引言

近10年来，我国高炉炼铁技术迅猛发展，连续10年成为世界第一产铁大国。高炉大型化、现代化、高效化、长寿化进程加快，并已取得了令人瞩目的技术成就。高炉长寿是现代大型高炉的重要技术特征，在我国大型高炉炼铁技术进步中，其作用尤为突出。

2　我国大型高炉长寿现状

据不完全统计，我国高炉容积大于1000 m^3 的大型高炉有50余座，2000 m^3 以上的大型高炉有25座，这些大型高炉的生产能力约占全国炼铁生产能力的50%以上。20世纪90年代，一批新建或大修技术改造的高炉采用了铁素体球墨铸铁冷却壁、铜冷却板、软水密闭循环冷却、陶瓷杯等现代高炉长寿技术，寿命已达到8~10年以上。其中我国自行设计建设的特大型高炉——宝钢2号高炉（4063 m^3）寿命已达到12年，武钢5号高炉（3200 m^3）、首钢4号高炉（2100 m^3）的寿命也相继达到12年。这些高炉至今未进行中修，仍在正常工作，预计高炉寿命将达到15年以上。表1列举了我国部分大型长寿高炉的实例。

表1　我国部分大型长寿高炉实绩

高炉	高炉有效容积/m^3	投产时间	停产时间	寿命/a
宝钢2号	4063	1991-06	尚在运行	13.0
武钢5号	3200	1991-10	尚在运行	12.8
首钢4号	2100	1992-05	尚在运行	12.1

本文作者：张福明，党玉华。原刊于《钢铁》，2004，39（10）：75-78。

续表 1

高炉	高炉有效容积/m^3	投产时间	停产时间	寿命/a
宝钢 1 号	4063	1985-09	1996-04	10.6
首钢 3 号	2536	1993-06	尚在运行	11.0
首钢 1 号	2536	1994-08	尚在运行	9.1
梅山 2 号	1250	1986-12	1997-09	10.8

我国新建或大修改造的大型高炉，遵循高效、长寿并举的原则，高炉一代炉役设计寿命 15~20 年，一代炉役平均利用系数大于 2.0t/(m^3·d)，一代炉役单位有效容积产铁量达到 10000~15000t/m^3。

3 高炉长寿技术的应用与发展

3.1 优化高炉炉型

我国炼铁工作者历来重视高炉炉型设计，通过研究总结高炉破损机理和高炉反应机理，优化高炉炉型设计的基本理念已经形成。

3.1.1 加深死铁层深度

实践证实，高炉炉缸炉底"象脚状"的异常侵蚀，主要是由于铁水渗透到炭砖中，使炭砖脆化变质，再加上炉缸内铁水环流的冲刷作用而形成的。加深死铁层深度，是抑制炉缸"象脚状"异常侵蚀的有效措施。死铁层加深以后，避免了死料柱直接沉降在炉底上，加大了死料柱与炉底之间的铁流通道，提高了炉缸透液性，减轻了铁水环流，延长了炉缸炉底寿命。理论研究和实践表明，死铁层深度一般为炉缸直径的 15%~20%。

3.1.2 适当加高炉缸高度

高炉在大喷煤操作条件下，炉缸风口回旋区结构将发生变化。适当加高炉缸高度，不仅有利于煤粉在风口前的燃烧，而且还可以增加炉缸容积，以满足高效化生产条件下的渣铁存储，减少在强化冶炼条件下出现的炉缸"憋风"的可能性。近年我国已建成或在建的大型高炉都有炉缸高度增加的趋势，高炉炉缸容积为有效容积的 16%~18%。

3.1.3 加深铁口深度

铁口是高炉渣铁排放的通道，铁口区的维护十分重要。研究表明，适当加深铁口深度，对于抑制铁口区周围炉缸内衬的侵蚀具有显著作用，铁口深度一般为炉缸半径的 45% 左右。这样可以减轻出铁时在铁口区附近形成的铁水涡流，延长铁口区炉缸内衬的寿命。

3.1.4 降低炉腹角

降低炉腹角有利于炉腹煤气的顺畅排升，从而减小炉腹热流冲击，而且还有助于在炉腹区域形成比较稳定的保护性渣皮，保护冷却器长期工作。现代大型高炉的炉腹角一般在 80°以内，本钢 5 号高炉（2600m^3）炉腹角已降低到 75.37°。

表2是近年我国部分建成或在建的大型高炉炉型参数。

表2 我国部分大型高炉炉型参数

项目	本钢5号高炉	鞍钢新1号高炉	唐钢2560m³高炉	首钢迁钢1号高炉	首钢3号高炉	济钢1750m³高炉
有效容积/m³	2600	3200	2560	2650	2536	1750
炉缸直径/mm	11000	12400	11000	11500	11560	9500
炉缸高度/mm		4900	4600	4200	4200	4200
死铁层深度/mm	1900	2400	2200	2300	2200	2000
炉腹角	75°22′12″	79°28′45″	79°59′31″	79°59′31″	78°02′36″	75°53′
高径比	2.244	2.204	2.402	2.268	1.985	2.327

3.2 炉缸炉底内衬结构

长寿炉缸炉底的关键是必须采用高质量的炭砖并辅之合理的冷却。通过技术引进和消化吸收，我国大型高炉炉缸炉底内衬设计结构和耐火材料应用已达到国际先进水平。

以美国UCAR公司为代表的"导热法"（热压炭砖法）炉缸设计体系已在本钢、首钢、宝钢、包钢、湘钢等企业的大型高炉上得到成功应用；以法国SAVOIE公司为代表的"耐火材料法"（陶瓷杯法）炉缸设计体系在首钢、梅山、宝钢、鞍钢等企业的大型高炉上也得到了推广应用。日本大块炭砖—综合炉底技术在宝钢、武钢等企业的大型高炉上也取得了长寿实绩。"导热法"和"耐火材料法"这两种看来似乎截然不同的设计体系，其技术原理的实质却是一致的。即通过控制1150℃等温线在炉缸炉底的分布，使炭砖尽量避开800~1100℃脆变温度区间。导热法采用高导热、抗铁水渗透性能优异的热压小块炭砖，通过合理的冷却，使炭砖热面能够形成一层保护性渣皮或铁壳，并将1150℃等温线阻滞在其中，使炭砖得到有效的保护，免受铁水渗透、冲刷等破坏。陶瓷杯法则是在大块炭砖的热面采用低导热的陶瓷质材料，形成一个杯状的陶瓷内衬，即所谓"陶瓷杯"，其目的是将1150℃等温线控制在陶瓷层中。这两种技术体系都必须采用具有高导热性且抗铁水渗透性能优异的炭砖。将两种设计体系组合在一起也不失为一种合理的选择，首钢1号高炉（2536m³）采用热压炭砖—陶瓷杯组合炉缸内衬技术，至今已安全运行10年，预计高炉炉缸炉底寿命可以达到15年。随着微孔炭砖、超微孔炭砖的相继问世，大块炭砖—综合炉底技术得到进一步发展，但采用此种结构的炉缸炉底须长期进行护炉操作。另一种值得关注的现象是高炉炉底和炉缸壁厚度都呈减薄趋势，个别大型高炉的炉底厚度已经减薄到2400mm，首钢首秦公司1号高炉（1200m³）炉缸采用热压炭砖，其炉缸壁厚度仅为800mm。

3.3 铜冷却壁

20世纪70年代末期，德国GHH公司和蒂森公司合作率先在高炉上应用了铜冷却壁，取得了令人满意的效果。高炉铜冷却壁具有高导热、抗热震、耐高热流冲击和长寿命等优越性能，越来越多地应用于国内外大型高炉的关键部位，为高炉高效长寿起到了重要的作用。

我国对铜冷却壁的研究始于20世纪90年代中期。广东汕头华兴冶金备件有限公司和首钢合作，于2000年1月设计研制出2块铜冷却壁，并安装在首钢2号高炉（1726m³）上试用，取得了显著的应用效果。2002年3月首钢2号高炉技术改造中，应用了该公司提供的120块铜冷却壁，这是我国高炉正式使用国产铜冷却壁，标志着铜冷却壁技术已经完全实现国产化。据不完全统计，目前我国已有20余座大型高炉采用了国产铜冷却壁[1]。

采用铜冷却壁的技术原理是依靠铜冷却壁优异的导热性、抗热震性和耐高热流冲击性，在其热面能够形成比较稳定的保护性渣皮。即使渣皮瞬间脱落，也能在其热面迅速地形成新的渣皮保护冷却壁，这种特性是其他常规冷却器所不能比拟的。实践证实，铜冷却壁是一种无过热冷却器，使用寿命可以达到20~30年。铜冷却壁在首钢、武钢、本钢、马钢、攀钢、湘钢等企业的大型高炉上已经得到了应用。

目前，我国已经研制出多种不同形式的铜冷却壁，有轧制铜板钻孔铜冷却壁、铜管铸造铜冷却壁、Ni-Cu合金管铸造铜冷却壁、铸造坯锻压钻孔铜冷却壁和连铸铜冷却壁等。轧制铜板钻孔铜冷却壁由于结构致密、组织缺陷少、冷却效率高，其应用范围最为广泛。

铜冷却壁是高炉长寿的关键技术之一，铜冷却壁的应用使高炉在不中修的条件下，寿命达到15~20年成为可能。铜冷却壁应使用在高炉热负荷最大的区域，即炉腹、炉腰和炉身下部，该区域是高炉异常破损严重且造成高炉短寿的关键部位，在此区域使用铜冷却壁对于延长高炉寿命具有重要作用。此外，在高炉炉缸（特别是铁口区）使用铜冷却壁也将会取得良好的应用效果。进一步优化铜冷却壁结构，降低造价是我国铜冷却壁技术发展的重要课题。

3.4 软水密闭循环冷却技术

高炉冷却系统对于高炉正常生产和长寿至关重要。20世纪80年代末期，我国高炉开始采用软水密闭循环冷却技术，经过不断地改进和完善，软水密闭循环冷却技术已日趋完善，并成为我国大型高炉冷却系统的主流发展模式。

软水密闭循环冷却技术使冷却水质得到极大改善，解决了冷却水管结垢的致命问题，为高效冷却器充分发挥作用提供了技术保障。该系统运行安全可靠，动力消耗低，补水量小，维护简便。

近年来，我国高炉软水密闭循环冷却技术进行了许多优化和改进：（1）根据冷却器的工作特点，分系统强化冷却，单独供水；（2）根据高炉不同部位的热负荷情况，在垂直方向上分段冷却，如炉缸、炉底设为一个冷却单元，炉腹、炉腰和炉身下部设为一个冷却单元；（3）为便于系统操作和检漏，采用圆周分区冷却方式，在高炉圆周方向分为4个冷却区间；（4）软水串联冷却，软水经炉底、冷却壁后，分流一部分升压再冷却风口、热风阀等。这种串联冷却系统具有占地省、投资低、动力消耗低的特点，在武钢1号高炉（2200m³）上已经得到应用。

3.5 薄壁内衬，砖壁一体化

高炉炉体破损机理的研究，使人们更加清楚地了解了高炉内衬和冷却器的工作条件；现代传热学理论的研究和运用，已将人们从传统的思维困惑中解脱出来，形成现代高炉长寿设计的基本理念，薄壁内衬技术就是在此条件下应运而生。所谓薄壁内衬就是对高炉内衬和冷却壁进行优化组合，形成砖壁一体化结构，解决炉腹、炉腰和炉身下部高热负荷区的短寿问题，使其寿命与高炉炉缸、炉底的寿命同步[2]。

我国已有数座大型高炉采用了砖壁一体化的薄壁内衬技术。冷却壁取消了凸台，消除了冷却壁破损最薄弱的部位，而且冷却壁热面全部采用耐火材料保护，即所谓全覆盖镶砖冷却壁。这种砖壁一体化的冷却壁是在第四代冷却壁的基础上优化演变而来的，其内衬厚度仅为150~250mm。大型高炉炉腹、炉腰、炉身下部采用铜冷却壁，炉身中部采用此种结构，炉身上部设2~3段C形光面水冷壁，这应是一种配置合理的长寿炉体结构。

3.6 耐火材料

我国大型高炉用新型耐火材料的开发与研究已取得显著进展。用于炉缸炉底的高导热半石墨炭砖、微孔炭砖已相继研制成功，并在大型高炉上使用。塑性相结合刚玉、微孔刚玉以及SIALON结合刚玉等新型陶瓷杯材料也陆续问世。SIALON结合刚玉、SIALON-SiC、高导热石墨砖、烧成微孔铝炭砖等一系列用于风口区以上的耐火材料也得到广泛应用。

3.7 自动化检测与控制

自动化检测是高炉长寿不可缺少的技术措施。炉缸炉底温度在线监测已成为监控炉缸炉底侵蚀状态的重要手段，也是建立炉缸炉底内衬侵蚀数学模型所必要的条件。炉腹、炉腰、炉身下部区域，温度、压力的检测为高炉操作者随时掌握炉况提供了有效的参考。通过对冷却水流量、温度、压力的检测，可以计算得出热流强度、热负荷等参数，而且还可以监控冷却系统的运行状况。炉喉固定测温、炉顶摄像、煤气在线自动分析、炉衬测厚等技术的应用使高炉长寿又得到了进一步的保障。我国宝钢、武钢、首钢、本钢、湘钢的大型高炉还引进了人工智能高炉冶炼专家系统，为延长高炉寿命创造了有利条件。

3.8 炉体维护技术

用含钛物料护炉，是由于在高温条件下还原生成TiC、TiN或Ti（C、N）等高熔点化合物，沉积在炉缸炉底，对其形成保护层。我国高炉已成功应用了含钛物料护炉技术，钒钛矿、含钛球团等护炉剂在高炉长寿实践中都取得了很好的效果，采用风口喷吹含钛物料、含钛精粉炮泥护炉也正在研究试验。应该指出，高炉炉役末期，采用含钛物料护炉是延长高炉寿命的主要技术措施，但由于采用炉缸炉底内衬结构不同，开始护炉的时间也存在差异。首钢3号高炉连续工作10年尚未进行护炉操作，这也从某种程度上证实了热压炭砖技术体系的合理性。

我国高炉炉体快速修补技术已经得到推广应用。炉衬遥控喷补、压浆等炉衬修补技术已成为现阶段延长高炉风口以上区域寿命的重要技术措施。微型冷却器、冷却壁水管再造等冷却壁修复技术也日渐成熟。

4 结语

我国大型高炉长寿技术已取得长足进步，已有10余座大型高炉寿命达到10年以上，正向15年迈进。新建或大修改造的高炉均采用了一系列高炉长寿综合技术，高炉设计寿命为15~20年。

（1）通过采用高炉综合长寿技术，我国部分大型高炉寿命已达到10年以上，正向国际先进水平迈进。

（2）高炉炉型的优化设计，为高炉高效长寿和高炉生产的稳定顺行奠定了良好基础。

（3）软水密闭循环冷却系统、合理炉缸内衬结构、铜冷却壁、薄壁内衬、自动化检测和炉体维护等高炉长寿综合技术是延长高炉寿命的重要技术保障。

（4）属于我国自行研制开发的高炉长寿原创技术仍显不足，这是我国炼铁工作者在21世纪所面临的新课题。

参考文献

[1] 张福明，姚轼. 铜冷却壁的设计研究与应用［C］//. 中国金属学会. 铜冷却壁技术研讨会论文集. 2003：39-42.
[2] 银汉. 现代薄壁高炉的特征［J］. 炼铁，2001（2）：7-10.

Present Situation and Development of Long Campaign Life Technologies of Large BF in China

Zhang Fuming　Dang Yuhua

Abstract: The present development situation of China large BF long campaign life technologies are narrated. In the design of large BF, the technologies like optimized BF profile, reasonable hearth lining, copper stave, soft water closed circulating cooling system and thin-walled lining etc. were applied to prolong BF campaign life. The BF campaign life can last over 15 years by applying the technologies like automation detection and control and proper maintenance etc.. In this thesis, the applications like hearth lining, copper stave, closed circulating soft water cooling system, thin-walled lining etc. are reviewed, also some suggestions for China large BF long campaign life technologies are brought out.

Keywords: blast furnace; long campaign life; design; copper stave

新型优质耐火材料在首钢高炉上的应用

摘 要：从1990年起，首钢4座高炉相继进行了重大技术改造。在高炉不同部位分别采用了Si_3N_4-SiC砖、热压小块炭砖、组合砖等新型耐火材料以及"陶瓷杯"技术。其中2、3、4号高炉采用了炭砖—高铝砖综合炉底，将NMA砖砌筑在炉底炉缸交界处"蒜头状"侵蚀区。2号高炉投产后炉底温度一直维持在100~200℃之间。1号高炉采用了"陶瓷杯"和热压炭块相互补充的炉衬结构。

关键词：高炉；耐火材料；长寿；设计

1 引言

从1990年起，首钢4座高炉相继进行了技术改造，高炉总容积由原来的4139 m^3扩大到8898 m^3。为使首钢高炉一代寿命（不中修）达到8~10年，采用了软水闭路循环冷却、第三代新型冷却壁、新型优质耐火材料、炉体监测控制等长寿技术。

2 耐火材料的选用

高炉炉衬的过早破损是导致高炉短寿的一个重要原因。首钢40多年的生产实践表明，高炉炉衬易损区域，一是炉底、炉缸；二是炉腹、炉腰和炉身下部。几座高炉大修改造设计中，在剖析上述两个区域炉衬破损机理的基础上，有针对性地选用了新型优质耐火材料。表1是首钢各高炉大修改造后的耐火材料使用情况。

表1 首钢4座高炉耐火材料使用情况

炉号	1	2	3	4
高炉有效容积/m^3	2536	1726	2536	2100
投产时间	正在施工	1991年5月15日	1993年6月2日	1992年5月15日
炉底、炉缸结构	陶瓷杯	综合炉底	综合炉底	综合炉底
炉底、炉缸	国产炭砖 NMA砖 莫来石砖 棕刚玉预注块	国产炭砖 NMA砖 高铝砖(1)	国产炭砖 NMA砖 高铝砖(1)	国产炭砖 NMA砖 高铝砖(1)
炉腹至炉身下部	NMD砖 $Si_3N_4 \cdot SiC$砖 高密度黏土砖(3)	$Si_3N_4 \cdot SiC$砖 高铝砖(2)	NMD砖 $Si_3N_4 \cdot SiC$砖 高密度黏土砖(3)	$Si_3N_4 \cdot SiC$砖 高密度黏土砖(3)

本文作者：张福明，刘兰菊。原刊于《炼铁》，1994，13（3）：22-25。

续表1

炉号	1	2	3	4
炉身中部	黏土砖(4)	黏土砖(4)	黏土砖(4)	黏土砖(4)
炉身上部	高铝砖(2)	高铝砖(2)	高铝砖(2)	高铝砖(2)
铁口组合砖	碳化硅-莫来石异形砖 高铝砖(1)	高铝异形砖(1)	碳化硅-莫来石异形砖 高铝砖(1)	莫来石异形砖 高铝砖(1)
风口组合砖	碳化硅-莫来石异形砖	高铝砖(1)	碳化硅-莫来石异形砖	莫来石异形砖

注：（1）$Al_2O_3 \geqslant 80\%$；（2）$Al_2O_3 \geqslant 65\%$；（3）ZGN-42；（4）GN-42。

2.1 炉底、炉缸区域

炉底炉缸区域的炭质耐火材料必须具有很高的抗渗透性、导热性、抗化学侵蚀性等，借助冷却壁的冷却作用，将1150℃等温线（铁水凝固线）尽量推向中心，在炭砖热面能够形成一层保护性的"渣皮"或"铁壳"，使炭砖免受渗透、化学、冲刷、热应力的破坏，以延长其使用寿命。其中最关键的就是要求炭砖具有优异的抗渗透性和高导热性。

美国UCAR公司开发的热压炭砖系列产品，其抗渗透性、热导率、抗碱性、机械强度等性能均能满足上述要求，全世界已有300多座高炉成功使用。继本钢5号高炉之后，首钢4座高炉都选择了热压炭砖。表2是UCAR公司NMA、NMD砖的理化性能。

表2 NMA、NMD热压炭砖理化性能

种类		NMA	NMD
体积密度/g·cm^{-3}		1.62	1.80
抗压强度/kPa		30500	31100
灰分/%		10	9.5
渗透性/mDa		9	8
导热系数 /W·(m·K)$^{-1}$	600℃	18.4	45.2
	800℃	18.8	38.1
	1000℃	19.3	32.2
	1200℃	19.7	28.5

NMA砖由于它的优良配方和特殊的成型工艺及构型，具有优异的导热性、极低的渗透性和优良的耐碱性，与传统的大型炭块相比，渗透率仅为后者的1/100~1/50，热导率高2~3倍[1,2]。另外，NMA砖形状较小，加上热固性C-34胶泥能提供膨胀补偿，从而能防止热应力断裂和收缩剥落；NMA砖的热导率高，可促使炉缸壁热面形成保护性"渣皮"或"铁壳"，防止气体、熔融渣铁、碱金属对炭砖的渗透、侵蚀和渣铁的冲刷磨蚀。为提高NMA砖的抗碱性，特意加入了SiO_2和石英材料，它们优先与碱发生化学反应生成安全而没有破坏性的化合物，从而缓解了由于化学侵蚀后体积膨胀对炭砖所产生的破坏。首钢的4座高炉都使用了NMA砖，将NMA砖砌筑在炉底炉缸交界处"蒜头状"侵蚀区。

2~4号高炉采用了炭砖-高铝砖综合炉底。2号高炉炉底炉缸砌筑结构如图1所示。综合炉底以上至铁口平面以下（即死铁层区域）全部采用NMA砖，前面砌一层高铝质保护砖，形成"简易陶瓷杯"结构，以避免开炉初期对NMA砖的各种破坏。高炉未使用NMA砖时，一般开炉9个月左右就要加入含钛炉料护炉，而2号高炉采用NMA砖之后，开炉18个月（1992年10月24日）才开始投入含钛炉料护炉，到1993年9月第二段冷却壁进出水温差平均0.3℃，75%冷却壁的热流强度为6970W/m²，炉底热电偶温度一直稳定在100~200℃之间，说明NMA砖的作用已经显示出来。

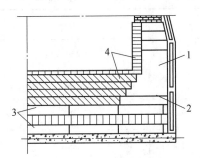

图1 2号高炉炉底炉缸砌筑结构
1—NMA砖；2—国产炭砖；3—3层满铺炭砖；4—高铝砖

2.2 炉腹至炉身下部区域

高炉护腹至炉身下部是目前高炉内衬最薄弱的环节，用于该区域的耐火材料应有较好的抗热震性、导热性和优良的抗化学侵蚀性、抗机械磨蚀性等。通过日本室兰3号高炉上的工业性对比试验发现：在高炉中部 Si_3N_4-SiC 砖的使用性能最为优异。我国从1985年在高炉上使用 Si_3N_4-SiC 砖以来，到目前已推广到20多座高炉上[3]。首钢的4座高炉在炉腹至炉身下部破损异常区，都使用了 Si_3N_4-SiC 砖。表3给出了 Si_3N_4-SiC 砖的理化性能。

表3 Si_3N_4-SiC 砖理化性能

项目	化学成分/%				物理性能					
	SiC	Si_3N_4	Fe_2O_3	f-Si	显气孔率/%	体积密度/g·cm⁻³	高温耐压强度/MPa	常温抗折强度/MPa	高温抗折强度/MPa(1400℃)	导热系数/W·(m·K)⁻¹	热膨胀系数/℃⁻¹(20~1500℃)
数值	>71	21~23	<2	<1	<18	<2.6	>150	>30	40	23	4.8×10⁻⁶

首钢高炉的炉腹至炉身下部是炉衬破损最严重的区域。Si_3N_4-SiC 砖具有良好的抗热震性、导热性、抗碱性和抗氧化性等特点，这种砖与合理的冷却形式和支撑结构配套使用，即可使炉内形成较稳定的渣皮来保护砌体，减缓其破损速度。为此采用了软水闭路循环冷却系统和第三代新型冷却壁，砖衬由冷却壁的凸台支撑。设计中采用了薄壁综合炉衬，紧贴冷却壁砌筑 Si_3N_4-SiC 砖，里层再砌一环高密度黏土砖（2号高炉用高铝砖），砖衬总厚度575mm。冷却壁的沟槽内采用SiC捣料填充。为防止 Si_3N_4-SiC 砖热膨胀造成的应力破损，在高度和圆周方向都预留了一定的膨胀缝，膨胀缝内填充耐热纤维片以吸收热

膨胀。高炉投产近3年来实践表明，该砖使用效果比较理想，与高铝质耐火材料相比，具有明显的优越性，但热震破损的问题仍没有完全解决。

在4号高炉大修时，炉腹有两块冷却壁的区域使用了UCAR公司推荐的NMD砖做工业性试验，生产4个月，温度稳定在70~150℃，NMD砖是热压半石墨制品，具有很高的热导率（比NMA还高）、极优的抗热震性和抗化学侵蚀性，适用于炉腹至炉身下部区域，表4给出了NMD砖和SiC砖性能对比。

表4 NMD砖和SiC砖的性能对比

材料		NMD砖	SiC砖
抗热冲击		最小250℃/min	最大50℃/min
热导率 /W·(m·K)$^{-1}$	600℃	45	21
	800℃	38	19
	1000℃	32	17
	1200℃	29	16
氧化临界温度/℃		800~900	800
抗碱侵蚀		卓越	优良

1992年在3号高炉设计中，炉腹至炉身下部紧贴第6~10段冷却壁砌筑了NMD砖，内层砌一环高密度黏土砖。在紧贴冷却壁处和冷却壁的沟槽内，填充了经过特殊处理的RP-20石墨质捣料，该捣料导热性好，易于填充，具有独特的膨胀性能。温度一旦超过200℃，捣料中石墨将发生破裂，使捣料的体积增加，这种"自动调节"作用为砖衬与冷却壁之间紧密接触提供了可靠的保证。3号高炉投产近半年来，此区域的使用情况一直较好，有希望在首钢的高炉上实现一代炉龄8~10年而不用中修的目标。

2.3 陶瓷杯技术在1号高炉上的应用

陶瓷杯的主要特点是利用低导热的陶瓷材料，将1150℃等温线阻滞在陶瓷层中，使炭砖避开800~1100℃的脆性断裂区，由于陶瓷杯的存在使铁水不直接与炭砖接触，从结构设计上缓解了铁水及碱金属对炭砖的渗透、冲刷破坏；同时所用的莫来石、棕刚玉等都是低导热的陶瓷材料，具有较高的抗渗透性和抗冲刷性[4,5]。陶瓷杯技术的优点在于：（1）防止铁水渗透；（2）减轻铁水环流冲刷（与合理的死铁层深度配合）；（3）提高铁水温度（18~25℃）和炉缸热稳定性。

首钢的设计、科研、生产、制造、施工等单位共同和法国专家进行了多次技术交流，经过论证，最后决定在1号高炉大修改造中应用陶瓷杯技术。与此同时吸取UCAR公司热压炭砖优点，将两种炉衬设计体系融为一体，互相补充。SAVOIE公司的设计经验认为，陶瓷材料和炭质材料的厚度之比应为1：(2~2.5)。1号高炉设计中采用了这一数据。图2是1号高炉陶瓷杯砌筑结构，表5是陶瓷杯材料的理化性能。

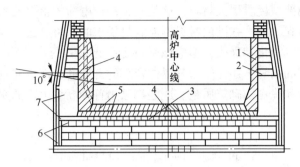

图 2　1 号高炉陶瓷杯砌筑结构

1—AMC66K 捣料；2—国产炭砖；3—国产莫来石刚玉砖；
4—棕刚玉预注块；5—莫来石 MS4；6—4 层国产炭砖；7—NMA 砖

表 5　陶瓷杯材料的理化性能

性能		莫来石 MS4	棕刚玉预注块
化学分析/%	Al_2O_3	70	88
	SiO_2		6
	TiO_2		3
	CaO		0.4
	Fe_2O_3	0.4	
	Na_2O+K_2O	0.6	
物理性能	体积密度/g·cm^{-3}	2.47	3.3
	气孔率/%	18.5	
	常温抗折强度/MPa	85	55
	0.2MPa 荷重软化点/℃	1650	
	永久线变化（1500℃，3h）/%	+0.5	1
	抗 CO 性/级	A	
	热膨胀系数/×10^{-6}·℃$^{-1}$	6.4	
	导热率/W·(m·K)$^{-1}$	2.2（1000℃）	4（1200℃）

2.4　风口、铁口组合砖的应用

高炉风口、铁口区过去一直采用常规高铝砖砌筑，使用寿命短。首钢几座高炉大修改造中都改为组合砖。投产以后该部位砖衬寿命明显提高。风口、铁口组合砖的材质是高级莫来石，为提高其热震稳定性，加入了一定量的 SiC，用振动成型工艺，使组合砖的外形尺寸能够满足设计要求。

3　结语

延长高炉寿命是一个系统工程，要从设计、制造、施工、生产、监测维护等方面入手，有的放矢地选用新型优质耐火材料，这是延长高炉寿命的重要技术措施。耐火材料的选用原则应以国内材料为主，在国内材料不能满足要求时，可有选择地引进国外有成功经验的耐火材料，并加以消化和吸收，但应该坚持走耐火材料国产化的道路。高炉设计中，在重视耐火材料选用的同时，还必须为其提供合理的有效的冷却，并且在炉腰、炉身区域确保砖衬的有效支撑和优良的砌筑质量。这些都已为首钢高炉生产的实践所证实。

参考文献

［1］Dzermejro A J. Blast furnace hearth design theory, materials and practice［J］. Iron and Steel Engineer, 1991, 68（12）: 23-30.
［2］Sjhinro J S, Barry J H. How the american hot-pressed-brick extruded-big-beam carbon hearth designs have eliminated salamander cutback and hazardous breakouts［J］. Steel Times, 1982, 210（7）: 363-368.
［3］伍积明. Si_3N_4 结合 SiC 砖的生产及其在炼铁工业上的应用［J］. 炼铁, 1993（1）: 47.
［4］Bauer J M. Ceramic cups to extend BF hearth life［J］. Steel Times, 1989（1）: 17-18.
［5］Bauer J M. Low conductivity blast furnace refractories［J］. Steel Times. 1993（1）: 49.

大型长寿高炉设计的探讨

摘 要：本文结合首钢高炉的大修改造和生产实践，分析探讨了高炉合理炉容与炉型、高炉冷却、耐火材料、炉体自动化检测等问题，提出了高炉设计中应注意的问题及大型长寿高炉的基本设计思想。分析表明，在条件允许的情况下，大型高炉炉容应设计大些，且确定合理的高径比，软水密闭循环冷却是高炉冷却技术的发展方向，高炉不同部位应采用不同的耐火材料，还要提高炉体自动化检测水平。

关键词：高炉；设计；寿命；冷却；耐火材料

1 引言

随着高炉的大型化和现代化，高炉寿命已成为重要的研究课题。长寿高炉应是炉龄达到 10 年以上而不用中修，一代炉役单位有效容积产铁量在 6000m³/t 以上。延长高炉寿命是一项系统工程，要从设计、制造、施工、操作、监测、维护等多方面考虑。本文仅从设计角度对这一问题进行分析探讨。

2 合理炉容与炉型

对于高炉生产全过程而言，其主要矛盾是煤气上升运动和炉料下降运动之间的矛盾[1]。高炉长寿是以高炉顺行为基础的，而合理的炉型设计对高炉顺行至关重要。

2.1 有效容积

有效容积的确定要考虑原燃料条件、操作条件、钢铁平衡等多方面的因素。新建或改建的大型高炉在确定炉容时要根据现有条件和今后发展情况而定。在条件允许情况下应将炉容设计得稍大一些，那种为追求个别先进指标而压小炉容是不可取的。因在相同的产铁量下，小炉容高炉的利用系数和冶炼强度均高于大炉容高炉，这样不但会增加高炉操作上的负担，且其顺行也会受到影响。

首钢 4 座高炉大修改造后，总容积由原来的 4139m³ 扩大到 8898m³，1 号、新 3 号和 4 号高炉有效容积都在 2000m³ 以上。

2.2 高径比（H_u/D）

目前高炉矮胖化的合理尺度尚无定论，但生产实践已证实了矮胖型高炉可提高冶炼强度且有利于高炉顺行。笔者统计了日本 15 座大型高炉的炉龄，其中寿命最短的 9 年零 8

本文作者：张福明。原刊于《首钢科技》，1994（5）：48-52，56。

个月,最长的13年零5个月(鹿岛3号高炉,H_u/D只有1.95,累计产铁$4.8×10^7$t,单位炉容产铁9535t/m³,也是目前世界上寿命最长的高炉)。这些高炉的高径比都在2.2以下,属矮胖型高炉。

矮胖型高炉的优点是煤气流速低,便于炉料和煤气的相对运动,有利于炉况稳定顺行和延长炉腹至炉身区域的寿命,矮胖型高炉的软熔带也相对较低,使熔渣对炉衬侵蚀区域相对减少。应该指出,矮胖型高炉只有同精料、炉料分布控制技术相结合,才能发挥其更大优势,进而达到延长高炉寿命的目的。

大型高炉有效高度可按下式验算[2]:

$$H_u = 5.6728 V_u^{0.2058} \tag{1}$$

不同级别高炉的高径比可参考下式:

$$H_u/D = 9.9803 V_u^{-0.1866} \tag{2}$$

2.3 死铁层深度

目前炉型设计中有加深死铁层的趋势。首钢新3号高炉(2536m³)将死铁层加深到2200mm,宝钢3号高炉(4350m³)将死铁层加深到2985mm。加深死铁层对于铁水流场及炉缸温度场的合理分布大有好处,它使炉缸中的死焦柱不直接坐在炉底而悬浮在铁水中,增加了铁水流通面积,铁水能顺利地流向铁口区,从而减少铁水环流对炉缸侧壁和炉底的冲刷。根据首钢的经验,死铁层深度为炉缸直径的20%左右较合适。加深死铁层后,炉底炉缸区应采用抗渗透性优异的耐火材料,以减少铁水的渗透侵蚀。

2.4 炉腹角和炉身角

目前高炉的炉腹角(α)与炉身角(β)趋于接近(α略小于β),且均有减小趋势。据首钢生产实践分析,α应在78°~81°之间,β不宜超过83°。减小α,一方面有利于炉缸煤气的顺利上升,另一方面也有利于在炉腹砖衬表面结上渣皮,减少对耐火材料的侵蚀,保护冷却设备,减小β,可使料柱对炉墙的侧压力和摩擦力减小,通过合理布料来抑制边缘煤气流的过分发展,减少热震对炉衬的破坏。α、β减小后须维持一定的炉腰高度,一般1800~3000mm较合理。

3 冷却系统

3.1 系统的选择

高炉冷却系统可分为工业水冷却、汽化冷却、软水密闭循环冷却三种,其中软水密闭循环冷却是高炉冷却的发展方向。

在冷却设备破损的诸因素中,冷却水管内壁结构造成导热效率下降、水管流通截面积减小、冷却强度下降,是冷却设备损坏的主要原因。而软水密闭循环冷却的最大优点即是可避免结垢。我国北方大部分地区的水质欠佳,硬度大,易结垢,因此新建或改建的大型高炉凡有条件的都应采用软水(或纯水)密闭循环冷却技术,当然也可采用经处理后的工业水冷却,以最大限度地降低水中CaO等物质的含量。北方一些钢铁厂的水质性能见图1和表1[3]。

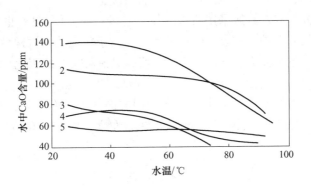

图1 各钢铁厂水质稳定性检验
1—邯钢；2—鞍钢；3—唐钢；4—首钢；5—本钢一铁

表1 冷却水中钙盐的变化　　　　　　　　　　　（mg/L）

项目	首钢	唐钢	鞍钢	本钢	邯钢
进水	69.75	75.28	118.83	80.94	141.20
出水	64.23	73.28	112.28	80.14	136.56

3.2 软水密闭循环冷却系统

在进行软水密闭循环冷却系统设计时，要确定水量、水压、水温、水温差、水速等参数，其中水速最为重要。一般对软水冷却系统，冷却壁的凸台管和前排管水速应设计为1.5~2.0m/s，后排管1.5m/s，软水进出水温差可控制在10℃以内；冷却水量要根据热负荷来确定，同时须满足水速的要求。国内近几年投产的大型高炉已有10余座采用了软水冷却技术。从使用情况看，软水冷却技术一方面可节水节能，另一方面可提高冷却效果。但也存在一些问题：(1) 管路相互串联，水量分布不均，水管烧坏后问题更为突出；(2) 高炉破损异常区（炉身下部）的冷却壁水管往往最先烧坏，为防止向炉内大量漏水而切断该路软水，则导致冷却效果下降，且目前软水检漏困难很多；(3) 国内设计的软水系统在高炉高度方向上都是从下向上串联，与热负荷分布不对应。这些问题可在今后使用过程中逐步解决。今后的软水系统设计，应分炉底炉缸、炉腹至炉身下部和炉身中上部三个区域冷却。国外高炉已采用这种冷却形式。软水回水支管和总管上都要设计排气装置，脱气罐、膨胀罐、稳压罐的设计要合理，以防冷却水管中形成气栓。

3.3 炉体冷却结构

炉体冷却结构目前有三种：(1) 冷却壁式，砖衬依靠冷却壁的凸台支撑；(2) 冷却板式；(3) 板壁结合式。对于砖衬来说，合理的冷却和有效的支撑至关重要。从国际上看，冷却壁的发展十分迅速，日本已发展到第四代，德国正致力于铜质冷却壁的开发和应用。就我国而言，可视具体情况选择冷却壁式或板壁结合式，如能解决砖衬的支撑问题，则冷却壁式最可行。首钢近几年改造的4座高炉全部采用冷却壁式，宝钢3号高炉也采用冷却壁式，使用区域从炉底一直到炉喉钢砖下端。冷却板式因铜质冷却板造价昂贵，炉壳开孔过多，强度下降，不利于高压操作，故不宜继续推广。

国外一些大型高炉的炉缸区采用喷水冷却，宝钢1号、2号高炉也采用喷水冷却。喷水冷却深度不够，只限于炉壳和硬质炉衬的冷却，对于陶瓷质内衬材料的冷却作用则显不足。另外喷水冷却水量消耗大，增加了冷却水净化处理等负担。因此，笔者认为炉缸不宜采用喷水冷却方式。

大型高炉炉底必须通水冷却。由于炉底圆周方向热流分布不均，炉底还应分区域冷却。炉底周边部位（象脚状侵蚀区）要加大冷却强度，中心部位则应保证适宜的冷却强度。

3.4 冷却设备

冷却壁今后将是高炉的主要冷却设备。对于铸铁冷却壁，应采用以铁素体为基的球墨铸铁。球墨铸铁的机械性能（如延伸率）优于其他材质，同时可消除相变时 Fe_3C 的分解开裂。不同材质的冷却壁性能见表2。有条件的厂家也可开发试验铜、铸钢等材质的冷却壁。

表2 不同材质的冷却壁主要性能对比

项 目	灰铸铁 HT15-33	低铬铸铁		高韧性球墨铸铁	
		FCH	GGL	FCD	GGG
抗拉强度 σ_b/kg·mm^{-2}	15	16	16	40	40
延伸率 σ_a/%	0	0	0	20	20
龟裂前热循环次数（300~800℃）	30~40	30~40	30~40	203~250	203~260
熔点温度/℃	约1150	约1150	约1150	约1150	约1150
导热系数 λ/W·(m·K)$^{-1}$	62.8	40.7~46.5	40.7~46.5	29.1~34.9	34.9~40.7

冷却壁内水管的布置要合理，冷却要均匀，不能存在冷却死区。根据首钢的经验，水管内径在30~48mm较合理，不宜过大；水管中心间距200~230mm；壁体不宜过厚，炉缸区壁厚100~150mm，炉腹至炉身下部应使用双排管冷却壁，壁厚300~350mm，炉身中上部壁厚200~260mm较合理，壁体越厚冷却效果越差，且加大了铸造难度，不易保证延伸率达到15%~20%，壁长应在1500~2000mm之间，宽度可根据每块冷却壁的水管（前排管）中心距确定，一般为800~1200mm；前排水管中心距热面的距离，镶砖冷却壁应为150mm（包括75mm的镶砖层），光面冷却壁应为65mm左右，此间距不宜过大；冷却壁凸台应对砖衬提供有效的支撑，目前普遍采用"Γ"形和鼻形两种，国外已开发出"C"形（双凸台）和下凸台型冷却壁，托砖效果很好。

镶砖冷却壁的沟槽不宜过大，高度75mm较合适；镶砖也不宜过厚，75mm即可。为提高镶砖的稳定性，可采用"燕尾"形结构。镶砖面积占冷却壁热面面积的40%最佳，不宜超过50%。镶砖方式目前有热铸法和冷镶法两种。根据首钢的生产实践，后者比前者效果好。镶砖用半石墨-碳化硅、半石墨炭砖等含碳导热型耐火材料为宜，炉身中上部可采用抗氧化性好的材料，冷却壁的沟槽也可采用碳化硅或半石墨质的捣料镶填。首钢新3号高炉大修时，炉腹至炉身下部冷却壁的沟槽内镶填了RP-20捣料。这是一种经过特殊处理的石墨质捣料，导热性好，易于填充，具有独特的膨胀性，其热膨胀可以弥补由于填充不实或其他原因造成的砖衬和冷却壁之间的空隙。当温度超过200℃时，捣料中经过特殊处理的石墨将发生破裂，从而使捣料体积增加，这种"自动调节"为砖衬与冷却壁紧密接触提供了可靠的保证。

对于采用板壁结合式的高炉，为节约投资，可用钢质冷却板，而且应为可更换式。为便于更换和检修，冷却板可用工业水冷却，冷却壁用软水冷却。

4 耐火材料的选择

高炉炉衬易损区域主要有两个：一是炉底、炉缸区域；一是炉腹、炉腰和炉身下部区域。这两个区域的工作条件十分恶劣，各种破损因素共同作用，造成了耐火材料的过早破损。

4.1 炉底、炉缸区

据生产实践分析，炉底炉缸产生象脚状侵蚀的原因主要有：（1）高温铁水及碱金属对炭砖的渗透侵蚀；（2）出铁时铁水环流对炭砖的机械冲刷；（3）高温热负荷造成的热应力破坏；（4）铁水熔蚀、氧化气体对炭砖的化学侵蚀等。

用于该区域的炭砖必须具有很强的抗渗透性、高导热性和优良的抗化学侵蚀性等，借助冷却壁的冷却作用将1150℃等温线（铁水凝固线）尽量推向中心，远离炉壳。

首钢4座大型高炉炉底炉缸交界处（象脚状侵蚀区）均采用美国联合碳化物公司（UCAR）的NMA热压小块炭砖。NMA具有优异的导热性、极低的渗透性和优良的抗碱性，比传统大块炭砖渗透率低50~100倍，热导率高2~3倍。首钢1号高炉采用NMA砖的同时还率先在国内采用了陶瓷杯技术。陶瓷杯由法国沙佛埃（Savoie）耐火材料公司提供。热压炭砖和陶瓷杯材料的理化性能见表3和表4。

表3 UCAR公司热压炭砖理化性能

项　目	假比重/t·m^{-3}	抗压强度/kPa	断裂模量/kPa	灰分/%	渗透性/mDa	硬度 R	导热系数/W·(m·K)$^{-1}$			
							600℃	800℃	1000℃	1200℃
NMA	1.62	30500	8100	10	9	93	18.4	18.3	19.3	19.7
NMD	1.60	31100	10100	9.5	8	85	45.2	38.1	32.2	28.5

表4 陶瓷杯材料的理化性能

种类	化学成分/%						物理性能							
	Al_2O_3	SiO_2	TiO_2	CaO	Fe_2O_3	Na_3+K_4O	体积密度/g·cm^{-3}	气孔率/%	常温抗折强度/MPa	0.2MPa荷重软化点/℃	1500℃ 3h后的永久线变化/%	抗CO性/级	热膨胀系数/℃$^{-1}$	导热系数/W·(m·K)$^{-1}$
莫来石 M_s	70	—	—	0.4	0.6	—	2.47	18.5	85	1650	+0.5	A	6.4×10^{-6}	2.2（1000℃）
棕刚玉 MONO-CORAL	88	6	3	0.4	—	—	3.3	—	55	—	1	—	—	4（1200℃）

大型高炉炉底炉缸宜采用综合结构，陶瓷质耐火材料与碳质耐火材料的厚度之比在1∶2~2.5之间较合适。炉底不宜过厚，设置炉底冷却装置后，厚度3000mm左右即可，首钢1号、新3号高炉的炉底厚度均为2800mm。炉缸侧壁应设计350~400mm的陶瓷质保护层，以减缓炉缸的过早破损。UCAR公司认为，如采用NMA砖，炉缸壁厚度有800mm即可，过厚不利于导热，使用大型炭砖时，更易造成温度梯度过大，而导致热应力断裂。

炉缸铁口区的设计十分重要，有资料表明，铁口深度应保持在炉缸半径的45%以上[4]，这样可减少铁水流动对铁口周围炉缸侧壁的冲刷（铁口炉缸侧壁也应适当加厚）。为延长铁口使用寿命，应采取角度为8°~15°的斜铁口，铁口直径在40~60mm之间为宜，出铁速度要控制在4~6t/min以内。大型高炉的风口、铁口应采用组合砖结构，风口可用碳化硅-莫来石等材料，铁口组合砖应与炉缸内衬材质一致，可采用陶瓷、碳化硅和碳质材料，并与炮泥材质配合使用。

4.2 炉腹至炉身下部

目前炉腹至炉身下部炉衬是最薄弱的环节。这个区域的破损机理十分复杂，众说不一，但就首钢来说，主要有以下原因：(1) 温度波动造成的热震应力对炉衬的破坏；(2) 高温热负荷的冲击；(3) 碱性物质、氧化性气体、CO气体及液态渣铁对炉衬的综合化学侵蚀；(4) 因上升煤气流与下降炉料对炉衬的机械冲刷而造成的磨损及崩料、坐料对炉衬的冲击等。

由于上述原因，用于该区域的耐火材料应有较好的抗热震性、导热性和优良的抗化学侵蚀性、抗机械磨损性等。首钢2号、4号高炉在此区域采用了Si_3N_4-SiC砖，1号、新3号高炉采用了NMD热压小块炭砖。这两种砖使用效果都比较好，只是Si_3N_4-SiC砖未能完全解决热震破损问题。

该区域应采用薄壁炉衬，砖衬厚度575~690mm即可。砖衬和冷却壁之间要紧密接触，使用Si_3N_4-SiC砖在圆周方向上要留有一定的膨胀缝以吸收热膨胀，膨胀缝内填充耐火纤维毡，Si_3N_4-SiC砖前面再砌一环黏土质保护砖。

Si_3N_4-SiC砖和NMD砖价格昂贵，并非所有厂家都能承受。最近国内开发了铝碳砖，完全可以推广应用到1000m³级的高炉上，高密度铝炭砖和焙烧铝炭砖比普通铝炭砖的性能好，应优先选用。根据首钢经验，炉腹至炉身下部区域应以导热型耐火材料为主，采用综合炉衬，紧贴冷却壁砌筑，借助冷却壁的冷却作用，在砖衬热面形成一层稳定的保护性渣皮，以减少砖衬和冷却设备的破损。

4.3 炉身中上部

该区域炉衬破损程度相对较轻，采用高铝砖和黏土砖即可，如选用含碳耐火材料，需具备优良的抗氧化性。炉身上部可不砌砖，直接用冷却壁工作，冷却壁应一直到炉喉钢砖下端。

以上是延长高炉寿命的几项主要措施。另外，提高炉体自动化检测水平也很重要，它包括砖衬的测温、测厚，冷却壁测温，冷却壁进出水温、水压、水量、水速的在线检测及软水检漏等项技术。综合运用这些技术，进行合理的高炉设计，才能为高炉长寿提供可靠的保证。

5 结语

延长高炉寿命是目前我国炼铁生产急需攻关的重要课题。影响高炉寿命的原因是多方面的，包括炉容及炉型的设计、冷却系统的选择、耐火材料的选用以及炉体自动化检测水平等。进行大型高炉设计时应综合考虑这些因素，从多方面入手，为高炉顺行、高产、长寿创造必要的条件。

参考文献

[1] 安朝俊. 首钢炼铁三十年 [M]. 北京：首都钢铁公司，1983：1.
[2] 高清志，石立平，高竞. 高炉内型设计与讨论 [J]. 炼铁，1990（4）：10-18.
[3] 顾飞，等. 高炉冷却水水质初探 [C]. 炼铁学术年会论文集，1993：344.
[4] 柴田耕一郎，等. 高炉炉缸铁水流动和控制 [J]. 国外钢铁，1992（10）：14.

延长大型高炉炉缸寿命的认识与方法

摘　要： 高炉炉缸是高炉的重要部位。近年来，许多大型高炉投产以后，频繁出现炉缸侧壁温度升高、局部过热、异常侵蚀等问题，甚至一些高炉还发生了炉缸烧穿事故。在当前原燃料条件和操作条件下，许多高炉寿命不足10年就被迫进行了炉缸大修。本文结合高炉生产实践和高炉破损调查情况，对炉缸侧壁异常侵蚀进行了分析研究，得出造成炉缸异常侵蚀的主要原因及其侵蚀机理，提出了延长高炉炉缸寿命的有效技术措施和途径。
关键词： 大型高炉；长寿；炉缸；内衬；铁水环流

1　引言

延长高炉寿命是现代高炉炼铁的一个重要的目标，也是炼铁技术的重要发展方向[1,2]。进入新世纪以来，我国约有120座容积2000m^3以上的大型高炉相继建成投产，这批高炉中包括8座5000m^3级特大型高炉和17座4000m^3级大型高炉。经过近20年来的生产实践，这批新世纪建成投产的高炉中，不少高炉寿命未能达到预期指标，有的甚至是投产不久就提前退役、被迫大修，还有一些高炉炉缸侧壁温度居高不降、局部过热，投产之后炉缸侧壁长期处于"高温运行"状态，炉缸烧穿事故时有发生[3~9]。

2　高炉炉缸功能与结构

炉缸是高炉的核心关键部位，是高炉冶炼进程的起始和终结。焦炭、煤粉等燃料在风口回旋区的燃烧，液态渣铁的聚集和贮存，铁及非铁元素的直接还原、铁水脱硫等一系列化学反应，液态渣铁在炉内分离及排放等都是在炉缸之中进行，因而炉缸在高炉冶炼过程中具有多重功能[10]。从功能解析的角度分析，高炉炉缸不仅是高炉贮存液态渣铁的熔池，还是硅、锰等元素直接还原以及炉渣脱硫、渣-铁界面反应的反应器，也是熔融渣铁的成分调节器，同时还是碳素燃烧反应、生成炉缸煤气的煤气发生器等。由此可见，高炉炉缸是一个集成多重功能于一体的复杂系统，是一个具有"自组织功能"的复杂结构[11]。典型的高炉炉缸结构如图1所示。

现代高炉炉缸是由炉壳钢板、冷却壁、炭砖等多层复合结构所构成，这种高炉结构是经过近200年的进化演变形成的，是高炉结构和功能不断发展演进的结果，图2是高炉炉缸传热过程结构解析。为了实现炉缸的多功能化，现代高炉炉缸设计结构的基本要素是：

（1）采用高强度、高性能的细晶粒炉壳钢板。结合现代高炉复杂的工况条件，对炉壳钢板的机械性能和热态力学性能提出了更高的技术要求。随着现代大型高炉炉内压力的

本文作者：张福明。原刊于《炼铁》，2019，38（6）：13-18。

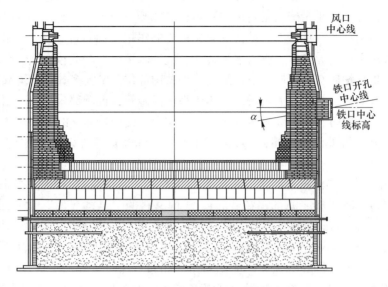

图 1 典型的高炉炉缸结构

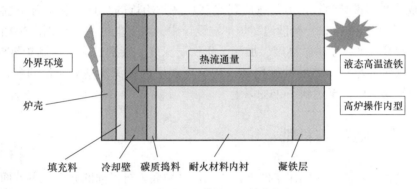

图 2 高炉炉缸传热过程结构解析

提高和运行工况的复杂多样,对钢板的冲击韧性、动态撕裂性、断裂韧性、应变时效敏感性、Z 向性能以及热疲劳性能等要求更加严苛。以 BB502、BB503、AG50、WSM50C、SM50B 等为代表的低合金高强度钢板是现代高炉炉壳常选材料。

(2) 采用冷却壁及纯水/软水密闭循环冷却系统。炉缸冷却经历了漫长的发展进程,从最初的炉壳喷水冷却,到夹套式冷却,再到冷却壁冷却,不同的炉缸冷却方式都有各自优势和缺陷。总体来看,进入新世纪以后,炉缸冷却采用冷却壁结构已成为主流模式,究其根本原因主要是为了节水,只有采用冷却壁才能实现闭路循环冷却,从而更加有效地调控冷却参数、降低水量消耗。因此,冷却壁+纯水/软水密闭循环冷却系统,已成为新世纪高炉炉缸冷却的显著特征[12]。近年来,为了强化冷却壁的冷却效果,一些高炉还在炉缸区域采用了铜冷却壁,其目的是利用铜冷却壁的高导热性,为炉缸炭砖提供更加可靠的冷却。炉底满铺炭砖之下,现代高炉均设置了冷却水管并联的水冷炉底结构,这是延长高炉炉底寿命不可或缺的重要保障。

（3）炉缸内衬采用高导热、高性能的优质炭砖。从20世纪中叶以来，近70年高炉生产实践表明，迄今为止炭砖仍是高炉炉缸内衬的重要功能材料、结构材料和首选材料，在高炉炉缸内衬结构中，具有不可替代的至关重要的作用。2000年以来，我国2000m³以上大型高炉普遍采用了导热性、抗铁水渗透性、抗铁水熔蚀性、抗碱侵蚀性等综合性能优良的炭砖，在炉缸炉底交界区域以及炉缸侧壁等关键部位，绝大多数高炉均采用了由国外引进的微孔大块炭砖、超微孔大块炭砖、热压小块炭砖等，也有部分高炉采用了全部国产的微孔炭砖。与此同时，在炉缸炭砖内衬设计结构上，遵循传热学的基本理念，以控制炉缸温度场合理分布为关注点，炭砖+综合炉底、炭砖+陶瓷杯等多种炉缸内衬结构均有应用。

（4）采用先进的界面技术和优质不定型耐火材料。在炉壳与冷却壁之间、冷却壁与炭砖之间、炭砖与陶瓷杯之间都存在着异质-异构材料的"结构界面"。炉缸结构设计优化的重要环节，就是合理处理好各种材料界面，使之成为一个有机的整体，使异质-异构的结构单元实现协同-匹配、耦合-集成。设计过程中需要运用传热学、热力学、材料力学和弹塑性力学的知识进行解析优化，优化界面"衔接、镶嵌"结构，采用优质不定型耐火材料（如碳素捣打料、自流浇注料、高强度碳质胶泥等）作为结构界面的功能材料和连接材料，是实现炉缸整体结构安全可靠的基本保障。

（5）优化炉型设计参数。为了适应现代高炉高效冶炼的特点，基于对炉缸冶金过程"三传一反"的解析和认识，新世纪高炉炉缸设计中，与上世纪的同级别高炉相比，适度扩大了炉缸直径、增加了炉缸高度、加深了死铁层深度。特别是为了抑制炉缸铁水环流，减少出铁时铁水环流对炭砖的冲刷侵蚀，高炉死铁层设计深度达到炉缸直径20%左右，有的高炉死铁层深度甚至达到了25%以上。

3 炉缸内衬侵蚀特征与机理

在当前高炉原料条件和操作条件下，造成炉缸炉底内衬侵蚀的原因是多方面的，不同的高炉也不尽相同、差异很大。从近年高炉炉缸破损调查情况分析，除了传统的侵蚀破损原因以外，还有一些新的侵蚀破损因素值得重点关注。

（1）炉缸内衬呈现"苹果状侵蚀"。当前，炉缸内衬侵蚀的形貌有所变异，已从传统的"象脚状侵蚀""锅底状侵蚀"，演变为"苹果状侵蚀"或"宽脸状侵蚀"。其侵蚀特征是：炉缸侧壁出现非均匀局部破损，炉缸异常侵蚀的最严重的部位出现在铁口中心线以下1.0~2.5m，铁口左右两侧0.5~2.0m的区域，呈现凹陷（凹坑）型异常侵蚀轮廓，非铁口区的炉缸壁侵蚀程度则相对较轻，传统的炉缸炉底交界处的"象脚状侵蚀"并不显著，炉缸侧壁异常侵蚀最严重的部位，已经从炉缸炉底交界处上移到死铁层中部的铁口下方两侧区域。从炉缸侵蚀剖面上观察，就如同一个苹果的外形。造成这种异常侵蚀形状的根本原因是炉缸死焦柱过于肥大、空隙度过低，小粒度碎焦和未燃煤粉堵塞柱体，造成渣铁滞流，炉缸透气性和透液性恶化。出铁时铁水环流加剧，从死焦柱下部的死铁层穿流而上的铁水流，在死焦柱根部向铁口区汇流，在铁口两侧炉缸侧壁形成涡流，铁水流黏性剪切力造成对铁口下方炉缸侧壁的冲刷磨蚀。沉浸在死铁层中的死焦柱由于根部肥大且透液性恶化，造成铁口区下方炉缸断面上的有效流动空间减少，在出铁强度高、出铁速度大时，直接导致在铁口区下方的两侧区域，铁水环流对炉缸侧壁造成严重冲刷磨蚀。即使在加钛补炉的条件下，由于铁水环流的扰动和连续不断的流动冲刷，极不利于在炉缸内衬破损严重区域形成相对稳定的凝铁层或 Ti（C，N）沉积保护层。而炉缸侧壁异常侵蚀最严重的部位，一般对应死焦柱根部沉浸在死铁层的位置。这是造成许多高炉尽管长期高强度加钛护炉，但炉缸侧壁温度仍然始终居高不降，最终被迫停炉大修的重要原因之一。

（2）炭砖选用不合理。2000 年以来，不少高炉炉缸炉底采用了大量高导热石墨炭砖。实践表明，炉缸炉底内衬与铁水直接接触的区域，或是炉役末期有可能与铁水接触的区域，都不宜选用石墨砖和石墨含量高的炭砖。石墨砖或石墨含量高的炭砖导热性高，但抗铁水熔蚀性能差，容易发生炭砖熔损，尤其是在铁水中 [C] 不饱和的情况下，更容易造成铁水对石墨炭砖的熔蚀。高炉设计时不仅要重视炭砖的热导性，也要重视炭砖的抗铁水渗透性和抗铁水熔蚀性，还要注重考查炭砖的气孔孔径、气孔率、透气度和气孔特性等综合指标。

（3）死铁层深度设计不合理。近期设计的高炉死铁层呈加深趋势，合理的死铁层深度可以有效缓解炉缸铁水环流的冲刷侵蚀，但过深的死铁层造成炉缸炉底承受更高的铁水静压力，铁水渗透、熔蚀的发生几率也会随之加大。在炉缸死焦柱呆滞、根部肥大、透气性和透液性恶化，死焦柱形状及位置难于控制的情况下，过深的死铁层实际上并不能改善炉缸流场分布，反而适得其反，会造成炉缸侧壁异常侵蚀位置上移。反之，过浅的死铁层使死焦柱直接沉坐在炉底上，同样会加剧铁水环流对炉缸侧壁的冲刷侵蚀。

（4）炉缸冷却结构设计与配置不合理。实践表明，炉缸区域的热负荷波动相对平稳，用于炉缸区域的冷却壁，其主要功能是为炉缸内衬提供足够的冷却，控制 1150℃ 等温线及温度场的合理分布，无论何种材质和结构的炉缸冷却壁，都不可能具备直接抵抗高温铁水的功能。因此，用于高炉炉缸的冷却壁与炉腹至炉身下部的冷却壁，其结构、功能和性能要求都不尽相同。炉缸冷却壁必须保持合理的冷却强度，使炭砖传递出来的热量能够通过与冷却水的热交换并顺畅导出，这是保障炉缸传热过程运行的基础。为了强化炉缸冷却，不少高炉在炉缸局部区域采用了铜冷却壁，但对铜冷却壁的设计结构、安装方式研究不够深入，其使用结果未达到预期效果。除此之外，铁口区冷却结构设计不合理，炉缸冷却壁与炉壳之间填料选用不当，炭砖与冷却壁之间的碳质捣料与炭砖的热导系数不匹配，冷却结构不合理等都会引发炉缸烧穿事故。

（5）高炉冶炼强化水平与炉缸炉底的可靠性和耐久性不匹配。2000 年以来，我国钢铁工业发展迅猛，钢铁产量持续攀升。一些钢铁企业以规模经济效益为中心，低水平粗放型发展。在高炉生产管理理念上，违背高炉炼铁的客观规律，盲目追求高产超产、高利用系数。新高炉投产后，片面追求所谓的快速达产、快速超产，以高产作为主要目标。在这种错误观念的主导下，不少高炉超出设备能力强化冶炼、超负荷生产，甚至以焦比和高炉寿命为代价，高炉投产 2~3 年就出现炉缸烧穿，代价巨大、教训惨痛。统计研究表明，国内外 50 余座长寿高炉的一代炉役期内的平均利用系数为 $2.0 \sim 2.3 t/(m^3 \cdot d)$，而出现炉缸烧穿高炉的利用系数大多数在 $2.5 t/(m^3 \cdot d)$ 以上，由此可见过高产量、超高利用系数是造成高炉短寿的"杀手"之一。

（6）高炉操作维护存在较大不足。1）为降低原燃料采购成本，违背高炉炼铁精料方针，大量使用品质差、有害元素含量高、价格低的低档原燃料。由于原燃料条件变化，造成钾、钠、铅、锌等有害元素在高炉内循环富集，与耐火材料发生化学反应生成化合物，使其体积膨胀，造成炉缸炉底内衬快速损坏。2）出铁制度不合理。有的大型高炉长期对侧出铁，造成炉缸流场偏析严重，铁水环流冲刷剧烈。尤其是炉缸直径13m以上多铁口的大型高炉，不采用交错轮流出铁的模式，而是长期采用固定对角的铁口出铁，形成"两用两备"的出铁模式，单铁口出铁负荷达到5000~6000t/d甚至更高，这样造成炉缸流场严重偏析，铁口区炉缸侧壁冲刷侵蚀异常严重，这是许多高炉炉缸侧壁破损的重要原因之一。3）铁口深度不足和出铁时铁口喷溅，铁水极易从铁口通道渗入砖缝，加速炭砖侵蚀；同时高温煤气也穿透到炭砖缝隙中，形成局部热点。加之盲目强化高炉冶炼，导致炉体破损加剧。4）风口等炉体冷却设备漏水，沿着炉壳、冷却壁渗漏到炉缸炉底，引起炭砖氧化、粉化，造成炉缸炭砖"非接触性"破损。5）含钛物料护炉加入量不够，错过最佳补炉时机且"时加时停"，对已经侵蚀的内衬修补不及时，不能形成稳定的保护性再生炉衬，炉缸修补效果不理想。6）炉缸压浆维护操作不当，压浆压力过高，泥浆的材质不合理，将已经很薄的残余砖衬压碎，或使泥浆从砖缝中压入炉内与高温铁水接触，出现不良后果，进而诱发炭砖渗铁或炉缸烧穿事故。

（7）炉缸炉底温度和热流强度在线监测措施缺乏。炉缸炉底内衬温度测量点少，热电偶测温点的设置也不尽合理；缺乏对冷却壁进出水温差、水流量、热流强度等参数的实时监测，造成不能及时准确发现炉缸炉底的异常状况，及时采取相应措施，往往是造成高炉炉缸烧穿事故的突发的原因。

4 延长炉缸寿命的技术措施

目前大型高炉普遍存在的炉缸内衬异常侵蚀，已成为炼铁行业共性技术问题。出现这种共性问题的原因是多方面的，既有普遍性，也有差异性。必须结合自身情况，有的放矢、因地制宜采取有效措施，加强在役高炉炉缸监测与维护。建立高炉一代炉役炉缸全生命周期操作维护理念[13]，对于延长高炉寿命至关重要（见图3）。

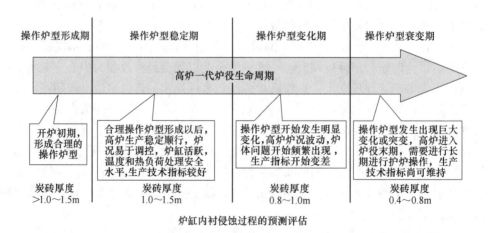

图3 高炉一代炉役炉缸全生命周期管控模型

4.1 优化炉缸结构设计

(1) 采用综合性能优异的炭砖。迄今为止，炭砖是构成高炉炉缸不可或缺的重要功能材料和结构材料，炭砖优异的综合理化性能是其他各类耐火材料所不能达到的。因此，选择综合性能优异的炭砖，是保障高炉长寿的基础，炭砖的品质、质量、性能及其稳定性，对高炉炉缸寿命的影响是至关重要和不可弥补的。2000年以来投产的许多大型高炉采用国外进口炭砖，如美国UCAR公司的热压小块炭砖，日本NDK公司的微孔大块炭砖和德国SGL公司的微孔大块炭砖。尽管如此，仍有多座高炉因炉缸侵蚀严重而被迫提前停炉大修，甚至发生了高炉炉缸烧穿事故，由此可见炭砖质量性能的稳定性尤为重要。

(2) 采用合理炉缸内衬结构。组合型的炉身质-异构炉缸炭砖的设计结构将可能在未来得到更广泛的应用。其结构特点是将高导热小块石墨质炭砖紧贴冷却壁砌筑，将其作为导热材料使用，其热面使用导热性、抗铁水渗透性和熔蚀性、抗碱性优异的大块微孔炭砖，形成异质-异构的组合型炭砖结构，采用优质碳质胶泥，减少炭砖砖缝。在炉缸内衬的砌筑结构上，避免炭砖砌体的横向贯通砖缝，防止铁水和碱金属的渗透侵蚀。与此同时，在铁口区及炉缸炉底交界部位，应适当增加炉缸侧壁炭砖厚度，以延长炭砖使用寿命。

(3) 强化炉缸冷却，提高冷却壁传热的比表面积，改善冷却壁冷却效率。在炉缸冷却壁材质、冷却水管直径、水管布置方式、壁体结构及厚度等方面需要结合高炉实际情况，进行认真研究和设计优化，从炉缸冷却壁的工作机制和所承受的最高热负荷分析，冷却壁的根本作用是为炭砖提供有效冷却，一旦炭砖侵蚀消失，冷却壁根本无法抵御高温铁水的熔蚀。因此炉缸冷却壁和炭砖是一对"生命共同体"，假如炭砖消失，冷却壁将直接面临铁水烧穿的危机。炉缸冷却壁结构宜采用冷却水管"四进四出"或"六进六出"的结构，均匀布置冷却水管，消除壁体冷却死区，提高冷却壁传热面积比和冷却效率，减少炉壳开孔数量，提高炉壳强度。因此，合理的冷却对于控制炭砖侵蚀和炉缸维护是不可或缺的技术保障。宝钢4号高炉大修时，将炉缸铜冷却壁改为铸铁冷却壁，其根本的出发点就是基于上述理念，而且炉役后期还可以采用高压水强化炉缸冷却。

(4) 进一步优化炉缸炉底结构设计。根据高炉生产的原燃料条件和操作条件，确定合理的高炉内型。特别是炉缸直径、炉缸高度、死铁层深度、炉腹角、炉身角等重要参数，要结合以往的高炉生产实践，认真对比分析、合理权衡选择，不可盲目照搬照抄。炉缸侧壁厚度、铁口区侧壁厚度、炉底厚度、铁口深度、铁口夹角、炭砖与陶瓷材料的结构匹配、炭砖与冷却壁的结构匹配等问题，在高炉顶层设计中都要充分研究，进行多方案的优化比选。炉缸区域炉壳宜采用带倾角的设计结构，以增强炭砖和冷却壁之间的摩擦力，提高砖衬砌体的结构稳定性，防止炭砖上浮或铁水钻透。

4.2 改善高炉操作

(1) 建立高炉维护期"安全、稳定、顺行"的生产管理理念。高炉炉役末期"带病运行"，存在较大的安全风险，必须高度重视炉缸内衬工作状况及其变化，树立"安全第一"的风险防范意识。与此同时，还必须重视高炉顺行，高炉顺行是高炉生产稳定的基础，也是规避异常风险、突发事故的保障，没有高炉顺行，就失去了高炉安全稳定运行的基础。

（2）提高鼓风动能，控制合理风速，适当加长风口长度，活跃炉缸工作，提高炉缸热量储备，控制死焦柱的形态和位置，改善炉缸中心的透气性和透液性，避免炉缸中心堆积。

（3）4000m^3级以上特大型高炉，可参考借鉴宝钢高炉操作三个控制警戒值的管理模式，风压不超过420kPa，透气性指数K值不超过2.6，炉身煤气流速不超过3.2m/s。

（4）煤气流控制要均衡合理，控制高炉炉顶降温喷水过多。要深入探索高炉顶压控制机制，炉顶压力和高炉压差不宜过高，不片面追求低炉腹煤气量。高炉应采取全风操作，疏松料柱、降低压差、提高高炉透气性。

（5）控制合理的理论燃烧温度，通过喷煤及加湿鼓风等措施进行调整，炉况失稳或煤气流不易控制时，应适当降低鼓风富氧量，维持合理的炉腹煤气量。

（6）高炉炉缸监测和管理始终采取"炉缸侧壁温度最高值"管理措施，对炉缸侧壁高温点区域以首次出现的最高温度作为监控温度和警戒值进行跟踪管理。

（7）要进一步探索高炉合理炉料控制与分布技术，合理控制中心与边缘煤气流的分布，不宜"边缘过重、中心过轻"。辩证认识和正确使用中心加焦技术，应适度开放边缘，疏导煤气上升途径，改善煤气运动及其分布，提高煤气利用率。

4.3 加强炉缸监测、运行与维护

（1）合理加钛护炉，控制好铁水成分。实践表明，高炉加钛护炉过程中，铁水中［Si］含量要达到0.40%以上，［S］含量降低到0.030%以下，［Ti］含量必须达到0.08%以上。护炉期间要将铁水中［Ti］含量提高到0.10%~0.15%；强化补炉期，要将铁水中［Ti］含量控制在0.15%~0.20%；铁水中［Si］+［Ti］的含量按照0.60%~0.70%控制，且铁水中［Ti］含量必须控制0.15%以上。

（2）以高炉稳定顺行、安全长寿为前提，合理控制高炉产量。高炉炉缸内衬异常破损，与高炉高强度冶炼具有直接关系。在强化护炉期间要按高炉正常产量减少20%控制产量，高炉护炉期间应采用全风操作，通过调整富氧率和炉温来控制冶炼强度。

（3）采取堵风口护炉措施。局部封堵炉缸侧壁高温区上方的风口，对高炉护炉是行之有效措施，实践表明，一般堵1~2个风口对炉况顺行不会产生明显影响，但封堵太多的风口对高炉炉缸均匀工作会带来不利。

（4）加强炉缸渣铁排放管理。高炉操作中，必须注意铁口深度的维护，保持足够的铁口深度和打泥量。同时必须采取多铁口轮流出铁制度，3~4个铁口比2个铁口交替轮流出铁，对于削弱炉缸出铁时，铁水环流及其对炉缸侧壁的冲刷侵蚀更为有利；出铁制度及铁口维护对于有效抑制炉缸异常侵蚀十分重要。

（5）慎重采取炉缸压浆修补措施。实践证实，炉缸区域采取外部压浆（灌浆）的方法来消除气隙的操作比较危险，操作中无法控制修补的程度和效果，操作不当还会适得其反，带来严重后果，不建议采取。

（6）增设炉缸监控设施。一是在炉缸局部过热区域增加热电偶测温或炉壳表面测温装置；二是可以应增加炉壳高温区红外热成像监控报警装置；三是增加热炉缸冷却壁热流强度/热负荷的监控系统。

（7）高炉操作应定期"排碱"，降低锌和碱金属的富集循环。有效降低炉料中锌和碱金属含量，对碱金属负荷高或超标的高炉，建议每月3~5天，提高炉温、降低炉渣碱度，使碱金属更多地进入炉渣、排出高炉，以减轻碱金属循环富集带来的危害和对炭砖的化学侵蚀。

（8）关于高炉开炉初期的炉缸维护。高炉投产以后，在15~30天内要保持较高炉温，不片面追求达产、高产、超产，应对炉缸炉底进行合理"养护"，有意识强化护炉，使铁水中石墨碳析出，有利于砖缝的填充和密封。

5 结语

高炉长寿是现代高炉炼铁的重要发展方向，也是保障高炉生产安全稳定顺行的重要基础。必须建立现代高炉长寿的工程理念，加强高炉全生命周期管理，实现设计–制造–建造–运行–维护–退役等全过程的多目标协同优化。特别是在高炉炉役末期，要厘清高炉操作主体思路，制定切实可行的技术方针，采取稳妥有效的技术措施，建立以高炉稳定顺行、安全长寿的高炉操作理念。对于炉缸内衬破损严重、长期持续高温带病作业、安全风险难以预测和控制高炉，必须及时进行炉缸内衬修复或大修，从根本上解决高炉炉缸存在的安全隐患。

参考文献

[1] 张寿荣，毕学工．中国炼铁的过去、现状和未来 [J]．炼铁，2015，34（4）：1-6．
[2] 张福明．面向未来的低碳绿色高炉炼铁技术发展方向 [J]．炼铁，2016，35（1）：1-6．
[3] 汤清华．关于延长高炉炉缸寿命的若干问题 [J]．炼铁，2014，33（5）：7-11．
[4] 王波，华建明．宝钢1号高炉炉缸侵蚀分析及对策 [J]．炼铁，2016，35（3）：30-32．
[5] 杨军强，张建良，焦克新，等．湘钢1号高炉炉缸破损调查研究 [J]．炼铁，2016，35（3）：50-52．
[6] 王红斌，李红卫，唐顺兵，等．太钢5号高炉控制炉缸侧壁温度升高的措施 [J]．炼铁，2015，34（2）：18-21．
[7] 郭天永，车玉满，王宝海，等．鞍钢新1号高炉炉缸内衬破损调查 [J]．炼铁，2015，34（4）：50-53．
[8] 霍吉祥，黄俊杰．首钢京唐2号高炉护炉措施 [J]．炼铁，2013，32（3）：14-16．
[9] 杨亚魁，陈畏林，张正东，等．武钢4号高炉炉缸温度异常升高的控制 [J]．炼铁，2015，34（1）：14-17．
[10] 张福明．基于自组织理论的高炉长寿技术 [C]//2017年全国高炉炼铁学术年会论文集．昆明：中国金属学会，2017：398-405．
[11] 张福明，赵宏博，钱世崇，等．高炉炉缸炉底温度场控制技术 [C]//第十届中国钢铁年会暨第六届宝钢学术年会论文集．上海：中国金属学会，2015：1-6．
[12] 张福明，程树森．现代高炉长寿技术 [M]．北京：冶金工业出版社，2012：340-356．
[13] 张福明．当代高炉炼铁技术若干问题的认识 [J]．炼铁，2012，31（5）：1-6．

Methods and Understandings on Prolong the Campaign Life of Large Blast Furnace Hearth

Zhang Fuming

Abstract: Blast furnace hearth is an important part of blast furnace. In recent years, after the commissioning of many large blast furnaces, there are frequent problems such as temperature rising of hearth side wall, local overheating, abnormal erosion and so on, and even some hearth breakout accidents occurred in blast furnaces. Under the current raw material and fuel conditions and operating conditions, a lot of blast furnaces have been forced to rebuild hearth under the condition of service life less than 10 years. Based on the blast furnace production practice and the investigation of blast furnace hearth breakage, this paper analyzes and researches the abnormal wear and erosion of hearth side wall, obtains the main causes of abnormal breakage of hearth and its erosion mechanism, and proposes some effective technical measures and approaches to prolong the service life of blast furnace hearth.

Keywords: blast furnace; long campaign life; hearth lining; hot metal circular flow; well death

热压炭砖—陶瓷杯技术在首钢1号高炉上的应用

摘　要：首钢1号高炉有效容积2536m³，于1994年8月9日送风投产。为延长高炉寿命，在高炉炉缸、炉底部位采用了美国UCAR公司的热压炭砖和法国SAVOIE公司的陶瓷杯技术，将"导热法"和"耐火材料法"两种炉缸内衬设计体系融为一体，相互补充，以期实现高炉长寿的目标。高炉投产16个月以来，生产稳步上升，炉缸、炉底内衬工作正常，热电偶温度变化平稳。铁水温度提高15~25℃，铁水质量提高。
关键词：高炉；内衬；耐火材料；设计；长寿

1　引言

1993~1994年首钢1号高炉进行了现代化新技术改造，高炉有效容积由576m³扩大到2536m³。设计要求高炉一代炉龄要达到10年以上，高炉单位容积产铁量要达到8000t/m³，经过调研和多次技术论证，决定在1号高炉上引进美国UCAR公司的热压炭砖和法国SAVOIE公司的陶瓷杯技术，将"导热法"和"耐火材料法"两种炉衬设计体系结合在一起，集二者之长，以期实现高炉长寿的目标。

2　炉缸、炉底内衬设计及特点

1号高炉炉缸直径11500mm，死铁层深度2200mm，3个铁口，铁口水平深度2200mm。炉缸、炉底内衬主要由6部分组成[1]：（1）炉底4层国产满铺大块炭砖；（2）炉底3层陶瓷垫（第1层为国产莫来石砖，第2、3层为SAVOIE公司的莫来石砖MS4）；（3）炉缸壁内侧5层陶瓷壁（全部为SAVOIE公司的棕刚玉预制块Monocoral）；（4）炉缸壁外侧铁口中心线以下、炉缸炉底交界处（即"象脚状"异常侵蚀区），紧贴冷却壁共砌筑26层UCAR公司的热压炭砖NMA；（5）炉缸壁外侧铁口中心线以上，共砌筑6层国产环形大块炭砖；（6）炉缸环形炭砖和陶瓷壁以上，风口组合砖以下，砌筑高铝砖（$Al_2O_3 \geq 80\%$）。

采用有限元法对炉缸、炉底内衬进行传热计算，得出了炉缸、炉底温度场分布曲线（设计值），如图1所示。

图1表明，这种炉缸、炉底内衬结构的温度场分布是合理的，1150℃等温线（铁水凝固线）处于炉底莫来石砖和炉缸壁棕刚玉预制块中，炉缸、炉底炭砖层避开了800℃的温度区间（炭砖脆化反应温度），使炭砖免受脆化断裂破损，可消除或减轻炉缸炭砖环裂和

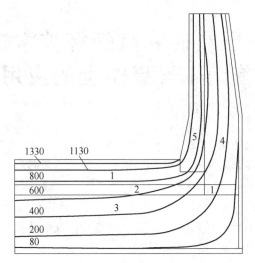

图1　1号高炉炉缸、炉底等温线分布（设计值）
1—莫来石砖 MS4；2—国产莫来石砖；3—国产大块炭砖；
4—热压炭砖 NMA；5—棕刚玉预制块 Monocoral

"象脚状"异常侵蚀。

1号高炉将"导热法"和"耐火材料法"炉衬设计体系融为一体，集中了热压炭砖和陶瓷杯的技术优点，二者相互补充，具有如下特点：

（1）防止铁水渗透。由于采用低导热的陶瓷质耐火材料，使1150℃等温线被阻滞在陶瓷层内，加上陶瓷杯特殊的设计结构和材料的热膨胀，使砖缝紧缩，最大限度减少了铁水对炭砖的渗透侵蚀。

（2）减轻铁水的流动冲刷。采用陶瓷杯应有合理的死铁层深度，一般约为炉缸直径的20%，使铁水在炉缸内的流动方向有所改变，而且在铁口区和炉缸炉底交界处均增加了陶瓷壁的厚度，因而可减轻铁水流动对炉底和炉缸壁的机械磨损。

（3）提高炉缸热稳定性，减少炉缸热损失。采用陶瓷杯后，能够提高铁水温度，改善铁水质量，可降低工序能耗，并为炼钢生产创造有利条件。

（4）有利于高炉操作。由于炉缸热储备量增加，使高炉易于操作，为高炉稳定顺行、活跃炉缸、冶炼低硅低硫生铁、复风操作等提供了良好的条件，并可减少炉缸堆积、风口灌渣等操作事故。

（5）大幅度延长炉缸、炉底寿命。高炉开炉以后，陶瓷杯对炭砖起到保护作用，使其不直接和铁水接触，从而使炭砖免受铁水的渗透、机械冲刷以及碱金属的化学侵蚀等破坏。而且陶瓷杯将800℃等温线阻滞在陶瓷材料中，使炭砖避开脆性断裂的温度区间。热压炭砖的优良配方和特殊的成型工艺及构型，使它具有优异的导热性、抗铁水渗透性和优良的抗碱性，可为陶瓷杯提供有效的冷却，从而延长陶瓷杯的寿命。由于NMA砖具有很高的导热性、抗铁水渗透性和抗化学侵蚀性，因此，即使在陶瓷杯出现破损，热压炭砖直接同铁水接触时，仍可依靠有效的冷却作用，使炭砖热面生成保护性渣皮或铁壳，最大限度地延长高炉寿命。

3 生产实践

1号高炉于1994年8月9日送风投产,至今已生产16个月。高炉投产以来,生产稳定,炉缸工作正常。表1是1号高炉与容积相同的3号高炉主要技术指标对比。通过一年多的生产实践和与3号高炉的分析对比证明:1号高炉热压炭砖-陶瓷杯技术的应用,取得了如下的初步成效。

表1 首钢1号、3号高炉主要技术指标对比 (1995年1~11月)

项 目	1号高炉	3号高炉
平均日产量/t	5259	5037
高炉利用系数/t·(m^3·d)$^{-1}$	2.094	2.008
入炉焦比/kg·t^{-1}	469.9	510.4
煤比/kg·t^{-1}	41.2	29.1
焦丁/kg·t^{-1}	35.5	34.3
燃料比/kg·t^{-1}	546.6	573.8
熟料比/%	97.38	97.89
入炉品位/%	57.10	56.38
渣量/kg·t^{-1}	386	431
生铁合格率/%	99.35	99.96
一级品率/%	90.04	92.77
风温/℃	952	872
炉顶压力/MPa	0.182	0.178
煤气利用率/%	41.186	39.132
焦炭负荷/t·t^{-1}	3.654	3.422
[Si]/%	0.428	0.49
[S]/%	0.02	0.018
铁水温度/℃	1496	1475
风口损坏/个	93	202
休风率/%	1.12	3.84

(1) 炉缸热量充沛,炉缸工作均匀活跃,为高炉稳定顺行创造了极为有利的条件。炉缸热损失减少,铁水温度比3号高炉提高15~25℃。高炉开炉以来,从未出现过炉缸堆积、风口灌渣等事故,风口损坏数量仅是3号高炉的1/2。

(2) 有利于降低燃料消耗。由于炉缸工作均匀活跃,热量充沛,为降低焦比、提高喷煤量创造了有利条件。1995年11月份,1号高炉入炉焦比415kg/t,煤比96kg/t,焦丁5kg/t,燃料比516kg/t,喷吹率18.6%。由于风温偏低(年平均仅为950℃)、鼓风没有富氧以及原燃料条件变差等不利因素的影响,限制了喷煤量的进一步提高。

(3) 有利于低硅低硫生铁的冶炼。由于炉缸物理热充沛,为冶炼低硅低硫生铁创造了条件。由表1可以看出1995年1~11月,1号高炉铁水含[Si]为0.428%,[S]为0.02%,而同样原燃料条件下,3号高炉铁水含[Si]为0.49%,[S]为0.018%。

(4) 缩短高炉复风时间。高炉休风后,由于炉缸内保持较高的热量储备,因此高炉复风迅速,易于恢复到正常的炉况状态。而且复风后渣铁温降小,从未出现过复风后渣铁排放困难的现象。

(5) 渣铁物理热提高,流动性改善,炉渣粘沟现象大大减少,降低了炉前工人的劳动强度,改善了劳动条件。

(6) 炉缸、炉底热电偶温度变化平稳(见图2),月平均温度变化为±0~10℃。开炉一年多来,热电偶平均升温速率仅为0.7~1.5℃/月。根据炉缸、炉底热电偶温度,采用计算机进行传热计算,得出了炉缸、炉底等温线分布曲线(计算值),如图3所示。同图1对比可以看出,1150℃等温线在陶瓷杯热面附近,陶瓷杯尚无明显侵蚀。推测在陶瓷杯热面可能存在一个凝固层,由于凝固层的存在,使1150℃等温线向炉缸中心推移。

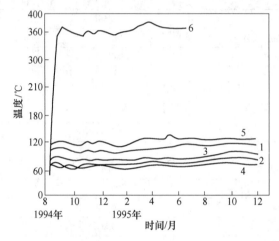

图2 1号高炉炉缸、炉底热电偶温度变化曲线
1—TC1793;2—TC1608;3—TC1769;4—TC1757;5—TC1709;6—TC1626(热电偶编号)

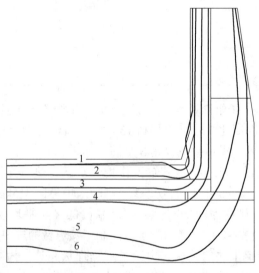

图3 1号高炉炉缸、炉底等温线分布(计算值)
1—1150℃;2—800℃;3—600℃;4—400℃;5—200℃;6—100℃

4 结语

热压炭砖-陶瓷杯技术在首钢1号高炉上的应用,经一年多来的生产实践初步表明:

(1) 炉缸热量充沛,炉缸工作均匀活跃,炉况稳定顺行,铁水温度提高15~25℃,炉缸热损失减少;

(2) 有利于降低燃料消耗和冶炼低硅低硫生铁,有利于提高铁水质量,缩短复风时间,减轻炉前工人的劳动强度;

(3) 炉缸、炉底热电偶测温及计算结果表明,一年多来1号高炉陶瓷杯并无明显侵蚀,炉缸、炉房耐火材料内衬工作状况正常;

(4) 1号高炉设计中将"导热法"和"耐火材料法"两种炉衬设计体系溶为一体,集二者之长,相互补充,相信会取得高炉长寿的实绩,但有待于高炉生产实践的长期检验。这可为国内外大型高炉炉缸、炉底内衬设计提供有效的借鉴。

致谢

参加此项工作的还有徐和谊、刘兰菊、刘泽长、张宗民、班润臣、尤新等同志。在本文成文过程中,得到了刘云彩、魏升明高工的大力支持和指导;承蒙首钢设计院副院长兼总工程师黄晋审阅了本文,并得到了李世良、赵志忠、王卫平、单洎华等同志的热情帮助,作者在此一并表示感谢。

参考文献

[1] 张福明,刘兰菊. 新型优质耐火材料在首钢高炉上的应用 [J]. 炼铁,1994 (3):22.

Application of Hot Pressed Carbon Bricks-Ceramic Cup Technology to No. 1 Blast Furnace at Shougang Co.

Zhang Fuming

Abstract: Hot pressed carbon, bricks of UCAR carbon company Inc. (U.S.A) and ceramic cup of Savoie Refractory Company (French) were introduced into the hearth and bottom of Shougang Co. No. 1 blast furnace ($2536m^3$) which blew in on August 9,1994. The "thermal solution concept" was combined with the "refractory solution concept" in the design of No. 1 BF in order Io meet the requirement for BF long campaign life. The BF has been working stably. the hearth and bottom lining working normally and the temperature of thermocouples varying steadily for 16 months since the BF blew in. The temperature of hot metal has raised 15~25℃ and the quality of hot metal has been improved.

Keywords: blast furnace; lining; refractory; design; long life

首钢1号高炉陶瓷杯砌筑质量管理

摘　要：首钢1号高炉（2536m³）炉底、炉缸部位采用了热压炭砖-陶瓷杯技术。在砌筑陶瓷杯过程中运用全面质量管理（TQC）对陶瓷杯耐火材料的验收、预组装以及砌筑等工序进行了严格的控制和管理，使陶瓷杯砌筑一次成功，达到了总体验收标准。

关键词：高炉；陶瓷杯；砌筑；管理

1　引言

1993~1994年，首钢1号高炉进行了技术改造，有效容积由576m³扩大到2536m³。为使高炉一代寿命达到10年以上，单位炉容产铁量达到8000t/m³，在炉底炉缸部位采用了美国UCAR公司的热压炭砖和法国SAVOIE公司的陶瓷杯技术。1号高炉陶瓷杯分为炉底陶瓷垫和炉缸陶瓷壁两部分。炉底陶瓷垫分3层砌筑而成，第1层厚度230mm，为国产莫来石砖；第2层和第3层厚度均为400mm，为Savoie公司的MS4莫来石砖。第1、2层莫来石砖采用十字形砌筑，沿砖列线方向砖的两个侧面均采用斜面咬砌结构；第3层莫来石砖由于与铁水接触，采用环形砌筑结构，每环砖在半径上也为斜面咬砌结构，炉底中心部位采用ϕ920mm的圆形中心砖，材质为棕刚玉预制块。炉缸陶瓷壁厚度400mm，铁口区厚度为840mm。材质为棕刚玉预制块。陶瓷壁总高度为4950mm，共5层，每层相邻的两预制块上下交错砌筑，每层之间没有通缝，为加强砌体整体稳定性，在陶瓷壁的冷面，在相邻两预制块之间采用锁砖连接。

2　陶瓷杯砌筑的质量控制与管理

陶瓷杯技术在我国是首次引进，首钢对陶瓷杯施工质量十分重视，曾多次邀请法方专家进行技术讲座，并派有关人员赴法国实地考察和培训。为确保施工质量和施工进度，运用全面质量管理（TQC）建立了统一协调的组织机构；并根据施工进度安排，将陶瓷杯砌筑划分为耐火材料检验（理化指标及外观尺寸）、陶瓷杯预组装等6个工序流程（见图1），严把每一工序的质量关。

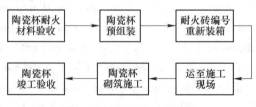

图1　陶瓷杯砌筑工艺流程

本文作者：张福明，李世良，刘兰菊，魏升明。原刊于《炼铁》，1996，15（5）：7-9。

2.1 建立专门的管理机构

成立陶瓷杯砌筑技术领导小组，由首钢公司一名副经理任组长，设计院、修建公司、炼铁厂、质检处等单位分别派有经验的技术人员参加，并请法国筑炉专家担任技术领导小组的顾问。技术领导小组负责陶瓷杯砌筑施工的全面工作（见图2），从耐火材料检验到砌筑竣工验收，均由技术领导小组负责技术监督和质量保证，每一工序环节完成以后，须经技术领导小组全体成员共同验收，验收合格后方可进行下一工序。

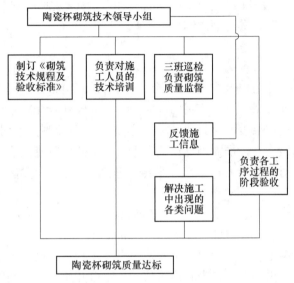

图2 技术领导小组业务流程

2.2 制订技术规程和验收标准

根据法方提供的技术文件和我国现行的《工业炉砌筑及验收工程规范》（GB/T 211—87）、《工业炉砌筑工程质量检验评定标准》（GB 50309—92），并结合首钢以往高炉砌筑施工的经验，以设计院为主，制订了《首钢1号高炉陶瓷杯砌筑技术规程及检验标准》。该规范包括对陶瓷杯耐火材料的检验、预组装、砌筑、验收等工序的技术要求和检验标准，并对施工操作方法、砌筑程序、泥浆的调配及使用、劳动保护、烘炉及开炉操作等作了详尽的说明。

2.3 施工人员的技术培训

要求各工序的施工人员熟练掌握《首钢1号高炉陶瓷杯砌筑技术规程和检验标准》及相关内容，并组织施工人员进行技术学习和实际操作培训，对施工人员进行笔试和现场操作两阶段考试，两项考试合格者，发给达标上岗证，方能上岗操作。

2.4 陶瓷杯耐火材料验收

陶瓷杯耐火材料分为炉底陶瓷垫和炉缸陶瓷壁两类，陶瓷垫第1层为国产莫来石砖；

第 2 层有 3 种砖型,即 H、J、K,总量为 3036 块;第 3 层有 8 种砖型,即 0C、2C～7C、8CR,总量为 3004 块,第 2、3 两层陶瓷垫的材质为 MS4 莫来石砖。陶瓷壁共有 21 种砖型,即 TK1～TK20、TK9T,总量为 294 块,材质为棕刚玉预制块。陶瓷杯耐火材料到货后,由技术领导小组组织技术供应处、炼铁厂、修建公司、质检处、设计院等部门对耐火材料进行验收,理化性能抽检结果表明,实物指标符合与法方签订的供货质量标准;耐火材料外观尺寸的检验是将砖放置在钢制平台上,每种砖型各取 10 块,用游标卡尺和靠尺进行测量,同时进行外观质量检验,经检验,各种砖型尺寸误差均在±1.00mm 以内,符合砌筑要求,扭曲四面检查误差也在 0.50mm 以内,均未发现有缺棱、缺角、熔洞、渣蚀、裂纹、大气孔以及疏松等外观缺陷。

2.5 陶瓷杯预组装

(1) 陶瓷垫预组装。在预组装钢制平台上(精度 0.10mm),用 90°靠尺测量砖的垂直度、高度、厚度、宽度等尺寸,其中 H、J、K 型砖各抽检 10 块,0C、2C～7C、8CR 型砖各抽检 1 块,抽检结果表明,第 2、3 层陶瓷垫砖。各项测量值均满足误差在±0.5mm 以内的要求,符合砌筑质量标准。因此,只在平台上对两层陶瓷垫进行四分之一圆的预组装,预组装完成后,对砌体半径、高度、砖缝、垂直度等进行了检验,结果达到了验收标准。

(2) 陶瓷壁预组装。陶瓷壁共有 21 种砖型,各种砖型抽检 1 块,检验方法与杯底相同,检测结果表明:砖的外观尺寸误差均在±0.5mm 以内,砖面之间的角度也很准确,所以决定只预砌陶瓷壁第 1 层。在预组装平台上用专用半径规按已确定的中心点画出陶瓷壁第 1 层内圆和外圆,画出 3 个铁口中心线,从 1 号铁口开始依次向两侧预砌,以内圆线为基准,调整杯壁砖热面与内圆线重合,并对砖缝进行适当调整,砖缝用 1～2mm 的纸板填充,第 1 层陶瓷壁预砌完成后,对砌体进行了测量,各类尺寸误差符合质量控制要求。预组装验收合格以后,将陶瓷杯材料逐一编号,重新装入集装箱,运往施工现场。装箱及运输过程中,为防止砖的意外损坏,采取了许多保护性措施。

2.6 陶瓷杯的砌筑及竣工验收

第 1 层陶瓷垫的顶面是第 2 层的基准面,要求其水平度准确,施工时对基准面用水平仪测量了 57 个点,对超出基准±1mm 的部位用铲平机进行处理,合格后再用 3m 平尺找平,并用塞尺检查平尺与砖之间的缝隙,保证其小于 2mm,使表面平整度和标高符合设计要求。基准面的各类尺寸经检验达到质量要求后,由技术领导小组全体成员签字认可,方可进行第 2 层陶瓷垫的砌筑。砌筑从中心开始,为减小砌砖误差,分 4 个扇区分别施工。砌筑过程中,砌砖、测量检验和统计记录工作分别由专人负责,各岗位各负其责,砌一块砖测量一次,记录一次,如果数据超标,则该砖必须重砌。技术领导小组成员同时跟班巡检,随时抽检,监督质量工作,如果抽验不合格,施工必须重新进行,施工小组也将受到考核。由于施工过程中严格管理,因此抽检结果完全合格,无一返工重砌。第 2 层陶瓷垫砌完后转入陶瓷壁砌筑。砌筑前对第 2 层陶瓷垫顶面用 3m 平尺找平,个别部位用磨平机处理,然后用水准仪观测,观测结果表明,观测的 20 个点中只有 20 号点误差为+4mm,超出规定标准,经铲磨处理后误差缩小到+1mm。陶瓷壁的砌筑难度较大,棕刚玉预制块共 303 块，重量最轻的 340kg/块，最重的 2056kg/块，总重量 328.4t。砌筑时使用法方提供的特殊吊装工具，从 1 号铁口开始向两侧砌筑。砌筑时要求铁口预制块中心线位置精确，用膨胀螺栓吊装调整砖缝和水平度，并用 0.5m 水平尺逐块找平，经检验，每块预制块的水平度、垂直度、砖缝均在±0.5mm 误差范围内。砌完两层后再用经纬仪配合校正每组砌体的垂直度。整个陶瓷壁砌完后，每组砌体的顶面水平误差均在±1.0mm 以内。第 2 层陶瓷垫和陶瓷壁砌完后，开始砌筑第 3 层陶瓷垫，第 3 层陶瓷垫与第 1 层陶瓷壁为斜面咬砌结构，故从最外环向中心砌筑，因为这样可保证杯底和杯壁之间的砖缝不超过 2.0mm。具体做法是：先在第 2 层陶瓷垫顶面标出高炉中心，用半径规控制各圆环砖半径，按 180°分成两个半圆工作面，每个 90°工作面从外圆向中心方向砌筑，每砌一环即用半径规检查上下半径，合格后再砌下一环、一个半圆完成后开始砌另一半圆，砌法同前。陶瓷杯砌筑施工完成后，技术领导小组负责对其进行了整体验收，对整个砌体的垂直度、水平度、半径、各面的标高、砌缝等进行了严格的检验，检验结果全部达标，陶瓷杯砌筑施工一次成功。

3 结语

首钢 1 号高炉于 1994 年 8 月 9 日建成投产，投产后，生产逐步上升，平均日产铁达 5600t，现在各项技术指标已达到或接近设计值。由于采用了陶瓷杯，炉缸热损失减少，铁水温度提高了 15~25℃，炉底、炉缸部位各测温点的温度值都很正常。

Quality Control to Ceramic Cup Laying on No. 1 BF at Shougang Co.

Zhang Fuming Li Shiliang Liu Lanju Wei Shengming

Abstract: The hot pressed carbon bricks-ceramic cup technology was introduced into the hearth and bottom of No. 1 BF at Shougang Co. TQC (Total Quality Control) was applied to the refractory acceptance, preliminary assembly and brick laying of ceramic cup. The ceramic cup was laid successfully and the laying quality reached acceptance standard.
Keywords: blast furnace; ceramic cup; lay; control

首钢 2 号高炉长寿技术设计

摘 要：首钢 2 号高炉有效容积 1726m³，于 1991 年 5 月 15 日扩容大修改造后投产。2002 年 3 月该高炉停炉进行现代化新技术改造，于 2002 年 5 月 23 日送风投产。在首钢 2 号高炉长寿技术设计中采用了软水密闭循环冷却、铜冷却壁、热压炭砖-陶瓷垫等一系列高炉长寿技术，经过 3 年的生产实践，取得了显著的效果。

关键词：高炉；长寿；铜冷却壁；设计

1 引言

首钢 2 号高炉有效容积 1726m³，于 1991 年 5 月 15 日扩容大修改造后送风投产。该高炉采用了并罐式无料钟炉顶、大型顶燃式热风炉、软水密闭循环冷却、第三代冷却壁及热压炭砖等技术。2002 年 3 月 6 日 2 号高炉停炉进行现代化新技术大修改造，本代炉龄 10 年 9 月，累计产铁达到 1428.7 万吨，平均利用系数为 2.122t/(m³·d)，单位容积产铁量为 8277t/m³。炉役期间该高炉进行了两次中修，同国内外长寿高炉先进水平相比仍存在着差距[1]。

2 号高炉 2002 年大修技术改造时，以"长寿、高效、低耗、清洁"为设计指导思想，积极采用现代高炉长寿综合技术，在不中修的条件下，高炉寿命要达到 15 年以上。本次高炉大修对上料、炉顶、出铁场、渣处理等系统进行了更新和完善，采用了高风温内燃式热风炉、铜冷却壁、人工智能高炉专家系统、软水密闭循环冷却、热压炭砖-陶瓷垫等多项先进技术。

2 高炉长寿设计理念

结合 2 号高炉 10 余年的生产实践和炉体维护过程中所暴露出的问题，对高炉长寿设计进行了研究和分析。生产实践表明，高炉炉缸、炉底和炉腹、炉腰、炉身下部是造成高炉短寿的两个关键部位。特别是炉腰和炉身下部，10 年间曾两次对该区域破损的冷却壁进行了更换和修复。因此有效地延长炉腹至炉身下部区域冷却壁的寿命无疑是保障高炉长寿的重要基础。为达到设计目标，确定了 2 号高炉长寿设计的理念。通过优化炉型，以确保高炉生产稳定、顺行、长寿；炉腹至炉身区域采用软水密闭循环冷却技术、铜冷却壁和优质耐火材料；炉缸、炉底部位采用优质高导热炭砖、合理的内衬结构和冷却系统；加强炉体自动化在线监测，为高炉操作提供准确的信息和参数。图 1 为首钢 2 号高炉长寿炉体设计结构。

本文作者：张福明，毛庆武，姚轼，钱世崇。原刊于《2005 年中国钢铁年会论文集》，2005：314-318。

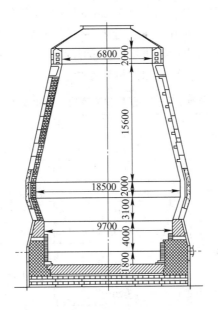

图 1　首钢 2 号高炉长寿炉体设计结构

3　优化高炉炉型

首钢历来重视高炉炉型的研究和探索。根据 2 号高炉生产运行实践和国内外高炉炉型设计发展趋势，设计中对高炉炉型进行了优化。加深了死铁层深度，以减轻铁水环流对炉缸内衬的冲刷侵蚀；适当加大了炉缸高度和炉缸直径，以满足高炉大喷煤操作和高效化生产的要求；降低了炉腹角、炉身角和高径比，使炉腹煤气顺畅上升，改善料柱透气性，稳定炉料和煤气流的合理分布，抑制高温煤气流对炉腹至炉身下部的热冲击，减轻炉料对内衬和冷却器的机械磨损。2 号高炉大修后的炉型主要参数为：有效容积 1780m³，炉缸直径 9.7m，炉腰直径 10.85m，死铁层深度 1.8m，炉缸高度 4.0m，有效高度 26.7m，炉腹角 79°29′31″，炉身角 82°36′14″，高径比 2.461。

4　高炉炉缸内衬结构

实践证实，炉缸、炉底是决定高炉一代寿命的关键部位。合理的炉缸内衬结构、高质量的炭砖和有效的冷却是延长炉缸、炉底寿命的有效措施。用于炉缸、炉底的炭砖，必须具有优异的导热性、抗铁水渗透性和优良的耐碱性。通过研究分析国内外大型高炉长寿技术发展趋势，结合首钢大型高炉生产特点和长寿实践，炉缸、炉底采用了热压炭砖-陶瓷垫结构。

炉底设 5 层满铺炭砖，第 1~4 层为国产半石墨质大块炭砖，第 5 层周边（靠近冷却壁）为热压炭砖 NMA，中间区域为国产微孔大块炭砖。综合炉底第 6、7 层中间部位为进口陶瓷垫，每层厚度为 500mm，周边为热压炭砖 NMA。炉底总厚度为 2800mm。

炉缸、炉底交界处至铁口中心线以上1528mm的炉缸壁全部为热压炭砖，靠近冷却壁为NMD砖，其内为NMA砖。热压炭砖之上设一层国产炉缸环形大块炭砖，其上是进口陶瓷质风口组合砖。炉缸壁垂直段热压炭砖厚度为1143mm，热压炭砖总高度为4676mm。表1是首钢2号高炉炭砖的理化性能和传热学参数。

表1 首钢2号高炉炉体炭砖理化性能和传热学参数

序号	性能指标		单位	炭 砖 品 种			
				国产半石墨质高炉炭砖	国产微孔炭砖	NMA	NMD
1	灰分		%	≤7	≤20	≤12	≤9
2	体积密度		g/cm³	≥1.60	≥1.60	≥1.61	≥1.82
3	显气孔率		%	≤18	≤16	≤18	≤16
4	常温耐压强度		MPa	≥31	≥36	≥33	≥30
5	抗折强度		MPa	≥8.0	≥9.0		
6	耐碱性			U（优）	U 或 LC		
7	氧化率		%	≤20	≤14		
8	透气度		mDa	≤70	≤9	≤11	≤5
9	导热系数	室温	W/(m·K)	≥6	≥6	≥20	≥60
		300℃		—	≥9	≥14	≥42
		600℃		≥14	≥14	≥14	≥38
10	平均孔半径		μm	—	≤0.5		
11	<1μm 孔容积率		%	—	≥70		
12	真密度		g/cm³	≥1.9	≥1.9		
13	铁水熔蚀指数		%	≤2	≤28		

5 炉体冷却结构

5.1 炉体冷却结构

高炉炉体全部采用冷却壁，共设17段冷却壁。炉缸第1~5段为光面冷却壁，厚度为120mm，材质为灰铸铁。炉腹下部（第6段）为双排管冷却壁，厚度为340mm，材质为球墨铸铁。炉腹上部、炉腰、炉身下部（第7、8、9段）为国产铜冷却壁，炉身下部（第10、11段）为双排管凸台冷却壁，壁体厚度为340mm，材质为球墨铸铁。炉身中、上部（第12~16段）为单排管凸台冷却壁，壁体厚度为265mm，材质为球墨铸铁。炉身上部、炉喉钢砖以下设一段"C"形水冷壁（第17段），水冷壁为光面结构，材质为球墨铸铁，水冷壁的热面不设砖衬。2号高炉炉体冷却壁主要技术性能见表2。

5.2 铜冷却壁

铜冷却壁具有高导热性、抗热震性、耐高热流冲击和长寿命等优越性能，在国外大型高炉的炉腹、炉腰和炉身下部区域得到成功应用，为延长高炉寿命起到了重要作用。

表2 首钢2号高炉炉体冷却壁主要技术性能

应用部位	结构形式	冷却壁材质	壁体厚度/mm
炉缸、炉底（第1~4段）	光面冷却壁	灰铸铁（HT-200）	120
风口区（第5段）	光面冷却壁	灰铸铁（HT-200）	120
炉腹下部（第6段）	双排管冷却壁	球墨铸铁（QT400-20）	340
炉腹上部（第7段）	国产铜冷却壁	无氧铜（TU$_2$）	140
炉腰和炉身下部（第8、9段）	国产铜冷却壁	无氧铜（TU$_2$）	140
炉身下部（第10、11段）	双排管中部凸台冷却壁	球墨铸铁（QT400-20）	340
炉身中部（第12~15段）	单排管中部凸台冷却壁	球墨铸铁（QT400-18）	265
炉身上部（第16段）	单排管上部凸台冷却壁	球墨铸铁（QT400-18）	265
炉身上部（第17段）	"C"形光面冷却壁	球墨铸铁（QT400-18）	170

2000年1月，首钢和广东汕头华兴冶金备件厂有限公司合作，开发研制出2块铜冷却壁，并安装在首钢2号高炉上试验，取得了明显的应用。本次2号高炉技术改造中，在炉腹上部、炉腰、炉身下部采用了3段国产铜冷却壁，共120块。这是我国高炉首次工业化应用国产铜冷却壁[2]。

2号高炉铜冷却壁采用轧制无氧铜板（TU$_2$）直接钻孔而形成冷却通道，提高了冷却壁的传热速率和效率，依靠铜的高导热性，使铜冷却壁本体保持较低的工作温度，促使铜冷却壁热面能够形成稳定的保护性"渣皮"，利用保护性渣皮代衬工作，这是铜冷却壁技术的基本设计思想。铜冷却壁的设计，经过传热计算，确定了冷却壁的外形尺寸、冷却通道数量、冷却通道直径、冷却通道间距、壁体厚度及燕尾槽尺寸等技术参数。铜冷却壁的设计至关重要，既要满足传热的要求，又要兼顾机械加工制造的可行性，还要优化设计，降低铜冷却壁造价。铜冷却壁主要技术参数见表3。

表3 首钢2号高炉铜冷却壁主要技术参数

项 目	第7段	第8段	第9段
冷却壁高度/mm	1420	1970	1700
冷却壁厚度/mm	140	140	140
燕尾槽深度/mm	37	37	37
冷却通道直径/mm	48	48	48
镶填料面积比	0.442	0.471	0.474

6 软水密闭循环冷却技术

高炉冷却系统对高炉正常生产和长寿至关重要。软水密闭循环冷却技术使冷却水质量得到极大改善，解决了冷却水管结垢的致命问题，为高效冷却器充分发挥作用提供了技术保障。该系统运行安全可靠，动力消耗低，补水量小，维护简便。

本次2号高炉技术改造对原有的软水密闭循环冷却系统进行了改进和完善。炉腹、炉腰和炉身区域的冷却壁（第6~16段）采用软水密闭循环冷却。软水循环量由原来的

2450m³/h 增加到 3410m³/h，设计供水温度 50℃，水压 0.6MPa。高炉软水密闭循环冷却系统主要传热学参数见表4。

表4 首钢2号高炉软水密闭循环冷却系统热传热学参数

冷却系统	最大热流强度 /kW·m⁻²	传热面积 /m²	系统热负荷 /MW	循环冷却水速 /m·s⁻¹	冷却水流量 /m³·h⁻¹	系统水温差 /℃
冷却壁前排管系统	46.52	491.01	19.65	1.83	1910	8.85
冷却壁后排管系统	17.45	141.51	2.20	1.50	300	6.29
冷却壁凸台Ⅰ系统	69.78	70.98	4.24	2.30	600	6.07
冷却壁凸台Ⅱ系统	63.97	73.87	3.98	2.30	600	5.71
合计		777.37	30.07		3410	7.58

本次2号高炉技术改造，软水密闭循环冷却系统进行了以下改进：（1）更换了3台水泵，正常生产时两用一备。其他辅助设施如柴油机事故泵、补水泵等基本完好，仅进行日常检修。（2）对原有12台空冷器进行更换，并增加了6台空冷器以提高冷却效率，提高系统工作可靠性。（3）对原系统的脱气、排气功能进行了完善，冷却支管布置采取自下而上串联，消除了冷却支管的水平或向下的折返，优化了管路布置，提高排气功能。重新设计了脱气罐、膨胀罐和稳压罐，提高整个系统的工作可靠性。（4）炉腹、炉腰、炉身下部区域的冷却壁（第6~11段）进出水管采用金属软管连接，减小冷却支管的阻力损失，使系统水量分配更加均匀。（5）根据冷却壁的工作特点和热负荷分布，强化了凸台管和前排管的冷却。将整个系统分为前排管、凸台管、后排管3个子系统。前排管和凸台管分别设2个供水环管，后排管设一个独立的供水环管，这种单独供水模式可以使冷却壁的各部位得到有效冷却，系统工作也更加稳定可靠。图2是2号高炉软水密闭循环冷却系统工艺流程图。

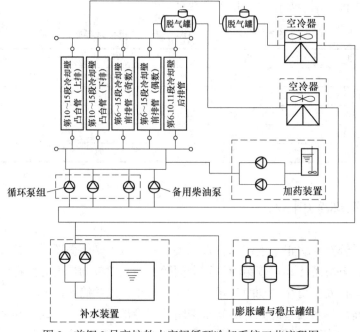

图2 首钢2号高炉软水密闭循环冷却系统工艺流程图

7 自动化检测与控制系统

自动化检测是高炉长寿不可缺少的技术措施。炉缸、炉底区域的温度在线检测是监控炉缸、炉底工作状况的重要手段。高炉炉底设两层热电偶，每层 25 个测点；炉缸壁共设 7 层热电偶，共计 92 个测点。通过热电偶的在线测温，可以随时掌握炉缸、炉底的工作状况，及时采取有效措施，抑制炉缸、炉底异常侵蚀过早出现。高炉炉腹至炉身部位共设 15 层热电偶，共计 224 个测点，用于检测冷却壁和砖衬温度。炉腰、炉身部位设 3 层炉体静压力检测装置，每层设 4 个测压点。根据炉体静压力检测值，得出炉内煤气分布状况，从而推断软熔带的位置和形状，随时掌握高炉工作状况。

炉喉设固定测温装置，共设 21 个测温点，可以在线检测炉喉煤气分布状况，从而推断煤气分布状态、炉料分布状况和煤气利用情况。炉顶设 2 台高温摄像仪，可以直观地观察布料情况和料面形状，为炉料分布控制提供有效信息。采用高炉煤气在线分析装置，可以在线分析煤气成分，计算得出煤气热值、煤气利用率等参数，为高炉操作和专家冶炼系统提供可靠数据。引进了芬兰罗德洛基公司人工智能高炉冶炼专家系统，为高炉操作提供预测和指导。该专家系统由技术计算、布料计算模型、炉缸仿真模型、炉缸侵蚀模型、炉身仿真模型、神经系统等组成。

为保证高炉冷却系统工作的可靠性，软水供回水支管、总管共设 33 个温度检测点，软水供回水支管、总管及补水泵出口共设 14 个压力检测点，稳压罐和膨胀罐设 3 点压力检测，膨胀罐设 4 点水位检测，供回水支管和总管共设 20 个流量检测点。

8 结语

首钢 2 号高炉于 2002 年 3 月 6 日停炉进行大修技术改造，经过 78 天的紧张施工，于当年 5 月 23 日送风投产，高炉开炉至今已经稳定生产 3 年。高炉各项技术经济指标不断提升，创造了国内同级别高炉的领先水平。在 2002 年 10 月，高炉月平均利用系数 $2.502t/(m^3 \cdot d)$、焦比 296.9kg/t、煤比 169.8kg/t、焦丁 14.3kg/t、燃料比 481kg/t、风温 1220℃。设计中所采用的高炉长寿技术发挥了显著作用，至今炉体工作状况稳定正常，实现了"高效、低耗、长寿、清洁"的设计目标。

参考文献

[1] 单泊华，等. 通过 2 号高炉破损调研探索首钢高炉长寿途径 [C] //2003 中国钢铁年会论文集. 北京：冶金工业出版社，2003：491-496.
[2] 张福明，等. 铜冷却壁的设计研究与应用 [C] //铜冷却壁技术研讨会论文集. 北京：中国金属学会、国际铜业协会（中国），2003：39-42.

Design of Long Campaign Life Technology for Shougang's No. 2 BF

Zhang Fuming Mao Qingwu Yao Shi Qian Shichong

Abstract: Shougang's No. 2 blast furnace with the effective volume 1726m^3 was blown in at May 15, 1991 after revamped with enlarging effective volume. This blast furnace was equipped with modern technology after blow down in Mar. 2002 and blow in at May 23, 2002. A series of long campaign life technologies, such as soft water obturation recirculation cooling system, copper stave, hot pressed UCAR carbon bricks, ceramic pad was adopted, which made the blast furnace attain prominent achievement in the past 3 years.

Keywords: BF; long campaign life; copper staves; design

首钢高炉炉缸内衬设计与实践

摘　要：20世纪90年代以来，首钢4座大型高炉相继进行了现代化新技术大修改造。为延长高炉寿命，设计中对高炉炉缸结构参数进行了优化设计，吸收导热法和耐火材料法设计体系的精华，采用国产优质耐火材料，关键部位引进国内尚不能生产的热压炭砖、陶瓷杯等材料。应用传热数学模型，计算炉缸、炉底等温线分布，通过结构和耐火材料的优化设计，控制炉衬中1150℃等温线的位置，抑制炉缸象脚状异常侵蚀的发生，延长炉缸、炉底寿命。近8年的生产实践表明，首钢高炉采用国外高炉长寿先进技术，取得了较好的实绩，创造了20世纪80年代以来首钢大型高炉炉缸工作8年无事故的纪录。目前，4座大型高炉炉缸工作正常，冷却水温差、冷却壁热负荷、炉衬温度等参数均在正常范围之内，高炉寿命有望达到10年以上。

关键词：高炉；长寿；设计；耐火材料

1 引言

　　20世纪80年代以来，日本和欧美等国都致力于高炉长寿化。到20世纪90年代末，世界主要产钢国的高炉寿命已达到或超过10~15年。其中日本川崎公司千叶厂6号高炉（4500m³）寿命已达到20年9个月。我国近年的高炉长寿化进程加快，宝钢、梅山等厂的高炉寿命均已达到10年左右。首钢4座高炉在"九五"期间相继进行了现代化新技术大修改造，高炉总容积由原来的4139m³扩大到9934m³。在高炉大型化的同时，设计中采用了多项高炉长寿新技术，力求使高炉寿命达到8~10年。在首钢高炉长寿化的进程中，最为突出的是高炉炉缸内衬设计结构的发展。迄今为止，首钢高炉寿命同20世纪80年代相比，已有了大幅度提高，大型高炉的炉缸已安全工作8年。

2 高炉炉缸侵蚀机理分析

　　从20世纪60年代起，首钢高炉开始采用炭砖-高铝砖综合炉底技术，使用情况一直较好。到20世纪80年代中期，随着高炉冶炼强度的提高，炉缸、炉底问题变得突出。4号高炉（1200m³）1983年6月进行大修，投产13个月，炉缸第2段冷却壁水温差大量超限，到1986年3月就发生了炉缸烧穿事故，寿命不足3年。3号高炉（1036m³）于1984年1月大修竣工投产，当年5月，炉缸铁口两侧冷却壁水温差达到1℃以上，第2段冷却壁改用1.3MPa高压水冷却，1989年11月炉缸烧穿，1990年3月停炉抢修，寿命仅6年。

　　首钢高炉炉缸、炉底的破损情况与国内外大体一致，即所谓"象脚状异常侵蚀"和

本文作者：张福明。原刊于《高炉长寿及快速修补技术研讨会论文集》，1999：164-169。

炉缸环裂。象脚状侵蚀最严重的部位是炉缸、炉底交界处,对应炉缸第 2 段冷却壁的位置,实测发现侵蚀最严重的区域距冷却壁不足 100mm。另外,炉缸环形炭砖均出现环裂现象,裂缝 80~100mm,如图 1 和图 2 所示。

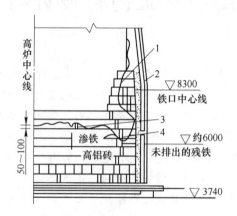

图 1　4 号高炉 1987 年 4 月抢修时炉缸破损情况
1—侵蚀线;2—裂缝;3—熔结物;4—焦炭

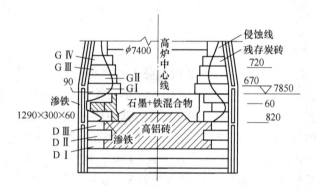

图 2　3 号高炉 1990 年 3 月抢修时炉缸破损情况

结合首钢的原燃料条件和生产条件,研究分析了首钢高炉炉缸侵蚀机理:

(1) 铁水对炭砖的渗透侵蚀。炉缸中铁水渗透到炭砖的气孔中,生成 Fe_3C 一类的脆性物质,形成脆化层,使炭砖热面脆化变质,理化性能下降。

(2) 铁水环流的机械冲刷。由于当时高炉的死铁层深度较浅,致使炉缸死焦柱直接坐在炉底上,炉缸透液性变差。出铁时,铁水沿炉缸周边形成环流,加剧了铁水对炭砖表面脆化层的冲刷磨蚀,使炭砖减薄。

(3) 熔融渣铁及 ZnO、K_2O、Na_2O 等碱金属对炭砖的化学侵蚀。

(4) 热应力对炭砖的破坏。由于炭砖两端存在着较大的温度梯度,产生热应力,造成炭砖断裂,使炉缸出现环裂。

(5) CO_2、H_2O 等氧化性气体对炭砖的氧化破坏。

由此可见,炉缸内衬的破损是一个综合侵蚀过程,有热化学侵蚀、机械磨蚀和热应力破损。因而,用于炉缸内衬的碳质耐火材料必须具有很高的抗渗透性、导热性和抗化学侵蚀性等。而且必须改进炉缸内衬的设计结构，采用合理的设计结构和新型优质耐火材料，抑制象脚状侵蚀和炉缸环裂，延长炉缸寿命。

3 高炉炉缸的内衬设计

3.1 炉缸参数的优化

日本福武刚和柴田耕一郎等对炉缸内铁水流动的研究[1,2]，使人们对炉缸渣铁排放现象有了更为深入的了解。图3是柴田耕一郎模拟试验中发现的炉缸铁水流场的分布。

我国学者刘云彩通过对炉缸死焦柱行为的研究，提出了炉缸上推力的作用及炉缸死铁层深度的推荐值[3]。

设计中研究分析了首钢40多年高炉生产实践，结合国内外最新研究成果，新设计的高炉适当加大了炉缸直径、炉缸高度和死铁层深度等参数，见表1。

表1 首钢高炉炉缸主要参数

炉 号	1	2	3	4
高炉有效容积 V/m^3	2536	1726	2536	2100
炉缸直径 d/mm	11560	9600	11560	10400
炉缸高度 l_1/mm	4200	3700	4200	4350
死铁层深度 h/mm	2200	1400	2200	1600
h/d	0.19	0.146	0.19	0.154
铁口角度	10°	10°	10°	10°
铁口深度 L_t/mm	2201	2250	2124	2116
L_t/R	0.38	0.47	0.37	0.41
风口数/个	30	26	30	28
铁口数/个	3	2	3	2

3.2 炉缸内衬设计体系

对于炉缸内衬的设计，世界上流行着两种不同的设计体系，即"导热法"和"耐火材料法"。导热法的基本原理是采用具有高抗渗透性和高导热性的碳质（或半石墨质）材料，紧贴炉壳或冷却壁砌筑，降低砖衬热面温度，使砖衬热面形成一层保护性"渣皮"或"铁壳"，这种保护层可防止气体、熔融渣铁、碱金属对炭砖的渗透侵蚀和渣铁的冲刷磨蚀，将1150℃等温线向炉缸中心推移，以延长炉缸寿命。耐火材料法（即陶瓷杯法）的技术原理是在炉底满铺炭砖之上砌筑1~2层的莫来石砖（陶瓷垫）；在炉缸壁环形炭砖热面砌一环棕刚玉预制块（陶瓷壁），由低导热的陶瓷质材料在碳质炉衬内形成一个杯状的陶瓷质内衬，即所谓陶瓷杯。利用低导热的陶瓷质材料，将1150℃等温线阻滞在陶瓷层中，使炭砖壁开800~1100℃的脆性断裂区间。由于陶瓷杯的存在，使铁水不直接与炭砖接触，从结构设计上缓解了铁水及碱金属对炭砖的渗透、冲刷等破坏；所采用的莫来石、棕刚玉等均是低导热的高级陶瓷质材料，具有较高的抗渗透性和耐冲刷性[4,5]。

首钢高炉炉缸内衬的设计思想，吸收了导热法和耐火材料法的精华，结合首钢高炉生产实践和经验，在水冷炭砖-高铝砖综合炉底技术的基础上，进行大胆的探索和实践。4座高炉在炉缸、炉底交界处至铁口平面以下的炉缸壁，紧贴冷却壁全部砌筑热压炭砖。1号高炉设计中，将导热法和耐火材料法结合一体，采用热压炭砖-陶瓷杯组合内衬技术。

3.3 热压炭砖技术的应用

美国 UCAR 公司是导热法体系的主要倡导者。其产品热压炭砖 NMA 是在高温加热条件下加压成型的，具有高导热性、低渗透性和优良的耐碱性。另外，NMA 砖形状较小，加上热固性胶泥 C-34 能提供热膨胀的补偿，从而能防止热应力断裂和收缩剥落。NMA 砖的高导热性，使 1150℃ 等温线向炉缸中心推移，促使炉缸壁热面能形成渣铁混凝物保护层，防止气体、熔融渣铁、碱金属对炭砖的渗透、化学侵蚀和渣铁的冲刷磨蚀（见图3）。为提高 NMA 砖的抗碱性，特意加入了 SiO_2，NMA 砖中的 SiO_2 优先与碱金属发生反应，生成层状化合物，从而缓解了由于化学侵蚀后体积膨胀对炭砖所产生的破坏[6]。

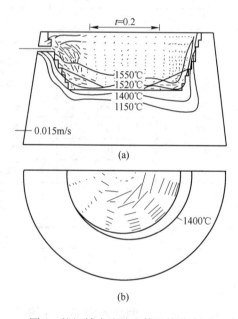

图3　炉缸铁水流动和等温线分布图

2、3、4 号高炉均采用了炭砖-高铝砖综合炉底，炉底总厚度 2800mm。炉底水冷管之上为厚度 150mm 的碳质找平层，炉底满铺 3 层国产炭砖，其上是 4 层综合炉底结构，综合炉底第 1 层周边是国产炭砖，中间部位立砌高度为 400mm 的高铝砖（$Al_2O_3 \geqslant 80\%$）；综合炉底第 2~4 层周边紧贴冷却壁为 NMA 砖，中间部位立砌高度为 350mm 的高铝砖，第 4 层综合炉底以上至铁口平面以下（即死铁层区域），炉缸壁全部采用 NMA 砖，NMA 砖之上，砌筑 3~4 层环形大块炭砖，炉缸壁垂直段 NMA 砖总厚度 1257.3mm。炉缸壁炭砖热面砌一环厚度 345mm 的高铝质保护砖，形成"简易陶瓷杯"结构，以避免开炉初期对 NMA 砖的各类破坏（见表1）。图4 是首钢 2 号高炉炉缸内衬结构图。

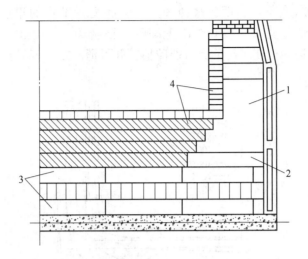

图 4　2 号高炉炉缸内衬结构图
1—NMA 砖；2—国产炭砖；3—满铺炭砖；4—高铝砖

3.4　陶瓷杯技术的应用

首钢 1 号高炉是最后一座进行扩容大修改造的高炉。设计中总结了前 3 座高炉炉缸内衬设计经验和生产实践，经过充分论证，决定在 1 号高炉上将导热法和耐火材料法结为一体，集二者之长，相互补充，以期实现高炉长寿的目标。

1 号高炉炉缸内衬主要由 6 部分组成（见图 5）[7]：（1）炉底 4 层国产满铺大块炭砖；（2）炉底 3 层陶瓷垫（第 1 层为国产莫来石砖，第 2、3 层为 SAVOIE 公司的莫来石砖 MS4）；（3）炉缸壁内侧 5 层陶瓷壁（全部为 SAVOIE 公司的棕刚玉预制块 MONOCORAL）；（4）炉缸壁外侧铁口中心线以下至炉缸炉底交界处（即象脚状侵蚀区），紧贴冷却壁共砌筑 26 层热压炭砖 MS4；（5）炉缸壁外侧铁口中心线以上，共砌筑 6 层国产环形大块炭砖；（6）炉缸环形炭砖和陶瓷壁以上、风口组合砖以下，砌筑高铝砖（$Al_2O_3 \geqslant 80\%$）。

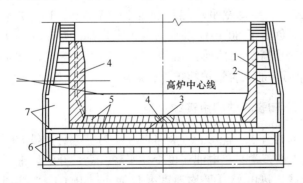

图 5　1 号高炉热压炭砖-陶瓷杯内衬结构
1—捣料；2—国产炭砖；3—国产莫来石刚玉；4—棕刚玉预制块 MONOCORAL；
5—莫来石砖 NMA；6—国产炭砖；7—NMA 砖

采用有限元传热模型对炉缸、炉底内衬等温线进行计算，得出了炉缸、炉底温度场分布曲线，如图6所示。计算表明，这种炉缸、炉底内衬结构的温度场分布是合理的，1150℃等温线处于炉底莫来石砖和炉缸壁棕刚玉预制块中，炉缸、炉底炭砖层避开了800~1100℃的温度区间，使炭砖免受断裂破损，并可消除或减轻炉缸炭砖环裂和象脚状异常侵蚀。

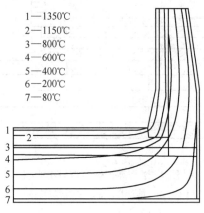

图6　1号高炉炉缸、炉底等温线分布

4　生产实践

4.1　热压炭砖的应用实践

首钢高炉未使用热压炭砖以前，一般开炉9个月左右就要加入含钛炉料护炉，炉缸第2段冷却壁改用高压水冷却，控制冷却壁热负荷在11630W/m² 以下。最早采用NMA砖的2号高炉，于1991年5月开炉，开炉18个月以后才开始投入含钛炉料护炉。直至目前，2号高炉炉缸已安全工作8年，累计产铁1058.8万吨，单位容积产铁量达到6050t/m³。第2段冷却壁水温差平均0.3℃，铁口区冷却壁水温差最高0.5℃（7根水管），80%以上的冷却壁水温差小于0.3℃。冷却壁平均热流强度8140W/m²。TiO_2 加入量7kg/t，铁水中[Ti] 含量为0.115%。创造了80年代以来，首钢大型高炉炉缸工作8年无事故的纪录。1993年6月投产的3号高炉，目前炉缸第2、3段冷却壁进出水温差稳定在0.2℃、0.3℃，高炉运行6年尚未加入含钛炉料护炉，炉缸工作安全稳定，累计产铁1157万吨。

4.2　1号高炉热压炭砖-陶瓷杯的应用实践

1号高炉1994年8月开炉，至今已生产4年9个月，累计产铁892万吨。高炉投产以来，生产稳定，炉缸工作正常，热电偶温度变化平稳。高炉开炉以来，对炉缸内衬的工作状况进行了长期的测试，热电偶月平均温度变化为±0~10℃，平均温升速率仅为0.7~1.5℃/月。目前，炉缸壁4层热电偶的平均温度分别为67℃、94℃、132℃、127℃，第4层铁口区热电偶平均温度186℃。炉缸第2、3段冷却壁85%的水温差稳定在0.2℃、0.3℃以下，铁口区冷却壁水温差0.4℃。该高炉投产至今未加含钛炉料护炉。近5年的生产实践表明，1号高炉炉缸热量充沛，炉缸工作均匀活跃，炉况稳定顺行。而且有利于降低燃料消耗和冶炼低硅低硫生铁，提高铁水温度。目前，1号高炉月平均利用系数 2.25t/(m³·d)，焦比350kg/t，煤比140kg/t，风温1080℃，焦炭负荷4.33t/t，取得了20世纪90年代首钢高炉生产的最佳技术指标。

5 结语

高炉炉缸、炉底寿命是决定高炉一代寿命的关键。炉缸内衬设计结构的不断改进和优化，是实现高炉长寿的基础和前提。

(1) 高炉炉缸内衬设计中，应充分重视炉缸参数的合理性，根据炉缸内衬侵蚀机理和炉缸工作过程的研究分析，确定合理的炉缸高度、死铁层深度、铁口深度等参数。

(2) 炉缸内衬结构和耐火材料的设计选择至关重要。用于炉缸、炉底的碳质耐火材料，特别是用于象脚状侵蚀区的炭砖，必须具有优异的抗铁水渗透性、高导热性和抗化学侵蚀性等。用于陶瓷杯的耐火材料，应根据不同的使用部位进行合理选择。炉底陶瓷垫材料应具备优异的抗铁水渗透性和抗铁水熔蚀性，炉缸陶瓷壁材料应具备优异的抗铁渣侵蚀性、抗碱性和抗CO侵蚀性。另外，设计中必须注重砌体的整体稳定性，碳质材料和陶瓷质材料的合理结合也是设计成败的关键。

(3) 长寿炉缸的设计，重点是炉缸内衬结构的设计，但绝不能忽视炉缸、炉底冷却系统的设计，无论采用何种设计体系，都要强调冷却系统的合理性和重要性。

(4) 首钢高炉实践表明，导热法和耐火材料法炉缸设计体系均能实现高炉长寿。炉缸设计体系和耐火材料的选择应根据具体的生产条件、建设投资等进行具体研究，实现优化合理配置。

参考文献

[1] 福武刚，等. 高炉回旋区和炉缸工作文集 [M]. 北京：冶金工业出版社，1986：1-19.
[2] 柴田耕一郎，等. 高炉炉缸铁水流动和控制 [J]. 国外钢铁，1992 (10)：14.
[3] 刘云彩. 高炉炉缸上推力的作用 [J]. 钢铁，1995 (12)：1-4.
[4] Bauer J M. Ceramic cups to extend BF hearth life [J]. Steel Time, 1989 (1)：18.
[5] Bauer J M. Low couductivity blast furnace refractories [J]. Steel Times, 1993 (1)：49.
[6] Albert J D. Blast furnace hearth design theory, materials and practice [J]. Iron and Steel Engineer, 1991 (12)：23-30.
[7] 张福明. 热压炭砖——陶瓷杯技术在首钢1号高炉上的应用 [J]. 炼铁，1996 (4)：12-15.

首钢高炉高效长寿技术设计与应用实践

摘　要：阐述了首钢高炉采用的多项高效长寿技术及取得的应用效果，对高炉内型、炉缸内衬结构、冷却体系、自动化检测、生产操作管理等技术进行了评述。

关键词：大型高炉；高效；长寿；设计

1　引言

20世纪90年代初，首钢总公司为充分发挥企业自身的潜力，将首钢建成大型钢铁联合企业，相继对2、4、3号高炉及1号高炉进行扩容和现代化新技术改造。随着北京奥运会的召开及首钢搬迁转移、战略性结构调整的需要，2号及4号高炉于2008年停产，1号及3号高炉分别于2010年12月18日及19日停产。1号及3号高炉至停产时，高炉运行状况良好，1、3号高炉及4号高炉炉龄分别达到16.4年、17.6年及15.6年，一代炉役单位炉容产铁量分别为13328t/m³、13991t/m³及12560t/m³，达到国内外高炉高效长寿的先进水平。

2　首钢高炉高效长寿技术设计

高炉高效长寿设计的关键是高炉内型、内衬结构、冷却体系、自动化检测的有机结合[1,2]。生产实践表明，目前高炉炉缸、炉底和炉腹、炉腰、炉身下部是高炉长寿的两个限制性环节，在设计中攻克这两个部位的短寿难题，将为高炉长寿奠定坚实的基础。首钢高炉炉体设计紧密围绕上述几个方面，通过炉型设计优化，选择矮胖炉型，为高炉生产稳定顺行、高效长寿创造有利条件；通过炉缸炉底的侵蚀机理分析研究，炉缸炉底部位采用"优质高导热炭砖-陶瓷杯"及"优质高导热炭砖-陶瓷垫"新型综合炉底内衬结构；炉腹至炉身区域采用软水密闭循环冷却技术、双排管铸铁冷却壁技术、倒扣冷却壁（"C"形冷却壁）技术，并实现了合理配置；有针对性地设计炉体自动化检测系统，加强砖衬侵蚀与冷却系统的检测、监控。通过这些现代高炉长寿技术的综合应用，以实现高炉高效长寿的目标。

2.1　首钢高炉矮胖炉型设计

我国炼铁工作者历来重视高炉炉型设计，通过研究总结高炉破损机理和高炉反应机理[3~5]，优化高炉炉型设计的基本理念已经形成。

在总结当时国内外同类容积高炉内型尺寸的基础上，根据首钢的原燃料条件和操作条

本文作者：毛庆武，张福明，姚轼，钱世崇。原刊于《炼铁》，2011，30（5）：1-6。

件,以适应高炉强化生产的要求,设计了矮胖炉型。首钢1号及3号高炉炉容、炉型相同,均为2536m³,高径比均为1.985,是当时同类级别高炉高径比最小的高炉,引起了国内外炼铁工作者的广泛关注和大讨论,引领了高炉矮胖炉型的发展,也为高炉矮胖炉型的设计奠定了坚实的基础。

实践表明,高炉炉缸炉底"象脚状"异常侵蚀的形成,主要是由于铁水渗透到炭砖中,使炭砖脆化变质,再加上炉缸内铁水环流的冲刷作用而形成的。加深死铁层深度,是抑制炉缸"象脚状"异常侵蚀的有效措施。死铁层加深以后,避免了死料柱直接沉降在炉底上,加大了死料柱与炉底之间的铁流通道,提高了炉缸透液性,减轻了铁水环流,延长了炉缸炉底寿命。理论研究和实践表明,死铁层深度一般为炉缸直径的20%左右。

高炉在大喷煤操作条件下,炉缸风口回旋区结构将发生变化。适当加高炉缸高度,不仅有利于煤粉在风口前的燃烧,而且还可以增加炉缸容积,以满足高效化生产条件下的渣铁存储,减少在强化冶炼条件下出现的炉缸"憋风"的可能性。近年我国已建成或在建的大型高炉都有炉缸高度增加的趋势,适宜的高炉炉缸容积应为有效容积的16%~18%。

铁口是高炉渣铁排放的通道,铁口区的维护十分重要。研究表明,适当加深铁口深度,对于抑制铁口区周围炉缸内衬的侵蚀具有显著作用,铁口深度一般为炉缸半径的45%左右。这样可以减轻出铁时在铁口区附近形成的铁水涡流,延长铁口区炉缸内衬的寿命。

降低炉腹角有利于炉腹煤气的顺畅排升,从而减小炉腹热流冲击,而且还有助于在炉腹区域形成比较稳定的保护性渣皮,保护冷却器长期工作。现代大型高炉的炉腹角一般在80°以内,国内E号高炉(2600m³)炉腹角已降低到75°57′49″。国内几座2500m³级高炉内型尺寸比较见表1。

表1 国内几座2500m³级高炉内型尺寸比较

项目	单位	首钢1、3号高炉	迁钢1、2号高炉	A号高炉	B号高炉	C号高炉	D号高炉	E号高炉
有效容积 V_u	m³	2536	2650	2580	2500	2560	2516	2600
炉缸直径 d	mm	11560	11500	11500	11400	11000	11200	11000
炉腰直径 D	mm	13000	12700	13000	12750	12200	12200	12800
炉喉直径 d_1	mm	8200	8100	8200	8100	8300	8200	8200
死铁层高度 h_0	mm	2200	2100	2004	2500	2200	2004	1900
炉缸高度 h_1	mm	4200	4200	4100	4500	4600	4500	4300
炉腹高度 h_2	mm	3400	3400	3600	3400	3400	3400	3600
炉腰高度 h_3	mm	2900	2400	2000	1800	1800	1900	2000
炉身高度 h_4	mm	13500	16600	17500	17000	17500	17400	17000
炉喉高度 h_5	mm	1800	2200	2300	2000	2000	2300	2000
有效高度 H_u	mm	25800	28800	29500	28700	29300	29500	28900
炉腹角 α		78°02′36″	79°59′31″	78°13′54″	78°46′15″	79°59′31″	81°38′02″	75°57′49″
炉身角 β		79°55′09″	82°06′42″	82°11′27″	82°12′44″	83°38′30″	83°26′34″	82°17′42″

续表1

项　　目	单位	首钢 1、3号高炉	迁钢 1、2号高炉	A号高炉	B号高炉	C号高炉	D号高炉	E号高炉
风口数	个	30	30	30	30	30	28	28
铁口数	个	3	3	3	3	3	2	3
渣口数	个	无	无	无	无	无	无	无
风口间距	mm	1211	1204	1204	1194	1152	1257	1234
H_u/D		1.985	2.268	2.269	2.251	2.402	2.418	2.258
V_1/V_u	%	17.38	16.29	15.16	17.29	16.96	17.10	15.31

2.2　炉缸炉底内衬结构设计

实践表明，高炉炉缸、炉底的寿命是决定高炉一代寿命的关键[6~8]，受到国内外炼铁工作者的高度重视。

从20世纪60年代起，首钢高炉开始采用炭砖—高铝砖综合炉底技术，使用情况一直较好。随着炼铁技术的发展，到20世纪80年代中期以后，高炉冶炼强度提高，炉缸、炉底问题变得突出。通过对10多次高炉停炉实测结果的研究分析[9]，总结得出了首钢高炉炉缸、炉底内衬的侵蚀是典型的"象脚状"异常侵蚀和炉缸环裂。象脚状异常侵蚀最严重的部位发生在炉缸、炉底交界处，对应炉缸第2段冷却壁的位置，实测发现侵蚀最严重的区域距冷却壁不足100mm。残余炭砖和高铝砖表面黏结有凝固的渣、铁及Ti（C、N）等高熔点凝结物。炉缸壁环形炭砖均出现环裂现象，裂缝80~200mm，裂缝中渗有凝固的渣、铁。

结合首钢高炉的原、燃料条件和操作条件，研究分析了首钢高炉炉缸、炉底内衬侵蚀机理，主要如下：（1）铁水对炭砖的渗透侵蚀；（2）铁水环流的机械冲刷；（3）熔融渣铁及ZnO、Na_2O、K_2O等碱金属对炭砖的熔蚀和化学侵蚀；（4）热应力对炭砖的破坏；（5）CO_2、H_2O等氧化性气体对炭砖的氧化破坏。

长寿炉缸炉底的关键是必须采用高质量的炭砖并辅之合理的冷却[10,11]。通过技术引进和消化吸收，我国大型高炉炉缸炉底内衬设计结构和耐火材料应用已达到国际先进水平。

以美国UCAR公司为代表的"导热法"（热压炭砖法）炉缸设计体系已在本钢、首钢、宝钢、包钢、湘钢、鞍钢等企业的大型高炉上得到成功应用；以法国SAVOIE公司为代表的"耐火材料法"（陶瓷杯法）炉缸设计体系在首钢、梅山、鞍钢、沙钢、宣钢等企业的大型高炉上也得到了推广应用；进口大块炭砖—综合炉底技术在宝钢、武钢、首钢京唐等企业的大型高炉上也取得了长寿实绩。"导热法"和"耐火材料法"这两种看来似乎截然不同的设计体系，其技术原理的实质却是一致的，即通过控制1150℃等温线在炉缸炉底的分布，使炭砖尽量避开800~1100℃脆变温度区间。导热法采用高导热、抗铁水渗透性能优异的热压小块炭砖NMA，通过合理的冷却，使炭砖热面能够形成一层保护性渣皮或铁壳。并将1150℃等温线阻滞在其中，使炭砖得到有效的保护，免受铁水渗透、冲刷等破坏。陶瓷杯法则是在大块炭砖的热面采用低导热的陶瓷质材料，形成一个杯状的陶瓷内衬。即所谓"陶瓷杯",其目的是将1150℃等温线控制在陶瓷层中。这两种技术体系都必须采用具有高导热性且抗铁水渗透性能优异的炭砖。

首钢2号（1726m³）、3号（2536m³）、4号（2100m³）高炉是"炭质炉缸-综合炉底"结构（见图1），1号高炉（2536m³）是"炭质-陶瓷杯复合炉缸炉底"结构（见图2），炉缸、炉底交界处即"象脚状"异常侵蚀区，均部分引进了美国UCAR公司的小块热压炭块NMA。这两种结构在首钢均得到成功应用，已取得了长寿业绩，首钢北京地区高炉炉龄统计见表2。特别是首钢1号和3号高炉炉容、炉型相同，在其他因素基本相同的条件下其炉龄基本是并驾齐驱，这也充分说明了当今炉缸炉底结构这两种技术主流模式基本成熟。

首钢高炉炉底陶瓷垫与炭砖的总厚度为2800mm。风口、铁口区域设计采用刚玉莫来石组合砖，提高其稳定性和整体性。

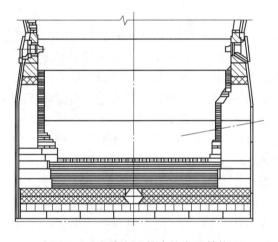 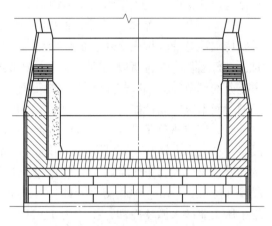

图1 "炭质炉缸-综合炉底"结构　　　　图2 "炭质-陶瓷杯复合炉缸炉底"结构

表2　首钢北京地区高炉炉龄统计

项目	容积/m³	开炉日期	停炉日期	炉龄	备注
1号高炉	2536	1994-08	2010-12	16年5个月	因北京市环保搬迁停炉
2号高炉	1726	1991-05	2002-03	10年10个月	
3号高炉	2536	1993-06	2010-12	17年7个月	因北京市环保搬迁停炉
4号高炉	2100	1992-05	2008-01	15年8个月	因奥运会停炉

2.3　高效长寿冷却技术的设计

（1）高炉冷却设备设计。20世纪90年代，高炉冷却主要有以下几种方式：1）炉腹至炉身下部全部采用铜冷却板；2）采用全部冷却壁；3）采用冷却壁与冷却板的组合方式。为使高炉寿命达到10~15年，首钢高炉全部采用冷却壁结构，在选择高炉各部位的冷却壁型式时考虑了以下因素：1）炉缸、炉底区域此部位的热负荷虽然较高，但比炉腹以上区域的热负荷要小，并且温度波动较小，在整个炉役中冷却壁前的炭砖衬能很好地保存下来，使冷却壁免受渣铁的侵蚀，因此在炉底、炉缸部位（包括风口带）均采用导热系数较高的灰铸铁（HT200）光面冷却壁，共设5段光面冷却壁。2) 高炉中部。这一区域跨越了炉腹、炉腰及炉身下部，是历来冷却壁破损最严重的部位。由于砖衬（渣皮）不能长期稳定地保存下来，冷却壁表面直接暴露在炉内，受到剧烈的热负荷作用和冲击、渣铁侵蚀、强烈的煤气流冲刷和炉料的机械磨损等，所以要求此区域的冷却壁有较高的热机械性能及较强的冷却能力。设计时采用了第三代双排管捣料型冷却壁，壁体材质为球墨铸铁（QT400-18），共设6段，炉腰及炉身下部冷却壁带凸台。2号高炉在2002年技术改造时第一次设计采用了3段国产铜冷却壁。3) 高炉中上部。此区域的冷却壁寿命主要受炉料的磨损、煤气流的冲刷及碱金属的化学侵蚀，并承受较高的热负荷，所以设计时采用了镶砖型带凸台冷却壁，壁体材质为球墨铸铁（QT400-18），镶砖材质为黏土砖，共设4段冷却壁。4) 在炉身上部至炉喉钢砖下沿，增加1段"C"形球墨铸铁水冷壁，水冷壁直接与炉料接触，取消了耐火材料内衬。

（2）高炉冷却系统的设计。根据首钢多年的实践得出采用先进的炉缸炉底结构的同时要特别注意炉缸炉底冷却，加强检测与监控。关键部位选用高导耐侵蚀的优质炭砖的同时，进行强化冷却，所以在冷却水量上要节约而不要制约，在冷却流量的设计能力上要考虑充分的调节能力，冷却流量控制应根据生产实践的实际情况实施，从而达到节能降耗的目的，而不能在设计能力上以冷却水量小说明设计先进，从而导致能力不足，在检测到炉缸炉底温度或热负荷异常时诸多措施难以实施。首钢高炉炉底水冷管、炉缸冷却壁（1~5段）、"C"形冷却壁、风口设备采用工业净水循环冷却，其中炉底水冷管、第1、4、5段冷却壁、风口大套采用常压工业水冷却，水压为0.60MPa（高炉±0.000m平面）；第2、3段冷却壁位于炉缸、炉底交界处即"象脚状"异常侵蚀区，故在此处进行强化冷却，采用中压工业净水循环冷却，压力为1.2MPa（高炉±0.000m平面）。风口中、小套采用高压工业净水循环冷却，压力为1.7MPa（高炉±0.000m平面）。炉腹以上冷却壁（"C"形冷却壁除外）采用软水密闭循环冷却（3号高炉除外）。

2.4　自动化检测与控制

自动化检测是高炉长寿不可缺少的技术措施。炉缸炉底温度在线监测已成为监控炉缸炉底侵蚀状态的重要手段，也是建立炉缸炉底内衬侵蚀数学模型所必要的条件。炉腹、炉腰、炉身下部区域，温度、压力的检测为高炉操作者随时掌握炉况提供了有效的参考。通过对冷却水流量、温度、压力的检测，可以计算得出热流强度、热负荷等参数，而且还可以监控冷却系统的运行状况。炉喉固定测温、炉顶摄像、煤气在线自动分析、炉衬测厚等技术的应用使高炉长寿又得到了进一步的保障。2号高炉在2002年技术改造时，引进了人工智能高炉冶炼专家系统，为延长高炉寿命创造了有利条件。

3　高炉高效长寿的生产管理

高炉的高效长寿离不开高炉冶炼技术的进步、高效长寿技术的研究与应用和生产操作的科学管理。

3.1 加强高炉的日常监测

（1）炉缸水温差自动监测。实现实时采集监测炉缸冷却水温差与热流强度变化，才能对炉缸工作状态进行正确判断，并据此做出相应的高炉上下部调剂、护炉措施及产量调节，以保证生产的顺利进行，延长高炉的使用寿命，达到长寿和高效的统一。

为实现实时采集监控炉缸第2段及第3段、铁口区域冷却壁水温差与热流强度变化，满足实时监控高炉炉缸运行状况的需要，开发在线冷却壁水温采集模块、数据处理模块及通信模块，建立炉缸冷却壁水温差在线采集通信系统，实时采集冷却壁水温，计算炉缸冷却壁的温差及热流强度，创建生产过程中冷却壁温差、炉缸热负荷的数据库，具备查看炉缸冷却水温差随时间变化曲线、查看炉缸热负荷随冷却壁变化的柱状图、炉缸热负荷圆周分布图、炉缸热负荷详细报表等功能，并实现炉缸热负荷超限的实时报警提示，为生产过程中高炉炉内状况和操作提供参考及指导。

（2）软水冷却系统检测与控制。除3号高炉炉腹以上冷却系统分为两段：即第6~12段冷却壁工业水冷却系统和第13~15段冷却壁软水密闭循环冷却系统外，1、2号及4号高炉炉腹以上冷却均为软水密闭循环冷却。

在软水供回水管路上均设有流量、压力、温度检测装置；冷却壁的每根回水支管上均设有压力表；炉体圆周冷却壁支管上间隔均布支管温度检测，以计算冷却壁的平均热负荷。膨胀罐上设有压力过低报警、水位过低报警及补水压力过低报警装置。以上检测数据除在水泵房有显示、记录外，还送入高炉主控室计算机，实现显示、存贮、记录、报警和打印等功能。软水系统的控制和调节，膨胀罐上的氮气稳压系统和补水系统均为自动控制。

（3）高炉专家系统。高炉专家系统拥有功能强大的数据库，利用检测设备直接测量及专家系统自动生成的数据，专家系统能够描绘高炉各项冶炼参数的变化趋势，尤其体现在十字测温、煤气成分、炉衬温度、冷却壁壁后温度的变化趋势等方面，专家系统提供高炉冶炼参数的变化趋势，用于分析判断炉况的变化，优化经济技术指标。

3.2 加强高炉的日常维护

（1）炉缸工作状态控制。高炉顺行稳定生产要求炉缸工作活跃，中心死料堆具有足够的透气性和透液性，炉缸环流减弱。若炉缸中心死料堆透气性和透液性差，铁水积聚在炉缸边缘，在出铁时易形成铁水环流导致炉缸内衬局部出现侵蚀，引发炉缸局部过热及炉缸烧出等事故。炉缸中心死料堆透气性和透液性差，大量渣铁滞留在死料堆中导致炉缸初始煤气难以渗透到中心，破坏炉内煤气分布，影响高炉炉内顺行及炉体长寿。因此，要采取活跃炉缸中心死料堆的措施，保持适当的炉缸炉底及侧壁温度，维持活跃的炉缸工作状态。

炉缸侧壁温度、炉缸炉底温度反映了炉缸内的温度场变化，随产量的提高，炉缸侧壁温度和炉底温度都呈升高趋势，随煤比的提高，炉缸侧壁温度呈升高趋势而炉底温度则呈下降趋势。炉缸工作活跃指数是监测炉缸工作状态的重要参数，为高炉长期高煤比生产下的冶炼参数调整提供依据，以达到高炉的顺行稳定生产。

提高原燃料质量，在高炉下部保持足够、稳定的鼓风动能的基础上，上部装料制度控制中心与边缘煤气的合理分配从而达到高炉顺行，这些措施有利于提高炉缸工作状态活跃性。通过对炉缸工作活跃指数的监测，及时调整各项高炉冶炼参数，保持指数在正常范围内，实现了高炉在高煤比下的顺稳生产，且炉缸侧壁温度保持在较低水平，实现了炉缸的长寿。

另外，炉缸压浆技术已成为现阶段延长高炉炉缸及铁口区域寿命的重要技术措施。出铁口区域一直是高炉压入维护的重点，是高炉寿命的薄弱环节。出铁口区域的砖衬往往受到来自泥炮、开口机对砖衬的反复冲击，砖衬易出现裂纹，形成高温煤气泄漏通道，需要进行铁口区域的压入维护，否则，不仅影响炉前工作的组织，而且造成铁口堵口困难等，也影响高炉顺行操作和高炉整体使用寿命。炉缸压浆技术的采用应着重注意不要损坏冷却壁及砖衬，在进行灌浆孔开孔时，注意避开冷却壁，防止损坏冷却壁本身；同时在压入过程中，掌握好入口压力和压入节奏，防止过高的压力冲击高炉炉缸砖衬（尤其是中后期高炉）。

（2）煤气流分布控制。合理煤气分布涉及高炉稳定顺行、节能降耗、长寿等问题，首钢高炉合理煤气分布目标：一是炉况的稳定顺行，二是煤气利用的提高、燃料比的降低，代表性的煤气分布形态为"中心煤气开、边缘煤气稳定"，中心煤气的"开"表现中心火柱窄而强，炉况顺行好，煤气利用率高、燃料比低，炉缸工作活跃。边缘煤气流的过分发展，不但会造成炉体热负荷升高，影响高炉长寿，而且煤气利用率变差，能量消耗高，影响高炉长期稳定顺行；边缘煤气流的稳定，有利于冷却壁的保护和渣铁保护层的稳定，中心煤气流对煤气利用、能量消耗、强化冶炼产生影响，也对边缘煤气流的稳定产生直接影响。高炉合理煤气分布的目标是实现高炉的稳定顺行，在此基础上提高煤气利用，实现高炉炼铁的节能降耗，实现高炉的高效长寿。

（3）操作炉型管理。高炉操作炉型管理涉及到炉型设计、冷却设备配置、耐火材料使用等设计因素，原料管理、炉体冷却、煤气分布、出铁管理、工长操作等使用因素，是高炉技术管理的综合体现，操作炉型是否能够长期稳定、合理也是高炉长寿的基础。高炉操作炉型管理应作为最重要的高炉生产日常管理制度，及时、准确了解高炉炉型的变化，量化分析得到的炉型变化信息，以判断、解决引起炉型变化的因素，维持正常的高炉操作炉型。

为减缓炉体的破损，首钢应用了高炉炉内遥控喷补造衬技术，喷补形成一个符合高炉冶炼规律的近似操作炉型，有利于维护炉墙冷却壁的使用寿命，延长高炉风口以上区域寿命，为高炉炉内煤气合理分布创造可靠的外围环境。

（4）炉前作业管理。高炉强化冶炼后，渣铁能否及时出净已成为高炉稳定、顺行的关键。出渣、出铁不及时易造成死焦堆中的渣铁渗透困难，破坏炉缸初始煤气分布，影响高炉操作炉型。铁口维护则直接影响铁口区域的操作炉型维护，铁口深度连续过浅，铁口区域炭砖易造成严重侵蚀，影响高炉炉缸的长寿。

量化分析高炉的出铁间隔、出铁时间、见渣时间、出铁量、铁口深度、打泥量，积极提高炉前操作水平，确保高炉不憋风，减少铁口冒泥，稳定铁口深度，提高炉前作业的稳定性。炮泥质量对出铁影响较大，要稳定炮泥质量，开发适应不同炉况和冶炼强度的炮泥，充分利用无水炮泥强度高、抗渣铁侵蚀性能好的特点，采用出铁次数少、出铁时间长、出铁间隔短的出铁方式是高炉炉前作业的趋势。

(5) 加钛护炉技术。现代高炉强化冶炼程度较高，尤其是处于炉役末期的高炉，含钛料的加入应成为炉缸维护的日常措施。长期连续加入含钛料，控制适宜的 TiO_2 加入量，这样一方面可在炉缸内部形成黏度较高的保护层，减缓铁水对炉缸的冲刷侵蚀；另一方面可在炉缸侵蚀处及时形成 TiC、TiN 及 Ti（CN）的聚集物，避免炉缸内部发生连续性侵蚀。高熔点的 TiC、TiN 及 Ti（CN）在炉缸生成、发育和集结，与铁水及铁水中析出的石墨等形成黏稠状物质，凝结在离冷却壁较近的被侵蚀严重的炉缸砖缝和内衬表面，进而对炉缸内衬起到保护作用。使用含钛料护炉时，炉渣中 TiC、TiN 在炉缸温度范围内不能熔化，以固态微粒悬于渣中，使炉渣流动能力恶化，TiC 和 TiN 越多，炉渣越黏，严重时失去流动性。首钢高炉在使用含钛料护炉时，合理控制入炉 TiO_2 加入量，合理利用炉缸温度梯度，可以较好的解决高炉炉缸维护与强化冶炼的矛盾。

4 生产指标

首钢注重高炉高效长寿的设计研究与生产操作的科学管理，实现了高炉高效长寿的目标。2010 年，是首钢搬迁转移、战略性结构调整的最后一年，1 号及 3 号高炉（2536m³）在炉役后期仍然保持安全稳定的生产，取得了较好的技术经济指标（见表3）。

表3 首钢1、3号高炉主要技术经济指标

项目	产量/t	利用系数/$t \cdot (m^3 \cdot d)^{-1}$	焦比/$kg \cdot t^{-1}$	煤比/$kg \cdot t^{-1}$	燃料比/$kg \cdot t^{-1}$	富氧/%	休风率/%	热风温度/℃	综合品位/%
1号高炉	2093035	2.338	340.4	144.9	505.2	1.12	2.11	1136	58.79
3号高炉	2026395	2.277	358.1	124.2	499.9	0.68	1.86	1076	58.85

5 结语

首钢高炉高效长寿技术水平虽有长足的提高，但与国际领先水平相比还有一定的差距，要真正达到世界领先水平仍需要继续努力。近年来，由于高炉强化冶炼，我国一些大型高炉出现炉缸烧出的情况，应引起炼铁工作者的高度重视。结合首钢及国内钢厂的实情，加强高炉高效长寿的设计研究与生产操作的科学管理，开发高炉高效长寿新技术，实现高炉高效长寿仍然是炼铁工作者重点研究的课题。当前，我国新建或大修改造的大型高炉，遵循高效、长寿并举的原则，高炉一代炉役设计寿命15～25年，一代炉役平均利用系数大于2.2，一代炉役单位有效容积产铁量达到12000～20000t/m³，相信在不久的将来，我国大型高炉长寿实绩将达到国际领先水平。

参考文献

[1] 张福明. 我国大型高炉长寿技术发展现状 [C]//2004 全国炼铁生产技术暨炼铁年会论文集. 无锡：中国金属学会炼铁分会, 2004：566-570.
[2] 张寿荣. 延长高炉寿命是系统工程高炉长寿技术是综合技术 [J]. 炼铁, 2002, 21 (1)：1-4.
[3] 宋木森, 邹明金, 等. 武钢4号高炉炉底炉缸破损调查分析 [J]. 炼铁, 2001, 20 (2)：7-10.

[4] 曹传根,周渝生,等.宝钢3号高炉冷却壁破损的原因及防止对策[J].炼铁,2000,19(2):1-5.
[5] 黄晓煜,孙金铎.鞍钢7号高炉炉身破损原因剖析[J].炼铁,2001(6):1-4.
[6] 傅世敏.高炉炉缸铁水环流与内衬侵蚀[J].炼铁,1995,14(4):8-11.
[7] 傅世敏.大型高炉合适炉缸高度的探讨[J].钢铁,1994,38(12):7-10.
[8] 傅世敏,周国凡,等.高炉炉缸结构与寿命[J].炼铁,1997,16(6):32-34.
[9] 单泪华,王颖生,等.通过2号高炉破损调研探索首钢高炉长寿途径[C]//2003中国钢铁年会论文集.北京:冶金工业出版社,2003:491-496.
[10] 钱世崇,程素森,张福明,等.首钢迁钢1号高炉长寿设计[J].炼铁,2005,24(1):6-9.
[11] 毛庆武,张福明,张建,等.迁钢1号高炉采用的新技术[J].炼铁,2006,25(5):5-9.

Design and Application of High Efficiency and Long Campaign Technology in Shougang

Mao Qingwu Zhang Fuming Yao Shi Qian Shicong

Abstract: The paper expounds several high efficient long campaign technologies used in blast furnace of Shougang and their application result, and gives comments to internal profile of blast furnace, structure of furnace hearth lining, cooling system, automation measurement and production management, etc..

Keywords: large sized blast furnace; high efficient; long campaign; design

铜冷却壁的设计研究与应用

摘　要：本文对铜冷却板和铸铁冷却壁进行了分析评述。讨论了铸铁冷却壁存在的技术缺陷，介绍了铜冷却壁设计研究及应用实践。

关键词：高炉；长寿；设计；铜冷却壁

1　引言

随着高炉炼铁技术进步，高炉正向大型化、高效化、现代化、长寿化方向发展，当前世界各主要产钢国都在竞相延长高炉寿命。到20世纪90年代末，日本和欧美等国的高炉寿命已达到或超过 10~15 年。我国高炉一代寿命也正向 10 年迈进，近年又有了新的突破。实践表明，高炉易损区域主要有两个，一是炉缸、炉底；二是炉腹至炉身下部。纵观近年国内外高炉长寿技术的发展历程，由于高炉炉缸炉底结构设计、耐火材料和冷却系统的不断改进和完善，影响高炉寿命的关键部位已集中到高炉炉腹至炉身下部区域。如何延长高炉炉腹至炉身下部区域的寿命，已成为国内外炼铁工作者关注的技术热点。

2　高炉冷却器的技术评价

高炉长寿技术的设计是高炉长寿的前提。长寿高炉设计的技术思想应是：（1）改善冷却介质，采用软水（纯水）密闭循环冷却系统；（2）采用新型高效冷却器；（3）采用优质耐火材料。这是高炉长寿的必要条件，当然，高质量的施工、合理的操作对于高炉长寿也同样重要。应该指出，新型高效冷却器对高炉长寿的作用不可忽视，国内外高炉长寿实践已充分证明了这一点。

高炉采用水冷却始于1884年，到1970年的近百年，高炉冷却主要是采用炉壳喷水冷却和冷却板冷却两种方式。20世纪70年代，原苏联设计开发了铸铁冷却壁，30年来，冷却壁技术发展迅猛，已成为高炉冷却结构的一种主要发展模式，方兴未艾。

2.1　冷却板与冷却壁

现代大型高炉炉腹至炉身下部的高热负荷区已很少采用炉壳喷水冷却，主要是冷却板与冷却壁冷却，同时还有两种技术组合的"板壁结合"结构。从实践结果分析，这三种冷却模式都有取得高炉长寿的实绩，同时又有明显的优点和缺陷（见表1）。冷却壁的主要优点是对高炉炉壳全部区域提供冷却（即所谓面式冷却）；而冷却板则只能对炉壳进行局部冷却（即所谓点式冷却）。但冷却板由于插入到砖衬中，可以提供高效的"深度"冷

本文作者：张福明，姚轼，苏维，黄晋。原刊于《铜冷却壁技术研讨会文集》，2003：39-42，50。

却,同时还可以对砖衬提供有效的支撑;而冷却壁只能对砖衬的冷面进行冷却,对砖衬的支撑作用也不如冷却板。冷却板损坏以后,可以方便地进行更换,而冷却壁损坏时则更换困难,在绝大多数情况下,更换冷却壁需要在高炉休风时进行。

表1 冷却板与铸铁冷却壁的对比

项目	冷却板	铸铁冷却壁
优点	(1) 对砖衬提供高效冷却; (2) 有利于支撑砖衬; (3) 更换简便、快捷; (4) 设计成多通道结构,提高冷却效率; (5) 采用密集式布置,增强冷却效果	(1) 冷却全部炉壳; (2) 高炉热损失少; (3) 冷却均匀,操作炉型合理; (4) 炉壳开孔少,减少炉壳热应力破损; (5) 双排管结构,强化凸台冷却(第三代)砖壁一体化,减薄砖衬厚度,使施工安装简化(第四代)
缺点	(1) 高炉热损大; (2) 不能对炉壳提供均匀、全部冷却; (3) 高温状态下易弯曲变形; (4) 炉壳开孔多,炉壳设计复杂; (5) 不利于形成稳定的操作炉型; (6) 要求匹配高级耐火材料(如半石墨、碳化硅等)	(1) 对砖衬支撑效果差; (2) 不易于更换维修; (3) 冷却壁边角及凸台部位易破损; (4) 在超过760℃时工作会出现相变,力学性能下降; (5) 水管与铸铁冷却壁体之间热阻大,传热效率低于铜冷却板

高炉采用何种冷却结构,与高炉所采用的炉料结构有重要关系。普遍认为,采用高球团矿率的高炉,炉墙热负荷和热震性波动较大,因而采用高导热性的耐火材料(如石墨、半石墨、碳化硅等)配合密集式铜冷却板较为适宜。而采用以烧结矿为主的高炉,炉墙热负荷较为稳定,温度波动较小,采用冷却壁更为合理。当然这还取决于工厂的传统和操作者的习惯。

2.2 铸铁冷却壁

铸铁冷却壁问世至今,已有30年的演变和发展史。日本已将铸铁冷却壁发展到第四代。其主要技术特征是采用力学性能优良的球墨铸铁,加强了壁体边角部位及凸台的冷却,采用独立供水的双排管冷却结构,壁体热面铸入SiC砖,形成砖壁一体化。法国FORCAST公司和澳大利亚BHP公司对铸铁冷却壁也进行了优化改进。采用改进后的铸铁冷却壁的高炉,在不中修的条件下,寿命已达到10~12年,但冷却壁破损严重。研究表明,铸铁冷却壁在高炉操作状态下,特别是热负荷和温度急剧波动的条件下,铸铁壁体的金相组织发生变化,出现热应力裂纹和龟裂。即使是第四代冷却壁,也不能完全克服这些缺陷,这些因素限制了冷却壁的寿命。

铸铁冷却壁出现破损的本质原因,应该说是由于铸铁冷却壁的制造工艺造成的。在冷却壁铸造过程中,为防止高碳铸铁中的石墨渗透到低碳的钢管中,必须在冷却水管表面进行防渗碳处理,这种防渗碳保护层的导热系数低,尽管厚度仅为0.1~0.3mm,但在操作状态下,所形成的热阻却相当大。引起壁体内温度梯度增大和热面温度升高,局部过热,这种作用也增加了裂纹生成的倾向。图1是铸铁冷却壁内的热负荷分布[1]。由此可见,长寿冷却壁应具备两个关键条件:一是冷却壁本体材料;二是要消除附加热阻。

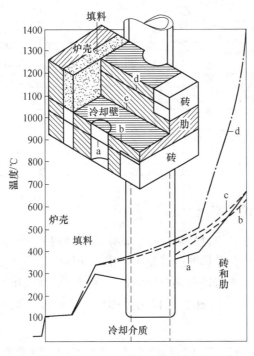

图 1 铸铁冷却壁热负荷分布

3 铜冷却壁的设计研究

为克服铸铁冷却壁的技术缺陷,新型高效冷却壁的开发成为必然。

3.1 技术思想

高效冷却壁的设计,关键是要求冷却壁本体无过热。只要将壁体热面温度控制在一定的温度范围,化学侵蚀、热应力破坏、机械磨损都将受到抑制,以延长其使用寿命。因此,采用优异力学性能和热学性能的材料,在设计结构上减少或避免附加热阻的形成是设计的关键。针对铸铁冷却壁的技术缺陷,德国 MAN-GHH 公司开发了铜冷却壁,并在蒂森公司的高炉上应用,获得成功[1]。即采用轧制铜板作为冷却壁本体,冷却通道直接钻孔而成,这样提高冷却水和壁体之间的传热速率,消除了铸入水管防渗碳保护层的附加热阻,大大增强了冷却壁的冷却效率。同铸铁相比,铜具有良好的热学性能和力学性能(见表2)。用轧制无氧铜(TU$_2$)板直接钻孔而形成冷却通道,提高冷却壁的传热速率和效率,依靠铜的高导热性,使铜冷却壁热面保持较低的工作温度,促使铜冷却壁表面能够形成稳定的保护性"渣皮",依靠保护性的"渣皮"代衬工作,这是铜冷却壁技术的基本设计思想。

表 2 铜、铸铁的物理和热学性能

项 目	铜	灰铸铁	球墨铸铁
抗拉强度 σ_b/MPa	196	160	400
屈服强度 σ_s/MPa			250

续表2

项　目	铜	灰铸铁	球墨铸铁
延伸率 δ_s/%	30	0	20
龟裂前热循环次数（300~900℃）		30~40	203~250
密度/g·cm^{-3}	8.9	7.0	7.2
熔点/℃	1083	约1150	约1150
导热系数/W·(m·K)$^{-1}$	360（400℃）	62.8	30~35
比热容/J·(kg·K)$^{-1}$	383	480	544

3.2 铜冷却壁的开发研制

铜冷却壁的设计，首先经过传热计算，确定冷却壁的外形尺寸、冷却通道数量、冷却通道间距、冷却通道直径、壁体厚度以及燕尾槽尺寸等热工参数。冷却通道的直径至关重要，要考虑通道内的冷却水流速和换热面积比，而且还要兼顾机械制造的可行性。

首钢2号高炉（1726m³）1999年12月检修时，在炉腰（第7段）安装了一块试验铜冷却壁（见表3和图2）。

表3　首钢2号高炉试验铜冷却壁主要技术参数

项　目	数　值
外形尺寸（长×宽×厚）/mm	1970×992×145
冷却通道数量/条	5
冷却通道间距/mm	210
冷却通道直径/mm	φ40
燕尾槽尺寸/mm	(65/75)×55
传热面积比	0.59
镶填料面积比	0.53

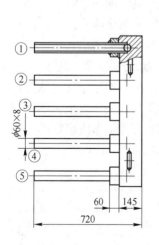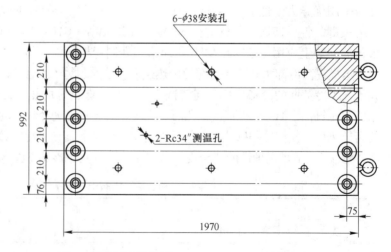

图2　铜冷却壁结构图

本次试验用铜冷却壁由北京首钢设计院设计，汕头华兴备件有限公司研制。铜冷却壁本体采用轧制的无氧铜（TU_2）铜板，设计中对铜冷却壁与炉壳的固定连接进行了特殊的处理，设计了壁本体与圆弧炉壳固定的特殊装置，保证冷却壁的良好固定。冷却壁与炉壳之间，注入自流浇注料，使冷却壁与炉壳之间具有良好的密封，防止煤气泄漏。

铜冷却壁组装完成以后，用1.3MPa的高压水进行试压检验，在30min内保持压力，检查壁体及所有焊缝处均无冒汗、渗漏，确认合格。然后在冷却壁沟槽内镶填SiC捣料。最后对冷却壁进行整体低温退火处理，以消除制造过程中所产生的内应力。

所采用的SiC捣料是与新型冷却壁配套使用的不定形耐火材料（见表4）。该捣料由高纯度的SiC和石墨组成，采用了以酚醛树脂为主的复合剂和抗氧化剂，形成微细的镶嵌结构，具有良好的化学稳定性、抗渣性和抗热震性，同时具有优良的导热性。使用该捣料有利于结成渣皮，可有效地保护冷却壁，延长冷却壁使用寿命。

表4 冷却壁用SiC捣料的理化性能

项目	数值
SiC/%	74.35
FC/%	12.94
Al_2O_3/%	2.44
SiO_2/%	6.01
Fe_2O_3/%	1.19
体积密度（200℃×3h）/g·cm^{-3}	1.22
气孔率（200℃×3h）/%	10
常温耐压强度（200℃×3h）/MPa	27.9
烧后线变化（200℃×3h）/%	-0.3

4 铜冷却壁的应用

铜冷却壁在高炉上安装以后，冷却壁热面砌一环Si_3N_4-SiC砖和一环高密度黏土保护砖，砖衬总厚度460mm。在铜冷却壁的壁体中安装了前、后两支热电偶，第一点插入壁体115mm，第二点插入壁体65mm，以在线监测冷却壁的温度变化和侵蚀状况。

为更加准确地监测铜冷却壁的工作状况，对冷却水量、水温差、热流强度进行了测试，结果见表5。

表5 铜冷却壁热流强度测试结果

序号	冷却壁水管编号	冷却通道对应的面积/m³	冷却水量/m³·h^{-1}	冷却水温差/℃	热流强度/kW·m^{-2}
1	73号	0.35657	11.61	0.9	33.993
2	74号	0.4137	11.99	4.1	137.840
3	75号	0.4137	11.25	3.0	94.634

续表5

序号	冷却壁水管编号	冷却通道对应的面积/m³	冷却水量/m³·h⁻¹	冷却水温差/℃	热流强度/kW·m⁻²
4	76号	0.4137	11.99	1.1	36.982
5	77号	0.35657	11.61	1.9	71.762

经过一年多的生产实践证实,首钢2号高炉钢冷却壁的应用已取得初步成功。2000年12月,汕头华兴备件有限公司、北京科技大学、北京钢铁设计研究总院等单位联合对首钢2号高炉试制的备用铜冷却壁进行了长期的热态实验和研究分析。结果证明,2号高炉的试验铜冷却壁设计参数和结构是合理的。铜冷却壁在1100℃高温状态下工作,壁体内部温度场分布均匀,壁体工作正常(图3)。

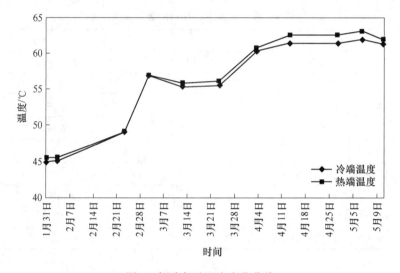

图3　铜冷却壁温度变化曲线

首钢2号高炉2001年检修设计中,在总结该高炉试验铜冷却壁的设计和应用的基础上,优化了铜冷却壁的设计。在炉腹、炉腰、炉身下部区域采用了3段共120块铜冷却壁。该冷却壁由北京首钢设计院设计,汕头华兴备件有限公司制造。

5　结语

新型高效冷却壁的开发和应用,是延长高炉寿命的重要技术措施。铜冷却壁的开发和试验,为推动高炉长寿技术进步,具有重要的技术经济价值和现实意义。

(1) 高炉采用冷却壁或板壁结合冷却结构,优于采用高密集纯铜冷却板的模式,更加适合我国国情。

(2) 铜冷却壁的本体采用轧制后的无氧铜(TU_2)板更为合理,减小壁体厚度和质量,降低工程造价,是铜冷却壁优化和改进的一个技术重点。

(3) 在投资允许的条件下,采用铜冷却壁是延长高炉寿命的一个有益措施。

致谢

参加本课题研究的还有汕头华兴备件有限公司余克事先生。并得到了北京钢铁设计研究总院副总工程师吴启常先生、首钢炼铁厂王颖生先生、王景智等先生的大力支持,作者在此一并表示感谢。

参考文献

[1] Heinrich P, et al. Copper blast furnace developed for multiple campaigns [J]. Iron and Steel Engineer, 1992 (2): 49-55.

The Design Research Work and Practice of Copper Staves

Zhang Fuming　Yao Shi　Su Wei　Huang Jin

Abstract: review about copper cooling plate and castiron stave was made. The diradvantags of castiron was considered and an introduction of copper stave design and practice was made.

Keywords: blast furnace; long campaign; design; copper stave

高炉炉缸炉底温度场控制技术

摘 要：通过对高炉炉缸炉底内衬侵蚀和温度过热现象的解析，阐述了当代高炉炉缸炉底温度场控制的理论。解析了高炉炉缸炉底温度过热的现象和形成机制，论述了高炉炉缸炉底工作过程中，内衬侵蚀破损的过程和机理。提出了"无过热-自保护"的炉缸炉底内衬设计理念，强调了通过设计合理的炉缸内衬结构、采用优质耐火材料和高效冷却系统，在线监测炉缸炉底温度和热流强度变化，控制炉缸炉底温度场合理分布，从而有效抑制炉缸炉底内衬侵蚀速率、延长高炉寿命。

关键词：高炉；炉缸；长寿；温度场；耐火材料

1 引言

近年来，国内新建或大修改造后的部分高炉相继出现了炉缸炉底过热、炉缸内衬异常侵蚀、甚至炉缸烧穿等事故，严重影响高炉正常生产和安全运行，还造成了巨大的经济损失。部分高炉炉缸炉底温度持续升高，为抑制炉缸炉底内衬侵蚀、延长高炉寿命，被迫采取强化护炉操作，使高炉稳定顺行生产受到影响。

据不完全统计，进入新世纪以来，我国已有数十座高炉发生炉缸炉底烧穿事故，除此之外，炉缸炉底出现局部温度过高的高炉数量呈现增加趋势[1,2]。因此，合理控制炉缸炉底温度、有效延长炉缸炉底寿命，已经成为当代中国炼铁工业面临的关键共性技术难题。

当前，诊断和判定炉缸炉底内衬出现异常最主要的方式，仍然是对炉缸炉底内衬和冷却系统温度场的监测。

2 炉缸炉底温度过热的表征与原因

炉缸炉底在高炉冶炼过程中，由于多种原因会导致其温度升高。炉缸炉底温度的在线监测和实时预警，是当前高炉维护、延长寿命的关键措施。炉缸炉底温度异常升高、过热，研究分析和实践证实主要由以下原因造成：（1）炉缸炉底内衬发生侵蚀和破损，必然导致其温度升高，致使温度场分布异常；（2）密闭的炉缸炉底内衬-冷却体系出现局部窜气或煤气渗漏造成温度升高；（3）炉缸炉底炭砖局部异常侵蚀，造成炭砖减薄、温度异常升高；（4）炉缸炉底炭砖砖缝胀裂，炭砖砌体局部渗铁、钻铁，其内充填铁水，造成温度升高；（5）冷却系统效能下降或失效，导致炭砖失去有效冷却，造成炭砖砖衬温度升高；（6）炭砖内部发生化学侵蚀，出现疏松、粉化甚至环状断裂，造成炭砖冷面温

本文作者：张福明，赵宏博，钱世崇，程树森。原刊于《第十届中国钢铁年会暨第六届宝钢学术年会论文集》，2015：1783-1788。

度升高；(7) 炭砖遇水氧化，局部出现腐蚀、缺陷或溶洞，最终导致炭砖温度升高；(8) 炉缸炉底炭砖在高温差条件下产生热应力，造成炭砖热应力破损，最终造成温度升高。

解析上述这些主要原因，可以看出，有一些是可以进行预防和治理的，属于影响高炉寿命的"非器质性病变"；而有一些就属于影响高炉寿命的"恶性重大疾病"，会直接威胁高炉正常生产，危及高炉寿命。

造成炉缸炉底温度异常升高的原因众多，从高炉炉缸炉底内衬侵蚀破损实践分析，凡是炉缸炉底内衬侵蚀破损，其最终结果必然会造成温度变化。

3 炉缸炉底工作条件与侵蚀机理

3.1 现代高炉冶炼技术特征

当代高炉在当前的资源、能源和技术条件下，生产技术指标仍保持稳定提升，这是炼铁工业技术创新、技术进步带来的效果。采用经济合理的炉料结构、提高风温、增加喷煤量、降低焦比和燃料比、提高鼓风富氧率、改善高炉操作、延长高炉寿命等综合技术措施，已成为当今高炉冶炼的重要技术特征[3]。

当代高炉以"高效率、低消耗、低成本、低排放"为主要技术发展理念[4]，所采取的高炉冶炼技术措施对于强化高炉生产无疑是正确的、合理的，但同时也带来了新的技术问题，主要表现在：(1) 高炉产量增加，生产效率提高，炉缸渣铁的排放通量和排放强度增加，铁口负荷增加，单铁口出铁负荷已达到 2500~3000t/d；(2) 高风温、高富氧、大喷煤强化了风口回旋区燃烧过程，回旋区结构发生了显著变化，高炉冶炼进程顺行难度增大；(3) 低焦比、高负荷、高煤比使高炉压差趋高、透气性变差，炉况稳定性衰减，高炉操作难度加大；(4) 炉缸死焦柱形态与结构发生变化，透气性与透液性趋于变差；(5) 铁水静压力增加，对出铁速度及对炉缸炉底炭砖的渗透作用产生影响。

3.2 炉缸炉底内衬工作条件

高炉在高风温、富氧大喷煤冶炼条件下，高炉冶炼进程发生变化，风口回旋区及料柱结构也随之发生变化。由于焦比降低、喷煤量增加，导致高炉透气性变差，操作难度增加；焦炭负荷提高使高炉死焦柱内部的粉焦增多，加上大量喷煤以后未燃粉煤量的增加，造成死焦柱内透气性和透液性恶化，高炉边缘气流发展，炉墙热负荷增高。

风口回旋区的结构变化，导致高炉冶炼进程出现新的变化，炉缸炉底工作条件趋于恶化：(1) 回旋区长度缩短、上翘，导致边缘气流发展；(2) 粉焦聚集在风口回旋区前端，形成"焦巢"结构，使死焦柱变得密实，使高炉透气性和透液性变差；(3) 死焦柱透气性与透液性恶化，气体、液体的顺畅运动受阻，对高炉顺行带来不利影响；(4) 死焦柱中心温度变低，炉缸工作活跃性下降，造成铁水环流加剧，炉缸炉底内衬冲刷侵蚀加剧。

3.3 炉缸炉底内衬侵蚀机理解析

造成炉缸炉底内衬侵蚀的原因众多，不同的高炉也不尽相同。除了通常的侵蚀破损原因以外，结合近年来高炉炉缸炉底的破损调查研究，下列原因也不容忽视：

（1）炉缸炉底温度在线监测措施缺乏。炉缸炉底内衬温度测量点少，热电偶测温点的设置也不尽科学合理；缺乏对冷却壁进出水温差、水流量、热流强度等参数的实时监测，造成不能及时发现炉缸炉底的异常状况，及时采取相应措施，结果往往是造成高炉炉缸烧穿事故的突发。

（2）炉缸冷却结构设计与配置不合理。用于炉缸炉底区域的冷却壁，其热负荷波动相对平稳，其主要功能是为炉缸炉底内衬提供足够的冷却，控制1150℃等温线的合理分布。用于高炉炉缸炉底的冷却壁与炉腹至炉身下部的冷却壁，其功能和性能要求也不尽相同。炉缸冷却壁要保持合理的冷却强度，使炭砖传递出来的热量能够顺畅与冷却水交换并导出，是保障炉缸炉底传热机制顺行的基础。为了强化炉缸冷却，不少高炉开始在炉缸局部区域采用铜冷却壁，但对铜冷却壁的设计结构、安装方式研究不够深入，其结果反而会适得其反。除此之外，铁口区冷却方式结构设计不合理，炉缸冷却壁与炉壳之间填料选用不当，炭砖与冷却壁之间的碳质捣料与炭砖的热导系数不匹配，冷却结构不合理等都会引发炉缸烧穿事故。

（3）炉缸炉底的可靠性、耐久性与高炉冶炼强化水平不匹配。21世纪初的10年间，我国钢铁工业发展迅猛，产量连年攀升。不少企业追求规模经济效益，以粗放扩张型发展获取经济利益。对于高炉生产而言，忽视高炉生产的科学规律，片面追求高产量、高利用系数。新高炉投产后，快速达产、快速超产，以效率最高为主要目标。在这种思想的主导下，不少高炉强化冶炼、超负荷生产，甚至不以焦比和高炉寿命为代价，高炉投产2~3年就出现炉缸烧穿，代价巨大、教训惨痛。统计研究表明，国内外50余座长寿高炉的一代炉役期内的平均利用系数为2.0~2.3t/(m^3·d)，而出现炉缸烧穿高炉的利用系数大多数在2.5t/(m^3·d)以上，由此可见过高产量、超高利用系数是造成高炉短寿的"杀手"之一。

（4）炭砖选用不合理。炉缸炉底内衬与铁水接触的部位或一代炉役末期要接触铁水的部位，不应选用石墨砖和石墨含量高的炭砖。石墨含量高的炭砖导热性高，但抗铁水熔蚀性差，容易发生炭砖熔损，不易黏结渣铁壳保护内衬。高炉设计时既要重视炭砖的热导性，也要重视炭砖的抗铁水渗透性和抗铁水熔蚀性，注重考查炭砖的气孔孔径、气孔率、透气度和气孔特性等综合指标。当前，新建高炉设计的死铁层呈现加深趋势，可以有效缓解炉缸铁水环流的侵蚀，但炉缸炉底要承受较高的铁水静压力，铁水渗透、熔蚀的发生几率也会随之加大。

（5）高炉操作维护存在不足。1）由于原燃料条件变化，造成钾、钠、铅、锌等有害元素在高炉内循环富集，与耐火材料发生化学反应生成化合物，使其体积膨胀，造成炉缸炉底内衬快速损坏；2）炉体冷却设备漏水，会沿着炉壳渗漏到炉缸，引起炭砖氧化、粉化，这是炉缸炭砖损坏的重要原因之一；3）铁口深度不够和出铁时铁口喷溅，铁水易从铁口通道进入砖缝，加速炭砖的侵蚀，同时高温煤气也穿透到炭砖缝隙中，形成局部热点；4）盲目强化高炉冶炼，导致炉体破损加剧；5）含钛物料护炉加入量不够，对已经侵蚀的内衬修补不及时，不能形成稳定的保护性再生炉衬；6）炉缸压浆维护操作不当，压浆压力过高，泥浆的材质不合理，将已经很薄的残余砖衬压碎，或使泥浆从砖缝中压入炉内与高温铁水接触，出现不良后果，进而诱发炭砖渗铁和炉缸烧穿事故。

4 炉缸炉底温度场的控制与管理

4.1 温度场控制的意义

炉缸炉底温度场控制与管理是当代高炉实现长寿的重要技术措施,是保障高炉生产稳定、安全的重要支撑技术。炉缸炉底的侵蚀过程是渣铁流场、温度场、应力场、化学侵蚀以及有害元素破坏等多要素耦合作用的结果,最终导致耐火材料内衬的侵蚀、破损、环裂、减薄等异常现象,这些都会直接快速地反映在温度场分布变化上。

温度场监控和管理是炉缸安全预警最直接的判断依据和监测手段。对于不同的高炉而言,炉缸炉底安全预警标准也不尽相同,科学合理的预警标准,应建立在对炉缸炉底温度场及侵蚀内型的实时计算监测的基础之上。

4.2 监测系统的硬件配置及性能

(1) 炉缸炉底测温电偶监测系统。为了在线监测炉缸炉底"象脚状"侵蚀区、铁口区的侵蚀状况,及时掌握炉缸活跃性的变化,优化炉缸炉底内衬热电偶布置,建议测温热电偶布置方案如图1所示[5]。

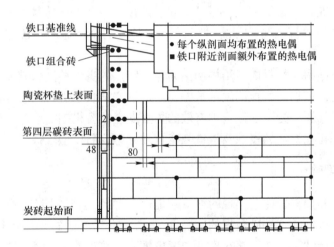

图1 炉缸炉底热电偶优化布置方案

(2) 炉缸冷却水温差与热负荷监测系统。为了保证冷却水温差变化对炉缸侵蚀及渣铁壳变化反映的敏感性和准确性,尤其是满足"隔热法"炉缸的监测需求,应采用高精度高分辨率的数字温度传感器,测温精度建议优于 0.05℃,分辨率优于 0.01℃(见图2)。实践证实这种方式的安全性、稳定性明显高于有线测温系统,且施工和维护简便。这种数字化无线热负荷监测系统,通过数十座高炉的成功应用,已证明了其稳定性和可靠性。

(3) "弱冷区"和监测"盲区"采用无线吸附式炉壳测温装置。高炉炉缸相邻冷却壁之间存在着一定的间隔,此间隔区域为传热上的"弱冷区",水温差监测对"弱冷区"侵蚀变化的敏感性较低,而一些高炉炉缸侵蚀严重甚至是烧穿部位恰处于"弱冷区"。此

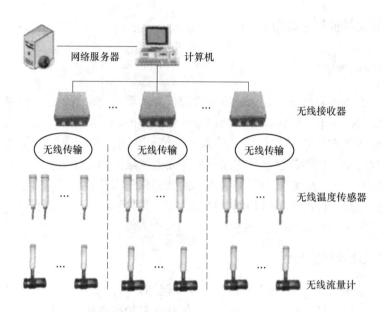

图 2　炉缸热负荷数字化无线采集通信系统拓扑图

外,炉缸炉底热电偶布置数目有限,尤其是到了炉役中后期如果砖衬内电偶损坏将难以恢复,即存在着监测的"盲区"。因此,为了实现对炉缸的全面监测,还应辅助炉壳表面温度监测。

如上所述,不同耐火材料内衬结构和不同操作特点的高炉炉缸,其安全预警标准存在着明显差异,因此仅依靠一次检测硬件数据,对炉缸安全状态进行判断存在着准确性差甚至可能造成误判的问题,为了建立合理有效的炉缸安全预警机制,应进一步依据传热学和炉缸炉底侵蚀机理建立专业的侵蚀及渣铁壳变化和异常诊断模型软件[6]。

4.3　智能诊断模型软件性能

智能诊断模型和预警软件应实现的如下功能[7]:(1)自动对基础硬件检测数据进行采集和滤波,保证侵蚀计算基础数据的准确性;(2)自动对炉缸炉底进行网格划分和三维非稳态温度场进行计算,并能够在模型中考虑铁水的凝固潜热对温度场和侵蚀的影响;(3)自动对炉缸炉底的不同横剖面、纵剖面的侵蚀内型进行图像重建和显示;(4)能够自动判断炉缸炉底可能出现的环裂、渗铁、气隙等异常;(5)能够对侵蚀加剧原因做出智能诊断和维护提示;(6)采取炉缸维护手段时能够自动计算并显示炉缸炉底渣铁壳的生成位置、厚度及形状变化;(7)对炉缸炉底侵蚀严重部位进行预警,防止炉缸烧穿事故的发生。

5　炉缸炉底温度过热的辩证治理

对炉缸炉底温度场进行在线监测管理的目的,是实现高炉全生命周期内的无过热和自保护。应当指出的是,炉缸炉底温度过热的治理标准并非一成不变的,而是在高炉整个生命周期的不同阶段，对于炉缸炉底的不同部位，无过热管理标准和对应的维护措施也要随之调整。

图 3 所示为高炉一代炉役生命周期内侵蚀内型的演变规律。不同类型的炉缸炉底虽然在不同阶段的持续时间会存在差异，但是基本都遵循这一演变进程，相应的在不同阶段，对炉缸炉底无过热的管理和自保护能力的变化也要区别对待。

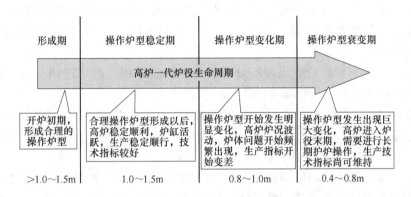

图 3　高炉一代炉役生命周期内侵蚀内型演变过程

表 1 为首钢高炉炉缸冷却壁热流强度的控制及采取的防控措施[8]。可见对于不同传热特性的炉缸，其安全管理标准也相应调整，同时在不同侵蚀阶段其对应的护炉措施和热流强度控制也逐渐变化。

表 1　首钢高炉炉缸冷却壁热流强度的控制及采取的防控措施

采用热压炭砖的炉缸冷却壁热流强度控制及措施		采用国产大块炭砖的炉缸冷却壁热流强度控制及措施	
控制值 /MJ·(m²·h)⁻¹	采取措施	控制值 /MJ·(m²·h)⁻¹	采取措施
≥41.86	加入含钛炉料护炉，保持[Ti]=0.08%~0.10%	≥33.49	加入含钛炉料护炉，保持[Ti]=0.08%~0.10%
≥50.23	提高含钛炉料加入量，使[Ti]≥0.10%	≥41.86	提高含钛炉料加入量，使[Ti]≥0.10%
≥54.42	封堵水温差超标的炉缸冷却壁上方的风口	≥46.03	封堵水温差超标的炉缸冷却壁上方的风口
≥62.79	高炉休风凉炉	≥54.42	高炉休风凉炉

基于不同类型的高炉实现炉缸安全长寿生产的本质都是"无过热-自保护"体系的建立，因此，在炉缸炉底温度场安全管理方面，进一步提出更加合理的残衬厚度管理及多级数字化预警机制，即安全预警标准应综合考虑热负荷、电偶温度、侵蚀厚度和渣铁壳，炉缸监测数据记录应分为实时值和历史最高值，并建立"工作标准"、"平衡标准"和"预警标准"三级预警指标，进而依据高炉生命周期的不同阶段的侵蚀特征，相应采取不同的炉缸维护手段及生产操作调节以实现高炉的安全高效生产。

6 结论

通过对高炉炉缸炉底温度过热现象的解析,提出必须建立基于炉缸炉底温度场控制为核心的高炉长寿技术体系,该体系的核心内容包括以下几个方面:

(1) 炉缸炉底过热现象的成因是炉缸内渣铁流场和砖衬温度场耦合作用的结果,其中炉缸铁水环流是造成炉缸过热、异常破损的主要原因,而炉缸炉底温度场则是内衬侵蚀状况最直接的体现。

(2) 为了实现对炉缸炉底温度场分布的全方位监测和是否"过热"进行科学判断,炉缸炉底精准检测硬件和三维温度场及侵蚀诊断模型软件是必备条件。

(3) 对于不同类型的炉缸炉底结构,在高炉一代炉役生命周期的不同阶段,对应不同部位,存在着不同的"无过热"判断标准和管理方法,高炉无过热-自保护体系的建立和维持也需依据其自身的传热特点及侵蚀特征因地制宜。

(4) 基于温度场、侵蚀内型及渣铁壳变化在线监测所得到的包括"工作标准"、"平衡标准"和"预警标准"的多级数字化预警机制是实现"自组织-自保护-无过热-永久性"炉缸炉底的科学方法和手段。

参考文献

[1] 汤清华,王筱留. 高炉炉虹炉底烧穿事故处理及努力提高其寿命[C]//2012年全国炼铁生产技术会议暨炼铁学术年会论文集(下). 无锡:中国金属学会,2012:89-94.

[2] 吴启常,王筱留. 炉缸长寿的关键在于耐火材料质量的突破[J]. 中国冶金,2013,23(7):11-16.

[3] 张福明. 当代高炉炼铁技术若干问题的认识[J]. 炼铁,2012,31(5):1-6.

[4] 张福明. 21世纪初巨型高炉的技术特征[J]. 炼铁,2012,31(2):1-8.

[5] Zhao H B, Cheng S S, Zhu J F, et al. Study on mechanics of "Elephant foot shaped" erosion of BF hearth [J]. The 5th ICSTI'09, Journal of Iron and Steel Research, International (Supply), 2009:10-14.

[6] 赵宏博,霍守锋,郝经伟,等. 高炉炉缸的安全预警机制[J]. 钢铁,2013,48(4):24-29.

[7] Zhao H B, Huo S F, Cheng S S. Study on the early warning mechanism for the security of blast furnace hearth [J]. International Journal of Minerals, Metallurgy and Materials. 2013, 20(4):345-353.

[8] 张福明,程树森. 现代高炉长寿技术[M]. 北京:冶金工业出版社,2012:503-520.

Control Technology on Temperature Field of Blast Furnace Hearth and Bottom

Zhang Fuming   Zhao Hongbo   Qian Shichong   Cheng Shusen

Abstract: The blast furnace (BF) hearth lining corrosion and the temperature overheat phenomenon are analyzed, the temperature field control concept of contemporary BF hearth is expounded. The phenomenon and formation mechanism of BF hearth and bottom temperature overheat are analyzed. The mechanism of hearth lining corrosion is analyzed under the BF working condition. The temperature distribution of hearth lining, and the hot metal flowing field distribution in the hearth are researched by CFD model. The design concept of "no overheat-self protection" of hearth lining is proposed. Reasonable measures such as structure design of hearth lining, high quality refractory and the high efficiency cooling system, hearth and temperature and heat flux intensity on-line monitoring, reasonable control of hearth bottom temperature field distribution, will be adopted to avoid the abnormal wear of hearth lining and prolong the BF campaign life. As a result, satisfied performance has been achieved in application.

Keywords: blast furnace; hearth; long campaign life; temperature field; refractory

基于自组织理论的高炉长寿技术

摘　要：目前，高炉炼铁仍是生产效能最高的炼铁工艺。在未来资源、能源、环境可承载的条件下，实现高炉炼铁的绿色化、智能化转型升级和可持续发展，必须建立新的发展理念。耗散结构和自组织理论是研究复杂开放系统演变过程机理及其规律的新理论，对于指导高炉炼铁协调发展与技术创新具有重要意义。本文阐述了系统论、耗散结构和自组织理论的核心观点，提出了高炉炼铁物质流通量参数、能量流通量参数、热流通量参数的概念及其物理意义。指出了未来高炉炼铁技术的发展理念、目标及方向，论述了运用自组织理论，通过高炉炉体功能、结构和效率的协同优化，实现高炉全寿命周期的动态自组织优化。提出并探讨了构建具有自组织特性的"自感知-自适应-自保护-自维护"炉体结构的方法及其路径。

关键词：耗散结构；开放系统；自组织；高炉；长寿；结构优化

1　引言

现代高炉炼铁已有近200年的历史，经历了多次重大的技术变革与创新，不断演进发展，渐趋至臻完善。当前，在全球范围内，高炉炼铁工艺流程占有绝对的主导地位。预计到21世纪中叶，高炉炼铁工艺流程的主导地位将不会发生根本性变化，高炉仍然是铁矿石还原和优质生铁的主流生产装置。进入21世纪以来，我国作为世界第一产铁大国，高炉炼铁技术进步及成就全球瞩目，在高炉大型化、装备现代化等方面取得重大进展举世公认；高炉生产技术指标不断提升，许多大型高炉都已进入国际先进行列[1]。应当看到，在高炉炼铁技术取得长足进步的同时，高炉寿命依然在不同程度上成为制约高炉正常生产的短板，一些现代化大型高炉投产不久就出现了炉缸烧穿或铜冷却壁大量损坏等严重事故，高炉安全生产受到严重威胁[2]。在研究解析高炉异常破损原因、采取有效措施延长高炉寿命的同时，应当建立起新的高炉长寿技术理念、理论和方法，构建自感知-自适应-自维护-自修复的高炉长寿技术理论体系，从根本上解决制约高炉寿命的关键技术难题，形成现代高炉长寿关键共性技术，并在工程实践中实施应用，在现有技术的基础上，进一步延长高炉寿命，实现高炉炼铁绿色化、智能化多目标协同发展。

2　高炉炼铁技术创新发展理念

未来钢铁工业的可持续发展，必须充分研究市场需求、产品竞争力和资源-能源-环境的承载能力，实现低碳绿色化发展[3]。对于具体的钢铁厂而言，要根据全厂的产量规模、

本文作者：张福明。原刊于《2017年全国高炉炼铁学术年会论文集（上）》，2017：398-405。

产品结构、总图布局，以实现物质流、能量流和信息流动态有序、协同连续、高效耦合运行为目标，合理确定高炉的座数和高炉容积[4]。在钢铁行业供给侧结构性调整、淘汰落后、转型升级的宏观经济形势下，不宜过分追求钢铁厂产能和产量，要转变以粗放型产能规模扩张的发展道路，坚持走集约化、绿色化、智能化、品牌化的可持续发展之路。

高炉大型化和装备现代化是我国钢铁工业发展的重要方向和路径，必须坚持钢铁厂流程结构优化前提下的大型化，以实现钢铁制造流程的动态有序、协同连续、集约高效、耗散优化为目标。未来我国钢铁工业的结构调整与优化，将以去产能、减量化发展为前提，淘汰落后、流程优化、结构调整、技术升级势在必行。因此在高炉设计建造过程中，必须以整个钢铁制造流程协同优化的视角，站在钢铁厂生存发展的全局性战略层次上，以构建钢铁制造全流程要素-结构-功能-效率协同优化为关注点，实现烧结、球团、焦化、炼钢、轧钢以及全厂能源系统的协同优化，同时注重流程网络的耗散优化、重构优化和时空关系，使物质流、能量流和信息流实现高效耦合"层流式"运行，提高运行效率和运行质量。总体看来，未来我国钢铁工业的发展，必须坚持以去产能、减量化为基本方针，持续开展钢铁产业结构调整和优化升级，通过淘汰落后、技术改造和装备升级，优化钢铁制造流程及其网络结构，构建钢铁制造流程耗散优化的自组织体系，实现市场-产品-质量与资源-能源-环境的和谐有序发展。

3 自组织理论

3.1 系统论

20世纪30~40年代，美籍奥地利理论生物学家L. V. 贝塔朗菲（L. Von. Bertalanffy）创立了系统论。其核心思想是系统的整体观念，强调任何系统都是一个有机的整体，不是各部分机械的组合和简单相加，系统的整体功能是各单元在孤立状态下所不具有的性质，亦即"整体大于部分之和"。系统的各单元不是孤立存在的，每个单元在系统中都起着特定的作用，单元之间相互关联、相互作用、协同匹配，从而构成了不可分割的有机整体。

系统论的出现，使人们的思维方式发生了深刻的变化，改变了传统的"还原论"的思维模式和方法，使人们从"孤立-静止、简单-抽象"的思维禁锢中解放出来，形成了分析复杂问题的系统观和整体观，建立了注重全局性、整体性的系统分析方法，为解决复杂问题提供了新思路和新方法。直至目前，系统论发展方兴未艾，与耗散结构理论、协同论、突变论、信息论、控制论、运筹学等新的科学理论相结合，成为人们认识事物及其发展变化规律的重要理论和方法。

3.2 耗散结构

20世纪60年代，比利时化学和物理学家伊利亚·普利高津（Ilya. Prigogine）经过多年的研究，建立了一种新的关于非平衡系统自组织的理论——耗散结构理论，因此而获得了1977年度诺贝尔化学奖。耗散结构理论是研究开放系统在远离平衡的非线性区域产生耗散结构的机理和规律的一种新理论。从热力学角度分析，耗散结构是一个远离平衡的、开放的不可逆系统，系统内部各子系统之间是非线性相互作用关系。系统通过与外界不断进行物质、能量和信息的交换，降低系统自身熵增，在系统内部的某个参量的变化达到一定阈值时，通过涨落，系统将出现非平衡突变，由原来的混沌状态转变成为一种在时间、空间或功能上的有序状态。这种有序结构，由于不断与外界交换物质和能量才能维持，消耗外界输入的物质和能量，同时也向外界释散出物质和能量，因此称之为"耗散结构"（Dissipative Structure）。

3.3 自组织理论

自组织理论（Self-organizing Theory）是20世纪60年代末发展起来的一种系统理论，是L. V. 贝塔朗菲系统论的发展。自组织理论主要研究的是复杂自组织系统的形成和发展机制问题，也就是在一定的条件下，系统如何自动地由无序到有序，从简单有序向复杂有序发展和演进。对于耗散结构而言，是开放且远离平衡的系统，与外界进行着物质、能量和信息的交换，系统内部各子系统之间是非线性相互作用，在一定的条件下可以自发地组织成时间、空间或功能上的有序结构，呈现出具有"生命特征"的自组织现象。自组织理论目前被广泛地应用于复杂工程问题和社会问题的研究，是将基础科学与哲学有机结合的一个新的科学理论。

复杂的、开放的、不可逆的、远离平衡的系统，都具有内在的自组织性，其作用机制源于系统具有多因子和单元性涨落，由于不同单元具有异质性，又可以具有相同的因子，通过单元之间的非线性耦合和/或系统与单元之间的非线性相互作用，从而使系统获得自组织性[5,6]。

4 高炉炉体的功能优化

4.1 高炉炼铁的物理本质

高炉炼铁是铁氧化物（烧结矿、球团矿、块矿等）以焦炭（及煤粉）作为主要燃料和还原剂，在高炉内经过一系列的物理-化学反应和冶金传输过程，生产出液态生铁的冶金工艺过程，焦炭是高炉炼铁工艺不可或缺的料柱骨架。其物理本质是铁素物质流在碳素能量流的驱动和作用下，按照设定的运行程序，沿着特定的流程网络动态-有序、协同-连续运行，实现铁素物质流和碳素能量流在整个流程范围内流动并转变/转换的过程。

由此可见，高炉炼铁区别于非高炉炼铁的重要工艺特征，主要表现在两个方面：一是采用焦炭作为主要燃料、还原剂、渗碳剂和料柱骨架；二是其产品为液态生铁。高炉冶炼进程是典型的竖炉逆流移动床过程，下降炉料与上升煤气流在相向运动过程中，经过一系列的物理-化学反应与传输过程，完成铁氧化物的还原和渗碳，最终形成液态生铁。

4.2 高炉炉体的功能解析

4.2.1 冶金反应器

从钢铁制造全流程的视野分析，高炉的基本功能应当解析为：（1）铁氧化物的还原和渗碳器；（2）液态生铁的发生器和连续供应器；（3）能源转换器；（4）冶金质量调控器[7]。对于高炉-转炉长流程钢铁厂而言，高炉在整个钢铁制造流程中的作用至关重要，是全流程物质流和能源流转变/转换的核心关键环节。因此，高炉冶炼过程要求是连续-稳定运行，稳定顺行成为高炉操作的核心要素。

高炉冶炼过程的冶金传输过程及反应和物质流、能量流的转变/转换都是在高炉炉体内完成的，高炉炉体是一个复杂的高温、高压、密闭冶金反应器，高炉冶炼则是多元-多相（态）复杂巨系统，是复杂的、开放的、远离平衡的不可逆过程，是有大量物质、能量和信息输入/输出的耗散结构。高炉炉体的本质功能，是保障高炉冶炼进程连续稳定运行的高效长寿冶金反应器，其生产效率和质量要满足钢铁厂对铁水供应的需求；其寿命周期要持续15~20年甚至更长。

4.2.2 核心流程结构

经过近200年的演进发展，现代高炉已经形成了炉喉、炉身、炉腰、炉腹、炉缸、死铁层"六段式"的炉体结构，炉体每个部位都有着不同的功能和作用。对于高炉冶炼过程而言，连续稳定运行是其重要的工艺特征。正是在这样一个特定的高温、高压、密闭的流程结构中，大通量物质流和能量流的输入/输出，并完成物质流、能量流的转变/转换。下降炉料从炉喉到炉缸，物质的形态、成分、温度、性能等因子发生了巨大变化；焦炭（煤粉）在风口前燃烧形成高炉煤气，煤气流与下降炉料逆行向上运动，经过炉缸、炉腹、炉腰、炉喉至炉顶，能量的形态、矢量、势量等因子也发生了巨大变化。基于耗散结构和冶金流程工程学理论，可以将高炉炉体视为一个物质流和能量流做动态-有序、协同-连续流动的流程路径（结构），进而言之，高炉炉体是高炉炼铁工序中核心关键的流程结构。在这个流程结构中，既有铁矿石、焦炭、煤粉、热风等物质或能量的输入，也有铁水、炉渣、煤气等物质或能量的输出。同时，在高炉冶炼过程中，伴随着物质和能量的输入/输出，信息也在不断地输入/输出。从微观尺度分析，高炉冶炼过程表现为质量传输、热量传输、动量传输和一系列物理-化学冶金反应工程的集成；从宏观尺度分析，高炉冶炼过程表现为物质流（特别是铁素物质流）、能量流（特别是碳素能量流）和信息流在高炉炉体特定的物理空间内做动态耦合运行。图1为高炉冶炼过程物质和能量输入/输出的耗散结构示意图[8]。

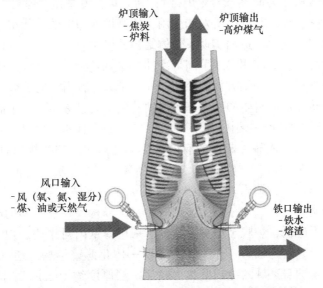

图1 高炉冶炼过程物质和能量输入/输出的耗散结构

因此，高炉炉体作为高炉冶炼的核心流程结构，从高炉生产运行角度看，高炉炉体结构直接影响高炉的稳定、顺行，进而影响高炉生产的高效、优质、低耗和长寿。从冶金流程工程学角度看，高炉炉体结构则直接影响整个钢铁制造流程的物质流、能量流和信息流动态耦合运行的效率和效果。由此可以推论，高炉长寿不仅是高炉生产稳定顺行的重要基础和前提，在整个钢铁制造流程的时空尺度上，高炉长寿也是维持冶金流程长期连续、协同稳定、动态运行的基础和前提。

4.3 高炉炉体的功能集成

由高炉炉壳、冷却器和耐火材料内衬等构成的高炉炉体，是高炉炼铁的核心关键工艺单元，既是高炉炼铁的冶金反应器，也是重要的能源转换器。铁氧化物（烧结矿、球团矿和块矿等）和部分非铁元素（硅、锰、硫、磷等）在高炉内被还原，形成液态生铁；焦炭（煤粉）在高炉内经过燃烧、气化、参与铁矿石的还原反应，形成了高炉煤气。铁素物质由固态氧化物被还原成液态生铁；绝大部分非铁氧化物由固态变成液态炉渣；固态焦炭（煤粉）变成了具有化学能、动能和热能的高炉煤气。在整个钢铁制造流程中，高炉炼铁工序是物质、能量和信息因子变化最多的多相态复杂系统，其物质流和能量流的转变/转换也最为复杂。因此，高炉炉体是集成多功能为一体的大型冶金反应器和能源转换器。

组成高炉炉体的各个子系统（单元），其功能和作用也不尽相同。炉壳是保障高炉炉体具有一定强度和刚度且具有承压密封功能的钢结构，其功能是固定炉体冷却器，支撑炉顶设备、煤气上升管，承受炉体附属设备的荷载，因此炉壳是维持高炉炉体的基本结构（骨架结构）。冷却对于现代高炉炉体长寿至关重要，炉体冷却结构经过不断演进发展，目前现代高炉已形成了以冷却壁为主流的炉体冷却模式，特别是在炉腹至炉身下部高热负荷区域采用铜冷却壁，使高炉炉腹至炉身区域从厚壁结构演进为薄壁结构，甚至演化为"无衬结构"，从而减少甚至摆脱了对耐火材料的依赖。依靠铜冷却壁的高效冷却作用，以快速形成保护性渣皮作为"动态自生炉衬"，为高炉长寿奠定了基础。应当说通过冷却壁材质和结构的优化，铜冷却壁的研发和应用，使高炉冷却壁的功能实现优化，炉体高热负荷区的冷却效率得到优化，进而实现了炉体结构的优化。

毋庸置疑，铜冷却壁的应用实现了高炉冷却壁结构-效率-功能的协同优化。冷却器的主要功能是为耐火材料内衬提供有效的冷却，降低耐火材料内衬的热应力和热面温度，促进在耐火材料内衬热面形成自生的、动态的保护性渣皮（渣铁壳），进而延长耐火材料内衬的使用寿命。对于炉腹至炉身中下部区域，当冷却壁内侧的炉衬侵蚀消失以后或取消铜冷却壁内侧炉衬结构时，依靠冷却壁的有效传热，能够在冷却壁热面直接形成动态性渣皮，即所谓"动态自生炉衬"。与此同时，冷却器还有保护炉壳、降低炉壳温度，从而延长炉壳使用寿命的作用。高炉内衬是由耐火材料构成的高炉冶金反应器的砖衬砌体（局部也可以是不定型耐火材料），一般是由若干种异质-异构、功能不同的耐火材料所组成。随着铜冷却壁的广泛应用，炉腹至炉身区域普遍采用薄壁内衬结构，炉衬厚度一般100~150mm，而且与冷却壁镶砖融为一体，成为砖壁一体化薄壁结构。还有不少高炉开炉之前，仅在高炉炉腹至炉身区域的冷却壁热面喷涂一层厚度约100mm的不定型耐火材料保护层，取消了砖衬砌体和冷却壁的镶砖结构。直至目前，高炉炉缸炉底耐火材料内衬仍是不可或缺且无法替代的结构，而且对炉缸炉底耐火材料内衬的材质、质量和结构要求更加严格，具有导热性、抗铁水渗透性和抗铁水熔蚀性优异的炭砖则是必不可少的功能材料。

高炉炉缸是高炉冶炼过程的起始和终结，也是高炉冶炼进程中冶金传输和物理-化学反应最为集中的区域，是典型的多元-多相态复杂系统。炉缸炉底还是高温液态渣铁积聚的区域，炉缸炉底内衬工作条件恶劣，承受着高温热应力、铁水渗透、化学侵蚀、机械冲刷磨损等各类破坏作用。而高炉炉缸炉底的使用寿命决定着高炉一代炉役寿命，是延长高炉寿命的核心关键环节。因此，延长高炉寿命的技术难点和技术重点就是要有效延长高炉炉缸炉底的使用寿命。

5 高炉炉体的结构优化

5.1 物质流和能量流通量参数

5.1.1 铁素物质流通量参数

高炉冶炼过程是大量物质、能量和信息输入/输出的耗散结构，是远离平衡、开放的不可逆过程。科学地表征高炉物质流的通量参数，不仅要精确阐释高炉冶炼的物质流流动特征，还应当具有可以精准测量、科学评价的特性。基于上述观点，笔者认为，单位时间内、单位高炉炉缸截面积的生铁产量（即高炉炉缸截面利用系数），其实物理本质是高炉冶炼过程中铁素物质流的流动速率，能够比较科学精准地描述和表征高炉冶炼过程铁素物质流的通量。式（1）给出了高炉铁素流通量参数的计算式。

$$F_\mathrm{m} = P/A \tag{1}$$

式中　F_m——铁素流通量参数（炉缸截面利用系数），$t/(m^2 \cdot d)$；

P——高炉日产生铁量，t/d；

A——高炉炉缸截面积，m^2。

铁素流通量参数结合高炉冶炼过程连续运行的工艺特征，从高炉冶炼过程动态连续运行的物理本质角度，是以炉缸截面作为铁素物质流转变/转换的流通截面，单位截面积的金属通量实质上就是钢铁制造流程铁素物质流的通量参数。要根据高炉原燃料条件、技术装备条件和操作条件，合理确定铁素流通量参数（即炉缸截面利用系数）。理论研究和生产实践证实，高炉炉缸截面利用系数可以客观反映高炉生产操作的优劣，是一个科学评价高炉冶炼效率的参数和指标。

5.1.2 碳素能量流通量参数

高炉冶过程得以连续进行的关键是碳素能量流的驱动，而炉缸风口前碳素（焦炭、煤粉等）燃烧生成炉缸煤气则是整个高炉冶炼进程的开始，由于焦炭、煤粉是高炉流程不可或缺的重要能源，因此高炉冶炼最主要的能量流是碳素能量流。同样，科学合理地表征高炉能量流的通量参数，要充分体现高炉冶炼过程能量流流动的物理本质和工艺特征，能够准确计算测量，具有普遍适用性且便于科学评价。基于上述观点，笔者认为，单位时间内、单位高炉炉缸截面积流通的炉腹煤气量（即高炉炉腹煤气量指数），其实质是高炉冶炼过程中碳素能量流的流动速率，能够比较科学准确地描述和表征高炉冶炼过程碳素能量流的通量。式（2）给出了高炉铁素流通量参数的计算式。

$$F_c = V_{BG}/A \tag{2}$$

式中 F_c——碳素流通量参数（炉腹煤气量指数），m/min；

V_{BG}——高炉炉腹煤气量，m^3/min；

A——高炉炉缸截面积，m^2。

碳素流通量参数（炉腹煤气量指数）可以解析为高炉煤气流通量参数，其物理本质是单位炉缸截面积流通的炉腹煤气量，也就是高炉炉缸截面上的高炉煤气空炉速度[9]。高炉炉腹煤气是炉缸风口回旋区内的焦炭、煤粉等碳素物质燃烧所生成的高炉煤气，经炉腹向上排升并穿透软融带，参与一系列物理-化学反应和冶金传输过程，是高炉冶炼过程碳素能量流的重要表现形态。理论上的高炉炉腹煤气量可以通过高炉工艺计算得出，实践中为了简化计算，一般以炉缸煤气代替炉腹煤气。高炉炉腹煤气与高炉原燃料条件、操作条件具有重要的非线性耦合关系，综合体现了燃料消耗、鼓风量、富氧量、鼓风温度、鼓风湿度等参数的影响，是目前评价高炉能源消耗、操作水平和炉况顺行程度的一个重要技术指标。

5.2 炉体热流通量参数与耗散优化

高炉冶炼过程中，热量的输入与输出是直接影响高炉冶炼稳定顺行的关键环节之一。将高炉视为一个耗散结构体系，在当前的原燃料条件下，理论上高炉冶炼过程中总的热量耗散约为4~6GJ/t，其中碳素物质（焦炭、煤粉）燃烧的化学热大约占高炉消耗总热量的60%，鼓风带入的物理热约占40%。热量输出项中除了液态渣铁和高炉煤气带出的物理热，炉体冷却水所带出的热量耗散约为8%~10%。未来高炉降低热量输入和消耗，特别是降低碳素燃料的消耗，高效回收炉渣显热，应是实现低碳绿色发展的重要途径。对于炉体冷却而言，在保证高炉炉体寿命的前提下，降低炉体热量耗散也应当给予关注。

高炉炉体各部位的功能不同，炉体不同部位的热流通量也存在较大差异。一般情况下，高炉炉腹至炉身下部的热流通量最高，其次是炉缸的中下部区域，特别是炉缸-炉底交界处，对应"象脚状"侵蚀区，炉缸内衬侵蚀严重、或炉役末期的热流通量可以达到60GJ/h甚至更高。高炉炉体热量耗散的热流通量参数一般采用热流强度表示，其物理意义是单位面积高炉冷却壁（热面）或炉壳（内表面）所传递的热量。因此，高炉炉体设计时，必须充分考虑高炉炉役末期，冷却器和冷却系统所能承受的最大热流通量。炉体冷却系统的功能设计不应片面追求冷却系统的节水、节能，应根据高炉冶炼过程动态变化和突变的工艺特征，留有一定的富裕能力，以适应高炉炉体热流强度和热流通量的波动、涌现、涨落和突变。高炉炉体顶层设计时，要依据炉体热流通量的变化和高炉操作数据的统计分析，经过归纳-综合、权衡-选择，确定合理的冷却水流量和压力等关键参数。当然，高炉炉体采用纯水（或软水）密闭循环冷却系统是现代高炉设计中的基本配置，工程设计的重点是构建热量耗散优化的冷却系统工艺流程、冷却参数选择、泵组-换热器-脱气罐-膨胀罐-稳压罐的配置、供回水管网及网络路径设置、智能化监测控制等。式(3)~式(5)是炉体热流通量和热流强度的计算式。

$$Q = c \times m \times \Delta t = c \times m \times (t_o - t_i) \tag{3}$$

$$q = Q/F = -\lambda \times \mathrm{grad}t = -\lambda \frac{\partial t}{\partial n} \qquad (4)$$

$$q = (t_\mathrm{h} - t_\mathrm{c}) \times \frac{\lambda}{S} \qquad (5)$$

式中 Q——热流通量，J；
　　c——水的比热容，J/(kg·K)；
　　m——冷却水量，m³/h；
　　Δt——冷却水进出水温差，℃；
　　t_o——冷却水出水温度，℃；
　　t_i——冷却水进水温度，℃；
　　F——传热面积，m²；
　　q——热流强度，W/m²；
　　$\mathrm{grad}t$——温度梯度，℃/m；
　　$\partial t/\partial n$——温度的方向导数；
　　S——内衬的厚度，m；
　　t_h——内衬热面温度，℃；
　　t_c——内衬冷面温度，℃；
　　λ——内衬的导热系数，W/(m·K)。

高炉炉体冷却耗散结构优化的设计方法和思维进路应当是：（1）采用纯水（或软水）密闭循环冷却技术；（2）参照高炉炉体热流强度进行传热学计算得出炉体总的最大热流通量，再计算出冷却水量、冷却水温差，进而设定出进水温度与出水温度；（3）根据冷却水的换热量，确定出冷却水热交换器的能力、型式和规格；（4）计算脱气罐、膨胀罐、稳压罐等罐组的容量及结构参数；（5）设计冷却系统工艺流程，初步确定供回水管网参数，计算管网系统的阻力损失；（6）根据计算管网系统阻力计算结果，确定循环泵组的优化配置（扬程、台数、工作制度及布置方式）；（7）设计系统各关键节点的流量、压力、温度监测控制系统，数据的自动采集与处理，构建信息流网络，使系统具备智能化调控的功能；（8）评估冷却系统的顶层设计，重点关注冷却系统的安全性和可靠性，确定备用电源供电、配置事故差油泵、供回水管道安全供水可靠性分析评价等[10]。

5.3　炉体结构优化

高炉炉体结构优化的核心关键，是要构建具有自组织特性的"自感知-自适应-自维护-自修复"的炉体长寿结构。高炉的结构特性和连续运行的工艺特征，使高炉炉体结构必须长期适应高炉冶炼过程的各种影响和破坏，具备抵抗恶性事故的可靠性、安全性和耐久性。遵循自组织理论，高炉炉体的功能拓展与功能集成是现代高炉区别于传统高炉的重要所在，现代高炉的多功能化和功能集成体现在应当具备优质铁水生产、高效能源转换和消纳废弃物并实现资源化的功能，因此，功能优化是现代高炉炼铁发展理念的重要创新。毋庸置疑，高炉炉体是实现高炉炼铁"高效、优质、低耗、长寿、安全"多目标优化的载体。

为适应高炉功能优化，炉体结构优化应当以全寿命周期的时间尺度，从高炉的设计、建造、运行、维护各个阶段都必须给予足够的重视。高炉炉体静态结构的优化重点是高炉设计和建造，尤其是高炉炉体结构设计，关乎到高炉的全寿命周期。高炉炉体结构设计的本质就是构建高效协同的高炉炉体自组织体系，应当建立耗散结构自组织体系优化的理念，遵循冶金学、传热学和材料科学的基础理论，采用概念设计-顶层设计-动态精准设计-仿真优化设计的系统设计方法，经过综合-权衡-选择-评估-决策的过程，以高炉全寿命周期和高炉炉体整体结构为关注点，注重顶层设计，通过空间结构设计优化实现炉体综合功能的协同优化。

炉体结构设计优化的思维进路和设计方法是：（1）以物质流和能量流通量参数为基础，设计合理的高炉内型，为高炉稳定顺行和高效长寿奠定基础；（2）高炉炉体结构的选择与确定；（3）高效冷却器参数与结构设计；（4）炉缸炉底的冷却和耐火材料结构设计优化；（5）采用传热学和材料力学数值仿真计算方法评估验证冷却器与炉缸炉底内衬的温度场、应力场等；（6）高炉炉壳传热学和弹塑性力学数值仿真计算、材质选择、结构设计。

5.4 高炉运行与维护的自组织

高炉建造过程中，要以工程设计为依据，科学组织、统筹管理，兼顾质量、进度、成本等要素，精细施工、精益管理，不宜片面追求工期进度或降低成本，应以质量为核心实现多目标协同优化。无数的案例证实，高炉施工建造过程的质量问题和隐患，是造成高炉短寿、出现恶性事故的直接原因，损失惨重、教训深刻，必须加强工程建造管理，提高施工水平，保证高炉一代炉役期间生产安全稳定。

高炉长寿的实质就是保持高炉一代炉役期间的合理操作炉型[11,12]。高炉投产以后，根据高炉炉型的演变进程可以划分为操作炉型形成期、操作炉型稳定期和操作炉型维护期三个阶段。高炉生产中，要通过精料、炉料分布控制、煤气流分布控制、炉体冷却与热负荷管理、渣铁流动控制等措施，保持高炉全寿命周期的合理操作炉型。高炉生产操作的调控，实质上就是对高炉冶炼过程的"他组织"，是物质、能量和信息的输入过程。高炉冷却系统的调控、含钛物料的加入等措施则主要是为了维护高炉炉体长寿，在高炉运行的状态下，通过高炉系统的自组织特性和自组织体系，促进形成动态的、自生的"保护性渣皮"，即所谓"自生炉衬"，通过自生炉衬的动态生成-涨落，实现高炉炉体具有自组织功能的自维护和自修复，从而延长高炉寿命。

未来高炉智能化的一个重要特征是要建立起自感知-自适应-自维护-自修复的炉体结构，在自动化、数字化、信息化的基础上，构建高炉炉体温度、压力、流量、应力/应变的精准监测和大数据分析处理，形成基于高炉冶炼-炉体长寿耦合的信息流管控体系，精准操作、精准护炉，实现与铁素物质流、碳素能量流和集成信息流高效耦合运行的协同管理。

6 结语

高炉炼铁要实现绿色化、智能化发展，必须建立新的技术发展理念，以适应经济社会发展的要求。在资源、能源和环境可承载的前提下，加大供给侧结构性调整，淘汰落后产能和工艺装备，推动技术进步和转型升级。运用耗散结构自组织理论，构建高炉炼铁动态有序、协同连续和耗散优化的流程体系，创新具有"自感知-自适应-自维护-自修复"功能的长寿炉体结构，积极采用高炉长寿创新理念、理论和方法，以高炉一代炉役全寿命周期为视角，注重高炉的概念设计、顶层设计和动态精准设计，促进物质流、能量流和信息流流程结构优化、实现高效耦合运行。高炉运行过程中，通过大数据、信息流的智能化管理，提高精准化智能操作水平，动态在线监测和调控炉体运行状态，增强高炉炉体的自组织性和自组织功能，保障高炉冶炼稳定顺行，进而实现高炉炼铁高效、优质、长寿、安全多目标协同优化。

参考文献

[1] 张寿荣，毕学工. 中国炼铁的过去、现状和未来 [J]. 炼铁，2015，34（4）：1-6.
[2] 汤清华. 关于延长高炉炉缸寿命的若干问题 [J]. 炼铁，2014，33（5）：7-11.
[3] 张福明. 面向未来的低碳绿色高炉炼铁技术发展方向 [J]. 2016，35（1）：1-6.
[4] 张福明，钱世崇，殷瑞钰. 钢铁厂结构优化与高炉大型化 [J]. 钢铁 2012，47（7）：1-9.
[5] 殷瑞钰. 冶金流程工程学（第2版）[M]. 北京：冶金工业出版社，2009：127-132.
[6] 殷瑞钰. 冶金流程集成理论与方法 [M]. 北京：冶金工业出版社，2013：50-54.
[7] 殷瑞钰. 冶金流程集成理论与方法 [M]. 北京：冶金工业出版社，2013：120-123.
[8] 马丁·戈德斯等著，沙永志译. 现代高炉炼铁（第3版）[M]. 北京：冶金工业出版社，2016：1-12.
[9] 项钟庸，王筱留，银汉. 再论高炉生产效率的评价方法 [J]. 钢铁，2013，48（3）：86-91.
[10] 张福明，程树森. 现代高炉长寿技术 [M]. 北京：冶金工业出版社，2012：340-356.
[11] 张寿荣，于仲洁. 武钢高炉长寿技术 [M]. 北京：冶金工业出版社，2010：227-231.
[12] 张福明. 当代高炉炼铁技术若干问题的认识 [J]. 炼铁，2012，31（5）：1-6.

涟钢 2800m³ 高炉无料钟炉顶齿轮箱水冷系统的设计

摘　要：介绍了涟钢2800m³高炉无料钟炉顶齿轮箱采用的闭路循环水冷系统，并与开路冷却模式的冷却效果、水量消耗、运行成本、管道系统等进行了对比分析，归纳了闭路循环冷却模式的特点及优势。通过热负荷计算，确定了冷却水量及水泵、热交换器等主要设备的参数，提出了闭路循环冷却系统在设计及操作过程中要注意的问题。

关键词：高炉；无料钟炉顶；齿轮箱；水冷系统

1　引言

布料溜槽控制器简称齿轮箱，是无料钟炉顶的重要设备之一，该设备运行情况，直接影响高炉能否正常布料，是维持高炉正常生产的关键[1]。齿轮箱的冷却系统对其使用寿命影响显著，是维持齿轮箱内部齿轮温度正常、实现高炉长寿生产的保障。因此，有必要对炉顶齿轮箱采用高效合理的冷却系统以保障高炉的生产顺行。

2　工程概况

涟源钢铁有限公司于2011年开始新建3号高炉，由首钢国际工程公司设计，其有效容积2800m³，年利用系数2.7t/(m³·d)，顶压0.25MPa，一代炉龄不少于15年，采用槽下分级入炉技术，并罐无料钟炉顶设备，陶瓷杯、铜冷却壁及全软水冷却长寿综合技术，高风温顶燃式热风炉与板式换热器组成的高温预热工艺，高炉煤气干法布袋除尘，大型静叶可调电动鼓风机及TRT发电等多项技术，其中炉顶齿轮箱采用氮气气封密闭循环水冷结构。

3　炉顶齿轮箱冷却系统的设计

3.1　齿轮箱冷却系统的必要性

涟钢新3号高炉采用自主改进型的并罐式炉顶，由换向溜槽、左右料罐、下部阀箱（内装调节阀及下密封）、中间斗、齿轮箱、布料溜槽等主要设备构成，其中齿轮箱传动系统采用平面二次包络蜗轮副传动技术，结构紧凑、效率高、承载力大、抗冲击能力强，同时使用自补偿石墨密封技术将气封间隙减到最小，大幅度降低密封氮气用量。

本文作者：孟祥龙，张福明，耿云梅，毛庆武。原刊于《钢铁研究》，2014，42（3）：36-38。

无料钟炉顶装料设备已是现代高炉的关键装备[2]。高炉炉顶设备从原来的大钟布料发展到无料钟布料后，布料溜槽控制器即齿轮箱成为重要设备，在布料过程中起到关键作用，其能否正常运行直接关系到高炉能否正常生产。齿轮箱工作环境恶劣，被高温高压及粉尘密集的炉顶煤气包围，煤气中含有大量的炉尘，炉尘进入到水冷齿轮箱内，会加速轴承、齿轮及各传动面的磨损，进而影响到水冷齿轮箱的寿命[3]。为保证齿轮箱内部件的稳定运行，在齿轮箱内设置了水冷装置和隔热装置，齿轮箱的冷却采用水冷方式，由冷却水交换传入齿轮箱内的热量来调节齿轮箱的温度，保证其工作环境长期处在正常温度下工作（30~40℃），最高不超过70℃。为防止炉顶荒煤气进入齿轮箱，需采用氮气密封防尘，向水冷齿轮箱内充入少量有压氮气，维持水冷齿轮箱对炉顶的微正压，在箱内运动件和静止件之间的间隙处形成动态气体密封。气封水冷同时作用来保证齿轮箱内部的零部件正常运行。

3.2 传统的开路冷却系统

当前国内高炉炉顶齿轮箱的水冷系统有开路和闭路两种模式。

在开路冷却模式中，冷却水通过冷却塔或冷却池中水的直接蒸发向环境散热实现冷却。开路冷却模式的原理如图1所示。

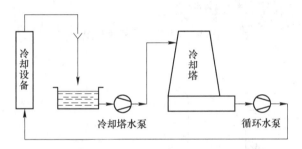

图1　开路冷却模式原理图

此种冷却方式冷却水与环境有一定时间的接触，容易混入杂质影响水质进而影响冷却效果，同时杂质也可能对冷却管路中的各部件产生不利影响，也增加了对管道的腐蚀。在蒸发过程中，冷却水有一定量的消耗，对于水资源相对匮乏的地区是个不小的压力。开路模式需设置冷却塔水泵，不仅增加工程投资，而且也增加了生产过程中的设备维护和能源消耗。开路冷却模式的管道系统复杂，由于没有稳压装置，在回水管路上需设置"U"形水封，防止炉况波动情况下，炉顶压力剧减造成冷却水回流，造成下水槽向高炉内漏水事故，因此在生产过程中要谨防"U"形水封发生击穿。图2为炉顶齿轮箱开路冷却模式流程图。

3.3 改进后的闭路循环冷却系统

鉴于炉顶齿轮箱开路冷却模式的缺陷，在涟钢2800m³高炉工程的设计中，使用了闭路循环的冷却模式。闭路循环冷却模式原理如图3所示。

闭路循环冷却模式与传统的开路冷却模式相比有如下优点：

（1）冷却可靠性高，冷却效率高。闭路循环冷却模式由于回路压力的增加，提高了水的沸点，可以维持较高的欠热度，同时降低了局部泡核沸腾的可能性，使整个系统的冷却效率显著提高。

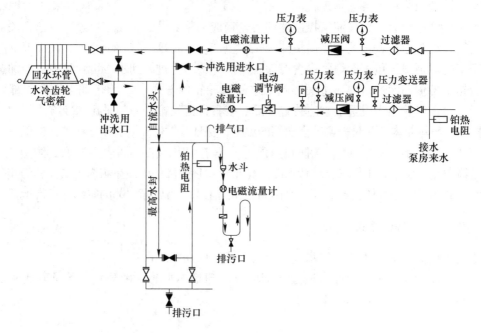

图 2 开路冷却模式流程图

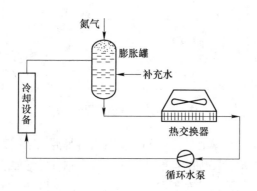

图 3 闭路循环冷却模式原理图

（2）水量消耗降低。闭路循环冷却模式中，整个系统与大气完全隔离，大幅度减少冷却水蒸发、系统泄漏等带来的损失，而正常补水量仅为总流量的 0.05% 左右，因此与开路冷却模式相比大幅度降低了水消耗量。

（3）运行成本降低。在开路冷却模式中，冷却塔的水泵设备要求其扬程要能克服管道系统阻力、供水点压力和剩余压力。而在闭路循环冷却模式中，冷却水的静压头可以得到充分利用，并设有膨胀罐，冷却系统的工作压力取决于膨胀罐内填充的氮气压力，因此不需要设置冷却塔水泵，只保留循环水泵，而且循环水泵的扬程只需克服管道系统阻力即可。

（4）管道系统简单。闭路循环冷却模式不需要设置"U"形水封，通过调整膨胀罐内氮气压力来保证冷却水不回流，使得冷却过程更加稳定。同时水泵、过滤器、热交换器等设备均设置于炉顶平台附近，整个冷却水回路行程可大大减小。

开路冷却模式与闭路循环冷却模式技术特点见表1。

表1 开路冷却模式与闭路循环冷却模式比较

项 目	开路冷却模式	闭路循环冷却模式
工况稳定性	易受炉况波动影响	工作稳定
灵活性	出现问题可通入少量氮气	出现问题时可通入大量氮气
冷却效率	低	高
能源消耗	大	小
管道系统	复杂	简单
适用性	受水封高度影响，改造项目适用性不强	适用性好

炉顶齿轮箱闭路循环冷却模式流程如图4所示。整个系统包括储水系统、补水系统、过滤排污系统、二冷水系统、齿轮箱冷却水系统及煤气压力平衡系统。所有参数的控制及显示均在主操作画面中完成，并成为独立系统，不与其他系统连锁。

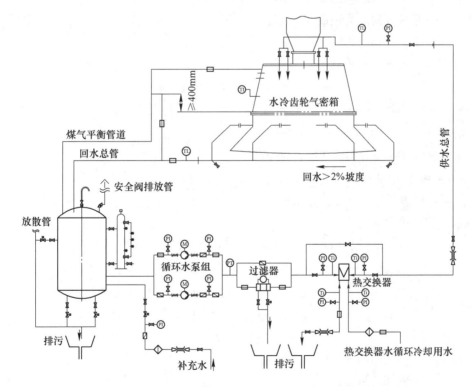

图4 闭路循环冷却模式流程图

由于考虑投资等原因，冷却水采用工业水，由于氮气密封能够大大减少炉尘进入循环水系统，同时配有过滤器设备，随时去除循环水中混入的杂质，可以很大程度上抑制管道的腐蚀及结垢现象。齿轮箱冷却用水量为 $10\sim15\,m^3/h$，反冲洗过滤器每隔一定时间自动冲洗一次，自动补水系统设上下限水位，补水量每天 $1\sim2\,m^3$。

3.4 闭路循环冷却系统参数选择

齿轮箱所承受的热负荷主要来源于炉料辐射、高炉煤气（主要是 CO_2 和水蒸气）的辐射、传热以及煤气的对流传热，这些传入齿轮箱的热量应由冷却水带走。齿轮箱是一个复杂的受热体，在进行热负荷计算时需做适当的简化处理。

所需冷却水流量为：

$$D_W = Q/(10^3 C_W \cdot \Delta t) \tag{1}$$

式中　D_W——冷却水流量，m^3/h；

Q——齿轮箱热负荷，J/h；

C_W——水的比热，$J/(kg \cdot ℃)$；

Δt——冷却水进出水温差，℃。

齿轮箱的热负荷可确定为：

$$Q = 3600(S_h \cdot q_h + S_r \cdot q_v) \tag{2}$$

式中　S_h——水平表面的受热面积，m^2；

q_h——水平表面的热流密度，W/m^2；

S_r——垂直表面的受热面积，m^2；

q_v——垂直表面的热流密度，W/m^2。

水平表面热流密度可确定为：

$$q_h = q_l + q_{CO_2} + q_{H_2O} + q_d + q_c \tag{3}$$

式中　q_l——炉料辐射热流密度，W/m^2；

q_{CO_2}——CO_2 辐射热流密度，W/m^2；

q_{H_2O}——水蒸气辐射热流密度，W/m^2；

q_d——煤气对流热流密度，W/m^2；

q_c——热传导热流密度，W/m^2。

各项热流密度可确定为：

$$q_l = \phi \cdot \varepsilon \cdot 4.96(T_1/100)^4 - (T_3/100)^4 \tag{4}$$

$$q_{CO_2} = \varepsilon_1 \cdot 3.5 \cdot \sqrt[3]{P_1 S_1} \cdot (T_2/1000)^{3.5} - (T_3/100)^{3.5} \tag{5}$$

$$q_{H_2O} = \varepsilon \cdot 3.5 P_2^{0.8} \cdot S_1^{0.6}(T_2/100)^3 - (T_3/100)^3 \tag{6}$$

$$q_d = 2.2(T_2 - T_3)^{1.25} \tag{7}$$

$$q_c = K(T_3 - T_4)/h \tag{8}$$

式中　T_1——炉喉料面的热点温度，℃；

T_2——炉喉煤气温度，℃；

T_3——炉内侧齿轮箱水平表面温度，℃；

T_4——齿轮箱内部水平表面温度，℃；

ϕ——辐射传热的形状系数；

ε——炉料辐射的吸收系数；

ε_1——气体辐射的吸收系数；

P_1——CO_2 分压，Pa；
P_2——水蒸气分压，Pa；
S_1——平均射线行程；
K——齿轮箱壁面导热系数，W/(m·℃)；
h——导热距离，m。

式（3）~式（8）可以确定出水平表面的热流密度，并应用类似办法可确定出垂直表面的热流密度。根据同级别其他高炉的实际生产参数，在 T_1 = 500℃、T_2 = 180℃、T_3 = 180℃、T_4 = 600℃ 的工况下确定出的炉顶齿轮箱正常冷却水流量为 8m³/h；在 T_1 = 900℃、T_2 = 500℃、T_3 = 550℃、T_4 = 600℃ 最大冷却水流量为 15m³/h。在设计中考虑到极限工况，实际选取的流量为 20m³/h。根据循环冷却水流量确定的其他各设备参数：水泵 2 台，水压 0.35MPa，流量 30m³/h；储水罐 1 个，5m³；过滤器 1 个，净水阻力 0.02MPa。

3.5 闭路循环冷却模式需注意的问题

闭路循环冷却模式优势明显，但在设计过程中仍有一些需要注意的问题：

（1）回水总管的水平段要保证一定坡度，一般按不小于 2% 确定，以保证回水管路畅通。

（2）密封氮气压力及流量必须严格控制，因为氮气压力过高、流量过大，吹动水面摇晃，导致水溢流，高炉炉顶压力与氮气压力差波动，也可以导致水溢流。

（3）要控制水泵的流量稳定，最好达到有压满流工况点。水靠重力向下流动时，若流量不足使管道达到有压满流，水必然携带气体进入管道内，在水冷板蛇形管内发生气水分离，转弯处立管正压力上升，如果聚集气体量大时，气体会沿管道返回上水槽，阻碍水向下流动。

（4）整套冷却系统与齿轮箱距离不可过远，除了能减少管道局部腐蚀后带来的检修工作，还能降低整个系统的阻损。

4 结语

涟源钢铁有限公司新 3 号高炉于 2013 年 3 月 22 日开炉，运行的 5 个月时间里，各项指标均达到设计要求，炉顶齿轮箱闭路循环水冷系统运行正常。

参考文献

[1] 潘幼清，姜文革．高炉炉顶齿轮箱冷却系统故障的诊断及处理 [J]．炼铁，2005，24（4）：44-46．
[2] 张福明，钱世崇，等．首钢京唐 5500m³ 高炉采用的新技术 [J]．钢铁，2011，46（2）：12-17．
[3] 蔡飞．齿轮箱冷却水循环系统补水快的原因分析 [J]．炼铁，2006，25（2）：44-45．

Design on Water Cooling System of Gear-box of Bell-less Top in a Blast Furnace (2800m³) at Liangang

Meng Xianglong Zhang Fuming Geng Yunmei Mao Qingwu

Abstract: The closed circuit water cooling system of gear-box of bell-less top in a blast furnace (2800m³) at Liangang was introduced. Comparative analysis of cooling effect, water consumption, running cost, pipe system and so on was carried out with opened circuit cooling system. The features and advantages of closed circuit cooling system were also discussed. Cooling water consumption and parameters of main equipments such as water pump and heat exchanger etc. were determined by heat load calculation. Some issues which should be paid attention in the design of closed circuit cooling system and operating process were put forward.

Keywords: blast furnace; bell-less top; gear box; water cooling system

Design and Operation Control for Long Campaign Life of Blast Furnaces

Zhang Fuming

Abstract: At the beginning of 1990s, Shougang blast furnace (BFs) No. 2, No. 4, No. 3 and No. 1 were rebuilt sequently for new technological modernization in succession. The campaign life of No. 1, No. 3 and No. 4 reaches 16.4, 17.6 and 15.6 years, respectively, and the hot metal output of one campaign reaches 3.38Mt, 3.548 and 2.637Mt, respectively; the hot metal output of BF effective volume of one campaign reaches 13328, 13991 and 12560t/m^3, respectively, which reaches the international advanced level of BF high efficiency and long campaign life. In BF designing, several advanced BF long campaign technologies were adopted, and closed circulating soft water cooling technology was applied in 4 BFs. Double row cooling pipe high efficiency cooling stave was developed which could prolong the service life of bosh, belly and stack. Hot pressed carbon brick and ceramic cup hearth lining structure were applied and optimized. BF operation was improved continuously to ensure stable and smooth operation of BF. Hearth working condition control was strengthened, burden distribution control technology was applied to achieve reasonable distribution of gas flow, and heat load monitoring was strengthened to maintain BF reasonable working inner profile. Proper maintenance at the end of BF campaign was enhanced. Hearth and bottom service life was prolonged by adding titaniferous material and enhancing hearth cooling. Gunning of lining was carried out periodically for the area above tuyere zone.

Keywords: blast furnace; long campaign life; hearth lining; cooling stave; operation control

1 Introduction

For a successful 2008 Olympic Games held in Beijing, Shougang shut down steel plants in Beijing region. Blast furnaces (BFs) No. 2 and No. 4 blew down in the year of 2008, and BFs No. 1 and No. 3 blew down respectively on December 18th and 19th in 2010. BFs No. 1 and No. 3 were in good operating conditions and performances when they shut down. The campaign life of BFs No. 1, No. 3 and No. 4 was 16.4, 17.6 and 15.6 years respectively, and the hot metal output of BF volume during one campaign was 13328, 13991 and 12560 respectively. The international advanced level is achieved in the high efficiency and long campaign life BFs.

本文作者：张福明。原刊于《钢铁研究学报》（英文版），2013, 20 (9): 53-60。

2 Design of BF Long Campaign Life

2.1 BF Inner Profile

The Chinese ironmaking engineers have always focused on the design of BF inner profile and such basic philosophy as optimized design of BF inner profile has been formed as a result of investigation and summarization of the BF breakage mechanism and BF reaction principle[1]. It is an effective technical method to restrain the elephant foot shaped abnormal wear by way of increasing the well-death depth properly. After increment of the well-death depth, a direct settlement of deadman onto the bottom is avoided, the access of hot metal flow between deadman and bottom is enlarged, the liquid and gas permeabilities in the hearth are improved, the peripheral flow of hot metal is lightened and the service life of hearth and bottom are effectively prolonged. It is confirmed by theoretical research and practice that the well-death depth is approximately 20% of the hearth diameter normally. A proper increment of hearth height is not only good for combustion of pulverized coal in front of the tuyere, but also can increase the hearth volume in order to store slag and hot metal under the condition of high efficiency production and reduce the potential irregular blasting under the condition of intensified smelting. A trend to increase the hearth height is on the rise for large BFs in China and an appropriate hearth volume should be 16%-18% of the effective volume of BF. The study shows that it has a remarkable effect to restrain the wear on the hearth lining around the taphole area by way of proper deepening of taphole with the taphole depth is approximately 45% of the hearth radius normally. Thus, it can reduce the hot metal whirlpool formed nearby the taphole area during tapping and prevent the scour on the hearth lining by the whirlpool, and prolong the service life of the hearth lining in the taphole area. Reduction of BF bosh angle is good not only for bosh gas smooth ascension, but also for easing heat flux attack on the bosh and good for formation of stable protective slag skull in the bosh area to protect cooler for a long term. The bosh angle of a modern large BF is within 80° normally. Table 1 shows the parameters of inner profile of some BFs of Shougang.

Table 1 Proper inner profile of BFs at Shougang

Item	No. 1, No. 3	No. 2	No. 4
Effective volume/m^3	2536	1726	2100
Diameter of hearth/m	11560	9700	10400
Diameter of belly/mm	13000	10850	11550
Diameter of throat/mm	8200	6800	8150
Height of well-death/mm	2200	1800	1600
Height of hearth/mm	4200	4000	4350
Height of bosh/mm	3400	3100	3400

Continued Table 1

Item	No. 1, No. 3	No. 2	No. 4
Height of belly/mm	2900	2000	2200
Height of stack/mm	13500	15600	13950
Height of throat/mm	1800	2000	2000
Effective height/mm	25800	26700	25900
Angle of bosh α	78°02′36″	79°41′42″	80°24′3″
Angle of stack β	79°55′9″	82°44′24″	83°3′07″
Numbers of tuyere	30	24	28
Numbers of taphole	3	2	2
Space between tuyeres/mm	1211	1270	1167
Height to diameter ratio	1.985	2.461	2.242

2.2 BF Hearth and Bottom Lining Structure

It has been proved by practice that the service life of a BF hearth and bottom is the key to determine BF campaign life and the ironmaking engineers have paid a great attention to effectively prolonging the hearth and bottom lining service life. Since 1960s, Shougang has been using such an integrated bottom technology as carbon bricks-high alumina bricks for the BFs that have been always in good conditions. With the development of ironmaking technology and since the middle period of 1980s, the BF production has been intensified and the problem of the BF hearth and bottom has been in the highlight. As a result of investigation and analysis by means of actual results of measurements of more than 10 BFs, when the BF blows down, the wear on the linings of Shougang's BF hearth and bottom is attributed to the typical and abnormal elephant-foot-shaped wear and annular cracks on the side wall of the hearth[2]. The position that the most serious erosion occurred due to the abnormal elephant-foot-shaped wear is located at the junction between the hearth and bottom, the position corresponding to the cooling stave on the 2nd part of the hearth and the most serious wear was less than 100mm only from the cooling stave hot face by way of actual measurements. The coagula with high melting point as solidified slag, hot metal and Ti (C, N) were coagulated on the surfaces of the residual carbon bricks and high alumina bricks, and annular cracks appeared on all the annular carbon bricks around the hearth. The cracks were 80-200mm and alkali metal, solidified slag, hot metal existed in the cracks.

The wear mechanism of hearth and bottom lining of Shougang's BFs was investigated and analyzed under consideration of the BF raw materials, fuel and operation, mainly including penetration and erosion of hot metal into and on the carbon brick, mechanical wear by peripheral flow of hot metal, corrosion and chemical erosion of such alkali metal as molten slag and hot metal, ZnO, Na_2O, K_2O, damage of carbon brick by thermal stress, oxidization and damage of carbon brick by oxide gas such as CO_2 and H_2O, etc. Application of high-quality carbon brick and rational cooling are the musts to prolong the service life of the BF hearth and bottom. The thermal conduction solution (hot-pressed carbon brick solution) of hearth lining design system, as a representative of UCAR in USA, has been successfully adopted for the large BFs. The refractory solution (ceramic cup solution) of hearth lining design system, as a representative of SAVOIE in France, has been popularly applied to the large BFs, too. Thermal conduction solution and refractory solution of hearth lining design systems have a same essence of the technical principle, i. e., control of distribution of 1150℃ isotherm in the hearth and bottom enables the carbon bricks to keep away from the 800-1100℃ embrittling temperature zone as possible as they can. Small size hot-pressed carbon bricks NMA with high thermal conductivity and excellent resistance to permeability of hot metal are used for the so-called thermal conductive solution. Through a rational cooling, a layer protective skull of slag or hot metal can be formed on the hot surfaces of the carbon brick and 1150℃ isotherm is restrained inside the protective skull. Therefore, the carbon bricks are effectively protected and prevented from the damages caused by penetration and wear of the hot metal. For the ceramic cup solution, the ceramic materials with low thermal conductivity are adopted on the hot surfaces of the carbon blocks so that a cup-shaped ceramic lining forms. The so-called ceramic cup that has a purpose to control the 1150℃ isotherm inside the ceramic cup. Both of these two technical systems must adopt the carbon bricks with high thermal conductivity and excellent resistance to permeability of hot metal[3].

Carbon brick-alumina brick comprehensive bottom lining structures are applied on BFs No. 2, No. 3 and No. 4 and the hot-pressed carbon brick-ceramic cup combined hearth and bottom lining structure is adopted for BF No. 1 (Fig. 1). Small size hot-pressed carbon brick NMA are applied to the junctions between the hearth and bottom of the four BFs. elephant-foot-shaped abnormal erosion areas[4]. Both of these two structures have been successfully applied and actual

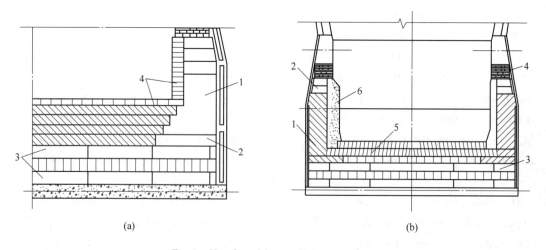

Fig. 1 Hearth and bottom lining structure

(a) Carbon brick-comprehensive bottom structure; (b) Carbon brick-ceramic cup combined structure

1—No metal area; 2—Carbon block; 3—Bottom carbon beam; 4—Alumina brick; 5—Ceramic pad; 6—Ceramic cup wall

achievements of BF long campaign life have been obtained. The campaign of Shougang's BFs in Beijing is indicated in Table 2. Especially for BFs No. 1 and No. 3 with the same volume and inner profile, under the similar conditions of raw materials, fuel and operation, the campaign of them exceeds the designed target that proves a long campaign life of the BFs can be realized by means of both of these two technical systems for hearth and bottom of the BFs.

Table 2 Campaign life of Shougang's BFs in Beijing

BF	Effective volume /m^3	Time of blowing in	Time of blowing down	Campaign life /d	Output during a campaign/Mt	Production perduring a campaign/t · m^{-3}
No. 1	2536	1994-8	2010-12	5990	33.80	13328
No. 2	1726	1991-5	2002-3	3950	15.40	8926
No. 3	2536	1993-6	2010-12	6415	35.48	13991
No. 4	2100	1992-5	2008-1	5715	26.38	12560

2.3 BF Cooling Technology

2.3.1 Cooling stave

In 1990s, there were 3 kinds of major cooling structure of BF: copper cooling plates were used for cooling from bosh to lower shaft; cooling staves were used for BF proper completely; cooling staves and copper cooling plates were used together for BF. To make the BF campaign life reach 10-15 years, the full cooling stave structure was adopted for Shougang's BFs. When the types of the cooling staves for all parts of the BFs were selected, the following factors were taken into consideration[5]:

(1) The heat load at the position of hearth is higher, but it is lower than the area above the bosh and the temperature fluctuation is smaller. The hot surface of the cooling staves can maintain the carbon bricks with a certain thickness to prevent the cooling staves from erosion by high-temperature molten slag and hot metal. Therefore, grey cast iron cooling staves with higher thermal conductivity are used in the hearth and bottom areas (including tuyere zone) and five rows of cast iron cooling staves are totally provided.

(2) The BF bosh, belly and lower shaft are the points that the cooling staves are most seriously damaged. Since the brick lining (or slag skull) can not be stably stored for a time, the surface of the cooling stave directly exposed to the BF and is impacted by severe heat load, eroded by molten slag and hot metal, and scoured by strong gas flow and due to the mechanical wear and tear of burden and so on, it is required that the cooling staves in the area should have a higher thermal mechanical property and a stronger cooling capability. In the design, the third generation of double-row-pipe cooling staves made of nodular graphite cast iron were applied. Six rows of brick-in-laid cooling staves were totally installed and flanged brackets were furnished for the

cooling staves at the belly and lower shaft to support the brick lining. Three rows of local copper cooling staves were adopted when No. 2 BF was rebuilt in the year of 2002.

(3) The middle part of shaft is mainly worn and torn by burden, scoured by gas flow, chemically eroded by alkali metal and bears a higher heat load. Therefore, the brick-inlaid cooling staves with beads and made of nodular graphite cast iron with inlaid clay bricks were used in design. Four rows of cooling staves were totally installed in middle part of shaft.

(4) One row of C shaped stave type cooler made of nodular cast iron was added between the upper part of shaft and lower edge of throat armour. The stave type cooler directly contact the burden and refractory lining was cancelled.

2.3.2 BF cooling system

It is proved by Shougang's BF production for many years that when the advanced hearth bottom lining structures are adopted, it shall specially take care of cooling of the hearth and bottom, and detection and monitoring shall be reinforced[6]. For the key positions, the high-quality carbon brick with high thermal conduction, resistance to penetration and anti-erosion shall be used and a rational cooling system shall be adopted for intensification of cooling. For design of the cooling water flowrate, a sufficient adjusting capacity shall be fully taken into account and control of cooling water flowrate shall be determined according to the actual conditions of the practice in order for energy saving and consumption reduction. It is not good in the design to unilaterally pursue a lower cooling water flowrate. In the middle and later stages of BF campaign, thinning of the lining will make the heat load and temperature rise up considerably, resulting in an insufficient cooling water flowrate and subsequently an inadequate capability of the cooling system, and an overheat or break-out will easily take place at the hearth in case the hearth and bottom temperature exceeds the normal range or the heat load is abnormal[7].

Industrial water circulating cooling system is used for bottom cooling pipes, hearth cooling staves (1-5 rows), C shaped stave type coolers and tuyeres of Shougang's four BFs. The bottom cooling pipes, 1st, 4th and 5th rows of cooling staves and tuyere coolers are cooled by industrial water with normal pressure of 0.60MPa. An intensified cooling system with medium pressure of 1.2MPa is carried out in the 2nd and 3rd rows of cooling staves located at the junction (abnormal "elephant-foot-shaped" erosion zone) between the hearth and bottom. Circulating industrial water cooling system with high pressure of 1.7MPa is used for the middle and noses jackets of tuyere. The closed circulating soften water cooling system is adopted in cooling staves above bosh. Fig. 2 shows the process diagram of closed circulating soften water cooling system for BF No. 2.

2.4 Automatic Detection and Control System

Automatic detection is an indispensable technical measure to prolong BF campaign life. An on-line monitoring of the hearth bottom temperature has become an important approach to control the status of erosion on the hearth bottom and also is a necessary condition to establish a mathematical model of erosion on the hearth bottom linings. Detection of the temperature and pressure in the parts of bosh, belly and lower shaft provides the BF operators with the reliable information to know

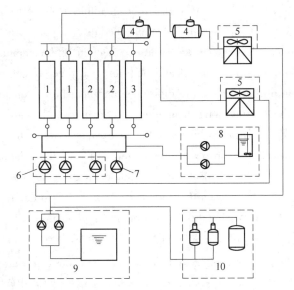

Fig. 2　Process scheme of closed circulating soften water cooling system for BF No. 2

1—Cooling unit for stave flanged bracket; 2—Cooling unit for stave body; 3—Cooling unit for stave snake type pipe;
4—Degasification tank; 5—Air cooler; 6—Circulating pump; 7—Stand-by diesel pump; 8—Supply unit of soften agent;
9—Unit of supply soft water; 10—Expansion tanks and pressure adjustable vessel

at any time the BF working conditions. By means of detection of the flowrate, temperature and pressure of the cooling water, such parameters as heat flux density and heat load can be calculated out and controlled, too[8]. Application of such technologies as stationary temperature probes of throat, top camera, automatic on-line analysis of gas, monitor of lining thickness and so on further ensures the long BF campaign life. To create favourable conditions for prolonging BF campaign life, an artificial intelligent expert system of BF ironmaking was introduced in the year of 2002 when BF No. 2 was rebuilt.

3　BF Operation Controls

3.1　Monitoring and Supervision

3.1.1　Automatic monitoring of hearth cooling system

The hearth working status can be correctly judged only a real-time collection and monitoring of the hearth cooling water temperature difference and variation of the heat flux density is realized, and on the basis of these, the corresponding burden distribution control and blast system adjustment are carried out, and measures for protection of the BFs and adjustment of the production capacity are conducted and taken to ensure a smooth and stable production, prolong the BF service life and reach the target of long life and high efficiency.

　To realize a real-time acquisition and monitoring of the cooling water temperature difference and variation of the heat flux density of the cooling staves at the 2nd and 3rd rows of the hearth and in the taphole area, and to satisfy the need of real-time monitoring of the hearth working status, an on-line cooling stave water temperature collection module, a data processing module and a communication module are developed; an on-line collection and communication system of the hearth water temperature difference is established for real-time detection of the stave water temperature, calculation of the temperature difference and heat flow strength of the hearth cooling staves; a database of the cooling stave temperature difference and hearth heat load during the production is established; a variation curve of the hearth stave water temperature difference, a variation curve of the hearth heat load, a circumference distribution diagram of the hearth heat load and a data report of the hearth heat load are automatically generated; a real-time alarming and warning for overrunning of the hearth heat load is realized in order to provide the reference and guidance about the BF internal status in the course of production.

3.1.2 Detection and control of soften water cooling system

A closed circulating soften water cooling system is used for cooling the areas above bosh of BFs No. 1, No. 2 and No. 4. The cooling system for bosh and above part of BF No. 3 is divided into 2 parts, industrial water is used for cooling stave of 1-12 rows and a closed circulating soften water system is used for cooling stave of 13-15 rows. Flowmeters, pressure gauges and temperature detectors are furnished in the soften water supply and return pipelines; pressure gauges are installed in each of the return water branch pipes and temperature detectors are arranged at certain intervals in the branch pipes of the cooling staves around the BF body to calculate the average heat load of the cooling staves. Alarming devices for detecting abnormally low pressure, abnormally low water level and abnormally low make-up water pressure are installed on the expansion tanks. All the above detected data, besides the displays and logs in the water pump house, are transmitted to the computers in the central control room for realization of such functions as display, storage, record, alarming and printing, etc. Control and regulation of the soften water system, nitrogen pressure stabilizing system on the expansion tanks and make-up water system are all automatically controlled.

3.2 Operation Controls

3.2.1 Hearth working status control

A stable and smooth BF production requires an active hearth working status. That is, the deadman at the hearth centre has sufficient gas permeability and liquid permeability and the peripheral flow of hot metal in the hearth is weakened. If the gas permeability and liquid permeability of the deadman at the hearth centre are bad, then, hot metal will accumulate at the hearth edge and the peripheral flow of hot metal easily formed during tapping will make the hearth lining partly eroded, make the hearth overheat locally and such accident as hearth break-out and the like will occur. And a large amount of slag and hot metal staying in the deadman will enable the initial gas to hardly penetrate the hearth centre and damage the gas distribution of the BF so that the smooth operation and long campaign life of the BF will be influenced. Therefore, measures shall be taken to activate the deadman in the hearth centre, maintain the hearth sidewall and bottom in appropriate temperature and keep the active hearth working situation.

The temperature of the hearth side wall and the bottom reflects the variation of the temperature field in the hearth and bottom. With the production increment, the temperature of the hearth side wall and bottom goes up and with the increment of pulverized coal injection (PCI); the temperature of the hearth side wall rises up while the bottom temperature drops down. The activity index of the hearth working is an important parameter to monitor the hearth working status and to provide a basis for adjustment of operating parameters under the condition of a long-time high PCR of BF so as to ensure a stable and smooth BF production.

Such measures as improving the quality of raw materials and fuel, control of the charging system on the rational distribution of the gas flow in the centre and peripheral based on a stable and sufficient blast kinetic energy will make the BF operation stable and smooth which are beneficial to the activity of the hearth working. By monitoring the activity index of the hearth working, all the BF operating parameters are timely adjusted and the activity index is maintained within a normal range in order to make the BF operation stable and smooth under the high PCR and the temperature of the hearth side wall is kept lower for safety and long life of the hearth and bottom.

3.2.2 Gas flow distribution control

Rational gas flow distribution is in relation with BF stable and smooth operation, energy saving, consumption reduction and long life. The Shougang's evaluation standards for BF gas rational distribution are: the first is to ensure stable and smooth operating condition; the second is to increase gas utilization and reduce fuel ratio, and ensure smooth furnace condition, high gas utilization, low fuel ratio, and hearth to work actively. Typical gas distribution is "central gas is open and peripheral gas is stable", the "open" of central gas means that central gas flow is narrow and strong. Over development of peripheral gas flow will not only result in the increase of furnace body heat load and affect BF campaign, but also make the gas efficiency worse, fuel consumption increasing and consequently affect the long and stable smooth operation of BF. The stable peripheral gas flow will benefit the protection of stave and stabilization of protective skull. Central gas flow has the effect on gas utilization, energy consumption, intensified smelting and has direct influence on the stabilization of peripheral gas flow. The target of rational gas distribution is to realize BF stable and smooth operation and then to increase gas efficiency, realize energy saving and consumption reduction for ironmaking and obtain BF long campaign life.

3.2.3 Working inner profile control

Management of BF working inner profile relates to design elements of profile design, configuration of cooler and usage of refractory and operating elements of raw material management, furnace body cooling, gas distribution, tapping management and BF operation. It is the comprehensive reflection of BF production management and foundation for long, stable and rational operating profile and also for long campaign life. Management of operating BF profile shall be taken as the most important regular management system of BF production. The variation of BF inner profile has to be timely and accurately monitored. The profile change information has to be quantitatively analyzed to judge and solve the problems that cause the profile change and to maintain normal BF working inner profile.

In order to reduce the damage to BF body, the remote controlled gunning technology for lining is adopted, the similar operating profile complying with BF smelting regulation is formed through gunning lining technology, which benefits to prolong the service life of stave and furnace body above the tuyere, and also provide a reliable condition for rational distribution of gas flow[9].

3.2.4 Casthouse operation control

After BF intensified smelting, in-time tapping of hot metal and slag becomes the key factor to BF stable and smooth operation. Behind-time tapping of slag and hot metal will result in difficult seepage of slag and hot metal in deadman, disturb the initial gas distribution in hearth and affect the BF working profile. Taphole maintenance will directly affect the maintenance of working profile in taphole area; taphole depth will be continuous over shallow, and carbon brick in taphole area will be severely corroded, affecting the long life of BF hearth. Quantitative analyses on BF tapping interval, tapping time, tapping quantity, taphole depth, clay quantity, enhance cast house operation level, ensure BF non-irregular blasting, reduce taphole excess clay, stabilize taphole depth, increase the stability of casthouse operation. The quality of clay will influent the tapping, which has to be stabilized; the taphole clay which adapts to different operating conditions and smelting intensity has to be developed; the features of anhydrous clay such as high strength and better anti-corrosion performance shall be fully utilized; it is the trend for BF casthouse operation to adopt tapping method with less tapping, long tapping time and short tapping interval[10].

3.3 Maintenance

In addition, the grouting technology for the hearth sidewall has become an important technical approach to prolong the service life of the BF hearth and taphole area. The taphole area is always the BF key position to be maintained and is a weak link of the BF campaign life, too. The brick lining in the taphole area is repeatedly impacted by the clay gun and driller, and cracks easily appear on it, forming the accesses to leak the high temperature gas. Therefore, a grouting operation is necessary for the taphole area. Otherwise, it is not only difficult to plug the taphole, but also affects the smooth BF operation and the whole BF campaign life. Care shall be taken when the grouting technology is adopted for the hearth so that the cooling staves and brick lining cannot be damaged and the grouting holes should keep away from the staves to prevent from damage if the holes are drilled while the inlet pressure and grouting rhythm should be well controlled during grouting and the BF hearth lining should be prevented from impact due to over-high pressure, especially in the middle and later stages of the BF.

For a modern BF with a higher intensity smelting, especially the BF at the late stage of the campaign, the addition of titaniferous material should become regular measure for maintenance of hearth and bottom, and long term and continuous addition of titaniferous material can keep proper addition of TiO_2, so that a protection layer with high viscosity shall be formed at the inside of hearth and bottom to slow down the corrosion on the hearth by flowing hot metal; on the other hand, accumulation of TiC, TiN and Ti (C, N) will be formed in time at hearth corrosion area to avoid continuous corrosion inside hearth. TiC, TiN and Ti (C, N) with high melting point are generated, grow, are accumulated in hearth and form sticky material together with hot metal and graphite from hot metal, and cemented on the surface of severely corroded hearth lining which is close to the stave; as a result, it will repair and protect the hearth lining. When using titaniferous material for maintenance, the TiC and TiN in the slag shall not be smelted under hearth temperature, suspended in the slag in the form of solid particle, and worsen the flowability of slag, and the slag becomes more sticky with more TiC and TiN, and will lost flowability at the worst. When titaniferous material is used for BF maintenance, the charged TiO_2 is rationally controlled, and the hearth temperature gradients is rationally utilized to solve the contradiction between BF hearth maintenance and intensified smelting. Table 3 shows the main parameters of BFs No. 1 and No. 3 in 2010.

Table 3 Main technical parameters of Shougang's BFs No. 1 and No. 3 in 2010

Item	No. 1	No. 3
Output/t	2093035	2026395
Productivity/t · (m^3 · d)$^{-1}$	2.338	2.277
Coke rate/kg · t^{-1}	340.4	358.1
PCR /kg · t^{-1}	144.9	124.2
Fuel ratio /kg · t^{-1}	505.2	499.9
Oxygen enrichment ratio /%	1.12	0.68
Blowing down ratio /%	2.11	1.86
Hot blast temperature /℃	1136	1076
w_{TFe} in burden /%	58.79	58.85

4 Conclusions

(1) BF long campaign life is the main technical developing trend of modern BF. BF design, operation and maintenance have important function on BF long campaign life. At early 1990's, a series of advanced long life technologies were adopted for Shougang's BF design. In BF production, the BF life is tremendously extended owing to adoption of these advanced long life technologies and better application effects are obtained.

(2) Optimization of the BF inner profiles, application of rational hearth bottom lining structure, adoption of the advanced staves and closed loop circulating soften water cooling system as well as deployment of perfect automatic monitoring system are the important foundations to prolong BF campaign life.

(3) During the BF operation, to intensify the on-line monitoring and maintenance is an

important measure to prolong BF campaign. The technical measures for Shougang's BF long campaign life are: on-line monitoring on hearth cooling water temperature difference and closed circulating soften water cooling system by using an automation model; optimization of BF operation by the expert system; to intensify the maintenance of BF body, to keep active hearth working status and rational gas flow distribution; to intensify the management of working inner profile and casthouse operation; adoption of gunning lining and titaniferous material to protect lining technologies, etc.

Acknowledgements

The author is grateful to colleagues of Mr. MAO Qingwu, Mr. YAO Shi, Mr. QIAN Shichong, Ms. NI Ping, et al who participate the technical study works on this project. Thanks are also given to Mr. LI Fu-lai, WEI Zhihai, et al for their help.

References

[1] Zhang Fuming, Dang Yuhua. Present situation and development of long campaign life technologies of large BF in China [J]. Iron and Steel, 2004, 39(10): 75-78.

[2] Zhang Fuming, Liu Lanju. Application of new type and high quality refractory materials for blast furnace in Shougang [J]. Ironmaking, 1994, 13(3): 22-25.

[3] Zhang Fuming, Wang Yingsheng. Design and practice of long campaign technology for blast furnace at Shougang Co. [J]. Iron and Steel, 1999, 34(Supplement): 251-254.

[4] Zhang Fuming. Design and application of combined hearth lining with hot pressed carbon brick and ceramic cup in Shougang's No. 1 blast furnace [J]. Iron and Steel, 2000, 35 (Supplement): 162-166.

[5] Mao Qingwu, Zhang Fuming, Yao Shi. Design and application of high efficiency and long campaign technology in Shougang [J]. Ironmaking, 2011, 30 (5): 1-6.

[6] Ma Hongbin, ZhangHeshun. Application of blast furnace long campaign life at Shougang [J]. Research of Iron and Steel, 2010, 38 (2): 38.

[7] Han Qing, Ding Rucai, Zhao Minge. Experience of prolong the campaign life of Shougang No. 4 BF [J]. Ironmaking, 2001, 20 (1): 3.

[8] Cheng Susen, Qian Shichong, Zhao Hongbo. Monitoring method for blast furnace wall with copper staves [J]. Journal of Iron and Steel Research, International, 2007, 14 (4): 1.

[9] Zhang Heshun, Wen Taiyang, Chen Jun. Application of maintenance to prolong blast furnace campaign life at Shougang [J]. Ironmaking, 2009, 28 (6): 21-24.

[10] Zhang Heshun, Wen Taiyang, Li Shuchun. Maintenance of taphole area for the long campaign of Shougang's BF [J]. Ironmaking, 2010, 29 (5): 14-17.

Design and Application of Blast Furnace Longevous Copper Cooling Stave

Zhang Fuming　Yin Guangyu

Abstract: Since 21st century, the application of copper cooling stave has been an important technical measure to prolong the campaign life of modern blast furnace. Combined with the application of blast furnace copper cooling stave at Shougang Qian'an Steel Plant, the optimum design of copper cooling stave heat transfer research and structure parameters are discussed. The heat transfer parameters, structure design and installation method of the copper cooling stave are introduced in this paper. The blast furnace engineering design and over 13 years operating practice of the copper cooling stave for No. 1 blast furnace at Shougang Qian'an Steel Plant are expounded.

Keywords: blast furnace; coppercooling stave; cooling system; campaign life extension; design optimuzation

1　Introduction

Since the 21st century, the blast furnace (BF) has made remarkable achievements in the aspects of enlargement, high efficiency, longevity, intelligence and so on. A series of BF longevity technologies have been developed and applied successively, and remarkable application results have been obtained. Compared with the traditional cast iron cooling stave, copper cooling stave has the advantages of high heat transfer efficiency, uniform and stable cooling, rapid formation of skull, long service life, etc. Using copper cooling stave can solve the problem of overheating damage of cooler in the area of high heat load from the bosh to lower stack of the BF proper. In order to prolong the BF campaign life effectively, copper cooling stave has been widely applied in BF, which is one of the most important technological progress of BF longevity technology at the beginning of 21st century. Since 2000, copper cooling stave has been popularized and applied in China. This technology has been applied in more than 200 BFs, and the application effect is very different. The excellent BF has been applied for more than 15 years without copper cooling stave damage, but copper cooling stave damage occurred in some BFs for 2 to 5 years[1]. Facing the future, the enlargement, longevity, greenization and intelligentization of BF are still the main developing tendency of the international ironmaking technology. In order to further prolong the BF campaign life and improve the production efficiency, the copper cooling stave is still one of the

---

本文作者：张福明，银光宇。原刊于 International Congress on the Science and Technology of Ironmaking, 8th, ICSTI, 2018。

important measures to realize the longevity of BF in the serious damaged area of the bosh, belly and lower stack of furnace body. The design, research and application of the copper cooling stave with high efficiency and longevity need to be carried out deeply[2].

2　Design of Copper Cooling Stave

2.1　Design Concept

With the rapid development and wide application of computation technology, the design of BF cooler should be based on the theory of heat transfer and numerical computation method to simulate and analyse the thermal condition of cooling stave. Therefore, in the design of cooling stave, the application conditions, temperature field and stress field of cooling stave should be researched. The purpose of this study is to optimize the cooling parameters and structural parameters of the cooler by applying this accurate and rapid method.

The design concept of modern BF cooling stave is to ensure that the maximum operating temperature of the cooling stave is not higher than the allowable operating temperature of the used material under the BF bad operating conditions, that is, when the furnace temperature or heat flux reaches the peak value. That is to say, the designed copper cooling stave is non-overheat and low stress under the BF operational condition.

2.2　Heat Transfer Model

It is obvious that the structure and cooling parameters of the cooling stave assembled in different areas of the BF are different because of the different heat transfer boundary conditions. However, for the closed circulating cooling system in series, the structure and cooling parameters of the cooling stave should be applied as the thermal calculation basis of the cooling stave with the thermal boundary condition of the maximum heat flux.

The temperature and peak heat flux in BF are the most important boundary conditions for temperature field analysis of cooling stave. During the whole production process of BF, the refractory lining in the front of the cooling stave hot face will be gradually erosion, wear or even disappear. The heat flux of the cooling stave will correspond to the calculated maximum heat flux. Because of the BF operating conditions change, the stuck skull may fall off for a short time. At this time, the heat flux of cooling stave will be greatly increased, which corresponds to the peak heat flux.

The practice of BF operation proves that although the temperature of gas in the bosh is the highest, but the thick and stable skull can be formed in this area, so the high heat load range is not in the bosh. Although the gas temperature in the lower stack is slightly lower, but the conditions for the skull formation are much worse than that in the bosh, and upward gas flow will appear transverse flow in the process of penetrating the cohesive zone coke window. The high temperature gas flow causes thermal shock to the lining and increases the heat load of the furnace wall, so the lower stack should be the key area of the research. The boundary conditions in different BF regions are different. Usually, the temperature in BF is 1600℃ as the boundary condition of the belly and the lower stack. The numerical computation based on heat transfer theory makes the heat transfer calculation of cooling stave structure more accurate, and the design of cooling stave can be optimized according to the calculation results.

2.3 Design Optimization of Copper Cooling Stave

It is proved by practice that the cooling structure of furnace body is the key factor to determine the BF campaign life. No matter the hearth and bottom, bosh, belly and lower stack, there is no reasonable cooling structure, even if any high-grade refractory is adopted, the BF will not be able to obtain longevity. Reasonable cooling structure is significant for prolong the BF campaign life, and even determines the length of BF campaign life in some sense. The reasonable cooling structure of the BF shall have the following conditions:

(1) The BF body adopts the full cooling structure, eliminates the cooling dead region, and constructs a complete cooling system according to the working conditions of each part from the bottom to the throat.

(2) In order to provide effective and reliable cooling for refractory lining, the furnace body structure, which is coordinated with cooling system and refractory lining, is constructed by optimizing the configuration of refractory lining structure. Coolers from the bosh to the stack should also provide effective support for the refractory lining.

(3) It can withstand high temperature heat load, and still has high efficiency heat transfer ability under the maximum heat flux of BF operating condition, so the cooler should not be overheated and damaged. The heat flux of the cooling stave in the hearth area should be 10-12kW/m^2, the heat flux should be 15kW/m^2 at the end of the BF campaign, and the heat flux should be 20-35kW/m^2 for the cooler in the bosh area, the heat flux in the lower stack should be 50-55kW/m^2, the heat flux in the middle stack is 30-40kW/m^2, and the heat flux in the upper stack is 15-20kW/m^2.

(4) For coolers in the BF bosh, belly and stack area, after the refractory brick lining has eroded or even disappeared, it is possible to rely on essential cooling ability to form a "self-protective permanent lining" based on the protective skull. Moreover, the BF operating inner profile formed by this approach of permanent lining should be beneficial to the BF operation stability.

(5) The reasonable BF cooling structure should be combined with the raw materials and fuel condition, operation condition and habit of BF, which is also an important factor in determining the BF cooling structure.

2.4 Design of Copper Cooling Stave

Shougang Qiangang is a new iron and steel union plant with capacity of 8.0 million tons crude steel, there are 3 BFs in Qiangang plant. The BF's technologies and equipments of Qiangang are shown in Table 1. Qiangang No. 1 BF and No. 2 BF are the same design and copper cooling stave have been applied in 6th to 8th ring of the furnace body. The number of cooling staves in each ring is 45 pieces. The structure of typical copper cooling stave is shown in Fig. 1[3].

Table 1 Main Technical specifications and equipment of Shougang Qiangang BFs

Plant	Qiangang (Qian'an Steel Plant)		
BF number	1	2	3
Effective volume/m^3	2650	2650	4000
Diameter of hearth/m	11.5	11.5	13.5
Design daily output/$t \cdot d^{-1}$	6000	6000	9200
Blew in/y-m	2004-10	2005-1	2010-1

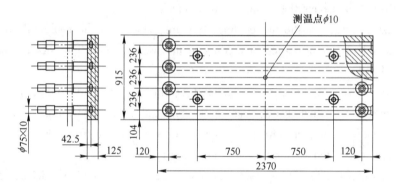

Fig. 1 Structure of belly copper cooling stave

There are 3 rings of copper cooling stave (6th-8th ring) in the bosh, belly and lower stack, 45 copper cooling staves in each ring, and 4 cooling channels in each cooling stave. The connection size of the pipe is DN 60 and the length is 30mm. The protective casing of water pipe is stainless steel pipe, and the cooling passage of copper cooling stave is compound flat pass pattern. The cooling passage is drilled at one end of the cooling stave and welded with copper plug. The hot face of the copper cooling stave is provided with a swallowtail groove with a width of 52mm, a bottom width of 66mm, a depth of 40mm and a distance between the slots of 100-114mm.

3 Application

3.1 Reasonable Cooling Water Volume Control

Reasonable cooling water volume is an important guarantee to realize the longevity of copper cooling stave. It is proved that some parts of BF copper cooling stave are damaged due to insufficient cooling intensity. Only if the cooling water is large enough can the cooling capacity of the copper cooling stave of BF be guaranteed, and the heat released from the furnace can be absorbed fully in time, the hot surface temperature of the cooling stave can be reduced to a reasonable level, and the safety and stability of the copper cooling stave can be realized.

The operational practice of Qiangang's BF shows that, the cooling water volume is the most important factor to determine the cooling capacity of cooling system. The design value of soften water volume of No. 1 and No. 2 BF is 4500m^3/h. In order to control the temperature of copper cooling stave below 80℃ in actual operation, increasing the amount of water to 5700m^3/h (including 9% bottom cooling water) improve the cooling efficiency of copper cooling stave. The actual water velocity of the cooling stave is more than 2.0m/s in order to facilitate the smooth discharge of bubbles in the cooling stave channel and enhance the cooling effect. In order to ensure that the water velocity of copper cooling stave is more than 2.0m/s, the soften water volume of No. 1 and No. 2 BF must be more than 5700m^3/h (including 10% bottom cooling water), and the soften water flow of No. 3 BF must be above 5600m^3/h.

The temperature change of copper cooling stave thermocouple is the most direct measure to monitor the working condition of copper cooling stave. The hot surface temperature of copper cooling stave should be lower than 150℃ for a long time, otherwise, the physical and thermodynamic properties of copper cooling stave will be changed greatly, and the service life will be shortened greatly.

3.2 Copper Cooling Stave Temperature Control

Fig. 2 shows the temperature change trend of copper cooling stave since Qiangang No. 1 BF blew in. It can be seen from Fig. 2 that the temperature of copper stave in the 6th ring of BF has been basically stable at 40-50℃ for 12 years since the BF blew in, and the temperature of copper cooling stave in 7th ring is basically stable at 50-80℃, and the temperature of copper cooling stave in 8th ring is basically stable in the temperature of copper stave of 40-60℃, and that of copper cooling stave in this area is basically stable at 40-60℃. The temperature stability of the copper cooling stave is the best in the 6th ring, and the temperature stability of the copper cooling stave in the 8th ring is common. The temperature stability of the 7th ring of the copper cooling stave located at the root of the cohesive zone is the worst, and the temperature of the copper cooling stave in this zone is generally controlled within the warning value.

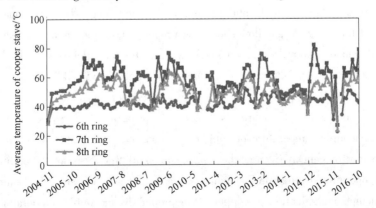

Fig. 2 Historical varying trend of average temperature of copper stave of No. 1 BF

The results show that the rolling copper plate cooling stave can bear the heat flux of $500kW/m^2$, and the hot face temperature of the copper cooling stave is less than 250℃. When the cooling water velocity of No. 1 BF copper cooling stave is higher than that of 1.5m/s, the heat flux duration is $81.41kW/m^2$ and the peak heat flux is $384kW/m^2$, which can last for 30 minutes.

The copper stave adopted in BF bosh, belly and lower stack, which are not only of high heat flux, but also less stable than the hearth and upper stack. The temperature of the cooling stave can reflect the mutative trend obviously. Copper cooling stave is applied in the middle part of BF with high heat load and molten slag and liquated iron. The purpose is to make use of its high thermal conductivity to quickly form a stable skull protective cooling stave and to maintain a stable and reasonable operating BF inner profile.

In order to ensure that the copper cooling stave hot face temperature and heat flux are always in safe working condition, according to the design temperature and heat flux of copper cooling stave, combined with the actual working condition of BF copper cooling stave, the normal temperature range of copper stave is controlled at 40-60℃, and the maximum temperature is not more than 130℃. The normal value of temperature difference of upper soften water is controlled at 1.5-3.5℃, and the maximum is not more than 4.5℃.

3.3 Gas Flow Distribution Control

In order to enhance the skull adhesion ability of copper cooling stave, the agglutinating skull stability was researched and investigated. By means of BF charging and blasting synthetically adjusting the gas flow distribution, controlling the content of harmful elements and regulating the cooling system of special furnace condition accurately, the local adhesion of BF wall or the large area falling off of skull can be avoided. A great deal of practical experience has been accumulated for the operation and maintenance of copper cooling stave.

The copper cooling stave is very sensitive to the instability of gas distribution in the furnace. How to stabilize the marginal gas flow to avoid frequent exfliation of skull is an important issue for the copper cooling stave. The unstable gas flow distribution easily causes the skull to fall off, endangering the service life of copper cooling stave. It is the key to ensure the copper cooling stave to stabilize the skull properly and to control the heat load. The mathematical simulation results show that the marginal gas flow should be properly developed and the heat load of copper cooling stave should be controlled reasonably. As a result, the thickness of skull in the front of copper stave hot surface is 20-30mm and the heat load range is $35-58kW/m^2$.

In practice, the skull adhesion on the hot surface of copper stave can be judged by the copper stave temperature detection. Compared with the thermocouple temperature of cast iron cooling stave, the thermocouple temperature of copper cooling stave can qualitatively reflect the general position of the root of cohesive zone, the varying of skull thickness and shedding of protective skull. The BF operational inner profile management in the area of bosh, belly and lower stack, mainly monitor the diversification of copper stave proper thermocouple temperature in this area, and focus on the uniformity and stability of the temperature along the circumferential direction to judge the rationality and stability of BF inner profile.

Qiangang has always attached great importance to the real-time monitoring of gas flow distribution, and has gradually established and perfected a variety of monitoring means. A BF expert system has been developed in Qiangang No. 3 BF to monitor and evaluate the temperature of copper stave.

3.4 Maintenance of copper cooling stave

During the BF operation, the refractory lining is corroded and damaged. Gunning maintenance for lining is an economical and effective measure to reduce the damage of cooling stave and prolong the BF campaign life. During BF long-time damping down, BF lining has been gunning repaired many times. In the process of gunning maintenance, through the combination of furnace body refractories, the thermocouple temperature change, continuously optimizes the material selection and gunning amount, strengthen the BF gunning control technology, and consummate the BF heating up temperature control.

A set of standardized, systematic and standardized technology for the application and maintenance of copper stave has been formed by combining the research and production practice. The reasonable maintenance of copper cooling stave has been achieved, and the accurate control of gas flow distribution in furnace has been realized. It effectively limits the load of harmful elements into the BF, maintains the reasonable operating furnace type, finally achieves the high efficiency and longevity of the copper stave, and avoids the reduction of the BF operational index and the long period of medium repair caused by the damage of the copper stave, the advantages of copper cooling stave are maximized. At present, Qiangang No. 1 BF has been put into operation for about 13 years (Fig. 3), No. 2 BF has been put into operation for 11 years, and No. 3 BF has been put into operation for about 8 years. None copper cooling stave damage has occurred. The service life of copper cooling stave of each BF is expected to reach the level of BF whole campaign life.

Fig. 3 Actual condition of No. 1 BF copper stave after 13 years operation

4 Conclusions

The copper cooling stave is an important technical measure to prolong the BF campaign life. The design of copper cooling stave should adopt heat transfer model and numerical calculation to determine the optimized structural parameters and thermal performance of copper cooling stave, determine the reasonable heat transfer ratio of cooling stave, diameter of cooling water pipe and flow velocity to meet the requirements of extreme heat load variation. In production and application, the operation management and maintenance of copper cooling stave should be strengthened, the cooling water volume should be adjusted reasonably. The heat load of furnace body and the working temperature of cooper stave should be controlled, so that the temperature of copper stave should be kept below 130℃. Improve the control of gas distribution and maintain the stability of furnace body heat load and skull. Strengthen furnace body maintenance, regular furnace body gunning repair and maintenance, to achieve BF work stability and longevity.

References

[1] Zhang F M, Cheng S S. Modern BF Long Campaign Life Technology [M]. Beijing: Metallurgical Industry Press, 2012: 321.
[2] Zhang F M. Campaign extension technology of BF based on self organization theory [C]//AISTech 2018 Proceedings. Philadelphia, 2018: 497.
[3] Wang J M, Gong W M, Ma G L, et al. The Improvement of ironmaking in Qian'an steel [C]//14th Annual Meeting of National Blast Furnace Ironmaking Proceedings. Jiuquan, 2013: 411.

Research and Practice on Inner Profile Design of Modern Blast Furnace

Zhang Fuming

Keywords: blast furnace; ironmaking; inner profile; long campaign life; design

1 Introduction

The inner profile of blast furnace (BF) is the geometric space of transfer of heat, mass and momentum, as well as metallurgical reaction during the blast furnace smelting process. Its physical essence is the geometry, structural dimension and relationship between them inside BF. BF inner profile is an important part of the BF proper. [1] For a long time, the design method with statistics and analogy is often used for design of BF inner profile. There is a few exploration and research theoretically on it. BF inner profile has a very important effect to stable and smooth operation, high efficiency and low consumption, and long service life of BF smelting. Reasonable BF inner profile is the foundation for BF stable operation and long service life. Therefore, the inner profile design of modern BF is to establish a new idea and method, following the metallurgical principle and rule, to carry out exploration and research of the BF inner profile.

2 Physical Nature and Technical Characteristics of BF Ironmaking

BF ironmaking is a large-scale continuous BF production process, with technological feature of high temperature, high pressure, closed and continuous mode. In the process of BF smelting, it is the coupling of multi-phase complex oxidation-reduction reaction coupling and the integration of metallurgical transfer process. The most important technical feature of BF smelting process is the opposite movement progress of BF burden lowering and gas flow rising, to complete transfers of heat, mass and momentum, as well as a series of physical-chemical action. This process includes heating, moisture evaporation, volatilization and decomposition; reduction of ferrous oxide and other nonferrous oxides; melting and slagging of the nonferrous oxides in the BF burden; carburization and melting of iron, and formation of pig iron; heat transfer between the lowing burden and rising gas, momentum transfer and mass transfer; carbon combustion reaction in the retrieving area before tuyere; flow and exhaust of slag and iron in BF hearth, etc.

本文作者：张福明。原刊于 Proceedings of the Iron & Steel Technology Conference, AISTech, 2018: 407-416。

For the process of BF metallurgy, continuous and stable operation is its important technological feature. It really is in this specific high temperature, high pressure and closed flow structure, and there is a large amount material flow and energy flow input/output, to complete transformation/conversion of material flow and energy flow. As to BF burden drop from throat to hearth, great change on material factors like state, composition, temperature, property, and so on, takes place; coke and pulverized coal burning before tuyere generates BF gas, the gas flow is moving upwards in retrograde way in comparison with the descending burden, passing through the hearth, bosh, belly, throat and top, and then the factors of energy state, vector, potential, etc., also have great change. Based on the engineering theory of dissipative structure and metallurgical process, the BF can be considered as a dynamic-order and collaborative-continuous flow structure of material flow and energy flow. In one word, BF body is the core and key flow structure in the BF ironmaking process.

In this specific flow structure, it not only has material or energy input of iron ore, coke, pulverized coal, hot blast, etc., but also material or energy output of hot metal, slag, gas, etc. And in the process of BF smelting, continuous input-output of information also exist, while input-output of material and energy go on. Analysis from microscale way, BF process represents a integration of mass transfer, heat transfer, momentum transfer, and a series of physical-chemical metallurgical reaction engineering; analysis from macroscale way, BF process represents a dynamic coupling operation in the specific physical space in the BF with characteristics of material flow (in particular flow of ferrous substance), energy flow (in particular flow of carbon energy) and information flow. Fig. 1 shows input-output process of material and energy of BF proper.[2]

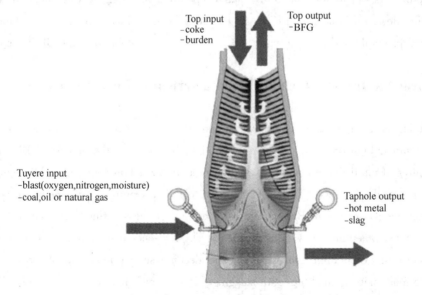

Fig. 1　Input-output process of material and energy of BF proper

3　Function and Optimization of BF Inner Profile

The production practice and research shows that BF inner profile has an important effect to metallurgical transfer and oxidation-reduction reaction during BF smelting process, especially, a great effect to the kinetics of the opposite movement process of burden lowering and gas flow rising. Meanwhile, there also have obvious effects to reasonable distribution of BF burden and gas flow, utilization of gas heat energy and chemical energy, etc. Overall, it can be summarized as follow:

There is effect to opposite movement of BF burden lowering and gas flow rising. The reasonable BF inner profile is benefit to stable and smooth BF operation, and it's better to improve the burden flow permeability. In one BF campaign, the long service life of BF is the basis for stable and smooth BF production. Construction of reasonable BF inner profile is the key links of stable and smooth BF production, and it is the core of the longevity of BF. Design of BF inner profile should ensure smooth opposite movement of solid burden, liquid slag/iron and gas flow, to realize high efficient transfers of heat, mass and momentum, and to improve the utilization efficiency of heat, mass and energy. The reasonable BF inner profile is propitious to smooth burden lowering and gas flow rising. BF, under normal smelting condition, has rational stack pressure difference and good permeability. Once pressure difference of the stack is at a high level, and the permeability goes worse, and the smooth BF condition will be suffered from problems of burden damage, hanging, breaking, over development of gas pipe and edge gas flow, etc., which result in impact of the hearth operation as well as a serious impact on the BF smelting process.

The design BF inner profile is the basis of BF operation inner profile. In one BF campaign, the BF inner profile is changed gradually, the radial dimension increases gradually with the lining erosion, and diameter of the hearth and belly is enlarged, in particularly, the belly diameter of the thick-wall BF. After erosion and disappearance of the brick lining, the belly diameter can be expanded by approx. 800-1200mm, and there is also outstanding expansion happened to the BF hearth diameter because of erosion of the heath side-wall. Due to different hearth erosion, expansion degree of the hearth diameter is not the same. Due to erosion of lining for the bosh and the lower part of the shaft, the belly diameter is increased, and bosh angle and shaft angle are decreased. Thanks to application of throat armour that the throat diameter basically remains the same or very little change. In height direction, the effective height of BF has not changed. But because of erosion of lining for hearth and bottom, depth of the well death is increased, heights for the hearth, bosh, belly and shaft have a certain change, and such changes have close relationship with erosion condition of the BF lining and cooler. A survey after shutdown of the BF found that after production of one BF campaign, BF bosh height is increased, belly height decreased, and shaft height increased. And some BFs are hard to distinguish the boundary of the bosh, belly and shaft. From the above, during the process from the design BF inner profile to BF operation inner profile, the most significant change is happened in the bosh, belly and lower stack. Major changes of the inner profile parameters represent on the belly diameter, belly height, bosh height and shaft height, so that they lead to smaller angle of bosh, angle of shaft and the height-to-diameter ratio accordingly. Change of BF inner profile is caused by erosion and damage of lining and cooler of BF during process of BF smelting, in turn, this change also has an effect on the BF smelting process.

The BF hearth is the beginning and the end of BF smelting process. BF output and efficiency depends on combustion capacity in the BF hearth tuyere raceway. And under condition of fixed tuyere combustion capacity, it depends on BF fuel ratio. Combustion capacity of BF tuyere depends on the hearth diameter and hearth section area. The larger the hearth diameter is, the larger the section area of the hearth is the more the number of the tuyere, and the larger of the BF combustion capacity would be. This is why the BF hearth diameter is used as the very basic parameter during design of the BF inner profile. This proves that the most fundamental core for decision of BF output and production efficiency is the hearth diameter and hearth section area. In recent years, a lot of iron making experts at home and abroad are based on the study of modern BF smelting technology to present the idea of measurement of BF production efficiency based on utilization of cross-sectional area of the BF hearth in order to have more scientific and reasonable evaluation of production efficiency of BF. From production practice of BF, guarantee of stable operation in the hearth tuyere raceway is the foundation for smooth and stable BF production.

The reasonableness of the relationship between the inner profile parameters of various parts of the BF can not be ignored in the BF smelting. There is a certain relationship between the radial and the high dimension of the BF. Such relationship is not only reflected in the statistical significance, but also has close correlation with transfer process of BF smelting. In the design of BF, whether the inner profile is reasonable, the inner profile parameters should be compared, and the relationship between various parameters should be paid to close attention, too.

4 Design Principle of BF Inner Profile

From the research and analysis of the development trend of BF inner profile of more than 200 BFs built at home and abroad over the past 30 years, summary of analysis and research of the metallurgical phenomenon, metallurgical transfer and numerical simulation study, as well as under conditions of material, fuel, operation and technical equipment of modern BF, design principle of reasonable BF inner profile can be as follow:

Design of BF inner profile should be propitious to the opposite movement of burden lowering and gas flow rising so as to reduce resistance to burden lowering and gas flow rising; be propitious to improve stack permeability and gas utilization so as to make heat energy and chemical energy of gas utilized sufficiently, and reduce fuel consumption; be propitious to smooth heat transfer and mass transfer during metallurgical reaction in BF; and be propitious to stable and smooth BF smelting process and create condition for longevity of BF.

Modern BF takes high blast temperature, oxygen enrichment and high PCI rate as the technical characteristics, and BF hearth should meet stable and smooth operation under condition of the high PCI rate. Parameter design of the inner profile in the hearth should be beneficial for the work in the tuyere raceway, be beneficial for improvement of the permeability and liquidity in the hearth area, be beneficial for the hearth more active with sufficient heat, and be beneficial for storage and emission of slag and iron. Determination of the reasonable hearth diameter and height is very important to the design of the entire BF inner profile parameters, and it impacts BF production and longevity directly.

Theory and Practice research proves that depth of the well death has heavy impact of the lining erosion of hearth and bottom. [3] Design of the reasonable well death depth has significant effect to the proper distribution of hot metal flow field and hearth temperature field in the hearth, which makes the dead coke column float in the hot metal, instead of settling down in the bottom, to increase hot metal circulating area, makes the hot metal have smooth flowing in the taphole area, and have effective control to flush and erosion to the hearth side-wall and bottom during hot metal circulating. Therefore, attempt to design reasonable well death is an important technical measure to suppress abnormal "elephant-foot-shaped" erosion of hearth and bottom, and also a significant principle for design of reasonable inner profile with long service life.

With development of BF cooling technology, application of efficient cooler and new quality refractory, design of modern BF inner profile must abandon the conventional concept of maintenance of BF campaign life by lining thickness increase. Design of BF inner profile should be beneficial for formation of protective slag skin at the hot face of cooler and lining, and the slag skin is used as the "permanent lining" in order to extend service life of cooler and lining. So reasonable designed angles of bosh and shaft play a vital role in service life extension of BF bosh and lower part of BF shaft.

There is reasonable relationship between the inner profile parameters of various parts of BF. Such relationship is not only the mathematical formula obtained from mathematical statistics of BF inner parameters, but also it is found that the proportion between the inner parameters of various BF parts has close relationship with transfer phenomenon of BF smelting process by deep research, and it is not simple linear mathematical relations.

5 Design Optimization of BF Inner Profile

5.1 BF Volume

Determination of quantity of BF and volume of BF in one iron and steel plant is a complicated engineering scientific issue. Selection of quantity and volume of BF must be considered as a top-level design in accordance with rationality, high efficiency and economy of the whole process structure in iron and steel works, and then rolling mill composition is considered comprehensively and reasonable productivity is evaluated, once more primary selection of corresponding BF number and volume is worked out. [3] Implementation of BF enlargement should be BF upsizing under precondition of whole process structure optimization in iron and steel works, neither promotion of too many BFs, nor blind pursuit of bigger BF in one iron and steel plant. BF number and capacity should be determined on the basis of process optimization principle of iron and steel plant with close attention to its rational position.

Selection of BF volume should also be based on conditions of material and fuel, operation condition and equipment condition, to determine BF production efficiency reasonably, and this is considered to set up the main parameters of BF for decision-making. The BF production capacity can be calculated based on Formula (1) or Formula (2).

$$P = A \cdot \eta_A \cdot \tau \tag{1}$$

In Formula (1):

P——Annual output of BF, tHM/a;

A——Cross-sectional area of hearth, m^2;

η_A——Utilization coefficient of hearth area, tHM/(m$^2 \cdot$ d);

τ——Annual working days of BF, day.

$$P = V \cdot \eta_V \cdot \tau \tag{2}$$

In Formula (2):

P——Annual output of BF, tHM/a;

V——Volume of BF, m^3;

η_V——Utilization coefficient of BF volume, tHM/(m$^3 \cdot$ d);

τ——Annual working days of BF, day.

Two indices, namely, BF volume utilization coefficient and smelting strength are generally used to evaluate the production efficiency of BF for a quite long time. Since BF hearth reaction is an extremely important metallurgical process in BF iron making, also the beginning and the end of BF smelting process. The utilization coefficient of BF hearth area reflects substantive characteristics of BF process. The utilization coefficient of the section area of BF hearth is used to measure BF production efficiency, which is more scientific and precise.

Table 1 shows volume utilization coefficient is different with utilization coefficient of BF hearth for different levels of BF. Fig. 2 is annual average volume utilization coefficient and annual average hearth area utilization coefficient for different levels of BF. It can be seen from Fig. 2 that volume utilization coefficient and hearth area utilization coefficient represent different change trends along with increase of BF volume.

Table 1 Production capacity and utilization coefficient for different levels of BF

BF Volume/m^3	1260	1800	2500	3200	4080	4350	5000	5500
BF volume productivity /t·(m$^3 \cdot$ d)$^{-1}$	2.5-2.7	2.4-2.6	2.4-2.6	2.3-2.5	2.2-2.4	2.2-2.4	2.1-2.3	2.1-2.3
Hearth area productivity /t·(m$^2 \cdot$ d)$^{-1}$	61.16-66.05	60.98-66.06	60.93-66.01	60.98-66.28	61.51-67.10	61.32-66.89	61.90-67.79	62.04-67.95
Annual operational period/day	350	350	350	355	355	355	355	355
Estimated annual yield /Mt·a^{-1}	110-119	151-163	210-227	261-284	312-340	339-370	372-408	410-449

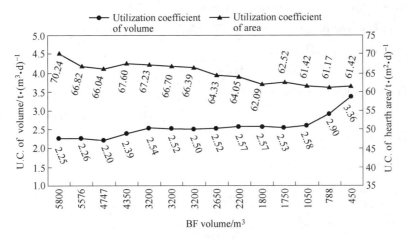

Fig. 2 Utilization coefficient of different levels of BF

5.2 Determination of Hearth Diameter

Hearth diameter is the most critical parameter in design of BF inner profile, and other parameters of BF inner profile have incidence relation with the hearth diameter directly or indirectly. Countries in Europe and America indeed adopt the hearth diameter to represent size of the BF instead of BF volume. The iron output by the unit area of the hearth is used to replace the volume utilization coefficient of BF, to show BF production efficiency, so that the importance of hearth diameter is proven.

Determination of the hearth diameter should follow the rule of BF smelting, which is beneficial for uniform distribution in all BF directions and radius direction in the hearth tuyere raceway, so as to create the prerequisites for smooth BF operation. It is advisable to make a preliminary estimation based on the empirical formula to determine a reasonable hearth diameter. At the same time, it can be checked by the parameters of the section utilization coefficient of the BF hearth with the same scale of production. Determination of the hearth diameter is the same with that of effective volume of BF, and it is a result with characteristics of comprehensive analysis, optimization and comparison, judgement and balancing, as well as reasonable selection. Fig. 3 is the relationship between BF hearth diameter and BF volume.

Determination method of conventional hearth diameter is in accordance with the combustion intensity of hearth section. Combustion intensity of the hearth section is defined as the combusted coke amount per square meter of the hearth section area every day (24h), and actually it is a measure of the degree of different strengthening smelting in BF.

$$I = \eta k = \frac{P}{V_u} \times k = \frac{Pk}{V_u} \tag{3}$$

$$J_A = \frac{P \times k}{A} \tag{4}$$

In Formula (3) and Formula (4):

I ——Smelting strength, normally 1.0-1.5 is taken, tHM/(m³·d);
η ——Utilization coefficient of effective BF volume, tHM/(m³·d);
k ——Coke ratio, kg/tHM;
P ——Daily output of hot metal, tHM/d;
V_u ——Effective volume of BF, m³;
J_A ——Combustion intensity of hearth section area, normally 24-40 is taken, tHM/(m²·d);
A ——Cross-sectional area of hearth, m².

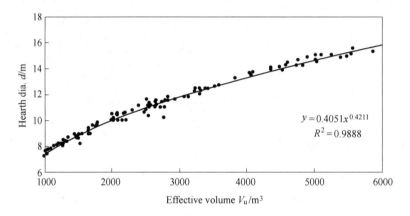

Fig. 3   Relationship between BF hearth diameter and effective volume of BF

After conversion by Formula (3) and Formula (4), Formula (5) and Formula (6) can be obtained:

$$IV_u = J_A A \qquad (5)$$

$$A = \frac{I \times U_u}{J_A} \qquad (6)$$

Section area of hearth:

$$A = \frac{\pi}{4}d^2 \qquad (7)$$

Then the hearth diameter can derived as:

$$d = \sqrt{\frac{4IV_u}{\pi J_A}}$$

$$= 1.128\sqrt{\frac{IV_u}{J_A}} \qquad (8)$$

In Formula (8):

d ——Hearth diameter, m.

Combining analysis and research of the metallurgical process in BF hearth, and current high-efficient, low consumption and low carbon development trend of BF ironmaking, determination of the hearth diameter should be based on high-efficient and collaborative operation concept of the material flow and energy flow to determine the hearth diameter in accordance with new idea and concept. Formula 9 is depended on parameters of the material circulation of BF ironmaking, and preliminary determination of the set-point of the hearth diameter is calculated based on Formula 9. Formula 10 is depended on parameters (gas index of bosh) of the energy circulation of BF ironmaking, and preliminary determination of the hearth diameter is calculated based on Formula 10.

$$d = \sqrt{\frac{4P}{\pi F_m}} \qquad (9)$$

$$d = \sqrt{\frac{4V_{BG}}{\pi F_c}} \qquad (10)$$

In Formula (11) and Formula (12):

    d——Hearth diameter, m;

    P——Daily output of hot metal by the BF, tHM/d;

    F_m——Utilization coefficient of section area of BF hearth, and 60-70 is taken, tHM/d;

    F_c——Gas index of BF bosh, 68-66 is taken, m/min.

Additionally, some scholars have pointed out that the hearth diameter can also be calculated based on gas flow rate of the bosh and gas index of the bosh (refer to Formula(11)):[4]

$$V_{BG} = AX_{BG} = \frac{\pi d^2}{4} X_{BG} \qquad (11a)$$

$$d = \sqrt{\frac{4V_{BG}}{\pi X_{BG}}} \qquad (11b)$$

$$d = \sqrt{\frac{4P v_{BG}}{\pi X_{BG}}} \qquad (11c)$$

In Formula (11):

    d——Hearth diameter, m;

    A——Cross-sectional area of hearth, m^2;

V_{BG}——Gas flow rate of bosh, m^3/min;

v_{BG}——Gas flow rate of the bosh per ton hot metal, m^3/tHM;

X_{BG}——Gas flow rate index of the bosh, 58-66 can be taken, m/min.

5.3 Design of Hearth Parameter

The BF hearth is the beginning and the end of BF smelting process. Parameter design of the hearth inner profile is also the foundation of parameter design of the BF. Importance of design of the hearth inner profile is to determine the reasonable hearth diameter, and this value is related to the rationality of the BF inner profile. Comprehensive consideration of BF smelting condition, hearth combustion strength, utilization coefficient of section area of the hearth, etc. should be carried out when the hearth diameter is determined, and refer to the inner profile parameter of the same level BF with similar material and fuel condition and smelting condition for preferential selection, weighing and decision.

Modern BF adopts the intensified smelting measures of high blast temperature, high oxygen enrichment oxygen enrichment and sufficient pulverized coal injection, etc. More and more attention is paid by domestic and foreign ironmaking workers to the hearth tuyere raceway structure, shape and its features. Control of the reasonable tuyere raceway is one of the importance of modern BF operation, and it makes gas flow be the fundamental to the appropriate distribution, and it is the key measure for control of the lower part of modern BF. For reasonable hearth design, we should not only focus on the determination of hearth diameter, but also give enough attention to following inner profile parameters of BF: hearth height, the ratio of effective volume and cross-sectional area of hearth, the ratio of hearth volume and effective volume of BF, depth of deadman, number of tuyeres, number of tapholes, etc.

For the determination of BF hearth diameter, Estimation is carried out according to the traditional empirical formula of hearth combustion intensity, and calculation is also carried out according to the utilization coefficient of BF hearth section area; under the condition of modern BF smelting, the utilization coefficient of hearth section for different levels of BFs is within the scope of 60-70tHM/($m^2 \cdot d$), the greater the BF, the higher utilization coefficient of hearth section will be, and comparing with the volume utilization coefficient, its change rule is just the opposite.

5.4 Useful Height of BF

BF is a complex reactor with the coexistence of gas, solid, liquid and powder. The gas flow with thermal and chemical energy completes momentum transfer, heat transfer and mass transfer during its backward motion with the solid charge flow and liquid slag iron flow, of which the mechanical process of multi-phase fluid with the characteristic of momentum transfer is the basis for smelting process, the metallurgy transmission process determines the stability of BF smelting and sufficient use of thermal and chemical energy, this is also the core or key point of BF smelting reinforcement.

The useful height of BF mainly depends on the effective volume of BF and metallurgical performance of coke, which is the main reason why the useful height is not linear increase after the volume expansion of BF. We must not intentionally increase the useful height by neglecting the coke quality for the use of gas thermal and chemical energy; and we must not artificially shorten the useful height for the so-called "smelting strengthening and stability promoting". Artificial increase or decrease of the useful height of BF will bring adverse effects on the production of BF, and there have been painful lessons in the design and production practice of both domestic and foreign BFs. In essence, the useful height of BF is the trip of lower furnace burden and rising gas in the BF. For 2000-5800m^3 BF, the useful height is basically maintained within the range of 24-32m. Fig. 4 shows the change trend of useful height of BF.

V_u/A, the ratio of effective volume of BF and cross-sectional area of hearth, is currently one parameter which is widely concerned. This value represents the corresponding effective volume of unit hearth area of BF. The larger the value, the larger effective volume of BF owned by unit hearth area of BF will be. If you assume that BF was a cylinder and cross-sectional area of hearth

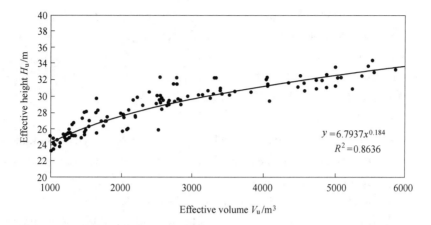

Fig. 4 The change trend of useful height of BF

was "equivalent cross-sectional area" for BF, namely the average cross-sectional area of BF, the physical meaning of the V_u/A is essentially the "equivalent height" of BF. The larger the BF volume, the larger the V_u/A will be, and for the 1000-5800m³ BF, its V_u/A is within the range of 22-32, which is very close to the useful height of BF. Fig. 5 shows the change trend of V_u/A along with the change of effective volume of BF.

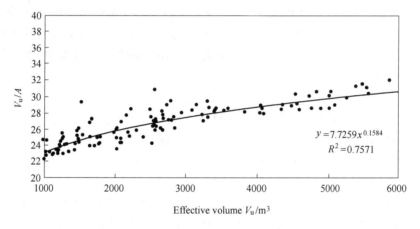

Fig. 5 Change trend of the ratio of effective volume of BF and cross-sectional area of BF hearth

The effective height of BF and physical analysis are based on the famous Ergun formula (Formula (12) and Formula (13)), which are the basic equation of motion of gas in the description of burden flow. It reflects the physical relation among effective height of BF, permeability, resistance of burden flow and gas velocity.

$$\frac{\Delta P}{H} = 150\frac{\mu v_0(1-\varepsilon)}{(\phi d_p)^2 \varepsilon^3} + 1.75\frac{\rho v_0^2(1-\varepsilon)}{\phi d_p \varepsilon^3} \tag{12}$$

Formula (12) can be simplified to:

$$\frac{\Delta P}{H} = 1.75\frac{\rho v_0^2(1-\varepsilon)}{\phi d_p \varepsilon^3} \tag{13}$$

In Formula (12) and Formula (13):

ΔP——Resistance loss of burden flow, N/m^2;

H——Height of burden flow, m;

μ——Viscosity of gas, m;

v_0——Apparent (uncharged) velocity of gas; m/s;

ε——Porosity of burden flow;

ϕ——Shape coefficient of burden particle means the rate of spheric surface area of the particles with equal volume and the surface area of the requested particle, and shape coefficient of the sphere is 1;

d_p——Average granularity of BF burden, m.

Ergun equation is the classic equation of to describe the resistance loss of the fluid through the movable filling bed, and it is the most popular kinetic equation for analysis of the uprising gas penetrated the BF burden during BF smelting process. Formula (13) shows that in addition to the physical property of the burden, resistance loss during gas rising is also influenced by burden porosity, average flow velocity gas seriously. And the resistance loss and the burden height is in positive correlation, namely, the higher the burden height is, the bigger of the resistance loss will be. This is also proven by BF production practice.

Apparent velocity of gas used for Ergun equation is also called the uncharged velocity V_0. This value means the average linear speed of the uprising gas in the BF section, and it can also be understood as the average velocity of gas under the uncharged BF status. Apparent velocity of gas can be calculated by Formula (14):

$$V_0 = \frac{V_{BG}}{60} \frac{P_0}{P} \frac{T}{T_0} \frac{1}{S} \qquad (14)$$

In Formula (14):

V_{BG}——Gas flow rate of bosh, m^3/min;

P_0——Absolute pressure under standard status, MPa;

P——Average absolute pressure inside BF, the average value of top pressure and blowing pressure in front of tuyere can be taken, MPa;

T_0——Absolute temperature under standard status, K;

T——Average temperature inside BF, the average value of top temperature and temperature in the raceway in front of the tuyere can be taken, MPa;

S——Equivalent section area of BF, $S = \varepsilon V_W / H_W$, m^2;

ε——Burden filling coefficient;

V_W——Working volume of BF, m^3;

H_W——Working height of BF, m.

6 Conclusions

BF inner profile is the basic dimension of metallurgical reactor of BF, and also the most fundamental process parameter of modern BF. Whether BF inner profile is reasonable, it directly impacts stable and smooth operation of metallurgical process of BF, meanwhile, it impacts efficiency and effectiveness of BF production. The effect of inner profile of BF on metallurgical transmission and metallurgical reaction of BF should be studied and analysed. In accordance with the development trend of modern BF inner profile, the reasonable BF inner profile should be designed to meet requirements of high efficiency, low consumption and long campaign life of modern BF ironmaking. The design of the inner profile of modern BF should meet the requirements of large scale, high efficiency and long life of the BF. As uninterrupted improvement of material and fuel condition as well as technical equipment level, design of the inner profile of modern BF should be based on the new design system and method which is established on the theory of dissipative structure.

References

[1] Zhang Fuming, Cheng Shusen, Long Campaign Technology of Modern Blast Furnace [M]. Beijing: Metallurgy Industrial Press, 2012: 340.

[2] Martin Gerdes, et al. Modern Blast Furnace Ironmaking (Third edition) [M]. Translated by Sha Yongzhi. Beijing: Metallurgical Industry Press, 2016: 1.

[3] Zhang Fuming, Qian Shichong, Yin Ruiyu. Structure optimization of iron & steel plant process and upsizing of blast furnace [J]. Iron & Steel, 2012, 47 (7): 1.

[4] Xiang Zhongyong, Wang Xiaoliu, Yin Han. Re-exploration of evaluation method on production efficiency of blast furnace [J]. Iron& Steel, 2013, 48 (3): 86.

Research and Application on Temperature Distribution Control Technology of Blast Furnace Hearth and Bottom

Zhang Fuming Zhao Hongbo Cheng Shusen

Keywords: blast furnace; hearth; long campaign life; temperature field; refractory

1 Introduction

In recent years, some blast furnaces, which were newly built or rebuilt and technical modernization, happened successively with accidents of overheat hearth and bottom, abnormal erosion of hearth lining, even hearth breakout, etc. These problems seriously affect normal and safe operation of BF, and result in enormous financial losses. Temperature of some BF hearth and bottom are rising continuously. In order to suppress erosion of the BF hearth and bottom, and prolong BF campaign life, BF reinforcement and protection measures must be taken into consideration so as to avoid influence of BF operational indexes.

According to the existing statistics, more than 10 BFs in China have accidents of breakout of hearth and bottom after entering the new century. Additionally, number of BF with local temperature overheat at heart and bottom shows a rising trend[1]. Therefore, reasonable control of BF hearth and bottom temperature as well as effective extension of BF hearth and bottom service life become a critical and common technical roadblock against the modern ironmaking industry in China.

At present, the uppermost approach to diagnosis and determination of abnormal lining of BF hearth and bottom is to monitor the temperature field of BF hearth/bottom lining and cooling system, and this is direct application of the heat transfer theory.

2 Technical Principle

2.1 Temperature Distribution of BF Hearth and Bottom

In the middle of 20th Century (1950s-1960s), with speed-up of progress of BF enlarge volume,

---

本文作者：张福明，赵宏博，程树森. 原刊于 Proceedings of the Iron & Steel Technology Conference, AISTech, 2015: 888-898。

increase of BF production efficiency, BF campaign life is the main obstacle to restrict the progress of BF ironmaking technology. Many technical researches and investigations were carried out for purpose of extension of BF campaign life at that time, and most significant technical progress ought to apply carbon brick to the places of BF hearth and bottom. Under reducibility condition in BF, carbon brick with prominent technical advantages of high temperature resistance, anti-erosion, resistance to wear, etc. became visible as upgraded and updated product, instead of ceramic material (high alumina brick, clay brick).

Application of carbon brick became reality for extension of BF campaign life undoubtedly, while a great change is realized to the design structure of BF hearth/bottom due to use of carbon brick. Different design structures of full laying of carbon brick at BF bottom, combined carbon brick-high alumina brick BF bottom, and so on have been popularly applied at BF at the time, and carbon block has been popularly used in the area from the BF hearth-bottom junction to the BF hearth side wall below the tuyere level. This design structure for BF hearth/bottom has lasted for nearly 30 years.

The original intention of carbon brick used for BF hearth/bottom is to replace ceramic material in order to resist various damages of flushing wear abrasion, chemical erosion and so on by high temperature slag and hot metal. Cooling process has to be considered in case of carbon brick used for BF hearth/bottom (in fact as early in 1852, BF cooling process existed and there was no application of carbon brick at that time). Whether air cooling, spray water cooling or cooling stave was used for BF hearth side wall, or air cooling, oil cooling and water cooling for BF bottom, in one word, cooling for carbon brick type BF is the technical measure with indispensable protection and support.

In the middle of 20th century, with the widespread use of carbon brick, "1150℃ Isothermal Theory" arose in response to the proper time and conditions, and became the design criteria of BF hearth/bottom with significant influence up to now. W. A. Archibald[2] and others pointed out a proposal of BF self-generated lining in 1957, and concept of long BF campaign life was firstly descibed based on cooling and heat balance. K. W. Cowling was the first to advance the "1150℃ Isothermal Theory" with approximate 1150℃ hot metal solidification temperature theory to observe and find that the erosion outline of BF hearth/bottom lining was the same with 1150℃ isothermal line. Hot metal is accumulated at the carbon brick lining of BF hearth bath. If only the carbon brick operating temperature is lower than this value, it is concluded that the carbon brick is under safe and working temperature, and its erosion would be suppressed.

2.2 Representation and Reason of Overheated BF Hearth/bottom Temperature

Basic theory of physics and thermodynamics shows that temperature is one important parameter for representation of heat. But high temperature does not mean high heat, vice versa. In heat transfer theory, the two physical quantity, namely temperature and heat, are combined, which is the outstanding Fourier's law of heat conduction (Formula (1) to Formula (3)).

$$Q = C \cdot m \cdot \Delta t \tag{1}$$

$$Q = q \times F \tag{2}$$
$$q = (t_1 - t_2) \times \lambda/S \tag{3}$$

Here:

  Q—— Quantity of heat transfer, J;
  C—— Specific heat capacity, J/(kg·K);
  m—— Quantity of cooling water, m³/h;
  Δt——Difference of temperature, K;
  F—— Heat transfer area, m²;
  q—— Rate of heat flow, W/m²;
  S—— Thickness of lining, m;
  t_1—— Temperature of lining hot surface, K;
  t_2—— Temperature of lining cold surface, K;
  λ—— Heat conductivity of lining, W/(m·K).

During smelting process of BF hearth/bottom, its temperature rises due to all kinds of reasons. On-line real time monitoring and warning of BF hearth/bottom temperature is the key measures of current BF maintenance and campaign life extension. Abnormal rising and overheat of BF hearth/bottom temperature are caused by the following reasons demonstrated by means of research analysis and practice[3].

(1) Erosion and damage on the hearth/bottom lining result in temperature rising necessarily so as to have abnormal temperature field distribution.

(2) Closed BF hearth and bottom lining-cooling system may have furnaces partial gas carry-over or gas leakage which result in temperature rising.

(3) Local abnormal erosion of BF hearth/bottom carbon brick creates carbon brick thinning and abnormal temperature rising.

(4) Spalling of carbon brick joint of BF hearth and bottom, local penetration of hot metal and hot metal drilling of carbon brickwork, as well as hot metal filling inside may result in temperature rising.

(5) Efficiency drop or failure of the cooling system may have effective cooling loss of carbon brick, which creates temperature rising to carbon brick lining.

(6) Chemical erosion, loose appearance, pulverization, even annular fracture inside of carbon brick can cause temperature rising at carbon brick cold surface.

(7) Carbon brick oxidation by water and vapour, local erosion, defect or karst cave occurred can lead to temperature rising of carbon brick.

(8) Under high temperature difference, BF hearth/bottom carbon brick result in breakage of carbon brick heat stress so as to cause temperature rising.

Through analysis of the above mentioned major reasons, it is easy to find out that some of them, as "non-organic disease" influencing longevity of BF, can be prevented and controlled; while some of them, as "malignant serious disease" influencing longevity of BF, will directly threat BF normal production and jeopardize BF campaign life.

There are plenty reasons which result in abnormal temperature rising to BF hearth/bottom. Practical analysis of erosion and damage of BF hearth/bottom lining shows that erosion and damage of BF hearth/bottom usually cause temperature rising. Some process of BF temperature rising takes long time with relative big on temperature change, and temperature rising sustains. Even if some BF protection measures are taken into consideration, yield results are little, and this means BF hearth/bottom has been seriously eroded, and it enters to "intensive care" stage. Some process of BF temperature rising takes short time with relative little on temperature change, and rate of temperature rising is less. Normally conventional BF protection measures are taken, afterwards, temperature rising can be controlled effectively, and the temperature may be lower than that before temperature rising. But temperature rising on some BFs lasts for a short time, and there is no obvious temperature rising sign. Temperature rises suddenly and rapidly, it is too late to take protection measures, and accident of BF hearth is break out. Although this phenomena is not normal, it must be drawn high and close attention, and it should be nipped in the bud.

3 Working Condition and Erosion Mechanism

3.1 Technical Feature of Modern BF Process

Significant changes have taken place to BF raw material and fuel condition, technical process equipment condition and production operation condition after entering the new century in comparison with that in the middle and late of last century. In recent 10 years, BF smelting process has a significant impact due to continuous rising of price of iron ore, gradual deterioration of material and fuel conditions, pressure increase of ecological environment, enlargement of BF process equipment, etc. there is no doubt that BF ironmaking process is restricted once again by natural sources shortage, insufficient energy supply, ecological environment protection, etc[4].

Realization of sustainable development of BF ironmaking becomes a severe task in the face of the iron making personnel. Production technical indexes of modern BF, under current resource, energy and technology, still maintain stable increase[5]. This is the result of technical innovation and technical progress in iron making industry. Application of integrative technical measures of economical and rational burden composition, improvement of blast temperature, increase of pulverized coal injection, decrease of coke rate and fuel ratio, improvement of BF operation, improvement of hot blast oxygen ratio, extension of BF campaign life, etc. are the main technical feature of the present BF smelting[6]. In recent years, based on technical policy of "high efficiency, low consumption, high quality, long campaign life, cleanness" for BF production, main operational and technical indexes of some large scale BFs in China have been achieved the international level[7,8]. BF productivity is 2.35-2.5tHM/($m^3 \cdot d$), average high blast temperature 1250±20℃, blasting oxygen enrichment ratio 5%-10%, coke rate 280-320kg/tHM, PCI rate 150-180kg/tHM, fuel ratio ≤500kg/tHM, top pressure 0.2-0.28MPa and coke load ≥5.0t/t.

At present, BF takes "high efficiency, low consumption and low cost" as the concept of main technical development. Technical measures used for BF smelting are absolutely correct and reasonable to intensify BF production, while new technical issues may occur in the following aspects:

(1) Through increase of BF output, improvement of productivity, increase of emission flux and discharge intensity of slag and hot metal in BF hearth as well as increase of hot metal taphole load, the individual taphole capacity can achieve to 2500-3000tHM/d.

(2) High blast temperature, high oxygen enrichment and high PCI rate have intensified combustion process in the raceway area of tuyere.

(3) Low coke rate, high load and high PCI rate make high BF pressure difference and bad gas permeability, attenuation of furnace condition stability, increase of BF operation difficulties as well.

(4) Dead man form and structure in BF hearth change, and gas flow permeability and liquidity go to worse.

(5) Increase of hot metal static pressure has influence to hot metal tapping speed and penetration of BF hearth/bottom carbon brick.

3.2 Working Condition of BF Hearth/Bottom Lining

Under smelting condition of high blast temperature and high PCI, BF process varies, and the tuyere raceway area and burden column structure vary also. Since coke rate decrease and PCI rate increase result in worse BF gas permeability, operation difficulty increases; improvement of coke load makes increase of coke breeze inside the BF dead man, in addition, increase of unburnt pulverized coal after high PCI causes worse gas permeability and liquidity inside the dead man, development of periphery gas flow and increase of heat load at BF wall. Fig. 1 shows the tuyere raceway structure under the high PCI operation condition. Fig. 2 shows the working conditions of BF hearth and bottom.

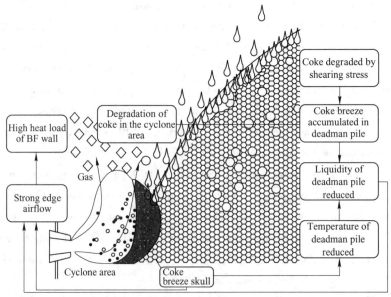

Fig. 1 Tuyere raceway structure under high PCI

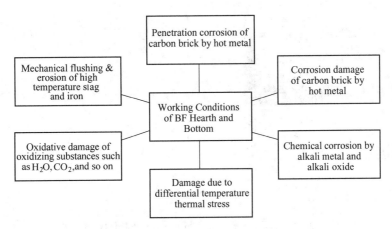

Fig. 2 Working conditions of BF hearth and bottom

3.3 Numerical Simulation Analysis Research for BF Hearth/Bottom Slag and Hot Metal

On one hand, erosional feature of BF hearth/bottom is affected by BF hearth/bottom lining structure and refractory properties, namely influence of temperature field, stress field, and performance of refractory melting slag and hot metal corrosion resistance, and selection of reasonability of BF hearth/bottom structure and refractory reasonable is whether or not mainly based on the original design proposal; on the other hand, after BF is put into operation, erosional feature of BF hearth/bottom is mainly affected by distribution of melting slag and hot metal field, i.e. BF operator improves distribution feature of melting slag and hot metal field in BF hearth by means of adjustment of raw material and fuel, as well as operation system, so as to suppress BF hearth/bottom erosion and prevent occurrence of safety accident. Therefore, it is utmost important to have numerical simulation analysis research for BF hearth/bottom melting slag and hot metal.

Calculation and analysis of hot metal flow inside the hearth show that hot metal circulation is the major reason for creation of serious erosion at BF bottom corner. Fig. 3 shows the distribution of flow field of hot metal in BF hearth. Fig. 3(a) is the longitudinal section of BF hearth field, and arrow size in the figure mean velocity, and it is visible that hot metal velocity at the junction of BF hearth and bottom is high; Fig. 3(b) is the transverse section at junction of BF heart and bottom, and serious circulation at BF hearth and bottom junction can be found. It is thus clear that hot metal circular flow at BF hearth is the arch-criminal to have BF hearth overheat and abnormal damage. And it is the most direct factor.

In order to reduce hot metal circular flow in BF hearth, dead man depth has to be increased to ensure floating status of dead man during BF operation all the time, in this way, there is no coke space existed in BF bottom. On the one hand, the circulation can be weakened (Fig. 4), on the other hand, in case of sufficient depth of well death at the hearth, hot metal velocity and temperature adjacent to BF bottom may be reduced so as to be beneficial to erosion reduction of BF bottom carbon brick.

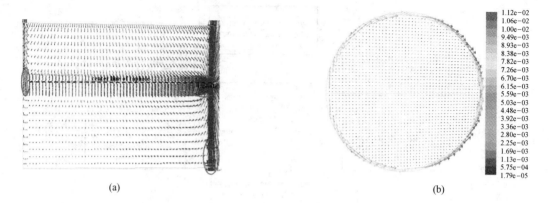

Fig. 3   Distribution of flow field of hot metal in BF hearth
(a) Flow field in vertical section of hearth; (b) Flow field of hot metal in cross section

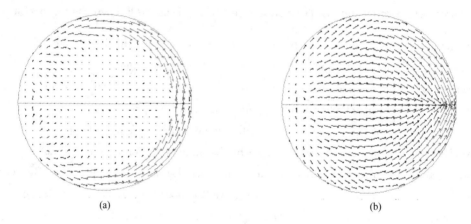

Fig. 4   Variety of hot metal circular flow inside BF hearth during dead man floating and settling
(a) Dead man sitting; (b) Dead man floating

In design of present BF, the well death depth is normally designed to have approximate 18%-25% of inner diameter of BF hearth. It is suggested to have further reasonable increase of design thickness of well death via load calculation on the dead man.

4   Control and Management Technology of Temperature Field

4.1   Practice of BF Hearth Break Out or Over Heat

Control and management for BF hearth/bottom temperature field is the important technical measures to have long campaign life of modern BF, important supporting technology for guarantee of stable and safe BF production. Although erosion progress of the entire BF heart/bottom is the combined result from destructive factors of melting slag and hot metal flow field, refractory temperature field, stress field, hazardous element, etc. Abnormality of final erosion, spalling or ring breakage, etc. of refractory can reflect at distribution variety of temperature field directly and rapidly. Fig. 5 shows variation curves of thermocouple temperature at BF hearth side wall before BF hearth break out accident of one local 3200m³ BF. It is possible to see obvious rapid rising of temperature adjacent to the thermocouple before break out. Since there is a certain distance from the thermocouple to the place burnt through, and its absolute temperature is not very high, it is failure to have enough attention. If the BF is provided with on-line computer monitor model for the BF hearth/bottom temperature field and erosion, the heat transfer modelling calculation can give more accurate warning to the break out place.

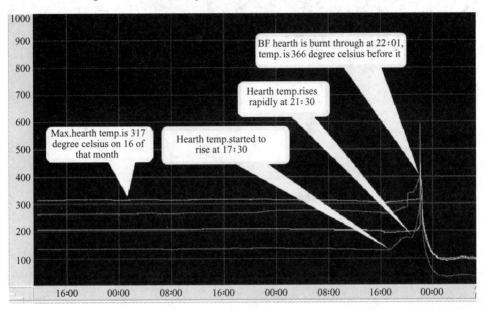

Fig. 5 Variation curves of temperature adjacent to thermocouple before BF hearth break out of one 3200m³ BF

Again, one local BF (1800m³) was built and put into production in February, 2004, super strength smelting was applied afterwards with productivity 2.5-2.8tHM/(m³·d) in long term, and daily output 4500-5000tHM/d. After 8 years since the put-into-production, temperature of the BF hearth side wall is overheat, with instantaneous temperature up to 800℃, and rate of temperature rising varies usually (Fig. 6). It was compelled to shut-down for overhaul. During inspection after dismantling, it is found that residual thickness at place where most serious erosion occurred to the NMA hot pressed carbon brick is approximate 300 mm, timely shut-down was carried out to avoid occurrence of BF hearth break out.

It can be known from the above examples, monitor and management of temperature field are the most direct judgement basis and monitor measure for safety warning of BF hearth. Simultaneously, as to different volume, different BF hearth/bottom structure, different production operation feature, there are different safety warning standard on BF hearth/bottom. Scientific and reasonable warning standard should be established on the basis of real time calculation and monitor of BF hearth/bottom temperature field and erosion profile. And accordingly, refer to the items as follows for detail of hardware, software composition and performance of BF hearth/bottom

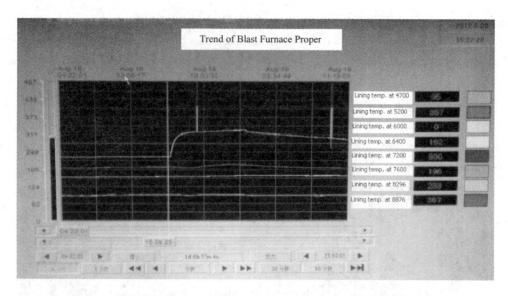

Fig. 6 Variation curves of thermocouple temperature at abnormal place before BF shut-down

temperature field monitor and safety warning system[9].

4.2 Performance Requirement for Accurate Detection Hardware

Selection and layout of thermocouples in hearth and bottom are listed as follows:

(1) The layout of thermocouples should be optimized to satisfy monitoring demands for the full campaign life, especially the key demands for taphole erosion, elephant's foot erosion at the corner of hearth, and hearth activity, as shown in Fig. 7.

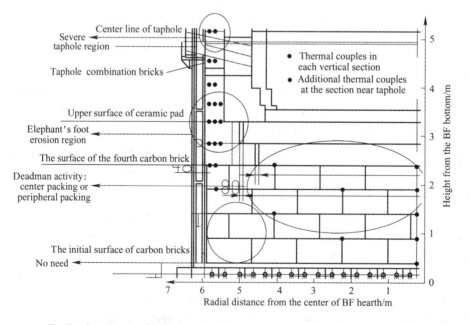

Fig. 7 Optimization for the layout of thermocouples in hearth and bottom of BF

(2) In order to reduce the damage of inbuilt thermocouples in brick lining and to ensure security and normal work of thermocouples during the whole campaign life, the flexible N-type thermocouples are selected, which have higher precision, wider range, and higher security.

(3) In order to ensure long-term normal and stable work as well as the security of data transmission under the BF onsite conditions of dusts and high temperature, Safety conduct is applied at the out-going end as protection measure, and anti-jamming highly integrated design is considered for data acquisition.

Selection and layout for monitoring system of water temperature difference and heat flux at hearth are listed as follows[10]:

(1) In order to guarantee the sensitivity and accuracy of water temperature difference to the changes of erosion and skull in hearth, especially for the monitoring requirements of "heat insulation method" hearth, the digital temperature sensors are selected, which have high precision and high stability. The measurement accuracy is recommended to be better than 0.05℃ because the maximum error of the water temperature difference is twice of the sensor precision. The accuracy of traditional temperature elements is only 0.1℃, and the error of water temperature difference is 0.2℃, corresponding to the error of heat flux above 2000W/m^2, while change of water temperature difference for "heat insulation method" hearth is always less than 0.1℃ during operation fluctuation.

(2) It is suggested that a wireless transmitter acquisition of all the measured point data be built with a single communication bus of high-temperature resistance and high security under the conditions with high moisture, high temperature, and lots of dusts on BF site, as shown in Fig. 8. Using practice of this wireless monitoring system in tens of blast furnaces has proved that the security and stability of this wireless system are significantly higher than the wired temperature measurement system. In addition, the construction and maintenance are much more convenient.

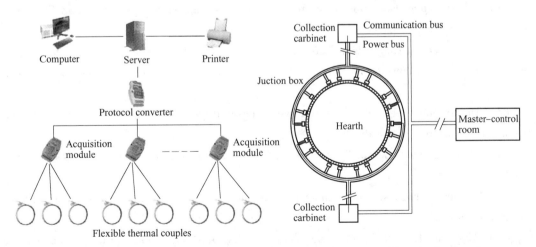

Fig. 8 Topography of wireless acquisition and communication system of heat load at hearth

Wireless surface temperature measuring device attached on BF's shell is used at "poor cooling" region and "monitoring blind" region of BF hearth. The "poor cooling" region is the certain space between adjacent cooling staves at BF hearth, where the sensitivity of water temperature difference to erosion change is low. Unfortunately, some BFs have severe erosion even burn-through just at this "poor cooling region". In addition, the number of thermocouples in hearth and bottom is usually limited, especially in the late period of BF campaign, the "blind zone" will form when thermocouples in this region are damaged. Therefore, temperature monitoring on BF shell temperature should be executed in order to achieve the overall monitoring of the hearth. Infrared temperature gun is mostly used to periodically detect shell temperature when the erosion of BF hearth is severe, but this method has disadvantages of low detecting frequency, large error, and potential safety hazard for BF operators. Therefore, it is suggested that high-precision wireless attached temperature measuring devices should be installed in the "poor cooling zone" and "monitoring blind zone". The accuracy of this device is 1℃. The shell temperature can be detected automatically, and data can be transmitted wirelessly every minute. This device can be used as an important auxiliary security monitoring tools for the security of BF hearth.

As mentioned above, safety warning standards are different for various hearths of different structures, refractory materials, and operational characteristics. Therefore, it is not reliable to solely rely on detection of hardware data to judge hearth security. In order to establish a reasonable and effective security warning mechanism of BF hearth, a professional diagnosing model should be established on the basis of erosion mechanism and heat transfer in hearth and bottom.

4.3 Performance Requirements for Intelligent Diagnosis Model and Warning Software

The intelligent diagnosing models and warning software should contain some functions listed as follows[11]:

(1) The real-time data transmitted from hardware is automatically acquired and filtered to ensure the accuracy of erosion calculation.

(2) The hearth and bottom are automatically meshed, and the 3D non-steady-state temperature field and erosion are calculated considering the influence of latent heat of hot metal solidification.

(3) The erosion images of different cross-sections and the vertical sections in the hearth and bottom are reconstructed and displayed.

(4) Abnormal conditions in BF hearth and bottom are diagnosed automatically, such as ring crack, iron penetration, and air gap.

(5) Intelligent diagnosis and maintenance measure are proposed based on different reasons for hearth erosion.

(6) The location, thickness, and shape of the skull layer in hearth and bottom are automatically calculated and displayed.

(7) The most severe eroded parts of hearth and bottom are alarmed to prevent hearth burn-through accidents.

5 Temperature Control and Dialectical Treatment

On-line monitoring management for temperature field of hearth bottom is to realize no-overheating and self-protection of BF in whole campaign life. It shall be noted that standards of treatment for overheating of hearth bottom are not consistent as always. Management standards of no-overheating and corresponding maintenance measures shall be adjusted depending on different stages of whole campaign life of BF and different parts of hearth bottom.

As shown in Fig. 9, it represents evolution trends of erosion in one campaign life of BF. Although duration of hearth bottom with different types may differ in different stages, this evolution steps are followed generally. In different stages, correspondingly, no-overheating management and self-protection capacity change of hearth bottom shall be conducted in differential ways.

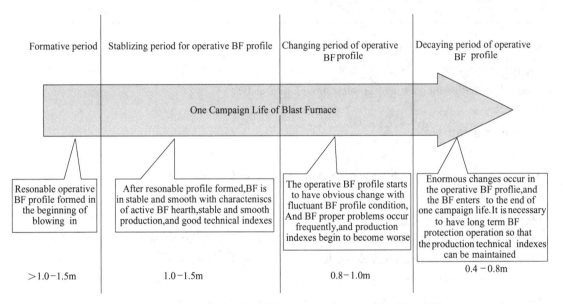

Fig. 9 Evolution process of profile erosion in one campaign life of BF

Table 1 shows control over thermal flow intensity of Shougang's BF hearth stave and related prevention measures, it can be seen from that safety management standards for hearths with different heat conductivity characters are adjusted accordingly. Corresponding measures and intensity of BF protection will change gradually in different erosion stages.

The essence of safe and long-time productions of different types BF is based on establishment of "no-overheating and self-protection" system. Thus, for safety management of hearth bottom temperature field, more reasonable remained lining thickness management system and multi-level digital early alarming mechanism shall be established, i.e safety alarming standards shall take thermal loads, couple temperature, erosion thickness and slag skull into consideration comprehensively and monitoring records of hearth shall be composed of actual value and maximum value. Three-level early alarming index of "work level" "balance level" and "early alarming

Table 1 Control over thermal flow intensity of Shougang's BF hearth stave and related prevention measures

\multicolumn{2}{c}{Heat flow intensity control and measure with application of cooling stave for hot pressed carbon brick type BF hearth}		Local carbon block is adopted for control over thermal flow intensity of BF hearth stave	
Control value /MJ·(m²·h)⁻¹	Applicable measure	Control value /MJ·(m²·h)⁻¹	Applicable measures
≥41.86	Addition of titaniferous material for BF protection, and maintained at [Ti] = 0.08%-0.10%	≥33.49	Addition of titaniferous material for BF protection, and maintained at [Ti] = 0.08%-0.10%
≥50.23	Increase of dosing amount of titaniferous material, maintained at [Ti] ≥0.10%	≥41.86	Increase of dosing amount of titaniferous material, maintained at [Ti] ≥0.10%
≥54.42	Block tuyere above hearth stave with transnormal water temperature difference	≥46.03	Block tuyere above hearth stave with transnormal water temperature difference
≥62.79	Damping down for BF cooling	≥54.42	Damping down for BF cooling

level" should be established. Then, based on erosion characters in different stage of one campaign life of BF, different BF protection measures and production operation adjustment can be adopted to realize the safe and efficient production of BF.

6 Conclusion

Through analysis of overheating phenomenon of BF hearth bottom, this paper concluded that BF long campaign life technical system with an orientation of temperature filed control of heath bottom must be established. Core content of this system is composed of the following four aspects:

The causes of overheating phenomenon is coupling effects between slag& hot metal flow field inside hearth and temperature field of brick lining. Among the causes, ring current of hot metal inside hearth is the main one to lead to overheating of hearth and abnormal breakage. While temperature field of brick lining is most direct representation of inner corrosion characteristics.

To realize all-dimensional monitoring of temperature field of hearth bottom and make judgements of "overheating" or not, precision inspection hardware, three-dimensional temperature field and corrosion detection modelling software for hearth bottom is necessary premises.

There are different judgemental standards and management methods for "no-overheating" as to different type's hearth bottom, in different stages of one campaign life of BF, and about different parts. Establishment and maintenance of "no-overheating and self-protection" system shall also based on heat conductivity and corrosion characteristics of itself in order to make a comprehensive judgement.

Multi-level digital early alarming mechanism, composed of "work level" "balance level" and "early alarming level", is established through on-line monitoring of temperature field, erosion condition and changes of slag skull, which is a scientific approach to realize "self-organized, no-overheating, self-protection, permanency" of BF hearth/bottom.

References

[1] Wu Qichang, Wang Xiaoliu. Key of long life of hearth is improvement of refractory material quality [J]. China Metallurgy, 2013, 23 (7): 11-16.

[2] Archibald W A, Brown T P, Leonard L A. Proposal for a self-lining blast furnace [J]. Iron & Steel, 1957, 30: 515-521.

[3] Zhang Fuming, Cheng Shusen. Long Campaign Life Technology of Modern Blast Furnace [M], Beijing: Metallurgical Industry Press, 2012: 30-49.

[4] Zhang Fuming. Research and innovation of the key technologies at modern large blast furnace [J]. Iron and Steel, 2009, 44 (4): 1-5.

[5] Zhang Fuming. Cognition of some technical issues on contemporary blast furnace ironmaking [J]. Ironmaking, 2012, 31 (5): 1-6.

[6] Zhang Fuming, Technical features of super large sized blast furnace in earlier 21st century [J]. Ironmaking, 2012, 31 (2): 1-8.

[7] Zhang Fuming. Developing prospects on high temperature and low fuel ratio technologies for blast furnace ironmaking [J]. China Metallurgy, 2013, 23 (2): 1-7.

[8] Zhang Fuming, Qian Shichong, Yin Ruiyu. Blast furnace enlargement and optimization of manufacture process structure of steel plant. Iron and Steel, 2012, 47 (7): 1-9.

[9] Zhao Hongbo. Hao Jiangwei, Teng Zhaojie. Development and application of the integral operating platform for blast furnace in China [J]. AISTech 2014 Proceedings of the Iron & Steel Technology Conference, Indianapolis, Ind., USA, 2014: 603-609.

[10] Zhao Hongbo, Cheng Shusen, Zhu Jinfeng, Pan Hongwei. Study on mechanics of "Elephant foot shaped" erosion of BF hearth [J]. The 5th ICSTI'09, Journal of Iron and Steel Research, International (Supply), 2009: 10-15.

[11] Zhao Hongbo, Huo Shoufeng, Cheng Shusen. Study on the early warning mechanism for the security of blast furnace hearth [J]. International Journal of Minerals, Metallurgy and Materials, 2013, 20 (4): 345-353.

高风温工艺与高效长寿
热风炉设计研究

我国大型顶燃式热风炉技术进步

摘 要：分析总结了我国大型顶燃式热风炉的设计开发和应用实践，对顶燃式热风炉的技术成就和发展方向进行了综合评述和探讨。顶燃式热风炉具有一定的技术优势，适合我国高炉扩容改造采用。

关键词：热风炉；高风温；燃烧器；技术进步

1 引言

顶燃式热风炉是由我国开发成功的一种新型高效长寿热风炉，在 300~2500m³ 高炉上得到了成功应用，取得了显著的经济效益和社会效益，备受世界炼铁工作者的关注。我国大型顶燃式热风炉的设计研究、综合技术开发和应用实践已处于世界领先水平。

2 顶燃式热风炉设计开发与应用

20 世纪 50 年代，我国高炉主要采用传统的内燃式热风炉。传统的内燃式热风炉存在着诸多技术缺陷，这些缺陷随着风温的提高日趋严重。为克服传统内燃式热风炉的技术缺陷，60 年代，出现了外燃式热风炉，将燃烧室与蓄热室分开，显著地提高了风温，延长了热风炉寿命。70 年代，荷兰霍戈文公司（现达涅利-康利斯公司）对传统的内燃式热风炉进行优化和改进，开发了改进型内燃式热风炉，在欧美等国得到应用，获得了成功。与此同时，以首钢为代表的我国炼铁工作者成功开发了顶燃式热风炉，并于 70 年代末在首钢 2 号高炉（1327m³）上成功应用。顶燃式热风炉自从问世至今已有近 30 年的发展历程。生产实践证实，顶燃式热风炉完全可以在 2500m³ 级大型高炉上应用，而且具有显著的技术优势。我国大型高炉采用顶燃式热风炉的实绩见表1。

表1 大型顶燃式热风炉应用实绩

厂名及高炉号	高炉容积/m³	布置形式	热风炉座数/座	投产时间
首钢2号	1327	矩形	4	1979年12月
首钢2号	1726	矩形	4	1991年5月
首钢4号	2100	正方形	4	1992年5月
首钢3号	2536	正方形	4	1993年6月
首钢1号	2536	正方形	4	1994年8月
邯钢4号	917	一列式	3	1997年7月

本文作者：张福明。原刊于《炼铁》，2002，21（5）：5-9。

3 顶燃式热风炉技术特点

顶燃式热风炉显著的技术特征是将燃烧室置于顶部，国外称之为"无燃烧室式热风炉"。其结构特点是完全取消了燃烧室，将燃烧器直接安装在热风炉的拱顶，以拱顶空间作为燃烧室。同内燃式、外燃式热风炉相比，具有如下特点[1]：

（1）同内燃式热风炉相比，取消了燃烧室和挡火墙，从根本上消除了内燃式热风炉的致命弱点，扩大了蓄热室容积，在相同热风炉容量条件下，蓄热面积增加25%~30%。

（2）同内燃式、外燃式热风炉相比，结构稳定性增强。钢壳结构均匀对称，气流分布均匀，传热对称均匀性提高。

（3）顶燃式热风炉采用大功率短焰燃烧器，直接安装在拱顶部位燃烧，使高温热量集中在拱顶部位，热损失减少，有利于提高拱顶温度。

（4）热效率提高。顶燃式热风炉燃烧期，高温烟气由上向下流动，烟气在流动过程中向蓄热室传热，在高度方向上形成了均匀稳定的温度场分布；热风炉送风期，冷风由下向上流动，温度由低变高，这是一种典型的逆向强化换热过程，提高了热效率。通过对首钢2号高炉实测表明，热风炉热效率为81%。

（5）耐火材料工作稳定性提高，热风炉寿命延长。顶燃式热风炉改善了耐火材料的工作条件，使温度均匀分布并与耐火材料工作条件相适应，延长了炉体寿命。

（6）布置紧凑，占地面积小，节约钢结构和耐火材料。实践表明，在相同高炉容积条件下，顶燃式热风炉比外燃式热风炉节约钢材约30%，节约耐火材料15%，节约投资20%，特别适合于我国钢铁企业现有高炉原地扩容大修改造采用[2]。

4 大型顶燃式热风炉设计实践

4.1 顶燃式热风炉工艺布置

顶燃式热风炉的结构特点，使其布置灵活多样，紧凑合理。顶燃式热风炉工艺布置形式有矩形布置（4座热风炉），一列式布置（3或4座热风炉），三角形布置（3座热风炉）。而内燃式、外燃式热风炉仅有一列式布置一种形式。首钢2号高炉顶燃式热风炉平面和立面布置如图1和图2所示。

4.2 顶燃式热风炉炉体结构

顶燃式热风炉炉体采用了将拱顶和大墙脱开的结构。拱顶砖衬单独支撑于焊在炉壳拱脚部位的托砖圈上。拱顶砖和大墙砖之间设有用耐火纤维填充的迷宫式滑移缝，可以吸收大墙受热产生的膨胀，使大墙与拱顶可以自由伸缩。这种设计结构增强了拱顶的稳定性，减少了拱顶及拱顶各孔口部位砖衬的膨胀量，减小了各孔口由于砖衬膨胀产生裂缝造成漏风和窜风的机会。热风炉拱顶结构是顶燃式热风炉的关键部位。典型的顶燃式热风炉采用半球型拱顶，在拱脚的垂直段上设置燃烧口，燃烧器的数量和布置方式是根据实验室实验和工业性试验确定的。拱顶是顶燃式热风炉的燃烧空间，为充分利用拱顶空间，使煤气燃

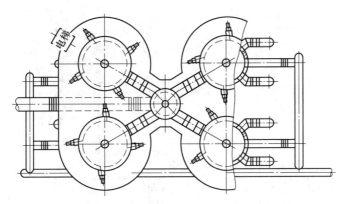

图 1 首钢 2 号高炉顶燃式热风炉平面布置示意图

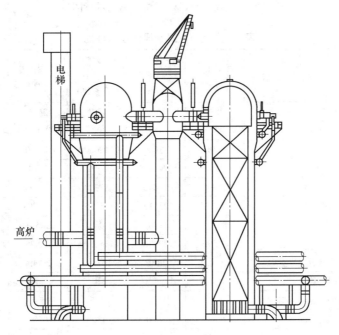

图 2 首钢 2 号高炉顶燃式热风炉立面布置示意图

烧完全，并使烟气在蓄热室格子砖内均匀分布，燃烧口需与拱顶大墙设一个切向角，而且为保证烟气的均匀分布，燃烧口设计成向上扩张的孔口，通过烟气流在拱顶空间内的循环来实现煤气的完全燃烧和烟气流的均匀分布，燃烧口结构如图 3 所示。

根据传热计算结果来确定蓄热室格子砖的高度和材质，确保蓄热面积设计合理，单位高炉有效容积加热面积一般为 $85\sim95m^2/m^3$，单位时间鼓风量的加热面积一般为 $34\sim40m^2/(m^3\cdot min)$。

大型顶燃式热风炉蓄热室格子砖一般分 3 段，高温区采用抗高温蠕变性能优异的硅砖或低蠕变高铝砖；中温区采用高铝砖；低温区采用黏土砖。这种耐火材料的优化配置与顶燃式热风炉特性相适应，可提高耐火材料的功能，降低工程投资。

大型顶燃式热风炉技术性能见表 2。

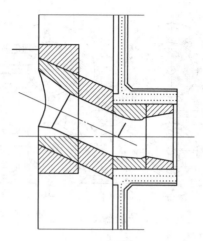

图 3 顶燃式热风炉燃烧口结构示意图

表 2 大型顶燃式热风炉技术性能

项　目	首钢1号高炉	首钢2号高炉（第5代）	首钢2号高炉（第6代）	首钢3号高炉	首钢4号高炉	邯钢4号高炉
高炉容积/m³	2536	1327	1726	2536	2100	917
热风炉直径/mm	8900	7000	7000	8900	8500	6980
热风炉全高/mm	48350	48680	48680	48350	45802	38256
蓄热室直径/mm	7594	5430	5792	7594	7194	5832
蓄热室全高/mm	36000	37840	37700	36000	34000	28000
蓄热室断面积/m²	45.293	23.16	26.35	45.293	40.65	26.713
拱顶燃烧空间/m³	365.1	190	203.2	365.1	306.7	186.3
一座热风炉蓄热面积/m²·座⁻¹	60596	29000	36687	60596	51260	27764
一座热风炉格子砖质量/t·座⁻¹	2314.2	1136	1378.4	2314.2	1974.3	1072
热风炉数量/座	4	4	4	4	4	3
单位炉容加热面积/m²·m⁻³	95.58	87.4	86.27	95.58	97.36	90.83
高炉最大风量/m³·min⁻¹	6500	3000	4300	6500	5800	2400
单位鼓风加热面积/m²·(m³·min)⁻¹	37.29	38.66	34.13	37.29	35.35	34.705
热风炉高径比	5.43	6.95	6.95	5.43	5.39	5.481

4.3 大功率短焰燃烧器

燃烧器是大型顶燃式热风炉最关键的设备。通过多年实验室实验和工业性测试研究，我国大型顶燃式热风炉采用短焰燃烧器，其特点是空气分割煤气，细流分割混合，交叉射流，强化空气、煤气混合。采用燃烧器、燃烧阀、燃烧口三位一体式设计结构，将热风炉炉墙内的燃烧口通道作为预混合段，使空气和煤气进入拱顶空间着火燃烧之前已充分混合均匀，达到短焰燃烧的目的。现役大型顶燃式热风炉短焰燃烧器能力已达 35000m³/h，在大功率短焰燃烧器开发成功以后，每个大型顶燃式热风炉可以配置 2~3 个燃烧器，进一步降低工程投资和设备维护量[3]。现役大型顶燃式热风炉燃烧器的技术性能指标见表3，短焰燃烧器结构如图4所示。

表3 大型顶燃式热风炉短焰燃烧器设计技术性能

项 目	首钢1号高炉	首钢2号高炉	首钢3号高炉	首钢4号高炉	邯钢4号高炉
每座热风炉的燃烧器数量/个	3	2/4	3	3	2
燃烧器额定能力/m³·h⁻¹	32000~35000	28000/15000	32000~35000	28000	25000
适用煤气	高炉、转炉煤气	高炉、焦炉煤气	高炉、转炉煤气	高炉煤气	高炉煤气
煤气温度/℃	0~80	0~80	0~80	0~80	40~45
助燃空气温度/℃	0~150	0~150	0~150	0~150	0~150

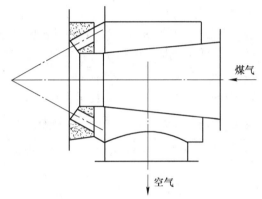

图4 短焰燃烧器结构示意图

5 大型顶燃式热风炉生产实践

5.1 顶燃式热风炉大型化

顶燃式热风炉在2500m³级高炉上已成功应用9年，完全满足大型高炉高效化生产的要求。顶燃式热风炉大型化的关键是大功率短焰燃烧器的开发成功和应用。实践表明，这种高效大功率燃烧器具有适应工况范围广、燃烧能力大、阻力损失小、空气过剩系数低（$\alpha \leqslant 1.05$）、混合充分、燃烧完全等优点，是实现热风炉大煤气量强化燃烧的核心设备。

5.2 顶燃式热风炉的寿命

首钢2号高炉（1327m³）第5代自1979年12月投产以来，其1号热风炉已正常工作22年3个月。炉体结构完整，拱顶、燃烧口、热风出口及格子砖仅出现轻微破损。实践表明，顶燃式热风炉是一种长寿型热风炉，完全可以满足两代高炉炉龄寿命的要求。

5.3 单一高炉煤气燃烧实现高风温

燃烧理论和生产实践证实，提高煤气热值是提高风温的有效措施。但是由于我国绝大多数钢铁企业的高热值煤气（焦炉煤气、转炉煤气）资源紧缺，用于热风炉的燃料为单一高炉煤气。由于近年我国高炉操作水平的提高，煤气利用率提高，使高炉煤气热值已降低到3300kJ/m³以下，理论燃烧温度仅能达到1280℃左右，为提高风温带来了一定难度。在采用单一高炉煤气烧炉的情况下，为实现高风温，近年来顶燃式热风炉进行了一系列技术完善和优化。

（1）采用煤气、助燃空气双预热技术。近年设计投产的顶燃式热风炉均设置了空气、煤气双预热系统。利用烟气余热，预热助燃空气和煤气，以提高理论燃烧温度和拱顶温度。首钢3号高炉采用煤气、空气双预热技术以后，风温提高了50~70℃。

（2）采用高炉煤气旋流脱水器，降低煤气含水量，提高风温。由于首钢大型高炉煤气清洗系统处理能力不足，造成煤气温度高、饱和水和机械水含量高、煤气热值低。首钢1、3号高炉在煤气管道上加设了旋流脱水装置，降低煤气含水量。实测表明，这项技术的实施，可提高风温15~20℃。

（3）采用冷风及烟气匹配技术。在首钢1、3、4号高炉热风炉上采用了冷风及烟气匹配技术，改善气流气场分布，使热风炉径向格子砖孔中气流速度梯度分布均匀化，提高热风炉热效率，综合提高风温20~25℃。

我国采用不同型式热风炉的部分大型高炉主要技术指标见表4。

表4 2000年我国部分大型高炉主要技术指标

项　　目	首钢1号	首钢3号	唐钢2560m³	马钢2500m³	武钢5号	鞍钢10号
高炉容积/m³	2536	2536	2560	2545	3200	2680
投产时间	1994年8月	1993年6月	1998年9月	1994年4月	1991年10月	1993年4月
热风炉型式	顶燃式	顶燃式	霍戈文内燃式	新日铁外燃式	霍戈文内燃式	新日铁外燃式
热风炉燃料	高炉煤气	高炉煤气	高炉煤气	高炉煤气	高炉煤气	高炉煤气
利用系数/t·(m³·d)$^{-1}$	2.119	2.309	1.6	1.9	2.185	2.134
入炉焦比/kg·t^{-1}	372.3	353.7	504	390	399	390
煤比/kg·t^{-1}	134.6	141.2	22	114	122	147
综合焦比/kg·t^{-1}	499	484.4	543	481	515	508
风温/℃	1100	1118	1032	1103	1102	1137

6　大型顶燃式热风炉发展和研究方向

（1）优化燃烧口设计结构和材质，进一步提高燃烧口寿命。由于首钢高炉煤气含水量高，煤气质量差，致使燃烧口出现过早破损，寿命4~5年。采用新型设计结构和优质耐火材料，提高燃烧口整体稳定性，延长燃烧口使用寿命。

（2）研究开发大型顶燃式热风炉自身预热技术。热风炉自身预热技术是由我国炼铁工作者率先提出的一种实现高风温的有效技术措施，在鞍钢10号高炉已取得良好的应用实绩[4]。实现高炉煤气烧炉情况下的高风温，研究开发大型顶燃式热风炉自身预热技术势在必行。

（3）采用烟气、冷风均配技术，进一步提高顶燃式热风炉的热效率。改善烟气、冷风的流场分布，提高烟气、冷风均匀分配程度，提高格子砖蓄热和加热能力，进一步提高热风炉热效率。

（4）开发新型顶燃式热风炉技术。应该指出，近年来，我国热风炉结构型式向多样化发展，内燃式、外燃式和顶燃式多种结构型式的热风炉并存发展。通过技术引进，使我国热风炉技术装备已达到国际先进水平。但值得注意的是，除宝钢以外，我国绝大多数大型高炉风温仅在 1000~1150℃ 左右，与国际先进水平相差甚大。当然，宝钢高炉达到 1200~1250℃ 高风温是以混烧高热值焦炉煤气作为高温热源条件的。采用单一高炉煤气烧炉实现高风温才是适合我国国情的技术发展途径。目前，不少钢铁企业从国外高价引进霍戈文改进型内燃式热风炉，这种发展趋势值得认真研究和探讨。改造型内燃式热风炉的应用还有待于生产实践的长期检验。另一种值得关注的发展趋向是我国 300~450m³ 级中小型高炉发展迅猛，其中顶燃式热风炉（或球式热风炉）的应用呈普及趋势。俄罗斯开发成功的顶燃式热风炉技术已向我国莱钢、济钢等厂输出，拟在 750m³、1750m³ 高炉上应用。目前首钢正在研究和试验新型结构的燃烧器，以适应顶燃式热风炉的发展需求。

7 结语

大型顶燃式热风炉是我国开发成功的一种高风温长寿热风炉，问世至今取得了显著的经济效益和社会效益。经过 20 余年的发展演变，顶燃式热风炉实现了大型化、高效化、长寿化，取得了令人瞩目的技术成就。顶燃式热风炉的成功应用是我国对世界炼铁工业的一个重要贡献。实践表明，顶燃式热风炉具有一定的技术优势，特别适合于我国钢铁企业现有高炉的扩容技术改造采用，今后应继续开发、研究、优化、完善顶燃式热风炉技术。

参考文献

[1] 黄晋，林起礽. 首钢大型顶燃式热风炉设计 [J]. 首钢科技，1999(2)：189.
[2] 倪苹，魏恩发，刘彦波. 邯钢 4 号高炉顶燃式热风炉的设计 [J]. 炼铁，1999(6)：35-36.
[3] 陈炳霖. 关于我国高炉高风温的探讨 [J]. 炼铁，1993(3)：13-18.
[4] 窦力威，陶欣，刘泉兴. 鞍钢高风温技术的开发与应用 [J]. 炼铁，2000(6)：15-18.

Technology Progress of Large Dome Combustion Hot Stove in China

Zhang Fuming

Abstract: The design research and practice application of large dome combustion hot stove in China were introduced. The technical achievements and development direction of large dome combustion hot stove is evaluated too. Practice shows that large dome combustion hot stove can be applied to BF transformation with volume expanded.

Keywords: hot stove; high temperature; burner; technique progress

高炉高风温技术发展与创新

摘　要：阐述了实现高风温的技术理念和关键技术，重点总结了燃烧低热值煤气条件下，获得高风温、高温热风稳定输送、优化热风炉操作，以及热风炉实现高效长寿的技术实践。
关键词：高炉；高风温；热风炉

1　引言

高风温是当代高炉炼铁的主要技术特征之一[1]。提高风温是当前钢铁行业发展循环经济、实现低碳冶炼、节能减排和可持续发展的重大关键共性技术，高风温对于提高高炉综合技术水平、减少 CO_2 排放、引领行业技术进步具有极其重要的意义。近年来，一系列高风温技术创新与应用实践取得了显著成效。新建或大修改造的高炉设计风温一般为 1250±50℃，当前提高风温乃是当代高炉炼铁技术发展的一个显著趋势。

长期以来，我国重点钢铁企业的高炉平均风温一直低于 1100℃。2001～2012 年，经过 10 余年的努力，我国重点钢铁企业高炉的平均风温从 1081℃ 提高到 1183℃，但尚未达到 1200℃，和国外先进水平相比差距甚大。2012 年，我国有 50 多座高炉年平均风温达到了 1200℃，少数先进高炉的风温达到了 1250℃ 以上。据不完全统计，我国容积 $1000m^3$ 以上的高炉约有 250 座，这些高炉的风温水平差距也很大，先进与落后高炉的风温相差 200℃ 以上，由此可见，提高风温仍是当代高炉炼铁的关键共性技术。必须指出，提高风温是当代高炉炼铁技术发展的重要途径，是引领炼铁工业实现可持续发展的重大关键共性技术，提高风温不是最终的目标，而是为高炉实现高效、低耗、低碳、低成本、低排放奠定基础，以提高风温为技术突破，进一步提高喷煤量、降低燃料消耗，从而构建高效率、低成本、低排放的当代高炉炼铁技术体系。以高风温为技术特征的当代高炉炼铁技术，在未来的发展进程中仍将具有其他炼铁工艺不可比拟的技术优势。

2　实现高风温的关键技术

2.1　燃烧高炉煤气实现高风温技术

目前，高炉煤气热值普遍低于 $3000kJ/m^3$，绝大多数钢铁厂高热值焦炉煤气和转炉煤气供给不足，热风炉以燃烧低热值高炉煤气为主，仅有少数企业具备高炉煤气富化条件。在燃烧低热值高炉煤气的条件下，如果不采用煤气和助燃空气预热技术，热风炉拱顶温度

本文作者：张福明。原刊于《炼铁》2013, 32 (6)：1-5。

仅能达到1200~1250℃，根本无法实现高风温，这是当前制约我国高炉提高风温最重要的原因之一。

面对能源供给短缺的现状，应当综合考虑钢铁厂高热值煤气的高效化利用途径，发挥高热值煤气的更大效能。如果将其掺入到高炉煤气中用于热风炉燃烧，尽管可以提高热风炉拱顶温度和风温，但从能源高效利用和长远发展的角度分析，无论是经济性还是技术先进性都不能认为是合理的技术方案。实现低品质高炉煤气的高效能源转换，研究开发燃烧单一低热值高炉煤气实现1280±20℃的高风温技术，才应当是我国当代高炉高风温技术创新的必由之路。实现这个目标的核心是采用煤气和助燃空气高效预热技术，通过提高煤气和助燃空气的物理热，从而有效提高热风炉拱顶温度。近年来，我国开发并应用了多种类型的煤气和助燃空气预热技术。2009~2010年，首钢京唐2座5500m³高炉相继建成投产，热风炉在燃烧单一高炉煤气条件下，为了实现高风温，开发并应用了煤气预热和助燃空气高效预热技术。采用热管换热器预热高炉煤气，使煤气预热温度达到200℃；同时采用2座预热炉强化预热助燃空气，将助燃空气温度预热到520~600℃，在燃烧单一高炉煤气条件下，热风炉拱顶温度可达1420℃，高炉风温可以达到1300℃[2]。首钢迁钢2号（2650m³）、3号（4000m³）的高炉上也采用了这种助燃空气强化预热技术，月平均风温达到1280℃。实践表明，这项利用低热值高炉煤气实现1300℃高风温技术值得借鉴和推广。首钢京唐5500m³高炉热风炉煤气和助燃空气预热工艺流程如图1所示。

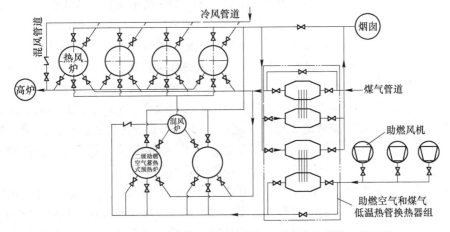

图1 首钢京唐5500m³高炉热风炉煤气和助燃空气预热工艺流程

2.2 采用合理的热风炉结构形式

高炉热风炉是加热鼓风的工艺装置，具有蓄热式加热的工艺技术特征。目前，高炉热风炉主要有内燃式、外燃式和顶燃式3种类型，并且这3种热风炉都有实现1250℃以上高风温的实践。通过对不同结构形式热风炉的燃烧、气体运动和传热等传输过程的研究解析，发现不同结构形式的热风炉在燃烧特性、气体流动与温度分布均匀性、能量转换效率等方面尚存在较大差异[3~5]。在选择热风炉结构形式时，值得关注的一种技术发展趋向是顶燃式热风炉呈现出蓬勃发展的态势。顶燃式热风炉在大型化进程中取得重大技术进步，在首钢京唐2座5500m³特大型高炉上应用并取得了显著成效，高炉月平均风温稳定达到 1250~1300℃。21 世纪初，我国自主开发并应用了多种类型的顶燃式热风炉，在生产实践中取得了较好的应用效果。未来一个时期，3 种结构的热风炉仍然会长期存在，但对于新建或大修改造的高炉，顶燃式热风炉则符合当前技术发展趋势。

针对外燃式和顶燃式热风炉的工艺特性、结构特点以及热风炉燃烧和换热过程进行了定量的分析对比，通过建立数学模型采用数值仿真计算，对以 DME 为代表的外燃式热风炉和顶燃式热风炉进行了传输过程的理论解析[5]。计算结果表明：

（1）顶燃式热风炉的高温烟气流速度分布均匀性优于外燃式热风炉。

（2）顶燃式热风炉格子砖表面温度分布均匀性优于外燃式热风炉。顶燃式热风炉格子砖表面的最大温差小于 50℃，而外燃式热风炉格子砖表面温度温差很大，受燃烧动力学和空间结构影响，最大温差大于 200℃。

（3）顶燃式热风炉流场分布和速度矢量场分布的均匀性优于外燃式热风炉。

2.3 采用高效格子砖

实践表明，采用高效格子砖可以强化蓄热室格子砖与气体之间的热交换，有效降低拱顶温度与风温的差值，显著提高风温。在维持格子砖活面积或格子砖重量恒定时，合理地减小格子砖格孔直径，可以增加格子砖的加热面积，有效改善气体与格子砖之间的传热过程，从而增加格子砖的换热量。具体而言，当热风炉燃烧时，燃烧后所形成的烟气可以将热量更多地传递给格子砖，热量利用更为充分，热风炉所排放的烟气温度也就更低；当热风炉送风时，高效格子砖有利于鼓风与格子砖之间的热交换，使鼓风可以从格子砖中获得更多的热量，有利于提高热风温度。但是过度缩小格子砖孔径，则会导致气体通过格孔通道的阻力增大，造成气体阻力损失增大，消耗更多的鼓风动能；而且过小的格子砖孔径还容易造成蓄热室中格子砖"通孔率"下降、格孔容易阻塞，导致蓄热室有效利用率下降。

毋庸置疑，通过采用高效格子砖来提高热风炉的热效率，是当前提高风温、降低热风炉燃烧煤气量、减少 CO_2 排放的关键技术之一。对于格子砖孔径和砖型的选用应统筹考虑，格子砖孔数越多、孔径越小并不意味着更加合理，需要综合考量蓄热室的热效率、有效利用率、气体运动阻力以及格子砖使用寿命等因素。

蓄热室格子砖在热风炉工作过程中受到热应力、机械荷载、相变应力和压力的综合作用，格子砖在高温下工作，蠕变变形是最突出的非弹性破坏。对蓄热室高温区域格子砖要求具有优良的高温体积稳定性、耐侵蚀性及抗蠕变性能。蓄热室高温区首选硅质格子砖，并且要求硅砖的残余石英必须低于 1%。根据蓄热室高度方向上的温度曲线划分温度区间，并依此而合理选用不同材质和理化性能的格子砖。首钢京唐 $5500m^3$ 高炉顶燃式热风炉蓄热室温度分布如图 2 所示，首钢京唐 1 号高炉 1 号热风炉温度变化曲线如图 3 所示。由图 3 可以看出，在风温达到 1300℃时，热风炉的温度变化平稳，表明热风炉蓄热室格子砖的配置是合理的。

2.4 改善热风炉操作

科学合理的热风炉操作对于提高风温至关重要，对于生产运行的热风炉，提高风温必须重视不断改进和优化操作。生产实践中，通过改进热风炉燃烧与送风的操作工艺，合理提高烟气温度（<500℃），缩小拱顶温度与送风温差值，是提高风温的重要操作措施。必

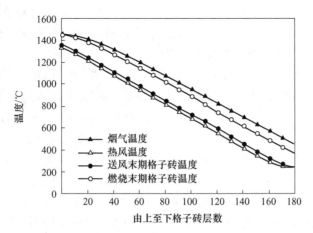

图2 首钢京唐5500m³高炉顶燃式热风炉蓄热室温度分布

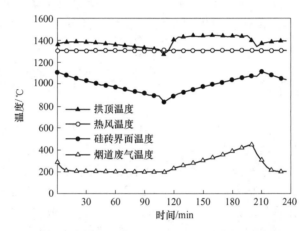

图3 首钢京唐1号高炉1号热风炉温度变化曲线

须注重优化热风炉燃烧操作,根据煤气条件和热风炉工况,设定合理的煤气与助燃空气比例,尽可能降低空气过剩系数,实现煤气的完全燃烧。在热风炉燃烧期,应尽可能降低烟气中残余的CO含量和O_2含量,以节约煤气消耗、提高烟气温度。热风炉燃烧初期宜采用强化燃烧模式,使拱顶温度快速达到设定值;燃烧过程中应合理调节煤气量和助燃空气量,在维持较低空气过剩系数条件下,使热风炉燃烧末期的烟气温度达到设定目标。热风炉燃烧操作时,要根据拱顶温度、烟气温度、烟气O_2含量等参数的变化,实时调节煤气量和助燃空气量,使拱顶温度和烟道温度在规定的燃烧时间内达到设定的目标。特别值得注意的是,在热风炉燃烧过程中,必须严格控制拱顶温度不能"超标",将热风炉整个燃烧期内的拱顶温度始终控制在1420℃以下,避免由于拱顶温度超高而造成NO_x大量生成,这是热风炉燃烧操作控制的关键环节之一。另外,不少钢铁企业由于低热值高炉煤气资源相对丰富,并不注重节约热风炉燃烧所消耗的高炉煤气,不仅增加了煤气消耗量,造成能源浪费、生产成本升高,而且也无助于提高风温,还增加了CO_2排放量。改进和优化热风炉燃烧操作不仅能够有效提高风温,而且还可以减少能源消耗,提高能源转换效率,降低热量耗散,减少CO_2排放。评价热风炉燃烧操作的主要因素有:在合理的燃烧周期内,热风炉拱顶温度和烟道温度满足设定参数的要求且不超标，空气过剩系数低，烟气中残余的 CO 和 O_2 含量低，煤气和助燃空气消耗量少。由此可见，在当前的技术条件下，改善和优化热风炉燃烧操作是必须给予高度重视。

与此同时，热风炉送风操作同样需要改进和完善。配置 4 座热风炉时，应采用"2 烧 2 送"交错并联送风操作模式，采用这种送风模式一般能提高风温 25℃ 左右。另外，要减少冷风混风量，混风量应根据送风温度变化进行在线调节，使高风温得到有效利用。3 座热风炉采用"2 烧 1 送"工作模式，可以合理地缩短送风周期、增加换炉次数。采用"2 烧 1 送"工作模式时，由于送风初期与末期的风温变化较大，为了保持恒定的风温，需要在热风中混入冷风，但也要注意减少冷风混风量。优化和改善热风炉操作是一项系统的工作，需要在工艺操作、设备管理、自动化控制等多方面协同优化。显而易见，采用精准可靠的热风炉操作数学模型，利用信息化、数字化、智能化措施提高热风炉操作水平是未来技术创新的重点课题。与此同时，开发应用热风炉"热交错并联"送风技术，减小风温波动，使高炉保持稳定的高风温。首钢京唐、迁钢高炉热风炉，采用热风炉精准操作技术，优化热风炉操作，在热风炉拱顶温度 1420℃ 的条件下，月平均风温稳定保持在 1250℃ 以上。

3 高温热风的安全输送

目前，制约高炉实现高风温的另一个重要环节是高温热风的安全输送。近期，随着风温的提高，许多热风炉的热风总管、热风环管、热风支管和送风装置等发生局部过热、发红、漏风，乃至管道烧损等事故[6]。因此，对于热风管道系统的设计和维护，必须给予足够重视，要系统地研究在高温、高压、高富氧送风条件下，实现 1280±20℃ 高风温所采取的可靠技术措施。在热风炉设计实践中，建立了"无过热-低应力"热风炉管道设计体系[7]，通过合理设置管道波纹补偿器、拉杆和管道支架，在有约束的条件下降低管道膨胀所产生的系统应力，特别是要合理解决管道系统在高压条件下所产生的"盲板力"。

热风管道承受着高温、高压和弹塑性变形作用，是热风炉系统中工况最恶劣的管道。随着高炉炼铁技术进步，热风管道工作压力已达到 0.35~0.55MPa，热风温度达到 1250℃ 以上，高温、高压的工况特性对于热风管道的设计提出了更高要求。近年来，我国高炉热风管道发生事故屡见不鲜，因此热风管道的设计必须予以足够的重视。计算表明，对管道直径为 2800mm、工作压力为 0.45MPa 的热风管道，其盲板力可达 2800kN，因此热风管道的设计必须保证在此工况下土建结构、管道钢壳、耐火材料砌体的工作稳定。通过多年的研究和探索，目前我国已经构建了一套完整的热风管道设计体系和方法，包括热风管道拉杆设计、波纹补偿器计算与选型、耐火材料设计计算、关键部位的有限元计算分析等。

热风管道设计应考虑工作温度、工作压力、耐火材料荷载、钢结构弹塑性变形、管道试压、环境温度的变化以及多工况耦合等因素，保证了热风管道的安全稳定运行。热风支管与热风总管的结合部采用三角形刚性拉梁的设计结构，利用了三角形稳定性的特点，将热风支管所产生的膨胀力和盲板力等作为内力，在三角形刚性拉梁的约束条件下克服在系统内部，以保证热风支管和热风总管三岔口的结构稳定，同时还可以抑制热风炉炉壳受热上涨对管道产生的影响。热风总管采用通长的一体化大拉杆结构并且合理设置波纹补偿器，要满足在环境温度变化的情况下，必须保持整个管道系统的结构稳定性，有效提高热风管道的安全性和可靠性，防止由于管道发生弹塑性形变和位移而导致耐火材料和钢结构的破坏。

长期以来，由于高炉风温、风压较低，热风管道尚可满足工况要求，但是当风温提高到1250℃以上时，热风管道异常损坏则经常发生，已成为影响提高风温和安全生产的重要因素。采用计算机数值仿真模拟技术优化热风管道设计，是当前应当大力推广的设计创新。通过采用管道受力分析计算软件CAESAR-II对热风炉所有管道系统进行有限元受力分析，使管道必须满足热风炉多种工况的要求，根据计算结果对管道进行详细的优化设计。对于结构复杂的热风管道，通过计算热风管道耐火材料内衬的温度场分布，合理确定耐火材料材质和厚度，从而优化耐火材料结构设计。高温、高压热风管道的设计是现代热风炉设计的重点之一，其重要性堪比拱顶和燃烧器。

合理的热风管道设计的关键内容主要包括3个方面：（1）管道钢结构、波纹补偿器、拉杆及管道支架的合理设计。全面解析整个管道系统的受力状态是优化设计的基础，建立完整的管道系统弹塑性力学分析数学模型，在有约束的条件下降低金属管道系统所产生的应力和应变，同时对于温变应力、膨胀应力和盲板力必须给予足够的重视，特别是管道系统的盲板力绝不可掉以轻心，必须采取稳妥可靠的解决措施。（2）热风管道内衬的优化设计。热风管道不同于其他的压力管道，其最大的差异在于输送介质为大通量条件下的高温、高压富氧空气。热风管道内衬的径向温度梯度很大，内外温差高达1150~1200℃，管壳和波纹补偿器的表面温度应控制在100℃左右。热风管道径向大梯度的温差，使耐火材料内衬结构在没有附加冷却的工况条件下，极易出现温差热应力而造成异常损坏。因此合理设计热风炉管道内衬结构、根据工况特点选择合理的内衬材质成为热风管道耐火材料内衬设计的关键。（3）合理解决热风管道内衬的热膨胀。热风管道内衬受热所产生的热膨胀，在热风管道设计中必须妥善处理。通过热风管道内衬温度场数值模拟，得出内衬的温度分布，根据耐火材料的热力学参数，精准计算出预留的膨胀缝大小、数量和位置。径向热膨胀一般采取在管道上部工作层砖衬与隔热层砖衬之间预留膨胀缝的措施，不应设置集中的径向膨胀缝，以避免热风管道上部砖衬的塌落或下沉。合理设置热风管道内衬轴向膨胀缝至关重要，应当在热风管道轴向方向上设置若干个独立的膨胀缝以吸收管道内衬所产生的热膨胀。轴向膨胀缝应采用迷宫式密封结构，提高膨胀缝的严密性，防止窜风、漏风，以提高热风管道的绝热保温效果，降低风温损失。热风管道波纹补偿器区域膨胀缝的设计结构尤为重要，这是热风管道中最为薄弱的环节，只有彻底攻克这项技术难题，才能真正实现高温热风的稳定、安全输送。

热风炉孔口部位的工况条件恶劣，孔口的异常破损也是影响热风炉长寿和提高风温的制约环节。热风炉孔口区域的耐火材料内衬不但承受高温、高压作用，还承受着气流收缩、扩张、转向运动所产生的冲击和振动作用。热风出口应当采用组合砖结构，组合砖宜采用具有双凹凸榫槽的锁砖结构，才能更好地抵御热风出口上部大墙砖衬对组合砖产生的压应力。除此之外，在热风管道上所有管道连接的"三岔口"部位也均应采用组合砖结构。

生产实践表明，热风管道的耐火材料砖衬在高温、高压热风的流动冲刷作用下，容易出现局部过热、窜风、甚至内衬脱落现象。对于运行中的热风管道应采用表面温度监测系统，可以在线监控热风管道关键部位的管壳温度，并可以进行数据处理和存储，实现信息化动态管理；同时为了监控热风管道受热膨胀而产生的变形情况，设置管道位移监测仪可以在线监测热风管道的膨胀位移。通过数字化在线监控装置，可以提高热风炉管道工作的可靠性，保障高温热风的稳定输送。该项技术已在首钢迁钢大型高炉上得到了成功应用[8]。

4　延长热风炉寿命

在当前技术条件下，热风炉拱顶温度应当控制在1420℃以下，其目的是为了降低热力型NO_x生成量，从而有效预防热风炉炉壳的晶间应力腐蚀。当代热风炉设计中，必须采取有效的预防晶间应力腐蚀技术措施，以延长热风炉使用寿命。生产实践中，通过控制合理的热风炉拱顶温度，防止热风炉燃烧和送风过程中NO_x的大量生成。热风炉炉壳应采用细晶粒钢板，提高炉壳自身的抗腐蚀性能；炉壳焊接安装以后必须对所有焊缝进行探伤检查和研磨，之后采取退火热处理措施，以消除炉壳制造、焊接过程所产生的残余应力。热风炉炉壳施工时，应采用高温区炉壳涂刷防酸涂料并喷涂耐酸喷涂料等综合防护措施，防止NO_x与冷凝水结合以后所形成的酸液对热风炉炉壳产生腐蚀破坏。为了有效防止热风炉燃烧和送风过程中水蒸气的冷凝，对于热风炉高温区炉壳还可以采取外保温等防护措施，使炉壳温度保持在水蒸气露点以上。总而言之，采取综合措施有效预防热风炉炉壳晶间应力腐蚀，是实现热风炉长寿、高效、高风温的重要技术保障。

影响当代热风炉寿命的关键因素除了拱顶和高温区炉壳的晶间应力腐蚀以外，还有热风炉炉体的耐火材料结构，包括拱顶、热风出口、热风管道以及蓄热室格子砖砌体，外燃式热风炉燃烧室与蓄热室连接部、内燃式热风炉的燃烧室与蓄热室的隔墙、陶瓷燃烧器本体等部位，这些部位的使用寿命仍是影响热风炉寿命的制约环节。当前值得重点关注的是热风炉热风出口、热风支管以及热风管道的寿命，应采取综合技术措施，采用"无过热-低应力"设计体系，着重解决高温、高压管道在复杂受力条件下的力学问题，与此同时，在耐火材料材质和设计结构上也应不断优化，以实现当代热风炉的高效长寿。

5　结语

（1）21世纪初，为实现高炉低碳冶炼、节能减排、循环经济，必须坚持"高效、低耗、长寿、清洁"当代高炉炼铁技术发展理念，不断提高风温和喷煤量，降低燃料消耗和生产成本，这将是当代高炉炼铁技术实现可持续发展的重要途径。

（2）实现高风温是提高喷煤量、降低焦比和燃料比的重要技术支撑和保障，是当前炼铁工业关键共性技术。在燃烧低热值高炉煤气条件下实现1280±20℃高风温，对于钢铁厂能源结构优化和高效能源转换意义重大。

（3）实现高风温是复杂的系统工程，要系统解决实现高风温的关键技术难题。采用合理的热风炉结构形式和高效格子砖，提高热风炉热效率；改善和优化热风炉操作，实现高温热风的安全输送；采取综合技术延长热风炉寿命。

（4）优化热风炉结构和内衬设计，重视热风管道的设计优化，构建并应用"无过热-

低应力"热风炉设计体系是实现热风炉高效、长寿、高风温的重要技术支撑。

致谢

衷心感谢首钢国际工程公司钱世崇、银光宇、毛庆武、李欣、梅丛华、倪苹、胡祖瑞、许云、范燕等同志对作者研究工作所提供的帮助；感谢首钢总公司王毅、马金芳、唐志强、郑敬先等领导和专家的支持与帮助。

参考文献

[1] 张福明.21世纪初巨型高炉的技术特征[J].炼铁,2012,31(2):1-8.
[2] 张福明,钱世崇,张建,等.首钢京唐5500m^{3}高炉采用的新技术[J].钢铁,2011,46(2):12-17.
[3] 胡祖瑞,程树森,张福明.霍戈文内燃式热风炉传输现象[J].北京科技大学学报,2010,32(8):1053-1059.
[4] 胡日君,程树森.考贝式热风炉拱顶空间烟气分布的数值模拟[J].北京科技大学学报,2006,28(4):338-342.
[5] 钱世崇,张福明,李欣,等.大型高炉热风炉技术的比较分析[J].钢铁,2011,46(10):1-6.
[6] 许秋慧,常建华,龚瑞娟,等.唐钢3200m^{3}高炉热风炉管道耐火材料塌落的处理[J].炼铁,2010,29(6):23-25.
[7] 张福明,梅丛华,银光宇,等.首钢京唐5500m^{3}高炉BSK顶燃式热风炉设计研究[J].中国冶金,2012,22(3):27-32.
[8] 马金芳,万雷,贾国利,等.迁钢2号高炉高风温技术实践[J].钢铁,2011,46(6):26-31.

Development and Innovation of High Blast Temperature of Blast Furnace

Zhang Fuming

Abstract: The paper expounds the technological concepts and key technologies of high blast temperature with respect to the current situation of blast furnace ironmaking technology and the challenges faced, and summarizes the activities in realizing high blast temperature, stable blowing of high temperature blast, optimized operation of hot stove and long campaign of hot stove under the condition of low caloric gas.

Keywords: blast furnace; high blast temperature; hot stove

首钢京唐 5500m³ 高炉 BSK 顶燃式热风炉设计研究

摘　要：介绍了首钢京唐钢铁厂 5500m³ 高炉 BSK 顶燃式热风炉的设计创新。优化集成了特大型顶燃式热风炉工艺；研究开发了助燃空气两级高温预热技术和顶燃式热风炉高效陶瓷燃烧器。根据顶燃式热风炉特性设计了合理的拱顶和陶瓷燃烧器结构；采用高效格子砖，优化了蓄热室的热工参数与结构，确定了合理的热风炉蓄热面积。优化热风炉炉体内衬设计；采用了有效的防止热风炉炉壳晶间应力腐蚀的技术措施。根据蓄热室传热计算，合理配置了热风炉炉体耐火材料，提高了耐火材料技术性能。优化热风管道系统耐火材料结构设计，使热风管道系统合理化并满足 1300℃ 高风温的要求。高炉投产后热风温达到设计水平，实现月平均风温 1300℃。

关键词：高炉；顶燃式热风炉；高风温；陶瓷燃烧器

1　引言

　　首钢京唐钢铁厂是我国在 21 世纪建设的具有国际先进水平的新一代钢铁厂。钢铁厂建设 2 座 5500m³ 高炉，年产生铁 898.15 万吨。这是我国首次建设 5000m³ 以上的特大型高炉，在全面分析研究了国际 5000m³ 以上的特大型高炉技术的基础上，积极推进自主创新，自主设计开发了无料钟炉顶设备、煤气全干法布袋除尘工艺、高炉高效长寿综合技术、顶燃式热风炉、螺旋法渣处理工艺等一系列具有重大创新的先进技术和工艺装备。

　　高风温是现代高炉炼铁的重要技术特征。提高风温可以有效地降低燃料消耗，提高高炉能量利用效率。设计中对改造型内燃式、外燃式、顶燃式 3 种结构形式的热风炉技术进行了研究分析，在首钢顶燃式热风炉技术和卡鲁金式顶燃式热风炉技术的基础上，综合两种技术的优势，设计开发了 BSK（Beijing Shougang Kalugin）型顶燃式热风炉技术，将顶燃式热风炉技术首次应用在 5000m³ 级特大型高炉上。

2　热风炉工艺技术研究

2.1　优化集成顶燃式热风炉工艺技术

　　高炉设计中对当时世界上已建成投产的 13 座 5000m³ 以上的特大型高炉工艺技术装备

基金项目：国家"十一五"科技支撑计划项目（2006BAE03A10）。
本文作者：张福明，梅丛华，银光宇，毛庆武，钱世崇，胡祖瑞。原刊于《中国冶金》，2012，22（3）：27-32。

和生产运行状况进行了综合研究分析。国内外 4000m³ 级的大型高炉主要采用外燃式热风炉，仅有个别高炉采用内燃式热风炉；5000m³ 以上的特大型高炉全部采用外燃式热风炉；全世界 4000m³ 以上的高炉尚无采用顶燃式热风炉的应用先例。

顶燃式热风炉将燃烧器置于拱顶部位，利用热风炉的拱顶空间进行燃烧，取消了独立设置的燃烧室，其结构对称、温度区间分明、热效率高、占地少，是一种高效节能长寿型热风炉，是热风炉技术的发展方向。

20 世纪 70 年代，首钢开始研究开发顶燃式热风炉技术。最初在 23m³ 的试验高炉上进行顶燃式热风炉工业化试验并获得成功，70 年代末期，将顶燃式热风炉技术应用在首钢 2 号高炉 (1327m³)，在世界上首次实现了大型高炉顶燃式热风炉的工业化应用。90 年代初期，首钢又将顶燃式热风炉技术相继推广应用到首钢 2 号 (1726m³)、4 号 (2100m³)、3 号 (2536m³)、1 号 (2536m³) 高炉上。历经 30 多年的持续研究创新，顶燃式热风炉技术已成为首钢具有完全自主知识产权的原始创新技术，在生产实践中取得了显著的技术经济效益[1]。

20 世纪 80 年代，前苏联冶金热工研究院开发了一种顶燃式热风炉，于 1982 年在下塔吉尔冶金公司的 1513m³ 高炉上建成应用。在这种顶燃式热风炉获得成功应用的基础上，该技术的创造者卡鲁金对这种顶燃式热风炉拱顶和燃烧器结构进行了技术改进和优化，形成了小拱顶结构的顶燃式热风炉，并将其命名为卡鲁金型顶燃式热风炉，这种卡鲁金型顶燃式热风炉技术在俄罗斯和乌克兰 10 余座 1386~3200m³ 高炉上得到应用[2]。

首钢京唐 5500m³ 高炉设计研究中，将首钢顶燃式热风炉技术和卡鲁金顶燃式热风炉技术体系结合一体，集成两种技术的优势，进一步优化创新，设计开发了适用于特大型高炉的 BSK（Beijing Shougang Kalugin）顶燃式热风炉技术，在国际上首次将顶燃式热风炉技术应用在 5000m³ 级特大型高炉上。

BSK 顶燃式热风炉兼具首钢型和卡鲁金型顶燃式热风炉的技术优势，其主要技术特征是：

（1）顶燃式热风炉的环形陶瓷燃烧器设置在热风炉拱顶部位，具有广泛的工况适应性，可以满足煤气和助燃空气多工况条件的运行，而且燃烧功率大、燃烧效率高、使用寿命较长。环形陶瓷燃烧器采用了特殊的旋流扩散燃烧技术，保证了空气和煤气的充分混合和燃烧，提高了理论燃烧温度和拱顶温度。

（2）利用拱顶空间作为燃烧室，取消了独立的燃烧室结构，加强了炉体结构的热稳定性。陶瓷燃烧器设在拱顶部位，高温烟气在旋流状态下分布均匀，有效地提高了高温烟气在蓄热室格子砖表面的均匀性和传热效率。

（3）蓄热室采用高效格子砖，适当缩小格子砖孔径，提高格子砖的加热面积，提高了热风炉传热效率。

（4）利用热风炉烟气余热预热煤气和助燃空气，助燃空气再经预热炉预热至 520℃ 以上，在采用单一高炉煤气作为燃料的条件下，可以使风温达到 1300℃。

（5）热风炉高温、高压管路系统采用低应力设计理念，通过管道体系和耐火材料结构优化设计，可以实现 1300℃ 高温热风的稳定输送[3]。

首钢京唐 5500m³ 高炉配置了 4 座 BSK 型顶燃式高风温长寿热风炉，设计风温为 1300℃，拱顶温度为 1420℃，热风炉高温区采用硅砖，设计寿命 25 年以上[4]。热风炉燃料为单一高炉煤气,采用烟气余热回收装置预热煤气和助燃空气,配置2座小型顶燃式热风炉单独预热助燃空气,可以使助燃空气温度达到520℃以上。热风炉高温阀门采用纯水密闭循环冷却,热风炉系统燃烧、送风、换炉实现自动控制。4座热风炉正常工作时,采用两烧两送交错并联送风模式,在使用高炉煤气作为燃料的条件下,风温可以达到1300℃;在三烧一送和两烧一送的工况条件下,风温也可以达到1250℃。BSK顶燃式热风炉的主要技术性能见表1。

表1 BSK顶燃式热风炉主要技术参数

项　　目	数　　值
热风炉座数/座	4
热风炉高度/m	49.22
热风炉直径/m	12.50
单座热风炉格子砖加热面积/m^2	95885
单位体积格子砖加热面积/$m^2 \cdot m^{-3}$	48
格子砖孔径/mm	30
格子砖高度/m	21.48
蓄热室截面积/m^2	93.21
送风温度/℃	1300
拱顶温度/℃	1420(格子砖上部最高温度1450)
拱顶耐火材料设计温度/℃	1550
烟气温度/℃	最大450,正常平均368
助燃空气预热温度/℃	520~600
煤气预热温度/℃	约215
冷风温度/℃	235
冷风风量/$Nm^3 \cdot min^{-1}$	9300
冷风压力/MPa	0.54
送风时间/min	60
燃烧时间/min	48
换炉时间/min	12
单位鼓风蓄热面积/$m^2 \cdot (m^3 \cdot min)^{-1}$	41.24
单座热风炉格子砖重量/t	2550
单位鼓风格子砖重量/$t \cdot (m^3 \cdot min)^{-1}$	1.097
单位高炉容积蓄热面积/$m^2 \cdot m^{-3}$	67.73

2.2 设计开发助燃空气高温预热技术

为使热风炉在燃烧单一高炉煤气条件下实现1300℃高风温,系统研究了提高热风炉理论燃烧温度、拱顶温度和热风温度的综合技术措施,在首钢助燃空气高温预热技术的基础上[5],设计开发了高效长寿型煤气、助燃空气两级预热技术[6]。

其主要技术原理是：采用分离式热管换热器，回收热风炉烟气余热预热煤气和助燃空气，经过预热后的煤气和助燃空气温度可以达到200℃左右，此过程被称为一级双预热；设置2座蓄热式助燃空气高温预热炉，用于预热助燃空气将其温度提高到520℃以上。助燃空气预热炉采用的煤气和助燃空气均经过热管换热器一级预热，预热炉拱顶温度可以达到1300℃，2座预热炉交替工作，用来加热热风炉燃烧使用的一部分助燃空气，助燃空气经过预热炉加热后温度可达1200℃，再通过与一级预热后的助燃空气混合，使混合后的助燃空气温度控制在520~600℃，此过程称为二级预热。这种工艺是一种自循环预热流程，显著地提高了助燃空气、煤气的物理热，可以使热风炉拱顶温度提高到1420℃甚至更高，从而可以有效地提高送风温度，热风炉系统总体热效率得到显著提高。

3 燃烧室结构优化研究

3.1 优化拱顶设计结构

拱顶是顶燃式热风炉的关键部位，拱顶结构设计的难点是要将陶瓷燃烧器和拱顶结构结合为一体，解决拱顶在热风炉燃烧、换炉、送风交替工作条件下的结构稳定性问题。BSK顶燃式热风炉炉体采用将拱顶和大墙砖衬脱开的自由膨胀结构，拱顶砖衬独立支撑在拱顶炉壳的托砖圈上。拱顶砖衬和大墙砖衬之间设有用陶瓷纤维填充的迷宫式滑移膨胀缝，可以吸收大墙受热产生的膨胀位移，使大墙与拱顶可以自由胀缩。这种自由滑动的设计结构增强了拱顶的稳定性，降低了拱顶及拱顶各孔口部位砖衬的热膨胀量，消除了各孔口砖衬由于热膨胀产生裂缝而造成的漏风和窜风。

3.2 环形陶瓷燃烧器结构

BSK顶燃式热风炉采用锥形拱顶，在拱顶的顶部中心区域设置环形陶瓷燃烧器。环形陶瓷燃烧器由煤气环道、助燃空气环道、煤气喷口、助燃空气喷口和预混室组成。环形陶瓷燃烧器与拱顶砌体采用相互独立的砌筑结构，环形陶瓷燃烧器的砌体由拱顶炉壳独立支撑，与拱顶砖衬砌体完全脱开，采用迷宫式密封结构，防止热膨胀应力破坏砖衬。陶瓷燃烧器采用高温综合性能优良的红柱石砖。图1为环形陶瓷燃烧设计结构。

3.3 陶瓷燃烧器几何结构优化

拱顶是顶燃式热风炉的燃烧空间，为充分利用拱顶空间，使煤气燃烧完全，并使高温烟气在蓄热室格子砖内均匀分布，采用了旋流扩散燃烧技术。环形陶瓷燃烧器的煤气、助燃空气喷口沿圆周切线方向布置，2环煤气喷口设置在燃烧器的上部并呈向下的倾角，2环助燃空气喷口设置在燃烧器的下部并呈向上的倾角，使喷出的气流以一定的速度在预混室内交叉混合并向下旋流，强化了煤气与助燃空气的扩散混合，以实现煤气完全燃烧。为实现烟气的均匀分布，设计了合理的环形陶瓷燃烧器预混室与锥形拱顶的几何结构，通过烟气流在拱顶空间内的收缩、扩张、旋流、回流而实现煤气的完全燃烧和高温烟气的均匀分布[7]。

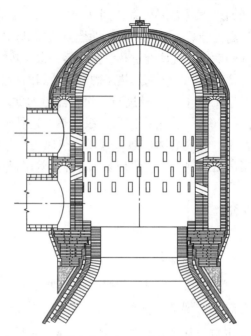

图1 BSK顶燃式热风炉环形陶瓷燃烧器设计结构

4 蓄热室结构设计优化

4.1 确定合理的热风炉工艺参数

BSK顶燃式热风炉设计中，根据设定的入炉风量、风温、煤气条件、助燃空气条件等初始条件和边界条件，建立了热风炉燃烧计算和传热计算数学模型，经过传热计算确定蓄热室格子砖的高度和材质，设计了经济合理的蓄热面积。为抑制热风炉炉壳的晶间应力腐蚀和NO_x的排放，送风温度1300℃时，热风炉拱顶温度控制在1420℃，设定拱顶温度与风温差值为120℃。由于设计开发了热风炉烟气余热回收利用工艺，改进了热风炉炉箅子和支柱的材质，提高了技术性能，在炉箅子和支柱允许工作温度下，烟气最高温度设定为450℃，平均温度为349℃，为热风炉操作提高烟气温度创造了条件。

4.2 优化配置热风炉数量

由于在国内外首次将顶燃式热风炉应用于5000m³以上特大型高炉，以热风炉系统稳定可靠运行为前提，设计配置了4座热风炉，采用交错并联送风模式，为提高风温创造了有利条件。国外生产实践证实，4座热风炉采用交错并联工作制度时，可以提高风温30℃[8]。为实现送风温度达到1300℃以上，结合不同的热风炉工作模式，设定了合理的热风炉工作周期，适当增加换炉次数，缩短送风时间，可以降低热风炉的蓄热量，在设定格子砖温度和温差的条件下减小蓄热室容积。在4座热风炉采用交错并联送风时，燃烧时间为60min，送风时间为48min，换炉时间为12min。

4.3 采用高效格子砖，优化蓄热室热工参数

经过研究计算和设计优化，每座热风炉蓄热面积为 95885m²，蓄热室断面积为 93.21m²，格子砖高度 21.48m，每座热风炉格子砖重量 2550t，单位高炉有效容积加热面积为 69.73m²/m³，单位鼓风的加热面积为 41.24m²/(m³·min)，热风炉总高度为 49.22m，炉壳最大内径为 12.5m，蓄热室格子砖砌体直径为 10.894m。

为了优化热风炉蓄热室的蓄热、加热性能，采用了直径为 30mm 的 19 孔高效格子砖，其加热面积为 48m²/m³，有效地实现了蓄热室断面的气流均匀分布，缩小了热风炉直径和外形尺寸，炉体结构简化，避免了大直径拱顶结构，提高了拱顶的结构稳定性和寿命。有效地提高了格子砖蓄热、加热能力，为实现稳定的高风温创造了条件。

BSK 热风炉设计中，由于优化了格子砖的热工参数和设计结构，格子砖加热面积比常规格子砖提高了 24%，大幅度提高了热风炉换热效率，蓄热室格子砖高度比 4000m³ 高炉外燃式热风炉降低了 13.5m，同时减少了热风炉直径，降低了热风炉整体高度，节约了工程建设投资。表 2 是高效格子砖的主要热工性能。

表 2 高效格子砖的主要热工性能

项 目	首钢京唐高炉 19孔格子砖	迁钢高炉 7孔格子砖	宝钢高炉 7孔格子砖	德国 7孔格子砖 I	德国 7孔格子砖 II
格子砖加热面积/m²·m⁻³	48	47.08	38.06	39.7	46.2
格子砖活面积/m²·m⁻³	0.36	0.33	0.409	0.298	0.405
填充系数/m²·m⁻³	0.641	0.67	0.591	0.7025	0.595
格子砖流体直径/mm	30	30	43	30	35
格子砖当量厚度/mm	34.6	28.5	31	35.4	25.8

4.4 蓄热室格子砖的优化配置

BSK 顶燃式热风炉设计中，经过热风炉燃烧期蓄热室温度场计算分析，得出蓄热室在高度方向的温度分布，根据传热计算结果和耐火材料的传热特性，对格子砖配置进行了设计优化，合理设置不同材质格子砖的使用区域。

顶燃式热风炉不同高度部位因所处的温度区间不同，所采用的格子砖材质也不同，由上至下依次为硅质、红柱石质、低蠕变黏土质、高密度黏土质。高温区采用抗高温蠕变性能优异的硅砖，中温区采用红柱石砖和低蠕变黏土砖，低温区采用高密度黏土砖。这种耐火材料的优化配置与顶燃式热风炉传热特性相适应，可提高耐火材料经济合理性，降低工程投资。

在蓄热室高度方向根据蓄热室温度分布、工作环境和耐火材料特性，从上至下共设 5 段不同材质的格子砖。蓄热室高温区采用高温体积稳定性、抗蠕变性和耐侵蚀性优异的硅砖。高温区采用硅砖，其工作温度区间控制在 800~1420℃，可有效防止硅砖的温度剧烈变化而引起的相变破损；第 2 段为红柱石砖；第 3 段为低蠕变黏土砖；第 2、3 段格子砖处于蓄热室高温区和低温区的过渡区间，温度变化比较敏感，因此采用抗热震性能优良的红柱石砖和低蠕变黏土砖；第 4 段处于蓄热室低温区，采用高密度黏土砖，以提高热风炉蓄热量；第 5 段采用抗压强度高、抗蠕变性能和抗热震性能优良的低蠕变黏土砖。

4.5 耐火材料内衬结构设计优化

设计研究中对热风炉用耐火材料及砌筑结构进行了综合分析。采用高温性能、高温结构强度、高温体积稳定性、热稳定性、耐侵蚀性优异的优质耐火材料，研究了耐火材料的结构整体性能和结构设计，运用低应力设计体系，采取有效措施消除或降低耐火材料体系的热应力、机械应力、相变应力和压应力。

根据热风炉各部位的工作温度、结构特点、受力情况及化学侵蚀的特点，分别选用不同性能的耐火材料，实现热风炉耐火材料功能性优化选择。

4.5.1 拱顶大墙结构

拱顶砖衬采用4层不同材质的耐火材料。由内向外分别为硅砖、轻质硅砖、轻质隔热砖、陶瓷纤维毡和硅钙板。为了防止拱顶区域炉壳发生晶间应力裂纹腐蚀，炉壳内表面涂刷防晶间应力腐蚀涂料，并喷涂耐酸喷涂料，防止含酸气体冷凝积聚腐蚀炉壳，抑制晶间应力腐蚀的发生，热风炉炉壳外部不设保温结构。

4.5.2 蓄热室大墙结构

蓄热室顶部与拱顶之间采用相互独立的砌筑结构，使拱顶与热风炉炉体上部砌体完全脱离。沿蓄热室筒体砖衬高度方向分为3段，上段由内向外分别采用硅砖、轻质硅砖、轻质黏土隔热砖、耐火纤维毡、硅钙板；中段为过渡段；下段由内向外分别采用黏土砖、轻质黏土隔热砖、耐火纤维毡、硅钙板。

5 热风管道内衬结构优化设计

5.1 热风管道

热风总管、热风支管、热风围管采用低蠕变莫来石砖和隔热砖组合砌筑结构。管壳内表面喷涂两层不定型耐火材料，上部砖衬与喷涂料之间设置一层陶瓷纤维棉。热风总管和热风支管采用低蠕变莫来石砖和两层隔热砖砌筑结构，管壳内表面喷涂两层不定型耐火材料，上部砖衬与喷涂料之间设置一层陶瓷纤维棉。为延长热风总管和支管的使用寿命，在热风管道钢壳内表面喷涂一层防晶间应力腐蚀的耐酸涂料。

5.2 孔口组合砖结构

热风炉各孔口的工作条件十分恶劣，强化热风炉燃烧和换热过程，耐火材料要承受高温、高压的作用，孔口耐火材料还要承受气流收缩、扩张、转向运动所产生的冲击和振动作用。热风炉各孔口在多种工况的恶劣条件下工作，是制约热风炉长寿和提高风温的薄弱环节。热风出口采用独立的环形组合砖构成，组合砖之间采用双凹凸榫槽结构进行加强，并在组合砖上部设有半环特殊的拱桥砖，以减轻上部大墙砖衬对组合砖产生的压应力。热风炉的热风出口、热风管道上的三岔口等部位均采用组合砖砌筑。

热风支管、热风总管、热风环管内衬的工作层的砖衬采用蠕变低、体积密度相对较低、高温稳定性优良的红柱石砖。

5.3 波纹补偿器内衬结构

根据对高温、高压热风管道热膨胀的计算结果,将波纹补偿器处的耐火材料砌体膨胀缝宽度设置在 50mm 以上,采用迷宫式密封结构,将热风管道的工作层和隔热层砖衬设计成相对独立的自由体系,采用限制性的定向膨胀结构,同一层砖衬允许轴向滑移,通过合理设置膨胀缝吸收砖衬的热膨胀。在工作层砖衬膨胀缝开口处设置了一环镶嵌式保护砖,可有效防止膨胀缝中填充的陶瓷纤维材料被气流冲掉,解决了热风管道窜风、发红的问题。

6 生产应用实践

首钢京唐 1 号高炉于 2009 年 5 月 21 日送风投产,高炉投产后生产稳定顺行,各项生产技术指标不断提升,高炉最高日产量达到 14245t/d,月平均利用系数达到 2.37t/(m^3·d),燃料比 480kg/t,焦比 269kg/t,煤比 175kg/t,风温 1300℃,达到了预期的设计水平。为充分发挥 BSK 顶燃式热风炉的技术优势,不断优化高炉操作,逐步提高风温。2009 年 12 月 13 日,高炉风温突破 1300℃,2009 年 12 月全月平均风温达到 1281℃;2010 年 1 月受焦炭质量影响,高炉风温有所减低,全月平均 1259℃;2 月中旬以后,随着焦炭质量的改善,高炉风温逐步攀升,恢复到 1305℃,2 月全月平均风温达到 1277℃;3 月全月平均风温达到 1300℃。实现了全高炉煤气条件下,高炉风温持续稳定达到 1300℃ 的设计指标,开创了特大型高炉高风温生产实践的新纪录,达到了国内外 5000m^3 以上特大型高炉高风温操作的领先水平。表 3 是首钢京唐 1 号高炉投产后 1 年的主要生产技术指标。

表 3 首钢京唐 1 号高炉主要生产技术指标

日期	平均日产量 /t·d^{-1}	利用系数 /t·(m^3·d)$^{-1}$	焦比 /kg·t^{-1}	煤比 /kg·t^{-1}	燃料比 /kg·t^{-1}	风温 /℃	工序能耗 /kgce·t^{-1}
2009 年 5 月	4840	0.88	551	83	634	914	799
2009 年 6 月	7425	1.35	503	62	565	998	538
2009 年 7 月	8525	1.55	483	49	532	1063	461
2009 年 8 月	11000	2.01	372	94	481	1166	409
2009 年 9 月	11660	2.12	354	101	483	1212	419
2009 年 10 月	12210	2.22	340	117	488	1262	414
2009 年 11 月	12500	2.27	299	145	484	1276	406
2009 年 12 月	12694	2.31	288	149	479	1281	393
2010 年 1 月	12657	2.30	307	137	482	1259	388
2010 年 2 月	12847	2.34	287	161	482	1277	375
2010 年 3 月	13035	2.37	269	175	480	1300	373
2010 年 4 月	12147	2.21	289	156	482	1275	381

7 结论

(1) 首钢京唐1号高炉投产后,在燃烧单一高炉煤气条件下,月平均风温突破1300℃,开创了特大型高炉高风温生产实践的新纪录,达到了国内外5000m³以上特大型高炉高风温操作的领先水平。

(2) 根据热风炉燃烧、气体流动、传输过程的理论研究,优化了热风炉燃烧器、燃烧室、蓄热室、炉体耐火材料内衬和管道结构设计,有效提高了热风炉的加热能力和工作效率,自主设计、研究开发了多项热风炉高效长寿技术。

(3) 优化热风炉拱顶和环形陶瓷燃烧器设计结构,将热风炉拱顶和陶瓷燃烧器作为整体进行设计优化。采用自由膨胀的无应力设计体系,延长拱顶和陶瓷燃烧器使用寿命。对热风炉的高温、高压管道进行了系统的设计优化,通过管系受力计算,合理设置波纹补偿器、拉杆和管道支架,实现了管道系统低应力设计,满足了高温热风稳定输送的要求。

(4) 为了有效抑制热风炉炉壳晶间应力腐蚀,热风炉高温区炉壳和热风管道内壁采取喷涂防酸涂料的综合防护措施。优化了热风管道的内衬设计结构,水平管道与垂直管道连接处采用各自独立的组合砖结构,使两者之间的热膨胀不互相干涉,解决了孔口砖衬热膨胀不均造成管道窜风的技术难题,满足了1300℃高风温的送风要求。

参考文献

[1] 张福明. 我国大型顶燃式热风炉技术进步 [J]. 炼铁, 2002, 21 (5): 5-10.
[2] Iakov Kalugin. High temperature shaftless hot air stove with long service life for blast furnace [C]. AISTech 2007 Proceedings-Volume I. Chicago: AIST, 2007: 405-411.
[3] 张福明, 钱世崇, 张建. 首钢京唐5500m³高炉采用的新技术 [J]. 钢铁, 2011, 46 (2): 12-17.
[4] Zhang Fuming, Qian Shichong, Zhang Jian, et al. Design of 5500m³ blast furnace at Shougang Jingtang [J]. Journal of Iron and Steel Research International, 2009, 16 (Supplement2): 1029-1034.
[5] 张福明. 现代大型高炉关键技术的研究与创新 [J]. 钢铁, 2009, 44 (4): 1-5.
[6] 梅丛华, 张卫东, 张福明, 等. 一种高风温长寿型两级双预热装置: 中国, ZL200920172956.4 [P]. 2010-07-14.
[7] 张福明. 长寿高效热风炉的传输理论与设计研究 [D]. 北京: 北京科技大学, 2010: 171-187.
[8] Peter Whitfield. The advantages and disadvantages of incorporating a fourth stove within an existing blast furnace stove system [C]. AISTech 2007 Proceedings Volume I. A Publication of the Association for Iron and Steel Technology, 2007: 393-403.

Design Study on BSK Dome Combustion Hot Blast Stove for Shougang Jingtang 5500m³ Blast Furnace

Zhang Fuming Mei Conghua Yin Guangyu
Mao Qingwu Qian Shichong Hu Zurui

Abstract: This paper introduces the technical innovation of BSK dome combustion hot stove for 5500m³ blast furnace at Shougang Jingtang Steel Plant. Ultra large dome combustion hot blast stove optimum technical process is integrated; high temperature 2 stage preheating technology of combustion air and dome combustion hot stove high-efficiency ceramic burner are developed; according to the characteristics of dome combustion hot blast stove, the rational structure of dome and ceramic burner are designed; high efficiency checker brick is adopted, the thermal specification and structure of regenerator are optimized; reasonable hot blast stove checker heating surface is determined. Hot blast stove proper lining design is optimized; the effective technical measures for hot blast stove shell inter-crystalline stress corrosion prevention is applied. According to the thermal conduction calculation of regenerator suitable hot blast stove refractory is configured; the refractory technical performance is improved additionally. Refractory structure design of hot blast pipe system is improved in order to meet the requirement of 1300℃ blast temperature. Blast temperature of design level is reached to 1300℃ monthly after blast furnace blow in.

Keywords: blast furnace; dome combustion hot blast stove; high blast temperature; ceramic burner

高风温内燃式热风炉的设计理论和应用实践

摘　要：高风温是现代高炉的重要技术特征。高风温内燃式热风炉具有工艺布置紧凑、结构合理、占地面积小、风温高、寿命长、操作灵活可靠的特点，在国内外近百座高炉上得到应用。本文通过对高风温内燃式热风炉设计理论的分析，总结了高风温内燃式热风炉的设计特点。结合首钢高炉高风温内燃式热风炉的设计实践和应用实践，对高风温内燃式热风炉的技术应用进行了评述。

关键词：热风炉；高风温；陶瓷燃烧器；设计理论

1　引言

高风温是现代高炉的重要技术特征。提高风温是增加喷煤量、降低焦比、降低生产成本的主要技术措施。提高风温可以通过提高煤气热值、优化热风炉结构、预热煤气和助燃空气、改善热风炉操作等技术措施来实现。理论研究和生产实践表明，优化热风炉结构、提高热风炉热效率、延长热风炉寿命是提高风温的有效技术途径。20世纪60年代末期，为降低生产成本，世界各主要产钢国开始研究不同结构形式的热风炉。70年代中期，荷兰霍戈文公司（Hoogovens，现称DCE）针对内燃式热风炉存在的诸多技术缺陷，研究开发了高风温内燃式热风炉。此种热风炉是在传统的内燃式热风炉的基础上进行设计优化，对内燃式热风炉进行了重大技术改进和创新，实现了热风炉高风温、高效和长寿，自问世至今在全世界百余座高炉上得到应用，特别是近8年内，全世界约有170座高风温内燃式热风炉投入使用。

2　高风温内燃式热风炉系统优化设计理论

通过对蓄热式热风炉技术的传热机理和传热过程的研究和解析，可以使热风炉系统的工艺流程和配置得到优化。高风温内燃式热风炉的问世，其创新的优化设计理念改变了人们传统的思维模式。

2.1　热风炉数量的确定

入炉风量和送风温度是热风炉设计的基本参数。在高炉容积和生产能力确定以后，热风炉数量和热风炉总加热面积等技术参数的设计是根据入炉风量和送风温度确定的。传统的设计模式通常是选择4座热风炉，热风炉加热面积设定为 $90\sim100m^2/m^3$。但由于现代热风炉系统的阀门和控制系统日益成熟可靠，而且设计建设大直径的热风炉技术条件已经

本文作者：张福明。原刊于《2008年全国炼铁生产技术会议暨炼铁年会论文集》，2008：767-770，774。

具备，因此设计3座热风炉已经完全可行。这样可以简化操作控制系统，降低工程投资，减少占地面积，而且可以根据工艺布置条件，预留第4座热风炉的位置，在热风炉系统大修和改造前，可以先行建设第4座热风炉，再对其他3座热风炉依次进行大修。3座热风炉的操作模式一般为"两烧一送"，风温的调节控制依靠混风实现。

2.2 降低加热面积

降低热风炉加热面积是 DCE 设计体系的显著特点。通过对热风炉传热过程的研究表明，提高热风炉加热面积，并非是提高风温的最佳选择。传统设计理念中，根据高炉容积确定的单位高炉容积加热面积也不能科学地表达热风炉的换热能力。适当降低热风炉的加热面积，可以有效地降低热风炉系统的工程投资。合理配置蓄热室格子砖，缩短热风炉燃烧周期和送风周期，可以实现低加热面积条件下的高风温。

2.3 缩短热风炉工作周期

在热风炉低加热面积的条件下，缩短送风时间可以提高热风炉加热效率，从而提高风温。送风时间的缩短，势必会影响热风炉整个工作周期。因此，热风炉要具有较强的燃烧能力，强化燃烧，可以在较短的时间内，迅速将热风炉加热到具备送风的条件。缩短换炉时间其目的是使3座热风炉交替工作流畅，减少在换炉期间的风温波动。

2.4 适当提高烟气温度

提高烟气温度可以有效地提高风温，但实际上多年以来，提高烟气温度一直不作为提高风温的主要措施。其原因主要是提高烟气温度以后，会产生一系列问题：（1）烟气温度提高，降低了热风炉本体的热效率，热风炉热能利用不充分。（2）烟气温度提高以后，热风炉下部温度升高，对热风炉支柱和炉箅子、烟道、阀门、烟囱等造成破坏。特别是烟气温度提高以后，使传统的热风炉支柱和炉箅子在高温热应力的作用下，破损严重。（3）烟气温度提高，热量不能得到充分利用，而且会因此加大煤气消耗量。当前的设计体系中，采取了许多综合技术措施：（1）采用耐高温 Si-Mo 合金铸铁热风炉支柱和炉箅子，提高格子砖支撑体系的抗高温能力；（2）支烟道、烟道、烟囱等采用内衬结构，支烟道、烟道设置波纹补偿器，避免由于烟气温度升高而造成的对烟气管路系统的破坏；（3）设置烟气余热回收系统，回收烟气余热预热煤气和助燃空气，提高热风炉理论燃烧温度，使烟气余热得到充分利用。

2.5 紧凑型工艺布置

高温内燃式热风炉通常采用一列式布置，3座热风炉呈一列式布置在高炉附近，占地面积小。热风炉的热风支管、热风总管；煤气支管、煤气总管；助燃空气支管、助燃空气总管布置在热风炉燃烧室一侧；热风炉支烟道、烟道；冷风总管、冷风支管布置在热风炉蓄热室另一侧。热风炉两侧的检修吊车框架、管架、平台为一体式结构，热风炉基础为整体筏式基础，热风炉炉壳和土建框架、管架、平台之间为独立结构，相互之间没有任何连接，以确保热风炉炉壳能够长期稳定工作。

3 高风温内燃式热风炉的结构特征

高风温内燃式热风炉是在传统的内燃式热风炉的基础上优化演变而来的。针对传统内燃式热风炉存在的技术缺陷，进行了卓有成效的设计优化和集成创新。

3.1 燃烧室

内置式燃烧室是内燃式热风炉区别于其他形式热风炉的主要技术特征。内燃式热风炉的燃烧室通常有圆形、苹果形、眼睛形三种结构型式。高风温内燃式热风炉选择了眼睛形燃烧室，这主要是取决于燃烧器的型式和火焰分布。传统的内燃式热风炉通常采用圆形燃烧室，燃烧室墙体稳定性差，容易产生裂缝、气流短路等现象。而且从燃烧室进入蓄热室的烟气气流分布不均匀，气流偏析严重，直接影响了蓄热室的蓄热能力和加热效率。DCE设计体系采用眼睛形燃烧室，缩小了燃烧室隔墙的弧长，使蓄热室的有效断面积加大；而且改善了烟气在蓄热室格子砖的分布，使蓄热室内格子砖的温度场分布更为均匀，提高了蓄热能力。

3.2 隔墙结构

隔墙是内燃式热风炉特有的结构，传统内燃式热风炉主要技术缺陷就是由于隔墙造成的。隔墙在垂直方向和水平方向上都存在着明显的温度梯度，隔墙受热膨胀所产生的位移量不同，造成隔墙开裂、倾斜。这样在热风炉燃烧期会使隔墙窜风、高温烟气短路，造成蓄热室下部格子砖和炉箅子、支柱过热，甚至烧坏炉箅子和支柱；在热风炉送风期，冷风会穿透隔墙裂缝，直接进入燃烧室，使蓄热室热效率下降，风温降低。

高风温内燃式热风炉的设计体系中，隔墙由多层结构组成。燃烧室中、下部，隔墙中间设有隔热层，采用低导热的耐火材料降低隔墙两侧的温度梯度，从而减小热应力，避免隔墙开裂或倾斜。为提高隔墙的密封性，在燃烧室下部区域的隔墙中间，设置一层耐高温合金钢板，用以消除热风炉送风期燃烧室和蓄热室下部因压力差而造成的窜气。隔墙砌体设置膨胀缝，用以吸收高温操作时，隔墙砌体所产生的热膨胀。隔墙为独立的板块式结构，与热风炉大墙采用滑动连接，使隔墙在高温条件下不受其他砌体的约束和干涉。为提高隔墙的稳定性，隔墙砌体采用带有凹凸锁砖结构。

3.3 矩形陶瓷燃烧器

燃烧器是热风炉的关键核心设备。DCE设计开发了矩形陶瓷燃烧器，与眼睛形燃烧室相匹配。矩形陶瓷燃烧器的断面呈矩形，中间是矩形煤气通道，外侧是助燃空气通道，在燃烧器顶端，助燃空气以一定的倾角向上喷出，与矩形煤气通道喷出的片状煤气流交叉混合，改善了煤气燃烧的传质过程，促进煤气燃烧。矩形陶瓷燃烧器是典型的长焰燃烧器，为使煤气能够充分燃烧，使热量分布均匀，燃烧器上方的火焰长度要达到20m以上。矩形陶瓷燃烧器的主要技术特点是：（1）具有很高的工作可靠性，适应多种不同工况条件。对于煤气和助燃空气不预热的常温状态、煤气和助燃空气均预热的高温状态以及煤气或助燃空气单预热的高温状态，陶瓷燃烧器均能适应。（2）燃烧功率大，燃烧能力调节范围广。燃烧器的工况范围可在额定能力的40%~135%内调节。（3）燃烧器运行平稳，有利于实现强化燃烧。由于采用矩形断面结构，燃烧器两侧的助燃空气与中心的片状煤气流交叉混合，改善了煤气燃烧的动力学条件。而且，中间矩形煤气通道内设有煤气导流板，通过冷态试验和调试，使导流板处于最佳位置，以保证中心矩形通道内煤气的均匀分布。实践证实，DCE开发的矩形陶瓷燃烧器结构合理，燃烧性能稳定热流分布均匀，无振动和燃烧脉动现象。正常条件下，烟气中的O_2、CO、NO_x等都达到设计标准。

3.4 悬链线拱顶

内燃式热风炉的拱顶结构有半球形、锥形和蘑菇形，悬链线拱顶属于蘑菇形。悬链线拱顶能够形成一种静态平衡形状，结构稳定性高，消除了向外的推力，能够承受高温、高压的变化。DCE设计的悬链线拱顶采用与热风炉大墙脱开的形式，拱顶砌体支撑在炉壳上，使大墙能够自由胀缩，不对拱顶砌体造成影响。拱顶砌体为3层独立式结构，工作层为硅砖，靠近炉壳的两层为隔热砖，硅砖为凹凸锁砖，以增加砌体的结构稳定性。在拱顶下部设铰接结构，即所谓"关节砖"，此种结构可以吸收拱顶砌体的热膨胀，对于消除拱顶裂纹具有积极作用。此外，拱顶砌体采用了板块式砌筑结构，每个模块之间都设置了膨胀缝，其内填充厚度不同的聚乙烯泡沫板或陶瓷纤维毡，用来吸收拱顶砌体所产生的高温热膨胀。DCE对悬链线拱顶在结构稳定性和热应力稳定性方面进行了许多研究和设计优化，使这种拱顶结构在流场和温度场分布也更加合理。

3.5 蓄热室和格子砖

蓄热室是蓄热式热风炉不可缺少的主要组成部分，蓄热室内的格子砖是加热冷风的主要传热载体。DCE设计的高风温内燃式热风炉，根据蓄热室内的温度场分布和格子砖的工作特性，在高度方向上，将蓄热室格子砖通常分为4段式结构。在蓄热室上部高温区，采用抗蠕变性能优异的硅砖；中温区上部采用低蠕变红柱石砖，下部采用低蠕变黏土砖；低温区采用黏土砖，在炉箅子上部砌筑若干层红柱石格子砖。根据蓄热室的温度场分布，采用不同材质和工作特性的耐火材料，使蓄热室的热量能够得到比较充分地利用，而且可以降低耐火材料的投资。格子砖与蓄热室大墙设有膨胀缝，使格子砖能够自由膨胀，消除格子砖膨胀所产生的热应力。格子砖上下接触面设有3个对应均布的凸台和凹槽，使上、下层格子砖能够啮合定位，消除格子砖的水平位移和旋转。

3.6 防止晶间应力腐蚀

热风炉炉壳晶间应力腐蚀是由于在很高的拱顶温度下，燃烧产物生成NO_x、SO_x而造成的，风温长期在1200℃以上时，热风炉炉壳会发生晶间应力腐蚀。研究表明，这些反应在温度超过1420℃时开始加剧。燃烧产物中的NO_x、SO_x与H_2O相遇生成硝酸和盐酸，导致炉壳腐蚀。特别是炉壳在高温应力状态下工作，晶粒之间的腐蚀更为严重。热风炉炉壳晶间应力腐蚀的产生是高温状态下形成了NO_x、SO_x，同时炉壳是在高温应力状态下工作，这两个不利的条件下对高风温的实现带来了挑战。

为了防止晶间应力腐蚀的发生，目前一般采用了如下技术：（1）在拱顶温度低于1350℃时，炉壳无需做特殊处理，但建议采用环氧材料保护炉壳内壁；（2）拱顶温度在

1350~1450℃时，炉壳内壁采用环氧涂层，但由于环氧材料不能完全起到防酸作用，最好在炉壳内层覆盖耐腐蚀的不锈钢钢箔或喷涂防喷酸涂料；(3) 当拱顶温度超过1450℃时，NO_x、SO_x将形成酸冷凝水，为消除这种现象，炉壳外部要进行保温绝热，使炉壳温度保持在120℃以上；(4) 热风炉炉壳采用细晶粒耐龟裂钢板，炉壳采用低应力设计，减少或消除炉壳焊接应力。

3.7 孔口和管道采用组合砖结构

热风炉的热风出口、点火孔、窥视孔、测温孔等均采用组合砖结构，以提高砌体的整体稳定性。热风出口由单独的环形组合砖组成，材质为红柱石，砖之间采用双凹凸榫槽进行加强，为减轻上部大墙对组合砖产生的压应力，在组合砖上部设有特殊的半环拱桥砖。热风支管、热风总管、热风环管内的工作层，均采用带凹凸棒槽的红柱石组合砖结构。热风支管与热风总管交汇处以及热风总管与热风环管交汇处，上部采用砌体均采用平拱吊挂结构，提高了砌体的结构稳定性。

4 高风温热风炉的应用实践

近年来，随着我国高炉大型化进程的加快，提高风温已成为现代高炉降低生产成本和能源消耗的重要技术措施。高风温内燃式热风炉在我国武钢、唐钢、首钢、鞍钢、太钢、湘钢等众多企业得到应用[1~3]。

2002年，首钢2号高炉（1780m³）技术改造时，采用了3座高风温内燃式热风炉。为提高风温，采用了助燃空气高温预热技术，利用2座热风炉预热助燃空气，使之达到600℃；利用热风炉烟气余热预热煤气，使煤气温度达到200℃，在使用单一高炉煤气作为燃料的条件下，使风温达到1250C。该高炉于2002年5月23日建成投产，同年10月利用系数就达到2.43t/(m³·d)，风温达到1220℃，焦比297kg/t，煤比170kg/t。

首钢迁钢1号、2号高炉（2650m³）也分别采用了3座高风温内燃式热风炉，两座高炉分别于2004年10月和2006年1月建成投产。2007年迁钢高炉主要技术指标见表1。

首钢采用的高风温内燃式热风炉主要技术性能参数见表2。

表1 2007年迁钢高炉主要技术指标

项目	利用系数 /t·(m³·d)$^{-1}$	焦比 /kg·t^{-1}	煤比 /kg·t^{-1}	焦丁 /kg·t^{-1}	燃料比 /kg·t^{-1}	风温 /℃	富氧率 /%	顶压 /MPa
1号高炉	2.468	302.9	156	29.8	488.7	1228	2.36	0.194
2号高炉	2.461	308.9	142.5	35.8	487.2	1214	2.37	0.195

表2 首钢高风温内燃式热风炉主要技术性能

项 目	首钢2号高炉	迁钢1号高炉	迁钢2号高炉
高炉容积/m³	1780	2650	2650
热风炉数量/座	3	3	3
热风炉操作方式	两烧一送	两烧一送	两烧一送

续表 2

项　目	首钢 2 号高炉	迁钢 1 号高炉	迁钢 2 号高炉
送风时间/min	45	45	45
燃烧时间/min	80	75	75
换炉时间/min	10	15	15
设计风温/℃	1250	1250	1250
拱顶温度/℃	1420	1400	1400
加热风量/$m^3 \cdot min^{-1}$	4200	5500	5500
炉壳内径/m	9.2	10.2	10.2
热风炉总高度/m	41.59	41.66	41.66
蓄热室断面积/m^2	35.8	44	44
燃烧室断面积/m^2	9.7	10.9	10.9
每座热风炉总蓄热面积/m^2	48879	64839	64839
单位高炉容积的加热面积/$m^2 \cdot m^{-3}$	82.38	73.4	73.4
格子砖高度/m	32.4	31.3	31.3
燃烧器长度/m	3.4	4.4	4.4

高风温内燃式热风炉在首钢高炉上的应用，已经取得初步成功。在工艺布置方面具有占地面积小，结构紧凑，布置灵活，管道走向通畅的特点。对于容积为 2650m^3 的高炉而言，高风温内燃式热风炉占地面积仅为 61m×38m。高风温内燃式热风炉的所有管道、波纹补偿器及管道支架的设置均经过详细的受力计算，特别是高温、高压管道的所出现的高温热膨胀位移和受压后产生的压力位移，在管道设计中均给予了充分的重视。波纹补偿器的合理设置、管道支架的优化选择为热风炉外部管道的稳定工作提供了可靠的保证。

高风温内燃式热风炉操作控制灵活可靠，能够实现全自动化联锁控制。对于流量、压力、温度、烟气成分的在线检测增强了热风炉的操控性能。

5　结语

高风温内燃式热风炉是在传统的内燃式热风炉基础上优化演变而来的，但克服了传统内燃式热风炉诸多的技术缺陷，在热风炉结构、管道布置、自动化控制等多面进行了技术创新，取得了显著的技术成就。

（1）先进的设计理念使热风炉设计更加优化，为热风炉实现高风温、长寿奠定了基础。

（2）采用合理的设计结构，对于传统的内燃式进行了有针对性的改进和技术创新。采用悬链线拱顶、复合式隔墙、矩形陶瓷燃烧器、眼睛形燃烧室，这些细节技术的改进和优化，使内燃式热风炉又焕发了生机和活力。

（3）工艺布置紧凑合理，降低了工程投资，成为现代大型高炉可以选择的一种经济实用可靠的热风炉型式，在 1000~3200m^3 高炉上已经得到推广应用。

（4）可靠的自动化检测与控制系统，使热风炉操作控制更加稳定，为实现高风温创造了有利条件。

致谢

北京首钢设计院毛庆武、姚轼、倪萍、李勇在本文成文过程中,为作者提供了部分资料和数据,作者在此表示感谢!

参考文献

[1] 张福明,毛庆武,张建,苏维.首钢2号高炉技术改造设计[J].设计通讯,2004(1):18-25.
[2] 文经国.武钢高风温内燃式热风炉技术的特征[J].炼铁,2001(S2):53-56.
[3] 舒军.高风温内燃式热风炉的发展特征[J].炼铁,1998(01):12-15.

当代高炉高风温技术理念与工程实践

摘　要：针对当前高炉炼铁技术面临的形势和挑战，提出了当代高炉炼铁技术的发展目标，阐述了高风温技术对高炉炼铁的重要意义和作用。分析和论述了实现高风温的技术理念和关键技术，重点阐述了燃烧低热值煤气条件下获得高风温、高温热风稳定输送、优化热风炉操作以及热风炉实现高效长寿的技术实践，指出了在高风温、低燃料比冶炼条件下，当代高炉炼铁技术具有广阔的发展前景。

关键词：高炉；炼铁；高风温；热风炉；低燃料比

1　引言

进入 21 世纪以来，高炉炼铁工艺再次受到自然资源短缺、能源供给不足以及环境保护等方面的制约，面临着较大的发展问题[1]。面对当前严峻的形势和挑战，21 世纪高炉炼铁工艺要实现可持续发展，必须在高效低耗、节能减排、循环经济、低碳冶金、清洁环保等方面取得显著突破，需要进一步提高风温、降低燃料比，以提高高炉炼铁技术的生命力和竞争力。面对当前国内外激烈的市场竞争环境，在资源短缺、能源供给不足、生态环境制约的条件下，为保障高炉炼铁工艺的可持续发展，实现低碳冶炼和循环经济，要着力构建高效率、低消耗、低成本、低排放的高炉炼铁生产体系[2]。

高风温是现代高炉炼铁的主要技术特征之一。提高风温是当前钢铁行业发展循环经济、实现低碳冶炼、节能减排和可持续发展的关键共性技术，高风温对于提高高炉综合技术水平、减少 CO_2 排放、引领行业技术进步具有极其重要的意义。

2　高风温的意义和作用

高炉冶炼所需要的热量，一部分是燃料在炉缸燃烧所释放的燃烧热，另一部分是高温热风所带入的物理热。高温热风所提供的热量占高炉冶炼所需要热量的约 20%，提高风温可以有效降低高炉冶炼过程的碳素消耗，热风带入高炉的热量越多，所需要的燃料燃烧热就越少，亦即燃料消耗就越低。实践证实，在风温 1000~1250℃ 的范围内，提高风温 100℃ 可以降低焦比约 10~15kg/t，由此可见，提高风温可以显著降低燃料消耗和生产成本。除此之外，提高风温还有助于提高风口前理论燃烧温度，使风口回旋区具有较高的温度，炉缸热量充沛，有利于提高煤粉燃烧率、增加喷煤量，还可以进一步降低焦比、提高高炉生产效率、显著降低 CO_2 排放，是引领当代高炉炼铁技术进步的关键共性技术。因

本文作者：张福明，钱世崇，银光宇，毛庆武，李欣。原刊于《2012 年全国高炉长寿与高风温技术研讨会论文集》，2012：24-31，54。

此，高风温是高炉实现大喷煤操作的关键技术，是高炉降低焦比、提高喷煤量、降低生产成本的重要技术途径，是高炉炼铁发展史上极其重要的技术进步[3]。

高风温技术是一项综合技术，涉及到整个钢铁厂物质流、能量流流程网络的动态运行和结构优化，应当在整个钢铁厂流程网络的尺度上进行研究[4]。高风温对于优化钢铁厂能源网络结构、降低生产成本和能源消耗、实现低品质能源的高效利用、减少 CO_2 排放等都具有重大的现实意义和深远的历史意义。进入21世纪以来，一系列高风温技术相继开发并应用在大型高炉上，高风温技术创新与应用实践取得了显著的技术成效。目前大型高炉的设计风温一般为1200~1300℃，提高风温已成为当代高炉炼铁技术发展的一个显著趋势。

当前，国内外高风温技术发展水平并不平衡，2010年中国重点钢铁企业的高炉平均风温为1160℃，先进高炉的风温已达到1250~1300℃，技术水平差距很大，进一步提高风温仍然是21世纪高炉炼铁技术的热点研究课题。应当指出，提高风温是当代高炉炼铁技术发展的重要途径，是引领炼铁工业实现可持续发展的重大关键共性技术，提高风温不是最终的目标，而是为高炉实现高效、低耗、低碳、低成本、低排放奠定基础，以提高风温为技术突破，进一步提高喷煤量、降低燃料消耗，从而构建高效率、低成本、低排放的当代高炉炼铁技术体系。以高风温为技术特征的当代高炉炼铁技术，在未来的发展进程中仍将具有其他炼铁工艺不可比拟的技术优势。

未来高炉炼铁的发展目标是：

(1) 燃料比不超过500kg/t，先进高炉燃料比应不超过480kg/t；入炉焦比应不超过300kg/t，先进高炉焦比应不超过280kg/t；煤比不小于180kg/t，先进高炉煤比应达到200~250kg/t，喷煤率达到45%~50%；

(2) 高炉有效容积利用系数达到2.0~2.3t/(m^3·d)，原燃料条件好、技术装备水平高的大型高炉应达到或超过2.5t/(m^3·d)；

(3) 高炉一代炉役寿命不小于15年，高炉一代炉役单位容积产铁量应达到10000~15000t/m^3；技术装备水平高、原燃料条件好的大型高炉，一代炉役寿命要力争达到20年以上，高炉单位容积产铁量达到15000t/m^3以上；热风炉寿命要大于或等于高炉的一代炉役寿命；

(4) 热风温度达到1200~1250℃，大型高炉风温应达到1250~1300℃；

(5) 高炉富氧率达到3%~5%，先进高炉富氧率应达到5%~10%。

3 获得高风温的关键技术

长期以来，中国高炉平均风温始终徘徊在1000~1080℃，2001~2010年的10年间，重点钢铁企业高炉平均风温由1081℃提高到1160℃，风温仅提高了79℃，可谓步履维艰，高炉风温是中国高炉和国外先进水平差距最大的技术指标。制约风温提高有许多因素，如何突破这些制约条件，达到1200℃以上的高风温乃是当今中国高炉炼铁的主要技术发展目标之一。在当前技术条件下，提高风温应着力解决利用低热值高炉煤气获得1250℃以上高风温的关键技术难题，通过技术创新实现热风炉高效率、低成本、低排放、长寿命的综合技术目标。

3.1 利用低热值煤气获得高风温技术

高炉煤气是高炉冶炼过程中产生的二次能源,热风炉燃烧大约消耗高炉煤气发生量的 40%~50%。随着高炉炼铁技术进步,大型高炉燃料比已降低到 520kg/t 以下,高炉煤气利用率提高到 45% 以上,煤气低发热值低于 3000kJ/m³。由于钢铁厂高热值的焦炉煤气和转炉煤气主要用于炼钢和轧钢等工序,高热值煤气供给不足,绝大部分热风炉只能燃烧低热值高炉煤气,在没有高热值煤气富化的条件下,导致热风炉理论燃烧温度和拱顶温度不高,进而难于实现 1200℃ 高风温,这是当前制约中国高炉提高风温最重要的原因。

面对能源供给短缺的现状,采用煤气、助燃空气高效双预热技术,不但可以回收热风炉烟气余热,减少热量耗散,还可以有效提高热风炉拱顶温度。在众多的预热技术中,要统筹考虑能量转换效率、技术可靠性以及设备使用寿命等因素,择优选用适宜可靠的煤气、助燃空气双预热技术。首钢京唐 5500m³ 高炉采用了煤气预热和助燃空气高温预热组合技术,利用分离式热管换热器回收烟气余热预热高炉煤气,设置 2 座小型热风炉预热助燃空气,使煤气预热温度达到 200℃,助燃空气预热温度达到 520~600℃,在燃烧单一高炉煤气条件下,热风炉拱顶温度可以达到 1420℃,高炉月平均风温达到 1300℃[5]。这项利用低热值高炉煤气实现 1300℃ 高风温技术值得推广。图 1 是首钢京唐 5500m³ 高炉热风炉煤气和助燃空气预热工艺流程。

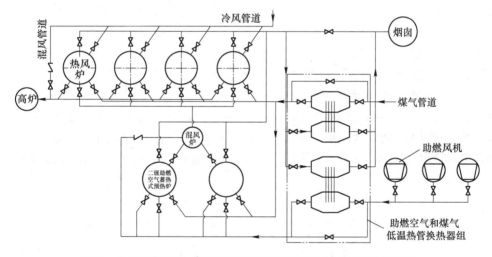

图 1 首钢京唐 5500m³ 高炉热风炉煤气和助燃空气预热工艺流程

为了提高煤气和助燃空气的物理热,一种新型煤气和助燃空气预热工艺也已问世并得到初步应用,即采用附加前置燃烧炉和高温板式换热器,可以将煤气温度预热到 200℃、助燃空气温度预热到 450℃ 以上,由此热风炉拱顶温度可以达到 1380~1400℃,风温可以达到 1250~1270℃。这种工艺在湘钢、水钢等高炉上应用,取得了初步的应用效果,关键是高温板式换热器的可靠性和使用寿命还有待于长期的生产检验。

另外一种值得关注的技术趋势是为了提高热风炉拱顶温度,采用热风炉富氧燃烧技术,即在助燃空气中进行富氧,这样可以降低助燃空气消耗量进而提高热风炉的理论燃烧温度,在一定条件下对于提高热风炉拱顶温度具有积极意义。西门子-奥钢联[6]和我国宝钢都开发了热风炉富氧燃烧技术，也取得了较好的应用效果。宝钢热风炉富氧燃烧实践表明，富氧率提高1%可以提高热风炉理论燃烧温度约20℃，可以降低空燃比，还可以使用低热值转炉煤气或高炉煤气代替高热值的焦炉煤气[7,8]。采用热风炉富氧燃烧技术的经济性值得结合具体情况研究论证[9]，同时应当注意控制热风炉燃烧过程NO_x的生成与排放，特别是在拱顶温度高于1420℃以上时，应着重控制NO_x的大量形成，从而有效抑制热风炉炉壳的晶间应力腐蚀和NO_x的排放。

3.2 选择合理的热风炉结构形式

高炉热风炉是典型的蓄热式加热炉，其工作原理不同于其他的冶金炉窑，是现代钢铁厂燃烧功率最大、能量消耗最高、热交换量最大的单体热工装置。尽管现有的内燃式、外燃式和顶燃式3种结构热风炉均有实现1250℃以上高风温的实绩，但不同结构的热风炉在燃烧工况适应性、气体流动及分布均匀性、能量利用有效性等方面仍存在差异[10]。综合考虑热风炉高效长寿和工况适应性，现代高炉采用顶燃式或外燃式热风炉是适宜的选择。值得指出的是，近年来顶燃式热风炉取得了重大技术进步，已推广应用于首钢京唐5500m³巨型高炉，并取得了显著的应用效果，月平均风温已稳定保持在1250~1300℃。目前，中国已开发并应用了各种结构的顶燃式热风炉，成为引领热风炉高风温技术创新的研究热点。同时应当看到，中国热风炉结构多样化将维持较长时期，3种结构的热风炉仍然会长期存在，但对于新建或大修改造的高炉，优先采用顶燃式热风炉则不失为一种明智之举。

文献［10］对外燃式和顶燃式热风炉的工艺特性、结构特点以及热风炉燃烧和换热过程进行了定量的对比，通过建立数学模型采用数值仿真计算，对以DME为代表的外燃式热风炉和顶燃式热风炉进行了传输过程的理论解析。计算结果表明：

（1）顶燃式热风炉炉内的高温烟气流速度以热风炉中心线为中心，呈对称式分布，这就保证了进入蓄热室的高温烟气是均匀的，进而有助于对格子砖的均匀加热。而外燃式热风炉燃烧产生的高温烟气无法以中心对称的方式进入蓄热室，也就无法使蓄热室同一平面的格子砖被均匀加热，降低了格子砖的利用效率。顶燃式和外燃式热风炉的高温烟气速度分布如图2所示。

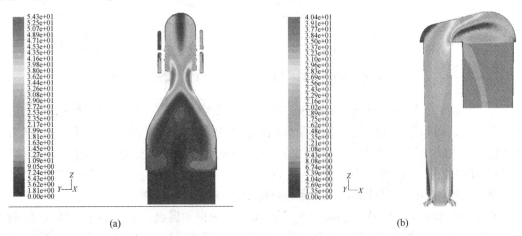

图2 顶燃式热风炉和外燃式热风炉速度场分布
(a) 顶燃式热风炉；(b) 外燃式热风炉

(2) 通过比较顶燃式热风炉和外燃式热风炉的格子砖表面温度分布图（见图 3）可以发现，顶燃式热风炉格子砖表面温度分布均匀，最大温差小于 50℃。而外燃式热风炉格子砖表面温度温差非常大，受燃烧效果和空间结构影响，最大温差大于 200℃。

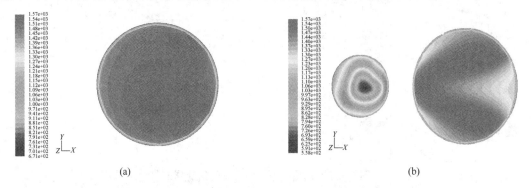

图 3　顶燃式热风炉和外燃式热风炉格子砖表面温度分布
(a) 顶燃式热风炉；(b) 外燃式热风炉

(3) 通过比较顶燃式热风炉和外燃式热风炉的流场分布和速度矢量场分布（见图 4 和图 5）可看出，顶燃式热风炉流场均匀有序，以热风炉中心线为轴心，均匀的旋流向下进入蓄热室格子砖，可以使格子砖受热均匀，热效率高。而外燃式热风炉受空间结构的影响，高温烟气无法均匀进入蓄热室格子砖，在进入格子砖之前有较大涡流存在，严重影响气流分布的均匀性，进而影响热风炉的利用效率和结构稳定性。

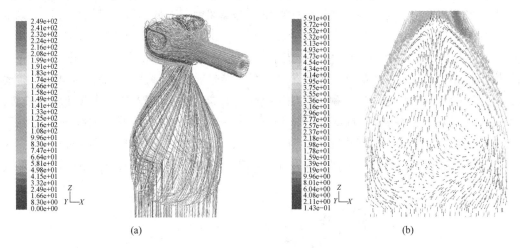

图 4　顶燃式热风炉流场与速度矢量分布
(a) 流线分布；(b) 矢量分布

3.3　采用高效格子砖

实践证实，缩小热风炉拱顶温度与风温的差值可以显著提高风温，其主要技术措施是强化蓄热室格子砖与气体之间的热交换。在保持格子砖活面积或格子砖重量不变的条件下，适当缩小格子砖孔径，可以增加格子砖加热面积、提高换热系数而增加热交换量。在热风炉燃烧期，高温烟气可以将更多的热量传递给格子砖，热量交换更加充分，使得烟气

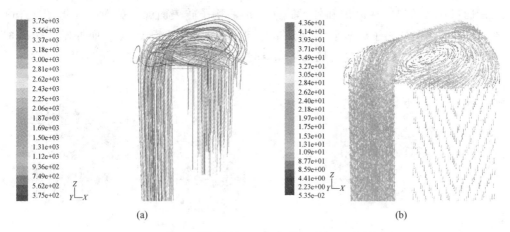

图 5　外燃式热风炉流场与速度矢量分布
(a) 流线分布；(b) 矢量分布

温度更低；在热风炉送风期，同样有利于鼓风与格子砖的热交换，使得热风温度更高，热风温降也更为平缓，在风温保持较高的状态下更加稳定。但是过度缩小格子砖孔径，则会导致气体通过格孔通道的阻力增大，造成气体阻力损失增大，消耗更多的鼓风动能；而且过小的格子砖孔径还容易造成格孔堵塞，导致蓄热室有效利用率下降。

对于格子砖砖型的选择需要综合考虑择优确定，并不是格子砖孔数越多、孔径越小就越有利，要综合考虑蓄热室热效率、蓄热室有效利用率和格子砖使用寿命等各种因素的影响。因此采用高效格子砖、提高热风炉热效率是提高风温、降低热风炉燃烧煤气量、减少CO_2排放的关键技术之一。

蓄热室格子砖在热风炉工作过程中受到热应力、机械荷载、相变应力和压力的综合作用，格子砖在高温下工作，蠕变变形是最突出的非弹性破坏。对蓄热室高温区域格子砖要求具有优良的高温体积稳定性、耐侵蚀性及抗蠕变性能。蓄热室高温区首选硅质格子砖，并且要求硅砖的残余石英必须低于1%。根据蓄热室高度方向上的温度曲线划分温度区间，并依此而合理选用不同理化性能的格子砖，图 6 是首钢京唐 $5500m^3$ 高炉顶燃式热风炉蓄热室温度分布。在热风炉高温区采用抗高温蠕变性能优异的硅砖，中温区采用低蠕变

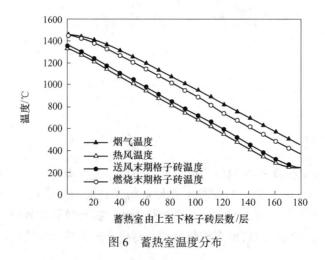

图 6　蓄热室温度分布

黏土砖，低温区采用高密度黏土砖。这种耐火材料的优化配置与顶燃式热风炉传热特性相适应，可提高耐火材料经济合理的功能性，降低工程投资。图7是首钢京唐1号高炉1号热风炉温度变化曲线，可以看出，在风温达到1300℃时，热风炉的温度变化平稳，证实了热风炉蓄热室格子砖配置的合理性。

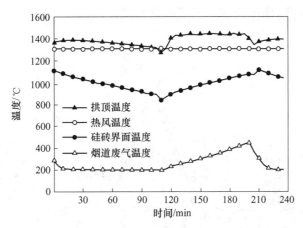

图7 首钢京唐1号高炉1号热风炉温度变化曲线

3.4 优化热风炉操作

优化热风炉燃烧、送风操作，缩小拱顶温度与风温的差值，适当提高烟气温度，是提高风温的主要操作措施。当前，热风炉燃烧操作应进一步优化，合理设定煤气与助燃空气配比，降低空气过剩系数，保证煤气的完全燃烧。热风炉烟气中残余 CO 含量和 O_2 含量应尽可能降低，燃烧初期应采取强化燃烧模式，使拱顶温度迅速达到设定数值；燃烧过程中合理调节煤气和助燃空气，在较低的空燃比下使烟气温度达到设定目标。根据拱顶温度、烟气温度、烟气 O_2 含量等参数在线调节煤气和助燃空气，优化热风炉燃烧操作是当前需要引起重视的问题。不少钢铁企业由于低热值高炉煤气资源相对丰富，并不注重节约热风炉燃烧所消耗的高炉煤气，不仅增加了煤气消耗量，造成能源浪费、生产成本升高，也无助于提高风温，而且还增加了 CO_2 排放量。

优化热风炉燃烧不但可以获得高风温，还可以降低能源消耗，提高能源转换效率，减少热量耗散，降低 CO_2 排放。热风炉送风操作也需要改进，4座热风炉应采用"两烧两送"交叉并联的工作模式，这样可以提高风温约25℃，同时应尽量减少冷风混风量，使高风温得到充分利用。3座热风炉采用"两烧一送"的工作模式，可以适当缩短送风周期、增加换炉次数，同样应减少冷风混风量。改善热风炉操作一方面应开发应用适宜的热风炉操作数学模型，采用信息化、数字化手段提高热风炉操作水平；另一方面应在交错并联送风技术的基础上，研究开发热风炉"热交错并联"送风技术，进一步减少冷风混风量，降低风温波动，使高炉获得更加稳定的高风温。

4 高温热风的稳定输送

风温提高以后，高温热风的稳定输送成为制约环节。近年来，不少热风炉的热风支管、热风总管和热风环管出现局部过热、管壳发红、管道窜风、甚至管道烧塌事故[11]，热风管道内衬经常出现破损，极大地限制了高风温技术的发展。因此，优化热风管道系统结构，采用"无过热-低应力"设计体系[12]，合理设置管道波纹补偿器和拉杆，妥善处理管道膨胀以降低管道系统应力，热风管道采用组合砖结构，消除热风管道的局部过热和管道窜风。图8是典型的热风炉热风管道平立面布置。

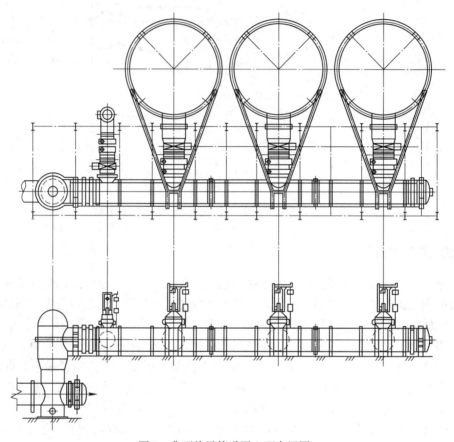

图8 典型热风管道平立面布置图

热风管道承受着高温、高压和弹塑性变形作用，是热风炉系统中工况最恶劣的管道。随着高炉炼铁技术进步，热风管道工作压力已经超过了煤气爆炸压力，工作温度达到了1250℃以上，这对于热风管道的设计提出了更高的要求。近年来我国高炉热风管道发生事故屡见不鲜，因此热风管道的设计必须予以足够的重视。以直径为2800mm、工作压力为0.45MPa的热风管道为例，其盲板力可达280t。因此热风管道的设计必须保证在此工况下土建结构、管道钢壳、耐火材料砌体的工作稳定。通过多年的研究和探索，目前已经构建了一套完整的热风管道设计体系和方法，包括热风管道拉杆设计、波纹补偿器计算和选型、耐火材料设计计算、关键部位的有限元计算分析等各个方面工作。

热风管道设计考虑了工作温度、工作压力、耐火材料荷载、钢结构受力变形、管道试压、外界温度随季节的变化以及多工况耦合等因素，保证了热风管道的安全稳定运行。热风支管采用三角形刚性拉梁的设计结构，利用了三角形稳定性的特点，保证了热风支管和热风总管三岔口的结构稳定,同时又能抑制热风炉炉壳受热上涨对管道的产生的影响。热风总管采用通长的大拉杆形式和合理的波纹补偿器设计,不但满足热风竖管的稳定、减小土建钢结构受力,同时满足在外界温度变化的情况下仍然保持整个管道系统的稳定,从而有效的提高了管道的安全性,避免了由于管道位移造成对耐火材料和钢结构的损坏。

确立热风管道设计方案之后,采用计算机软件CAESARⅡ对管道进行有限元分析(见图9),对热风管道各个部位进行详细分析,确保管道能够满足热风炉多种工况的要求,对管道进行详细的优化设计。对于结构复杂的热风管道,不同部位采用了不同的管道壁厚,不但节约了工程投资,更有效地提高了管道的安全性。通过热风管道耐火材料温度场分布的计算,不但可以确定耐火材料总厚度,而且详细分析各层耐火材料之间的温度,充分利用不同耐火材料的理化性能特点,优化耐火材料选型和设计,提高管道的绝热保温效果。

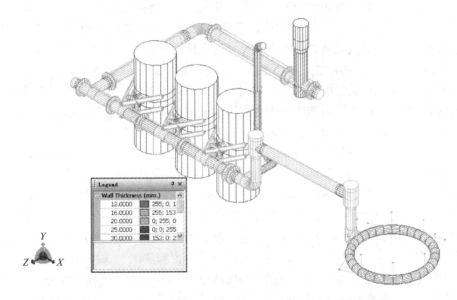

图9 复杂热风管道的CAESARⅡ有限元受力分析

通过对热风管道的受力分析、钢壳热膨胀、盲板力影响的定量分析,耐火材料保温效果的准确计算,为热风管道耐材结构设计提供完整的原始工艺参数。耐火材料砌筑设计还需要在此基础上考虑耐火砖自身受热膨胀、制造成本等因素。通过上述多方面综合考虑、完整准确的计算、有效的计算机仿真分析,在热风管道设计方面取得了长足的进步。

热风炉各孔口在多种工况的恶劣条件下工作,也是制约热风炉长寿和提高风温的薄弱环节。热风炉各孔口耐火材料要承受高温、高压的作用,还要承受气流收缩、扩张、转向运动所产生的冲击和振动作用。热风出口应采用独立的环形组合砖构成,组合砖之间采用双凹凸榫槽结构进行加强,以减轻上部大墙砖衬对组合砖所产生的压应力。热风炉的热风出口、热风管道上的三岔口等关键部位均应采用组合砖结构。热风管道内衬的工作层,应采用抗蠕变、体积密度较低、高温稳定性优良的耐火材料;绝热层应采用体积密度低、绝热性能优良的耐火材料,还应注重合理设计耐火材料膨胀缝及其密封结构。

值得指出的是,热风管道耐火材料内衬在高温高压热风的流动冲刷作用下,容易出现局部过热、窜风、甚至内衬脱落现象。对于已运行的热风管道应采用表面温度监测系统，可以在线监控热风管道关键部位的管壳温度，并可以进行数据处理和存储，实现信息化动态管理；同时为了监控热风管道受热膨胀而产生的变形情况，设置激光位移监测仪可以在线监测热风管道的膨胀位移。通过数字化在线监控装置，可以提高热风炉管道工作的可靠性，保障高温热风的稳定输送。该项技术已在首钢迁钢大型高炉上都得到了成功应用[13]。

5 延长热风炉寿命

研究表明，当热风炉拱顶温度达到1420℃以上时，燃烧产物中的NO_x生成量急剧升高，燃烧产物中的水蒸气在温度降低到露点以下时冷凝成液态水，NO_x与冷凝水结合形成酸性腐蚀性介质，对热风炉高温区炉壳造成晶间应力腐蚀，降低了热风炉使用寿命，成为制约风温进一步提高的主要限制因素。因此，热风炉一般将拱顶温度控制在1420℃以下，旨在降低NO_x生成量从而有效抑制炉壳晶间应力腐蚀。当代热风炉设计中应采取有效的晶间应力预防措施，延长热风炉使用寿命。通过合理控制热风炉拱顶温度，抑制热风炉燃烧及送风过程NO_x的大量生成；采取高温区炉壳涂刷防酸涂料和喷涂耐酸喷涂料等防护措施，加强对热风炉炉壳的隔离保护；采用细晶粒炉壳钢板，采用热处理措施消除或降低炉壳制造过程的焊接应力；对于热风炉高温区炉壳采取保温等综合防护措施，预防热风炉炉壳晶间应力腐蚀的发生，以延长热风炉寿命达到与高炉寿命同步。

影响当代热风炉寿命的关键因素除了拱顶和高温区炉壳的晶间应力腐蚀以外，还有热风炉炉体的耐火材料结构，包括拱顶、热风出口、热风管道以及蓄热室格子砖砌体，外燃式热风炉燃烧室与蓄热室连接部、内燃式热风炉的燃烧室与蓄热室的隔墙、陶瓷燃烧器本体等部位的使用寿命仍是影响热风炉寿命的制约环节。长期以来，各种结构形式热风炉的拱顶结构设计均进行了许多卓有成效的改进，改进型内燃式热风炉采用了悬链线形拱顶结构和板块式砌筑结构，在热风炉拱顶底部设置可以自由膨胀的"关节砖"，有效地避免了热风炉拱顶由于高温热应力影响而出现的异常破损；新日铁外燃式热风炉和顶燃式热风炉由于采用锥球形小拱顶结构，也较好地解决了拱顶在受热、受力条件下的异常破损问题。当前值得重点关注的是热风炉热风出口、热风支管以及热风管道的寿命，应采取综合技术措施，构建"无过热-低应力"体系，着重解决高温、高压管道在复杂受力条件下的力学问题，与此同时，在耐火材料材质和设计结构上也要不断优化，以实现当代热风炉的高效长寿。

6 结语

（1）新世纪初，当代高炉炼铁技术面临着资源短缺、能源供给不足和生态环境的制约，实现高炉炼铁工艺的可持续发展，必须采用高风温、高富氧大喷煤技术，降低燃料消耗和生产成本，这是当代高炉炼铁技术发展的必由之路。

（2）高风温是综合技术，是降低燃料比、提高喷煤量的重要技术保障。在当前条件下，利用低热值高炉煤气实现1250℃以上高风温，是实现低品质能源高效利用和高效能源转换最优化的技术措施。要系统解决高风温的获得、高温热风的稳定输送和高效利用等关键技术问题，采用高效长寿热风炉及高效格子砖，优化热风炉操作，保障高温热风的稳定输送，延长热风炉寿命，使高温热风得到高效利用。

（3）到21世纪中期，当代高炉采用高风温、高富氧大喷煤、精料、长寿、环境保护等综合技术，进一步降低燃料比，具有广阔的发展前景。

致谢

衷心感谢吴启常大师、银汉教授级高级工程师、杨天钧教授、程树森教授等专家，对高效长寿热风炉技术研究的大力支持和帮助；衷心感谢张建、梅丛华、倪苹、胡祖瑞、许云、范燕等设计人员的辛勤工作；感谢首钢京唐公司、迁钢公司、首秦公司的领导和专家的支持与帮助。

参考文献

[1] 张寿荣. 进入21世纪后中国炼铁工业的发展及存在的问题 [J]. 炼铁, 2012, 31 (1): 1-6.
[2] 张福明. 当代高炉炼铁技术若干问题的认识 [J]. 炼铁, 2012, 31 (5): 1-6.
[3] 张福明. 21世纪初巨型高炉的技术特征 [J]. 炼铁, 2012, 31 (2): 1-8.
[4] 张福明, 钱世崇, 殷瑞钰. 钢铁厂流程结构优化与高炉大型化 [J]. 钢铁, 2012, 47 (7): 1-9.
[5] 张福明, 钱世崇, 张建, 等. 首钢京唐5500m^3高炉采用的新技术 [J]. 钢铁, 2011, 46 (2): 12-17.
[6] Peter M Martin, Mark Geach, Kim Michelsson, et al. Improved blast furnace stove operation with the use of oxygen enriched combustion air [C]. 6th International Congress on the Science and Technology of Ironmaking-ICSTI 2012 Proceedings. Rio de Janeiro, RJ, Brazil, 2012: 870-880.
[7] 林成城, 徐宏辉. 热风炉高风温新技术开发 [J]. 钢铁, 2005, 40 (11): 9-12.
[8] 刘振均, 李海波, 冯新华. 宝钢热风炉富氧燃烧技术的应用 [J]. 炼铁, 2009, 28 (6): 14-15.
[9] 孟凡双, 金国一. 热风炉富氧燃烧特性与操作措施研究 [J]. 鞍钢技术, 2011 (3): 18-21.
[10] 钱世崇, 张福明, 李欣, 等. 大型高炉热风炉技术的比较分析 [J]. 钢铁, 2011, 46 (10): 1-6.
[11] 许秋慧, 常建华, 龚瑞娟, 等. 唐钢3200m^3高炉热风炉管道耐火材料塌落的处理 [J]. 炼铁, 2010, 29 (6): 23-25.
[12] 张福明, 梅丛华, 银光宇, 等. 首钢京唐5500m^3高炉BSK顶燃式热风炉设计研究 [J]. 中国冶金, 2012, 22 (3): 27-32.
[13] 马金芳, 万雷, 贾国利, 等. 迁钢2号高炉高风温技术实践 [J]. 钢铁, 2011, 46 (6): 26-31.

Technological Philosophy and Application on High Blast Temperature Technology of Contemporary Blast Furnace

Zhang Fuming　Qian Shichong　Yin Guangyu　Mao Qingwu　Li Xin

Abstract: For the current situation and challenges facing the blast furnace (BF) ironmaking technology, the contemporary BF ironmaking technological development targets are pointed. The importance and function of high blast temperature technology on the BF ironmaking are expounded. The technical philosophy and key technology for high temperature are analyzed and discussed, focused on attaining the high blast temperature under the condition of the low heat value gas combustion, high-temperature hot blast stable conveying, optimize the operation of hot blast stove, as well as achieve the high efficiency and long campaign life of hot blast stove. The contemporary BF ironmaking technology has a broad prospect for development under the conditions of high blast temperature and low fuel ratio smelting.

Keywords: blast furnace; ironmaking; high blast temperature; hot blast furnace; low fuel ratio

大型高炉热风炉技术的比较分析

摘　要：主要针对5000m³级别大型高炉的高风温热风炉技术进行技术比较分析，选择5000m³级别大型高炉的设计实例，在风温、风量、燃烧介质等热风炉设计参数相同的同口径条件下，对Didier外燃式热风炉和顶燃式热风炉进行本体表面积和表面积散热比较，同时通过数值模拟分析，比较这两种热风炉的高温烟气速度分布、高温烟气流场分布、格子砖顶面温度分布，为大型高炉热风炉形式的合理选择提供建设性建议。通过比较分析，顶燃式热风炉的本体结构技术、流场热传输技术较其他形式热风炉具有明显优点，顶燃式热风炉技术是目前高风温热风炉技术发展的趋势，对于大型高炉采用顶燃式热风炉技术可以取得可观经济效益。

关键词：大型高炉；热风炉；技术分析

1　引言

近20年来，高炉炼铁技术在大型、高效、长寿、低耗、环保方面取得进步，高炉长寿、高风温、富氧喷煤等单项技术的进步突出。高炉寿命提高10年，提出25年目标；热风炉高风温提高100℃达到1250℃，提出1300℃目标；燃料比降低到550kg/t以下，提出了490kg/t以下的目标；煤比达到150kg/t以上，提出200kg/t目标。所有技术的进步均体现在能耗和效率两个方面，并与高炉大型化发展密不可分，高炉大型化是高炉炼铁技术发展的必然趋势，我国高炉大型化应该追溯到1985年9月宝钢的投产开始，然而真正走大型化应该是从2004年。不完全统计，至2011年3月底我国在役和正在建设的4000m³以上高炉累计18~20座，其中5000m³以上高炉3~4座，我国高炉大型化，进入了钢铁企业结构优化的新阶段，同时大型高炉的热风炉技术选择成为关注的热点。目前4000m³级别高炉热风炉技术有内燃式、外燃式、顶燃式三种。5000m³级别高炉热风炉技术有外燃式和顶燃式两种。5000m³级别以上高炉热风炉技术统计见表1。

表1　5000m³级别以上高炉热风炉技术统计

国家	公司/炉号（代）	有效容积/m³	投产时间	炉缸直径/m	热风炉形式	座数	设计风温/℃
乌克兰	克里沃罗格/9号（3代）	5026	2003年11月	14.7	Didier	4	1300
俄罗斯	切列波维茨/5号（3代）	5580	2005年9月	15.1	Didier	4	1300
德国	施韦尔根/2号（1代）	5513	1993年1月	14.9	Didier	3	1250
日本	NSC君津/4号（3代）	5555	2003年5月	14.5	NSC式	4	1250

本文作者：钱世崇，张福明，李欣，银光宇，毛庆武。原刊于《钢铁》，2011，46（10）：1-6。

续表 1

国家	公司/炉号（代）	有效容积/m³	投产时间	炉缸直径/m	热风炉形式	座数	设计风温/℃
日本	君津/2 号（不详代）	5245	2004 年 5 月		NSC 式	4	1250
日本	NSC 大分/1 号（4 代）	5775	2009 年 8 月	14.9	NSC 式	4	1250
日本	NSC 大分/2 号（3 代）	5775	2004 年 5 月	14.9	NSC 式	4	1250
日本	NSC 名古屋/1 号（3 代）	5443	2007 年 4 月	14.5	NSC 式	4	1250
日本	住金鹿岛/1 号（2 代）	5370	2004 年 9 月	15	Koppers	4	1250
日本	神户加古川 2 号（2 代）	5400	2007 年 5 月	不详	不详	4	1250
日本	JFE 千叶/6 号（2 代）	5153	1998 年 5 月	15	不详	4	1250
日本	JFE 京滨 2 号（2 代）	5000	2004 年 3 月	不详	不详	4	1250
日本	JFE 仓敷/4 号（3 代）	5005	2002 年 1 月	14.4	不详	4	1250
日本	JFE 福山 4 号（4 代）	5000	2006 年 5 月	14.0	不详	4	1250
日本	JFE 福山 5 号（代）	5500	2005 年 3 月	不详	不详	4	1250
中国	首钢京唐 1 号（1 代）	5576	2009 年 5 月	15.5	BSK 顶燃	4	1300
中国	首钢京唐 2 号（1 代）	5576	2010 年 6 月	15.5	BSK 顶燃	4	1300
中国	江苏沙钢 2 号（1 代）	5800	2010 年 10 月	15.3	DME 外燃	3	1250
南韩	浦项光阳/4 号（代）	5500	2009 年 10 月	15.6	Koppers	4	1250
南韩	现代唐津厂 1 号（1 代）	5250	2010 年 4 月	14.8	DME 外燃	3	1280
南韩	现代唐津厂 2 号（1 代）	5250	2011 年	14.8	DME 外燃	3	1280

2 外燃式、顶燃式热风炉本体结构的比较

外燃式热风炉有地得式（Didier）、柯柏式（Koppers）、马琴式（Martin and Pagenstecher）、新日铁式（NSC）外燃式和 DME 外燃式。外燃式热风炉的构思是在 1910 年由达尔（F. Dahl）提出，1928 年首先在美国卡尔尼基钢铁公司建成，1938 年 Koppers 提出专利，柯柏式 1950 年用于高炉，其特点是燃烧室拱顶和蓄热室拱顶由各自不同半径的半球形砌体构成（见图 1）[1]；1959 年出现了地得式（Didier）热风炉，其特点是拱顶由近似半个卵形拱顶连接（见图 1）。1965 年德国蒂森（Thyssen）公司使用马琴（Martin and Pagenst）式外燃热风炉，其特点是蓄热室顶部具有圆锥形的缩口，使蓄热室拱顶与燃烧室拱顶由两个半径相同的 1/4 球形和大半个圆柱体所组成（见图 1）；新日铁（NSC）式热风炉，于 20 世纪 60 年代末综合了柯柏式和马琴式外燃热风炉的特点，首先在日本八幡制铁所高炉上使用，其特点是蓄热室顶部具有圆锥形的缩口，蓄热室顶部直径与燃烧室直径相同，拱顶由两个半径相同的半球形拱顶和 1 个圆柱体的联络管所组成；DME 外燃式是在地得式（Didier）的基础上进行燃烧器和燃烧室支撑结构的改进，其特点与（Didier）外燃热风炉基本相同。发展演变到现在，外燃式热风炉主要有地得式和新日铁式两种类型。

顶燃式热风炉早在 20 世纪 20 年代哈特曼（Hartmann）就提出了设想，但未受到人们

重视,从 20 世纪 70 年代开始在我国的中小型高炉上应用,1979 年首钢 2 号高炉 (1327m³) 使用了顶燃式热风炉,开创了大型高炉使用顶燃式热风炉的先河,并经过技术发展,形成目前的新型顶燃式热风炉,其特点是采用锥形与球形结合的拱顶结构(见图1),拱顶作为燃烧室,拱顶砌体受力均匀,上部拱顶的直径较小,而下部又不承受推力,具有很好的稳定性,拱顶和管道出入口均采用异形砖砌筑,以提高砌体的整体性。目前应用于 5000m³ 级别高炉的热风炉结构形式主要有两种:BSK 新型顶燃式和外燃式热风炉。针对这两种热风炉的典型形式进行比较,详见表2。

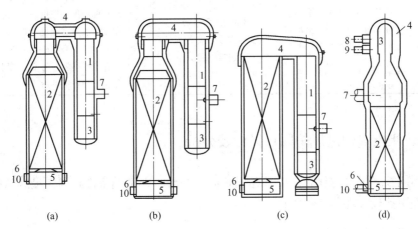

图 1 外燃式、顶燃式热风炉结构示意图
(a) 新日铁式;(b) 马琴式;(c) 地得式;(d) 顶燃式
1—燃烧室;2—蓄热室;3—燃烧器;4—拱顶;5—炉箅子及支柱;6—冷风入口;
7—热风出口;8—煤气入口;9—助燃空气入口;10—烟气出口

表 2 顶燃式与外燃式热风炉的比较

型号	NSC 外燃式	Didier 外燃式	顶燃式
布置	只能一列式	只能一列式	布置灵活,一列式、矩形、菱形
拱顶特点	拱顶由两个半径相同半球顶和一个圆柱形连接管组成,连接管上设有膨胀补偿器。拱顶对称,尺寸小,结构稳定性良	拱顶由两个不同半径的接近1/4球体和半个截头圆锥组成。整个拱顶呈半卵形整体结构。拱顶尺寸大,结构复杂,不对称,稳定性差	燃烧室在热风炉顶部,弥补内燃式和外燃式不足,燃烧和蓄热有机统一的结构形式,拱顶直径小,结构完全对称,稳定性很好
燃烧器	三孔套筒式燃烧器,燃烧稳定;但火焰相对长,结构复杂对阀门等设备要求严格,易产生脉动。有火井,寿命可达 15 年	栅格式燃烧器,火焰短,空气过剩系数小;但结构复杂,空煤气预热温度要求大致相等,空煤气预热条件具有局限性,对燃烧室掉砖掉物敏感,热风阀漏水易造成损坏。有火井,寿命可达 15 年	顶燃式热风炉的陶瓷燃烧器设置在热风炉拱顶部位,旋流扩散环形燃烧器,环形多孔,气流出口呈旋流喷射,煤气和空气在进入蓄热室之前充分燃烧,燃烧稳定,火焰短,空气过剩系数小,具有广泛的工况适应性;适应各种空煤气预热条件。无火井,寿命可达 25 年
气流	气流分布相对较好	气流分布相对较差	气流分布相对非常好

续表 2

型号	NSC 外燃式	Didier 外燃式	顶燃式
优势与存在的问题	结构稳定性较好；拱顶温度高超过 1420℃，外形较高，占地面积大，砖形复杂，拱顶联络管的砌砖和波纹补偿器问题，混风室与燃烧器之间联通管出现问题，晶间应力腐蚀问题。以上问题难以解决	外形低。拱顶结构庞大稳定性较差，占地面积大，砖形复杂。拱顶温度高超过 1420℃，晶间应力腐蚀问题；对煤气和助燃空气温度差敏感性问题；对燃烧室掉砖和热风阀漏水易造成燃烧器损坏问题。以上问题难以解决	结构均匀对称。消除本体结构和传热的不对称性。布置灵活多样，紧凑合理占地面积小，节约钢材和耐材，高温烟气分布好，格子砖的热效率提高。由于其燃烧器的结构特点，热风炉的最高温度并不在拱顶处出现，而在格子砖上表面，从而有效地降低了拱顶的温度，使拱顶更加稳定，拱顶温度一般不会超过 1100℃，可有效地减轻拱顶壳产生晶间应力腐蚀。不足在于燃烧器位置高，带来管系设计复杂，设计能够解决

3 5000m³ 级别高炉热风炉同口径比较分析

对于不同技术模式的比较，必须建立在同一平台，在基础条件相同、最终目标和效果一致的条件下比较才能具有科学性。

3.1 热风炉基本设计参数

对于热风炉同口径比较的基础就是基本设计参数，如燃料、操作制度、鼓风流量、送风温度等。两种热风炉的基本设计参数比较见表 3。

表 3 热风炉基本设计参数表

名　称	单位	3 座工作（1 座检修）	4 座热风炉工作
热风炉操作制度		单炉送风 2 烧 1 送	冷交叉并联 2 烧 2 送
鼓风流量（设计）	Nm³/min	正常 8270	
热风温度（围管处）	℃	1250	1300
风压（工作）	MPa	0.53	
冷风温度	℃	240，最大 250	
冷风湿度	g/m³	10~40	
热风炉拱顶温度（操作）	℃	1420~1450	
送风时间	min	30	60
换炉时间	min	10	
燃烧时间	min	50	50
助燃空气、煤气预热温度	℃	设计能力 200℃，按照 180℃计算	
格孔直径	mm	30（19 孔格子砖）	
热风炉烟气温度（最高/平均）	℃	400/325	
热风炉炉壳工作温度	℃	80~120	

续表3

名称	单位	3座工作（1座检修）	4座热风炉工作
格子砖蓄热面积	m²/m³	48.6	
单位鼓风蓄热面积	m²/m³	41.19	
单位鼓风格子砖质量	t/m³	1.06	

3.2 同口径条件下热风炉基本结构设计参数

根据表3中热风炉基本设计参数，通过热工燃烧计算和换热技术计算，结合目前正在运行的热风炉结构参数，进行两种不同形式热风炉的基本结构设计，基本结构设计参数见表4。

表4 热风炉基本结构设计参数

顶燃式热风炉结构参数			Didier外燃式热风炉结构参数		
热风炉总高度	m	约51.5	热风炉总高度	m	约40.790
热风炉直径	mm	10350/10900/11920	蓄热室直径/燃烧室直径	mm	9500/5900
蓄热室截面面积	m²	67.9	蓄热室截面积/燃烧室截面积	m²	57.3/15.77
格子砖高度	m	24.84	格子砖高度	m	31.44
格子砖总重量	t	2120	格子砖总重量	t	2182

3.3 外燃式和顶燃式热风炉表面热损失比较

同口径条件下，根据热风炉基本结构设计参数相比，按照热风炉不同区域、不同砖衬厚度计算炉壳表面温度。顶燃式热风炉表面积与表面热损失计算表详见表5，Didier外燃式热风炉表面积与表面热损失计算表见表6。

表5 顶燃式热风炉表面积与表面热损失计算

区域	直径/m	高度/m	面积/m²	壳体/℃	环境/℃	Δt/℃	炉壳散热总热损失/W
燃烧室顶部	7.88	3.10	79.38	52.00	25.00	27.00	33413.42
燃烧室	7.88	9.66	234.63	59.00	25.00	34.00	124367.98
拱顶锥段	7.04/11.92	5.82	201.14	85.00	25.00	60.00	188146.36
拱锥直段	11.92	5.10	190.98	81.00	25.00	56.00	166733.18
蓄热室上部	11.92	1.56	58.42	78.00	25.00	53.00	48270.69
蓄热室中部	11.92	2.86	104.14	75.00	25.00	50.00	81177.13
蓄热室中部	10.90	7.39	252.54	75.00	25.00	50.00	196854.93
蓄热室下部	10.35	12.00	390.14	48.00	25.00	23.00	139892.50
烟气室	10.35	3.35	108.28	44.00	25.00	19.00	32073.62
合计		50.84	1619.65				1010929.81
α/W·(m²·℃)⁻¹	15.59（大气流动情况复杂，计算按照大气流速为4m/s时的综合经验值）						

续表5

炉壳散热热损失流量/kJ·h⁻¹	3639347.31	热风流量/m³·h⁻¹	456000.0
4座炉壳总热损失流量/kJ·h⁻¹	14557389.24	4座炉壳总热损失比率/%	1.71
热风带出的热量/kJ·m⁻³	1864.62	设计风温水平/℃	1300.0
影响风温/℃		22.26	

表6 Didier外燃式热风炉表面积与表面热损失计算

区域	直径/m	高度/m	面积/m²	壳体温度/℃	环境温度/℃	Δt/℃	炉壳散热总热损失/W
燃烧室顶部	3.57	3.70		85.00	25.00	60.00	
拱顶过渡直段	3.57/5.32	8.25	293.94	85.00	25.00	60.00	274951.48
蓄热室拱顶	5.32	5.36		85.00	25.00	60.00	
燃烧室火井上部	5.90	15.50		78.00	25.00	53.00	178545.71
燃烧室火井中部	5.90	8.50	540.22	75.00	25.00	50.00	168439.35
燃烧室火井下部	5.90	6.70		59.00	25.00	34.00	57269.38
燃烧室火井底部	5.90	1.30	62.90	52.00	25.00	27.00	26476.50
蓄热室上部	9.50	16.00		81.00	25.00	56.00	422481.68
蓄热室中部	9.50	9.12	1136.91	75.00	25.00	50.00	215013.00
蓄热室下部	9.50	9.12		48.00	25.00	23.00	98905.98
烟气室	9.50	3.35		44.00	25.00	19.00	30012.23
合计		86.9	2033.97				1472095.31
α/W·(m²·℃)⁻¹	15.59（大气流动情况复杂，计算按照大气流速为4m/s时的综合经验值）						
炉壳散热总热损失/kJ·h⁻¹	5299543.10	热风流量/m³·h⁻¹					456000.0
4座炉壳总热损失流量/kJ·h⁻¹	21198172.4	4座炉壳总热损失比率/%					2.49
热风带出的热量/kJ·m⁻³	1864.62	设计风温水平/℃					1300.0
影响风温/℃		32.41					

3.4 比较讨论

从上述同口径热风炉本体炉壳散热比较，可以显示：（1）外燃式热风炉本体炉壳散热量较顶燃式热风炉高45.62%；（2）因本体表面热损失，风温影响相差10℃；（3）按照每提高风温100℃降低焦比15kg计算[2]，提高10℃降低焦比1.5kg，2座高炉900万吨，年节约成本2430万元（焦炭价格按照1800元/吨计算）。

4 外燃式和顶燃式热风炉数值模拟计算定性比较分析

4.1 顶燃热风炉与外燃热风炉模型建立

针对上述同口径设计参数,根据设计所配置的不同热风炉形式的结构参数进行建模、计算和分析。该模型由煤气管道、空气管道、煤气环道、空气环道、半球形炉顶、喷嘴、燃烧室和蓄热室等部分组成,图 2 是外燃式和顶燃式热风炉模型与网格划分示意图,表 4 是热风炉基本结构设计参数。

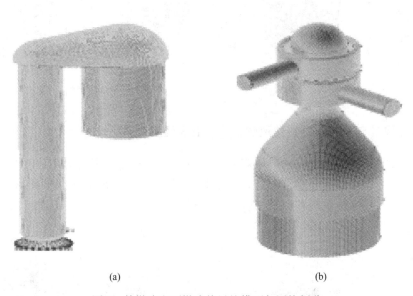

图 2 外燃式和顶燃式热风炉模型与网格划分

两种热风炉在燃烧方式上都可看作是非预混燃烧,因此在数学模型的选择上是类似的。实际中的燃烧过程是湍流和化学反应相互作用的结果,解决这一过程目前最常用的数学模型是 $k\text{-}\varepsilon\text{-}g$ 模型[3]。边界条件、煤气和空气入口均为质量流量边界,壁面设置为非滑移边界,传热采用第二类边界条件且壁面绝热。流体密度采用 PDF(概率密度函数)混合物模型给出,比热容采用混合定律给出。辐射模型为 Discrete Ordinates 模型[4]。

4.2 顶燃热风炉与外燃热风炉速度分布比较

比较顶燃式热风炉与外燃式热风炉的速度分布图(见图 3)可清楚地发现,顶燃式热风炉炉内的速度以热风炉中心线为中心,呈对称分布,这就保证了进入格子砖的烟气是均匀的,进而保证格子砖可均匀受热。而外燃式热风炉燃烧产生的高温烟气无法以中心对称的方式进入格子砖,也就无法使同一平面的格子砖被均匀加热,降低了格子砖的利用效率。

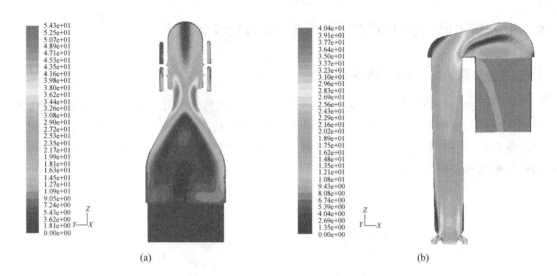

图 3 顶燃式热风炉和外燃式热风炉速度场分布比较

4.3 顶燃热风炉与 Didier 外燃热风炉格子砖表面温度分布比较

比较顶燃式热风炉与外燃式热风炉的格子砖表面温度分布图（见图4）可发现，顶燃式热风炉格子砖表面温度分布均匀，最大温差小于 50℃。而外燃式热风炉格子砖表面温度温差非常大，受燃烧效果和空间结构影响，最大温差大于 200℃。

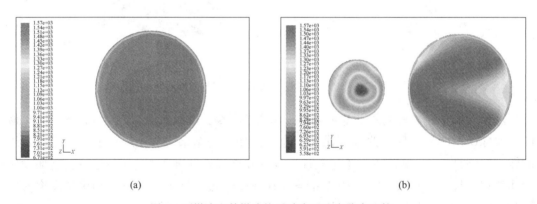

图 4 顶燃式和外燃式格子砖表面温度分布比较

4.4 顶燃热风炉与外燃热风炉流场比较

比较顶燃式热风炉与外燃式热风炉的流场分布和速度矢量场分布（见图5和图6）可看出，顶燃式热风炉流场均匀有序，以热风炉中心线为中心，均匀的旋流向下进入格子砖，确保格子砖受热均匀，热效率高。

而外燃式热风炉受空间结构的影响，高温烟气无法均匀进入格子砖，在进入格子砖之前有较大涡流存在，严重影响气流分配均匀性，进而影响热风炉的使用效率和结构稳定性。

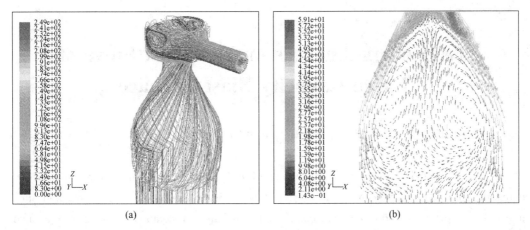

图 5 顶燃式热风炉流场与速度矢量分布

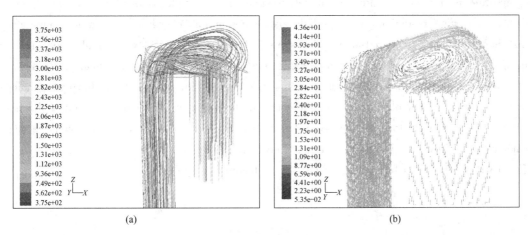

图 6 外燃式热风炉流场与速度矢量分布

5 结论

顶燃式热风炉技术是目前高风温热风炉技术发展的趋势，顶燃式热风炉的本体结构技术、流场热传输技术较其他形式热风炉具有明显优点，对于大型高炉采用顶燃式热风炉技术可以取得可观经济效益，仅热风炉本体炉壳散热比较，2 座 5000m³ 级别高炉，年产 900 万吨铁水，采用顶燃式热风炉年节约成本 2430 万元左右。

参考文献

[1] 项钟庸，王筱留. 高炉设计-炼铁工艺设计理论与实践 [M]. 北京：冶金工业出版社，2009：462.
[2] 周传典. 高炉炼铁生产技术手册 [M]. 北京：冶金工业出版社，2003：358.
[3] 王应时，范维澄，周力行，等. 燃烧过程数值计算 [M]. 北京：科学出版社，1986：72.
[4] 郭敏雷. 顶燃式热风炉传热及气体燃烧的数值模拟 [D]. 北京：北京科技大学，2008.

Technology Analysis on Hot Blast Stove of Large Capacity Blast Furnace

Qian Shichong   Zhang Fuming   Li Xin   Yin Guangyu   Mao Qingwu

Abstract: Technologies of hot blast stove for 5000m^3 grade blast furnace were compared and analyzed. Taking an example of 5000m^3 grade blast furnace, the body surface area and surface area heat dissipation of Didier external-combustion stove and top-combustion stove were compared under the same conditions of such design parameters as the blast temperature, blast volume, combustion medium etc. Through simulation analysis, the high temperature flue gas velocity distribution, high temperature gas flow field distribution and checker brick top surface temperature distribution for Didier external-combustion stove and top-combustion stove were also compared to provide constructive suggestions for reasonable selection of hot blast stove technology. Because of obvious advantages in the body structure technology and the flow field heat transferring technology over other types of hot blast stove, top-combustion stove was the development trend of the high temperature hot blast stove technology. The application of top-combustion stove on large scale capacity blast furnace can achieve significant economic benefits.

Keywords: large blast furnace; hot blast stove; technology analysis

高炉热风炉高温预热工艺设计与应用

摘　要：首钢 2 号高炉技术改造设计采用了高温预热及高风温长寿热风炉技术。助燃空气高温预热炉及煤气热管换热器投入使用后的第一个月，助燃空气预热到 630℃，高炉煤气预热到 200℃，高炉月平均利用系数 2.502t/(m³·d)、焦比 296.9kg/t、煤比 169.8kg/t、焦丁 14.3kg/t、燃料比 481kg/t、风温 1220℃。实现了"高效、低耗、长寿、优质、清洁"的设计目标。

关键词：高炉；热风炉；高温预热；高风温

1　引言

随着高炉大型化和高效化，高炉炼铁技术正向高效、长寿、高风温、大喷煤量方向发展。提高风温、增大高炉喷煤量是降低生铁能耗和成本最有效的技术措施，可以带动炼铁技术的全面提高，不仅可以代替昂贵的焦炭，节约宝贵的炼焦煤资源，降低炼焦所产生的环境污染；同时还可以促进高炉稳定顺行，强化高炉冶炼，因此受到世界各国的普遍重视。

2　国内外高炉提高风温的技术措施

热风炉是为高炉加热鼓风的设备，是现代高炉不可缺少的重要组成部分。在高炉实际生产中提高风温所受到的限制因素主要是热风炉承受高风温的能力及燃料的发热值。由于炼铁技术和工艺的不断发展，热风炉的结构形式和使用的耐火材料等项技术日臻完善，热风炉已能承受 1250℃ 或更高的风温[1]。

为了提高热风温度，目前在热风炉上采用了多项提高风温的技术，基本有以下几种。

2.1　采用富化煤气技术

国外热风炉基本上采用了富化高炉煤气的措施，常用的富化气为焦炉煤气（也有用转炉煤气及天然气等进行富化）。

2.2　利用换热器对煤气、助燃空气进行预热的技术

这是目前普遍采用的技术。换热器的形式主要有回转式换热器、固定板式换热器、热管换热器及管式换热器等。热管换热器的热管内热循环媒质主要为纯水、热煤油等。目前国内大多数厂家采用以纯水为热循环媒质的热管换热器。

本文作者：毛庆武，张福明，黄晋。原刊于《2005 中国钢铁年会论文集》，2005：449-453。

2.3 热风炉自身预热技术

这项技术是通过设置四座热风炉和配套的管道、阀门等设施，利用其中一座热风炉送风后的余热预热助燃空气，供给其他热风炉燃烧用，采用"两烧一送一预热"的工作制度。

2.4 助燃空气高温预热技术

助燃空气高温预热技术是在总结现有用于提高热风温度的各项技术的基础上，从而提出一种能够满足热风温度大于1250℃的需要，工艺简单，工作可靠，适应各种形式的热风炉使用的助燃空气高温预热装置。

助燃空气高温预热的工艺原理是：设置两座助燃空气高温预热炉，通过燃烧低热值的高炉煤气将预热炉加热后，再用来预热热风炉使用的助燃空气。预热炉燃烧温度在1000℃以上，助燃空气可以被预热到600℃以上。由于供热风炉使用的助燃空气的物理热被显著提高，热风炉的加热能力也被相应提高。同时利用热风炉烟气余热预热高炉煤气到200℃，使采用全烧高炉煤气的热风炉风温达到1250℃以上。助燃空气高温预热工艺流程图如图1所示。

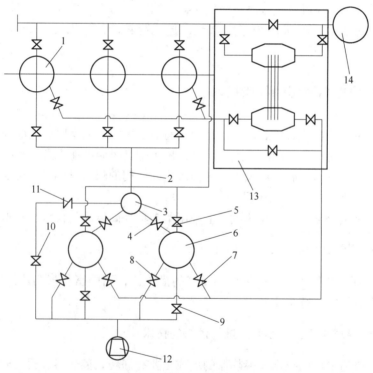

图 1 高温预热工艺流程图

1—热风炉；2—温度检测点；3—混合室；4—热助燃空气切断阀；5—烟气切断阀；
6—助燃空气高温预热炉；7—煤气切断阀；8—空气切断阀；9—冷助燃空气切断阀；10—助燃空气旁通阀；
11—助燃空气调节阀；12—助燃风机；13—煤气预热器；14—烟囱

助燃空气高温预热装置具有温度调节简便灵活；操作成熟，可靠；提供的高温预热助燃空气温度稳定，可达到600℃以上，具备通用的余热预热设备无法达到300℃以上预热温度的优点。而且助燃空气高温预热装置可以是各种形式的热风炉或加热炉，在高炉扩容大修改造时，其原有热风炉可改造成助燃空气高温预热炉，助燃空气高温预热装置的使用寿命可达到高炉热风炉的使用寿命。同时，能广泛适应高炉各种生产状况的操作需要。

3 工艺开发与设计

3.1 工艺技术方案的确定

2000年，根据首钢总公司的发展规划，决定于2002年对首钢2号高炉进行一次现代化新技术改造，要求热风温度达到1250℃。首钢2号高炉的热风炉系统为1979年建成的4座顶燃式热风炉，已经使用了20多年，也经历了两代高炉炉役，无论是从炉壳结构强度、拱顶内衬、管道设备均已无法适应现代高炉使用风温1250℃、高炉一代炉役大于15年（无中修）的要求。因此，热风炉系统必须进行新建或大修改造，才能实现热风温度1250℃的要求。如果在原地进行大修改造，将对高炉的正常生产产生影响，延长高炉停炉后大修工期，而且受场地限制，很难进行彻底完善的大修改造；而采用原有顶燃式热风炉高温预热助燃空气，要求的拱顶温度和压力不高，可以安全稳定的生产；新建热风炉必须满足全烧高炉煤气的条件，既要适应助燃空气高温预热及高炉煤气低温预热，又要适应助燃空气及高炉煤气不预热的条件，同时要求高的换热效率及较小的占地面积。基于以上原因，结合2号高炉的现场情况，决定在首钢2号高炉停炉前，先在南面场地上新建3座霍戈文改造型高风温内燃式热风炉，并相应配套建设助燃风机、煤气热管换热器、烟囱等设施；高炉停炉后，4座顶燃式热风炉拆除2座（1号及2号），改造2座热风炉（3号及4号）作为高温预热助燃空气的预热炉，达到合理利用原有设施、减少投资的目的。通过采用助燃空气高温预热技术，在全烧高炉煤气的条件下使热风温度达到1250℃。

3.2 新建热风炉系统工艺设计

新建3座霍戈文改造型高风温内燃式热风炉呈一列式布置。热风炉主要技术参数及不同条件下的计算结果比较详见表1及表2。

表1 热风炉主要技术参数

项目	设计值	项目	设计值
热风炉炉壳内径/mm	$\phi 9200/\phi 10468$	每座热风炉总蓄热面积/m²	48879
热风炉总高度/mm	41590	燃烧器长度/m	3.4
热风炉高径比	4.521	每1m³高炉容积具有的加热面积/m²·m⁻³	82.38
热风炉蓄热室断面积/m²	35.8	每1m³/min高炉鼓风的加热面积/m²·(m³/min)⁻¹	34.9（按风量4200m³/min）
热风炉燃烧室断面积/m²	9.7		

表 2 计算结果比较

序号	项目	单位	高温预热	煤气富化	煤气预热	不预热
1	热风炉座数	座	3	3	3	3
2	风量	Nm³/min	4200	4200	4200	4200
3	热风温度	℃	1250	1250	1150	1050
4	拱顶温度	℃	1420	1420	1310	1220
5	火焰温度	℃	1442	1442	1325	1235
6	湿度	g/Nm³	14	14	14	14
7	焦炉煤气用量	%	0	11.4	0	0
8	送风时间	min	45	45	45	45
9	燃烧时间	min	80	80	80	80
10	格子砖高度	m	32.4	32.4	32.4	32.4
11	助燃空气温度	℃	600	20	20	20
12	高炉煤气温度	℃	200	45	200	45
13	高炉煤气中含水	%	10.98	10.98	10.98	10.98
14	高炉煤气热值	kJ/m³	3368	3130	3368	3130
15	冷风温度	℃	170	170	170	170
16	烟气操作温度	℃	398	398	383	330
17	烟气最高温度	℃	398	398	400	400
18	一座热风炉煤气最大流量	Nm³/h	73840	61231	79160	75340
19	一座热风炉助燃空气最大流量	Nm³/h	56190	71070	51770	49870
20	一座热风炉烟气最大流量	Nm³/h	121400	123700	121700	116420

3.3 热风炉预热工艺设计

热风炉的预热系统设计了低温预热及高温预热两套系统，通过预热高炉煤气及助燃空气，在全烧高炉煤气的条件下，使热风温度达到1250℃。

3.3.1 低温预热工艺设计

利用热风炉烟气余热，通过分离式热管换热器对热风炉用高炉煤气进行预热。预热后，高炉煤气温度可达200~250℃。分离式热管换热器的烟气、煤气两个箱体分散布置，通过外联管传输水媒质换热介质。

3.3.2 高温预热工艺设计

2号高炉原有4座首钢式顶燃热风炉于1979年建成投产，在1991年2号高炉大修时，高炉扩容为1726m³，仅更换了3座热风炉的内衬、格子砖、燃烧器、部分工艺管道及设备，其中1号热风炉的格子砖仍为5孔格子砖。此次2号高炉技术改造，充分发挥顶燃热风炉具有投资省，占地面积少，耐火材料结构合理的技术优势，以节省投资、缩短建设工期、满足工艺要求为目的，仍然采用顶燃热风炉作为助燃空气高温预热炉，充分利用了4座热风炉矩形布置的优势[2]，重点进行了助燃空气高温预热工艺的改造设计，主要有以下方面：(1) 拆除已经运行23年多的1号和2号两座旧顶燃热风炉，利旧改造3号、4号顶燃热风炉作为高温预热助燃空气的预热炉。(2) 原有热风竖管作为预热炉的混风室，原有倒流休风管改为冷助燃空气入口，原热风出口改造为热助燃空气出口。(3) 降低原3号、4号顶燃热风炉的标高，设计合理的高径比，热风炉中下部（标高22.800m以下）的内衬及格子砖利旧，保护性拆除热风炉中上部10.098m炉壳、内衬及格子砖。拆下的格子砖进行筛选，重新砌筑。热风炉内衬及格子砖材质设计上部为低蠕变高铝质、中部为高铝质、下部为黏土质。(4) 按照助燃空气不低于600℃进行设计计算燃烧器的能力，每座热风炉由4套燃烧器改为2套燃烧器，在减少设备数量的同时，保证加热燃烧的均匀性。(5) 在原综合楼一层新建2台预热炉用助燃风机。改造后的助燃空气高温预热炉技术参数详见表3。

表3 改造后的助燃空气高温预热炉技术参数

项目	设计值	项目	设计值
热风炉炉壳内径/mm	φ7000	蓄热室格子砖段数/段	3
蓄热室直径/mm	φ5792	热风炉座数/座	2
热风炉总高度/mm	38882	每个燃烧器能力/$m^3 \cdot h^{-1}$	21000
热风炉高径比	5.555	热助燃空气温度/℃	≥600
热风炉蓄热室断面积/m^2	26.35	拱顶温度/℃	1200
每座热风炉总蓄热面积/m^2	26691	烟道温度/℃	300

4 生产实践与应用

首钢2号高炉于2002年5月23日点火投产，助燃空气高温预热炉于2002年9月8日投入运行。助燃空气高温预热炉及煤气热管换热器投入使用后，运行良好，达到了设计要求。预热炉及煤气热管换热器的投入运行后的高炉技术经济指标详见表4。

表4 首钢2号高炉2002年10月主要技术经济指标

项目	指标	项目	指标
高炉有效容积/m^3	1780	渣量/$kg \cdot t^{-1}$	292
利用系数/$t \cdot (m^3 \cdot d)^{-1}$	2.502	入炉风量/$Nm^3 \cdot t^{-1}$	1108
焦比/$kg \cdot t^{-1}$	296.9	富氧率/%	0.3
煤比/$kg \cdot t^{-1}$	169.8	送风压力/MPa	0.304
焦丁/$kg \cdot t^{-1}$	14.3	风温/℃	1220
燃料比/$kg \cdot t^{-1}$	481	顶温/℃	204
综合焦比/$kg \cdot t^{-1}$	444.2	顶压/MPa	0.166
综合冶炼强度/$t \cdot (m^3 \cdot d)^{-1}$	1.111	热助燃空气温度/℃	630
综合入炉矿品位/%	59.79	高炉煤气预热温度/℃	200
熟料率/%	89.26	热风炉拱顶温度/℃	1370

通过2号高炉的生产实践证明，只燃烧低热值的高炉煤气，再好的热风炉也无法获得高风温，对助燃空气和煤气进行预热是获得高风温的有效手段和途径。由于首钢2号高炉（1780m³）助燃空气高温预热技术的成功应用，因此在首秦1号高炉（1200m³）也采用了助燃空气高温预热技术，而且将应用于首秦2号高炉（1780m³）及迁钢2号高炉（2650m³）。

5 结语

大型高炉助燃空气高温预热技术工艺开发与设计在首钢2号高炉上得到了成功应用，该工程投产后解决了首钢风温不足的矛盾，2号高炉最高煤比达到190kg/t，煤比得到了大幅度的提高，在没有富氧的条件下，最高月平均煤比达到了169.8kg/t。该项工程投产后，节约了焦炭等资源消耗，实现了清洁化生产，改善了劳动条件，取得了显著的经济效益、社会效益和环境效益。

大型高炉助燃空气高温预热技术的建成投产，为首钢进一步提高风温及喷煤量，扩大喷吹煤种，创造了必要的先决条件。助燃空气高温预热完全采用国产化技术和设备，对我国高炉助燃空气高温预热技术改造提供了有益的借鉴和参考。

参考文献

[1] 吴启常，张建良，苍大强. 我国热风炉的现状及提高风温的对策 [J]. 炼铁，2002（5）：1-4.
[2] 黄东辉，韩向东. 首钢2号高炉利用旧热风炉预热助燃空气的实践 [J]. 炼铁，2004（2）：15-18.

Design and Application of High Temperature Preheating Process for Hot Blast Stove of BF

Mao Qingwu　Zhang Fuming　Huang Jin

Abstract: The technology of high temperature preheating and high temperature and long campaign life hot blast stove had been adopted during the Shougang's No. 2 BF technology rebuild design. The one month after high temperature preheating combustion air stove and gas preheating equipment were used, combustion air was heated to 630℃, BF gas was heated to 200℃, monthly average productivity 2.502t/(m³·d), coke ratio 296.9kg/t, coal ratio 169.8kg/t, coke nut ratio 14.3kg/t, fuel ratio 481kg/t, blast temperature 1220℃. Achieved design target of "high efficiency, low consumption, long campaign life, high quality, cleaning".

Keywords: BF; hot blast stove; high temperature preheating; high temperature blast

首钢京唐钢铁厂 5500m³ 高炉热风炉系统长寿型两级双预热系统设计

摘 要：对首钢京唐钢铁厂 5500m³ 高炉热风炉系统长寿型两级双预热系统进行了阐述，该系统的设计和实施满足了热风炉系统在采用单一高炉煤气的情况下，能够实现1300℃风温。整个系统既能高效使用高炉煤气，又能够长期可靠运行，为高炉可靠、经济、环保地运行打下了坚实的基础。

关键词：高炉；热风炉；单一高炉煤气；两级双预热；高风温；热效率；长寿

1 目前国内热风炉系统预热系统的现状

从全世界范围来看，大家都在努力提高热风温度，从而为提高喷煤量、节约宝贵的焦炭而创造条件。热风温度在1150℃以上时，提高100℃风温，可以节约焦炭 8~15kg/tFe[1]。少数钢铁企业由于拥有多余的高热值煤气（焦炉煤气或转炉煤气），故获得1250℃及以上高风温比较容易实现。但是由于各钢铁厂高热值煤气越来越紧张，这就迫使人们考虑在采用单一高炉煤气的情况下如何获得1250℃以上的风温（对于大型高炉要求1300℃及以上的风温）。目前国内许多钢铁厂在这方面做了大量的工作，涌现出了各种各样的工艺流程。实践证明，各种各样的工艺流程，都有其特点，为推动热风炉风温水平的提高做出了一定的贡献。但是仔细分析目前各厂采用的工艺流程，又或多或少的存在一定的缺陷。在新设计的热风炉或者旧热风炉改造时，应该仔细研究，从而推动我国热风炉高风温技术的进一步发展。

目前我国高炉热风炉系统为了获得高风温所采取的措施中存在一些缺陷：

（1）热风温度偏低。对于仅采用低温热管换热器而预热高炉煤气和助燃空气的热风炉系统，在采用单一高炉煤气的情况下，其风温水平一般不会超过1200℃，故这种工艺流程已经不能满足高风温的要求。

（2）预热系统的寿命不能与热风炉本体同步，且在大型高炉热风炉系统上应用有一些缺陷。为了达到1250℃热风温度，目前国内普遍采用的是附加燃烧炉加中温换热器的组合方式，将助燃空气和煤气预热到较高温度，从而满足1250℃的高风温。中温换热器主要有两种类型，一种为扰流子中温管式换热器（烟气温度一般不超过600℃）；另外一种为强制油循环中温管式换热器（此流程目前在台湾中钢采用，使用较少）。前者由于工艺特点及制造水平限制，目前运行结果表明不能和高炉寿命同步，更不能和热风炉寿命同步，其使用寿命约为10年左右；当高炉大型化以后，这种换热器的体积也变得非常庞大，节省投资的优势不明显；而且一般只能将煤气和空气预热到300℃，风温最高一般只能达

本文作者：梅丛华，张福明，毛庆武，银光宇，倪苹。原刊于《第十届全国大高炉炼铁学术年会论文集》，2009：564-567。

到1250℃而无法达到1300℃及以上,且一旦系统出现故障,对热风温度的影响较大。由于没有充分回收热风炉废烟气余热,故系统尚有一定改进余地。

目前国内一些高炉上使用了高风温组合换热系统。该系统从原理上讲是对附加燃烧炉加中温换热器的组合方式的一种改进,提高了整个热风炉的热效率,充分回收了热风炉废烟气余热,且热风炉助燃空气采用了两级预热。该系统关键部件为空气二级扰流子换热器,如何提高其使用寿命是该工艺需要重点解决的问题。

(3) 对热风炉系统总体热效率重视不够。首钢2号高炉率先利用旧的热风炉,加以适当的改造,用来预热助燃空气,以获得600℃的助燃空气。国内很多高炉也采用了类似的系统,从而获得1250℃的风温。但是在采用这样的系统时,往往对热风炉系统的总体热效率重视不够。在国内已经投产的高炉的热风炉系统中,在配置蓄热式热风炉预热助燃空气的情况下,一般只配置助燃空气低温热管或者煤气低温热管换热器,而不是两者同时配置。这样做的缺点有两条:一是根据热风炉系统的热平衡计算,在只配置一种介质低温预热的情况,将导致大量烟气余热排向大气,不仅导致环境的恶化,也降低了整个热风炉系统的总体热效率,一些观点认为,由于现在的高炉都在提高喷煤比,故喷煤系统需要一部分热风炉烟气用来制粉。但是通过整个喷煤系统和热风炉系统烟气平衡的计算,喷煤制粉系统需要的热风炉烟气量占热风炉系统的烟气只是很小一部分,根本无法全部消化。还有一部分人认为,部分没有换热的热风炉烟气可以直接用来制粉,取消制粉系统的烟气炉。实践证明这样的系统存在一些缺陷,因而未得到钢铁厂的普遍采用(只有个别钢铁厂采用,国内某钢铁厂采用的条件,是热风炉距制粉距离较近)。第二个缺点是目前普遍的做法是只进行空气的低温热管换热,而不进行煤气的低温热管换热。这样做其实并不好,因为,为了环保和节能,各钢铁厂目前普遍采用了高炉煤气干法除尘系统,不仅在中型高炉上得到了大量使用,而且在国内大型高炉上($3200 \sim 5500m^3$) 也获得了应用。在实践过程中发现,采用干法除尘系统煤气的低温冷凝液具有很强的酸腐蚀性,在许多钢铁厂出现了管道和设备的点腐蚀。而进行煤气的低温热管换热,既提高了热风炉系统总的热效率,又减少甚至消除了煤气的低温冷凝,从而为热风炉系统的长期安全生产创造了条件。

2 首钢京唐钢铁厂 $5500m^3$ 高炉热风炉系统长寿型两级双预热系统设计

综合国内各种预热工艺流程的优缺点,并且结合首钢在该领域积累的大量工程实践,在首钢京唐钢铁厂 $5500m^3$ 高炉热风炉预热系统设计中,采用了具有我公司自主知识产权的长寿型两级双预热系统。

2.1 工艺流程

热风炉系统长寿型两级双预热系统设工艺流程如图1所示。

二级助燃空气蓄热式预热炉的助燃空气和煤气分别通过其助燃空气燃烧阀和煤气燃烧阀进入预热炉混合燃烧,完成对二级助燃空气蓄热式预热炉蓄热。二级助燃空气蓄热式预热炉蓄热完成后,将其切换到送风状态,这时来自低温空气总管的空气(约190℃)通过二级助燃空气蓄热式预热炉冷风管道和冷空气阀进入二级助燃空气蓄热式预热炉,与其内

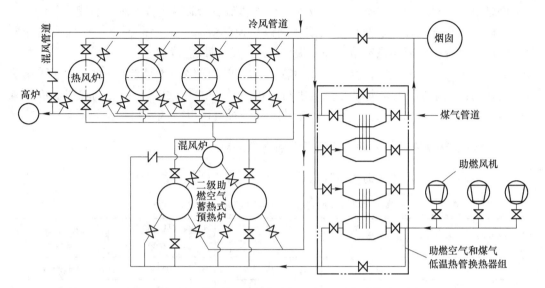

图 1 热风炉系统高风温的长寿型两级双预热系统流程图

的格子砖换热后，产生的高温空气经过热空气阀进入混风炉，与流经冷空气混风管道和冷空气混风阀进入混风炉的低温空气（约 190℃）混合，产生的中温助燃空气（450～700℃）通过中温助燃空气管道输送到热风炉，用于热风炉的燃烧。

来自管网的高炉煤气（包括所有的热风炉和二级助燃空气蓄热式预热炉燃烧需要的所有高炉煤气），经过煤气低温热管换热器预热到约 200℃ 后，然后通过两根煤气管道分别输送到二级助燃空气蓄热式预热炉和热风炉，与各自的助燃空气混合燃烧，完成各自的燃烧过程。

二级助燃空气蓄热式预热炉和热风炉各自燃烧产生的高温烟气，其热量被各自蓄热室的格子砖吸收后，通过各自烟道阀的废烟气最高温度被控制在 450℃，然后通过各自的烟气管道汇总到烟气换热器入口，这样所有的废烟气都尽可能参与助燃空气和煤气的低温换热。完成换热后的废烟气最终通过共用烟囱排入大气。

2.2 工艺流程的特点

在配置二级助燃空气蓄热式预热炉预热助燃空气的情况下，同时配置助燃空气低温热管和煤气低温热管换热器，且二级助燃空气蓄热式预热炉用助燃空气也利用废烟气预热，大大提高了整个热风炉系统的热效率，整个预热系统的使用寿命与热风炉本体同步，满足了现代高炉长期高风温稳定运行的要求。国内现有的预热系统追求的只是最高 1250℃ 的风温水平，而本工艺流程满足了 1300℃ 及以上的风温水平要求。煤气采用低温热管换热器进行预热的目的，不仅利用废烟气的余热，提高热风炉系统总的热效率，而且根据目前国内的生产实践，煤气采用低温热管换热器进行预热后，其温度大大超过煤气结露温度，消除煤气低温冷凝，减少了煤气管道的酸腐蚀，从而为热风炉系统的长期安全生产创造了条件。

热风炉和二级助燃空气蓄热式预热炉所需要的所有助燃空气都由集中助燃风机提供，而不是将两者分开，使设备功能集中，减少了设备的台数，简化了工艺流程，为热风炉和二级助燃空气蓄热式预热炉所有助燃空气都经过助燃空气低温热管换热器创造了条件。

在煤气和空气低温热管换热器长期使用以后换热效率变低而需要重新充填换热介质时，热风炉系统仍然能够稳定地为高炉提供1250℃的热风温度。

而这种预热系统的配置，反过来又为热风炉本体的优化设计创造了条件。由于采用了这种完善的预热系统，热风炉的废气温度可以提高到450℃，可以有效提高热风温度，减少热风炉本体的格子砖使用量，热风炉本体的建设投资可以降低5%。

2.3 煤气低温热管换热器和助燃空气低温热管换热器配置

煤气低温热管换热器和助燃空气低温热管换热器的设计参数见表1。

表1 煤气和助燃空气低温热管换热器设计参数

名 称	单位	烟气	煤气低温热管换热器	助燃空气低温热管换热器
流 量	Nm³/h	700000	505349	319000
进口温度	℃	350	45	20
出口温度	℃	约165	215	200
流动阻力	Pa	≤500	≤550	≤600
最高管内蒸气温度	℃		260	248
换热面积	m²		6811+14224（热侧）	
回收热量	kW		21461+35373=56833	

由于低温换热器区域管道直径较大，低温换热系统的各种阀门均选择了液动阀门，并且具有调节功能，简化了阀门选型和工艺操作的需要。

2.4 二级助燃空气蓄热式预热炉设计参数

二级助燃空气蓄热式预热炉采用顶燃式热风炉，其工艺设计参数见表2。

表2 二级助燃空气蓄热式预热炉参数

名 称	单 位	参 数
空气蓄热式预热炉座数	座	2
周期	h	1.50
送风时间	h	0.75
拱顶温度	℃	1315
热空气温度	℃	1200
空气入口温度	℃	190
助燃空气温度	℃	190
煤气温度	℃	190
废烟气平均温度	℃	340
废烟气最高温度	℃	450

续表2

名称		单位	参数
介质消耗量	加热空气量	m³/h	129149
	煤气消耗量	m³/h	99083
	助燃空气消耗量	m³/h	68888
	烟气量	m³/h	161866
	格子砖重量	t	768.8
	加热面积（一座预热炉）	m²	35865
	煤气热值	kJ/m³	3012.5
	热效率	%	76.40

3 结语

在首钢京唐钢铁厂 5500m³ 高炉热风炉预热系统设计中，总结了国内相关工艺流程的优缺点，结合首钢的工程实践，提出了长寿型两级双预热系统。该系统已于2009年5月投入运行，从运行情况看，系统能够满足在使用单一高炉煤气的情况下，获得1300℃的热风温度。由于高炉投入运行时间不长，故该系统的能力还没有完全发挥出来；目前高炉风温水平已经达到1180℃，技术优势已初步呈现出来。长寿型两级双预热系统优点体现在：

（1）提高热风炉烟气温度，缩小热风炉尺寸，从而减少热风炉本体投资。

（2）助燃空气采用二级预热，即助燃空气预热由一级热管换热器和二级蓄热式预热炉组合而成；煤气采用热管换热器。

（3）获得了较高的助燃空气温度和适当的煤气预热温度，从而在使用单一高炉煤气的情况下获得1300℃以上的热风温度。

（4）在获得1300℃以上的热风温度的情况下，整个热风炉系统的热效率仍保持在很高的水平。

（5）助燃空气二级预热采用蓄热式热风炉，与热风炉本体寿命保持同步。

（6）在助燃空气和煤气低温预热系统失效而进行检修时，仍然能够获得较高的热风温度，从而保证高炉始终在高风温水平下操作。

参考文献

[1] 项钟庸, 王筱留. 高炉设计——炼铁工艺设计理论与实践 [M]. 北京：冶金工业出版社, 2007.

Double-Preheating Longevity System of Hot Blast Furnace in 5500m³ BF of Shougang Jingtang Steel Plant

Mei Conghua Zhang Fuming Mao Qingwu Yin Guangyu Ni Ping

Abstract: Double-preheating longevity system of hot blast furnace is introduced in 5500m³ BF of Shougang Jingtang steel plant, hot blast furnace system can get 1300℃ blast temperature on all BF gas condition. This system may utilize BF gas with high efficiency, run stably, make BF work smoothly.

Keywords: BF; hot blast stove; all BF gas; double-preheating; high blast temperature; thermal efficiency; longevity

高效长寿顶燃式热风炉燃烧技术研究

摘　要：本文结合首钢京唐 5500m³ 特大型高炉顶燃式热风炉，论述了顶燃式热风炉的技术优势和燃烧器燃烧的技术特征，并利用数学仿真重点研究了顶燃式热风炉燃烧器燃烧机理，解析了顶燃式热风炉燃烧室内的燃烧过程，分析了炉内的速度、温度以及浓度分布。通过对燃烧器结构的改进优化，改善了燃烧室内的燃烧状况。

关键词：顶燃式热风炉；燃烧器；燃烧机理；优化设计

1　引言

顶燃式热风炉是一种针对于内燃式和外燃式热风炉的不足而发展起来的新型热风炉结构。其结构特点是取消了燃烧室，将燃烧器直接布置在热风炉的拱顶，以拱顶空间作为燃烧室[1]。其结构设计充分吸收了内燃式、外燃式热风炉的技术优点。与内燃式、外燃式热风炉相比，顶燃式热风炉具有如下特点：（1）取消了燃烧室和隔墙，扩大了蓄热室容积，在相同的容量条件下，蓄热面积增加 25% ~ 30%[2]。（2）结构稳定性增强。（3）采用大功率短焰燃烧器，直接安装在拱顶部位燃烧，使高温热量集中在拱顶部位，热损失减少，有利于提高拱顶温度。（4）其工作过程是一种典型的逆向强化换热过程，提高了热效率。（5）耐火材料稳定性提高，热风炉寿命延长。（6）布置紧凑，占地面积小，节约钢材和耐火材料。

2　顶燃式热风炉燃烧机理和燃烧特性

顶燃式热风炉将环形预燃室置于炉顶，在预燃室下部布置了几十个小口径陶瓷燃烧器，环形空气和煤气通道以及陶瓷烧嘴布置在燃烧室的壳体内。煤气和助燃空气以一定切向角度由喷嘴喷出，混合气体在燃烧室内旋转燃烧，使煤气和助燃空气混合均匀。从喉口到扩张段，横截面积突然增大，使得该区域出现较大的回流，该回流区的产生是顶燃式热风炉的关键技术所在。回流的存在一方面能够起到稳定火焰的作用，使得新补充的煤气和空气始终能和高温气体接触，连续燃烧；另一方面扩张段中心的回流切断了火焰的中心发展，而使火焰沿着锥角空间扩张，从而缩短了火焰长度，使火焰不会对蓄热室格子砖产生冲击。生产实践表明，燃烧室内煤气能实现完全燃烧，而且在所有工作制度下都不会出现脉动燃烧现象，由于燃烧火焰不直接接触耐火材料砌体，砌体不会出现局部过热现象。

燃烧室布置在蓄热室上部，与蓄热室在同一中心轴线上，可保证烟气均匀进入格子

本文作者：张福明，胡祖瑞，程树森，毛庆武，钱世崇。原刊于《2010 年全国炼铁生产技术会议暨炼铁学术年会》，2010：610-614。

砖，据测定其不均匀度为±3%~5%，大大提高了热风炉蓄热室的利用率，并使得温度沿拱顶、格子砖、内衬和炉壳均匀分布，减少了温差热应力，提高了热风炉的寿命。

3 首钢京唐钢铁厂1号高炉顶燃式热风炉燃烧数学仿真计算

由于热风炉炉内温度高达1200℃以上，燃烧器结构复杂，很难用实际测量的方法获悉内部的速度、温度以及浓度分布状况，因此目前国内外对热风炉炉内燃烧过程的研究普遍采用仿真计算的方法。数值模拟仿真计算[3]是通过建立数学模型把实际的物理、化学过程简单化、数学公式化，根据过程的特点用数学方程加以描述。这种方法的特点是：准确可靠，灵活多变，速度快，费用少。仿真计算能够避免现场实测的困难及物理模型和原型不尽相似的缺陷，提供更为接近实际的计算数据；能够较容易地改变模型结构和操作条件对实际过程进行深入研究，在较短时间内得到大量信息；不需要购买成套的实验设备，能在短期得到理想的计算结果。因此，数值模拟仿真计算在现代燃烧技术方面具有极大的优越性和发展前景。随着现代计算机技术的高速发展，计算机仿真模拟的优势也越来越明显。

3.1 物理模型

本文针对首钢京唐钢铁厂1号高炉所配置的BSK型顶燃式热风炉进行建模、计算和分析。该模型由煤气管道、空气管道、煤气环道、空气环道、半球型炉顶、喷嘴、燃烧室和蓄热室等部分组成，图1是BSK顶燃式热风炉的示意图。

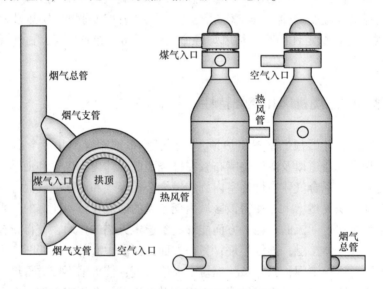

图1 BSK顶燃式热风炉示意图

3.2 边界条件

首钢京唐5500m³高炉BSK顶燃式热风炉正式投入运行之后，对其三个阶段的运行参数进行了采样分析和计算。表1和表2分别是三个时间高炉煤气的成分和相关操作参数的采样表。从表1可以看出，随着生产的进行，高炉煤气利用率提高，进入热风炉燃烧的煤气热值降低。该热风炉采用了助燃空气和煤气双预热技术，其煤气和空气管道入口处两种气体的相关参数见表2。

表1　煤气成分采样　　　　　　　　　　（%）

日　　期	CO	CO_2	N_2	CH_4	H_2
2009年6月20日	23.23	19.59	53.93	0.4	2.85
2009年7月30日	21.80	18.93	56.32	0.02	2.93
2009年9月24日	20.87	23.2	52.02	0.05	3.86

选取表2中9月24日的数据给出数学模型所对应的边界条件，煤气和空气入口均为质量流量边界，空气入口的质量流量为50.36kg/s，湍流强度为10%，入口温度839K，平均混合分数和平均混合分数均方差都为0；煤气入口的质量流量为70.72kg/s，湍流强度为10%，入口温度为441K，平均混合分数为1，平均混合分数均方差为0。出口为压力出口边界，设定压强为大气压即表压为0，设定回流湍流强度10%，平均混合分数和平均混合分数均方差都为0。对于壁面边界，壁面为非滑移边界，传热采用第二类边界条件且壁面绝热。流体密度采用PDF混合物模型给出，比热容采用混合定律给出。辐射模型采用P1模型。

表2　顶燃式热风炉操作参数

日　　期	2009年6月20日	2009年7月30日	2009年9月24日
煤气入口流量/$Nm^3 \cdot h^{-1}$	187163	133352	190000
煤气入口温度/℃	27	163	168
煤气入口压力/kPa	8.5	2.09	4.34
理论需要空气量/$Nm^3 \cdot h^{-1}$	123340	78811	112860
空气入口流量/$Nm^3 \cdot h^{-1}$	150000	101779	140534
空气入口温度/℃	373	360	566
空气入口压力/kPa	12.5	12.6	10.7
拱顶温度实测值/℃	1280	1360	1390
拱顶温度仿真计算值/℃	1293	1345	1400

3.3　计算结果及分析

3.3.1　速度分布

图2给出了燃烧室空间 $Y=0$（左）和 $X=0$（右）截面的速度分布云图。从图中可以看出，整个空间内速度基本沿中心呈对称分布。随着高度的下降，整个预燃室煤气和空气的混合气在径向方向逐渐均匀，从煤气喷嘴层面到空气喷嘴层面燃烧室的中心低速区逐渐减小，到达预燃室底部时，低速区基本已经消失。

在喉口区域由于气体运动空间大幅度压缩，靠近炉墙的气体速度迅速升高，中心位置的速度大小基本不变。靠近炉墙处气体速度很大，而且旋流非常强烈，对炉墙造成了强烈的冲刷，因此喉口处的炉墙在设计时需考虑到这一因素。

另外，从图 2 中还能看到，当气体通过喉口以下的扩张段时，气体速度迅速减小，而且依然保持中心速度低、边缘速度高的分布规律，而且扩张段中心存在一个较大的低速区。图 3 是扩张段至燃烧室底部的速度矢量图，从图中明显看出由于从喉口到扩张段，横截面积突然增大，使得在该区域出现了一个较大的回流区，这个回流的产生是 BSK 顶燃式热风炉的关键技术所在。一方面回流的存在能够起到稳定火焰的作用，使得新补充的煤气和空气始终能和高温气体接触；另一方面扩张段中心的回流切断了火焰的中心发展，而使火焰沿着锥角空间扩张，从而缩短了火焰长度，使火焰不会对蓄热室格子砖产生冲击。当气体通过扩张段到达燃烧室底部时，竖直向下的速度基本占主导，这有利于高温烟气的热量利用。燃烧室底部中心与边缘的速度差异缩小，均匀性较好。

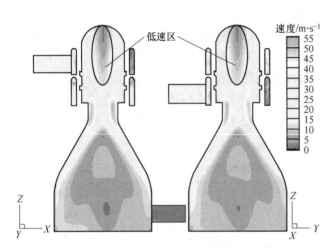

图 2　燃烧室速度分布云图

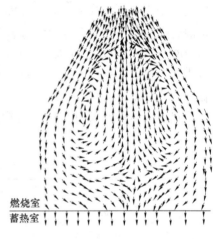

图 3　扩张段速度矢量图

3.3.2　浓度分布

从图 4 可以看出，预燃室上部包括半球拱顶内均被煤气充满，一氧化碳浓度和煤气入口浓度一致。在预燃室下部，助燃空气从两层空气喷口喷出，包裹煤气，并依靠旋流作用切割煤气燃烧。由流场计算结果与分析可知，此处气流速度中等，混合强度一般，因此预燃室下部中心区域的一氧化碳浓度依然很高。

在喉口区域由于靠近炉墙处的速度大于中心区域，即外围空气的速度远大于中心煤气速度，因此在喉口处，空气以较大速度切割中心煤气，发生强烈混合和燃烧，从图中也能看到在喉口处一氧化碳浓度急剧下降。在扩张段混合空间的增大使得煤气和助燃空气进一步混合、燃烧，扩张段中心的回流也切断了煤气的中心流动，使煤气向边缘流动，更好地与空气混合。图 4 左图中 $Y=0$ 平面扩张段底部，一氧化碳浓度分布略微不对称，右侧一氧化碳浓度要稍高于左侧，主要原因是上部各喷嘴流量不均匀，本文将在第 4 部分对喷嘴结构进行优化。总体来说浓度分布较为合理，煤气燃烧充分。

3.3.3　温度分布与火焰形状

燃烧室的温度分布与浓度分布密切相关，对于图 4 中一氧化碳浓度很高的区域，由于没有发生燃烧，温度很低。例如预燃室上部包括半球拱顶和预燃室下部的中心区域。

喉口区域由于煤气和助燃空气发生了较为强烈的混合和燃烧,因此从喉口处温度便迅速升高,甚至中心区域的煤气也从原来的215℃逐渐被加热到600℃。

扩张段向下,更多的煤气发生燃烧反应,炉内温度迅速升高,在扩张段底部温度基本趋于稳定。图 5 是燃烧室底部的温度分布云图,由于燃烧充分,燃烧室底部均处于1350℃以上的高温状态,另外一氧化碳浓度存在轻微的不均匀,使得左右两侧温度不是完全均匀,通过对底部截面温度进行面积分计算,求得底部截面的平均温度为1449.7℃,整体温度分布状况良好。

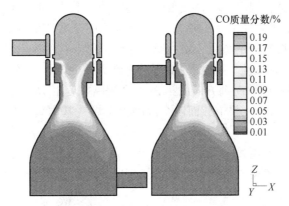

图 4　燃烧室浓度分布

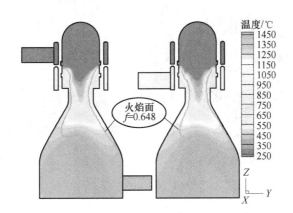

图 5　燃烧室温度分布和火焰面形状

根据燃烧 k-ε-g[4] 模型,需引入混合分数的概念对温度场和浓度分布进行求解。顶燃式热风炉的燃烧是非预混燃烧,非预混燃烧可以看作简单化学系统,温度与混合分数呈线性关系。计算得出当平均混合分数 $f<0.648$ 时,温度随着平均混合分数的增大线性递增;当 $f>0.648$ 时,温度随着平均混合分数的增大线性递减;仅当 $f=0.648$ 时,温度达到最大的1487℃, $f=0.648$ 对应的区域即为火焰面。图 5 中线条即表示火焰面轮廓线。

火焰面的形状与炉内煤气和助燃空气的混合、燃烧情况、温度场、浓度分布都是相吻合的。在预燃室的下部,火焰面反映出煤气与助燃空气的交界面,由于该处产生的少量高温烟气将周围的助燃空气和煤气加热了近200℃,因此此处温度大约为1250℃。在扩张段底部沿火焰面高度方向从上向下,气体温度呈一定温度梯度逐渐上升;靠近热风出口附近，一氧化碳浓度高于燃烧室另一侧，因此火焰向热风出口侧发生了轻微偏斜，火焰面以下温度基本保持稳定。

4 燃烧器结构的设计优化

燃烧室底部煤气浓度和烟气温度分布不均匀主要是燃烧器空气喷嘴流量不均匀导致的，根据相关文献对类似问题的研究[5]可知，调节空气喷嘴的截面积能够调节空气的流量，在调整个别喷嘴的情况下，气体质量流量会随着喷嘴截面积的增大而增大。

图6（a）是优化前空气第一层喷嘴一氧化碳浓度分布，根据其浓度分布情况分析，需要减小第三象限助燃空气的流量，增大第二象限助燃空气的流量。各喷嘴大小在设计时完全一样，宽度均为160mm，现将第三象限顺时针方向的第2~4个喷嘴宽度由160mm减小到140mm，将第二象限顺时针方向的第3~5个喷嘴宽度由160mm增大到180mm。优化后的结果如图6（b）所示。整个截面四个象限上空煤气基本呈对称分布，混合的均匀性得到大幅度提高。

图7是优化后燃烧室温度分布和火焰形状，对比图5中优化之前的温度场和火焰形状，不仅火焰偏斜的现象得到改善，而且扩张段以下的温度分布变得更加均匀。

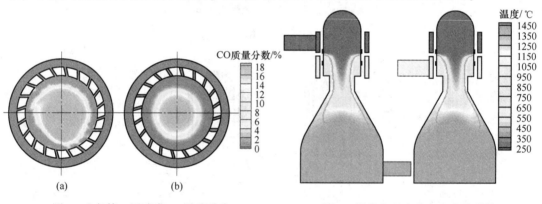

图6 空气第一层喷嘴CO浓度分布　　　　图7 燃烧室温度分布和火焰形状

5 结论

（1）首钢京唐1号高炉BSK型顶燃式热风炉燃烧室内的速度分布情况较为合理，喉口处的高速气流加速了空煤气的混合，扩张段中心的回流区域对稳焰和缩短火焰长度起到了重要作用，蓄热室表面边缘速度略高于中心，整齐均匀性良好。

（2）空气从喷嘴喷出后以较大速度切割中心煤气，发生混合、燃烧。虽然热风出口一侧一氧化碳浓度要略高于另一侧，但总体来说浓度分布较为合理，煤气燃烧充分。

（3）燃烧室底部均处于1350℃以上的高温状态，一氧化碳浓度轻微的不均匀性，使得左右两侧温度不是完全均匀，通过对底部截面温度进行面积分计算，求得底部截面的平均温度为1449.7℃，整体温度分布状况良好。火焰面向热风出口轻微偏斜，基本达到短焰扩散燃烧的目的。

(4) 通过燃烧器喷嘴结构优化能大幅提高空气、煤气混合的均匀性,改善燃烧室内浓度、温度分布以及火焰形状。

致谢

本研究受到国家科技支撑计划项目"新一代可循环钢铁流程工艺技术——长寿集约型冶金煤气干法除尘技术的开发"(2006BAE03A10)的支持。

参考文献

[1] 项钟庸,郭庆第. 蓄热式热风炉 [M]. 北京:冶金工业出版社,1988:1-11,210-233.
[2] 张福明. 大型顶燃式热风炉的进步 [J]. 炼铁,2002,21 (5):5-10.
[3] Li Baokuan. Metallurgical Application of Advanced Fluid Dynamics [M]. Beijing:Metallurgical Industry Press,2004:1-6.
[4] 王应时,范维澄,周力行,等. 燃烧过程数值计算 [M]. 北京:科学出版社,1986,(1):72.
[5] 吴狄峰,程树森,赵宏博,等. 关于高炉风口面积调节方法的探讨 [J]. 中国冶金,2007 (12):55-59.

Research of Combustion Technology on High Efficiency and Long Campaign Life Top Combustion Hot Blast Stove

Zhang Fuming Hu Zurui Cheng Shusen Mao Qingwu Qian Shichong

Abstract: The technology advantage and characteristics are introduced in this paper with the oversized top combustion hot blast stove equipped in Shougang Jing tang 5500m^3 BF. The Combustion Mechanism of burners are analyzed by numerical methods, which discussed the combustion process in the top combustion stove as well as the velocity, temperature and concentration distribution. The combustion status is improved by optimize the construction of burner.

Keywords: dome combustion hot blast stove;burner;combustion mechanism;optimal design

Dome Combustion Hot Blast Stove for Huge Blast Furnace

Zhang Fuming Mao Qingwu Mei Conghua Li Xin Hu Zurui

Abstract: In Shougang Jingtang 5500m³ huge blast furnace design, dome combustion hot blast stove (DCHBS) technology is developed. DCHBS process is optimized and integrated, reasonable hot blast stove (HBS) technical parameters are determined. Mathematic model is established and adopted by computational fluid dynamics (CFD), the transmission theory is studied for hot blast stove combustion and gas flow, and distribution results of HBS velocity field, CO density field and temperature field are achieved. Physical test model and hot trail unit are established, verified the numeral calculation result through test and investigation. Computer 3D simulation design is adopted, HBS process flow and process layout are optimized and designed, combustion air two-stage high temperature preheating technology is designed and developed, two sets of small size DCHBSs are adopted to preheat the combustion air to 520-600 ℃. With the precondition of BF gas combustion, the hot blast stove dome temperature can reach over 1420 ℃. According to DCHBS technical features, reasonable refractory structure is designed, effective technical measures are adopted to prevent hot blast stove shell inter-crystalline stress corrosion, hot blast stove hot pipe and lining system are optimized and designed. After blow in, the blast temperature keeps increasing, the monthly average blast temperature reaches 1300 ℃ when burning single BF gas.

Keywords: dome combustion hot blast stove; high blast temperature; ceramic burner; high temperature combustion air preheating; CFD

1 Introduction

Hot blast temperature is an important technical feature and development orientation of modern blast furnace (BF) iron making. Increasing blast temperature can reduce fuel consumption and improve blast furnace energy utilization efficiency effectively. If the blast temperature can be increased by 100 ℃, the coke rate can be saved by 15-20kg/t, and beneficial conditions can be exerted on PCI.

In Shougang Jingtang 5500m³ BF design, three kinds of HBS technologies including internal HBS, external HBS and DCHBS are studied and analyzed. On the basis of Shougang DCHBS technology [1] and Russia Kalugin DCHBS [2], the advantages of two kinds of technologies are integrated, and BSK (Beijing Shougang Kalugin) new DCHBS technology is developed and designed. It is the first time worldwide to apply DCHBS in 5000m³ grade huge BF [3].

2 Study on of DCHBS

2.1 Optimized Integration of Technical Process for DCHBS

In the design of Shougang Jingtang 5500m³ BF, the technological equipment and operation conditions of existing 13 sets of huge BF above 5000m³ worldwide are studied and analyzed. The 4000m³ grade large BF worldwide mainly uses external HBS, with several exceptions which use internal HBS. The 5000m³ grade huge BFs all use external HBS. There is no previous application of DCHBS on BFs above 4000m³ worldwide.

Through modern research methods including HBS combustion, gas and air flow and heat transmission theory, computational fluid dynamics (CFD) simulation, physical model flow field analyze, pilot trial of cold phase and hot phase etc., a series of design and study is carried out on HBS, and BSK DCHBS technology suitable for huge BF is designed and developed.

BSK DCHBS combines the technical advantages of Shougang type and Kalugin type DCHBS. The main technical features include: (1) The circular ceramic burner of DCHBS is arranged at dome position which has wide operating condition adaptation, can fulfill the multi-operating conditions of gas and combustion air with large combustion power, high combustion efficiency and long service life. Circular ceramic burner uses cyclone pervasion combustion technology, which can ensure the complete mixing and combustion of combustion air and gas, increase the flame temperature and dome temperature. (2) Dome space is used as combustion chamber, independent combustion chamber is deleted, and heat stability of stove proper is improved. Ceramic burner is arranged at dome position, high temperature fume is distributed evenly under cyclone condition, the evenness and heat transmission efficiency of high temperature fume on regenerative chamber checker brick surface can be effectively increased. (3) Regenerative chamber uses high efficiency checker brick, checker brick channel diameter is relatively decreased, the heating surface of checker brick is increased, hot blast stove heat transmission efficiency is improved. (4) HBS fume waste heat is reused to preheat the gas and combustion air. The combustion air is preheated above 520℃ by preheating stove. The blast temperature can reach 1300℃ using BF gas only. (5) HBS high temperature and high pressure pipe system adopts low stress design philosophy. By pipe system and refractory material structure optimized design, stable transferring of 1300℃ high temperature hot blast can be achieved [4].

2.2 Determination of the Numbers for DCHBS

Most of 5000m³ grade huge BFs worldwide adopt 4 HBSs to improve the dependability of HBS system stable operation. Approved by oversea operation practice, blast temperature can be increased by 30℃ when 4 HBSs are operated under staggering parallel operating condition [5].

Considering the worldwide 5000m³ huge BF HBS operation practice, in order to achieve long service life, high efficiency and stable operation of BF, Shougang Jingtang 5500m³ BF is configured with 4 DCHBSs, adopting "two burning, two blasting" staggering parallel burning-blasting process. According to different HBS operation modes, reasonable HBS operating period is designated which can reduce the HBS heat accumulation. High blast temperature of 1300℃ can be achieved using BF gas only. When 4 stoves are under staggering parallel blasting, HBS burning time is 60min, blasting time is 48 min and change-over time is 12min.

2.3 Technical Specifications

The design blast temperature of Shougang Jingtang 5500m³ BF BSK DCHBS is 1300℃ (max. 1310℃), dome temperature is 1420℃ (max. 1450℃), and the design campaign life is above 25 years. The HBS fuel is BF gas only. HBS fume waste heat recovery unit is used to preheat the gas and combustion air. Two sets of small size stoves are provided to preheat combustion air, which can be preheated above 520℃. The HBS high temperature valves are cooled by demineralized water closed circulating cooling system. The HBS system burning, blasting and change-over can be controlled full-automatically. Staggering parallel burning-blasting mode is adopted, and blast temperature of 1300℃ can be achieved using BF gas as fuel only. The blast temperature can reach 1250℃ when using other burning-blasting modes [6]. Table 1 shows the main technical specifications of BSK DCHBS.

Table 1 Main technical specifications of BSK DCHBS

Item	Parameters
HBS number/set	4
HBS height/m	49.22
HBS diameter/m	12.50
Regenerative chamber section area/m²	93.21
Checker brick height/m	21.48
Checker brick channel diameter/mm	30
Design blast temperature/℃	1300 (max. 1310)
Dome temperature/℃	1420 (checker brick top max. temperature 1450)
Fume temperature/℃	Average368 (max. 450)
Combustion air preheating temperature/℃	520-600
Gas preheating temperature/℃	215
Blast flow volume/Nm³·min^{-1}	9300
Blast time/min	60
Burning time/min	48
Change-over time/min	12

Continued Table 1

Item	Parameters
Checker brick heating area/$m^2 \cdot m^{-3}$	48
Heating area of each stove/m^2	95885
Unit BF volume checker area/$m^2 \cdot m^{-3}$	69.73
Unitblast flow volume checker area/$m^2 \cdot (m^3 \cdot min)^{-1}$	41.24

3 Study on Transfer Theory of DCHBS

3.1 Combustion Feature CFD Study

Based on hydrodynamics mass conservation, momentum conservation and energy conservation theory, governing equations are built, original conditions and boundary conditions are set, numerical calculation is carried out through CFD numerical model on BSK DCHBS, transmission theory of BSK DCHBS burning process is studied, distribution of velocity flow field, concentration field and temperature field inside combustion chamber of BSK DCHBS is analyzed. The design structure of HBS and burner is optimized according to study result and the distribution of HBS high temperature fume inside combustion chamber is more even. Fig. 1 to Fig. 4 are the simulation calculation results of velocity field, flow curve field, CO concentration field and temperature field inside combustion chamber of BSK DCHBS [7].

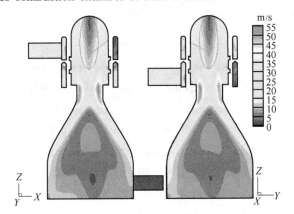

Fig. 1 Flow velocity field distribution of DCHBS   Fig. 2 Flow curve distribution of DCHBS

3.2 Physical Model Test

In order to verify the correctness of numerical calculation result, according to original design model parameters, original conditions and boundary conditions of numerical calculation, plexiglass physical model of HBS burner is built following similarity principle, physical simulation test unit of high efficiency cyclone pervasion DCHBS is built in 1 : 10 scale of geometry size.

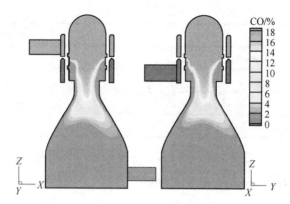

Fig. 3 CO concentration field distribution of DCHBS

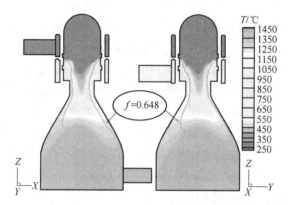

Fig. 4 Temperature field distribution of DCHBS

Through physical model test study, the following targets can be achieved: (1) the velocity, pressure and temperature etc. of physical model test unit can be measured, and compared with CFD numerical calculation result in order to verify the correctness of numerical calculation results. (2) accordingto similarity principle, normal temperature air is used to simulate BF gas, heated air is used to simulate combustion air, the flow field of physical model is put into second simulation zone by adjusting the flow rate of combustion air and gas, Euler Dimensionless Number is basically unchanged, actual pressure loss of BSK HBS is measured according to measured pressure loss in order to compare and verify the flow field numerical calculation result. (3) through measuring the gas flow rate at different injection position of circular burner, the evenness of gas distribution is verified. (4) by adjusting the structure type and angle of combustion air and gas nozzle, the gas flow rate is evenly distributed; actual ceramic burner of HBS design is optimized. The numerical calculation results are verified by physical model simulation test.

3.3 Hot State Simulation Trial

The HBS hot state simulation test is to investigate in more details the transmission theory of DCHBS combustion, fluid dynamic and heat transmission. Two hot trial stoves are built for

Shougang Jingtang 5500m³ BF HBS in 1 : 10 scale, "one stove burning-one stove blasting" mode is adopted to hot simulate the actual working conditions of HBS.

At two hot trial stoves, 289 temperature measuring points and corresponding gas flow measuring points are equipped, and several HBS hot trials are carried out. By online monitoring the regenerative chamber checker brick temperature change under burning condition and blasting condition in trial HBS, considering the pressure measuring and fume element measuring, temperature field distribution and burning condition inside trial stove can be analyzed. The HBS hot trial can verify directly the numerical calculation results of DCHBS burning and heat transmission process, and can provide theoretical and test basis for actual HBS optimized design and reasonable checker brick configuration.

4 Research and Design

4.1 3-D Simulating Design

In the design, HBS process arrangement is studied, and new rectangular process arrangement is designed and developed considering existing Shougang rectangular layout. The new HBS rectangular process layout has compact process layout, smooth and short flow process, reducing the hot blast main length significantly. 4 stoves are arranged in non-symmetrical rectangular type, and hot blast vertical pipe is arranged outside the rectangular area of 4 stoves. Independent HBS frame is arranged to support pipe and valves. One bridge crane can fulfill all the equipments maintenance requirement in HBS area.

As DCHBS burner is arranged at dome position, the installation positions of all pipes and valves are relatively higher. Influenced by the HBS heat expansion, the pipe system stress and heat expansion shifting is more complicated, and the requirement on pipeline system design and stress calculation is more accurate. Especially the pipeline system design in HBS area, reasonable pipe stress, easy installation, operation and maintenance shall be considered generally. In order to improve design accuracy and arrange pipe reasonably, 3-D computer aid optimized design is adopted, compact arrangement and reasonable flow of HBS is achieved. Fig. 5 is the 3-D simulation design of DCHBS for No. 1 BF at Shougang Jingtang.

4.2 Combustion Air High Temperature Preheating Technology

In order to achieve 1300℃ blasting temperature with only BF gas, general technical measures are studied to increase HBS flame temperature and dome temperature. On the basis of combustion air high temperature preheating technology developed by Shougang[8], high efficiency long service life gas and combustion air two stage preheating technology is designed and developed[9].

Its main technical theory is as follows: separated hot tube heat exchanger is used to recover the HBS fume waste heat and preheat the gas and combustion air. The temperature of gas and combustion air can reach about 200℃ after preheating. This cycle is called 1st stage double

Fig. 5 3-D simulating design of DCHBS for No. 1 BF at Shougang Jingtang

preheating. Two sets of small size stoves for combustion air high temperature preheating are used to preheat the combustion air to above 520℃. The gas and combustion air used in combustion air preheating stoves are from hot tube heat exchanger 1st stage preheating. Two preheating stoves can be operated alternatively in order to preheat some combustion air used for HBS burning. The combustion air temperature can reach 1200℃ after preheating which is mixed with combustion air from 1st stage preheating. The mixed combustion air temperature is controlled at 520-600℃. This process is called 2nd stage preheating. This process is a self-recycling preheating process, the physical heat of combustion air and gas can be increased significantly, the HBS dome temperature can be increased to 1420℃ and over, the blast temperature can be increased effectively, general heat utilization of HBS can be improved remarkably. Fig. 6 is the combustion air and gas preheating system process flow scheme.

4.3 Circular Ceramic Burner

In order to fully utilize the dome space, completely burn the gas and distribute evenly the high temperature fume in regenerative chamber, cyclone pervasion combustion technology is adopted. The nozzles of circular ceramic burner gas and combustion air are arranged along tangent to periphery. Two rings of gas nozzles are arranged at the burner top with downward inclination, two rings of combustion air nozzles are arranged at burner bottom with upward inclination, the injected flow can be mixed inside pre-mixing chamber with certain velocity and revolve downwards, intensify the pervasion mixing of gas and combustion air, in order to achieve complete combustion of gas. In order to achieve the even distribution of gas, reasonable circular ceramic burner pre-mixing chamber and cone dome geometry structure is designed. Through fume flow shrinking, expansion, cyclone and return flow in dome space, the gas can be burned completely and high temperature fume can be distributed evenly. Fig. 7 is the circular ceramic burning design structure.

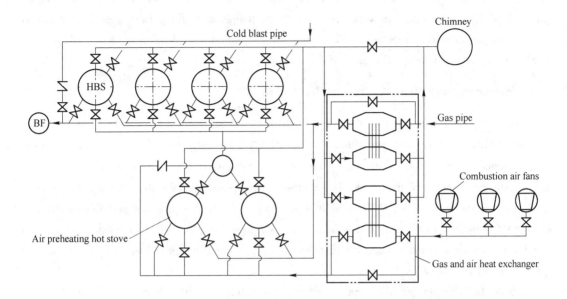

Fig. 6 Process flow sheet of preheating for combustion air and gas

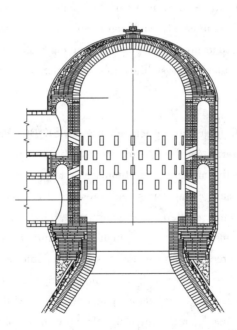

Fig. 7 Design construction of circle ceramic burner for BSK DCHBS

BSK DCHBS circular ceramic burner is completely suitable with the combustion condition of preheating combustion air to 600℃, and preheating gas to 200℃. The burner adopts cyclone pervasion burning type, has big regulation range of combustion air and gas flow rate, complete and even mixing. Before high temperature fume enters regenerative chamber checker brick, the combustion is complete, which can restrain effectively the section high temperature zone in dome and reduce NO_x emission. After burning, the high temperature fume flow field is distributed evenly and properly. The results of CFD numerical calculation and physical model simulation test verify that flow field inside HBS can fulfill the design requirement.

4.4 Intercrystalline Stress Corrosion Prevention

Intercrystalline stress corrosion exists widely in high blast temperature HBS. The main reason is that when HBS dome temperature reaches above 1400℃, the formation NO_x of increases quickly and its concentration reaches 350×10^{-6} or above. NO_x combines with the condensed water at HBS inner wall and forms corrosive acid. The corrosive acid penetrates deeply, expands and cracks along the crystal lattice from shell position with stress. The pulse pressure and tiredness stress during HBS operation enhances corrosion and cracking process. The HBS high temperature area shell intercrystalline stress cession becomes the restriction condition of influencing HBS campaign life and improving blast temperature.

In BSK DCHBS design, the max. design temperature of HBS dome is controlled within 1450℃, and normal operating temperature is controlled within 1420℃ in order to ensure safe temperature operation of HBS, control NO_x formation and prevent intercrystalline stress corrosion effectively. Also low alloy microlite corrosion proof steel plate is adopted, welding stress generated during shell manufacturing is eliminated, corrosion proof painting is applied at inner wall of HBS high temperature area shell, one layer of acid proof gunning material is used at corrosion proof painting in order to increase the intercrystalline stress corrosion proof ability of HBS shell.

4.5 Refractory Material

In order to fulfill the 1300℃ high blasting temperature of BSK DCHBS and prolong its campaign life, hot blast refractory material and bricklaying structure is designed and optimized. Low stress design is adopted, and effective measures to eliminate or reduce refractory material system heat stress, mechanical stress, phase change stress and pressure stress is used. According to working conditions of different HBS area, different refractory material is used to achieve the optimized selection of HBS refractory material. In HBS high temperature area, silicon bricks with high temperature proof and good creep performance are used. In middle temperature zone, and alusite bricks and low creep clay bricks are used. In low temperature zone, high density clay bricks are used. The optimized configuration of refractory material adapts with the working characters of DCHBS, improves the economic rationality of refractory material and reduces the project investment.

Low creep mullite bricks and insulation bricks composite bricklaying structure is adopted for hot blast main, hot blast branch and hot blast bustle. Two layers of indefinite form refractory material are used at internal surface of pipe shell. One layer of ceramic fiber felt is filled between pipe upper bricks and gunning material. The expansion joints at hot blast pipe bricklaying are filled with ceramic fiber material which can resist 1420℃ high temperature and absorb the heat expansion of refractory material lining stably. In order to prolong the service life of hot blast main and branch, one layer of acid proof painting to prevent intercrystalline stress corrosion is applied at internal wall of hot blast pipe shell.

As working under several bad operational conditions, all holes and openings of HBS are the weak points for HBS long life and blast temperature improving. In order to prolong the service life of HBS holes and openings, composite brick structure is applied at hot blast outlet, T-section of hot blast pipe, gas inlet and combustion air inlet etc. The hot blast outlet is composed of independent ring type composite brick. Among composite bricks, biconcave tenon groove structure is used for intensifying. Above composite bricks, semi-ring special arch bricks are used to reduce the pressure stress generated by upper stove wall brick lining on composite bricks.

4.6 Hot Blast Pipe

The hot blast branch, main and bustle of HBS is high temperature and high pressure pipe, and is the most complicated system of HBS pipeline system. Hot blast pipe is the important tache to achieve high blast temperature, is the technical guarantee of stable transferring high temperature blasting. In the design, through HBS piping system stress calculation analyse, the expansion displacement caused by operating temperature, environmental temperature and operating pressure is considered, low stress pipe design philosophy is adopted, hot blast pipe is optimized in order to fulfill the 1300℃ high temperature blasting requirement. Different structure bellow compensator and tie-rod are arranged reasonably. The bellow compensator position of hot blast branch is arranged between hot blast valve and hot blast main. At the end of hot blast main, pressure equalizing type bellow compensator is arranged to optimize the pipe support and provide technical assurance for hot blast pipe stable operation.

5 Application

Shougang Jingtang No.1BF was blown-in and start-up on May 21st, 2009. After blow-in, the production is stable and smooth, and all operation technical parameters keep improving. The max. daily output is 14245t/d, and the monthly average productivity is 2.37t/($m^3 \cdot d$), fuel ratio is 480kg/t, coke rate is 269kg/t, PCI rate is 175kg/t, blast temperature is 1300℃, all of which have reached the expected design level. In order to fully exert the technical advantages of BSK DCHBS。 BF operation keeps optimizing and blast temperature is increased steadily. On Dec. 13th 2009, the BF blasting temperature exceeds 1300℃, average blast temperature in December 2009 reached 1281℃. In March 2010, the average blast temperature reached 1300℃. The design value of BF blasting temperature at 1300℃ continuously and stably with only BF gas is achieved, and new record of huge BF high blast temperature operation practice is created, and reaches the advanced level of worldwide 5000m^3 grade huge BF high blasting temperature. Table 2 is the Main operating technical parameters of No. 1 BF at Shougang Jingtang.

Table 2 Main operating technical parameters of No. 1BF at Shougang Jingtang

Time (year-month)	Average daily output /tHM·d⁻¹	Productivity /tHM·(m³·d)⁻¹	Coke rate /kg·tHM⁻¹	PCI rate /kg·tHM⁻¹	Fuel rate /kg·tHM⁻¹	Blast temperature /℃	Process energy consumption /kgce·t⁻¹
2009-5	4840	0.88	551	83	634	914	799
2009-6	7425	1.35	503	62	565	998	538
2009-7	8525	1.55	483	49	532	1063	461
2009-8	11000	2.01	372	94	481	1166	409
2009-9	11660	2.12	354	101	483	1212	419
2009-10	12210	2.22	340	117	488	1262	414
2009-11	12500	2.27	299	145	484	1276	406
2009-12	12694	2.31	288	149	479	1281	393
2010-1	12657	2.30	307	137	482	1259	388
2010-2	12847	2.34	287	161	482	1277	375
2010-3	13035	2.37	269	175	480	1300	373
2010-4	12147	2.21	289	156	482	1275	381

6 Conclusions

(1) In Shougang Jingtang 5500m³ huge BF, DCHBS technology is integrated and innovated, the design blast temperature is 1300℃. Numerical calculation, physical model and hot simulation test are adopted, and theoretical study on HBS combustion, gas flow and heat transmission process is carried out. 3-D simulation design is adopted, HBS process flow and process layout is optimized, several HBS high efficiency long campaign life technologies are studied and developed.

(2) DCHBS proper and circular ceramic burner design structure is optimized. High performance refractory material is configured reasonably. In order to effectively restrain the HBS shell intercrystalline stress corrosion and control reasonable dome temperature, acid and corrosion proof insulation measures are taken at HBS high temperature area shell and hot pipe inner wall.

(3) After blow-in, the monthly average blast temperature of Shougang Jingtang No. 1 BF reaches 1300℃ burning only BF gas, creating new record of huge BF high blasting temperature operation and reaching international advanced level of 5000m³ grade huge BF high blasting temperature operation.

Acknowledgements

Mr. Qian Shichong, Mr. Yin Guangyu, Ms. Ni Ping, Mr. Han Xiangdong, et al. From Beijing Shougang International Engineering Technology Co., and Professor Cheng Shusen from University of Science and Technology Beijing also take part in this issue. Sincere thanks for the great support from Shougang Jingtang Company and Zhengzhou Aneca Company. Thanks for the help from Mr. Wu Qingsong.

References

[1] Zhang Fuming. Technology progress of large dome combustion hot stove in China [J]. Ironmaking, 2002, 21 (5): 5-10.

[2] Iakov Kalugin. High temperature shaftless hot air stove with long service life for BF [C]//AISTech 2007 Proceedings-Volume I. Chicago: AIST, 2007: 405-411.

[3] Zhang Fuming, Qian Shichong, Zhang Jian, et al. Design of 5500m^3 blast furnace at Shougang Jingtang [J]. Journal of Iron and Steel Research International, 2009, 16 (Supplement2): 1029-1034.

[4] Zhang Fuming, Qian Shichong, Zhang Jian, et al. New technologies of 5500m^3 blast furnace at Shougang Jingtang [J]. Iron and Steel, 2011, 46 (2): 12-17.

[5] Peter Whitfield. The advantages and disadvantages of incorporating a fourth stove with in an existing blast furnace stove system [C]//AISTech 2007 Proceedings Volume I. A Publication of the Association for Iron and Steel Technology, 2007: 393-403. Indianapolis, Indiand, U.S.A

[6] Mao Qingwu, Zhang Fuming, Zhang Jianliang, et al. Design and research of new type high temperature top combustion hot stove for super large sized blast furnace [J]. Ironmaking, 2009, 29 (4): 1-6.

[7] Zhang Fuming. Transfer theory and design research on long campaign life and high efficiency in hot blast stove [D]. Beijing: University of Science and Technology Beijing, 2010: 171-187.

[8] Zhang Fuming. Research and innovation of the key technologies at modern large blast furnace [J]. Iron and Steel, 2009, 44 (4): 1-5.

[9] Mei Conghua, Zhang Weidong, Zhang Fuming, et al. A high blast temperature and long service life facility of two stage preheating for combustion air and gas: China, ZL200920172956.4 [P]. 2010-07-14.

Research on High Efficiency Energy Conversion Technology for Modern Hot Blast Stove

Zhang Fuming　Li Xin　Hu Zurui

Abstract: High blast temperature is one of the important technical characteristics of modern blast furnace (BF), which is also an important technical approach for the green development of ironmaking. Increase blast temperature can improve and promote the BF operation smooth and stable, reduce coke rate, fuel consumption and CO_2 emission. Top combustion hot blast stove technology has been applied in Shougang Jingtang's 5500m^3 BF for 8 years. Under the condition of burning single BF gas, high efficiency energy conversion and over 1250℃ high blast temperature have achieved. The combustion and heat transformation process were researched by numerical simulation technology to optimize the structure design. The high efficiency annular ceramic burner and checker brick were developed and applied, the flue gas waste heat was recovered and reused to preheat the combustion air and gas. Prominent achievement of high blast temperature under the condition of single BF gas burning has been realized.

Keywords: blast furnace; hot blast temperature; top combustion hot blast stove; waste heat recovery; combustion; energy transformation

1　Introduction

Reducing energy consumption and cost of ironmaking process are very important to improve the market competitiveness of iron and steel industry,[1] because energy consumption of ironmaking process accounts for about 70% of the total energy consumption of iron and steel manufacturing, and cost of pig iron accounts for about 50% the total cost of iron and steel manufacturing. High temperature blast provides about 20% of the heat in BF smelting, and high blast temperature is an important technical characteristic of modern BF ironmaking, which is also an important way to realize high efficiency energy conversion. High blast temperature has the following important meaning:[2]

(1) Reducing fuel consumption, saving coke, reducing the production cost.

(2) Increasing the temperature of tuyere raceway, increasing quantity of pulverized coal injection, further reducing coke rate.

(3) Increasing the hearth heat, which is helpful for the BF operational performance.

本文作者：张福明，李欣，胡祖瑞。原刊于 2019 TMS Conference Proceeding（Energy Technology 2018-Carbon Dioxide Management and Other Technologies）. The Minerals, Metals & Materials Society, Phoenix, 2019：133-152.

(4) Improving production efficiency, reducing CO_2 emissions.

Chinese steel workers have studied the key technologies of hot blast stove from their respective needs. The research they have made are as follows: The research for the reason caused the piping broken and optimizing design[3], Study on standards of refractory used for hot blast pipes between blast furnace and hot blast stove[4], The optimization of hot blast stove quantity and operation[5], The optimization of hot blast stove control system[6], High radiative coating used in checker bricks[7].

Acquisition, transmission and utilization of high blast temperature is a systematic project, which is composed of a series of key technologies, including low heat-value gas utilization technology, high efficient and clean combustion technology, high efficient heat transfer technology in regenerator, low stress piping system, etc. This paper focused on joint research which combined numerical simulation and experimental research, and also focused on analyzing and verifying the influence of checker brick on heat transfer process. The purpose of the research is to increase the energy conversion efficiency of hot blast stove.

2 Promote Blast Temperature by Low Heat Value BF Gas

2.1 Technological Philosophy

Shougang Jingtang steel plant is an annual production capacity of approximate 10 million tons of large iron and steel union enterprises, product positioning for high grade plate and strip, cold rolled products ratio is 54.6%, the steel products processing facilities complete, so the high calorific value of coke oven gas, converter gas to supply in steelmaking, rolling process, hot blast stove only the low calorific value of BF gas as fuel. With the progress of BF operational technology, the utilization rate of gas is continuously increasing, and the calorific value of BF gas is decreasing gradually, which is about 3000-3200kJ/m^3. The top combustion hot blast stove can only reach 1200-1300℃ with the single BF gas, and so it is difficult to reach the high blast temperature above 1250℃.[8]

Shougang Jingtang's 5500m^3 BF in the design process, in order to realize the high blast temperature reaches 1250-1300℃[9], which fuel of hot blast stove only the BF gas, a large amount of research, comprehensive analysis and demonstration of various technical measures of high temperature at home and abroad have been applied, the design and development of BF gas and combustion air, preheating process with innovation.[10,11]

2.2 Theoretical Combustion Temperature and Dome Temperature

During the hot blast stove burning, the main source of heat is the chemical heat of the gas, and the preheated temperature of the combustion air and gas is fixed. When all the heat is used to heat the combustion products, no other heat losses, the temperature which can reach called the theoretical combustion temperature of hot blast stove.

During the operation of the hot blast stove, actually the dome temperature of hot blast stove is lower than the theoretical combustion temperature 30-50℃ because of heat dissipation of stove wall and incomplete combustion. Practice has proved that the theoretical combustion temperature is the main contradiction to limit the increase of blast temperature in hot blast stove, and the formula is shown in equation (1).

$$t_T = \frac{Q_A + Q_G + Q_{DW}}{C_P V_P} \tag{1}$$

In equation t_T——Theoretical combustion temperature, ℃;

Q_A——Combustion air sensible heat, kJ/m³;

Q_G——Gas sensible heat, kJ/m³;

Q_{DW}——Low calorific value of gas, kJ/m³;

C_P——Heat capacity of combustion product at t_T, kJ/(m³·℃);

V_P——Volume of combustion product, m³.

It can be seen from equation (1), that the theoretical combustion temperature of the hot blast stove can be improved by increasing the physical heat (sensible heat) of the combustion air and gas. Increase the preheat temperature of combustion air and gas can improve their physical heat effectively. Fig. 1 shows the theoretical combustion temperature under the condition of preheating combustion air and gas.

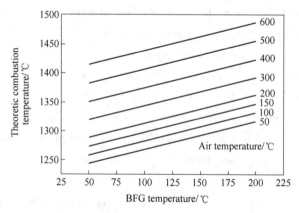

Fig. 1 The theoretical combustion temperature of hot blast stove under the condition of combustion air and gas preheating

During the blasting period, the cold blast absorbs heat from checker brick, then flows into the hot blast pipeline and blast into the BF through the tuyere. The hot blast temperature which is received by the BF is called the blast temperature. During the whole blasting period, the hot blast temperature at the hot blast outlet is gradually decreased with the increase of the heat exchange between the cold blast and the checker brick. BF operation requires constant blast temperature, so we usually mix cold blast into the hot blast pipeline system, and maintain the constant blast temperature by adjusting the mix cold flow at different stages of blasting.

2.3 High Efficiency Preheating Technology

It is a worldwide problem to realize high temperature only by burning BF gas. The blast temperature of Japan, Europe and other advanced BFs have reached 1250℃, were blended with some high calorific value gas (such as natural gas, coke oven gas, converter gas and so on). Research shows that, the dome temperature of hot blast stove is depending on the theoretical combustion temperature, according to theoretical combustion temperature calculation formula can be found that improve the physical heat of gas and air are the effective technical measures to improve the flame temperature at the same gas calorific value. According to the combustion calculation, the BF gas should be preheated to 200℃ and the combustion air should be preheated to over 500℃ to achieve high blast temperature at 1300℃.[12]

At the beginning of 21st Century, the combustion air preheating technology was adopted to increase dome temperature and achieve industrial application success in Shougang. The existing 4 sets of Shougang's type top combustion hot stove which belong to Shougang's No. 2 BF were put into production in 1979. No. 2 BF rebuilt in 2002, focusing on the innovative design of high temperature combustion air preheating process, mainly in the following aspects:

(1) Remove two old top burning hot stoves, No. 1 and No. 2, which have been running for more than 23 years. No. 3 and No. 4 top combustion hot blast stoves were used as preheating furnace for high temperature preheated combustion air.

(2) The three high temperature internal combustion hot stoves introduced from Corus were applied, which combustion air was preheated to 600℃ by the original Shougang's top combustion stove. The blast temperature reaches 1250℃ only by burning BF gas. Fig. 2 is a photo of Shougang's No. 2 BF after renovation.

Fig. 2 Photo of Shougang's No. 2 BF hot blast stove

2.4 Preheating Process Integration

The comprehensive advantages and disadvantages of various preheating process, and combined with a large number of engineering practice in the field of accumulation, in the design of hot blast stove preheating system of 5500m³ BF of Shougang Jingtang steel plant, the longevity of independent intellectual property rights of the two stage double preheating system. The technology is developed and designed successfully on the basis of high temperature preheating technology of combustion air in shougang. The main technical principle is: the separated type heat pipe heat exchanger, using hot BF flue gas waste heat preheating gas and combustion air is preheated gas and combustion air temperature can reach 200℃, this process is called a double preheating. Two sets of regenerative combustion air temperature preheating stove used to preheat the combustion air, the temperature is raised to 550℃. The combustion air preheating furnace gas and combustion air through the heat pipe heat exchanger preheating, by burning through a preheated BF gas to preheat furnace after heating, the dome temperature can reach 1300℃, and then used for combustion air heating hot blast stove use, after combustion air temperature after heating temperature can reach 1200℃ for export and then, after a mixed combustion air preheating, the combustion air to control the temperature of 550-600℃, this process is called two preheating. This is a process of self-circulating preheating process, significantly improve the physical heat of air and gas, the dome temperature of hot blast stove also improved, which can effectively improve the air temperature, hot blast stove system overall efficiency is significantly improved. Fig. 3 shows the two stage dual preheating system with high blast temperature in the hot blast stove system.

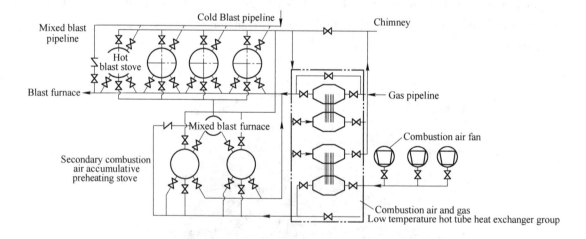

Fig. 3 Process flow chart of two stage double preheating system for hot blast stove

With the development of plate heat exchanger, it has shown many advantages, such as longer service life, less fouling, and gradually began to replace the traditional tubular heat exchanger.

The system has been put into operation in May 2009, and the practice has proved that the system can satisfy the thermal air temperature of 1250-1300℃ under the condition of using a single BF gas.[13]

2.5 Comparison of Efficiency of Hot Blast Stove

The two processes of two stage double preheating and only heat exchanger preheating are calculated and compared:

(1) The thermal efficiency of the main body of the hot blast stove.

The two process gas temperature is basically the same, the two stage double preheating process of blast temperature is 1300℃ higher than the low temperature preheating 1200℃, so the body slightly high thermal efficiency of hot blast stove. The thermal efficiency of the two stage double preheating process was calculated to be 77.01%, and the thermal efficiency of the hot blast stove was 74.84% at the low temperature preheating process, and the former was 2.17% higher than that of the latter. [14]

(2) The cooling, heat dissipation of the shell, pipe and valve of the hot blast stove system. Table 1 shows the comparison of the heat dissipation of two kinds of preheating process.

Table 1 Heat dissipation of different preheating process

Process	Shell/kJ · (Nm^3 blast)$^{-1}$	Pipe/kJ · (Nm^3 blast)$^{-1}$	Valve cooling /kJ · (Nm^3 blast)$^{-1}$
Two stage high temperature preheating	52.58	71.23	17.71
Low temperature preheating	44.42	57.12	8.51
Different/%	15.5	26.2	51.9

(3) Comparison of thermal efficiency of hot blast stove system.

Overall, the two stage high temperature double preheating process has improved 100℃ blast temperature, reducing coke rate 15kg/tHM, but low temperature preheating process to consume 34313 Nm^3/h gas, the thermal efficiency of the hot blast stove system is reduced by 0.29%, but for the entire system of BF, the energy consumption is reduced (Table 2).

Table 2 Comparison of different preheating process

Process	Blast temperature /℃	BF gas flow /Nm^3 · h^{-1}	Air flow /Nm^3 · h^{-1}	Coke rate /kg · tHM^{-1}	Thermal efficiency of hot stove/%	Thermal efficiency of the hot blast stove system/%
Two stage high temperature preheating	1300	338089	218406	290	77.01	83.76
low temperature preheating	1200	303776	196240	305	74.84	84.05
Different/%	100	34313	22166	-15	2.17	-0.29

3 Numerical Simulation and Experimental Study

Burner is very important as the core equipment of hot blast stove. During the research and design of huge high temperature top combustion hot blast stove, hot blast stove accurate design system have been established, which is based on the theoretical research, experimental investigation, industrial trial, 3-D precision design. The system integrated application of a variety of advanced research methods, and developed several calculation programs independent. Research group not only analyzed and calculated the hot blast stove theoretically, but also did the cold/hot test and industrial test to verify the calculation results. The 3-D design method also was used to enhance design efficiency and precision, and to optimize design scheme.

3.1 Numerical Simulation of Top Combustion Hot Blast Stove

The combustion type in the dome combustion hot blast stove is turbulent non-premixed combustion, which is simulated by the mixed fraction k-ε-g model. The control equations include:

Continuity equation:

$$\frac{\partial \rho}{\partial t} + \nabla \cdot (\rho U) = 0 \tag{2}$$

Momentum equation:

$$\frac{\partial \rho U}{\partial t} + \nabla \cdot (\rho U \times U) - \nabla \cdot (\mu_{\text{eff}} \nabla U) = - p' + \nabla \cdot [\mu_{\text{eff}} (\nabla U)^T] + B \tag{3}$$

In the equation, ρ is density of main stream; U is average velocity of main stream; μ_{eff} is effective viscosity, which is defined as: $\mu_{\text{eff}} = \mu + \mu_T$, μ_T is eddy viscosity, B is body force.

k-ε equation:

$$\frac{\partial \rho k}{\partial t} + \nabla \cdot (\rho U k) - \nabla \cdot \left[\left(\mu + \frac{\mu_T}{\sigma_k}\right) \nabla k\right] = P + G - \rho \varepsilon \tag{4}$$

In the equation, k is the turbulent kinetic energy, ε is the turbulent dissipation rate, P is shear generated, G is volume generated, σ_k and σ_ε respectively represent Prandtl number of k and ε. In the calculation, $\sigma_k = 1.0$, $\sigma_\varepsilon = 1.3$. C_1, C_2 and C_3 are empirical constant, $C_1 = 1.44$, $C_2 = 1.92$, $C_3 = 0.09$.

Considering that combustion in hot blast stove is diffusion combustion, a non-premixed PDF model is used. The main fuel of hot blast stove is BF gas, which produce CO_2, H_2O and N_2 after combustion. CO_2 and H_2O belongs to polar molecules gas, within a certain wave length radiation and energy absorption. In common engineering combustion equipment, radiation heat transfer is the main method to transfer heat, up to 90%. So radiation model must be used when we simulate the combustion in hot blast stove. For radiation heat flux q_r:

$$- \nabla q_r = aG - 4a\sigma T^4 \tag{5}$$

The expression of $-\nabla q_r$ can be directly introduced to the energy equation, thereby obtaining the heat source caused by the radiation. The absorption coefficient and scattering coefficient of a given mixture are required for the use of this radiation model. Because the mixture contains CO_2 and H_2O, the absorption coefficient is calculated by using the gray gas weighted average model (WSGGM). Boundary conditions: set flow, temperature and composition of air and gas inlet, resistance coefficient of checker brick.

Through iterative calculation, pressure distribution, velocity distribution, temperature distribution, CO concentration distribution and flow field distribution of the dome combustion hot blast stove are obtained, as shown in Fig. 4.

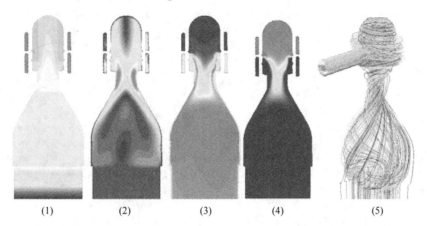

Fig. 4   Results of CFD simulation for top combustion hot blast stove
(1) Pressure distribution; (2) Velocity distribution; (3) Temperature distribution;
(4) CO concentration distribution; (5) Flow field distribution

It is found by calculation that, after flowing into the premix chamber, velocity of air and gas increase quickly when they down through the throat because of flowing area decreases which is caused by shrinkage throat. A mixture of air and gas here is greatly enhanced, and nonuniformity caused by upper stream is also effectively reduced. Through the strengthening mixture of the throat, velocity distribution, temperature distribution and concentration field distribution in the combustion chamber are axisymmetric. CO is fully burned in the combustion chamber before flowing into the checker brick, avoid the occurrence of secondary combustion within the checker brick and to extend the service life of checker brick. The uniform swirling flow field in dome combustion hot blast stove avoids the production of local high-temperature zone, which can produce a large number of NO_x.

3.2 Cold Blast Distribution Uniformity of Top Combustion Hot Blast Stove

During blasting period, the cold blast flow into the bottom of the stove through the branch pipe, and then flow into the regenerator chamber. After exchanging heat in regenerator chamber, the cold blast become hot blast, and leave the hot blast stove from hot blast outlet. In order to improve the efficiency of the use of regenerator chamber, and make the heat in regenerator chamber to be taken away evenly and fully as much as possible, cold blast should be distributed evenly into the chamber, avoid nonuniformity.

Three schemes are designed: (1) Single cold blastinlet pipe; (2) Single cold blast inlet pipe with guide plate; (3) Using the two waste gas branch pipes as cold blast inlet pipes. After analyzing simulation results of three schemes, it is found that if used single cold blast inlet, whether guide plate used or not, there are wide range of high speed area and low speed area near the inlet. If two waste gas branch pipes are used as cold blast inlets, it could decrease the high speed area and low speed area obviously. The distribution of cold blast is more uniform, which is beneficial to improve the heat exchanging efficiency of the regenerator chamber. Fig. 5 shows the velocity distribution of cold blast in the blasting chamber.

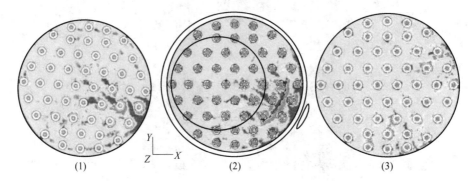

Fig. 5 Velocity distribution of cold blast in the blasting chamber
(1) Single cold blast inlet pipe; (2) Single cold blast inlet pipe with guide plate;
(3) Using the two waste gas branch pipes as cold blast inlet pipes

3.3 Industrial Cold Test of Top Combustion Hot Blast Stove

The cold industrial test for hot blast stove of Shougang Jingtang's 5500m³ huge BF was carried out to verify the numerical simulation. The testing content included air jet velocity, gas jet velocity (medium was air), velocity in throat, regenerator chamber upper velocity distribution, regenerator chamber lower velocity distribution, flow field inside the stove, etc.

Test results show that the distribution of air and gas among nozzles are not uniform. The speed of nozzles which are near the branch pipe is higher than others. The gas jet velocity is more uniform than air's. The test results are consistent with the numerical simulation, so we believe that the simulation is reliable. Because air and gas swirl down through the throat, and flow near the wall, the velocity near the wall is higher, and velocity in center of the combustion chamber is lower, which covers an area of about 10% of the regenerator chamber upper area.

3.4 Hot Condition Simulating Trial

For further study of transfer theory of combustion, gas flow and heat transfer process of dome combustion type hot blast stove, 2 hot blast stoves for hot condition trial based on prototypical of dome combustion hot blast stove of Jingtang's 5500m³ BF are established. The "one stove burning and one stove blasting" working mode is applied to simulate real working process of the hot blast stove.

The 2 hot blast stoves for hot condition trial are provided with 289 temperature meters and corresponding gas flow meters for many times of hot condition test of hot blast stove. On-line detection is used to test temperature variety of checker brick in the regenerator chamber at combustion and blasting statuses of hot blast stove, and with combination of pressure detection and fume composition detection to analyse temperature field distribution status and combustion status in the experimental hot blast stoves. The results of hot condition test of hot blast stove prove that, the results of the flow field numerical simulation of combustion chamber and the numerical simulation of regenerator chamber are correct. The simulation and the test provide theoretic and experimental basis for optimal design of the hot blast stove and configure checker brick reasonably.

4 Research on Regenerator Chamber

Regenerator chamber has the characteristics of periodic heat absorption and exothermic, and is another key part of hot blast stove. The design of checker brick and channel regenerator chamber, and selection of refractory material are emphasized during the design of hot blast stove.

4.1 Optimization of Checker Brick

The checker brick should meets the following conditions:
(1) Having large heatin gsurface, can satisfy the requirements of the heat transfer rate;
(2) Having enough filling rate, can satisfy the heat storage capacity;
(3) Having enough compressive strength, can support regenerator chamber;
(4) Checker bricks in high temperature zone have excellent creep resistance;
(5) Easy to be manufactured, low cost, quality is stable.

Fig. 6 shows the three kinds of checker bricks currently are used in a wide range. Table 3 shows the specification comparison of these three checker bricks.

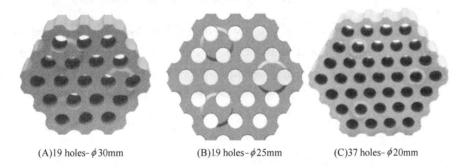

(A)19 holes-φ30mm (B)19 holes-φ25mm (C)37 holes-φ20mm

Fig. 6 Three types checker brick

Table 3 Specification comparison of three checker bricks

Items	A	B	C
Diameter of hole/mm	30	25	20
Numbers of hole	19	19	37
Filling rate	0.612	0.627	0.655
Active area	0.388	0.373	0.345

Continued Table 3

Items	A	B	C
Thickness of brick/mm	120	120	120
Heating area per volume/$m^2 \cdot m^{-3}$	48.6	56.1	64.7
Equivalent depth/mm	25.2	22.4	20.2
Thickness between holes/mm	17.3	15.2	13.5

Compare with checker brick A, checker brick C has smaller hole and living area. Under the same simulation condition, the pressure loss will increase with smaller hole. If we use checker brick A (ϕ30mm) as the benchmark, when the hole of checker brick decrease to 25mm, the pressure loss will increase 50%, and when the hole of checker brick decrease to 20mm, the pressure loss will increase 150%, as shown in the Fig. 7.

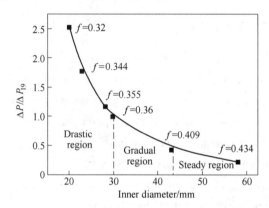

Fig. 7 Relationship between checker brick hole and pressure loss

In order to reduce pressure loss, a series of method should be used after using small hole checker brick, like increasing the diameter of regenerator chamber and flow area, and lowering gas velocity. The convective heat transfer coefficient of the fluid has a great relationship with the flow condition, which is showed by the Reynolds number. The calculation equation of the Reynolds number is:

$$Re = \rho v d / \mu \qquad (6)$$

In equation (6), ρ is density of the fluid, v is velocity, d is diameter of checker brick hole, μ is dynamic viscosity. Table 4 shows regenerator chamber specification comparison with 3 different checker bricks.

Table 4 Specification comparison with 3 different checker bricks

Items	Option A	Option B	Option C
Effective volume of BF/m^3	5500	5500	5500
Blast volume/$Nm^3 \cdot min^{-1}$	9300	9300	9300
Blast temperature/℃	1300	1300	1300
Diameter of regenerator chamber/m	10.89	10.0	10.89

Continued Table 4

Items	Option A	Option B	Option C
Height of regenerator chamber/m	17.48	22.5	21.48
Ratio of height to diameter	1.61	2.25	1.97
Diameter of checker brick hole/mm	20	25	30
Average Reynolds number	1504	2036	1980
Pressur eloss of regenerator chamber/Pa	616	783	446
Heatin garea of hot blast stove/$m^2 \cdot HBS^{-1}$	104452	96524	95885
Weight of checker bricks per hot blast stove/$t \cdot HBS^{-1}$	2164	2175	2435
Heating area of checker brick per volume of blast/$m^2 \cdot (m^3 \cdot min)^{-1}$	47.48	41.52	41.24
Height of hot blast stove/m	46.2	50	50.2

Option A uses checker brick with diameter 20mm hole. In this situation, the height of regenerator chamber should be less than 18m, in order to control the pressure loss, and avoid blocking the hole. So the diameter of the regenerator chamber should be increased to get enough checker brick weight for storing heat. Average Reynolds number of option A is just 1504, which is less than option B, when they have the same pressure loss. This is not good for heat convection.

If the checker brick diameter 20mm like option C is adopted, and keep the heating area as same as option A and B, the weight of regenerator chamber will increase about 12%. The Reynolds number is same as option B. So we can get balance easily between heating area and weight, and we also can get higher using efficiency of regenerator chamber when we use diameter 30mm checker brick like option B.

During the hot blast stove repair, it was founded that on the upper surface of the wall and inner surface of upper silicon checker brick there was a thick and loosened layer of residue which was formed by reaction of silicon bricks and dust under high temperature condition. Colors of the residue were earth yellow and dark gray, and the original color of silicon brick was pale yellow. After cooling to room temperature, the residue began to crack and peel from the checker brick. It was clean on the top surface of regenerator chamber and no residue, because the gas velocity here was too high to keep dust staying. But silicon checker brick was eroded because of being long-term exposed under high temperature and gas condition. The relatively weak part of the silica brick was eroded seriously, and surface of the weak part became honeycomb. The damaged checker brick details were shown in Fig. 8.

The internal erosion of hot blast stove is mainly determined by the chemical composition of BF gas and the surface temperature of refractory lining. After testing of BF gas gravity dust, it is found that the main composition of BF gas dust were MgO, CaO, Fe_2O_3, K_2O, Na_2O and ZnO, which is related to the reaction taking place inside the BF. There will be some small particles that have not been removed as the gas flows into the hot blast stove. When the BF gas react in hot blast stove, the products of combustion will adhere to the surface of the refractory lining under

 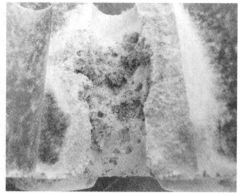

Fig. 8 Erosion of upper silicon checker brick in hot blast stove

high temperature condition. This will cause damage to the silica and aluminium siliceous refractory which always happen in BF.

Hot blast stove always works under the condition of high temperature and high pressure. For long-term used hot blast stove which is designed to use for 25-30 years, it has big risk to use smaller hole (like diameter 20mm) checker brick.

Considering the above factors, it is advisable to use diameter 25mm checker brick which has large heating area, a moderate amount of pressure loss and unblocked characteristic, to achieve high efficiency, long service life and low cost.

4.2 Numerical Simulation and Optimization of Regenerator Chamber

The hot blast stove is a typical heat exchanger, which is made of a large number of checker bricks. Due to the periodic characteristics of the hot blast stove, the heat storage and heat release of the hot blast stove have the characteristic of hysteresis and nonlinearity. The calculation of hot blast stove regenerator chamber has been based on several empirical formulas for a long time.[15]

The hot blast stove regenerator chamber calculation program developed by BSIET, is based on the German H. Hausen theory to calculate the vertical temperature field of regenerator chamber. After practical use and continuous optimization and improvement of the program for several years, it can accurately calculate the temperature distribution in regenerator chamber, and provide effective guidance for hot blast stove precise design. Fig. 9 shows the temperature distribution curve of regenerator chamber of hot blast stove of Tonggang's new No. 2 BF. It was calculated by the program. With the guidance of the calculation result, we could choose different refractory material for different part of the regenerator chamber. For example, the height of the silicic checker brick eventually identified as $70 \times 120mm = 8400mm$ high, which was considered with both the calculation result and the requirement of the temperature range 950-1200℃. With the understanding of the temperature change along the height of the regenerator chamber in different time, we could reasonably choose the refractory materials of different parts to ensure the long-term stable operation of the hot blast stove.

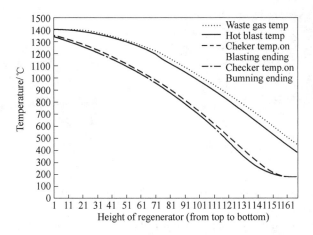

Fig. 9 Temperature distribution curve in regenerator chamber

5 Design of Low Stress and No Overheat System for Hot Blast Pipeline

5.1 Static Force Analysis of Pipeline

The hot blast pipe, main pipe and bustle pipe in hot blast stove system are high temperature and high pressure pipes, and they are the most complicated piping design of hot blast stove. Hot blast pipeline is an important link to achieve high hot blast temperature, and is the guarantee of stable transportation of high temperature hot blast.

In the design through the flexible structure with rigid structure should be designed with a combination of low power pipeline, pipe system using professional analysis software design and calculation of pipeline (Fig. 10), reasonable design of pipe, corrugated expander, pipe rack, pipe rod, so that the hot blast pipeline into a low stress system. Effectively guarantee the normal service life of the pipe system. The hot blast duct and three fork pipe fixing bracket, a fixed bracket on both sides of duct axial compensator for thermal displacement, absorption duct, is arranged between the fixed bracket and compound compensator is used to heat the hot blast stove, hot blast stove and the hot blast rising displacement absorption branch. The hot blast duct with rigid rod body, overcome the expansion force, the whole pipeline system of thermal pressure force, the end of the hot blast duct is used to absorb the hot blast duct compensator, and the elongation of the rod thermal expansion displacement, to overcome the blind plate force hot blast duct, hot blast duct of the low stress drop. Through the optimization design of refractory in the pipeline, a no overheating system of hot blast pipe was established.

5.2 Design Optimization of Hot Blast Branch Pipe

The hot blast duct and three fork branch intersection is the position of stress concentration. Through the pull beam structure is arranged in a hot blast pipe on the triangle layout, the

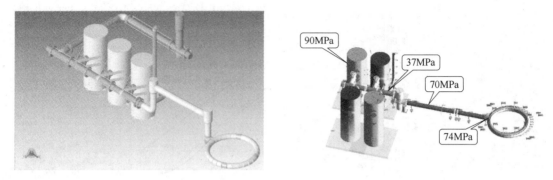

Fig. 10 Force analysis of hot blast pipe system

characteristics of triangle stability, improve the stability of steel pipe/hot blast duct three forks (Fig. 11). Special method should be used because the hot blast outlet exists larger vertical displacement after heating up the hot blast stove. A vertical displacement between hot blast main pipe and hot blast outlet should be kept when hot blast outlet is installed (Fig. 12). The hot blast outlet will rise up with the heating of hot blast stove. When elevation of hot blast outlet and hot blast valve is same, the hot blast valve should be installed. This method could decrease stress of hot blast outlet.

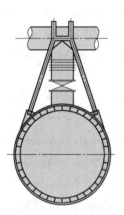
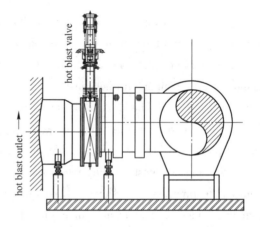

Fig. 11 Triangular tension beam structure Fig. 12 Staggered installation of hot blast valve

5.3 Design Optimization of Corrugated Compensator

Hot blast pipe compensator stiffness is relatively small, high temperature and high pressure and harsh working conditions, refractory heavy load makes the compensator has become a weak link in hot blast pipe system. The load of the compensator is improved by adding the structure of the ripple compensator. The expansion joint of traditional double axial compensator has wide width and is prone to overheat and even wind. By optimizing the form of compensator structure and adopting the compound form of hinge and axial, the width of expansion joint of compensator is reduced, and the tightness of refractory material is improved (Fig. 13).

Fig. 13　Optimization of hot blast pipe system compensator

5.4　Optimization of Refractory Structure Design

The hot blast pipeline adopts a heavy brick, three layer of heat insulation brick, spraying a layer of unshaped refractory. The upper part of the pipeline section 120 degree heavy brick with locking structure, improve the structural stability of lift. The expansion joint is designed reasonably in the axial direction of the pipe, and the expansion joint is protected by heavy bricks. Hot blast valve located near the expansion joints, convenient maintenance for the valve, avoid change valve, damage the overall stability of the pipeline refractories (Fig. 14).

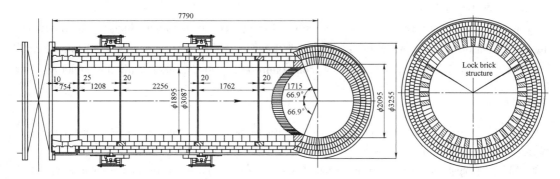

Fig. 14　Optimization of refractory structure for hot blast pipe

6　Application

Top combustion hot blast stove and efficient energy conversion technology has applied to the Shougang Jingtang's 5500m^3 huge BF, Qiangang's No.1 BF (2650m^3), the new No.2 BF (2650m^3) of TISCO, No.8 BF (3052m^3) of Liangang, No.2 BF (1950m^3), the new No.3 BF (1800m^3) of Xiangtan Iron & Steel Co. Which is the first BF of Shougang Jingtang's 5500m^3 the application of the technology huge BF, after the commissioning of monthly average hot blast temperature reached 1300℃, the combustion condition of single BF gas, BF hot blast temperature stable design index reached 1300℃, has created considerable economic benefits and good social benefits for the user.

The new technology in Qiangang's No. 2 BF (2650m^3) application, the production index steadily, gas preheating temperature 180-200℃, combustion hot blast preheating temperature 350-400℃, hot blast stove burning BF gas, can provide 1250℃ hot blast temperature to improve the stability of BF.

7 Conclusions

With the using of two stage double preheating process, hot blast stove of Shougang Jingtang 5500m^3 hot blast furnace could get hot blast as high as 1300℃ with fully using of blast furnace gas.

With the using of joint research combined with numerical simulation, cold condition test and hot condition experiment, the flow field in combustion chamber and cold blast distribution in blasting chamber are optimized.

It is advisable to apply diameter 25mm checker brick which has large heating area, a moderate amount of pressure loss and unblocked characteristic, to achieve high efficiency, long service life and low cost.

Pay attention to design optimization of hot blast pipeline, no overheat and low stress hot blast pipeline system is applied with structure optimization, refractory optimization and using of advanced design system.

Top combustion hot blast stove and high efficient energy conversion technology has been applied to several hot blast furnaces, which scale from 1800m^3 to 5500m^3. The practical application prove that the technology can increase the blast temperature efficiently, economically and steadily.

References

[1] Zhang Shourong. Development and problems of ironmaking industry after entering the 21st century in China [J]. Ironmaking, 2012, 31 (1): 1-6.

[2] Zhang Fuming. Cognition of some technical issues on contemporary blast furnace ironmaking [J]. Ironmaking, 2012, 31 (5): 1-6.

[3] Tang Yao, Zhao Ruihai, Luo Zhihong. The study of design concept of hot blast piping of high temperature hot blast stove [J]. Ironmaking, 2016, 35 (4): 30-33.

[4] Xu Guotao, Xiang Wuguo, Wang Xibo. Study on standards of refractory used for hot blast pipes between blast furnace and hot blast stove [J]. Research on Iron &Steel, 2017, 45 (1): 58-62.

[5] Tang Wenquan, Wei Cuiping. The comparation and optimization of quantity of hot blast stove [J]. Metallurgical Economy & Management, 20156: 37-39.

[6] Zhu Lihong, Huang Han, Wei Jie. Intelligent optimal control method study of burning process for hot blast stove [J]. Computer Measurement & Control, 2016, 24 (5): 74-77.

[7] Tian Fengjun, Sun Chuansheng, Liu Changfu. The applied research and progress of high radiative coating

technology [J]. Energy for Metallurgical Industry, 2017, 36: 20-22.
[8] Zhang Fuming. Technical features of super large sized blast furnace in earlier 21st century [J]. Ironmaking, 2012, 31 (2): 1-8.
[9] Zhang Fuming. Developing prospects on high temperature and low fuel ratio technologies for blast furnace ironmaking [J]. China Metallurgy, 2013, 23 (2): 1-7.
[10] Zhang Fuming, Qian Shichong, Yin Ruiyu. Blast furnace enlargement and optimization of manufacture process structure of steel plant [J]. Iron and Steel, 2012, 47 (7): 1-9.
[11] Zhang Fuming, Qian Shichong, Zhang Jian. New Technologies of 5500m^3 blast furnace at Shougang Jingtang [J]. Iron and Steel, 2011, 46 (2): 12-17.
[12] Wei Hongqi. Application and practices of 1300℃ air temperature in shougang jingtang united iron and steel Co., Ltd [J]. Ironmaking, 2010, 29 (41): 7-10.
[13] Zhang Fuming, Hu Zurui, Cheng Shusen. Combustion technology of BSK dome combustion hot blast stove at Shougang Jingtang [J]. Iron and Steel, 2012, 47 (5): 75-81.
[14] Mao Qingwu, Zhang Fuming, Zhang Jianliang, et al. Design and research of new type high temperature top combustion hot stove for super large sized blast furnace [J]. Ironmaking, 2010, 29 (4): 1-6.
[15] Zhang Fuming, Mei Conghua, Yin Guangyu. Design study on BSK dome combustion hot blast stove of Shougang Jingtang 5500m^3 blast furnace [J]. China Metallurgy, 2012, 22 (3): 27-32.

Study on Heat Transfer Process of Dome Combustion Hot Blast Stove

Zhang Fuming Hu Zurui Cheng Shusen

Abstract: The technology advantages and heat transfer characteristics in the regenerator are introduced in this paper with the huge dome combustion hot blast stove configured in 5500m^3 blast furnace at Shougang Jingtang. Heat transfer theory is discussed, and the heat transfer process in the regenerator checkers is analyzed especially. The 2-dimensional model for heat transfer and calculation under different shaped checkers is established. The result of study shows proper reduction of the channel diameter of checkers can promote the heat transfer efficiency of regenerator, and benefit to enhance the blast temperature. The number and diameter of checker's channels are determined through synthetic consideration. The reasonable thickness of the silica checker in regenerator can prolong the blasting sequence period and enlarge the adjustable range of blast temperature to guarantee the smooth and efficient working status of silica checkers. The correctness of investigation on heat transfer characteristics of checkers and rationality of optimum design on regenerative structure for huge dome combustion hot blast stove are proved by production practice.

Keywords: dome combustion hot blast stove; high blast temperature; heat transfer theory; checker brick; regenerator

1 Introduction

High blast temperature is one of the important technology character for modern blast furnace. For present iron and steel industry, improve hot blast temperature is the key-common technology which can develop circulating economy, achieve low-carbon and green metallurgy, energy saving and low emission. It's very important to improve blast furnace ironmaking technology, decrease carbon dioxide emission and make industry technology progress. In recent years, with the continuous innovation of dome combustion hot blast stove technology, a series of key technologies of high efficiency, high blast temperature and long campaign life have been successfully developed, which makes the technical and economic advantages more prominent[1]. Compare to the internal combustion hot blast stove and external combustion hot blast stove, significant technical advantages of dome combustion hot stove has been widely recognized at home and

---

本文作者：张福明，胡祖瑞，程树森。原刊于 Proceedings of the Iron & Steel Technology Conference, AISTech, 2017：835-843。

abroad[2]. Since entering the 21st century, the dome combustion hot blast stove technology developing rapidly, Shougang Jingtang for the first time in the world applied the dome combustion hot blast stove technology in the 5,500m³ huge blast furnace, under the condition of burning the full BF gas, the design hot blast temperature is 1,300℃[3]. In order to study the heat transfer process and technical characteristics of large dome combustion hot stove, through the establishment of mathematical model and theoretical computation, get the results of relationship between the checker brick type and channel diameter, and regenerator reasonable segmentation of checker brick. The research results provide effective technical support and reference for engineering design optimization and practical production.

2 Technical Characteristics

Dome combustion hot blast stove is a new type hot blast stove at the end of 1970s successfully developed, its main characteristic is to cancel the independent combustion chamber structure, using dome space as the combustion chamber, and the burner is arranged in the stove dome. Dome combustion hot blast stove can operation in various gas and combustion air conditions, and the large combustion power, high combustion efficiency, long campaign life, is a kind of high efficiency energy-saving and long campaign life hot blast stove, which is the developing trend of modern hot blast stove technology. Dome combustion stove adopts annular ceramic burner with swirling diffusion combustion, the combustion air and gas fully mixed, contribute to complete combustion of gas. In lower air excess coefficient, the burner can effectively improve the flame temperature and the dome temperature[4]. The unique structural characteristics of the dome combustion hot blast stove can enhance the stability of the stove proper structure, reduce the thermal stress caused by the structural design and prolong the service life of the hot blast stove. Ceramic burner is configured in the dome of stove, high temperature flue gas is uniformly distributed in the vortex state, effectively improve the high temperature flue gas in the regenerator checker brick surface distribution uniformity and heat transfer efficiency, thereby greatly improving the utilization rate of the regenerator, the temperature distribution of the dome circumference is uniform, it can reduce the thermal stress in the lining and shell, to prolong the campaign life of hot blast stove.

Dome combustion hot blast stove technology is applied in Shougang Jingtang 5,500m³ huge BF, which is the world's first dome combustion hot blast stove technology applied in huge blast furnace of 5,000m³ grade, the highest design blast temperature is 1,300℃. Under the condition of full BF gas combustion, using heat recycled from waste gas generated by hot blast stove can preheat BF gas and combustion air to 180-215℃, then using two combustion air preheaters which are two small hot blast stoves, to preheat the combustion air to 520-600℃ with second stage. The gas and air preheating technology makes stove dome temperature can reach 1,420℃, which is the foundation for the realization of the 1,300℃ high temperature.

There is no doubt that it is no precedent for reference in the international to use dome combustion hot blast stove in the 5,500m³ huge blast furnace, at that time that is an international technical issues. Therefore, it is great significance to study engineering science problem such as the transfer theory of combustion, gas motion and heat transfer, optimize engineering design, realize system integration, and satisfy production requirements. In engineering design, the hot blast stove regenerator heat transfer computational model is established, using the developed simulation software, the relationship between thermal parameters and heat transfer process is studied systematically. The thermal parameters contains checker brick materials, brick type, channel dimension and thickness and so on. Focus on the analysis of the effect of aperture and height of silica bricks on the heat transfer process of regenerator, provides reliable technical support for implementation of precision design engineering.

3 Heat Transfer Calculational Model for Regenerator

3.1 Physical Model

During the engineering design of Shougang Jingtang 5,500m^3 blast furnace dome combustion stove, through the investigation of home and aboard large blast furnace design and application, based on the existing technology, carried out a feasibility study to dome combustion stove. In the feasibility study, the 19 channels checker brick is adopted in the regenerator, the diameter of channel is 30mm, and the checker brick is a regular hexagon. According to the temperature field distribution and heat transfer characteristics of hot blast stove, the regenerator is divided to 3 sections, high density clay bricks applied in low temperature zone, low creep clay bricks applied in middle temperature zone, silicic bricks adopted in high temperature zone which with excellent creep resistance. Table 1 shows the Shougang Jingtang 5,500m^3 blast furnace dome combustion hot blast stove main technical performance parameters. Table 2 shows the distribution and physical parameters of each kind brick.

Table 1 Main technical specifications of dome combustion hot blast stove

Items	Data
Number of stove	4
Height/m	49.22
Regenerator diameter/m	10.90
Regenerator area/m^2	93.21
Checker brick height/m	21.48
Checker brick channel diameter/mm	30
Design hot blast temperature/℃	1300 (max. 1310)
Dome temperature/℃	1420
Waste flue temperature/℃	Normal average 349 (max. 450)
Combustion air preheating temperature/℃	520-600
Gas preheating temperature/℃	215
Blast volume/Nm^3 · min^{-1}	9300

Continued Table 1

Items	Data
Checker brick heating surface/$m^2 \cdot m^{-3}$	48
Total heating surface of one stove/m^2	95885
Heating surface of BF/$m^2 \cdot m^{-3}$	69.73
Heating surface of blast volume/$m^2 \cdot (m^3 \cdot min)^{-1}$	41.24

Table 2 Physical parameters and distribution of checkers

Checker brick	Material	Height/m	Heating capacity /$J \cdot (kg \cdot K)^{-1}$	Density /$t \cdot m^{-3}$	Thermal conductivity /$W \cdot (m \cdot K)^{-1}$
First section	High density clay	9.12	836.8+0.234T	2.15	0.84+0.00052T
Second section	Low creep clay	1.92	836.8+0.234T	2.10	0.84+0.00052T
Third section	Silicic	10.44	794.0+0.251T	1.80	0.93+0.0007T

In order to check and verify the feasibility study, structure design and optimization of regenerator checker brick in hot blast stove, a regenerator heat transfer mathematical model is established to optimize the design. It's many kind of calculation model about research of hot blast stove regenerator. The method of the fluid channel section shape checker to simplify the complex pipeline with a certain thickness is adopted in this paper. The hot blast stove regenerator is regarded as a space is consisted of many pipelines, with a single brick channel as the research object, thereby a mathematical model is established, such as shown in Fig. 1. The fluid channel diameter is the channel inner diameter, outer diameter of channel can be calculated by the active area of the regenerator (i.e. brick effective channel cross-sectional area). The fluid channel thickness can be calculated by the inner and outer diameter of the channel[5,6].

$$\frac{\pi d^2}{4} = \frac{\pi D^2}{4} \times s \qquad (1)$$

In equation: d is inner diameter, D is outer diameter, and s is active area. According to equation (1), $D = 0.05$m.

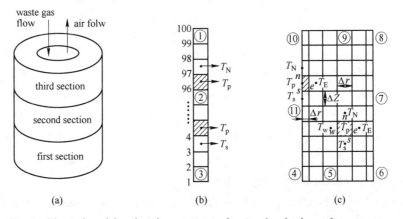

Fig. 1 Physical model and gird compartmentalization for checkers of regenerator
(a) Subsection of checker chamber; (b) Grid of gas; (c) Grid of checker brick

3.2 Mathematic Model

Because the gas flow and heat transfer in the regenerator is a very complex process, some reasonable assumptions should be defined to simplify the model. According to the engineering design and operational application, the mathematic calculation using the following assumptions:

(1) The assumption of the regenerator in the same cross section, gas distribution uniformity, heat transfer process of gas pipelines is the same, gas pipelines between each other are heat insulation.

(2) Assumes that the density of the gas is the function of temperature and pressure, the thermal properties of bricks is only temperature function is not affected by phase transformation of bricks.

(3) Assuming the gas velocity and temperature distribution uniformly at the same height on the cross section. In the combustion period and blast stage, gas temperature and gas flow rate to maintain the stable entrance. It should be pointed out that the basic assumptions are based on blast stove operation in practice, with reference to the assumption of heat transfer model of hot blast stove regenerator in the literature[7-11], and supplement and perfected the assumption appropriately, to make the assumption closer to the actual working condition. In this paper based on the assumption of the model, the temperature field of gas is calculated by a one-dimensional unsteady convection heat transfer equation[12], checker brick uses the two-dimensional unsteady heat transfer equation[13], coupling the temperature of gas and checker brick.

One-dimensional unsteady convection heat transfer equation:

$$\rho_g C_{p,g} \left(\frac{\partial T_g}{\partial t} + v_g \frac{\partial T_g}{\partial z} \right) + v_g \frac{\partial P}{\partial z} = \frac{4\alpha}{d}(T_w - T_g) \qquad (2)$$

Two-dimensional unsteady heat transfer equation in checker brick:

$$\rho_s C_{p,s} \frac{\partial T_s}{\partial t} = \frac{1}{r} \frac{\partial}{\partial r}\left(r\lambda_s \frac{\partial T_s}{\partial r}\right) + \frac{\partial}{\partial z}\left(\lambda_s \frac{\partial T_s}{\partial z}\right) \qquad (3)$$

In equations: ρ_g, ρ_s is the density of gas and checker brick; $C_{p,g}$, $C_{p,s}$ is the heat capacity of gas and checker brick; v_g is the velocity of gas; P is the pressure of gas; λ_s is the thermal conductivity of checker brick; α is convective heat transfer coefficient; d is inner diameter of checker brick channel; T_w is the temperature of wall of checker brick channel; T_g is gas temperature; T_s is checker brick temperature.

Fig. 1 (b) and (c) show the grid of gas and brick, then calculate using the integral discrete and integral solve coupling method, the channel calculation process written into a set of simulation software[14].

4 Thermal Characteristics of Checkers

4.1 Computational Scheme

In order to compare and analyze the heat transfer performance of different brick type, 4 kinds of checker brick with different channel diameters were selected, which main thermal parameters were shown in Table 3. The calculation based on the developed simulation software, respectively, the 4 kinds of thermal parameters of checker input calculation program, according to the research results of large dome combustion hot blast stove combustion process[15], set on the average temperature of regenerator top surface is 1,400℃, flue gas flow is 267,000Nm³/h, the blast flow is 9,300Nm³/min. The temperature change curve of checker chamber with different checker brick can be calculated.

Table 3 Thermal parameters of different types of checkers

Type	Hydraulics diameter /mm	Flowing area /m²·m⁻²	Fill coefficient /m³·m⁻³	Heating surface /m²·m⁻³	Thickness /m
5 channels	58.3	0.434	0.566	29.78	0.038
7 channels	43.0	0.409	0.591	38.05	0.031
19 channels	30.0	0.360	0.640	48.60	0.023
37 channels	20.0	0.354	0.646	70.80	0.018

4.2 Calculation Results and Discussion

Fig.2 shows the curves of flue gas temperature with time on burning. As can be found from Fig.2, the flue gas temperature gradually increased, at the same time, the more the number of channels and the smaller the grid-hole diameter, then the flue gas temperature lower. When around 30 minutes, the flue gas temperature of the 7 channels, the 19 channels and the 37 channels checker brick are nearly the same, and the flue gas temperature of the 5 channels checker brick is still high, which indicates that the flue gas heat can't be utilized effectively. Fig.3 shows the curves of blast temperature with time on blasting. See from Fig.3, during 60 minutes of on blast, blast temperature decreased gradually with the time, and the temperature drop rate is different for different type checker brick, brick channel number more, smaller grid-hole diameter, blast temperature drop rate is smaller. During blasting period, the hot blast temperature of the 7 channels, the 19 channels and the 37 channels checker brick are always higher than that of 5 channels.

From the comparison of 4 different types of brick can be seen in the regenerator structure, stove changing cycle, blast flow, flue gas flow under the same conditions, during the burning period and blasting period, the flue gas temperature and blast temperature change with time are different for checker brick with different channel number and diameter. For the 7 channels, 19 channels

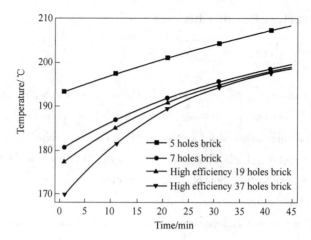

Fig. 2　Temperature curves of flue gas in burning sequence period

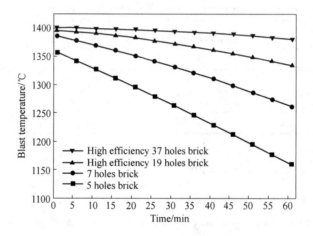

Fig. 3　Temperature curves of hot blast in blasting sequence period

and 37 channels brick, the flue gas temperature there is small difference in prophase of on gas, flue gas temperature change is the same after 30 minutes. However, the blast temperature change is different obviously from begin to end on blast. The temperature drop of 7 channels, 19 channels and 37 channels brick are 140℃, 65℃ and 20℃ respectively.

The calculated results show that the brick of more channels, as the grid-hole diameter is smaller, the heating area and heat transfer coefficient increase, which is helpful to increase the heat exchange capacity of 5 channels regenerator. For the 7 channels, 19 channels and 37 channels bricks, high temperature flue gas can make more heat exchange with bricks than 5 channels, high temperature heat can be fully transferred to the bricks, therefore the temperature of flue gas is low in prophase of combustion period; while blasting period, due to the checker brick heat reserve expressively, not only it can get higher blast temperature, but also ensure the blast temperature change gently, and the blast temperature is more stable.

4.3 Resistance Loss

In order to study the resistance loss of different checker brick type, the relationship between the resistance loss of the checker brick with different channel diameter and the physical parameters of the checker brick is deduced by means of Formula 4. Calculating found that the gas of hot blast stove in a laminar or transitional zone, the ratio of resistance coefficient is equal to the heating area ratio, pressure loss is inversely proportional to the active area and to the square of channel diameter. Thus it can be seen that the smaller active area of the checker brick, the smaller channel diameter, cause to the greater resistance loss. Based on the 19 channel checker brick, the ratio of the resistance loss of other brick type and 19 channel checker brick is calculated, as shown in Fig. 4. As can be seen from Fig. 4, the maximum resistance of the 37 channel checker brick is about 2.5 times of the 19 channel, and the resistance loss of the 5 channel and the 7 channel checker brick is about 1/4 and 1/2, respectively, of 19 channel.

$$\frac{\Delta P_n}{\Delta P_m} = \frac{\dfrac{\xi_n L_n \rho v_n^2}{2 d_n}}{\dfrac{\xi_m L_m \rho v_m^2}{2 d_m}} = \frac{\xi_n v_n^2 d_m}{\xi_m v_m^2 d_n} \approx \frac{F_n}{F_m}\left(\frac{\phi_m}{\phi_n}\right)^2 \frac{d_m}{d_n} = \frac{\phi_m}{\phi_n}\left(\frac{d_m}{d_n}\right)^2 \tag{4}$$

In equation: P is the resistance loss of gas flowing through the bricks; ξ is the resistance coefficient for gas flow through bricks; L is bricks height of gas flows through; v is gas velocity flow checker brick; d is checker brick channel inner diameter; F is heating area of checker brick; ϕ is active area of brick.

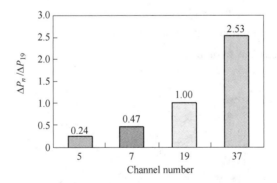

Fig. 4 The resistant pressure ratio between of 19 channel and the others checker brick

4.4 Summary

Thus, to some extent, reducing the grid-hole diameter of the checker brick is helpful to improve the thermal efficiency of the regenerator and improve the heat transfer process of hot blast stove. But excessive narrow bricks will result in increased the resistance of gas flow through regenerator, loss of gas resistance, and cost of blast energy. Moreover, small brick channel is not conducive to ensure regenerator chamber of hot blast stove "through channel ratio", but also easy to cause the channel plug, resulting in effective utilization of regenerator the rate of decline. Therefore, the choice of the checker brick type is not the more number and the smaller of channels more favorable, but also to take into account the thermal efficiency, effective utilization of the regenerator and the use of life and other factors.

According to the calculation results, in the engineering design of Shougang Jingtang 5,500m³ large blast furnace of dome combustion hot stove, the 19 channel checker brick is applied, which grid-hole diameter is 30 mm, maintain the feasibility study in the design scheme. The bricks can moderately increase heating area, the amount of heat exchange, and improve the heat transfer coefficient, in the same active area or heat storage quality of checker brick. In the period of combustion, high temperature flue gas can be fully heat transfer to brick works, the flue gas temperature is lower. Also, in the period of blasting, it is helpful to the heat exchange between blast and bricks, make hot stove obtain high temperature, which is the technical foundation for the realization of the 1,300℃ ultra-high blast temperature.

5 Regenerator Checker Brick Configuration

Shougang Jingtang 5,500m³ blast furnace dome combustion hot stove, the total height of regenerator is 21.48meter, the obvious temperature difference in the height direction on the regenerator. After the temperature field calculation of stove regenerator[16], obtains the temperature distribution of regenerator in the height direction at the end of combustion period. According to heat transfer calculation results and refractory performance, optimize the allocation of regenerator, reasonable use of bricks of different materials in different temperature range. Fig. 5 shows the temperature distribution of the flue gas, the hot blast and the checker brick at the end of combustion period and blast period.

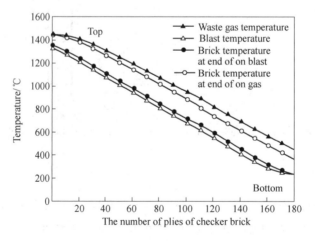

Fig. 5 Temperature distribution of regenerator

The checker brick of the regenerator is subjected to the combined action of thermal stress, mechanical load, phase transformation stress and pressure in working. Creep is the most seriously non-flex destroy for checker brick in high temperature. The high temperature bulk stability, corrosion resistance and creep resistance of the checker brick in the high temperature zone of the regenerator are required. The regenerator of high temperature zone preferred siliceous bricks, and residual quartz must be less than 1%. The temperature range is divided according to the temperature curve in the height direction of the regenerator, and the different physical and chemical properties checker bricks are reasonably selected. In high temperature area, using silicic bricks which with excellent creep resistance, high density clay bricks applied in low temperature area, and low creep clay bricks using in middle temperature. The optimal configuration of the refractory is in accordance with the heat transfer characteristics of dome combustion hot blast stove, which can improve the economic and reasonable function of the refractory and reduce the investment of the project[17].

6 Application

No. 1 blast furnace of Shougang Jingtang put into production in May 2009, blast furnace production stable, continuously improve the production of technical parameters, the highest daily production reached 14,245t/d, the monthly average productivity reached 2.37t/(m^3 · d), fuel ratio is 480kg/tHM, the coke rate is 269kg/tHM, the coal ratio is 175kg/tHM, April 2010 the average blast temperature reaches 1,300℃. Monthly average blast temperature steadily reached 1,300℃ only the BF gas using as fuel, creating a new record in the production practice of large blast furnace with high temperature, reached the advanced level at home and abroad more than 5,000m^3 huge blast furnace with high temperature applications. Fig. 6 shows the No. 1 blast furnace in blast temperature is 1,300℃ operating conditions, combustion period and blasting period dome temperature, silicon brick interface temperature, ceramic burner loop channel temperature and flue gas temperature curve with time. It can be found from Fig. 6, at the end of

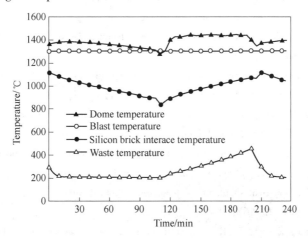

Fig. 6 Temperature curve of No. 1 hot blast stove of Shougang Jingtang's No. 1 BF

burning period, the temperature of top of regenerator reached 1,420℃, the flue gas temperature reached 400℃, the blast temperature reached 1,300℃ stable. At the end of blasting period, the temperature of top of regenerator reduced to 1,300℃, silicon brick interface temperature controlled at 800℃, to achieve efficient and stable work of hot blast stove.

7 Conclusions

In the same active area or heat storage quality of checker brick, moderately reduce the grid-hole diameter of checker brick can increase heating area and the amount of heat exchange, also improve the heat transfer coefficient. During the burning period, high temperature flue gas can be fully heat transfer to bricks, the flue gas temperature is lower. Also, during the blasting period, it is helpful to the heat exchange between blast and bricks, obtain higher blast temperature, which will be more stable and change gently.

The selection of the checker brick type should be considered synthetically. It is not more favorable for the more channels and smaller grid-hole diameter. It must consider the thermal efficiency of the regenerator, the effective utilization of the regenerator and the service life of the checker brick and other factors.

After Shougang Jingtang huge dome combustion hot stove put into production, monthly average blast temperature reached 1,300 degree only the BF gas using as fuel. Hot blast stove stable work with high heat exchange efficiency, practice has proved that the correctness of research on heat transfer characteristics of large dome combustion hot stove checker brick and the rationality of the regenerator structure of design and optimization.

References

[1] Zhang Fuming. Technological progress of large dome combustion hot stove in china [J]. Ironmaking, 2002, 21 (5): 5-10.
[2] Qian Shichong, Zhang Fuming, Li Xin, et al. Comparatively analysis of large blast furnace hot blast stove technology [J]. Iron and Steel, 2011, 46 (10): 1-6.
[3] Zhang Fuming, Qian Shichong, Zhang Jian, et al. New technologies of 5500m^3 blast furnace at Shougang JIngtang [J]. Iron and Steel, 2011, 46 (2): 12-17.
[4] Zhang Fuming, Hu Zurui, Cheng Shusen, et al. Combustion technology of BSK dome combustion hot blast stove at Shougang Jingtang [J]. Iron and Steel, 2012, 47 (5): 75-81.
[5] Guo Minlei. Numerical simulation of heat transfer and gas combustion of dome combustion hot blast stove [D]. Beijing: University of Science and Technology Beijing, 2008: 61-63.
[6] Zhang Fuming. Study on transmission theory and design of long campaign Life and high efficiency hot blast stove [D]. Beijing: University of Science and Technology Beijing, 2010: 171-187.
[7] Chen Lan, Tang En. Numerical simulation of temperature field distribution in regenerator of hot blast stove [J]. Ironmaking, 2004, 23 (6): 38-40.
[8] Zheng Zhong, Huang Zhenyi. Study on heat transfer and operating system of hot blast stove based on FLUENT [J]. Industrial Heating, 2008, 37 (5): 37-41.

[9] Zhang Yin, Liu Zhongxing, He Youduo. Calculation of heat transfer process in hot blast stove regenerator [J]. Journal of Baotou University of Iron and Steel Technology, 2001, 20 (1): 4-7.
[10] Luo Sheng, Lu Jian, Chen Yisheng, et al. Study on operation system of regenerative hot blast stove [J]. Journal of Baotou University of Iron and Steel Technology, 2005, 23 (2): 107-112.
[11] Zhang Lilin, Zheng Chuguang, Wang Hai. Simplified model of temperature field in hot blast stove [J]. Metallurgical Energy, 2004, 23 (2): 23-26.
[12] Muske K R, Howse J W, Hausen G A, et al. Model-based control of a thermal regenerator. Part 1: Dynamic Model [J]. Computers and Chemical Engineer, 2000, 24 (11): 2519-2531.
[13] Muske K R, Howse J W, Hausen G A, et al. Hot blast stove process model and model-based controller [J]. Journal of Iron and Steel Institute, 1999, 76 (6): 56-62.
[14] Guo Minlei, Cheng Shusen, Zhang Fuming, et al. Development and application of simulation software for heat transfer calculation of checker brick in hot blast stove [C]//Chinese Society of Metals, 2008 Chinese Ironmaking Technology Conference & Ironmaking Conference Proceedings, Ningbo, 2008: 822-833.
[15] Zhang Fuming, Hu Zurui, Cheng Shusen, et al. Combustion technology of BSK dome combustion hot blast stove at Shougang Jingtang [J]. Iron & Steel, 2012, 47 (5): 75-81.
[16] Guo Minlei, Cheng Shusen, Zhang Fuming, et al. Calculation of temperature distribution of checker brick on blasting period [J]. Iron &Steel, 2008, 43 (6): 15-21.
[17] Zhang Fuming, Mei Conghua, Yin Guangyu, et al. Study on design of BSK dome combustion hot blast stove for 5500m^3 blast furnace at Shougang Jingtang [J]. Chinese Metallurgy, 2102, 22 (3): 27-32.

Research and Application on Waste Heat Recycling and Preheating Technology of Ironmaking Hot Blast Stove in China

Li Xin   Zhang Fuming   Yin Guangyu   Cao Chaozhen

Abstract: China is the country with the largest steel output in the world. During the process of iron-making, the blast furnace consumes a lot of energy. The hot blast stove system is an important system for providing high temperature hot air to the blast furnace, and its energy utilization efficiency and has an important influence on energy conservation and emission reduction. This paper analyzed and compared the use effect, thermal efficiency, technical characteristics and development prospects of different waste heat recycling technology, based on the research and application of different waste heat recovery and utilization methods of hot-blast stove system in different projects, such as tubular heat exchangers, heat pipe exchangers, plate heat exchangers, preheating furnaces and combined preheating methods. Furthermore, this paper gave some recommendations for the future development path of waste heat recycling technology of hot blast stove system.

Keywords: hot blast stove; ironmaking; energy recycling; heat exchanger; preheating

1 Introduction

China is the world's largest steel producer. According to data released by the World Steel Association, the world's crude steel output in 2017 was 1.691 billion tons, of which China's crude steel output was 832 million tons, accounting for 49.2%.

During the steel production process, the energy consumption of the iron-making process accounts for 60%-70% of the total energy consumption of steel production. The production cost of blast furnace (BF) iron-making accounts for 60%-70% of the steel manufacturing cost[1]. As an important part of blast furnace system, hot blast stove produces a large amount of waste gas at a temperature of 200-300℃. If this energy is fully recovered, the energy consumption of the iron-making process can be reduced by about 4%-5%[2]. Therefore, research and application of safe, efficient, long-lived waste heat recycling technology is increasingly important for blast

---

本文作者：李欣，张福明，银光宇，曹朝真。原刊于TMS2019 Conference Proceedings (10th International Symposium on High Temperature Metallurgical Processing). The Minerals, Metals & Materials Society, San Antonio, 2019: 33-46。

furnace iron-making.

On the other hand, hot blast provides the second largest heat for blast furnace. The first one is coke. Increasing the hot blast temperature by 100℃ can reduce the coke ratio by 10-15kg/t. With the advancement of modern steel technology, in iron and steel enterprises, the high calorific value gas like coke oven gas and converter gas are mainly used in steel making and steel rolling processes, or in the preparation of H_2 or other high value-added products. Blast furnace gas (BFG) has gradually become the only fuel for hot blast stoves. However, with the advancement of blast furnace smelting technology, the gas utilization ratio in the blast furnace is continuously improved, and the calorific value of the BFG is continuously reduced. The calorific value of the BFG of the advanced large blast furnace is only 3000kJ/Nm^3 (About 2230kJ/kg). If gas and combustion air (CA) preheating technology are not used, the dome temperature of the hot blast stove can only reach 1200-1250℃ [3].

During the hot blast stove burning, the main source of heat is the chemical heat of the gas, and the preheated temperature of the CA and gas is fixed. When all the heat is used to heat the combustion products, no other heat losses, the temperature which can reach called the theoretical combustion temperature of hot blast stove.

During the operation of the hot blast stove, actually the dome temperature of hot blast stove is lower than the theoretical combustion temperature 30-50℃ because of heat dissipation of stove wall and incomplete combustion. Practice has proved that the theoretical combustion temperature is the main constraint to limit the increase of blast temperature in hot blast stove, and the formula is shown in equation 1.

$$t_T = \frac{Q_A + Q_G + Q_{DW}}{C_P V_P} \tag{1}$$

In formula t_T——Theoretical combustion temperature, ℃;

Q_A——Combustion air sensible heat, kJ/m^3;

Q_G——Gas sensible heat, kJ/m^3;

Q_{DW}——Low calorific value of gas, kJ/m^3;

C_P——Heat capacity of combustion product at t_T, kJ/(m^3·℃);

V_P——Volume of combustion product, m^3.

It can be seen from formula 1, that the theoretical combustion temperature of the hot blast stove can be improved by increasing the physical heat (sensible heat) of the combustion air and gas. Increase in the preheat temperature of CA and gas can improve their physical heat effectively. Fig. 1 shows the theoretical combustion temperature under the condition of preheating CA and gas.

Considering the decrement from the dome temperature to the hot blast temperature, the hot blast temperature will be lower than 1200℃. Therefore, a variety of high temperature preheating technology combined with waste gas heat recycling technology have been developed.

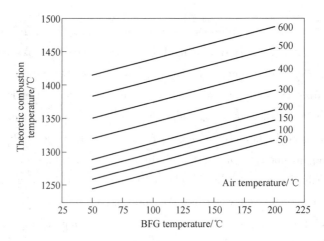

Fig. 1 The theoretical combustion temperature of hot blast stove under the condition of combustion air and gas preheating

2 Tubular Heat Exchanger

2.1 The Principle of Tubular Heat Exchanger

Both the tubular heat exchanger and the plate heat exchanger use the inner wall to separate the flow spaces of the two different fluids. Heat is transformed through methods of wall conduction and convection.

The tubular heat exchanger is also called tube type heat exchanger or shell-and-tube heat exchanger. The heat exchange element is thin-walled tube welded on the tube plate. The inner diameter of the tube is 50-100mm, and the wall thickness of tube is 3-4mm, the length is determined by the heat exchange area. The heat exchange tube is required to have corrosion resistance and good thermal conductivity.

During working, high-temperature and low-temperature fluids flow on both sides of the tube. The flow inside the tube is called the tube process, and the flow outside the tube is called the shell side. The surface area of the tube bundle is the heat exchange area. In order to increase the flow rate of fluid inside the tube, the entire tube is often divided into several groups, so that the fluid travels back and forth within these tube groups, which is called multi-tube process. In order to increase the flow rate of fluid outside the tube, a certain number of baffles perpendicular to the tube bundle are installed to increase the fluid flow rate and turbulence intensity, and improve heat exchanging.

2.2 The Characteristics of Tubular Heat Exchanger

The advantages of the tubular heat exchanger are simple structure, low manufacturing difficulty and strong adaptability, especially in chemical industry with high temperature and high pressure conditions. However, the tube bundles are easily deformed or broken under the thermal stress and strain, which will cause fluid leakage and heat exchange efficiency reduction. The service life and reliability of heat exchanger will be seriously affected. From a safety point of view, tubular heat exchangers are not suitable for BFG preheating.

The tubular heat exchanger has been used successfully in European steel enterprise. It can preheat the CA to 350℃, and the BFG to 420℃. The hot air stove can get 1550℃ dome temperature, and can provide hot blast with 1275-1300℃ to blast furnace[4]. However, this type of heat exchanger is relatively rare in China. In addition to safety consideration, the waste gas heat exchanger is easy to accumulate ash, which will cause heat exchanger blockage, heat exchange efficiency reduction. This default also affects the promotion and application of the tubular heat exchanger in China.

2.3 The Application of Tubular Heat Exchanger

The TISCO No. 3 blast furnace (1800m^3), designed by BSIET, is equipped with four internal combustion hot air stoves. The design hot blast temperature is 1200℃. It was put into operation on June 31, 2007.

The hot blast stove system uses tubular heat exchanger to preheat the CA and gas. The preheating parameters are shown in Table 1.

Table 1 The parameters of TISCO No. 3 BF tubular heat exchanger

Volume of blast furnace/m^3	1800
Type of hot blast stoves	Internal combustion HBS
Number of hot blast stoves	4
Fuel of hot blast stoves	Blast furnace gas
Waste gas flow/Nm$^3 \cdot h^{-1}$	250,000
BFG flow/Nm$^3 \cdot h^{-1}$	155,000
Combustion air flow/Nm$^3 \cdot h^{-1}$	105,000
Waste gas temperature/℃	220-250
BFG inlet temperature/℃	60
Combustion air inlet temperature/℃	20
BFG outlet temperature/℃	220-250
Combustion air outlet temperature/℃	220-250

This tubular heat exchanger has been used for 11 years, and still can preheat the air and gas to about 300℃. It proves that the tubular heat exchanger can achieve the goal of long-life and reliability with rationally designing the flow area and strictly controlling the product quality.

3 Heat Pipe Heat Exchanger

3.1 The Principle of Heat Pipe Heat Exchanger

The heat pipe is a heat transfer element with high thermal conductivity. It is a sealed pipe which inside it air is evacuated and an appropriate amount of heat medium is filled with. Heat is transferred through phase change (evaporation and condensation) of the internal heat medium [5]. Fig. 2 shows the working principle of the heat pipe. The heat pipe includes evaporator, condenser, and connecting section (also referred to as a transmission section). When it is working, the heat from high temperature medium is transferred to the heat medium in the evaporator, and the heat medium is evaporated to steam by the heat. After reaching the condenser through the connecting section, the steam is condensed into liquid, and the heat released by the phase change is transferred to the low temperature fluid. The liquid heat medium return to the evaporator by gravity or capillary action to complete a whole cycle.

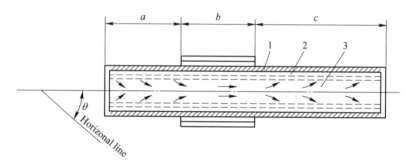

Fig. 2 The principle of heat pipe exchanger
1—Outer shell; 2—Suction core; 3—Steam space

Under the continuous circulation of the heat medium, heat is continuously transferred from the high temperature medium to the low temperature medium. Because it relies on the latent heat of phase change to transfer heat, the heat conductivity of heat pipe is far higher than good conductor of heat like silver and copper.

3.2 The Characteristics of Heat Pipe Heat Exchanger

The heat pipe heat exchanger for hot blast stove mostly uses pure water as the heating medium, and adopts carbon steel-water heat pipe technology, in which the steel pipe is handled with passivation treatment to improve the service life. The advantages of the heat pipe heat exchanger include: high heat transfer efficiency with the use of phase change heat transfer, large heat transfer area per unit of volume by welding fins on the heat pipe, low running cost because of no additional power requirements, completely relying on phase change and gravity (or capillary), and high safety.

The heat pipe heat exchanger has two types: integral type and separate type. The integral heat pipe heat exchanger directly arranges the gas heat exchanger and the air heat exchanger on the top of waste gas heat exchanger. So it has the advantages of small cover area and less investments. The disadvantages is hard to repair. If one heat pipe leaks, it will affect the use of the entire heat exchanger. The separate type divides the waste gas, air and gas into different units, and the units are connected by external pipes to realize the phase change flow between the different units. This type is more flexible to arrange. There is no way to leak between different media such as waste gas and gas. The heat medium can be replenished and vented on site, and the maintenance is convenient. This type can meet the requirements of large-scale blast furnace, because after separating, each one is not too large. But the investment of this type is higher than integral type.

3.3 The Application of Heat Pipe Heat Exchanger

Heat pipe heat exchangers have been used in the field of hot blast stoves since the 1980s to recover heat from waste gas, and reduce the temperature from 300℃ to 150℃.

The Shougang Jingtang No. 1 blast furnace (5500m^3), designed by BSIET, is equipped with four dome combustion hot blast stoves. The design hot blast temperature is 1300℃. It was put into operation in May 2009.

The hot blast stove system of No. 1 blast furnace adopts two-stage double preheating technology, in which the low temperature preheating process adopts a separate heat pipe heat exchanger, and the preheating parameters are shown in Table 2.

Table 2 The parameters of Shougang Jingtang No. 1 BF heat pipe heat exchanger

Volume of blast furnace/m^3	5500
Type of hot blast stoves	Dome combustion HBS
Number of hot blast stoves	4
Fuel of hot blast stoves	BFG
Waste gas flow/$Nm^3 \cdot h^{-1}$	700,000
BFG flow/$Nm^3 \cdot h^{-1}$	505,000
Combustion air flow/$Nm^3 \cdot h^{-1}$	319,000
Waste gas temperature/℃	350
BFG inlet temperature/℃	45
Combustion air inlet temperature/℃	20
BFG outlet temperature/℃	215
Combustion air outlet temperature/℃	200

This heat exchanger has been used for 9 years. In 2013, the gas heat exchanger was overhauled and some gas pipe bundles were replaced. At present, the gas can only be preheated to 80℃, and the air can only be preheated to 90-100℃. Heat exchange efficiency has fallen dramatically.

Shougang Qiangang 4000m³ blast furnace also uses the heat pipe heat exchanger to recover the heat of waste gas of the hot blast stove. The heat exchanger was put into use in 2010. After two years, the preheating effect began to decline, and leakage problem began to occur[6]. These failure or leakage phenomenon also appeared in other blast furnaces of China. Fig. 3 shows severe corrosion of the outer and inner heat pipe bundles after disassembly.

Fig. 3 Severe corrosion of the outer and inner heat pipe bundles

4 Plate Heat Exchanger

4.1 The Principle of Plate Heat Exchanger

The plate heat exchanger works in the same way as the tubular heat exchanger. It uses the wall to separate the two fluids and transfer heat through the wall. The difference is that plate heat exchanger uses the plate to separate the high temperature fluid and low temperature fluid into many narrow flow area, which are wide about 7-10mm. The length and number of the flow area are determined by the heat exchange area. In order to increase the turbulence intensity and heat transfer coefficient, the sheets are often pressed into rugged shape, such as horizontal flat corrugated shape, herringbone corrugated shape, oblique corrugated shape, etc. Corrugated sheets used in the field of hot blast stoves are elliptical or sine wave forms etc. [7-9] The sheets are made of corrosion-resistant alloy steel with thickness of 0.8-1.2mm.

4.2 The Characteristics of Plate Heat Exchanger

The plate heat exchanger has the advantages of high heat transfer efficiency, small size, not easy to accumulate dust, and easy cleaning. The heat transfer coefficient of the plate heat exchanger is higher by 1-3 times than the tubular heat exchanger. But the disadvantage of the plate heat exchanger is too much welding spots, which are more likely to cause leakage. The other disadvantages are higher pressure loss than heat pipe heat exchanger and high investment.

In recent years, due to the improvement of welding technology and sealing technology, and the improvement of corrugated sheet material and the price drop, the application of plate heat exchangers in hot blast stoves has gradually increased. Now it has been used in various grades of blast furnace, which include 1000-5000m^3.

4.3 The Application of Plate Heat Exchanger

The blast furnaces designed by BSIET recent years, include Liangang No. 8 blast furnace (2800m^3), Xianggang No. 3 blast furnace (1800m^3) and Qinggang No. 2 blast furnace (1800m^3), all used plate heat exchangers. The BSIET also completed the reconstruction projects of Qiangang No. 2 blast furnace (2650m^3) and No. 3 blast furnace (4000m^3), in which to replace the heat pipe heat exchangers by plate heat exchangers.

The Liangang No. 8 blast furnace, designed by BSIET, is equipped with three dome combustion hot blast stoves. The design hot blast temperature is 1250℃. It was put into operation in March 2013. The hot blast stove system adopts a combined preheating technology, and both the gas and the CA are preheated by plate heat exchanger, and the preheating parameters are shown in Table 3.

Table 3 The parameters of Liangang No. 8 BF plate heat exchanger

Volume of blast furnace/m^3	2800
Type of hot blast stoves	Dome combustion HBS
Number of hot blast stoves	3
Fuel of hot blast stoves	BFG
Waste gas flow/Nm$^3 \cdot h^{-1}$	355,000
BFG flow/Nm$^3 \cdot h^{-1}$	255,000
Combustion air flow/Nm$^3 \cdot h^{-1}$	155,000
Waste gas temperature/℃	325
BFG inlet temperature/℃	70
Combustion air inlet temperature/℃	20
BFG outlet temperature/℃	220
Combustion air outlet temperature/℃	450

Now this plate heat exchanger has been used for more than 5 years. The gas and CA can be preheated to 180℃. If the pre-burning furnace is put into use, the air preheating temperature can be further improved. It shows that this type of heat exchanger can work well under the conditions of HBS waste gas and BFG. It can meet the long-term and stable use requirements of the hot blast stove, as long as the anti-corrosion and welding problems are handled well. Fig. 4 shows the plate heat exchanger installed on site.

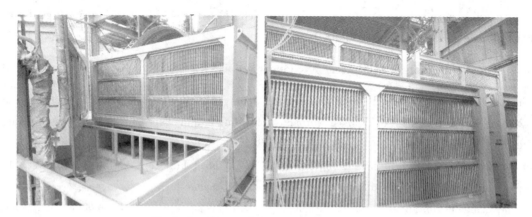

Fig. 4   Installation of plate heat exchange on site

5 Preheating Furnace

5.1 The Characteristics of Preheating Furnace

The method is preheating CA with preheating furnace, which is equivalent to a small hot blast stove. The internal regenerator is used to transfer heat through periodic combustion (heat storage) and air supply (heat release). The advantages of preheating furnace are high heating capacity and high heating temperature, which can preheat the CA to above 500℃ or higher. The disadvantages are that the investment is high and additional fuel is consumed. So this technology is often used in reconstruction projects, or projects need high blast temperature (>1250℃).

5.2 The Application of Preheating Furnace

Shougang No. 2 blast furnace had four Shougang type dome combustion hot blast stoves, which were put into production in 1979. In 2002, when the No. 2 blast furnace was reconstructed, three internal combustion hot blast stoves were built. At the same time, the old No. 3 and No. 4 dome combustion hot blast stoves were rebuilt, and they were used as preheating furnaces to preheat CA. Under the condition of fully using BFG, the high temperature of 1250℃ can be provided to the blast furnace. The technical parameters are shown in Table 4.

Table 4   The parameters of Shougang No. 2 BF CA preheating furnace

Volume of blast furnace/m³	1726
Type of hot blast stoves	Inner combustion HBS
Number of hot blast stoves	3
Type of preheating furnace	Dome combustion HBS
Number of preheating furnace	2
Fuel of hot blast stoves	BFG

Continued Table 4

Volume of blast furnace/m^3	1726
Consumption of preheating furnace BFG/Nm$^3 \cdot$ h^{-1}	31,000
Preheating temperature of CA/℃	600
Waste gas temperature/℃	290

Shougang No. 2 blast furnace both used preheating furnace and heat exchanger, which recycling the waste gas heat to preheat BGF. With this method, the hot blast stoves could provide hot blast of 1250℃ to the blast furnace[10].

6 Combined Preheating Technology

6.1 The Background of Combined Preheating Technology

This preheating technology is mainly developed in response to the development of steel production technology. At present, iron and steel associated enterprises' high-calorific value gas such as coke oven gas and converter gas mainly use for steelmaking and steel rolling processes. Meanwhile, increasing hot blast temperature is of great significance for operating BF smoothly, reducing coke ratio, increasing output and improving efficiency. Therefore, it is necessary to develop new technology with which the hot blast stoves can continuously and steadily provide 1250℃ hot blast to the blast furnace under the condition of fully using BFG.

On the other hand, with the application of large-scale blast furnace, fine material technology, high blast temperature technology, high-pressure operation technology, and oxygen-enriched pulverized coal injection technology, the fuel ratio of blast furnace is continuously reduced, and the chemical energy of BFG is fully utilized. The calorific value of BFG is only about 3000-3200kJ/m^3. Using BFG, the dome temperature of HBS only could be 1200-1300℃, and the hot blast temperature only could be 1100℃, which totally can't meet the requirements of modern blast furnace.

If only low-temperature preheating technology is adopted, the CA and gas can be preheated to 200℃, the hot blast temperature can be raised to 1150℃. It is still difficult to stably provide 1200℃ hot blast, not to mention 1250℃ or higher.

Therefore, combined preheating technology has been developed and used by Chinese steel companies. This technology uses the low-temperature heat exchanger simultaneously with combustion furnace or preheating furnace. The CA and gas are preheated to about 200℃ by the low-temperature heat exchanger, and then the CA is further preheated to 450℃ or higher by preheating furnace.

6.2 The Characteristics of Combined Preheating Technology

A typical technology of combined preheating technology is two-stage double preheating technology. The low-temperature preheating section of this technology can use heat pipe heat exchanger, tubular heat exchanger or plate heat exchanger. The function of this part is to recycle the residual heat of the waste gas of the HBS system. The temperature of waste gas is reduced from 300℃ to 150℃, and the CA and BFG are preheated to about 200℃.

The high-temperature preheating section uses the preheating furnace to further preheat the CA to 450℃ or higher. Old hot blast stoves or new small preheating furnaces can be used in this part. Two preheating furnaces work with one burn and one delivery working system, alternately provide high temperature CA to hot blast stoves.

The advantages of this technology are following.

(1) Higher CA temperature and suitable gas preheating temperature are obtained. Hot blast stoves can continuously and steadily provide 1300℃ hot blast to the blast furnace under the condition of fully using BFG.

(2) The preheating furnace has long service life and works stably without failure problem.

(3) When the low temperature preheating device fails or is repaired, the HBS system can still provide higher than 1200℃ hot blast to BF by increasing the preheating temperature of CA. This is very important to ensure that the blast furnace is always operated at a high blast temperature level, and to avoid fluctuations of the BF working condition, and to reduce the coke ratio.

The disadvantage of this technology is that the investment is high, especially for new projects. Because two new preheating furnaces need to be built. But the economic benefit is still significant because lots of coke could be saved and output could be increased. It is also benefit for BF smoothly operating with steady high temperature hot blast.

6.3 The Application of Combined Preheating Technology

In the design of hot blast stove preheating system of $5500m^3$ BF of Shougang Jingtang steel plant, two stage double preheating system has been used. This technology is developed and designed successfully on the basis of high CA preheating temperature of Shougang[11].

The separated type heat pipe heat exchanger is used to recycle the residual heat of waste gas, and preheat CA and BFG to about 200℃. This process is called stage one double preheating. Two sets of regenerative preheating furnace are used to preheat the CA to 550℃.

The BFG and CA of preheating furnace are preheated by heat pipe heat exchangers. The dome temperature of preheating furnace can reach to 1300℃. The temperature of CA for HBS can reach to 1200℃ after it going through the preheating furnace. Then it will be mixed with the CA preheated by heat pipe heat exchanger, and the final temperature after mixing is 550-600℃. This process is called stage two. This is a process of self-circulating preheating process. It significantly improve the physical heat of air and gas. The dome temperature of hot blast stove also is improved. With this process, the thermal efficiency of HBS system is improved significantly. Fig. 5 shows the two stage double preheating process of high blast temperature HBS system.

Shougang Jingtang No. 1 BF was put into operation in May 2009. Under the condition of fully using BFG, the HBS system stably supply 1300℃ hot blast to the BF. In recent years, with the

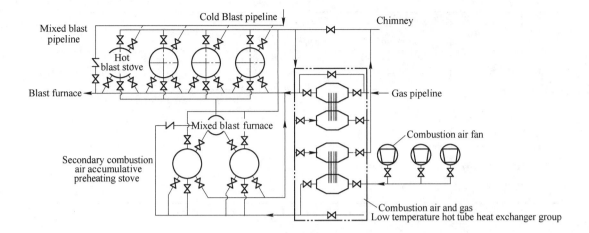

Fig. 5 Two stage double preheating process of high blast temperature HBS system

heat pipe heat exchanger gradually failing, BFG and CA can only be preheated to 80-100℃. But the HBS system still can provide 1200-1250℃ hot blast to the BF, which is benefit for the BF to reduce coke ratio.

Through quantitative comparison between low temperature double preheating and two-stage double preheating process under the design conditions of Shougang Jingtang 5500m^3 blast furnace, it is proved that the two-stage double preheating process consumes 34,313Nm^3/h BFG more than the low temperature double preheating process. However, the hot blast temperature increases from 1200℃ to 1300℃. In advantage of hot blast increased 100℃, the BF can save 15kg coke per ton of hot metal, and can save 143,700 tons per year. The economic benefits are significant.

Shougang Qiangang No. 3 blast furnace, which was put into operation in January 2010, also applied two-stage double preheating technology. The low-temperature preheating section used heat pipe heat exchanger to recycle the residual heat of waste gas, and preheat BFG to 180℃. The high-temperature preheating section used preheating furnace to preheat CA to 650℃. After putting into production, the monthly average blast temperature was 1280℃, and the effect of saving coke was remarkable [12].

By summarizing the characteristics and application of the five preheating technologies, the advantages and disadvantages of each of them are shown in Table 5.

Table 5 The comparison of five preheating technologies

Type of preheating	Preheating temperature/℃	Hot blast temperature/℃	Problems	Application range
Tubular heat exchanger	200-250	1150	Low thermal efficiency, easy to be blocked	Small
Heat pipe heat exchanger	200-250	1150	Easy to be blocked, the heat medium is easy to fail	Gradually decreasing.

Continued Table 5

Type of preheating	Preheating temperature/°C	Hot blast temperature/°C	Problems	Application range
Plate heat exchanger	200-250	1150	Two much welding spots, easy to leak	Gradually increasing.
Preheating furnace	>450	1250	Do not recycle the residual heat of waste gas, low thermal efficiency	small
Combined preheating technology	BFG: 200-250; CA: >450	>1250	High investment	Often used in the BF larger than 3000m³

7 Conclusion

With suitable heat exchange area and flow area, tubular heat exchanger can work long and steadily. However, the heat exchange efficiency of tubular heat exchanger is lower than plate heat exchanger's. Currently, it is used less in the field of blast furnace hot blast stoves in China.

The heat pipe heat exchanger has high heat exchange efficiency. It is not easy to leak, and has good safety. However, it can fail after 2-3 years of use. In recent years, the application of heat pipe heat exchanger gradually reduced in the field of blast furnace hot blast stoves in China.

With the advancement of the manufacturing process of plate heat exchangers and the reduction of manufacturing costs, plate heat exchangers, which were expensive and had high leakage risks, are being used more widely in Chinese steel enterprises. It can be combined with pre-burning furnaces. Through this combination, the preheating temperature of CA will be increased to 400°C, which meets the hot blast temperature requirements of 1200-1250°C.

For large blast furnaces above 3000m³, considering the factors of investment, longevity, reliability and stability, only the two-stage double preheating technology can meet the requirements.

Reference

[1] Zhang Fuming. BF ironmaking technological development trend characterized by green and low carbon emission [J]. Ironmaking, 2016, 35 (1): 1-6.

[2] Zhen Changliang, Cheng Cuihua. Review on technique and device of recycling hot blast stove residual heat [J]. Energy for Metallurgical Industry, 2017, 36 (3): 47-50.

[3] Zhang Fuming. Development and problems of ironmaking industry after entering the 21st century in China [J]. Ironmaking, 2013, 32 (6): 1-5.

[4] Shu Jun. An approach to method achieving high blast temperature with low calorific value BF gas [J]. Iron and Steel, 1997, 32: 399-401.

[5] Zhang Hong, Zhuang Jun. Research and application of heat pipe technology in industry [J]. Chemical Equipment and Technology, 2001, 22 (6): 11-14.

[6] Qi Lidong. Technical innovation and application of gas preheater for blast furnace hot blast stove [J]. Metallurgical Equipment, 2017: 41-56.

[7] Wu Zhiqiang. A corrugated sheet for double preheating device of blast furnace hot blast stove [P]. CN201120266285.5, 2012-02-29.

[8] Fujian Lixin. Heat Exchange Equipment Manufacture Corporation. A slab that is used for evaporative cooling or condensation [P]. CN201220389408.9, 2012-08-08.

[9] Shanghai Lalleen Engineering Technology Co., Ltd. Heat transfer plate for gas phase plate preheater [P]. CN201020195319.1, 2011-01-19.

[10] Huang Donghui, Han Xiangdong. Application practice of burning technique of high temperature air hot stove for shougang's No. 2 BF [J]. Ironmaking, 2004, 23 (2): 15-18

[11] Zhang Fuming, Li Xin, Hu Zurui. Research on high efficiency energy conversion technology for modern hot blast stove [C]. TMS2018, Energy Technology 2018 (Carbon Dioxide Management and Other Technologies), 2018: 133-152.

[12] Ni Ping. Study and application of high temperature hot blast technology on Qiangang 4000m^3 blast furnace [C]. 2011 CSM Annual Meeting Proceedings, 2011.

高炉富氧喷煤
与煤气干法除尘设计研究

浅谈首钢高炉喷煤技术的发展方向

摘　要： 首钢高炉喷煤始于1964年。讨论了进一步发展首钢高炉喷煤技术应该采取的一些措施，如提高精料水平，提高热风温度、改用分配器系统的喷吹方式、实现富氧大喷煤、喷吹烟煤、采用煤粉浓相输送技术，提高喷煤自动化控制水平等。

关键词： 高炉；煤粉喷吹；炉料；高温鼓风；烟煤

1　引言

1964年4月30日首钢1号高炉开始喷煤，迄今已30年有余，30年来首钢一直采用双罐双系列多管路喷煤工艺，改造后的2号、4号高炉仍沿用了该工艺。3号高炉易地大修时，炉容扩大到2536m³，设计煤比150~200kg/t，喷煤量为48~64t/h，传统的双罐双系列工艺已不能满足此设计要求。为进一步提高喷煤量，更好地解决输煤和喷煤两系统之间的矛盾，设计中采用三罐三系列多管路喷煤工艺。在中间罐和喷吹罐之间设置了4根刚性拉杆，以克服煤粉倒罐时因浮力产生的称量误差，但并未完全实现连续计量。1号高炉的喷煤系统设计与3号高炉相同。从图1和表1不难看出，4座高炉改造投产后，燃料比和焦比有所上升，而喷煤量和喷煤率大幅度下降，其原因是多方面的，但笔者认为以下几点不容忽视：（1）由于高炉生产规模扩大，原燃料消耗量增加，原燃料条件变差，为高炉生产带来了许多不利因素；（2）高炉投产以后，对高炉喷煤技术的重要性和必要性认识不足，片面追求高强度冶炼，忽视了高产、低耗、长寿的统一关系；（3）高炉容积扩大以后，在新的冶炼条件下高炉喷煤操作规律尚需进一步摸索；（4）喷煤工艺设计存在许多不足之处，基本上都是在原有基础上加以改进，没有大的突破和创新。

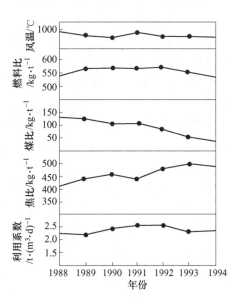

图1　近几年首钢高炉主要生产指标的变化

然而时代在进步，技术在发展。近10多年来，高炉喷煤技术发展迅速，取得了引人注目的成就。西欧从1980年第一座高炉开始喷煤算起，到1991年已有42座高炉喷煤[1]。

本文作者：张福明。原刊于《炼铁》，1995，14（6）：16-20。

表 1 首钢 4 座高炉大修改造后喷煤技术指标对比

项 目	1 号高炉		2 号高炉		3 号高炉		4 号高炉	
	改造前	改造后	改造前	改造后	改造前	改造后	改造前	改造后
时间/年	1991	1994	1989	1994	1991	1994	1989	1994
有效容积/m^3	576	2536	1327	1726	1036	2536	1200	2100
利用系数/$t \cdot (m^3 \cdot d)^{-1}$	3.009	1.8232	2.258	2.351	2.609	2.41	2.589	2.381
风温/℃	1051	910	1057	931	1047	999	1058	990
风量/$m^3 \cdot min^{-1}$	138	4478	2612	3998	2190	5231	2579	4824
焦比/$kg \cdot t^{-1}$	406.5	539.8	408.6	487.8	444.8	458.7	395.6	490.2
煤比/$kg \cdot t^{-1}$	113.8	28.6	131.6	51.0	119.7	49.8	143.3	53.3
燃料比/$kg \cdot t^{-1}$	520.3	568.4	540.2	538.8	564.5	508.5	538.9	543.5
喷煤率/%	21.87	5.0	24.36	9.47	21.20	9.79	26.59	9.8
富氧率/%	3.01	0	3.24	2.36	0.81	0	3.68	0.25

近几年，西欧高炉喷煤技术发展很快，目前已处于世界领先地位。1991 年英国克里夫兰厂 4 号高炉（600m^3）成功地进行了高富氧大喷煤工业试验，在鼓风含氧 40% 的条件下，喷煤量达到 300kg/t，并且实现了半焦半煤操作（焦比 270kg/t，煤比 270kg/t）[2]，创出了世界上高富氧大喷煤高炉冶炼的最新水平。

在日本，从 1981 年新日铁大分厂 1 号高炉建成第一套喷煤装置到 1992 年末，33 座高炉中有 27 座喷煤。1991 年日本平均喷煤量超过 100kg/t 的高炉已有 9 座，最高月平均喷煤量为 187.8kg/t（神户厂 3 号高炉 1991 年 9 月），小仓厂 2 号高炉 1991 年喷煤量 148kg/t，焦比 357kg/t，燃料比 505kg/t。表 2 列出了国外部分高炉主要喷煤技术指标。

表 2 国外部分高炉喷煤技术指标

国别	厂名	炉号	高炉容积/m^3	最高月平均			时间	全年平均			时间	喷吹方式
				焦比/$kg \cdot t^{-1}$	煤比/$kg \cdot t^{-1}$	喷煤率/%		焦比/$kg \cdot t^{-1}$	煤比/$kg \cdot t^{-1}$	喷煤率/%		
德国	施韦尔根	1	4670	307	201	39.5	1992 年 11 月	320	179	35.9	1992 年	分配器
	汉伯恩	4	2243					350	145	29.3	1992 年	分配器
	鲁劳特	6	2151					360	146	28.9	1992 年	分配器
英国	斯肯索普	Vict	1791	314	201	39.0	1991 年 10 月	332	165	33.2	1991 年 7 月～1992 年 6 月	多管路
		Anne	1791	330	161.7	32.9	1992 年 6 月	356	142	28.5	1991 年 7 月～1992 年 6 月	多管路

续表2

国别	厂名	炉号	高炉容积/m³	最高月平均			时间	全年平均			时间	喷吹方式
				焦比/kg·t⁻¹	煤比/kg·t⁻¹	喷煤率/%		焦比/kg·t⁻¹	煤比/kg·t⁻¹	喷煤率/%		
法国	敦刻尔克	4	4497	287	194	40.3	1992年5月14日~6月14日	309	170	35.5	1992年	多管路
荷兰	艾莫依登	6	2678	240	204	42.5	1992年11月	307	175	36.3	1992年	分配器
日本	神户	3	1845	313	187.8	37.5	1991年9月	323	175	35.7	1991年	多管路
	大分	1	4158		141.5		1992年6月	360.9	128.3	26.2	1992年	分配器
	小仓	2	1850					357	148	29.3	1992年	分配器
韩国	浦项	3	3795					387.5	102.3	20.9	1992年	分配器
	光阳	3	3800					373.2	118.2	24.1	1992年	分配器

在我国，自1989年以来，冶金部大力推动喷煤技术已取得了明显的进展。高炉喷煤量不断提高，全国高炉喷煤总量1988年为203.75万吨，1989年为215万吨，1994年为354万吨。高炉喷吹烟煤技术已过关。高炉富氧喷煤技术取得了进展，高炉喷煤工艺装备全面进步。此外，煤粉倒罐时连续计量技术、高炉风口煤粉分配技术、煤粉单支管计量、氧煤枪、煤粉浓相输送等项技术也取得了初步成功。国内近几年大修改造和新建的大型高炉，如宝钢2号高炉（4063m³）、3号高炉（4350m³）、武钢5号高炉（3200m³），鞍钢11号高炉（2580m³），马钢新1号高炉（2545m³）等在喷煤系统设计中引进和吸收了国外一些先进技术，为首钢喷煤技术的继续发展提供了良好的外部条件。总而言之，首钢的高炉喷煤技术与国内外先进水平相比还有一定差距，许多技术问题尚未解决，随着喷煤量的增加，下列问题会更加突出。如：各风口之间煤量分配差值较大，喷吹效果不佳，不利于提高喷煤量；煤粉在风口区燃烧的问题，随着煤比的进一步提高，如果不采用新的喷煤工艺，煤粉的利用率也不会很好；喷煤的计量技术及自动化控制水平更有待于提高。本文就首钢高炉喷煤技术的发展方向谈几点看法。

2 首钢高炉喷煤技术的发展方向

2.1 提高精料水平

精料是高炉实现大喷煤量、高风温、高顶压、高煤气利用率及高炉长寿的基础。尤其是高炉大量喷煤时，对原燃料的要求更加严格。国外实现大喷煤量的高炉都有很好的原燃料条件。烧结矿 TFe 57%~58%，FeO 小于10%，碱度（CaO/SiO$_2$）大于1.6，TI（+6.3mm）75%~80%，RDI 小于30%，入炉烧结矿中小于5mm的粉末含量低于10%，渣量250~300kg/t，焦炭灰分小于10%，M_{40} 80%~87%，M_{10} 6%~7%，反应性小于28%，SCR大于60%，大部分焦炭粒度40~80mm，平均粒度大于50mm，这对于高炉大量喷煤和其他先进技术的应用非常有利。近几年，随着首钢高炉生产规模的扩大，原燃料消耗量大幅度增加，烧结矿和焦炭供不应求，有50%左右的焦炭需要外购，由于产地较多，焦炭成分和理化性能波动很大，原燃料条件变差（见表3和表4）。提高原燃料成分及理化性能的稳定性，减少波动，这是保证高炉稳定顺行的基本条件。因此，首钢在今后一个时期内应该下大力气提高精料水平，不断改进烧结矿质量，提高品位，降低FeO含量，改善烧结矿的还原性和高温性能，加强入炉前的筛分，减少入炉粉末，有条件要采用分级入炉技术；降低焦炭灰分（<10%），提高焦炭强度（M_{40}>80%，M_{10}<7%），提高入炉焦炭的平均粒度（>40mm），同时将10~25mm的焦丁与烧结矿混装入炉，提高料柱的透气性。

表3　近几年首钢自产烧结矿化学成分及性能　　　　　　　　　　　　（%）

年份	TFe	FeO	SiO_2	CaO	MeO	MnO	S	残C	CaO/SiO_2	TI
1988	57.79	11.08	6.00	8.98	2.47	0.18	0.012	0.08	1.51	87.17
1989	57.91	11.28	5.73	8.80	2.50	0.15	0.012	0.05	1.54	88.00
1990	57.77	10.73	5.66	9.08	2.52	0.15	0.013	0.05	1.69	88.36
1991	56.64	10.65	5.84	9.87	2.86	0.26	0.013	0.05	1.69	87.75
1992	56.37	10.57	5.74	10.35	3.03	0.16	0.017	0.05	1.80	88.68
1993	56.26	10.54	5.84	10.64	3.12	0.17	0.020	0.05	1.82	88.48
1994	54.58	10.22	6.11	12.15	3.37	0.19	0.030	0.05	1.90	89.03

表4　近几年首钢高炉焦炭化学成分及性能　　　　　　　　　　　　（%）

年份	水分	灰分	挥发分	S	M_{40}	M_{10}
1988	3.90	12.91	1.09	0.82	77.0	8.20
1989	3.90	13.36	1.17	0.76	76.90	8.90
1990	3.90	12.42	1.17	0.69	76.80	8.20
1991	4.40	12.61	1.16	0.66	76.70	8.30
1992	4.40	12.26	1.17	0.63	78.5	7.90
1993	4.70	12.43	1.37	0.64	78.1	7.90
1994	5.00	12.63	1.44	0.64	79.30	7.70

2.2 提高热风温度

近几年首钢高炉的风温水平一直偏低，年平均尚未达到1000℃，这严重限制了喷煤量的提高。造成风温偏低的主要原因有：（1）由于干法布袋除尘系统尚未投入使用，高炉煤气净化一直采用湿法处理系统，煤气的含尘量和水分较高，发热值低，加之焦炉煤气大部分供应民用，热风炉只燃烧单一高炉煤气，拱顶温度仅有1300℃左右；（2）1号、3号高炉的热风炉烟气余热回收预热助燃空气系统建成以后未投入使用；（3）个别高炉热风炉的加热面积较小，如2号高炉每立方米炉容的加热面积仅为84.7m²/m³；（4）热风炉的装备水平和结构形式不能满足高风温要求。为了在首钢的高炉上实现1150~1200℃的高风温,需要设计、生产等单位多方配合,不懈努力。高炉煤气干法除尘系统要尽快投产,为热风炉提供合格的煤气,有条件应采用高炉、转炉混合煤气作为烧炉燃料;热风炉助燃空气预热系统也要尽快投产,一般可提高风温80~150℃;今后设计中要增设热风炉自身预热系统,对于顶燃式热风炉的结构形式要进行改进和完善,充分满足高风温的要求。

2.3 改进喷吹方式

目前高炉喷吹方式主要有两种:一种是多管路系统,另一种是分配器系统,煤粉分配器是一项很重要的喷煤新技术。首钢5号高炉（1036m³）1995年2月中修时对喷煤系统进行了改造,将原有多管路系统改为分配器系统,大大简化了操作,设备故障率降低,减轻了工人的劳动强度。该设备运行5个月就显示出明显的技术优势,在原燃料条件基本相同的条件下,5号高炉的煤比一直在全厂处于领先地位（见表5）。应在其他高炉上推广分配器技术。

表5　首钢各高炉1995年4~8月喷煤量对比　　　　（kg/t）

月份	1号高炉	2号高炉	3号高炉	4号高炉	5号高炉	全厂平均
4	6.9	60.6	1.1	41.7	57.3	28.1
5	6.2	25.7	10.7	60.5	63.1	27.9
6	7.4	27.2	16.5	38.3	66.1	26.7
7	21.7	60.7	20.5	34.7	66.1	34.6
8	52.1	83.4	11.8	42.6	61.5	45.9

2.4 完善串联罐系统的连续计量

准确的连续计量,不仅可以随时检测全过程的喷煤量,而且便于监测倒罐过程,防止煤粉存积,除有利于安全喷吹外,还可为实现自动化控制喷煤量创造必要的前提条件。解决连续计量问题的关键是消除煤粉倒罐时因波纹补偿器产生的浮力对称量造成的误差。包钢1号高炉、宝钢2号高炉等都较好地解决了这个问题。首钢要从根本上解决连续计量问题,应采用新型结构（压力补偿型）波纹补偿器,以消除倒罐时产生的浮力,并用计算机程序校正计量误差,实现喷煤全过程的连续计量。

2.5 实现富氧大喷煤

富氧大喷煤是指导我国高炉生产的重要技术方针。富氧能够提高煤粉燃烧率,大喷煤量必须同富氧相结合,喷煤量达到150kg/t时,富氧率应达到2%~4%,喷煤量达到200kg/t时,富氧率应达到5%~8%。目前首钢高炉用氧都是炼钢余氧,供氧能力严重不足。要建设高炉专用的制氧机,使高炉鼓风含氧量达到25%。在喷煤量达到150kg/t以上时,应采用氧煤枪喷吹,提高煤粉的燃烧率。

2.6 采用喷吹烟煤和煤粉浓相输送技术

首钢曾在高炉上喷吹过烟煤并获得了成功,但因安全问题未能坚持下来,几十年来一直喷吹无烟煤。喷吹烟煤是高炉喷煤技术的发展方向，而实现喷吹烟煤技术的关键则是解决工艺过程的安全问题。低氧浓度操作是保证安全的关键，在整个系统中（包括制粉、输煤、喷煤）均应设氧浓度监控点，监控值通常为5%～10%，并设置上限、极限报警。同时要防止输煤喷煤过程中的煤粉存积，消除火源，防止静电，杜绝火花，设置温度、压力、CO浓度监控及氮气灭爆系统、消防灭火系统、泄爆装置等。实际上喷吹烟煤技术目前已很成功，完全可以在首钢高炉上推广应用。

煤粉浓相输送可减少压缩空气的消耗，减轻管道磨损，降低静电产生的能量，有利于烟煤的安全输送，而且输送能力大，输送距离长，混合比可达90～100kg/kg，也应推广应用。

2.7 提高喷煤自动化控制水平

首钢高炉喷煤自动化控制水平有待于进一步提高，在喷吹烟煤时的监控技术，单支管计量、风口进风流量计量、喷煤量调节装置的联锁控制以及连续计量的自动校正等方面要全部实现计算机管理，实现喷煤全过程的自动化在线控制。

2.8 探索大喷煤量下的高炉操作技术

高炉喷煤操作实践表明：大喷煤量时，必须维持合理的理论燃烧温度，一般为2050～2350℃。要通过提高风温（1150～1200℃）、增加富氧率（4%）及其他调剂手段，维持合理的理论燃烧温度；充分发挥无料钟炉顶的技术优势，积极采用炉料分布控制技术，在焦炭负荷加大以后，炉内焦炭层的厚度仍保持不变，并通过上部调剂，控制好煤气流的合理分布，提高煤气利用率；风口要广喷、均喷，不轻易停枪，维持炉缸周围工作的均匀；开展高炉富氧大喷煤冶炼技术的基础研究和工业试验工作，借鉴国内外的成功经验，不断探索现代化大型高炉的冶炼规律。

3 结语

首钢高炉喷煤技术以前曾居世界先进水平，近些年来与世界先进水平有了一定的差距。提高高炉喷煤技术是一项系统工程，涉及范围广，需各方面互相配合。首钢高炉喷煤技术起步早，有很好的技术基础，只要领导重视，科技人员及广大炼铁工作者努力，首钢高炉喷煤技术重返世界先进行列是大有希望的。

参考文献

[1] 唐文权，国外高炉喷煤技术 [J]. 炼铁，1993（6）：45.
[2] Campbell A, et al. Oxygen-coal injection at cleveland ironworks [J]. Ironmaking and Steelmaking, 1992 (2): 120.

On Technical Measures to Further Advance the PCI Technology Progress of Shougang Corporation

Zhang Fuming

Abstract: The injection of coal into blast furnace in Shougang Co. was started in 1964. Some technical measures to further advance the PCI technology progress of Shougang Co. are described, such as improving the quality of burden material, raising hot blast temperature, adopting injection system with distributors, carrying out high pulverized coal injection with oxygen enrichment to the blast, injecting bituminous coal, conveying the pulverized coal to dense flow, improving the automatic control for coal injection and so on.

Keywords: blast furnace; pulverized coal injection; burden material; high temperature blast; bituminous coal

关于高炉富氧喷煤技术问题的探讨

摘　要：简介了高炉煤粉燃烧过程和燃烧机理，探讨了提高煤粉燃烧率及喷煤量的一系列技术问题。分析认为，煤粉在风口区的燃烧是一复杂的物理化学反应过程，其在风口回旋区内的燃烧率一般只有70%～80%，未燃烧煤粉对高炉顺行的破坏作用不容忽视。

关键词：高炉；喷煤；燃烧效率；富氧鼓风

1　引言

发展高炉富氧喷煤技术是我国"九五"期间钢铁工业结构优化的重点。高炉喷煤已不仅仅是一个以煤代焦、降低能耗的技术问题，它关系到整个钢铁工业主流程的结构优化，具有深远的战略意义。目前，高炉炼铁技术正面临着直接还原、熔融还原等非高炉炼铁技术的挑战，实现富氧大喷煤是其最终出路。富氧喷煤技术的关键是提高煤粉燃烧率，这是实现高炉大喷煤（煤比200kg/t以上）的限制性环节。本文就提高煤粉燃烧率及喷煤量的一系列技术问题进行了探讨。

2　煤粉燃烧过程和燃烧机理

煤粉在由喷枪进入直吹管、风口、回旋区的过程中，快速进行着热量、质量和动量的传输以及一系列物理化学反应。研究表明，煤粉由喷枪进入直吹管后，首先被热风快速加热（煤粉温度一般为50～80℃，热风温度1000～1250℃），预热时间一般只有10ms左右，随后进行脱气和快速热分解（即煤的热分解和挥发分的二次热分解），然后着火，挥发物进行燃烧反应，最后是残炭（或称半焦）与氧化性气体进行燃烧的多相反应，该反应占煤粉燃烧全过程50%以上的时间（见图1和图2）。

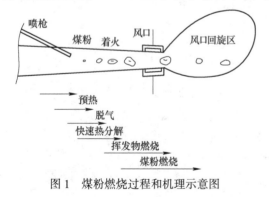

图1　煤粉燃烧过程和机理示意图

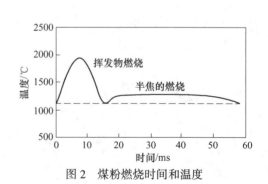

图2　煤粉燃烧时间和温度

本文作者：张福明。原刊于《首钢科技》，1996（5）：52-59。

关于煤粉的燃烧过程和机理，近年来又取得了一些新的研究成果：

（1）煤粉在预热阶段，辐射传热是其主要的传热过程。在此阶段，粒度74μm的煤粉加热速率为 $10^5 \sim 10^8$ ℃/s。煤粉在直吹管内的传热和运动状态对脱气和热分解非常重要[1]。

（2）高速加热条件下，煤粉的脱气和快速热分解几乎同时进行，燃烧与气化的第一步受煤粉热分解过程的制约[2]。

（3）煤种对快速热分解及燃烧气化过程有决定性影响。烟煤的热分解及燃烧气化速度快于无烟煤，但热分解后形成大量冷凝物和烟炭；气氛对于高速加热快速热分解影响较小，而对其后的多相燃烧气化反应影响较大[3]。

（4）残炭的燃烧气化为扩散反应。燃烧动力学研究表明，在高炉生产条件下，其反应速率主要受氧向残炭表面传质速度的限制，残炭燃尽所需时间取决于燃烧区的温度、气相中的氧浓度、煤粉粒度及其化学性质。在粒度、温度相同的条件下，煤粉燃烧速度随气相中氧浓度成正比地增加[4]。

上述关于煤粉燃烧过程和机理的研究，有助于采取有效措施，提高煤粉燃烧率和喷煤量。

3 煤粉燃烧率和未燃煤粉

文献 [4] 认为，喷入的煤粉在回旋区内未能完全燃烧，煤粉燃烧率最高只有70%～80%（见图3），因而残炭或半焦的最终气化是在回旋区以外进行的。

有关专家将煤粉燃烧率定义为：

$$\eta = \frac{1 - A_0/A_1}{1 - A_0} \times 100\% \quad (1)$$

式中　η——煤粉燃烧率；
　　　A_0——煤粉燃烧前的灰分，%；
　　　A_1——煤粉部分燃烧后的灰分，%。

理论和实测结果都表明，煤粉在风口回旋区不可能100%燃烧，必然有一

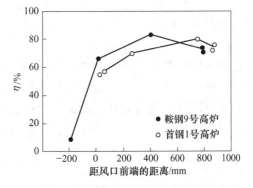

图3　风口区煤粉燃烧率变化规律

部分未燃煤粉存在。未燃煤粉在高炉内的行为主要是与 CO_2、H_2O 等氧化性气体发生气化反应；参与铁及非铁元素的直接还原和生铁渗碳等反应，并有少量未燃煤粉进入炉渣或随煤气逸出炉外。

一般说来，高炉喷煤后风口回旋区的结构将发生变化（见图4）。风口回旋区内尚未燃尽的煤粉（残炭）将在煤气流的带动下沉积在软熔带根部和炉缸死焦柱内（见图5）。由此可知，如果未燃煤粉大量增加，则势必会造成焦窗堵塞、炉缸中心堆积、料柱透气性和透液性变差。日本和国内一些高炉，在喷煤量提高以后出现了边缘气流过分发展、炉墙热负荷增加的现象，就可能是未燃煤粉过多所致。

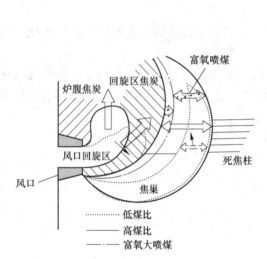

图 4 高炉喷煤时风口回旋区结构

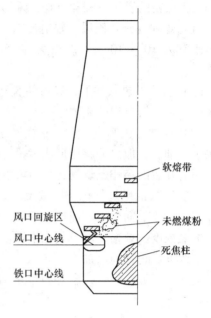

图 5 未燃煤粉在高炉内的沉积位置

研究表明,未燃煤粉活性较高,在高炉内比焦炭更易于与二氧化碳发生气化反应(见图6)。从这个意义上讲,部分未燃煤粉吸附在焦炭表面,可优先与 CO_2 发生气化反应,从而保护焦炭,减少焦炭的熔损。

目前高炉能接受的最大未燃煤粉量尚不清楚,但可做如下推测:在高炉条件下,碳素溶解反应温度区间为 600~1200℃,目前我

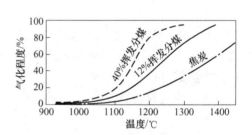

图 6 焦炭和未燃煤粉的气化率

国焦炭反应性约为30%,如果发生碳素溶解反应的焦炭有70%由未燃煤粉代替,高炉喷煤时焦比400kg/t,则高炉可接受的未燃煤粉量为 400×30%×70%=84kg/t。

1966年,首钢1号高炉在大喷煤条件下煤粉利用率高达97%以上,但这并不说明未燃煤粉量无论多大都可保证有较高的煤粉利用率,因为该高炉较高的煤粉利用率是在风温1050~1150℃、富氧率2%~4%、理论燃烧温度大于2050℃、风口前煤粉燃烧率达70%~80%的条件下获得的。笔者认为,片面强调未燃煤粉对焦炭的保护作用,甚至想利用高反应性未燃煤粉保护焦炭,从而降低焦比、提高喷煤量的观点是值得商榷的。这是因为高炉焦炭在下降过程中,其粒度和理化性能将发生变化,尤其在块状带中下部,焦炭和 CO_2 气体发生碳素溶解反应($C_{焦}+CO_2=2CO$),使焦炭粒度变小,强度变差。在此区域内保护焦炭,减少焦炭的碳素溶解损失,虽然会使焦炭在软熔带、滴落带、风口带以及死焦柱区具有适宜的粒度和较高的强度,改善料柱的透气性和透液性,从而有利于高炉顺行,但问题是未燃煤粉能否随煤气流穿透软熔带到达块状带,以及在块状带内能否有效地保护焦炭,目前尚无定论。而且未燃煤粉过多,也势必给高炉操作带来困难,并降低煤粉利用率。因此,提高风口前煤粉燃烧率才是提高喷煤量的最有效的途径。

4 提高煤粉燃烧率的途径

4.1 提高鼓风含氧量

煤粉在高炉风口前的燃烧主要受氧的扩散传质限制,因此,提高煤粉燃烧区气相中的氧浓度可以促进煤粉燃烧。有关专家证实,热风温度 600~1100℃时,煤粉燃烧率几乎与风温成正比,将鼓风含氧量从21%增至25%,效果比风温从1000℃提高到1100℃约好4倍[5]。鼓风含氧量与煤粉燃烧率的关系如图7所示,风温与煤粉燃烧率的关系如图8所示[6]。从图7、图8可以看出,增加鼓风含氧量比提高风温对煤粉燃烧率的影响更大。

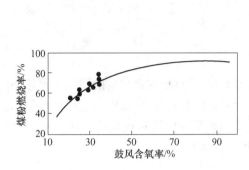

图7 鼓风含氧率与煤粉燃烧率的关系

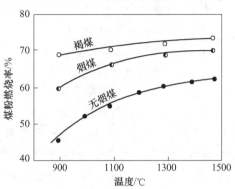

图8 风温与煤粉燃烧率的关系

目前富氧鼓风已成为提高煤粉燃烧率和喷煤量的有效措施。卢森堡PW公司曾根据欧洲高炉的喷煤实践,对喷煤量和富氧率提出了这样的定性关系:煤比小于120kg/t时,不必富氧;煤比在120~160kg/t时,需鼓风富氧2%~3%;煤比大于160kg/t时,除鼓风富氧外还可采用氧煤枪。

国内外一些高炉鼓风富氧与喷煤量的关系见表1,喷煤量与富氧率的关系如图9所示。

表1 国内外一些高炉鼓风富氧与喷煤量的关系

项目	炉容/m³	时间	煤比/kg·t⁻¹	富氧率/%
英国钢铁公司斯肯索普厂维多利亚女王号高炉	1534	1992年	144.2	7.32
荷兰霍戈文公司艾莫伊登厂6号高炉	2327	1994年	177	4.9
德国蒂森公司施韦尔根厂1号高炉	4337		220	3.24
美国钢铁公司盖瑞厂13号高炉	2953		283.5	6~7
日本NKK公司福山厂4号高炉	4288	1994年10月	218	3
鞍钢3号高炉	831	1995年8~11月	203	3.24
首钢原1号高炉	576	1966年	225	2.1

喷煤量与富氧率的关系可用风口前氧过剩系数(EXO)进行定量分析。假定煤粉中可燃成分碳、氢完全燃烧。生成 CO_2 和 H_2O,则可导出 EXO 的表达式:

$$EXO = \frac{实际供氧量}{理论耗氧量} = \frac{V_0[(1-f)W + 0.5f] \times 60^5/n}{\left[\left(\frac{32}{12}P_{cr} \cdot C^y + \frac{16}{2}P_{cr} \cdot H^y\right) \times \frac{22.4}{32 \times 1000}\right]/n_1}$$

整理得：

$$EXO = \frac{V_b[(1-f)W + 0.5f] \times 10^5/n}{(3.11 P_{cr} \cdot C^y + 9.33 P_{tt} \cdot H^y)/n_1}$$

式中　V_b——入炉风量，m³/min；

　　　f——鼓风湿度，%；

　　　W——鼓风含氧量，%；

　　　P_{cr}——喷煤量，kg/h；

　　　C^y——煤粉中碳含量，%；

　　　H^y——煤粉中氢含量，%；

　　　n——送风风口数目，个；

　　　n_1——喷煤风口数据，个。

根据首钢高炉富氧喷煤实践，当 EXO 大于1.15时，煤粉在风口前的燃烧率可达70%以上。文献[7]研究发现，EXO 同煤粉置换比也有密切关系（见图10）。从中可看出，EXO 越高，置换比也越高，大喷煤时须提高鼓风含氧量以保证有较高的 EXO（≥1.45），从而获得较高的煤粉燃烧率（>80%）和置换比（>0.8）。

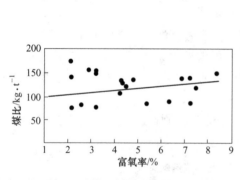

图9　喷煤量与富氧率的关系

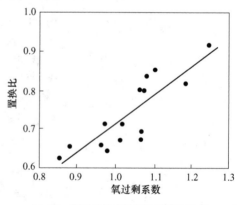

图10　置换比与氧过剩系数的关系

4.2　选择适宜的煤粉粒度

文献[8]指出，直径为 d_p 的固体碳燃烧时，若受气体扩散的控制，则煤粉燃烧时间与其粒径的平方有如下关系：

$$t_b = B_0 \times d_p^2$$

式中　t_b——煤粉燃烧时间，ms；

　　　B_0——当气氛中氧浓度一定时为常数，ms/m²；

　　　d_p——煤粉粒径，m。

由此可知，煤粉粒度越细，比表面积越大，则反应能力越强，燃烧时间越短，因此细粒煤粉有利于提高煤粉燃烧率。20 世纪 60 年代，喷吹细粒煤粉（小于 200 目占 80% 以上）曾是首钢高炉大喷煤的经验之一。文献［9］对不同煤种的粒径与燃烧率的关系进行了更深入的研究，认为当煤粉可燃基挥发分小于 20% 时，粒径越细燃烧率越高；但当可燃基挥发分在 25%～40% 时，煤粉燃烧率并不随粒径变细而增加，而是在小于 200 目占 60% 时燃烧率最高；当可燃基挥发分大于 40% 时，煤粉粒径越细，燃烧率反而下降。一般说来，高变质程度的煤种，粒度小于 200 目的应控制在 70% 以上；中低变质程度的煤种，粒度小于 200 目的可扩大到 60% 左右。因此，要综合考虑煤种、煤质、加工成本、燃烧率等因素，并通过试验选择合适的煤粉粒度。

英国钢铁公司斯肯索普厂喷吹粒煤，其中粒径小于 2mm 的占 95%，小于 74μm 的占 10%～30%，平均径粒 0.6mm。美国 Davy 公司和德国 Aachen 大学曾对该厂喷吹的粒煤进行了研究，认为这是一种含有较高结晶水的褐煤，高温下会发生爆裂，而且会大幅度降低风口前理论燃烧温度，喷吹此种粒煤，需用较高的风温和鼓风富氧（富氧率 7%～10%）来补偿理论燃烧温度的下降。目前我国钢铁工业用氧十分紧缺（尤其是高炉用氧），所以笔者认为目前我国高炉应以喷吹粉煤为主，不宜提倡喷吹粒煤。

4.3 提高热风温度

提高热风温度可以促进煤粉的预热、脱气、热分解等反应，提高煤粉燃烧率。煤粉预热、脱气、热分解需要热量补偿，补偿温度可通过下式计算：

$$t_c = \frac{Q_c + Q_{1500}}{V_b + C_b};$$

式中　t_c——喷吹煤粉时的补偿温度，℃；

Q_c——煤粉热分解吸收的热量，kJ/kg；

Q_{1500}——煤粉温度提高到 1500℃ 时所需的物理热；

V_b——入炉风量，m³/tHM；

C_b——热风在 t 风温时的热容，kJ/(cm³·℃)。

煤粉燃烧动力学的研究结果表明，在高炉风口区对于提高煤粉燃烧率而言，提高风温的作用不及鼓风富氧的作用显著；但从热力学角度分析，煤粉燃烧前的预热、脱气、热分解对煤粉燃烧过程的影响也很大，而这一过程是需要热量（温度）来补偿的。提供补偿热量最有效的途径就是依靠热风的物理热，这比鼓风富氧产生的化学热更易获得，其作用也更明显。

综合鼓风富氧、提高风温等因素对煤粉燃烧的影响，可用风口前理论燃烧温度进行定量分析，其表达式为：

$$t_f = \frac{Q_t + Q_1 + Q_f - Q_x}{V_g \cdot C_g}$$

式中　t_f——风口前理论燃烧温度，℃；

Q_t——入炉碳素生成 CO 产生的热量，kJ；

Q_1——焦炭带入风口区的热量，kJ；

Q_f——鼓风带入的物理热，kJ；

Q_x——煤粉预热到 1500℃ 时及热分解所吸收的热量，kJ；

V_g——炉缸煤气量，m³/t；

C_g——炉缸煤气的平均热容，kJ/(m³·K)。

国内外经验表明，在大喷煤条件下，理论燃烧温度一般为 2050~2350℃。也可按日本新日铁公司提出的经验公式进行简易计算：

$$t_r = 1550 + 0.839 t_b - 6.033 W_m - (2.37 \sim 2.75) W_c + 4.973 V_{O_2}$$

式中　t_r——风口前理论燃烧温度，℃；

　　　t_b——鼓风温度，℃；

　　　W_m——鼓风湿度，g/m³；

　　　W_c——每立方米鼓风的喷煤量，kg/m³；

　　　V_{O_2}——每立方米鼓风富氧量，m³/m³。

由上式可知，提高风温和富氧量，理论燃烧温度升高；提高喷煤量，理论燃烧温度降低。因此，喷煤时须通过提高风温、富氧鼓风等措施维持合理的理论燃烧温度，保证高炉顺行。

4.4　确定合理的喷枪位置

理论和实践都表明，喷枪插入位置对煤粉燃烧率有重要影响（见图11）。据文献 [10] 介绍，在喷枪出口距直吹管前端 720mm 时，煤粉在直吹管内的燃烧率约为 1.92%~8%。煤粉呈流股状态由喷枪进入直吹管内，被速度为 200m/s 左右的热风包围，很快被吹散成"煤粉云"，并快速进行预热、脱气、热分解和挥发物燃烧等一系列反应。在喷枪出口处，煤流是连续稳定的（流化不好的则在喷枪出口处呈脉动煤流），煤粉在直吹管及风口的停留时间很短（一般 10ms 左右）。尽管煤粉在直吹管内的燃烧率不超过 20%，但这一过程对煤粉的预热、脱气、热分解、着火等反应非常重要。煤粉在高速热风的扰动下与热风混合良好，有利于传热和氧的传质，为煤粉在回旋区内燃烧创造了有利条件。但喷枪距风口前端距离过长，则可能会出现直吹管和风口内壁结渣、喷枪结焦等现象，这时需适当调整喷枪位置及插枪角度，以减少煤流对风口的磨蚀（见图12）。在喷煤量提高时，会出现风压升高和直吹管的微压振动，这时若将喷枪出口位置靠近风口前端，则因直吹管内的煤粉燃烧量减少而使炉壁侧焦炭消耗量和下降速度增加，炉壁热负荷降低，因此需确定一个高燃烧率的最佳喷枪位置，使煤粉在直吹管内形成均匀燃烧区，减少"煤粉云"在直吹管内的随机波动，使煤粉分散均匀，提高燃烧率。

日本田村建二等通过数学模型计算，得出了喷枪最佳位置与喷煤量、煤粉粒度之间的关系，并回归成下式：

$$L \leq \frac{52 d_P^2 - 53 d_P + 17.5}{P_{cr} - 30}$$

式中　L——喷枪距风口前端位置，m；

　　　d_P——煤粉粒度，mm；

　　　P_{cr}——煤比，kg/t。

在煤比为 80~180kg/t、粒度为 0.074mm 时，由上式可以得出 $L \leq 0.277 \sim 0.092$m。

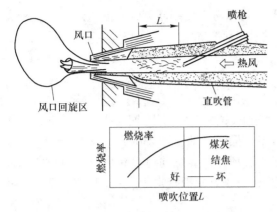
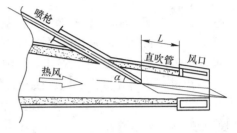

图 11　煤粉燃烧率与喷枪位置的关系　　　　图 12　最佳喷枪位置的选择

4.5　喷吹混煤与填加煤粉促燃剂

国内外研究结果表明，在煤粉中填加少量促燃剂（催化剂），如生石灰、石灰石、白云石、$KMnO_4$ 溶液等，对改善煤粉的燃烧性能、提高煤粉燃烧率有一定作用。这是由于煤粉燃烧过程主要受氧的扩散控制，在煤粉中填加少量含氧促燃剂，可提高燃烧区氧或氧化性物质的浓度，改善煤粉燃烧时的供氧条件，促进煤粉燃烧，从而提高煤粉燃烧率。德国蒂森公司是通过向煤粉中添加 $CaCO_3$ 粉来提高煤粉燃烧率的。

另外，煤种对煤粉燃烧率也有较大影响。实验表明，煤的种类对煤粉的快速热分解和挥发物燃烧有决定性影响。挥发分高的煤，挥发物燃烧迅速，残炭表面有连通裂纹和微孔，比表面积大，有利于氧的扩散和碳-氧反应的进行，因此，在相同条件下，烟煤的燃烧率高于无烟煤。但烟煤的 C/H 低，置换比也低于无烟煤。将烟煤和无烟煤混合喷吹，其效果优于喷吹单一煤种。一方面，在无烟煤中加入烟煤对无烟煤有助燃作用，可降低着火温度，提高燃烧率；另一方面，通过配煤可以保证有较高的置换比，而且同喷吹高挥发分的烟煤相比，其安全性也更有保障。德国蒂森公司认为混煤的挥发分控制在 20%～25% 最佳，这样既可保证煤粉在风口前有较高的燃烧率，又可得到较高的置换比。

4.6　关于氧煤枪的应用

为提高煤粉在风口回旋区的燃烧率，各种结构型式的氧煤枪（氧煤燃烧器）相继问世。其主要特点是：在煤粉燃烧区局部富氧，增加该区域的局部氧浓度，促进氧的扩散，强化煤粉燃烧过程，从而提高煤粉燃烧率。高炉风口处的风速约 180～260m/s，煤粉由喷枪出口到回旋区的停留时间很短，在高速热风的扰动下，煤流周围的氧浓度迅速发生变化。这就要求煤粉着火后，"煤粉云"周围仍能保持较高的氧浓度（要大于鼓风中的氧浓度），以发挥氧煤枪局部富氧促进煤粉燃烧的作用。如果煤粉着火后，"煤粉云"周围的氧浓度与鼓风富氧时的氧浓度相差无几，则氧煤枪富氧便失去了其实用意义。从国内外高炉喷煤实践看，煤比在 150kg/t 以上的高炉大多采用鼓风富氧的方式，仅在德国、英国、日本、瑞典等国家的部分高炉上试用氧煤枪，氧煤枪的应用尚不普遍。另外，氧煤枪的供氧系统较复杂，安全供氧不可忽视。因此笔者认为，当煤比在 200kg/t 以下时，采用鼓风富氧更为经济可行；当煤比在 200kg/t 以上时，可以考虑使用新型结构的氧煤枪。但氧煤枪的结构、寿命、安全供氧等问题还有待于进一步研究。

5 结论

（1）高炉富氧喷煤是我国钢铁工业结构优化的重点，"九五"期间我国部分钢铁企业的煤比将达到 200kg/t，并最终实现高炉半煤半焦操作（即煤比 250kg/t，焦比 250kg/t）。

（2）进一步研究煤粉在风口回旋区的燃烧机理，以燃烧动力学和热力学的角度掌握煤粉燃烧过程的动量、热量、质量的传输机理，特别是模拟直吹管及风口回旋区条件的实验研究更具有实际意义。

（3）大喷煤量操作应有较高的煤粉燃烧率，这是获得较高置换比和稳定顺行的前提，不宜过分夸大未燃煤粉在高炉内的积极作用。未燃煤粉在高炉内的行为还有待于进一步研究。

（4）富氧、高风温是进行大喷煤量操作时维持高炉稳定顺行重要的前提条件，可以用理论燃烧温度进行定量分析和控制。

（5）煤种、煤粉粒度及喷枪位置对煤粉燃烧率、置换比及喷煤量都有重要影响。

（6）当煤比在 200kg/t 以下时，采用鼓风富氧比较经济可行；当煤比大于 200kg/t 时，应考虑使用新型氧煤枪。

参考文献

[1] 丁玉龙，等. 高炉直吹管条件下粉煤热解前的传热分析 [J]. 北京科技大学学报，1993，2：145.
[2] Bortz S, et al. Experiments on Pulverized coal combustion under condition simulating blast furoace environments [J]. Ironmaking and Steelmaking, 1983, 5: 222.
[3] 杨天钧. 煤粉高速加热条件下的快速热分解 [J]. 北京钢铁学院学报. 1988, 3: 279.
[4] 沈颐身，等. 氧煤燃烧器的研究与开发 [J]. 钢铁，1991，1：11.
[5] 杨永宜，等. 进一步提高高炉喷煤量而不降低煤利用率的可行性研究 [J]. 北京钢铁学院学报，1986. 2: 1.
[6] 周渝生，等. 高炉氧煤强化炼铁工艺的开发 [J]. 北京科技大学学报，1993. 3: 387.
[7] 魏升明，等. 影响煤焦置换比因素的分析 [J]. 首钢科技，1981. 3: 69.
[8] SPalding D B. 燃烧与传质 [M]. 北京：国防工业出版社，1969：208.
[9] 顾飞，等. 高炉喷吹煤粉的煤种及粒径研究 [J]. 钢铁，1996. 4: 11.
[10] 安朝俊. 高炉生产——首钢炼铁三十年 [M]. 北京：首都钢铁公司，1983：351.
[11] 田村建二，等. 高炉 V-スラエイ部ての微粉炭の燃烧量限界と吹入位置の适正化 [J]. 铁と钢，1991，6：775.

大型高炉紧凑型长距离喷煤技术

摘　要：首钢2号、3号高炉制粉喷煤系统技术改造中，采用自行设计开发的紧凑型长距离直接喷煤技术。制粉系统采用大型中速磨、高效袋式煤粉收集器、大倾角胶带机；喷煤系统采用双系列串联罐、总管-分配器、自动给煤机、流化喷吹、压力平衡型波纹补偿器、直接喷吹等技术，在生产实践中获得了成功。

关键词：高炉；喷煤；直接喷吹；中速磨

1　引言

我国是煤炭资源丰富的国家，但我国的炼焦煤资源匮乏。随着我国钢铁工业的迅猛发展，炼焦煤供需矛盾日益突出。面对有限的炼焦煤资源，大力发展高炉喷煤技术已经成为必然。对于地处首都、需大量高价采购外地焦炭的首钢而言，提高喷煤量具有更特殊的战略、环境和经济意义。

1999年，首钢根据长远发展规划的要求，决定建设一套新的高炉制粉喷煤系统，结合首钢现役高炉的生产状况和技术装备，决定对2号（1726m^3）、3号（2536m^3）高炉的喷煤系统进行全面新技术改造。该项目的实施，可以降低首钢高炉生产成本，降低能源消耗，实现清洁化生产，淘汰能耗高、污染大的落后工艺和设备，提高高炉喷煤整体技术装备水平和综合技术经济指标。

2　技术方案的确定

由于首钢是国有特大型老企业，制粉喷煤系统工艺设备老化，场地拥挤，工艺落后，存在诸多问题。实施制粉喷煤系统的新技术改造存在着相当大的困难，技术观点上也存在着分歧。首钢现有制粉喷煤系统存在的突出问题是：（1）制粉能力严重不足，只能维持高炉喷煤量110kg/t；（2）工艺技术落后，设备老化；（3）安全措施不完善，不能喷吹烟煤和混煤；（4）场地狭小，建设新的制粉喷煤车间存在很大难度；（5）原有煤场储量偏小，很难适应大喷煤量的要求。

为解决上述问题，在技术方案设计时，对目前国内外现行的各种高炉制粉喷煤技术进行了分析研究和技术对比，吸收国际上最新的设计理念和设计思想，结合我国和首钢的实际情况，对于大型高炉采用短流程中速磨制粉、长距离直接喷吹技术进行了重点技术研究和开发创新。

结合首钢的实际条件和国内外高炉制粉喷煤技术的发展趋势，经过充分的技术研究和

本文作者：张福明。原刊于《2004年全国炼铁生产技术暨炼铁年会文集》，2004，729-734。

论证,决定在首钢原四制粉车间附近,利用原有的储煤场,新建一个高炉制粉喷煤车间,将制粉、喷煤系统合建在一个厂房内。采用大倾角胶带机、大型中速磨煤机、高效袋式煤粉收集器、双系列串联罐组、总管-分配器长距离直接喷吹新工艺。首钢新型高炉喷煤技术工艺流程如图1所示。

3 工艺设计与技术特点

新的高炉制粉喷煤系统由北京首钢设计院设计,采用长距离直接喷吹、大型中速磨、引射式燃烧炉、低压脉冲高效袋式煤粉收集器、大倾角胶带机、自动可调式给煤机以及煤粉分配器等国产化先进技术和设备,并且设计了两套喷煤系统之间煤粉的互相供给和分配装置。

3.1 工艺总体布置

首钢是具有80多年历史的老厂,由于生产能力的不断扩大,使首钢厂区十分拥挤,场地非常紧张。在满足生产和工艺要求的前提下,充分利用炼铁厂现有场地,利用原有四制粉车间的储煤场,新建的制粉喷煤车间位于首钢原四制粉车间北侧,采用紧凑型工艺布置,将制粉、喷煤系统合建在一个厂房内。制粉喷煤主厂房占地$12 \times 45 m^2$,其他附属设施布置在车间附近。

3.2 上煤系统

由于利用首钢原有四制粉车间的储煤场,场地拥挤、狭窄,无法采用普通胶带机上煤,使原煤的运输成为一个重要的技术问题,设计开发了采用大倾角胶带机(75°)上煤工艺。设计中对原四制粉车间储煤场进行了改造,适当扩大了煤场的存储能力,在不影响原有生产设施的条件下,采用2条大倾角胶带机,将原煤提升至高度40m的原煤仓。设计2个双曲线原煤仓,单仓有效容积$350m^3$,可存储原煤7h。

采用大倾角胶带机上煤的新工艺,特别适用于场地拥挤的老企业制粉喷煤系统新技术改造,而且占地面积少,投资省,设备简化。

3.3 干燥气供应系统

设计中对不同的干燥气供应工艺进行了研究分析,结合首钢的实际条件,采用混合干燥气供应工艺。但由于新建的制粉喷煤车间距离2号、3号高炉较远,因此采用距离较近的4号高炉($2100m^3$)热风炉烟气。2002年,在首钢2号高炉技术改造过程中,又新建了一套由2号高炉热风炉抽引热风炉烟气的干燥气供应系统,以实现干燥气的稳定供应。

由于热风炉烟气温度较低,而且温度不稳定,先将热风炉烟气抽送至制粉车间的燃烧炉内,与燃烧炉所产生的高温烟气充分混合后,再进入中速磨中。

采用的全封闭引射式燃烧炉为圆筒形双层结构,燃烧炉尾部为燃烧室,前端为环状引射式混合室。全封闭引射式燃烧炉具有结构紧凑、调节范围大、占地面积小、便于控制、节约能源等特点。引射式燃烧炉主要技术性能见表1。

表1 引射式燃烧炉主要技术性能

项　目	参　数
最大流量/$Nm^3 \cdot h^{-1}$	75000
额定流量/$Nm^3 \cdot h^{-1}$	68000
烟气入口温度/℃	130~150
混合后温度/℃	240~280
出口处压力设定/Pa	-200~-100
高炉煤气热值/$kJ \cdot m^{-3}$	2973
燃烧器长度/mm	1630
炉体长度/mm	7400
炉体直径/mm	3200
管道直径/mm	1600

3.4 磨煤制粉系统

3.4.1 中速磨选型

我国从20世纪80年代中期开始和国外中速磨制造厂商合作，通过技术引进和消化吸收，我国在中速磨设计、制造技术上已经取得突破，在大型电厂和高炉上都已能够采用我国自行设计、制造的中速磨。设计中对中速磨制造企业及其用户均进行了技术考察，对中速磨的设备结构特点、运行特点、使用情况等进行了客观的分析对比，最终选用了MPS型中速磨。

3.4.2 中速磨生产能力的确定

3.4.2.1 煤种的选择

近30年来，首钢高炉一直喷吹京西无烟煤，由于京西煤矿已开采多年，煤质条件不断恶化，煤质硬、灰分高且价格较贵。因此首钢专门成立了煤种调研小组对今后拟采用的煤种进行了考察调研，并委托有关部门对所选煤种进行了全面检验分析和评价。表2是根据原煤检验分析数据，经过优选以后作为中速磨设备选型的煤质数据。

表2 中速磨设备选型的煤质数据

项　目	数　据
全水分 M_f/%	≤10
空气干燥基水分 M_{ad}/%	≤1
干燥无灰挥发分 V_{daf}/%	≤25
灰分 A_{ad}/%	12
硫分 $S_{o.ar}$/%	≤0.8
低发热值 $Q_{net.v.ar}$/$MJ \cdot kg^{-1}$	29.57
可磨系数 HGI/%	255
冲刷磨损指数 K_e/%	3.5

3.4.2.2 中速磨生产能力的确定

根据所确定的煤种和煤质条件，经过计算分析，采用了两台MPS212型中速磨，配置SLS225型动-静态煤粉分离器。MPS212型中速磨标准出力为67.7t/h，在$HGI \geqslant 55$，$R_{90} \leqslant 20\%$，$W^Y \leqslant 10\%$的设计条件下，其实际出力可以达到40t/h，可以满足两座大型高炉正常生产时煤比达到200kg/t的要求。MPS212型中速磨主要技术性能见表3。

表3　MPS212型中速磨主要技术性能

项　目	参　数
标准出力（$HGI=80$，$R_{90}=16\%$，$W^Y=4\%$）/t·h^{-1}	67.7
实际出力（$HGI \geqslant 55$，$R_{90} \leqslant 20\%$，$W^Y \leqslant 10\%$）/t·h^{-1}	40
原煤粒度/mm	≤40
一次风量（标准状态）/kg·s^{-1}	21.65
磨盘工作直径/mm	2120
磨辊直径/宽度/mm	1650/560
磨辊数量/个	3
磨盘转速/r·min^{-1}	24.8
传动速比	39.7
主电机功率/kW	560
主电机转速/r·min^{-1}	985
主电机电压/V	6000
主电机防护等级	IP54
中速磨阻损/Pa	6700
分离器型号	SLS225
分离器电机功率/kW	37
分离器电机电压/V	380
分离器旋转速度/r·min^{-1}	11~115
密封风流量/kg·s^{-1}	0.57

3.5　煤粉收集系统

随着技术进步，煤粉收集的工艺流程和主要设备不断地更新和改进。20世纪90年代以后，我国吸收国外成功经验，成功开发研制了低压脉冲高效煤粉收集器。中速磨设备自身已经配置了粗粉分离器，而高效袋式煤粉收集器的允许入口浓度已经达到500~1000g/Nm^3，出口排放浓度低于30mg/Nm^3。设计中采用高效袋式煤粉收集器一级收粉短流程工艺，仅在煤粉收集器后设一台抽风机，取消了排粉风机、细粉分离器等设备，优化了工艺流程。抽风机流量103100m^3/h，全压14100Pa，电机功率630kW。高效袋式煤粉收集器的主要技术性能见表4。

表 4 GMZ4200 高效袋式煤粉收集器技术性能

项　目	参　数
过滤面积/m²	2061
处理风量/m³·h⁻¹	102000
过滤风速/m·min⁻¹	0.823
滤袋尺寸/mm	ϕ120×6000
滤袋数量/个	1824
滤袋材质	抗静电针刺毡 NOME×(500g/m²)
入口浓度/g·m⁻³	≤1000
出口浓度/mg·m⁻³	≤30
入口温度/℃	≤120
喷吹压力/MPa	0.25
氮气消耗量/m³·min⁻¹	6~10
设备耐压/Pa	14000
设备阻力/Pa	≤1200

3.6 喷煤系统

3.6.1 喷煤工艺流程

通过对串罐式和并罐式、总管-分配器和多管路系统的技术研究和分析对比，结合首钢高炉的生产工艺特点，采用了串罐式、总管-分配器喷煤工艺。设两个喷煤系统分别对 2 号、3 号高炉进行直接喷吹，均采用串联罐双系列总管-分配器的喷吹工艺。喷煤系统的主要工艺参数见表 5。

每座高炉的喷煤系统设有 A、B 两个喷吹系列，每个系列由煤粉仓、储煤罐、喷煤罐、给煤机、喷煤总管、分配器、喷煤支管及喷枪等主要设备组成。煤粉仓和储煤罐之间采用 DN400 球阀和 DN400 波纹器连接；储煤罐和喷煤罐之间采用 DN400 代钟式球阀和 DN400 压力平衡式波纹器连接。喷煤罐下部设有 DN250 下煤球阀，煤粉通过下煤球阀进入自动可调煤粉给料机，在经喷煤总管输送至炉前分配器中，再由分配器分配到各个喷煤支管，经喷枪、风口喷入高炉。自动可调式给煤机、煤粉分配器、压力平衡型波纹补偿器、高温合金喷枪等全部为国产设备。

高炉喷吹气源采用压缩空气，由已投产的喷煤专用空压站供应，气源压力为 1.2MPa。储煤罐、喷煤罐的充压和流化、煤粉收集器的脉冲反吹、煤粉仓的流化与惰化等，全部采用氮气。

表5 喷煤系统主要工艺参数

项　目	2号高炉	3号高炉
高炉有效容积/m^3	1726	2536
利用系数/$t \cdot (m^3 \cdot d)^{-1}$	2.25	2.25
日产铁量/$t \cdot d^{-1}$	3884	5706
煤比/$kg \cdot t^{-1}$	200	200
热风压力/MPa	0.35	0.35
喷煤量/$t \cdot h^{-1}$	32.37	47.55
喷煤罐有效容积/m^3	30	30
气源压力/MPa	1.2	1.2
气固比/$kg \cdot kg^{-1}$	25	25
喷煤总管直径/mm	100	100
喷煤支管直径/mm	25	25
输送距离/m	452	358

3.6.2 煤粉交叉供给

正常生产中，一台中速磨对应一个高炉喷煤系统。由于两座高炉的生产能力不同，对煤粉量的需求量也不一样，因此设计了煤粉交叉互给装置。在每个煤粉收集器下设有两个收煤斗，分别对应每座高炉的两个系列。在收煤斗与煤粉仓之间设置交叉溜管，这样任意一台中速磨生产的煤粉可以向任意一座高炉的喷煤系统输送，使两座高炉的煤粉可以互相补充和分配，在一台中速磨出现故障时，也可以由一台中速磨供应两座高炉喷煤。

3.7 煤粉长距离直接喷吹工艺

由于新建的高炉制粉喷煤车间距离2号、3号高炉较远，设计中对煤粉的输送方式进行了研究和论证。采用传统的输煤工艺可以解决煤粉长距离输送的问题，但是要增加输煤设备、喷煤设施等，煤粉输送到高炉附近还要进行再次收集，然后再经喷煤系统喷入高炉。这样设备复杂，工艺环节多，占地面积大，工程投资高。采用长距离煤粉直接喷吹工艺可以简化工艺流程，节约工程投资和占地，但是大型高压高炉煤粉直接喷吹距离长达450m的工艺在我国尚无先例，而且在国际上也不多见，设计中重点对煤粉长距离直接喷吹工艺进行了可靠性计算和研究，对煤粉直接喷吹管路系统进行了优化，实现了煤粉长距离直接喷吹。其中2号高炉喷煤总管长度达到452m，3号高炉喷煤总管长度达到358m。

3.8 安全防火防爆系统

设计中严格执行《高炉喷吹烟煤系统防爆安全规程》（GB 16543—1996）的有关规定。在中速磨入口烟气管道上、煤粉收集器入口和出口管道上分别设置了氧浓度分析仪，在线监测系统的氧浓度；在每个煤粉仓上设置了CO浓度分析仪；一旦O_2、CO含量超标均能自动报警、充氮直至停机；对系统的温度和压力进行严格控制；干燥炉装有火焰监测器及熄火保护装置;喷吹气体在紧急状态下,可以由压缩空气切换为氮气。整个车间厂房的设计,完全按照有关规程要求配备了火灾报警、消防泵自动启动以及相应的消防设备。制粉、喷煤系统按照喷吹烟煤设计,在整个系统的设计中采取了防静电措施。

4 生产实践与应用

首钢新型高炉制粉喷煤工程于2000年11月投入运行(图1)。工程投产以后,解决了首钢制粉能力不足的矛盾,制粉能力达到了80t/h,2号、3号高炉喷煤量得到了较大幅度的提高。3号高炉最高月平均煤比达到165kg/t,2002年该高炉年平均利用系数达到2.3t/($m^3 \cdot d$),入炉焦比为351.1kg/t,煤比为138.5kg/t。2号高炉于2002年3~5月停炉进行大修技术改造,投产后的2号高炉生产指标不断攀升,2002年10月,高炉利用系数达到2.5t/($m^3 \cdot d$),入炉焦比为296.8kg/t,煤比为170kg/t,喷煤率达到36.4%。

首钢2号、3号高炉制粉喷煤技术改造工程采用自行设计开发的紧凑型长距离直接喷煤技术,经过3年的生产运行实践,大幅度提高了喷煤量,节约了焦炭、电力等资源消耗,实现了清洁化生产,有效地控制了粉尘排放,降低了环境污染,改善了劳动条件,取得了巨大的经济效益和显著的社会效益、环境效益。生产实践证实,首钢大型高炉紧凑型长距离制粉喷煤工艺技术的设计是成功的,达到了国际先进水平,该项技术成果于2003年12月通过了北京市科技成果鉴定。

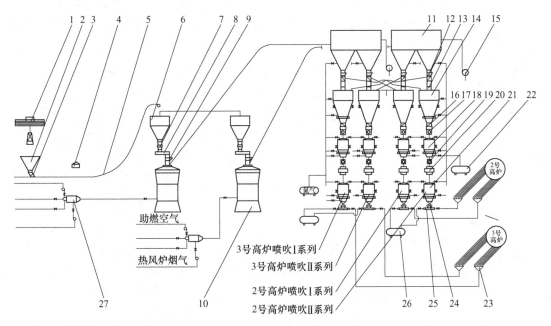

图1 首钢大型高炉制粉喷煤工艺流程图

1—抓斗吊;2—受煤斗;3,7—闸门;4—除铁器;5—上煤胶带机;6—原煤仓;8—电子胶带称给煤机;9—锁气器;10—中速磨;11—高效袋式煤粉收集器;12—插板阀;13—软连接;14—煤粉仓;15—排风机;16,17—球阀;18—储煤罐;19—代钟式球阀;20—压力平衡型波纹补偿器;21—喷煤罐;22—充压气包;23—煤粉分配器;24—下煤球阀;25—自动可调式给煤机;26—喷吹气包;27—燃烧炉

5 结语

大型高炉紧凑型长距离制粉喷煤系统的建成投产，为首钢进一步提高喷煤量、扩大喷吹煤种、节能降耗，创造了必要的先决条件。新型高炉制粉喷煤工艺，使首钢喷煤技术发生了质的飞跃。高炉制粉喷煤系统完全采用国产化技术和设备，整体设计理念先进，工艺布局合理，特别是采用紧凑型短流程工艺，占地面积小，工程投资低，对我国高炉制粉喷煤技术改造提供了有益的借鉴和参考，成为我国老企业高炉制粉喷煤技术改造的示范，具有较高的推广应用价值和广阔的推广应用前景。

（1）采用国产化技术和设备，自行设计开发了大型高炉紧凑型长距离制粉喷煤技术。首钢新型高炉制粉喷煤技术完全依靠我国自有技术和工艺设备，自行设计开发成功的。

（2）采用紧凑式工艺布局。在首钢现有的场地条件下，采用紧凑式布置，节约了占地和工程投资，也使新型高炉制粉喷煤技术在首钢得以实现。

（3）采用长距离直接喷煤工艺，取消了煤粉输送工艺环节，优化了喷煤工艺流程。2号、3号高炉喷煤总管总长分别达到452m、358m，使首钢高炉喷煤技术实现了质的飞跃。

（4）采用自动化检测与过程控制技术，实现了喷煤自动控制和喷煤量调节。采用O_2浓度、CO浓度、温度、压力、流量、质量等参数的在线监测技术，全部实现计算机联锁控制，使整体技术装备和控制水平达到国际先进水平。

（5）大型高炉紧凑型长距离制粉喷煤技术为我国高炉制粉喷煤技术的设计和应用提供了有益的参考和借鉴，在国内外高炉制粉喷煤系统中具有广泛的推广应用价值和推广应用前景。

引进技术在迁钢3号高炉喷煤系统中的应用

摘 要：阐述了引进 Danieli Corus 公司先进技术在迁钢3号高炉喷煤系统中的应用实践，采用中速磨煤机制粉，一级大布袋收粉，三罐并列，主管加分配器以及浓相长距离直接喷吹工艺，其中喷吹罐的工作周期及能力配置、流态化措施及气力输送系统等关键技术与高炉能力紧密匹配，实现功能最大化利用及节能减排，整套工艺达到国际先进水平。

关键词：特大型高炉；喷煤；引进技术

1 引言

高炉喷煤是现代高炉炼铁生产广泛采用的技术之一，喷吹煤粉从当年简单的以煤代焦和提供热量的思路出发，到如今已经成为调剂炉况热制度、改善炉缸工作状态、降低燃料消耗以及节能减排等方面的重要技术措施[1]。高炉大型化已经成为趋势，喷煤工艺各环节应做出相应改进，以满足高炉大型化的需求，其中装备水平、煤粉流态化措施、长距离浓相输送以及风口均匀喷吹技术等方面已经引起重视[2]。迁钢3号高炉（4000m^3）于2010年1月8日点火投产，其配套的制粉喷煤系统在喷吹部分的设计过程中引进了荷兰 Danieli Corus 公司的优势技术，希望能对今后国内的高炉喷煤设计工作起到参考作用。

2 工艺设计方案

2.1 工程概况

3号高炉设计利用系数2.4，日产铁量9600t/d，设计喷煤比190kg/t，小时喷煤量达到77t/h。配套制粉喷煤工艺设备能力达到250kg/t，按照喷吹烟煤和无烟煤混合煤设计，采用中速磨制粉、一级大布袋收粉、并罐直接喷吹、单主管单分配器工艺，喷煤罐充压及流化采用氮气，煤粉输送采用压缩空气。基本设计情况见表1。

表1 基本设计情况

名 称	迁钢3号高炉
高炉有效容积/m^3	4000
高炉产量/$t \cdot d^{-1}$	9600
高炉利用系数/$t \cdot (m^3 \cdot d)^{-1}$	2.4

本文作者：孟祥龙，张福明，王维乔。原刊于《炼铁》，2013，32（2）：34-37。

续表1

名　　称	迁钢3号高炉
风口数量/个	36
煤比/kg·t^{-1}	190~250
喷煤量/t·h^{-1}	77~108
中速磨煤机台数/台	2

2.2　工艺流程

原煤由皮带机从原料场储煤筒仓运至喷煤主厂房，并向原煤仓加煤，原煤仓中的原煤经称重式给煤机进入中速磨煤机，干燥剂采用热风炉废烟气与混风炉燃烧高炉煤气产生的高温烟气的混合气，排粉风机形成负压将干燥剂吸入磨煤机，配合原煤在磨煤机中进行粉碎和干燥。粒度较大的颗粒经粗粉分离后重新磨制，合格煤粉沿管道进入布袋收集器，收集后的煤粉经收集器灰斗后进入煤粉仓，煤粉仓下设置3个高压并列式喷吹罐，煤粉经下煤管道进入喷吹罐中，并作为喷吹罐操作周期的一部分。离开喷吹罐的单一煤粉流在补气器中与输送气体混合，然后将煤粉送到高炉附近。煤粉与输送气体在补气器中混合，并通过输煤管线进入分配器，分配器将该单流平均分配到各喷吹管线，各喷吹管线再将煤粉送到高炉各风口。

3号高炉喷煤系统采用的主管路加分配器的喷吹模式，是目前国内高炉喷煤工艺广泛采用的一种喷煤形式，喷吹部分简要流程如图1所示。

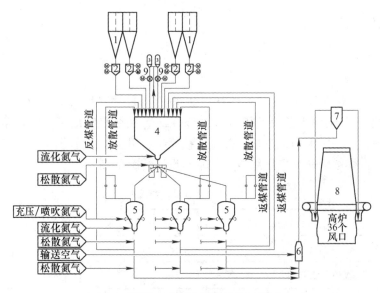

图1　迁钢3号高炉喷煤系统流程图
1—煤粉收集器；2—煤粉振动筛；3—仓顶除尘器；4—煤粉仓；
5—喷吹罐；6—补气器；7—分配器；8—高炉；9—旋转给料机

3 引进的先进技术

3.1 输煤总管缩径设计与末端方管设置

在与 Danieli Corus 合作中,输煤主管的设计有两点与以往思路不同:

(1) 输煤主管的管内径沿着煤粉运行方向是逐级递增的(见表2)。在喷煤主厂房补气器出口处使用的是内径119.7mm圆管,到末端已经逐级递增至内径133.3mm。

表2 迁钢3号高炉输煤主管各段

管 段	长度/m	内径/mm	壁厚/mm	备注
第1段	60	119.7	10	圆管
第2段	46	122.7	8.5	圆管
第3段	91	128.3	20	圆管
第4段	51	133.3	17.5	圆管
第5段	6	126×126	12	方管
合 计	254	—	—	—

输煤主管输送煤粉和压缩空气时,由于压损,气体体积增大,引起速率逐渐增大。采取输送管线内径逐级增大的措施,可以补偿增大的输送速率所带来的管道腐蚀、压降和气耗增大等一系列问题。但在计算时不能无限地扩大内径,这是由于在输送过程中,要时刻保证输送速率大于沉积速率,才能使煤粉被输送气体顺利输送。

(2) 在输煤主管末端进入分配器之前加入了1段长6m的方形管道,垂直布置,如表2中的第5段管道。

由于煤粉和压缩空气两相流在沿着一定长度的圆管运行后,会出现压力场和速度场不均匀的现象,主要体现为其运行轨迹沿着管道中心线呈螺旋状前进。这种不均匀性会造成管线内出现固态沉积,进入分配器后的煤粉不能均匀地被分配到各支管中,从而引起各支管的喷煤量差异,更进一步地影响整个炉况。因此,在进入分配器入口,输送管线竖直段的最后一段采用方形管,避免这种不均匀现象,以保证分配器平稳分配煤粉。

3.2 煤粉仓和喷吹罐的能力配置

煤粉仓和喷吹罐是喷吹环节储存煤粉的重要装置,其几何参数的设计会对喷煤过程的周期操作产生直接影响,3号高炉的煤粉仓及喷吹罐设计参数见表3。

表3 迁钢3号高炉煤粉仓及喷吹罐参数

名 称	煤粉仓	喷吹罐
数量/个	1	3
有效容积/m³	1025	93
储存(喷吹)时间/min	432(正常)	27.1(正常)
高径比(h/d)	1.68	3.21
下锥体角度/(°)	60	36.4

三罐并列式喷吹形式是1、2、3号罐之间进行轮流交替喷吹,罐与罐切换之间存在倒罐周期。与两罐并列式不同的是,三罐并列式倒罐周期基本就是倒罐操作时间,一般不专门考虑等待时间。倒罐周期由几个特定的工作阶段组成,包括单罐喷吹时间 t、喷吹罐减压放散时间 t_p、装煤时间 t_z、充压时间 t_c、保持时间 t_d。国内经验当满足 t 小于 $1.25\sim1.4(t_p+t_z+t_c+t_d)$ 时,应设置3罐并列式的布置形式。

从表4中喷吹罐工作周期各阶段的时序来看,满足 t 小于 $1.25\sim1.4(t_p+t_z+t_c+t_d)$,因此选择了3罐并列的方式,这与国内以往的经验也吻合。由于保持时间 t_d 内喷吹罐要求不断补充气体以保证工作压力,因此工作周期的精准设计不但能够节约能源,而且对实际操作具有指导意义,应引起重视。

表4 迁钢3号高炉喷吹罐工作周期

喷吹时间 t	放散时间 t_p	装煤时间 t_z	充压时间 t_c	等待时间 t_d
21.7	7.5	7	5.5	1.7

3.3 煤粉仓与喷吹罐的煤粉流态化措施

原煤经过磨煤机磨制后进入煤粉仓和喷吹罐中后,由于自重或者环境压力的原因,颗粒间相互接触,形成固定床状态,而且在靠近容器底部的出粉口有被压实效果,即使在制粉环节已被处理成含有少量水分的细小颗粒,仍然很难靠自重流出。在煤粉仓和喷吹罐的出粉口设置流化装置将解决这一问题。

煤粉仓内属于非高压环境,但由于一般情况下煤粉仓容积较大,料柱也较高,内部煤粉的重力将最终作用于下部出料口。3号高炉采用了流态化板设备来保证煤粉顺利出仓。

煤粉仓在其底部下封头处设置流态化板装置,板上形成若干个圆孔(见图2),流态化气体不间断地从下到上流过流态化板,在流态化板上部空间煤粉出口管道处形成连续流态化的煤粉,流态化的煤粉通过下煤管道被输送至喷吹罐。流态化气体使用氮气,保持煤粉时刻处于惰性环境下,防止燃烧和爆炸。

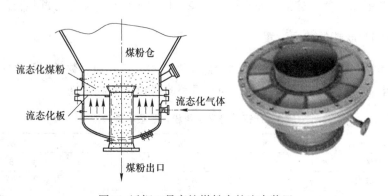

图2 迁钢3号高炉煤粉仓的流态装置

喷吹罐属于压力容器,内部气体及煤粉处于高压状态,而且通常其高径比也较煤粉仓大,在其下部出粉口位置煤粉颗粒紧密连接。喷煤罐的流态化效果将直接影响到能否为高炉快速均匀地输送煤粉,需要更加精确地控制流化气体流量及压力,实际流态化气体的操作要满足大于煤粉流态化速度而小于悬浮速度。

3号高炉的喷吹罐与其配套的煤粉仓采用了相同的流态化措施,即在下锥体封头处设置了流态化板,流态化板的安装位置和结构与煤粉仓基本相似,并采用下出料形式。

流态化气体的流量及压力控制需同时考虑喷吹罐内环境压力及煤粉自重的影响,经过详细计算并结合生产实践方可确定。

3.4 煤粉的浓相输送技术

以往国内高炉的喷煤系统一般采用稀相输送方式,通常固气比仅为10~15kg/kg,输送能力低,能源消耗大,而且高速行进的煤粉对管道及设备的磨损大,增加检修负担。3号高炉喷煤主管中的固气比设计为不小于35kg$_{煤粉}$/kg$_{气体}$,相比较稀相是个不小的提升。

在浓相输送技术的运用中,输煤主管道内输送气体的速度与喷吹罐正常工作时的压力是设计要点。喷煤总管道输送气体速度v_s直接影响输送管路中的固气比和输送速度,应控制在煤粉的临界流态化速度v_c和悬浮速度v_x之间。

实际生产中保证$v_c \leq v_s \leq v_x$是依靠调整输送气体与煤粉在输煤总管中的比例来实现的,设计中在喷煤主管道上设置了二次补气器来调整煤粉输送浓度,图3为本次采用的补气器的结构形式。

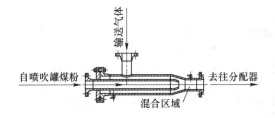

图3 迁钢3号高炉输煤主管补气器

喷吹罐正常工作压力也同样制约着喷煤浓度。国内某研究通过试验和理论计算的结果表明,一般情况下,喷煤浓度随着喷吹罐工作压力增加而降低。

当喷吹罐压力为1.4MPa时,最大输送浓度只能达到42kg$_{煤粉}$/kg$_{气体}$,再加上二次补气,则只能达到30kg$_{煤粉}$/kg$_{气体}$左右;当喷吹罐压力为0.3MPa时,最大输送浓度能达到148kg$_{煤粉}$/kg$_{气体}$,加上二次补气也可达到90kg$_{煤粉}$/kg$_{气体}$[3]。

煤粉在不同的罐压下能够达到的最大质量浓度不同,但实际体积浓度却相同。这是因为喷吹罐压力增大后,气体随着煤粉输送过程中,煤粉颗粒间会压缩进更多质量的气体,喷吹罐的压力越大,能够达到的质量浓度越低。当然喷吹罐的压力不可无限降低,其要求应满足能够克服喷煤系统两相流输送管路总阻损后将煤粉顺利输送至高炉风口。两相流输送管路总阻损包括两相流输送管道阻损和各个部件的局部阻损,具体包括喷煤罐出口流量调节阀、二次补气器、输煤主管、煤粉分配器、喷煤支管、缩径喷嘴和喷煤枪等阻损之和,要通过详细计算才能确定。3号高炉喷吹罐的最大设计工作压力为1.42MPa。

3.5 煤粉的均匀喷吹技术

煤粉进入分配器后经过各喷煤支管到达风口喷枪继而喷入高炉，其均匀性将直接影响到炉缸热状态。喷吹的均匀性是衡量喷吹质量的重要指标，这也是引进的 Danieli Corus 喷煤技术的主要优点之一。

3 号高炉采用了"瓶式"分配器（见图 4），并对其内部结构进行了调整，解决了以往此类型分配器，煤粉和输送气体在其内部产生涡流，阻力大易积粉的问题，效果比较理想，能够达到生产要求。

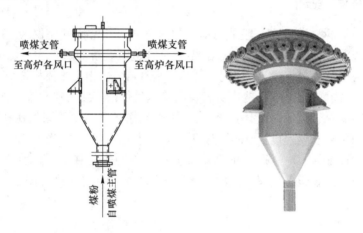

图 4 迁钢 3 号高炉煤粉分配器

要保证各支管喷吹均匀，除了确保在分配器中煤粉均匀分配外，在喷煤支管的设计上，利用每个管道里阻损均等的原理，使煤粉均匀分配到每个风口，实现了各风口喷吹煤粉的均匀性控制在 4% 的误差范围内。国内常用以下经验公式计算喷煤支管阻损 ΔP_b：

$$\Delta P_b = P_b - P_f \qquad P_b = \sqrt{P_f^2 + 8.85(1+\mu)^{0.75} G_b^{1.75} L_b D_b^{-4.75}}$$

式中 P_b——喷煤支管起点的压力，MPa；

P_f——喷煤支管终点的压力，MPa；

μ——固气比，$kg_{煤粉}/kg_{气体}$；

G_b——喷煤支管载气流量，kg/s；

L_b——喷煤支管当量长度，m；

D_b——喷煤支管内径，cm。

由于喷煤支管起点的压力 P_b、喷煤支管终点的压力 P_f、固气比 μ 对于各个支管相同，同时要求各支管载气流量 G_b 和喷煤支管内径 D_b 相同，在 3 号高炉喷煤系统的设计中将分配器置于炉顶平台（标高 57m），增加喷煤支管的垂直段比例，减小阻损以提高煤粉分配均匀程度，同时使各支管当量长度 L_b 相等，实现各支管等阻损。

4 系统运行情况

本系统自 2010 年 1 月投产，输送及喷吹在短期内即达到设计要求。经过 2 年多的运行,该系统可靠、平稳、顺利,各项指标均良好(2012年部分生产指标见表5)。

表5 迁钢3号高炉部分生产指标

月份	利用系数/t·(m³·d)⁻¹	煤比/kg·t⁻¹	焦比/kg·t⁻¹	燃料比/kg·t⁻¹
1	2.39	157	303	510
2	2.08	148	315	513
3	2.45	156	299	504
4	2.34	164	304	511
5	2.41	164	305	509
6	2.24	155	307	505

5 结语

(1)随着高炉不断大型化,喷煤工艺技术及装备水平应不断提升以满足炉容扩大后的要求。

(2)煤粉仓与喷吹罐的设计,包括外形尺寸与工作周期之间的相互配合,对节能减排和指导操作具有积极意义。

(3)煤粉流态化与浓相输送等技术与国际先进水平尚存在差距,仍有进一步提升空间。

参考文献

[1] Wu K, Ding R C, Han Q, et al. Research on unconsumed fine coke and pulverized coal of BF dust under different PCI rates in BF at Capital Steel Co [J]. ISIJ Int, 2010, 50 (3): 390.
[2] 周春林, 沙永志, 等. 高炉喷煤工艺优化及系统改进 [J]. 钢铁, 2009, 7 (44): 20-23.
[3] 吉永业, 等. 关于太钢浓相喷煤技术的优化 [J]. 钢铁, 2010, 45 (8): 20-24.

首钢高炉喷煤技术发展与创新

摘　要：首钢最早在20世纪60年代就在国内率先开始了高炉喷煤的工业尝试，在随后几十年时间里，先后经历了几个具有不同时期特点的发展阶段。进入21世纪后，伴随着高炉不断大型化的趋势，与大型高炉相配套的制粉喷煤系统相继在首钢的迁钢和京唐等地获得应用。如今，整合了浓相长距离输送、风口均匀喷吹、全自动控制等技术的，具有首钢特色的大型高炉喷煤系统已在首钢广泛应用。

关键词：高炉；喷煤；首钢；发展

1　引言

高炉喷吹燃料的思想起源于1840年法国马恩省炼铁厂喷吹木炭屑，目前已经发展为各种不同类型的喷吹燃料，如煤粉、焦炉煤气、除尘灰及其他还原性物料[1~3]。在众多的喷吹燃料中，由于煤粉较低的成本和较好的技术优势而被国内大部分高炉采用。喷煤工艺作为高炉炼铁的配套设施，已经成为现代高炉炼铁的显著技术特征[4]，是调剂炉况热制度、改善炉缸工作状态、降低燃料消耗以及节能减排等方面的重要技术措施[5]。首钢作为我国高炉喷煤技术的先行者，50年来一直通过不断地完善喷煤装备与工艺来满足高炉炼铁技术发展过程中的不同需求。

2　首钢喷煤技术发展历程

我国的高炉喷煤技术开发和应用起步较早，早在20世纪60年代就开始了高炉喷煤技术的研究。首钢是我国高炉喷煤技术的开创者和先行者之一，在喷煤领域的探索及发展过程大致可分为4个阶段。图1简要地描述了首钢喷煤五十年的发展历程中，不同阶段所具有的不同技术特点。

阶段一：起步。1963年6月，世界上首套现代化工业性煤粉喷吹装置在美国AK钢公司的Bellafonte高炉上投入使用。首钢在同年开始对喷煤技术进行了系统的研究，并在首钢试验厂$18m^3$高炉上进行试验，于1964年投产建成当时国内第一套喷煤系统，比西欧及日本的第一套喷煤装置早大约十五六年[6]。

阶段二：探索。20世纪60年代末到90年代，首钢在喷煤工艺优化方面做了包括喷吹煤种的改变、富氧喷吹的尝试及探索与炉型相适应的喷煤操作等尝试，实现在煤种挥发分高达32.5%的条件下，煤比123kg/t；富氧鼓风25.2%的情况下月平均煤比控制在246kg/t；针对矮胖高炉进行精料及喷煤改造、调整装料及改进送风制度等一系列工作，

本文作者：孟祥龙，张福明，曹朝真。原刊于《中国冶金》，2014，24（3）：21-26。

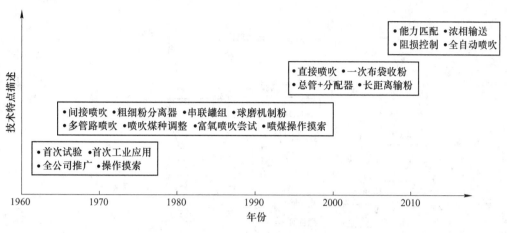

图 1 首钢喷煤发展历程

使年平均煤比提高了 28.9kg/t[7]。

阶段三：变革。到 20 世纪 90 年代末，喷煤技术已在国内大部分高炉上应用，相应的操作也日趋稳定，此时工艺及装备水平成了制约喷煤量进一步提高的障碍[8]。同期首钢开发了一系列新的工艺及装备，直接喷吹、以中速磨制粉、一次布袋煤粉收集、单管路长距离输送等技术获得应用。该时期的喷煤系统全面采用当时国内的先进技术，同时整合了一些自行开发的技术和装备，在首钢喷煤技术的发展历史上是一个重要转折点。

阶段四：提升。进入 21 世纪后，国内新建高炉大型趋势明显，旧有的系统已不能满足需求。首钢在迁钢与京唐大型高炉的建设过程中，投产了一系列适用于大型高炉的喷煤系统。新的喷煤系统更加注重与大型高炉相匹配，同时生产操作也更加精细化，不再追求过高的喷煤比，而是更加强调喷煤保障高炉顺行的作用。在这一阶段，首钢的喷煤技术装备力量和生产操作水平较以往有了更大提升。

3 首钢喷煤工艺的改进

3.1 间接喷吹向直接喷吹的转变

20 世纪 90 年代末之前，首钢高炉喷煤一直采用传统的"集中制粉、间接喷吹"工艺。整个工艺流程分为三个部分，即煤粉制备、煤粉输送和煤粉喷吹。当时的首钢采用 3 个集中制粉车间承担所有高炉的煤粉制备，每座高炉设有 2~3 台仓式泵，仓式泵交替工作以保证煤粉的连续输送。在煤粉喷吹环节，"双重罐、多管路、高压喷吹"是当时首钢传统高炉喷煤技术的特点。每座高炉附近都设有 1 座喷煤塔，喷煤塔由两个独立的系列组成，每个系列由上、下两个重叠罐构成，其中上罐为储煤罐，接受远距离输送来的煤粉，而下罐为喷煤罐，内部的煤粉在高压气体和喷送器的共同作用下，通过喷煤支管、喷枪从风口喷入高炉。90 年代初期，首钢 1、3 号高炉扩容大修改造时，为满足大型高炉喷煤的需要，在"双系列、双罐串联、多管路喷吹"技术的基础上，又开发了"三系列、三罐串联、多管路喷吹"的喷煤工艺。当时首钢的间接喷吹工艺流程如图 2 所示。

由于在间接喷吹工艺中，完成磨制的合格煤粉要从制粉车间输送至喷吹车间，不仅流

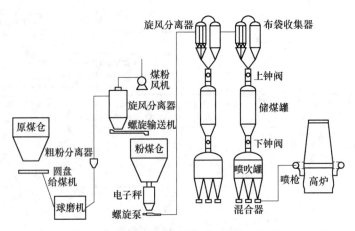

图 2　首钢早期的间接喷吹喷煤系统流程

程复杂，建设投资高，动力消耗大，而且增加了喷吹易燃易爆的烟煤时的不安全因素，同时首钢旧的制粉车间中的球磨机设备运行噪声大，能耗高，设备老化严重，运行成本增加，制粉能力不足，难以支撑日益增长的喷煤量。首钢于 1999 年底筹划建设新的喷煤系统，其流程如图 3 所示。

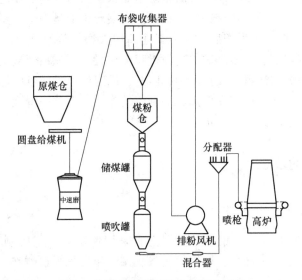

图 3　首钢中速磨制粉直接喷吹流程

该套喷煤系统与以往相比有所创新：首先，首次采用中速磨制粉一级大布袋收集的直接喷吹工艺，减少建设及运行成本的同时更加注重系统安全性；其次，成功实现长距离煤粉输送，该套喷煤系统同时给当时的首钢 2、3 号高炉提供煤粉，其中 2 号高炉输送距离 358m，3 号高炉输送距离 452m，是当时国内煤粉输送距离之最；另外，自行开发的可调给煤机、锥式分配器、点式流化器、喷煤支管测堵装置、高温合金喷枪等新装备也获得应用[9]。该系统于 2000 年 11 月投入使用，已经具备当前喷煤工艺的雏形，是首钢喷煤发展过程中的一个里程碑。

3.2 串联罐组向并联罐组的改进

2007年之前,包括2000年投入使用的首钢四制粉车间在经过历次改造后,虽然已经采用了直接喷吹模式并使用总管加分配器技术,但是在喷吹罐组的布置形式上一直延续使用串联罐组。由于在倒罐过程中,波纹补偿器产生浮力影响,所以连续计量不够准确一直是串联罐组布置形式存在的问题。这不仅影响到喷煤量的计量,而且影响倒罐过程中对煤粉存积的检测,直接威胁到安全生产。另外,进入21世纪后,国内部分高炉喷煤系统已经开始采用并联罐组的布置形式,其建设投资小、能源消耗少等优势逐渐被业内所认可。首钢于2007年1月4日投产的迁钢2号高炉,其配套的喷煤系统采用了并罐喷吹模式,是首钢与荷兰Danieli Corus公司合作开发,也是首钢在喷煤工艺上首次引进国外先进技术。

该套喷煤系统除了在喷吹罐布置形式上采用了并列罐组,同时也对当时的一些新技术和设备进行了尝试。首先在煤粉仓与喷吹罐锥体底部设置了板式流化器代替以往使用的点式流化器,以实现煤粉流态化,煤粉输送浓度达到18~20kg/kg,比以往首钢旧的工艺提高了3倍;其次,采用了均匀喷吹技术,对分配器设备及喷煤各支管阻损都进行了严格设计;另外最重要的是,实现了无人值守全自动喷吹。之前首钢曾经多次组织对迁钢1号高炉和北京老厂区高炉进行自动化喷吹试验,全部以失败告终,原因是串罐喷吹形式造成上下罐相互影响称量,造成重量叠加,导致重量补偿累计到一定数值后,出现程序紊乱。本次的全自动喷吹以中速磨为核心,进行连锁控制,同时实现了对下煤量、煤温、喷煤量、充压稳压自动调节及自动倒罐。在自动控制及计量和调节精度方面,按照高炉要求自动调节,喷煤量计量精度可以控制在1%误差范围内,各风口喷吹煤粉的均匀性控制在4%的误差范围内;不但节约人工成本,提高生产效率,而且在减少了人为操作失误的同时稳定生产,减少设备维修率。

迁钢2号高炉喷煤系统全自动喷吹的实现过程并非一帆风顺。投产初期就遇到预设喷吹风流量曲线错误、称量错误、提前倒罐、充压阀自动打开等一系列问题,必须人工干预才能进行正常生产,而且严重影响高炉喷煤系统的稳定运行及喷煤量的提高。为了解决上述问题,首钢进行了部分技术改造,针对问题对程序进行研究、分析、修正,解决了充压氮气供气能力不足、罐压过高限制喷煤量、喷煤量波动、系统不稳定等一系列问题,最终实现了全自动喷吹。

3.3 特大型高炉喷煤系统的应用

进入21世纪后,首钢在搬迁调整的过程中,高炉大型化的趋势十分明显。2010年1月投产的迁钢3号高炉其有效容积为4000m^3,2009年5月和2010年6月投产的京唐1、2号高炉有效容积更是达到了5500m^3,这种超大型的高炉对配套的喷煤系统提出了更高的要求。

3.3.1 迁钢3号高炉喷煤系统

迁钢3号高炉的喷煤系统结合了荷兰Danieli Corus公司的技术,设计煤比190kg/t,设备能力达到250kg/t,采用中速磨制粉、一级大布袋收粉、三罐并列直接喷吹、单主管单分配器的全自动喷吹工艺,支管数量36个,喷煤罐充压及流化采用氮气,煤粉输送采用压缩空气，输煤总管采用逐渐缩径设计，煤粉仓与喷煤罐的煤粉流态化采用板式流化器设备，煤粉下出料方式，输送设计固气比为不小于35kg/kg。表1为该喷煤系统的基本设计参数。

表 1　迁钢 3 号高炉喷煤系统基本设计参数

名　称	迁钢 3 号高炉
高炉有效容积/m³	4000
高炉产量/t·d⁻¹	9600
高炉利用系数/t·(m³·d)⁻¹	2.4
风口数量/个	36
设计煤比/kg·t⁻¹	190~250
设计喷煤量/t·h⁻¹	77~108
中速磨煤机台数/台	2
单台磨热风量/m³·h⁻¹	120000
布袋收尘器过滤面积/m²	3926

均匀喷吹是引进的 Danieli 喷煤技术的优势之一，采用"瓶式"分配器，并对其内部结构进行了调整，解决了以往此类型分配器煤粉和输送气体在其内部产生涡流，阻力大易积粉的问题，同时将"瓶式"分配器置于炉顶平台，增加喷煤支管的垂直段比例，减小阻损以提高煤粉分配均匀程度，同时使各支管当量长度相等，达到各支管等阻损，从而实现各风口均匀喷吹。该喷煤系统喷吹部分简要流程如图 4 所示。

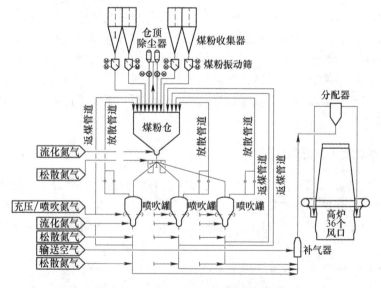

图 4　首钢迁钢 3 号高炉喷煤系统流程

表 2 为迁钢 3 号高炉及喷煤系统 2012 年下半年的运行指标。

表2 迁钢3号高炉喷煤系统运行指标

时间	利用系数/t·(m³·d)⁻¹	煤比/kg·t⁻¹	焦比/kg·t⁻¹	燃料比/kg·t⁻¹
2012年7月	2.37	142.03	315.66	501.38
2012年8月	2.23	153.35	308.58	504.64
2012年9月	2.50	182.34	285.83	509.56
2012年10月	2.21	152.29	309.69	502.19
2012年11月	2.04	154.61	317.28	507.64
2012年12月	2.08	166.48	307.19	518.43

3.3.2 京唐1、2号高炉喷煤系统

京唐1、2号高炉的喷煤系统结合了德国Kuttner公司的技术，设计煤比220kg/t，设备能力250kg/t，采用中速磨制粉、一级大布袋收粉、三罐并列直接喷吹、双主管双分配器的全自动喷吹工艺，支管数量42个，喷煤罐充压、流化及煤粉输送均采用氮气，煤粉仓煤粉流化采用点式流化器，而喷煤罐采用流化罐装置，煤粉上出料，输送设计固气比为不小于40kg/kg，这也是首钢历史上输送固气比的最高值。表3为京唐1、2号高炉喷煤系统基本设计参数。

表3 京唐1、2号高炉喷煤系统基本设计参数

名称	京唐1、2号高炉
高炉有效容积/m³	5500
高炉产量/t·d⁻¹	12650
高炉利用系数/t·(m³·d)⁻¹	2.3
风口数量/个	42
设计煤比/kg·t⁻¹	220~250
设计喷煤量/t·h⁻¹	116~138
中速磨煤机台数/台	2
单台磨热风量/m³·h⁻¹	116000
布袋收尘器过滤面积/m²	3038

在保证均匀喷吹方面，使用"碗式"分配器，同时喷煤支管采用迂回缠绕的方式实现支管当量长度相等，而且在每个支管上设置了拉瓦尔阻损管，克服了高炉各风口压力差异等因素对支管流量的影响，从而进一步提高喷吹均匀性。该系统喷吹部分简要流程如图5所示。

表4为京唐1号高炉及喷煤系统2012年下半年的运行指标。

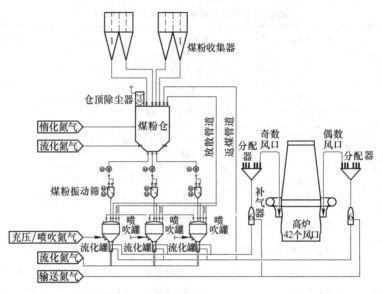

图5 首钢京唐1、2号高炉喷煤系统

表4 京唐1号高炉喷煤系统运行指标

时间	利用系数/t·(m³·d)⁻¹	煤比/kg·t⁻¹	焦比/kg·t⁻¹	燃料比/kg·t⁻¹
2012年7月	2.3	146.90	294.60	471.25
2012年8月	2.12	158.95	312.08	494.23
2012年9月	1.77	123.26	389.81	535.71
2012年10月	2.18	159.00	302.31	487.85
2012年11月	2.25	153.39	299.80	480.68
2012年12月	2.29	143.84	307.25	479.45

新的喷煤系统特点可概括为：（1）中速磨等设备能力选择与实际煤种特性相结合；（2）喷吹工作循环周期与高炉喷煤能力配合；（3）实现浓相输送；（4）支管阻损精确控制；（5）无人值守全自动喷吹；（6）国外先进技术及装备的应用。另外京唐高炉喷煤实际指标与设计值之间存在一定差距，是由于喷煤量受到精料水平、热补偿、富氧率、炉料和煤气流分布等制约[10]，同时目前对该级别的高炉操作及管理尚在探索阶段，在实际生产中，要考虑到原煤质量、焦炭性能、精料水平、炉渣量及炉渣黏度、富氧率及热风温度等因素的影响，需要进一步摸索合理的喷煤比[11]，这也体现了首钢在新的时期更加注重喷煤保障高炉顺行的作用。

4 结语

从国内的第一套喷煤装置开始，首钢在喷煤领域不断探索，取得了历史最高煤比333.1kg/t、首先在矮胖高炉上进行高富氧多煤种喷吹、当时最远距离煤粉输送及单项设备的顺利应用等成绩；同时在大型高炉建设过程中，通过对国外先进技术的消化吸收，掌握浓相输送、均匀分配及全自动喷吹等技术，并已成功应用于生产实践。

参考文献

[1] 张立国,王再义,等. 鞍钢高炉喷吹除尘灰的研究与应用[J]. 中国冶金,2012,22(10):47.
[2] 陈永星,王广伟,等. 高炉富氧喷吹焦炉煤气理论研究[J]. 钢铁,2012,47(2):12.
[3] 郭同来,柳正根,等. 高炉喷吹焦炉煤气风口回旋区的数学模拟[J]. 东北大学学报,2012,33(7):987.
[4] 王连昌. 喷煤促进高炉冶炼进步[J]. 莱钢科技,2005(6):83.
[5] 张福明. 高风温低燃料比高炉冶炼工艺技术的发展前景[J]. 中国冶金,2013,23(2):1.
[6] 温大威. 中国高炉喷煤史回顾[J]. 宝钢技术,2005(1):6.
[7] 由文泉,温仕湛. 首钢矮胖高炉喷煤生产实践[J]. 炼铁,1997,16(4):14.
[8] 陈茂熙. 高炉喷煤的几个问题探讨[J]. 钢铁技术,1996(1):9.
[9] 张进副,李伟广. 新技术在首钢喷煤系统的应用[J]. 炼铁,2004,32(5):6.
[10] 张寿荣,毕学工. 关于大量喷煤高炉的某些理论问题的思考[J]. 钢铁,2004,39(2):8.
[11] 周渝生,项钟庸. 合理喷煤比的技术分析[J]. 钢铁,2010,45(2):1.

Technical Development and Innovation of PCI System in Shougang

Meng Xianglong Zhang Fuming Cao Chaozhen

Abstract: The first industrial test of PCI technology in domestic was began by Shougang as early as 1960's, in the following decades, the PCI technology went through some stages with different characteristic. In the 21th century, the volume of blast furnace become larger and larger, and PCI system matching large sized blast furnace were put into practice in Qiangang and Jingtang in Shougang. Now, PCI system with Shougang feature for large sized blast furnace, which combined with the technology of dense phase long distance conveying, equal injection and full-automatic control, has been widely applied.

Keywords: blast furnace; PCI; Shougang; development

大型高炉煤气干式布袋除尘技术研究

摘　要：对高炉煤气干式布袋除尘技术的发展现状进行了总结，阐明了干式布袋除尘技术的工艺流程及低压脉冲喷吹干式布袋除尘技术的原理，重点阐述了研究开发的高炉煤气温度控制、煤气含尘量在线监测、煤气管道系统防腐及除尘灰浓相气力输送等关键技术。

关键词：高炉；煤气干式布袋除尘；温度控制；含尘量监测

1　引言

高炉煤气干式布袋除尘技术是 21 世纪高炉实现节能减排、清洁生产的重要技术创新。与传统的高炉煤气湿式除尘技术相比，干式布袋除尘技术提高了煤气净化程度、煤气温度和热值，可以显著降低高炉生产过程中的新水消耗和动力消耗，还可以提高炉顶煤气余压发电量和二次能源的利用效率，减少环境污染，是钢铁工业发展循环经济、实现可持续发展的重要技术途径，已成为当今高炉炼铁技术的发展方向。

2　高炉煤气干式布袋除尘技术发展现状

20 世纪 80 年代我国炼铁工作者总结了高炉煤气干式布袋除尘技术的开发经验和应用实践，自主开发了高炉煤气低压脉冲喷吹干式布袋除尘技术，在我国 300 m^3 级高炉上进行试验并取得成功，使这项技术实现了工业化应用。经过几年的发展，高炉煤气低压脉冲喷吹干式布袋除尘技术在我国中小型高炉上得到迅速推广应用，21 世纪初新建的 1000 m^3 级的高炉相继采用此项技术，目前已推广应用到数十座 2000~5500 m^3 级大型高炉上。

目前，我国自主开发的高炉煤气干式布袋除尘技术在设计研究、技术创新、工程集成及生产应用等方面已取得突破性进展，该项技术全面取代了传统的高炉煤气湿式除尘工艺，而且为数众多的大型高炉采用了煤气"全干式"除尘工艺，完全摆脱了传统的煤气湿式除尘备用系统。我国已研究开发了低压脉冲喷吹清灰、煤气温度控制、煤气含尘量在线监测、煤气管道系统防腐及除尘灰浓相气力输送等关键技术，使大型高炉煤气干式布袋除尘技术日臻完善。我国高炉煤气干式布袋除尘技术的发展进程如图 1 所示，我国部分大型高炉煤气干式布袋除尘系统技术参数见表 1。

本文作者：张福明。原刊于《炼铁》，2011, 30（1）：1-5。

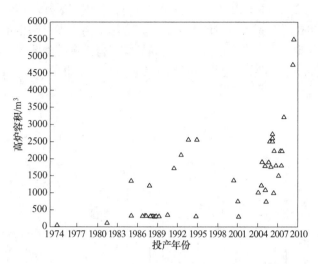

图1 我国高炉煤气干式布袋除尘技术的发展进程

表1 我国部分大型高炉煤气干式布袋除尘系统技术参数

项 目	高炉容积 /m³	煤气量 ×10⁴/m³·h⁻¹	炉顶压力 /MPa	煤气温度 /℃	箱体数量 /个	箱体直径 /mm	单箱滤袋数量 /个	滤袋规格 /mm	总过滤面积 /m²	单箱过滤面积 /m²	标况过滤速度 /m·min⁻¹	工况过滤速度 /m·min⁻¹	净煤气尘含量 /mg·m⁻³
首秦1号高炉	1200	23	0.17	100~250	10	4000	248	130×6000	6080	608	0.63	0.40	≤5
首秦2号高炉	1800	35	0.25	100~250	14	4000	248	130×6000	8512	608	0.69	0.32	≤5
迁钢2号高炉	2650	50	0.25	100~250	14	4600	250	160×7000	12320	880	0.68	0.31	≤5
唐钢3200m³高炉	3200	55	0.25	120~260	15	6000	468	130×7000	20070	1338	0.46	0.21	≤8
迁钢3号高炉	4000	75	0.28	100~250	13	6200	409	160×7000	18616	1432	0.67	0.28	≤5
首钢京唐1号高炉	5500	87	0.28	100~250	15	6200	409	160×7000	21586	1439	0.59	0.23	≤5

注：滤袋规格为直径×长度。

3 高炉煤气干式布袋除尘技术原理

3.1 高炉煤气干式布袋除尘技术工艺流程

布袋除尘技术最早应用于环境除尘领域[1,2]。高炉煤气是易燃、易爆的有毒气体，而且处理流量大、温度波动大、系统压力高、粉尘含量高,采用布袋除尘技术具有很高的技术难度和风险。布袋除尘技术基于纤维过滤理论,将其应用在高炉煤气除尘系统,代替了高炉煤气用水清洗的湿式除尘技术,迄今已发展成为高炉煤气低压脉冲喷吹干式布袋除尘技术。

高炉煤气干式布袋除尘系统工艺流程如图 2 所示。

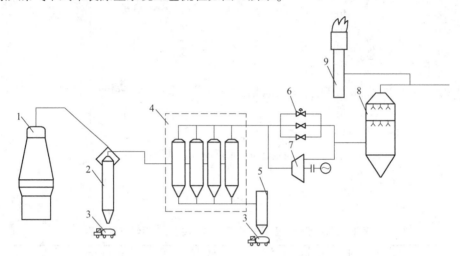

图 2　高炉煤气干式布袋除尘系统工艺流程
1—高炉；2—重力除尘器（或旋风除尘器）；3—罐车；4—干式布袋除尘器；
5—集中灰仓；6—减压阀组；7—余压发电装置（TRT）；8—氯化物脱除装置；9—煤气放散塔

高炉冶炼过程中产生的煤气经导出管、上升管和下降管进入重力除尘器（或旋风除尘器），经过重力除尘后的荒煤气经荒煤气总管及支管进入各个干式布袋除尘器。经干式布袋除尘器净化处理后的净煤气经净煤气支管进入到净煤气总管,再通过余压发电装置（TRT）或减压阀组减压后进入煤气管网,供钢铁厂作为二次能源利用。

3.2　高炉煤气低压脉冲喷吹干式布袋除尘技术原理

（1）高炉煤气布袋除尘过滤机理。高炉煤气经过重力除尘器或旋风除尘器进行粗除尘以后,煤气中大颗粒粉尘被捕集,煤气含尘量一般可以降低到 $10g/m^3$ 左右。经过粗除尘以后的煤气还需要进行净化处理,使煤气含尘量降低到 $10mg/m^3$ 以下。未经净化处理的高炉煤气中悬浮着形状不一、大小不等的微细粉状颗粒,是典型的气溶胶体系。高炉煤气布袋除尘的过滤机理基于纤维过滤理论,其过滤过程可以分为两个阶段：当含尘煤气通过洁净布袋时,在扩散效应、直接拦截、重力沉降及筛分效应的共同作用下,首先进行的是布袋纤维对粉尘的捕集,起过滤主导作用的是纤维,然后是阻留在布袋纤维中的粉尘与纤维一起参与过滤,此过程称为"内部过滤"；当布袋纤维层的粉尘达到一定容量以后,粉尘将沉积在布袋纤维层表面,在布袋表面形成一定厚度的粉尘层,布袋表面的粉尘层对煤气中粉尘的过滤将起到主要作用,此过程称为"表面过滤"。在布袋除尘的实际运行中,表面过滤是主导的过滤方式,对高炉煤气布袋除尘技术具有重要意义。

（2）脉冲喷吹清灰机理。布袋除尘器工作时,其阻力随布袋表面粉尘层厚度的增加而加大,当阻力达到规定数值时,就必须及时清除附着在布袋表面的灰尘。清灰的基本要求是从布袋上迅速均匀地剥落沉积的粉尘,而且又要求在布袋表面能保持一定厚度的粉尘层。清灰是保证布袋除尘器正常工作的重要因素,常用的清灰方式有机械清灰、反吹清灰和脉冲清灰。用于高炉煤气布袋除尘的方式主要是反吹清灰和脉冲清灰,两者的主要技术特征如下:

1)反吹清灰是利用与过滤气流相反的气流使布袋变形造成粉尘层脱落的一种清灰方式,日本高炉煤气干式除尘系统均采用布袋反吹清灰方式。我国在20世纪80~90年代由日本引进的太钢3号高炉(1200m³)、首钢2号高炉(1327m³)、攀钢4号高炉(1350m³)煤气干式布袋除尘系统均采用此种工艺。反吹清灰采用大规格布袋,布袋直径为300mm,长度为10m。含尘荒煤气由箱体下部进入布袋内部过滤后到达箱体上部,即所谓"内滤式布袋除尘",除尘后的净煤气进入箱体顶部的净煤气支管,再汇集到净煤气总管。为了进行布袋清灰,采用净煤气加压反吹清灰工艺,由净煤气管道引出净煤气,经反吹风机加压后进入除尘器箱体,煤气流反向流动,由布袋外部进入布袋,将沉积在布袋内壁的灰尘吹落,完成清灰过程,反吹清灰一般采取离线清灰方式。

2)我国从20世纪80年代开始,将脉冲清灰技术应用于高炉煤气布袋除尘工艺,并使这项技术得到迅速推广应用。与反吹清灰工艺不同,脉冲清灰采用小规格布袋,布袋直径一般130~160mm,长度为6~7m。含尘荒煤气由箱体下部进入箱体,经布袋外部过滤后进入布袋内部,即所谓"外滤式布袋除尘"。除尘后的净煤气由布袋内部进入箱体上部的净煤气支管,再汇集到净煤气总管。脉冲除尘的布袋内部设有专用的骨架结构,以支撑布袋在工作时始终保持袋状,不致被压扁而失效。脉冲清灰是利用加压氮气或煤气(压力为0.15~0.60MPa)在极短的时间内(不大于0.2s)由布袋袋口高速喷入布袋内,同时诱导大量煤气,在布袋内形成气波,使布袋从袋口到袋底产生急剧膨胀和冲击振动,具有很强的清灰作用。脉冲喷吹清灰冲击强度大,而且其强度和频率都可以调节,提高了清灰效果,系统阻力损失低,动力消耗少,还可以实现布袋过滤时在线清灰,在处理相同煤气量情况下,布袋过滤面积比反吹清灰要低。京唐1号高炉煤气干式布袋除尘系统清灰原理如图3所示。

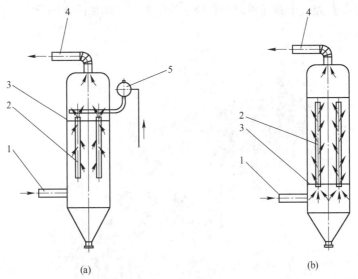

图3 京唐1号高炉煤气干式布袋除尘系统清灰原理

(a)脉冲清灰;(b)反吹清灰

1—荒煤气管道;2—布袋;3—花板;4—净煤气管道;5—脉冲喷吹装置

4 关键技术的研究与开发

高炉煤气干式布袋除尘技术是一项集成技术，涉及冶金、机械、化纤纺织、燃气、自动化检测与控制等多个工程技术领域。为提高系统运行的可靠性，不断优化工艺装备，对关键技术进行了研究开发，解决了一系列工程设计、设备制造、施工建设及生产操作过程中出现的问题，提高了整体技术装备水平和控制水平，实现了向特大型高炉推广应用的技术突破。

4.1 工艺流程优化和技术参数的确定

高炉煤气干式布袋除尘工艺流程的合理设计是确保系统稳定运行的基础，要遵循流体设计的基本原则，合理布置除尘器和煤气管道。煤气管道的布置和设计直接影响到除尘器内气流分布和阻力损失的均匀性，对进入各除尘器的煤气量及粉尘量要均匀分配。除尘器应采用双排并联布置方式，含尘煤气和净煤气总管设置在两排除尘器中间，通过支管与除尘器连接。煤气管道按等流速原理设计，使进入每个除尘器的煤气量分配均匀，而且整个系统应做到工艺布置紧凑、流程短捷顺畅、设备检修维护便利。除尘器采用低过滤速度设计理念，通过计算流体力学（CFD）仿真计算，研究分析除尘器内流场分布。根据数学仿真计算结果，确定合理的气流速度和气流方向，在除尘器内设置导流板，优化除尘器结构，使煤气在除尘器内的流场均匀分布，保证除尘器内的布袋在煤气流动均匀平稳的工况下工作，这是大型高炉煤气布袋除尘器设计的关键技术。合理确定除尘器过滤面积、过滤速度、气流上升速度、阻力损失及清灰周期等技术参数，特别是对于炉顶压力较高的大型高炉，煤气工况流量、压力、温度和过滤速度等参数的合理设计是保证系统可靠运行的关键要素。技术研究和生产实践表明，大型高炉煤气布袋除尘的工况过滤速度一般应控制在 0.5m/min 以下。首钢京唐 1 号高炉（5500m^3）煤气干式布袋除尘系统三维仿真设计图如图 4 所示。

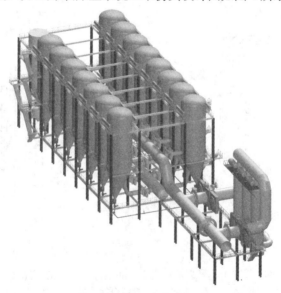

图 4 京唐 1 号高炉煤气干式布袋除尘系统三维仿真设计图

4.2 滤布的选择

滤布是布袋除尘过滤粉尘的介质，滤布的材质和性能对高炉煤气干式布袋除尘系统运行的稳定性和可靠性具有重要影响。由于高炉煤气温度高且不稳定，煤气中含有水分，煤气的相对湿度随原燃料条件和高炉操作条件变化较大，煤气中还含有腐蚀性介质。因此，滤布的选择要适合高炉煤气特点，要求滤布除尘效率高、耐高温和耐腐蚀性好、耐水解性好且使用寿命长。目前适用于高炉煤气干式布袋除尘系统的典型滤布有玻璃纤维、薄膜复合芳香聚酰胺针刺毡（NOMEX）、FMS复合针刺毡、聚酰亚胺（P84）及聚四氟乙烯（PTFE）复合针刺毡等，实际应用中要综合煤气工况和粉尘的特点，合理选择性能优良的滤布。几种典型滤布的理化性能见表2。

表2 几种典型滤布的理化性能

项目	材 质	质量/g·m^{-2}	厚度/mm	透气度/m^3·(m^2·min)$^{-1}$	断裂强度/N 径向	断裂强度/N 纬向	断后伸长率/% 径向	断后伸长率/% 纬向	连续使用温度/℃	短时使用温度/℃	处理方式
玻璃纤维复合针刺毡Ⅰ	玻璃纤维、P84/玻璃纤维基布	≥800	2.0~3.2	10~20	≥1800	≥1800	<10	<10	260	350	PTFE处理
玻璃纤维复合针刺毡Ⅱ	玻璃纤维、芳纶聚酰胺/玻璃纤维基布	≥800	2.0~3.2	8~20	≥2000	≥2000	<10	<10	220	260	PTFE处理
NOMEX	芳纶聚酰胺	≥550	2.0~2.4	10~20	≥800	≥1000	<20	<40	204	240	热定型、烧压
聚四氟乙烯针刺毡	PTFE/PTFE长丝	≥600	1.1	8~10	≥800	≥1000	<15	<15	240	260	热定型

注：测试透气度的压力为127Pa；断裂强度测试试样尺寸为50mm×200mm。

4.3 煤气温度控制技术

煤气温度控制是干式布袋除尘技术的关键要素，正常状态下，煤气温度应控制在80~220℃，煤气温度过高或过低都会影响系统的正常运行。当煤气温度达到250℃时，超过一般布袋的安全使用温度，布袋长期在高温条件下工作，会出现异常破损甚至烧毁。由于煤气中含水，当煤气温度低于露点温度时，煤气中的水蒸气发生相变凝结为液态，出现结露现象，造成布袋黏结。因此采用煤气干式布袋除尘技术，高炉操作要更加重视炉顶温度的调节控制[3]。

高炉炉顶温度升高时可采取炉顶雾化喷水降温措施，同时在高炉荒煤气管道上设置热管换热器，用软水作为冷却介质，通过热管换热使软水汽化吸收高温煤气的热量，有效降低煤气温度。实践表明，此项技术措施可有效解决煤气高温控制的技术难题，煤气温度能够降低50~90℃。

如果煤气温度过低则要采取综合措施，提高入炉原燃料质量、降低入炉原燃料水分、加强炉体冷却设备监控、合理控制炉顶温度和荒煤气管道保温等技术措施都能取得成效。

特别在高炉开炉、复风时要注重煤气温度的控制，降低煤气的含水量，将煤气温度控制在露点温度以上 20~30℃。目前已成功开发出煤气低温状态的高效快速加热装置，利用蒸汽作为煤气加热介质，通过热管换热将蒸汽热量传递给煤气，可以使煤气温度提高到露点温度以上，抑制水分凝聚灰尘黏结布袋。这种煤气温度控制装置可以有效控制煤气高温和低温的异常状况，使高炉煤气干式布袋除尘系统可以适应多种工况条件，提高了系统的适应性和可靠性。京唐 1 号高炉煤气温度控制装置的工艺原理如图 5 所示。

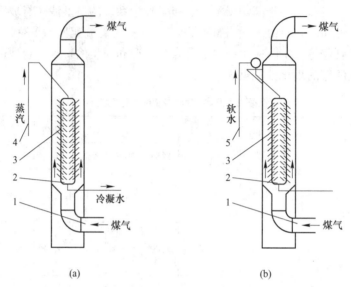

图 5　京唐 1 号高炉煤气温度控制装置的工艺原理
（a）升温装置；（b）降温装置
1—煤气管道；2—软水罐；3—热管；4—蒸汽管道；5—软水管道

4.4　煤气含尘量在线监测技术

煤气含尘量在线监测系统是监控高炉煤气干式布袋除尘系统、炉顶煤气余压发电系统稳定运行的重要检测设备。高炉煤气含尘量在线监测系统采用电荷感应原理，在流动的高炉煤气中，粉尘颗粒因摩擦、碰撞产生静电荷，形成静电场，静电场的变化即可反映煤气含尘量的变化。煤气含尘量在线监测系统通过测量静电荷的变化，从而推断出煤气中含尘量的数值，以此判定布袋除尘系统的运行是否正常。当布袋出现破损时，净煤气管道中含尘量增加，静电荷量强度增大，电荷传感器可以及时检测到电荷量值并输出到变送器，实现煤气含尘量的自动在线监测。

4.5　煤气管道系统防腐技术

采用高炉煤气干式布袋除尘技术后，净煤气冷凝水中的氯离子含量显著升高，这主要是由于高炉原燃料中的氯化物在高炉冶炼过程中形成气态的 HCl。当煤气温度达到露点温度时，气态 HCl 与冷凝水结合，形成酸性水溶液而引起酸腐蚀。在潮湿的中性环境中，煤气中的氯离子也会对煤气管道和不锈钢波纹补偿器产生点腐蚀、应力腐蚀和局部腐蚀。通过检验分析采用干式布袋除尘技术的净煤气冷凝水发现，冷凝水中氯离子含量高达

1000mg/L，煤气冷凝水的 pH 值低于 7，有时甚至达到 2~3，对煤气管道和波纹补偿器具有强腐蚀性，会造成煤气管道系统异常腐蚀。

为了抑制煤气管道系统的异常腐蚀，对煤气管道波纹补偿器的腐蚀机理进行了分析研究，采取了以下措施：

（1）对不锈钢波纹补偿器的材质及结构进行了改进，将波纹补偿器材质由奥氏体不锈钢 316L（00Cr17Ni14Mo2）改进为耐氯离子腐蚀的不锈钢 Incoloy825，提高了材质的抗酸腐蚀性能。

（2）在煤气管道内壁喷涂防腐涂料，使金属管道与酸性腐蚀介质隔离，抑制管道异常腐蚀。

（3）为了脱除高炉煤气中的氯化物，开发了氯化物脱除装置，应用化学和物理吸附原理，有效脱除高炉煤气中的氯化物。在净煤气管道上设置喷洒碱液装置，使碱液与高炉煤气充分接触，降低煤气中的氯化物含量。

4.6 除尘灰浓相气力输送技术

高炉煤气布袋除尘灰的收集输送是影响系统正常工作的关键因素，传统的机械式输灰工艺存在着诸多技术缺陷。除尘灰气力输送技术利用氮气或净煤气作为载气输送除尘灰，将每个布袋除尘器收集的除尘灰通过管道输送到灰仓，再集中抽吸到罐车中运送到烧结厂回收利用，实现了除尘灰全程密闭输送，解决了传统机械输灰工艺的技术缺陷，优化了工艺流程，降低了能源消耗，减少了二次污染，攻克了输灰管道磨损等技术难题。

5 结语

大型高炉采用煤气干式布袋除尘技术是炼铁技术的发展趋势，是实现炼铁工业高效低耗、节能减排、降低水资源消耗、发展循环经济的重要支撑技术。通过系统研究和技术集成，我国已经掌握了大型高炉煤气干式布袋除尘的关键技术，并在 1000~5500m³ 大型高炉上得到成功应用。我国自主开发的高炉煤气低压脉冲喷吹干式布袋除尘技术在工艺流程和除尘器结构优化、煤气温度控制、管道系统防腐、除尘灰气力输送及数字化控制系统等方面均取得突破，在生产实践中取得了显著的经济效益、社会效益和环境效益。

致谢

衷心感谢高鲁平、张建、毛庆武、郑传和、侯建、章启夫、韩渝京、陈玉敏、任绍峰、钱世崇、李欣、于玉良、李林等专家的辛勤工作。本项研究得到国家科技支撑计划项目"新一代可循环钢铁流程工艺技术——长寿集约型冶金煤气干法除尘技术的开发"（2006BAE03A10）的资助，特此感谢。

参考文献

[1] 向晓东. 烟尘纤维过滤理论、技术及应用 [M]. 北京：冶金工业出版社，2007：62-65.

[2] 张殿印，王纯，俞非漉. 袋式除尘技术 [M]. 北京：冶金工业出版社，2008：43-50.

[3] 张福明. 现代大型高炉关键技术创新 [J]. 钢铁，2009（4）：1-4.

Study on Dry Dedusting Technology by Bag Filter for Large Sized Blast Furnace Gas

Zhang Fuming

Abstract: The paper presents the current development and the process flow of dry dedusting technology by bag filter for blast furnace gas, and states the philosophy of LP pulse bag filter technology. The author particularly focuses on BFG temperature control, on-line monitoring technology of dust content in gas, corrosion proof of gas pipeline system and dense phase ash pneumatic conveying technology, etc.

Keywords: blast furnace; dry bag filter of gas; temperature control; dust content monitoring

大型转炉煤气干法除尘技术研究与应用

摘　要：转炉煤气干法除尘技术具有高效能源转换、节约新水、节能减排、清洁环保的技术优势，可以大幅度降低水消耗、高效回收蒸汽和煤气，减少环境污染，是当代转炉冶炼实现高效能源转换的关键技术。针对首钢京唐"全三脱"转炉冶炼工艺过程和技术特征，解析了转炉煤气发生泄爆的工艺机制，研究开发了控制煤气泄爆的安全技术措施。采用CFD数值仿真技术，对蒸发冷却塔内煤气流动过程进行了研究解析，优化了蒸发冷却塔的设计、雾化喷嘴的布置及其流量控制。研究开发了300t转炉在"全三脱"冶炼条件下，合理控制转炉煤气成分、温度的关键技术，有效地提高了煤气、蒸汽的回收率，大幅度降低了煤气泄爆，实现了工艺稳定运行、能源高效回收、排放显著降低的目标。生产实践表明，"全三脱"冶炼条件下全年回收煤气达到85m^3/t以上，回收蒸汽达到110kg/t以上，煤气泄爆率控制在0.03%以下，保证了炼钢生产的安全稳定运行，取得了显著的经济效益和生态环境效益。

关键词：炼钢；转炉；煤气干法除尘；蒸发冷却器；静电除尘器；能源回收

1　引言

　　首钢京唐钢铁厂是按照循环经济理念建设的新一代可循环钢铁厂，具备"优质产品制造、高效能源转换、消纳废弃物并实现资源化"的三重功能[1]。以构建高效率、低成本洁净钢生产体系为目标[2]，大力降低资源和能源消耗，实现高效能源转换和回收利用；降低水资源消耗，实现污水和废弃物的"零排放"。并开发了"全三脱"铁水预处理工艺，集成创新并应用了"2+3"300t转炉洁净钢冶炼工艺和转炉煤气干法除尘技术。

　　转炉煤气干法除尘工艺具有除尘效率高、节水效果好、能源消耗和运行费用低、使用寿命长、维护维修少的优点。特别是在降低新水消耗、能源消耗方面具有显著优势，可将转炉煤气含尘量降低到15mg/m^3以下，大幅度降低粉尘排放；同时还可实现污水零排放，含铁粉尘经压块处理后可直接供转炉使用，实现废弃物的资源化回收利用[3]。

　　转炉煤气干法除尘技术于20世纪60年代末期开发成功，在欧洲钢厂率先得到应用，取得了较好的应用效果。1969年，德国萨尔茨吉特钢厂对3座转炉煤气除尘系统进行改造，将原有的湿法除尘工艺改造为干法除尘工艺，粉尘排放量降低了65%[4]。1997年，我国宝钢二炼钢厂在国内首次引进转炉煤气干法除尘技术，将其应用在2座250t转炉上。但由于引进国外成套技术装备投资巨大，加上煤气泄爆、系统稳定运行等技术问题在较长时期内未能得到有效解决，导致转炉煤气干法除尘技术在我国未能得到推广应用[5~7]。

　　进入新世纪以来，随着转炉煤气干法除尘技术的日渐成熟以及我国对烟气排放浓度的

本文作者：张福明，张德国，张凌义，韩渝京，程树森，昌占辉。原刊于《钢铁》，2013, 48 (2)：1-9。

要求日趋严格，2006年以后，国内大型钢厂相继引进了转炉煤气干法除尘技术，部分中型钢厂也陆续采用了我国自主开发的转炉煤气干法除尘装置，在降低粉尘排放、提高煤气回收等方面均取得了较好的应用效果。国内许多钢厂为解决煤气干法除尘系统泄爆问题，还开发应用了许多行之有效的煤气泄爆控制技术[8]，而且引进设备的国产化程度也逐渐提高，100~150t级转炉煤气干法除尘装备基本已经实现全面国产化。首钢京唐钢铁厂采用铁水"全三脱"预处理工艺生产洁净钢，由于这种新型冶炼模式具有其工艺特殊性，导致在这种冶炼条件下，转炉煤气干法除尘系统煤气泄爆几率比转炉常规冶炼时大幅度增加，安全生产隐患更为突出。因此研究特大型转炉在"全三脱"冶炼条件下，煤气干法除尘技术的高效稳定运行技术具有极其重要的意义。

2 "全三脱"洁净钢生产工艺技术

所谓"全三脱"就是对全量铁水进行脱硫、脱硅、脱磷预处理，以此构建高效率、快节奏、低成本的洁净钢生产系统。该冶炼工艺于20世纪80年代由日本最早提出并实施，取得了很好的应用效果[9]。铁水在装入脱碳转炉之前首先对铁水进行脱硫、脱硅、脱磷"全三脱"预处理，以最大限度地降低钢水中的硫、磷含量，为生产高品质洁净钢奠定基础。与转炉常规冶炼工艺相比，由于铁水采用"全三脱"预处理工艺，转炉冶炼的主要功能被简化为脱碳和升温，因此转炉冶炼周期缩短，可由常规冶炼的36~38min缩短到30min以下，可以实现转炉的高效快速冶炼和少渣冶炼。因此可以有效地降低转炉生产成本，提高生产效率，还可以有效地降低钢水中硫、磷含量，实现优质洁净钢生产[10]。图1是典型的"全三脱"条件下转炉炼钢工艺流程。

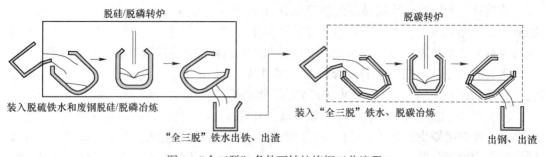

图1 "全三脱"条件下转炉炼钢工艺流程

经过"全三脱"预处理后的铁水装入脱碳转炉之前，铁水中硅、锰、磷的含量已经很低，铁水装入脱碳转炉以后，由于没有硅、锰、磷等元素氧化的"前烧期"，氧枪降枪吹炼以后立即发生碳-氧反应并生成大量的CO，同时转炉内温度迅速升高，生成含大量CO的煤气进入静电除尘器与其内滞留的空气混合，遇静电火花而极易发生泄爆事故。在转炉常规冶炼条件下，由于可以形成稳定的硅、锰、磷等元素氧化的"前烧期"，"前烧期"生成的烟气主要为惰性混合气体（以CO_2和N_2为主），在轴流风机的抽吸作用下，残留在除尘管道和静电除尘器内部的空气随烟气一起被抽出，相应降低了泄爆几率。因此，转炉煤气干法除尘工艺在欧洲各钢厂均用于转炉常规冶炼。日本虽然采用"全三脱"冶炼工艺，但是转炉煤气除尘采用的是传统的湿法除尘工艺（OG），没有"全三脱"冶炼条件下煤气干法除尘的应用实绩。

高效率、快节奏、低成本生产洁净钢是 21 世纪钢铁工业发展的重要目标。随着市场对高性能、高质量洁净钢需求的日益增长，采用常规转炉冶炼工艺生产洁净钢存在较大难度。因此，以铁水"全三脱"预处理为代表的转炉冶炼新流程，成为高效率、低成本、稳定生产洁净钢的先进工艺之一。铁水经过"全三脱"预处理以后，硅、锰、磷元素的含量大幅度降低，因此转炉脱碳冶炼过程与常规冶炼相比，发生了很大的改变，转炉冶炼的供氧制度、造渣制度、温度控制、钢铁料装入等工艺操作制度也需要进行相应调整。这种冶炼工艺的改变又影响了转炉煤气除尘系统的工艺技术的选择和应用。

3 转炉煤气干法除尘工艺

3.1 工艺流程

与传统的转炉煤气湿法除尘工艺（OG）相比，转炉煤气干法除尘技术具有节水、节电、环境清洁等优势，而且除尘灰经压块后可以直接作为转炉原料回收利用。转炉煤气（1400~1600℃）由烟罩收集后导入汽化冷却烟道，并在进入蒸发冷却器前通过热交换将高温煤气热量回收，使转炉煤气温度降低到 1000℃ 以下，进入蒸发冷却器进行转炉煤气的二次降温和粗除尘；经过蒸发冷却器冷却后的煤气温度降低到 210~230℃，再进入到干式静电除尘器中进行煤气精除尘，经过静电除尘器净化的转炉煤气由轴流风机加压后，合格煤气经煤气冷却器再次降温后进入转炉煤气柜中，作为二次能源回收利用。首钢京唐转炉煤气干法除尘工艺流程如图 2 所示。

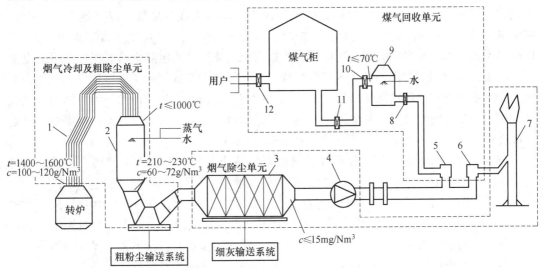

图 2 转炉煤气干法除尘工艺流程

1—汽化冷却烟道；2—蒸发冷却器，$\phi=6m$；3—干式电除尘器，$\phi=12.6m$；
4—ID 风机，$Q=192000m^3/h$，$P=1980kW$，$\Delta p=8.5kPa$；5—回收杯阀，DN2600mm；
6—放散杯阀，DN2000mm；7—放散烟囱；8—高温眼镜阀，DN2600mm；9—煤气冷却器，$\phi=6.9m$；
10—回收眼镜阀，DN2200mm；11—气柜入口切断阀，DN2600mm；12—气柜出口切断阀，DN2600mm；
t—烟气温度；c—烟气粉尘浓度

3.2 "全三脱"冶炼条件下转炉煤气泄爆机制

采用转炉煤气干法除尘工艺,最大的技术难题是静电除尘器内容易出现煤气泄爆,不仅会造成安全事故,还会影响正常生产。静电除尘器发生煤气泄爆的根本原因是静电除尘器内煤气中的CO(或H_2)与O_2的体积分数达到一定比例后,遇到电场中高压电弧火花发生煤气爆炸。理论研究表明,转炉煤气除尘系统产生煤气爆炸的一般条件是:煤气成分达到燃烧爆炸范围;煤气温度达到其燃烧温度(610℃)以上;煤气除尘系统中存在具有一定能量的"火种"。实践表明,静电除尘器容易在下列情况下发生煤气泄爆现象:

(1)装入脱碳转炉的铁水是经过"全三脱"预处理以后的铁水。在这种条件下,由于脱碳转炉开吹后没有硅、锰的氧化期,氧气直接参加脱碳反应,因此,转炉吹炼以后碳-氧反应十分剧烈,迅速生成CO。如果产生的CO在炉口没有被完全燃烧而进入静电除尘器,与吹炼前烟道中残留的空气混合就会产生爆炸,使泄爆阀开启,被迫中断吹炼,造成生产故障,影响正常生产。

(2)实际生产中,由于设备或生产组织等各种原因,有时会出现转炉吹炼断吹、停吹,经过一段时间后再恢复吹炼的情况。在转炉停止吹炼以后,空气进入到煤气除尘系统中,当重新降枪吹炼时,容易导致系统中CO和O_2的体积分数达到临界值,在高压电场的作用下发生煤气泄爆。

(3)转炉吹炼后期,当钢水碳的质量分数降至0.1%以下时,转炉内产生的煤气量明显减少。此时会有部分空气被吸入烟道和煤气除尘系统,使除尘系统内部O_2的体积分数升高至临界值,造成煤气泄爆。

(4)转炉采用"全三脱"冶炼工艺时,一般在脱碳转炉冶炼过程中不加入废钢,使脱碳转炉热量消耗少,熔池温度高,脱碳速度快,煤气生成量增加,煤气速度也相应提高。由于煤气中CO含量快速增加,在炉口又未能充分燃烧,使煤气中含有大量CO,难以稳定形成以CO_2为主体的惰性气体流。当这部分气体进入静电除尘器时就极有可能发生煤气泄爆。与此同时,由于转炉内煤气生成量迅速增加,使煤气除尘系统风机抽吸能力不足,此时不得不采取提前降低罩裙以减少煤气燃烧,这样就会造成以CO_2为主体的惰性气体生成量不足,使气流进入静电除尘器内置换时间过短、混匀时间不足而产生爆炸,这是造成转炉"全三脱"冶炼条件下安全生产的最大隐患。

(5)首钢京唐地处海滨,海洋性的气候条件造成各种物料含水量较高。在转炉冶炼过程中,煤气中H_2的体积分数容易达到临界值,增加了煤气"氢爆"的发生概率。

研究表明,将转炉冶炼过程产生的煤气的体积分数控制在$\varphi(CO) \leq 9\%$、$\varphi(O_2) \leq 6\%$、$\varphi(H_2) \leq 1\%$的范围内,煤气除尘系统就不会发生煤气泄爆。发生煤气泄爆的条件如图3所示。转炉煤气中爆炸性气体的体积分数与转炉操作具有直接关系。煤气泄爆主要发生在转炉加料开吹阶段、吹炼后期的提枪再下枪点吹以及溅渣护炉等阶段。与转炉常规冶炼不同的"全三脱"冶炼工艺,在脱碳转炉吹炼开始阶段,存在突出的煤气泄爆隐患。脱碳转炉与常规转炉冶炼过程的主要区别在于:常规转炉冶炼具有稳定的硅、锰氧化期,在硅、锰氧化期间由于开罩作业,容易产生大量以N_2为主要成分的惰性气体流。在轴流风机的抽吸作用下,滞留在管道和静电除尘器内部的空气随惰性气体一同由放散烟囱排出,从而有效避免了煤气泄爆。在"全三脱"冶炼条件下,脱碳转炉吹炼前期没有硅、锰氧化期,吹炼以后立即产生含有大量 CO 的煤气,无法生成惰性气体柱塞流。同时,由于脱碳转炉不加废钢,导致熔池温度快速上升,含有大量 CO 的煤气迅速产生。当这些煤气进入静电除尘器后,遇到静电火花时与存留的空气(含氧气体)混合而发生泄爆。因此,控制脱碳转炉在前烧期内煤气完全燃烧,确保在煤气管道内能够稳定形成一段"非泄爆"的惰性气流变得极其困难。

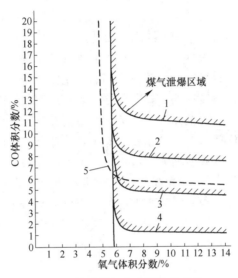

图 3　转炉煤气产生爆炸的条件

1—$\varphi(H_2)=0$, 200℃; 2—$\varphi(H_2)=1\%$, 200℃; 3—$\varphi(H_2)=2\%$, 200℃;
4—$\varphi(H_2)=3\%$, 200℃; 5—$\varphi(H_2)=1\%$, 400℃

4　蒸发冷却器 CFD 仿真研究

4.1　蒸发冷却器的功能

转炉煤气具有温度高、粉尘多、CO 含量高的特点。约 1500℃ 高温的转炉煤气经过汽化冷却烟道被冷却至 900~1000℃ 后,进入蒸发冷却器。蒸发冷却器利用雾化水蒸发冷却原理,将高温转炉煤气进行快速冷却,以达到进入电除尘器的温度要求。蒸发冷却器的作用是对煤气进行再次降温和粗除尘。在蒸发冷却器的上部均匀设置有多个雾化喷嘴,煤气在蒸发冷却器内与逆流的雾化液滴进行充分换热以后,煤气温度由 900℃ 降低到 200℃ 左右。降温后的煤气再进入静电除尘器进行精除尘。

冷却水量的精确控制和雾化喷嘴的布置方式是蒸发冷却器高效工作的技术关键。冷却水量取决于煤气流量和温度,如果冷却水量不足、煤气冷却效果不佳、煤气温度偏高,就会严重影响静电除尘器的除尘效果和使用寿命,增加煤气泄爆的几率。冷却水量过大,会造成烟尘结块,形成污泥,影响蒸发冷却器的工作效率。雾化喷嘴的布置方式应根据转炉煤气在蒸发冷却器内的流动特点进行相应地调整,从而达到冷却均匀、高效的目的。另外,由于转炉吹炼过程中煤气量和温度是变化的,因此必须建立可靠的控制模型,对冷却水量进行精确调节。

为实现精确控制煤气温度和冷却水量,降低由于转炉煤气温度过高而引起的泄爆,采用CFD仿真模拟研究转炉煤气在蒸发冷却器内的流动特性,建立了蒸发冷却器内煤气流动、水滴雾化以及传热数学模型,得出最优化的蒸发器结构参数和喷水装置布置参数。

4.2 蒸发冷却器内煤气流场分布

图4为蒸发冷却器内煤气速度场分布。可以看出,煤气在汽化冷却烟道内的速度较高,且由于烟道形状的不规则性,在蒸发冷却器的 X-Z 截面上,煤气的高速区主要集中在烟道弯管的外弧侧部分,而沿烟道内弧侧流动的煤气速度则较低。当煤气进入蒸发冷却器后,速度有所降低,大部分煤气沿着蒸发冷却器的左侧流动,但在蒸发冷却器内侧却出现了一个小范围的低速区,速度在6m/s以下。在蒸发冷却器的 Y-Z 截面上,由于蒸发冷却器底部弯管的作用,煤气在此截面上的高速区稍微向右偏移,而在蒸发冷却器的左侧同样也存在着一个小范围的低速区,煤气主要沿着蒸发冷却器的右侧运动。在蒸发冷却器底部的出口弯管处,煤气的速度有所增大,但仍出现了速度分布不均的现象,煤气在管道弯曲处的内弧侧速度较大。图4(c)为煤气在汽化冷却烟道、蒸发冷却器不同高度横截面上的速度分布。从图4(c)中可以明显地看到,煤气在管道横截面上的速度始终分布不均。

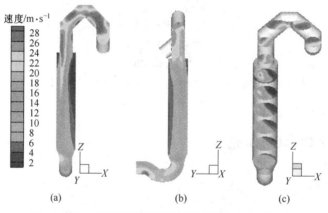

图4 蒸发冷却器内不同截面的速度分布
(a) X-Z 截面;(b) Y-Z 截面;(c) 管道横截面

在汽化冷却烟道内,煤气在烟道末端弯管段横截面上的高速区主要集中在弯管段的外弧侧部分。进入蒸发冷却器以后,煤气的速度有所降低,但冷却烟道的不规则性对蒸发冷却器内的煤气流动仍有较大影响,煤气的高速区仍然集中在蒸发冷却器的一侧,即烟道的外弧侧方向,从而导致蒸发冷却器另一侧的煤气速度始终较低,一直延伸到蒸发冷却器的底部。随着煤气向蒸发冷却器下部流动,低速区的范围开始变小,煤气在管道横截面上的速度分布开始变得均匀,维持在12m/s左右。

4.3 蒸发冷却器内温度分布

图5是蒸发冷却器内的温度场分布。将煤气在模型入口处的温度设定为1500℃,由

于汽化冷却烟道的冷却作用，煤气在靠近管道壁面处的温度较低，而管道中心的温度则较高。当煤气进入蒸发冷却器后温度有所降低，约为900~1000℃。在汽化冷却烟道内，由于煤气在管道的外弧侧速度较高，高温煤气更新较快，导致煤气在管道横截面上的高温区向烟道的外侧偏移。进入蒸发冷却器后，煤气高温区的中心开始向蒸发冷却器右侧偏移，而在蒸发冷却器左侧的温度则相对较低，这是由于蒸发冷却器内部回流的影响。蒸发冷却器右侧回流的存在，使得高温煤气向蒸发冷却器右侧方向的传热强烈，导致该区域的煤气温度相对较高。随着煤气向蒸发冷却器下部的流动，煤气温度稍微有所降低，而且蒸发冷却器横截面的温度分布也逐渐变得均匀，蒸发冷却器底部温差保持在100℃以内。

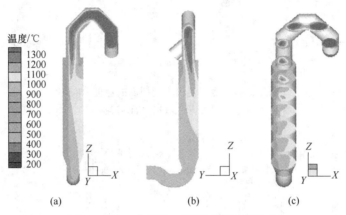

图5　蒸发冷却器内不同截面的温度分布
（a）X-Z截面；（b）Y-Z截面；（c）管道横截面

通过蒸发冷却器内速度场和温度场的计算分析可以得出，在未喷水的情况下，煤气在蒸发冷却器内的温度较高，平均温度约为900℃。汽化冷却烟道的不规则性对蒸发冷却器内的煤气流动影响较大。煤气进入蒸发冷却器后主要沿着蒸发冷却器的-X方向（即烟道的外弧侧方向）向下运动，而在蒸发冷却器的+X方向煤气速度则较低，且在此处产生了一个范围较大的回流区，大大减小了蒸发冷却器的有效利用体积。随着煤气向蒸发冷却器下部的流动，煤气在蒸发冷却器横截面上的速度分布变得越来越均匀。

因此，应根据转炉煤气在蒸发冷却器内的流动特征，合理优化蒸发冷却器的设计。可在蒸发冷却器喷嘴前设置不同形状的煤气导流板，达到使速度和温度均匀分布的目的。

4.4　蒸发冷却器设计优化

利用CFD模拟了蒸发冷却器内的煤气流场和温度场。为使煤气和水蒸气均匀分布，对喷水装置的安装位置及断面分布、煤气流速进行了设计优化。对蒸发冷却器内粉尘的运动进行了数学模拟，得到粉尘在蒸发冷却器内的运动轨迹和沉降位置，并提出相应的粉尘沉积位置。

研究表明，转炉汽化冷却烟道布置方式与蒸发冷却器具有重要的相关性。汽化冷却烟道的布置及走向极大地影响了烟道内煤气流动状态。当烟道布置具有较少的转向时，可在很大程度上改善烟道内气体流动不均匀的状况，有利于烟道内煤气流均匀地进入蒸发冷却器。与此同时，在蒸发冷却器煤气入口处设置煤气导流板，可以有效地调节煤气流分布状态,改善蒸发冷却器温度场分布。在相同的初始条件下增加煤气导流板,可使蒸发冷却器煤气出口温度降低 5~8℃。图 6 和图 7 分别给出蒸发冷却器设置导流装置前后不同截面的速度分布和粉尘分布。

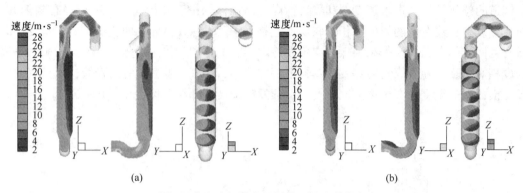

图 6　蒸发冷却器内不同截面的速度分布
(a) 未设导流装置;(b) 增设导流装置

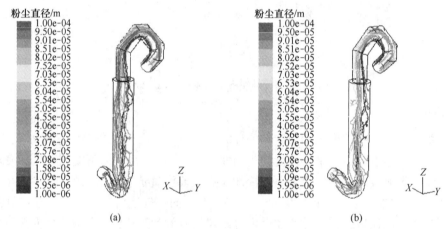

图 7　蒸发冷却器内不同截面的粉尘分布
(a) 未设导流装置;(b) 增设导流装置

5　应用实践

5.1　系统优化集成

为构建高效率、快节奏、低成本的洁净钢生产体系,首钢京唐炼钢厂采用"2+3" 300t 转炉"两步法"冶炼工艺,同时采用转炉煤气干法除尘工艺,在世界上首次将两种当今国际先进的工艺技术优化集成为一体[11,12]。

工程设计研究中,根据大型转炉"全三脱"冶炼工艺机制和技术特征,通过对转炉煤气形成过程的理论和试验研究,对转炉煤气干法除尘系统进行了工艺设计优化。改进了煤气管道的布置,使煤气管道布置顺畅,而且使煤气在蒸发冷却器内的流场、温度场分布均匀，有效提高了降温效果。优化设计了蒸发冷却器内双流雾化喷嘴的布置方式，使蒸发冷却器内冷却效率提高，解决了冷却不均和灰尘结块的问题。自主开发了除尘灰链带输送工艺和设备，优化了系统控制功能，大幅度减少了除尘灰输灰系统的设备故障。自主设计了除尘灰回收、压块工艺，将回收的除尘灰在炼钢厂附近进行资源化处理，降低了除尘灰转运过程的二次环境污染，取得了显著的经济和环境效益。工程中采用了自主研制开发的高温眼镜阀等国产化关键设备，大幅度降低了工程投资。

5.2 应用研究与技术创新

通过基础理论研究和工业试验，初步探索了"全三脱"冶炼条件下的转炉煤气干法除尘的工艺操作制度和关键控制技术。在基础研究和工业试验的基础上进一步深入研究，制定了一整套满足首钢京唐300t转炉"全三脱"冶炼条件下的煤气干法除尘工艺操作制度，为转炉快节奏、高效化冶炼提供可靠的技术保障。开发了"全三脱"冶炼条件下转炉煤气的综合防爆技术，在生产实践中取得成功应用：

（1）转炉煤气干法除尘系统发生煤气泄爆主要集中在转炉吹炼前期、吹炼中断后复吹和吹炼后期3个阶段。$\varphi(CO)/\varphi(O_2)$ 或 $\varphi(H_2)/\varphi(O_2)$ 达到临界值后，在静电除尘器的高压电场作用下就会发生煤气泄爆，因此，防止煤气泄爆的关键在于转炉煤气中 $\varphi(CO)/\varphi(O_2)$ 或 $\varphi(H_2)/\varphi(O_2)$ 必须避开爆炸临界值区域。

（2）优化转炉开吹、停吹后再开吹过程的供氧流量及罩裙位置控制。为了严格控制转炉冶炼过程中氧化反应速率，保证吹炼初期产生的CO能在炉口完全燃烧后变成CO_2，制定了氧气流量和罩裙位置的优化控制方案。转炉开吹时，氧枪降至最低位，同时控制氧气流量在较小的范围内，罩裙完全打开，使生成的CO完全燃烧生成CO_2。CO_2为非爆炸性气体，利用CO_2气体在整个煤气除尘系统内形成一股以CO_2为主的惰性气体柱塞流，推动煤气管道中残余的空气由放散烟囱排出。将其后形成的含有大量CO的转炉煤气，利用非爆炸性的惰性气体流与空气中的氧气隔离，将CO与O_2的体积分数控制在爆炸临界值区域之外。

（3）合理控制供氧强度、罩裙位置、风机转速。氧枪位置在脱碳转炉冶炼初期应控制在1.8~2.5m，吹氧量为81000~90000m^3/h的50%，且应保持30~35s，随后按20%~30%比例增加到100%，但其总时间应控制在90s内。活动罩裙在转炉吹炼初期处于初始位置，然后再按照30%~35%的比例降低活动罩裙高度，90s后降低到活动罩裙与炉口距离为50~100mm。除尘风机在活动罩裙高位、氧枪高位、低氧流量时采用高转速运行（1400~1500r/min）；活动罩裙降至低位后，除尘风机进入可控流量程序运行。经过氧气流量、活动罩裙高度、氧枪位置、风机转速等多项工艺参数的联合调控，形成在静电除尘器内长度方向上的气体组分不同且具有明显范围的气体流动特征（$\varphi(CO) = 0~4\%$；$\varphi(CO_2) = 16\%~32\%$；$\varphi(O_2) = 0~20\%$），要保证以$CO_2$为主体组分形成的惰性气体区间，在静电除尘器内的停留时间，与其相邻的气体组分区间间隔60s。实践证实，采用上述控制技术可以有效地消除或减少静电除尘器内煤气泄爆。

5.3 应用实绩

自2009年3月首钢京唐炼钢厂投产3年以来，经过不断地探索和研究，形成了一整套适应300t大型转炉"全三脱"冶炼与转炉煤气干法除尘相互匹配的生产操作技术和工艺制度。目前，转炉煤气干法除尘系统整体运行状况安全稳定，煤气泄爆问题已经基本杜绝，煤气和蒸汽回收量均已超过预期指标。

5.3.1 转炉煤气和蒸汽回收实绩

2010年，转炉常规冶炼时，转炉煤气回收量均达到100m³/t以上，全年平均为103.3m³/t，最高月份达到107.8m³/t。"全三脱"冶炼时，转炉煤气回收量均达到80m³/t以上，全年平均为85.0m³/t，最高月份达到86.6m³/t。2010年两种工况的平均煤气回收量达到96.6m³/t，煤气热值保持在7620~8374kJ/m³之间；2011年两种工况的平均煤气回收量达到97.9m³/t，煤气热值达到7536kJ/m³。

2010年，转炉常规冶炼时，蒸汽回收量均达到80kg/t以上，全年平均为81.8kg/t，最高月份达到86.5kg/t。"全三脱"冶炼时，蒸汽回收量均达到110kg/t以上，全年平均为113.4kg/t，最高月份达到122.0kg/t。2010年两种工况的平均蒸汽回收量达到96.6kg/t，2011年两种工况的平均蒸汽回收量为98.9kg/t，蒸汽温度为220℃，蒸汽压力为2.9~3.8MPa。实践证明：转炉"全三脱"冶炼时，煤气回收量和蒸汽回收量均超过设计指标（煤气回收量设计值为80m³/t，蒸汽回收量设计值为100kg/t）。图8~图10给出2010年和2011年各月煤气和蒸汽回收情况。

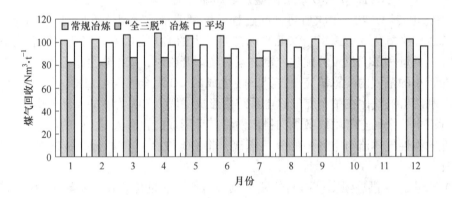

图8 2010年转炉煤气回收情况

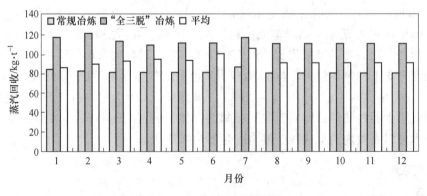

图9 2010年蒸汽回收情况

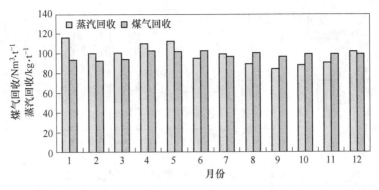

图 10　2011 年转炉煤气和蒸汽回收情况

5.3.2　煤气含尘量和泄爆情况

转炉煤气干法除尘系统投产以后，精细操作，注重系统稳定运行，尽可能减少非冶炼状态下产生的粉尘。2010～2011 年逐月对转炉煤气含尘量进行了检测。检测结果表明，2010 年和 2011 年转炉煤气平均含尘量分别为 4.3mg/m³ 和 3.8mg/m³，优于设计指标 15mg/m³ 的要求，2010 年和 2011 年转炉煤气含尘量见图 11。

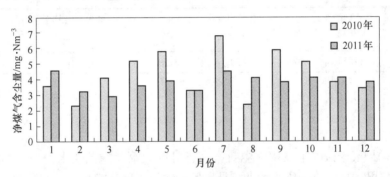

图 11　2010 年和 2011 年转炉煤气含尘量

经过系统的理论研究，制定了脱碳转炉煤气干法除尘系统防止转炉开吹过程煤气泄爆的基本控制原则。由于脱碳转炉开吹后没有硅、锰氧化期，氧气直接参与脱碳反应，煤气中 CO 的体积分数高于转炉常规冶炼。根据"全三脱"冶炼工艺特性和煤气产生泄爆的机制研究，防止脱碳转炉开吹期煤气泄爆的技术原则是：

（1）在脱碳转炉开吹后 1～2min "前烧期"内，应采用较小供氧量（不超过常规冶炼开吹供氧量的 70%），避免"前烧期"内煤气生成量超过风机能力；

（2）脱碳转炉"前烧期"内，烟罩处于一定的高位，同时配合适宜的风机转速，使煤气完全燃烧，持续时间约为 1.5min，从而形成比较稳定的惰性气体流。

根据上述原则和理论研究，自主开发了基于理论研究的计算软件，通过设计脱碳转炉开吹操作条件，制定了在 1.5min 内可以稳定形成惰性气体流的操作制度，并绘制了在此操作模式下的转炉煤气曲线，用于指导转炉冶炼。全年煤气泄爆率已经降低到 0.03%～0.04%，并且控制状态稳定。图 12 给出 2010 年和 2011 年煤气泄爆次数。

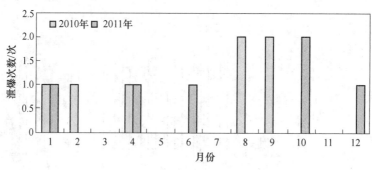

图12 2010年和2011年煤气泄爆次数

5.3.3 降低水消耗

与煤气湿法除尘工艺相比，转炉煤气采用干法除尘工艺以后，水、电消耗量显著降低。在同等条件下可以减少新水消耗$0.248m^3/t$（约降低80%）；减少电力消耗$5.06kW·h/t$（约降低50%）。表1给出首钢京唐炼钢厂和首钢迁钢一炼钢厂转炉煤气采用不同除尘工艺时的水、电消耗对比。

表1 转炉煤气干湿法除尘水、电耗量对比

炼钢厂	转炉座数/座	转炉容量/t	煤气除尘工艺	耗水量/$m^3·t^{-1}$	耗电量/$kW·h·t^{-1}$
首钢迁钢一炼钢	3	210	湿法除尘	0.284	8.54
首钢京唐	2（脱磷）+3（脱碳）	300	干法电除尘	0.036	3.48

6 结语

（1）转炉煤气干法除尘技术除尘效率高，可以大幅度降低新水消耗，实现二次能源的高效回收利用，有效减少环境污染，是现代钢铁厂实现节能减排、发展循环经济的重要技术途径。

（2）"全三脱"洁净钢冶炼工艺具有高效率、快节奏、低成本的技术优势，是生产高质量、高性能洁净钢最具有发展前景的工艺技术。由于"全三脱"冶炼工艺的特性，脱碳转炉冶炼过程中不存在"前烧期"，氧气直接与铁水中的碳元素发生氧化反应，快速生成大量CO。故在采用煤气干法除尘工艺时极易出现煤气泄爆。

（3）蒸发冷却器是转炉煤气干法除尘的关键设备。采用CFD数学仿真研究了蒸发冷却器内煤气流场和温度场的分布。通过合理布置汽化冷却烟道、设置煤气导流装置、优化喷嘴布置等措施，使蒸发冷却器内煤气速度和温度分布均匀，保证了煤气干法除尘系统稳定、可靠地运行。

（4）系统研究了转炉"全三脱"冶炼工艺煤气爆炸特性和机制。基于理论研究结果，设计开发了安全防爆技术和操作控制技术。通过工业试验和生产实践，研究了300t大型转炉"全三脱"冶炼条件下氧枪枪位、供氧量、风机转数、活动罩裙的最佳工艺控制参数，探索了"全三脱"冶炼条件下防止煤气泄爆的工艺规律和操作控制技术。

（5）基于对转炉"全三脱"冶炼工艺和煤气干法除尘技术的研究，开发了大型转炉

"全三脱"冶炼条件下,煤气干法除尘工艺预防煤气泄爆技术,保证了转炉冶炼和煤气干法除尘工艺的稳定、高效、安全运行。首钢京唐300t大型转炉在"全三脱"冶炼条件下,煤气回收量年平均达到85m³/t以上,煤气热值达到7536kJ/m³以上,蒸汽回收量达到110kg/t以上,蒸汽温度为220℃,蒸汽压力为2.9~3.8MPa,煤气含尘量控制在5mg/m³以下,全年煤气泄爆次数控制在0.03%~0.04%的水平。新水和电消耗与湿法除尘工艺相比显著降低。

参考文献

[1] 殷瑞钰. 论钢厂制造过程中能量流行为和能量流网络的构建[J]. 钢铁, 2010, 45(4): 1.

[2] 殷瑞钰. 关于高效率低成本洁净钢平台的讨论——21世纪钢铁工业关键技术之一[J]. 炼钢, 2011, 27(1): 1-10.

[3] 张春霞, 殷瑞钰, 秦松, 等. 循环经济中的中国钢厂[J]. 钢铁, 2011, 46(7): 1.

[4] Fingerhut W. 转炉干法除尘技术最新进展及市场动态[J]. 李晓强, 译. 世界钢铁, 2008(5): 12-16.

[5] 潘秀兰, 常桂华, 冯士超, 等. 转炉煤气回收和利用技术的最新进展[J]. 冶金能源, 2010, 29(5): 37-42.

[6] 张堃. 转炉煤气干法除尘回收技术展望[J]. 冶金动力, 2009, 133(3): 71-75.

[7] 王永刚, 王建国, 叶天鸿, 等. 转炉煤气干法除尘技术在国内钢厂的应用[J]. 重型机械, 2006(2): 1-3.

[8] 马宝宝, 刘飞. 济钢三炼钢转炉干法除尘系统泄爆控制的实践[J]. 机械与电子, 2010(13): 87-88.

[9] Kawamoto M. Recent development of steelmaking process in sumitomo metals[J]. Journal of Iron and Steel Research International, 2011, 18(S2): 28-35.

[10] 殷瑞钰. 冶金流程工程学[M]. 北京: 冶金工业出版社, 2005: 325-328.

[11] 张福明, 崔幸超, 张德国, 等. 首钢京唐炼钢厂新一代工艺流程与应用实践[J]. 炼钢, 2012, 28(2): 1-6.

[12] Zhang Fuming, Cui Xingchao, Zhang Deguo. Construction of high-efficiency and low-cost clean steel production system in Shougang Jingtang[J]. Journal of Iron and Steel Research International: 2011, 18(S2): 42-51.

Research and Application on Large BOF Gas Dry Dedusting Technology

Zhang Fuming   Zhang Deguo   Zhang Lingyi   Han Yujing
Cheng Shusen   Yan Zhanhui

Abstract: Basic oxygen furnace (BOF) gas dry dedusting technique is the key technology of contemporary BOF steelmaking to achieve high efficient energy conversion, which is provided with high efficient energy conversion, water saving, energy saving and lower emission, environment clean and friendly. It can significantly reduce water consumption, efficient recycling of steam and gas, to reduce environmental pollution. According to the process and technical feature of BOF steelmaking under the condition of "full three removal" at Shougang Jingtang, the technical mechanism of BOF gas explosion was analyzed, and the safe technical measures to control the BOF gas explosion were researched and developed. The gas flow process in evaporative cooling tower was investigated and analyzed by means of computational fluid dynamics (CFD) numerical simulation technology, and the layout and flow control of nozzles in evaporative cooling tower were optimized. A number of key technologies are researched and developed to control the gas composition and temperature rationally of 300t converter under the condition of "full three removal" smelting. The recovery efficiency of gas and steam have been improved effectively; the gas explosions were reduced significantly. The targets are achieved which include stable operation, efficient energy recovery, and significantly reduced emissions. It is proved by application, the annual recovery of gas and steam under the condition of "full three removal" has been reached practically more than $85m^3/t$ and $110kg/t$ respectively, and annual numbers of gas explosion are controlled less than 0.03%. BOF dry dedusting technology has ensured the safe and stable operation of steelmaking production; it has created significant economic and ecological environment benefit.

Keywords: steelmaking; BOF; gas dry deduting; evaporative cooler; electrostatic precipitator; energy recovery

冶金煤气干法除尘技术创新与成就

摘　要：高炉炼铁和转炉炼钢过程中所产生的高炉煤气和转炉煤气是钢铁厂重要的二次能源，冶金煤气干法除尘是钢铁制造绿色发展的关键共性技术。首钢京唐钢铁厂采用了高炉煤气和转炉煤气干法除尘技术，攻克了 $5500m^3$ 特大型高炉和 300t 转炉煤气干法除尘国际性技术难题，实现了世界首套 $5500m^3$ 特大型高炉煤气纯干法除尘、首套铁水"脱硫-脱硅-脱磷"工艺条件下 300t 转炉煤气干法除尘重大技术突破和工业化稳定应用。本文结合首钢京唐钢铁厂工程设计研究，论述了特大型高炉和转炉煤气干法除尘的技术创新及其成就。

关键词：高炉；转炉；煤气干法除尘；煤气余压发电；节能减排；循环经济

1　引言

　　钢铁冶金过程产生的大量的炼铁高炉煤气和炼钢转炉煤气是重要的二次能源，必须经过除尘净化后才能回收利用。长期以来，冶金煤气普遍采用湿法除尘，消耗了大量水资源和能源并且产生二次污染。冶金煤气干法除尘是国家重点支持发展的绿色关键技术，其除尘效率和能源回收效率高、节水降耗、节能减排的经济效益和社会效益突出，是实现钢铁绿色制造、低碳清洁生产、能源高效转换、节能减排、循环经济的重大关键共性技术[1]。

　　21 世纪初，为疏解非首都功能、改善北京环境质量，国家决定首钢迁离北京，在河北唐山曹妃甸建设新的钢铁基地——首钢京唐钢铁厂。首钢京唐钢铁厂是基于新一代可循环钢铁制造流程理念，自主设计建造的现代化特大型钢铁项目，也是我国首个临海靠港建设且全部生产薄带材的千万吨级钢铁厂，具有先进钢铁产品制造、能源高效转换和消纳、处理废弃物并实现资源化的"三大功能"[2]。工程设计与技术研发过程中，攻克了 $5500m^3$ 高炉和 300t 转炉煤气干法除尘的国际性关键技术难题，实现了世界首套 $5500m^3$ 高炉煤气纯干法除尘、首套铁水"脱硫-脱硅-脱磷"预处理条件下 300t 转炉煤气干法除尘的重大技术突破，项目整体技术和关键指标均达到国际一流水平。

2　高炉煤气纯干法除尘技术

2.1　项目背景

　　2005 年以前，全球仅有 13 座 $5000m^3$ 以上特大型高炉，其中 12 座高炉采用煤气湿法除尘工艺，只有日本大分 2 号高炉采用煤气干法布袋除尘和湿法除尘并用工艺，干法除尘

本文作者：张福明。原刊于《第十一届中国钢铁年会论文集》，2017：261-267。

采用净煤气加压离线反吹清灰技术,干法除尘作业率约为 90%,尚无法实现全量高炉煤气干法除尘。针对 5500m³ 高炉煤气采用干法布袋除尘所面临的煤气流量大（87 万 Nm³/h）、炉顶压力高（0.28~0.3MPa）、温度波动大（100~350℃）等国际性关键技术难题,开展了大量研究并取得一系列工程创新成果。

2.2 设计创新

工程设计研究过程中,提出了 5500m³ 高炉煤气采用自主研发的纯干法脉冲布袋除尘技术原创性设计思想[3],系统研究了煤气流量、温度、压力、流速、初始含尘量等工艺参数的变化规律和特征及其对布袋除尘系统的影响。建立了数值仿真模型,通过仿真优化和精准设计,提出了控制煤气流均匀分布的等流速设计原理和方法,创建了特大型高炉煤气彻底取消湿法除尘备用的"纯干法"脉冲布袋除尘高效集约的工艺流程[4],实现了 5500m³ 高炉开炉、休风、复风操作时,煤气流量、压力、温度、成分等工艺参数发生复杂变化条件下,多工况模式的安全可靠稳定运行,煤气干法除尘作业率达到 100%。

2.3 技术研究

2.3.1 理论研究

创新应用动态精准设计体系和方法,创建了三维仿真动态协同优化设计平台,形成了基于 CFD、FEM、3D-CAD 多模型仿真研究、多场耦合优化的工程动态精准设计体系,研究探索了保证全系统稳定可靠运行的设计关键要素。提出了基于流体力学、粉体工程学和纤维过滤理论的恒流速控制、布袋-灰膜耦合过滤、在线脉冲激波清灰、除尘器流场分布均匀化等设计理论和准则,设计研制了大容量高效脉冲布袋除尘器。

建立了基于布袋除尘机理[5]和布袋滤料评价的物理实验和工业性试验平台,通过实验研究、试验测试和应用研究,发现了高炉煤气在高压工况下,粉尘的过滤/分离过程、布袋除尘工艺机理、运行规律以及布袋异常破损的原因,建立了高温、高压、腐蚀工况条件下布袋选型的优化模型和评价体系。针对高过滤负荷条件下布袋脉冲清灰的技术难题,通过布袋脉冲清灰工业性试验,研究发现了脉冲频率、脉冲压力、脉冲阀型式、喷嘴结构、压力波传导等对布袋清灰过程和效果的影响及规律,开发研制出高效双侧在线脉冲清灰装置。

2.3.2 关键技术研发

（1）针对煤气温度对布袋寿命和耐用性的影响,开展了数值仿真、实验和试验研究,研发出基于高效热管换热原理的大功率煤气温度智能化调控技术、系统和方法。研制出高炉煤气大功率（15.8MW）热管换热技术及装置,利用热管相变吸收高炉煤气的热量,从而有效控制高温煤气对布袋的损坏[6,7]。当高炉煤气温度超过 250℃ 以上时,可以降低高炉煤气温度 50~100℃,使进入布袋除尘器的煤气温度控制在 250℃ 以下安全工作温度区间,避免了由于煤气温度超高造成的布袋烧损,攻克了煤气温度变化影响布袋寿命的关键技术难题,布袋使用寿命达到 3 年以上。

（2）针对氯化物对高炉煤气余压发电设备（TRT）和煤气管道的腐蚀破坏,系统研究了高炉煤气中氯化物的生成原因、生成规律及其有效抑制措施,发明了高炉煤气喷碱脱氯装置和管道系统综合防腐方法，有效解决了高炉煤气采用干法除尘技术以后，TRT 叶片结垢、波纹补偿器和煤气管道异常腐蚀破损的关键技术难题[8]。

(3) 研究解析了高炉煤气全干法除尘与全干式 TRT 高效耦合运行的难题和规律，攻克了 36.5MW 超大功率 TRT 发电机组的稳定运行、智能控制等难题，形成高精度炉顶压力稳定控制技术，实现 TRT 作业与高炉同步，年平均发电量达到 54kW·h/t 以上，实现了清洁能源高效转换。

(4) 针对除尘灰颗粒细微、流动性差、易燃易爆等物理化学性能和难于收集、输送的技术难题，基于气力输送原理，研发并应用了高炉煤气除尘灰全程密闭气力输送工艺。设计研发出流态化气力输送装置，采用净化后的高炉煤气或氮气作为载气，将除尘灰输送至储灰仓并且回收煤气。攻克了传统机械式除尘灰输送二次污染、故障率高等技术难题，实现了除尘灰全程密闭输送与收集。

2.4 工程创新

(1) 首钢京唐 5500m³ 高炉采用煤气纯干法除尘技术及装备并取得了成功应用，全面解决了 5500m³ 高炉煤气流量大、系统压力高、结构尺度大、复杂程度高、安全稳定运行等一系列国际技术难题。构建了高效长寿集约型工艺流程，自主研发出基于物理化学吸附的煤气管道防腐技术和基于高效换热原理的大功率煤气温度智能控制技术；创新研发出与超大容量全干式 TRT 耦合匹配工艺；形成了拥有完整自主知识产权的技术及装备，建成了世界首例 5500m³ 高炉煤气纯干法除尘示范工程[9]。表 1 为首钢京唐 5500m³ 高炉煤气脉冲布袋纯干法除尘工艺的技术参数。

表 1 首钢京唐 5500m³ 高炉煤气脉冲布袋纯干法除尘技术参数

项　目	参　数
高炉容积/m³	5500
炉缸直径/m	15.5
日产铁量/t·d⁻¹	12650
煤气处理量/Nm³·h⁻¹	780000（最大 870000）
炉顶压力/MPa	0.28~0.3
煤气温度/℃	100~250
除尘器数量/个	15
除尘器直径/mm	6200
单体除尘器布袋数量/条	409
布袋规格 $\phi \times L$/mm	160×7000
单体除尘器过滤面积/m²	1439
总过滤面积/m²	21585
标准状态过滤速度/m·min⁻¹	0.59
工作状态过滤速度/m·min⁻¹	0.23
粗煤气含尘量/mg·m⁻³	≤10000
净煤气含尘量/mg·m⁻³	≤5

（2）开发研制出煤气含尘量在线监测装置和信息管理系统，实现了粗煤气和净煤气中含尘量在线精准监测，并以此精准监控布袋工作及破损状况。研发出布袋除尘器除尘灰位自动化监测与信息管理系统，依此精准判定布袋除尘器积灰状态并实现除尘灰自动输送。设计研发出全过程数字化控制系统，实现了布袋除尘器在线、离线多模式全过程自动清灰。

（3）建立了基于大数据管理和数据挖掘的高炉煤气智能化管理和动态精准控制系统，实现了煤气流量、成分、温度、压力、热值、露点、含尘量、氯离子含量等多参数自动化监测、数据采集、信息传输和管理，使高炉操作、炉顶压力调控、煤气温度控制、布袋除尘控制、喷碱脱氯、TRT 运行控制、热风炉燃烧、煤气管网输送、2×300MW 机组燃烧等高炉煤气能量流网络实现全流程高效协同运行和动态精准控制。

3 转炉煤气干法除尘技术

3.1 项目背景

"全三脱"是指铁水在装入脱碳转炉之前，对全量铁水进行脱硫、脱硅、脱磷"三脱"预处理，采用 KR 脱硫工艺和专用转炉进行铁水脱硅—脱磷预处理，以提高钢水洁净度，构建高效率、低成本洁净钢生产平台[10]。转炉煤气干法除尘效率高，降低水耗、节能环保优势突出，转炉净煤气含尘量显著降低[11]，除尘灰经压块或喷吹可直接供转炉使用，实现固体废弃物资源化高效回收利用。针对 2 座 300t 脱磷转炉和三座 300t 脱碳转炉全量煤气采用干法除尘技术的世界性技术难题，深入开展了理论、技术和应用研究工作。图 1 为铁水"全三脱"预处理条件下转炉炼钢工艺流程图。

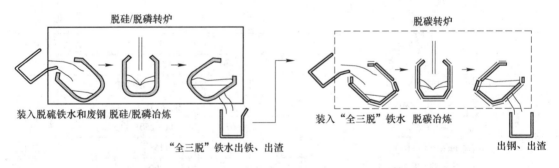

图 1　铁水"全三脱"预处理条件下转炉炼钢工艺流程

3.2 技术研究

3.2.1 理论研究

（1）针对转炉煤气干法除尘工艺过程及其技术特征，研究解析了铁水"全三脱"条件下，转炉煤气生成的热力学及动力学规律，建立了基于控制转炉煤气成分和生成速率的热力学及动力学仿真模型。研究发现了铁水"全三脱"条件下，脱碳转炉的供氧强度、氧枪枪位等操作参数的控制规律，形成了完整的动态精准操作控制技术理论体系。

(2) 针对转炉煤气干法除尘频繁发生泄爆的关键技术难题，建立数值仿真模型，研究解析了转炉煤气形成的冶金传输过程及其机理。研究发现了转炉煤气成分、温度、闪络等原因造成煤气泄爆的条件、规律及其工艺机理，提出了转炉冶炼过程中煤气体积分数的安全控制范围，制定了转炉煤气安全防爆控制准则，研究开发出有效控制煤气泄爆的安全控制技术[12]。

(3) 研究解析了蒸发冷却器的功能和传输过程，建立了蒸发冷却器内煤气流动、水滴雾化、粉尘沉降以及传热数学模型，应用多相态复杂传输过程CFD多场耦合数值仿真方法，研究了转炉煤气在蒸发冷却器内的流动特性，发现了影响蒸发冷却器工作效能的关键要素，得出了最优化的蒸发器结构参数和雾化喷嘴参数，确定了冷却水量精准控制模型和雾化喷嘴的优化布置方式，实现精确控制煤气温度和冷却水量，降低由于转炉煤气温度过高而引起的泄爆，实现了蒸发冷却器温度场、速度场和粉体场的均匀分布，有效提高了蒸发冷却器的冷却和除尘效率。

3.2.2 关键技术研发

(1) 建立了泄爆阀模拟泄爆、煤气管道压力释放等多个物理模型、实验研究和工程化试验平台，通过实验/试验研究深入研究解析了铁水"全三脱"预处理条件下，转炉冶炼与煤气干法除尘耦合匹配的关键技术难题，验证了仿真研究结果。

(2) 在包钢和首钢股份210t转炉上进行了模拟铁水"全三脱"冶炼工业试验，验证了脱磷-脱碳转炉冶炼工艺的理论研究和仿真研究结论，证实了转炉煤气生成特性及其成分和温度的变化规律，探索研究了"全三脱"条件下，脱碳转炉的氧枪枪位、供氧强度、风机转速、活动罩裙的最佳控制参数，得出了铁水"全三脱"预处理条件下，有效抑制转炉煤气泄爆的工艺操作方法、控制规则和技术诀窍。建立了基于智能化控制的转炉冶炼和煤气干法除尘高效耦合运行体系，创新实践了脱碳转炉冶炼初期控制煤气稳定生成的关键技术和方法，形成了自主开发的核心操作控制技术，实现了全系统安全可靠稳定运行。

(3) 自主开发研制出一系列转炉煤气干法除尘核心关键设备，主要包括：基于CFD仿真模拟和构建数字化样机，设计研制了转炉煤气蒸发冷却器的核心关键设备——高效双流雾化喷枪，实现了煤气冷却、预除尘和煤气改质，解决了多相态复杂传输的关键技术问题。设计研制了煤气泄爆阀，通过设备结构优化设计和数值模拟，建立了试验平台进行工况爆炸试验，实测了煤气爆炸压力等关键参数，完成了实体样机的研制和工业应用。采用FEM、3D-CAD等先进设计方法，研发出结构优化、功能完善的煤气高温眼镜阀数字化样机，完成了实体样机研制并进行了工业化应用。上述设备总体性能达到引进水平，替代进口产品。

3.3 工程创新

(1) 集成创新了国际首套铁水"脱硫-脱硅-脱磷"预处理工艺条件下，300t转炉煤气干法除尘技术及关键设备并取得成功应用。国内外首次针对转炉煤气全干法除尘技术，攻克了300t脱磷-脱碳转炉冶炼工艺智能控制、煤气成分控制和安全稳定运行等难题，集成创新并成功应用了转炉煤气干法除尘技术并取得重大技术突破，实现了整体工艺技术的集成创新和关键工艺设备的自主创新。

(2) 针对除尘灰的物化特性，发明了除尘灰资源化、高值化回收利用新工艺，研发出转炉煤气除尘灰压块、喷吹等直接用于炼钢循环利用的先进工艺及装备。除尘灰压块后替代矿石作为转炉原料再次利用，还可作为脱磷剂喷吹用于脱磷转炉，将除尘灰经过顶吹脱磷氧枪喷入熔池，提高了资源回收利用效率，实现了固体废弃物在炼钢工序内的高效消纳处理和固体废弃物资源的高值转化。

（3）研究探索了铁水"全三脱"条件下，脱磷-脱碳转炉冶炼工艺控制规律和操作特征，提出并成功应用了脱碳转炉煤气处理采用干法除尘工艺电除尘不泄爆的方法、干法除尘条件下半钢冶炼方法、双联冶炼工艺脱碳转炉的自动控制方法、脱磷转炉的自动控制方法等多项技术发明，实现了铁水"全三脱"条件下，转炉煤气干法除尘系统安全可靠稳定运行，全年基本实现"零泄爆"。

（4）基于大数据信息化管理，建立了铁水预处理-转炉冶炼-煤气干法除尘多工序耦合运行动态精准控制体系和流程网络，构建了转炉炼钢过程物质流、能量流和信息流的高效协同运行程序和规则，实现了"一键式"智能化炼钢和全过程煤气、蒸汽回收智能化调控，实现了转炉冶炼与煤气干法除尘动态有序、高效协同、耦合匹配的能量回收智能化动态精准控制。

4 应用实践

（1）自主设计研制出关键系统和核心装备，实现集成创新。特大型高炉煤气和转炉煤气干法除尘技术装备总体性能达到国际最优：自主设计研发出国际首创的特大型高炉煤气全干法脉冲布袋除尘技术，彻底取消湿法备用系统；在国际上首次创新应用300t脱磷-脱碳转炉全量煤气干法除尘技术；设计研制出国内外单体处理能力最大的煤气布袋除尘器；实现了智能监测控制系统系列关键技术突破。

（2）自主研发了物质流-能量流-信息流耦合运行的智能化精准控制系统与方法。设计研制出煤气含尘量自动监测系统与装置；研发出转炉冶炼控制与煤气生成-回收智能化控制系统与方法；研发出煤气温度、流量、压力多系统智能化调控与耦合匹配技术，实现与36.5MW超大容量全干式TRT系统高效耦合运行；实现了钢铁冶金过程与冶金煤气能量流网络全流程的智能化控制。

（3）为了实现大流量高炉煤气和转炉煤气的收集、除尘和能量高效回收利用，针对高炉煤气和转炉煤气干法除尘不同的工艺技术特性，创新采用自动化、数字化、信息化和智能化先进技术和方法，构建了冶金煤气工艺参数的自动化检测、关键参数的监测和全过程信息流网络。创新建立了物质流-能量流-信息流高效协同运行的流程网络，实现了全流程智能化控制。

（4）首钢京唐1号、2号高炉相继于2009年5月和2010年6月建成投产，高炉煤气纯干法除尘技术已在5500m^3高炉上成功应用8年。实践证实，高炉煤气纯干法布袋除尘技术运行稳定可靠，经历了各种复杂工况的考验，年平均净煤气含尘量达到2~4mg/Nm^3，煤气温度比湿法除尘提高80~100℃，布袋除尘系统阻力损失小于2000Pa，年平均TRT发电量达到54kW·h/t，布袋使用寿命达到3年，取得了重大的经济和环境效益。图2和图3分别是首钢京唐两座5500m^3高炉近年来高炉煤气含尘量和煤气余压发电量的情况。

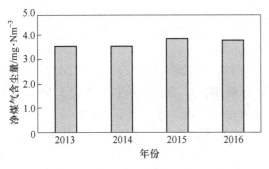

图 2　首钢京唐 5500m³ 高炉净煤气含尘量

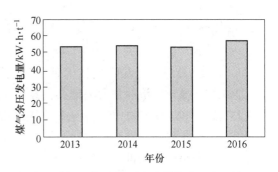

图 3　首钢京唐 5500m³ 高炉煤气余压发电情况

(5) 首钢京唐两座脱磷转炉和三座脱碳转炉，投产至今已稳定运行 8 年，年平均转炉净煤气含尘量为 3~6mg/Nm³；年平均回收转炉煤气达到 110Nm³/t、回收蒸汽达到 97kg/t；转炉煤气干法除尘系统年泄爆次数降低到 3 次（万分之一），基本实现"零泄爆"。图 4~图 7 分别为近年来转炉煤气回收、蒸汽回收、转炉净煤气含尘量、系统泄爆次数的实绩。

(6) 首钢京唐采用高炉煤气和转炉煤气干法除尘技术，节约新水 697 万吨/年，节电 8790 万千瓦时/年，TRT 发电 4.9 亿千瓦时/年，节约能源 16.2 万吨标煤/年，减少 CO_2 排放约 40 万吨/年。

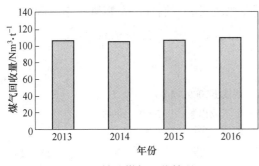

图 4　转炉煤气回收情况

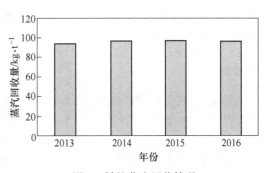

图 5　转炉蒸汽回收情况

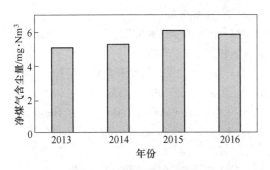

图 6　转炉净煤气含尘量

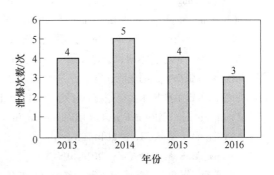

图 7　转炉煤气干法除尘系统泄爆情况

5 结论

（1）冶金煤气干法除尘技术是实现钢铁绿色制造、低碳清洁生产、能量高效转换、节能减排、循环经济的重大关键共性技术。首钢京唐钢铁厂工程创新中，攻克了5500m^3特大型高炉和300t转炉煤气干法除尘国际性技术难题，实现了世界首套5500m^3特大型高炉煤气全干法除尘、首套铁水"全三脱"条件下300t转炉煤气干法除尘重大技术突破，关键技术和指标达到世界一流。

（2）项目研究系统解决了冶金装备大型化以后，冶金煤气复杂工况条件下非线性变化的技术难题。针对煤气温度控制、管道系统防腐、能量高效转换、除尘灰资源化利用、智能化协同控制等关键技术难题，研发出一系列有效技术措施，保障了特大型高炉和转炉煤气干法除尘技术的稳定运行和成功应用。

（3）项目实施形成了一整套具有完全自主知识产权的关键技术工程集成创新成果和产业化示范，形成了集设计-研发-制造-建造-运行一体化的工程实施平台和产业链，有力推动了我国工程设计、科研开发、冶金、材料、装备制造、工程建造、信息化等相关产业的科技进步和产业协同创新。

致谢

衷心感谢北京首钢国际工程公司和首钢京唐公司冶金煤气干法除尘项目团队成员的辛勤工作，感谢毛庆武、张德国、闫占辉、颉建新、张凌义、章启夫、曹朝真、李欣等专家的大力支持和帮助。

参考文献

[1] 张福明. 现代大型高炉关键技术的研究与创新 [J]. 钢铁, 2009, 44 (4)：1-5.

[2] 殷瑞钰. 冶金流程工程学（第2版）[M]. 北京：冶金工业出版社, 2009：109-113.

[3] 张福明. 大型高炉煤气干式布袋除尘技术研究 [J]. 炼铁, 2011, 30 (1)：1-6.

[4] 张福明, 钱世崇, 张建, 等. 首钢京唐5500m^3高炉采用的新技术 [J]. 钢铁, 2011, 46 (2)：12-17.

[5] 向晓东. 现代除尘理论与技术 [M]. 北京：冶金工业出版社, 2004：145-147.

[6] Zhang Fuming. Study on dry type bag filter cleaning technology of BF gas at large blast furnace [J]. Journal of Iron and Steel Research International, 2009, 16 (Suppl. 2)：608-612.

[7] 张福明, 毛庆武, 章启夫, 等. 首钢京唐5500m^3高炉煤气干法除尘技术开发与应用 [C] //第三届中德（欧）冶金技术研讨会论文集. 北京：中国金属学会, 2011：70-77.

[8] 章启夫, 张福明, 毛庆武, 等. 京唐5500m^3特大型高炉煤气干法袋式除尘系统设计 [C] //2010年全国炼铁生产技术会议暨炼铁年会论文集. 北京：中国金属学会, 2010：175-179.

[9] Zhang Fuming. Research and Application on Blast Furnace Gas Dry Cleaning Technology of Huge Blast Furnace [C] //AISTech 2014 Proceedings of the Iron & Steel Technology Conference. Indianapolis, Indiana, U. S. A.：AISTech, 2014. 195-204.

[10] 张福明, 崔幸超, 张德国, 等. 首钢京唐炼钢厂新一代工艺流程与应用实践 [J]. 炼钢, 2012, 28 (2)：1-6.

[11] 张德国,魏钢,张宇思. 欧洲转炉干法除尘技术调研及京唐"三脱转炉"应用分析 [C] //2007年中国钢铁年会论文集. 北京:中国金属学会,2007:17-25.
[12] 张福明,张德国,张凌义,等. 大型转炉煤气干法除尘技术研究与应用 [J]. 钢铁,2013,48(2):1-9.

Achievement and Innovation of Metallurgical Gas Dry Dedusting Technology

Zhang Fuming

Abstract: Metallurgical gas includes blast furnace gas (BFG) and basic oxygen furnace gas (BOFG), produced by blast furnace ironmaking and basic oxygen furnace steelmaking. Metallurgical gas is an important secondary energy in steel plant, the BFG and BOFG dry dedusting process are the key common technology of steel manufacturing green development. The BFG and BOFG dry dedusting technologies were applied in Shougang Jingtang steel plant, the international technological difficult problems of $5500m^3$ huge blast furnace and 300t converter metallurgical gas dry dedusting technology have been solved. The major breakthrough and stable industrialization application of the world's first $5500m^3$ huge blast furnace gas dry dedusting technology and the first example under the conditions of hot metal desulfurization - desilication - dephosphorization pretreatment 300t converter gas dry dedusting technology have been achieved. In this paper, combine with the study on engineering design of Shougang Jingtang steel plant, the technological innovation and achievement of BFG and BOFG dry dedusting process are discussed.

Keywords: blast furnace; converter; gas dry dedusting; top gas recovery turbine; energy saving and emission reducing; circulating economy

首钢京唐1号5500m³高炉煤气干法除尘自动化控制系统的创新设计与实现

摘　要：本文介绍了首钢京唐1号5500m³高炉煤气干法除尘自动控制系统的结构和特点，以及自动化新技术在该系统中的应用。并针对类似工程在运行中出现的问题，对关键仪表的选型提出了改进意见，重点论述了料位计和含尘量在线监测系统的优化设计等。详细阐述了基于现场总线、可编程控制器、工业微机和工业以太网的综合自动化技术在该系统中的应用，以及高炉煤气全干式布袋除尘器自动控制的全过程。该自动控制系统结构合理，技术先进，已正常稳定运行，值得在特大型高炉煤气干法除尘中推广和借鉴。

关键词：干法除尘；自动控制系统；可编程控制器

1　引言

高炉煤气干法除尘技术是21世纪高炉实现节能减排、清洁生产的重要技术创新，不仅可以显著降低炼铁生产过程的水消耗，而且可以提高二次能源的利用效率、减少环境污染。高炉煤气干法除尘可以使高炉煤气含尘量降低到$5mg/m^3$以下，煤气温度提高约100℃且不含机械水，煤气热值提高约$210kJ/m^3$，提高炉顶煤气余压发电量35%以上，因此高炉煤气采用干法除尘已成为当今高炉炼铁技术的发展方向，也是国家钢铁行业当前首要推广"三干一电"（高炉煤气干法除尘、转炉煤气干法除尘、干熄焦和高炉煤气余压发电）节能技术中的一项，属于冶金工业的绿色环保技术[1]。

本文重点论述了自动控制系统的硬件、软件和监控系统的创新设计和在以往的实际生产中遇到的一些问题及相应采取的解决措施。比如：料位计采用分体式、耐高温并抗震的料位计；含尘量在线监测传感器表面采用特殊涂敷材料的专利技术等。

2　自动化控制系统硬件创新设计

首钢京唐1号高炉煤气干法除尘系统要求采用具有高性能、高可靠性并经济实用的可编程控制器（PLC）。笔者在总结首钢首秦、首钢迁钢、济钢、重钢、宣钢高炉煤气干法除尘自动化控制系统的成功设计经验后，将首钢京唐1号高炉煤气干法除尘自动控制系统PLC选用AB 1756系列可编程控制器，主机架选用10槽结构，CPU选用1756-L63 8M模块，PLC主机架与远程I/O的通过ControlNet网相联。为进一步提高了系统的可靠性，主PLC和ControlNet网络都采用冗余结构。本系统共有1个主PLC柜和多个远程I/O箱。

本文作者：任绍峰，张福明，刘燕，付海峰，刘飞飞，周为民。原刊于《第七届（2009）中国钢铁年会论文集》，2009：348-354。

PLC 与上位工控机采用标准工业以太网连接。上位系统共有 3 台工控机用于系统监控。其结构示意图如图 1 所示。

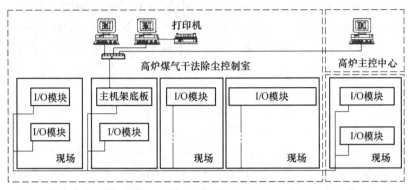

图 1 首钢京唐高炉煤气干法除尘自动化监控系统结构示意图

PLC 主机架与远程 I/O 和智能电气柜是通过 ControlNet 网相联的方式，节省大量电缆的同时也保证了信号的可靠性。

3 自动化控制系统软件和上位组态软件的优化设计

首钢京唐高炉煤气干法除尘自动化控制系统的上位机操作系统选用 Windows 2000 Professional 中文版。上位组态软件选用 9701-VWSTZHE。PLC 编程软件选用 9324-RLD300NXZHE，其具有以下特点：

（1）支持多种操作系统平台：Windows NT/2000 等；
（2）符合 IEC 1131-3 标准的多种编程模式；
（3）强大的在线帮助功能，界面友好，信息量大，极大方便应用开发人员的使用；
（4）提供的软件存取保护，防止非法访问，安全、可靠。

4 自动化控制系统的先进技术

首钢京唐 1 号高炉煤气干法除尘工程设备包括换热器系统、布袋除尘系统、卸灰系统。自动化控制系统对其进行控制，是工程最重要的组成部分之一。

首钢京唐 1 号高炉煤气全干法除尘工艺流程图如图 2 所示。

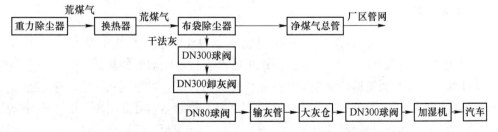

图 2 首钢京唐 1 号高炉煤气全干法除尘工艺流程图

4.1 换热器系统的自动控制

包括 3 个 DN2800 电动蝶阀、2 个 DN2800 眼睛阀和自动补水装置。

3 个 DN2800 蝶阀启闭根据炉顶四点平均温度（t_1）、旋风除尘器前温度（t_2）及旋风除尘器后温度（t_3）三者之一进行控制，荒煤气总管温度只作显示而不参与控制。在显示器画面中，可手动选取 t_1、t_2、t_3 之一作为 3 个蝶阀的自动连锁控制参数，并在规定范围内输入。

自动方式时，当荒煤气总管温度大于 260℃ 或旋风除尘器后荒煤气温度大于 280℃ 的时候，程序会自动把换热器管道上的 2 号和 3 号 DN2800 蝶阀打开并在开到位后才关闭 1 号 DN2800 蝶阀；当旋风除尘器后荒煤气温度小于 250℃ 或者荒煤气总管温度小于 200℃ 的时候打开主管道 1 号 DN2800 蝶阀并在开到位后才关闭 2 号和 3 号 DN2800 蝶阀。生产时保证从旋风除尘器出口的煤气通向干法除尘系统的一条通道是打开的，不能同时关闭两条通道。

首钢京唐高炉煤气全干法除尘换热器系统流程图如图 3 所示。

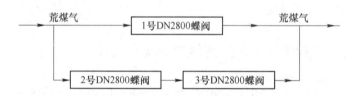

图 3　首钢京唐高炉煤气全干法除尘换热器系统流程图

4.2 脉冲反吹的自动控制

脉冲反吹系统是布袋除尘系统的关键，共 16 个箱体（包括大灰仓），除尘箱体每个箱体上 38(19+19) 个脉冲阀，分两侧布置，大灰仓上 19 个脉冲阀。24V 直流电接通后第一个脉冲阀启动，接通时间 0.1~0.3s（时间间隔可调）。向一排滤袋喷射氮气，完成一排滤袋的反吹清灰，第一个脉冲阀喷吹后 5~20s 第二个脉冲阀动作（时间间隔可调），直到全部脉冲阀动作完毕，完成一个箱体的反吹工作，再自动进行第二个箱体的反吹，直至所有工作箱体（1~15 个）完成反吹。

若先关闭净煤气支管蝶阀反吹，称为离线反吹；也可以不关闭荒、净煤气蝶阀，边过滤边反吹，称为在线反吹。全部操作由 PLC 完成。设计同时考虑离线反吹、在线反吹。

4.3 自动卸灰系统

通过对 2 个 DN150 电动球阀、2 个 DN100 电动球阀以及除尘器箱体上 15 个 DN300 球阀、15 个 DN300 放灰阀、15 个 DN80 球阀和 30 个仓壁振动器的控制，实现干法除尘系统的自动卸灰。当除尘器箱体的灰位达到高灰位时开始卸灰，当达到低灰位时停止卸灰。每次只能操作一个箱体，如果运行中有 2 个或 2 个以上的箱体同时到达高灰位，这时需要人工干预选择，保证同一时间只能对一个或两个箱体进行卸灰，防止灰量过大，堵塞输灰管道。布袋除尘灰由气力输送至大灰仓。大灰仓的灰由罐车运输。

4.4 首钢京唐1号高炉煤气干法除尘上位监控系统

本系统有除尘器本体画面、大灰仓画面、历史曲线画面、系统报警及高炉指令画面等。所有重要的测量参数有自动记录曲线，并有历史记录。首钢京唐1号高炉煤气干法除尘主画面如图4所示。

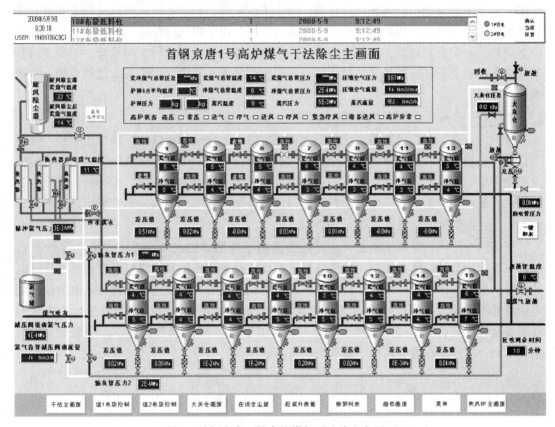

图4 首钢京唐1号高炉煤气干法除尘主画面

5 关键仪表设备的改进

5.1 料位计

料位计性能优劣直接影响到干法除尘卸灰系统的稳定与否，该料位计选用分体式、耐高温并抗振的射频导纳料位计，消除了高温的除尘灰及仓壁振动器对料位计的影响，从而大大减少了料位计信号的误报，消除了卸灰系统不稳定的隐患。

5.2 含尘量在线监测系统

含尘量在线监测装置是检验高炉煤气干法除尘效果的重要检测设备，也是保证后续TRT系统和热风炉系统长期稳定运行的重要设备之一，本系统含尘量在线监测装置的传感器表面采用特殊涂敷材料,避免由于高炉煤气中含有水分导致传感器表面黏结灰尘,从而提高了装置的稳定性[2]。

5.2.1 问题与分析

在高炉煤气全干法除尘后煤气中粉尘含量的在线监测,一直是个技术难题。因为煤气中不仅含有粉尘,而且还含有水分,煤气温度高且变化较大,当系统工况不稳时,会造成干法除尘器中的布袋结露或非正常爆裂。粉尘浓度升高,如不及时采取措施将会导致后续炉顶煤气压差发电或热风炉系统不能稳定运行。同时高炉煤气毒性很大,且干法除尘作为整个高炉的一个重要环节,一旦投入运行很难随时停止,不易进行实时检修。笔者连同有关人员共同攻克了该技术难题,设计出一套高炉煤气干法除尘含尘量在线监测系统,并得到推广应用。

5.2.2 含尘量在线监测系统原理

高炉煤气含尘量在线监测系统采用电荷感应原理。在流动粉体中,颗粒与颗粒、颗粒与管壁、颗粒与布袋之间因摩擦、碰撞产生静电荷,形成静电场,其静电场的变化即可反映粉尘含量的变化。含尘量在线监测系统就是通过测量静电荷的变化,来判断布袋除尘系统的运行是否正常。插入箱体输出管道中的传感器及时检测到电荷量值并输出到变送器。

5.2.3 含尘量在线监测系统设计与实现

下面结合附图和具体实施对含尘量在线监测系统做详细说明。在图5中:传感器1,包括布袋除尘器各箱体净煤气出口传感器、荒煤气总管传感器和净煤气总管传感器检测电荷信号,根据箱体数量的增加,传感器的数量也相应地增加;变送器2进行从电荷信号到电压信号的转换,并进行硬件补偿,根据传感器数量的增加,变送器的数量也相应地增加;安全栅3具有本安防爆功能;系统主板4用于A/D转换及数据补偿;供电单元5为系统供电;液晶显示器6以棒图格式显示各布袋净煤气出口。

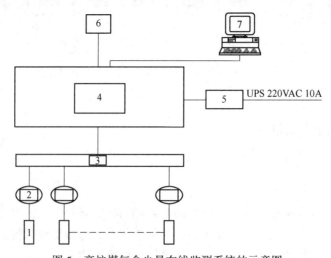

图5 高炉煤气含尘量在线监测系统的示意图

1—传感器;2—变送器;3—安全栅;4—系统主板;5—供电单元;6—液晶显示器;7—PLC系统

由传感器测得电荷信号进入到变送器,经过电荷信号到电压信号的转换并进行补偿后,进入安全栅阵列进行防爆隔离,然后进入系统主板进行A/D转换及补偿,最后以4~

20mA 标准电流信号输出。

当任一箱体有布袋破裂时，会使该箱体含尘量值上升，同时净煤气总管的含尘量值也会略有上升，此时系统即可在监视盘和上位机上显示故障状态，并发出报警信号，易于维护和操作人员及时采取相应的措施。

5.2.4 含尘量在线监测系统技术进步及创新点

该高炉煤气全干法除尘含尘量在线监测系统创新点在于：采用电荷感应原理，通过测量静电荷的变化，来判断布袋除尘系统的运行是否正常；传感器表面采用特殊涂敷材料；接地网的制作采用传感器端悬空而在变送器端接地的连接方式；硬件及软件补偿[3,4]。

首钢京唐1号高炉煤气全干法除尘含尘量系统填补了国内的空白，技术达到国内领先水平，为冶金行业重点推广应用该技术做出了积极的贡献[5]。

6 结语

为将首钢京唐公司建设成为能源循环型的钢铁联合企业，首钢京唐1号5500m^3高炉采用全干式低压脉冲布袋除尘技术对高炉煤气进行除尘，取代原有湿法除尘系统，并实现全过程的自动监控，有着明显的社会和经济效益。首钢京唐1号高炉煤气干法除尘采用了先进的自动控制系统后，与湿法除尘相比，全干法除尘系统有着明显的优越性，显著优点如下：

（1）节水且还可节约运行费用，省掉了湿法除尘建设大型的水洗塔和沉淀池等投资和占地，杜绝大量污泥、污水的产生及对环境的污染。

（2）全干法除尘工艺在运行中通过脉冲反吹布袋除尘技术，能实现自动连续除尘，显著减少粉尘外排量，干的粉尘可充分回收。

（3）全干法系统排出的煤气压力损失小，温度高，比湿法高出约 100～170℃。经干法除尘后的煤气，热值高、水分低，煤气的理论燃烧温度高，应用领域扩大。热风炉采用干式热煤气可提高热风温度50℃左右，相应降低炼铁焦比 8kg/tFe。同时，高炉煤气全干法除尘投资省、占地少、建设周期短、运行成本低。

首钢京唐1号5500m^3高炉煤气全干法除尘现已实现了远程集中监控，自动化控制系统稳定运行，这为推动我国特大型高炉向节能、环保、高效方向的发展做出了贡献。

参考文献

[1] 张福明. 现代大型高炉关键技术的研究与创新 [J]. 钢铁，2009，44（4）：1-5.
[2] 任绍峰，张福明，等. 高炉煤气干法除尘含尘量在线监测装置 [P]. ZL 2005 2 0112339.7. 2006-10-18.
[3] 任绍峰，马维理. PLC 在迁钢 2650m^3高炉煤气干法除尘控制系统中的应用[J]. 钢铁增刊（2）：2006 中国金属学会青年学术年会论文集，2006.
[4] 任绍峰，马维理，胡国新. 济钢 1750m^3高炉煤气全干法除尘自动控制系统[J]. 冶金自动化增刊（2）：中国计量协会冶金分会 2007 年会论文集，2007.
[5] 吴建常. 建设资源节约、环境友好型钢铁企业——在中国钢铁工业协会环保与节能工作委员会上的讲话 [R]. 2007.

The Innovative Design and Realization of Automation Control System for Shougang Jingtang 5500m³ Blast Furnace Dry Dusting System

Ren Shaofeng Zhang Fuming Liu Yan Fu Haifeng
Liu Feifei Zhou Weimin

Abstract: In this paper, we introduce the configuration of automation system for dry dusting system in 1# Shougang Jingtang 5500m³ blast furnace. And we make suggestions concerning the selection of main instruments, considering the problems which have occurred in other similar projects for which those instruments are used. We emphasize on the optimum application of level sensing device and online monitoring system for dust content. We expound emphatically the application of integrated automatic technology based on field bus, PLC, industrial computer and Ethernet network. And automatic control process of dusting system in blast furnace etc. The control system has reasonable structure and advanced technology. It runs normally now, is worth promoting.

Keywords: dry dusting; automation control system; PLC

Research and Application on Blast Furnace Gas Dry Cleaning Technology of Huge Blast Furnace

Zhang Fuming

Keywords: blast furnace; bell-less top; dome combustion hot stove; bag filter dedusting

1 Introduction

Blast furnace gas (BFG) dry bag filter cleaning technology is an important technical innovation in 21st century to realize the energy saving, reduction of emission and clean production for blast furnace (BF). Compared with traditional BFG wet cleaning technology, it has increased gas cleanness, gas temperature and heat value, obviously reduced fresh water and power consumption during iron making process and also can increase efficiency of secondary energy source, reduce environment pollution. It is an important technical solution for cycling economy and sustainable development of iron and steel industry and become the developing orientation of BF iron making technology.[1]

The project of Shougang Jingtang iron and steel plant is a key project for restructuring of Chinese iron and steel industry and promotion of overall technological equipment level for iron and steel enterprises. The annual capacity of the iron and steel plant is 9.7 million tons, two 5500m^3 BFs have been built up with annual capacity of 8.98 million tons of hot metal (HM). It is the first in China to build huge BF above 5000m^3. The characteristics of design and technical equipment of huge BF above 5000m^3 in the world was studied and analysed during design. All independent innovation was performed. A series of innovative advanced technologies and process equipments were designed and developed independently such as bell-less top, top combustion hot blast stove, gas fully dry bag filter cleaning, long campaign life technology and improved screw type slag granulation process.[2]

2 Technological Principle

2.1 Process Flow of BFG Dry Bag Filter Cleaning Technology

At the beginning the bag filter cleaning technology was adopted in environment cleaning field.

本文作者：张福明。原刊于 Proceedings of the Iron & Steel Technology Conference, AISTech, 2014: 195-204。

BFG is a flammable, explosive and toxic gas, and the throughput is large, temperature fluctuation is great, system pressure is high and dust content is high so it has high technical difficulty and risk to adopt bag filter cleaning technology. Bag filter cleaning technology is based on fibre filtering theory, the application in BFG cleaning system changed the wet cleaning technology of BFG cleaning with water and developed BFG fully dry low pressure pulse injection bag filter cleaning process.

BFG generated during BF operating will enter cyclone catcher through eduction pipe, ascension pipe and down comer, the crude gas after cyclone cleaning will enter into each bag filter catcher through crude gas main and branch. The clean gas purified by bag filter shall enter clean gas main through clean gas branch and enter gas piping system through top gas recovery turbine (TRT) or pressure reducing valves block and then is supplied to iron and steel plant as secondary energy resource. Fig. 1 shows the process flow diagram for gas dry bag filter cleaning of Jingtang's No. 1 BF.

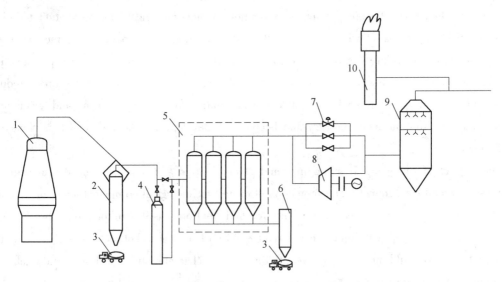

Fig. 1 Process flow diagram of BFG dry bag filter cleaning

1—BF; 2—cyclone catcher; 3—tank car; 4—heat exchanger; 5—bag filter dust catcher; 6—centralized dust bin; 7—pressure reducing valve block; 8—TRT; 9—chloride removing tower; 10—gas bleeding tower

2.2 Technical Principle of Pulse Injection Bag Filter Cleaning Process

After primary cleaning by cyclone catcher, large particles in BFG shall be collected, dust content in gas is generally reduced to about $10g/m^3$. BFG has to be purified after primary cleaning to reduce the dust content to below $10mg/m^3$. Superfine powder particles suspend in the BFG not purified which is a typical aerosol system. The filtering mechanism of BFG bag filter cleaning is based on fibre filtering theory, its filtering process can be divided into two stages: under the joint function of dispersing effect, direct intercept, gravity deposition and screening effect when the gas go through clean bag, first the bag fibre shall arrest the dust, it is the fibre who plays main function, and then the dust arrested in bag fibre together with the fibre join the filtering, this process is called "internal filtering"; when the dust in bag fibre layer is up to certain volume, the dust will deposit on the surface of bag fibre layer and form a dust layer with certain thickness, dust layer on bag surface will play main function on filtering of gas dust, this process is called "surface filtering". In actual operation of bag filter cleaning, surface filtering is the main filtering method, having an important significance on BFG bag filter cleaning technology.

In bag filter operation, the resistance will be increased with thickness increase of dust layer on bag surface, when the resistance had reached the set value; the dust on bag surface has to be removed in time. The basic requirement on dust removing is to rapidly and evenly remove the dust deposited on bag surface while a certain thickness of dust on bag surface has to be maintained. Dust removing is an important factor in normal operation of bag filter, dust removing method commonly used are mechanical dust removing, back blown dust removing and pulse dust removing, back blown and pulse dust removing are mainly used in BFG bagfilter cleaning.

Reverse blown dust removing is one method to remove dust in which the gas flow counter to filtering gas flow will be applied to deform the bag and make the dust layer fall down; all Japanese BFG dry bag filter cleaning processes adopted bag reverse blown dust removing method. Pulse dust removing technology was applied to BFG bag filter cleaning process in 1980's in China and has been rapidly promoted. Pulse injection dust removing is to inject pressurized nitrogen or gas (pressure is within 0.15MPa to 0.6MPa) in a very short time ($\leq$0.2s) with high velocity into bag, and at the same time induce huge amount of gas to form an gas wave inside the bag, generate a rapid expansion and impacting shock from the bag opening to bag bottom, having a strong dust removing effect. The impacting strength of pulse injection dust removing is strong, and its strength and frequency can be adjusted, dust removing effect is increased, system resistance loss is low, power consumption is less and on-line dust removing during bag filtering process can be realized, bag filtering surface is lower than that of reverse blown dust removing with same throughput of gas. Fig. 2 illustrates the principle of dust removal of low pressure pulse injection bag filter cleaning technology for BFG.

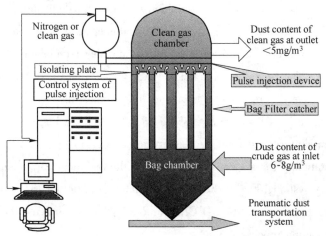

Fig. 2 Principle of low pressure pulse injection bag filter cleaning for BFG

3 Research and Development of Key Technologies

3.1 Process Overview

BFG fully dry bag filter cleaning technology is an integrated technology, relating to many engineering technology fields such as metallurgy, machinery, chemical fibre and textile, automatic detection and control. In order to increase the reliability of system operation and optimize process equipment, research and development were conducted on key technologies, a series of problems arose in engineering design, equipment manufacture, construction and operation were resolved, overall technical equipment level and control level were increased and technical breakthrough were made in application on huge BF.[3,4]

3.2 Optimization of Process Flow and Determination of Technical Parameters

Rational design of BFG dry bag filter cleaning process is the foundation for stable operation of the system. Deduster and gas pipeline are rationally arranged following the basic principle of fluid design. The layout and design of gas pipeline will directly influence the gas flow distribution inside the deduster and uniformity of resistance loss, gas volume and dust amount that enter into each deduster should be evenly distributed. BFG bag filter dedusters adopt double row in parallel arrangement, dust containing gas and clean gas main are arranged between two rows of deduster and connected with deuster through branch. Gas pipeline is designed according to equal velocity principle and evenly distributes gas volume that enters into each deduster, the process layout of whole system is compact, process flow is short and smooth and equipment maintenance is convenient. The deduster adopts low filtering velocity design concept, flow filed distribution inside the deduster is researched and analysed by computational fluid dynamics (CFD), rational gas flow velocity and gas flow direction are determined based on mathematical simulation calculation results, deflector is provided in deduster, deduster structure is optimized, and the gas flow field is evenly distributed inside deduster to ensure that the bag inside the deduster work under even and stable gas flow, this is the key technology in the design of large BF gas bag deduster. The technical parameters of deduster filtering surface, filtering velocity, gas flow ascending speed, resistance loss and dust removing cycle should be rationally determined, for super large BF with top pressure up to 0.28-0.3MPa, the rational design of gas flow, pressure, temperature and filtering velocity under working condition is the key factor to ensure reliable operation of the system. Table 1 shows the technical parameters of Jingtang No.1 BF gas dry cleaning system, Fig.3 shows the result of flow field CFD simulating calculation inside the deduster, Fig.4 shows the 3D CAD simulating design drawing for Jingtang No.1 BF gas dry bag filter cleaning process.

Table 1 Technical parameters of Jingtang's No.1 BF gas dry cleaning system

Items	Parameters
BF effective volume	5,500m³
Diameter of the hearth	15.5m
Daily Production	12,650 tHM/d
Gas volume	760,000 (maxiume 870,000) Nm³/h
Top pressure	0.28-0.30MPa
Gas temperature	100-250℃
Number of deduster	15 sets
Diameter of the deduster	6,200mm
Number of bag per single deduster	409
Specification of bag $\phi \times L$	160×7,000mm
Total filtering surface	21,586m²
Filtering area of single deduster	1,439m²
Filtering velocity under standard condition	0.59m/min
Filtering velocity under working condition	0.23m/min
Dust content in crude gas	≤10,000mg/m³
Dust content in purified gas	≤5mg/m³

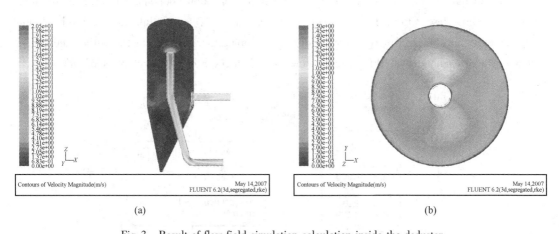

Fig. 3 Result of flow field simulation calculation inside the deduster
(a) Comprehensive velocity distribution at vertical section inside;
(b) Comprehensive velocity distribution at horizontal section inside the deduster

3.3 BFG Temperature Control Technology

BFG temperature control is the critical factor for stable and reliable operating of bag filter system. At normal status, gas temperature shall be controlled in 80-220℃. Normal running of this system can be impacted by too high and too low gas temperature. When gas temperature reaches to 250℃, and this exceeds applicable safety temperature of ordinary bags, abnormal damage, even

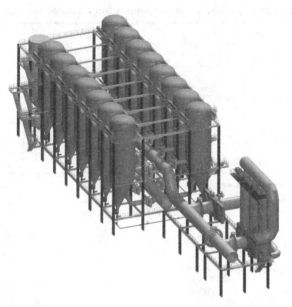

Fig. 4　3D CAD simulating design for Jingtang's No. 1 BF gas dry cleaning system

burnt bags can be happened in case of long term high temperature working condition of bags. Due to water in gas, and when gas temperature is lower than the dew-point temperature, phase transition can happen to water steam so as to appear condensation phenomenon and bag sticking. Therefore, BFG bag filter cleaning technology is applied, regulating and control of top temperature should be paid more attention during BF operation. Fig. 5 illustrates the mechanism scheme of heat exchanger of BFG, Fig. 6 shows the photograph of heat exchanger of BFG.

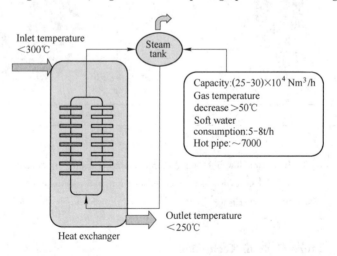

Fig. 5　Mechanism scheme of heat exchanger of BFG

When top temperature rises, mist water spraying at the top is used as temperature lowering measure. 3 groups of water spraying device for temperature lowering are provided at the top of BF with 5 nozzles each group. On/off of the 3 groups of water spraying device are controlled

Fig. 6 Photograph of heat exchanger of BFG

automatically as per the top temperature. 3 grade control system is considered for spraying water according to the top temperature. When the top temperature reaches to 255℃, the system gives alarm and the 1st water spraying device is turned on automatically; When the temperature rising is keeping on to 270℃, the 2nd water spraying device is turned on automatically for temperature lowering purpose; When the top temperature reaches to 290℃, the 3rd water spraying device is turned on automatically for water spraying and temperature lowering; When the top temperature is lowered to 220℃, the top water spraying devices for temperature lowering are closed.

High-efficient hot tube heat exchanger is a kind of gas temperature lowering device developed based on principle of hot tube heat exchange for efficient control of BFG temperature. Hot tube heat exchanger is provided at the BF crude gas pipeline, and soft water is used as cooling medium. BFG heat is exchanged via the hot tube to make soft water evaporate and absorb gas heat so as to lower gas temperature efficiently. Under normal condition of gas temperature less than 220℃, the crude gas after the cyclone dust catcher enters to the bag dust catcher directly for purification treatment; when temperature of the crude gas through the cyclone dust catcher is higher than 220℃, the system opens the valve of the hot tube heat exchanger automatically. The high temperature gas enters to the hot tube heat exchanger to lower the gas temperature around 200℃, and then it comes to the bag dust catcher. When gas temperature lowers to 180℃, temperature lowering system is closed automatically. At beginning of No.1 BF blow in, temperature of the crude gas reaches to 260℃ for times, even to 300℃. 3 sets of high efficient hot tube heat exchanger provided at the crude gas pipeline is used to lower gas temperature above approximate 70℃ to meet requirement of safe and reliable running of the bag filter cleaning system. It is proven from practice that this technical measure can solve technical problem on high temperature control of gas.

Comprehensive measures should be taken to control too low gas temperature, for instance, technical measures of improvement of input quality of raw material and fuel, decrease of moisture of input raw material and fuel, intensification of monitor of furnace proper cooling device, rational control of top pressure, temperature insulation of the crude gas pipeline, etc. to achieve high efficiency. Especially during BF blow-in and BF re-blow, control of gas temperature should be drawn to attention in order to reduce moisture in gas, and control gas temperature 20-30℃ above the dew-point.

3.4 Anticorrosion Technology Applied to Gas Pipeline System

Due to application of BFG dry dust catcher technology, chloride ion in condensate of the purified gas is obvious high. Main reason for this is that chloride exists in BF raw material and fuel. Gaseous HCl is formed during BF smelting process. When gas temperature reaches to the dew point, gaseous HCl is combined with condensate so as to generate acid watery solution, which leads to acid corrosion. At damp neutral environment, chloride ion in gas also produces dot corrosion, stress corrosion and local corrosion to the gas pipeline and stainless steel bellow compensator. Test and analysis is taken to condensate of the purified gas after dry bag deduster, it is found that chloride ion in the condensate reaches to 1000mg/L, pH in the gas condensate is lower than 7, sometimes even to 2-3. Heavy corrosion is happened to the gas pipeline and bellow compensator and abnormal corrosion is occurred to the gas pipeline. In order to suppress abnormal corrosion of the gas pipeline system, analysis and study on corrosion mechanism of bellow compensator of gas pipeline is carried out to improve material and structure of stainless steel bellow compensator. Material of bellow compensator is improved to stainless steel Incoloy-825, which can prevent corrosion due to chloride ion, instead of austenite stainless steel 316L (00Cr17Ni14Mo2); polyethylene resin anti-corrosion paint is sprayed at the inner wall of the gas pipeline to separate metal pipeline and acid corrosion medium in order to suppress abnormal corrosion; For elimination of chloride in BFG, the chloride removing device is developed with application of chemical and physical adsorption principle to remove chloride in BFG efficiently. Lye sprinkler is provided at the purified gas pipeline to make lye contact with BFG sufficiently in order to reduce chloride content in gas. Since the blast furnace was put into production, pH of condensate in gas is stabilized at around 6, with chloride content 400mg/L, and there is no abnormal corrosion happened to the gas pipeline and bellow compensator. Fig. 7 illustrates the process diagram of de-chloride facility of BFG, Fig. 8 shows the photograph of chloride removal facility of BFG.

3.5 On-line Monitor Technology of Dust Content in BFG

On-line monitor device of dust content in BFG is the critical detection device for stable operation of monitor of BFG bag cleaning system, TRT. BFG dust content on-line monitor system researched and developed adopts charge induction principle. In running BFG, dust particles

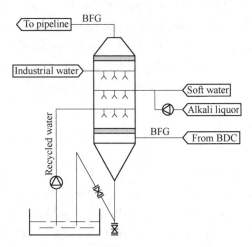

Fig. 7 Process diagram of chloride removal facility of BFG

Fig. 8 Photograph of chloride removal facility of BFG

produce static charge due to friction and collision so that static field is formed. Change of its static field reflects change of BFG dust. BFG dust on-line monitor system deduces BFG dust figure via measurement of change on static charge. This is used to judge whether the bag dust catcher runs normally. When the bag has damaged, dust in the purified gas pipeline increases, and strength of static charge increases, too. Charge sensor can monitor charge volume timely and sent it to the transformer so as to realize automatic monitor of BFG dust.

3.6 Dense-phase Pneumatic Transportation Technology for Collected Dust

Collecting and transportation of dust from the BFG bag filter dedusting catcher is the critical factor, which impacts normal operation of the system. There are many technical defectives on conventional mechanical dust conveying process. Dust pneumatic conveying technology has been developed with nitrogen or purified gas as carrying gas for dust conveying. Dust from every bag of the dust catcher is collected and transferred to the dust bin via pipeline, and then they are centrally taken to the tank car to the sintering plant for recovery and utilization. Airtight conveying in the whole course is realized, technical detectives on conventional mechanical dust conveying process is solved, the process flow is optimized, energy consumption is reduced, secondary pollution is decreased and technical difficulties like wear-off of dust conveying pipeline and so on are overcome.

4 Application in Production

Jingtang's No. 1 BF adopts independent developed BFG fully dry bag filter cleaning system with complete elimination of wet dust catcher system as standby. Technical breakthrough of gas fully dry bag filter cleaning system is realized in huge BF firstly all over the world. The No. 1 BF was blown in and put into production on May 21 of 2009, No. 2 BF was blown in and commissioning on June 20 of 2010. Fig. 9 shows the photograph of No. 1 BF after commissioning, Fig. 10 Shows the photograph of BFG bag filter cleaning facility of No. 1 BF, Fig. 11 Shows the photograph of BFG bag filter cleaning facility of No. 2 BF. After four years' uninterrupted production and application, BFG dry cleaning system has been proven by various operating conditions. It runs stably and reliably, and various parameters have fully achieved or exceeded the design level. From June of 2009 till now, dust in the purified gas is stabilized in 2-4mg/m^3, minimum dust content in the purified gas only approximate 1mg/m^3, annual average 3.74mg/m^3, which is better than the design parameter 5mg/m^3 average gas temperature is 140℃, and approximate 100℃ gas temperature is improved in comparison with that from wet cleaning system, with high gas cleanness and low moisture. Table 2 shows main operating parameters of BFG dry cleaning system of No. 1 BF in 2010.

Fig. 9 Photograph of Jingtang's No. 1 BF after commissioning

BFG calorific value and hot blast stove flame temperature are improved accordingly. Under

Fig. 10 BFG bag filter cleaning facility of Jingtang's No. 1 BF

condition of full application of BFG for BF hot blast stove, hot blast temperature can be stabilized in range of 1250-1300℃ in long term. BFG fully dry cleaning process is applied to realize "zero consumption of fresh water, zero discharge of waste water", eliminate large quantity of poison sewage and slurry produced from wet cleaning process. Issues on fresh water consumption, secondary water pollution and slurry treatment are resolved from the springhead.

Fig. 11 BFG bag filter cleaning facility of Jingtang's No. 2 BF

Table 2 Main operating parameters of BFG dry cleaning system of No. 1 BF

Date	Crude gas temperature/℃	Purified gas temperature/℃	Dust in crude gas /mg·m^{-3}	Dust in purified gas/mg·m^{-3}
2010-01	319	109	7320	4.07
2010-02	191	112	7320	4.44
2010-03	228	120	10930	4.73
2010-04	127	122	9170	1.47
2010-05	138	135	7130	4.97
2010-06	141	137	8470	4.67
2010-07	154	150	8060	4.86
2010-08	174	167	7730	4.73
2010-09	156	152	9000	4.96
2010-10	172	166	8840	4.65
2010-11	177	170	7800	4.83
2010-12	169	163	7950	4.78

Shougang Jingtang's BF adopts fully dry TRT system which is matched with BFG fully dry cleaning system. Fully dry axial type adjustable static blade turbine is applied with generator power 36.5MW, and TRT power generation is promoted greatly. TRT power generation is stably and continuously improved since the BF commissioning. Average monthly power generating capacity achieve more than 52kW·h/tHM in 2010, maximum average monthly power generating capacity reached 64.9kW·h/tHM. Annual average power generating capacity which is improved by approximate 45% in comparison with the gas wet cleaning process. Fig. 12 shows the top pressure of Jingtang's BFs in 2013. Fig. 13 shows the top temperature of Jingtang's BFs in 2013. Fig. 14 shows the TRT power generation performance of Jingtang's BFs in 2013.

Up to now, the service life of the bag can reach to 18 months, stable and reliable performance of operation is achieved. There is no abnormal corrosion appeared at gas pipeline, bellow compensator, etc. Satisfactory result and good performance on anti-corrosion measure of the gas pipeline have been achieved since BF commissioning.

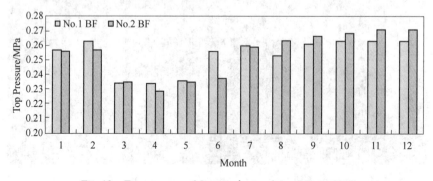

Fig. 12 Top pressure of Jingtang's two huge BFs in 2013

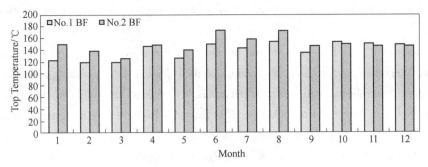

Fig. 13 Top temperature of Jingtang's two huge BFs in 2013

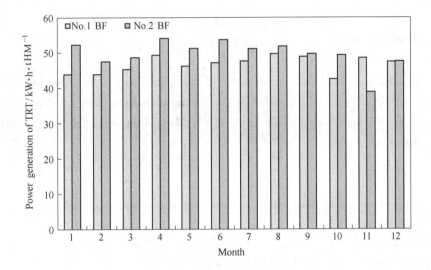

Fig. 14 Power generation performance of TRT of Jingtang's two huge BFs in 2013

5 Conclusins

BFG dry bag filter cleaning technology applied on large BF is the development orientation of iron making technology, and is an important support technology for realization of high efficiency and low consumption, energy saving and emission control, reduction of water source consumption, improvement of energy utilization efficiency, development of recycle economy for iron making industry.

By system research and technical integration, the important breakthrough is realized on BFG fully dry bag filter cleaning technology of Jingtang two huge BFs, with complete elimination of wet cleaning system as standby. The system has been operated stably for 4 years with major technical parameters of dust content in purified gas, gas temperature, TRT power generating performance, etc. has achieved or exceeded the design level, and huge technical and economic efficiency has been obtained. China has already been mastered the core and key technology on BFG dry bag filter cleaning in huge BF, and it has been successfully applied in the Jingtang's two 5500m³ huge BFs.

Breakthrough progress has been made in aspects of design, study, technical innovation, engineering integration, production, application of BFG dry bag filter cleaning technology, which is independent researched and developed in China. A series of key and core technologies have been researched and developed to make the BFG fully dry bag filter cleaning technology on huge BF be perfect day by day, such as the low pressure impulse injection and dust cleaning technology of BFG dry bag filter cleaning, gas temperature control technology, gas dust online detection and digital control system, dense-phase pneumatic conveying of collected dust, anti-corrosion of pipeline system, etc.

6 Acknowledgements

The author would like to thank for my colleagues of Prof. Gao Luping, Mr. Hou Jian, Mr. Zhang Jian, Mr. Zhang Qifu, etc. who participate the technical research and develop works on this issue. Thanks for the help from my colleagues of Mr. Meng Xianglong, Dr. Cao Chaozhen, Dr. Li lin, and Dr. Guo Yanyong.

References

[1] Zhang Fuming. Study and innovation of the key technologies on modern large blast furnace [J]. Iron and Steel, 2009, 44 (4): 1-5.

[2] Zhang Fuming, Qian Shichong, Zhang Jian. New technologies of 5500m³ blast furnace at Shougang Jingtang [J]. Iron and Steel, 2011, 46 (2): 12-17.

[3] Zhang Fuming. Study on dry type bag filter cleaning technology of BF gas at large blast furnace [J]. Journal of Iron and Steel Research International, 2009, 16 (2): 608-612.

[4] Zhang Fuming. Study on dry cleaning technology by bag filter for large sized blast furnace gas [J]. Ironmking, 2011, 30 (1): 1-5.

Technical Development of Low Carbon and Greenization Blast Furnace Ironmaking

Zhang Fuming   Meng Xianglong   Hu Zurui

Abstract: Blast Furnace (BF) ironmaking is a very important key process of iron and steel conventional manufacture system, which is the core and key element of ferrous substance flow conversion, the core unit of energy conversion in the iron and steel manufacture process and the central link of energy flow network. Since stepping on 21st century, BF ironmaking process has been faced with many major sustainable development issues on restriction again by aspects of natural resources shortage, undersupply of energy resources, ecological environment protection, and so on. Under the economic situation of market downturn and weak demand, and in the face of increasingly severe competition and challenge, it is a must to have remarkable breakthroughs in terms of high efficiency and low consumption, energy conservation and emission reduction, circular economy, clean and environment protection, low carbon smelting, intelligent-intensivism, green development, etc., taking low carbon and green development is regarded as the leading factor to have further optimization of process flow, such as, improve the quality of raw materials, optimize the BF operation, increase the hot blast temperature, reduce the coke rate and fuel ratio, in order to improve vitality and competitiveness of BF ironmaking technology. In the future, BF ironmaking technology developing philosophy should be low carbon, green, high efficiency, low energy consumption, intelligent and integration. The contemporary BF ironmaking "three functions" of the hot metal production, energy conversion and waste disposal will be achieved. The new generation ironmaking process design and optimization will be the core and key issue of the BF in the future. The development trend of BF ironmaking technology under the concept of circular economy is discussed, the technical development route of the future BF is expounded, and the key common technical innovation of the development of low carbon and green ironmaking is pointed out in this paper.

Keywords: blast furnace; ironmaking; high blast temperature; PCI; low CO_2 emission

1 Developing Concept

1.1 Main Technical Orientation of Modern BF Ironmaking

After the first industrial revolution, the BF ironmaking process, which had experienced nearly 200 years of development and evolution, has achieved significant technical progress in respects of enlargement, high efficiency, long campaign life, intensification, intelligence, etc. It has been developed into the ironmaking process with the highest iron oxide reduction efficiency. BF ironmaking has been transmuted into engineering science, which is becoming perfection gradually. BF main function is evolving all the time in pace with its course of technical development [1].

During BF production, huge energy consumption and energy conversion are executed with characteristics of physical sensible heat and chemical energy of BF products (hot metal, liquid slag and BF gas) from conversion of fossil mass energy such as coke, pulverized coal, natural gas and so on, so that BF has function of energy converter [2]. With progress of energy saving of BF ironmaking, as well as implementation of technologies of top gas pressure recovery turbine (TRT), recovery of other waste heat and energy, it is more and more prominent for BF as an energy converter. BF ironmaking is maximum energy conversion process unit in iron and steel manufacturing process at present, and also the central link for establishment of energy flow network in iron and steel plant.

Modern BF ironmaking should establish production system of dynamic precision and high efficient operation. High efficiency, low cost and less emission are considered as the goals to improve production efficiency, decrease energy consumption and reduce environmental pollution so as to achieve the goal of collaborative development of modern BF ironmaking with characteristics of high efficiency, low consumption, long campaign life, low carbon and greenization. Main technical characteristics of new century BF ironmaking technology are improvement of comprehensive technical equipment level, development of circular economy and achievement of low carbon and green manufacture [3]. BF operation is controlled rationally, and the scientific and reasonable index assessment system are established to achieve multi-objective collaboration optimization of BF ironmaking. BF ironmaking technology system with characteristics of high efficiency, low cost, low consumption and less emission should be built.

Technology and equipment of large BF in 21st century pay more attention to technical innovation and low carbon green development. Key common technologies should be advanced technologies for new BF ironmaking with major breakthrough, collaborative optimization and intensiveness innovation. Great technical progress should be obtained in many fields such as BF high efficiency and long campaign life, high quality and low consumption, energy conservation and emission reduction, circular economy, clean and environment protection, low carbon and greenization, and so on [4].

1.2 Functional Prolongation of BF

BF iornmaking, taking coke as the main fuel and reductant, is the process unit which is used for reducing iron oxide into hot metal. It belongs to typical shaft type reactor and reverse industrial reaction device with huge multiphase complex system. During the opposite movement process of burden lowering and gas flow rising, the thermal transmission, mass transmission, momentum transmission and a series of physical-chemical reactions can be achieved during BF smelting process. Premise for smooth achievement of BF high temperature reduction process is to have coke column (namely skeleton effect) in BF, which is irreplaceable. For this reason, different degree of carburizing process of hot metal always exists during BF reduction process.

The most important function of BF is to provide high quality hot metal to converter, and it is the foundation of converter steelmaking process. BF reduces ferrous substance of solid iron oxide mineral into liquid pig iron, so that it has function of melter. And continuous reduction and production of hot metal by BF shows BF is a consecutive hot metal supplier. Meanwhile, BF ironmaking process has significant functions of regulating and controlling the composition and quality of hot metal. And temperature, composition and deviation of hot metal can be stabilized by means of BF operation, especially in controlling of sulphur, silicon, and so on in hot metal in reasonable ranges.

Besides metallurgical functions, BF ironmaking process also has functions of solid waste disposal and changing waste into resources, including in coking, sintering and pelletizing which are matched for BF. BF ironmaking system is an important link in achievement of circular economy in iron and steel plant. For instance, coking process and pulverized coal injection (PCI) of BF are applied to treat waste plastics, sintering or pelletizing process are considered to briquette dust from the iron and steel plant for recovery of secondary resources, collected dust is injected into BF, and coke oven is used to pyrolyze and handle municipal solid waste, etc..

2 Pulverized Coal Injection and Oxygen Enrichment

In the 70s to 80s in 20th century, the outburst in succession of two oil crisis helped BF PCI technology dramatically developed and become the key core technology of metallurgical engineering technology progress. Up to now, the BF PCI has been still the important energy saving, emission controlling and low carbon green technology for modern BF, which is the major technology that must be kept on developing for modern BF ironmaking. Under the condition of lower fuel ratio, improving the PCI quantity and reducing coke rate is an important technical approach to reduce energy consumption and CO_2 emission of BF in the future. This method not only reduces the consumption of coke, but also reduces the environmental pollution during coking process.

Currently, there are 170 BFs with the inner volume over 2000m^3 in China, and 22 large BFs over 4000m^3. The large BF has higher production efficiency, less heat dissipation, lower energy consumption and lower production cost. Practice shows that the large BF fuel ratio and coke rate decreased significantly, the fuel ratio of Shougang Jingtang and Baosteel large BFs is less than 500kg/tHM, the coke rate is lower than 300kg/tHM, the coal ratio reaches 180-200kg/tHM.

Oxygen-enriched blasting is one of the effective technical measures for modern BF to increase production efficiency. Increasing oxygen enrichment percentage is significant in modern BF ironmaking and should deserve high attention. BF oxygen-enrichment, PCI and high blast temperature integrated coupling technology can effectively improve the work of hearth tuyere raceway zone, increase pulverized coal burning rate and PCI rate, effectively reduce bosh gas volume, improve BF permeability, promote BF stable and smooth operation and increase gas utilization ratio so as to effectively reduce fuel consumption and CO_2 emission.

BF oxygen-enrichment, PCI and high blast temperature are the important supporting technologies to reduce fuel consumption and CO_2 emission. Under current condition, coal injection rate shall be further increased to make the PCI rate up to 200kg/tHM and even higher, fuel ratio reduced to below 500kg/tHM, by which the foundation of modern BF will keep vitality and competitiveness.

Doubtlessly 21th century modern BF must strive to promote and apply PCI, oxygen-enrichment and high blast temperature integrated technology in new ways, which will become the important technical route for BF low carbon green development in future. Taking the BF low carbon green development and reduction of BF fuel consumption as the technical target, taking the increment of blast temperature, oxygen-enrichment ratio and PCI rate as the technical route and taking the beneficiated burden technology, long campaign life technology and optimized operation as the technical support, the oxygen-enrichment, PCI, high blast temperature technology are integrated to have a broad development prospect [5]. Table 1 shows the large BF (the BF inner volume over 4000m^3) operational parameters of China in first half of 2016 [6].

Table 1 Operational parameters of large BF of China in first half of 2016

Item	Data	Item	Data
Productivity	2.085t/(m^3·d)	Silicon content in hot metal	0.42%
Fuel ratio	517.1kg/tHM	Sulphur content in hot metal	0.03%
Coke rate	349.4kg/tHM	Temperature of hot metal	1502.6℃
PCI rate	159.8kg/tHM	Total ferrous content in burden	59.01%
Oxygen enrichment ratio	3.36%	Utilization efficiency of BF gas	48.21%
Blast temperature	1211℃	Slag volume	297.7kg/tHM
Blast pressure	404.6kPa	Energy consumption	11.27GJ/tHM
Top pressure	172.3kPa	Volatile matter of coal	18.33%

3 High Blast Temperature

High blast temperature technology is an important technical approach for BF to low down coke rate, increase PCI quantity and improve energy conversion efficiency. Using BF gas as the main fuel, the waste heat of flue gas can be recovered to preheat the combustion air and gas, and the dome temperature and hot blast temperature of the hot blast stove can be improved. Increasing the hot blast temperature is beneficial to reduce the fuel consumption by 10-15kg/tHM, and can improve the BF operation at the same time, so that the waste heat and secondary energy of BF can be efficiently recycled and reused. As a result, the CO_2 emission can be reduced significantly. At present the designed blast temperature of large BF is generally 1250-1300 degree celsius, to increase blast temperature is one of the important technical features of the 21 century BF ironmaking.

In early 21 century, super-large BF above 5000m^3 commonly adopted external combustion hot blast stove, NSC and Didier external combustion hot blast stove are the most representative. Shougang Jingtang 5500m^3 BF developed and applied dome combustion hot blast stove, which is the first in the world to adopt dome combustion hot blast stove in super-large BF above 5000m^3, making a breakthrough in super-large BF hot blast stove structure type [7, 8].

The structure type of modern large BF hot blast stove takes on diversified development, the main technical developments are applied. Increase blast temperature to 1250 degree celsius and higher by applying combustion air and gas low temperature double preheating or enriched gas. By adopting preheating stove preheating combustion air, the blast temperature can realize 1250-1300 degree celsius under condition of burning full BF gas only. Increase the uniformity of gas flow distribution by optimization of combustion process, study of gas flow movement rule and the study of regenerator heat transfer, use high efficiency checker brick, increase heat transfer area, intensify heat transfer process and shorten the different value between hot blast stove dome temperature and blast temperature. Optimize hot blast pipeline system structure, adopt non-over-heat and low stress design system, rationally configure hot blast pipeline bellows expansion joint and tie rod structure, deal effectively with pipeline expansion to reduce pipe system stress, hot blast pipeline adopt composite brick structure, eliminate local over-heat and air leakage of hot blast pipeline. Take effective measures to prevent the intergranular stress in stove shell to prolong the service life of the hot blast stove. Optimize hot blast stove operation, rationally set up hot blast stove working cycle, increase hot blast stove heat exchange efficiency. Optimize combustion process, reduce fuel consumption. Effectively reduce NO_x and CO_2 emission to realize energy saving and emission control and low carbon environment protection.

4 BF Gas Dry Dedusting and TRT

BF gas dry dedusting technology is the important technical innovation for 21 century BF to realize high efficiency, low consumption, energy saving and emission control and clean production. The BF gas dry dedusting technology increases the purity of gas, gas temperature and calorific value, not only remarkably reduces the fresh water consumption and energy consumption in ironmaking procedure but also increases the utilization rate of secondary energy and reduces environment pollution. BF gas dry dedusting and TRT coupling technology is the important key technology for BF ironmaking to realize energy saving and emission control, low carbon smelting and high efficiency energy conversion, and it becoming the important technical developing direction for modern ironmaking industry to perform cycling economy and realize low carbon green development.

At present the BF gas dry bag dedusting technology which has been innovated independently by China have achieved great breaking progress in term of design research, technical innovation, engineering integration and production application, it has been used in super-large 5500m^3 BF for 8 years more, the system operates safely and stably, the dust content in purified BF gas is reached 2-4mg/m^3, bag service life is reached 36 months, and electric energy production of TRT is increased by 54kW·h/tHM, more than 45% compared with wet dedusting technology. The key core technologies such as impulse injection dust removal technology, gas temperature control technology, on-line detection of gas dust content, dense phase conveying of collected dust and anti-corrosion for pipeline system have been researched and developed to make the large BF gas fully dry bag dedusting technology become better and better day by day [9].

5　Conclusion

The key common technologies of BF oxygen enrichment, PCI, high blast temperature, BF gas dry dedusting technique, TRT, and high efficiency long campaign are still the important technical foundation, support and guarantee for BF low carbon green development in future. Facing the future, inheritance innovation, integration innovation and re-innovation, on the basis of existing technologies, continuously improve, modify and pursue integration effect and collaborative innovation of comprehensive technologies must always persisted on.

By execution of comprehensive technologies of circular economy, low carbon metallurgy and green development, innovated application of beneficiated burden, long campaign, high blast temperature, oxygen enriched PCI, energy saving and emission control, low carbon smelting and environment protection, further lower down resources and energy consumption, insisting in high efficiency, low cost, low emission and green development concept, promote the sustainable development of BF ironmaking, the future BF ironmaking still has a broad developing prospect.

Acknowledgements

The authors sincerely thank the Beijing Scholars Program for supporting.

References

[1] Zhang Fuming, Cao Chaozhen. Development orientation of low carbon and greenization BF ironmaking technology [C]. 2016 AISTech proceedings volume I (AIST). Pittsburgh, Pennsylvania, USA, 2016: 155-163.

[2] Yin Ruiyu. Metallurgical process flow engineering (second edition) [M]. Beijing: Metallurgical Industry

Press, 2009: 274-277.
[3] Zhang Fuming. Technical features of super-large blast furnace in early 21th century [J]. Ironmaking, 2012, 2: 1-8.
[4] Zhang Fuming. Understandings on some iron making technologies by modern blast furnace [J]. Ironmaking, 2012, 5: 1-6.
[5] Zhang Fuming, Cheng Shusen. Long Campaign Life Technology of Modern Blast Furnace [M]. Beijing: Metallurgical Industry Press, 2012: 579-581.
[6] Zhang Shourong, Jiang Xi. Production and development of large blast furnaces in China [J]. Iron and Steel, 2017, 2: 1-4.
[7] Zhang Fuming. The development and innovation of BF high blast temperature technology [J]. Ironmaking, 2013, 6: 1-5.
[8] Zhang Fuming. The development prospects of high blast temperature low fuel ratio of blast furnace smelting technology [J]. China Metallurgy, 2013, 2: 1-7.
[9] Zhang Fuming. Study on large blast furnace gas dry bag dedusting technology [J]. Ironmaking, 2011, 1: 1-5.

Pulverized Coal Injection Technology in Large-Sized Blast Furnace of Shougang

Meng Xianglong   Zhang Fuming   Li Lin   Cao Chaozhen

Abstract: Pulverized coal injection (PCI) is the magnificent technological feature in modern large sized blast furnace. The technical development of blast furnace relies on the advancement of PCI technology. The advanced technologies were adopted in PCI designed by BSIET recent years, such as DANIELI from Holland and KUTTNER from Germany, which is compared with domestic independent technology in this paper. The analysis of each advantage is a reference of design promotion. Some key technologies of PCI are investigated such as pulverized coal silo and vessel configure, pulverized coal fluidizing technology, pneumatic conveying system optimization. As a result, modern PCI facility must meet the demand of large sized blast furnace. PCI technology will improve rapidly to international advanced level by combining the domestic experience with the adopted technology.

Keywords: large sized; blast furnace; iron-making; pulverized coal injection

1 Introduction

Fuel injection in blast furnace (BF), which originates from Mame iron-making works charcoal scraps injection in France in 1840, has been used in industrial production since 1960s, by continuously study and experiment, most BFs in the world have adopted kinds of injection fuel like coal, dead oil, natural gas, coke oven gas, dedusting ash and other reducing material by now [1-5]. Because of lower cost and richer reserves, pulverized coal is adopted by majority of the BFs, and more than 90% iron are produced in the blast furnace using pulverized coal injection (PCI) technology [6]. Shougang is a traditional steel-making enterprise in China, and began the first PCI industrial test as early as 1960's, by several decades' development, a complete PCI process has been established. Especially the volume of BF becomes larger and larger in recent years, and PCI system matching large sized BF has been put into practice in Shougang widely [7,8]. Qiangang 4000m^3 BF and Jingtang 5500m^3 BF have been built by Shougang in recent years, and the two PCI systems were cooperated with Danieli and Kuttner. PCI technique characteristics of Shougang large sized BF are introduced by analyzing technological process, equipment capacity, pulverized coal fluidizing technique, dense phase conveying, even injection and automatic control etc.

本文作者：孟祥龙，张福明，李林，曹朝真。原刊于 AISTech-Irom and Steel Technology Conference Proceedings, 2014：805-813。

Moreover, suitable PCI rate for Shougang large sized BF is discussed with practical production. As a result, modern PCI facility must meet demand of large sized blast furnace.

2 PCI in Large Sized BF be in Shougang

PCI technology in BF is widely adopted in China because of the influence on improving economic benefit, reducing the cost of iron, regulating heat system and reducing environmental pollution etc. Shougang is one of the inaugurator and forerunner in BF iron-making PCI field in China, whose exploration on PCI technology began from 1963. Until now, PCI technology is utilized in all the BFs in Shougang. Qiangang 4000m^3 BF and Jingtang 5500m^3 BF, which were built in the early 21th century, adopted advanced technology in the PCI system. Medium speed mill pulverizing, once bag filter collection, parallel vessel dense phase conveying and automatic equal injection process was applied, and the whole process could meet the demand of large sized BF.

2.1 Coal Pulverizing System

Because of more complicated operation, the requirements to each system in large sized BF will increase including PCI system. Table 1 shows the production and PCI indexes of large sized BF in Shougang.

Table 1 Production and PCI indexes of large sized BF in Shougang

Item	Jingtang 5500m^3 BF	Qiangang 4000m^3 BF
Effective volume of BF/m^3	5500	4000
Daily iron production/t · d^{-1}	12650	9600
Productivity/t · (m^3 · d)$^{-1}$	2.3	2.4
Tuyere quantity/No.	42	36
PCI rate/kg · t^{-1}	220	190
PCI quantity/t · h^{-1}	116	77
Pulverized coal size/μm	<74 occupy 80wt%	<74 occupy 60wt%
Pulverized coal moisture/wt%	<1.3	<1.5
Pulverized bulk density/t · m^{-3}	0.5-0.6	0.68

According to both the pulverized coal requirements for BF and technical advancement and reliability, Medium speed mill pulverizing process was adopted in both Jingtang 5500m^3 BF and Qiangang 4000m^3 BF. As the core equipment in coal pulverizing system, stable and efficient running of medium speed mill plays an important role in BF production. Raw coal conditions of the two BFs are different, especially in the raw coal size and total moisture which have a great influence on mill efficiency. The raw coal conditions are showed in Table 2.

Table 2 Raw coal conditions of the two BFs

Item	Jingtang 5500m³ BF	Qiangang 4000m³ BF
Total moisture Mt/wt%	≤10	≤13
Ash content/wt%	≤13	≤12
Fugitive constituent/wt%	≤38	≤35
HGI	55	55
Size/mm	≤40	≤60

Pulverized coal demand is bigger in Jingtang 5500m³ BF, and the raw coal quality is better than that in Qiangang 4000m³ BF. Considering about the influence of each item to equipment efficiency, Jingtang 5500m³ BF adopted two MPS235 HP-II vertical miller type mills, whose basic output is ≥75t/h, capacity factor is 8.1-10kW · h/t$_{(coal)}$, primary gas quantity is 98.3-131.1t/h, inlet temperature is ≤350℃ and outlet temperature is ≤90℃. Qiangang 4000m³ BF adopted two HPS1203 vertical miller type mills, whose basic output is ≥75t/h, capacity factor is 9.5kW · h/t$_{(coal)}$, primary gas quantity is 162t/h, inlet temperature is ≤400℃ and outlet temperature is ≤110℃.

Drying gas for pulverized coal of the two BFs is mixed gas, which is composed of hot blast stove exhaust and high temperature gas that is produced by horizontal type flue furnace. According to the drying gas quantity for mill, Jingtang BF flue furnace capacity is $1.16 \times 10^5 Nm^3/h$, which is $1.2 \times 10^5 Nm^3/h$ in Qiangang BF. Blast furnace gas and coke oven gas mixing burning is adopted.

Qualified pulverized coal from mill is as gas-powder state. Once bag filter collector, which is compact and safe and more suit for bituminous coal separating compare with traditional multiple stage collection process, is adopted for gas and pulverized coal separating. Table 3 shows the bag filter collectors' parameters of the two BFs.

Table 3 Parameters of bag filter collectors

Item	Jingtang 5500m³ BF	Qiangang 4000m³ BF
Filtering velocity/m · min⁻¹	0.87	0.85
Filtering area/m²	3038	3926
Temperature of gas-powder/℃	≤120	≤120
Inlet dustiness/g · Nm⁻³	400-600	400-600
Outlet dustiness/mg · Nm⁻³	≤20	≤30
Resistance pressure/Pa	≤1500	≤1500
Gas leakage factor/%	≤3	≤2

2.2 Injection System

Parallel vessels arrangement and main pipe with distributor direct injection process is adopted in both Jingtang BF and Qiangang BF. This kind of process, which has advantages of less investment, higher efficiency and accurate measurement and suits for high volatile coal injection, is used widely in China. The technological processes are showed in Fig. 1.

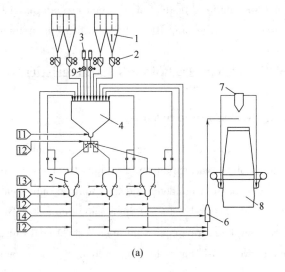

(a)

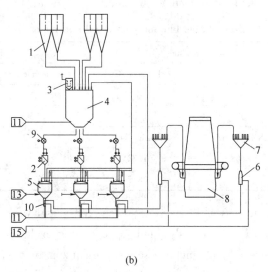

(b)

Fig. 1 Flow chart of PCI system
(a) Qiangang BF PCI system; (b) Jingtang BF PCI system
1—bag filter collector; 2—vibrating screen; 3—pressure relief filter; 4—coal bin; 5—feed vessel; 6—mixing tee;
7—distributor; 8—BF; 9—rotating feeder; 10—fluidizing tank; 11—fluidizing N_2; 12—fluffing N_2;
13—pressurizing N_2; 14—conveying air; 15—conveying N_2

All in all, three vessels arranged parallel, pulverized coal fluidizing and pressurizing by inert gas, monitoring temperature and oxygen content online are applied in the two PCI systems both. But there are still some differences in detail, including: (1) vibrating screens of Qiangang BF are set between bag filter collectors and coal bin, which are set between coal bin and feed vessels in Jingtang BF; (2) plate type fluidizing devices are used in Qiangang BF, instead of multi-point type fluidizing devices and fluidizing tanks in Jingtang BF; (3) single main pipe with single distributor is adopted in Qiangang BF, which are doubled in Jingtang BF, and Qiangang BF distributor is set on furnace top platform instead of cast house platform in Jingtang BF; (4) the conveying gas is compressed air in Qiangang BF, which is nitrogen in Jingtang BF.

3 Large Sized BF PCI System Technique Characteristics

3.1 Equipment Capacity Match Large Sized BF

Accompany with the increasing size of blast furnace, PCI technology and facility should be improved to meet the demand of volume expansion. Besides the capacity of medium speed mill, main exhaust fan, pulverized coal filter collector and injection line number, the storage volume of coal bin and feed vessel should be suitable for the increasing volume of BF.

PCI quantity, mill efficiency and raw coal conditions restrict coal bin and feed vessel storage time, which is controlled between 5-8 hours for coal mill examination and repair, at the same time, the injection time of feed vessel matches the increasing volume of BF too (Table 4).

Table 4 Coal bin and feed vessel parameters

Item	Coal bin		Feed vessel	
	Qiangang BF	Jingtang BF	Qiangang BF	Jingtang BF
Quantity/No.	1	1	3	3
Effective volume/m³	1025	1108	93	80
Storing/feeding time/min	Normal 432	Normal 312	Normal 21.7	Normal 22.8
Height to diameter (h/d)	1.68	1.78	3.21	2.475
Angle of cone/(°)	60	38.7	36.4	49.8

For the parallel vessels process of PCI system, each feed vessel works alternately in cycle to supply pulverized coal for BF continuously, and the operating cycle consist of several working stages including feeding time (T) of single vessel, depressurizing time (T_p), filling time (T_z), pressurizing time (T_c) and holding time (T_d). The operating cycle times are showed in Fig. 2.

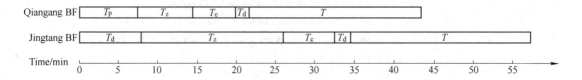

Fig. 2 Operating cycle of feed vessel/min

Each stage of operating cycle should meet the requirement of material flow, valve action and so on. Moreover, the time of each stage should be adjusted at any time in production process, especially for the holding time accuracy controlling, for the reason that high pressure nitrogen is supplied continually for holding working pressure in feed vessel in T_d, which results in energy waste.

3.2 Application of Single Technique and Equipment

Ore and gas move oppositely with longer distance inside large sized BF which has longer smelling period, it will cost much more time to adjust by operators when the condition of BF is unstable. As a result, stable running is significant to large sized BF. Pulverized coal burning in the tuyere area has direct influence on hearth heat condition, and Shougang BF adopted several techniques and equipment in PCI system to ensure pulverized coal supplying.

Firstly, pulverized coal fluidizing technique is applied. Pulverized coal fluidizing could increase conveying density, prevent spontaneous ignition and hanging in feed vessel. The feed vessel which was filled with N_2 and pulverized coal with high pressure is a pressure vessel. Because of gravity and N_2 pressure, coal grain contacts each other and forms the state of fixed bed, especially at outlet of vessel's lower part where the pulverized coal was extruded firmly. It could not flow out itself by gravity although the coal has been processed into tiny grain with little moisture at milling process; it is why the fluidizing device should be used in feed vessel.

A fluidizing tank was set below feed vessel of Jingtang 5500m³ BF, as is showed in Fig. 3. Fluidizing N_2 flows up into buffer for rectification, and passes the sintered metallic plate evenly with high speed. As a result, pulverized coal will loosen, and flows into the transport main pipe through trumpet-shape outlet.

As it's shown in Fig. 4, a plate type fluidizing device with several holes was set in lower head of feed vessel in Qiangang 4000m³ BF. Fluidizing N_2 flows through the plate from the bottom up continuously, so pulverized coal at outlet above the plate was fluidized and filled into feed vessel through pipe. N_2 also ensures the inert environment in feed vessel to prevent coal from burning and exploding.

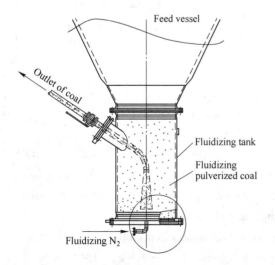

Fig. 3 Fluidizing tank of Jingtang 5500m³ BF

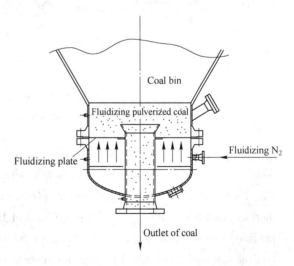

Fig. 4 Plate type fluidizing device of Qiangang 4000m³ BF

Supplying pulverized coal for BF duly and evenly responds to fluidizing effect. Considering the ambient pressure and coal powder gravity, it's necessary to control fluidizing N_2 flow and pressure accurately as well as to make sure that the coal power speed is between critical fluidizing velocity and suspending velocity. In the production practice, fluidizing N_2 flow is controlled around $120Nm^3/h$ in Jingtang BF and $130Nm^3/h$ in Qiangang BF, which gives good result and prepares the ground for dense phase conveying.

Secondly, dense phase conveying technique is applied. Qualified pulverized coal was injected into BF's tuyeres through distributor from feed vessel by pneumatic conveying. Shougang Jingtang and Qiangang BF adopted dense phase conveying technique which depends on increasing the proportion of conveying gas and pulverized coal in transport main pipe by adjusting feed vessel working pressure and additional gas quantity. Under the different feed vessel pressure, the actual volume concentration is the same, while the maximum mass concentration of pulverized coal is not. With increasing pressure of feed vessel, more gas will be compressed between pulverized coal granules during the conveying process. Higher pressure in feed vessel brings lower mass concentration. Nowadays, Qiangang BF feed vessel working pressure is less than 1.42MPa, which is 1.2MPa in Jingtang BF's with the solid-gas rate is more than $35kg_{(coal)}/kg_{(gas)}$ and $40kg_{(coal)}/kg_{(gas)}$. Fig. 5 shows the mixing tee equipment of the two BFs.

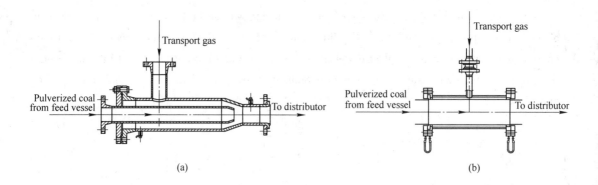

Fig. 5 Mixing tee in transport main pipe
(a) Mixing tee of Qiangang BF; (b) Mixing tee of Jingtang BF

Thirdly, equal distribution and injection technique is applied. Pulverized coal gets into distributors, reaches the injection lances at tuyere through injection lines, and finally enters into BF. Uniformity, one of the significant measurement of injection quality, has a direct effect on the thermal condition of BF hearth. Injection lines of Jingtang and Qiangang BF are designed as equal pressure loss. Distributor device of Qiangang BF is set on the furnace top platform to ensure the equivalent length, and Laval nozzles are adopted in each injection line in Jingtang BF to achieve the pressure-loss requirement. Besides, different distributors are used in the two BFs for equal injection as it is shown in Fig. 6.

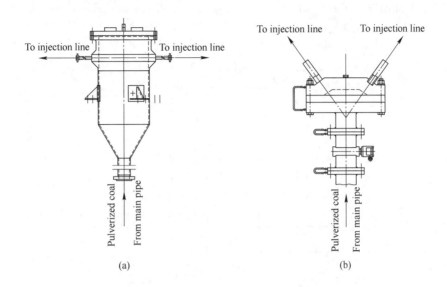

Fig. 6 Different distributor devices of the two BFs
(a) Distributor of Qiangang BF; (b) Distributor of Jingtang BF

4 Full Automatic Injection Technique

PCI system, as an important adjusting method of BF condition, should react rapidly when BF is unstable, including PCI rate automatic regulation, PCI undulating self-correcting in normal conditions and automatic start and stop in accident conditions. The PCI processes of Shougang Jingtang and Qiangang BFs, which are modern large sized BF, realize full-automatic unmanned on duty injection. Chain controlling with PLC is adopted in the whole process including caw coal storage and transportation, pulverized coal preparation, conveying and even injection. The communication between PLC and host computer is via industrial Ethernet. Procedural parameters are controlled to be around set point through PID loop. PV tracking is used in all PID loops which are operated in full-automatic, semi-automatic and manual mode. In full-automatic mode, valves are regulated by PLC. In semi-automatic mode, valves are regulated by workers' operating on control screen. In manual mode, valves are regulated on set by operators and opening will be shown on control screen. The control system consists of raw coal storage and transportation system, pulverized coal preparation system, drying gas system and injection system. The injection system automatic control is core of the whole control system, including feed vessels automatic alternating, PCI rate automatic metering and adjusting, even distribution control, feed vessels pressure automatic regulating, fluidizing N_2 pressure automatic control, failure warming and removal.

In the injection system automatic control, strict start and stop conditions should be set in each stage of feed vessel operating cycle, the feed vessel cannot be put into operation and the operating cycle will be stopped to end the automatic injection unless all the start conditions are met. Feed vessel sequential diagram is shown in Fig. 7.

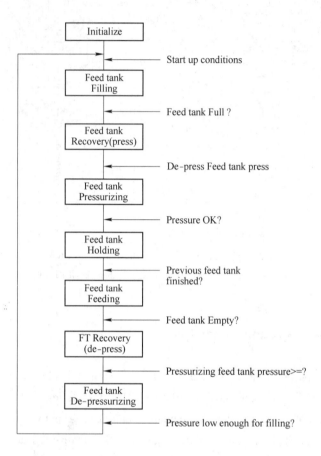

Fig. 7 Feed vessel sequential diagram

PCI rate accuracy control and automatic regulation are significant parts of full-automatic injection function. In productive process, PCI rate is affected by differential pressure in transport pipe, additional gas flow, pulverized coal size and humidity and coal quantity in feed vessel etc. In Shougang, several mathematic models are established as N_2 pressure — PCI rate, feed vessel working pressure — PCI rate, pulverized coal quantity in feed vessel — PCI rate, feed vessel pressurizing — feed vessel weighing, injection lance pressure — PCI rate, and accuracy control and automatic regulation was realized by vessel pressure — additional gas flow control module and feed vessels alternating revise module. Fig. 8 shows the PCI rate accuracy control principle.

PCI rate relative error standard deviation could reach 0.5% and max relative error is controlled within 2% by adopting PCI rate accuracy control technique. Full automatic injection technique ensures the whole PCI system running stably and safely for BF smooth operation and coke saving.

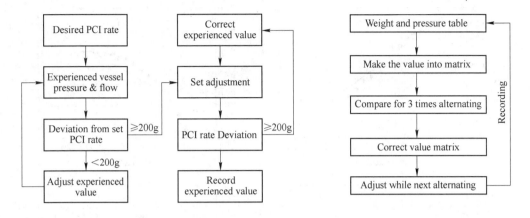

Fig. 8 PCI rate accuracy control principle

5 Exploration of Suitable PCI Rate

In the past few decades, for throughput demand and weak awareness of energy conservation and environment protection, BF strengthening smelting concept was prevailing in China to give rise to the situation of excessively paying attention to throughput, ignoring energy conservation and high fuel ratio[9-11]. Nowadays, steel capacity is increasing rapidly to approach saturation, energy conservation and environment protection awareness is strengthened, moreover, adverse effects of strengthening smelting on BF long campaign are realized, Shougang began to explore reasonable smelting intensity including controlling suitable PCI rate.

Raw coal quality, coke performance, burden condition, slag quantity and viscosity, oxygen enrichment ratio and blast temperature should be considered to decide suitable PCI rate. Coke will be pulverized and heavily loaded under excess PCI rate which should match the coke quality. Fig. 9 shows the relation between PCI rate and coke quality in Shougang large sized BF.

The average coke quality of Jingtang BF in 2012 was that the ash content was 11.98%, M_{40} was 91.08%, M_{10} was 5.55%, CRI was 20.84%, CSR was 70.95%, average grain size was 55.21mm, which in Qiangang BF was that the ash content was 12.06%, M_{40} was 89.12%, M_{10} was 5.64%, CRI was 20.95%, CSR was 67.99%, average grain size was 49.76mm. The experience show that, PCI rate should be kept between 150kg/t and 180kg/t to meet the demand of gas permeability under the conditions that M_{40} is 86%-87% and M_{10} is 6%-8%.

PCI rate should match burden condition and slag ratio. Gas and liquid permeability of hearth will decrease accompanying with the PCI rate increasing, moreover, adverse effect on lower part gas distribution will result in more sluggish of BF center, slag and liquid iron will gather around hearth edge to cause flooding. The average burden grade of Jingtang BF in 2012 was 58.94%,

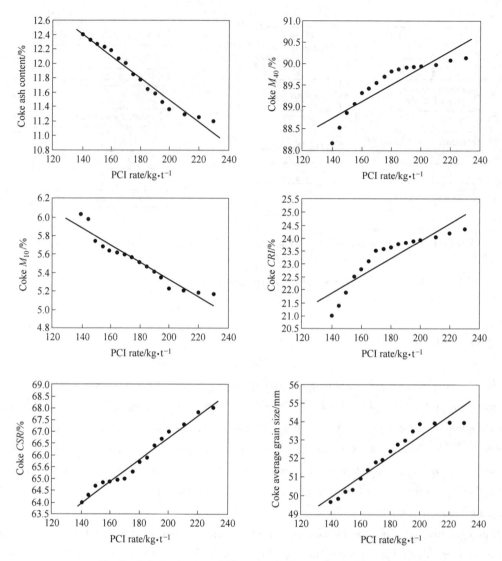

Fig. 9 Relation between PCI rate and coke quality in Shougang

sinter SiO_2 content was 5.49% and slag ratio was 289.08kg/t; the average burden grade of Qiangang BF in 2012 was 57.9%, sinter SiO_2 content was 5.46% and slag ratio was 321.5kg/t, and the experiment showed that the BF was running stably when the PCI rate was controlled in 130-180kg/t.

There is an effect of blast condition on PCI rate too. Suitable theoretical combustion temperature and gas distribution should be kept to ensure hearth thermal conditions. Theoretical combustion temperature is influenced by PCI rate and pulverized coal burning rate, and blast temperature and oxygen enrichment ratio influences the pulverized coal burning rate directly. So PCI rate should match the blast temperature and oxygen enrichment ratio too. The average blast temperature of Jingtang BF in 2012 was 1242.38℃ with the oxygen enrichment ratio of 3.87%; The average blast temperature of Qiangang BF in 2012 was 1245℃ with the oxygen enrichment ratio of 4.84%. The experiment showed that PCI rate of the two BFs should be controlled in 140-170kg/t to keep BF running stably under these blast conditions.

In practical production, the actual PCI rate of Jingtang BF in 2012 was 150-160kg/t, which was 150-170kg/t in Qiangang BF, the corresponding average fuel ratio was 484.54kg/t and 507.76kg/t respectively, the average utilization coefficient was $2.21t/(m^3 \cdot d)$ and $2.29t/(m^3 \cdot d)$ respectively.

6 Conclusions

Accompany with the increasing size of BF, PCI technological process and facilities should be improved to meet the demand of volume expansion. Parallel vessels arrangement, main pipe with distributor direct injection process is suitable for large-sized BF. The capacity of medium speed mill, main exhaust fan, pulverized coal filter collector as well as the number of injection line and the storage volume of coal bin and feed vessel should adapt to the increasing volume of BF.

Stable running is significant to large-sized BF. The techniques of pulverized coal fluidizing, dense phase conveying, equal injection and automatic control could ensure the PCI system supplying pulverized coal stably and evenly for BF which should be adopted in large-sized BF widely.

Excessive smelting intensity results in energy waste, environment pollution and high fuel ratio. Suitable PCI rate should be explored for reasonable smelting intensity which is beneficial to keep BF running stably and long campaign.

Burden condition, coke performance, slag ratio, oxygen enrichment ratio and blast temperature, which have complex effect on BF running, should be considered to decide suitable PCI rate.

References

[1] Zhang Liguo, Wang Zaiyi. Study and application of dust injection into BF [J]. China Metallurgy, 2012, 32 (3): 40-48.

[2] Chen Yongxing, Wang Guangwei. Theoretical analysis of injecting coke oven gas with oxygen enriched into blast furnace [J]. Iron and Steel, 2012, 47 (2): 12-16.

[3] Bi Xuegong, Rao Changrun. Development status of simultaneous injection of agricultural and forestry residues and its perspective [J]. Henan Metallurgy, 2012, 20 (3): 1-5.

[4] Guo Tonglai, Liu Zhenggen. Numerical simulation of blast furnace raceway with coke oven gas injection [J]. Journal of Northeastern University (Natural Science), 2012, 33 (7): 987-991.

[5] Gao Jianjun, Guo Peimin. Numerical simulation of injection of coke oven gas with oxygen enrichment to the blast furnace [J]. Iron Steel Vanadium Titanium, 2010, 31 (3): 1-5.

[6] Hou Xing. Current situation and development for coal powder injection of blast furnace at home and abroad [J]. Jingxi Metallurgy, 2012, 32 (3): 40-48.

[7] Ma Jibo, Fang Yiliu. Development of pulverized coal injection technology in China [J]. Shandong Metallurgy, 2009, 31 (1): 9-16

[8] Meng Xianglong, Zhang Fuming. Application of imported technology on pulverized coal injection in Qiangang 4000m^3 blast furnace [J]. Ironmaking, 2013, 32 (2): 34-37.

[9] Zhai Xinghua. Approach of pulverized coal injection system design [J]. Ironmaking, 2003, 5 (22): 5-8.
[10] Xiang Zhongyong, Yin Han. Discussing on energy saving and emission reduction target and combustion intensity of BF [J]. Iron & Steel Technology, 2013, 1: 2-9.
[11] Zhou Yusheng, Xiang Zhongyong. Technical considerations on reasonable pulverized coal injection rate for blast furnace process [J]. Iron and Steel, 2010, 45 (2): 1-8.

非高炉炼铁工程设计研究

气基竖炉直接还原技术的发展现状与展望

摘　要：阐述了气基竖炉直接还原工艺的技术特点和发展现状，分析了在我国资源和能源条件下气基竖炉直接还原技术发展所面临的主要问题。基于气基竖炉直接还原工艺的特点，对该工艺的原料、还原气等进行了分析研究。指出非常规天然气资源的有效开采和加压煤制气工艺投资、运行成本的显著降低，将是未来气基竖炉直接还原技术发展的主要推动力，同时利用钢铁企业过剩的煤气资源和我国局部地区相对丰富的天然气资源生产直接还原铁，是今后我国气基竖炉直接还原技术发展的重要方向。参照唐山地区的原料和能源价格，对年产量为80万吨/年的直接还原铁装置的生产成本和技术经济可行性进行了分析，分析结果表明：原燃料价格波动对DRI成本影响显著，其中还原气成本约占DRI生产成本的10%~25%；若按DRI替代转炉废钢计算效益，要求天然气价格低于1.8元/m^3。
关键词：炼铁；直接还原；竖炉；球团；煤制气

1　引言

当前，由于高炉炼铁工艺赖以依存的炼焦煤资源日益稀缺，造成高炉炼铁生产成本居高不下，高炉炼铁工业可持续发展受到资源供给不足、能源稀缺、生态环境的制约。直接还原工艺是以气体燃料、液态燃料或非焦煤作为能源，将球团或铁矿石还原成直接还原铁（DRI）的炼铁工艺。气基竖炉直接还原工艺是最具有代表性的直接还原工艺之一，近年来在墨西哥、伊朗、沙特、阿联酋、印度等国家发展迅速，近期投产的单套生产装置产能已达到200万吨/年以上[1]。

2　竖炉直接还原工艺

2.1　发展现状

竖炉直接还原工艺起源于20世纪50年代，随着天然气的大量开采和利用，推动了竖炉直接还原技术的发展，相继出现了Midrex、HYL-I、HYL-III、HYL/Energiron-ZR等以气基竖炉为还原反应器的直接还原工艺。2010年全世界DRI产量首次突破7000万吨，气基竖炉直接还原工艺生产的DRI产量占直接还原工艺总产量的74%以上，成为主要的直接还原工艺[2]。2011年全球DRI产量为7330万吨，气基竖炉直接还原工艺的DRI产量达到76%以上，其中Midrex工艺的总产量为4440万吨，HYL工艺的总产量达到1110万吨。2012年全球DRI总产量达到7402万吨，气基竖炉直接还原工艺生产的DRI产量达到77%，全球DRI/HBI的贸易量达到1475万吨，约占全球DRI总产量的20%。

本文作者：张福明，曹朝真，徐辉。原刊于《钢铁》，2014, 49 (3)：1-10。

2.2 工艺流程

2.2.1 Midrex 工艺

从 20 世纪 60 年代问世至今，Midrex 工艺一直是气基竖炉直接还原工艺的主流技术。在近 50 年的发展进程中，Midrex 工艺本身没有出现革命性的技术变革，其技术进步主要体现在竖炉大型化、提高竖炉生产效率、还原气重整工艺优化等方面。2000 年全球 DRI 总产量达到 4320 万吨，其中采用 Midrex 工艺生产的 DRI 产量达到全球 DRI 总产量的 68%，到 2011 年采用 Midrex 工艺生产的 DRI 产量仍维持全球 DRI 总产量的 60%以上，足见该工艺仍是当今世界上最具代表性和主导性的气基竖炉直接还原工艺。图 1 是典型的使用天然气为燃料的 Miderex 工艺流程。

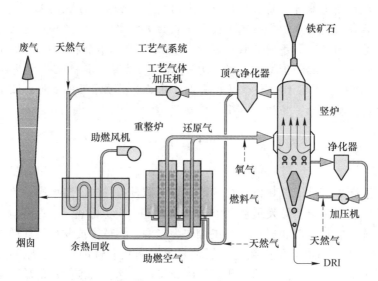

图 1 典型的 Midrex 生产工艺流程（天然气）

为了使天然气中的 CH_4 转化成还原气，Miderex 工艺将天然气与富含 CO_2 的炉顶回收气混合后，送入独立的高温重整炉进行催化重整，在回收竖炉顶气 CO_2 的同时，在镍质催化剂的催化反应条件下将 CH_4 转化为 H_2 和 CO，得到 H_2+CO 体积分数约为 90%、H_2 与 CO 的体积分数比为 1.5~1.7、温度为 900~950℃ 的还原气，直接供给还原竖炉使用。由于采用 CO_2 重整工艺，该工艺无需设置单独的 CO_2 脱除系统，与水蒸气重整工艺相比，还原气中的 CO 含量较高，为避免 CO 还原过程放热造成局部炉料高温黏结，需在还原竖炉高温区下部设置破碎装置。由于 Miderex 工艺竖炉反应温度较低，炉料在还原带的停留时间约为 6h，在竖炉内的停留时间约为 10h，竖炉内的操作压力约为 0.2~0.3MPa。

2.2.2 HYL/Energiron-ZR 工艺

自 1957 年第一个 HYL 气基直接还原铁厂建成以来，HYL 直接还原技术先后经历了一系列的重要变革。1980 年将固定床反应罐改为竖炉实现了连续化生产；1986 年采用炉顶气脱除 CO_2 工艺，可以在还原过程中有选择地脱除氧化性气体（H_2O 和 CO_2），实现节能减排。1995 年引入了氧气喷入技术，采用特殊设计的氧气喷枪可以将高温区控制在一个较小的范围内，通过部分还原气不完全燃烧将气体温度提高至1085℃。氧气喷入技术的采用，摆脱了气体加热炉材质对还原气温度进一步提高的制约，使还原气温度可以达到1000℃以上，使竖炉内CH_4自重整的热力学条件得以实现，为竖炉实现自重整工艺奠定了基础。1997年取消了天然气的催化裂解工序（即取消了独立的重整炉），以自重整技术取代重整炉催化裂解工艺。1999年HYLSA公司开发了DRI热送技术（HYL-HYTEMP），并在墨西哥蒙特雷钢厂4M型HYL生产装置上投入工业化运行。该工艺可以将还原竖炉生产的热态DRI（温度约为700℃），利用N_2或热态还原性气体，以脉动柱塞流的方式直接送到电炉车间电炉上方的料仓，再通过给料管直接加入到电炉中，实现了竖炉与电炉生产工序铁素物质流的热态连接，可以显著提高电炉的冶炼效率、降低能耗。

图2是典型HYL/Energiron-ZR自重整工艺流程图。HYL/Energiron-ZR工艺采用竖炉内自重整工艺取代了独立的重整炉，设置了还原竖炉顶气CO_2脱除装置和水蒸气加湿装置，可以选择性去除还原气中的氧化性气体成分，对还原气成分进行灵活调节。由于采用水蒸气重整工艺，重整后还原气中H_2与CO体积分数比较高（可达5.6），还原气中较高的H_2含量可以使还原尾气中CO_2含量降低，减少CO_2排放，降低环境负荷，还可以降低球团在竖炉内还原过程中的膨胀率[3]，有利于降低竖炉还原区内出现炉料高温黏结的几率。HYL还原竖炉操作采用较高的工作压力（0.6~0.8MPa），通过提高系统压力可以减小还原气体体积，降低炉内的还原气流速，使相同产能的竖炉容积减小，提高生产效率、减少粉尘带出量。由于自重整工艺无需对还原气进行预重整，因此简化了还原气体回路的工艺流程，可以使用天然气、煤制气、焦炉煤气等多种气源，且无需对工艺装置进行较大的改造。此外，还可以通过调节CO_2脱除量、加湿量和吹氧量，实现对DRI中碳含量的灵活调节（1%~3.5%），从而满足电炉冶炼的工艺要求[4]。

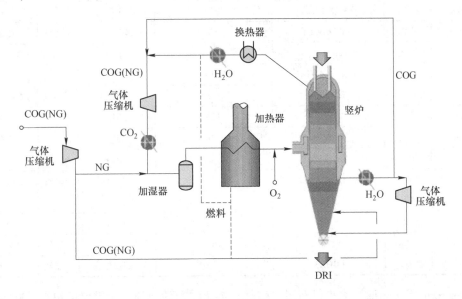

图2 典型的HYL/Energiron-ZR工艺流程

2.3 生产规模

进入新世纪以来,直接还原技术发展迅猛。伊朗、印度、埃及、阿联酋等国家,相继建成了多座气基竖炉直接还原生产装置。目前,单套竖炉生产装置的设计规模最高可达275万吨/年,大型化成为气基竖炉直接还原工艺在本世纪初最显著的技术特征,单套装置的产能和生产效率是目前非高炉炼铁工艺中最大的,甚至可以和高炉炼铁工艺相媲美。表1给出了近期建成投产的气基竖炉直接还原装置。

表1 近期建成投产的气基竖炉直接还原装置

工艺	国家	工厂	生产能力/万吨·年$^{-1}$	投产时间	还原剂	产品
HYL-Ⅲ	委内瑞拉	Matesi	2×75	2004年	天然气	HBI
Energiron-ZR	印度	Welspun Maxsteel-2	60	2009年	天然气	CDRI
Energiron-Ⅲ	阿联酋	Emirates Steel 1 (GHC)	160	2009年	天然气	HYEMP/CDRI
HYL-ZR	阿联酋	Gulf Sponge iron	20	2010年	天然气	CDRI
HYL-Ⅲ	委内瑞拉	Sidor	80	2011年	天然气	CDRI
Energiron-Ⅲ	阿联酋	Emirates Steel 2 (GHC)	160	2011年	天然气	HYEMP/CDRI
Energiron-ZR	埃及	Suez Steel	195	2013年	天然气	HYEMP/CDRI
Energiron-Ⅲ	埃及	Ezz Rolling Mills	190	2013年	天然气	CDRI
Energiron-ZR	美国	Nucor Louisiana	250	2013年	天然气	CDRI
Energiron-ZR	印度	Jindal Steel & Power	250	2014年	合成气/COG	HYEMP/CDRI
Midrex	伊朗	HOSCO Ⅰ & Ⅱ	2×80	2009/2010年	天然气	CDRI
Midrex	印度	Essar 钢铁Ⅳ	150	2010年	天然气	CDRI
Midrex	伊朗	呼罗珊干贴Ⅰ	80	2010年	天然气	CDRI
Midrex	伊朗	IMPADCO	80	2011年	天然气	CDRI
Midrex	阿曼	Jindal Shadeed	150	2011年	天然气	HDRI/HBI
Midrex	伊朗	Mobarakeh G-H	2×150	2012年	天然气	CDRI
Midrex	巴基斯坦	Tuwairqi	128	2012	天然气	HDRI/CDRI
Midrex	印度	JSW	120	2013	天然气	HDRI/CDRI
Midrex	伊朗	IGISCO	80	2013	天然气	CDRI
Midrex	伊朗	Arfa 钢铁	80	2013	天然气	CDRI
Midrex	印度	Jindal Steel & Power	180	2013	天然气	HDRI/CDRI
Midrex	埃及	ESISCO	176	2013	天然气	HDRI/CDRI
Midrex	伊朗	Saba	150	2013	天然气	CDRI
Midrex	巴林	SULB	150	2013	天然气	HDRI/CDRI
Midrex	俄罗斯	LGOK3	180	建设中	天然气	HBI

气基竖炉直接还原工艺可以摆脱对烧结、炼焦工艺的依赖,无需使用焦炭,减少了工艺生产流程和单元工序,是典型的短流程生产工艺。竖炉直接还原工艺使用的原料为球团(或配加一定比例的块矿),天然气作为燃料和还原剂,因此该工艺在天然气资源丰富的

地区发展迅速。气基竖炉生产的 DRI 由于采用高品质球团和洁净的天然气生产，因此 DRI 中 S、P 等杂质元素含量低，化学成分稳定，金属化率高，粒度均匀，将 DRI 进行热压形成热压铁块（HBI），更便于运输和贸易。因此，DRI/HBI 成为可以替代废钢的重要金属原料，在利用电炉生产高纯净度的合金钢时，DRI/HBI 作为杂质含量低的洁净铁源，普遍受到下游用户的青睐。

气基竖炉直接还原装置的大型化和生产效率的提高是带动该工艺快速发展的源动力。回顾气基竖炉直接还原工艺的发展历程可以看出，技术创新和工艺优化是这种工艺保持技术领先的基础，主要表现在：

（1）提高还原气温度。20 世纪 70 年代，竖炉还原气的入炉温度约为 780℃，为了提高生产效率，改善还原反应热力学条件，到 20 世纪 90 年代，还原气的入炉温度提高到 850℃，由此生产效率提高了 13%。为了防止竖炉内球团在高温条件下出现黏结，开发了原料球团的喷涂工艺，在原料球团表面喷涂一层防止黏结的喷涂材料，还原气入炉温度进而提高到 900℃，竖炉生产效率在原有的基础上又提高了约 11%。

（2）在高温还原气中加入氧气，进一步提高还原气温度。在经过加热后还原气的温度可达 930℃，在还原气进入竖炉之前再兑入氧气，还原气与氧气发生部分氧化反应，释放热量，可将还原气温度提高到 1085℃；同时还原气与氧气发生部分氧化反应，还有助于还原气的改质反应。由于提高了还原气的入炉温度，有利于竖炉内还原反应的进行，生产效率提高了 20%，降低能源消耗约 5%。

（3）还原气改质（重整）工艺日益成熟。长期以来，采用天然气作为气基竖炉的还原气体，由于天然气的主要成分是 CH_4，必须在还原气入炉之前进行改质处理，将 CH_4 转化成为具有还原能力 H_2 和 CO。以 Midrex、HYL 为代表的气基竖炉直接还原工艺，对于还原气体的改质工艺都进行了许多改进和创新。尽管两种工艺采用的天然气改质工艺不尽相同，改质后进入竖炉的还原气成分、温度也差异较大，但天然气改质工艺的日益成熟可靠，是气基竖炉直接还原工艺扩大产能、提高生产效率、实现大型化的一个重要技术支撑。

2.4 还原剂与还原工艺

2.4.1 还原剂

用于竖炉直接还原工艺的还原剂一般为具有还原性的气体，如天然气、人工煤气等。按照铁氧化物还原的冶金原理，根据冶金物理化学和铁冶金学的学术定义，气基竖炉直接还原工艺的还原反应实质上属于典型的间接还原反应，即以 H_2 或 CO 气体作为铁氧化物的还原剂，而将其称为直接还原工艺则主要是为了区别于高炉炼铁工艺的学科分类。从某种意义上讲，气基竖炉直接还原工艺的单元过程与高炉软熔带以上块状区的单元过程具有较大的相似性。典型的气基竖炉直接还原工艺是以天然气或人工煤气为主体能源，在竖炉内的还原剂为 H_2 和 CO 气体。天然气是一种多组分的混合气态化石质燃料，主要成分为烷烃，其中 CH_4 含量可以达到 90% 以上。采用天然气作为竖炉直接还原的主体能源时，必须对天然气进行改质，将其裂解为以 H_2 和 CO 为主要成分的还原性气体。反应式（1）、式（2）为天然气裂解反应。

$$CH_4 + H_2O \Longrightarrow 3H_2 + CO \tag{1}$$

$$CH_4 + CO_2 = 2H_2 + 2CO \qquad (2)$$

2.4.2 天然气改质工艺

将天然气改质转化成为具有还原性的 H_2 和 CO 气体,是竖炉直接还原工艺的重要技术特征之一。由于 CH_4 的裂解反应为吸热反应,需要在 900~1000℃ 时才能得到质量优良的还原气体和较高的天然气转化率[5],同时可以获得较高的能源利用率。不同的竖炉直接还原工艺其天然气改质工艺路线也不尽相同[6],Midrex 工艺主要采用镍基催化工艺,利用富含 CO_2 的竖炉顶气与天然气重整而实现天然气的改质,以式(2)为主要反应;而 HYL 工艺则主要采用天然气与水蒸气的重整工艺,以式(1)为主要反应。由反应式(1)和式(2)可以看出,基于反应式(1)的 HYL 天然气改质工艺,重整反应产物中 H_2 的体积分数为 CO 的 3 倍,H_2 与 CO 体积分数比理论值约为 3,还原气中 H_2 含量高,而 CO 含量低;而基于反应式(2)的 Midrex 天然气改质工艺,重整反应产物中 H_2 与 CO 的体积分数大体相同,H_2 与 CO 体积分数比理论值约为 1。

2.4.3 竖炉还原反应特征

在竖炉还原进程中,以 H_2 和 CO 作为还原剂实现对铁氧化物的逐级间接还原。H_2 还原氧化铁的动力学条件要优于 CO,特别是在传质过程中,H_2 的还原速率要明显高于 CO 的还原速率。由于水煤气置换反应的存在(见式(3))[7],在温度高于 810℃ 时,H_2 的还原能力要高于 CO,基于该反应所提出的低碳氢冶金工艺已成为竖炉直接还原技术和高炉喷吹焦炉煤气技术新的研究开发热点。此外,由于 CO 还原氧化铁的反应是放热反应,而 H_2 还原氧化铁的反应是吸热反应(见式(4)和式(5)),因此,不同的 H_2 与 CO 体积分数比对于竖炉内的温度分布和还原反应的影响存在较大区别[8]。还原气中较高的 H_2 含量意味着更高的还原反应热负荷,需要更高的还原气入炉温度,如 HYL 工艺中还原气的入炉温度要求要比 Midrex 工艺高约 50℃,而 Midrex 工艺中由于还原气中的 CO 含量较高,炉料在炉内出现局部高温黏结的趋势要比 HYL 工艺更加明显。

$$CO_2 + H_2 = CO + H_2O(g) \qquad \Delta G^{\ominus} = 30459.5 - 28.14T, \text{ J/mol} \qquad (3)$$

$$1/4Fe_3O_4(s) + H_2 = 3/4Fe(s) + H_2O(g) \qquad \Delta G^{\ominus} = 35550 - 30.40T, \text{ J/mol} \qquad (4)$$

$$1/4Fe_3O_4(s) + CO = 3/4Fe(s) + CO_2 \qquad \Delta G^{\ominus} = -9832 + 8.58T, \text{ J/mol} \qquad (5)$$

3 原料

目前,用于竖炉直接还原的原料以高品质球团为主,可以适当配加至 30% 的块矿或高炉用球团。由于竖炉直接还原工艺中,炉料在加热、还原过程中始终处于固态,因此要求原料具有良好的理化性能与冶金性能。表 2 给出了竖炉直接还原用原料和高炉球团的理化性能指标,其中直接还原竖炉球团要求全铁(TFe)大于等于 67%,转鼓指数(+6.3mm)大于 93%,低温还原粉化率(+6.3mm)大于 80%,膨胀指数小于 10%,这些指标比 3000m³ 级大型高炉球团的质量要求更为严格[9]。竖炉直接还原工艺对球团品位没有特殊要求,但球团的全铁含量会对 DRI 的质量产生直接影响,这一影响主要体现在对炼钢工序造渣剂消耗和能耗上,因此实际生产中要求用于直接还原竖炉球团的含铁品位越高越好。直接还原竖炉的原料结构同高炉相比差异很大,其中球团的比例很高(大于等于70%，甚至达到100%），因此直接还原竖炉工艺对球团膨胀指数的要求也明显高于高炉。此外，低温还原粉化率也是考核直接还原竖炉球团技术性能的重要指标之一。TiO_2含量主要影响球团的低温还原粉化率，碱金属含量、粒度组成、冷态强度等指标对直接还原竖炉的影响与高炉类似。

表2 竖炉和高炉原料理化性能对比

项目	球团	块矿	高炉球团（3000m³级高炉）
TFe/%	≥67	≥66	≥64
FeO/%	1.0	—	<1.5
Na_2O+K_2O/%	0.1	≤0.1	≤0.1
TiO_2/%	0.2	≤0.2	—
LOI/%	—	≤1.5	—
粒度分布（+31.8mm）/%	—	≤5	—
粒度分布（+15.9mm）/%	≤5	—	—
粒度分布（−9.5mm）/%	≤15	—	≥85（9~18mm）
粒度分布（−6.3mm）/%	≤1	≤8	≤5（−6mm）
粒度分布（−3.2mm）/%	—	≤2	—
气孔率/%	>20	—	—
压溃强度/kN·球⁻¹	>2000	—	>2000
转鼓指数（+6.3mm）/%	>93	>90	>90
落下强度（+6.3mm）/%	>95	>90	—
还原指数（800℃）/%	>3	>3	—
还原指数（950℃）/%	>4	>4	—
膨胀指数/%	<10	—	<15
低温还原粉化率（+6.3mm）/%	>80	>70	>85（+3.15mm）
低温还原粉化率（−3.2mm）/%	<10	<20	—
未破损球团/%	>60	—	—

理论研究和生产实践证实，用于直接还原竖炉的球团，其综合性能要高于大型高炉的球团，对于低温还原粉化等关键指标的要求也高于高炉球团。为了进行气基直接还原竖炉工艺的可行性研究，首钢曾将3种不同产地的球团委托有关研究单位进行了对比实验，其理化性能检验结果见表3。A、B、C三种球团中，受低温还原粉化率指标的影响，球团A和C适用于竖炉直接还原生产，而球团B不能单独作为竖炉的原料使用。因此，鉴于气基竖炉对原料的苛刻要求，在开展国内气基竖炉建设之前，一定要重视对原料条件的检测和论证，特别是低温还原粉化率、膨胀指数、强度、含铁品位等主要性能指标要满足竖炉直接还原工艺的要求。

表 3 首钢球团理化性能分析

项 目		A	B	C
化学成分 /%	TFe	65.4	64.8	67.9
	FeO	0.77	0.84	0.39
	Fe_2O_3	92.70	91.73	96.63
	S	0.009	0.002	0.002
	P	0.015	0.013	0.031
	SiO_2	4.18	5.09	1.25
	Al_2O_3	0.31	1.01	0.43
	CaO	0.47	0.30	0.81
	MgO	0.92	0.46	0.14
	Na_2O	0.144	0.020	0.001
	K_2O	0.064	0.001	0.044
	TiO_2	0.030	0.130	0.020
转鼓指数 (+6.3mm)/%		95.0	95.1	93.2
耐磨指数 (-0.5mm)/%		4.3	3.5	6.0
还原指数 (800℃)/%		3.6	3.4	4.5
还原指数 (950℃)/%		5.4	5.4	5.7
膨胀指数 (800℃)/%		4.0	-1.0	4.7
膨胀指数 (950℃)/%		6.4	5.9	11.8
低温还原粉化率 (+6.3mm)/%		89.2	39.7	90.4
低温还原粉化率 (-3.2mm)/%		7.2	54.0	7.6
还原后球团未破碎率/%		75.9	33.6	77.4

4 还原气的选择

世界上已经建成的气基直接还原竖炉装置，基本上都是以天然气作为还原气，主要集中在伊朗、委内瑞拉、墨西哥等天然气资源丰富的地区。为了使气基直接还原工艺摆脱对天然气资源的依赖，探索和采用新的还原气已成为该技术的发展方向。近年来，以使用煤制气、焦炉煤气和合成气等为能源的气基竖炉直接还原技术的开发逐渐受到重视，并取得了一定的进展，未来有望在印度和我国率先得到实施。

4.1 天然气

囿于资源禀赋条件，我国常规天然气资源相对贫乏，2012 年我国天然气的对外依存度达29%，天然气进口量达400 多亿立方米，而且购气成本还在不断升高，以常规天然气为能源的气基直接还原技术在我国发展缓慢。由于世界能源消费的不断攀升，包括页岩气在内的非常规能源越来越受到重视，随着开采技术的渐趋成熟和进步，除常规天然气外，对煤层气和页岩气的开发和利用逐渐成为可能，这也为我国气基竖炉直接还原技术的发展提供了新的契机。

4.1.1 页岩气

页岩气是以吸附或游离状态存在于泥岩和页岩层中的天然气资矿，全球资源总量约为 456 万亿立方米[10]。长期以来，受储集层渗透率低、采收率低、开采难度较大等因素的影响，页岩气开采难以取得发展。近年来，美国在页岩气开采方面取得了较大进展，从 2005 年到 2010 年美国的页岩气产量增长了近 6 倍，达到 1378 亿立方米，实现了对页岩气的大规模商业化开采，页岩气的大量开采压低了天然气价格，为美国的能源市场带来了革命性巨变。受美国成功开发页岩气的影响，全球页岩气勘探开发呈现快速发展态势。

我国的页岩气储量丰富，居世界首位，约为 15 万亿~30 万亿立方米，主要分布在松辽盆地、渤海湾盆地、江汉盆地、四川盆地、黔北和柴达木盆地等地区，具有较好开发前景[11]。页岩气的开发为扩大我国天然气应用提供了较大空间，但同时也存在诸多困难。我国的页岩气资源条件差，表现为藏储层埋深大、丰度低、开采难度大，开采成本约为美国的 2~4 倍。页岩气作为一个新兴的非常规能源，其开发需要大量投入，而我国页岩气的开发才刚刚起步，开采经验不足，对开采技术的掌握还非常有限。目前，国际上页岩气的开采主要采用水平井钻井法和水力压裂法，即通过将岩石层压裂，释放出其中的天然气，美国超过一半的天然气都是通过压裂法开采获得，这项技术需要消耗大量水资源，而页岩气储区一般都为缺水地区，而且巨大的耗水量不仅有可能造成水资源的浪费，还有可能造成地下水污染，甲烷大量溢出还有可能带来温室效应，因此，页岩气的开发有可能会对环境带来负面影响。上述这些因素都制约着我国页岩气的发展，预计未来以页岩气作为竖炉还原气的进程仍很漫长。

4.1.2 煤层气

煤层气俗称"瓦斯"，是另一种重要的非常规天然气，指吸附在煤矿内壁上以甲烷为主要成分的烃类气体，属于煤的伴生矿产资源，其热值与天然气相当。全球埋深浅于 2000m 的煤层气资源约为 240 万亿立方米，是常规天然气探明储量的两倍多，我国浅煤层气资源量预计达 36.8 万亿立方米，居世界第 3 位，主要分布在晋陕内蒙古、新疆、冀豫皖和云贵川渝等四个含气区，其中 80% 以上储量和产量来自山西省[12]。煤层气是煤矿瓦斯爆炸事故的根源，当空气浓度达到 5%~16% 时，遇到明火就会发生爆炸，如果直接排放到大气中，其温室效应约为 CO_2 的 21 倍，对生态环境破坏性极强。而作为一种高效、洁净的能源，煤层气的商业化应用则具有巨大的经济效益，因此在采煤之前需要先开采煤层气。煤层气的开采一般采用地面钻井开采和井下瓦斯抽放系统抽出两种方式。

煤层气在我国已经开采了 20 多年，相比页岩气的开发技术更加成熟，我国的煤层气产业已积累了一定的政策和技术基础，但是由于我国煤层气资源具有"低压、低渗、低饱和"及地质构造复杂的特点，构造煤、超低渗、深部煤层气资源等难采资源量占我国煤层气资源总量的 70%~80%，造成单井产气量低，成本效益倒挂，使大部分煤层气开采企业的经济效益并不显著。另外，受我国天然气管网建设落后的影响，目前煤层气的"商品气"只占到不足开采量的一半。2012 年我国煤层气产量达到 125 亿立方米，其中地面开发煤层气产量为 25.7 亿立方米，煤矿瓦斯产量为 99.4 亿立方米，利用总量为 52 亿立方米，总利用率仅为 41.53%，煤矿瓦斯抽采利用率只有 32%。按照我国煤层气"十二五"发展规划，2015 年我国煤层气产量目标为 300 亿立方米，但受煤层气管道建设滞后、供应范围相对局限、矿权重叠等因素的制约，当前，煤层气的发展面临一系列难题[13]。

综上所述，非常规天然气的发展为我国开展气基竖炉直接还原工艺提供了新的可能和契机，但页岩气和煤层气的开发都面临一定的困难，短期内无法改变天然气供应短缺的局面。在我国天然气资源相对丰富的局部地区，气基竖炉直接还原技术可能会有一定的发展空间。

4.2 人工煤气

我国油气资源短缺，但煤炭资源丰富，这为开展煤制气—直接还原铁生产奠定了能源基础。此外，气基直接还原技术的发展，扩大了还原气体的适用范围，使得以煤制气、焦炉煤气等人工煤气生产直接还原铁成为可能。

4.2.1 煤制气

以动力煤为原料制取竖炉用还原气属于成熟技术。目前世界上主要的煤气化工艺有鲁奇法（Lurggi）、恩德法（Ende）、德士古法（Texaco）和壳牌法（Shell），以上工艺均可用于制取竖炉用还原气，但不同的煤制气工艺其原料要求、还原气成分、运行效果、成本和投资等具有较大差别[14]，表4对不同种类的煤气化工艺进行了比较。鲁奇法投资较低，热效率高，作业率高，设备无需备用，还原气 H_2 与 CO 体积分数比和 CH_4 含量较高，适宜作为竖炉还原气合成气使用，但需要对煤气进行煤气净化处理，净化系统复杂，需要使用块煤，对原料条件要求较高。恩德法为粉煤沸腾床高温气化炉工艺，对煤的适用性较高，生产效率较高，建设投资相对较低，但煤气中 $CO+H_2$ 的含量较低，碳转化率偏低，煤气净化处理难度较大，主要适用于化肥行业。德士古法是目前业绩最多的第二代气流床煤气化工艺，已在我国应用近 20 年，因采用水煤浆高压泵送，运行压力高，技术成熟，国产化率高，但要求对煤气进行脱碳处理和煤气转换，通常需要备用装置。壳牌法无需对煤粉制浆，煤制气还原势高，有效成分含量高，无需脱碳处理，近年来在技术和操作上取得了较大改进，但需要进行煤气转换，主体设备和技术需要引进，废热锅炉投资较高。上述煤制气工艺中，鲁奇法煤制气工艺比较适宜制备竖炉还原气，但其技术先进性相对不足；壳牌法技术先进，发展形势较好，但其设备和技术需要引进，投资成本过高。

表4 煤制气工艺比较

项 目	鲁奇法	恩德法	德士古法	壳牌法
气化炉	固定床	沸腾床	气流床	气流床
煤种	褐煤、烟煤	褐煤等	烟煤、无烟煤	全部
煤粉粒度	5~75mm	<10mm	75%>75μm	90%<100μm
进料方式	块煤	粉煤	60%水煤浆	干煤粉
气化压力/MPa	2.0~3.0	2.5~3.0	2.6~8.4	2.0~4.0
气化剂	氧气+水蒸气	氧气+水蒸气	氧气	纯氧+蒸气
氧耗/$m^3 \cdot (1000 \cdot m^3)^{-1}$	140~240	270	380~430	330~360
煤耗/$kg \cdot (1000 \cdot m^3)^{-1}$	720	678	640	600
气化温度/℃	900~1050	950	1400~1600	1300~1400
碳转化率/%	96~98	90~95	95	99.5

续表4

项 目	鲁奇法	恩德法	德士古法	壳牌法
煤气中 CO+H_2 含量（体积分数）/%	约60	39~59	约80	约90
煤气中 CH_4 含量（体积分数）/%	9~12	<1	0.01	0.01
煤气中 H_2/CO（体积分数）	1.7~2.3	0.9	0.4~0.5	0.4~0.5
净化煤气温度/℃	250~350	300	200~260	350
冷煤气效率/%	80	72~76	70~76	80~85
单炉气化能力/t·d^{-1}	1800	—	2000	3000
关键技术问题	碳氢化合物的利用	碳的转化	粗煤气冷却	粗煤气冷却
50万吨/年直接还原厂配套煤制气装置投资/亿元	3.0~3.5	2.0~2.3	8~10	10~12

目前，制约煤制气—竖炉直接还原技术发展的关键因素是加压煤制气技术投资和运行成本过高的问题，高额的技术引进费用和投资成本限制了其市场发展。随着煤制气工艺的技术进步和市场竞争，这一现状有望未来将被打破。印度 Jindal 公司采用鲁奇法＋Energiron-ZR 工艺，建设世界首套煤制气竖炉直接还原装置，预计将2014年投产运行，其成功应用将对煤制气竖炉直接还原技术的发展起到积极的示范和推动作用。我国内蒙古、江苏、贵州等地区也正在开展煤制气—直接还原铁生产的研究论证工作。

4.2.2 焦炉煤气

焦炉煤气（COG）是炼焦过程的副产品，与天然气相比，其 H_2 和 CO 含量高且 CH_4 含量低，更适宜作为竖炉用还原气生产 DRI。长期以来，焦炉煤气的高效利用未受到足够的重视，大量焦炉煤气尤其是独立焦化厂产生的焦炉煤气无法全部利用而被大量放散，钢铁联合企业的焦炉煤气尽管利用率较高，但主要用作加热炉燃料，其化学能未被充分利用[15]。2012年我国焦炭产量约为4.43亿吨，其中钢铁企业自产焦炭为1.3亿吨，自产焦炉煤气546亿立方米，独立焦化厂过剩焦炉煤气量约为200亿~260亿立方米。以低品质煤制气或高炉煤气置换出焦炉煤气，以焦炉煤气作为还原气，利用直接还原竖炉生产 DRI，也将是我国钢铁企业二次能源高效利用、节能减排和实现可持续发展的有效途径之一，同时这也是气基竖炉直接还原技术未来发展的方向之一。

Midrex 工艺和 HYL 工艺目前都已开发出了使用焦炉煤气的气基竖炉流程[16]。Midrex 工艺首先将净化预热后的焦炉煤气在热反应器内进行重整处理（见图3），同时在还原气体回路中增加 CO_2 脱除装置，可用于生产冷态或热态 DRI。HYL-ZR 工艺无需对还原气体回路进行改造，但需要将焦炉煤气首先通入冷却回路（见图2），因而只能生产冷态 DRI。研究表明，对焦炉煤气利用的难点在于，焦炉煤气中含有少量的 BTX（苯、甲苯、二甲苯混合物）、焦油和萘等杂质，而 BTX 在还原气体通过加热器时会产生析碳，造成管道堵塞。此外，对于 HYL-ZR 工艺，竖炉内的工作压力约为0.6MPa，而焦化厂输出的煤气压力约为5kPa，因此需要对焦炉煤气进行加压，煤气中焦油和萘在增压过程中会大量析出，也会堵塞设备和管道。如何消除 BTX 等杂质对还原工艺的危害，是以焦炉煤气为燃料的气基竖炉直接还原工艺需要突破的技术关键[17]。

综上所述，我国煤炭资源丰富，煤制气—竖炉直接还原技术将是未来我国直接技术发展的重要方向，进一步降低煤制气工艺的投资和运行成本是该技术发展的关键。非常规天

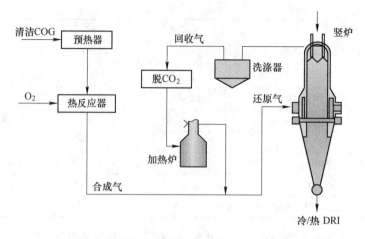

图 3 采用焦炉煤气的 Midrex 生产工艺流程

然气作为天然气资源的补充未来将具有较好的发展前景,但受资源条件和开采技术的制约,短期内难以取得较大进展。在我国天然气资源相对丰富的局部地区,天然气气基竖炉直接还原技术或将存有一定的发展空间。此外,利用钢铁厂富余的焦炉煤气或焦炉煤气和天然气的混合气生产 DRI,为气基直接还原技术的发展提供了更多选择,同时也是钢铁企业实现二次能源高效转化和利用的有效途径,有望未来实现工业化应用。

5 生产运行成本

为了研究分析气基竖炉直接还原工艺的技术可行性,以生产规模为 80 万吨/年的直接还原工厂为模型,参照 2013 年上半年河北唐山地区的资源和能源平均价格,对其生产运行成本进行了分析测算,测算结果见表 5。技术经济分析结果表明,原燃料价格波动对 DRI 成本影响显著,其中还原气成本约占 DRI 生产成本的 10%~25%。使用焦炉煤气生产 DRI,其成本比使用天然气降低约 244.09 元/吨。

表 5 年产 80 万吨/年直接还原铁工厂生产成本测算

项 目	单价/元	单耗 NG（COG）	天然气生产成本 /元·吨$^{-1}$（DRI）	焦炉煤气生产成本 /元·吨$^{-1}$（DRI）
原辅材料			1538.51	1538.51
球团矿/t·t^{-1}	1100	1.39	1529.00	1529.00
MDEA 溶液/kg·t^{-1}	36.96	0.04	1.48	1.48
涂层水泥/kg·t^{-1}	0.62	4.00	2.48	2.48
其他化学品和消耗品/kg·t^{-1}	55.44	0.10	5.54	5.54
燃料动力			591.24	349.59
水/m^3·t^{-1}	7.60	1.30（1.90）	9.88	14.44
电/kW·h·t^{-1}	0.38	90.0（148.3）	34.20	56.35

续表5

项 目	单价/元	单耗 NG（COG）	天然气生产成本 /元·吨$^{-1}$（DRI）	焦炉煤气生产成本 /元·吨$^{-1}$（DRI）
天然气（焦炉煤气）/m^3·t^{-1}	2.0（0.46）	264.44（580.48）	528.89	267.02
氮气/m^3·t^{-1}	0.15	20.00（25.00）	3.00	3.75
氧气/m^3·t^{-1}	0.29	50（25）	14.50	7.25
压缩空气/m^3·t^{-1}	0.0769	10.00	0.77	0.77
工资及附加			8.1	8.1
修理费			24.00	24.00
折旧			76.00	76.00
其他费用			21.84	19.40
经营成本			2183.68	1939.59
总成本			2259.68	2015.59

表6列出了使用不同还原气生产DRI的测算成本，若按DRI替代转炉废钢计算效益，2013年11月河北地区重型废钢不含税价约为2200元/t，则使用焦炉煤气生产DRI的效益为84.41元/t；使用天然气生产DRI，当天然气价格为1.8元/m^{3}时，DRI成本与废钢价格持平，若天然气价格进一步升高则DRI成本将高于废钢。按照鲁奇法或壳牌法煤制气工艺的制气成本测算，目前采用煤制气生产DRI的成本要高于废钢。

表6 采用不同还原气生产DRI成本测算

还原气种类	焦炉煤气	天然气				煤制气
单价/元·m^{-3}	0.46	1.80	2.00	2.50	3.00	1.6
总成本/元·t^{-1}	2015.59	2206.26	2259.68	2393.24	2526.80	2377.61

6 结论

（1）气基竖炉直接还原工艺是当前世界上具有主导地位的非高炉炼铁工艺，近年来在装备大型化、生产高效化、燃料多样化等方面取得长足技术进步，技术发展成效显著。预计在未来一定时期内，仍将成为引领直接还原工艺发展的主流技术。

（2）气基竖炉直接还原工艺由于其工艺特征，对含铁原料的要求较为苛刻，我国的气基竖炉直接还原装置的建设应结合资源条件进行充分的研究论证。页岩气、煤层气等非常规天然气的开采和应用，为气基竖炉直接还原工艺提供了更为广阔的发展空间，但由于资源条件、开采技术和开采成本等因素的制约，我国短期内仍难以改变天然气资源短缺的局面，利用天然气作为还原气生产DRI仍将不会取得显著进展。

（3）依托我国丰富的煤炭资源，开发基于煤制气工艺的气基竖炉直接还原技术，将是我国直接还原技术发展的主要方向。开发新型加压煤制气技术，有效降低煤制气的工程投资和运行成本是煤制气—直接还原技术发展的关键。对于我国众多独立焦化厂为了实现焦炉煤气的高效化综合利用，可以探索利用焦炉煤气作为还原气生产DRI；对于钢铁联合企业，利用钢铁厂富裕的低品质煤气置换出富含 H_2 的焦炉煤气，将其作为还原气生产 DRI，是实现钢铁企业二次能源高效转换的有效途径之一，有望未来实现工业化应用。

（4）结合唐山地区的资源和能源条件，以年产 80 万吨/年直接还原厂为模型进行的技术经济可行性分析结果表明，利用焦炉煤气生产 DRI 具有一定的经济效益；如果使用天然气生产 DRI，则要求天然气价格低于 1.8 元/m^3；目前使用煤制气生产 DRI 的成本要高于废钢价格。

参考文献

[1] Pablo Duarte, Thomas Scarnati. The JSPL DR project: A practical example of the flexibility of energiron DR technology for using any energy source without requiring scheme change [C] // 7th China International Steel Congress Proceedings. Beijing, 2012.

[2] World DRI Statistics [EB/OL]. [2013-9-26]. http://www.midrex.com.

[3] 王兆才, 陈双印, 储满生, 等. 煤制气-竖炉生产直接还原铁浅析 [J]. 中国冶金, 2013, 23 (1): 20.

[4] Martinis A, Patrizio D, Volptti A. Energiron direct reduction technology integration with electric steel making technology and blast furnace technology [C] // 7th China International Steel Congress Proceedings. Beijing, 2012.

[5] 曹朝真, 张福明, 毛庆武, 等. 焦炉煤气二氧化碳重整热力学规律研究 [C] //2011 年全国冶金节能减排与低碳技术发展研讨会论文集. 唐山, 2011.

[6] 齐渊洪, 钱晖, 周渝生, 等. 中国直接还原铁技术发展的现状及方向 [J]. 中国冶金, 2013, 23 (1): 9.

[7] 黄希祜. 钢铁冶金原理 [M]. 北京: 冶金工业出版社, 2002.

[8] George Tsvik. Impact of H_2/CO ratio on syngas-based direct reduction shaft furnace [C] // 6th Europe coking and Ironmaking Conference Proceedings (ECIC). Dusseldorf, 2011.

[9] 项钟庸, 王筱留, 等. 高炉设计——炼铁工艺设计理论与实践 [M]. 北京: 冶金工业出版社, 2009.

[10] 姜福杰, 庞雄奇, 欧阳学成, 等. 世界页岩气研究概况及中国页岩气资源潜力分析 [J]. 地质前缘, 2012, 19 (2): 198.

[11] 张金川, 姜生玲, 唐玄, 等. 我国页岩气富集类型及资源特点 [J]. 天然气工业, 2009, 29 (12): 109.

[12] 李景明, 巢海燕, 李小军, 等. 中国煤层气资源特点及开发对策 [J]. 天然气工业, 2009, 29 (4): 9.

[13] 邹才能, 董大忠, 王社教, 等. 中国页岩气形成机理、地质特征及资源潜力 [J]. 天然气资源勘探与开发, 2010, 29 (12): 641.

[14] 周渝生, 钱晖, 齐渊洪, 等. 煤制气生产直接还原铁的联合工艺方案 [J]. 钢铁, 2012, 47 (11): 27.

[15] 王太炎. 焦炉煤气开发利用的问题与途径 [J]. 燃料与化工, 2004, 35 (6): 1.

[16] Peter Diemer, Hans Bodo Lungen, Martin Reinke. Utilization of coke oven gas for the production of DRI [C] // 6th Europe coking and Ironmaking Conference Proceedings (ECIC). Dusseldorf, 2011.

[17] 曹朝真, 张福明, 毛庆武, 等. 利用焦炉煤气生产直接还原铁关键技术研究 [C] // 2011 年全国冶金节能减排与低碳技术发展研讨会论文集. 唐山, 2011.

Current Status and Prospects of Gas-based Shaft Furnace Direct Reduction Technology

Zhang Fuming Cao Chaozhen Xu Hui

Abstract: The technical characteristics and current status of gas-based shaft furnace direct reduction process have been expounded and the major problems faced by gas-based direct reduction technology under China's resource and energy condition have been analyzed. Based on the characteristics of the gas-based shaft furnace direct reduction process, analysis of raw materials and reduction gases of this technology have been undertaken. It has been pointed out that a significant reduction in investment and operating costs of pressurized coal gasification technology and the effective exploitation of unconventional natural gas resource is the main driving force of the future development of gas-based shaft furnace direct reduction technology; meanwhile it is still another important direction of current gas-based shaft furnace development to produce DRI by taking advantage of the excess gas resources of steel plants and natural gas resource in some areas where the natural gas resource is relatively abundant. According to the raw material and energy prices in Tangshan region, the production costs of DRI has been calculated and the technical and economic feasibility has been analyzed under the scale of 0.8Mt/a. The results show that the effect of price fluctuations of raw materials on the DRI costs is significant, and the reducing gas accounts for about 10%-25% of DRI total costs. If considering the economic benefits of using DRI instead of scrap in converter, the natural gas price should be lower than 1.8 yuan RMB/m^3.

Keywords: iron-making; direct reduction; shaft furnace; pellet; coal gasification

我国首座 HIsmelt 工业装置的设计优化与技术进展

摘　要：简要阐述了澳大利亚 Kwinana 厂 HIsmelt 工业装置 3 年多的生产实践，指出了生产过程中暴露出的一系列问题，认为 HIsmelt 工艺技术是可行的，但需要系统优化。结合我国首座 HIsmelt 工业装置的设计和建设实践，从炉缸耐材、矿粉预热、煤气净化、矿煤喷吹等方面，提出了 HIsmelt 工艺优化的主要技术路径。从 SRV 大型化、全氧冶炼、钒钛矿冶炼等方面，简要阐述了 HIsmelt 技术的最新进展。
关键词：HIsmelt 工艺；熔融还原炉（SRV）；设计优化

1　引言

绿色清洁炼铁是钢铁行业实现可持续发展的必然选择。熔融还原炼铁工艺与高炉工艺相比，在资源适用性、低碳绿色等方面具有显著优势，是低碳炼铁新技术发展的一个重要方向。而 HIsmelt 工艺是一种经过 30 多年的不断发展，并经过工业化生产验证，可以完全摆脱焦炭的熔融还原炼铁技术。

2012 年，山东墨龙公司与力拓公司签订协议，将澳大利亚 Kwinana 厂的 HIsmelt 工业装置搬迁至我国。目前，该项目已经完成设计和建设工作，正在进行热试车调试工作。本文在总结 Kwinana 厂生产实践的基础上，重点阐述墨龙 HIsmelt 工业装置设计优化的特点。

2　Kwinana 厂的生产实践

Kwinana 厂于 2003 年 1 月开始建设，2004 年 9 月进行设备调试，2005 年 4 月进行系统热试车，2005 年 9 月开始正式试生产，试生产一直延续到 2008 年 12 月。生产过程中，最大小时产量达到 80t，日最高产量达到 1834t，周最高作业率达到 99%，最长连续操作时间达到 68d，实际设备利用率达到设计能力的 75%～80%。随着生产效率的提高，煤比逐渐降低，日最低煤比达到 810kg/t 铁水，理论最低煤比有望接近 700kg/t 铁水（干基）。Kwinana 厂 3 年多的生产实践表明，HIsmelt 工艺技术是可行的。

但是，作为一种新的炼铁工艺，HIsmelt 工艺没有经过长期的生产实践验证，还存在不少问题。由于 Kwinana 厂的开发理念是尽量采用成熟的技术组合，核心开发内容集中在熔融还原炉（SRV）及其关键设备上，而对生产过程中暴露出一系列的问题估计不足[1~3]，主要表现在：

（1）作业率不高，没有达到设计要求。外围设备问题较多，如预热器系统能力长期达不到要求，热矿输送系统双螺旋给料器磨损严重、耐材脱落、链斗输送机故障等。矿煤

本文作者：曹朝真，张福明，毛庆武，李欣。原刊于《炼铁》2016，35（5）：59-62。

喷枪多次出现破损、磨漏现象,需要频繁更换。还原煤气热值低造成火焰不稳定,给余热锅炉蒸气压力和温度的控制带来问题。

(2) 辅助系统设计能力不合理。五大辅助系统中,预热器系统和热风系统的运行能力只达到 80%~90%,而煤粉系统、烟气脱硫系统和余热回收系统则达到 110%~120%。

(3) 炉缸耐火材料侵蚀严重。由于 HIsmelt 采用较高的二次燃烧率,炉渣中 FeO 含量较高,使得熔融还原炉的炉缸耐火材料尤其是渣线部位,受熔渣侵蚀、磨损相当严重,最初的炉缸耐材设计寿命是 18 个月,但第一代炉役只维持了不到 3 个月,3 年间曾 4 次更换炉缸内衬。

(4) 操作经验不足。第一次热装时铁水在前置炉缸出现冻结,重新打通前炉和熔池非常困难;因熔池渣温不足和加矿过量,造成碳平衡失控,出现了泡沫渣,炉内的铁大部分转化成了钢,在水冷壁上结成钢壳,对供水管路产生巨大应力,导致大量漏水;此外,由于缺乏操作经验,对炉缸关键区域的耐材状态和炉壳温度重视不足,造成炉缸内衬过度侵蚀,出现跑铁事故。

3 墨龙 HIsmelt 工业装置的设计优化

结合 Kwinana 厂生产过程中出现的问题,在墨龙 HIsmelt 工业装置的设计和建设过程中,围绕炉缸耐材结构优化、矿粉预热、煤气净化及余热余能利用和固体料喷枪改造等几个方面,对 HIsmelt 工艺进行了系统优化。

3.1 炉缸耐材结构

SRV 是 HIsmelt 工艺的核心反应器,是整个工厂的核心部分。SRV 的耐材砌筑属于 HIsmelt 工艺的核心技术,直接影响到整个工艺过程的运行效果。由于 HIsmelt 工艺采用高二次燃烧率操作,使得炉渣中的 FeO 含量高达 4%~5%,给炉缸耐材带来了严重的侵蚀,尤其是渣线部位的侵蚀更为明显。Kwinana 厂最初的炉缸耐材砌筑方案如图 1(a)所示,炉缸侧壁的耐材内表面与炉底垂直,顶部耐材断面为三角形,希望通过形成挂渣层使炉衬得到保护,从而延长耐材的使用寿命。使用中发现由于液态渣铁在高速气流作用下剧烈波动,渣铁界面难以保持稳定,渣线波动范围内耐材侵蚀严重;一旦侧壁渣线部位的耐材被侵蚀形成凹陷,上部的耐材就会失去足够的支撑,在液态渣铁的冲刷和浮力作用下,炉缸侧壁的顶部耐材很快就会剥落,从而造成炉缸耐材寿命过短,甚至造成烧穿事故。为了克服上述问题,2008 年 10 月,Kwinana 厂在炉缸渣线部位沿圆周方向安装了渣区铜冷却壁,如图 1(b)所示,取代了渣线部位的耐材,冷却壁直接与渣液接触,通过强制冷却实现渣皮保护;同时,铜冷却壁还可以从上部压住炉缸侧壁的耐材,从而缓解由于铁水浮力对耐材造成的不利影响。在使用渣区铜冷却壁 2 个多月的时间里,炉缸耐材情况良好,根据耐材的侵蚀速度测算,耐材的使用寿命有望达到 2 年以上。

在墨龙 HImelt 工业装置的建设过程中,对炉缸耐材的方案设计给予了高度重视,并提出了一系列的优化方案。其中主要的优化措施之一,就是进一步改善炉缸炉底工作层的耐材质量,使用高铬质耐火砖取代原有的镁铝质耐火砖,从而进一步提升工作层耐材抵抗渣铁侵蚀的能力;另一方面,从优化炉缸砌筑结构入手,结合数值模拟和炉缸传热模型计

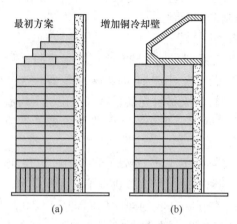

图1 Kwinana厂SRV炉缸侧壁耐材结构及渣线铜冷却壁安装示意
（a）最初方案；（b）增加铜冷却壁

算结果，充分考虑耐材热膨胀等因素的影响，合理配置耐材的砌筑方案。通过上述改进工作，有望进一步提高炉缸耐材的炉役寿命，从而显著提高SRV的作业率。

3.2 矿粉预热

矿粉预热系统的作用是将铁矿粉（<6mm）在装入熔融还原炉之前进行预热，可以为下一步在SRV炉中的熔融还原提供部分能量，同时也是提高熔融还原炉反应效率和增加HIsmelt工艺产量的有效措施之一。Kwinana厂的矿粉预热系统采用循环流化床工艺，矿粉在循环流化床反应器中完成预热和煅烧，除部分铁矿粉被还原外，最高可将铁矿粉预热至850℃。矿粉预热系统工艺流程如图2所示。

Kwinana厂的矿粉预热系统在运行过程中，先后出现螺旋给料机、链斗输送机故障、耐材脱落、排料口堵等问题，给工厂的正常生产造成了很大影响。为此，在新流程的设计中，结合国内的生产实践，从选择成熟可靠的工艺技术出发，将矿粉预热系统进行了改造，如图3所示。

改造后的矿粉预热系统采用两级回转窑预热工艺，首先将矿石导入烘干回转窑，在烘干回转窑内将铁矿粉干燥，烘干回转窑出料温度80~100℃，排烟温度80~150℃；经干燥后的矿粉再导入预还原回转窑，在预还原窑内使矿粉部分还原，回转窑烟气出口温度为600~650℃，运行过程中回转窑采用50%煤+50%SRV煤气加热，同时进一步加热铁矿粉达到750℃以上，满足物料的还原度达到15%以上；矿粉经回转窑出口排出，进入热矿斗式提升机，通过斗式提升机将矿粉送入热矿喷吹系统的热矿料斗。回转窑在我国被广泛应用于赤铁矿的磁化还原焙烧选矿技术，具有工艺简单、技术成熟等特点。

3.3 高温低热值煤气利用

在冶炼过程中，煤粉经喷枪被高速地喷入熔池，并与铁水接触，煤中的挥发分被裂解形成CO和氢气，碳素熔于铁液并被炉料中的铁氧化物氧化形成CO，熔池中快速释放的CO、H_2和用作载气的N_2，在SRV上部与热风中的富氧（35%）二次燃烧生成煤气，煤

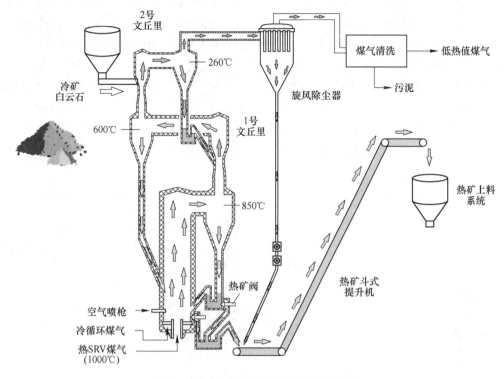

图 2 Kwinana 厂矿粉预热系统工艺流程

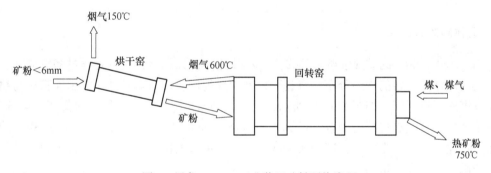

图 3 墨龙 HIsmelt 工业装置矿粉预热流程

气温度一般为 1450~1650℃，压力为 0.08~0.1MPa，热值为 1800~3000kJ/Nm³。Kwinana 厂对 SRV 高温煤气的净化采用湿法除尘工艺，工艺过程为高温煤气首先进入水冷烟罩，温度降至 800~1000℃，然后通过煤气环缝洗涤塔对荒煤气进行除尘和降温处理，煤气温度降至 100℃以下，煤气含尘量降至 10mg/Nm³ 以下，再经煤气冷却塔进一步降温后，使煤气质量达到后续工序的使用要求。以上除尘工艺可以满足对煤气的除尘要求，但对于 1000℃以下的煤气物理热无法进行回收利用，而且需要增加污水处理系统；此外，由于 SRV 煤气热值低，在供给燃气锅炉燃烧时会造成火焰不稳定，需要与高热值煤气混合使用。

为实现对 SRV 炉煤气的净化和余热余能的高效利用，针对煤气处理系统提出了一套全新的工艺方案（见图4）。首先将来自SRV炉的1450~1650℃的煤气通过水冷烟罩降温至800~1000℃，然后进入旋风除尘器进行粗除尘，煤气含尘量由20~100g/Nm³降至8~20g/Nm³；再通过空气-煤气换热器，将煤气温度降至150~200℃，同时将空气加热至350~450℃；净化后的煤气一部分与高热值煤气混合后通过煤气支管进入燃气锅炉燃烧用于发电，另一部分通过煤气支管进入热风炉作为燃料，用于给SRV生产热风，来自热风炉和燃气锅炉的烟气经烟囱一起排放。该工艺通过水冷烟罩和空气-煤气换热器回收煤气的物理热，同时通过燃气锅炉和热风炉对煤气的化学热进行回收，其中热风炉以直接向SRV供给热量的形式回收煤气余热，有利于提高能量的利用效率，减少燃料消耗量。

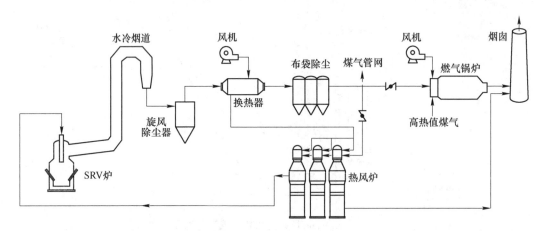

图4 墨龙HIsmelt工业装置煤气净化和能量回收工艺流程

3.4 固体料喷枪改造

Kwinana厂SRV的矿煤喷吹采用多点、小喷枪的设计思路，目的是通过多点喷吹，实现炉缸内渣铁和物料的均匀分布，有利于提高炉缸的反应动力学条件。共由4只矿粉喷枪和4只煤粉喷枪组成，如图5（a）所示，位于炉缸耐材上部，沿圆周方向交替均匀分布。实际生产中，多次出现喷枪破损、磨漏现象，需要频繁更换喷枪，严重影响了SRV的正常运行；此外，生产中发现，减少喷枪数量、增大单只喷枪的喷吹能力同样可以达到活跃炉缸和满足矿煤喷吹量的要求。

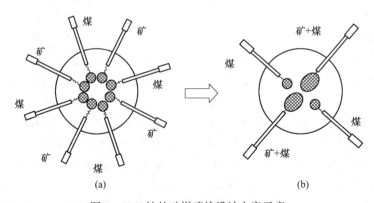

图5 SRV炉的矿煤喷枪设计方案示意

在新工艺的设计过程，采用了简化的喷枪设计方案，将矿煤喷枪由原来的8支减少为4只，如图5(b)所示，同时可以实现矿煤混喷，简化了喷吹系统的工艺流程，减少了潜在的设备故障点，有利于进一步提高SRV的作业率。

4 HIsmelt技术的最新进展

4.1 HIsmelt大型化开发

为了进一步推进HIsmelt技术的发展，开展了HIsmelt工业装置大型化的开发工作，完成了生产能力为200万吨/年，SRV炉缸内径为8m的工艺设计方案。在新方案的设计中，保留了原有前置炉缸、水冷盘管冷却壁和主体耐材砌筑结构的设计方案；改进了热风喷枪设计，提高了喷吹效率，简化了喷枪结构，提高了使用寿命；为了便于加工、制造和运输，采用了双水冷烟罩和煤气处理系统；同时，采用巨型喷枪和矿煤混喷的方式，进一步简化固体料喷吹流程。

4.2 HIsarna熔融还原技术开发

HIsarna工艺融合了力拓的HIsmelt技术和旋风炉（CCF）技术，被认为是HIsmelt技术未来的重要发展方向。HIsarna工艺继承了HIsmelt的技术思想，可以彻底摆脱焦炭，直接使用矿粉和煤粉冶炼。该技术的一项重要突破是实现了全氧冶炼，煤气在炉内基本被全部燃烧，可以使用高挥发分的低品质煤，还可以用于处理钒钛磁铁矿和高磷矿，结合CCS技术（CO_2捕获和封存技术），可以显著降低CO_2排放。

2008年，HIsarna工艺作为唯一一项熔融还原炼铁技术入选欧洲ULCOS（超低CO_2炼钢）项目。从2010年开始，在荷兰的艾默伊登厂开始筹建试验工厂，2011年，在荷兰的艾默伊登厂建成了年产6万吨/年，炉缸直径2.5m的HIsarna试验装置，目前已经完成了四次试验，通过试验验证了赤铁矿和中低品质煤的冶炼效果，达到了8t/h铁水的设计产能，煤比接近750kg/t铁水，二次燃烧率大于90%，铁水温度控制在1450℃，碳含量在4%~4.5%，工艺的技术可行性得到了检验，取得了预期效果[4]。2016年，欧盟决定提供740万欧元资金支持，计划进行为期半年的第五次试验，主要用于确定该项技术能否长期稳定生产铁水。如果项目试验成功，后期还将建造工业化规模生产厂，据估算成本在3亿欧元左右。未来HIsarna装置试验的目的将主要侧重于收集设计参数和进行规模放大。

4.3 HIsmelt工艺处理钒钛磁铁矿技术

钒钛磁铁矿是一种极具综合利用价值的复合矿石，我国在高炉冶炼钒钛磁铁矿方面积累了丰富的经验，但对其资源化的综合利用目前仍然面临较多难题。一方面，原料中的二氧化钛含量会降低烧结矿质量，给高炉冶炼带来不利影响；另一方面，由于高炉内的强还原性气氛，冶炼过程中会生成钛的低价化合物和TiC（TiN），它们是胶体态高度弥散的固相物，与熔渣有很好的润湿性，会导致泡沫渣严重、炉渣黏稠、铁水粘罐、渣铁挂渣多、渣铁分离不理想等一系列问题，而高炉的软熔带要求炉渣黏度不能太高。目前，高炉工艺中钒的回收率约为50%，无法对钛进行回收[5]。

HIsmelt 工艺在处理钒钛矿方面与高炉相比具有显著优势,其独特的强氧化性气氛,使渣中的氧势较高,足以有效抑制 TiO_2 还原和高熔点 TiC (TiN) 生成,同时不影响钒的回收和在高温下还原钛铁矿。铁矿石在熔融状态下发生直接还原,反应速度快,炉渣黏度较低,渣内不会残留大量气体,所以炉内不会产生严重的泡沫渣[6]。而且,冶炼钒钛矿还有利于 SRV 炉水冷壁的挂渣和形成冷凝渣皮保护耐材。目前,力拓公司正在开展利用 HIsarna 试验装置进行钒钛矿冶炼的前期准备工作。

致谢

力拓公司 Neil Goodman 先生、Jacques Pilote 先生以及首钢国际工程公司炼铁所姚轼、章启夫、梅丛华、孟玉杰、王维乔、刘永、李林、孟祥龙等同事,参加了本项目的技术研发和设计研究工作,作者在此一并表示感谢。

参考文献

[1] Goodman N, Dry R. HIsmelt 炼铁工艺 [J]. 世界钢铁, 2010 (2): 1-5.

[2] Bates P, HIsmelt A C. The future in ironmaking technology [C] // French Iron and Steel Technical Association. 4th European Coke &Ironmaking Conference. Paris: French Iron and Steel Technical Association, 2000: 597-602.

[3] Meng X L, Li L, Zhang F M, et al. Application on HIsmelt smelting reduction process in China [C] // Association for Iron and Steel Technology. AISTech2015 Proceedings. Cleveland: Association for iron and steel technology, 2015: 1135-1145.

[4] Peeters T, et al. Development of HIsarna iron making technology for low CO_2 steelmaking [C] // Steel Institute VDEh. Düsseldorf: 4th VDEh-CSM-Seninar on Metallurgical Fundamentals, 2014: 66-72.

[5] 杜鹤桂. 高炉冶炼钒钛磁铁矿原理 [M]. 北京: 科学出版社, 1996: 25-30.

[6] 邝亚丽, 王华, 卿山. 钛铁矿和高磷铁矿混合矿氧气顶吹熔融还原炼铁的工艺条件 [J]. 过程工程学报, 2011, 11 (6): 1024-1029.

转底炉直接还原技术的进展

摘　要：转底炉直接还原工艺可以利用各种铁矿和含铁原料以及钢铁厂粉尘生产直接还原铁，是一种具有技术前景的煤基直接还原工艺。本文介绍了国内外转底炉直接还原工艺的发展历程，对 Inmetco、DRYIron、Fastmet、Itmk3 等多种转底炉直接还原工艺进行了综合评述，分析了其技术原理和工艺特点。对钢铁厂含铁、含锌固体粉尘的资源化综合处理提出了解决方案，阐述了转底炉直接还原技术在我国应用的技术前景。

关键词：炼铁；转底炉；直接还原；熔融还原

1　引言

转底炉（Rotary Hearth Furnace）作为直接还原装置生产直接还原铁（DRI）始于20世纪60年代。转底炉最初是用于轧钢的环形加热炉，经过多年的技术发展和创新，人们利用转底炉处理钢铁生产过程中产生的粉尘和固体废弃物，目前已成为煤基直接还原技术的主要工艺流程之一。由于采用了压块技术、能源利用和环境保护方面的最新技术以及合理的转底炉设计，这种工艺还克服了传统的煤基直接还原过程的粉化、黏结、金属化率低及硫含量高等缺陷。工艺过程的反应基于煤基直接还原动力学，炉料在转底炉内反应迅速、还原周期短，操作控制灵活简便，特别适用于钢铁厂含铁、含锌等粉尘和固体废弃物的资源化回收利用，对于实现循环经济和钢铁绿色制造具有重要的作用和显著的技术优势。

2　转底炉直接还原工艺的发展历程

2.1　Heat Fast 工艺

将转底炉用于直接还原的最初设想是在20世纪60年代由美国的 Midland 公司（现在 Midrex 公司的前身）提出的。1965~1966年，在美国明尼苏达州的 Cooley 建成了第一座直径为 6.1m 的直接还原转底炉，通过中试获得成功，该转底炉工艺被称为 Heat Fast 工艺。该工艺是将铁矿粉与煤粉混合、造球，在1台链箅机上进行生球干燥，然后在转底炉上进行还原，产品在竖窑中进行冷却。

由于 Midland 公司同期正在进行气基直接还原工艺的开发（即现今的 Midrex 工艺），而且成效显著，因此对于 Heat Fast 工艺的工业化规模开发研究未能继续进行。

本文作者：张福明。原刊于《华西冶金论坛第26届（厦门）会议——全国能源与工业炉热工学术研讨会论文集》，2011：357-363。

2.2 Inmetco 工艺

加拿大的国际镍集团（INCO. Ltd）于 1974 年开始对处理不锈钢厂的废料进行研究，并在安大略科尔伯恩港（Port Colborne）建成 1 座直径为 6.8m 的转底炉进行半工业试验，开发了一种利用转底炉处理工业废料生产直接还原铁的工艺，将煤和工业废料混合，进行造球，用转底炉进行还原，然后用料罐将还原后的金属化球团热装入电炉进行熔炼。其后在美国成立 INCO 公司，并在该公司下成立国际金属回收公司（INMETCO）。1978 年 INMETCO 公司为处理利用冶金废弃物，在美国 Ellwood 市建成了世界上第 1 座具有生产规模直径为 16.7m 的转底炉，用于回收合金钢冶炼废料中的镍、铬和铁。此装置用合金钢厂回收的废料生产直接还原铁，然后热装入电炉生产合金铁水。在此装置上采用铁精矿粉生产直接还原铁也取得很好的效果，被命名为 Inmetco 工艺。1983 年底，德国 SMS-Demag 公司获得该工艺的经营权。

2.3 DRYIron 工艺

DRYIron 工艺是由美国 MR&E（Maumee Research and Engineering）公司的前身 Midland Ross 公司在 20 世纪 60 年代开发的 Heat Fast 工艺演变发展而来的。MR&E 公司曾为美国 Ameri Steel 公司的电炉除尘灰处理系统以及美国 Rouge Steel 公司的钢铁厂除尘灰和污泥处理系统提供 DRYIron 的相关设备。DRYIron 工艺基于两项专利技术：利用氧化铁粉与炭粉混合物成型的干式压块技术和采用特殊振动传送装置的装料技术。该工艺采用高压成型机压块、能源利用和环境保护的新技术，对 Heat Fast 工艺进行了改进和创新，在降低能源消耗、简化工艺流程、降低投资与生产成本、减少环境污染等方面取得良好效果[1,2]。DRYIron 工艺流程如图 1 所示。

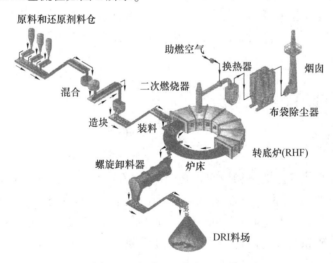

图 1 DRYIron 工艺流程示意图

2.4 Fastmet 与 Fastmelt 工艺

1982 年，Midrex 公司结合其开发成功的气基直接还原 Midrex 技术进行了系统分析和研究，认为 Midrex 工艺的还原气体必须使用天然气，因此该装置只能在天然气资源充沛和廉价的地区使用，很难在全球范围内大面积推广应用。开发利用资源充沛、分布广泛、储量充足且廉价的煤炭作为还原剂的直接还原工艺势在必行。在系统研究了当时全部已有和正在开发的煤基直接还原技术以后，认为 20 世纪 60 年代开发的 Heat Fast 工艺具有巨大的市场前景，随即开始了大规模的深入研究。

日本神户公司与 Midrex 公司合作，在 Heat Fast 工艺的基础上，共同研究开发了使用铁矿粉和煤粉生产直接还原铁的 Fastmet 工艺。为了对 Fastmet 工艺进行验证，在神户公司的加古川制铁所建设了年产 2 万吨的试验装置，从 1985 年 12 月开始进行了大约 3 年的生产工艺验证，并结合钢铁厂粉尘处理进行系统研究。1995 年 8 月在加古川制铁所建成 Fastmet 示范工厂，2000 年向日本新日铁公司广畑厂出售了第一套 Fastmet 工业化设备，处理钢铁厂废弃物 19 万吨/年。2001 年第 2 套 Fastmet 商业化设备在日本神户公司加古川厂建成投产，处理钢铁厂废弃物 16 万吨/年[3]。

由于用煤作为还原剂存在灰分和硫分的波动，对直接还原铁的质量造成影响，因此 Fastmelt 工艺被开发出来。这种工艺是在 Fastmet 工艺的基础上发展演变的，是将 Fastmet 生产出来的直接还原铁在高温状态下用熔炼炉熔化，实现渣-铁分离，并将残留的 FeO 还原，同时进行脱硫，控制铁水中的含碳量，这一冶金过程被称为"熔分"。直接还原铁熔化所用的能源可以是电或煤，熔化炉产生的气体主要成分 CO 可作为转底炉的燃料使用。这种由 Fastmet 工艺生产出的直接还原铁后，再用专用的电炉或转炉熔炼成铁水的工艺被称为 Fastmelt 工艺，或可将其理解为这是一种"两步法"的熔融还原工艺。Fastmet 与 Fastmelt 工艺流程如图 2 所示，典型的 Fastmet 与 Fastmelt 产品化学成分见表 1。

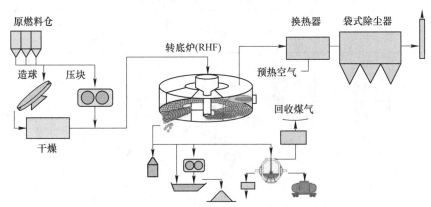

图 2　Fastmet 与 Fastmelt 工艺流程示意图

表 1　典型的 Fastmet 与 Fastmelt 产品化学成分　　　　　（%）

项　目	C	Si	S	P	Fe	FeO	脉石	TFe	金属化率
Fastmet 直接还原铁	4		0.15		78.2	11.3	6.35	86.9	90
Fastmelt 铁水	2.0~4.5	0.1~0.6	<0.05	<0.04	95.5~98				

2.5　Itmk3 工艺

在 20 世纪 90 年代中后期，日本神户公司和 Midrex 公司联合开发的 Fastmet 工艺获得突破性进展，并迅速得到工业化应用。人们发现，在 1350~1400℃ 的高温条件下，转底炉中的直接还原铁出现轻度熔化，生成粒铁（Nuggets），同时脉石也出现熔化，形成渣-铁初步分离。为了检验和证实工艺原理，并使其实现工业化，开发者从 1996 年开始，在加古川厂的直径为 3.2m、设计能力为 350kg/h 的中试转底炉上进行中试[5]，在技术原理和工艺得到验证以后，在美国明尼苏达州建成了 2.5 万吨/年的试验装置，并于 2004 年完成了试验工作。

这种采用矿粉和煤粉利用转底炉生产粒铁的工艺被命名为第三代炼铁工艺——Itmk3，这种工艺也是在 Fastmet 工艺的基础上发展演变的，是将含碳球团布在转底炉的炉床上，在 1400℃ 高温条件下迅速加热，在氧化铁快速还原的同时，实现海绵铁的渗碳，在反应的最后阶段使铁渣熔融，分离出脉石成分，熔融铁粒在转底炉内凝结成粒铁，经冷却后排除，使粒铁和炉渣在转底炉外实现分离。采用这种工艺所得到的粒铁的金属铁含量达到 96%~97%，且不含脉石成分。

Itmk3 工艺的开发者认为，如果把当今占主导地位的高炉炼铁流程作为第一代炼铁工艺；以 Midrex 为代表的直接还原流程作为第二代炼铁工艺；Itmk3 工艺则是第三代炼铁工艺。应该指出，这种观点未必会得到广泛的技术共识。但从 Itmk3 工艺的技术特征可以看出，在转底炉内实现了渣-铁分离且工艺流程简单，同其他直接还原工艺相比，在技术和产品质量等方面显现出优势和潜力。Itmk3 工艺生产的成品粒铁的化学成分见表 2，Itmk3 的工艺流程如图 3 所示。

表 2　Itmk3 工艺生产的成品粒铁的化学成分　　（%）

成分	金属铁	C	P	S
含量	97	2.0~2.5	0.01~0.02	0.07~0.11

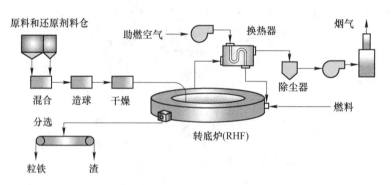

图 3　Itmk3 工艺流程示意图

3　转底炉直接还原工艺的技术原理

3.1　转底炉直接还原的工艺过程

转底炉直接还原工艺是以煤炭作为还原剂，将铁矿粉、钢铁厂粉尘、煤粉等混合后制成球团，再将球团装到转底炉的炉床上，在炉床上形成 1~3 层球团，在炉床旋转的条件下，球团被转底炉内的辐射热快速加热，球团内的铁氧化物与球团内的碳在 1200~1400℃ 的温度条件下发生还原反应，生成直接还原铁，整个还原过程需要 6~12min，在温度约为 900~1000℃ 时，生产的直接还原铁从转底炉的卸料口连续排出，炉料在炉内的停留时间约为 20min。在转底炉内按照炉料运行的进程和炉内温度一般分为装料区、加热区、还原区和卸料区 4 个区间，其温度分布如图 4 所示，转底炉内冶金传输原理的解析如图 5 所示。

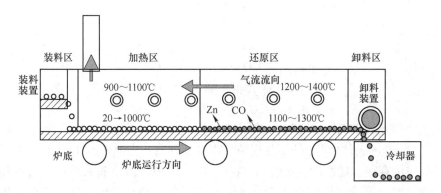

图 4　转底炉内温度分布

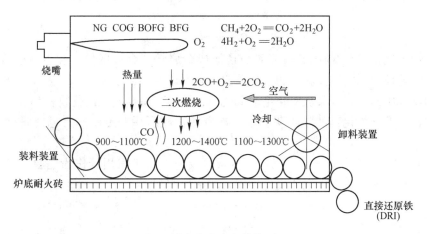

图 5　转底炉内冶金传输原理

转底炉炉底的旋转方向与反应生成的气流方向相反，冶金传输过程和化学反应在球团与气流的相向运动进程中完成。炉内的温度和气氛由设在炉墙上的烧嘴进行精准控制，从而获得最佳的还原热力学和动力学条件。加热区为氧化性气氛，热量的主要来源是通过烧嘴燃烧产生的；还原区为还原性气氛，球团还原生成的 CO 在此区间内二次燃烧释放热量，球团与气流之间主要通过辐射传热。还原反应生成的 CO 被用作转底炉内的燃料，仅需加入相当于 15% 必要能量的燃料进行补充。转底炉的烟气从装料端顶部排出，经过换热器预热助燃空气，将助燃空气预热到 500~600℃，用于烧嘴助燃。废气经废热锅炉燃烧产生蒸气后，再经过除尘净化，由烟囱排出。

3.2 转底炉直接还原的基本反应

转底炉直接还原的基本反应包括球团中铁的氧化物在转底炉内被还原成金属铁、球团中的氧化锌在转底炉内被还原成金属锌,碳在转底炉内的氧化反应和碳素溶解反应等。

(1) 球团中的铁氧化物在转底炉内被还原为金属铁。发生的还原反应如下:

$$3Fe_2O_3 + C = 2Fe_3O_4 + CO$$
$$Fe_3O_4 + C = 3FeO + CO$$
$$FeO + C = Fe + CO$$
$$3Fe_2O_3 + CO = 2Fe_3O_4 + CO_2$$
$$Fe_3O_4 + CO = 3FeO + CO_2$$
$$FeO + CO = Fe + CO_2$$
$$C + CO_2 = 2CO$$

铁氧化物的还原是逐级进行的,转底炉直接还原工艺也遵循这个规律。温度低于900℃时,气相成分受烟气的影响而使还原反应受到抑制;当温度为900~1000℃时,由于碳素溶解反应的存在,还原反应接近 Fe-FeO 平衡,反应速率受碳素溶解反应控制;当温度高于1100℃时,碳素溶解反应活跃,气相中只要存在 CO_2,就立即与 C 反应生成 CO,反应速率受碳素溶解和还原反应共同控制。

(2) 球团中所含的氧化锌在转底炉内被还原成金属锌。发生的还原反应如下:

$$ZnO + C = Zn + CO$$
$$ZnO + CO = Zn + CO_2$$

转底炉内氧化锌在 950~1000℃时被还原成金属锌,金属锌的气化温度也在此区间,因此氧化锌的还原反应与金属锌的气化反应是相继进行的连锁反应。金属锌气化后挥发到转底炉的尾气中,金属锌又被氧化生成氧化锌,然后与碱金属和卤化物一起被收集在布袋除尘器中,将富含氧化锌的除尘灰作为二次资源回收利用,可用作有色金属冶炼厂的粗原料。

(3) 碳在转底炉内的氧化反应和碳素溶解反应。由于采用碳作为还原剂,含碳球团中的碳在转底炉内还会发生氧化反应和碳素溶解反应如下:

$$C + O_2 = CO_2$$
$$C + CO_2 = 2CO$$

碳素溶解反应是转底炉直接还原反应得以合成的关键反应。此外,在转底炉内还有燃料燃烧反应和二次燃烧反应,燃烧反应热为球团的加热和还原反应提供热量,驱动冶金反应进程的顺利完成。

4 转底炉直接还原工艺的技术特点

转底炉直接还原工艺的技术特点如下:

(1) 转底炉直接还原工艺是典型的煤基直接还原,以非炼焦煤作为还原剂,可以完全不用焦炭,与气基直接还原工艺相比,摆脱了天然气资源的制约,工艺的适应性更加广泛。

（2）采用含碳球团使还原动力学条件得到明显改善，较好地解决了传质与传热的矛盾，加快了还原反应进程。这是由于含碳球团中矿粉和煤粉之间紧密接触，还原反应气体扩散距离短，因而整个还原反应的时间不取决于气体扩散而取决于传热速率，高温状态的还原反应动力学条件为快速还原提供了良好条件。

（3）对原料适应性更加广泛，既可以使用铁矿粉和精矿粉，也可以利用钢铁厂的含铁、含锌固体废弃物生产直接还原铁，同时可去除废弃物中的锌，生成副产品——粗氧化锌，采用转底炉工艺，直接还原铁的金属化率可以达到90%，脱锌率可达到95%以上，转底炉工艺的金属化率与脱锌率如图6所示。

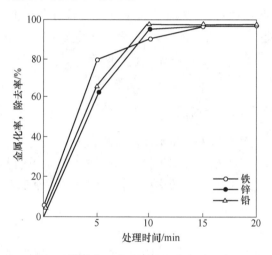

图 6　转底炉工艺的金属化率与脱锌率

（4）还原反应的温度比其他直接还原工艺的温度高 300~700℃，还原热力学条件充沛。该工艺是薄料层在高温中加热，加热速率快，还原快速，球团在转底炉内停留时间短（一般 20min 左右），且操作控制灵活、简便。

（5）转底炉直接还原工艺对炉料的强度要求不高，造球和压块都可以满足工艺要求，对原料的适应性也更加广泛。这是由于炉底上的料层很薄，且随炉底一起转动，球团不承受料层的压力，又与炉底之间处于相对静止状态，球团破损率很低，不会出现炉料黏结、结圈和结瘤等问题。

（6）转底炉直接还原工艺设备简单，建设投资低于其他直接还原工艺，技术较成熟，操作控制灵活，生产效率高，能源消耗低，可以实现规模生产。

5　转底炉直接还原工艺的应用与发展前景

5.1　转底炉直接还原工艺的应用

转底炉直接还原工艺经过30余年的发展，目前已经实现工业化生产，在美国和日本得到推广应用。1978年美国 INMETCO 公司建成了世界上首座用于钢铁厂固体粉尘处理的转底炉直接还原生产线，投产以来生产正常，成为美国中部钢铁厂废弃物处理中心；1997年美国 MR&E 公司在 Ameri-Steel 建成了产能为2.2万吨/年的 DRYIron 生产线；1999年美国动力钢公司（Steel Dynamics, Inc.）的子公司动力铁公司（Iron Dynamics）建成投产了世界上生产规模最大的 IDP（Iron Dynamics Process）转底炉直接还原生产线，生产规模为 50 万吨/年；2000 年 4 月日本新日铁广畑厂建成世界上第一条 Fastmet 生产线，生产能力为 14 万吨/年；2000 年 5 月日本新日铁君津厂建成生产能力为 13 万吨/年的 Fastmet 生产线；2001 年日本新日铁光厂引进 MR&E 开发的 DRYIron 技术，建成了处理能力为 6 万吨/年的生产线。

20 世纪 80 年代末期，我国已开始转底炉直接还原技术的研究工作。1992 年北京科技大学与舞阳钢铁公司合作，借鉴 Fastmet 和 Inmetco 工艺，建成了直径为 3m 的转底炉直接还原试验装置。1994 年北京科技大学对敞焰加热含碳球团直接还原工艺进行改进，提出了名为 COF-R 的新工艺，在四川阿坝设计建设了一座直径为 7.4m 的试验装置[5]。2000 年北京科技大学与山西明亮钢铁公司合作，在山西翼城建设了我国第一座生产直接还原铁的转底炉，2006~2007 年，进行了 7 个月的连续生产[6]。

5.2 转底炉直接还原工艺的发展前景

转底炉直接还原工艺具有显著的技术优势和发展潜力，将成为煤基直接还原的主要工艺流程，特别适用于钢铁厂含铁、含锌等粉尘和固体废弃物的资源化回收利用，对于实现循环经济和钢铁绿色制造，具有重要的作用，转底炉直接还原工艺的关键技术将在未来的发展中不断得到优化、创新。

5.2.1 球团配碳技术

采用含碳球团是转底炉直接还原工艺的重要技术特征。含碳球团中煤粉或焦粉与铁氧化物紧密接触，改善了铁氧化物与固定碳的固相-固相还原反应的动力学条件，高温状态下加快了反应进程。球团中的含碳量决定了直接还原铁的金属化率和碳含量，也影响还原反应进程和工艺操作制度，其技术关键是还原剂和铁氧化物的均匀混合。研究表明，球团中固定碳与铁氧化物的理论摩尔比为 1.5∶1，实际生产中固定碳与铁氧化物的重量比约为 1∶6，即球团中固定碳含量一般为 11%~15%。

5.2.2 转底炉内气氛的控制

转底炉内气氛的控制是转底炉操作的关键技术，必须精准地控制转底炉各区间的气氛、煤气流速、煤气成分和温度，确保球团的加热和还原，避免球团中的固定碳在加热区的过度烧损和直接还原铁在排料区的氧化。合理设定球团含碳量的配比、优化烧嘴设计、燃烧控制及炉气中 CO/CO_2 比率的控制是转底炉工艺热工操作的核心。

5.2.3 机械设备的改进效果

机械设备改进后的主要效果如下：

（1）转底炉转动时，沿半径方向上球团运行的角速度相等而线速度不等，因此装料装置要实现球团在炉底径向和圆周方向的均匀分布，避免球团的破碎和堆积。

（2）提高直接还原铁卸料装置的可靠性和使用寿命。

（3）提高炉底传动系统运行的可靠性。

（4）提高炉体设备的强度和耐高温性能，减少炉体结构的热应力变形。

（5）改进了炉体耐火材料内衬结构和材质，延长了炉体寿命。

(6) 炉底水封结构设计可靠，阻止了固态残留物堵塞水槽。

机械装备的不断改进和创新将有力推动转底炉工艺的技术进步和推广应用。

5.2.4 处理钢铁厂固体废弃物

采用转底炉工艺处理钢铁厂粉尘可以实现固体废弃物资源化合理利用，是实现节能减排、循环经济的有效技术措施。对钢铁生产过程中产生的含铁、含碳、含锌粉尘进行集中处理，采用转底炉工艺可以生产直接还原铁并脱除氧化锌，实现铁素资源和非铁元素的充分再利用。这是21世纪现代化钢铁厂发展循环经济、实现钢铁绿色制造的重要途径。

5.2.5 直接还原铁的应用

转底炉生产的高温直接还原铁热装到埋弧电炉（SAF）或熔炼炉（KDP）进行熔炼、分离，可生产合格铁水，供转炉或电炉使用，也可以将直接还原铁热装入电炉替代废钢；日本将转底炉直接还原铁作为金属化炉料装入高炉，加入100kg直接还原铁可以降低焦比约23kg/t[7]。

6 结语

转底炉直接还原工艺经过30多年的发展，已经实现了工业化生产，可以利用各种铁矿粉和含铁原料以及钢铁厂粉尘生产直接还原铁，是一种具有技术发展前景的煤基直接还原工艺。转底炉工艺技术成熟、对原燃料适应性强及操作控制灵活、简便，且可以处理钢铁厂粉尘，比其他直接还原工艺更具有技术优势和发展潜力。采用转底炉工艺处理钢铁厂粉尘可以实现固体废弃物资源化合理利用，是实现节能减排、循环经济的有效技术措施。我国天然气资源较少，而煤炭资源丰富，因此研究开发新一代煤基直接还原技术是发展我国非高炉炼铁技术的重要技术途径。

参考文献

[1] Valdis R Daiga, Deane A Horne, James A Thornton. Steel mill waste process on a rotary hearth furnace to recover valuable iron units [A]. Ironmaking Division of the Iron & steel Society. 2002 Iornmaking Conference Proceedings [C]. Nashville, Tennessee：Ironmaking Division of the Iron & steel Society, 2002：655.

[2] Rodney A Apple, et al. Waste oxide reduction facility at rouge steel [A]. Ironmaking Division of the Iron & steel Society . 2002 Iornmaking Conference Proceedings [C]. Nashville, Tennessee：Ironmaking Division of the Iron & steel Society, 2002：693.

[3] James M McClelland, Jr P E. Fastmet proven process for steel mill waste recovery [A]. Ironmaking Division of the Iron & Steel Society. 2002 Ironmaking Conference Proceedings [C]. Nashville, Tennessee：Ironmaking Division of the Iron & Steel Society, 2002：667.

[4] Osamu Tsuge, Shoichi Kikuchi, Koji Tokuda, et al. Successful iron Nuggets production at ITmk3 pilot plant [A]. Ironmaking Division of the Iron & Steel Society. 2002 Ironmaking Division of the Iron & Steel Society, 2002：511.

[5] 王尚槐, 冯俊小, 姚夏漪, 等. 环形转底炉海绵铁生产方法 [J]. 冶金能源, 1996, 15 (6)：7.

[6] 孔令坛, 郭明威. 我国第一台金属化球团转底炉 [A]. 中国金属学会. 2007 中国钢铁年会论文集 [C]. 北京：冶金工业出版社, 2007.

[7] Tetsuharu Ibaraki, Hiroshi Oda, et al. Dust recycling technology by the rotary hearth furnace [A]. The Iron and Steel Institute of Japan. 2006 Asia Steel International Conference Proceedings [C]. Fukuoka, Japan: The Iron and Steel Institute of Japan, 2006: 432.

The Evolution of Direct Reduction Technology of Rotary Hearth Furnace

Zhang Fuming

Abstract: The direct reduction technology of rotary hearth furnace (RHF) is a coal-based direct reduction technology, which has a great prospect. With this process, direct reduction iron could be produced by varied kinds of iron ores, ferrous contained material and the dust collected from iron & steel plant. In this paper, the domestic and abroad development of rotary hearth furnace direct reduction technology is introduced, in which the technique mechanism, characteristics and comprehensive evaluation of Inmetco, DRYIron, Fastmet and Itmk3 are introduced. Furthermore, based on this technology, the solution of integrated processing of the solid dust containing ferrous and zinc is put forward, as well as the technical outlook for the application of rotary hearth furnace direct reduction technology in China is introduced.

Keywords: ironmaking; rotary hearth furnace; direct reduction; smelting reduction

焦炉煤气二氧化碳重整热力学规律研究

摘 要：本文分析了富氢煤气还原对气体成分的要求；通过热力学模拟计算对焦炉煤气二氧化碳重整的热力学规律进行了研究。结果表明：富氢煤气还原时其 H_2/CO 应为 $1\sim2$。对焦炉煤气进行 CO_2 重整，其气体产物的 H_2/CO 较低，当 CO_2 配入量大于20%时，气体产物的 H_2/CO 可以降低到2.0以下；提高反应温度和降低反应体系压力有利于焦炉煤气 CO_2 重整反应的进行，综合考虑 CH_4 转化率、气体产物的还原势等因素，焦炉煤气 CO_2 重整反应的温度应大于1000℃；焦炉煤气重整后的 H_2/CO 主要决定于 CO_2 的配入量，通过调节水蒸气和 CO_2 加入量可以对还原气体成分进行有效控制。

关键词：焦炉煤气；自重整；CO_2；DRI

1 引言

焦炉煤气是炼焦过程中的副产品，主要成分是氢气和甲烷，其中氢气和一氧化碳含量约占到60%。因此，焦炉煤气除可作为燃料外还是很好的还原剂。目前，随着我国焦化行业的快速发展，焦炉煤气出现大量过剩，过剩的焦炉煤气未加利用而被直接排放掉了，这不仅造成资源的极大浪费，也对环境造成严重污染[1]。如何对焦炉煤气资源进行合理的利用，已经引起冶金工作者的广泛关注[2]。对焦炉煤气的利用存在着燃料化和资源化两种途径，由于氢气的发热值仅为 $10.8MJ/m^3$（CH_4 热值为 $37.6\sim45.2MJ/m^3$），因此将焦炉煤气用于加热和发电在经济上是不合理的。受天然气资源的制约，我国海绵铁生产以煤基还原为主，由于存在能耗高、污染严重、生产效率低下等问题，一直发展缓慢。如果将焦炉煤气经过重整后，用作还原剂生产直接还原铁，不仅可以缓解我国直接还原铁需求，而且可以使氢气优良的还原性能得到利用，此外由于氢气是清洁能源，发展焦炉煤气还原有利于减少钢铁行业的二氧化碳排放（与煤基海绵铁生产过程相比，吨铁 CO_2 排放量可以从2000kg降低到400kg以下），实现清洁生产，符合低碳可循环经济的发展需求[3]。

焦炉煤气重整是指焦炉煤气经过净化和脱硫后，通过加入氧气、二氧化碳或水蒸气等转化剂，在重整炉或反应炉中将甲烷转化为氢气和一氧化碳的过程。目前，在以天然气为能源的气基直接还原工艺中，以 Midrex 工艺和 HYL 工艺应用最为广泛，均采用水蒸气法（$CH_4+H_2O = CO+3H_2$）对天然气进行重整。HYL 工艺近年来发展很快，HYL-ZR 自重整工艺取消了独立的重整炉，在竖炉中就可以实现重整和还原过程，是目前最先进的天然气重整工艺。焦炉煤气重整制取甲醇的生产实践表明，对焦炉煤气进行水蒸气重整，其重整产物中的 H_2/CO 较高（可达4.6）[4]。由于氢气还原氧化铁的反应是吸热过程，过高的 H_2/CO 对于还原反应的供热提出了更高要求，很难适应气基直接还原反应的需要。如果

本文作者：曹朝真，张福明，毛庆武，徐辉。原刊于《第八届（2011）中国钢铁年会论文集》，2011：357-363。

向重整气中配入一定量的二氧化碳,通过二氧化碳重整反应（$CH_4+CO_2=2CO+2H_2$）使气体产物中的 CO 含量明显提高,则可以实现对气体产物 H_2/CO 的有效调节,从而为焦炉煤气生产直接还原铁的实现创造条件。

本文从热平衡角度分析富氢煤气还原氧化铁时,H_2/CO 对气体利用率的影响,并通过热力学模拟计算对焦炉煤气二氧化碳重整反应的过程进行模拟,分析其重整反应的热力学规律,对焦炉煤气水蒸气重整和二氧化碳重整的效果进行对比。

2 富氢煤气还原技术分析

使用氢气还原氧化铁时其气体产物是水,因此,氢气不仅是优良的还原剂,也是一种清洁能源。用氢气取代焦炭还原氧化铁,开展富氢（纯氢）冶金技术研究,是炼铁技术发展的新方向。氢气还原氧化铁的动力学条件要好于 CO,特别是在传质范围内,氢气的还原速度要明显高于 CO 的还原速度,这从氢气和 CO 还原氧化铁的热力学平衡图中可以看出[5]。当反应温度大于 818℃时,氢气与氧的结合力要大于 CO,因此还原能力也更强。富氢（纯氢）煤气与 CO 相比,由于还原气中氢气的存在,使得还原反应的动力学条件得以改善。

$$1/4Fe_3O_4(s) + H_2 = 3/4Fe(s) + H_2O(g) \quad \Delta G^{\ominus} = 35550 - 30.40T, \text{J/mol} \quad (1)$$

$$1/4Fe_3O_4(s) + CO = 3/4Fe(s) + CO_2 \quad \Delta G^{\ominus} = -9832 + 8.58T, \text{J/mol} \quad (2)$$

由式（1）和式（2）可以看出,CO 还原氧化铁的反应是放热反应,而氢气还原氧化铁的反应是强吸热反应,因此,如何持续向反应区供给热量是富氢直接还原技术的难点。图 1 给出了反应温度为 900℃,含铁原料预热,还原气物理热供热时,不同温降条件下气体用量和利用率与氢气含量之间的关系。

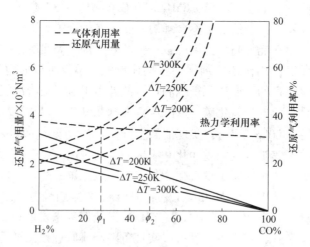

图 1 还原气用量及气体利用率与 H_2%关系图（含铁原料预热）

图 1 中可以看出,在反应温度为 900℃时,富氢气体的热力学利用率随氢含量的增加而提高,因此提高 H_2/CO 有利于提高还原气的综合利用率。但由于氢气还原反应为吸热反应,提高煤气的 H_2/CO 吨铁还原需要的热量增加,同时由于炉内热收入主要靠还原气

物理热来提供，因此，要增加炉内的供热量就必须加大还原气量，而加大气量又会造成气体利用率的进一步降低，这使得反应炉内的气体成分和气体利用率很难得到最优化的协调统一，即反应炉内热平衡和化学平衡之间的矛盾决定了富氢气体一次利用率极限的存在。在现有的气基直接还原工艺中（如 HYL、Midrex），由于受气体加热炉换热器材质的制约，最高加热温度只能达到 970℃，为进一步提高气体温度常采用吹氧燃烧部分还原气的方法来实现，气体温度可提高至约 1085℃。为维持一定的反应温度，还原气的温降就具有一定局限性。如图中所示，取气体温降为 250℃，当 $H_2/CO=1.67$ 时，还原气利用率接近热力学利用率，约为 34.5%；当 $H_2/CO>1.67$ 时，还原气的利用率迅速降低，如当 $H_2/CO=4$ 时，还原气利用率降低至 26.5%。因此，在现有气基直接还原工艺还原气温降水平下，为获得尽可能高的气体利用率，适宜的 H_2/CO 应为 1~2。

3 焦炉煤气二氧化碳重整热力学模拟计算

由上述分析可知，过高的 H_2/CO 不利于富氢气体还原反应的进行，而焦炉煤气水蒸气重整气体产物的 H_2/CO 达到 4 以上，不能满足富氢煤气还原的经济性要求。若对焦炉煤气进行 CO_2 重整，可以在不引入 H_2 的前提下，增加 CO 的量，有利于重整产物 H_2/CO 的降低。为此本文对焦炉煤气 CO_2 重整过程进行热力学模拟计算，并对不同工艺条件下重整产物 H_2/CO 的变化规律进行分析。

3.1 计算方法

计算中采用的焦炉煤气成分见表 1。

表 1　净化后的焦炉煤气成分（体积分数）

焦炉煤气	H_2	CO	CO_2	CH_4	N_2	H_2O	C_2H_6
含量/%	54.5	6	2.3	26.2	6.5	2.0	2.5

对于 CH_4-C_2H_6-H_2-H_2O-CO-CO_2-O_2 体系，选取式（3）~式（6）作为独立反应。

$$CH_4 + CO_2 = 2CO + 2H_2 \quad (3)$$

$$C_2H_6 + 2CO_2 = 4CO + 3H_2 \quad (4)$$

$$CO + H_2O = CO_2 + H_2 \quad (5)$$

$$2H_2 + O_2 = 2H_2O \quad (6)$$

则有：

分压总和方程：

$$P_{CH_4(平)} + P_{H_2(平)} + P_{H_2O(平)} + P_{CO(平)} + P_{CO_2(平)} + P_{O_2(平)} = P_{总} \quad (7)$$

独立反应的平衡常数方程：

$$K_1 = P_{CO(平)}^2 P_{H_2(平)}^2 / P_{CH_4(平)} P_{CO_2(平)} P_{(总)}^2 \quad (8)$$

$$K_2 = P_{CO(平)}^4 P_{H_2(平)}^3 / P_{C_2H_6(平)} P_{CO_2(平)}^2 P_{(总)}^4 \quad (9)$$

$$K_3 = P_{H_2(平)} P_{CO_2(平)} / P_{H_2O(平)} P_{CO(平)} \quad (10)$$

$$K_4 = P_{H_2O(平)}^2 / P_{H_2(平)}^2 P_{O_2(平)} P_{(总)} \quad (11)$$

元素原子摩尔量恒定方程：

$$n_{H(初)} = n_{H(平)} \tag{12}$$

$$n_{C(初)} = n_{C(平)} \tag{13}$$

$$n_{O(初)} = n_{O(平)} \tag{14}$$

联立方程式（7）~式（14），可以对不同初始条件下的平衡态气体组分含量进行计算[6]。

3.2 计算结果及分析

3.2.1 CO_2 配加量的影响

以表1中焦炉煤气成分为例，取反应温度为1050℃，体系压力为0.5MPa，焦炉煤气量为100kmol，配氧量为1kmol，改变 CO_2 配入量进行热力学平衡模拟计算，计算结果如图2和图3所示。

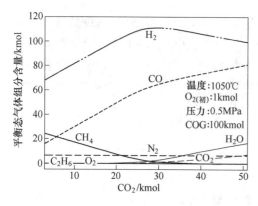

图2 二氧化碳含量对平衡态组分的影响

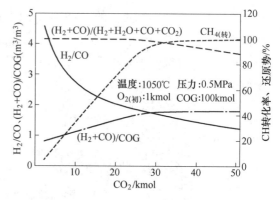

图3 不同 CO_2 配加量对 H_2/CO 的影响

图2为焦炉煤气二氧化碳重整平衡态组分随二氧化碳配入量变化曲线，计算表明，在该计算条件下，平衡态时 C_2H_6 和 O_2 消失；而且随着 CO_2 配入量的增加，CH_4 量逐渐减少，当 CO_2 约为30kmol时，CH_4 消失，此时，H_2 量达到最大值，CO增加速度开始变慢。继续增加 CO_2 配入量，平衡体系中开始出现 H_2O 和 CO_2。考虑到平衡系统总气量的变化，图3中对气体产物的还原势等进行了计算，结果表明，当 CO_2 配入量小于25%时，平衡态气体的还原势均在99%以上，当 CO_2 配入量大于30%时，气体的还原势开始下降，此时，CH_4 转化率达到95%以上；还原气体的 H_2/CO 随 CO_2 配入量的增加而迅速降低，当 $CO_2 >$ 20kmol时，$H_2/CO<2$。

3.2.2 反应温度的影响

对于焦炉煤气 CO_2 重整反应来说，除上述四个独立重整反应外，在反应温度较低时还有可能产生析碳现象，因此，在对不同温度下的重整反应进行热平衡计算时考虑析碳反应（$2CO = C(s) + CO_2$）。取焦炉煤气的量为100kmol，配氧量为1kmol，CO_2 配入量为30kmol，反应体系压力为0.5MPa，改变反应体系的温度，进行热力学平衡模拟计算，计算结果如图4和图5所示。

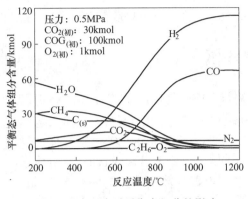
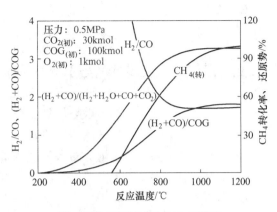

图 4　反应温度对平衡态组分的影响　　　　图 5　反应温度对 H_2/CO 的影响

图 4 为平衡态组分含量随温度变化曲线，从图中可以看出，不同温度下平衡体系中均不含 C_2H_6 和 O_2；随着反应温度的提高，体系中 H_2 和 CO 量迅速增加，CO 含量在 900℃以上趋于稳定，H_2 含量则在 1000℃以上趋于稳定；H_2O、CH_4 和 CO_2 在 1000℃以上基本消失；对于析碳反应，在 600℃以下 C 含量较高并维持一稳定值，从 600℃开始，C 含量迅速降低，约在 930℃时全部消失。图 5 表明，气体产物的 H_2/CO 随温度的升高而迅速降低，当反应温度大于 900℃时达到一个稳定值约为 1.7，CH_4 转化率、气体产物的还原势以及还原气体总量均在 1000℃以上达到最大值并趋于稳定，可见 CO_2 重整反应温度应大于 1000℃。

3.2.3　反应压力的影响

为考察反应体系压力对 CO_2 重整反应的影响，分别取系统压力为 0.1~0.8MPa，焦炉煤气的量为 100kmol，配氧量为 1kmol，CO_2 配入量为 30kmol，反应温度为 1050℃，进行热力学平衡模拟计算，计算结果如图 6 和图 7 所示。

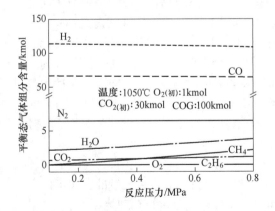
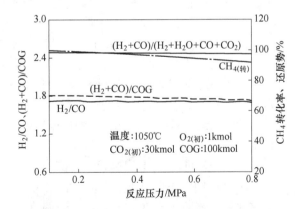

图 6　反应体系压力对平衡态组分的影响　　　　图 7　反应体系压力对 H_2/CO 的影响

上图中计算结果表明，增大反应体系的平衡压力，H_2 和 CO 量有所降低，H_2O、CH_4 和 CO_2 含量则不断提高，系统中没有 C_2H_6 和 O_2；气体产物的还原势基本稳定在 98% 左右，

H_2/CO 基本不变约为 1.71，CH_4 的转化率则略有下降。因此，反应体系的压力对 CO_2 重整反应的影响不大，低压下更有利于 CH_4 的转化，但系统压力对还原气体总量影响显著。

4　焦炉煤气二氧化碳重整与水蒸气重整效果比较

上述计算表明，以 CO_2 为重整剂对焦炉煤气进行重整，气体产物中的 H_2/CO 可以达到 2.0 以下，能够满足富氢气体还原的需要。为比较焦炉煤气水蒸气重整与 CO_2 重整的重整效果，分别对不同 CO_2 含量下的水蒸气重整和不同 H_2O 含量下的 CO_2 重整气体产物的 H_2/CO 变化规律进行了比较。取焦炉煤气的量为 100kmol，配氧量为 1kmol，反应体系压力为 0.5MPa，反应温度为 1050℃，进行热力学平衡模拟计算，计算结果如图 8 和图 9 所示。

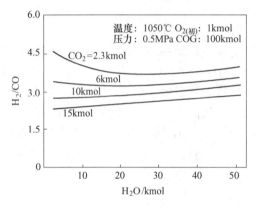 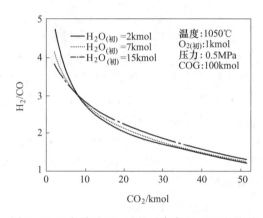

图 8　CO_2 含量对加湿重整平衡态 H_2/CO 的影响　　图 9　H_2O 含量对 CO_2 重整平衡态 H_2/CO 的影响

图 8 中分别取 CO_2 的初始含量为 2.3kmol、6kmol、10kmol 和 15kmol，改变 H_2O 的加入量，计算结果表明，当 CO_2 的初始配入量较低时（<6%），重整后气体产物的 H_2/CO 随 H_2O 加入量的增加呈现先降低后增加的趋势，并且总体变化幅度不大；继续提高 CO_2 配入量，气体产物的 H_2/CO 随 H_2O 加入量的增加略有增加，当 CO_2 的初始配入量大于 15kmol 时，重整后气体产物的 H_2/CO 可以达到 2 以下。图 9 中分别取 H_2O 的初始含量为 2kmol、7kmol 和 15kmol，改变 CO_2 的加入量，计算发现，水蒸气的配入量对气体产物的 H_2/CO 影响并不十分显著，重整后气体产物的 H_2/CO 随 CO_2 配入量的增加而迅速降低，而且水蒸气的初始配入量越小，其降低的速度越快，图中三条曲线在 CO_2 配入量约为 8.5kmol 时相交。

综上分析可知，焦炉煤气重整过程中，水蒸气配入量对其重整产物的 H_2/CO 影响较小，H_2/CO 的大小主要决定于 CO_2 的配入量，通过改变水蒸气和 CO_2 配入量的方法可以实现对其重整效果（H_2/CO 等）的有效调节，从而对还原气体成分进行合理控制。

5　结语

通过对富氢煤气还原技术进行分析和对 CH_4-C_2H_6-H_2-H_2O-CO-CO_2-O_2 体系进行热力

学平衡模拟计算，主要结论如下：

（1）综合考虑富氢气体还原时的热平衡和化学平衡，富氢煤气中 H_2/CO 的适宜值应为 1~2。

（2）对焦炉煤气进行 CO_2 重整可以显著降低气体产物的 H_2/CO，当 CO_2 配入量大于 20% 时，气体产物的 H_2/CO 可以降低到 2.0 以下。

（3）提高反应温度有利于焦炉煤气 CO_2 重整反应的进行，综合考虑 CH_4 转化率、气体产物的还原势和 H_2/CO 等因素，焦炉煤气 CO_2 重整反应的温度应大于 1000℃。

（4）反应体系的压力对焦炉煤气 CO_2 重整效果影响不大，降低压力更有利于 CH_4 的转化，但会造成还原气体积总量增大。

（5）焦炉煤气重整后 H_2/CO 的大小主要决定于 CO_2 的配入量，通过调节水蒸气和 CO_2 加入量可以对还原气体成分进行合理控制。

参考文献

[1] 王太炎. 焦炉煤气开发利用的问题与途径 [J]. 燃料与化工，2004，36（6）：1-3.
[2] 苏亚杰. 焦炉煤气生产直接还原铁试验与研究新进展 [J]. 煤化工，2006，2：10-12.
[3] 张春霞，胡长庆，严定鎏，等. 温室气体和钢铁工业减排措施 [J]. 中国冶金，2007（1）：7-12.
[4] 曹朝真，郭培民，赵沛，等. 焦炉煤气自重整炉气成分与温度变化规律研究 [J]. 钢铁，2009（1）：11-15.
[5] 方觉，等. 熔融还原与直接还原 [M]. 沈阳：东北大学出版社，1996：57-62.
[6] 黄希祜. 钢铁冶金原理 [M]. 北京：冶金工业出版社，1990：80-100.

Study on Thermodynamic Laws of Reforming of Coke Oven Gas by Carbon Dioxide

Cao Chaozhen　Zhang Fuming　Mao Qingwu　Xu Hui

Abstract：The requirements of iron ore reduction by hydrogen rich gas was analysed by computation of heat balance and the thermodynamic laws of reforming of coke oven gas was studied by thermodynamic calculation in this paper. The results showed that the optimal H_2/CO range was 1 to 2. It would be less than 2.0 that the value of H_2/CO of gaseous product of reforming of coke oven gas by CO_2 while the addition of CO_2 is more than 20 percent, which is lower than by H_2O. Higher temperature and lower pressure is favorable to the reforming by CO_2 and the reforming temperature should be more than 1000℃ considering the factors that the rate of reforming and reduction potential of reducing gas. The value of H_2/CO which is mainly depend on the addition of CO_2 can be adjusted effectively by adjusting the amount of CO_2 and H_2O.

Keywords：coke oven gas；self-reforming；CO_2；DRI

利用焦炉煤气生产直接还原铁关键技术分析

摘 要：本文对焦炉煤气和天然气生产直接还原铁方案进行了对比，并对焦炉煤气利用过程中的甲烷自重整、BTX 裂解、DRI 热态输送等技术进行了分析。如何对焦炉煤气进行精制是焦炉煤气利用的技术关键，具有发展前景的焦炉煤气处理工艺应该能够合理有效地利用二次热源，同时应具有最低的动力费用消耗，在此基础上本文提出了几种可行的焦炉煤气净化方案。

关键词：焦炉煤气；自重整；BTX；DRI

1 引言

近年来，我国钢铁行业发展迅速，钢产量已连续多年居世界第一位，但我国钢铁工业结构不合理，低附加值的钢材产量过剩，而高档钢材每年仍需进口。随着我国钢铁工业结构的调整和对钢铁产品质量要求的提高，电炉钢短流程必然会得到较快发展。由于我国废钢资源不足，每年的废钢进口量都在 1000 万吨以上，而且废钢中杂质元素的不断积累会对优质钢的生产造成不利影响，直接还原铁作为废钢的重要替代品是电炉炼钢的理想原料，它具有纯净度高、成分稳定等优点，是发展钢铁生产短流程的基础。此外，发展直接还原铁生产既可以改变长期以来传统炼铁工艺对焦煤的依赖，同时可以减少二氧化碳排放量，符合钢铁工业可持续发展的技术要求，是钢铁行业实现节能减排的有效途径。

直接还原铁生产根据使用还原剂的种类不同，可以分为煤基还原和气基还原两种。目前，全世界直接还原铁生产中，约 90%以上是通过气基直接还原工艺生产的，典型工艺有 Midrex 和 HYL-Ⅲ。气基竖炉是当今世界公认的直接还原铁生产的主流技术，它具有生产能力大、技术成熟、操作稳定、生产效率高等优势。我国直接还原铁技术的发展应适应钢铁行业主体设备大型化的趋势，优先发展具有较大生产能力的直接还原工艺，因此我国直接还原技术应把气基还原作为主要的发展方向。传统的气基还原需要有丰富的天然气资源做保障，而我国天然气资源缺乏，而且，天然气裂解工艺复杂、投资较大，因此，采用天然气作为原料气的气基直接还原工艺在我国很难得到发展。目前国内气基直接还原技术的发展方向主要包括：(1) 煤气化（水煤浆气化、粉煤气化）配竖炉工艺生产还原铁技术方案；(2) 天然气转化生产还原气配竖炉工艺生产直接还原铁技术方案；(3) 焦炉煤气转化生产还原气配 HYL 工艺生产还原铁技术方案。以上工艺方案中，天然气方案受资源条件限制，发展困难；煤制气竖炉方案在技术上是可行的，具有较好的发展前景，但目前主要受到煤制气技术发展的制约，煤制气竖炉流程的发展有赖于加压压煤制气技术的突破；焦炉煤气作为一种富氢气源可以用作优质还原气生产直接还原铁，其在冶金工业中的

本文作者：曹朝真，张福明，毛庆武，徐辉。原刊于《第八届（2011）中国钢铁年会论文集》，2011：3158-3163。

应用前景越来越受到人们的重视，我国焦化企业每年产生大量过剩的焦炉煤气，这为开展焦炉煤气竖炉法生产直接还原铁提供了可能，此外，使用低热值的煤制气加热焦炉，从而置换出部分焦炉煤气用于直接还原铁生产，也是一种可以扩大焦炉煤气来源的可行的技术方案。焦炉煤气生产直接还原铁方案是目前国内直接还原技术领域研究的热点，并有望最快实现工业化。

2 焦炉煤气自重整直接还原铁生产工艺

传统的气基竖炉直接还原铁生产工艺以 Midrex 和 HYL 法为代表，主要以天然气为原料生产直接还原铁。天然气的主要成分是甲烷，而甲烷无法直接参与还原反应，需要首先将其分解为 H_2 和 CO，因此，在天然气进入竖炉前首先要经过气体重整炉，在金属催化剂的作用下与水蒸气反应发生分解。HYL 法自 1997 年开始率先完全取消了天然气重整炉，实现了天然气在竖炉内的自重整（Self-reforming），在不增设重整炉的前提下提高了竖炉的生产效率，并且开发出可使用焦炉煤气、煤制气等多种气源的气基还原工艺，即 HYL-ZR 工艺。本文以 HYL-ZR 工艺为例，对以焦炉煤气和天然气为还原气的自重整工艺进行对比分析，其工艺流程图如图 1 所示。

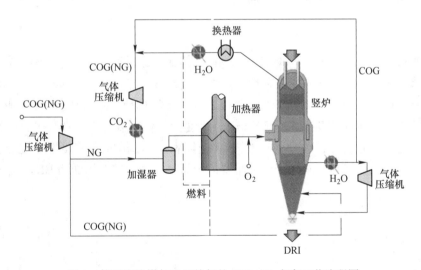

图 1 使用焦炉煤气和天然气的 HYL-ZR 方案工艺流程图

从图 1 中可以看到，在 HYL-ZR 工艺方案中，还原气存在两个回路：还原回路和冷却回路。使用天然气作为还原气时，天然气加压后进入还原回路，经加湿后进入气体加热炉，被加热到 930℃以上，再经加氧部分燃烧的方式使还原气温度进一步升高到 1085℃，然后沿竖炉中部的环形管道通入炉内。在竖炉内甲烷在热金属铁的催化作用下与水蒸气发生重整反应（$CH_4 + H_2O = CO + 3H_2$），生成 CO 和 H_2 参与还原反应并与铁矿石间进行热交换，反应后的气体经炉顶排除炉外，经换热、脱水后除一部分送入气体加热炉作为燃料气外，大部分顶气经加压、脱除 CO_2 后继续沿还原回路参与反应。在该工艺方案中，是否添加冷却回路取决于产品的用途，通过添加冷却回路可以实现对直接还原铁的冷却和渗碳，并可以实现对直接还原铁含碳量的灵活调整，从而得到稳定态的直接还原铁；如果用于生产热态直接还原铁则无需冷却段。使用焦炉煤气作为还原气时，焦炉煤气必须首先经过冷却回路，在竖炉下部与温度大于650℃的活性金属铁作用，使焦炉煤气中的BTX发生裂解，同时直接还原铁被焦炉煤气冷却并进行渗碳。焦炉煤气通过竖炉冷却段后经脱水后与炉顶气汇合，一部分送入气体加热炉用作加热还原气的燃料，剩余的顶气则沿还原回路进行循环。在还原回路和冷却回路之间可以通过压力控制来防止串气。

通过以上对比可以看到，使用焦炉煤气与使用天然气生产直接还原铁最大的区别在于是否需要设置冷却段，使用焦炉煤气还原时必须设置冷却回路，使煤气中的杂质分解以满足还原回路对还原气体的要求，因此，使用焦炉煤气的HYL-ZR方案无法得到热态的直接还原铁，从而无法实现与后续电炉间的含铁料的热态输送。

3 利用焦炉煤气生产直接还原铁关键技术

目前，世界上超过80%的直接还原铁生产是以天然气为原料的，天然气竖炉技术已经非常成熟。焦炉煤气与天然气相比甲烷含量较低，氢气含量高，更宜作为还原气使用，但由于焦炉煤气中杂质较多，因此如何净化焦炉煤气就成为其利用的关键。以下将从几个方面对其利用过程中涉及的关键技术问题进行分析。

3.1 甲烷自重整技术

甲烷的自重整是指无需专设重整炉，原料气体在输送过程中或在竖炉内在高温活性金属铁的催化下进行重整。气基直接还原一般要求还原气氛中 $CO+H_2>90\%$，$(CO_2+H_2O)/(CO_2+H_2+CO+H_2O)<5\%$。焦炉煤气中含有20%以上的 CH_4，在低温条件下，直接用焦炉煤气还原铁矿石会发生渗碳反应，降低还原速度；在较高温度下，CH_4 作为惰性气体存在，同样会降低还原速度；此外，从充分利用能源的角度出发也需要将甲烷转化加以利用，因此使用焦炉煤气与天然气一样都需要进行原料气体重整。图2为 CH_4 重整热力学平衡图。

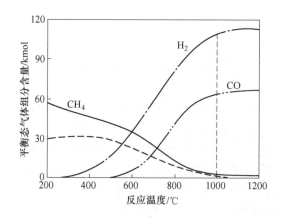

图2 CH_4 重整热力学平衡图

热力学计算表明[1]，提高温度有利于甲烷重整（水蒸气重整）反应的进行，当温度大于1000℃时，甲烷的重整比较彻底，因此要想实现竖炉内的自重整，反应气的温度应大于1000℃。在气基直接还原工艺中，受加热炉材质的影响，还原气最高只能加热到970℃，从1995年开始HYL工艺引入了O_2气喷入技术，吨铁用氧量为12～20m³，该技术使用了特殊设计的燃烧器，借助天然气的部分氧化使还原气的温度达到1085℃，氧气喷入技术的发展最终使得自重整工艺得以实现。第一套采用HYL-ZR自重整技术的生产装置，于1998年4月在墨西哥的蒙特雷TERNIUM HYLSA薄板厂建成投产，该装置的运行实践证明了甲烷自重整技术的可行性。

3.2 焦炉煤气BTX裂解技术

前已述及，以焦炉煤气为原料的HYL-ZR工艺与使用天然气工艺相比存在较大区别，这是由于焦炉煤气中除含有CH_4、H_2、CO等有用成分外，还有少量的BTX（苯、甲苯、二甲苯混合物）、焦油和萘等杂质，BTX对HYL-ZR工艺本身最大的影响就是当气体通过加热器时会产生析碳，从而造成管道堵塞；此外，在HYL-ZR工艺中，竖炉内的工作压力为0.6MPa，而焦化厂送出的煤气压力约5kPa，因此需要在使用前对焦炉煤气进行加压，由于煤气中焦油和萘的存在，在增压的过程中会大量析出，容易造成设备和管道的堵塞。如何消除BTX等杂质对还原工艺的危害，是以焦炉煤气为原料的气基直接还原工艺能否打通的技术关键。

来自焦化厂的焦炉煤气中，一般含有焦油不超过50mg/Nm³，萘不超过500mg/Nm³，BTX≤4000mg/Nm³，为了降低焦油和萘对煤气加压机的影响，焦炉煤气应首先进行脱萘、脱焦油处理，可通过变温吸附法将焦炉煤气中的萘降至50mg/Nm³以下，焦油降至10mg/Nm³以下。对于BTX的裂解方案，HYL公司曾对BTX含量为0.5～20g/Nm³的焦炉煤气进行了裂解实验，结果表明，通过将焦炉煤气通入竖炉冷却段的方式，可使BTX的分解率达到95%以上。在此实验基础上，HYL公司提出了HYL-ZR焦炉煤气生产直接还原铁工艺方案，即增加一个冷却回路，使焦炉煤气首先全部经过竖炉冷却段进行BTX裂解，循环一定时间后再通入还原回路参与重整和还原反应。该工艺方案目前仅通过了实验室规模的验证，尚未用于工业规模的实际生产，因此，其实际使用效果有待进一步验证。

3.3 直接还原铁热态输送技术

直接还原铁热态输送技术（HYTEMP）是HYL-ZR工艺除O_2喷入技术以外的另一项主要技术，该技术通过一个专用的气体加热装置，将运载气体（工艺气或N_2）加热到600℃左右，并使用运载气体将约700℃的直接还原铁直接输送至电弧炉内，通过这种方式可以减小直接还原铁的温降。该装置已于1998年在墨西哥的蒙特雷TERNIUM电炉厂投入工艺生产，目前已累计输送热态直接还原铁800多万吨。通过直接还原铁热送技术可以有效回收直接还原铁的显热，不仅能够获得显著的经济效益，同时可以减少温室气体排放。电炉使用热态（约700℃）高碳DRI可节约电能130kW·h/t钢水，电炉的供电冶炼时间可减少30%，电炉的生产能力可提高20%。在HYL-ZR焦炉煤气生产直接还原铁工艺方案中，由于必须设置冷却段，因此只能生产冷态的直接还原铁，这使得后续电炉的直接还原铁热送无法实现，增加了电炉的冶炼周期和生产成本。如何使用焦炉煤气生产热态直接还原铁，目前仍是焦炉煤气使用过程的一个技术难点，有待于冶金工作者的进一步攻关。

4 BTX 裂解技术路线探讨

HYL-ZR 焦炉煤气生产直接还原铁工艺方案，采用在不改变原有工艺框架（NG 方案）的基础上，使焦炉煤气在竖炉下部与热金属铁接触裂解 BTX 等杂质的方法，生产冷态直接还原铁，该方法已经得到实验室结果的支持，是一种可行的 BTX 处理方法，但由于该方案无法使直接还原铁的物理热得到有效利用，与天然气方案相比，其技术经济性有待提高。新的有前途的焦炉煤气处理工艺应该能够合理有效地利用二次热源，同时具有最低的动力费用消耗。考虑到焦炉煤气中的 BTX 对还原工艺的危害，主要体现在加热和加压过程中的析碳和焦油析出，因此，应在焦炉煤气进入直接还原系统之前进行煤气精制。

焦炉煤气从炭化室出来经上升管、桥管到集气管，到达集气管处煤气的温度约为 650~700℃，目前焦化厂对此处焦炉煤气的物理热并未进行利用，如果可以对来自焦炉的热的焦炉煤气直接进行净化处理，则既可以使煤气的余热得到回收又可以避免由于煤气升温带来的问题。此外，焦油、BTX 等杂质属于高分子聚合物，如苯基重烃等，由于其碳链较长，因此其热稳定差，如能将焦炉煤气升高至更高温度（比焦化温度更高）则可将其中的杂质裂解，基于以上分析可以采用高温裂解炉对焦炉煤气进行处理，如图 3 所示。该焦炉煤气净化工艺只需在焦炉旁安装一个焦炉煤气裂解炉，无需对现有焦炉进行改造，同时不需要使用催化剂。将从焦炉出来的红焦由炉体上部装入裂解炉，并在炉内二次加热至 1200~1250℃，同时，来自焦炉的 650℃ 左右的焦炉煤气则由裂解炉下部进入炉内，在高温焦炭的作用下，焦炉煤气中的焦油可高效转化为合成气体，通过这种方法可以有效地使焦油、BTX 等杂质分解，用于生产洁净焦炉煤气，目前日本正在进行该技术的开发。

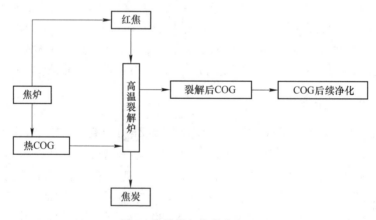

图 3　焦炉煤气净化方案一

除高温裂解法以外，以金属铁为催化剂，以焦炉煤气为冷却介质，在熄焦炉内对焦炉煤气进行处理也是一种煤气净化的思路，如图 4 所示。从焦炉出来的红焦配加一定量的直接还原铁一起装入息焦炉内，冷的焦炉煤气则由下部通入炉内，与焦炭进行热交换后，煤气由炉顶排出，一部分炉顶气经换热、脱水和加压后重新进入炉内循环，另一部分则作为净煤气排出；通过换热焦炭被冷却，同时焦炉煤气中的杂质在直接还原铁的作用下被分解。与传统的干熄焦工艺相比，由于焦炉煤气中的杂质和甲烷裂解吸热，使得其具有更高的传热效率，此外，煤气中的氢气还可以与焦炭中的硫化物和有机硫反应生产硫化氢，从而显著降低焦炭内的硫含量，达到脱硫效果[2]。

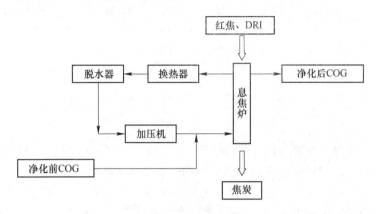

图 4 焦炉煤气净化方案二

传统的天然气竖炉工艺中，通常使用水蒸气重整法对甲烷进行裂解，制取还原气，而焦炉煤气中的杂质与甲烷相比，热稳定性差，更容易发生重整反应，图 5 中计算了 800℃ 下，热力学平衡态时 C_6H_6 和 CH_4 的转化效率。由图中可以看到，相同条件下 C_6H_6 优先与水发生重整反应，而且其重整开始温度也比 CH_4 更低。因此可以考虑采用重整法对焦炉煤气进行净化处理。

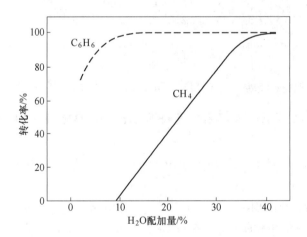

图 5 平衡态下水蒸气配加量与 CH_4、C_6H_6 裂解率的关系

图 6 是焦炉煤气重整裂解净化方案，与方案一相似，该方案直接利用来自焦炉的热焦炉煤气，采用管道喷氧或裂解炉加热的方式将焦炉煤气由 650℃ 左右升高至 800~1000℃，焦炉煤气与水蒸气在裂解炉内催化剂的作用下发生重整反应，除去焦油和 BTX 等杂质，裂解后的焦炉煤气从裂解炉排出后继续进行脱 S、脱氰、脱 NO_x 等净化处理。此处的焦炉煤气重整是半重整，通过配入过量的水蒸气使焦炉煤气中的焦油等大分子化合物优先裂解

而大部分甲烷并未分解。通过添加重整裂解炉，使苯、萘、焦油等杂质在一个反应器内被去除，从而有可能使原有的焦炉煤气净化工艺得到简化。

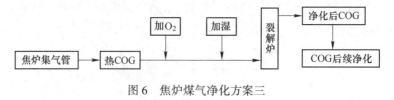

图6 焦炉煤气净化方案三

5 小结

以长流程为主导，合理利用焦炉煤气资源，发展气基直接还原，走氢冶金短流程的道路是我国钢铁行业发展的一个重要方向。利用焦炉煤气生产直接还原铁的关键是如何对煤气进行深度净化以满足还原工艺的需要，HYL-ZR工艺采用在反应塔下部使用焦炉煤气循环冷却热直接还原铁裂解BTX的方案，无法生产热还原铁，从而无法实现竖炉与电炉间的热态输送。为了利用焦炉煤气生产热态直接还原铁，可以考虑改变现有的HYL-ZR技术方案，即在煤气进入直接还原系统之前进行煤气精制。具有发展前景的焦炉煤气处理工艺应该能够合理有效地利用二次热源，同时应具有最低的动力费用消耗，在此基础上本文提出了几种可行的焦炉煤气净化方案。

参考文献

[1] 黄希祜．钢铁冶金原理［M］．北京：冶金工业出版社，1990：80-100.
[2] 郭占成，黄孝文．用焦炉煤气干熄焦和焦炭脱硫的方法［P］．200410078284.2.

Key Technology Analysis of DRI Production by Using Coke Oven Gas as Reduction Agent

Cao Chaozhen   Zhang Fuming   Mao Qingwu   Xu Hui

Abstract: The schemes of DRI production by using coke oven gas and nature gas as reduction agents was compared and the technologies of CH_4 self-reforming and BTX cracking and hot DRI conveying were studied in this paper. It is the key technology of coke oven gas utilization that how to get rid of the influence of the impurities of coke oven gas. The process of coke oven gas purification which has a bright future should make the secondary source of heat be utilized effectively and properly and at the same time it should have a minimum power consumption. Based on the above viewpoint several feasible schemes of coke oven gas purification were proposed.

Keywords: coke oven gas; self-reforming; BTX; DRI

低碳绿色炼铁技术的发展前景与展望

摘　要：第一次工业革命以来，以碳素为基础的钢铁冶金工艺技术不断演进发展，成就斐然。近 200 年来，无论是焦化、铁矿粉造块、高炉炼铁和以碳素为还原剂与发热剂的非高炉炼铁工艺，都是以碳素作为驱动工艺过程的能量，最终排放出大量的 CO_2。面向全球碳减排和碳中和的发展态势，炼铁工业和传统工艺要最大限度减少对碳素能源的依赖，减少碳素消耗和 CO_2 排放，将成为到本世纪中叶的主要发展命题。本文概述了国内外氢冶金和低碳炼铁技术的发展现状，对主要关键工艺技术进行了比对研究和分析评价。结合我国炼铁工业的资源和能源条件，对低碳冶金、氢冶金和碳-氢耦合冶金技术进行了探讨，提出了焦化+烧结+球团+高炉长流程实现低碳冶金和碳-氢耦合冶金的技术路线。针对全氧高炉、氢基竖炉直接还原铁和熔融还原炼铁技术的发展提出了建议，研究分析了其工艺过程的物理本质和关键技术。指出了至 2050 年我国炼铁技术的主要发展理念和技术路径。

关键词：低碳；炼铁；高炉；氢冶金；碳减排

1 引言

在全球积极推进"碳减排和碳中和"的背景下，国内外为数众多的钢铁企业开始研究探索低碳冶金、超低碳冶金、非碳冶金或氢冶金的前沿技术课题。以日本、瑞典、德国、韩国等工业发达国家为代表，相继开展此项研发工作：有的是在原有高炉炼铁工艺的技术上进行改进创新，力图实现碳-氢耦合冶金（日本 COURSE50）[1]；有的是创建全新工艺流程，以"绿氢"为能源，生产钢铁材料，完全摆脱碳冶金的技术路线[2~4]；有的是在已有技术的基础上进行再创新[5,6]，从传统的煤基碳素还原转变为气基氢还原或碳-氢耦合还原。我国宝武、河钢、酒钢等企业或研究机构，也相继开展了低碳冶金或氢冶金的试验探索和工程研究工作，氢-氧高炉、全氧高炉、焦炉煤气-合成气竖炉直接还原，甚至是基于 HIsmelt 铁浴法的熔融还原工艺装置，都在开展工业化应用的前期试验或者已经实现工业化初步应用。

据有关机构测算，我国钢铁工业碳排放量约占碳排放总量的 15%，是制造业中碳排放量最高的行业。面对日益严峻的生态环境保护形势，钢铁工业面临着巨大的碳减排压力，迫切需要开发能够显著降低 CO_2 排放的突破性低碳冶金技术[7]，以满足"碳达峰"和"碳中和"的政策要求[8]。与此同时，我国作为世界上最大的粗钢生产国和消费国，2020 年粗钢产量达到 10.65 亿吨，占全球粗钢产量的 57%，碳排放量占全球钢铁碳排放总量 60% 以上，钢铁工业碳减排的重要性和急迫性刻不容缓。我国目前钢铁制造流程，

本文作者：张福明，程相锋，银光宇，曹朝真。原刊于《2021 年全国高炉·非高炉炼铁技术学会议论文集》，2021：1-11。

仍是以高炉+转炉的传统"长流程"工艺结构占主导地位。2020 年高炉生铁产量达到 8.88 亿吨,转炉生产粗钢比率约占 90% 左右[9];钢铁制造流程能源结构"高碳化",煤、焦炭等碳素消耗占能源总量近 90%。

全球范围内,目前已有许多国家和钢铁企业对低碳冶金技术,特别是氢冶金技术进行了战略布局,而且呈现出迅猛发展态势。我国应积极探索和统筹规划适合我国国情的低碳冶金和节能减排技术路线图,对未来技术研发方向和发展目标进行系统科学的战略谋划,面对世界百年未有之大变局,承担起钢铁大国的责任担当。

2 国外相关技术发展现状

2.1 日本 COURSE50 项目

2.1.1 项目研发概况

COURSE50 项目是由日本钢铁联盟发起,并得到日本新能源和产业技术开发组织(NEDO)的支持,由神户制钢、JFE、新日铁、住友金属、日新制钢和新日铁工程公司共同研发的突破性炼铁工艺技术[10]。日本根据现代钢铁材料应用现状、性能质量和生产技术的发展趋势,预测到 2050 年以后,仍然需要大量使用铁矿石通过高炉冶炼工艺,制造高品质、高性能的钢铁材料。因此,COURSE50 项目是利用天然铁矿石,采用高炉炼铁工艺实现 CO_2 大幅度减排的新技术研发项目。

COURSE50 项目的核心关键目标,旨在开发减少高炉炼铁工艺 CO_2 排放、包括从高炉煤气中分离和回收 CO_2 的新型高炉炼铁工艺技术,最终减少约 30% 的 CO_2 排放量[11]。项目计划在 2008~2030 年完成全部技术研发工作;2030~2050 年完成全部技术推广应用工作,实现新研发技术的工程化和产业化。按项目研发最终目标值测算,2050 年以后,日本炼铁工序的吨铁 CO_2 排放量将从目前 1.64t 降低到 1.15t。图 1 为 COURSE50 项目工艺流程。

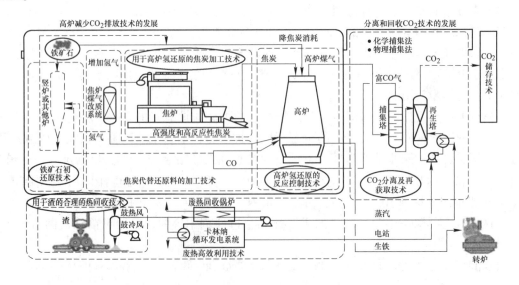

图 1 COURSE50 项目工艺流程

日本研究者认为，减少高炉炼铁工艺 CO_2 排放的关键技术，是开发出高炉工艺氢还原铁矿石的新技术，其目标是实现 CO_2 减排约 10%。强化高炉内氢还原铁矿石过程需要两项支撑技术：一是提高 H_2 体积分数的焦炉煤气重整/改质技术；二是高强度、高反应性焦炭的生产技术。COURSE50 项目高炉煤气中 CO_2 的分离和回收技术，其研发目标是实现 CO_2 减排约 20%。高炉富氢还原和炉顶煤气中的 CO_2 分离和回收，合计可降低 CO_2 排放约为 30%。同时，为了避免高炉煤气中 CO_2 脱除回收过程新增能源的 CO_2 排放，还需要开发出钢铁厂未能全量利用的新型余热回收技术，从而支撑高炉煤气中 CO_2 的分离、回收的能量输入。

2.1.2 项目研发进展

COURSE50 项目在 2018 年先后进行了 5 次试验（从 2018 年 10 月 29 日到 11 月 27 日，共计 30 天），通过改变还原气体和原材料配比，试验了氢气对炼铁过程的改善效果。第二阶段的研究重点是在高炉煤气中分离回收 CO_2 和余热，并计划于 2022 年度进行实际高炉的放大试验测试。

2018 年，项目研发者采用三维高炉数学模型，研究了氢气喷吹技术对 CO_2 排放减排的影响；2022 年度将在 2 座高炉上进行工业试验，目前已开展高炉工业性试验的前期准备工作。

2.2 瑞典钢铁公司的"突破性氢能炼铁技术"（HYBRIT）

2.2.1 项目研发概况

2016 年，瑞典钢铁公司（SSAB）、LKAB（欧洲最大的铁矿石生产商）和 Vattenfall（大瀑布公司，欧洲最大的能源公司之一）联合创建基于氢能冶金的 HYBRIT 项目[12]。目标是使用非化石质燃料所生产的电力和氢气替代炼焦煤等传统工艺，以实现无化石能源的炼钢技术。2018 年夏该项目进入中试阶段，在瑞典吕勒奥的 SSAB 工厂开工建设了全球第一个无化石能源的钢铁制造中试工厂。在 HYBRIT 中试工厂中，采用电解水产生氢气，试验氢气直接还原技术的可行性，采用该技术生产直接还原铁（DRI），然后将 DRI 与废钢等一起用于炼钢生产。

2019 年夏，在瑞典 Malmberget 的 LKAB 工厂建造无化石燃料的球团生产线，采用生物质燃料替代化石燃料，以实现基于无化石燃料工艺的铁矿石球团制造技术。另外，计划在 Svartberget 的 LKAB 工厂建造一个位于地下 25~35m 的氢气存储中试装备，该项目靠近吕勒欧的 HYBRIT 中试工厂。预计该存储中试装备将于 2022~2024 年投入运行。

2020 年 8 月 5 日，LKAB 在球团生产工艺中使用无化石燃料工业化试验获得成功。新一代球团厂的开发正在进行中，LKAB 对 HYBRIT 的主要研发工作是开发无化石燃料的球团生产工艺，因此需要在球团焙烧过程中改进和创新加热工艺。LKAB 在 Malmberget 一个现有的球团厂，正在进行全面的试验研发工作，例如用生物油代替石油，在试验期内，该项目的 CO_2 排放量减少了 40%，该项试验将持续到 2021 年。除此之外，以无化石燃料电力和生物燃料为基础的替代燃料也进行了试验，其中包括采用氢气和等离子体等。

2.2.2 项目未来研发计划

(1) 建设工业级的示范线。HYBRIT 工业化示范线将于 2023 年开工建设，计划于

2025年示范工厂建成投产。该示范工厂产能将达到100万吨/年，约为LKAB在Malmberget总产能的20%，约占SSAB吕勒欧厂高炉产能的50%。与此同时，SSAB奥克斯松兰德厂的高炉将进行改造，计划将于2026年实现工业化无化石能源的铁矿石冶炼工艺突破。SSAB最早将于2025年之前，有望将CO_2排放量减少25%。

（2）示范工厂工业化试验成功之后，将进行传统工艺流程的彻底改造。计划于2030~2040年将其全部的传统高炉流程转变为电炉流程。同时，SSAB开始逐步淘汰全公司的轧钢厂和热处理厂的化石燃料燃烧加热工艺，2045年全面实现无化石能源的钢铁制造。

高炉流程和HYBRIT流程的对比如图2所示。

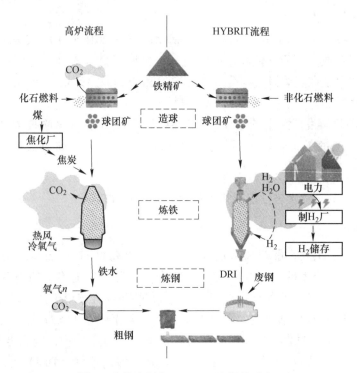

图2 高炉流程和HYBRIT流程的对比

2.3 高炉喷吹富氢还原气体

2.3.1 德国蒂森克虏伯高炉喷吹氢气试验项目

德国蒂森克虏伯计划将在2050年前实现碳中和的战略目标，实现温室气体的"净零排放"。为实现这个战略目标，蒂森克虏伯制定了两个关键性中期目标：一是到2030年，生产和能源供应系统的CO_2排放量比2018年减少30%；二是使用该公司产品和技术客户的CO_2排放量比2017年减少16%（例如汽车制造企业、家电制造企业等）。2018年蒂森克虏伯CO_2排放量约为2400万吨，约占德国CO_2排放总量的3%。

蒂森克虏伯作为国际先进的钢铁制造企业，采用传统高炉炼铁工艺，其焦比、煤比和燃料比分别约为300kg/t、200kg/t和500kg/t，在国际上处于先进行列，特别是蒂森克虏伯的高炉喷煤技术，近30年来一直处于领先水平。面向本世纪中期的碳中和战略，蒂森克房伯认为以氢能的利用是实现碳中和战略目标的关键。

2019 年 11 月，杜伊斯堡厂 9 号高炉（容积为 1833m³，炉缸直径为 10m，日产量约为 4600t/d），开始进行喷吹氢气炼铁试验，在 1 个风口上喷吹氢气代替煤粉，取得初步成效，计划下一阶段逐步在高炉 28 个风口上全部喷吹氢气，并将从 2022 年起开始在北莱茵-威斯特法伦州的全部 3 个高炉上喷吹氢气，预计可减少 20% 的 CO_2 排放。与此同时，在实现氢能稳定供应的条件下，蒂森克房伯已经开始规划建造利用氢气进行直接还原的冶金工厂。

2.3.2 安赛洛米塔尔纯氢炼铁技术研发

在全球积极推动碳中和的背景下，安米集团计划投资 6500 万欧元，在德国汉堡厂进行氢直接还原铁矿石的项目研究，项目研发技术路线与瑞典 HYBRIT 类似，并计划在未来几年建设中试工厂。目前安米汉堡厂采用天然气生产直接还原铁，计划未来几年，在汉堡厂进行利用氢直接还原铁矿石的工艺试验，中试厂的规模为 10 万吨/年。该研究项目的氢气制备将采用变压吸附工艺，从安米汉堡厂的冶金煤气中分离氢气，使其纯度达到 95% 以上，待未来有足够数量绿氢（来自可再生能源的氢）时，将采用绿氢生产直接还原铁。

安米在德国不莱梅厂的研究项目是通过电解槽制备氢气，并将氢气喷入高炉风口。该项目将可以减少高炉炼铁过程中所需的碳素消耗，从而减少 CO_2 排放。

安米在法国敦刻尔克厂，正在开发将钢铁生产过程中产生的副产煤气转化成合成气体喷入高炉的工艺，包括向高炉喷吹直接还原副产煤气技术，以及向高炉风口喷吹合成气体技术。开展炼铁高炉-直接还原竖炉耦合工艺流程的规模性工业试验，利用合成气体替代化石质燃料将有助于大幅降低 CO_2 排放，在可以大规模获取"绿氢"时，将向高炉喷吹"绿氢"，进而实现炼铁过程的 CO_2 净零排放。安米公司计划在欧洲板材生产基地应用这项技术，将不同来源的气体（如富氢焦炉煤气）喷入高炉，这种高效、经济的技术方案，可以有效减少钢铁生产过程中的 CO_2 排放。安米阿斯图里亚斯厂拥有先进的焦炉煤气处理工艺，可以向高炉喷吹"灰氢"，即从天然气和焦炉煤气中提取氢气，该项目将于 2021 年启动。

2.3.3 德国迪林根和萨尔钢公司富氢炼铁技术

德国迪林根（Dillinger）和萨尔钢铁公司（Saarstahl）计划投资 1400 万欧元，从 2020 年开始，用于研究将钢铁厂的富氢焦炉煤气，在萨尔钢铁公司的 2 座高炉上喷吹，从而实现利用氢气替代部分碳素燃料的工艺技术，这项研究所涉及的设备及基础设施不影响高炉运行。

在钢铁生产中使用氢能是减少 CO_2 排放的一个关键技术要素，高炉喷吹富氢焦炉煤气是生产绿色钢材和合理利用资源/能源的重要措施。在高炉生产过程中，氢气代替碳素作为还原剂和能量载体，可以有效降低在高炉生产过程 CO_2 排放。

2020 年 8 月，迪林根（Dillinger）钢铁公司发布，该厂建成并投产了德国首家在高炉常规运行中，使用氢气作为还原剂的氢基钢铁生产厂，该厂的钢铁生产新工艺由 Dillinger 和 Saarstahl 钢铁公司共同开发和投资。该工艺通过向高炉中喷吹富氢焦炉煤气，进一步减少 CO_2 排放，目前已经取得在钢铁生产中使用氢气的工程经验，未来计划在 2 座高炉上喷吹纯氢。两家钢铁公司的目标是到 2035 年将使 CO_2 排放量减少 40%，除此之外，两家公司还在进一步推进大型能效提升项目，并对现有系统的优化进行可持续投资。

2.4 氢能竖炉直接还原技术

2.4.1 安米氢基 DRI-EAF 工艺流程研究

安米欧洲公司在汉堡厂拥有欧洲唯一的直接还原-电炉（DRI-EAF）工艺装置，计划在汉堡厂进行氢还原铁矿石生产直接还原铁（DRI）的工业化规模试验，以及电炉炼钢流程使用无渗碳 DRI 进行工业试验。除此之外，安米法国敦刻尔克厂已经启动一项研究工作，建造一座大型 DRI 工厂并配置 EAF。最初考虑使用天然气竖炉工艺生产 DRI，结合自身在 DRI 生产方面的独特经验，以及汉堡厂在氢基 DRI 项目的工业试验结果，目前考虑将全部使用氢生产 DRI。

2.4.2 氢基 Midrex 工艺

研究认为，气基竖炉直接还原技术利用天然气、页岩气、煤层气、焦炉煤气以及合成气等多种氢基气体燃料，生产 DRI 可以显著降低 CO_2 排放[13~16]。神户制钢全资子公司米德雷克斯的 Midrex 工艺已经得到商业化应用，而且已被证实为 CO_2 排放量最低的炼铁工艺路线，这是因为该工艺采用的天然气的氢气含量高于燃煤。典型 Midrex 工厂所使用的还原气体中 H_2 和 CO 的体积百分数分别为 55% 和 36%，而高炉使用的还原性气体中 CO 体积百分数约为 90% 以上。因此，与电炉配合使用的 Midrex 工艺吨钢 CO_2 排放量约为高炉-转炉流程的 50%。截至 2018 年底，全球共有 79 座 Midrex 工厂投入运行，年产量约为 6000 万吨。

目前采用 Midrex 工艺生产的直接还原铁（DRI）产量超过全球产量的 60%。在建工厂投产后，预计 2020 年将达到年产 7500 万吨/年的规模。为显著减少钢铁工业的碳足迹，该公司正在开发一种利用低碳能源制备纯氢气并利用所制氢气生产直接还原铁的技术，被命名为"MIDREX H_2"，该技术在现有或新的直接还原铁厂都具有相当的应用潜力。

3 常规高炉低碳炼铁技术的发展前景

3.1 优化高炉炉料结构

新中国成立 70 多年来，我国钢铁工业发生了翻天覆地的变化。由于铁矿石资源禀赋和技术传承等多种原因，我国高炉炉料一直以烧结矿为主，配加少量酸性球团或块矿。进入 21 世纪以后，我国球团技术取得飞跃性发展，大型链算机-回转窑和大型带式焙烧机球团生产线相继建成投产，高炉球团矿入炉比率不断提高（见图 3），为降低铁前工序的能耗、CO_2 和污染物排放，提高高炉入炉矿品位，促进高炉高效生产发挥了重要作用[17]。

目前我国大多数钢铁厂都配有球团生产线，在现有铁前系统工艺装备和技术条件下，应充分发挥现有球团产线的生产效能，进一步加大球团矿入炉比率，提高高炉入炉矿综合品位，降低燃料比和渣量，进而降低 CO_2 排放[18]。

研究表明，带式焙烧机使用富氢焦炉煤气生产球团矿时，吨矿 CO_2 排放量仅为 60kg/t；而采用碳素固体燃料为主生产烧结矿时，吨矿 CO_2 排放量约为 155kg/t。显而易见，烧结工序的 CO_2 排放约为球团工序 2.5 倍，因此发展球团工艺、提高球团矿入炉比率，非常有利于降低铁前系统的 CO_2 排放。

以首钢京唐为例，目前 3 座 5500m³ 高炉球团矿入炉比率已稳定达到 55% 左右，高炉

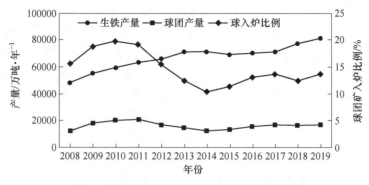

图 3 我国高炉生铁产量和炉料结构变化趋势

渣比约为 220kg/t,高炉燃料比为 480~490kg/t[19]。由于高炉使用超低硅碱性球团矿,综合入炉矿品位提高(61.5%~62%),高炉燃料比和渣比显著降低,极其有利于降低 CO_2 排放。首钢京唐球团设计产量为 1200 万吨/年,烧结矿产量为 1100 万吨/年,如果充分发挥现有球团产线的生产效能,每年增产球团矿 100 万~150 万吨/年,可将单座高炉球团矿入炉比率提高到 70% 以上。

基于现有条件,完全可以在单座高炉上试验 70% 以上球团矿冶炼,积极探索通过炉料结构优化和改进实现碳减排的发展路径。理论计算表明,首钢京唐单座高炉球团矿比率从 55% 提高至 70%,即每年用 100 万吨/年球团矿替代烧结矿,球团矿替代烧结矿而带来的降低 CO_2 排放量约为 12.9 万吨/年。同时高炉球团矿入炉比率达到 70%,高炉渣量将降低至 200kg/t 以下(约为 180kg/t),燃料比可降低到 470kg/t(接近常规高炉碳基冶金的理论燃料比 465kg/t),力争实现常规高炉超低燃料比冶炼,降低燃料比而带来的 CO_2 排放降低量为 12.1 万吨/年,因此单座高炉如果球团矿入炉比率达到 70%,每年降低 CO_2 排放量可达 25 万吨。由此推论,3 座 5500m³ 高炉提高球团矿入炉比率达到 60%,可以降低 CO_2 排放约 30.3 万吨/年。

3.2 优化高炉操作

3.2.1 提高并稳定风温

高风温是现代高炉冶炼的热能基础,高炉鼓风所带入的物理热从某种意义上讲,是一种"清洁低碳能源",是回收利用高炉冶金过程伴生煤气而获得的高温热能[20]。现代高炉鼓风所带入的物理热一般占能量输入的 18%~25%,是高炉冶金过程"三传一反"的热力学基础,也是冶金过程基元反应的重要热量来源。

提高风温的现实意义在于:有效降低碳素消耗和碳排放、维持合理的理论燃烧温度、为风口喷吹燃料提供能量基础和保障。现代高炉风温应达到 1200℃ 以上,有条件的要达到 1250℃,技术装备领先的高炉要力争达到 1280±20℃。

3.2.2 改善高炉送风操作

高炉透气性是未来高炉超高强度冶炼、超低碳冶炼的最关键的制约环节。在极低焦炭消耗的条件下,改善高炉透气性将成为高炉炼铁技术发展最重要的研究课题,也是高炉炼铁工艺生存和发展的重大技术命题。

面向未来,随着可再生能源发电技术的推广应用,我国电力结构将发生根本性转变(目前煤基火力发电量约占 70%),基于可再生能源产生的"绿电"占比将逐步提高。提高鼓风富氧率(5%~10%)、降低吨铁鼓风量(<800m³/t)将成为现实,"绿电"将作为一种清洁"零碳"能源将被广泛应用于冶金工业。随着富氧率的提高,炉腹煤气量降低,有助于提高 CO 和 H_2 的还原势,对于改善高炉透气性,促进间接还原过程、提高生产效能都将发挥重要作用,这将成为未来高炉实现低碳冶炼的一个重要途径。

应当指出,未来随着社会废钢资源的增长,电炉流程的逐步推广应用将成为不争的事实。预计未来在高炉数量减少、高炉生铁总量减少的条件下,单座高炉的生产效能还将会进一步提高,采用高富氧-大喷煤和氧-煤强化高炉冶炼将成为主要技术途径之一。

高风温与高富氧技术耦合匹配,是未来高炉大量喷吹富氢(氢基)燃料的重要基础。无论高炉喷吹天然气、焦炉煤气还是"电解制备的绿氢",高风温和富氧都是不可或缺的重要支撑技术。2020 年我国 300 余座高炉风温与燃料消耗、富氧率与燃料消耗及利用系数的关系如图 4~图 7 所示。

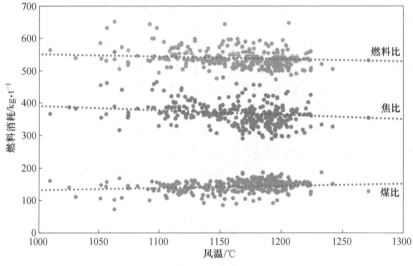

图 4 高炉风温与燃料消耗的关系

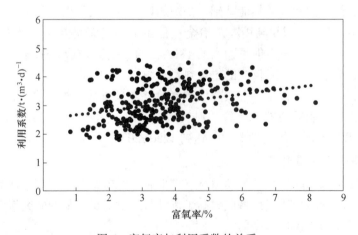

图 5 富氧率与利用系数的关系

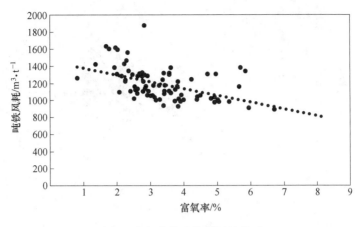

图 6 富氧率与吨铁风耗的关系

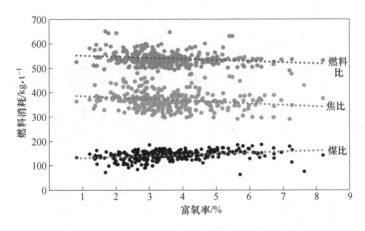

图 7 富氧率与燃料消耗的关系

3.2.3 提高炉顶压力

提高炉顶压力实质上是压缩了煤气体积、降低了煤气流速、延长了煤气在高炉内的停留时间，进而改善间接还原动力学条件，从而有助于煤气和铁矿石之间的气-固反应，增强了煤气的扩散、穿透能力，强化了煤气吸附、界面反应等还原过程。在宏观上提高顶压可以有效提高高炉透气性，促进高炉稳定顺行。近年来，我国高炉顶压提高幅度较大，一些先进高炉的顶压已经达到 280kPa、接近 300kPa，有效提高了煤气利用率、降低了燃料消耗。图 8 和图 9 是 2020 年全国 300 余座高炉炉顶压力和煤气利用率及燃料消耗的关系。

3.3 发展理念与目标

3.3.1 工程理念

面向未来，以高炉为中心的炼铁系统协同优化和动态有序、协同连续、精准高效运行，是炼铁系统流程结构优化和技术发展的重点。必须加强以高炉稳定顺行为基础的工程运行理念，建立系统性、全局性的工程思维模式，以技术、经济多目标整体优化为导向，

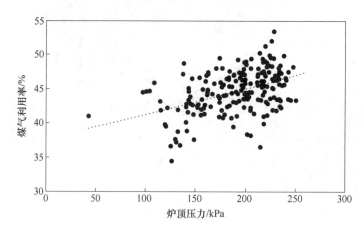

图 8　高炉炉顶压力和煤气利用率的关系

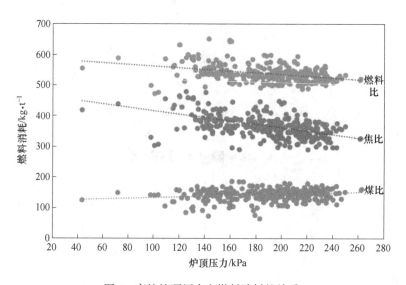

图 9　高炉炉顶压力和燃料消耗的关系

不片面追求所谓的"超高利用系数""极低成本运行"以及个别技术指标的"领先",摒弃不讲客观、不顾原燃料条件、不论技术装备条件的盲目攀比和"比大比小"。

遵循钢铁制造流程的基本规律[21],科学认识高炉冶炼过程的动态运行规律,不断总结提升,加强知识管理,做好卓越炼铁工程师的培养,造就基础扎实、经验丰富、求真务实、视野开阔的领军人才及团队,形成具有企业特色的现代化高炉炼铁生产、运行管理的工程思维和工程理念。

3.3.2　发展目标与路径

(1) 在常规高炉使用固体碳素燃料(焦炭+煤粉)的条件下,降低碳素燃料消耗是减少 CO_2 排放最直接、最有效的技术措施。因此,到 2030 年前后,我国高炉燃料比应普遍降低到 520kg/t 以下(达标水准);一大批装备精良的高炉燃料比应低于 500kg/t (平均先进水平);部分先进高炉应将燃料比降低到 480kg/t 以下(低碳);领先高炉燃料比应降低到 470kg/t 甚至更低（超低碳）。

（2）进一步加强精料技术研究，探索并构建以球团矿为主的新型炉料结构体系，降低整个炼铁流程的碳素消耗和污染物排放，进而降低 CO_2 排放。

（3）继续推进高风温、富氧、喷煤、高顶压等关键技术的再创新，进一步降低高炉燃料消耗，提高煤气利用率，促进高炉稳定顺行，减少 CO_2 排放。

（4）加强大型高炉操作规律的研究，建立动态有序、协同连续、精准高效的现代高炉运行理念，以高炉生产长期稳定顺行为基础，不断改善、优化、提升大型高炉的动态精准操控水平。

（5）进一步加强高炉、热风炉等主体系统运行的过程智能化监测和维护，采取有效技术措施延长高炉寿命，奠定高炉高效低耗低排放运行技术装备基础，提高高炉生产效能[22]。

（6）构建料场、烧结、球团、焦化、高炉炼铁系统一体化集中智能管控平台，着重解决不同工序的界面技术优化，实现物质流、能量流和信息流的高效协同运行，通过炼铁工序全流程智能化动态管控，以高炉为中心构建信息物理系统（CPS）和集成控制中心，提高智能化精准控制水平。

4 探索碳-氢耦合冶金技术

4.1 烧结过程富氢气体燃料喷吹技术

烧结过程在料面喷吹天然气或焦炉煤气，可以有效降低固体燃料消耗，改善料层传热机制，进而还可以降低烟气中 CO 等有害物质的排放[23]。目前铁前工序的 CO_2 排放约占全流程的 80%，其中每吨烧结矿碳素固体燃料消耗约为 50~60kg/t，CO_2 排放约为 150~170kg/t，所以降低烧结工序的碳素燃料消耗对降低碳排放意义重大。当前国内不少烧结机正在开展"烧结烟气双循环"和烧结料面喷吹天然气（或焦炉煤气）的技术升级改造。根据测算，烧结实现烟气高效循环利用和"热风烧结"，固体碳素燃料消耗可降低约 2kg/t；喷吹天然气可降低固体碳素燃料约 2.5kg/t，两项技术集成应用可以降低碳素固体燃料消耗约 4.5kg/t。因此如果烧结实现烟气循环和喷吹天然气技术，每吨烧结矿的 CO_2 排放可降低约 12kg/t，因此建议有条件的企业要加快推进烧结工序节能减排工艺系统改造。

4.2 高炉喷吹富氢气体燃料

结合未来能源供给条件，具备良好能源基础和技术基础的企业，在可以在高炉上开展喷吹天然气（或焦炉煤气）工业性试验及应用。理论研究和生产实践表明，通过高炉风口向高炉喷吹天然气，有利于提高高炉产量和降低燃料比。俄罗斯和美国等天然气资源充沛的国家，高炉从 20 世纪 70 年代起就开始喷吹天然气，取得了提高产量、降低焦比和燃料比的生产效果。高炉喷吹天然气技术较为成熟，2019 年美国安塞洛米塔尔公司印第安纳港 7 号高炉（4800m^3）天然气喷吹量为 50kg/t，高炉燃料比 480kg/t[24~28]。根据理论推算，高炉喷吹 1m^3 天然气可替代 1.25kg 碳素。天然气富含甲烷和氢，高炉喷吹以后甲烷在高温条件下经过裂解变成 H_2 和 CO，可以提高炉腹煤气中 H_2 和 CO 体积百分数，提高高炉煤气还原势。在高风温（1200~1250℃）的基础上，匹配富氧鼓风，通过提高富氧率，可以保持合理的风口理论燃烧温度、降低炉腹煤气量，进而促进间接还原、降低碳素燃料消耗，在现有基础上进一步降低 CO_2 排放。因此，常规高炉喷吹富氢/含氢气基燃料（还原剂）是长流程钢铁工艺开展氢冶金的有效途径之一。

以 5500m³ 高炉为例，根据理论计算，喷吹天然气 30kg/t，高炉每天需要 412.5t/d 天然气（按高炉利用系数 2.5t/(m³·d)、日产量 13750t/d 测算），折合 53 万立方米/天。由于固体碳素燃料消耗的降低，理论计算吨铁可降低 CO_2 排放约 33~36kg/t，高炉预计每年可减排 CO_2 约为 14.5 万吨/年。如果喷吹焦炉煤气，由于 H_2 的体积百分数更高（>50%），理论上 CO_2 的减排效果会更加显著。

因此，有条件的高炉喷吹天然气、焦炉煤气等氢基燃料，是减少碳素燃料消耗和 CO_2 排放的一个新的技术途径。在当前技术条件下，"绿氢"的大量制备和使用仍处于研究探索阶段，规模化、商业化的应用还有待时日，与此同时经济性也需要结合碳减排政策制度量力而行，不可一哄而上。

4.3 高炉炉顶煤气循环技术

高炉炉顶煤气中，CO 和 H_2 的体积分数分别约为 20%~28% 和 1%~6%，高炉煤气低发热值约为 3000~3300kJ/m³。传统工艺中一般将高炉煤气作为气体燃料使用，主要用于热风炉、焦炉、CCPP、发电机组的燃烧，因此煤气中的 CO 经过燃烧后，最终生成 CO_2 排放。高炉煤气顶气脱除 CO_2（体积分数约为 18%~23%）以后，不仅可以实现 CO_2 的分离捕集（CCS），还提高了炉顶煤气的热值。因此俄罗斯和日本将炉顶煤气脱除 CO_2 以后再经过风口或炉身喷口喷入高炉，实现顶气的循环再利用[29]。这种顶气脱除 CO_2 再循环工艺，在气基竖炉直接还原工艺中（如 HYL-ZR 和 Midrex 等）已广泛应用，属于成熟可靠技术，只不过气基竖炉的顶压比高炉更高（约 0.5~0.6MPa）。

对于喷吹富氢燃料和高富氧的高炉，由于炉腹煤气量的大幅度降低，风口前理论燃烧温度又不可能大幅度提高，加之 H_2 还原 FeO 的过程为吸热反应（见式（1）），因此高炉"三传一反"冶金过程将可能会出现供热不足的情况，因此从炉身喷吹炉顶循环煤气，其技术本质就是实现碳-氢耦合冶金，因为 CO 还原 FeO 的过程是放热反应（见式（2）），这样可以把传热、传质和动量传输的"三传过程"，在新的炉料结构和燃料结构条件下，建立起新的动态耗散结构，并实现过程优化，达到铁素物质流、碳素能量流、氢素能量流和热量的动态耦合和协同运行。

$$FeO + H_2 = Fe + H_2O \quad \Delta H_{298}^{\ominus} = 28010 J/mol \quad (1)$$

$$FeO + CO = Fe + CO_2 \quad \Delta H_{298}^{\ominus} = -13190 J/mol \quad (2)$$

4.4 气基竖炉直接还原

我国高炉—转炉长流程占 90% 以上，由于高炉无法完全摆脱对焦炭的依赖，所以高炉—转炉长流程工艺实现碳中和的提升空间有限[30]。研究表明采取富氢喷吹、炉顶煤气脱除 CO_2 后循环利用等综合技术以后，大约可以降低 CO_2 排放约 30%。基于绿氢气基直接还原铁+电炉冶炼+绿电是实现碳中和最为有效的技术途径，目前国外不少钢铁企业计划通过氢气（绿氢）直接还原铁（DRI）+电炉（EAF）技术路线实现 CO_2 排放大幅度降低。

以首钢京唐为例，目前球团矿产量为 1200 万吨/年，采用秘鲁高品位铁矿粉（TFe=69.8%），铁矿粉中 SiO_2 质量百分数低于 1.5%，可以制备出高品位低硅优质球团矿。由于球团工序具有一定的富余产能，可以生产 50 万吨/年直接还原用球团矿。可以预见，我国部分钢铁企业具备生产 DRI 的球团制备和供给能力，同时具备较为充沛的天然气和焦炉煤气资源，具备装备先进、流程合理的竖炉+电炉（DRI+EAF）制造流程和良好的资源、能源供给条件，以气基竖炉替代高炉、以电炉替代转炉，配加大宗废钢，或采用全废钢冶炼，下游工序可以和现有长材或薄板坯连铸连轧产线流程匹配衔接，进而实现钢铁制造流程重构优化和集成优化。预计未来，钢铁制造流程结构调整和重构优化，将是 2030 年以前我国钢铁工业发展的一个重点课题。

面向未来，我国不少钢铁联合企业具备高炉-转炉长流程（BF+BOF）和竖炉-电炉短流程（DRI+EAF）"长短结合"耦合匹配的基础条件，可以形成长短两种流程的集成优化和动态协同，以长流程为主、以短流程为补充，面向不同的产品需求和产线配置，采取不同的工艺技术路线，进一步加大废钢资源的使用比率，从而降低 CO_2 排放总量和吨钢排放量。因此开展流程耦合集成的前期技术储备，做好气基直接还原竖炉+电炉流程工艺的概念研究和顶层设计，具有十分重要的现实意义和长远意义。

因此利用焦炉煤气、煤制气、天然气以及绿氢，建设规模适度的 DRI 直接还原装置，热装热送加入电炉（EAF），再配加一定量的废钢，可以生产以长材或薄规格带材产品（薄板坯连铸连轧 ESP/CSP），可以有效降低钢铁制造全流程的 CO_2 排放。

5 结语

（1）在全球碳达峰、碳中和的背景下，钢铁工业作为重工业的代表性产业，应顺应时代潮流，在已有技术的基础上降低化石质碳素能源的消耗，改进工艺流程、创新关键技术，大幅度降低碳素消耗和 CO_2 排放。

（2）当前技术条件下，以常规高炉冶炼为主流的炼铁工艺，喷吹富氢燃料替代固体碳素燃料、炉顶煤气脱除 CO_2 后循环再利用、以高炉流程为基本工艺路线的碳-氢耦合还原；以绿氢为代表的气基竖炉直接还原和以生物质能源为驱动的球团矿制备等新兴技术，将在未来取得工程化应用和推广。

（3）发挥既有资源优势，开展低碳绿色炼铁新技术的探索研究。依托国内外资源，优化炉料结构、提高球团矿入炉比率，积极推进造块工序有效降低碳素燃料消耗和碳减排工作。在采用复合喷吹氢基燃料和采用顶气脱除 CO_2 后循环等先进技术以后，先进高炉燃料比降低到 470kg/t 以下，进而实现 CO_2 排放减少 30% 以上。

参考文献

[1] Shigeaki Tonomura, Naoki Kikuchi, Natsuo Ishiwata, et al. Concept and current state of CO_2 ultimate reductionin the steelmaking process (COURSE50) aimed at sustainability in the Japanese steel industry [J]. J. Sustain. Metall, 2016 (2): 191-199.

[2] Ahmed I Osman, Mahmoud Hefny, Maksoud M, et al. Recent advances in carbon capture storage and utilisation technologies: A review [J]. Environmental Chemistry Letters, 2020, 19 (1).

[3] 张京萍. 拥抱氢经济时代全球氢冶金技术研发亮点纷呈 [N]. 世界金属导报, 2019-11-26 (1).

[4] 胡俊鸽, 王再义, 陈妍, 等. 日本钢铁业研究的减排 CO_2 炼铁新技术 [J]. 冶金丛刊, 2011, 193 (3): 47-50.

[5] 严珺洁. 超低二氧化碳排放炼钢项目的进展与未来 [J]. 中国冶金, 2017, 27 (2): 6-11.

[6] 郑少波. 氢冶金基础研究及新工艺探索 [J]. 中国冶金, 2012, 22 (7): 1.

[7] 赵沛, 董鹏莉. 碳排放是中国钢铁业未来不容忽视的问题 [J]. 2018, 53 (8): 1-7.

[8] 张贤. 碳中和目标下中国碳捕集利用与封存技术应用前景 [J]. 可持续发展经济导刊, 2020 (12): 22-24.

[9] 王维兴. 2020 年我国炼铁技术发展评述 [J]. 世界金属导报, 2021-4-6 (B02).

[10] Kota Moriya, Koichi Takahashi, Akinori Murao. Effect of large amount of co-injected gasesous reducing agent on combustibility of pulverized coal analyzed with non-contact measurement [J]. ISIJ International, 2020, 60 (8): 1662-1668.

[11] Murao Ryota, Michitaka Sato, Tatsuro Ariyama. Design of innovative blast furnace for minimizing CO_2 emission based on optimization of solid fuel injection and top gas recycling [J]. ISIJ International, 2004, 44 (2): 2168-2177.

[12] 罗晔. HYBRIT 项目将进入中试阶段 [J]. 世界金属导报, 2018-08-28 (B01).

[13] 于恒, 周继程, 郦秀萍, 等. 气基竖炉直接还原炼铁流程重构优化 [J]. 中国冶金, 2021, 31 (1): 31-35, 45.

[14] 张福明, 李林, 刘清梅. 中国钢铁产业发展与展望 [J]. 冶金设备, 2021 (2): 1-6, 29.

[15] 张福明, 曹朝真, 徐辉. 气基竖炉直接还原技术的发展现状与展望 [J]. 钢铁, 2014, 49 (3): 1.

[16] 曹朝真, 张福明, 毛庆武, 等. 利用焦炉煤气生产直接还原铁关键技术分析 [C] //第八届中国钢铁年会论文集. 北京: 中国金属学会, 2011: 267.

[17] 张春霞, 上官方钦, 胡长庆, 等. 钢铁流程结构及对吨钢 CO_2 排放的影响 [J]. 钢铁, 2010, 45 (5): 1.

[18] 张福明. 中国高炉炼铁技术装备发展成就与展望 [J]. 钢铁, 2019, 54 (11) 1-8.

[19] 张福明. 首钢绿色低碳炼铁技术的发展与展望 [J]. 钢铁, 2020, 55 (11) 11-18.

[20] 张福明, 银光宇, 李欣. 现代高炉高风温关键技术问题的认识与研究 [J]. 中国冶金, 2020, 30 (12): 1-8.

[21] 张福明. 钢铁冶金从技艺走向工程科学的演化进程研究 [J]. 工程研究-跨学科视野的工程, 2020, 12 (6): 527-537.

[22] 张福明. 我国 5000m^3 级高炉技术进步与运行实绩 [J]. 炼铁, 2021, 40 (1): 1-8.

[23] 张俊杰, 裴元东, 周晓冬, 等. 550m^2 烧结机喷吹天然气工艺实践 [N]. 世界金属导报, 2021-02-02 (B02).

[24] Meng Xianglong, Zhang Fuming, Wang Weiqiao, et al. Analysis of pulverized coal and natural gas injection in 5500m^3 blast furnace in Shougang Jingtang [C] //AISTech 2015 Proceedings, AIST, Cleveland, 2015: 946-958.

[25] Halim K S A. Effective utilization of using natural gas injection in the production of pig iron [J]. Materials Letters, 2007, 61 (14-15): 3281-3286.

[26] Michal J Wojewodka, James P Keith, Stephen D Horvath, et al. Natural gas injection maximization on C and D blast furnace at arcelorMittal burns harbor [C] //AISTech 2014 Proceedings, AIST, Indianapolis, 2014: 767-779.

[27] Armin K Silaen, Tyamo Okosun, Zhao Jiaqi, et al. Study of injection natural gas into blast furnace [C] // AISTech 2016 Proceedings, AIST, Pittsburgh, 2016: 595-602.

[28] Megha Jampani, Chris P Pistorius. Increased use of natural gas in blast furnace ironmaking: Methane reforming [C] //AISTech 2016 Proceedings, AIST, Pittsburgh, 2016: 573-580.

[29] Hiroshi Nogami, Jun-ichiro Yagi, Shin-ya Kitamura, Peter Richard Astin. Analysis on material and energy balances of ironmaking systems on blast furnace operations with metallic charging, top gas recycling and natural gas injection [J]. ISIJ International, 2006, 46 (12): 1759-1766.

[30] 李峰,储满生,唐珏,等. 基于LCA的煤制气—气基竖炉—电炉短流程和高炉—转炉流程环境影响分析 [J]. 钢铁研究学报, 2020, 32 (7): 577-583.

Development Prospect and Perspective on Low Carbon Green Ironmaking Technology

Zhang Fuming   Cheng Xiangfeng   Yin Guangyu   Cao Chaozhen

Abstract: Since the first industrial revolution, carbon-based iron and steel metallurgical technology has been continuously evolving and developing with remarkable achievements. In the past 200 years, no matter coking, iron ore fine agglomeration, blast furnace ironmaking, and non-blast furnace alternative ironmaking processes using carbon as reducing agent and heating agent, carbon is always used as the energy to drive the process with a large amount of CO_2 ultimately emitted. Facing the development trend of global carbon emission reduction and carbon neutral, the ironmaking industry and traditional processes aims to minimize the dependence on carbon energy, and reduce carbon consumption and CO_2 emission, which will become the main development proposition by the middle of this century. The development status of domestic and foreign hydrogen metallurgy and low-carbon iron-making technologies, and carries out comparative research, analysis and evaluation of the main key process technologies are summarized. Based on the resources and energy conditions of China's ironmaking industry, low-carbon metallurgy, hydrogen metallurgy and carbon-hydrogen coupling metallurgy technologies are discussed. The technical route of long course process (coking + sintering + pelletizing + blast furnace) is proposed to realize low-carbon metallurgy, hydrogen metallurgy and carbon-hydrogen coupling. Suggestions for the development of direct reduction based on hydrogen shaft furnace, full oxygen blast furnace, and smelting reduction process are put forward, and the physical nature and key technologies of the process are investigated and analyzed. The main development philosophy and technical approach of iron-making technology in China until 2050 are pointed out.

Keywords: low-carbon; ironmaking; blast furnace; hydrogen metallurgy; carbon emission reduction

Application on HIsmelt Smelting Reduction Process in China

Meng Xianglong   Li Lin   Zhang Fuming   Cao Chaozhen

Abstract: HIsmelt smelting reduction process is an important branch of alternative ironmaking. Molong HIsmelt project is the first commercial application of HIsmelt smelting reduction process in China. Characteristics of HIsmelt process were introduced. Raw material and fuel, charge structure, reduction mechanism, chemical constituent of hot metal and slag, energy utilization efficiency and environmental protection were analyzed comparing with BF process. Production process, equipment capacity and operation features of HIsmelt was introduced combine with Molong HIsmelt project.

Keywords: alternative ironmaking; smelting reduction; HIsmelt

1 Introduction

Coke is essential for blast furnace (BF) ironmaking process, although the coking coal is becoming scarcity. Therefore there have been a great number of attempts over the past 75 years to develop alternative processes [1,2]. The HIsmelt represents one of the main alternative ironmaking processes to the blast furnace, which has a relatively long history and having been conceived in the early 1980's. A 20-year development phase, involving multiple pilot plants, cumulated in a 2.7m ID vertical smelter being built in 1996 and commissioned in 1997. After that, a HIsmelt plant which has a 6m ID smelter with the production capacity of 0.8Mt/a pig iron was built in Kwinana, Western Australia in 2005 [3-5].

A HIsmelt plant is being built in Molong Company for iron source at present in Shandong, China and expected to be commissioned at the end of 2015. Production process, equipment capacity, operation features and design optimizations of HIsmelt were introduced; thermodynamic conditions of smelting were analyzed combined with Molong HIsmelt project.

2 HIsmelt Process and Technical Features

2.1 HIsmelt Process

HIsmelt consists of raw materials handling, coal injection, ore injection, stoves, smelt reduction

---

本文作者：孟祥龙，李林，张福明，曹朝真。原刊于 Proceedings of the Iron & Steel Technology Conference, AISTech, 2014: 1135-1145。

vessel (SRV), cast house, slag handling, offgas cleaning and cooling, waste heat recovery, oxygen plant, turbine blower and waste water collection etc. HIsmelt process is presented in Fig. 1.

(1) Coal system: wet coal reclaimed from the stockpile will be kept in storage bin and fed into a coal mill to be crushed and dried. Dry, granulated coal with the average grain size of about 3mm and moisture content of less than 2% will be injected into the SRV metal bath by N_2 via separate injection systems.

(2) Ore system: ore will be transported to preheat plant from stock yard and be heated to 600-800℃. After that, hot ore will be fed into injection system and injected into the SRV metal bath via water-cooled ore injection lances by N_2.

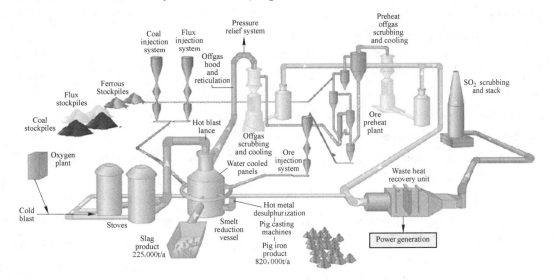

Fig. 1 HIsmelt process

(3) Hot blast system: hot air blast, whose typically oxygen enrichment is 35% and temperature is 1200℃, will be injected from the top of SRV and burn with CO in the relatively oxidizing region in upper section of the vessel. Heat produced by post-combustion will transfer between the upper region and the lower region by spitting molten slag and hot metal through convection or radiant to maintain the necessary thermal balance.

(4) Hot metal and slag handing system: after ore reduction, hot metal will be tapped continuously via a forehearth with the temperature of 1450-1500℃, transported to desulphurization unit and casted into pig iron by pig casting machine. Slag will be tapped periodically per 2-3 hours via a water-cooled notch.

2.2 Technical features of HIsmelt process

Comparing with BF process, HIsmelt process has the advantages of

(1) More flexible raw material. Pulverized iron ore (grain size<6mm) injection is adopted in HIsmelt process, besides, some ferruginous material like small granularity waste, iron scale powder in rolling process, return fines and others, could be used in HIsmelt process. Lower grade

ore and coal, which is difficult to be used in BF process, may be accepted in HIsmelt process.

(2) Lower investment. Coking, sintering and pelletizing plants are not necessary in HIsmelt process, which will reduce the investment comparing with BF process. Besides, smaller material yard results in less occupied area of the whole plant.

(3) Lower operation cost and higher efficiency. The raw material includes lower grade pulverized ore, non-coking coal and ferruginous waste without any coke, sinter or pellet, which will lead to operation cost decreasing. In HIsmelt process, carbon of coal will dissolve in hot metal immediately and react with FeO to produce offgas, slag and more hot metal. The violent production of the offgas will produce a fountain to agitate molten metal bath violently and promote material fusion and reduction reaction. The rate of FeO reduction by dissolved carbon is faster than by solid carbon, which will result in high efficiency.

(4) Less environmental pollution. Pollutants generated in coking, sintering or pelletizing process as SO_x, NO_x or dioxin, which may result in environmental pollution, will be reduced. Volatile matter in coal will be split, so there is almost no CH_x in offgas. Sulphur in coal could also be absorbed by molten iron and little sulphur component in offgas.

There also has some defects in HIsmelt process including

(1) Lower heat transfer efficiency. Heat in SRV is transferred based on splashing and dropping of molten metal and slag moving between regions. The designed heat transfer efficiency in SRV is about 75%-85%; it is lower than that in fixed bed which could be more than 90%. Moreover, slag above the molten metal may prevent convection or radiant heat transfer between the upper region and the lower region. Lower heat transfer efficiency may result in higher coal consumption.

(2) Sulfur couldn't be removed. Smelting experiment with high phosphorus ore in Kwinana HIsmelt plant shows that, most of the phosphorus in ore could be removed to slag, but sulfur content in hot metal was up to about 0.2% ± 0.08%. In China, high sulfur raw material is common, so desulphurization unit should be adopted before steelmaking process. However, desulphurization process has developed rapidly to deal with high sulfur content in hot metal.

(3) High FeO content in slag. Ore and coal is injected into molten metal through solids injection lances. Lower temperature for reduction reaction results in a great quantity of FeO and solid iron in slag. Part of molten metal will be spattered with slag into the upper region to be oxidized once again. Slag rate has reached 460kg/t in Kwinana HIsmelt plant, while the mass fraction of FeO and solid iron in slag reached to about 5.0% both.

Table 1 shows the comparison of elements content on HIsmelt and BF process. Main parameters of HIsmelt and BF process are presented in Table 2.

Table 1 Comparison of elements content on HIsmelt and BF   (wt%)

Elements content	C	Si	Mn	P	S
BF	4.5	0.5±0.3	0.4±0.2	0.09±0.02	0.04±0.02
HIsmelt	4.4±0.15	<0.01	<0.02	0.02±0.01	0.1±0.05

Table 2 Comparison of main parameters on HIsmelt and BFs

Indexes	HIsmelt	Angang 2580m³ BF	WISCO 3200m³ BF	Bao steel 4350m³ BF	Jingtang 5500m³ BF
Raw material	Ore powder	Sinter, pellet, lump ore	Sinter, pellet, lump ore	Sinter, pellet, lump ore	Sinter, pellet, lump ore
Sinter/wt%	0	78	66	73	70
Pellet/wt%	0	19	25	11	20
Lump ore/wt%	0	3	9	16	10
TFe/wt%	—	59.46	59.25	60.08	59.1
Reductant	Non-coking coal	Coke	Coke	Coke	Coke
Pre-reduction device	Circulated fluidized bed	Null	Null	Null	Null
Coke ratio/kg·t^{-1}	235	368	318	300	290
Coal ratio/kg·t^{-1}	900-950	130	163	172	160
Fuel ratio/kg·t^{-1}	1135-1185	550	520	492	490

3 Thermodynamic Analysis of High Phosphorus Ore Smelting in HIsmelt Process

Hot metal produced in HIsmelt process contains less silicon, manganese and phosphorus, which has great relationship with gas component, reaction temperature, oxygen enrichment rate and slag component. Thermodynamic conditions for high efficiency dephosphorization as high oxidizing, high alkaline, good flowability and lower temperature could be achieved in HIsmelt process. The SRV has functions of iron oxides reduction, dephosphorization and desiliconization, and the thermodynamic condition is intermediate between later stage of BF reduction reaction and earlier stage of converter oxidation reaction. Professor BI Xuegong researched on dephosphorization capability of HIsmelt by analyzing distribution of phosphorus between molten metal and slag[6]. Dephosphorization capability of slag could be indicated by two indexes named phosphate capacity and phosphorus distribution ratio, which could be converted by each other. The main reactions between molten metal, slag and gas phase during reduction of high phosphorus ore can be written as

$$\frac{1}{2}P_{2(g)} + \frac{5}{4}O_{2(g)} + \frac{3}{2}(O^{2-}) = (PO_4^{3-}) \tag{1}$$

$$\frac{1}{2}P_{2(g)} = [P] \tag{2}$$

$$CO_{(g)} + \frac{1}{2}O_{2(g)} = CO_{2(g)} \tag{3}$$

$$Fe + [O] = (FeO) \tag{4}$$

$$\frac{1}{2}O_{2(g)} = [O] \tag{5}$$

$$[C] + [O] = CO \tag{6}$$

Dephosphorization capability of slag indicated by reaction (1) is defined as phosphate capacity:

$$C_{PO_4^{3-}} = \frac{(\%PO_4^{3-})}{P_{P_2}^{\frac{1}{2}} \cdot P_{O_2}^{\frac{5}{4}}} = \frac{K_1 \cdot a_{O^{2-}}^{\frac{3}{2}}}{\alpha \cdot \gamma_{PO_4^{3-}}} \quad (7)$$

Where,

K_1: equilibrium constant of each reaction;

$(\%PO_4^{3-})$: mass percent of PO_4^{3-} in slag;

P_{P_2}: partial pressure proportion of phosphorus in gas phase;

P_{O_2}: partial pressure proportion of oxygen in gas phase;

α: conversion coefficient of phosphate radical from mole percent to mass percent;

$a_{O^{2-}}$: activity of oxygen ion in slag;

$\gamma_{PO_4^{3-}}$: activity coefficient of phosphate radical.

Equation (7) shows that, phosphate capacity relate to activity of oxygen ion and activity coefficient of phosphate radical in slag, which is depended on slag components. Moreover, reaction temperature also has influent on phosphate capacity. Combined with reaction (2) and (5), dephosphorization capability could be indicated as

$$C_{PO_4^{3-}} = \frac{(\%PO_4^{3-})}{\left(\frac{f_P[\%P]}{K_2}\right) \cdot \left(\frac{f_O[\%O]}{K_5}\right)^{\frac{5}{2}}} \quad (8)$$

Where,

$[\%P]$: mass percent of phosphorus in molten metal;

$[\%O]$: mass percent of oxygen in molten metal;

f_P, f_O: activity coefficient of phosphorus and oxygen in molten metal

Phosphorus distribution ratio (L_P) could be indicated as

$$L_P = \frac{(\%P)}{[\%P]} \quad (9)$$

Where, $(\%P)$ is mass percent of phosphorus in slag.

Two equations of phosphorus distribution ratio could be derived based on equation (9) as

$$L_P = \frac{\delta}{\alpha \cdot \beta} \cdot f_P \cdot \frac{K_1}{K_2 \cdot K_4^{\frac{5}{2}} \cdot K_5^{\frac{5}{2}}} \cdot \frac{a_{O^{2-}}^{\frac{3}{2}}}{\gamma_{PO_4^{3-}}} \cdot \gamma_{FeO} \cdot (\%FeO)^{\frac{5}{2}} \quad (10)$$

$$L_P = \frac{\delta}{\alpha \cdot \beta} \cdot f_P \cdot \frac{K_1}{K_2 \cdot K_3^{\frac{5}{2}}} \cdot \frac{a_{O^{2-}}^{\frac{3}{2}}}{\gamma_{PO_4^{3-}}} \cdot \left(\frac{P_{CO_2}}{P_{CO}}\right)^n \quad (11)$$

Analysis on equation (10) and equation (11) shows that, L_P depends on activity of oxygen ion in slag ($a_{O^{2-}}$), activity coefficient of phosphate radical ($\gamma_{PO_4^{3-}}$) and ratio of partial pressure by CO_2 and CO (P_{CO_2}/P_{CO}). Activity of oxygen ion in slag is indicated by activity of basic oxide. Dephosphorization capability could be judged by $a_{O^{2-}}$, $\gamma_{PO_4^{3-}}$ and P_{CO_2}/P_{CO}, which are the rationale of researching high phosphorus ore smelting and developing operation parameters in HIsmelt process[7,8].

Research on dephosphorization of HIsmelt process was developed by professor Guo Hanjie based on molecular theory. Reactions of (1)-(6) shows all the reactions which have influence on dephosphorization in SRV. Actually, three reactions could indicate the dephosphorization process as (3), (4) and (12) shown below by synthesizing the six reactions above.

$$[P] + \frac{5}{2}[O] + \frac{3}{2}(O^{2-}) = (PO_4^{3-}) \tag{12}$$

Reaction (12) shows the essence of dephosphorization. According to phase diagram, $3CaO \cdot P_2O_5$ is a kind of stable compound as the resultant of reaction instead of $4CaO \cdot P_2O_5$. The reaction formula of (12) could be written as (13).

$$2[P] + 5[O] + 3(CaO) = 3CaO \cdot P_2O_5 \tag{13}$$

Moreover, reaction (13) could be written as

$$2[P] + 5(FeO) + 3(CaO) = (3CaO \cdot P_2O_5) + 5[FeO] \tag{14}$$

The standard free energy of reaction (14) could be calculated as

$$2[P] + 5[O] = P_2O_5 \qquad \Delta G_1^\ominus = -742,032 + 532.71T \tag{15}$$

$$3(CaO) + P_2O_{5(g)} = (3CaO \cdot P_2O_5) \qquad \Delta G_2^\ominus = -684,335 + 63.6T \tag{16}$$

$$(FeO) = [Fe] + [O] \qquad \Delta G_3^\ominus = 121,010 - 52.35T \tag{17}$$

Standard free energy of reaction (14) is calculated by the equation:

$$\Delta_r G^\ominus = \Delta G_1^\ominus + \Delta G_2^\ominus + \Delta G_3^\ominus = -821,317 + 334.56T \tag{18}$$

Equilibrium constant of reaction (14) could be calculated by equation (18) as

$$\ln K^\ominus = \frac{98,787}{T} - 40.24 \tag{19}$$

Equilibrium constant of reaction (14) also could be written as

$$K^\ominus = \frac{a_{3CaO \cdot P_2O_5}}{[\%P]^2 \cdot a_{FeO}^5 \cdot a_{CaO}^3} \tag{20}$$

According to simultaneous equation (19) and (20), mass percent of phosphorus in molten metal could be calculated as

$$[\%P] = a_{3CaO \cdot P_2O_5}^{\frac{1}{2}} \cdot a_{FeO}^{-\frac{5}{2}} \cdot a_{CaO}^{-\frac{3}{2}} \cdot e^{-\frac{49,391}{T} + 20.12} \tag{21}$$

Thermodynamic conditions for high efficiency dephosphorization as high oxidizing, high alkaline, high flowability and lower temperature could be analyzed by equation (21).

(1) High oxidizing. Phosphorus content in molten metal grows in inverse proportion to the 2.5 power of the activity of FeO in slag. According to equation (17), high content of FeO in slag results in high content oxygen in molten metal.

Equilibrium constant of reaction (17) is

$$K^\ominus = L_O = \frac{[\%O]}{a_{FeO}} \tag{22}$$

When $T = 1600℃$, equilibrium constant of reaction (17) is 0.23, which indicate that oxygen content in molten metal $[\%O] = 0.23a_{FeO}$. It shows that Oxygen content in molten metal will increase when the activity of FeO in slag increases, meanwhile, dephosphorization capability of slag will also increase.

(2) High alkaline. Phosphorus content in molten metal grows in inverse proportion to the 1.5th power of the activity of CaO in slag, which indicates that lime injection plays a key role in smelting. The relationship between CaO which should be excess and acidic oxide is shown in equation (23).

$$n_{CaO}^E = (n_{CaO} + n_{MgO} + n_{MnO}) - 2n_{SiO_2} - 3n_{P_2O_5} - 3n_{Al_2O_3} - n_{Fe_2O_3} \quad (23)$$

(3) Good flowability. For dynamic requirement, CaO and FeO in slag and phosphorus in molten metal should diffuse to metal-slag interface quickly to promote the reaction.

(4) Decrease the reaction temperature as low as possible provided keeping the slag liquid. According to molecular theory, under the condition of slag component fixed, the value of $a_{3CaO \cdot P_2O_5}^{\frac{1}{2}} \cdot a_{FeO}^{-\frac{5}{2}} \cdot a_{CaO}^{-\frac{3}{2}}$ could be considered as constant which is not sensitive to temperature. Let $F(T) = e^{-\frac{49,391}{T} + 20.12}$, the value of $F(T)$ was calculated and shown in Table 3.

Table 3 Temperature and equilibrium constant of dephophorization

Reaction temperature/K	1473	1573	1673	1773	1873	1973
Equilibrium constant	4.467×10^{11}	6.286×10^9	1.473×10^8	5.269×10^6	2.691×10^5	1.857×10^4
Value of $F(T)$	1.499×10^{-6}	1.263×10^{-5}	8.252×10^{-5}	4.362×10^{-4}	1.930×10^{-3}	7.347×10^{-3}

Value of $F(T)$ will increase rapidly when the temperature increases, which indicates that the temperature has a great effect on phosphorus content in molten metal [9,10]. The phosphorus content could be calculated once the components of slag and reaction temperature are given.

4 Core of HIsmelt Process

4.1 SRV System

In SRV, iron ore, or any appropriate ferrous feed material, coal and fluxes are injected into a molten iron bath. Coal is used as the reductant and energy source for iron ore reduction, gangue removes into slag, gases produced by the smelting process are released out of vessel. The SRV consists of a cylindrical lower section, and a top cone. Principle of Reaction on SRV of HIsmelt Process is presented in Fig. 2.

SRV system is the core of HIsmelt process, which consists of SRV proper, forehearth, mixing solids injection lances, coal injection lances and hot air blast (HAB) lance, etc. SRV proper was designed as pressure vessel, whose lower section is barrel and top cone connect to the offgas chamber with a welded connection. The lower region of barrel contained a refractory hearth, sidewalls and slag zone for containment of the molten metal, while the upper region had penetrations for the water cooled panels to protect the shell of the SRV from the process materials

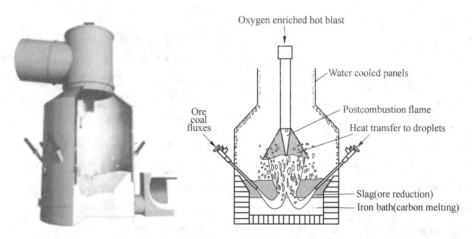

Fig. 2 Principle of reaction on SRV of HIsmelt process

(molten iron, slag and hot gases) by maintaining a protective slag coating on. Water cooled HAB lance was mounted from the lid of the vessel offgas chamber. Two mixing solids injection lances and two coal injection lances were equally spaced around the vessel circumference, while two fuel injection lances were assembled symmetrically above. The two fuel injection lances will be used for heating hearth when blow-in start.

In normal operation, preheated iron ore, coal and fluxes will be injected by nitrogen through mixing solids injection lances into the molten metal bath located in the refractory lined hearth of the SRV. The end of the mixing solids injection lances were above the surface of molten metal. These raw materials injected into deep of molten metal will react with the hot metal to produce offgas, slag and more hot metal. Volatile matter in coal will split into CO and H_2, while carbon will be dissolved in hot metal. Iron oxides of ore will be reduced by dissolved carbon into CO and molten metal, and gangue, fluxes and ash of coal will mix together to form molten slag. The violent production of the offgas will produce a fountain of molten slag and hot metal. Gases produced by the smelting process will be post-combusted in the vessel top-space by hot oxygen enriched air and produce a large amount of heat. Molten slag and hot metal in the fountain will absorb heat produced by gases post-combustion through convection or radiant and pass it back to molten metal. Heat transfer between the upper region and the lower region could maintain the necessary heat for smelting reaction. Part of slag in the fountain will be attached to the vessel walls and injection lances to reduce heat loss. In addition, the level of the hot metal in the SRV will be kept constant via an external forehearth which will be directly connected to the hearth of the SRV as a barometric leg for the hot metal. The hot metal will overflow from the forehearth to the hot metal ladles via a series of refractory lined fixed and tilting runners. The slag will be periodically tapped from the SRV via a water-cooled slag notch into slag pots. The operation pressure of smelting process in the SRV is about 0.08-0.1MPa.

4.2 SRV Refractory

According to the operation temperature, erosion characteristics and masonry strength requirement of different parts, kinds of refractories were adopted in SRV.

Refractory lining of hearth consists of working lining and permanent lining. The working lining sidewalls were constructed by corundum-spinel brick RESISTAL KSP95, while alumina-chrome brick Cr50-RK was adopted at serious erosion region near slag level to prevent infiltration and erosion by molten slag and metal. The thickness of working lining sidewalls is about 1200mm, with the service life of about four years. The thickness of the permanent lining sidewalls is about 440mm, in which 350mm thick pourable high-alumina refractory was casted against the steel shell and 90mm thick compressible high-alumina refractory was rammed between working lining sidewalls and pourable refractory. The working lining sidewalls refractory expansion could be absorbed by compressible high-alumina refractory.

The working lining bottom was constructed by 500mm thick corundum-spinel brick RESISTAL KSP95. The permanent lining bottom was constructed by 600mm thick high-alumina brick and rammed using 50mm thick compressible high-alumina refractory.

Steel cooling panels were covered with corundum pourable refractory. Copper cooling panels set at slag region were sprayed with a little pourable refractory. In normal operation, molten slag will be sprayed on copper cooling panels to form crust, which could reduce heat loss. End tapholes were constructed by bauxite prefabricated element SAINT SP7. Slag drain taphole was constructed by Cr30-RK and blocked by Al_2O_3-SiC-C.

The working lining sidewalls of forehearth were constructed by 230mm thick corundum-spinel brick RESISTAL KSP95, while the permanent lining sidewalls were constructed by 170mm thick high-alumina brick RESISTAL B80 and 66mm thick light weight fireclay brick from the inside out. 40mm thick compressible high-alumina refractory was rammed between the working lining and high-alumina brick; 50mm thick high-alumina refractory was rammed between high-alumina brick and light weight fireclay brick.

The working lining bottom of forehearth and the connecting part between forehearth and SRV were constructed by bauxite prefabricated element SAINT SP7, while the permanent lining bottom of forehearth was constructed by 230mm thick high-alumina brick RESISTAL B80 and 150mm thick high-alumina ramming refractory.

4.3 SRV Water Cooling System

The SRV water cooling system consists of panels cooling system and core plant cooling system (Table 4).

Table 4 Parameters of SRV water cooling system

Indexes		Water flow /t·h^{-1}		Supply pressure /MPa		Return pressure /MPa		Temperature differential /℃	
		Design	Normal	Design	Normal	Design	Normal	Design	Normal
Panels circuit cooling system		6000	5580	1.5	0.8	0.6	0.5	23	10
Core plant circuit cooling system	High pressure	800	720	1.4	1.0	1.35	0.235	48	12
	Low pressure	1861	—	0.8	0.63	0.8	0.247	6	2

(1) Panels circuit cooling system

The panels circuit cooling system was recirculate closed circuit water to the SRV panels and will cool the water via wet surface air coolers. There were 15 rows of panels to protect the shell of the SRV. The 1th row was copper cooling panels set at slag region of shell, and the 2th to 15th rows were steel cooling panels which were designed according to the sharp of inside walls. Barrel and roof steel panels had 2 layers of tubes to reduce SRV heat losses, and other steel panels were of a single layer tube design. Typical pipe diameter was DN80 ($\phi 89mm \times 8mm$) with the material of 12Cr1MoVG. Emergency cooling water system has been designed to keeping cooling water supply.

(2) Core plant circuit cooling system

The core plant circuit cooling system was recirculates closed circuit water to the HAB lance, SRV relief valves, solids injection lances, coal injection lances and slag notch, and cool the water via plate and frame heat exchangers. High pressure cooling water will be used for HAB lance while low pressure water for the others. Flow and heat load monitoring devices were set at each cooling water inlet and outlet.

4.4 SRV Core Plants

SRV core plants includes HAB lance, solids injection lances, fuel injection lances and slag notch, etc. The solids injection lances of Kwinana HIsmelt plant consists of four coal injection lances and four ore injection lances. In Molong project, two mixing solids injection lances and two coal injection lances were adopted. Another four holes blocked up by water cooling cap were reserved for adding lances later.

4.4.1 HAB lance

The HAB lance, as complete set of equipment, was mounted by a flanged connection from the top of the lid of the vessel offgas chamber. Refractory lining was sprayed onto the inner wall of water-cooled HAB lance. Considering the effects of the HAB lance position on heat transfer efficiency in vessel, the height of HAB lance could be regulated in three grades as 0m, 0.6m and 1.2m. Normal operating pressure of HAB is 0.17MPa, the temperature is 1200℃, and oxygen enrichment rate is 40%.

4.4.2 Solids injection lances

Four solids injection lances were designed. Preheated ore, coal and fluxes will be conveyed near the SRV, mixed and injected by two mixing solids injection lances. Another two lances were for coal injection, which will be used to heating hearth refractory when blow-in start and keeping hot metal temperature while SRV off-wind temporarily.

4.4.3 Fuel injection lances

For preheating hearth refractory and keeping temperature in SRV, two water-cooled natural gas injection lances were assembled. The two fuel injection lances will be cooled by nitrogen continuously in normal operation to prevent blocking by molten slag. Design nitrogen flow is 2000Nm^3/h, and pressure is 1MPa.

4.4.4 Discharge ports

Four discharge ports were designed in SRV as slag notch, slag drain taphole, SRV end taphole and forehearth end taphole. Molten slag produced by smelting will be periodically tapped via water-cooled copper slag notch which was inlayed with steel lining in the center and installed in SRV shell. The slag notch will be blocked up by clay and cap while none of slag is tapped. In normal operation, the slag will be tapped 45 min and 80t per time every two hours. Slag drain taphole was set at higher part of hearth, with the function of molten slag draining before examining and repairing. Slag will be drained 30 min and 120t per time at early stage of SRV blow-in, and may be 220t per time after hearth refractory begin to be eroded. The SRV and forehearth have an end taphole for each to enable the SRV and forehearth to be fully drained at an end tap.

4.4.5 Offgas chamber

Offgas produced by smelting will be collected in offgas chamber in the top part of SRV and delivered through offgas duct to offgas hood which will collect, partially cool, and deliver the offgas from the SRV to the gas cleaning plant. The hood will be cooled by a combination of forced and natural convection cooling of boiler quality water, which will raise steam for use in other areas of the plant. The offgas chamber has a flanged connection with a diameter of 5000mm to the offgas duct. In normal operation, the offgas flow will reach 236,000Nm^3/h, the temperature will be 1500℃, and the pressure will be 0.08MPa.

4.5 Shell of SRV and Forehearth

Gas pressure inside SRV will be about 0.08-0.1MPa, so the SRV and forehearth shell were designed as pressure vessel using Q345R. The SRV bottom plate is supported on welded beams on concrete foundation. The forehearth, whose base is on sliding supports component of cast iron and steel plates to allow for expansion of the SRV and the forehearth, is of a fabricated construction and sufficient strength to prevent distortion from refractory expansion. CN140 coating was sprayed on bottom plate and support beams to prevent thermal stresses. The internal diameter of barrel lower region and offgas chamber are 9280mm and 5260mm respectively.

5 Conclusions

HIsmelt smelting reduction process is an important branch of alternative ironmaking. Practice in Kwinana in Australia has proved there is absolutely no doubt that the core process works. As a developing technology, the economic value of HIsmelt process has not been reflected in Kwinana plant. Molong HIsmelt project is the first commercial application of HIsmelt smelting reduction process in China. Some key technologies as operation cost, raw material range, heat transfer efficiency and refractory service life are still the focus of HIsmelt process.

(1) The HIsmelt process represents one of the alternative ironmaking processes to the blast furnace. It has an ability to treat various steel plant revert materials directly including lower grade ore and coal which would be difficult to use in BF process. The HIsmelt process creates conditions

for strong migration of phosphorous from metal to slag, which shows the potential of high phosphorus ore smelting.

(2) Heat produced by post-combustion transfers between the upper region and the lower region to maintain the necessary thermal balance, which has lower efficiency and results in higher coal consumption. The crucial element for replacing blast furnace process by HIsmelt is whether the heat utilization efficiency would be improved further.

(3) Some design improvements were adopted in Molong HIsmelt project, including solids injection lances number, refractory structure, HAB lance height and water cooling panels arrangement. The rational of design improvements will be proved after Molong HIsmelt plant is commissioned.

Acknowledgements

The author would like to thank Mr. Neil Goodman, Mr. Mao Qingwu, Mr. Zhang Qifu, Ms. Meng Yujie, Ms. Liu Ran, etc. for their support and assistance. Thank experts from Molong Company for their support of information.

References

[1] Guo Peimin, Zhao Pei, Pang Jianming, Cao Chaozhen. Technical analysis on smelting reduction ironmaking process [J]. Iron Steel Vanadium Titanium, 2009, 30 (3): 1-8.

[2] Sha Yongzhi, Teng Fei, Cao Jun. Progress of ironmaking technology in China [J]. Ironmaking, 2012, 31 (1): 7-11.

[3] Neil Goodman, Rod Dry. HIsmelt plant ramp-Up [C] //Proceedings of the 5th International Congress on the Science and Technology of Ironmaking, Shanghai, 2009.

[4] Bates P, Coad A. HIsmelt, the future in ironmaking technology [C] //4th European Coke & Ironmaking Conference, Paris, 2000.

[5] Dry R J, Bates C P, Price D P. HIsmelt-the future in direct ironmaking [C] //58th Ironmaking Conference Proceedings, Chicago, 1999.

[6] Bi Xuegong, Zhou Jindong, Huang Zhicheng. Analysis on high phosphorus ore smelting of HIsmelt process [C] //2008 Ironmaking Conference, Ningbo, 2008.

[7] Zhang Ben, Gu Xuzhong, Guo Zhancheng. Relation between stickingand metallic iron precipitation on the surface of Fe_2O_3 particles reduced by CO in the fluidized Bed [J]. ISIJ International, 2001, 51 (9): 1403-1409.

[8] Pawlik C, Schustre S. Reduction of iron ore fines with CO-rich gases under pressurized fluidized bed conditions [J]. ISIJ International, 2007, 47 (2): 217-225.

[9] Li Lin, Peng Jie, Meng Xianglong, Guo Hanjie. Experimental research on influencing factors of metal-slag melting separation after titanomagnetite prereduction[J]. Mining and Metallurgical Engineering Supplement, 2014, 34: 378-382.

[10] Li Lin, Li Yongqi, Guo Hanjie. Experimental research on influencing factors for pre-reduction of titanomagnetite pellets[J]. Mining and Metallurgical Engineering, 2014, 34 (3): 80-89.

现代钢铁冶金工程设计与应用研究

冶金工程设计的发展现状及展望

摘 要：冶金工程设计对促进冶金工业的可持续发展具有特殊的重要作用。我国已经具备了现代大型钢铁厂的流程设计、工艺设计、设备设计以及系统集成能力，创建并应用动态-精准设计体系，建立了现代冶金厂工程设计理念、理论和设计方法，设计建设了具有国际先进水平的新一代可循环钢铁厂，在流程高效化、生产集约化、装备大型化等方面取得显著成效。与此同时，冶金关键单元技术和工程化集成技术得到重点突破，在关键、共性、重大技术领域取得一系列科技成果，对冶金工程设计的理论与实践提供了有效支撑。未来我国冶金工程设计应以冶金流程工程学和动态-精准设计理论与方法为核心，构建冶金工程设计创新理念、理论及设计方法的完整体系。

关键词：冶金流程；工程设计；设计理论；设计方法；技术创新

1 引言

我国冶金工程设计体系在20世纪50年代由苏联引入，长期以来基本沿用苏联的设计理论和设计方法。20世纪80年代以后，随着宝钢工程的设计建设，我国冶金工程设计又相继引入了日本和欧洲的设计方法，但仍属传统的"静态-分割"设计方法，即静态的半经验、半理论的设计方法。进入21世纪以后，我国著名冶金学家殷瑞钰院士创立了冶金流程工程学[1]，继而结合新一代钢铁制造流程的系统研究和首钢京唐钢铁厂工程的设计建设，提出了基于动态-精准设计体系的新一代冶金工程设计理论和方法[2~4]。我国冶金工程设计经历了近60年的发展，目前正在形成以冶金流程工程学、冶金流程集成理论与方法[5]、工程哲学[6]等理论为基础的具有一定国际影响的学科体系。

近年来，在冶金流程工程学理论研究、冶金厂动态-精准设计研究等领域取得了具有创新性的研究成果，以冶金流程工程学、冶金流程集成理论和方法为理论指导的冶金工程设计实践也取得了显著的应用成果。基于先进的冶金工程设计理论、采用动态-精准设计体系设计的首钢京唐钢铁厂工程，已成为新一代可循环钢铁制造流程的示范工程。

2 冶金工程设计体系演化与发展历程

2.1 传统冶金工程设计体系

我国冶金工程设计的历史可以追溯到19世纪后期的洋务运动，其后经历了半封建、半殖民地时期，再到建国以后学习原苏联时期和改革开放初期。从总体上讲，这一时期从

本文作者：张福明，颉建新。原刊于《钢铁》，2014，49（7）：41-48。

国外引进并学习了通用的设计方法、步骤及其工具，但实际上仍处于一种缺乏现代设计理论体系支撑的状态，无论是流程制造业的设计还是离散型制造业的设计大体都是如此。

直到20世纪末期，我国钢铁工业的发展基本上是以产能扩张、基建投资为主。冶金工程的科研开发工作也侧重于局部领域的理论、材料或单体技术的研究，局限在细节技术、单元操作和单体工艺，较少从钢铁制造全流程优化的角度去分析研究问题。虽然取得了一定程度的发展，但不少优秀的研究成果不能"固化"于工艺、装备的升级换代上，不能稳定、有效地"融合"在钢铁厂生产流程之中，难以对钢铁工业或钢铁企业的整体结构优化产生根本性的推动作用，造成了我国一些钢铁厂追求产能扩张、低水平重复建设，存在整体流程结构混乱、能源消耗高、制造成本高、生产效率低、环境污染严重等诸多问题。

究其深层次的原因，主要还是由于长期以来对钢铁制造流程的研究重视不够，在工程设计方面局限于单元工序，忽略工序间匹配优化和流程结构优化的整体协同效应。而在理论研究方面主要是建立在反应解析、过程解析的"三传一反"基础上，使得研究微观课题较多，注重追求单元工序或单一装置的强化，而缺乏从整体上研究钢铁制造流程的理念。实践证实，单元工序或装置的优化仅能解决钢铁厂生产过程中的局部问题，对全局和整体结构不能产生根本性的影响；而流程的优劣、合理与否则综合影响产品的成本、质量、生产效率、投资效益、过程排放与环境效益等技术经济指标，直接关系到企业的生存与发展。

2.2 传统设计体系的弊端

改革开放之前，囿于我国当时的国情和钢铁工业粗放、简单的发展方式，我国冶金工程设计一直延续着原苏联的设计理念和设计模式，即对不同工序装备的能力进行静态估算，加上工序之间的匹配连接，形成一种堆砌起来的、粗放的生产流程。其特点是只从单元工序的局部出发（即停留在本工序范围内提出静态要求，很少提出上、下游工序之间动态、有序、协调、集成运行方面的要求），分别预留出不同的富裕能力；各工序装备能力的"富裕系数"取决于设计人员（或不同工序用户）的主观需求而不同，而各工序之间的连接方式则是堆砌性的静态连接，缺乏动态运行的计算，采用这种设计方法构建出来的钢铁厂生产流程和工艺装备，在实际运行过程中则会出现前后工序的能力不匹配，功能不协调，信息不顺畅且难于调控。

这种"静态估算、简单堆砌"的设计理念和方法，缺乏构建钢铁厂物质流、能量流、信息流网络的理论和方法，未能形成整体流程的设计理念和方法，未有成熟的整体设计分析工具。传统的经验设计方法在当今的工程设计中仍占据主导地位，由此带来的后果是工程设计在方法上难以有突破性的创新，只能是停留在传统设计方法的基础上进行有限的局部改进。采用传统的工程设计方法，难以真正实现钢铁厂的运行效率、产品质量、投资效益的最佳化；而且，往往是流程建成之时即是技术改造之始。流程一旦建成以后，只能在不合理流程的基础上改进完善，许多问题甚至无法从根本上得到解决。

2.3 工程设计理念的创新与初步成效

20世纪90年代以后，我国冶金学家开始研究钢铁制造流程结构及其运行规律[7,8]，结合国内外钢铁工业发展趋势,对我国钢铁工业技术进步战略的判断、选择和有序推进做了大量工程技术和理论研究工作。在理论上提出并阐述了钢铁制造流程的多因子物质流控制、钢铁制造流程解析与集成、钢铁厂结构优化和发展模式、钢铁工业与绿色制造等一系列观点[9~13];促进了一大批钢铁厂工艺流程结构的优化,有力推动了我国钢铁工业的持续快速发展。这一时期,冶金工程设计新的理论基础已经初步形成。

世纪之交,我国钢铁工业发展突飞猛进,钢铁产量连续多年保持全球第一,一大批新建或新技术改造的钢铁厂在新世纪之初相继投产。在钢铁产能规模扩张的同时,钢铁生产工艺技术装备水平也得到了大幅度提升,冶金工程创新取得了长足进步。冶金工艺技术装备大型化、现代化成为这一时期的显著特征,一批容积 2500m^3 以上的大型高炉、容量 200t 以上的大型转炉、宽度 2.0m 以上大型宽带钢热轧-冷轧生产线、面积 260m^2 以上的大型烧结机、炭化室高度 6.0m 以上大型焦炉等具有国际先进水平的冶金工艺和技术装备相继建成投产。与此同时,一批新建的钢铁厂也纷纷建成投产,集成了原料场、烧结、焦化、炼铁、炼钢-连铸、热轧、冷轧、能源、环境等全流程工序,这些新建钢铁厂工程设计不仅在技术装备现代化、大型化方面成效显著,而且在工艺优化、流程设计、系统集成等方面也取得长足进步。

3 冶金工程设计发展现状

3.1 冶金工程设计总体进展

进入 21 世纪以后,在我国冶金工艺技术装备大型化、现代化的同时,冶金流程工程理论研究也取得重大突破。在冶金制造流程结构优化研究和多项冶金领域关键、共性、重大技术研究成果的基础上,创建了以整个钢铁制造全流程中物质流、能量流和信息流实现"动态-有序""高效-协同""连续-紧凑"运行为核心的冶金流程工程学。冶金流程工程学和冶金流程集成理论与方法已作为现代冶金工程设计学科的重要理论基础,成为指导和构建冶金厂流程设计、总图布置、工艺装备选择、流程结构优化的核心思想体系。

目前,我国冶金工程设计已经具备了现代大型钢铁厂的流程设计、工艺设计、设备设计以及系统集成能力。然而我国钢铁工业正在经历从单纯追求数量到主要追求市场竞争力的过程,围绕工艺流程创新、产品开发、节能降耗、环境保护等领域提出的设计要求越来越高,同时在海外市场与国外大型设计公司的竞争日趋激烈,继续沿用传统的设计理念、理论和方法难以满足钢铁产业结构调整升级的需求,也难以在国际竞争中占有相对优势,这就需要从更高、更广阔的视野以及工程本质的研究上提出新的设计理论,从而提升整个行业的设计水平。

3.2 冶金工程设计发展特征

3.2.1 冶金工程设计的理论创新

基于冶金流程工程学、工程哲学等理论,冶金工程设计建立了新的工程理念和技术理念。设计是工程的原工程,工程设计是工程建设的灵魂,是对项目建设进行全过程的详细策划和表述项目建设意图的过程,是科学技术转化为生产力的关键环节,是体现技术和经济双重科学性的关键要素,是实现项目建设目标的决定性环节。

没有现代化的设计,就没有现代化的建设。科学合理的工程设计,对加快工程项目的建设速度、提高工程建设质量、节约建设投资、保证项目顺利投产并取得较好的经济效益、社会效益和环境效益具有决定性作用。钢铁厂的竞争和创新表象上体现在产品和市场,其根源却来自于设计理念、设计过程和制造过程,工程设计正在成为市场竞争的始点,设计的竞争和创新关键在于工程复杂系统的多目标群优化[14]。

3.2.2 冶金工程设计实现静态设计向动态-精准设计的转变

传统的工程设计方法,即采取静态的半经验、半理论的工程设计方法。动态-精准设计是从动态-协同运行的总体目标出发,对先进的技术单元进行判断、权衡、选择,再进行动态整合,把各有关的单元技术通过在流程网络化整合和程序化协同,使"物质流""能量流""信息流"的流动/流变过程在规定的时-空边界内动态-有序化运行,形成一个动态-有序、连续-紧凑的工程整体集成效应,达到多目标优化的目的。

而现代工程设计方法是要运用冶金流程工程学动态-精准设计的理论开展钢铁厂工程设计创新工作,对于钢铁制造流程设计而言,运用冶金流程工程学理论尤为重要。冶金流程工程学的核心是通过研究钢铁制造流程的功能优化、结构优化、效率优化、耗散最小化,实现钢铁制造流程的有序化、协调化、高效化、连续化生产。其基本方法是采用工序功能集合的解析-优化、工序之间关系集合的协调-优化、流程工序集合的重构-优化的方法,开展钢铁厂工程设计创新工作。

动态-精准设计方法是建立在动态-有序、协同-连续/准连续地描述物质/能量的合理转换和动态-有序、协同-连续/准连续运行的过程设计理论的基础上,并实现全流程物质/能量的动态-有序、协同-连续/准连续的运行过程中各种信息参量的设计;甚至进一步推进到计算机虚拟现实。其目的是提高钢铁厂的市场竞争能力,实现可持续发展。

3.2.3 形成对钢铁制造流程动态运行过程物理本质的新认识

钢铁制造流程动态-有序运行的基本要素是"流""流程网络"和"程序"。其中"流"是制造流程运行过程中的物质性主体,"流程网络"同"节点"和"连接器"构成是"流"运行的承载体和时间-空间边界,而"程序"则是"流"的运行特征在信息形式上的反映。物质流(主要是铁素流)在能量流(主要是碳素流)的驱动和作用下,按照设定的"程序",沿着"流程网络"做动态-有序的运行,实现钢铁厂"3个功能"的多目标群优化。

3.2.4 工程设计方法的创新

应用动态-精准设计理论、方法和先进设计手段,采用三维动态计算机辅助设计、模拟仿真技术、计划网络控制技术和有限元计算分析软件等现代设计方法,提高工程设计动态-精准设计程度。时间管理对于制造流程中多因子"物质流"运行的紧凑-连续性具有决定性的影响,在钢铁制造流程中,各工序、装置运行过程在时间上的协调至关重要。时间在钢铁制造流程中的表现形式是以时间点、时间域、时间位、时间序、时间节奏、时间周期等形式表现出来。

解析时间的表现形式,要建立有效的钢铁厂信息调控系统。时间在钢铁制造流程中具有既是自变量又是目标函数的两重性,研究钢铁制造流程整体运行的本质和运行规律,使钢铁制造流程中的物质流、能量流、信息流实现协同优化，使其"层流"运行，减少"紊流"运行状态，从而提高钢铁制造流程的运行效率。

3.2.5 设计人员设计理念和设计思维的转变

长期以来，由于工程设计人员既受到传统教育模式的影响，又受到工作背景和经历的影响，一般工程设计人员对基础科学和技术科学知识积累较多，而对工程科学的理论研究相对不足，参加工作后又长期从事钢铁生产制造流程某一生产工序的工程设计工作，习惯于从某一具体生产工序的角度考虑问题，其设计思路具有一定的局限性，很难从钢铁生产制造整个流程的全局高度去分析、解决本工序的工程设计问题。因此，开展21世纪钢铁厂工程设计创新工作，工程设计人员其思维方式的转变，思维层次的提高更具有战略意义，也是解决工程设计持续创新的关键。

4 21世纪以来取得的主要成就

4.1 冶金工程设计总体成效

进入21世纪以后，我国冶金工程公司开始应用"新一代钢铁厂动态-精准设计理论与方法"进行现代化钢铁厂设计。通过对钢铁制造流程的工程科学问题的研究探索，对冶金工程设计理念、理论和方法提出了新的认识，并将创新研究成果应用于工程实践，推动了钢铁厂结构调整和流程优化，为钢铁企业节省了工程投资，为钢铁制造流程的整体调控和智能化高度集成提供了理论依据。与此同时，冶金关键单元技术和工程化集成技术得到重点突破，在关键、共性、重大技术领域取得一系列科技成果，对冶金工程设计理论与实践提供了有效的支撑。

首钢京唐、首钢迁钢、鞍钢鲅鱼圈等钢铁厂，都是近年来我国自主设计建设的国际一流的高效化钢铁厂，通过对首钢京唐、首钢迁钢、鞍钢鲅鱼圈等大型钢铁厂全流程高效快节奏生产工艺与质量控制技术集成的研究，探索了全流程高效快节奏生产工艺与运行规律、研究开发出各工序间的工艺衔接与界面匹配技术、炼钢-精炼-连铸-连轧高效快节奏运行模式、建立全流程信息化集成系统和大批量、低成本洁净钢生产体系与质量保证体系，并取得了重大科技成果。冶金关键单元技术和工程化技术集成在诸多方面得到了突破。

4.2 构建先进的工艺流程

冶金厂工程设计兼具流程制造业与装备制造业双重属性。制造业分为流程制造业与装备制造业两大类，钢铁生产工艺过程的研究属流程制造业，而钢铁生产装备技术的研究则属于装备制造业。因此，21世纪钢铁厂工程设计理念和设计方法的运用有所侧重，针对钢铁生产工艺过程的研究主要是在冶金-材料学科基础科学和技术科学研究的基础上，运用工程科学的理论——冶金流程工程学分析解决问题；而针对钢铁生产装备技术的研究则主要是在满足钢铁生产工艺要求的前提下，运用现代设计方法分析解决问题。

按照冶金流程工程学理论，在解析各工序（单元）"功能-结构-效率"的基础上对钢铁制造全流程进行系统优化和集成，为构建新一代钢铁厂先进工艺流程探索了理论并积累了实践经验。图1所示为首钢京唐钢铁厂钢铁制造工艺流程。

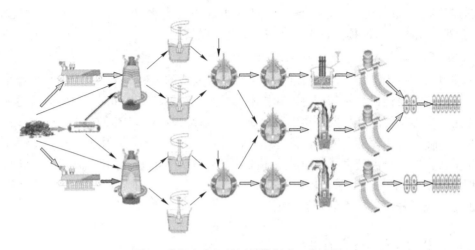

图1 首钢京唐钢铁厂钢铁制造工艺流程

4.3 建立连续-紧凑的生产布局

新一代钢铁厂的功能已拓展为先进钢铁产品制造、高效能源转换和消纳废弃物并实现资源化的"3个功能"。钢铁厂钢铁产品制造功能是在尽可能减少资源和能源消耗的基础上，高效率地生产出成本低、质量好、排放少且能够满足用户需求不断变化的钢材，供给社会生产和居民生活消费。能源转换功能与钢铁制造功能相互协同耦合，即钢铁生产过程同时也伴随着能源转换过程。以高炉-转炉-热轧流程为代表的钢铁联合企业，其实质是冶金-化工过程，也可以视为是将煤炭通过钢铁冶金制造流程转换为可燃气、热能、电能、蒸汽甚至氢气或甲醇等能源介质的过程。废弃物消纳-处理和再资源化功能，即钢铁制造流程中的诸多工序、装备可以处理、消纳来自钢铁厂自身和社会的大宗废弃物，改善社会环境负荷，促进资源、能源的循环利用。

以"流"（物质流、能源流、信息流）为核心，建设最优化的"物质流、能源流、信息流"耦合的生产布局，实现物质-能量-时间-空间的相互协调，促进钢铁生产整体运行稳定、协调，实现高效化、集约化、连续化，为新一代钢铁厂集约化的"联合-集中"布局建立了示范。

4.4 开发高效-短捷的界面技术

界面技术是指主体工序之间的衔接-匹配、协调-缓冲技术及相应的装置（装备）。主要体现在要实现生产过程物质流（包括流量、成分、组织、形状等）、生产过程能量流、生产过程温度、生产过程时间等基本参数的衔接、匹配、协调、稳定等方面。界面技术在传统的冶金厂工程设计中一直被忽略，而从动态-精准设计的角度分析，界面技术则是优化钢铁生产流程、促进生产流程整体运行的稳定、协调和高效化、连续化的关键环节。高炉—转炉生产流程中重要的界面技术包括：高炉-转炉区段的界面技术、炼钢炉-连铸机区段的界面技术和连铸机-热连轧机区段的界面技术。

在高效化生产流程中生产过程日趋连续化，将炼铁、炼钢和轧钢三个单元工序有机地结合为一体，整体进行生产调度安排。为保证流程的连续性，采用"连续-紧凑"较刚性的连接减少系统缓冲环节，避免缓冲造成过多的时间延误和温度损失。在炼铁-炼钢界面采用"一罐到底"铁水运输工艺，减少铁水倒罐过程；在连铸-热轧界面采用连铸坯在线"热装热送"工艺，提高连铸坯热装温度，提高生产效率，实现节能减排。图 2 所示为钢铁制造流程中典型的界面技术。

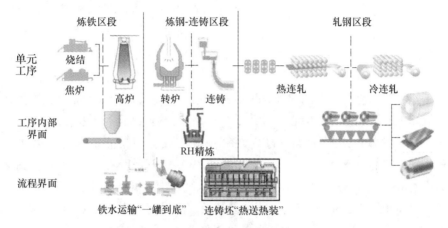

图 2　钢铁制造流程中典型的界面技术

4.5　自主研发先进的大型工艺装备

科学合理的工艺技术装备大型化是钢铁工业技术发展的主导方向，也是提高生产效率、实现节能减排和降本增效的根本措施。以首钢京唐钢铁厂为例，设计并采用了处理量 2500 万吨/年大型原料场，7.63m 焦炉和 260t/h 大型干熄焦，400 万吨/年球团带式焙烧机，500m^2 烧结机，5500m^3 高炉，"2+3" 大型转炉"全三脱"冶炼工艺及多功能 RH 二次精炼装置，2150/1650mm 双流板坯连铸机，2250/1580mm 热连轧机组，以及 2230mm、1700mm、1420mm 冷连轧机组等国内外先进的大型冶金技术装备，使全流程的生产效率显著提高、生产运行成本明显降低，为新一代钢铁厂工艺技术装备大型化、高效化积累了经验。

4.6　关键、共性单元工艺技术取得重点突破

近年来，我国成功开发了利用劣质煤生产高品质焦炭技术和低品质矿的综合利用技术；开发了大型高效、节能环保的烧结工艺技术，显著提高了烧结矿质量和生产效率；自主设计了具有国际先进水平的特大型高炉，掌握了实现特大型高炉冶炼稳定顺行的综合技术；设计开发了基于"全三脱"冶炼工艺的高效率、低成本洁净钢生产工艺流程和技术装备，实现了转炉炼钢高效化生产与无缺陷连铸坯高效连铸生产的耦合匹配；开发了高效、低成本的精准轧制技术，设计并应用了新一代高性能、高质量钢铁产品制造工艺；开发并应用了沿海钢铁厂大型海水淡化工艺技术；开发并应用了高效能源转换技术，利用冶金二次能源使钢铁厂自供电率达到 96% 以上；按照循环经济理念，建立了冶金资源、能源循环和绿色制造技术体系。

4.7 应用实践

首钢京唐钢铁厂将新一代可循环钢铁流程工艺技术所开发的各项重大单元技术成果进行系统集成，成为具有 21 世纪国际先进水平的钢铁厂，建成了新一代可循环钢铁流程的示范基地，实现了我国钢铁工业的新飞跃。

在设计开发 5500m³ 特大型高炉无料钟炉顶设备[15]、1300℃高风温顶燃式热风炉[16]、特大型高炉煤气全干法除尘[17]、"2+3"全三脱洁净钢生产平台[18,19]、"一罐到底"的多功能铁水罐直接运输[20]等重大单元技术的基础上，通过工序功能-结构-效率的解析、优化和系统集成，构建了 2 座高炉、1 个炼钢厂、2 套热连轧机的流程结构，实现了全流程的精准匹配和动态协调。与国内外先进钢铁厂相比，系统运行效率显著提高。通过建立全厂生产指挥中心、能源管控中心，构建了全厂流-程序-网络运行系统[21,22]。

通过研究开发炼铁-炼钢界面标准轨距"一罐到底"多功能铁水直接运输技术[23]，将高炉与炼钢之间的距离缩短为 900m[24,25]，减小铁水温降 30~50℃，使铁水进入 KR 脱硫站的平均温度达到 1387.5℃。由于取消了铁水倒罐站，取消了炼钢厂房铁水倒罐跨，减小炼钢厂房面积约 1150m²，节约倒罐站设备及厂房投资 4000 万元以上，每年可减少烟尘排放约 4700t，减少岗位定员约 100 人，每年降低生产运行成本 6000 万元以上。自主研究开发了利用汽轮发电机低温、低压乏汽进行低温多效海水淡化新工艺，实现了"热-水-电三联产"，系统热效率提高到 82.23%。系统可以进行 3 种工况切换运行，充分利用不同品质的富裕蒸汽资源，减少蒸汽放散。设计并应用了 4×1.25 万吨/天的低温多效蒸馏技术的海水淡化装置，每年可替代淡水资源 1750 万吨，为社会制碱厂提供浓盐水约 1000 万吨/年，减少了浓盐水的直接外排，实现了资源的循环利用。其中，2 套日产水量 1.25 万吨的低温多效海水淡化装置的 2 套装机容量 25MW 前置发电机组，每年发电 3.5 亿千瓦时，可大幅度减少外购电量。采用上述先进工艺技术，每年可创造经济效益约 9800 万元，社会效益和环境效益显著。

构建了大型转炉"2+3"洁净钢生产平台，与常规的复吹工艺相比，通过提高基元反应效率，建成以铁水"全三脱"少渣冶炼为工艺特色的洁净钢生产新流程，实现钢材洁净度提高 1 倍，硫、磷、氢、氮、氧杂质元素总含量可降低到 0.007%以下，炼钢石灰消耗量降低到 25kg/t，炼钢渣量减少到 50kg/t，转炉生产效率提高 1 倍，与传统转炉生产普通钢相比，吨钢生产成本降低 39.6 元。

研究并采用新一代钢铁厂精准设计和流程动态优化技术，通过建立动态—有序运行的理论框架和物理模型及仿真模型，对钢铁厂各工序（系统）从原燃料的消耗、产能匹配、各项工艺参数的确定、能源动力的消耗到能源设施的布局、工序之间的衔接等进行了深入的解析研究，在温度、物质的成分品位、运行节奏、能源的输入和输出等方面均进行了精准的计算和优化配置。通过运用精准设计理论，构建了首钢京唐钢铁厂动态有序、连续紧凑、精准协调的生产运行体系。实践表明，从高炉出铁进入炼钢厂脱硫、脱硅-脱磷、脱碳；炉外精炼、连铸；到热轧工序形成热轧产品，通过对全过程的运行节奏、温度调节、成分控制、作业组织进行动态、精准的优化，可将整个时间目标控制在 400min 以内。

5 冶金工程设计发展趋势及展望

5.1 冶金工程设计发展方向

目前，在冶金工程设计领域，我国已经具备了现代钢铁厂的流程设计、工艺设计、设备设计和系统集成能力，建立了新一代钢铁厂功能-结构-效率优化和系统集成、流-程序-网络的运行体系、高效紧凑的界面技术、动态精准设计、洁净钢生产平台等冶金工程设计学科的基础理论，通过首钢京唐、鞍钢鲅鱼圈、马钢新区、首钢迁钢、邯钢新区、重钢新区等重大项目的实践，使我国冶金工程设计的理论研究和实践得到了快速的发展。

我国冶金工程设计企业的优势在于现代特大规模钢铁联合企业的设计和建设，以及将新产品、工艺、装备技术为一体的集成创新[26~29]。未来的发展方向是将冶金流程工程学理论及其动态-精准设计方法，紧密结合我国新建或改建钢铁厂工程建设项目推广应用。以实施工程设计理论工程化集成创新重大专项为载体，验证冶金工程设计创新理念、理论及设计方法的应用效果，通过工程化应用不断完善工程设计创新理念、理论及设计方法，并通过生产实际应用构建新一代可循环钢铁制造流程，转变我国钢铁工业经济发展方式，提高我国钢铁企业市场竞争力和可持续发展能力，以此推动我国由钢铁大国向钢铁强国转变。

5.2 冶金工程设计发展思路

未来一个时期，提高我国冶金工程设计创新能力的总体战略思路应当是：以冶金流程工程学和动态-精准设计理论和方法为核心，构建我国冶金工程设计创新理念、理论及设计方法的完整体系。重视顶层设计和需求引导，结合我国新建或改建的钢铁厂工程建设项目，以实施冶金工程设计创新理念、理论及设计方法工程化技术集成创新为载体，结合钢铁制造过程的动态运行，完善新一代钢铁制造流程，全面提升我国冶金工程设计创新能力，以此带动我国钢铁制造流程实现全面转型，转变我国钢铁行业经济发展增长方式，提高我国钢铁企业的市场竞争力和可持续发展能力。

5.3 冶金工程设计发展目标

今后一个时期，应着力建立和完善现代钢铁厂的"3个功能"，现代钢铁厂的设计目标不应局限于单一的钢铁制造功能，更应具有多重目标的属性，即：满足市场需求，提高企业核心竞争力和可持续发展能力，建设资源节约型、环境友好型钢铁厂，大力发展循环经济，实现与社会和生态环境的和谐发展。

建立新一代钢铁厂工程设计建设的基本目标是：建立高效率、低成本、稳定生产洁净钢的生产体系；高效率、低排放的能源转换体系和余热余能回收利用系统以及大宗社会废弃物的处理-消纳-再资源化系统。

以实现流程动态-有序运行的动态-精准设计设计方法为工具，设计的新一代钢铁制造流程应具有如下主要特征：流程高效化、生产集约化、工艺现代化、装备大型化、产品洁净化、资源循环化、环境友好化、效益最佳化，并且与社会和生态环境和谐发展。体现在新一代钢铁厂应具有高质量、高效率、低成本、清洁化的生产运行体系。

应继续保持现有的理论探索优势，缩小在信息化技术应用开发、数字化设计方法，特别是在三维动态仿真设计等领域与国外先进水平的差距。

6 结论

（1）冶金工程设计是将冶金科学技术转化为生产力的关键环节，是体现技术和经济双重科学性的关键要素，是实现冶金工程建设目标的决定性环节，直接影响钢铁制造流程的竞争力和可持续发展，具有特殊的重要性。

（2）21世纪以前，我国冶金工程设计沿用苏联的设计理论和设计方法，属于"静态-分割"和"半经验、半理论"的传统设计体系，存在着诸多弊端，难以真正实现钢铁厂的运行效率、产品质量、投资效益的最佳化。

（3）当前，我国已经具备了现代大型钢铁厂的流程设计、工艺设计、设备设计以及系统集成能力，建立了现代冶金工程设计理念、理论和设计方法，创建并应用动态-精准设计体系，设计建设了具有国际先进水平的新一代可循环钢铁厂，在流程高效化、生产集约化、装备大型化等方面取得显著成效。

（4）我国冶金工程设计的优势在于集成创新，与国外先进水平相比，在核心技术产品的设计研制、数字化三维仿真设计和设计方法创新等方面仍存在差距。

（5）未来一个时期，我国应以冶金流程工程学和冶金流程集成理论与方法为指导，以动态-精准设计理论和方法为核心，构建冶金工程设计创新理念、理论及设计方法的完整体系。

致谢

衷心感谢中国工程院殷瑞钰院士、徐匡迪院士、干勇院士、张寿荣院士，以及中国金属学会仲增庸教授、李文秀教授、洪及鄙教授和宝钢王喆教授等院士专家的悉心指导、大力支持与帮助。

参考文献

[1] 殷瑞钰. 冶金流程工程学（第2版）[M]. 北京：冶金工业出版社，2009.
[2] 殷瑞钰，汪应洛，李伯聪. 工程演化论[M]. 北京：高等教育出版社，2011.
[3] 殷瑞钰. 高效率、低成本洁净钢"制造平台"集成技术及其运行[J]. 钢铁，2012，47（1）：1-8.
[4] 殷瑞钰. 关于新一代钢铁制造流程的命题[J]. 上海金属，2006，28（4）：1-5，13.
[5] 殷瑞钰. 冶金流程集成理论与方法[M]. 北京：冶金工业出版社，2013.
[6] 殷瑞钰，汪应洛，李伯聪. 工程哲学[M]. 北京：高等教育出版社，2007.
[7] 殷瑞钰. 中国钢铁工业的崛起与技术进步[M]. 北京：冶金工业出版社，2004.
[8] 中国金属学会，中国钢铁工业协会. 2006年-2020年中国钢铁工业科学与技术发展指南[M]. 北京：冶金工业出版社，2006.
[9] 殷瑞钰. 冶金工序功能的演进和钢厂结构的优化[J]. 金属学报，1993，29（7）：289-315.
[10] 殷瑞钰. 钢铁制造流程的多维物流控制系统[J]. 金属学报，1997，33（1）：1-5.
[11] 田乃媛. 钢铁制造流程多维物流管制研究的进展[J]. 钢铁研究，2002，30（5）：1-4.

[12] 刘青,田乃媛,殷瑞钰.炼钢厂的运行控制[J].钢铁,2003,38(9):14-18.
[13] 刘茂林,田乃媛,徐安军.高炉-转炉区段物流过程解析[J].北京科技大学学报,2000,22(1):8-11.
[14] 李喜先.工程系统论[M].北京:科学出版社,2007.
[15] 张福明,钱世崇,张建,等.首钢京唐5500m³高炉采用的新技术[J].钢铁,2011,46(2):12-17.
[16] 张福明,梅丛华,银光宇,等.首钢京唐5500m³高炉BSK顶燃式热风炉设计研究[J].中国冶金,2012,22(3):27-32.
[17] 张福明.大型高炉煤气干式布袋除尘技术研究[J].炼铁,2011,30(1):1-5.
[18] 殷瑞钰.关于高效率、低成本洁净钢平台的讨论[J].中国冶金,2010,20(10):1-10.
[19] 张福明,崔幸超,张德国,等.首钢京唐炼钢厂新一代工艺流程与应用实践[J].炼钢,2012,28(2):1-6.
[20] 李湘臣,田乃媛.铁水包功能综合化的边界条件及实施的可行性分析[J].钢铁研究,2006,34(3):51-53.
[21] 张春霞,殷瑞钰,秦松,等.循环经济社会中的中国钢厂[J].钢铁,2011,46(7):1-6.
[22] 顾里云.首钢京唐钢铁公司能源管控系统建设的理论与实践[J].冶金自动化,2011,35(3):24-28.
[23] 邱剑,田乃媛.首钢炼铁—炼钢界面模式的研究[J].钢铁,2004,39(4):74-78.
[24] 张福明,钱世崇,殷瑞钰.钢铁厂流程结构优化与高炉大型化[J].钢铁,2012,47(7)1-9.
[25] 尚国普,向春涛,范明浩.首钢京唐钢铁厂总图运输系统的创新及应用[J].中国冶金,2012,22(8):1-6.
[26] 张寿荣.关于21世纪我国钢铁工业的若干思考[J].炼钢,2002,18(2):5-10.
[27] 殷瑞钰.中国钢铁工业的成就、命题和发展[J].冶金管理,2003(5):4-11.
[28] 徐匡迪.20世纪——钢铁冶金从技艺走向工程科学[J].上海金属,2002,24(1):1-10.
[29] 殷瑞钰.节能、清洁生产、绿色制造与钢铁工业的可持续发展[J].钢铁,2002,37(8):1-8.

Current Status and Expectations of Metallurgical Engineering Design

Zhang Fuming Xie Jianxin

Abstract: The Chinese metallurgical engineering design plays an important role for promoting sustainable development of metallurgical industry. China has already possessed the expertise of process design, plant design and equipment design as well as the technical know-how of system integration for modern large-scale steel plants, and it has created and put into application a dynamic-precision engineering system. A new generation of circulating steel plant, advanced by international standard, is designed and built. These metallurgical plants feature significantly high-efficient process, intensive production, large-scale equipment, etc. With the development of these metallurgical plants, a system of engineering philosophy, theory and method for metallurgical plant is established. And at the same time, significant breakthroughs in key-unit metallurgical technology and engineering integration are made; a series of scientific-technological achievements in various technical fields of key technologies, common technologies and significant technologies are acquired, providing effective support to the theory development and practice of the studies of metallurgical plant design. In the coming period, the metallurgical process engineering and the theory and method of dynamic-precision should be centered, to establish a comprehensive system of philosophy, theory and design method for metallurgical plant design innovation.

Keywords: metallurgical process; engineering design; design theory; design method; technological innovation

新一代钢铁厂循环经济发展模式的构建

摘　要：本文分析了现代钢铁工业发展中存在的问题，提出了现代钢铁工业发展循环经济的基本模式。新一代可循环钢铁厂要具备钢铁产品制造、能源转换和消纳社会废弃物的三种基本功能，循环经济将是未来钢铁厂生存和发展的必由之路。本文结合首钢京唐钢铁厂的设计论述了新一代钢铁厂的高效可循环的设计理念，阐述了"减量化、再利用、再循环"的原则在新一代钢铁厂设计中的应用。

关键词：循环经济；钢铁工业；节能减排；设计理念

1　钢铁工业发展循环经济理念的提出

在工业化发展的进程中，环境破坏和生态恶化的根源是工业系统所采用的以"高开采-低利用-高能耗-高排放"为特征的单向发展模式造成的。钢铁行业是高消耗、高能耗型行业，也是对环境污染比较严重的行业，钢铁行业的能耗约占全国总能耗的10%。钢铁生产的末端治理需要大量资金，回收的资源得不到有效的利用，影响企业自身的可持续发展、企业的经济效益和当地的环境质量。

随着我国经济社会的发展以及人民对生存环境日益提高的要求，国家对钢铁行业提出越来越严格的资源、能源与环境要求。保护地球环境与实现可持续发展，是21世纪钢铁工业面临的最大挑战。钢铁制造实现绿色化，是21世纪钢铁产业发展的奋斗目标[1]。

循环经济是一种可持续发展的生态经济，是按照自然生态系统的物质循环、能量守恒和生态规律利用自然资源和环境容量重构经济系统，目的旨在使经济系统和谐地纳入到自然生态系统的物质循环过程中。"减量化（Reduce）、再利用（Reuse）、再循环（Recycle）"（简称"3R"原则）是循环经济最重要的实际操作原则。20世纪90年代以来，发展循环经济已成为国际社会的一大趋势。

传统经济是由"资源-产品-污染排放"所构成的物质单向线性流动的开环式经济过程。循环经济是一种建立在物质不断循环利用基础上的经济发展模式，它要求经济活动按照自然生态系统的模式，组织成一个"资源-产品-资源再生"的闭环反馈式循环过程，以期实现"最佳化的生产、最适度的消费、最少量的废弃"。循环经济倡导的是一种与环境、社会和谐的经济发展模式。

钢铁材料作为最重要的结构材料和工程材料，在未来相当长的时期内是不可替代的基础性和功能性材料，同时也是最易于回收和可再生的资源。钢铁材料能够大规模、低成本地制造，循环往复地利用，并做到与环境友好。钢铁企业通过实施清洁化生产、节能减排和物质循环，能够生产出更多性能优良、使用寿命长、效率高和对环境影响符合要求的优

本文作者：张福明。原刊于《2008年冶金循环经济发展论坛论文集》，2008：57-61。

质产品，实现经济效益、环境效益和社会效益的协调统一，促进工业经济的可持续发展。

贯彻落实科学发展观，发展循环经济的直接目的是通过发展产品深加工，延伸产品链，优化整体物流链，扩展物质的循环利用领域；通过提高资源和能源利用效率，实现"资源效率提高-原材料燃料消耗降低-环境改善-钢铁产品成本降低-企业市场竞争力提高"的良性循环。目的就是要最大限度地提高能源、水资源和矿产资源的利用效率；最大限度地提高环境空间利用效率；最大限度地提高钢材产品的使用效率；最根本的目的是要开发出高档次、高质量的满足社会需要的钢材产品。

2 发展循环经济的基本思路

2.1 指导思想

按照科学发展观的要求，钢铁企业以市场为基础，以持续提高企业的核心竞争力为目标，按照循环经济的"3R"原则，通过实施可持续发展战略，促进金属资源节约、能源高效利用、水资源节约、清洁生产、节能减排以及资源回收与综合利用，追求生态环境和经济效益的最佳化。发展具有市场潜力和比较优势的钢铁制造系统，延长产品生产链，生产高档次、高附加值的精品板材，以获取更高的资源利用价值和钢材使用价值。构建资源和能源节约、生产与管理高效、环境清洁的运行管理模式，使21世纪钢铁厂成为与环境友好的能源和资源节约型绿色冶金工厂。

2.2 基本思路

循环经济是把清洁生产、资源综合利用、可再生能源开发、产品的生态设计和生态消费融为一体，运用生态学规律来指导人类社会经济活动的模式。从而达到"生态工业系统"与"自然生态系统"相耦合的自然循环型。

钢铁制造流程经过不断的技术开发和集成优化，可以使整体功能得到拓展，不断地延伸其制造链、经营链形成工业生态链。由于工业生态链的构成，铁素资源形成循环经济中的重要一环。同时，钢铁厂的社会经济职能也通过生态化转型得到进一步扩展[2]。

从钢铁企业的未来社会经济角色来看，钢铁厂生产流程应该具有三种主要功能[3]：
(1) 钢铁产品制造功能；
(2) 能源转化功能；
(3) 社会部分大宗废弃物的处理——消纳功能。

在此认识前提下，实施清洁生产、绿色制造、生态化转型等技术进步措施，解决好钢铁厂与生态环境的和谐发展，承担起应负的社会责任，即在获得良好的市场竞争力的同时，实现可持续发展。

建设循环型企业的目标是"零排放"，企业内部建立资源循环、能源循环、水循环、冶金渣和固体废弃物循环的四个循环体系。

钢铁企业将面向市场，调整产品结构；依靠技术进步，促进产业升级；加大力度，实现可持续发展，拓展企业的钢铁生产功能，使它除充分发挥钢铁生产功能外，还具有能源转换、固体废弃物处理和为相关行业提供原料等功能，实现物质和能源的大、中、小循环。

小循环——以铁素资源为核心的上下生产工序之间的循环；

中循环——各生产分厂之间的物质和能量循环；

大循环——钢铁企业与社会之间的物质和能量循环。

图1是新一代钢铁厂循环经济模式的示意图，图2是以钢铁厂为核心的循环经济产业链。

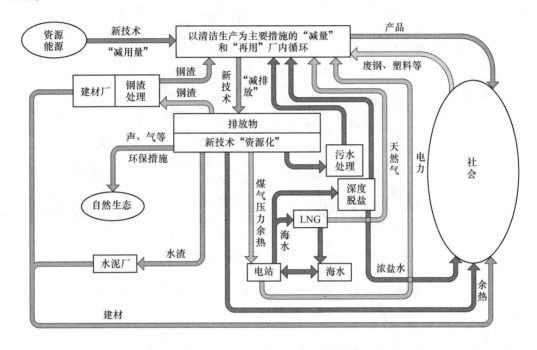

图1 首钢京唐钢铁厂循环经济模式示意图

首钢京唐钢铁厂发展循环经济的基本思路是：

(1) 以循环经济理念为指导，按照"减量化、再利用、再循环"的原则统筹规划，使资源充分、合理利用，有效控制和消除环境污染；使用较少的资源、原材料和能源投入，生产出更多市场需要的高性能、高附加值产品，使钢铁生产的资源利用更合理、更充分，产品生产成本更低、市场竞争力更强，实现钢铁生产经济效益和环境效益共赢的目标。

(2) 全面推广采用先进的节能、节水、资源回收利用和环保技术，走新型工业化道路，实现"少投入、多产出、低污染、零排放、高效益、可持续发展"的战略目标，使企业具有很强的核心竞争力。

(3) 强化能源与环境的科学管理。包括建立完善的能源调度和环境监测管理体系，对各种能源实行集中管理和统一调配，把科学、完善的节能与环境监测管理体系纳入生产管理之中。

(4) 面向市场，调整和优化产品结构，提高产品的加工深度，发展精品板材产品。

(5) 贯彻全面协调、可持续发展的科学发展观。一方面，钢铁企业为相关行业提供工业原料，同时把钢铁产品生产过程中产生的二次能源用于城市生活，改善城市的环境质

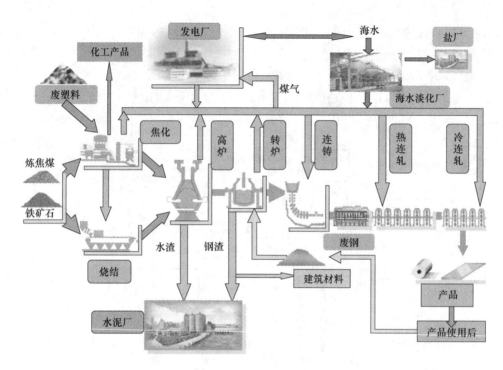

图 2 以钢铁厂为核心的循环经济产业链

量;另一方面,把钢铁厂作为社会废弃物无害化处理中心,实现企业和社会、人与自然和谐发展,实现企业效益、社会效益和环境效益相统一。

3 首钢京唐钢铁厂发展循环经济的实施方案

3.1 减量化

通过大力推广应用大型化技术装备、先进的工艺技术、现代管理理念,最大限度降低原燃料、原材料及各种资源消耗,从而降低生产过程中废渣、废料、废次材及污染物的产生,节约资源、降低成本、减少排放、提高效益。

(1) 烧结采用 500 m^2 大型烧结机和厚料、低 SiO_2、低 FeO 烧结工艺,提高烧结矿强度、质量和利用率,烧结矿品位达到 58% 以上,转鼓指数达到 78% 以上,高炉返矿率由传统的 12% 降低到 8%。为高炉生产提高精料水平、降低原料消耗创造了良好的条件。烧结利用废气余热作为原料干燥的热能和用于点火助燃空气,降低能耗,燃料用量由传统工艺的 50kg/t 降低到 43kg/t。

(2) 球团采用大型带式焙烧机,物料在整个焙烧过程中处于静止状态,其生产工艺和链箅机-回转窑工艺相比,单机产能大,原料消耗少,铁矿粉耗量由传统工艺的 1010kg/t 降低到 990kg/t 以下。

(3) 焦化采用 7.63m 特大型焦炉和 2×260t/h 特大型干熄焦工艺,提高焦炭强度,降低高炉返焦率,提高入炉焦成品率,减少炼焦煤用量。采用干熄焦工艺使焦炭强度 M_{40} 提高 3%~8%，M_{10} 降低 0.3%~0.8%。M_{40} 增加 1%，铁水可降低燃料消耗 3~5kg/t，M_{10} 降低 0.2%，铁水可降低燃料消耗 4~7kg/t，有利于降低炼铁工序的燃料消耗。煤气脱硫采用真空碳酸钾配制酸工艺生产硫酸，减少硫铵生产所需硫酸的外购量 2.96 万吨/年。

（4）炼铁采用 5500m³ 超大型高炉，采用精料技术和合理炉料结构，熟料率提高到 90%，综合入炉品位提高到 61% 以上，吨铁入炉矿降低到 1600kg/t 以下。通过优化炉料结构，采用并罐式无料钟炉顶装料设备和无中继站直接上料工艺，采用高风温和富氧、大喷煤技术，焦比由传统的 380kg/t 降低到 270kg/t 以下，节约焦炭约 100 万吨/年。渣比由传统的 310kg 降低到 250kg/t 以下，每年减少水渣约 54 万吨。大幅度降低了炼铁生产过程的燃料消耗和工序能耗。

（5）炼钢采用"2+3" 300t 大型转炉"全三脱"洁净钢冶炼工艺技术、铁水 100% 预处理措施、"少渣法"冶炼、副枪及智能化科学炼钢、顶底复合吹炼等综合技术，吨钢入炉钢铁料消耗由传统的 1085kg/t 降低到 1068kg/t，每年减少钢铁料消耗约 17 万吨。

采用活性石灰、挡渣出钢等技术，吨钢石灰、白云石等原材料消耗由传统的 70kg/t 和 20kg/t 分别降低到 40kg/t 和 15kg/t，吨钢渣量由传统的 110~130kg/t 降低到 70~90kg/t，每年减少转炉渣排放 40 万吨。

总图布置上缩短了炼铁与炼钢的距离，缩短铁水运输距离，采用"一罐到底"铁水运输工艺，减少运输过程的温降和金属损失；提高连铸机作业率和连浇炉数，钢水成坯率由传统的 97% 提高到 97.5%。

（6）轧钢减少氧化铁皮和废钢的产生，提高金属成材率。采用节能型加热炉，减少烧损量；连铸坯热送热装率达到 75%，减少二次加热烧损；采用连续化生产工艺，综合成材率提高到 93.02%。

3.2 再利用

3.2.1 含铁固体物的处理及综合利用

为实现企业循环经济的目标，对烧结、球团、炼铁、炼钢、轧钢等各主要生产工序含铁尘泥充分回收，并在厂内循环利用，回收炼钢钢渣中的渣铁，炼钢、轧钢产生的废钢全部回收作为炼钢入炉料。

通过充分回收及利用钢铁厂生产过程中的铁素资源，使铁素资源得到 100% 利用。钢铁厂生产系统铁素资源循环流程如图 3 所示。

3.2.2 非含铁固体物的处理及综合利用

3.2.2.1 主要非含铁固体物的处理及综合利用

贯彻循环经济原则，除含铁资源循环利用外，所有固体废弃物，如高炉水渣、转炉渣等，均须综合利用，既要节约成本，又要保护环境。固体废弃物具体利用，大部分采用地区经济协作的方式进行有效处理。钢铁厂非铁固体废弃物资源循环流程如图 4 所示。

3.2.2.2 其他非含铁固体物的处理及综合利用

（1）脱苯系统再生器产生的残渣混入焦油中回收；焦油渣、生化处理污泥均掺入炼焦煤中再利用，其总量约 1 万吨/年；

（2）锌渣约 1.2 万吨/年，全部送冶炼厂重熔使用；

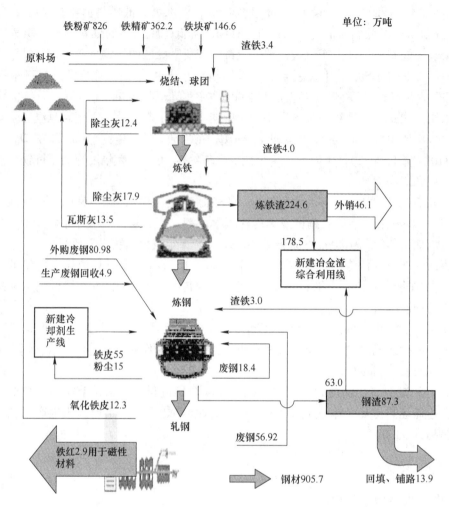

图 3 钢铁厂生产系统铁素资源循环流程

(3) 冷轧产生的废盐酸 1.2 万吨/年，设置废酸再生站进行再生处理利用；

(4) 自备电站锅炉粉煤灰渣约 26 万吨/年，用于生产建筑材料。

3.2.3 能源循环利用

3.2.3.1 能源循环利用的设计原则

在现代化大生产中循环经济体现在生产的各个环节，充分利用生产过程中产生的余压、余热、余气进行能源转化，最大限度降低能源消耗，提高资源的利用率，改变传统的"高消耗-低利用-高排放"的生产模式，通过对废弃物进行回收利用、无害化及再生的方式，以达到资源的永续利用，促进社会经济的可持续发展，体现"低消耗-高利用-低排放"的现代新型工业化生产特征。

3.2.3.2 能源循环利用措施

利用高炉炉顶煤气余压、干熄焦显热和富裕高炉煤气进行发电，每年自发电 55.04 亿千瓦时，电力自供率达到 94%。利用热风炉废气对助燃空气、煤气进行预热和作为干燥剂干燥喷吹煤粉。利用烧结环冷机高温废气和干熄焦烟气通过余热锅炉，产生蒸气，回收

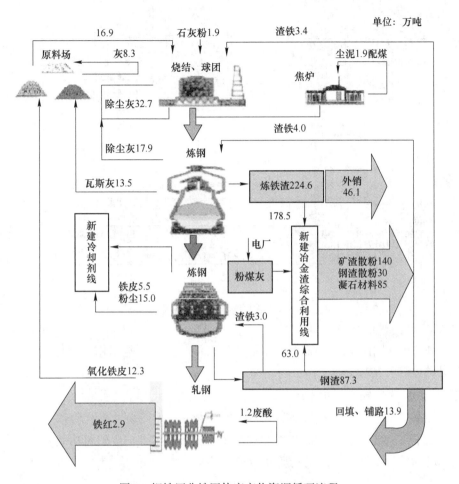

图 4 钢铁厂非铁固体废弃物资源循环流程

转炉和加热炉气化冷却产生的蒸汽用于其他工序生产需要,每年通过回收余热、余汽增加蒸汽约 312 万吨。图 5 是钢铁厂煤气循环利用示意图。

3.3 再循环

实现企业与社会资源大循环,实现钢铁废物社会资源化,社会废物钢铁资源化,将钢铁厂优势转化为社会优势,使钢铁厂成为循环型社会的重要部分。

3.3.1 钢铁厂废物社会资源化

(1) 每年产生高炉水渣 224.6 万吨,可直接用于水泥厂作为水泥原料使用,每年生产钢渣 87.3 万吨,一部分钢渣经加工细磨后可用于高活性水泥掺和料,另一部分经磁选后的钢渣可加工用于道路基层材料和生产钢渣砖,用于建筑材料。由于利用钢铁厂水渣、钢渣制造水泥,每年减少水泥行业石灰石开采量约 320 万吨,减少水泥行业 CO_2 排放约 220 万吨,减少标准煤消耗约 22 万吨,减少粉尘排放约 7 万吨。

(2) 焦化每年生产硫铵、焦油、轻苯、重精苯约 27.8 万吨,为社会提供优质化工原料。

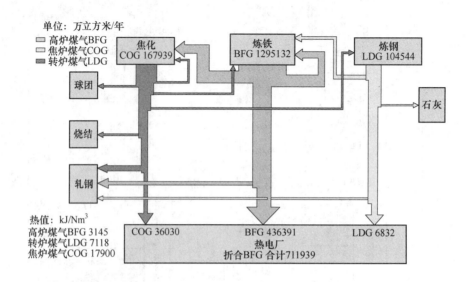

图 5　钢铁厂煤气循环利用流程

（3）钢铁厂利用厂内余热进行海水淡化，将淡化后高含盐水用于当地盐厂进行制盐，为社会提供良好的制盐资源。

3.3.2　社会废物钢铁厂资源化

（1）钢铁厂每年消纳社会废钢资源约 100 万吨。

（2）焦化工序设计预留了废塑料添加系统，具备消纳社会废塑料功能。

图 6 是钢铁厂与社会之间的物质和能量循环示意图。

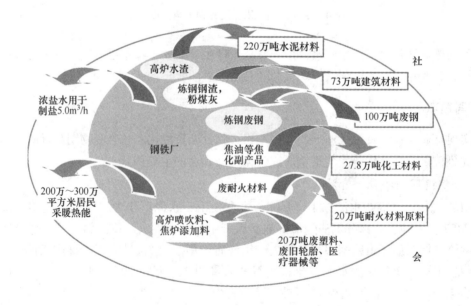

图 6　钢铁厂与社会之间的物质和能量循环示意图

4 结语

21世纪新一代钢铁厂实施循环经济战略，促进金属资源节约、能源高效利用、节约水资源、实现清洁化生产、节能减排、保护生态环境，达到企业与社会、人与自然之间的高度和谐发展，实现可持续发展。钢铁厂设计中煤气100%循环利用，铁素资源100%循环利用，工业废水100%回收处理循环利用，固体废弃物100%循环利用，水循环利用率可以达到97.5%，废水基本实现"零排放"，吨钢耗新水为3.84m^3/t，电力自供率达到94%，吨钢粉尘排放量0.3kg/t、SO_2为0.25kg/t，吨钢综合能耗为669kg/t，吨钢可比能耗为649kg/t，达到21世纪国际先进水平，基本实现了废水、固体废弃物"零排放"的目标。为构建新一代可循环钢铁生产流程、发展循环经济进行了卓有成效的探索。

参考文献

[1] 张寿荣. 21世纪前期钢铁工业的发展趋势及我国面临的挑战 [C] //2007中国钢铁年会论文集，北京：冶金工业出版社，2007，11：1-107.

[2] 刘铁男. 钢铁产业发展政策指南 [M]. 北京：经济科学出版社，2005，12：185-191.

[3] 殷瑞钰. 冶金流程工程学 [M]. 北京：冶金工业出版社，2005，9：381.

The Construction of Recycling Economic Advanced Mode for New Type Iron and Steel Factory

Zhang Fuming

Abstract: New briefing the open question of developing iron and steel industry recently and represent the advanced recycling economic mode for iron and steel industry. It is equipped with 3 kind fundamental functions such as products manufacture, energy transfer and sale satisfied the social wastes. The recycling economic is only way for survive and developing of iron and steel industry. A high efficiency design involved the principle of "reduce, reuse and recycle" already applied in Shougang Jingtang iron and steel factory.

Keywords: recycling economic; iron and steel industry; energy saving and decrease discharge; design idea

钢铁厂工程设计的创新与实践

摘　要：本文分析了首钢京唐钢铁厂工程建设的背景，结合首钢京唐钢铁厂工程设计的实践，论述了新世纪钢铁厂工程设计创新的理念、主要技术路线、主要研究方法和内容，并提出了新世纪钢铁厂工程设计创新所赋予的新内涵，对新世纪国内、外新建或改建钢铁厂工程设计具有重要的引领示范作用。

关键词：钢铁厂；工程设计；创新；实践；新内涵

1　引言

钢铁工业是我国国民经济的重要基础产业和实现新型工业化的支撑产业。钢铁产业涉及范围广、产业关联度高、消费拉动大，在经济建设、社会发展等方面发挥着重要作用。新世纪我国钢铁工业面临全球化市场激烈竞争、资源制约、环境生态问题等日益严峻的挑战，不单是产品质量、性能的问题，而是面临着成本、质量、性能、效率、过程排放、过程综合控制、环境、生态等多目标群的挑战。我国应对这些挑战的总体战略是：用较低的GDP钢产量完成我国的工业化进程；以较低的资源、能源消耗生产满足我国工业化发展所需要的铁产品；以发展板带材集成技术为载体，构建新一代可循环钢铁制造流程；钢产量的增加不以增加环境负荷为代价。

首钢京唐钢铁厂是我国第一个自主设计、建造的具有21世纪国际先进水平的千万吨级钢铁厂。该项目为优化我国钢铁产业布局和产业结构、推进华北地区钢铁工业优化升级、促进环渤海地区经济协调发展、构建具有21世纪国际先进水平的新一代可循环钢铁制造流程具有重大意义。首钢京唐钢铁厂工程是我国首次自主设计的千万吨级沿海钢铁项目，工程设计是实现项目建设目标的重要关键环节，工程设计理念、理论和方法的创新是项目建设成功的重要保障，首钢京唐钢铁厂工程设计的创新与实践对21世纪国内外新建或改扩建钢铁厂具有重要的引领和示范作用。

2　工程设计理念创新

首钢京唐钢铁厂是新世纪初设计建造的钢铁厂，也是我国首次在沿海地区通过填海造地建设的千万吨级现代化钢铁厂。工程设计按照循环经济理念，遵循冶金流程工程学的原理，以构建高效率、低成本洁净钢生产工艺为核心，努力实现资源和能源的高效可循环利用，大幅度减少环境和生态污染，实现清洁生产和节能减排，将传统钢铁厂单一的产品制造功能拓展为具有优质产品制造、高效能源转换和消纳废弃物实现资源化的三项功能，构建物质流、能量流和信息流协同耦合运行的新一代可循环钢铁厂。

本文作者：张福明，颉建新，周颂明。原刊于《2011中国钢铁工业科技与竞争战略论坛论文集》，2011：116-140。

2.1 新世纪钢铁厂的发展目标

结合 20 世纪中后期国内外钢铁产业的发展趋势，新世纪钢铁厂发展的目标不再追求单一的生产规模，而是生产规模、生产效率、制造成本、产品质量、环境保护等多方面综合的目标集合，其最终目标是使钢铁企业具有市场核心竞争力并且能够实现可持续发展。现代钢铁厂应着力构建高效率、低成本的洁净钢生产体系，以满足市场对洁净钢的需求；同时应按照循环经济理念，建设资源节约、能源高效利用、生态环境友好的绿色钢铁制造体系，提高钢铁企业的核心竞争力和可持续发展能力，实现企业与社会、企业与环境的和谐发展，建立新一代可循环的钢铁制造流程，大力发展循环经济、低碳冶金和绿色制造。

2.2 新世纪钢铁厂的功能

针对"十一五"期间我国钢铁工业可持续发展受到资源、能源和环境的束缚，单纯采用传统钢铁流程难以保证我国钢铁工业可持续健康发展。为了解决这一重大问题，我国冶金专家经过多年研究，提出新一代可循环钢铁制造流程的工艺思想。

按照工程哲学和工程系统论的观点，从物理角度分析，由性质不同的诸多单元工序组成的钢铁制造流程运行的本质是一种多维"流"按照一定的"程序"在复杂的流程网络结构中流动运行的现象。钢铁制造流程是一种开放的、不可逆的、远离平衡的耗散过程结构[1]。传统钢铁厂的功能只是单纯制造钢铁产品，其特点是物质流单向线性流动的开放式生产过程，忽视钢铁制造过程中能源的高效转换、回收和综合利用，除利用一部分废钢资源之外，也没有消纳城市生活污水、废塑料等社会大宗废弃物的功能，钢铁企业与社会之间没有建立起发展循环经济的框架。新一代可循环钢铁制造流程把清洁生产、资源综合利用和可再生能源开发融为一体，成为闭环式生产过程，在原有功能的基础上拓展，具备先进钢铁产品制造、高效能源转换和消纳社会废弃物并实现资源化的三重功能[2,3]。

图 1 所示的新一代可循环钢铁生产流程的技术思想，首钢京唐钢铁厂的生产规模为 970 万吨/年，每年大约消耗铁矿石 1344 万吨，煤炭 714 万吨，生产粗钢 970 万吨，钢材 905 万吨。同时可消纳社会废钢 160 万吨，废塑料 20 万吨，利用余热余能发电 55 亿 kW·h，钢铁厂自发电率为 94%。通过炉渣回收处理生产水泥超微粉混合料约 300 万吨，降低 CO_2 排放，有利于改善生态环境，同时可以减少生产水泥的资源消耗，提高资源利用率。

2.3 新世纪钢铁厂的主要特征

新一代钢铁厂主要特征是：装备大型化、生产集约化、工艺现代化、产品精品化、资源循环化、流程高效化、效益最佳化和企业与社会和谐发展，构建高质量、高效率、低成本、清洁化的生产运行体系。

3 工程设计主要技术路线

3.1 钢铁厂工程设计主要特点

（1）钢铁厂工程设计是围绕质量、性能、成本、投资、效率、资源、环境等多目标

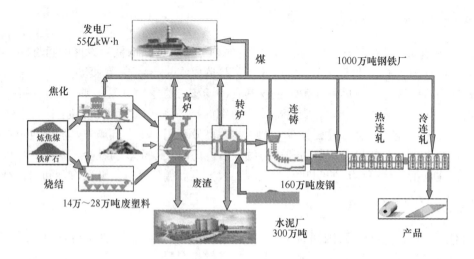

图 1　新一代可循环钢铁生产流程示意图

群进行选择、整合、互动、协同等集成和优化、进化的过程；

（2）钢铁厂工程设计是一个在实现单元工序优化的基础上，通过集成和优化，实现全流程的系统优化的过程；

（3）钢铁厂工程设计是一个实现钢铁制造全流程动态-精准、连续-高效运行的过程。

3.2　钢铁厂工程设计创新与实践的主要技术路线

（1）按照新一代钢铁厂功能、结构、效率的优化和系统集成理论，在对各工序装置功能进行解析优化的基础上，通过采用各工序间动态匹配、协同紧凑的流程结构和界面技术进行衔接，实施各工序单元功能的系统集成，从而构建钢铁厂的总体功能和效率，实现钢铁制造流程的有序化、协调化、高效化、连续化生产[4,5]。

（2）通过对首钢京唐钢铁厂物质流（铁素流）、能量流（碳素流）、信息流的"流、程序、网络"的系统构建，研究钢铁制造流程中输入端（如原料、辅料、能源等）、输出端（如产品、副产品、各类排放物等）、制造过程（如生产效率、能源利用率、物料收得率等）的多目标群的集成优化，实现物质流、能源流耗散的最小化。

（3）围绕钢铁厂发展循环经济的实施措施，按照"减量化、再利用、资源化"的原则，构建首钢京唐钢铁厂循环经济的运行体系。

（4）围绕钢铁厂环境保护的实施措施，按照"源头消减、过程控制、末端治理"的原则，构建首钢京唐钢铁厂环境保护的运行体系。

4　工程设计主要研究方法和内容

首钢京唐钢铁厂产品定位于为汽车、机电、石油、家电、建筑及结构、机械制造等行业提供热轧、冷轧、热镀锌、彩涂等高端精品板材产品；是我国第一个利用天然深水港条件、通过围海造地靠海建设的大型钢铁联合企业；第一个高端板带材专业生产基地、第一个运用动态-有序、连续-紧凑的精准设计体系建设的新一代钢铁厂;第一个集成应用了国内外先进技术建设的钢铁厂;第一个自主创新和循环经济示范基地。

首钢京唐钢铁厂总体目标是:建设集高效率、低成本、循环经济、精品制造于一体的多目标群优化的新一代可循环现代化钢铁厂,实现"高起点、高标准、高要求"和"产品一流、管理一流、环境一流、效益一流"的发展目标,如图2所示。

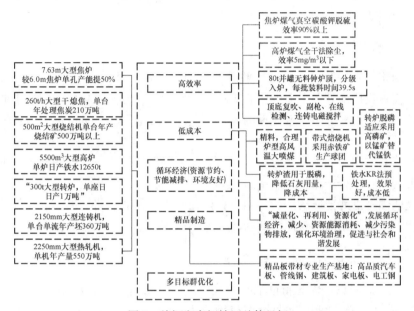

图2 首钢京唐钢铁厂总体目标

首钢京唐钢铁厂工程设计运用冶金流程工程学的理论,以构建动态-有序、连续-紧凑的运行体系,从钢铁厂的功能-结构-效率优化和系统集成出发,运用动态-精准设计理论,从产品的定位和生产规模,功能和运行流程的结构、设施配置和现代技术的应用、循环经济和节能减排、环境保护到企业与社会的协调和谐发展等方面均按照新世纪钢铁厂的理论和要求进行设计。

4.1 在生产产品定位方面

新世纪我国钢铁工业发展的主线是动态地适应市场发展趋势,以调整结构为主要途径,以增强市场竞争力和注重可持续发展为总体目标。我国钢铁工业应是以首先满足内需为主的产业结构。

新世纪我国钢铁工业产品发展的重点是发展板材,特别是薄板,包括冷轧薄板、镀锌薄板、彩涂薄板、冷轧硅钢片、不锈钢薄板等深加工、高附加值产品。

以市场需求为导向,首钢京唐钢铁厂产品定位是以发展新世纪先进的钢材产品为目标,通过建立高效率、低成本的洁净钢生产平台,构建我国第一个精品板材专业生产基地。

生产产品定位为满足汽车、机电、石油、家电、建筑及结构、机械制造等行业对热轧、冷轧、热镀锌、彩涂等高端精品板材产品的需求,其中以高强度、高塑性、高成型性的汽车板为标志产品,弥补我国市场空缺,替代进口。其中:

2250mm 热轧生产主要钢种有：碳素结构钢、优质碳素结构钢、锅炉及压力容器用钢、造船用钢、桥梁用钢、管线用钢、耐候钢、IF 钢、双相（DP）和多相（MP）及相变诱导塑性钢（TRIP）、超微细晶粒高强钢等。

1580mm 热轧生产主要钢种有：碳素结构钢、优质碳素结构钢、低合金结构钢、汽车结构用钢、集装箱及车厢用钢、耐候钢、IF 钢、双相钢（DP）、多相钢（MP）、相变诱导塑性钢（TRIP）等。

4.2 在生产规模定位方面

钢铁厂未来的发展模式，将逐步形成城市周边区钢厂和海港工业生态带钢厂两种模式。首钢京唐钢铁厂地处环渤海的中心部位——曹妃甸，对于优化我国钢铁工业空间布局将发挥重要作用，可充分利用曹妃甸的港口、土地、交通运输、资源供应、环境容量等综合优势。

根据所确立的产品大纲，结合未来现代板带型钢厂的模式。大型联合企业钢厂结构主要有：全板带型、综合产品型、扁平材及其延伸产品型、混合流程型、薄板超大型五种结构模式，吸取国外沿海建厂的经验，结合曹妃甸码头、外部铁路、供水供电等外部配套条件的承载能力，确立京唐钢铁厂为薄板超大型结构模式，生产规模约为 970 万吨/年钢。

4.3 功能—结构—效率的优化和系统集成

（1）首钢京唐钢铁厂作为新世纪的钢铁厂，其功能不仅是先进的钢材产品制造功能，同时还具备高效能源转化和消纳社会废弃物的功能。

（2）首钢京唐钢铁厂能源转换功能，体现在能源流"流-流程网络-程序"的构建上，在洁净钢生产和产品精准轧制生产系统的基础上，设计了高效率能源转化的大型炼焦系统、高效率的煤气、热量回收和发电设施、预留了消纳社会废塑料的设施等，京唐钢铁厂总体能源转换构架-能源流网络如图 3 所示。

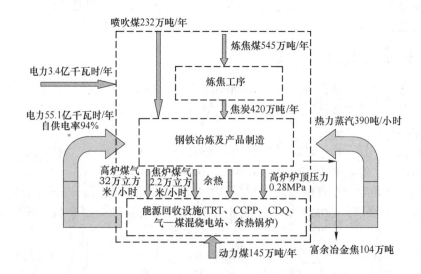

图 3 首钢京唐钢铁厂总体能源转换构架（能源流网络）

（3）在钢铁制造流程运行结构合理化方面，对一个约970万吨/年钢规模的联合钢铁厂，以2座高炉、1个炼钢厂、2个热轧厂是比较合理的流程结构，有利于构筑高效率、低成本的生产运行体系。

首钢京唐钢铁厂在运行流程的结构上，遵循各种不同钢材的合理生产规模和先进的制造流程、制造流程中前后工序/装备的协调优化、生产不同类型钢材时兼顾铸机—轧机之间物流的横向兼容优化、铸机—轧机之间的整数对应优化、高炉座数不宜过多（一般以2~3座为宜）、炼钢厂不宜超过2个（以利于优化铁路输送网络）的原则。

设计了2座5500m³高炉、一个由2座300t脱磷转炉+3座300t脱碳转炉构成的具有"全三脱"洁净钢生产功能的炼钢厂、2250mm和1580mm 2个热轧生产线，其结构模式体现了新世纪钢铁厂合理的流程结构；京唐钢铁厂流程结构优化如图4所示。

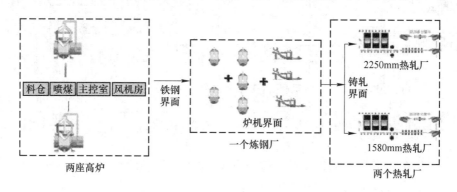

图4　京唐钢铁厂京唐钢铁厂流程结构优化

（4）京唐钢铁厂建立以连铸为中心的精准匹配流程如图5所示，以此为基础，构建京唐钢铁厂生产工艺流程、各生产工序产能匹配—物质流网络如图6所示。

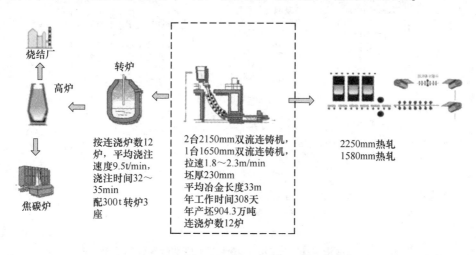

图5　建立以连铸为中心的精准匹配流程

（5）首钢京唐钢铁厂各生产工序功能-解析优化和系统集成：京唐钢铁厂轧钢、连铸生产工序功能-解析优化和系统集成如图7所示，京唐钢铁厂炼钢、炼铁生产工序功能-解析优化和系统集成如图 8 所示，京唐钢铁厂烧结、焦化生产工序功能-解析优化和系统集成如图 9 所示。

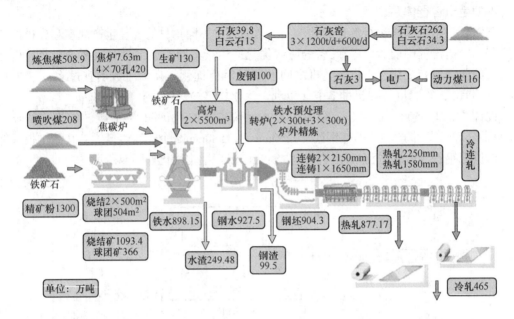

图 6　首钢京唐钢铁厂生产工艺流程及主要生产工序产能匹配-物质流网络图

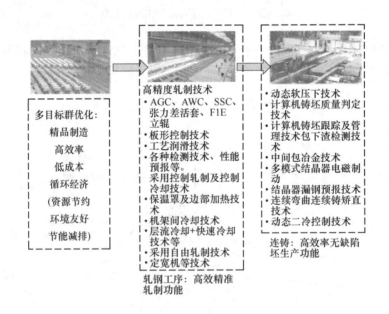

图 7　首钢京唐钢铁厂轧钢、连铸生产工序功能优化和系统集成

（6）首钢京唐钢铁厂钢铁制造流程"界面技术"的研究与应用。生产流程工序间主要界面关系包含焦化和烧结球团工序与高炉炼铁工序间界面技术，高炉炼铁工序与转炉炼钢工序间界面技术，连铸工序与热轧工序间界面技术等（见图 10）。

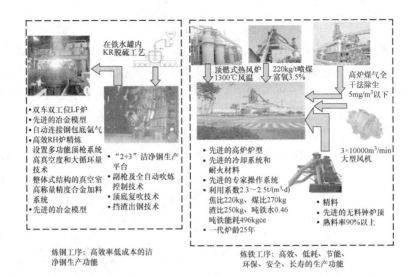

图 8　首钢京唐钢铁厂炼钢、炼铁生产工序功能优化和系统集成

图 9　首钢京唐钢铁厂烧结、焦化生产工序功能优化和系统集成

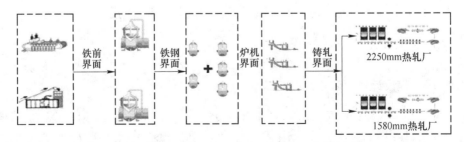

图 10　首钢京唐钢铁厂生产流程工序间主要界面关系

4.4 在资源循环化方面

以资源、能源高效利用和循环利用为核心,以"减量化、再利用、资源化"为原则,以低消耗、低排放、高效率为特征,大力实施循环经济,充分回收利用生产过程中产生的二次能源和废弃物,对生产中产生的余热、余压、余气、含铁物质及其他废弃物、污水等要进行充分的回收和利用,以降低钢铁厂能源、资源的消耗和废弃物排放。

通过实施循环经济,京唐钢铁厂做到富余煤气利用率 100%、铁素利用率 100%、吨钢新水消耗量 3.84t、水循环利用率 97.5%,固体废弃物和污水基本零排放。钢铁厂每年可提供 1800 万吨浓盐水用于制盐,330 万吨高炉水渣、转炉钢渣、粉煤灰等用于生产建筑材料,同时,钢铁厂可以回收处理消化大量废钢铁、废塑料等社会废弃物,实现了钢铁厂对社会废弃资源的循环利用(见图 11)。

图 11 消纳社会废弃物功能—首钢京唐钢铁厂消纳社会废弃物

5 新世纪钢铁厂工程设计创新所赋予的新内涵

5.1 流程制造业的设计理念创新

制造业分为流程制造业与制品(装备)制造业两大类,钢铁生产工艺过程的研究属流程制造业,而钢铁生产装备技术的研究则属于制品(装备)制造业,钢铁厂工程设计具有流程制造业与制品制造业两重性,因此,新世纪钢铁厂工程设计理念和设计方法的运用要有所侧重。

针对钢铁生产工艺过程的研究主要是在冶金—金属材料学科基础科学和技术科学研究的基础上,运用工程科学的理论—冶金流程工程学去分析解决问题。

针对钢铁生产装备技术的研究主要是在满足钢铁生产工艺要求的前提下,运用现代设计方法去分析解决问题。

5.2 钢铁厂工程设计作用和意义的新认识

工程设计是工程建设的灵魂,是对项目建设进行全过程的详细策划和表述项目建设意图的过程,是科学技术转化为生产力的关键环节,是体现技术和经济双重科学性的关键因素，是能够实现项目建设目标的决定环节。没有现代化的设计，就没有现代化的建设。做好设计工作，对加快工程项目的建设速度，节约建设投资，保证项目投产取得较好的经济效益、社会效益和环境效益都起着决定性作用。

钢铁厂的竞争和创新表面上凸显在产品和市场，其根源却来自于设计理念、设计过程和制造过程，设计正在成为市场竞争的始点，设计的竞争和创新关键在于工程复杂系统的多目标群优化。

5.3 新世纪钢铁厂功能的新认识

新一代钢铁厂功能拓展为三大功能：

（1）钢铁产品制造功能。钢铁厂的产品制造功能主要是在尽可能减少资源和能源消耗的基础上，高效率地生产出成本低、质量好、排放少且能够满足用户不断变化要求的钢材，供给社会生产和居民生活消费。

（2）能源转换功能。钢铁生产过程同时也伴随着能源转换过程。以高炉-转炉-热轧流程为代表的钢铁联合企业，实际是铁-化工过程，也可以看出是将煤炭通过钢铁冶金制造流程转换为可燃气、热能、电能、蒸汽甚至氢气或二甲醚等能源介质的过程。

（3）废弃物消纳-处理和再资源化功能。钢铁厂制造流程中的诸多工序、装备可以处理、消纳来自社会的大宗废弃物、改善社会环境负荷，促进资源、能源的循环利用。

新一代钢铁厂工程设计要从单纯的钢铁产品制造功能拓展为以上三大功能，从而构建可循环的钢铁制造流程，即高效化的生产制造流程、高效能源转换的流程和循环经济的制造流程。

5.4 新世纪钢铁厂工程设计要实现静态设计向动态-精准设计的转变

传统的工程设计方法，即采取静态的半经验、半理论的工程设计方法；而现代工程设计方法是要运用冶金流程工程学动态—精准设计的理论开展钢铁厂工程设计创新工作，对于钢铁生产制造流程设计而言，运用冶金流程工程学理论尤为重要。

冶金流程工程学的核心是通过研究钢铁制造流程的功能优化、结构优化、效率优化、耗散最小化，实现钢铁制造流程的有序化、协调化、高效化、连续化生产；其基本方法是采用工序功能集合的解析优化、工序之间关系集合的协调-优化、流程工序集合的重构-优化的方法，开展钢铁厂工程设计创新工作；其目的是提高钢铁厂的市场竞争能力，实现可持续发展。

5.5 钢铁生产制造流程动态运行过程物理本质的新认识

钢铁制造流程动态-有序运行的三个基本要素是："流""流程网络"和"程序"，其中"流"是制造流程运行过程中的物质性主体，"流程网络"同"节点"和"连接器"是构成"流"运行的承载体和时间-空间边界，而"程序"则是"流"的运行特征在信息形式上的反映。

物质流（主要是铁素流）在能量流（主要是碳素流）的驱动和作用下，按照设定的"程序"，沿着"流程网络"作动态—有序的运行，实现钢铁厂三种功能的多目标群优化。

铁素流运行功能——钢铁产品制造功能。

能量流运行功能——能源转换功能以及与剩余能源相关的废弃物消纳-处理。

铁素流—能量流相互作用过程的功能——实现过程工艺目标以及与此相应的废弃物消纳-处理。

5.6 钢铁生产流程中"界面技术"的新认识

"界面技术"：指主体工序之间的衔接-匹配、协调-缓冲技术及相应的装置（装备）；主要体现在要实现生产过程物质流（应包括流量、成分、组织、形状等）、生产过程能量流、生产过程温度、生产过程时间等基本参数的衔接、匹配、协调、稳定等方面。优化钢铁生产流程应注意研究开发"界面技术"促进生产流程整体运行的稳定、协调和高效化、连续化。钢铁生产流程中"界面技术"主要类型：

(1) 高炉-转炉区段的"界面技术"；
(2) 炼钢炉-连铸机区段的"界面技术"；
(3) 连铸机-热连轧机区段的"界面技术"。

5.7 应用动态-有序、连续-紧凑的设计理论、方法和先进设计手段的新认识

采用三维动态、模拟、仿真技术和有限元等分析软件等现代设计方法，提高工程设计动态-精准设计程度。

时间对于制造流程中多因子"物质流"运行的紧凑-连续性具有决定性的影响。在钢铁制造流程中，各工序、装置运行过程在时间上的协调至关重要。时间在钢铁制造流程中的表现形式是以时间点、时间域、时间位、时间序、时间节奏、时间周期等形式表现出来。解析时间的表现形式，要有效建立起钢铁厂信息调控系统。时间在钢铁制造流程中具有既是自变量又是目标函数的两重性。

研究钢铁制造流程整体运行的本质和运行规律，以协调钢铁制造流程中物质流、能量流、信息流的优化，使之层流运行，减少紊流运行状态，从而提高钢铁制造流程的运行效率。

5.8 新世纪钢铁厂工程设计面临工程设计人员思维方式的转变

由于现有工程设计人员既受传统教育方式的影响，又受工作背景的影响，一般工程设计人员对基础科学和技术科学知识积累较多，而对工程科学的理论研究相对不足，参加工作后又长时期从事钢铁生产制造流程某一生产工序的工程设计工作，其思路往往容易从某一生产工序的角度考虑问题，较难站在钢铁生产制造整个流程角度去分析、解决本生产工序的工程设计问题。

因此，开展新世纪钢铁厂工程设计创新工作，工程设计人员其思维方式的转变、思维层次的提高更具有战略意义，也是解决工程设计持续创新的关键。

5.9 新世纪钢铁厂设计建设工作流程产生新的变化

钢铁厂设计建设的工作流程包括：设计前期工作、设计工作、项目实施三个阶段。其中，设计前期工作包括：项目建议书、厂址选择、可行性研究报告、环境影响报告书、设计任务书等内容；设计阶段工作包括：初步设计、施工图设计。

运用冶金流程工程学理论开展钢铁厂工程设计创新工作主要对设计前期工作和设计工作阶段产生重大影响，特别是对项目的建设规模，产品方案，厂址，工艺流程，装备水

平，交通运输，供水、供电条件等的指导思想、设计原则产生了深刻变化和重大影响，钢铁厂基本建设工作流程所赋予的新内涵。

参考文献

[1] 殷瑞钰. 冶金流程工程学（第2版）[M]. 北京：冶金工业出版社，2009.
[2] 北京首钢国际工程技术有限公司. 冶金工程设计理念的创新与实践 [M]. 北京：冶金工业出版社，2010：7.
[3] 殷瑞钰，汪应洛，李伯聪等. 工程哲学 [M]. 北京：高等教育出版社，2007：7.
[4] 袁熙志. 冶金工艺工程设计 [M]. 北京：冶金工业出版社，2003：2.
[5] 郭鸿发，等. 冶金工程设计 [M]. 北京：冶金工业出版社，2006：6.

The Innovation and Practice of Steel Factory Engineering Design

Zhang Fuming Xie Jianxin Zhou Songming

Abstract: This paper analyses background of Shougang Jingtang steel factory engineering constructing, Integrating with practice of Shougang Jingtang steel factory engineering design. It discusses idea, main technology route, main research means and content. The new meaning of new century steel factory engineering design has been put forward. There is an obvious illustrative for new or rebuild steel factory engineering design in country or overseas at new century.

Keywords: steel factory; engineering design; innovation; practice; new meaning

首钢京唐炼钢厂新一代工艺流程与应用实践

摘　要：研究分析了现代化炼钢厂洁净钢生产技术，提出了以生产效率、制造成本和产品性能为核心的洁净钢生产技术理念。通过对新一代炼钢厂高效率、低成本、高质量钢铁产品制造功能的解析与集成，结合首钢京唐钢铁厂炼钢—连铸工艺的设计研究，应用动态精准设计体系，优化配置铁水预处理、转炉冶炼、二次精炼、连铸等单元工序，构建了基于动态有序生产体系的高效率、低成本、高质量洁净钢生产平台。

关键词：炼钢；连铸；铁水预处理；二次精炼；流程

1　引言

首钢京唐钢铁联合有限责任公司（以下简称"首钢京唐"）钢铁厂项目（曹妃甸）是我国"十一五"规划的重大工程，产品定位于精品板材，设计生产规模为927.5万吨/年。建设有两座5500 m^3 高炉，炼钢厂配置两座300t脱磷转炉、三座300t脱碳转炉和三台板坯连铸机，建设2250mm和1580mm两条热连轧生产线，2230mm、1700mm和1550mm三条冷轧生产线，冷热轧转换比为57.2%，涂镀板比为65.7%。

热轧主导产品为高品质汽车板，重要产品为管线钢、压力容器钢和造船用钢。高强度钢抗拉强度最高可达1200MPa，管线钢级别为X100。冷轧产品包括固溶型、析出型、烘烤硬化型、双相钢（DP）及相变诱发塑性钢（TRIP）等高强度钢，最高强度级别为825MPa，热镀锌产品最高强度级别为590MPa，彩色涂层产品最高强度级别为440MPa。

首钢京唐钢铁厂的设计，遵循"工艺现代化、流程高效化、效益最佳化"的设计理念，炼钢—连铸工序采用铁水"全三脱"预处理模式（全量脱硫、脱硅、脱磷预处理），应用动态精准设计体系，优化"高炉铁水运输—铁水预处理—转炉冶炼—二次精炼—连铸"各单元工序的流程，构建了基于动态精准生产体系的高效率、低成本、高质量洁净钢生产体系。

2　洁净钢生产体系的构建

在钢铁工业的发展进程中，洁净钢的生产加快了工艺流程的优化和产品质量的提高。构建一种全新的、大规模、高效率、低成本生产洁净钢的生产体系，对我国现代钢铁工业的发展具有重要意义。洁净钢生产体系的构建不仅是单纯的脱硫、脱磷、脱碳、脱氧等工艺技术和品种质量问题，应该包括工艺、设备、技术管理和生产运行等诸多因素，实现高效、优质和低成本的目标。洁净钢生产体系必须采用高效、稳定的运行模式。炼钢—连铸

本文作者：张福明，崔幸超，张德国，罗伯钢，魏钢，韩丽敏。原刊于《炼钢》，2012，28（2）：1-6。

制造流程中整个系统的产能不仅取决于单元工序的产能,还取决于单元工序之间物质流的流通能力和效率,因而通过解析各单元工序的功能,改变传统的单元工序静态生产能力核算的设计理念,建立动态精准设计体系,是实现炼钢-连铸工艺流程优化的重要方法,也是构建高效率、低成本洁净钢生产体系的基本理念[1,2]。

通过对炼钢—连铸各工序功能的解析与集成,按照铁水"全三脱"的设计理念,首钢京唐炼钢厂建立了动态有序、紧凑连续、高效低耗的洁净钢生产体系[3]。建立了"铁水短流程运输—铁水预处理—高效转炉冶炼—钢水二次精炼—高效连铸"协同优化的生产工艺流程,主要单元工序配置见表1,工艺流程如图1所示。

表1 首钢京唐炼钢厂主要工序配置

单元工序	数量
铁水KR脱硫装置	4套
300t铁水脱磷、脱硅转炉	2座
300t顶底复吹脱碳转炉	3座
CAS精炼装置	2套
双工位LF	1座
双工位RH	2台
2150mm双流板坯连铸机	2台
1650mm双流板坯连铸机	1台

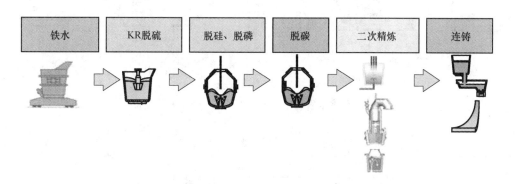

图1 首钢京唐钢铁厂炼钢—连铸生产工艺流程

2.1 炼铁-炼钢界面技术

铁水运输采用300t铁水罐直接运输技术,取代了常规的鱼雷罐运输工艺。采用铁水直接运输技术可以减少铁水倒罐操作,缩短工艺流程,降低烟尘污染、提高铁水温度、有利于铁水脱硫处理和转炉多加废钢、降低铁水消耗和能源消耗、降低生产运行成本。实践表明,采用铁水"一罐到底"直接运输技术,铁水温度比采用鱼雷罐运输提高约30~40℃。表2为2010年高炉出铁温度和KR处理前的铁水温度。

表 2 2010 年铁水温度

月 份	1月	2月	3月	4月	5月	6月	7月	8月	9月	10月	11月	12月
高炉出铁温度/℃	1513	1520	1518	1517	1523	1514	1517	1505	1507	1506	1514	1512
KR处理前铁水温度/℃	1409	1421	1413	1410	1413	1411	1394	1370	1378	1381	1382	1387

2.2 铁水预处理工艺

为构建洁净钢生产体系,铁水预处理采用铁水脱硫、脱磷、脱硅的"全三脱"预处理工艺,配置4套KR脱硫装置用于铁水脱硫预处理,采用2座300t转炉进行铁水脱磷、脱硅预处理。

2.2.1 铁水脱硫预处理

铁水采用KR机械搅拌脱硫装置,可以高效稳定地满足高品质板材对硫含量的要求,KR脱硫工艺流程如图2所示,主要技术参数见表3。KR脱硫工艺的主要技术特点是:(1)具有良好的动力学条件,脱硫效率高且稳定,脱硫率可达到90%~95%;(2)脱硫剂采用石灰及少量萤石,脱硫剂价格低廉,生产成本低;(3)采用活性石灰套筒窑生产的石灰粉末,通过气力输送,可实现资源综合利用,降低生产成本;(4)KR脱硫工艺采用二次扒渣,处理周期为36~40min,操作时间与脱磷转炉冶炼周期相匹配。

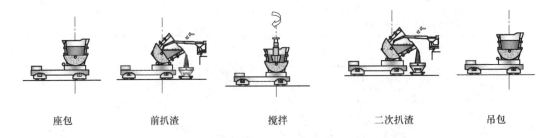

座包　　　前扒渣　　　搅拌　　　二次扒渣　　　吊包

图 2 首钢京唐炼钢厂 KR 铁水脱硫处理工艺流程

表 3 首钢京唐炼钢厂 KR 脱硫工艺主要技术参数

项 目	参 数
处理铁水量/万吨·年$^{-1}$	898.15
每罐铁水平均处理量/t	287
每罐铁水处理时间/min	36~40
处理铁水能力/万吨·年$^{-1}$	1100
处理前铁水硫含量/%	≤0.07
终点目标硫含量目标/%	≤0.002 占 20 0.0021~0.005 占 50 0.0051~0.01 占 30
脱硫剂消耗/kg·t^{-1}	6~10

2.2.2 铁水脱硅脱磷预处理

通过对转炉冶炼功能的解析研究，采用转炉分阶段冶炼的技术理念，将传统转炉脱硅、脱磷、脱碳集成一体的功能优化为采用专用的转炉进行铁水脱磷、脱硅预处理，顶底复合吹炼转炉则专用于脱碳升温，改变了传统转炉冶炼的操作模式，原来一座转炉的冶炼功能由2座转炉采取串联作业来实现[4]。"全三脱"冶炼的操作模式是采用2座转炉前后串联作业，即用于铁水预处理的转炉，主要进行铁水脱磷、脱硅操作，称其为脱磷转炉；用于脱碳的转炉承接来自脱磷转炉处理后的"半钢铁水"，主要完成脱碳操作。这种优化了的"全三脱"冶炼模式缩短了转炉冶炼周期，提高了转炉冶炼效率和钢水洁净度，特别对于生产低磷钢和超低磷钢，采用"全三脱"转炉冶炼模式具有显著的技术优势[5]。首钢京唐炼钢厂配置2座300t脱磷、脱硅转炉和3座300t脱碳转炉，可实现铁水"全三脱"预处理和工艺流程优化，主要技术特点是：（1）转炉内脱磷反应空间大，能够实现大气量底吹搅拌，加速脱磷反应，创造良好的脱磷动力学条件，生产运行成本低，可以经济地获得低磷铁水。（2）优化了转炉入炉原料，可实现精料操作。（3）脱磷时间短，简化了转炉冶炼工艺，高速吹炼，实现快节奏生产[4]。（4）转炉分阶段冶炼，有利于脱碳转炉采用锰矿，减少Fe-Mn合金的消耗，降低生产成本。（5）脱碳转炉精炼渣可作为脱磷剂使用，降低生产成本，实现资源合理利用。（6）转炉少渣冶炼，减少钢渣处理量，节能环保，实现绿色生产。（7）高炉可以适度利用高磷铁矿，利于降低原料的生产成本。

因此，采用专用转炉进行铁水脱硅、脱磷预处理，不仅有利于低磷钢的生产，还可以优化工艺流程、提高生产效率、降低运行成本，体现了现代化炼钢厂发展循环经济、减量化生产的发展方向，是钢铁厂经济运行的一个系统化工程，有利于提高产品的市场竞争力。

铁水预处理工序采用4套KR脱硫装置和2座脱硅脱磷转炉，采取2对1的操作模式，即2套KR与1座脱硅脱磷转炉匹配，铁水总处理能力约为1100万吨/年，满足转炉年产927.5万吨/年钢水的要求。根据不同钢种的要求，铁水经过"全三脱"预处理以后，铁水中的硅、磷、硫含量可以达到表4的质量控制目标。

表4 首钢京唐炼钢厂铁水预处理后的控制目标   （%）

铁水	Si	P	S
普通铁水	0.05~0.23	<0.015	<0.01
低磷、低硫铁水	0.05~0.23	<0.015	<0.005
超低磷铁水	0.05~0.23	<0.01	<0.01

2.3 转炉冶炼工艺

转炉冶炼工序配置3座300t脱碳转炉，由于采用铁水"全三脱"预处理工艺，转炉工序的主要任务是脱碳升温，冶炼周期缩短，可由常规冶炼的36~38min缩短到30min以下，实现转炉的高效冶炼和少渣冶炼。为保证钢水的洁净度，采用顶底复合吹炼、副枪、挡渣出钢、钢包渣改质处理等技术。根据不同钢种的要求，转炉冶炼终点钢水成分可以达到表5的控制水平。

表5 首钢京唐炼钢厂转炉冶炼终点钢水的控制情况 （%）

钢水	C	Mn	P	S
普通钢水	0.06	0.6	<0.01	<0.01
超低硫钢水	0.06	0.8	<0.01	<0.004
超低磷钢水	0.03	0.6	<0.005	<0.01

2.4 精炼工艺

根据热轧和冷轧产品的质量要求以及不同精炼装置的功能，精炼工序配置2座RH、1座LF、2座CAS，按照产品的质量要求，各种钢水二次精炼装置可以单独使用或采取双重精炼处理工艺。

2.4.1 RH精炼工艺

多功能RH真空精炼装置特别适用于现代转炉冶炼和板坯连铸生产。2台RH真空处理装置，可单独或与CAS、LF进行串联作业，实现脱碳、真空脱氧、脱氢和脱氮，用以大规模生产低碳优质钢种，如超低碳IF钢、硅钢等。处理低碳钢、超低碳钢和对气体含量控制要求较高的钢种，如DQ、DDQ、EDDQ系列钢板，可通过RH真空自然脱碳或强制脱碳、真空脱氧、脱气处理。

RH真空处理装置采用双工位，配置多功能顶枪，通过顶枪吹氧生产超低碳钢；加铝吹氧进行化学升温；顶吹燃气和氧气为真空槽补充加热，以减少RH处理时钢水温降，消除真空槽内冷钢，避免钢种之间污染。经过多功能RH真空处理装置处理后的钢水成分可达到C<0.0015%，H<0.0002%，N<0.003%，O<0.002%，RH主要技术参数见表6。

表6 首钢京唐炼钢厂RH主要技术参数

项目	参数
公称容量/t	300
处理周期/min	23~55（平均31）
钢水罐升降方式	液压缸顶升
真空泵能力/kg·h^{-1}	1250（0.5torr）
钢水循环率/t·min^{-1}	最大250
处理能力/万吨·年$^{-1}$	768
冶金效果/%	C<0.0015，H<0.0002，N<0.003，O<0.002

2.4.2 LF精炼工艺

LF具有如下精炼功能：（1）利用钢水加热功能，可协调炼钢和连铸工序生产节奏，保证多炉连浇的顺利进行；（2）在还原性气氛下造碱性渣，对钢水进一步进行脱硫处理，有利于冶炼超低硫钢；（3）可以加入合金及渣料进行钢水脱氧、脱硫及合金化处理，控制钢水成分，提高钢水质量；（4）采用底吹氩搅拌工艺均匀钢水温度和成分。

对要求生产低氧、低硫的钢种，如低合金钢、低牌号管线钢等，均可采用LF处理。首钢京唐炼钢厂配置双工位、电极旋转式LF精炼装置1台，其主要技术参数见表7。

表7 首钢京唐炼钢厂LF主要技术参数

项 目	参 数
公称容量/t	300
平均处理周期/min	40
变压器额定容量/MV·A	45
电极调节方式	电液比例阀
钢水平均升温速度/℃·min^{-1}	≥4.5
处理钢水能力/万吨·年$^{-1}$	333
脱硫率/%	≥60

2.4.3 CAS精炼工艺

CAS精炼装置除不具备脱硫功能以外，可以实现LF的大部分功能，可作为LF精炼工艺的并列或替代工艺。对于普通热轧产品，如SS400、SM490钢等，可以单独采用CAS精炼工艺。首钢京唐炼钢厂两台CAS精炼装置配置顶枪、具有加热功能，公称容量300t，处理周期28~40min，处理钢水能力666万吨/年。

2.5 连铸工艺

连铸配置3台高效板坯连铸机，设计年产904.3万吨坯。在保证钢水洁净度、提高铸坯的表面和内部质量、提高连铸生产率及可靠性、实现板坯高温热送等方面采用了30余项先进技术，体现了当今国际连铸技术的发展趋势，表8为板坯连铸机的主要技术参数。

表8 首钢京唐炼钢厂板坯连铸机的主要技术参数

项 目	参 数	
	2150mm板坯连铸机	1650mm板坯连铸机
机型	直弧形、连续弯曲、连续矫直	直弧形、连续弯曲、连续矫直
台数×流数	2×2	1×2
基本弧半径/m	9	9.5
浇注断面/mm	230×(1100~2150)	230×(900~1650)
切割定尺长度/m	9~11	8~10.5
拉坯速度/m·min^{-1}	1.0~2.5	1.2~2.5
冶金长度/m	48	48
连浇炉数	10~12	10~12
连铸机产量/万吨·年$^{-1}$	624.3	280

（1）直弧形连铸机采用分节密排辊列、连续弯曲、连续矫直的连铸机机型，满足高拉速下连铸坯内部洁净度的要求，减小连铸坯的弯矫变形，提高了连铸坯的内部质量。

（2）采用结晶器钢水电磁制动技术。结晶器钢水电磁制动技术特别适合2.0m/min以上的高拉速浇注。电磁制动技术可以控制钢水的流速和方向，使结晶器内的钢水流场分布始终保持在合理状态，避免钢水卷渣，保证高拉速条件下连铸坯的表面质量和内部质量，可以有效提高连铸坯的洁净度。

(3) 配备结晶器液压振动装置。结晶器液压振动可以在浇注过程中改变振幅和振频,实现正弦和非正弦振动,有效地减少连铸坯振痕深度,特别适合高拉速条件下保护渣的有效供给,提高连铸坯的表面质量。

(4) 采用铸坯动态轻压下技术。通过建立连铸二冷水控制模型,实时判断连铸坯内部的液芯位置,在铸流导向段的适当位置,控制系统自动调整扇形段的辊缝开度,从而对连铸坯实施动态轻压下。可以有效地改善连铸坯内部的中心偏析和中心疏松,从而获得良好的连铸坯内部质量,在消除连铸坯中心偏析方面具有显著效果。

3 生产运行实践

3.1 工艺流程优化

首钢京唐炼钢厂采用优化的工艺流程,构建了高效率、低成本洁净钢生产体系,集成"铁水直接运输—'全三脱'铁水预处理—转炉炼钢-钢水二次精炼-高效连铸"一体化的短流程工艺,整个工艺流程紧凑合理,KR 脱硫装置独立设置,脱磷转炉与脱碳转炉分跨设置,工艺流程紧凑连续、物质流运行顺畅、运行高效稳定。3 台连铸机有明确的产品和产量分工,分别与各自对应的"KR 脱硫—脱磷转炉—脱碳转炉—精炼装置—连铸"保持层流运行,生产组织运行稳定。表 9 为典型生产过程各工序的生产时间。

表 9 首钢京唐炼钢厂典型低碳钢各工序的生产时间

项 目	参 数
KR 脱硫处理周期/min	20(2 套 KR 处理节奏)
脱磷转炉处理周期/min	20
脱碳转炉冶炼周期/min	30
RH 处理时间/min	25
单炉浇注时间/min	30
铸坯规格/mm	230×1500
工作拉速/m·min^{-1}	1.9
连浇炉数	10

3.2 生产运行实践

3.2.1 KR 脱硫系统

铁水经 KR 进行脱硫预处理后,铁水 S<0.002%的比例为 93.1%,S<0.005%的比例为 99.5%。KR 处理后铁水硫含量大幅度降低,可为高附加值产品提供优质铁水。目前,处理过程铁水平均温降为 29℃,脱硫剂消耗稳定在 7kg/t 左右,综合指标达到国内先进水平。表 10 为 2011 年 1~9 月的 KR 实际运行指标。

表 10 首钢京唐炼钢厂 KR 脱硫实际生产指标

项目	数值
脱硫剂消耗/kg·t^{-1}	7.2
钢铁料消耗/kg·t^{-1}	1004
铁水进站温度/℃	1412
铁水脱后温度/℃	1383
生产周期/min	31

3.2.2 转炉系统

300t 脱磷转炉采用了专用脱磷氧枪、加大底吹流量、使用返回渣、"一键式"脱磷冶炼等先进技术。目前，脱磷转炉终点半钢平均 C = 3.4%；平均 Si = 0.036%，脱硅率为 90.8%；平均 P = 0.033%，脱磷率为 63.3%；平均半钢温度为 1334℃。脱磷转炉冶炼时，每炉使用约 5t 脱碳转炉返回渣作为脱磷转炉造渣剂，经济效益显著。目前，铁水"全三脱"比例达到 85% 以上；脱碳转炉终点钢水月平均 O = 0.054%；石灰消耗降低到 16.5kg/t 以下；脱碳转炉终点钢水 P 平均为 0.007%，终点温度月平均为 1670℃。脱碳转炉采用副枪自动化冶炼工艺，由于冶炼模型和操作模式与常规转炉冶炼不同，具有一定的特殊性，需要在长期的生产实践中进一步积累经验。表 11 和表 12 分别为 2011 年 1~9 月脱磷转炉和脱碳转炉的生产指标；表 13 是脱碳转炉终点碳含量及温度命中率的精度范围。

表 11 首钢京唐炼钢厂脱磷转炉生产指标

项目	参数
"一键式"脱磷比例/%	97
平均终点钢水 P/%	0.033
平均终点钢水 C/%	3.4
平均半钢温度/℃	1334
终渣 TFe/%	12.2
钢铁料消耗/kg·t^{-1}	1036
石灰消耗/kg·t^{-1}	15.1
轻烧白云石消耗/kg·t^{-1}	0.9
矿石消耗/kg·t^{-1}	17.2
萤石消耗/kg·t^{-1}	1.5
使用返回渣炉次比例/%	92.4

表 12 首钢京唐炼钢厂脱碳转炉生产指标

项目	参数
自动炼钢比例/%	100
终点碳含量及温度双命中率/%	85.6
低碳钢终点 O≤0.085% 比例/%	85.3
终渣 TFe/%	17.2

续表12

项　　目	参　　数
平均出钢温度/℃	1671
钢铁料消耗/kg·t^{-1}	1041
石灰消耗/kg·t^{-1}	16.5
轻烧白云石消耗/kg·t^{-1}	12.2
矿石消耗/kg·t^{-1}	2.9
萤石消耗/kg·t^{-1}	0.6
留渣炉次比例/%	37.3

表13　首钢京唐炼钢厂脱碳转炉终点碳含量及温度命中率的精度范围

C/%	C控制精度/%	温度控制精度/℃	C-T双命中率/%
0.02~0.05	±0.015	±12	92.00
0.05~0.10	±0.02	±12	86.86

3.2.3　洁净钢生产控制水平

铁水采用"全三脱"预处理、转炉精料操作、多功能钢水精炼、连铸中间包/结晶器钢水冶金，为高效率、低成本、批量化生产杂质含量低的洁净钢生产创造了有利条件，构建了高效率、低成本的洁净钢生产体系。转炉工序采用全自动吹炼和全流程计算机监控，精炼工序配置LF、RH、CAS精炼站，在洁净钢批量生产的基础上减小钢水成分与温度的波动，稳定产品性能，提高产品质量。铁水"全三脱"预处理可以降低白灰、合金料消耗，脱碳转炉渣回收用作脱磷转炉的脱磷剂，低成本生产洁净钢，实现资源循环利用，降低生产成本。项目投产后，经过近两年的探索和实践，目前首钢京唐洁净钢生产已初见成效。

(1) 碳含量。RH真空脱碳处理后可使钢水C降低到0.0008%；成品C达到0.0012%。

(2) 硫含量。从控制原辅材料中的硫含量着手，由铁水KR脱硫开始加强各个工序控制，实现最低成品S达0.0001%以下。

(3) 磷含量。转炉采用"全三脱"冶炼工艺，最低成品P达到0.003%以下。

(4) 氮含量。炼钢采用低氮模式生产，钢水N降低到0.0008%，精炼过程N<0.0005%，连铸采用全保护浇注，ΔN<0.0002%，最低成品N达到0.0012%。

(5) 氢含量。钢水经过RH真空处理后，H<0.0002%，最低成品H达0.00006%。

(6) 全氧含量。IF钢T.O<0.003%，成品最低T.O达0.002%；X80管线钢、SQ700MCD钢T.O<0.0012%，成品最低T.O达0.0005%。

4　结语

(1) 首钢京唐炼钢厂采用铁水"全三脱"预处理设计理念，通过工艺流程、技术装备的优化，构建了基于动态有序、紧凑连续、高效稳定的洁净钢生产体系，具有高质量、高效率、低成本、可循环的洁净钢生产技术特征。首钢京唐钢铁厂洁净钢生产体系的构建，提出了21世纪一种高效率、低成本、可循环生产洁净钢的技术发展模式和方向。

（2）提高钢材洁净度是未来钢铁工业的重点课题。洁净钢的生产是一项复杂的系统工程，是建立在工艺流程、技术装备、生产操作和质量管理基础之上的技术体系。新一代钢铁厂应构建洁净钢的生产平台，通过优化工艺流程、提高技术装备、改善生产操作、提高质量水平，实现高效率、低成本大批量生产用户需要的洁净钢材。

（3）首钢京唐炼钢厂投产后的生产运行实践证实，基于新一代"全三脱"冶炼模式的洁净钢生产系统，可以提高生产效率、降低生产成本，有效提高洁净度。通过对铁水预处理、转炉冶炼、二次精炼和连铸工序的技术装备和工艺流程进行优化，各工序布置紧凑合理，层流运行稳定，目前已初步达到了预期的目标。

致谢

本文的研究工作得到了殷瑞钰院士的悉心指导，首钢京唐公司王天义总经理、王毅常务副总经理、杨春政副总经理和炼钢作业部闫占辉等相关技术人员给予了大力支持和帮助，谨在此致以衷心的感谢！

参考文献

[1] 殷瑞钰. 关于高效率低成本洁净钢平台的讨论——21世纪钢铁工业关键技术之一 [J]. 炼钢, 2011, 27 (1): 1-10.
[2] 殷瑞钰. 冶金流程工程学 [M]. 北京: 冶金工业出版社, 2005: 325-328.
[3] Zhang Fuming, Cui Xingchao, Zhang Deguo. Construction of high-efficiency and low-cost clean steel production system in Shougang Jingtang [J]. Journal of Iron and Steel Research International, 2011, 18 (Suppl. 2): 42-51.
[4] 刘浏. 中国转炉炼钢技术进步 [J]. 钢铁, 2005, 40 (2): 1-5.
[5] 刘浏. 转炉炼钢技术的发展 [J]. 中国冶金, 2004, 14 (2): 7-11.

New Generation Steelmaking Plant Process Flow and Its Application Practice at Shougang Jingtang

Zhang Fuming   Cui Xingchao   Zhang Deguo   Luo Bogang
Wei Gang   Han Limin

Abstract: The technology for clean steel production in a modern steelmaking plant is analyzed and the technological philosophy of the clean steel production with efficiency, manufacturing cost and product performance as its core is also put forward. By way of reviewing and integrating the manufacturing functions of the high efficiency, low cost and high quality steel produced by the steelmaking plant of new generation and by optimizating allocation of the operating procedures such as hot metal pretreatment, converter smelting, secondary refining and continuous casting with the help of the dynamic and precise design system in connection with the design and research of the steelmaking-continuous casting process at Shougang Jingtang Iron & Steel Plant. A clean steel production platform with dynamic and orderly production system as its base is built up.

Keywords: steelmaking; continuous casting; hot metal pretreatment; secondary refining; process flow

首钢京唐 2250mm 热轧生产线采用的先进技术

摘　要：首钢京唐钢铁有限公司 2250mm 宽带钢热轧是按照动态精准设计体系设计，设计产能为 550 万吨/年，产品抗拉强度可达 1000MPa，设计中采用了当今国际轧钢技术领域的 20 多项先进技术，整体工艺技术装备达到国际先进水平。本文介绍了首钢京唐 2250mm 热轧设计特点和采用的先进技术。工程设计中采用步进式加热炉、热装热送技术、定宽压力机、二辊可逆式 R1 粗轧机、四辊可逆式 R2 粗轧机、7 架 4 辊精轧机、CVC^{plus} 板形控制技术、具有快速冷却功能的层流冷却技术（TMCP）等；自主设计开发了托盘式钢卷运输设备、交-直-交传动控制、自动化检测与控制系统及完善的除尘环保系统。该生产线投产后，主要技术经济指标达到设计水平，实现了高效、优质、节能的目标。

关键词：热轧；宽带钢轧机；工艺与装备；先进技术

1　引言

首钢京唐 2250mm 宽带钢热连轧生产线采用了当今国际先进的新技术、新设备，产品质量、能源消耗、生产效率等方面达到国际先进水平。

2　产品大纲与产品规格

该生产线生产的主要钢种有碳素结构钢、优质碳素结构钢、锅炉及压力容器用钢、造船板、桥梁钢、管线钢、耐候钢、IF 钢、双相钢（DP）、多相钢（MP）、相变诱导塑性钢（TRIP）、超微细晶粒高强钢等。年产热轧钢卷为 550 万吨，成品材 546 万吨。其中钢板为 45 万吨，平整卷为 100 万吨，管线钢为 20 万吨，供冷轧用料 301.6 万吨，商品卷 79.4 万吨，年需连铸板坯量 561.2 万吨。带钢厚度为 1.2~25.4mm，带钢宽度为 830~2130mm，钢卷内径为 762mm，最大钢卷外径为 2200mm，最大钢卷重量为 40t，最大单位卷重为 24kg/mm，产品最高抗拉强度为 1000MPa。热轧产品及产量分配见表 1，产品品种及规格见表 2。

表 1　热轧产品及产量分配

产品用途	钢　种	代 表 钢 号（JIS 标准）	热轧成品年产量 万吨/年	%
冷轧原料卷	深冲及超深冲带卷（含 IF）	SPCC~SPCE	156.6	28.68
	结构用钢	Q195~Q345（GB 标准）	97.6	17.88

本文作者：张福明，颉建新。原刊于《轧钢》，2012，29（1）：45-49。

续表1

产品用途	钢种	代表钢号（JIS标准）	热轧成品年产量	
			万吨/年	%
冷轧原料卷	340~590MPa级高强度钢	HSS-CQ、HSS-DQ、HSS-DDQ	42.8	7.84
	590MPa以上级高强度钢	HSS-BH、DP、TRIP	4.6	0.84
	小计		301.6	55.24
热轧商品卷	碳素结构钢	SPHC、SPHD、SPHE	86.0	15.75
	结构钢	SS330~SS540、SM400~SM570	25.4	4.65
	汽车结构用钢	SAPH310~SAPH440、DP、TRIP	28.0	5.13
	锅炉和压力容器用钢	SB42、SB49	8.0	1.47
	造船板	AH32、EH36（GB标准）	37.0	6.78
	管线钢	X42~X80（API标准）	20.0	3.66
	焊接气瓶用钢板	SG295、SG325	6.0	1.10
	高耐候性结构钢	09CuPCrNi-A、09CuPTiRe（GB标准）	34.0	6.22
	小计		244.4	44.76
	合计		546.0	100

表2 产品品种及规格

产品品种	产品规格			年产量/万吨·年$^{-1}$	
	厚度/mm	宽度/mm	卷重/t	热轧钢卷	热轧成品
供2230mm冷轧原料卷	2.0~6.0	1000~2130	最大38	231.6	231.6
供1700mm冷轧原料卷	1.8~6.0	850~1580	最大34	70.0	70.0
商品钢板	5.5~25.4			47.0	45.0
平整分卷	1.2~12.7	850~2100	5~38	102.0	100.0
管线钢	6~19	1050~2100	最大38	20.0	20.0
热轧商品卷	1.2~25.4	850~2130	最大38	79.4	79.4
合计				550.0	546.0

3 2250mm热轧生产工艺流程优化及采用的先进技术

现代热轧带钢生产工艺有常规流程轧制工艺和短流程连铸连轧工艺[1]。根据首钢京唐2250mm生产线生产高品质及高附加值产品和生产规模大型化的要求，设计采用常规流程轧制工艺。其主要优势是工艺技术成熟可靠，可以生产以汽车板、管线钢、DP钢等为代表的优质高档产品；可以形成规模效益、年产量可以达到550万吨，能够充分发挥设备潜力，实现高效化生产；热轧与连铸工序紧凑布置，可以实现连铸坯的热送热装和直接轧制，热送热装率可达75%以上，与薄板坯连铸连轧工艺相比，同样可以降低生产成本、节约能源消耗。

常规热连轧机组根据粗轧机的工艺布置形式主要分为全连续式、3/4 连续式、半连续式三种形式[2,3]。全连续式和 3/4 连续式布置的热连轧机组设备多、轧线长、投资高，目前国内外钢铁企业已不采用；而半连续式布置的连轧机组是经济型布置形式，依据轧机规格和产品方案，其年生产能力可达 200 万~600 万吨。半连续式热连轧机的粗轧机组主要有单机架和双机架两种配置工艺，一般年生产规模超过 400 万吨的热连轧机组大多采用双机架的工艺布置方式。

首钢京唐 2250mm 热轧主轧线工艺平面布置如图 1 所示，主要设备及其性能见表 3。

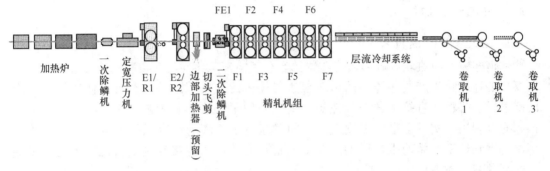

图 1　首钢京唐 2250mm 热轧生产线工艺流程

表 3　首钢京唐 2250mm 热轧生产线主要工艺技术装备

项　目	数量/台（套）	技术性能及参数
加热炉	4	板坯加热能力 350t/h
定宽压力机	1	侧压力 22000kN，最大减宽量 350mm，42 次/min
E1/R1	1	轧制压力 38000kN
E2/R2	1	轧制压力 55000kN
保温罩/废钢推出机	1	液压翻转式，内衬隔热材料
中间坯边部加热器	1	预留
切头飞剪	1	剪切力 12000kN，剪切速度 0.3~1.75m/s
F1~F7	1	F1~F4 最大压力 50000kN，F5~F7 最大压力 45000kN
层流冷却	1	冷却段长度 103360mm，水压 0.07MPa
卷取机	3	3 助卷辊全液压卷取机，卷取温度 100~850 ℃
托盘运输系统	1	1 卷/min

3.1　工艺流程

连铸坯出铸机后经辊道直接运送到热轧板坯库，直接热装的板坯送至加热炉，不能直接热装的板坯下线进入保温坑保温后再送至加热炉加热，并留有直接轧制的可能。

出炉后的板坯，经高压水除鳞后，经辊道输送到定宽压力机，按照工艺要求将板坯宽度调整到设定的尺寸，然后由粗轧机组轧制 4~8 道，达中间坯厚度要求。中间坯经带保温罩的中间辊道输送到切头飞剪进行切头。飞剪前预留了中间坯边部加热器。切头飞剪配有中间坯头尾形状检测仪及剪切优化控制系统，可实现优化剪切，减少中间坯切头切尾的金属损失。切头后的中间坯经精轧前高压水除鳞装置清除二次氧化铁皮,进入精轧机组,轧制成 1.2~25.4mm 厚的带钢。

出精轧机组的带钢在输出辊道上由高效层流冷却系统进行冷却,冷却到设定温度后由全液压卷取机卷取,卷取后的钢卷经卸卷、打捆、称量、标记后由运输系统送到钢卷库内存放和冷却。需要检查的钢卷送到钢卷检查线进行检查和取样。在钢卷库内冷却后的钢卷按下工序加工工艺要求分别送至平整分卷机组、钢板横切机组、冷轧生产线或按销售计划发货,钢卷运输采用自主设计开发的钢卷托盘式运输装置[4,5]。

3.2 采用的先进技术

3.2.1 板坯热装热送技术

炼钢连铸厂与热轧厂毗邻紧凑布置,有利于实现工序间工艺流程的直接连接和一贯制管理的连续化生产。连铸机的出坯辊道与轧机的上料辊道重合,可以使连铸机生产的高温无缺陷板坯直接运送至加热炉。由于采用了热装轧制工艺,因此板坯自身余热得到充分利用,实现了最大限度节能;同时减少了板坯的库存量及库容,大幅度缩短了由炼钢到轧钢产品的生产周期。在工艺布置上预留了直接轧制的可能,为进一步降低能耗预留了条件。

3.2.2 粗轧机组先进技术

(1)采用板坯定宽压力机,可连续进行板坯侧压,运行时间短、效率高、板坯温降小,侧压后板坯头尾形状规整,板坯减宽侧压有效率可达 90% 以上。采用定宽压力机的主要技术优点:1)调宽能力大,最大侧压量可达 350mm,可大幅度减少板坯的宽度种类,充分发挥连铸机的产能,稳定连铸生产操作,提高板坯质量;2)便于热轧厂生产计划组织,提高板坯热装热送比例,节能降耗;3)改善中间坯的头尾形状,降低切损,提高金属收得率;4)由于减少了板坯宽度的种类,便于板坯库管理,相应提高了板坯库的利用率[6~8]。

(2)粗轧机组采用二辊可逆式轧机 R1+四辊可逆式轧机 R2 的半连续式轧制工艺,以降低设备投资,提高粗轧机组的利用率,缩短轧线长度,减少轧件的热量损失。粗轧机组能力大、轧制道次灵活,R1 轧机工作辊辊径大,可实现大压下轧制,对定宽压力机大幅度减宽后产生的板坯厚度增厚快速减薄;R2 粗轧机工作辊辊径小、轧机横向刚度好,对中间坯进一步减薄,可向精轧机组提供横向厚差小的薄规格中间带坯。

(3)粗轧机组采用 SSC 短行程控制和 AWC 宽度自动控制。经立辊宽度压下和水平辊厚度压下后,板坯头尾部将发生失宽现象,根据其失宽曲线采用与该曲线对称的反函数曲线,使立辊轧机的辊缝在轧制过程中不断变化,这样轧出的中间坯再经水平辊轧制后,头尾部失宽量减少。短行程法可减少切头损失率 20%~25%,还可显著提高头尾部的宽度精度。

(4)R1、R2 机架前后采用强力侧导板,可以有效防止板坯产生镰刀弯;提高了一次除鳞机喷嘴压力和对板坯的打击力,进一步提高了产品表面质量;设置了 E1、E2 立辊轧机,进一步提高产品宽度精度和控制能力;R1 和 R2 粗轧机牌坊断面积和电机功率较大,从而提高了粗轧机轧制能力;在中间辊道上设置有可分段开启的保温罩并预留了电感应边部加热器,可以减少中间坯在中间辊道上的热量损失、中间坯头尾温差和中间坯中部与边部的温差,提高中间坯横向和纵向温度均匀性,有利于稳定精轧机的操作,满足多品种产品的生产要求。

3.2.3 精轧机组先进技术

(1) 精轧机采用全液压压下和 AGC 系统,厚度控制效果显著。采用先进的弯辊加连续可变凸度控制(CVC^{plus})的板形控制技术,轧机凸度控制可达 1000μm,适用于轧制薄规格、低凸度宽带钢产品。

(2) 精轧机组采用 7 架四辊不可逆轧机,采用液压低惯量活套,有利于精轧机组的速度和带钢张力的控制,在精轧机出口设置上下表面质量检查仪,可在线检查和分析带钢表面质量。精轧机机架间采用抑尘喷水装置,可以有效减少粉尘生成量,在 F4~F7 机架间设置了除尘排烟系统,保证了清洁的生产环境和高标准的大气排放要求,实现了绿色生产。

(3) 设置了 F_1E 立辊轧机,可对中间坯进行对中和导向,同时其微量压下的作用可防止带钢边裂,并进一步提高产品宽度精度和控制能力;提高了精轧除鳞机喷嘴压力和对板坯的打击力,进一步提高产品表面质量;采用曲柄式飞剪和最优化控制系统,提高了剪切能力,可剪切 X80~X100 等钢种。

3.2.4 层流冷却先进技术

采用高效节能型层流冷却装置,可满足双相钢及多相钢等高强钢的生产要求。层流冷却系统包括精调区和修整区,可严格控制带钢冷却速率和最终冷却温度;层流冷却系统增加了边部遮挡系统,使带钢温降更为均匀。

3.2.5 卷取机先进技术

全液压助卷辊卷取机卷取温度达到 100~850℃,具有位置控制(即踏步控制)、最终压力控制和连续打开控制的功能,可防止卷曲过程中产生带钢压痕,卷曲钢卷紧密、齐边卷取质量好。

3.2.6 托盘钢卷运输先进技术

首钢国际工程公司自主开发研制了双排式托盘钢卷运输系统[4,5],满足了主轧线钢卷运输 1 卷/min 的工序节奏要求,运输方式、速度匹配灵活,实现了一条主轧线对应多条后部工序、工位的钢卷运输,全线采用卧式钢卷运输方式,可防止钢卷边部受损,确保了钢卷质量和倒运次数。

3.2.7 电气传动与自动化控制先进技术

3.2.7.1 电气传动

供配电系统在 110kV、35kV、10kV、0.4kV 侧均未设置高次谐波滤波装置,可以节约能源、减少对电网的污染,而且优化了整个车间布置,减少占地面积,节省工程投资,减少设备维护量。

所有辊道电机和辅助传动均采用 660V 电压等级的电机,变频装置采用公共直流母线方式,逆变器采用 690V 电压等级,节约投资、节省能源。主电机风机采用变频风机,根据主电机的 RMS 值(电机负载方均根值)通过二级数学模型,计算电机发热量,给出调速风机转速设定值,再通过所采集的电机实际温度值及进出口风温的实际值对调速风机进行闭环控制,节能降耗。主传动系统采用 TEMIC 电气公司生产的 TMD-70 双 PWM 控制的交直交变频器,工作电压为3300V,电气元件为IEGT,其优点是可从公共电网上吸收清洁的交流电源,具有本征四象限能力,电动与发电状态可以实现平滑转换,具有100%能量再生能力且无换相失败,可实现功率因数调节;这种双PWM控制的交直交变频器具有低能耗、低电磁和谐波干扰、轧钢过程中不发生无功冲击、无需任何补偿设备特点。

辅传动变频调速采用带公共直流母线的结构方式,使电机制动的反馈能量与电动状态运行的电机进行能量交换,使整流变和整流单元容量减小,降低工程投资。低压MCC柜采用智能性开关柜,实现自动操作集中监视,提高自动化水平,减少维护工作量;能源介质系统的风机水泵,均采用交流变频调速以实现节能降耗。

3.2.7.2 自动化控制

自动化控制系统包括电气传动级(L_0级)、基础自动化级(L_1级)、过程控制级(L_2级)生产管理级(L_3级)。L_0级主传动设备采用AFE结构(有功前端)的交直交变频调速系统;L_1级采用TEMIC电气公司生产的V系列控制器硬件设备;L_2级采用PC服务器3台,型号是美国生产的Stratas/FT3300,具有双机热备功能。L_1级和L_2级、L_2级和L_3级通过以太网(TCP/IP协议)进行通信,L_1级之间通过TC-NET100LAN网进行通信,L_1级和L_0级之间通过TOSLINE-20进行数据收集和快速通信。

4 生产应用实践

首钢京唐2250mm热轧工程于2008年12月10日建成投产,一次调试成功,经过半年的生产实践,迅速达到设计能力。吨钢成品消耗设计指标为1.031t/t,实际指标为1.0257t/t;热轧工序设计综合能耗为1195MJ/t,实际生产指标达到1195MJ/t;实际生产板坯热装率达到75%,综合技术经济指标达到国内先进水平,产品质量指标见表4。

表4 产品质量指标

项 目	设计指标	实际生产指标	国内先进指标
带钢平直度/IU	±30	±20	±30
带钢厚度/μm	±50	±30	±50
带钢宽度/mm	0~20	5~15	0~20
带钢板形/μm	±15	±15	±15
钢卷塔型/mm	最大50	40左右	最大50

5 结语

首钢京唐2250mm热轧生产线集成应用了当今国内外先进、成熟的工艺及装备,如采用了热装热送技术、定宽压力机、二辊可逆R1粗轧机、四辊可逆R2粗轧机、7机架四辊精轧机、CVC^{plus}板形控制技术、层流冷却技术等,开发研制了托盘式钢卷运输设备,并优化设计了电气传动和自动化控制系统。

2250mm热轧生产线设计中,按照动态精准设计理论,优化工艺流程,构建流程紧凑、运行高效、系统协同的生产体系,使连铸—热轧—冷轧工序的衔接匹配合理;整体工艺流程顺畅，工艺布置合理，设备选型和技术经济指标先进，节能环保设施齐全。

2250mm 生产线投产后快速达产，生产能力达到 550 万吨/年，产品质量达到国内外先进水平，主要技术经济指标达到或超过设计水平，实现了优质、高效、低耗的目标。

参考文献

[1] 陈应耀. 我国宽带钢热轧工艺的实践和发展方向 [J]. 轧钢, 2011, 28 (2): 1-8.

[2] 钱振伦. 我国宽带钢热连轧机的最新发展及其评析（一）[J]. 轧钢, 2007, 24 (1): 33-36.

[3] 钱振伦. 我国宽带钢热连轧机的最新发展及其评析（二）[J]. 轧钢, 2007, 24 (2): 32-34.

[4] 韦富强. 首钢京唐公司热轧钢卷运输系统研究 [J]. 轧钢, 2007, 24 (6): 36-40.

[5] 韦富强. 托盘式钢卷运输的冶金流程工程学分析及其应用 [J]. 轧钢, 2009, 26 (2): 32-35.

[6] 中国金属学会热轧板带学术委员会. 中国热轧宽带钢轧及生产技术 [M]. 北京: 冶金工业出版社. 2002: 53-87.

[7] 曲家庆. 热轧宽带钢生产工艺比较 [J]. 研究与探讨, 2007 (2): 38-41.

[8] 黄波. 我国热轧宽带钢轧机建设情况综述 [J]. 轧钢, 2009, 26 (1): 47-52.

Advanced Technologies of 2250mm Hot Strip Rolling Line at Shougang Jingtang

Zhang Fuming　Xie Jianxin

Abstract: 2250mm broad strip hot rolling line of Shougang Jingtang Iron & Steel Plant is designed according to dynamic exactness design system. Rolling mill designed product capacity is 5.5Mt/a. The tensile strength of production reach to 1000MPa. More than 20 items advanced technologies in today international steel rolling technical field have been adopted. The whole process and equipment reach to the international advanced level. This paper introduces design peculiarity and advanced technologies of 2250mm hot rolling line at Shougang Jingtang. In engineering design, heating furnace by step, casting-hot charge rolling, sizing press, two-high R1 reversing roughing mill, four-high R2 reversing roughing mill, 7 stands four-high finshing mill, CVC^{plus} plate shape control technique, laminar strip cooling system with speediness cooling process (TMCP) and so on have been adopted. The coil pallet transportation system has been researched independent. The engineering have been optimized which include of AC-DC-AC drive control technique, automation measure and control system, perfect dusting system. After the engineering project commission, the main technical economy parameters have achieved design target, and the goal of high quality, high efficiency, low cost has been reached.

Keywords: hot rolling; wide strip rolling mill; process and equipment; advanced technology

大型带式焙烧机球团技术创新与应用

摘　要：本文介绍了首钢京唐504m²带式焙烧机球团工艺和技术装备。为提高球团矿质量和性能，满足2座5500m³巨型高炉的高效低耗生产的需求，采用合理的炉料结构，配置了年产400万吨/年带式焙烧机球团生产线。该项目采用了一系列先进技术与创新，如球团工程的流程设计、工艺布局、设备开发、节能减排、自动化控制系统等。开发与应用了一系列先进技术和设备，如大型干燥窑、直径为7.5m的造球盘、梭式布料器、大型带式焙烧机、高效燃烧器等。工程投产后，建立了基于配料研究、造球、布料、焙烧温度控制等多工序的综合控制管理技术体系，球团生产运行和球团技术指标达到先进水平，取得了显著的经济效益和环境效益。

关键词：球团；带式焙烧机；原料；节能；工程设计

1　引言

首钢京唐钢铁厂是"十一五"国家重大工程，是按照循环经济理念建设的新一代可循环钢铁厂[1]。为了实现5500m³巨型高炉的稳定生产、优化炉料结构、提高精料水平[2]，设计并应用了国内首条504m²带式焙烧机球团生产线，构建了"高效率、低成本、节能环保"球团生产线，满足了2座5500m³巨型高炉高效低耗生产对炉料的要求。

本项目设计研究针对带式焙烧机球团生产工艺过程和技术特征，通过冶金过程工艺理论研究、工艺流程和功能解析、数值仿真设计优化、工业试验研究和关键设备开发研制，系统地研究了大型带式焙烧机球团技术的工艺优化、关键设备国产化和精准控制体系，为该项技术的成功应用提供了重要的技术保障。

2　工程设计研究与创新

2.1　优化工艺流程

工艺设计中基于冶金流程工程学理论[3]和首钢京唐钢铁厂的建厂理念，以流程优化、功能优化、结构优化、效率优化为目标，以动态精准设计为指导思想，注重工艺系统的顶层设计和物质流、能量流、信息流的协同高效运行，构建了以"流""流程网络"和"运行程序"为基本要素的新一代球团厂动态精准运行体系[4]。注重原料准备工序的精准设计，为精准配料及高质量、多品种球团提供了多种调节方式，突出了造球、预热、焙烧、

本文作者：张福明，王渠生，韩志国，黄文斌。原刊于《第十届中国钢铁年会暨第六届宝钢学术年会论文集》，2015：1737-1743。

冷却、工艺风机等主工序的功能解析和高效集约化配置，为高效、低耗、优质、清洁的绿色球团生产奠定了基础。

设计中对工艺流程、功能集成、集约高效、布局紧凑方面进行了大量研究。对预配料、干燥、辊压、熔剂与燃料制备、配料、混合、造球、焙烧、冷却到成品分级等多工序进行紧密衔接，最大限度地缩短物流运距、减少物料转运，将功能相同或相近的建筑物联合设置，以减少占地面积，实现了连续紧凑。

2.2 优化工艺技术

采用国际先进的带式焙烧机工艺，焙烧机全长为126m，台车宽度为4m，台车面积504m^2，料层厚度为400mm，球团生产能力为400万吨/年。

2.2.1 提高热风利用率

带式焙烧机集7个工艺区段于一体，通过台车的循环运行，台车上的生球依次完成干燥、预热、焙烧、均热和冷却过程。干燥段分为鼓风干燥段和抽风干燥段，鼓风干燥段主要是利用二冷段热风对生球进行脱水干燥，抽风干燥段是通过回热风机抽取的预热段和焙烧段的热风对生球进行干燥。在预热段和焙烧段利用冷却1段热风和外部热源加热，在均热段不再使用外部热源，主要是利用球团自身放热和冷却1段的热风。合理的回热风利用系统最有效地实现热的再利用，高效短捷的流程将散热面积降到最低，有效降低了热耗、提高了能源循环利用率[5,6]。

2.2.2 合理利用钢厂二次能源

本项目不同于国外采用重油或天然气为热源，焙烧机采用焦炉煤气作为外部热源。在固定的炉罩上设置32组烧嘴，通过流量的灵活调节实现工艺温度的灵活调节，最大限度地适应不同原料的需求。与此同时可以实现钢铁厂副产煤气的高效化利用，使高热值煤气得到最优化的利用，实现能源高效转换。

2.2.3 系统调控灵活

焙烧工序依靠工艺风机系统平衡焙烧机各个工艺段的温度和压力。工艺风机系统采用变频调速风机及自动控制阀门，实现各个工艺段温度的灵活调节和热量的合理分配。通过先进可靠的造球系统保证生球入炉强度，减少烟气含尘量，有效延长风机使用寿命。

2.2.4 提高台车作业率

带式焙烧机工艺采用铺底料循环系统保护台车，以延长台车耐热件的使用寿命。头部设置更换台车装置，可以实现在5min之内台车的在线快速更换，离线维修，有效提高了台车的作业率。

2.3 采用内配固体燃料工艺

为了提高球团质量，设计了熔剂与燃料制备工序，配置了内配固体燃料、内配白云石工序，预留根据不同原料调整燃料配比和生产熔剂性球团的手段。内配燃料工艺可以增加球团的孔隙率和还原性，这种球团用于高炉生产能提高生产率和降低焦比。这种高气孔率、高还原性的球团还能有效降低燃料消耗、降低焙烧机算床温度以及风机的电耗。

在工艺设计中，首次将熔剂与燃料制备系统布置在配料室旁边，将熔剂与燃料收集器置于配料室料仓的顶部，直接将熔剂粉、燃料粉输送到配料料仓中，通过优化流程有效简化了设备配置。

熔剂与燃料制备系统中采用热废烟气自循环新工艺，以降低系统热耗，降低热风炉设备规格及投资，减少废气排放量，有利于节能环保。

2.4 开发球团往复式布料技术

研究开发了新型球团专用布料胶带机，将往复式布料器与造球盘下的集料皮带集成为一条皮带，通过控制布料器头轮直径及高度，将生球落料高度控制到最小；通过布料器往复行走，实现生球单行程布料。有效减少生球的转运次数和落差，提高生球粒度合格率，布料均匀两侧无堆积。先进的布料胶带机+宽胶带+双层辊筛布料工艺，保证在带式焙烧机上生球料层均匀一致，具有较好的透气性。

2.5 开发应用大型矿粉干燥窑

为了满足大型带式焙烧机的正常生产，联合设备制造商共同开发大型球团干燥窑技术，设计研制了国内特大型 $\phi 5m \times 22m$ 矿粉干燥窑，满足控制物料水分的要求。本项目相比常规的球团生产工艺配置增加了原料干燥、辊压工序，确保球团精矿粉水分稳定、粒度及比表面积的均匀，提高其成球性，进而提高生球强度和质量。

首次开发了我国最大规格的 $\phi 5m \times 22m$ 矿粉干燥窑。根据矿粉干燥特点，专门设计了长径比为 4.4 的短粗型回转式干燥窑，既能强化干燥效果，也能降低出料口气体流速，从而减小除尘器的负荷，确保粉尘排放达标。

首次在干燥机上采用液压马达传动装置，实现无级变速技术，确保矿粉水分能按照生产要求进行灵活调整。设计了尾气自动补热系统，确保尾气温度在露点以上，以防止尾气由于温度过低而产生结露。

2.6 开发研究大型造球系统

球团生产的稳定顺行，最重要的是要保证生球质量，同规格的造球盘能力受物料成球性影响较大。为确保造球效果，在国内首次与制造厂共同开发最大规格的 $\phi 7.5m$ 新型造球盘与大型带式焙烧机相配套。与常规 $\phi 6m$ 造球盘相比，单机造球能力大幅度提高；圆盘直径加大，增加球团滚动次数，改善造球效果，提高生球强度。同时通过规模化造球，造球盘数量减少，厂房占地减小，工程造价降低。

在设计中采用造球盘盘面角度调整机构和球盘转速变频调速装置，以满足造球工序对水分和物料变化的要求，此外特别考虑采用全新形式的固定刮刀及造球盘的支撑结构来适应设备大型化的需要。

2.7 集成创新应用大型技术装备

为与 504m² 的大型带式焙烧机相匹配，首次采用了 $\phi 1700mm$ 辊径的辊压机、600m² 电除尘器及叶轮直径为 $\phi 3.6m$ 的耐热风机，其中 600m² 电除尘器是国内单台除尘面积最大的电除尘器；主要工艺风机实现国产化，解决了以前球团大型风机由国外引进的问题。所有工艺风机均采用高压变频调速，控制调节方便，有效降低电耗。

2.7.1 台车运动力学研究

台车是带式焙烧机核心设备中的最关键的设备，台车总重量约占整个焙烧机总重的17%。由于难于通过直接测量，检测台车在工况下温度和应力分布变化，只能通过已知条件建立模型、编制专门的程序，模拟台车在整个循环中的温度和应力分布情况。在此基础上，使用 ANSYS 软件进行有限元应力分析，分析台车的失效形式。根据台车热工条件，通过三维设计和热应力有限元分析，合理确定台车和箅条的结构（见图1和图2）。

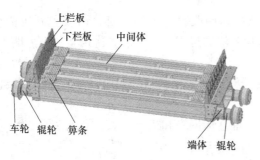
图1 台车三维设计结构图

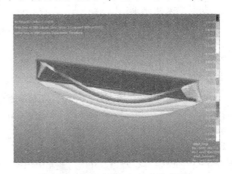
图2 台车有限元应力分析图

计算分析表明台车在一个运行周期中承受交变机械应力和脉动热应力，其中热应力占主导因素；尤其进入高温段，温度上升快，温度梯度大。根据计算结果，一方面，要选择即满足强度要求同时又经济合理的材料，并对材料的化学成分、组织状态、铸造和加工质量等各方面提出特别要求，有效保证主梁热强度；另一方面，通过传热学研究，设计开发新型我国箅条，优化箅条结构，减少箅条对台车梁的直接热传导，降低台车梁的热负荷。通过对箅条形状的优化，能对风流进行合理引导，有效减少箅床通风阻力，减少堵料。

2.7.2 创新采用自润滑辊套技术

借鉴近年来在烧结台车上的成熟的自润滑轴承技术，台车辊轮轴承采用自润滑轴承结构，具有制造成本低、制造周期短、使用寿命长的特点，可以有效地降低维护成本，提高了焙烧机的作业率。

自润滑轴承机体上嵌入由二硫化钼和二硫化钨组成的固体润滑剂，在摩擦过程中润滑剂微粒吸附在金属表面，起到润滑作用。在使用过程中无需添加油脂，实施免维护，而且使用寿命长，能够有效降低设备维护费用。

2.7.3 开发风箱端部及隔断密封技术

在生产实践中，为了进一步改善系统密封性，通过对风箱端部及隔断密封工作条件的深入研究，优化设计了新型带式焙烧机的风箱端部及隔断密封装置，有效地提高密封效果，降低了系统漏风率。

在生产实践的基础上，对带式焙烧机炉罩与台车间的密封效果进行分析，采取针对性改进措施，开发了新型炉罩与台车间的密封技术，进一步提高密封效果。

3 节能环保技术研究开发

节能环保是现代冶金工厂的重要标志。首钢京唐球团秉承节能减排、绿色环保的先进理念，采用带式焙烧机球团生产工艺，在节能减排、清洁生产和循环经济等方面创新采用了多项先进技术，节能环保技术优势显著。主要表现在：

（1）带式焙烧机在一个密闭炉罩内完成干燥、预热、焙烧、冷却的全部工艺，同时带式焙烧机整体布置在一个封闭厂房内，有效减少粉尘泄漏。

（2）炉罩高温热风直接回用，管道距离短，回风管道面积大，热量利用效率高。炉罩采用多层耐热材料和保温材料，有效减少表面散热，降低热耗。工艺风机全部采用高压变频调速，降低电耗。

（3）全部除尘灰均采用浓相气力输灰系统返回配料室使用，充分回收和利用资源。输送设备和管道实现全密封，避免了传统除尘灰输送过程中产生的二次扬尘，极大地改善了工作条件和厂区环境。

（4）首次开发的矿粉干燥窑上采用专门研制的高效扬料板，采用高温旋风配风装置，提高热交换效率，增强干燥效果，减低热耗。

设计研究中，建立了基于能源高效转换的能量流网络和运行体系，实现系统节能。以"减量化、再循环、再利用"为用能准则，建立能量输入-输出模型，实现优化用能。主要技术创新包括：

（1）利用钢铁厂自产的焦炉煤气作为球团焙烧的燃料，优化燃料结构，取代了高热值天然气，使钢铁厂能源流结构得到优化，焦炉煤气的效能得到高效化的利用。

（2）开发研制适用于焦炉煤气的燃烧器，设计合理的燃烧器结构，实现低空气过剩系数调控，保证燃料燃烧完全和足够的火焰温度，降低 CO_2 和 NO_x 的大量产生和排放，实现高效清洁燃烧，减少污染物排放。

（3）合理设置燃烧器安装位置和数量，以满足球团焙烧工艺为前提，实现燃烧和温度场精准控制，减少能源消耗、降低污染物排放。

（4）设计开发了高效的"风流回热利用系统"，将带式焙烧机焙烧过程产生的各梯级热能高效利用，实现球团干燥、预热、焙烧等不同工艺过程能量的合理匹配和高效利用。

（5）开发研制了球团内配碳技术，可以将钢铁厂碳素粉尘及焦化粉末作为能源[7]，实现固体废弃物的资源化、高效化综合利用，还可以提高球团焙烧质量、降低工序能耗。

4 生产应用

首钢京唐球团厂于 2010 年 8 月投产以来，研究探索了大型带式焙烧机球团生产工艺规律，研究开发了不同原料条件下稳定生产高品质、高性能球团的关键技术[8,9]。建立了基于配料研究、造球、布料、焙烧控制等多工序的综合控制管理技术体系。经过近 5 年生产实践，设备运行状况良好，球团矿质量和性能达到了设计指标。

4.1 原料适应性

球团原料主要采用秘鲁高硫磁铁矿，投产以来还试用过各种不同原料，如使用多种铁精矿粉和辅料，高比例使用赤铁矿，也曾全部使用赤铁矿，先后生产过普通酸性氧化球团矿、蛇纹石球团矿、白云石球团矿以及低硅镁质球团矿，在不同品种球团矿生产转换过程中，对带式焙烧机炉罩上 32 个烧嘴焦炉煤气使用量进行灵活调节，使整个工艺过程的焙烧强度和焙烧温度曲线得以控制，实现平稳过渡，充分显示出工艺操作制度适应性高的强大优势。

4.2 球团矿质量

优化的布料工艺，紧凑连续的工艺流程，极大保障了生球强度。焙烧机自身短距离的高效热风循环系统方便精确调节鼓风干燥与抽风干燥，有效控制生球不会破裂。在保证干球强度的条件下，减少膨润土添加量，提高球团矿含铁品位。投产以后，通过对造球工艺的探索和设备改进完善，进一步减少湿返料量。生产近5年来，焙烧机透气性一直处于优异水平，没有出现球团黏结现象。2014年平均膨润土消耗为15.07kg/t，球团矿品位达到65.95%，球团矿强度达到3000kN/球，达到国内领先水平。

4.3 应用实践

带式焙烧机工艺只有一台主体设备，设备集中、管理集约，烧嘴操作灵活，自动化水平高，使得生产稳定顺行，有效降低劳动强度，减少岗位定员，降低人工成本，提高劳动生产率。建立了基于智能化生产控制系统，使其具有以下功能：

（1）设计完善的自动化检测和控制系统，使整个工艺全过程的信息实现在线监控。

（2）构建优化的数学模型，开发布料闭环控制、温度闭环控制、风系统闭环控制、铺底料平衡控制和煤气安全系统，实现球团生产的最优化操作。

（3）实现设备远程遥控和无人值守，劳动生产率大幅度提高。

为保证生产稳定顺行，生产工艺的优化非常重要。主要在以下方面进行了创新：

（1）生产过程采用计算机集中控制和调节，主要工艺环节如风机系统、料厚控制系统、铺底料循环系统、烧嘴系统采用工业电视监控和专家控制系统管理，自动化水平高，生产稳定，煤气消耗低[10]。

（2）投产后对造球系统优化，从料仓给料方式到造球盘给料皮带、造球盘底衬等进行多项改造，同时设置生球破碎机，改善返料成球性能，进一步提高成品球合格率。

（3）开发设备管理周期软件，对易损部件定期维护更换和保养，根据使用经验及时调整检修计划，对具备条件的设备尽可能做到离线维修，将问题解决在发生之前，有效保证作业率。

（4）根据生产经验调整岗位定员，提高劳动生产率，年作业率达到98%以上。

该项目集成创新了国内外先进工艺和技术，优化工艺流程和关键单元技术，使其工艺流程合理，工艺配置完备，工厂布局紧凑，生产运行高效。工程设计中特别注重原料准备阶段和焙烧热量调节的精细化，可以适应多种原料条件，为生产多品种、高质量球团矿创造了有利条件。生产实践证实，带式焙烧机球团生产工艺，可以高效率、低成本、大规模稳定生产高品质、高性能的球团矿。

首钢京唐400万吨/年球团厂投产运行以后，主要原料为秘鲁高硫磁铁矿，碱金属含量较高。由于带式焙烧机技术性能先进可靠、工艺适应性强、控制调节灵活，可以很好地适应多种物料，因此生产运行稳定顺行，技术指标先进。采用多种原料稳定生产优质球团，完全满足京唐两座5500m³巨型高炉高效低耗生产的精料要求。球团生产能耗、球团矿质量以及设备作业率等主要指标都达到国际先进水平，与国内外先进球团生产线的指标对比见表1。

表 1　球团主要生产技术经济指标

项　目	首钢京唐球团厂（带式机）	国内 A 球团厂（带式机）	国内 B 球团厂（回转窑）	首钢矿业球团厂（回转窑）	巴西 CVRD（带式机）	伊朗 GMI（带式机）
设计能力/万吨·年$^{-1}$	400	210	500	200	750	500
作业率/%	98.32	93.34	95.86	90.84	96.44	82.20
球团矿 TFe/%	65.95	57.8	63.26	65.2	65.5	66.78
抗压强度/N·球$^{-1}$	3000	2427	2584	2782	3300	2700
筛分指数/%	0.35	2.82	2.31	0.73	2.47	3.35
工序能耗/kgce·t^{-1}	17.11	35.12	25.51	17.55	30.78	19.10

4.4　运行效果

（1）风机寿命延长。工艺风流系统互联互通，烧嘴调整灵活。球团静止的焙烧过程，使粉料量大幅度降低。最终表现为热循环利用风中粉尘量一直保持在极低水平，平均在 0.5% 以下，灰尘粒度更细，耐热风机转子使用寿命延长，在没有耐热多管除尘器的条件下，已经使用 4 年以上，没有进行任何修复和更换。

（2）作业率提高。耐热件消耗是球团生产最大的维护成本。采用铺底铺边保护台车，效果明显。设备运行稳定，检修量很少。运行 3 年时台车没有更换过甚至中间体都没有翻转过，更换的箅条量也仅占总体箅条数量的 2.15%。台车可以实现在线快速更换，目前可以做到在 5min 内更换一个台车，热工况条件保持稳定，有效提高作业率，年工作日达到 350d 以上。

（3）耐火材料寿命延长。由于焙烧炉是静止的，耐火材料也是固定的，没有机械振动、变形及球团磨损和结圈造成的损坏。台车的快速更换，减少了因事故造成的焙烧炉和焙烧设备的急冷急热，延长设备和耐火衬的使用寿命。使用 4 年，还未进行过任何修补和更换。

5　结语

针对我国首台产量 400 万吨/年大型带式焙烧机球团生产线的设计开发、技术装备集成以及生产应用研究，开发并形成了大型带式焙烧机球团高效低耗生产技术，主要技术创新如下：

（1）首次自主设计、自主集成建造了我国第一个产量 400 万吨/年以上大型带式焙烧机球团生产线，在工艺功能优化、流程优化、设计优化、效率优化等多方面具有创新，拥有多项自主知识产权，实现了高效低耗生产。

（2）开发研制并应用了往复式布料设备、台车密封装置、大型干燥窑及大型造球盘等一系列具有自主知识产权的大型化技术装备，大型技术装备实现了全面国产化，实现了自主设计创新与装备制造创新。

（3）研究探索了大型带式焙烧机球团生产工艺规律，开发了不同原料条件下稳定生产高品质、高性能球团的关键技术；建立了基于配料研究、球团造球、焙烧控制等多工序的综合控制管理技术体系。

(4) 首钢京唐球团厂于 2010 年 8 月投产以来，经过 4 年多的运行，已经完全达到设计水平，年作业率达到 98% 以上，实现了球团品位达到 65.95%，球团抗压强度不小于 3000kN，工序能耗 17.11kgce/t，达到同类技术的国际先进水平，环境效益和社会效益显著。

致谢

首钢京唐公司和首钢国际工程公司的夏雷阁、利敏、张卫华、张全申、李祥、赵宏森等参加了本项目的研究和设计开发工作，作者一并表示衷心感谢。

参考文献

[1] Zhang Fuming, Qian Shichong. Design, construction and application of 5,500m^3 blast furnace at Shougang Jingtang [C]. AISTech 2014 Proceedings of the Iron & Steel Technology Conference, 2014, Indianapolis, Ind., USA: 753-765.

[2] Zhang Fuming, Cao Chaozhen, Meng Xianglong. Technological development orientation on ironmaking of contemporary blast furnace [J]. Advanced Materials Research, 2014: 875-877, 1138-1142.

[3] 殷瑞钰. 冶金流程工程学（第2版）[M]. 北京：冶金工业出版社, 2009: 152-158.

[4] 殷瑞钰. 冶金流程集成理论与方法 [M]. 北京：冶金工业出版社, 2013: 160-168.

[5] 李国玮, 夏雷阁, 青格勒, 等. 京唐带式焙烧机原料方案及热工制度研究 [J]. 烧结球团, 2011, 36 (2): 20-24.

[6] Englund D J, Davis R A. CFD model of a straight-grate furnace for iron oxide pellet induration [J]. Minerals & Metallurgical Processing, 2014, 31 (4): 200-208.

[7] Umadevi T, Kumar Prachethan P, Kumar Prasanna, et al. Investigation of factors affecting pellet strength in straight grate induration machine [J]. Ironmaking and Steelmaking, 2008, 35 (5): 321-326.

[8] 夏雷阁, 苏步新, 李新宇, 等. 焙烧温度与球团矿强度和还原性的关系 [J]. 烧结球团, 2014, 39 (1): 21-24.

[9] 夏雷阁, 苏步新, 李新宇, 等. 首钢504 m^{2}带式焙烧机热工制度的试验研究 [J]. 矿冶工程, 2014, 34 (3): 69-75.

[10] 夏雷阁, 刘文旺, 黄文斌. 大型带式焙烧机在首钢京唐球团的应用 [C] //2011年度全国烧结球团技术交流年会论文集, 2011: 116-120.

Innovation and Application on Pelletizing Technology of Large Travelling Grate Machine

Zhang Fuming Wang Qusheng Han Zhiguo Huang Wenbin

Abstract: The pelletizing process and technical equipment of 504m^2 travelling grate machine at Shougang Jingtang steel plant are introduced in this paper. In order to improve the pellet quality and performance, meet the requirement of two 5500m^3 huge blast furnaces high efficiency - low consumption operation, adopt reasonable burden composition, therefore a large scale induration travelling grate machine pelletizing plant with annual output 4.0 million tons is configured at Shougang Jingtang. In this project, a number of advanced technologies and innovations are implemented, such as the pelletizing process design, general layout, equipment development, energy saving and emission control, automatic operation system, etc. A series of advanced technologies and large scale equipment are developed and applied, such as the large drying rotary kiln, the balling disc with diameter 7.5 meter, the shuttle type green pellet distributor, the large travelling grate machine and the seal facility, the high efficiency burner and so on. After commissioning, the comprehensive operation system is established based on charging mixture investigation, balling, green pellet distribution, and baking temperature control. As a result, the operational parameters and pellet quality have achieved advanced technical level; huge operational benefit and magnificent environmental benefit have been obtained.

Keywords: pellet; induration travelling grate machine; raw material; energy saving; engineering design

首钢京唐钢铁公司绿色低碳钢铁生产流程解析

摘　要：我国钢铁行业面临着资源能源短缺、市场盈利下降和生态环境制约三个方面的压力与挑战，在低碳循环经济成为当前社会发展的主趋势的情况下，钢铁企业需要向低碳、绿色钢铁生产的方向转型升级。以京唐钢铁公司2015年上半年的生产调研为基础，计算京唐钢铁公司生产过程的能耗、CO_2排放情况，并对钢铁生产过程能量流运行、废弃物处理与能源转换情况进行分析，为其他钢铁企业减少二氧化碳排放、节约能源、保护环境提供一个参照。研究表明，京唐钢铁公司吨钢综合能耗604.5kgce，吨钢CO_2排放2.165t，与产品结构相类似的宝钢和武钢相比，能耗水平位于行业先进水平；基于钢铁生产的能量流分析表明，京唐钢铁公司通过采用干熄焦发电、煤气干法除尘、TRT余压发电等先进技术，实现余热余能回收136.26kgce/t钢，余热余能回收率48.31%，高出全国平均水平10.89%；京唐钢铁公司基于循环理念，建立高效的煤气-电能转换中心和余热蒸汽回收利用中心，基本实现了煤气资源、固体废弃物资源、水资源的循环利用与废水零排放，生产过程中SO_2、NO_2和粉尘的全面达标排放。

关键词：绿色低碳生产流程；吨钢综合能耗；CO_2排放；余热余能回收；废弃物处理

1　引言

　　钢铁工业一直在我国扮演着经济基础产业与工业化支撑产业的角色，然而，钢铁工业的高速发展，也给我国的环境保护、资源和能源的供应造成了巨大的压力。作为资源、能源的密集型产业，我国钢铁工业的能源消耗占全国总能耗的16%左右，CO_2排放量占全国工业排放总量的9.3%[1]，钢铁工业的废水、粉尘、固废和SO_2排放量分别占全国工业排放总量的8.53%、15.18%、17.0%和3.7%[2,3]。《国家十三五发展规划纲要》明确要求："十三五"期间，单位GDP生产能耗下降16%，CO_2排放降低17%；工业生产过程中SO_2和化学需氧量各削减8%，并将进一步完善相关能源与环境的法律法规的制定，以产品能耗与能效标准限额，倒逼企业改革与发展，通过推广重点工程领域的关键节能技术，推进节能减排、能源管理工作[4]。

　　近年来，关于钢铁行业转型发展问题引起了社会的广泛关注，在绿色低碳循环经济已成为当前社会经济发展主旋律的情况下，钢铁行业同样应大力发展绿色低碳循环经济。钢铁行业建设和发展绿色低碳循环经济的主要特征是："低排放、低污染、高效率"，旨在最大程度提高能源的利用效率，并大力提升清洁能源比例，改善能源消费结构[5,6]。钢铁行业的发展与转型不能仅局限于冶金-材料制备层面，作为流程制造业的钢铁企业，还应当开展和完善能源转换与大宗废弃物消纳与处理功能，钢铁企业的发展道路应该通过绿色制

本文作者：刘宏强、张福明、刘思雨、付建勋。原刊于《钢铁》，2016，51（12）：80-85。

造过程走向生态化转型的道路上去[7,8]。新一代钢铁厂的发展需要解决的问题是在生产先进产品的同时,如何实现钢铁生产过程的节能与减排。

首钢京唐钢铁联合有限责任公司(以下简称"京唐公司")作为国家"十一五"规划的重点工程,是按照循环经济理念建立的钢铁联合企业。其设计思想是打造新一代可循环钢铁流程,核心思路是将钢铁企业由单一的钢材生产功能,转变成为具有洁净钢生产、高效能源转换、大宗废弃物消纳处理与综合利用在内的三大主要功能的平台,以"减量化、资源化、再利用"为原则,在生产高品质、高附加值的钢铁产品的同时,消纳社会和自身产生的废弃物,实现生产过程高效率、低污染、低排放[9,10]。本文依托于京唐公司开展研究,以京唐公司2015年上半年的生产数据为样本,开展钢铁生产流程的研究,意在摸清京唐公司钢铁生产过程中能源消耗与CO_2排放情况,探讨钢铁生产过程余热余能的利用水平,分析钢铁生产过程中能源转换、废弃物消纳处理以及环保运行水平,依此为其他钢铁企业减少二氧化碳排放、节约能源、保护环境提供一个参照。

2 京唐钢铁厂的设计理念与生产情况

京唐公司坐落于河北唐山曹妃甸地区,一期项目建成于2010年底,形成了生铁898万吨、钢坯970万吨、材912万吨的生产规模。产品定位是面向汽车、家电、食品包装等领域各种强度级别的低碳、超低碳冷轧产品以及各种高强度结构钢、管线钢类热轧产品。目前主要产品包括高端汽车板、高端家电板、建筑板以及食品包装用电镀锡板。

京唐公司钢铁生产流程遵循构建新一代钢铁流程的思想理念,采用"2+1+2"的流程结构,建设2座$5500m^3$高炉、1座转炉炼钢厂和2条热连轧生产线(2250mm和1580mm)以及相应的冷轧、涂镀生产线,钢铁生产的全流程按年产900万吨精品钢的生产路线进行设计。图1是京唐公司钢铁生产的流程示意图。在以工序能力匹配为前提下,京唐公司坚持装备大型化、高效化、连续化、专业化的原则,选用先进、节能、高效的生产设备,代表性设备有7.63m大型焦炉、$500m^2$大型烧结机、$504m^2$带式焙烧机、$5500m^3$大型高炉等大型化工艺装备,为降低资源、能源消耗,提高劳动生产率奠定了基础。京唐公司在建设过程中采用了国际、国内先进技术达220余项[11]。具有代表性的生产工艺包括:高效能源转换的大型焦炉炼焦技术;精料、高风温、富氧大喷煤等高炉强化冶炼技术;"一包到底"炼铁-炼钢界面技术;高效率、低成本的洁净钢生产线等[12~15]。

京唐钢铁厂的布置创新性的采用了"一"字形总图布置,物流短捷顺畅、工序衔接紧密。各生产工序之间、工序内部设施之间,在满足物料传输工艺要求的前提下,尽量紧凑布置,缩短物料传输距离。在工序界面层次,生产单元按工艺顺序采用紧凑化布局,从原料场到高炉全皮带运输,总长度仅42km;炼铁-炼钢界面采用铁水罐"多功能化"技术,取消了炼钢厂的倒罐站;炼钢-热轧界面通过辊道实现铸坯"红送",选择标准轨铁路机车输送方式,最大限度地减少厂内运输。京唐公司目前采用"铁水全三脱"冶炼工艺,轧制过程中冷轧与热轧的转换比约为70%左右,图2所示为京唐公司钢铁生产流程示意图及2015年上半年的主要产品的产量。京唐公司能源工序包括供电、供水、供气以及热电联产部门,在能源利用方面,京唐公司通过采用大型化的装备、减少工序环节、减少装置数量、优化工序界面,以最大限度地减少能量耗散,提高能源的利用效率;通过采用

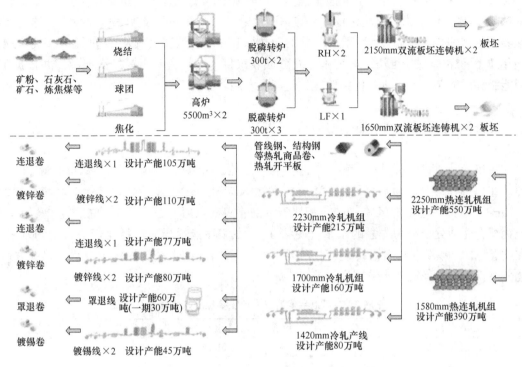

图 1 京唐公司钢铁主体生产装置及其生产流程示意图

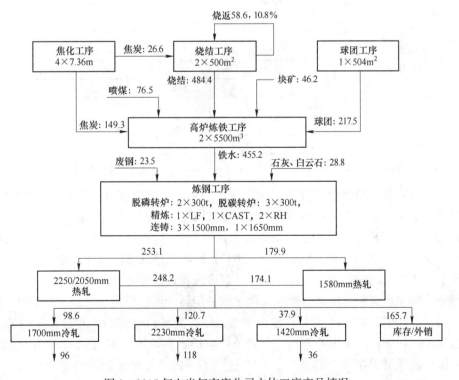

图 2 2015年上半年京唐公司主体工序产品情况

全烧高炉煤气热风炉、全干法除尘匹配 TRT 炉顶余压发电装置、烧结机余热回收技术、炼铁-炼钢界面铁水罐多功能化技术、转炉干法除尘、余热回收及煤气回收技术、连铸-热轧界面的铸坯热送热装技术、轧钢加热炉余热回收技术、低温蒸汽及发电后乏汽用于海水淡化技术、煤气-煤粉混烧发电技术等充分回收煤气、余热、余压等二次能源；并根据不同能级和用能装置特性进行优化配置，使能源利用效率提升。

3 京唐公司钢铁生产能耗分析

2015 年上半年，京唐公司铁水产量为 455 万吨，粗钢产量 441 万吨，钢材产量 423 万吨。2015 年上半年，京唐公司吨钢综合能耗为 604.5kgce/t，京唐公司吨钢新水消耗量为 3.29m³，吨钢耗电 692.9kW·h/t，相比于宝钢、武钢以薄、带材产品为主的企业，目前吨钢综合能耗处于较为先进的水平（见表1）；在工序能耗层面，相比于全国平均水平，焦化、球团、炼铁、热轧、冷轧工序能耗均低于全国平均水平，相比于宝钢，目前在焦化、炼铁工序方面能耗水平较为先进。

表 1 京唐 2015 年上半年钢铁生产主要经济技术指标及比较

指标名称	单位	京唐	宝钢	武钢	行业平均水平
吨钢综合能耗	kgce/t	604.5	610.2	609.5	580.4
吨钢耗电	kW·h/t	692.9	650.2	615	470.2
吨钢耗新水	m³/t	3.29	3.15	3.58	3.45
焦化能耗	kgce/t	80.26	100.30	89.8	101.5
烧结能耗	kgce/t	51.0	48.3	46.2	48.2
球团能耗	kgce/t	19.5	—	—	27.6
炼铁能耗	kgce/t	369.6	371.0	386.2	389.0
炼钢能耗	kgce/t	2.1	-5.9	—	-4.2
套筒窑能耗	kgce/t	170.4	144.0	—	—
热轧能耗	kgce/t	45.6	52.2	57.9	50.8
冷轧能耗	kgce/t	64.3	61.9	137.7	68.2
冷轧比	%	63.56	67.34	—	—

注：宝钢、武钢及行业平均水平数据来源于中国钢铁工业协会 2015 年企业能耗数据统计。

京唐公司是全部生产薄、带材产品的钢铁企业，冷轧与热轧转换比约为 65%，钢铁生产过程耗电量明显高于生产线棒材产品的钢铁企业，吨钢耗电量高于行业平均水平约 220kW·h/t，在钢铁生产结构上，京唐公司采用铁水"三脱"预处理工艺，废钢的加入量相对较少，2015 年上半年，铁钢比为 1.032，铁前工序的能耗约占总能耗的 81.3%，相对行业平均水平高约 3%~5% 左右；另外，京唐公司通过采用海水淡化设备生产除盐水，替代部分外购淡水资源，根据对海水淡化过程的能耗分析，海水淡化能耗折合 7.5kgce/t，京唐公司目前吨钢新水消耗量 3.3t 左右，吨钢外购新水 1.6t 左右，除盐水使用量 1.7t 左右，相比于其他钢铁厂使用外购水，京唐公司吨钢水资源能耗多出 12.24kgce/t 钢；同时京唐采用长流程高炉-转炉炼钢模式，其能耗高于国内部分企业采用的短流程电炉炼钢模式。上述几个原因导致京唐公司综合能耗高于行业平均吨钢综合能耗。

4 京唐公司余热余能回收利用分析

钢铁生产过程属于高温生产,绝大部分生产工序均需要在高温（>1000℃）进行生产,同时会伴随着大量的余热余能产生。提高钢铁生产过程中产生的余热余能的回收利用率,对实现钢铁低碳、绿色生产有着重要意义。张春霞等人基于对国内主要的大中型钢铁企业生产实况的调研,分析得到目前钢铁生产过程的余热资源总量约为8.44GJ/t钢,钢铁企业余热资源回收利用率约26%,主要集中在高温余热资源的回收利用,中温、低温余热资源的回收利用相对较少[16]。

京唐公司通过为实现钢铁生产过程的余热余能资源的充分回收利用,建立并采用了包括干熄焦余热发电、烧结环冷烟气余热回收、TRT余压发电、一罐到底等多种装置和技术,实现了对干熄焦余热、烧结矿显热、高炉煤气余压、铁水显热等多种余热余能资源的综合回收利用。表2是基于京唐公司2015年上半年生产数据,在对各工序能量流研究的基础上,折算的京唐公司吨钢可供回收利用的余热余能资源及其回收利用情况汇总见表2,并与全国重点企业[16~18]的余热余能回收利用情况进行了对比分析。

表2 京唐公司吨钢余热余能回收利用情况汇总 （GJ/t）

工序	余热余能名称	资源总量	京唐回收情况	全国平均水平[16~18]
焦化	红焦显热	0.573	0.495	0.420
	焦炉烟气显热	0.195	0	0
	荒煤气显热	0.475	0.02	0.02
	焦化工序小计	1.243	0.515	0.440
烧结	烧结烟气	0.373	0	0
	烧结矿显热	0.913	0.616	0.254
	烧结工序小计	1.286	0.616	0.254
炼铁	高炉煤气显热	0.312	0.052	0.048
	高炉煤气余压能	0.298	0.186	0.129
	热风炉废气显热	0.557	0.206	0.230
	高炉渣显热	0.537	0	0
	高炉冷却水	0.734	0	0
	铁水显热	1.418	1.291	1.180
	炼铁工序小计	3.711	1.735	1.587
炼钢	转炉煤气显热	0.558	0.338	0.212
	转炉渣显热	0.142	0	0
	转炉工序小计	0.700	0.338	0.212
热轧	连铸坯显热	0.597	0.240	0.240
	加热炉烟气显热	0.534	0.411	0.212
	加热炉气化冷却显热	0.195	0.139	0.149
	热轧工序小计	1.326	0.790	0.601
吨钢	汇总	8.267	3.994	3.094

研究表明,京唐公司吨钢可供回收的余热余能资源为8.267GJ,目前实现回收

3.994GJ，余热余能回收率为48.31%，全国平均水平为3.094GJ，余热余能回收利用率为37.42%。在京唐公司的各生产工序中，余热余能回收利用率最高的工序是热轧工序，余热余能回收利用率为59.57%，热轧加热炉通过采用加热炉烟气换热技术，回收了76.9%的加热炉烟气显热；余热余能回收利用率最低的工序为焦化工序，主要是焦炉烟气与荒煤气显热均未被有效回收利用；炼铁工序吨钢余热余能资源总量最高，为3.711GJ/t，目前实现回收利用了46.75%，炼铁工序主要未被回收利用的能量为高炉渣显热，京唐公司采用水冲渣工艺，将高炉渣显热转换为热水的显热，但目前尚未对冲渣水的显热资源进行回收与利用，目前，公司正在积极研发相关冲渣水余热资源回收利用技术。

总体来说，通过采用先进的生产工艺与技术装备，目前京唐公司余热余能回收利用水平处于全国先进水平。在能源回收利用方面，基于热力学第一、第二定律，对余热余能回收利用进行能级匹配、逐级回收与分级利用，建立了能源综合利用体系，对余热余能进行分级回收与综合利用。实现了对钢铁生产过程中众多余热余能资源的高效率回收利用。

5 京唐公司钢铁生产CO_2排放分析

碳排放计算在方法论上可分为生命周期评价（Life Cycle Assessment，LCA）方法以及投入产出（Input-Output，I-O）评价两种方法[19~23]。目前，国内主要采用的是投入产出计算方法的较多，国家发改委制定的《钢铁生产企业温室气体排放核算方法与编制指南》（以下简称《钢铁指南》）是基于投入产出计算方法，通过研究钢铁生产的整体系统边界，计算企业层面的温室气体排放情况，计算方法简单便捷，将钢铁企业看作一个整体，计算温室气体排放量，不考虑系统内部各物质之间的转换，适合钢铁企业层面的温室气体排放计算[24]。

图3是京唐公司的碳排放计算边界，京唐公司的碳排放计算范围包括：（1）化石燃

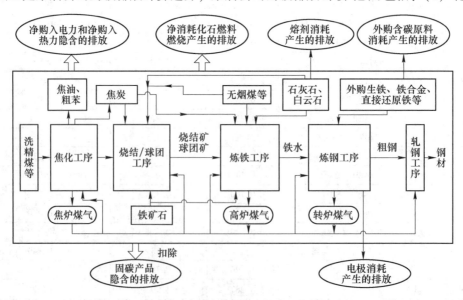

图3 京唐公司企业温室气体排放及其核算范围

料燃烧产生的 CO_2 排放 E_{com}；（2）工艺生产过程产生的 CO_2 排放 E_{pro}，包括烧结、球团、炼铁、炼钢过程中熔剂和含碳原料带入而产生的 CO_2 排放；（3）外购电力、蒸汽排放产生的 CO_2 排放 E_{ind}；（4）副产品煤气外销抵扣的 CO_2 排放 E_{ded}；（5）外销钢铁产品隐藏碳的抵扣 E_{ind}，主要是指钢材、输出的焦炭、煤气、焦油中的碳含量抵扣的 CO_2 排放；钢铁企业温室气体排放总量 E_{total}，可按照公式 $E_{total} = E_{com} + E_{pro} + E_{ind} - E_{ded} - E_{hid}$ 进行计算。

京唐公司 2015 年上半年主要外购燃料（洗精煤、动力煤、无烟煤、汽油与柴油），净购入电力以及含碳产品的输入输出情况分析见表 3。京唐公司 CO_2 排放情况见表 4。计算结果显示，2015 年上半年京唐公司 CO_2 排放总量为 954.59 万吨，吨钢温室气体排放 CO_2 为 2.165 吨。

表 3　京唐公司 2015 年上半年含碳物质消耗情况

类　　别	用　　量
洗精煤/万吨	241.74
动力煤/万吨	60.78
无烟煤/万吨	76.67
电力/万千瓦时	-13767
石灰石/万吨	24.7
白云石/万吨	5.5
电极/万吨	1772
废钢/万吨	23.46
焦油（外售）/t	66520
粗苯（外售）/t	16630
气、柴油/t	1408

表 4　京唐公司 2015 上半年 CO_2 排放情况　　　　　　　　（万吨）

E_{com}	E_{pro}	E_{ind}	E_{hid}	E_{total}
978.23	18.80	-16.76	25.68	954.59

京唐公司与不同国家和地区的主要钢铁企业 CO_2 温室气体排放情况的对比见表 5[25]，在同口径仅计算直接与间接温室气体排放情况下，比韩国 POSCO 温室气体排放高约 30kg/t；比日本钢铁 JFE 高出 30~50kg/t 左右。京唐钢铁生产过程中吨钢温室气体排放与世界先进企业差距较小。

表 5　京唐与几个主要钢铁企业 CO_2 排放强度分析

来源	CO_2 排放强度/$t \cdot t^{-1}$	备　　注
日本住友金属	1.95	能源消耗
	2.09	能源消耗+熔剂消耗
日本 JFE	2.00	能源消耗
	2.07	能源消耗+熔剂消耗

续表5

来源	CO_2排放强度/$t \cdot t^{-1}$	备注
日本新日铁	1.82	能源消耗
韩国POSCO	2.19	包括：2.12直接排放和0.17间接排放
中国台湾中钢	2.18	未说明
国际钢铁协会	1.70	国际钢铁工业可持续发展指标（未说明）

6 京唐公司能源高效转换体系分析

开展生产过程的余能、余热、废水以及固废进行循环利用，是京唐公司实现清洁、低碳、节能生产重要措施之一。近几年来京唐公司全流程吨钢综合能耗不断降低，从2009年的671kgce/t降低至604kgce/t，降幅达9.98%。同时，京唐公司通过优化二次能源的利用方式，发展热电联产、余热余能发电，使得企业的余热余能回收水平、企业自发电率、余热余能自供电率不断提升，2015年上半年实现自发电量富余，发电富余外售共计1.85×10^8kW·h，余热余能自供电率提升至46.7%。

6.1 京唐公司煤气-电能转化

钢铁公司一次能源以煤炭资源为主，一次能源煤炭在入厂之后，经过能源转换成为二次能源，洗精煤在焦炉内经过焦化反应转换为焦炭，焦炉煤气能源、喷吹煤、焦炭在高炉炼铁过程部分转换为高炉煤气资源，动力煤在辅助能源工序用于发电和产生蒸汽资源。2015年上半年，京唐公司的二次能源主要包括焦炭（54%）、副产品煤气（40%）、余热蒸汽（4%）和余热发电2%。京唐公司的煤气资源占二次能源比例的40%左右，京唐公司吨钢煤气资源回收总量为1690m^3，其中焦炉煤气141m^3/t、高炉煤气1451m^3/t、转炉煤气98m^3/t，吨钢煤气资源折合297.93kgce，相当于吨钢综合能耗的49.30%。提高钢铁企业的煤气资源利用率，降低煤气资源的放散率，这对降低钢铁生产过程的吨钢综合能耗意义重大。

在正常生产情况下，京唐公司各生产工序焦炉煤气资源消耗量为81.78%，高炉煤气资源消耗量为79.7%，转炉煤气资源消耗量72.7%，钢铁生产各工序消纳煤气资源共计237.48kgce/t钢，富余煤气资源量为56.22kgce/t钢，为了最大程度地使用煤气二次能源，降低煤气的放散，京唐公司建立了2×300MW发电机组，既可以全烧燃煤发电，又可以掺烧煤气发电，最高可掺烧45000m^3/h焦炉煤气和36000m^3/h高炉煤气。现阶段正常生产情况下，焦炉煤气零放散，高炉煤气在热风炉等大用户调整情况下有少量放散，在自备电站等用户检修时，高炉煤气有较大量放散。

图4所示为京唐公司余能发电与煤气发电情况分析。2015年上半年，京唐公司焦化工序通过采用高温高压（540℃，9.8MPa）锅炉并配备2×30MW蒸汽发电机组，利用干熄焦工艺回收红焦显热发电，共计发电2.06亿千瓦时，吨焦发电量为109.3千瓦时，并实现外供蒸汽29.67万吨，炼铁工序5500m^3高炉采用TRT发电回收利用高炉煤气余压能，TRT发电总量为2.28亿千瓦时，吨铁发电量达50.15千瓦时，同时2×300MW燃煤-燃气混烧发电机

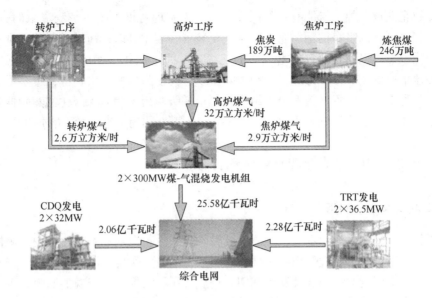

图 4 京唐余能发电、煤气发电分析

组回收利用富余煤气资源,实现发电 25.58 亿千瓦时。2015 年上半年共计发电 31.90 亿千瓦时,企业实际自供电率达 104.51%,实现富余电力外供 1.85 亿千瓦时。

6.2 余热蒸汽与海水淡化

京唐公司蒸汽发生系统主要包括热轧汽化冷却蒸汽、炼钢余热蒸汽、烧结余热蒸汽、干熄焦余热蒸汽抽汽,以及能源辅助系统中 2×35t 锅炉、2×130t 锅炉和 2×300MW 发电机组蒸汽抽汽。图 5 是烧结、焦化、热轧以及炼钢余热蒸汽回收情况。钢铁生产过程各工序回收的蒸汽全部外送至综合管网统一调度使用,实现对余热蒸汽资源的综合利用。

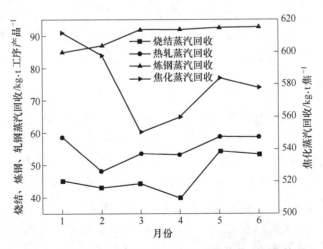

图 5 2015 年京唐公司主体生产工序余热蒸汽回收情况

京唐公司通过建立不同压力的蒸汽管网,将余热蒸汽资源进行分级综合利用。一般来说,钢铁企业的低温余热蒸汽资源往往难以得到充分利用,尤其是在夏季,低温蒸汽资源大多被钢铁企业放散掉，未能有效的回收利用。为实现对钢铁企业蒸汽资源的充分利用，京唐公司建立了全国唯一的钢铁企业海水淡化装置，利用其毗邻海域的优势，利用钢铁生产过程的低温蒸汽以及企业自发电以后的乏气进行海水淡化。2015年上半年京唐公司海水淡化除盐水产量809万立方米，并利用除盐水替代了部分的淡水资源，缓解了水资源不足的压力。海水淡化前置汽轮发电机组实现能量梯级利用，用海水淡化的蒸发器替代了汽轮机的凝汽器，全系统能量利用率82.23%，实现煤气-热-水-电多能源介质协同优化。

7 京唐公司废弃物消纳处理及环保运行分析

7.1 固体废弃物回收利用

京唐公司每年产生高炉水渣、钢渣、各类除尘灰、粉煤灰以及氧化铁皮等固体废总量约467万吨，京唐公司通过采用各种循环利用加工工艺，实现对固体废弃物的资源化再利用。通过对各类固废的综合处理回收利用，京唐公司极大提高了资源的回收利用率，降低了生产过程中煤炭、矿石等资源的消耗总量，并与建筑行业、水泥行业等形成产业链。

7.2 水资源综合利用

京唐公司毗邻海域，建有海水淡化工程，生产除盐水替代部分淡水资源，用以降低水资源缺乏的压力。京唐公司在钢铁生产过程中，按照循环经济理念，在最大限度减少新水的使用量的同时，尽可能延长生产过程中新水的循环利用率，努力实现废水的零排放综合化处理，降低废水排放量，实现水资源的再生循环利用。从2009年至今，京唐公司吨钢新水资源消耗量、吨钢外购水资源消耗量逐年降低，其中：吨钢新水消耗量从 $4.45m^3$ 降低至 $3.3m^3$，吨钢外购新水量从 $2.46m^3$ 降低至 $1.5m^3$，除盐水替代外购淡水比例从44.72%提升至51.52%。在废水处理方面，京唐公司建立了2座污水处理站，每座处理站处理生产废水 $2.4\times10^4m^3/d$，生活污水处理能力达 $2400m^3/d$，污水深度脱盐处理能力达 $600m^3/h$，京唐公司将处理后的二次水用水泵送至厂区生产-消防给水管网回用，基本实现了生产、生活废水"零"排放。

7.3 环保运行情况

京唐公司一期工程环保总投资约76亿元，占工程总投资的11.21%，包括废水处理设施8套、固废处理设施5套、废气处理设施128套。通过采用先进的环保工艺技术，布袋除尘、电除尘后粉尘排放浓度分别小于 $20mg/m^3$ 和 $30mg/m^3$，电厂海水脱硫小于 $50mg/m^3$，脱硝小于 $300mg/m^3$。2015年吨钢烟粉尘排放量0.407kg，吨钢 SO_2 排放量0.389kg，大气环境质量 SO_2 平均 $23\mu g/m^3$，优于国家二级标准年均值 $60\mu g/m^3$；NO_2 浓度 $18\mu g/m^3$，优于空气质量二级标准；人工监测降尘为 $15.9t/(km^2 \cdot 月)$。京唐公司相关空气污染物排放指标与国家要求数据对比见表6。

京唐公司在创建低碳、环保循环经济钢铁企业发展模式下，通过采用先进的生产设备与技术，实现了钢铁企业余热余能的高效率回收利用，对钢铁生产过程的废水、废渣等进行了资源化回收、再处理与利用，基本实现了废弃物的零排放，同时建立了高效的环保运行机制，钢铁企业的粉尘排放、SO_2、NO_2排放均达到并优于国家标准排放。

表6 京唐公司主要污染物排放情况

污染物	排放情况	国 家 要 求
吨钢烟粉尘	0.407kg/t钢	环评要求：0.430kg/t钢
吨钢SO_2排放	0.389kg/t钢	环评要求：0.410kg/t钢
吨钢NO_2排放	0.304kg/t钢	环评要求：0.410kg/t钢
厂区空气SO_2检测	23μg/m³	国家Ⅰ级：20μg/m³ 国家Ⅱ级：60μg/m³
厂区空气NO_2检测	18μg/m³	国家Ⅰ级：20μg/m³ 国家Ⅱ级：40μg/m³

8 结论

（1）京唐公司采用"高炉—转炉"长流程生产模式，生产装备上实现了大型化、高效化、连续化、专业化，具有众多高效节能的生产设备，烧结、球团、焦炭100%自产，企业实际自供电率100%，水资源采用海水淡化+外购部分新水模式。基本实现对煤气资源、固体废弃物资源、水资源的零排放综合利用和SO_2、NO_2等污染物的达标排放，目前钢铁生产过程吨钢综合能耗为604.5kgce/t，吨钢CO_2排放量为2.165t，余热余能回收率48.25%。

（2）京唐公司定位于全部生产薄、带材钢铁产品，冷轧与热轧的转换比约为70%左右，目前吨钢综合能耗为604.5kgce/t，相比于同类型的宝钢和武钢，目前能耗水平处于行业领先地位；在工序能耗层面，焦化、球团、炼铁、热轧等工序能耗均比行业平均水平数值低。

（3）京唐公司吨钢余热余能总量为8.27GJ/t，通过采用干熄焦工艺、TRT余压发电等余热余能回收技术，实现余热余能回收136.13kgce/t；余热余能回收率48.25%，高出全国平均水平11.26%。

（4）京唐公司2015年上半年CO_2排放总量为954.59万吨，吨钢CO_2排放量为2.165t，焦化、烧结、球团、炼铁、炼钢、套筒窑、热轧、冷轧工序的吨钢CO_2排放量分别为169.95kg/t、229.43kg/t、19.37kg/t、1294.45kg/t、37.60kg/t、46.06kg/t、205.31kg/t、125.66kg/t。

（5）京唐公司对钢铁生产过程中的固废、水资源进行了充分的循环利用，基本实现了钢铁生产过程各类固废的全部循环利用，通过海水淡化及废水处理实现了水资源的循环利用与废水零排放；生产过程SO_2、NO_2、粉尘的排放全面达标。

参考文献

[1] 王维兴．钢铁工业能耗现状和节能潜力分析［J］．中国钢铁业，2011（4）：19．
[2] 王维兴．科学评价中国钢铁工业能耗现状与国内外对标［J］．四川冶金，2009（4）：1．
[3] 杨建新，刘炳江．中国钢材生命周期清单分析［J］．环境科学学报，2002（4）：519．

[4] 《国民经济和社会发展第十二个五年规划纲要（草案）》（摘编）[N]. 人民日报, 2011-03-06 (006).
[5] 柳克勋. 关于钢铁企业发展低碳经济的思考 [J]. 再生资源与循环经济, 2010 (4): 11.
[6] 张春霞, 王海风, 张寿荣, 等. 中国钢铁工业绿色发展工程科技战略及对策 [J]. 钢铁, 2015 (10): 1.
[7] 殷瑞钰. 钢铁制造流程的本质、功能与钢厂未来发展模式 [J]. 中国科学 (E辑: 技术科学), 2008 (9): 1365.
[8] 廖中举, 李喆, 黄超. 钢铁企业绿色转型的影响因素及其路径 [J]. 钢铁, 2016 (4): 83.
[9] 杨春政. 新一代钢铁流程运行实践 [J]. 钢铁, 2014 (7): 30.
[10] 张福明, 颉建新. 冶金工程设计的发展现状及展望 [J]. 钢铁, 2014 (7): 41.
[11] 北京首钢国际工程技术有限公司. 冶金工程设计理念的创新与实践 [M]. 北京: 冶金工业出版社, 2010.
[12] 张贺顺, 任全烜, 郭艳永, 等. 首钢京唐公司炼铁低成本冶炼实践 [J]. 中国冶金, 2015 (9): 27.
[13] 张福明, 崔幸超, 张德国, 等. 首钢京唐炼钢厂新一代工艺流程与应用实践 [J]. 炼钢, 2012 (2): 1.
[14] 张福明, 张德国, 张凌义, 等. 大型转炉煤气干法除尘技术研究与应用 [J]. 钢铁, 2013 (2): 1.
[15] 李金柱, 王飞, 杨春政. 首钢京唐铁水包多功能化应用实践 [J]. 炼钢, 2014 (4): 61.
[16] 蔡九菊, 王建军, 陈春霞, 等. 钢铁企业余热资源的回收与利用 [J]. 钢铁, 2007 (6): 1.
[17] 仇芝蓉. 我国钢铁企业余热资源的回收与利用 [J]. 冶金丛刊, 2010 (6): 47.
[18] 吴春华. 钢铁企业余热余能资源利用现状分析 [J]. 冶金能源, 2014 (2): 54.
[19] 王微, 林剑艺, 崔胜辉, 等. 碳足迹分析方法研究综述 [J]. 环境科学与技术, 2010 (7): 71.
[20] International Organization for Standardization. Life Cycle Assessment Principles and Framework [M]. ISO: Geneva, Switzerland, 2006.
[21] BSI. Specification for the Assessment of the Life Cycle Greenhouse Gas Emissions of Goods and Services (PAS 2050) [R]. London: British Standards Institution, 2008.
[22] 张玥, 王让会, 刘飞. 钢铁生产过程碳足迹研究——以南京钢铁联合有限公司为例 [J]. 环境科学学报, 2013, 33 (4): 1195.
[23] 高成康, 陈杉, 陈胜, 等. 中国典型钢铁联合企业的碳足迹分析 [J]. 钢铁, 2015 (3): 1-8.
[24] 刘宏强, 付建勋, 刘思雨, 等. 钢铁生产过程二氧化碳排放计算方法与实践 [J]. 钢铁, 2016 (4): 74.
[25] 张春霞, 上官方钦, 张寿荣, 殷瑞钰. 关于钢铁工业温室气体减排的探讨 [J]. 工程研究-跨学科视野中的工程, 2012 (3): 221.

Green Low-Carbon Analysis of Iron and Steel Manufacturing Process of Shougang Jingtang Iron and Steel Company

Liu Hongqiang　Zhang Fuming　Liu Siyu　Fu Jianxun

Abstract: Resources and energy shortages, market declining profitability and ecological and environmental constraints have become three aspects of the pressures and challenges in China's steel industry. Low-carbon and circular economic became the main trends in current social development, and iron and steel enterprises should upgrade to the direction of low-carbon and green steel production. This paper was based on the production survey in the first half of 2015 year in Shougang Jingtang Company. Energy consumption and CO_2 emissions in the steel production process were calculated. The energy flow operation, waste treatment and energy conversion in iron and steel production process were analyzed. Also, references were provided for other steel companies to reduce carbon dioxide emissions, save energy and protect environment. Studies had shown that comprehensive energy consumption and CO_2 emissions for per ton of steel were 604.5kgce and 2.165t. Compared with Baosteel and Wuhan Iron and Steel with similar product structure, energy level located in the advance level. Based on the energy flow analysis of steel production, Shougang Jingtang Company used the CDQ, TRT and other advanced technology. The waste heat and energy recovery achieved 136.26kgce/t steel, and the recovery rate of waste heat and energy was 48.31%, which was 10.89% higher than the national average value. Based on the cycle thought, Shougang Jingtang Company established the efficient gas-electric energy conversion center and steam heat recycling center, and basically realized zero emissions and comprehensive utilization of coal gas, solid waste and water resources and discharged SO_2, NO_2 and other pollutants.

Keywords: green and low-carbon steel manufacturing process; energy consumption for per ton of steel; CO_2 emissions; waste heat and energy recovery; waste disposal

钢铁流程固废资源化利用逆向供应链体系探讨

摘　要： 以"减量化-再利用-再循环"的3R原则为指导，以实现钢铁流程固体废弃物再利用为目标，设计了钢铁流程固体废弃物资源化利用逆向供应链体系。结合各类固体废弃物的成分特性，设定有价元素为新资源，构建了钢铁流程固体废弃物资源化利用逆向供应链组合方案，并配置了流程自循环、外部利用和产物升级的梯级实施路径。为了达到逆向供应链操作过程的动态有序、协同连续和集约高效，建议引入固体废弃物全生命周期绿色化管理体系，并贯穿到固体废弃物资源化利用逆向供应链组合方案的初步规划与具体实施中，这将有利于降低企业固体废弃物环境管理投入成本和取得优良的钢铁全流程绿色化效果。

关键词： 钢铁流程；固体废弃物；逆向供应链；3R原则；全生命周期

1　引言

我国钢铁企业每年产生的固体废弃物超过4亿吨，约占工业固体废弃物年总排放量的1/4，且多以低值化利用和工程堆存填埋为主，如果能充分利用好钢铁流程中的固体废弃物，将对钢铁产业的节能减排、低碳生产和循环经济产生重要意义。现有的研究表明，以"资源-产品-污染排放"为主的物质单向流动模式难以解决钢铁企业经济盈利与环境保护产生的冲突，限制了企业的绿色可持续发展。而以3R原则为指导，积极开拓"资源-产品-再生资源循环"的新型物质流动模式，使钢铁流程从炼铁、炼钢到轧钢过程产生尽可能少的固体废弃物并且消纳一定的固体废弃物，这是钢铁流程绿色可持续循环发展的新模式[1~3]。

目前，已经使用的钢铁固体废弃物众多，如日本根据钢铁渣的成分特性，分别把钢铁渣用作水泥原料、铺路材料、土建用材料、制造农业肥料及土壤改良剂等，部分炉渣还用于海洋人工海藻礁[4]。韩国用高炉渣生产耐火专用隔热材料的矿棉和Pos-Ment水泥以及用于修复生态系统的人工鱼礁Triton新型环保产品；利用镍铁炉渣替代混凝土专用的细骨料产品；以高炉渣、转炉渣和镍铁渣等为原料，以干式熔融法制得仿玄武岩Pos-Pipe和Pos-Tile产品；以炼铁过程的煤沥青为原料直接合成电子元件基板石墨烯工艺等[5]。我国以高炉矿渣制造水泥混合材料，重矿渣用作混凝土骨料和道路材料，钢渣粉用作道路工程材料、工程回填材料等[6]。综上所述，固体废弃物利用的方法各异，处理后得到的产物也不同，这就使得钢铁流程固体废弃物从产生到消化过程的管理相对分散，能系统地指导钢铁全流程固体废弃物高效利用的科学方法较少。

近年来，已有一些钢铁流程固体废弃物科学回收决策系统、管理绩效评价、收集和运

本文作者：刘清梅、张福明、李海波、骆振勇。原刊于《钢铁》，2019，54（10）：117-124。

输成本等相关领域的研究。如 Froling M 等用集成规划方法为含锌固体废弃物回收与运输计划创建决策支撑系统[7]；冀巨海等研究了钢铁企业绿色供应链管理绩效关联评价的相关因素[8]；Giannetti B 等考虑了国外企业在废钢铁逆向物流网回收的环境效益和经济效益[9]；鲁春林等通过建立收集和运输模型来实现降低成本[10]；钟磊钢等从绿色物流的理念出发，分析钢铁行业逆向物流过程中存在的回收控制不严、企业规模小和市场不规范等问题[11]。但是，这些研究多以某一种或一类固体废弃物为研究对象来开展相关工作，而针对全流程的总体钢铁固体废弃物统筹规划和科学路径实施的研究不多，而且以减量化为目标的固体废弃物资源化应用方法也没有相对统一的观点。

本文以循环经济"减量化–再利用–再循环"原则为指导，引入逆向供应链体系，设计"正向生产–逆向缩减"的钢铁流程固体废弃物资源化利用逆向供应链体系，以长流程钢铁企业生产线为生产单元，设定固体废弃物的有价元素为新资源，构建全流程固体废弃物资源化利用的新组合方案以及相应的梯级实施路径，并提出钢铁流程固体废弃物全生命周期绿色化管理的科学方法。

2 逆向供应链体系设计

钢铁流程固体废弃物资源化利用逆向供应链体系以拉动用户消费为设计思路，实现回收产物的成本比购买新产物或者处置旧产物的成本低，以此获得经济激励，并得到后续客户的满意认可。钢铁流程固体废弃物资源化利用逆向供应链体系如图 1 所示。

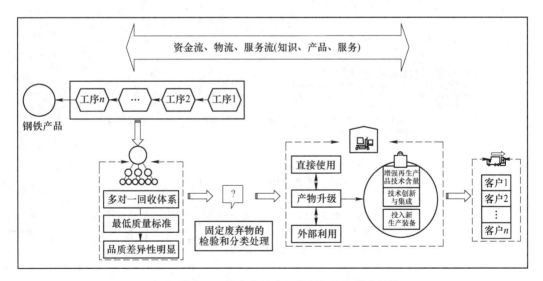

图 1 钢铁流程固体废弃物资源化利用逆向供应链体系

由图 1 可见，固体废弃物资源化利用逆向供应链涵盖了以下 4 个主题：

（1）多对一回收体系。多对一回收体系是在钢铁流程稳定生产的基础上，持续稳定地获得大量固体废弃物，这是企业投资资源化利用逆向供应链的前提，在目前的钢铁生产中，各个流程的固体废弃物种类繁多，要建立符合自身条件的多对一回收体系，以方便统一管理。

（2）固体废弃物的检验和分类处理。固体废弃物的检验和分类处理可以评价出固体废弃物的质量水平，并为资源化利用逆向供应链后续的各种产物定制提供恰当的处理方法。

（3）以直接使用、产物升级和外部利用3种路径进行处理，三者之间对应的固体废弃物存在协作与竞争的关系，无论是哪种路径均能获得价值增值，并且寻求最大增值的处理方式，实现从"直接使用或者报废处理"向"产物升级"的方式转变。

（4）采取多组分销形式满足不同客户需求，进入资源化利用逆向供应链的产物可以将仍有利用价值部分分离出来再使用，因为从废物流中回收产物有高度不确定性，所以这些产物的可用性与质量存在不可预测性，对应的客户也不相同。

在整个固体废弃物资源化利用逆向供应链体系中，关键是创造固体废弃物的高附加值，需要采用优化的组合模式，部分还需要引入新的制造技术，使最终产物全部或部分再次进入行业的供应链循环。

3 逆向供应链组合方案与实施路径

3.1 钢铁流程固体废弃物成分分析

以年粗钢产量900万吨的单体长流程钢铁企业为生产单元分析全流程固体废弃物的种类、产生量和主要成分。钢铁流程生产线如图2所示，钢铁流程固体废弃物种类与产生量见表1[12]。

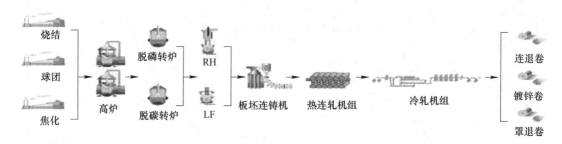

图2 钢铁流程生产线示意图

表1 钢铁流程固体废弃物种类与产生量

种类	产量/万吨·年$^{-1}$	种类	产量/万吨·年$^{-1}$
焦油渣与生化污泥	1.20	炼钢一次灰	1.20
湿熄焦粉	4.70	炼钢二次灰	3.99
焦化环境灰	5.04	渣钢	19.40
烧结电场灰	1.34	氧化铁皮	1.89
烧结脱硫灰	4.44	热轧油泥	1.82
烧结环境灰	9.54	含油污泥	0.12
球团环境灰	1.02	冷轧酸再生泥饼	0.36

续表1

种　类	产量/万吨·年$^{-1}$	种　类	产量/万吨·年$^{-1}$
粉煤灰炉渣	2.89	锌渣	1.59
高炉炉前灰	3.86	粉煤灰	20.00
高炉干法灰	4.81	筛下焦末	48.60
高炉旋风灰	13.20	钢渣尾渣	77.50
脱硫钢渣	3.90	高炉渣	257.00
脱碳钢渣	2.80		

钢铁流程固体废弃物产生量约为505万吨/年，其中铁前与炼铁流程产生总量约为380万吨/年，主要固体废弃物有高炉渣、高炉干法灰、高炉旋风灰、高炉炉前灰、烧结电场灰、烧结脱硫灰、烧结环境灰、球团环境灰、粉煤灰、粉煤灰炉渣、焦油渣、生化污泥、焦化环境灰、湿熄焦粉和筛下焦末等，其中高炉渣占比约为68.1%；炼钢与轧钢流程产生量约为125万吨/年，主要固体废弃物有钢渣尾渣、渣钢、脱硫钢渣、脱碳钢渣、炼钢一次灰、炼钢二次灰、氧化铁皮、热轧油泥、冷轧酸再生泥饼、含油污泥和锌渣等，其中钢渣尾渣占比约为62.4%。

按照产物的铁和碳等有价元素进行划分，高炉炉前灰、高炉干法灰、烧结环境灰、球团环境灰、高炉旋风灰、烧结电场灰、热轧油泥、氧化铁皮、渣钢和钢渣尾渣等全铁质量分数相对较高，其中的氧化铁皮、高炉炉前灰、烧结环境灰和高炉旋风灰的全铁质量分数分别达到74.00%、60.17%、49.49%、41.25%，具有极高的铁元素再利用价值；焦化环境灰、湿熄焦粉、筛下焦末、焦油渣和生化污泥等碳质量分数达到60%以上，具有极高的碳元素再利用价值。典型的含铁类、含碳类和其他类固体废弃物主要成分分别见表2~表4[13]。

表2　含铁类固体废弃物主要成分（质量分数）　　　　　　　　　（%）

种类	TFe	SiO_2	CaO	MgO	Al_2O_3	K_2O	Na_2O	ZnO	Cl
炉前灰	60.17	1.83	1.08	0.28	0.85	0.21	0.22	0.19	0.09
烧结环境灰	49.49	5.50	9.10	1.84	3.23	0.39	0.54	0.031	0.38
烧结电场灰	36.37	2.56	6.03	0.66	1.58	12.96	6.27	0.20	15.03
高炉旋风灰	41.25	4.48	2.15	0.53	2.92	0.50	0.32	0.41	0.75
高炉干法灰	34.97	2.22	0.92	0.22	1.48	0.64	0.58	2.60	2.85
炼钢二次灰	28.57	16.10	19.69	6.12	3.84	1.79	1.05	1.84	1.39
钢渣	21.71	14.39	38.12	7.17	3.14	0.049	0.18	0.007	0.13
氧化铁皮	74.00	0.34	0.10	0.16	0.084	0.002	0.019	0.009	0.012

表3　含碳类固体废弃物主要成分（质量分数）　　　　　　　　　（%）

种类	C	灰分	S
焦化环境灰	65.68	7.87	0.81
焦油渣	64.39	8.58	0.73
湿熄焦粉	86.55	12.45	0.87
筛下焦末	87.77	11.19	0.72

表 4　其他类固体废弃物主要成分（质量分数） （%）

种类	SiO_2	CaO	MgO	FeO
高炉渣	34.28	40.10	8.38	0.68
粉煤灰	50.60	2.80	1.20	7.10

3.2　逆向供应链组合方案

钢铁流程固体废弃物的资源化遵守循环经济固体废弃物处理 3R 原则，该原则如图 3 所示[14]。

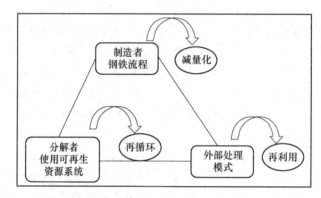

图 3　循环经济固体废弃物处理的 3R 原则示意图

首先，减量化（Reducing）即减少资源利用量和废物排放量；其次，再利用（Reusing）是对固体废弃物进行充分回收利用；然后，再循环（Recycling）是应对物料进行再生循环利用。根据全流程固体废弃物的成分分析，确立不同类别固体废弃物的工艺来源、产量以及成分特征，在此基础上构建资源化利用逆向供应链组合方案，如图 4 所示。

在整个组合方案中，各生产阶段生成的多种类固体废弃物组成了资源化利用逆向供应链的相关模块，在交互过程中涉及资源、组织、服务价值的动态配置，并共同构成供应链模型。随着生产工序的成熟和下游客户提出新需求或者产生的固体废弃物不断解散和重组，企业可以在本组合方案的基础上不断优化组合过程，使全流程资源化利用逆向供应链日趋完善。

3.3　实施路径

3.3.1　流程自循环

根据钢铁流程固体废弃物资源化利用逆向供应链体系设计（见图 1），以"直接使用"为主要实施路径结合固体废弃物的成分属性进行设计。在钢铁流程固体废弃物资源化利用逆向供应链组合方案中（见图 4），将焦化环境灰和筛下焦末配入混匀矿使用，焦化干熄焦灰供烧结使用或喷吹高炉，焦油渣和生化污泥配入焦化型煤使用；烧结环境灰配入混匀矿使用，高炉炉前灰和高炉旋风灰配入烧结回用。在炼钢和轧钢流程，炼钢一次灰返到冷固球团造球后供炼钢使用，炼钢产生的氧化铁皮、脱硫渣进入混匀矿或炼钢回用，渣钢为炼钢回用；热轧氧化铁皮配入混匀矿使用，热轧油泥干燥后铁质量分数达 50% ~

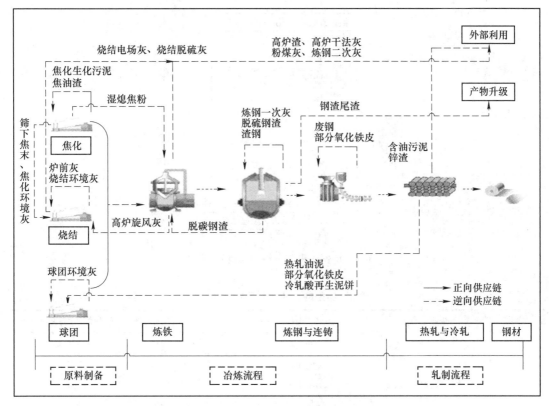

图 4 钢铁流程固体废弃物资源化利用逆向供应链组合方案

60%，供球团使用；冷轧酸再生泥饼进入混匀矿后达到球团直接配用要求使用。

采用自循环路径要建立独立的资源化利用逆向供应链循环，以成为钢铁流程最直接高效和经济的运行模式，为了扩大自循环固体废弃物的使用量，可以开发新的流程技术。如在炼钢流程开发液态钢渣回收利用途径，利用脱碳钢渣较高碱度和脱磷转炉冶炼半钢炉渣低碱度的生产特点，将脱碳转炉渣用于脱磷转炉造渣以替代部分熔剂[15]。炼钢流程液态钢渣回收利用途径如图 5 所示。

通过钢铁流程自循环实施路径，可以最直接地获得各类固体废弃物的信息，及时了解应用渠道的合理性，并结合钢铁工艺流程完善资源化利用逆向供应链的组合模式。

3.3.2 外部利用

钢铁企业不参与固体废弃物的回收过程，通过资源外部管理实现客户化定制。在钢铁流程固体废弃物资源化利用逆向供应链组合方案中（见图 4），设计烧结电场灰、烧结脱硫灰、高炉干法灰、高炉渣、粉煤灰、炼钢二次灰、含油污泥和锌渣等固体废弃物为外部利用的组合模式，可以引入钢铁流程以外的成熟工艺，如从固体废弃物的物性分析可知，高炉干法灰、烧结电场灰、炼钢二次灰的碱金属质量分数高，通过钢铁流程内部自循环困难，此时可以采用转底炉或竖炉的外部利用方法，将钢铁流程中多种类的固体废弃物灵活地组织起来，实现外部再利用。典型的工业化转底炉和竖炉的应用情况分别见表 5 和表 6。

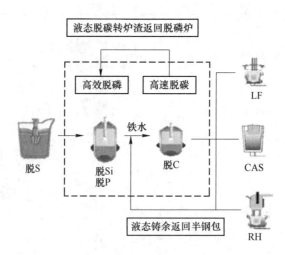

图5 炼钢流程液态钢渣回收利用途径

表5 典型的工业化转底炉应用情况

序号	企业名称	工艺类型	投产时间	处理能力/万吨·年$^{-1}$
1	日本新日铁住金君津	Inmeteo	2000年5月	18.0
		DRyIron	2002年12月	13.0
		DRyIron	2008年3月	31.0
		Fastmet	2000年4月	19.0
		Fastmet	2005年2月	19.0
2	日本新日铁住金广畑	Fastmet	2008年12月	19.0
		Fastmet	2011年10月	22.0
3	日本神户制钢加古川厂	Fastmet	2001年4月	1.4
4	日本JFE钢铁西日本厂	Fastmet	2009年4月	19.0
5	美国ITmk3厂	ITmk3	2010年1月	50.0
6	韩国浦项厂	DRyIron	2009年9月	20.0
7	韩国光阳厂	DRyIron	2009年12月	20.0
8	中国台湾中钢	DRyIron	2007年12月	26.0
9	中国马钢	DRyIron	2009年5月	20.0

表6 典型的工业化竖炉应用情况

序号	企业名称	处理能力/t·h^{-1}	所用原料	投产年份
1	墨西哥Sicartsa	80	压块型砖和废钢等	1998
2	德国蒂森克虏伯	25~50	压块型砖和废钢等	2004
3	日本新日铁住金	60	含碳球团和废钢等	2005
4	日本JFE钢铁公司	80	压块型砖和废钢等	2008
5	中国太钢	502	碳钢和不锈钢固体废弃物压块型砖	2011

3.3.3 产物升级

该模式需要把新技术、新工艺的研发放在突出位置，依靠冶金技术、信息技术和工程技术集成来处理固体废弃物，以提升再生产物的技术含量。在钢铁流程固体废弃物资源化利用逆向供应链组合方案中（见图4），设计对钢渣尾渣进行新产物的工艺开发，根据现状采用钢铁流程自循环的实施路径，如在烧结矿中配加脱碳钢渣代替白灰等熔剂，有利于回收氧化铁和氧化钙等成分，但存在磷富集而导致铁水磷质量分数升高的困难。采用外部利用为实施路径已经有一些方法，如将钢渣应用在道路的填筑材料、透水混凝土制品和制作硅肥等领域。以应用道路填筑材料为例，按照 JTJ 034—2000《公路路面基层施工技术规范》、CJJ 35—1990《钢渣石灰类道路基层施工及验收规范》，钢渣以一定的配合比应用在路基填筑材料中。但是，钢渣中含有微量的铅、汞和铬等重金属离子，其应用于道路工程后经过长期雨水冲刷和浸泡，重金属离子会浸出而污染土壤和地下水，造成生态环境的破坏。

因此，在现有模式下，无论是部分应用还是工程堆存，与循环经济固体废弃物处理的 3R 模型预设的结果均存在差距。根据含铁类固体废弃物主要成分的检测结果（见表1），钢渣铁质量分数达到 20%，另外，熔融钢渣排放时温度达到 1100 ℃ 以上，富含铁以及丰富的热能和化学能，均为产物升级路径提供了资源条件。

在实际工艺开发中，采用此路径构建钢铁流程固体废弃物资源化利用逆向供应链存在新产物可用性与质量不可预测的技术难题，还需要积极建立和完善此类废弃物资源化利用标准体系，实现多项技术集成和可复制性，以保证技术可靠性后的推广应用价值。

4 逆向供应链组合方案的实施建议

钢铁流程固体废弃物资源化利用逆向供应链的优势如下：

（1）全流程固体废弃物利用采用"多对一回收体系"，为统一管理提供了科学依据。

（2）采用多种方案相结合的原则，设计了"直接使用-产物升级-外部利用"相互转化的实施路径，促使了全流程 28 类固体废弃物利用方案的动态优化。因此，整个资源化利用逆向供应链组合方案是"变废为宝"，这就需要更为科学的固体废弃物可持续性的管理方案，本文建议运用固体废弃物全生命周期绿色化管理体系。目前，国际上很多制造业均开展了诸如产物制造、废弃物处置的全生命周期绿色化管理，如瑞典 Volvo 汽车公司和芬兰 Neste Oil 石油公司均通过全生命周期绿色化管理对产物研发、生产、废弃过程中所产生的环境影响进行了分析，并制定相应方案。

在钢铁流程固体废弃物资源化利用逆向供应链方案实施过程中，加强对固体废弃物的全生命周期绿色化管理，包括方案的最初项目建议书、可行性研究报告、初步规划设计、详细规划设计、施工图设计、施工招投标、施工建设到后期的使用效果，整个固体废弃物的全生命周期均考虑资源化利用逆向供应链实施过程的减量化、再利用和再循环，有利于将项目实施整个过程中所产生的环境影响降至最小，减少固体废弃物的产生量、降低环境风险以及钢铁企业在环境管理的投入成本。

5 结语

钢铁流程绿色化发展需要开拓"资源-产品-再生资源循环"的新型物质流动模式，以"减量化-再利用-再循环"的3R原则为指导，设计了"正向生产-逆向缩减"的资源化利用逆向供应链体系，并涵盖"多对一回收体系""固体废弃物检验和分类处理""直接使用、产物升级和外部利用方案"和"多组分销"等4个主题，达到了统一管理、动态实施、高附加值利用的目的。

清晰地指出了钢铁流程固体废弃物资源化利用逆向供应链组合方案的"流程自循环""外部利用"和"产物升级"的资源化利用逆向供应链操作路径，以有价元素碳为新资源，设计出焦化环境灰、筛下焦末、焦化干熄焦灰、焦油渣和生化污泥等的利用方式；以有价元素铁为新资源，设计出烧结环境灰、高炉炉前灰、高炉旋风灰、炼钢一次灰、氧化铁皮、渣钢、热轧油泥和冷轧酸再生泥饼等的利用方式。通过加强外部成熟工艺的应用，为流程内部自循环困难的高炉干法灰、烧结电场灰和炼钢二次灰等设计出有效的利用方式。

为了追求固体废弃物资源化利用逆向供应链实施效果最大化，建议引入固体废弃物的全生命周期管理，有利于资源化利用逆向供应链组合方案有计划地逐步推进，促进钢铁流程绿色化与固体废弃物排放的减量化。

致谢

感谢冶金工业规划研究院李新创院长对作者学术观点的建立和完善给予的启发和帮助。

参考文献

[1] 殷瑞钰. 钢厂模式与工业生态链——钢铁工业的未来发展模式[J]. 钢铁, 2003, 38 (10): 1.

[2] 张福明. 面向未来的低碳绿色高炉炼铁技术发展方向[J]. 炼铁, 2016, 35 (1): 1.

[3] 李新创. "六位一体"促进钢铁绿色发展[J]. 冶金经济与管理, 2016, 8 (2): 4.

[4] 马军, 邹真勤. 国内外钢铁企业固体废弃物资源化利用及技术新进展[J]. 冶金经济与管理, 2006, 4 (2): 32.

[5] 罗晔, 吴瑾, 王超. 国内外钢铁企业固体废弃物资源化利用及技术新进展[J]. 中国冶金, 2017, 27 (10): 76.

[6] 孙书晶. 钢铁工业固体废弃物资源化途径探析[J]. 科技创新导报, 2017, 17 (2): 138.

[7] Froling M, Schwaderer F, Bartusch H. Integrated planning of transportation and recycling for multiple plants based on process simulation [J]. European Journal of Operational Research, 2010, 69 (2): 958.

[8] 冀巨海, 郭忠行. 钢铁企业绿色供应链管理绩效灰色关联评价[J]. 太原理工大学学报, 2012, 43 (2): 224.

[9] Giannetti B, Bonilla. An emergy-based evaluation of a reverse logistics network for steel recycling [J]. Journal of Cleaner Production, 2012, 37 (2): 48.

[10] 鲁春林, 霍佳震. 钢铁企业内废钢逆向问题的研究[J]. 钢铁研究, 2017, 35 (1): 1.

[11] 钟磊钢, 崔阳. 我国钢铁行业绿色逆向物流分析[J]. 冶金经济与管理, 2007, 44 (6): 44.

[12] 刘宏强,张福明,付建勋.首钢京唐钢铁公司绿色低碳钢铁生产流程解析[J].钢铁,2016,51(12):81.

[13] 裴元东,赵志星,安钢.首钢京唐烧结利用钢铁流程废弃物的研究与实践[C]//第十届中国钢铁年会.上海:中国金属学会,2016.

[14] 殷瑞钰.流程制造业与循环经济[J].再生资源与循环经济,2009,2(12):3.

[15] 王莉,曹盛,吉立鹏.炼钢厂含铁固体废弃物资源化回收利用[J].中国冶金,2018,28(3):73.

Reverse Supply Chain System of Solid Waste Resource Utilization in Iron and Steel Process

Liu Qingmei   Zhang Fuming   Li Haibo   Luo Zhenyong

Abstract: Guided by the 3R principle of "reducing-reusing-recycling" and aiming at realizing the reuse of the solid waste in the iron and steel process, the reverse supply chain system of the solid waste resource utilization in the iron and steel process was designed. Combining with the composition characteristics of various solid wastes, valuable elements were set as new resources and a reverse supply chain combination scheme for the solid waste resource utilization in the iron and steel process was constructed, which configured the cascade implementation path of the process self-circulation, external utilization and product upgrading. In order to achieve the dynamic order, collaborative continuity and intensive efficiency of the operation process of the reverse supply chain, it was suggested to introduce the green life cycle management system for solid wastes, and run through the preliminary planning and implementation of the reverse supply chain combination scheme for the solid waste resource utilization, which will be conductive to reducing the cost of the solid waste environment management and achieving a better green effect on the whole iron and steel process.

Keywords: iron and steel process; solid waste; reverse supply chain; reducing-reusing-recycling principle; life cycle

首钢京唐烧结厂降低生产工序能耗的实践

摘 要：从首钢京唐烧结厂设计实际出发，详细介绍了京唐烧结厂节能的具体措施；采取厚料层烧结，控制原、燃料和白云石粒度，稳定烧结生产，推行低温烧结工艺，生石灰强化烧结，堵漏风，控制点火温度等一系列降低固体燃耗、电耗、煤气消耗以及水耗的措施，同时介绍了烧结余热回收措施。自2009年5月投入生产以来，生产逐步走上正轨，烧结料层厚度达780mm、固体燃耗由投产初的50.11kg/t降低到当前的42.65kg/t，取得了较好的经济效益。

关键词：烧结；能耗；固体燃耗；余热；节水

1 引言

为了满足高炉对烧结矿产量和质量的要求，钢铁厂烧结机逐渐向大型化发展。首钢京唐钢铁联合有限公司烧结一期工程在曹妃甸新建两台烧结机规格500m^2的烧结厂，烧结机的设计以"高效节能"为原则，而烧结工序能耗主要包括生产用的固体燃耗、煤气燃耗、电耗、水、蒸气、压缩空气等消耗。其中，固体燃料消耗约占75%，电耗15%，煤气消耗约占6%。设计首钢京唐钢铁公司烧结厂，烧结机面积500m^2，设计过程中采取多项技术措施；两台烧结机分别于2009年5月和12月建成，投产以来，不断优化烧结工艺，烧结工序能耗逐月降低，取得了较好的效果。

2 降低固体燃耗措施

2.1 厚料层烧结

在主抽风机能力和原料条件不变的情况下，厚料层烧结的关键在于改善料层透气性，它要求料层具有较强的氧化气氛。在一定范围内，料层越厚，自动蓄热能力越强，越有利于节约燃料，同时厚料层烧结可增加低价铁氧化物氧化放热、减少高价氧化物分解吸热，大幅度降低燃料消耗，改善烧结矿质量[1]。京唐公司烧结机台车栏板设计高为800mm，料层厚度设计750mm、最大可达800mm；采用梭式布料、辊式布料器联合布料，该布料系统不会破坏混合料颗粒，有利于提高料层透气性。同时，采用粒度为12~20mm的冷烧结矿作为铺底料，铺底料厚为20~40mm。为配合厚料层烧结，采取了以下改善料层透气性的措施。

2.1.1 强化混合制粒

烧结料配完料后，经带式输送机运送至混合工序，设计了一次混合和二次混合，混合

本文作者：贺万才，张福明，李长兴，王代军。原刊于《中国冶金》，2012，22（1）：20-24。

机采用圆筒混合机,混合机安装的关键点是倾斜角度。混合的目的:一是实现混合料均匀,获得化学成分均一的混合料;二是对混合料加水润湿和制粒[2]。具体内容如下:

(1) 一次混合机安装倾角为2.5°,二次混合机安装倾角为1.5°。

(2) 在一、二次混合机筒体内安装带花纹的含油尼龙橡胶衬板,衬板上的花纹,形成一定比例的料磨料内衬,对混合料产生一定的摩擦力。混合机工作时靠这种摩擦力将烧结混合料带到一定的高度,使混合料在混合机内形成规则的滚动,对混合料造球十分有利,减少混合机粘料,以改善制粒。

(3) 在一、二次混合机内按合理加水曲线安装雾化水喷头,实现雾化水造球。

(4) 混合料混合时间为2.8min;制粒时间为5.0min,从而提高料层的透气性。

通过这4项措施的实施,第1台烧结机于2009年5月投入生产,对混合料中+3mm粒级进行测定,测定结果为+3mm粒级达75%左右。

2.1.2 蒸汽预热混合料

为了达到烧结必要的料温,降低过湿层阻力,设计了二次混合后蒸汽预热混合料的措施。

在二次混合机和烧结机前的混合料矿槽分两次用蒸汽预热混合料,尽可能提高料温到65℃左右。该措施的实施,投产后效果明显,极大地增强了料层透气性,促进烧结顺利进行。实践证明:料温的提高,一方面能减少过湿层厚度,改善透气性;另一方面其热量能代替部分燃料燃烧热,降低燃耗。通过逐步提高烧结料层厚度,烧结矿强度得到改善,固体燃耗逐步降低。

2.1.3 采用梭式布料、辊式铺料

布料作业是将混合料和铺底料均匀布在烧结机台车上,为获得透气性良好的烧结料,在设计烧结机布料时采取以下措施:

(1) 在混合料矿槽上部安装了梭式布料器,以保证烧结机宽度方向上均匀布料。

(2) 在混合料矿槽下部的泥辊安设九辊布料器,实现了混合料沿台车高度方向的合理偏析,使大颗粒混合料布到台车中下部,小颗粒混合料布到中上部。

(3) 为了使烧结机料面平整,还在泥辊横梁上设计了可调整高度和角度的刮料板。因此,在保证料面平整的同时,还可根据生产需要调整料层厚度,并适当提高台车两侧的料层,使料层断面略呈"锅底"状,以减少边缘效应和漏风率。

2.2 控制原、燃料和白云石粒度

2.2.1 合理控制原、燃料粒度

(1) 生产实践和研究表明,燃料最适宜的粒度为0.5~3mm,一般控制在3mm以下。为此,燃料破碎间布置电磁除铁器吸起碎焦中的铁杂物,同时进行燃料预筛分,筛除大焦块,保证进入破碎机的燃料粒度均匀且无铁杂物。精心操作,确保燃料0.5~3mm粒级在(85±2)%范围内,确保在焙烧过程中燃料分布均匀,燃烧完全,热效率提高,消耗量低。

(2) 原料粒度过大,不利于造球,熔化温度高,耗热量大。一般控制其上限,即+8mm粒级不大于10%。

2.2.2 控制白云石粒度

根据高炉需要,京唐公司烧结矿MgO含量为1.9%。为此,根据原料MgO含量的不

同，京唐烧结厂配加 3.5%~6.5% 的白云石。白云石粒度过粗，其分解热耗增大，在烧结过程中不易完全矿化。据生产经验一般控制 0~3mm 粒级含量大于 90% 为宜。

2.3 调整低价铁氧化物含量，配加含碳固体废弃物

（1）由于京唐所用原料主要为进口矿粉，熔化温度较高，液相形成所需要的热量也多。而磁铁矿因含有低价铁氧化物，在烧结过程中发生氧化反应，放出大量的热，能显著降低能耗。

（2）炼铁、焦化工序在生产过程中产生的高炉灰、焦化除尘灰等冶金工业废弃物含有较高的碳，将高炉灰、焦化除尘灰应用于烧结生产，在烧结过程燃烧放热，可替代部分固体燃料，从而达到降低固体燃料消耗的目的。因为高炉灰、焦化除尘灰粒度较细，控制配比在 1%~2%，既不影响透气性和烧结矿产质量，又可降低固体燃耗 4.5~5.5kg/t。

2.4 稳定烧结生产，降低返矿循环量

返矿主要来源于台车表层、两侧、底部等部位。返矿量的增加，主要是烧结过程波动造成的，其结果是烧结矿产量降低，燃耗上升。因此，稳定烧结生产是降低返矿量、降低燃耗的主要措施之一。

2.4.1 提高混匀料成分稳定率

烧结配料是将各种烧结料按配比和烧结机所需要的给料量，准确进行配料。配料系统采用自动质量配料法，各种原料均自行组成闭环定量调节，再通过总设定系统与逻辑控制系统，组成自动质量配料系统。其特点是设备运行平稳、可靠，配料精度高达 0.5%~1%，使烧结矿合格率、一级品率均有较大幅度提高，同时可减少烧结燃料耗量。配料室料仓底设计装有电子皮带秤，并定期对电秤进行校正或整改，标定给料圆盘的下料量，达到配料有效计量。

2.4.2 优化料仓给料设备

一条烧结生产线，配料室设计有 17 个料仓，根据不同原料性质采用了不同仓型和仓下给料形式。在生石灰料仓的锥段部分安装了流化装置，避免蓬仓、堵仓带来的下料不畅，实现了配料仓下稳定连续给料，保证配料的准确性，保证烧结矿质量指标，使之碱度稳定率为 99%、品位稳定率 100%、合格品率 100%、一级品率 92%。

2.4.3 优化燃料用量计算方法

传统的计算燃料用量方法极易导致生产中燃料波动，改进方法是考虑返矿残碳量，把混匀料和返矿分开计算燃料配比。计算公式为：

$$燃料用量 = 混匀料量 \times m\% + 返矿量 \times n\%$$

式中，$m\%$ 为混匀料的燃料配比；$n\%$ 为返矿的燃料配比。

方法改进后，烧结配碳稳定，燃料用量降低。同时，完善和应用自动配料技术，提高生产稳定性。

2.5 低温烧结工艺

低温烧结工艺是通过降低烧结温度，发展性能优良的黏结相来固结烧结料，其关键在

于控制烧结温度和气氛。在配矿结构相同的原料里,烧结矿FeO质量分数能表明烧结过程中温度水平高低和气氛。可见,推行低温烧结工艺主要是通过控制和降低FeO质量分数来控制烧结过程温度水平。因此,在保证烧结矿强度前提下尽可能降低烧结矿FeO质量分数。生产实践统计表明,在现有原料条件下,FeO质量分数每降低1%,固体燃耗下降1.35kg/t。对配矿结构相同的料堆,确定燃料用量后,加大FeO稳定率考核力度,提高烧结矿FeO控制水平,防止烧结过程热水平大起大落。而配矿结构不同的料堆,尽可能采取低FeO质量分数操作,降低固体燃耗。采取上述措施后,FeO稳定率提高,含量明显下降(见表1)。

表1 投产以来烧结矿FeO质量分数

时间	2009-05-08	2009-09-12	2010-01-04	2010-05-08
质量分数/%	8.90	8.34	7.86	7.25

2.6 生石灰强化烧结

生产实践表明:配加生石灰能强化烧结,降低燃耗。配加生石灰主要作用在于:能强化制粒效果,提高制粒增强小球强度,改善混合料的粒度组成,提高料层透气性,降低固体燃耗;减少碳酸钙分解吸热,而且生石灰消化放热可提高料温10℃左右;生石灰粒度细,比表面积大,极易形成低熔点物质,降低液相形成温度,大幅度降低燃耗[3]。根据碱度的不同,生石灰配比为5.5%~7.5%,生石灰加水系统设计两次加水,并采用蒸汽加热,使水温达到50℃左右。本措施提高了生石灰消化速度,使生石灰充分润湿消化,强化烧结的作用得到增强,混合料粒度组成得到改善,从而促进了烧结料层透气性的改善和固体燃耗的降低,降低燃耗约1.25kg/t。

固体燃耗占烧结工序能耗的75%,降低固体燃耗是烧结节能的关键。设计烧结厂时采取上述6项措施,自第1台烧结机投入运行以来,设备运转稳定,烧结生产逐步走上正轨,固体燃耗稳步降低。固体燃耗数据见表2。

表2 烧结矿强度指标和固体燃耗

时间	料层厚度/mm	转鼓指数/%	固体燃耗/kg·t^{-1}	筛分指数/%
2009年5~8月	700	76.7	50.11	7.48
2009年9~12月	740	79.4	47.33	7.11
2010年1~4月	770	80.8	44.56	6.89
2010年5~8月	780	81.1	42.65	6.70

3 降低电耗

主抽风机容量占烧结厂总装机容量的30%~50%,减少抽风系统的漏风率,增加通过料层的有效风量和降低烧结负压对节约电耗意义重大。

3.1 堵漏风

由于烧结料层越厚,阻力越大,风箱负压越高,漏风率也相应增加,这给堵漏风工作增加了难度。烧结机抽风系统漏风主要体现在：台车在高温下变形磨损，风箱密封装置磨损、弹性消退，机头机尾处的风箱隔板与台车底部间间隙增大，台车滑板与风箱滑板密封不严，相邻台车之间接触缝隙增大，抽风管道穿漏等。因此，有必要对烧结机滑道系统及机头、机尾密封板等部位进行优化设计，加强密封，改进台车、首尾风箱隔板、弹性滑道的结构；同时，加强对整个抽风机系统的维护检修，及时堵漏风，将漏风率降至最低程度。设计了烧结机集中润滑系统，健全润滑机制；要求生产者及时更换紧固台车挡板、箅条；安装箅条压辊和台车边箅条；台车端部加衬板；将台车挡板4块改为2块；箅条压钉改为螺栓固定；风箱内部涂抹耐磨耐高温材料；经常检查焊补抽风系统连接法兰和膨胀节等易漏风部位；利用烧结机检修和待开的时机，对诸如头、尾部密封盖板、水封漏斗、风箱上下滑道密封等部位进行检修，使漏风率大大降低。

3.2 降低烧结负压

吸取其他烧结工程卸灰漏斗的教训，卸灰漏斗因密封不严，在较大负压作用下，大烟道内积灰无法在正常生产时清放，导致大烟道内积灰严重时占据大烟道容积的一半以上，烟道负压升高，有效风量降低，电耗升高。大烟道下部设计接大容量灰斗，可贮存较多灰尘，减少烟道内积灰，灰斗下接双层卸灰阀。此措施能有效降低负压，从而降低电耗。

资料显示[4]：漏风率减少10%，可增产5%~6%，每吨烧结矿可减少电耗2kW·h，成品率提高1.5%~2.0%。日本新日铁大分厂2号烧结机采用降低漏风率措施后，漏风率降低了12.5%，电耗降低1.96kW·h/t，相当于每降低10%的漏风率，电耗降低1.56kW·h/t；梅山烧结厂将漏风率从71.14%降至42.99%后，电耗降低4.33kW·h/t，相当于降低10%的漏风率，电耗降低1.54kW·h/t。京唐烧结通过实施堵漏风和大烟道下部设计接大容量灰斗以降低负压，漏风率由投产初的50%降低到41%，主抽风系统电耗稳定在13kW·h/t，比设计值小1.5kW·h/t。

3.3 减少风机空转时间

根据生产实际情况合理控制主风门的开度，如停机10min以上，就及时关闭风门，很好地降低了抽风机功率，降低电耗；其次，由于每台烧结机都有2台功率11000kW的大型抽风机，电耗相当惊人，风机启动采用先进的变频软启动技术，风机可以做到随时启动。因此，对抽风机的停止和启动做出规定：如停机1h以上必须停抽风机。这一举措的实施，很好地控制了抽风机空转的时间，降低了烧结矿的吨矿电耗。当前，每台风机的功率在9700kW。

4 降低煤气消耗

4.1 严格控制点火温度和点火时间

点火的目的是补充烧结料表面热量的不足，点燃表面烧结料中的燃料，使表层烧结料烧结成块。点火温度的高低和点火时间的长短应根据各厂的具体原料条件和设备情况而定，达到点火的目的即可。无烟煤和焦炭的着火温度在700~1000℃，因此点火温度达到

1000℃即可，甚至更低就可以把燃料点着，满足点火的要求，同时节约了煤气消耗；又因料层厚度增加，与点火器距离缩短，火焰长度缩短，火焰高温部分较容易到达料面。根据实际情况，将点火温度控制在（1050±50）℃，点火时间在1min以内。

4.2 应用新型节能型点火器

采用矮炉膛线形点火技术，在台车横向上，通过烧嘴不同间距的布置达到在台车表面均匀点火的效果。为提高表层烧结矿质量，降低烧结矿燃耗，在点火炉前设置预热段，在点火炉后面设有较大的保温炉。点火温度可达到1100℃以上。该点火炉炉衬由耐火层和保温层组成，保温效果好，炉墙采用耐火浇注料预制块拼装而成，炉顶采用吊挂预制块结构。烧结机点火器炉衬采用耐火预制块拼装结构，施工维修十分方便。高温火焰带宽适中，温度均匀，高温点火时间和机速匹配良好，采用的烧嘴流股混合良好，火焰短，燃烧完全。

通过应用新型节能点火器，严格控制点火温度和点火时间；京唐烧结厂平均点火温度为1030℃，投产初煤气消耗约4.02m³/t（烧结矿）；通过几个月的运行调整，生产工艺流程越来越顺，煤气消耗降低到2.87m³/t（烧结矿），煤气消耗降低约15MJ/t（烧结矿）。

5 降低水耗

5.1 热水热媒采暖

京唐烧结厂建于唐山唐海县，属于采暖区域；配料室、烧结主厂房、主控楼、成品筛分间及大部分附属设施需要采暖。按以往设计经验一般考虑采用蒸汽作为热媒采暖，鉴于节能减排政策，本工程采用热水采暖。水作为热媒，热水相比于蒸气具有以下优点：

（1）热水供应系统的热能利用效率高，由于在热水供热系统中没有凝结水和蒸汽泄漏以及二次蒸发气的热损失，因而热效率比蒸汽供热系统高，按经验一般可节省燃料20%~40%。

（2）运用按质调节的方法进行调节，既节约热量，又能较好地满足卫生要求。

（3）热水蓄热能力高，由于系统中水量多，水的比热容大，因此在水力工况和热力工况短时间失调时不至于显著引起供热状况的波动。

（4）可以进行远距离输送，热损失较小，供热半径大。

（5）在热电厂供热情况下，充分利用低压抽气，提高热电厂的经济效益。

（6）热水采暖采用集中换热站供热，冷凝水采用回收装置集中回收；蒸汽采暖系统冷凝水则直接排放。京唐烧结厂总供热负荷约2700kW，按照经验值计算，热水采暖相对于蒸气采暖可节水4.5t/h，一个采暖季按120天计算，整个采暖季节水约12960t。

5.2 烧结节水措施

京唐烧结厂水系统设计时从节水角度出发，在设备选型时合理地控制了设备的水量和水压。工艺设备冷却水采用厂区净环水循环；冲洗地坪等采用工业水，排水回收处理后再进入管网；湿式除尘和配料室排水采用沉淀池技术处理后再供湿式除尘使用，多余部分溢流入排水系统回收。

6 余热回收

烧结系统的耗能约占在钢铁工艺总耗能的 10%～12%，而烧结机与冷却机废气带走的热约占烧结能耗的 40%～50%；因此，回收和利用这些余热极为重要，余热回收主要在烧结矿成品显热及环冷机的排气显热两个方面。

烧结生产时烧结矿从烧结机尾部落下，经单辊破碎后通过导料溜槽，再经板式给矿机给到环冷机台车上，在溜槽部分烧结矿温度高达 700～800℃，以辐射热形式向外散热，落到环冷机后其温度仍在 600℃ 以上，环冷机上设计有冷却密封罩，密封罩内通过鼓风机使冷却风强制穿过矿层，经烧结矿加热使冷却风温度升高到 300℃ 以上，这样的冷却风可利用使其显热来产生蒸汽[5]。当前，烧结余热利用多采取制蒸汽；马钢、济钢烧结厂已建设余热发电设施，经过考察，余热发电系统运行不够稳定，尚未达到预期设想。京唐公司又建有海水淡化，需要消耗大量蒸汽；综合衡量，最终决定采取余热产蒸汽方案，弥补厂区蒸汽供应不足。其流程如图 1 所示。

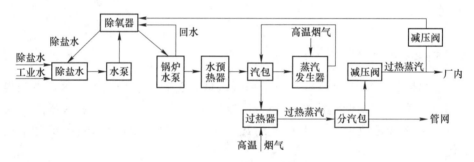

图 1 余热蒸汽发生系统流程

除盐水经过除氧器蒸汽除氧，产生热水；通过锅炉给水泵加压至预热器，利用环冷机上烧结矿产生的热废气，加热除盐水进入汽包。除盐水通过汽包下降管进入热管蒸汽发生器，在此处热废气横向冲刷热管受热侧，热管通过相变传热至上联箱来的饱和水，饱和水吸热变成汽水混合物由上联箱通过总上升管进入汽包，汽水分离后，饱和水通过下降管回至下联箱，再次受热蒸发，如此反复循环，将烟气热量传入水侧产生饱和蒸汽。烟气经过过热器、热管蒸汽发生器利用后，通过烟筒排入大气。饱和蒸汽从汽包出口管道进入过热器，产生 260℃ 过热蒸汽，经分气缸分配蒸汽，一部分蒸汽经减压阀减压至 0.8MPa 进入烧结生产主管道供烧结厂内使用；另一部分蒸汽设计压力约在 1.3MPa 并入京唐厂区综合管网。

7 节能效果

京唐烧结厂在设计时采取各种新工艺以及生产科学管理等一系列措施，使烧结工序能耗逐步下降，达到国内先进水平（见表3）。

表3 京唐烧结工序能耗对比　　　　　　　　　　（kg(标煤)/t）

厂名	京唐烧结	武钢股份烧结	宝钢股份烧结	鞍钢烧结	太钢烧结
工序能耗	50.61	58.12	55.66	56.2	54.7

8 结语

（1）京唐烧结厂自2009年5月投产，一年多来，该厂设备的平稳运行表明国内已可自主设计500m^2烧结机。通过各种措施的运用，目前固体燃耗约为42.65kg/t。

（2）新型节能型点火器的应用，点火温度控制在1030℃，从投产初煤气消耗约4.02m^3/t（烧结矿）降低到2.87m^3/t（烧结矿），煤气消耗降低约15MJ/t（烧结矿）。

（3）采取堵漏风和降低烧结负压的措施，烧结漏风率由40%~50%降低到32%，主抽风机负荷减小，主抽风系统电耗稳定在13kW·h/t，比设计值小1.5kW·h/t。

（4）热水作供暖热媒，一个采暖季节水约12960t；利用环冷机高温废气热量产过热蒸汽，经分气缸分配蒸汽，一部分蒸汽经减压阀减压至0.8MPa进入烧结生产主管道供烧结厂内使用；另一部分并入京唐厂区综合管网。

（5）京唐烧结工序能耗为50.61kg（标煤)/t，低于国内其他几大钢铁集团烧结工序能耗。

参考文献

[1] 王悦祥. 烧结矿与球团矿生产 [M]. 北京：冶金工业出版社，2006.
[2] 夏铁玉，李政伟，颜庆双. 鞍钢三烧车间600mm厚料层烧结生产实践 [J]. 烧结球团，2005，30 (1)：45.
[3] 魏建新. 武钢降低炼铁系统能源消耗的实践 [J]. 可持续发展，2007 (5)：19.
[4] 王中林. 降低烧结工序能耗的研讨 [J]. 节能与环保，2004 (11)：42.
[5] 卓阿诚. 韶钢六号烧结机余热回收利用的设计特点及生产实践 [J]. 广东科技，2008 (14)：216.

The Practice of Reducing Energy Consumption of Production Processes of Shougang Jingtang Sinter

He Wancai　Zhang Fuming　Li Changxing　Wang Daijun

Abstract: The paper sets out actual design of Shougang Jingtang sinter plant. The specific measures of saving energy are taken, consisting of decreasing solid fuel consumption, electricity consumption, gas consumption and water consumption such as adopting deep bed sinter, controlling the particle size of raw material, fuel and dolomite, stabilizing sintering process, promoting low temperature sintering process, intensifying sintering by calcium oxide, preventing air leakage, controlling ignition temperature, introducing recycle the heat of sintering at the same time. Since commissioning in May 2009 sinter bed thickness has reached about 780 mm. The solid fuel consumption is lowered from 50.11 kg/t to 42.65kg/t and the obvious economical benefit is obtained.

Keywords: sinter; consumption; solid burn consume; residual heat energy; saving water

首钢冶金石灰技术创新与工程实践

摘　要：本文分析了当前我国石灰生产技术发展现状，介绍了石灰生产工艺的产能规模和主要技术装备。结合首钢冶金石灰技术发展创新历程，系统阐述了套筒窑工艺原理、工艺流程、技术特点和发展趋势，论述了首钢套筒窑的工程设计特点、采用的新技术以及生产应用实践，全面总结分析了当代冶金石灰生产工艺工程设计与工程创新成就，对套筒窑工艺技术研究与设计开发的技术前景和技术发展方向进行了研究探讨。
关键词：活性石灰；套筒窑；工程设计；节能减排

1　我国石灰技术发展现状

进入21世纪以来，我国经济持续高速稳定发展，2011年全国工业用石灰产量超过了1.2亿吨[1]，冶金行业石灰需求量占总量70%以上。这带动了石灰行业的发展，石灰生产工艺技术装备水平得到大幅度提升[2]。目前，我国自主集成开发的回转窑、双膛窑、套筒窑、梁式窑、新型气烧竖窑、新型焦炭竖窑、中石立窑和马式窑得到了广泛推广。

石灰是钢铁生产过程不可或缺的重要辅助原料，对铁水预处理、炼钢、钢水精炼等关键工序影响重大。冶金石灰是炼钢生产主要的造渣剂，其产品质量直接影响到钢水质量。在氧气顶吹转炉炼钢生产中使用活性石灰可以改善氧气顶吹转炉炼钢生产条件，加快成渣速度，提高脱磷、脱硫效率，稳定操作，缩短冶炼时间，提高炉龄，从而提高炼钢生产的经济效益，并且为精品钢材的生产提供优质钢水。活性石灰用于烧结生产，则可以改善烧结矿质量，强化烧结过程，提高生产效率。随着钢铁清洁生产技术进步，石灰还被广泛地用于钢铁厂烟气脱硫、脱硝等环境污染治理。因此，对冶金用石灰的品质性能提出了新的更高要求。在推广先进活性石灰生产工艺时，应做到因地制宜，根据产能需求、原燃料供应条件、石灰产品用户的实际需求等因素合理选用。

套筒窑技术是国内外先进的石灰生产技术之一，其技术优势明显，发展前景广阔。20世纪90年代初，首钢在国内率先引进德国贝肯巴赫套筒窑技术，二十多年来，在消化吸收原技术的基础上创造出具有首钢特色的套筒窑新技术，产能包括300t/d、500t/d、600t/d。其中，600t/d活性套筒窑技术是在长期技术研究基础上自主开发出来的，是国内外最大产能的套筒窑技术，在工艺系统构成、热工系统煅烧原理及工艺特点、窑本体结构特征、内衬结构、内套筒、燃烧系统设备、换热器技术、布料和出灰装置等方面进行了设计研究和技术创新。首钢套筒窑技术相继应用于首钢京唐、迁钢、首秦、长钢、江苏淮钢、山西晋钢等10余项工程，并在生产实践中取得了成功。

本文作者：周宏，张福明。原刊于《第九届中国钢铁年会论文集》，2013：852-858。

2 首钢套筒窑工艺技术特点

2.1 工艺流程和系统组成

2.1.1 工艺流程

套筒窑石灰生产工艺流程包括原料储备、上料、装料、石灰煅烧、成品处理等工艺单元，具有典型的竖炉生产特征。石灰石在储料跨存储，采用抓斗吊将石灰石装入料仓，由电振给料机卸料，石灰石通过溜槽，经振动筛筛分后加入称量斗；需向窑内装料时，称量斗液压闸门开启，石灰石卸入料斗中，用卷扬提升装置将料斗提升至窑顶，加入窑顶中间料仓，在料钟关闭的条件下，打开中间料仓闸门将石灰石卸入旋转布料器，旋转布料器旋转到某一特定位置后，在中间料仓闸门关闭的条件下开启料钟将料卸入窑内；石灰石料通过在窑内的预热、煅烧、冷却过程变成石灰，再通过液压出灰机将石灰卸入窑下石灰仓，由窑下电振给料机将石灰卸到窑下输灰胶带机上。

2.1.2 系统组成

套筒窑主要由原料储存（原料跨或原料场）、原料准备、窑本体、卷扬机房、液压站、主控楼、风机房、废气风机房、成品石灰储运系统、除尘系统组成。根据项目实际情况设置高低压配电室、煤气加压站、给水泵站、空压机等设施。窑本体系统由窑体、上料装置、出料装置、燃烧室、换热器、喷射器、耐火材料内衬以及风机系统等构成，单座套筒窑系统如图1所示。

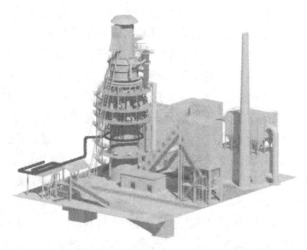

图1 套筒窑三维仿真工艺流程

2.2 工艺技术特征

首钢设计开发的套筒窑技术具有合理的内衬结构和气流分配方式，煅烧活性石灰具有显著的技术优势。其主要技术特征是：(1) 采用环形结构，热气流分布均匀，边缘效应减小，有利于石灰煅烧，套筒窑窑体内部设计结构如图2所示；(2) 采用逆流+并流的先进煅烧工艺,石灰在并流区域烧成,石灰成品品质高、质量稳定,活性度大于360mL;(3)采用废气预热驱动空气、冷却内套筒后的热空气作为燃烧一次风,降低燃料消耗,燃料消耗为3887~4096kJ/kg;(4)采用全负压操作,有效防止粉尘外溢,特别适合于现代化钢铁厂环境保护和清洁生产的要求;(5)操作简单,可实现全自动化控制,采用精确的燃料和助燃风分配技术,提高燃烧效率,降低能耗,有效改善热量分配、温度分布和煅烧效果;(6)采用先进的分料技术,物料在套筒窑内多次重新分料,可以有效减小物料粒度不均对煅烧效果的影响,并降低窑内气体阻力;(7)耐火材料内衬设计结构合理,使用寿命长,大修周期在5年以上,年作业率高达96%。

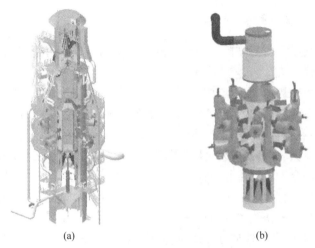

图2 套筒窑窑体内部设计结构
(a)三维设计结构;(b)套筒窑燃烧器设置

2.3 窑体设计结构

套筒窑窑体由内、外套筒组成。外筒为整体结构,内筒又分上内套筒、下内套筒两个独立部分。外筒是窑体的主要承载结构,由钢板构成并衬以耐火材料,与内套筒同心布置,形成一个环形空间,石灰石就在该环形区域内煅烧。内套筒为双层结构,夹层内通入空气冷却,防止其高温变形。筒体内外两侧砌有耐火砖。部分高温废气可通过上部内套筒输出用来预热驱动空气;热气流通过下内筒内部形成循环气流,改变窑内下部热气流方向,产生并流煅烧带。套筒窑设有上下两层燃烧室,燃烧室通过耐火材料砌筑的拱桥与内套筒相连。不同规格套筒窑窑体技术参数见表1。

表1 不同规格套筒窑窑体技术参数

项 目	技 术 参 数		
产能规格/t·d^{-1}	300	500	600
窑体总高/m	49.08	49.8	50.8
窑体有效高度/m	22.5	24.3	24.3
窑壳外径/m	6.7	8.0	9.0

续表1

项　目	技　术　参　数		
窑壳内径/m	5.6	6.9	7.9
内套筒外径/m	2.7	3.8	4.8
火桥跨度/m	1.55	1.55	1.55
上烧嘴数量/个	5	6	7
下烧嘴数量/个	5	6	7

2.4　装料、煅烧及出灰设备

上料系统由称量斗及密封闸门、单斗提升机、中间料仓及密封闸板、旋转布料器、料钟及料位检测装置等组成。

窑体煅烧系统由换热器（使窑内废气与驱动空气实现热交换）、燃烧系统（含上燃烧器、下燃烧器）、喷射器（用驱动空气将窑内部分气体带出形成再循环气流，产生窑内并流区域）、上内套筒和下内套筒等设备组成。

出灰系统将经预热、煅烧和冷却生成的石灰，在冷却带底部由抽屉式出灰机直接卸入窑下部灰仓，然后经仓下振动给料机排出。

内衬结构根据内外套筒设计要求[3]，结合热工系统要求，采用合理的耐火材料内衬设计结构；火桥底部拱桥镁铝尖晶石砖采用3层结构；内套筒和窑壳采用托砖圈技术；采用低温空烘窑-带料烘窑二步烘窑方法或带料烘开窑技术等。

2.5　风机与除尘系统

套筒窑风机系统主要由废气风机、驱动风机、内套筒冷却风机组成。废气风机采用高压风机，作用是将窑内废气抽出，使窑保持负压；驱动风机采用罗茨风机，向喷射器供给驱动空气；内套筒冷却风机采用离心风机，向内套筒供应冷却空气；套筒窑除尘系统主要由高温布袋除尘器、除尘风机组成。

2.6　燃烧系统设计创新

套筒窑可使用多种燃料，如天然气、焦炉煤气、转炉煤气、高焦混合煤气等。燃烧过程通过烧嘴在燃烧室内进行。燃烧室一般设置在窑体中部窑皮外侧，燃烧室分为上、下两层，每层燃烧室的数目因套筒窑产能不同而不同。同一层燃烧室均匀布置，上、下两层交错布置。每个燃烧室与下内套筒之间均由耐火砖砌筑而成的拱桥相连，燃烧产生的高温烟气通过拱桥下的空间进入石灰石料层。燃烧器技术实现完全国产化，采用CFD数值仿真计算、三维设计建模等数字化设计方法优化了燃烧器设计结构，使煤气气流速度场、温度场分布更为均匀合理，确保燃烧器具有广泛的适应性，可以根据设计要求调整燃烧器工况适用性能。

2.7　自动化控制系统设计创新

系统控制采用以PLC为核心的计算机控制系统。系统构成为工程师站、操作站、PLC

系统三个部分。网络采用工业以太网和设备网，通过网络进行数据通信；通过人机操作界面（HMI）完成工艺流程动态画面显示、传动系统运行状态显示、工艺参数设定、操作方式的选择、生产报表统计、打印，以及故障显示等功能。其主要特点是：操作方式具有现场、手动和自动三种方式；上料、布料和出灰系统可全自动进行，参数设定简单而方便；采用"以产量为目标，确定出灰速度，调整窑内温度，控制上料批次"的控制理念；采用"小闭环，大连锁"的控制方针，确保窑况稳定安全运行；操作和报警信息自动记录存储；采用先进而实用的数据报表系统。

3 工程创新与应用实践

3.1 首钢京唐套筒窑工程

首钢京唐钢铁厂是按照循环经济理念设计建造的新一代可循环钢铁厂，设计了基于高效率、低成本洁净钢生产的新一代钢铁工艺流程。为满足"全三脱"洁净钢生产要求，设计建造了4座500m³活性石灰套筒窑，用于高品质、高性能活性石灰和轻烧白云石的生产。在工程设计中，本着自主集成、优化设计的理念，在首钢已有套筒窑技术的基础上进行再次创新，通过对其热工系统的改进优化，采用全自动化控制系统，使京唐套筒窑可适应较小粒度的石灰石原料。一方面可以解决原料供应问题，扩大资源的可利用率；另一方面可使产品粒度适当减少，更适用于300t大型转炉"全三脱"冶炼的需要，形成一整套具有首钢京唐技术特色的套筒窑新技术。

为满足套筒窑产品用户对产品粒度、成品质量的多样性要求，在工程设计中，独创了全新的石灰成品处理系统，套筒窑的成品石灰和轻烧白云石通过这套深加工系统可以高效地完成筛分、破碎、制粉、储存、运输等复杂的深加工过程，满足烧结、铁水"全三脱"预处理、转炉冶炼、钢水精炼等各工序对石灰成品品质、性能和粒度的要求，图3所示为首钢京唐套筒窑工程三维仿真设计模型和建成后的工程实景。

(a)   (b)

图3 首钢京唐套筒窑工程全景

(a) 三维仿真设计模型；(b) 建成后的套筒窑

工程投产一年的连续生产实践表明，套筒窑窑顶废气温度低，出灰温度容易控制，燃料消耗为单位石灰3953kJ/kg，石灰活性度月平均达到381mL（采用高钙石灰石时为417mL，4N Cl，5min）。首钢京唐500t/d套筒窑工程设计和新技术集成是合理的，主要技术经济指标对比见表2。

表 2　京唐主要技术经济指标比较

项　目	设计值	2012 年生产平均
石灰石耗量/t·t^{-1}	1.80	1.65
燃料消耗/kJ·kg^{-1}石灰	3971	3953
电耗/kW·h·t^{-1}石灰	54.00	51.85
活性度/mL(4mol/L HCl)	≥350	417（高钙灰） 381（普通灰）

3.2　首钢迁钢套筒窑工程

首钢迁钢共建设 3 座套筒窑，为转炉冶炼、LF 精炼和 KR 铁水脱硫预处理工序提供高品质活性石灰。1 号、2 号套筒窑日产 500t 活性石灰，为迁钢一炼钢厂供应石灰产品；3 号套筒窑日产 600t 活性石灰，为迁钢二炼钢厂供应石灰产品。同时为了满足二炼钢厂 KR 铁水脱硫对石灰粒度和混料的要求，设计了相应的脱硫剂制粉工艺系统。

1 号套筒窑于 2010 年 5 月进行了扩容改造和自动化控制系统升级改造，适当增加风机风量，增加换热器能力，并用于高 CaO 含量的活性石灰生产，为实现 210t 转炉少渣冶炼、降低石灰消耗创造了有利条件；2 号套筒窑目前用于生产轻烧白云石。

2009 年，为满足二炼钢厂石灰供应，设计建造了 3 号套筒窑。采用自主研发的 600t/d 套筒窑技术，填补国内 600t/d 国产化套筒窑技术的空白，使首钢套筒窑工艺的产能规格实现系列化。设计开发了全国产化的 600t/d 套筒窑窑体结构和内衬结构，集成采用一系列先进技术，如并流煅烧工艺、旋转布料技术、冷却气热量回输和废气换热技术、负压操作技术、托板出灰技术、带料烘开窑技术、增加托砖圈结构、增加窑壳冷却梁技术、优化拱桥耐火材料结构及燃烧室结构、合理确定环形空间、降低附壁效应、优化竖向工艺布置等。该技术的研发成功，实现了 600t/d 套筒窑的完全自主设计，达到了建设投资省、适应能力强、生产稳定顺行的目标，整体技术装备达到国际先进水平。图 4 所示为迁钢 600t/d 套筒窑三维仿真设计模型和工程实景。

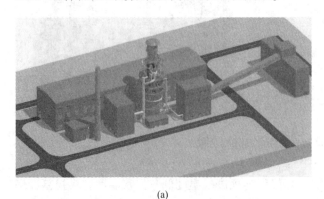
(a)

(b)

图 4　迁钢 600t/d 套筒窑工程全景
(a) 三维仿真设计模型；(b) 建成后的套筒窑

首钢迁钢600t/d套筒窑，2009年12月16日建成投产。投产2周后实现连续稳定运行，连续24天达到日产600t设计能力，最高日产可达660t，各项经济技术指标均达到或超过设计水平。2010年石灰产品平均活性度高达380mL（4mol/L HCl，5min）。2011年开始，采用高CaO石灰石原料生产活性石灰，石灰活性度均高于380mL。转炉煤气热值7106kJ/N·m³，600t/d套筒窑实际生产过程中转炉煤气平均耗量560m³/h，实际热耗为3980kJ/kg石灰，低于设计指标。首钢迁钢600t/d套筒窑主要技术指标见表3。

表3 迁钢600t/d套筒窑主要技术指标

项 目	设计指标	2012年平均生产指标
石灰石耗量/t·t^{-1}	1.8	1.7
燃料消耗/kJ·kg^{-1}石灰	4096	3980
本体电耗/kW·h·t^{-1}石灰	28.0	27.0
活性度/mL(4mol/L HCl)	≥350	398

4 石灰生产工艺技术展望

4.1 套筒窑技术的持续创新

我国石灰产业发展迅猛，造成优质石灰石资源日渐不足，在生产工艺选择时也必须考虑提高石灰石的利用率，提高工艺对原料的适应性，减少资源消耗，降低生产成本。套筒窑石灰生产工艺具有诸多技术优势，已成为主流技术发展模式。但套筒窑的单体产能与回转窑相比仍不具备优势，同时由于工艺固有的特点，对原料条件的要求相对苛刻。

未来套筒窑技术一方面要加强技术研究与工程创新，设计开发更大产能的特大型套筒窑，实现工艺技术装备的大型化、高效化；另一方面要提高原料适应性，通过改善原料结构、粒级匹配、布料控制、窑体结构优化等综合技术措施，实现资源高效利用，构建高效率、低成本的生产工艺流程。与此同时，还应适用多种燃料条件，开发研究利用低热值高炉煤气和蓄热式燃烧的新型燃烧工艺和燃烧器，降低高热值煤气的消耗和燃料消耗，有效降低燃烧过程CO_2排放。

4.2 构建基于循环经济的石灰生产工艺

石灰生产是利用自然资源进行矿物加工的工艺过程，石灰石煅烧过程产生大量的CO_2，增加了温室性气体的排放。目前，国内外石灰生产过程产生的CO_2尚未得到有效的收集和脱除，这是未来石灰生产工艺需要研究解决的课题。另外石灰生产过程产生的粉尘尚未进行分类处理，大多和钢铁厂的其他粉尘混合用于烧结，二次资源尚未得到高效化综合利用。

应开发研究石灰生产工艺CO_2减排技术，设计研发生产副产品工艺技术。如将回收的CO_2作为化工产品，用于食品加工、化工等行业；将回收的CO_2气体作为惰性气体用于钢铁厂等。石灰粉尘和筛下细粉应实现分类处理，提高二次资源的利用效率，实现资源减量化和再利用。进一步研究套筒窑余热高效利用技术，实现低品质余热的高效化利用，进一步降低燃料消耗，减少 CO_2 排放。开发研究适用于套筒窑工艺的助燃空气预热、煤气高效蓄热式燃烧器等，实现生产高效、节能减排、低碳生产、循环经济。

5 结论

(1) 首钢套筒窑石灰生产工艺技术研究与创新历经二十多年，在技术研究、工艺优化、工程设计、系统集成、生产应用等方面取得显著成就。基于自主集成创新首次设计开发了国内最大的 600t/d 套筒窑，系统地解决了窑体结构稳定性、燃烧和传热均匀性、工艺与设备可靠性等重大关键技术难题，整体工艺技术装备达到国际先进水平。

(2) 采用三维仿真设计、CFD 数值模拟计算、试验研究、设计结构优化等现代化系统研究方法，在已有技术基础上，集成创新、协同优化，设计开发了燃烧器、预热器等关键设备，并实现大型化工程应用。

(3) 生产实践证实，自主设计开发的大型套筒窑工艺取得了优异的生产实绩，实现了石灰生产高效、低耗、优质、清洁。特别是石灰成品的质量、性能满足了洁净钢生产的要求，为洁净钢生产、转炉少渣冶炼创造了有利条件。

(4) 冶金石灰是钢铁制造过程不可或缺的重要辅助原料，是实现高效率、低成本洁净钢生产的重要支撑条件。在现有技术的基础上进一步实现套筒窑大型化、提高原料适用性、降低能源消耗、减少 CO_2 排放、发展循环经济将是未来套筒窑技术的主要发展趋势。

参考文献

[1] 中国石灰协会技术专家组. 2012 年中国石灰窑技术发展报告 [C] //中国石灰协会编. 2012 年中国石灰工业技术交流与合作大会论文汇编. 浙江嘉兴，2012：1-13.

[2] 周宏，等. 关于活性石灰生产工艺装备选择的探讨 [J]. 石灰，2006，83 (4)：29-37.

[3] 张德国，等. 关于套筒窑砌筑设计的几点改进 [J]. 耐火材料，第五届国际耐火材料学术会议论文集，2007，41 (增刊275)：342-346.

Technology Innovation and Engineering Practice of Metallurgical Lime in Shougang

Zhou Hong   Zhang Fuming

Abstract: This paper analyzes the current situation of lime production technology in China, and introduces the lime production capacity and the main technical equipment. Combined with Shougang Metallurgical lime technology development and innovation process, system elaborates the principle of sleeve kiln process, process flow, technical characteristics and trends, discusses the Shougang sleeve kiln engineering design features, the new technologies adopted and production application practice, comprehensively summarizes and analyzes the contemporary metallurgical lime production process engineering and achievements of engineering innovative , sleeve kiln technology research and design development prospects and technological development direction have been discussed.

Keywords: active lime; annular shaft kiln; engineering design; energy conservation and emission reduction

现代钢铁冶金工程设计方法研究

摘　要： 本文论述了钢铁材料在经济社会发展中的重要作用和地位，分析了我国钢铁工业的发展历程与现状，论述了科技进步对产业发展的推动作用和重要意义，指出了我国钢铁工业未来发展方向。本文回顾了我国钢铁冶金工程设计的发展演变历程，分析了工程方法论与钢铁冶金工程设计方法的关系以及钢铁冶金工程设计的地位和作用，阐述了钢铁制造流程的物理本质和现代钢铁制造流程的"三大功能"。从钢铁制造流程的物理本质、特征以及动态运行的特征要素、设计方法的路径、动态-有序运行过程中的动态耦合等方面，对钢铁冶金工程设计的问题进行识别与定义。阐述了现代钢铁冶金工程的概念设计、顶层设计、动态-精准设计的方法体系，以及现代钢铁制造流程的系统集成与结构优化。

关键词： 钢铁冶金；流程制造；结构优化；工程设计；设计方法

1　钢铁工业的发展历程与现状

1.1　钢铁材料的地位与作用

从全球经济、社会发展和我国现代化进程的角度分析，在可以预见时间范围内，钢铁仍然是世界上重要的基础材料和结构材料，也是世界上消费量最大的功能材料。钢铁由于资源丰富、成本相对低廉、材料性能优越、易于加工且便于循环利用，因此，钢铁作为一种重要的结构材料和功能材料的地位不会发生重大变化。钢铁仍是"必选材料"，同时也是可循环利用的材料，是经济社会发展过程中必需的基础性材料[1,2]。

1.2　中国钢铁工业的发展进程与方向

从历史发展来看，我国的工业化进程落后于西方工业发达国家。我国最早的近代新式钢铁企业"汉冶萍公司"于1890年在汉阳兴建，1893年投产，建厂时间与日本八幡制铁所属于同一时期。由于清政府的腐败和经济发展落后，我国钢铁工业长期落后。1949年，中华人民共和国成立时，当年钢产量仅为15.8万吨。新中国成立后，国家把优先发展重工业作为经济建设的重点，而发展钢铁工业则是重中之重。发展钢铁工业必须建设钢铁厂，而建钢铁厂首先要开展钢铁厂设计。

新中国成立至20世纪80年代，是我国钢铁工业奠定基础的阶段[3]，新中国成立后经济建设以发展重工业为中心，1959年我国年钢产量达到1387万吨。1958年全国性的大炼钢铁开始，钢铁工业建设成为群众运动。"文化大革命"期间，国民经济濒临崩溃的边缘。由于当时经济工作的口号是"以钢为纲"，钢铁产量的趋势是增长的，钢铁生产规模在不断扩大。新中国成立后经济建设的主导思想是全面学习苏联，大量引进苏联工程技术和管理技术。旧中国没有中国自己的钢铁工业技术。通过引进苏联的钢铁工业技术在我国钢铁厂的建设过程中，形成了中国钢铁工业的技术队伍，1953年建立了中国钢铁工业的设计院。"文革"期间新增的许多钢铁项目是我国科技人员独立设计完成的。可以认为，在学习苏联引进技术的基础上，我国钢铁工业设计力量已基本形成。

20世纪70~80年代，我国改革开放以后，开始宝钢工程的建设，由日本新日铁公司主要承担项目的规划设计和主要设备供货，是项目的技术总负责方。宝钢钢铁联合项目的全面引进，对推进我国钢铁工业的发展起了重大作用。此前的技术引进以单项技术和部分专业技术引进为主，设备引进以国内不能制造的部分为主，基本以引进硬件为主，对软件引进重视不够。宝钢工程的全面引进，包括硬件和软件，为我国学习国外钢铁工程的设计提供了难得的机遇。宝钢钢铁联合企业采用的技术属于国际20世纪70~80年代水平，比以前引进的苏联技术要领先约20年，大大提高了中国钢铁工业的总体水平。

20世纪80年代以后，我国地方钢铁企业取得了很大发展。不仅生产规模扩大，在技术方面已赶上或接近重点钢铁企业水平。应当肯定，宝钢引进技术的消化吸收在地方钢铁企业发展中发挥了重大作用。

长期以来，钢铁技术的引进均以专项技术引进方式进行。宝钢工程的建设，使我国钢铁界在专项技术消化吸收基础上逐渐形成对当代钢铁工业技术的整体理解。钢铁工业属于流程工业，对流程工业认识的不断加深，使钢铁界对当代钢铁工业技术提升到一个新的高度。

经过改革开放，特别是20世纪90年代的技术结构调整，新世纪以来，在市场需求拉动和大量投资拉动的双重作用下，我国粗钢产量快速增长，2015年我国粗钢产量达到8.04亿吨，约占世界粗钢产量的50%。当前，我国钢铁产能过剩，在市场、资源、环境的多重压力下，我国钢铁工业正面临结构调整、压缩过剩产能的严峻挑战和考验[4]。

20世纪90年代是我国钢铁工业的崛起过程，通过对连铸、高炉喷煤、高炉长寿、连续轧制、转炉溅渣护炉和综合节能等6项关键共性技术的开发、集成和推广，促进了我国钢铁制造流程结构的优化，实现了节能降耗，提高了生产效率，为我国钢铁工业快速发展奠定了基础。由此可以看出科技进步对我国钢铁工业发展的推动作用与重要意义[5,6]。

新世纪以来，首钢京唐钢铁厂的设计和建设基于冶金工程技术基础科学、技术科学和工程科学，运用现代钢铁冶金工程设计理念、理论和方法进行工程决策、规划、设计、建设、运行等，构建了新一代可循环钢铁制造流程，引领了我国钢铁工业科技进步的发展方向。

回顾我国钢铁工业的发展历程，可以更加深刻地认识产业发展的客观规律，从中感悟出经济-社会-技术-工程非线性耦合的交互作用和重要影响。20世纪钢铁工业快速发展的启示是：经济增长是钢铁工业规模扩大的拉动力，科技进步是钢铁工业发展的推动力，资源、能源和环境约束则是钢铁工业发展的限制力，市场-价格-质量-生态则是对钢铁企业的筛选力。

经过几十年的努力，我国钢铁工业的设计队伍正在成长为与世界钢铁强国要求相匹配的技术力量。能够设计现代化的联合钢铁企业及各种类型的钢铁厂，提供国际一流水平的钢铁产品，设计水平已进入国际一流。总体来看，我国钢铁工业未来发展必须依靠技术创新提高技术水平。

我国钢铁工业的科学、健康、可持续发展，必须在这几种力量的平衡协调中寻求发展——走钢铁流程绿色化和智能化的新型工业化道路[7]。我国钢铁工业发展的主线应是动态地适应市场发展趋势，继续以调整结构为主要途径，以增强市场竞争力和注重可持续发展作为总体目标。我国钢铁工业应建立以内需为主的结构。从钢铁企业未来社会、经济角色来看，钢铁制造流程应具有三个主要功能：钢铁产品制造功能、能源高效转换功能、大宗废弃物的消纳-处理-再资源化功能，如图1所示。

图1 钢铁制造流程功能的演变

2 钢铁冶金工程设计方法发展历程与演进

2.1 传统钢铁冶金工程设计方法

冶金工程设计是运用冶金工程技术基础科学、技术科学、工程科学的研究成果进行集成与应用，并实现工程化的一门综合性学科[8]。制造业主要分为流程制造业与装备制造业两大类，钢铁制造属于典型的流程制造业，而服务于钢铁制造流程的工艺装备技术则属于装备制造业。就制造业而言，产品设计主要面向的是消费者用户，工程设计主要面向生产产品的企业用户。因此，冶金工程设计同时具有流程制造业与装备制造业的双重属性。

我国冶金工程设计理论体系在20世纪50年代由前苏联引入，长期以来基本沿用前苏联的设计方法（经验型设计方法）。20世纪80年代以后，随着宝钢工程的设计建设，我国冶金工程设计又相继引入了日本和欧洲的设计方法（半经验-半理论型设计方法），但仍属传统的"静态-分割"设计方法，即静态的"半经验-半理论"的设计方法。

传统的"静态-分割"设计方法，主要存在以下问题：在工程理念方面，传统钢铁冶金工程设计方法遵循的是"征服自然"工程理念；在工程思维方面，传统钢铁冶金工程设计方法思维模式是还原论思维模式；在工程系统观及工程系统分析方法方面，传统钢铁冶金工程设计方法基本上没有形成现代工程系统观及工程系统分析方法，或者说是以模糊整体论与机械还原论为基础的分析方法。

传统钢铁冶金工程设计方法还存在着内容的缺失：一是采用基础科学（解决原子、分子尺度上的问题）和技术科学（解决工序、装置、场域尺度上的问题）的思维方式来解决工程科学（解决制造流程整体尺度、层次和流程中工序、装置之间关系的衔接、匹配、优化问题）问题，使得建设项目在工程设计的思维方式上存在着先天不足；二是采用从传统设计方法发展到现代设计方法的产品设计方法来替代冶金工程设计方法，主要是针对钢铁制造流程装备技术设计方法的研究，而钢铁冶金工程设计具有流程制造业与制品制造业两重性，忽略了对钢铁制造流程工艺流程设计方法的研究。从系统观、整体观方面看，传统冶金工程设计方法强调了子系统（工序）的设计，忽视了系统-子系统、子系统-子系统之间的动态运行关系的设计。

2.2 现代钢铁冶金工程设计方法的形成

20 世纪 80 年代以后，我国钢铁工业的迅猛发展促进了科技工作者对冶金工程设计理论及方法的深入研究，新的研究成果不断涌现，我国已初步建立起现代钢铁冶金工程设计的知识体系并且不断完善。进入 21 世纪以来，《工程哲学》（第 1 版）[9]、《工程哲学》（第 2 版）[10]、《工程演化论》[11]、《冶金流程工程学》（第 1 版）[12]、《冶金流程工程学》（第 2 版）、《冶金流程集成理论与方法》[13]等重要学术专著相继出版，标志着我国冶金工程学科从基础科学、技术科学到工程科学的知识体系已经建立并不断完善。

工程理念是工程建造、工程运行的灵魂[14]。承载工程理念的工程设计则是对工程项目建设进行全过程的总体性策划和表述项目建设意图的过程，是科学技术转化为生产力的关键环节，是实现工程项目建设目标的基础性、决定性环节。没有现代化的工程设计，就没有现代化的工程，也不会产生现代化的生产运行绩效。科学合理的工程设计，对加快工程项目建设速度、提高工程建设质量、节约建设投资、保证工程项目顺利投产并对取得较好的经济效益、社会效益和环境效益具有决定性作用。钢铁厂的竞争力和创新看似体现在产品和市场，究其根源却来自于设计理念、设计过程和制造过程，工程设计正在成为市场竞争的始点，工程设计的竞争和创新的关键在于工程复杂系统的多目标总体优化。

首钢京唐钢铁项目是我国自主设计建设的第一个特大型钢铁联合企业。以钢铁流程工业的指导思想进行设计建设。目前已投产的铁前系统和炼钢系统，实际运行水平已处于国际先进。首钢京唐钢铁厂工程设计与建设，运用现代钢铁冶金工程设计方法进行工程决策、规划、设计、构建、运行等并获得成功，进一步验证了现代钢铁冶金工程设计方法推广应用的重大理论价值。

3 工程方法论与现代钢铁冶金工程设计方法

工程方法论是研究工程方法的共同本质（结构化、功能化等）、共性规律（要素选择优化、程序化、协同化等）和一般价值的理论（和谐化、效率化、效益化、优质化等），是阐明正确认识、评价和指导工程活动的一般方法、途径及其规律，其核心和本质是研究各种工程方法所具有的共性特征和工程方法所应遵循原则和规律。

工程方法论对现代钢铁冶金工程设计方法的指导作用主要是建立起现代工程思维、工程理念、工程系统观及工程系统分析方法等，并与钢铁冶金工程的决策、规划、设计、建造、运行等过程有机地结合起来，并发挥指导作用。

钢铁冶金工程设计是建立在基本要素、原理、工艺技术、设备（装置）、程序、管理、评价基础上集成、建构的过程。从方法论上看，要研究从对钢铁冶金工程要素的合理选择、集成出发，建构出结构合理、功能优化、效率卓越的可运行、有竞争力的工程实体的方法问题。因而，宏观上工程方法论主要是围绕着工程整体的结构化、功能化、效率化和环境适应性等维度展开的。

钢铁冶金工程设计的最主要的命题是解决好整体流程的结构、功能和动态运行过程上的多目标优化；解决好流程中工序/装置之间动态-有序、协同-连续/准连续问题，并形成物质流、能量流网络和信息流网络；解决好工序、装备和信息控制单元本身的结构-功能-效率问题。因此，钢铁冶金工程设计理论是一门复杂的学问，工程设计理论需要通过工程设计、工程运行和工程管理等实践中的范例和失败教训为基本素材，利用基础科学、技术科学特别是工程科学的最新成就加以研究、总结，概括出新的认识——新的工程设计理论和方法。

在钢铁冶金工程设计方法中，工程方法论的"二阶性"主要体现在：体系结构化及其方法，协同化及其方法，程序化及其方法，功能化及其方法，和谐化及其方法。上述内容体现了工程方法论的共性原则和规律，是指导钢铁冶金工程设计的基本理论和方法。

4 现代钢铁冶金工程设计方法

4.1 钢铁冶金工程设计问题的识别与定义

4.1.1 钢铁制造流程的物理本质及其特征

钢铁制造流程的物理本质是物质、能量和信息在不同的时-空尺度上流动/流变的过程。也就是物质流在能量流的驱动下，按照设定的"程序"，沿着特定的"流程网络"作动态-有序的运行，并实现多目标的优化。优化的目标包括产品优质、低成本，生产高效-顺行，能源利用效率高，能耗低，排放少，环境友好等。演变和流动是钢铁制造流程运行的核心。

钢铁制造流程是由各单元工序串联作业，各工序协同、集成的生产过程。一般前工序的输出即为后工序的输入，且互相衔接、互相缓冲-匹配。钢铁制造流程具有复杂性和整体性特征，复杂性表现"复杂多样"与"层次结构"两个特点。

4.1.2 钢铁制造流程动态运行的特征要素

钢铁制造流程动态运行的特征要素是"流""流程网络""运行程序"，其中"流"是制造流程运行过程中的动态变化的主体，"流程网络"（即"节点"和"连接器"构成的图形）是"流"运行的承载体和时-空边界，而"运行程序"则是"流"的运行特征在信息形式上的反映。

4.1.3 钢铁制造流程运行的特点

由此可以推论出，钢铁制造流程运行的物理本质是一类开放的、远离平衡的、不可逆的、由不同结构-功能的相关单元工序过程经过非线性相互作用，嵌套构建而成的流程系统。在这一流程系统中，铁素流（包括铁矿石、废钢、铁水、钢水、铸坯、钢材等）在能量流（包括煤、焦、电、汽等）的驱动和作用下，按照一定的"程序"（包括功能序、时间序、空间序、时-空序和信息流调控程序等）在特定设计的复杂网络结构（如生产车间平面布置图、总平面布置图等）中的流动运行现象。这类流程的运行过程包含着实现运行要素的优化集成和运行结果的多目标优化。

以钢铁制造流程整体动态-有序、协同-连续运行集成理论为指导，钢铁冶金工程设计的核心理念是：在上、下游工序动态运行容量匹配的基础上，考虑工序功能集（包括单元工序功能集）的解析优化，工序之间关系集的协调-优化（而且这种工序之间关系集的协同-优化不仅包括相邻工序关系，也包括长程的工序关系集）和整个流程中所包括的工序集的重构优化（即淘汰落后的工序装置、有效"嵌入"先进的工序/装置等）。

4.1.4 钢铁冶金工程设计方法的路径

基于对钢铁制造流程动态运行物理本质的认识，钢铁冶金工程设计的重要目的，就是通过选择、综合、权衡、集成等方法，构建出符合钢铁制造流程运行规律和特点的先进流程，可以归纳概括为：

（1）钢铁制造流程具有复杂的时-空性，复杂的质-能性、复杂的自组织性、他组织性等特点，并体现为多因子、多尺度、多层次、多单元、多目标优化。

（2）钢铁冶金工程设计是围绕质量/性能、成本、投资、效率、资源、环境等多目标群进行选择、整合、互动、协同等集成过程和优化、进化的过程。

（3）钢铁冶金工程设计是在实现单元工序优化基础上，通过集成和优化，实现钢铁冶金全流程系统优化的过程。

（4）钢铁冶金工程设计是在实现全流程动态-精准、连续（准连续）-高效运行的过程指导思想统领下，对各工序/装置提出集成、优化的设计要求。

（5）钢铁冶金工程设计创新要顺应时代潮流，从单一的钢铁产品制造功能提升到实现钢铁厂"三个功能"的过程。

因而，钢铁冶金工程设计方法的路径是建立在描述物质/能量的合理转换和动态-有序、协同-连续运行过程设计理论的基础上，并努力实现全流程物质流/能量流运行过程中各种信息参量的动态精准，并进一步发展到计算机虚拟现实。

4.2 钢铁制造流程的动态运行与界面技术

4.2.1 钢铁制造流程动态-有序运行过程中的动态耦合

研究钢铁制造流程动态-有序运行的非线性相互作用和动态耦合是现代钢铁冶金设计方法重要内涵，主要体现在钢铁制造流程区段运行的动态-有序化、界面技术协同化和流程网络合理化。不同工序/装置在其动态-有序运行过程中的动态耦合是流程形成动态结构的重要标志。

4.2.2 钢铁制造流程界面技术协同化

所谓"界面技术"是相对于钢铁制造流程中炼铁、炼钢、铸锭、初轧（开坯）、热轧等主体工序之间的衔接-匹配、协调-缓冲技术及相应的装置（装备）。"界面技术"不仅包括相应的工艺、装置，还包括平面图等时-空合理配置、装置数量（容量）匹配等一系列的工程技术，如图2所示。

"界面技术"主要体现实现生产过程物质流（应包括流量、成分、组织、形状等）、生产过程能量流（包括一次能源、二次能源以及用能终端等）、生产过程温度、生产过程时间和空间位置等基本参数的衔接、匹配、协调、稳定等方面。

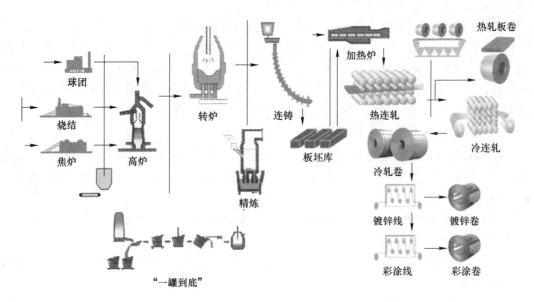

图 2　现代钢铁制造流程的界面技术

"界面技术"是在单元工序功能优化、作业程序优化和流程网络优化等流程设计创新的基础上，所开发出来的工序之间关系的协同优化技术，包括了相邻工序之间的关系协同-优化或多工序之间关系的协同-优化。"界面技术"的形式分为物流-时/空的界面技术、物质性质转换的界面技术和能量/温度转换的界面技术等。

4.3　现代钢铁冶金工程的概念设计

4.3.1　建立现代工程思维模式——概念设计

概念设计是工程科学层次上的问题，首先要从创建流程的耗散结构、耗散过程出发，突出流程应该动态-有序、协同-连续运行的概念。在新一代钢铁制造流程的设计研究中，概念设计研究要建立起系统研究分析钢铁制造流程物理本质和动态运行特征的工程思维模式，采用解析与集成的方法，整体研究钢铁制造流程动态运行的规律和设计、运行的规则。因此，对新一代钢铁制造流程的研究首先应从研究整体流程的动态运行本质开始，进行流程层次上整体动态运行的概念研究。对流程动态运行进行理性抽象的方法是系统地思考生产流程动态运行的物理本质，用解析与集成的方法，整体研究流程动态运行的规律和设计运行的规则。

4.3.2　现代钢铁制造流程两类基本流程的选择

结合市场需求和资源供给能力，针对现代钢铁制造流程已演变成的两类基本流程进行选择。一种是以铁矿石、煤炭等天然资源为源头的高炉—转炉—精炼—连铸—热轧—深加工流程或熔融还原—转炉—精炼—连铸—热轧—深加工流程；另一种是以废钢为再生资源和电力为能源的电炉—精炼—连铸—热轧—深加工流程。研究表明，无论何种流程结构，流程动态运行系统本身都是一种耗散结构，必须构建一个优化的耗散结构，使物质流得以动态-有序、协同-连续地持续运行。

4.4 现代钢铁冶金工程的顶层设计

钢铁工业未来发展，必须在充分理解钢铁制造流程动态运行过程物理本质的基础上，进一步拓展钢铁厂的功能，以新的模式实现绿色化、智能化转型，融入循环经济社会。通过对钢铁制造流程动态运行过程物理本质的研究，可以推论出现代钢铁制造流程应当具有"三个功能"：

（1）铁素物质流运行的功能——高效率、低成本、洁净化钢铁产品制造功能。

（2）能量流运行的功能——能源合理、高效转换功能以及利用过程剩余能源进行相关的废弃物消纳-处理功能。

（3）铁素流-能量流相互作用过程的功能——实现过程工艺目标以及与此相应的废弃物消纳-处理-再资源化功能。

现代钢铁制造流程工程顶层设计以概念设计为基础，并确立钢铁制造流程中"流"的动态概念，强调以动态-有序、协同-连续运行的观念，形成集成的、动态-精准运行的工程设计观。在顶层设计中突出流程结构优化和流程功能的拓展，以"三个功能"为设计的总体目标，强调以要素选择、结构优化、功能拓展和效率卓越为顶层设计的原则。在方法上强调从顶层（流程整体）决定底层（工序/装置），形成从上层指导、规范下层的思维模式。

4.5 现代钢铁冶金工程的动态-精准设计

现代钢铁冶金工程设计方法体系形成了基于冶金流程工程学、冶金流程集成理论与方法的钢铁制造流程动态-精准设计方法，从开始设计就以"流"和"动"的概念为指导，将分割-粗放的传统设计方法进化到动态精准设计方法，这是建立在钢铁制造流程动态运行物理本质基础上，特别是钢铁制造流程动态-有序运行中的运行动力学理论基础上的工程设计方法。

以先进的概念研究和顶层设计为指导，运用图论、排队论和动态甘特图等工具为手段，研究高效匹配的界面技术实现动态-有序、协同-连续的物质流设计、高效转换并及时回收利用的能量流设计及以节能减排为中心的开放系统设计，从而在更高层次上体现钢铁制造流程的"三个功能"。动态精准设计方法是建设项目工程设计顶层设计阶段的进一步深化，是宏观尺度下工程设计动态精准设计的具体方式和方法。

5 结语

（1）现代钢铁冶金工程设计应当从要素-结构-功能-效率集成优化的观点出发，在工程设计中体现出动态有序、协同连续的流程运行优化，这是动态精准设计和钢铁厂实际生产运行过程的理论核心。

（2）在冶金工程相关的基础科学、技术科学的基础上，以冶金流程工程学理论和方法为核心，构建钢铁冶金工程设计创新理念、理论及方法的完整体系。深入开展对钢铁制造流程动态运行物理本质的研究，探索物质流、能量流、信息流的协同耦合和集成理论研究，重视钢铁制造流程动态运行中"流""流程网络"和"运行程序"的研究，以及对钢铁企业结构、功能、效率的影响。

（3）重视以网络化整合、程序化协同为重要手段（集成创新），重视概念设计、顶层设计和动态精准设计，提高钢铁制造流程的设计水平和生产过程的运行效率。重视新工艺、新技术、新装备的开发、设计制造并通过多层次、多尺度、多因子集成优化，有效、动态地"嵌入"到钢铁生产流程中。同时要高度重视全厂性、全流程层次上能量流研究及其网络化整合，提高能源效率，进一步从流程上推进节能减排。

（4）从工程哲学的视角来看，钢铁冶金工程设计过程和生产运行过程长期以来集中注意的是局部性的"实"，而往往忽视了贯通全局性的"流"。未来工程设计、生产运行和过程管理中既要注重解决具体的、局部的"实"，更应集中关注贯通全局的"流"。

（5）面向未来，我国钢铁工业应当以绿色化、智能化发展作为产业转型升级的主要方向，重视产业结构的调整升级和企业结构的概念研究、顶层设计，推动冶金工程动态精准设计及其动态精准运行，以工程哲学的思维和战略思考，解决产业层面、企业层面的复杂性命题，获得新的市场竞争力和可持续发展能力。

致谢

感谢张春霞、徐安军等专家的宝贵意见和建议；感谢曹朝真、李欣、周颂明、张旭孝、张文皓、上官方钦等青年学者的参与和帮助。

参考文献

[1] 徐匡迪. 20世纪——钢铁冶金从技艺走向工程科学 [J]. 上海金属，2002（1）：1-10.
[2] 殷瑞钰. 冶金流程工程学（第2版）[M]. 北京：冶金工业出版社，2009：1-10.
[3] 张寿荣. 论21世纪中国钢铁工业结构调整 [J]. 冶金丛刊，2000（1）：39-44.
[4] 张寿荣，于仲洁. 中国炼铁技术60年的发展 [J]. 钢铁，2014，49（7）：8-14.
[5] 徐匡迪. 中国钢铁工业的发展和技术创新 [J]. 钢铁，2008，43（2）：1-13.
[6] 张寿荣. 钢铁工业与技术创新 [J]. 中国冶金，2005，15（5）：1-5.
[7] 张春霞，王海风，张寿荣，殷瑞钰. 中国钢铁工业绿色发展工程科技战略及对策 [J]. 钢铁，2015，50（10）：1-7.
[8] 张福明，颉建新. 冶金工程设计的发展现状及展望 [J]. 钢铁，2014，49（7）：41-48.
[9] 殷瑞钰，汪应洛，李伯聪，等. 工程哲学（第1版）[M]. 北京：高等教育出版社，2007.
[10] 殷瑞钰，汪应洛，李伯聪，等. 工程哲学（第2版）[M]. 北京：高等教育出版社，2013.
[11] 殷瑞钰，李伯聪，汪应洛. 工程演化论 [M]. 北京：高等教育出版社，2011.
[12] 殷瑞钰. 冶金流程工程学（第1版）[M]. 北京：冶金工业出版社，2004.
[13] 殷瑞钰. 冶金流程集成理论与方法 [M]. 北京：冶金工业出版社，2013.
[14] 李喜先. 工程系统论 [M]. 北京：科学出版社，2007.

Research on Design Method of Modern Iron and Steel Metallurgy Engineering

Yin Ruiyu Zhang Shourong Zhang Fuming Xie Jianxin

Abstract: This paper expounds the important role and status of steel materials in the economic and social development, analyzes the development history and current situation of China's iron and steel industry, and discusses the important role in promoting the development of science and technology progress for steel industry, points out the future development orientation of China's steel industry. Also this paper reviews the development course of Chinese metallurgical engineering design evolution, analyzes the relationship between engineering methodology and metallurgical engineering design method, metallurgical engineering design status and the function, and elaborates the physical essence of steel manufacturing process and modern steel manufacturing process of the "three functions". From the physical essence of steel manufacturing process, characteristics and dynamic operating characteristics factors, path design method, dynamic and orderly operation of dynamic coupling process, etc., the identification and definition of metallurgical engineering design problem are proposed. The methodological system of concept design, top-level design, dynamic precision design for modern iron and steel metallurgy engineering method, as well as the systematic integration of modern steel manufacturing process and structure optimization, economic and social assessment are proposed.

Keywords: iron and steel metallurgy; process manufacture; structure optimization; engineering design; design method

钢铁冶金工程知识研究与展望

摘　要：从工程集成、工程建构、工程运行和工程管理的视角，结合钢铁冶金工程讨论了工程知识。从知识论的角度看，钢铁冶金过程是物理知识、化学知识、技术知识、管理知识等诸多知识范畴的集成，包括了工程科学、工程技术、工程设计、工程管理、工程哲学等方面的知识。分析了钢铁材料对经济社会发展的重要作用，阐述了钢铁冶金工程流程运行的主要特征，解析了钢铁冶金流程动态运行的物理本质。系统梳理了冶金工程知识的结构，揭示了冶金工程知识系统的复杂性、协同性、整体性、集成性等特征；论述了钢铁冶金工程知识的形成和发展过程，以及技术进步对冶金工程知识创新的推动作用；讨论了"界面技术"对冶金制造流程协同—集成的重要性，以及信息技术对钢铁冶金工程智能化的推动作用。在工程哲学的视野下，探讨了宏观动态冶金学和钢铁冶金工程知识面向智能化和绿色化的发展方向和目标。

关键词：钢铁冶金；工程知识；工程哲学；冶金流程；界面技术

1　引言

　　钢铁材料是支撑经济社会发展的重要的基础材料，钢铁冶金工程是重要的流程制造业和基础产业。现代钢铁冶金工程知识，是将微观基础冶金学的知识、专业工艺冶金学的知识和动态宏观冶金学的知识相互嵌套、综合集成起来，所构建的一个集成化、工程化的知识体系。既面向工程科学、工程技术、工程设计，又面向工程决策、工程管理、工程评估甚至关联到工程哲学等。这是观察研究冶金科学与工程所必须拥有的视野。

2　钢铁冶金工程概述

2.1　钢铁冶金的地位和特点

　　钢铁由于资源丰富、成本相对低廉、材料性能优越、易于加工且便于循环利用，因此，仍然是当前世界上重要的结构材料，也是世界上消费量最大的功能材料（例如电工钢、不锈钢等）[1]。

　　从钢铁制造流程分析，它属于典型的流程制造过程。即整个制造流程是由若干个相关的异质、异构且时-空过程各异的单元工序/装置所组成，上下游工序之间紧密衔接，上游工序是下游工序基础，上游工序的输出即为下游工序的输入。钢铁制造流程是由原料造块

本文作者：殷瑞钰，张福明，张寿荣，颉建新。原刊于《工程研究-跨学科视野中的工程》，2019，11（5）：438-453。

(烧结或球团)、炼铁、炼钢、轧钢以及钢材深加工等若干个工序组成的制造流程。从流程制造的特征分析，钢铁制造流程是由多工序协同、相互连接形成的整体系统，具有显著的层次性、过程性和复杂性[2]。典型的钢铁制造流程如图1所示。

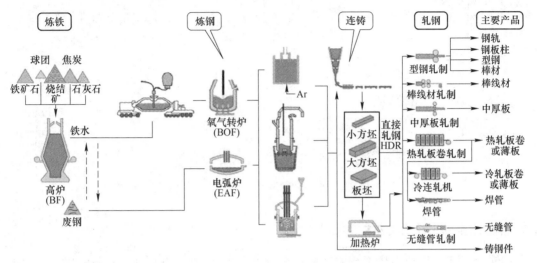

图 1　典型的钢铁冶金制造流程

2.2　钢铁冶金制造流程的认识深化

制造业通常可分为流程制造业与装备制造业两类。一般而言，流程制造业为装备制造业提供产品或原材料，而装备制造业则是利用流程制造业生产出来的产品作为原材料制造出最终工业品或消费品。流程制造业一般是指原料经过一系列以改变其物理、化学性状为目的的加工-变性处理，获得具有特定物理、化学性质或特定用途产品的工业，钢铁冶金工业属于典型的流程制造业。

流程工业的工艺流程中各工序（装置）加工、操作的形式是多样化的，包括化学变化、物理转换等；其作业方式则包括连续化、准连续化和间歇化等形式。动态运行是流程制造业的本构特征，构成流程制造动态运行过程的要素包括"流""流程网络"以及"运行程序"，这也是流程制造业共有的特征；而流程制造动态运行过程中的"流"一般包括"物质流""能量流"和"信息流"，同时还有与之相应的"流程网络"以及"运行程序"。

通过对钢铁制造流程追求动态-有序、协同-连续/准连续运行的物理本质研究，可以清晰地推论得出：在未来绿色化可持续发展进程中，钢铁制造流程（特别是高炉-转炉长流程）将主要实现"三个功能"，如图2所示。现代钢铁制造流程应具有三个功能，即钢铁产品制造功能、能源高效转换功能和大宗废弃物的处理-消纳和再资源化功能[3]。进而，从现代钢铁制造流程的三个功能的演化，可以得出钢铁产业的可持续发展必须遵循绿色化、智能化、减量化和品牌化的发展规律，不宜盲目、无序地扩大生产规模，粗放式发展，而应当在资源能源可获取、生态环境可承载、市场竞争可容纳的前提下，科学合理地确定不同类型钢厂、不同钢铁制造流程的市场定位、产品方案、工艺流程以及装备水平，进而使现代钢厂融入到循环经济社会中[4]。

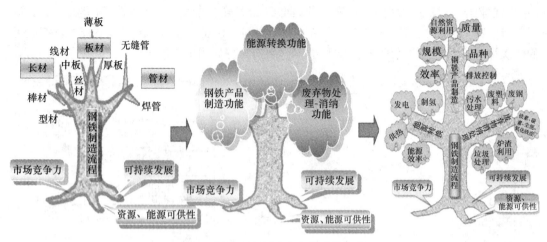

图 2　钢铁冶金制造流程的功能演进

2.3　钢铁冶金的工程本质及其动态运行规律

钢铁冶金流程动态-有序运行的基本要素是"流""流程网络"和"运行程序"。其中"流"是制造流程运行过程中的动态变化性主体,"流程网络"(即"节点"和"连接器"构成的图形)是"流"运行的承载体和时间-空间边界,而"运行程序"则是"流"的运行特征在信息指令形式上的反映。

从热力学角度分析,钢铁冶金流程则是一类开放的、非平衡的、不可逆的、由不同结构-异质功能的单元工序通过非线性相互作用、动态耦合等机制所构成的复杂系统,其动态运行过程的本质是耗散过程。这一认识是基于钢铁冶金流程宏观动态运行规律及其物理本质深入解析的结论,建构了冶金流程工程科学研究的理论基础和体系,特别是突破了经典热力学关于"孤立系统""物料平衡与热平衡"等传统观点的束缚,建立起了开放、动态、有序、协同、集成等观念,显著提升了现代冶金学在工程科学层次上的理论深度与广度,有利于现代冶金学在工程科学(例如冶金流程工程学等)的理论指导下走向绿色化、智能化。

通过对流程制造过程的深入解析,可以从流程制造的表象特征进而研究得出更深层次的运行规律。以钢铁冶金为例,可以清晰地阐释钢铁制造流程的物理本质是:在一定外界环境条件下,物质流(主要是铁素流)在能量流(主要是碳素流)的驱动和作用下,按照设定的"运行程序",沿着特定(设定的)的"流程网络"做动态-有序的运行,并且实现多目标优化——这就是所谓的"三流一态"。

通过对钢铁冶金工程本构特征及其运行规律的深入研究,使冶金学从孤立的局部性研究走向开放的动态系统研究,从间歇-等待-随机组合运行的流程走向准连续-协同-动态-非线性耦合的动态-有序、协同-连续流程。

与此同时,现代钢铁制造流程的功能应当拓展为"三个功能",再通过"三个功能"的拓展获得新的产业结构和产业经济增长点,从而建构具有循环经济特征的产业集群和产业生态链,成为绿色发展理念下循环经济社会的一个组成部分[5]。

现代钢铁厂的钢铁产品制造功能是在尽可能减少资源/能源和时间/空间消耗的基础

上，高效率地生产出成本低、质量好、排放少且能够满足用户不断变化需求的钢材，供给社会生产和居民生活消费。能源转换功能与钢铁制造功能相互协同耦合，即钢铁生产过程同时也伴随着能源转换过程。以高炉-转炉-热轧流程为代表的钢铁联合企业为例，其实质是冶金-化工过程，也可以视为是将煤炭通过钢铁冶金制造流程转换为可燃气、热能、电能、蒸汽甚至氢气或甲醇等能源介质的过程。废弃物消纳-处理和再资源化功能，即钢铁厂制造流程中的诸多工序、装备可以处理、消纳来自钢铁厂自身和社会的大宗废弃物、改善区域环境负荷，促进资源、能源的循环利用。

因此，现代钢铁冶金的工程科学研究及其知识体系，要在微观基础冶金学和专业工艺冶金学的基础上，上升到物质流、能量流、信息流及其动态运行过程（"三流一态"）的宏观动态冶金学层次上，并使这三个层次的冶金学嵌套集成起来，扩充成新的冶金学体系。这是工程科学领域研究范畴以及工程知识体系结构的拓展和创新。

3 冶金学和冶金工程知识体系及其发展

3.1 冶金学学科知识的形成与发展

冶金学有着悠久的历史发展过程，至今仍在不断发展。从青铜器时代、铁器时代开始，冶金术开始发展，直至现代，从 1925 年英国法拉第学会在伦敦召开"炼钢过程中的物理化学"会议开始，冶金学开始跨入了现代科学的发展进程。然而，冶金学又是有别于物理学、化学、生物学、地学、天文学、数学等以研究自然物理现象为主要目标的基础科学的。从根本上看，冶金学是属于研究人工物世界的工程科学、技术科学范畴，是重在研究发展现实生产力的工程知识、技术知识和工程科学知识。其知识的来源是多元化、多层次、集成、综合性的，不仅是只来源于基础科学。冶金学的知识重在对各类要素、各类知识集成起来，并能转化为现实的、直接的生产力，发明、集成、综合、转化是其特征。这是新工科发展过程中应该给予重视的。

经过近百年的探索、研究发展进程，当代冶金学（冶金科学与工程）已逐步构成了由三个层次的知识集成构建而成的框架体系，即原子/分子层次上的微观基础冶金学，工序/装置层次上的专业工艺冶金学和全流程/过程群层次上的动态宏观冶金学。可见，随着不同层次科学问题研究的深入，学者们的研究目标、研究领域不断扩宽，认识问题的视野发生了层次性跃迁，并进而嵌套集成为一个新的知识结构，即不囿于经典热力学孤立系统观念，跨入探索冶金企业全流程的过程群的集成优化、结构优化，研究的对象发生了变层次、变轨的跃迁。同时，扩大了研发领域，既引导企业全流程中过程和过程群的自组织结构以及他组织调控过程中共同形成的耗散结构和耗散过程优化的研究，又引导新的工程设计、工程运行的理论和方法。当代冶金学发展的战略目标也跟着时代的发展，发生了战略性的变化。当代冶金学的战略目标除了制造新一代产品以外，已经聚焦于冶金工厂的绿色化（绿色、低碳、循环发展）和智能化（智能化设计、智能化生产、智能化服务、智能化管理等）。

有鉴于冶金生产的实质是在开放、动态的流程系统中，通过输入/输出"流"在耗散结构中动态-有序、协同-连续运行并实现多目标优化，冶金科学与工程的知识必须要将微观基础冶金学的知识、专业工艺冶金学的知识和动态宏观冶金学的知识相互嵌套、集成综合起来,形成一个集成化、工程化的知识体系。这个知识体系,既要面向工程科学、工程技术、工程设计,又要面向工程决策、工程管理、工程评估甚至关联到工程哲学等。这是观察研究当代冶金科学与工程应有的视野和思路。

当代冶金科学与工程的知识体系层次结构和工程视野的解析如图3所示。从图3中可以看出,微观基础冶金学包括冶金过程物理化学、冶金原理、金属学、传热学等知识,主要研究原子/分子尺度的微观基础理论问题。专业工艺冶金学包括炼铁学、炼钢学、金属压力加工学、冶金反应工程学等,主要研究工序/装置尺度的技术科学问题。由于钢铁冶金企业生产运行的实质是开放、动态的输入/输出过程(流)中实现多目标优化,因而动态宏观冶金学主要是冶金流程工程学,主要研究整体钢铁制造流程及钢铁冶金企业(钢铁厂)尺度的工程科学问题。

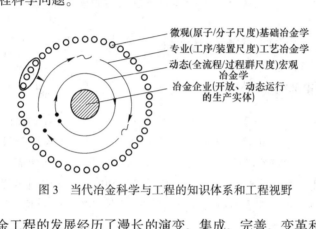

图3 当代冶金科学与工程的知识体系和工程视野

现代钢铁冶金工程的发展经历了漫长的演变、集成、完善、变革和创新的过程。其中理论体系的形成、建立和发展,技术的发明、开发、应用和革命,生产工艺流程的组合、集成、演变和完善,在第一次工业革命以后大约200年的历史进程中,不断交替出现和相互促进。理论的形成、发展和不断完善,是指导技术发明和技术创新,以及工程集成和工程创新的重要动力。

回顾冶金学科的发展演进历程不难看出,由于钢铁冶金工程包括矿物开采与加工、高温冶金过程、凝固-成形过程,金属塑性变形过程与材料性能控制过程,具有工序繁多、功能各异、过程复杂、流程结构多样等特点,因而钢铁冶金工程基础理论的形成、建立和发展,是多领域、多学科的理论研究和相互交叉发展的过程,呈现出一种典型的解析-组合-再解析-再组合的不断发展和不断完善的过程。

3.2 钢铁冶金工程科学知识的形成与发展

钢铁冶金工程科学主要体现为钢铁制造流程的整体结构优化和动态运行过程优化,初期阶段是要解决好多维的过程物流管制系统,即物质状态转变、物质性质控制和物质流管制三方面的融合、贯通、协调/控制;确定制造过程的基本参数和派生参数;进而发展到钢铁冶金过程物质流、能量流和信息流的综合调控等宏观系统的问题[6]。

现代钢铁冶金生产,一般是在一个联合企业内包括烧结、焦化、炼铁、炼钢、连铸、热轧、冷轧、钢材深加工等多个相关工序组成的制造流程,即从矿物加工直到钢材深加工都在一个钢铁厂完成。所以钢铁冶金流程是涉及多组元、多过程、多工序、多层次、多领域的流程集成系统。在这种条件下有时会出现这样的现象：即对某个具体的冶金工序而言是最优化的，却不一定能使整个冶金流程整体得到优化，也就是局部最优化不等于全流程系统优化。因此，随着技术进步、生产发展，必然需要在更高层次和更大的时-空尺度上研究冶金过程整体性、集成性和协同性的工程科学，也就是冶金流程工程学——宏观动态冶金学。

冶金流程工程学属于宏观层次上的工程科学的范畴，主要研究冶金生产流程的物理本质、本构特征和整体行为，旨在厘清冶金生产流程中相关的物质（和能量）流动（与储运）的驱动力，研究冶金制造流程中所涉及的有关结构、功能、效率的问题，包括流的行为、网络构建、程序指令和环境条件约束等命题，乃至空间和平面布置、时间和时序安排与控制、流程过程中排放与消纳（或者再循环）的控制和优化等一系列学问。

从工程哲学角度分析，在冶金制造流程中，不仅要研究"孤立""局部"的"最佳"，更重要的是要解决制造流程整体动态运行过程的最佳；因此，不能用机械论的拆分方法来解决相关的、异质功能的而又往往是不易同步运行工序/装置的组合集成问题。重要的是要研究多因子、多尺度、多层次的开放系统动态运行的过程工程学问，要分清工艺表象和物理本质之间的表里关系、因果关系、非线性相互作用和动态耦合关系，并探索出其内在规律。表 1 阐释了冶金工程科学知识体系的结构层次与特征。

表 1 冶金学科知识体系的结构层次与特征

科学分类	研究尺度	研究方法		层次	系统特征	系统控制
		白箱	黑箱			
基础科学	原子/分子	原子/分子	系统背景	微观	孤立系统：与外界没有物质、能量交换，可逆过程——平衡状态	PLC
技术科学	场域/装置	场域/装置	分子/流程	中观	开放系统：与外界有某些物质、能量和信息交换	PLC/MES、EMS
工程科学	流程/复杂系统	流程/工序关系	分子/场域	宏观	开放系统：与外界不断有物质、能量和信息交换，不可逆过程——远离平衡态	ERP/MES/PCS

冶金工程科学之冶金流程工程学的知识体系包括：认知科学；工程逻辑——系统集成指导下的钢铁厂结构分析和结构演进；工序功能集的解析和优化；工序功能集、工序关系集在流程中的协调与优化；流程工序集的重构与优化；过程工程与信息技术的结合等。冶金流程工程学作为工程科学，追求通过制造流程的整体优化来解决流程系统的功能优化、结构优化和效率优化，进而在更大的尺度上解决企业的模式-结构、产品的市场竞争力、资源与能源的可供性以及企业所面临的环境问题和生态协调性等问题。

4 工程演化、技术进步与冶金工程知识创新

4.1 钢铁冶金工程科学知识的形成与发展

经过近百年的探索、研究和发展，当代冶金学（冶金科学与工程）已逐步形成了由

三个层次的知识集成构建而成的框架体系，即原子/分子层次上的微观基础冶金学，工序/装置层次上的专业工艺冶金学和全流程/过程群层次上的动态宏观冶金学。可见，随着不同层次科学问题研究的深入，学者们的研究目标、研究领域不断扩展，认识问题的视野发生了层次性跃迁，并进而嵌套集成为一个新的知识结构，即不囿于经典热力学孤立系统观念，跨入探索冶金企业全流程的过程群的集成优化、结构优化，研究的对象发生了变层次、变轨的跃迁。同时，扩大了研究领域，既引导研究制造流程中过程和过程群的自组织结构以及他组织调控过程中共同形成的耗散结构和耗散过程优化的研究，又引导新的工程设计、工程运行的理论和方法。当代冶金学发展的战略目标也跟着时代的发展，发生了战略性的变化；当代冶金学的战略目标除了制造新一代产品以外，已经聚焦于冶金工厂的绿色化（绿色、低碳、循环发展）和智能化（智能化设计、智能化生产、智能化服务、智能化管理等）。

从钢铁冶金工业的发展演进历程可以看出，钢铁冶金工业现在面临的挑战是多方面的，要解决这些复杂环境下的复杂命题，就必须从战略层面上来思考钢铁厂的要素-结构-功能-效率问题，实质上这是全厂性的生产流程层面上的问题，就必须从生产流程的结构优化及其相关的工程设计等根源着手。进而还可以清晰地认识到，这样的系统性、全局性、复杂性问题，不是依靠技术科学层次上的单元技术革新和技术攻关所能解决的，而是需要以工程哲学的视野，在全产业、全过程和工业生态链等工程科学层次上解决。表2列出了冶金学科不同尺度及其对比。

表2　冶金学科不同尺度及其对比

层次	学科分支	时空尺度	基础研究内容	研究方法特征	吸引子	物理基础
微观	冶金物理化学	原子、分子、离子	热力学函数能量关系；冶金反应化学亲和势	实验室测定平衡；相图研究；反应速率和机理测定	平衡态	经典热力学，物质构造学说
介观	冶金反应工程学（"三传一反"）	反应器、场域	反应器内介质浓度、温度、停留时间分布；颗粒液滴气泡弥散体系	数学和物理模拟；场域条件下参数测量；比拟放大	非平衡稳定态	传输现象理论，线性非平衡热力学
宏观	冶金流程工程学（"三流一态"）	冶金流程整体（全厂性）和区段流程（车间性）	多因子物质流控制；流程解析和集成；物质流-能量流-信息流协调运行	工程设计和模拟运行，流程动态运行优化和信息化表征、调控	动态-有序、连续-紧凑、耗散优化	牛顿力学（动力学），非线性、非平衡、开放系统热力学，耗散结构理论

4.2　钢铁冶金工程理念的转变

工程理念是工程哲学的核心概念之一，是源于客观世界而表现在主管意识中的哲学概念，它是人们在长期、丰富的工程实践的基础上，经过深入的理性思考而形成的对工程发展规律、发展方向和有关的思想信念、理想追求的集中概括和高度升华，在工程活动中，工程理念发挥着根本性的作用。工程理念在工程活动中发挥着最根本性的、指导性的、贯穿始终的、影响全局的作用。

长期以来，传统的钢铁冶金工程理念基本上是以"还原论"的思维模式处理问题，也就是将钢铁制造流程分割为若干工序、装置，再将工序、装置解析成某种化学反应过程或是传质、传热和动量传输的过程，再在工序之间进行简单拼接、叠加处理就算形成了制造流程，其时间/空间问题涉及较少，动态运行过程中的相互作用关系和协同连接的界面技术往往被忽视。

毋庸置疑，工程理念是工程决策、工程设计、工程建构和工程运行的灵魂。工程理念是对工程项目建设进行全过程的总体性策划和表述项目建设意图的过程，是实现工程项目建设目标的基础性、决定性环节。没有与时俱进的工程理念，就没有现代化的工程设计，就没有现代化的工程，也不会产生现代化的生产运行绩效。先进的工程理念、科学合理的工程设计，对加快工程项目的建设速度、提高工程建设质量、节约建设投资、保证工程项目顺利投产并对取得较好的经济效益、社会效益和环境效益具有决定性作用。钢铁厂的竞争力和创新看似体现在产品和市场，其根源却来自工程理念、设计过程和制造过程，工程理念和工程设计正在成为市场竞争的始点，工程设计的竞争和创新关键在于工程复杂系统的多目标群优化。

新一代钢铁冶金制造流程，使冶金学从孤立的局部性研究走向开放的动态系统研究，从间歇-等待-随机组合运行的流程走向准连续-协同-动态-非线性耦合的动态-有序、协同-连续流程。

钢铁冶金工程的演化和新一代可循环钢铁制造流程的构建，实际上是工程理念、工程思维模式的转变和创新，这是从"还原论"思维模式所暴露出的缺失中，探索到了整体集成优化的新理念、新思路。

从工程哲学角度来看，在钢铁冶金工程中，不仅要研究"孤立""局部"的"最佳"，在当代，更重要的是要解决整体动态运行过程的最佳；不能用机械论的拆分方法来解决相关的、异质功能的而又往往是不易同步运行工序/装置的组合集成问题。重要的是要研究多组元、多因子、多尺度、多层次的开放系统动态运行的过程工程学问，要厘清工艺表象和物理本质之间的表里关系、因果关系、非线性相互作用和动态耦合关系，并探索出其内在规律。

4.3 技术进步与关键共性技术对冶金工程知识的推动

4.3.1 技术创新推动了钢铁工业的发展

20世纪90年代是我国钢铁工业迅速崛起时期，期间通过对连铸、高炉喷煤、高炉长寿、连续轧制、转炉溅渣护炉和综合节能等六项关键共性技术的开发、集成和全国性推广，促进了我国钢铁制造流程结构的优化，实现了节能降耗，提高生产效率和产品质量，为我国钢铁工业快速发展奠定了基础[7]。

21世纪以来，首钢京唐钢铁厂的设计建造，基于冶金流程工程学、冶金流程工程集成理论与方法，创新并实践了现代钢铁冶金工程设计的新理念、新理论和新方法，从工程决策开始，直到规划、设计、建造、运行和管理等过程，构建了新一代可循环钢铁制造流程，引领了我国钢铁工业科技进步的发展方向。

1856年以来，纵观钢铁工业的发展史，从技术层面上讲，实际上就是钢铁制造工艺的技术创新史[8]。其共同遵循的规律是工序功能集合的解析-优化，工序间关系集合的协同-优化和流程工序集合的重构-优化，既体现着原始创新，更多的是集成创新。

4.3.2 技术创新推动了钢铁工业的发展

钢铁冶金的科技进步和工程演化，既受到专业技术的渐进性、突变性进步的影响，还受到相关支撑技术的渐进性、突变性进步的影响。因此冶金工业科技进步的方式不仅以单体技术进步的形式出现，而且以互动的、协同的、网络化集成形式出现。

关键共性技术的突破、集成，对冶金制造流程的整体结构优化产生了重大影响，做好关键共性技术的开发、应用和推广，既要注重对关键共性技术的判断和正确选择，也必须结合不同类型钢铁厂特点进行深化和创新，还必须正确把握这些关键共性技术研发、投资的时序安排和互相关系。

所谓关键共性技术就是指在冶金生产流程中技术关联度大，对企业结构影响力大，并且在整个冶金行业领域具有共性的技术。关键共性技术的选择、集成，必须对整个冶金制造流程工艺、装备和产品结构之间的关系进行深入研究和总体认识，并且理性分析和适当排序，寻找对不同类型冶金企业影响面大的共性技术和对整体生产流程关联度大的关键技术，在深化认识的基础上，确立这些关键共性技术对冶金工业不同发展阶段所起的战略主导地位。进而分步、有序地推进，并相继将其集成起来，促进冶金企业生产流程整体结构的优化。可见，生产流程的结构优化成为冶金企业技术创新的重要命题。

20世纪90年代，我国冶金工业科技发展逐步转变为以生产流程结构优化为主线，以技术改造、工艺和装备升级换代为支撑，研究方向和投资重点逐步转向关键共性技术及其工艺、装备研究开发和生产流程的集成优化方面。技术改造讲究冶金制造流程的结构优化和产品的结构优化，出现了不同类型的钢厂模式。

面向新世纪，我国冶金工业发展的主流应是主动动态地适应市场发展趋势，继续以结构优化为主要途径，进一步强化节能、环境治理和生态协同发展，同时加速薄板产线的建设与产品研发，以增强市场竞争力和可持续发展能力作为总体目标，在钢铁产量减量化的前提下，积极推动冶金工业的绿色化、智能化和品牌化发展。

4.4 "界面技术"对冶金制造流程协同、集成的重要性

氧气转炉、连续铸钢、大型高炉、大型宽带连轧机等关键共性技术对钢铁冶金工程的演化和技术进步具有重要的推动作用，这些关键共性技术一般是单元工序的功能集合的优化创新和技术演进，由于单元工序工艺、设备、装置的创新、优化和演进，从而引起了钢铁制造流程中一系列"界面技术"的演变和优化，甚至还出现了不少新的"界面技术"，这些界面技术与单元工序形成了新的组合和流程结构，因而对于界面技术的研究及有关界面技术的工程知识，成为钢铁冶金工程知识的重要创新内容。

界面技术是与钢铁制造流程中相关的炼铁、炼钢、铸锭、初轧（开坯）、热轧等主体工序之间的衔接-匹配、协调-缓冲技术及相应的装置（装备）。界面技术不仅包括相应的工艺、装置，还包括平面图等时-空合理配置、装置数量（容量）匹配等一系列的工程技术，如图4所示。

界面技术主要体现实现生产过程物质流（应包括流量、成分、组织、形状等因子）、生产过程能量流（包括一次能源、二次能源以及用能终端等）、生产过程温度、生产过程时间和空间位置等基本参数的衔接、匹配、协调、稳定等方面。在很大程度上体现了工序之间关系集合的协同-优化。

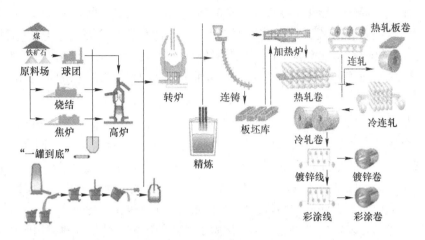

图 4　现代钢铁制造流程的界面技术

界面技术是在单元工序功能优化、作业程序优化和流程网络优化等流程设计创新的基础上，所开发出来的工序之间关系的协同优化技术，包括了相邻工序之间的关系协同-优化或多工序之间关系集合的协同-优化。界面技术形式分为物流-时/空的界面技术、物质性质转换的界面技术和能量/温度转换的界面技术等。

现代钢铁冶金工程的界面技术主要体现在：

（1）简捷化的物质流、能量流通路（如流程图、总平面图等）；

（2）工序/装置之间互动关系的缓冲-稳定-协同（如动态运行 Gantt 图等）；

（3）制造流程中网络节点功能优化和节点群优化以及连接器形式优化（如装备数量、装置能力和位置合理化、运输方式、运输距离、输送规则优化等）；

（4）物质流效率、速率优化；

（5）能量流效率优化和节能减排；

（6）物质流、能量流和信息流的协同优化等。

4.5　信息技术对钢铁冶金工程智能化的推动

4.5.1　钢铁冶金制造流程的运行特征

钢铁冶金制造流程由异质、异构、相关协同的工序构成，钢铁企业以不可拆分的制造流程整体协同运行的方式存在，适合于连续、批量化生产。钢铁制造流程中存在着许多复杂的物理、化学过程，甚至往往出现气、液、固多相共存的连续变化，物质/能量转化过程复杂，难以全部实现数字化。钢铁冶金制造流程是复杂的大系统，输入的原料/燃料组分波动，外界随机干涉因素多，难以直接实现数字化。组成钢铁制造流程的单元工序/装置的功能是不同的，钢铁冶金制造流程属于异质、异构单元组合的集成体。钢铁冶金单元工序/装置之间的关系属于异质、异构单元之间非线性相互作用、动态耦合过程，匹配、协同的参数复杂多变，难以实现数字化。产品性能、质量、生产效率取决于工艺流程设计优化，各个工艺过程的优化和全流程运行的整体优化。钢铁冶金工程的智能化主要体现在制造流程运行过程的智能化。

根据钢铁冶金制造流程的技术特征，其智能制造的涵义应该是以钢铁企业生产经营全过程和企业发展全局的智能化、绿色化、产品质量品牌化为核心目标研发出来的生产经营全过程的数字物理融合系统。其关键技术是生产工艺/装置技术优化、工艺/装置之间的"界面"技术优化和制造全过程的整合-协同优化，以此为基础嵌入数字信息技术，以"三流"协同、"三网"融合为切入口，从而构成体现智能特色的数字物理融合系统——CPS[9]。

4.5.2 钢铁冶金工程"界面技术"的物理系统优化和信息数字系统的构建

钢铁冶金制造流程是由一系列相关的、异质-异构的工序/装置以及它们之间的"界面技术"构成的。"界面技术"承载着物质流/能量流/信息流的沟通、传递功能；工序/装置之间的功能衔接、匹配；工序/装置之间物质流、能量流的承接、缓冲功能等。"界面技术"鲜明地影响着制造流程的结构，尤其是动态运行结构和运行效率。

长期以来对单元装置的自动化、无人化做了大量的研发，并较好地应用于生产制造过程，解决了局部的自动控制问题。应该作为钢铁冶金制造流程智能化的基础。然而这只是实现钢铁冶金制造流程全盘智能化中的一部分。

"界面技术"的物理系统优化和数字化建模也是流程制造业智能化的核心技术之一，至今研究较少，应该引起关注，深入探索。应该认识到"界面技术"同样也是由物理系统硬件和信息数字软件构成的，应该建立在扎实的工程科学基础上。似可按如下方法探索：

（1）以图论为工具构建制造流程的静态网络；

（2）以排队论-Gantt图等方法描述"界面"过程的运行动力学；

（3）制定钢铁冶金流程动态运行规则（设定运行程序）；

（4）以软件仿真手段研究"界面"过程动力学及其比较优化；

（5）解决物质流/能量流/信息流协同优化的自感知、自学习、自决策、自执行、自适应的数字信息系统，在此问题上要解决好物质流网络、能量流网络和信息流网络之间的"三网"联结、融合问题，构建起基于物理系统过程耗散优化的CPS系统，实现工程化的全流程智能化；

（6）将钢铁生产过程和能源系统联系起来，实现钢铁企业的物质流-能量流之间的关联。

4.5.3 钢铁冶金制造流程的智能化

智能化是钢铁工业的重要发展方向之一，必须高度重视，不能错失时机，但也不会在短时内一蹴而就，要经历一个探索、研发、积累、集成、创新的过程。钢厂智能化要与数字化物理融合系统的概念相对接，突出"流""流程网络"和"运行程序"的概念，特别是优化的物质流网络、能量流网络和信息流网络之间的协同运行，实现全厂性动态运行、管理、服务等过程的自感知、自学习、自决策、自执行、自适应[10]。钢铁冶金制造流程物理系统优化是钢铁厂智能化的重要基础性前提，要充分认识制造流程的运行特点，不宜盲目搬用离散型制造的某些概念和方法。在钢铁冶金工程中，"界面技术"的研发是信息物理系统建构中的一个缺失环节，"界面技术"优化对于"三流"协同、"三网"融合具有重要价值，应作为解决智能化钢厂的重要内容之一，必须予以高度重视。

4.6 钢铁冶金工程知识与工程管理知识的集成与融合

如前所述，现代钢铁冶金工程是由多个由不同结构-异质功能的单元工序通过非线性相互作用、动态耦合等机制所构成的复杂系统，具有整体性、复杂性和多层次性。而钢铁冶金工程的管理活动则成为一项多领域的复杂管理活动。

钢铁冶金工程管理知识，不但具有一般工程管理知识的普遍性，还具有产业/专业的特殊性，遵循"实践-认识-再实践-再认识"的辩证路径，工程管理应该是能打开"工程黑箱"的管理，应该是钢铁冶金工程知识与管理工程知识在工程活动中必须有效集成、深度融合在一起，能在工程活动的全生命周期内更好地发挥作用，指导和引领整个钢铁冶金工程活动。

钢铁冶金工程活动中，无论是工程的规划、决策、设计、建造，还是工程运行、维护、退役，都存在大量的矛盾冲突，这些矛盾冲突不仅仅表现在工艺方案的选择、技术装备的选型和工程建造的实施等方面，并不是基础科学、技术科学和工程科学知识所能涵盖的，往往涉及到工程与社会、工程与经济、工程与生态环境、工程与文化伦理的和谐共融发展的问题。钢铁冶金工程作为人工造物活动，无论是钢铁冶金工程的建造还是钢铁产品的生产，都必须要考虑到整体经济效益的最优化，降低工程建造/生产运行成本、减少资源/能源消耗、减少污染物排放，还要兼顾考虑整体与局部的经济效益、社会效益和生态环境效益等方面。钢铁冶金工程建造/运行过程中，既要保证工程的安全与质量，还要满足成本与时间等要求。

从钢铁冶金工程全生命周期的视角来看，包括了规划决策、勘察设计、工程建造、运行维护、工程退役等几个阶段，每个阶段的工程活动中，其重点核心和关键内容都不尽相同。而工程管理是贯穿于整个工程活动的综合性工作，包括工程活动的全过程和各专业领域的管理，涉及了经济、社会、政治、人文、伦理、生态等诸多领域。

因此，钢铁冶金工程管理知识融合了工程技术、经济、管理、法律/法规、人文、伦理等诸多领域，结构复杂、交叉融合，具有更强的综合性、集成性、融合性、协同性、和谐性、跨学科"泛专业性"等特征。

因此，钢铁冶金工程活动过程中，工程管理知识是不可或缺的工程知识，在工程思维、工程决策、工程设计、工程建造、工程运行等环节中，工程管理知识更是发挥重要的作用，是工程活动过程中重要的"斡件"，特别是工程决策过程中，工程管理知识发挥无可替代的重要作用，是工程决策者进行综合、选择、权衡、协调、集成的重要知识基础和决策依据[11]。

4.7 冶金工程知识对工程创新与产业进步的引领

纵观钢铁冶金工程的演化历程不难看出，理论源自于实践，理论是通过对实践活动的观察、思考、感悟、试验验证、总结归纳、整理提炼等一系列步骤升华而成的，反过来再用理论来指导工程实践活动，因而，理论是一种最具代表性的知识。从工程哲学的角度分析，对于钢铁冶金工程学而言，其理论基础不仅包括基础科学理论、技术科学理论，还包括工程科学理论（冶金流程工程学、冶金流程工程集成理论与方法）。因而，冶金工程学理论是建构在冶金基础科学理论、冶金技术科学理论，以及冶金工程科学理论基础之上的知识体系，是经过试验研究凝练、升华和验证的冶金工程的知识精华，对于钢铁冶金工程，特别是全流程和全厂性的工程问题具有普适性。

在研究钢铁制造全流程或全厂性的工程设计及动态运行全过程中，遇到的问题往往是工程与工程科学问题。因为工程体现了相关的功能不同的异质技术的集成，不但要充分考量技术要素和工程要素的合理配置，而且还必须考虑到资源、能源、土地、资金、生态环境、市场、劳动力等基本经济要素的有效配置，所以对工程和工程科学的思维方式必然是一种开放的、动态的、集成优化的复杂思维模式。在工程模型的构造过程中，其思维逻辑和工程知识一般应包括：

（1）确立正确的工程理念（这方面在工程决策、规划、设计等过程尤为重要），而且正确工程理念则是来自于对工程知识的学习、实践、认知、领会和运用；

（2）建立符合时代需要的集成理论和方法（即在规划、设计和生产运营过程中，要重视集成性优化和进化性创新的理论、方法和知识）；

（3）在冶金工程设计与构建过程中展开、落实，并使工程理念和理论逐步物质化（即获得要素-结构-功能-效率协同优化的工程系统，建立动态有序、协同连续、耗散优化的钢铁制造流程及其网络结构）；

（4）在钢铁制造动态运行与管理中具体实施，获得预期的多维目标（即在钢铁冶金工程系统的实际动态运行过程中，必须注意多目标优化及其选择与权衡）；

（5）开展生命周期评估（即从自然资源的源头开始，经过流程系统的运行、加工制造、消费、社会废弃物的资源化-能源化消纳-处理-再资源化等过程来评价钢铁冶金流程工程系统的价值及其合理性）；

（6）深化钢铁冶金工程系统对自然-社会环境的适应性、进化性以及价值评估的认识（也就是要拓展钢铁冶金流程的正面影响，避免负面影响）。

由于钢铁冶金工程知识体系复杂庞大，涉及的专业领域广泛，知识门类众多，是多学科知识的汇聚和交融，因而钢铁冶金工程知识是一种综合知识、集成知识，是一个工程知识群，并非某个单一基础学科的知识。钢铁冶金工程知识（群）具有系统集成性。工程是开放、复杂系统，集成是工程知识的基本特征，钢铁冶金工程知识（群）的"集成性"主要体现在：

（1）现代工程目标是多目标性的，需要集成性知识；

（2）现代工程知识是复合性、链接性的，需要集成性知识；

（3）现代工程的技术复杂性要求有序性、协同性——集成性知识；

（4）现代工程的创新集成性和环境和谐性也需要集成性知识。

从钢铁冶金工程演化和工程创新的发展历程可以看出，工程知识创新不仅体现于单元性、局部性创新，而且更体现于集成创新，这是由于钢铁冶金工程的基本属性和工程特征所决定的，因而工程集成创新是经常出现的、重要的工程知识创新的活力，工程知识的重要活力之一是持续进行着的因时因地的集成创新。

钢铁冶金工程知识群的系统集成性还包括集成软件的硬件化、硬件的软件化、核心原理知识与构成性知识之间结构化、功能化集成、多环节、多结点、多层级的接口（界面）知识——非线性、动态耦合集成。集成度的内涵包括层次性集成度、链接性集成度、相关性、涌现性等。

钢铁冶金工程知识（群）集成的对象和要素，主要包括：要素集成、方法集成、过程集成、装备集成、目标集成以及环境-生态集成等。在工程功能集成创新方面，表现在功能优化、功能拓展和新功能涌现。

钢铁冶金工程知识集成的一般机制是：钢铁冶金工程知识集成-驱动工程要素整合集成-通过工程设计和建构-形成工程体系实物结构-工程实体运行转化为功能-体现工程的现实生产力作用。即要素群选择-有序、整合、协同-工程设计-工程建构-工程运行-体现功能-转化为现实生产力。

21世纪初，我国自主设计建造的首钢京唐钢铁厂工程，基于新一代可循环钢铁制造流程工程理念，按照冶金流程工程学理论与方法，自主设计建设的、具有21世纪国际先进水平的大型钢铁项目，也是我国建设的真正临海靠港的全部生产薄带材的大型现代化钢铁厂[12]。首钢京唐钢铁厂具有"三大功能"，即：不仅注重优质、高效的钢铁产品制造功能，同时注重高效、清洁的能源转换功能和消纳、处理大宗社会废弃物并实现再资源化的功能。首钢京唐厂一期设计规模为年产粗钢870万~920万吨/年，该工程于2007年3月开工建设，2010年6月26日全面竣工，顺利投产。构建了以2座高炉、1个炼钢厂、2条热连轧生产线和4条冷轧生产线为物理框架的高效洁净钢铁生产工艺流程。自主设计建造了$5500m^3$高炉、高效率低成本洁净钢生产平台、5万吨/天海水淡化装置等一批具有代表性的先进工艺与装备，采用先进技术220余项，其中自主创新和集成创新达到了三分之二。自主设计建造的具有21世纪国际先进水平的新一代钢铁制造基地，成为临海靠港建设大型钢铁厂的工程示范，引领了我国钢铁工业的发展方向。众所周知，首钢京唐钢铁厂是在冶金工程理论和知识的指导下设计建造的大型钢铁工程，对冶金工程创新与钢铁产业技术进步的引领与推动作用显著、意义重大。

5 工程哲学对冶金学、冶金工程思维进程的引领性

5.1 哲学视野下的冶金工程

从工程哲学的视角来看，钢铁冶金工程设计过程和生产运行过程长期以来集中注意的是局部性的"实"，而往往忽视了贯通全局性的"流"。未来工程设计、生产运行和过程管理中既要注重解决具体的、局部的"实"，更应集中关注贯通全局的"流"。流程工程如果脱离了"流"的动态概念，就等于失去了"灵魂"。钢铁冶金工程设计和工厂生产运行都要"虚""实"结合，必须首先确立理念——"虚"；构建并形成动态有序、协同稳定、连续紧凑的开放系统——优化的耗散结构体系，即通过工程设计和动态的生产运行付诸于实践——"实"，实现复杂系统动态运行过程的多目标优化，追求流程运行过程的耗散"最小化"。总而言之，应当从要素-结构-功能-效率集成优化的工程理念出发，在工程设计中和生产运行过程中体现出动态有序、协同连续的本质要求，这是动态精准设计和钢铁厂实际生产运行过程的理论核心。

面向未来，钢铁工业的发展方向是构建绿色化、智能化、质量-品牌、集成化、多元化的产业体系，即通过加强资源和能源可持续性供应、人才、资本、效率、环境等要素的支撑，拓展市场、功能、服务；提升效率、改进质量、扩大品种、塑造品牌、延伸产业价

值链以及环境-生态链,从而增强企业的综合竞争力和可持续发展能力[13]。

具体而言,在经过一段时间的快速增长以后,当前,我国钢铁工业要压缩过剩产能、淘汰落后产品、落后工艺和装备、落后的生产线,必须以绿色化、减量化作为产业升级的主要方向,重视资源、能源的可获取性和生态环境的可承载能力,坚决制止以高消耗、高能耗、高污染为代价,破坏生态环境,盲目扩张钢铁产能、提高产量的粗放式发展方式,着力构建资源、能源节约型和生态环境友好型钢铁企业。重视能源的高效利用、节能减排、清洁生产,特别是能量流网络化的高效利用,进一步提高钢铁冶金全流程的能源效率。重视所有钢铁产品的高效率、低成本洁净钢生产平台的构建;重视钢铁产品的质量、性能与功能的时代适应性,研发战略性新兴产业所需要的钢材,特别是能源、交通、动力、海洋用钢等,贴近客户服务,满足客户需求,倡导冶金工程与材料工程相结合的产品研发思路和方法,在扩大钢材品种的同时,更要重视产品品牌的塑造。构建钢铁冶金工程基于"互联网+智能设计+智能制造"体系,实现物质流、能量流、信息流高效耦合协同运行;基于新一代信息技术实现钢铁制造过程和能源转换过程的智能化调控,不断提高系统运行效率、钢铁产品质量,实现钢铁制造流程的智能化运行;同时,高度重视智能化服务和市场拓展。重视人力资源综合素质的提高,强调"德才兼备",善于学习、不断创新,培养具有团队创新能力的复合型梯队人才队伍,特别是重视战略型高端人才的培养和领军人才的培养。重视市场容量的演变及其区域经济的变化趋势,防止区域内同行业的恶性竞争。

5.2 冶金工程知识的研究思路和发展方向

在冶金工程相关的微观基础冶金学和专业工艺冶金学的基础上,进一步发展宏观动态冶金学,以此三个层次冶金学集成理论作为研究思想进路,构建钢铁冶金工程设计创新理念、形成理论及方法体系。在深入认识钢铁制造流程动态运行物理本质和本构特征的研究,探索物质流、能量流、信息流的协同耦合和集成理论研究,重视钢铁制造流程动态运行中"流""流程网络"和"运行程序"的研究,以及对钢铁企业结构、功能、效率的影响。

"三流"的集成,离不开"三网"的物理框架融合和信息贯通。因此,要重视以网络化整合、程序化协同为重要手段(集成创新),重视概念设计、顶层设计和动态精准设计,提高钢铁制造流程的设计水平和生产过程的运行智能化水平。重视新工艺、新技术、新装备的开发、设计制造并通过多因子、多层次、多尺度集成优化,有效地、动态地体现到钢铁生产流程中。

高度重视全厂性、全流程层次上能量流研究及其网络化整合,提高能源利用和转化效率,进一步从流程结构、网络合理化、网络延伸上推进节能减排和排放过程的源头削减。

高度重视信息流在钢铁制造流程中贯通,提高与物质流、能量流优化集合的信息有效调控水平。重视用于钢铁冶金工程设计的计算机软件、硬件的开发与应用,推动三维仿真设计在钢铁冶金工程建设项目中的全面推广应用。采用信息化、数字化、智能化设计手段和方法,实现设计参数结构优化和仿真模拟,模拟工程建造、运行、管理、维护的过程,形成虚拟现实的仿真体系。逐渐构建钢铁冶金工程基于"互联网+智能设计+智能制造"体系,推动智能化冶金三维仿真设计和钢铁冶金生产过程的三维动态模拟仿真,实现钢铁冶金工程的智能化、绿色化发展。

高度重视具有综合素质的精英人才、卓越工程师、战略科学家的培养、训练和使用。高度重视关于生态环境保护、气候变化等时代责任和社会伦理命题的战略性对策研究。高度重视工程管理知识的凝练、传播、学习和运用，自觉应用先进的工程管理理念、知识和方法进行钢铁冶金工程项目管理、质量管理、运营管理，不断提高管理效率和效益。

综上所述，通过冶金工程知识一系列深入研究，中国钢铁工业应当以绿色化、智能化发展作为产业转型升级的主要方向，重视产业结构的调整升级和企业结构的顶层研究、顶层设计，解决产业层面、企业层面的复杂性命题，获得新的市场竞争力和可持续发展能力。

5.3 冶金学和冶金工程知识的发展展望

当今，信息-智能、环境-生态、资金-金融的影响力日益深入，同时产业、企业的成本、质量竞争压力越来越大，冶金学不应局限在小尺度问题的思考上，冶金学需要拓展和创新。

回顾起来，1925年英国法拉第学会（Faraday Society）在伦敦召开的"炼钢过程物理化学"讨论会，深刻地影响着当时的冶金学与冶金工业的发展，然而，迄今已经将近100年，难道不应该有新的思考、创新发展？初步看来新冶金学应该在微观基础冶金学（冶金物理化学、金属学等）、工艺冶金学（炼铁学、炼钢学等）的基础上，进一步扩展提升，建立起宏观动态冶金学（冶金流程工程学等）。

冶金科学的命题已经或是正在发生演变：

（1）由原子尺度扩展到流程尺度；

（2）由平衡态研究推进到耗散结构研究；

（3）由局部的孤立系统研究集成到全流程的开放系统研究；

（4）由特定点的"质-能"衡算发展到开放系统的输入流-输出流研究。

（5）宏观动态冶金学体现着整体性、自组织性、结构化、绿色化、智能化的内涵和特征。

宏观动态冶金学——冶金流程工程学是大时-空尺度、跨时-空尺度的、兼涉物质-能量-时间-空间-信息的整体性学问，并且体现着以动态-有序、衔接-匹配、协同-连续为特征的结构化知识，构建合理的耗散结构和耗散过程。冶金流程工程学是直接面向下列方向的：

（1）面向绿色化、智能化的冶金学；

（2）信息自组织与他组织融合的冶金学；

（3）开放-动态、集成一体化的冶金学；

（4）多因子-多层次嵌套、集成、协同的冶金学。

钢铁冶金工程创新和产业进步，不仅依赖于冶金学科的创新，更有赖于高素质工程技术人才的培养和造就，其前提是冶金工程知识的不断创新、完善，使其体系化、集成化、理论化，成为冶金工程技术人员就职前教育和继续工程的持续获取新的工程知识的重要来源和途径。

卓越的工程师不但要有精深的专业知识，还要有宽阔的知识视野，系统的工程思维，不能成为"专业分工的奴隶"，不能局限于"碎片化"的知识，卓越工程师要具有综合集成的创造力，还应有高瞻远瞩的工程理念、卓越非凡的创新精神、强烈的社会责任感和历史使命感。

时代呼唤着新冶金学，21世纪应该有新冶金学问世，对此中国冶金学人应当仁不让，不应沿着老路迷失在细节之中，不应在老框框内打转转，要看到大潮流、大方向，要看到更上一层楼的上升通道，建立新的冶金工程知识体系，开辟学科发展新路径。

6 结语

（1）钢铁是经济社会发展的重要基础材料、结构材料和功能材料，仍是当前的首选材料。钢铁冶金工程是重要的基础产业和流程制造业，是经济社会发展的重要产业基础。

（2）钢铁冶金工程的物理本质是一类开放的、远离平衡的、不可逆的、由不同结构-功能的单元工序过程经过非线性相互作用，动态耦合嵌套构建而成的流程系统。

（3）钢铁冶金工程知识是由三个层次的知识集成构建而成的框架体系，即原子/分子层次上的微观基础冶金科学知识，工序/装置层次上的专业工艺冶金学技术知识和全流程/过程群层次上的动态宏观冶金学工程知识。

（4）在冶金工程相关的微观基础冶金学和专业工艺冶金学的基础上，应进一步发展宏观动态冶金学，以此三个层次冶金学集成理论作为研究思想进路，构建钢铁冶金工程的创新理念，并形成理论及方法体系，进一步发展钢铁冶金工程知识。

致谢

感谢中国工程院工程知识论项目研究资助，感谢项目组专家学者提出的宝贵意见建议。

参考文献

[1] 殷瑞钰. 冶金流程工程学 [M]. 2版. 北京：冶金工业出版社，2009：1-10.
[2] 殷瑞钰. 冶金流程集成理论与方法 [M]. 北京：冶金工业出版社，2013：14-16.
[3] 殷瑞钰. 冶金流程工程学 [M]. 1版. 北京：冶金工业出版社，2004.
[4] 张春霞，殷瑞钰，秦松，等. 循环经济社会中的中国钢厂 [J]. 钢铁，2011，46（7）：1-6.
[5] 张春霞，王海风，张寿荣，等. 中国钢铁工业绿色发展工程科技战略及对策 [J]. 钢铁，2015，50（10）：1-7.
[6] 殷瑞钰. 过程工程与制造流程 [J]. 钢铁，2014，49（7）：15-22.
[7] 徐匡迪. 中国钢铁工业的发展和技术创新 [J]. 钢铁，2008，43（2）：1-13.
[8] 张寿荣. 论21世纪中国钢铁工业结构调整 [J]. 冶金丛刊，2000（1）：39-44.
[9] 殷瑞钰. 关于智能化钢厂的讨论——从物理系统一侧出发讨论钢厂智能化 [J]. 钢铁，2017，52（6）：1-12.
[10] 殷瑞钰. "流"、流程网络与耗散结构——关于流程制造型制造流程物理系统的认识 [J]. 中国科学：技术科学，2018，48（2）：136-142.
[11] 殷瑞钰，汪应洛，李伯聪. 工程方法论 [M]. 北京：高等教育出版社，2017：45-60.
[12] 张福明，颉建新. 冶金工程设计的发展现状及展望 [J]. 钢铁，2014，49（7）：41-48.
[13] 殷瑞钰，张寿荣，张福明，等. 现代钢铁冶金工程设计方法研究 [J]. 工程研究-跨学科视野中的工程，2016，8（5）：502-510.

Research on Iron & Steel Metallurgical Engineering Knowledge

Yin Ruiyu   Zhang Fuming   Zhang Shourong   Xie Jianxin

Abstract: Engineering knowledge combined with iron and steel metallurgical engineering is discussed in this paper from the perspectives of engineering integration, construction, operation and management. From the viewpoint of knowledge theory, iron and steel metallurgy processes require the integration of many knowledge categories, such as physical, chemical, technical, management knowledge in engineering science, technology, design, management, as well as engineering philosophy knowledge. In this paper, the important role of iron and steel materials in economic and social development is analyzed. The main characteristics of iron and steel metallurgical engineering process operation are expounded, and the physical essence of dynamic operation of iron and steel metallurgical process are analyzed. The structure of metallurgical engineering knowledge is systematically combed, and the features of complexity, cooperation, integrity and integration of metallurgical knowledge system are revealed, and the developing prospect of accurate application of contemporary metallurgical engineering knowledge to the intelligence of metallurgical engineering is revealed. The structure system, formation and development process of iron and steel metallurgical engineering knowledge are discussed. The promoting effects of engineering evolution and technological progress on the innovation of metallurgical engineering knowledge are expounded. The importance of "interface technology" in the coordination and integration of metallurgical manufacturing process and the promoting effect of information technology on the intellectualization of iron and steel metallurgical engineering are discussed. From the perspective of engineering philosophy, the developing direction and goal of intellectualization and green macroscopic dynamic metallurgy and metallurgical engineering knowledge are discussed in this paper.

Keywords: iron and steel metallurgy; engineering knowledge; engineering philosophy; metallurgical process; interface technology

协同设计管理在三维工厂设计中的研究与应用

摘　要：首钢京唐钢铁厂是国家"十一五"规划的重点建设项目，是具有21世纪国际先进水平的现代化钢铁厂。北京首钢国际工程技术有限公司是该工程的总承包设计单位，在该工程设计中采用协同设计管理和三维工厂设计技术，通过构建协同设计管理环境开展三维工厂设计项目，解决设计应用中数据和软件资源整合与共享，下游专业要能够方便地利用上游专业的设计成果，在不同专业软件之间双向数据流动和进行协同设计的管理及设计流程的变革。

关键词：冶金；工程设计；协同设计管理；三维工厂设计

1　引言

首钢京唐钢铁厂是我国"十一五"规划的重点项目，是我国钢铁工业落实科学发展观和钢铁产业发展政策，为促进我国钢铁产业布局调整、优化产业结构、提升我国钢铁工业国际竞争力而实施的重大项目，坚持"高起点、高标准、高要求"的建厂方针，实现钢铁厂"钢铁产品一流、工艺技术一流、生态环境一流、经济效益一流"和"钢铁产品精品制造基地、循环经济和自主创新示范基地"的建厂目标。

在此背景下开展首钢京唐钢铁厂工程设计难度和深度是前所未有的。北京首钢国际工程技术有限公司采用新的技术创新设计，通过理念新颖、紧密贴合京唐钢铁厂目标需求的协同设计管理三维工程设计来完成创新的技术项目。这是国内冶金企业首次通过在协同设计管理平台构建适合于开展协同设计管理工作的环境，解决设计应用中数据和软件资源整合与共享、下游专业方便地利用上游专业的成果、不同专业软件之间双向数据流动来进行协同设计的管理及设计流程的变革。在提供三维模型的同时提供施工图纸，并达到准确无误，从而大大提高设计精准度，提高工作效率并节约费用。

2　构建协同设计的总体技术思路

所谓协同设计，就是应用先进设计管理技术，进行资源性、功能性整合，并进行设计管理流程再造，实现不同的设计单位、设计人、设计专业之间进行协调和配合。

三维工厂设计打破了分专业"出图"的制约，具有多专业在一个模型中共存，建立的是一个全面的数字化工厂的特点，协同设计管理在三维工厂设计上具有独特的优势，也与三维工厂设计的发展趋势是相符的。

本文作者：张福明，张严，王建涛。原刊于《中国科技成果》，2010（5）：18-20。

2.1 设计流程的再造与优化

以往的设计流程为主工艺专业首先开展工作，再以提出资料的方式要求诸如土建等配套专业开始设计。然后各专业在做完一定阶段的设计后，再次以互提资料的形式交叉沟通，反映本专业的设计情况和对其他各个专业的设计要求。一旦接到资料的配套专业发现所提出的设计要求难以实现，必然通知相关专业修改设计或重新设计。

在这种模式下的设计是一种多专业依次开始、渐次反馈、修正（甚至推倒重来）、螺旋式展开的串行工作模式。完全依靠设计完成后的阶段性设计成品会审来避免专业之间的各类冲突。

按照现代项目管理的基本理论和网络计划控制技术，对设计流程进行了再造和优化，借助先进的IT技术和管理软件，在协同设计环境下，各专业则是同时开始工作，每个专业、每个人及专业管理人员几乎能看到对方实时的设计，协同设计管理实现并行设计，保证各个专业有序而及时的沟通。

2.2 优化设计审核模式

传统的设计审核是分专业、分阶段，由少量专门人员承担设计检查。

在协同设计环境中所有合作设计人员可以随时进行相互的设计审核，甚至可异地参与，不受时间、空间和专业的限制，实现所见即所得，极大地提高设计的质量和工作的效率。

从三维模型直接生成的二维图中的颠覆性问题、专业冲突性的问题大大减少，二维图设计审核投入的人力少，节省人力资源，提高了设计质量和设计效率。

2.3 提高设计的精准性

通过自动完成的碰撞检查，保证了设计的精确性，避免了因设备、结构、管道"碰撞"造成的损失，减少了建设过程中施工返工，提高了工程建设的工作效率，也使工程设计更趋优化，更加科学与合理。

2.4 提高整体设计效率

当采用协同设计管理开展三维工厂设计，可以整合不同文档格式和应用软件。建立统一设计标准，上游专业输入的信息，下游专业只能使用而不用修改，工程人员能够迅速、准确地找到所需的工程信息，解决多专业配合过程中出现很难找到所需信息、找到的信息不能直接引用的问题和信息重复输入、信息存在不唯一、信息标准不统一的问题。面对设计变更造成的同一个设计文件具有多个版本，可以保证只有最新的版本在被共享。

2.5 缩短设计工期

在协同设计管理整个设计过程中，实现并行设计、设计审核提前、高效碰撞检查、三维模型提取二维图纸和材料表在设计阶段完成，以及整体设计的提高，在高质量的前提下缩短设计工期。

2.6 支持工程设计向工程总承包演进

三维工厂设计的重要特性是管理多专业元素,如管道、设备、建筑结构等。在协同设计管理的基础上,多专业的模型可进行碰撞检查;利用接口输出数据进行应力分析、流体分析;根据模型可生成任意平断面图、轴测图并自动标注尺寸,项目各个阶段即时材料用量报表输出,现场吊装分析把安装的碰撞也消灭在设计的过程中。这将非常有利于设计单位逐步过渡到开展设备加工制造和现场建设、设备安装、试车、交钥匙工程的组织管理。

3 协同设计管理在三维工厂设计中的实践应用

3.1 建立协同工作规则

根据首钢京唐钢铁厂 5 个项目的专业特点,以项目为核心,把所有三维工厂设计内容集中在同一个项目数据源中,定义目录结构。该目录结构充分考虑了冶金专业进行设计的特点,界面采用最流行的 Windows 资源管理器工作界面,用户不必关心数据的物理位置,无需进行更多的应用培训,极易上手。

在协同设计环境统一命名文件,三维设计文件名与该项目二维设计图相对应,便于三维设计模型中与施工图对照。

3.2 实现数据共享

在统一的环境下分区建立模型,所有模型参考到一起,有效地避免不同设计人之间的主观误差或产生的设计错误。在设计阶段就避免了类似项目大量出现的碰撞。经统计,仅管道碰撞点就检测出 90 多处。

如图 1 所示,首钢京唐钢铁厂制氧站分子筛纯化系统是设备、管道、应力计算等相对集中和复杂的区域,三维工厂设计提高了设计精度,准确客观地反映了设计中的碰撞。

图 1 首钢京唐钢铁厂制氧系统储罐区 3D 设计图

数据共享还实现了自动提取材料与设备清单，极大避免了二维设计时材料设备量统计等繁琐的设计工作，简化了设计流程，为设计人员节省了大量时间和精力，提高了设计效率，降低了设计成本。

3.3 实现文件控制和查询的管理

协同设计管理具有完备的用户管理机制及权限继承体系，按照不同的用户角色设置不同的安全访问权限，使不同的人所能浏览和操作的文件及文件内容有所不同，同时还可以进行日志文件管理。

整个项目完成后统一导出进行了归档。当再有此类的新项目时，可以直接导入修改三维模型文件，极大地提高了设计效率。

3.4 实现了远程异地设计和沟通

协同设计实现了在首钢京唐钢铁厂施工现场如同在设计公司本地设计一样方便，公司本地随时配合施工现场发现的问题修改三维模型。

项目的各个参与方（设计单位、施工单位、建设方等）也在一个统一的平台上工作和交流，增强了项目的协同性，提高了工作效率。京唐钢铁厂和设计公司不在同一地区，实现远程、异地设计就显得尤为重要。

4 应用的特点

（1）方便项目资料的发布和各专业、部门之间的设计配合。可以快速找到其他专业的图纸和文档，方便引用。修改文件后可以附加相应的注释，使其他设计人员及时了解文件的设计状况。

参与项目的工艺专业负责人深有感触说：以前各专业每周要开一次碰头会，最起码要2~3h，协商解决设计中出现的问题，由于意见不统一常常达不到效果。而在协同设计下规范了设计人的工作，能够及时发现专业间或设计文件出现的问题，通过公司内部网络信息很容易沟通解决。

（2）有良好的检入/检出机制，同一时间同一个文件只能一个用户编辑，保证了文档的唯一性。可以创建多个文件版本，但只可以在最新版本上进行修改编辑；任何历史版本都可以回溯，避免发生设计错误。

（3）集成各种设计软件（MicroStation、AutoCAD 等）以及 OFFICE 办公软件，可以将系统中的属性信息自动写入到设计文件图框中的内容中，减少了设计人员工作量。

（4）借助 Web 客户端，管理人员工作电脑上可以不安装任何工程软件（AutoCAD/MicroStation/Office），只用 IE 浏览器就可以查看各种 DGN/DWG 图纸，光栅影像文件和 OFFICE 文件。

5 结束语

在首钢京唐钢铁厂设计中采用三维工厂协同设计管理，北京首钢国际工程技术有限公司大力开展了各专业骨干人员培训，掌握了协同设计和三维工厂设计的精髓，通过实践更将学习到的知识转化到实际工程项目上。经过不断探索，总结形成了公司的管道标准，解决了因冶金管道应用没有统一标准的三维建库等问题，有效地推动了项目的开展。建立的行之有效的规章制度，确保了协同设计管理的规范化、标准化。协同设计管理的应用填补了冶金行业在这一领域的空白，为今后协同三维工厂设计提供了宝贵的经验，也必将推动相关行业工程设计水平的不断提升。

钢铁制造流程智能制造与智能设计的研究

摘　要：本文分析了钢铁制造流程智能制造和智能设计的主要内容。钢铁制造流程智能制造包括：智能化流程设计、智能化流程生产运行、智能化流程管理、智能化流程供应链、智能化流程服务体系。钢铁制造流程智能设计包括：基础科学研究，采用CAE三维仿真设计分析计算技术；技术科学研究，采用机械三维仿真设计技术；工程科学研究，采用数字化三维仿真工厂设计技术。实现钢铁制造流程智能制造，智能设计既是智能制造的关键环节，又发挥着先导和引领作用，没有先进的智能设计，很难有先进的智能制造，必须高度重视钢铁制造流程智能设计的作用和意义，智能设计的先进程度决定着智能制造的先进水平与未来。

关键词：钢铁制造流程；智能制造；智能设计；研究

1　引言

　　制造业是国民经济的主体，是立国之本、兴国之器、强国之基。打造具有国际竞争力的制造业，是我国提升综合国力、保障国家安全、建设世界强国的必由之路。改革开放以来，我国制造业持续快速发展，建成了门类齐全、独立完整的产业体系，有力推动工业化和现代化进程，显著增强综合国力，支撑世界大国地位。

　　但是，同世界先进水平相比，在自主创新能力、资源利用效率、产业结构水平、信息化程度、质量效益等方面还存在明显差距，我国制造业转型升级和跨越发展的任务紧迫而艰巨。面对新的国际、国内环境，我国立足国际产业变革大势，通过实施《中国制造2025》，全面提升我国制造业发展质量和水平，改变我国制造业"大而不强"的局面，将我国建成具有全球引领和影响力的制造强国。

　　智能制造就是落实《中国制造2025》的重要举措。智能制造是基于新一代信息通信技术与先进制造技术深度融合，贯穿于设计、生产、管理、服务等制造活动的各个环节，具有自感知、自学习、自决策、自执行、自适应等功能的新型生产方式。实现制造的数字化、网络化、智能化发展，数字化是基础，网络化是关键，智能化是方向。加快发展智能制造，是培育我国经济增长新动能的必由之路，是抢占未来经济和科技发展制高点的战略选择，对于推动我国制造业供给侧结构性改革，打造我国制造业竞争新优势，实现制造强国具有重要的战略意义[1~4]。

　　虽然智能制造与智能设计的手段多种多样，但是数字化三维仿真设计技术却是实现智能设计唯一的先进手段，没有数字化的设计，智能设计就没有了实现智能的数字化基础，更无法实现虚拟仿真设计。智能制造与智能设计是先有智能设计，后有智能制造，是二者

本文作者：颉建新，张福明。原刊于《第十二届中国钢铁年会论文集》，2019：228-234。

的内在关系,因此,必须优先发展智能设计,再带动智能制造的发展。

制造业包括流程型制造业和离散型制造业,钢铁制造流程是典型的流程型制造业,实现钢铁制造流程的智能制造,是实现制造业智能制造的重要组成部分,而智能设计是钢铁制造流程智能制造关键、重要的环节。

2 钢铁制造流程智能制造的主要内容

2.1 钢铁制造流程智能制造的内涵及组成

钢铁制造流程智能制造包括:智能化流程设计、智能化流程生产运行、智能化流程管理、智能化流程供应链、智能化流程服务体系等。钢铁制造流程智能制造内涵应是围绕钢铁制造流程"产品制造、能源转换、废弃物消纳处理与资源化"三个功能的价值提升,基于物质流网络、能量流网络、信息流网络的关联和协同集成,将物联网、大数据、云计算等信息手段与钢铁制造流程的设计、运行、管理、服务等各个环节深度融合,实现信息深度自感知、智能化自决策、精准控制自执行、自适应等功能,可有效优化钢铁制造流程结构,特别是动态运行结构,提升全流程运行过程智能化控制和管理水平[5~9]。其主要组成有:

(1) 以提升产品质量为目标,通过生产运行参数的"窄窗口"优化,提高大宗钢材及关键钢材质量的稳定性、可靠性和适用性,实现产品质量的品牌化。

(2) 以降低资源、能源消耗,减少过程排放为目标,实现清洁生产、环境友好,推动绿色化发展。

(3) 形成以采购、生产、销售、物流、用户服务等多目标优化的智能化经营模式。

(4) 以提高资金利用效率为手段,加快企业资金流动及进一步增值。

(5) 以延伸产业链为手段,促进低碳经济、循环经济发展。

(6) 加快钢铁工业转型升级。

2.2 钢铁制造流程智能制造的具体内容

钢铁制造流程智能制造是针对钢铁企业的整体解决方案,其具体内容有:

(1) 钢铁制造流程智能设计。它是数字物理系统的先导,是建立先进的物理系统(钢铁制造流程)的起点,同时,这个物理系统及所构成的硬件网络将有利于数字信息便捷、高效地与之相融合。智能设计是全方位研究物质流智能化、能量流智能化、信息流智能化的综合方案。

(2) 智能化流程生产运行。它涉及从焦化、烧结开始到轧钢生产过程动态运行的、起伏变化的物质流、能量流和各类信息流,即动态变化的"三流"要通过智能化制造平台实现自感知、自决策、自执行和自适应,生产过程需采用先进传感器件、仪器仪表、各类自动化控制、制造平台、可视化等关键共性技术。

(3) 智能化流程管理。它是战略性的高层次综合判断、决策、执行系统,涉及经营战略目标、财务资金运筹、客户服务,并延伸到与经济、社会相适应等高层次目标的决策与执行,延伸产业链,促进低碳经济、循环经济发展,推动钢铁工业转型升级。

（4）智能化流程供应链。它涉及物料采购、储存、运输和合理分配，涉及产品订单、发运、储存和销售的高效合理化，它是企业物流输入和输出的智能化问题。

（5）智能化流程服务体系。它涉及采购、生产、销售、物流、用户服务等在内的智能化经营模式，目的是提高资金利用效率，加快企业资金流动和进一步增值。

实施钢铁制造流程智能制造，应在深刻理解流程动态运行过程物理本质的基础上，从流程运行要素及优化运行网络、运行程序的物理模型入手，构建出全流程网络化、层次化信息流模型。通过工序功能集合的解析优化、工序之间关系集合的协同优化和流程工序集合的重构优化，最终实现钢铁制造全流程结构优化、功能拓展、工业生态园区的构建和有效运行[5~9]。

3 钢铁制造流程智能设计的主要内容

智能设计是指应用现代信息技术，采用计算机模拟人类的思维活动，提高计算机的智能水平，从而使计算机能够更多、更好地承担设计过程中各种复杂任务，成为设计人员的重要辅助工具[10]。钢铁制造流程智能设计包括三方面：一是基础科学（研究尺度：分子原子尺寸），研究科学的合理性，采用CAE三维仿真设计分析计算技术，模拟仿真研究冶金工艺过程、温度场、流场等；二是技术科学（研究尺度：工序或装置尺寸），研究技术的可行性，采作机械三维仿真设计技术，模拟仿真研究冶金装置的优化；三是工程科学（研究尺度：制造流程或车间尺寸），研究工程项目的整体性、系统性、协同性优化，采用数字化三维仿真工厂设计技术，模拟仿真研究钢铁制造全流程的优化，实现设计、施工、运维的一体化[11]。

也就是说，在钢铁制造流程物理系统建成之前，通过智能设计先虚拟出了钢铁制造流程物理系统，从基础科学研究、技术科学研究、工程科学研究三个层次进行虚拟仿真研究，提出改进、优化、创新的钢铁制造流程物理系统，并为钢铁制造流程实现智能制造提供所有数字化基础数据，这样，才能真正实现先进钢铁制造流程物理系统的智能制造。

3.1 钢铁制造流程全流程智能设计

（1）研究构建动态精准设计体系。以钢铁制造流程中的焦化、烧结球团、炼铁、炼钢、轧钢、工业炉、冶金装备、电气与自动化、能源与环境、建筑工程设计及关键共性技术为研究对象，以工程科学"三流一态"、耗散理论为指导，通过运用先进的三维仿真设计技术，实现钢铁制造流程数字化三维工厂设计和关键共性技术的突破。研究方向是以动态精准三维仿真设计为基础，构建基于循环经济理念的钢铁制造流程，创建具有钢铁产品制造、高效能源转化、消纳废弃物并实现资源化"三大功能"的现代钢铁制造流程，基于动态精准设计理论和数字化三维设计体系，构建三维仿真设计、仿真研究、虚拟现实的高效精准设计平台。

（2）研发现代钢铁制造流程工艺技术和装备。通过对焦化生产、烧结球团生产、炼铁生产、炼钢生产、轧钢生产过程进行数值模拟仿真，实现钢铁制造流程工艺技术和装备技术的精准计算和设计优化，进而提高钢铁生产产品质量、降低产品成本，全面提升钢铁企业市场竞争力和可持续发展能力。研发基于蓄热式燃烧、分段加热、低氮氧化物排放的现代大型捣固焦炉系统技术，实现煤炭资源的高效化利用，减少资源和能源的消耗；研发现代球团工艺，实现高效、低耗、低排放，降低粉尘、SO_x、NO_x、二噁英等污染物的排放；应用三维仿真技术，研究熔融还原、直接还原的冶金过程和"三传一反"现象，建立数学模型应用CFD数值模拟技术，解析气体运动、燃烧、传热、传质等传输过程，开发具有工程化应用可行性和可靠性的冶金装置；通过应用冶金流程集成理论和方法，建立数学模型，以非线性数学规划、动态甘特图等为研究方法，优化钢铁制造流程铁素物质流和能量流的运行，研发可靠适用的界面技术，提高钢铁厂运行效率和能源转换效率。

（3）打造先进的冶金三维仿真设计技术应用研发平台，推动三维仿真设计技术在钢铁行业全面推广应用，为实现数字化、智能化钢铁制造流程奠定基础。

3.2 钢铁制造流程各工序智能设计

3.2.1 焦化工序智能设计

基于蓄热式燃烧、分段加热、低氮氧化物排放的现代大型捣固焦炉系统技术，实现煤炭资源的高效化利用，减少资源和能源的消耗。实现焦炉数字化三维仿真工程设计、焦炉生产过程三维仿真分析计算等，焦炉主体设备、炉体砌砖、管道等全面实现三维仿真设计。对焦化过程能源转换、焦炉加热、设备运行等工艺过程进行仿真研究，建立基于甘特图的动态精准运行体系，实现动态运行管理。

3.2.2 烧结球团工序智能设计

研发现代球团工艺，实现高效、低耗、低排放，降低粉尘、SO_x、NO_x、二噁英等污染物的排放。采用三维仿真技术，解析烧结、球团工艺的燃烧、传热、传质和动量传输，控制合理的冶金反应过程，研究开发低碳厚料层烧结工艺和球团焙烧新工艺；对球团造球、预热、焙烧等工艺过程进行数学解析，并与实验结果进行比对研究，提高烧结球团的生产效率、降低能耗；建立三维仿真模型和物理模型，开发研究新型烧结球团干法脱硫、脱硝新工艺；开发烧结球团数字化三维仿真工程设计、烧结球团生产过程三维仿真分析计算等。

3.2.3 炼铁工序智能设计

基于三维仿真设计技术，研究熔融还原、直接还原的冶金过程和"三传一反"现象，建立数学模型应用CFD数值模拟技术，解析气体运动、燃烧、传热、传质等传输过程，开发具有工程化应用可行性和可靠性的冶金装置和设备。在高温低氧热风炉、煤气干法除尘-TRT、无料钟炉顶、炉前设备、炉渣处理及热能回收、高炉长寿、高炉高富氧大喷煤等多项技术领域开展三维仿真设计和CFD、FEM等仿真研究，建立数学模型和物理模型，研究高炉关键设备设计，开发设计软件，实现精准设计，完成炼铁数字化三维仿真工程设计、炼铁生产过程三维仿真分析计算等。

3.2.4 炼钢工序智能设计

基于冶金流程集成理论和方法，建立数学模型，以非线性数学规划、动态甘特图等为研究方法，优化钢铁制造流程铁素物质流和能量流的运行，开发可靠适用的界面技术，提高钢铁厂运行效率和能源转换效率。建立仿真模型，研究铁水脱硫、脱磷、脱硅预处理过程的热力学和动力学机理，提高处理效率，降低资源和能源消耗；完成炼钢-精炼-连铸工艺的数字化三维仿真工程设计、炼钢生产过程三维仿真分析计算等。

3.2.5 轧钢生产工序智能设计

以高效连铸-连轧为研究重点，实现铁素资源的高效利用和能源高效利用。研发大型宽带钢热连轧工程三维仿真设计，建立三维仿真设备信息管理体系，为生产高效运行管理提供支撑。基于仿真模型研究，研发高性能、高质量钢材的加热、轧制及热处理仿真过程，研究轧钢过程形变、相变及温度变化规律，开发研究现代基于控轧控冷的轧钢新技术，不断完善棒材、线材高效轧机仿真设计，开发具有多功能的新型轧机，完成轧钢数字化三维仿真工程设计、轧钢生产过程三维仿真分析计算等。

3.2.6 冶金装备智能设计

冶金装备设计属产品设计和工业设计范畴，与工程设计具有区别和差异。以大型关键冶金装备三维仿真设计开发为研究内容，建立三维模型，应用 FEM 等分析软件进行计算分析，开发数字化样机，研究设备在工作状态的力学和热力学变化，建立应力场、应变场和温度场等多场耦合的三维仿真研究模型，结合物理样机的设计研制及其试验测试，完成现代冶金装备的工程化应用。在三维仿真机械设计的基础上，将三维设计的成果延伸到设备制造，在 3D-CAD 的基础上形成 3D-CAM，进一步推动数字化设计、数字化制造多产业的技术进步。开展紧凑型高炉无料钟设备、高炉炉前液压设备、炼钢转炉悬挂及驱动装置、高效板坯及方坯连铸机、连铸坯动态冷却与结晶控制、高效棒线材轧机、钢材深加工设备等具有自主知识产权的现代冶金装备，在设备可靠性、经济性、耐久性、节能性等方面形成突破，替代国外同类产品并输出国外，参与国际竞争。

3.2.7 工业炉工艺及装备智能设计

基于三维仿真模型，系统研究燃料燃烧、气体运动和加热过程，通过 CFD 数值仿真计算，解析蓄热式燃烧、高温低氧燃烧、全氧燃烧等不同燃烧工艺的特征和优势，研究基于高温低氧燃烧的工艺过程，降低燃料消耗、减少 NO_x、CO_2 等污染物的排放，开发现代燃烧加热装置，大幅度降低燃料消耗、降低污染物排放、提高能源转换效率和热能利用效率。运用三维仿真设计技术，开发新型高温环形加热炉、转底炉、高温低氧加热炉等热工炉窑，完成三维仿真设计及其附属设备的三维数字化样机的开发，研究钢铁厂粉尘处理的转底炉直接还原新工艺开发，形成由配料、造球、焙烧、冷却、气体处理等全工艺流程的数字化三维虚拟工厂运行，以此为示范，在烧结球团、焦化、炼铁、炼钢、轧钢等工序推广应用。进行工业炉数字化三维仿真工程设计、工业炉钢坯加热过程三维仿真分析计算等研究工作。

3.2.8 能源环境系统智能设计

创建系统节能体系，以减少能量消耗、实现能量高效转换为目标，建立数学模型，科学评价能量的输入、转换和输出过程，以数学仿真计算为研究手段，建立基于碳素能量流为核心的冶金能量流网络，减少过程耗散，降低 CO_2 排放，实现能源高效利用。冶金能源技术领域的重点研究和开发内容：建立冶金厂能量流网络模型，采用数值仿真研究科学合理用能，提高能源转换和利用效率；建立钢铁制造流程三维仿真能源管网体系，合理配置能源、减少耗散；完成制氧、余热锅炉、热电厂、CCPP 等单元工序的工厂和管网的三维仿真设计；建立余热利用-发电-海水淡化等多工序耦合的能源转换网络，完成三维仿真设计，建立能源多效转换的数学仿真模型；深入开展钢铁厂水资源高效利用课题研究，建立数学仿真模型和流程网络，实现"一水多用""串级使用""高效回收"，实现污水"零排放"；进行冶金能源工程数字化三维仿真设计与三维仿真分析计算等。

环境保护技术领域重点研究和开发内容：加强钢铁厂粉尘、烟尘、污水、SO_x、NO_x、CO_2等多种污染物及温室性气体的动态数值化研究，建立数模模型，运用数学分析方法，研究各类污染物的产生、排放、收集、处理和资源化的有效途径；采用三维仿真设计，开发研究新型高效除尘器、高效脱硫脱硝装置，实现三维样机开发；采用CFD数值仿真计算，研究粉尘、烟尘的产生及其运动规律，开发新型捕集和气力输送系统；采用三维仿真和数值模拟开发高效率长寿命除尘器；进行冶金环境工程数字化三维仿真设计与三维仿真分析计算等。

3.2.9 电气与自动化系统智能设计

基于现代冶金流程能源流和信息流流程网络理论，建立钢铁制造流程三维供配电网络管理系统，在数学优化的基础上，合理设计钢铁制造流程供电、配电和发电设施，加强余热、余能、副产煤气的高效发电，实现能源高效转换，减少电力消耗，实现钢铁厂电力能源的自给自足。实现变电站、发电厂、配电所、供电管网等三维仿真设计；不断完善自动化模型开发研究，以信息化、数字化促进技术进步；开展冶金过程设备运行与工艺控制的模型开发；建立冶金电气与自动化工程数字化三维仿真设计与冶金生产过程二级模型控制仿真计算分析研究等工作。

3.2.10 建筑工程系统智能设计

以建筑工程三维仿真设计为重点，进行大型高炉、烧结球团、炼钢、轧钢等主要冶金工序的三维土建结构设计。通过对麦达斯等有限元设计软件进行二次开发，不断完善大型冶金设备基础、柱基础、土建结构、网架结构的三维仿真设计优化，实现建筑减量化、轻量化、绿色化，减少钢材消耗；进行冶金建筑工程数字化三维仿真设计与建筑结构三维仿真分析计算等研究工作。

4 钢铁制造流程智能制造与智能设计的相互关系

（1）设计是竞争的起点，是工程的元工程。任何产品的竞争虽然最终是通过产品的质量、性能、成本、服务等因素表现出来，但追溯到其根源，将会深层次地延伸到产品的设计层面，任何产品决定其竞争力的质量、性能、成本等要素都是通过产品的设计来体现，设计是产品竞争力的起点，未来的市场竞争中，设计是决定产品竞争力的根源要素。

（2）对一个工程项目来说，项目设计的好坏在很大程度上决定着该工程的好坏和成败。对于一个行业来说，这个行业对设计工作的整体重视程度和整体设计水平往往会成为反映和决定该行业兴旺程度的关键指标和因素[12]。

（3）从更大范围和更长远尺度看，设计工作的"整体状况"必然深刻影响到国家的经济社会发展状况乃至人类文明的进步。人们必须以踏踏实实、细心细致的态度认识和从事设计工作，更必须以宏观和战略的眼光认识和从事设计工作。设计工作本身只是全部工程活动中的一个部分和一个环节，却是一项具有全局性意义和必然影响全局的工作[12]。

（4）设计面向未来，设计引领未来，战略设计引领战略未来[13]。实现钢铁制造流程制能制造，智能设计既是关键环节，又发挥着先导和引领作用，设计决定着工程项目原始物理系统的先进程度，没有先进的智能设计，很难有先进的智能制造。因此，必须高度重视钢铁制造流程智能设计作用和意义，从一定意义说，智能设计的先进程度决定着智能制造的先进水平和未来。

5 结论

（1）智能制造是基于新一代信息通信技术与先进制造技术深度融合，贯穿于设计、生产、管理、服务等制造活动的各个环节，具有自感知、自学习、自决策、自执行、自适应等功能的新型生产方式。实现制造的数字化、网络化、智能化发展，数字化是基础，网络化是关键，智能化是方向。制造业包括流程型制造业和离散型制造业，钢铁制造流程是典型的流程型制造业，实现钢铁制造流程的智能制造，是实现制造业智能制造的重要组成部分，而智能设计是钢铁制造流程智能制造关键、重要的环节。

（2）智能设计是指应用现代信息技术，采用计算机模拟人类的思维活动，提高计算机的智能水平，从而使计算机能够更多、更好地承担设计过程中各种复杂任务，成为设计人员的重要辅助工具。钢铁制造流程智能设计包括三方面：一是基础科学（研究尺度：分子原子尺寸），研究科学的合理性，采用CAE三维仿真设计分析计算技术，模拟仿真研究冶金和轧制工艺过程、温度场、流场等；二是技术科学（研究尺度：工序或装置尺寸），研究技术的可行性，采作机械三维仿真设计技术，模拟仿真研究冶金装置的优化；三是工程科学（研究尺度：制造流程或车间尺寸），研究工程项目的整体性、系统性、协同性优化，采用数字化三维仿真工厂设计技术，模拟仿真研究钢铁制造全流程的优化，实现设计、施工、运维的一体化。

（3）任何产品的竞争虽然最终是通过产品的质量、性能、成本、服务等因素表现出来，但追溯到其根源，将会深层次地延伸到产品的设计层面，任何产品决定其竞争力的质量、性能、成本等要素都是通过产品的设计来体现，设计是产品竞争力的起点，未来的市场竞争中，设计是决定产品竞争力的根源要素。

（4）设计面向未来，设计引领未来，战略设计引领战略未来。实现钢铁制造流程制能制造，智能设计既是智能制造的关键环节，又发挥着先导和引领作用，没有先进的智能设计，很难有先进的智能制造。因此，必须高度重视钢铁制造流程智能设计作用和意义，从一定意义说，智能设计的先进程度决定着智能制造的先进水平与未来。

参考文献

[1] 国家制造强国建设战略咨询委员会. 中国制造2025蓝皮书（2017）[J]. 北京：电子工业出版社，2017，6.

[2] 李新创. 智能制造助力钢铁工业转型升级[J]. 中国冶金，2017（2）：1-5.

[3] 郑忠，黄世鹏，等. 钢铁智能制造背景下物质流和能量流协同方法[J]. 工程科学学报，2017（1）：115-124.

[4] 孙彦广. 钢铁工业数字化、网络化、智能化制造技术发展路线图[J]. 冶金管理，2015（9）：4-8.

[5] 殷瑞钰. 冶金流程工程学（第1版）[M]. 北京：冶金工业出版社，2004.
[6] 殷瑞钰. 冶金流程工程学（第2版）[M]. 北京：冶金工业出版社，2009.
[7] 殷瑞钰. 冶金流程集成理论与方法[M]. 北京：冶金工业出版社，2013.
[8] 殷瑞钰. 关于智能化钢厂的讨论——从物理系统一侧出发讨论钢厂智能化[J]. 钢铁，2017（6）：1-12.
[9] 殷瑞钰. "流"，流程网络与耗散结构——关于流程制造型制造流程物理系统的认识[J].《中国科学：技术科学》，2018（2）：136-142.
[10] 颉建新. 关于工程公司三维设计技术应用若干问题的分析与研究[J]. 首钢科技，2012（4）：49-54.
[11] 许秀斌. 智能概念设计综述[J]. 中国科技信息，2011（12）：94，96.
[12] 徐灏，邱宣怀，蔡春源，等. 机械设计手册（第1册）[M]. 北京：机械工业出版社，1991.
[13] 李喜先，等. 工程系统论[M]. 北京：科学出版社，2009.

Research of Intelligent Manufacturing and Intelligent Design on Iron andSteel Manufacturing Process

Xie Jianxin   Zhang Fuming

Abstract：The main content of intelligent manufacturing and intelligent design on iron and steel manufacturing process are analyzed in this paper. The intelligent manufacturing on iron and steel manufacturing process includes：the smart process design, the smart process production run, the smart process management, the smart process supply chain, the smart process service system. The intelligent design on iron and steel manufacturing process includes：the basic scientific research, adopting CAE three-dimensional simulation design analysis computing technology；the technical science research, adopting mechanical engineering three-dimensional simulation design technique；the engineering science research, adopting digitization three-dimensional simulation plant layout technique；realizing intelligent manufacturing on iron and steel manufacturing process, the intelligent design is not only key link of intelligent manufacturing, but also give play to forerunner and guide affect. Without having advanced intelligent design, there will be no advanced intelligent manufacturing, so we must pay high attention to the effect and significance of intelligent design on iron and steel manufacturing process. The advanced degree of intelligent design decides the advanced level and future of intelligent manufacturing.

Keywords：steel manufacturing process；intelligent manufacturing；intelligent design；research

冶金流程工程学的典型应用

摘　要：钢铁是经济社会发展不可或缺的重要基础材料，钢铁产业是国民经济的重要基础产业。分析研究了钢铁制造流程的物理本质和主要特征，以冶金流程工程学的观点阐述了现代钢铁制造流程功能拓展的重大意义。阐述了钢铁冶金工程知识体系及其发展演化进程，论述了基础科学、技术科学和工程科学知识体系的形成与发展历程，指出冶金流程工程学是宏观冶金学，是面向钢铁冶金全流程和产业层次的工程科学。研究分析了在冶金流程工程学指导下现代钢铁冶金工程设计体系和方法的创新，阐释了概念设计、顶层设计和动态精准设计的理念、内涵和方法；介绍了首钢京唐钢铁厂的工程设计创新及其应用实绩。

关键词：钢铁冶金；流程工程学；工程科学；工程设计；耗散结构

1　引言

从 20 世纪 90 年代开始，我国著名冶金学家、工程哲学家和战略科学家殷瑞钰院士，基于长期对钢铁制造流程及其物理本质的深入思考研究，以钢铁制造全流程和工程哲学的视野，对钢铁制造流程的物理本质及其动态运行的本构特征，进行了长达 10 余年的研究分析，做出了精辟的阐释，并创立了冶金流程工程学[1]。冶金流程工程学的创建，使现代冶金学跨入到工程科学层次，对钢铁冶金工程规划、设计、决策、建造、运行等都具有十分重要的指导意义，已成为现代钢铁冶金工程实现可持续发展的基本理念、理论和方法，在工程实践中取得了显著的应用成效。

2　钢铁冶金工程的物理本质和主要特征

钢铁乃是目前人类社会应用最广泛的结构材料，还是全球消费量最大的功能材料，甚至可以说是"必选材料"或是"首选材料"。钢铁产业是经济社会发展的重要基础产业和流程制造业，是各门类产业发展的重要基础，是经济社会发展不可或缺的支撑。

钢铁制造流程其实质上是典型的耗散结构体系，即为开放、不可逆、远离平衡的系统，是由异质、异构的单元工序通过非线性相互作用、动态耦合等机制所构成的复杂系统，其动态运行过程的本质是耗散过程[2]。

钢铁制造流程，是由烧结（球团）、炼铁、炼钢、轧钢以及产品加工等若干个不同的工序串联/并联所组成的流程系统。其上下游工序之间衔接紧密，上游工序乃是下游工序基础，上游工序的输出就是下游工序的输入。从流程制造的特征分析，钢铁制造流程是由多工序协同、相互连接形成的整体系统，具有显著的层次性、过程性和复杂性。

本文作者：张福明，颉建新。原刊于《钢铁》，2021, 56 (8): 10-19。

3 现代钢铁制造流程的功能拓展

21世纪初，在我国钢铁工业快速发展时期，殷瑞钰院士提出：新一代可循环钢铁厂的功能，应当摆脱产品制造单一功能的传统思维，将其拓展为钢铁产品制造、能源转换，以及处理废弃物、实现资源化的"三大功能"[3]。以实现三大功能为核心，发展循环经济，降低CO_2排放，与其他行业乃至社会实现生态链接，从而实现钢铁产业自身、上下游产业链之间，以及产业循环经济区域的绿色化、可持续、协同发展[4~6]。

现代钢铁制造流程功能拓展[7]，是冶金流程工程学理论和实践的重大创新之一。使人们以全流程和全产业链的角度，从自然生态、制造流程、产业链接、经济社会的工程哲学视野，通过对钢铁制造流程功能解析与集成[8]，揭示了物质/能源转化/转变的本质特征，也就是说钢铁产品的制造过程，是在能量驱动下的物质转化/转变过程，物质、能量和信息是相伴相行的耦合过程，也是协同运行、流动、流变的过程。在此规律认识的基础上，将传统钢铁产品制造的单一功能，拓展到能源转换/转化功能，以及在能量的驱动下，实现钢铁制造过程自身排放的废弃物和社会大宗废弃物的消纳、处理并实现资源化的功能。

应当指出，现代钢铁制造流程的三大功能，是区别于传统钢铁制造流程最重要的特征之一，也是冶金流程工程学最重要的理论创新之一。图1为现代钢铁制造流程的三大功能。

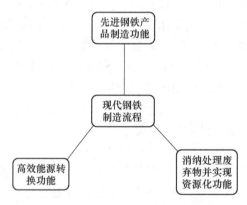

图1　现代钢铁制造流程的功能拓展

4 现代钢铁冶金的知识体系及其发展演进

4.1 现代钢铁冶金基础科学的形成与发展

18世纪末至19世纪初的第一次工业革命以后，高炉开始使用焦炭、机械动力和热风炼铁，开启了近代高炉冶炼的新纪元；1856年，英国工程师、发明家亨利·贝塞麦发明了转炉炼钢工艺，极大地推动了炼钢技术进步。应当说第一次工业革命，特别是蒸汽机的发明和应用，使社会生产力发生了巨大的变化，从而使人类社会步入了工业文明的新时代。

现代冶金学的起源是第一次工业革命之后，随着自然科学的发展，而逐渐形成的一门研究由自然界矿石之中提取金属或制备合金的学问，除了高等数学、普通化学、普通物理等普适性的基础自然科学之外，人们开始运用物理的方法来研究化学问题，例如研究冶金反应过程进行的可能性、最大反应限度及其反应速率和生产效率等问题。冶金物理化学其核心，是冶金过程热力学和冶金过程动力学，这是冶金物理化学最重要的两大组成部分。冶金物理化学是现代钢铁冶金重要的基础理论科学知识，现代钢铁冶金学是建构在冶金物理化学基础科学之上的学科，甚至可以说没有冶金物理化学也谈不上现代冶金科学。

冶金过程热力学主要研究冶金反应的可能性（热力学第二定律）、方向和反应条件（反应温度、吸热还是放热）等，这是热力学研究最主要的内容。冶金过程动力学，是研究冶金反应进行的机理及其反应速度（微观动力学），对于宏观冶金过程，包括对于高温多相/多态耦合反应的传热、传质及化学反应进程中的动力学，属于冶金传输原理的研究范畴（宏观动力学）。因此冶金过程动力学，是主要研究冶金化学反速率、反应限度和反应极限等基础科学问题。

20世纪的百年间，乃是钢铁冶金工程发展成为科学的重要时期[9]。20世纪20年代之后，化学热力学理论被应用于冶金领域中，逐渐发展为冶金过程物理化学[10]。1925年，英国法拉第学会在伦敦召开了"炼钢过程物理化学"讨论会，这次会议对钢铁冶金学的形成和发展意义重大，使钢铁冶金从技艺上升到科学研究的层次。化学反应热力学，揭示了冶金化学反应过程的本质和规律，具有划时代的科学意义。

这些冶金过程基础理论的研究，通过合理的简化、假设和典型化，采用"机械还原论"的研究方法，将发生在原子/分子间的冶金反应过程假设成为"孤立系统"进行研究，再将研究结果"还原"到实际冶金反应过程中。

4.2 现代钢铁冶金技术科学的形成与发展

钢铁冶金技术科学，主要是研究解决工序、装置、场域尺度上的技术问题，例如如何提高单元工序和单元过程的生产效率，降低工序消耗，减少单元工序的过程耗散，强化工艺过程的热量、质量和动量传输，以及冶金反应过程"三传一反"等技术层次的科学问题。总体来看，钢铁冶金技术科学不但包括炼铁学、炼钢学、金属压力加工学等专业工艺冶金学，还包括冶金反应工程学。

冶金反应工程学是在化学反应工程学的基础上而发展起来的一门新兴学科。20世纪60年代以后，世界钢铁技术发展取得长足进步，冶金反应工程学因此应运而生。1972年，日本鞭岩和森山昭教授，合著出版了《冶金反应工程学》[11]，创建形成了化学反应工程学在冶金过程的应用科学。冶金反应工程学的主要研究特点，是在宏观动力学的基础上，充分考虑冶金反应过程操作条件、初始条件、边界条件和反应器结构优化，从而提高冶金过程的反应效率和效能优化。例如反应器内基本现象解析、建立数学模型、数值仿真模拟、单元过程最优化、动态特性最优化和物理模拟等。

冶金反应工程学主要研究某些典型冶金反应器的工艺特性及其功能改进，如高炉冶炼过程解析模拟，钢水连铸的中间罐冶金、钢包冶金等装置优化等。这些研究主要是应用数学模型化的方法，首先建立物理模型，进而建立数学模型，给定初始条件和边界条件，选用或开发适用的计算软件（如CFD、FEM模型软件），借助于编程和计算机数值计算，解析研究某一类冶金反应器及其系统操作过程的现象、特性和规律，从而通过对比研究和参数优化，得到优选的解决方案、工艺参数或结构尺寸；与此同时对新工艺、新装置、新设备的开发过程中，还可以预测其功能的特性，或者进行数值化的动态运行（仿真模拟），进而指导生产操作工艺的改进以及装置的优化设计。对于生产运行的冶金设备和装置的技术改造和操作优化，采用冶金反应工程学进行仿真优化研究，能够获得较好的技术解决方案和装置的效能提升。

4.3 现代钢铁冶金工程科学的形成与发展

冶金工程科学属于宏观冶金学范畴，是由殷瑞钰院士在本世纪初创建的冶金流程工程学。冶金流程工程学、冶金流程集成理论与方法，主要研究钢铁制造流程整体和联合钢铁企业尺度的工程问题。冶金流程工程学的理论基础，包括牛顿力学、普利高津耗散结构理论，以及非线性、非平衡、开放系统热力学，主要研究冶金制造流程动态运行的物理本质及其本构特征，全流程物质流-能量流-信息流协同耦合，如何实现动态有序、协同连续运行状态（即所谓"三流一态"）等工程层次的科学问题。

冶金流程工程学的创建，使现代钢铁冶金学形成了基础科学（微观）、技术科学（中观）和工程科学（宏观）三个层次的完整知识体系，使现代冶金学在系统全局的高度上，将科学-技术-工程融会贯通、集成一体[12]。应当看到，当前钢铁工业所面临的挑战是多方面的，解决复杂环境下的命题必须从战略层面上，系统思考钢铁厂要素-结构-功能-效率问题，实质上是全厂性生产流程层面问题，应当从生产流程结构优化及其相关工程设计等根源着手。从钢铁冶金工程知识体系可以清晰地认识到，钢铁制造全流程的系统性、全局性、复杂性问题，不仅仅依靠技术科学层次上的单元技术革新和技术攻关所能解决，而更需要以工程哲学的视野，创新思维理念和工程方法，在全产业、全过程和工业生态链等工程科学层次上不断推进和完善。现代钢铁工业应当建构出与自然环境、经济社会和相关产业之间和谐友好、相互依存、相融共生、绿色低碳可持续发展的新格局（见图2）。

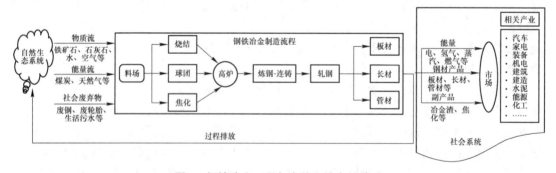

图 2　钢铁冶金工程与自然和社会的关系

冶金流程工程学基于钢铁制造全流程层次的系统视野，揭示了钢铁制造流程的物理本质和本构特征，创新提出了物质流、能量流和信息流的基本概念，首创提出了钢铁制造流程的三个要素（即"流""流程网络"和"运行程序"）[13,14]，使工程思维由传统的关注单元操作或单元过程，跃迁到注重钢铁制造流程全局性多目标的集成优化。表1阐释了现代钢铁冶金工程知识体系的结构及特征。

表 1 现代钢铁冶金工程知识体系的结构及特征

层次	学科分支	时空尺度	基础的研究内容	研究方法特征	吸引子	物理基础
微观	冶金物理化学	原子、分子、离子	热力学函数能量关系；冶金反应化学亲和势	实验室测定平衡；相图研究；反应速率和机理测定	平衡态	经典热力学，物质构造学说
中观	冶金反应工程学（三传一反）	反应器、场域	反应器内介质浓度、温度、停留时间分布；颗粒液滴气泡弥散体系	数学和物理模拟；场域条件下参数测量；比拟放大	非平衡稳定态	传输现象理论，线性非平衡热力学
宏观	冶金流程工程学（三流一态）	冶金流程整体（全流程）和区段流程（工序或系统）	多因子物质流控制；流程解析和集成；物质流-能量流-信息流协同运行	工程设计和模拟运行，流程动态运行优化和信息化表征、调控	动态有序、连续紧凑、耗散优化	牛顿力学（动力学），耗散结构理论，非线性、非平衡、开放系统热力学

5 现代钢铁冶金工程设计理论与方法创新

现代钢铁冶金工程设计的核心内容，是构建出符合钢铁制造流程动态运行规律的物质流、能量流和信息流顺畅的流程网络，以及高效协同的运行程序，形成"三流耦合、三网协同"的信息物理系统，采用数字化、信息化和智能化设计手段及方法，仿真模拟钢铁制造过程单元操作、单元过程和全流程动态运行现实的过程虚拟，再通过解析、综合、分析、选择、权衡、集成等工程方法[15]，从而构建出符合钢铁制造流程动态运行规律的先进流程。钢铁冶金工程设计方法，其实质就是构建以全流程动态有序、协调连续运行为目标的设计方法，形成动态有序、连续运行系统的实体（硬件）+虚体（软件）集成，即构建钢铁制造流程信息物理系统（具有自感知、自适应、自学习、自决策、自执行等功能）、实现智能运行的设计方法。

在冶金流程工程学理论指导下，现代钢铁冶金工程设计涵盖概念设计、顶层设计和动态精准设计三个层次。动态精准设计是以构建钢铁制造流程信息物理系统（CPS）为目标，以数字化、信息化和智能化设计为手段，面向钢铁冶金全流程动态仿真和虚拟现实，实现多平台数字化交互协同设计和设计产品的数字化信息交付，进而通过工程设计、建造、运行、维护和退役等全生命周期的信息数字化系统的构建，实现钢铁冶金工程设计的智能化升级。

5.1 概念设计

概念设计是工程科学层次上的问题，是钢铁冶金工程设计最根本的立足点和出发点。概念设计或称之为概念研究，其核心内容是建立起现代钢铁冶金工程动态运行的思维理念和工程系统观，深刻理解并认识钢铁制造流程动态运行的物理本质、运行规律及其本构特征，从而建立起钢铁制造流程耗散结构优化的基本思路、目标和思维方法，制定出钢铁制造流程层次的重大工程方案研究，并进行科学决策。例如钢铁制造流程是选择高炉-转炉

"长流程",还是选择废钢-电炉"短流程"的重大工程问题,就是需要在工程概念设计阶段通过研究分析、综合权衡来确定。

5.2 顶层设计

顶层设计是基于概念设计的结果,从钢铁制造流程动态运行过程物理本质出发,对钢铁制造流程进行"赋能",拓展钢铁厂功能,实现绿色化、智能化发展。顶层设计实质上是钢铁制造流程静态物理框架的构建过程,包括全流程层次、单元工序层次和工序之间的设备/装置层次。首先要构建出运行高效顺畅合理的铁素物质流及其流程网络框架结构;其次就是在物质流网络框架结构的基础上,构建出与物质流协同运行的能量流网络框架结构;再次是构建出信息流网络框架结构。从钢铁制造流程智能化出发,物质流和能量流网络框架的设计,实质上是钢铁制造流程"实体"物理系统的构建,而信息流网络框架的设计及其运行程序的设计,就是"虚体"信息系统的构建。此外,顶层设计还包括钢铁厂三大功能的拓展与构建,从单一目标转化为多目标集成优化等。这些设计内容都是从钢铁制造全流程的层次/角度开展的,是以钢铁制造流程宏观层次为设计对象,所形成的系统性、全局性、自上而下的工程总体设计。

5.3 动态精准设计

现代钢铁冶金工程的动态精准设计,是在概念设计、顶层设计的基础上开展的工程详细设计,从而形成了从宏观到微观、从顶层到底层完整的工程设计体系,也就是工程方法论的"程序化"[16]。其核心思想理念是:以钢铁制造流程的要素为设计关注点,充分体现出不同工序内部物质和能量的高效转换,注重不同工序之间的界面技术优化,重视工艺流程设计、主要设备配置和生产装置的选择和选型,在工艺流程图、工艺布置图、设备安装图、工程总平立面图的设计中,注重时间-空间的协调关系,注重物料运行距离、运行轨迹、运行方式等对钢铁制造流程效率、成本和时间的影响因素的评价和优化。

具体而言,空间维度反映了流程运行的网络结构、总图布局、运行轨迹,体现在动态运行的集约高效、紧凑顺畅;时间维度则反映出流程的连续性、协调性和动态耦合性,以及运输过程、等待过程中的能量耗散。动态精准设计的核心内容,就是构建出精准运行的时间-空间协同框架,构建动态有序、协同连续的物理系统,是奠定钢铁制造流程信息物理系统(CPS)的基础,进而实现"三流耦合"和"三网协同"[17]。

在动态精准设计中,不同工序之间的界面技术优化是非常重要的环节,应采用运筹学、排队论和图论等现代理论,以及甘特图(Gantt Chart)、关键路径(CPM)、程序运行分析评价(PERT)等设计工具,对钢铁制造流程的工序界面,进行时间和空间的耦合研究分析和设计优化。动态精准设计过程中,还要重视工程的集成优化和工程创新,这是工程设计从抽象的理念、概念,转变成具象的现实工程实体极其重要的过程[18]。此外,动态精准设计中,还应当重视全流程整体动态运行的稳定性、可靠性和高效性,制定流程高效稳定运行的规则和程序。在精准设计中,通过建立运行过程的仿真模型,对不同工序进行深入解析研究,以实现效率优化、结构优化和功能优化为前提,合理确定设备/装置的数量和能力。现代钢铁冶金工程动态精准设计流程如图3所示。

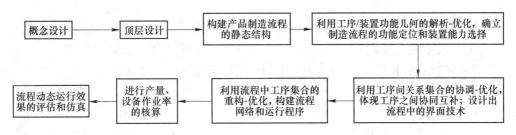

图 3 现代钢铁冶金工程动态精准设计的程序化框图

6 首钢京唐钢铁厂的工程设计及实践

6.1 工程概念设计

21世纪初设计建设的首钢京唐钢铁厂工程，是在冶金流程工程学理论和新一代可循环钢铁制造流程理念指导下，自主设计的新一代可循环钢铁制造流程工程示范。概念设计中所确立的设计目标是实现流程高效、装备大型、系统集成、循环经济。通过多方案对比分析和综合权衡，选择了生产效能高、成熟可靠、运行稳定的高炉-转炉长流程工艺。产品结构确定为汽车、机电、石油、家电、能源所需的高级板带材，一期工程设计生产能力为870万~920万吨/年，其中冷轧产品比例设计达到60%以上。

6.2 工程顶层设计

顶层设计中针对工程重大工艺技术方案，进行了全方位的研究分析和权衡综合，结合资源能源供给条件、产品市场竞争力以及产业链衔接情况，做出了重大工程设计方案决策。例如：针对轧机工艺装备设计选型，进行了多方案的研究对比分析。如果建设1条常规热轧+1套中厚板生产线，生产能力大约为700万吨/年；如果建设2条常规热轧生产线，生产能力可达900万吨/年，为充分发挥沿海临港的建厂优势，设计选用了2250mm和1580mm两条常规热连轧生产线。针对炼钢流程设计，通过对洁净钢生产前沿技术的研究分析，没有采用常规的转炉复合吹炼工艺，决策采用脱硅/脱磷-脱碳转炉冶炼工艺，构建高效率-低成本洁净钢生产工艺流程[19]。炼铁流程设计中，坚持流程结构优化下的高炉大型化设计理念，围绕高炉数量、容积的选择进行了深入的对比研究分析，设计选用2座5500m³高炉，而没有采用3座4000m³高炉的工艺方案[20]。针对炼铁-炼钢界面技术，采用了铁水罐多功能化技术，即铁水"一罐到底"直接运输工艺[21]。根据能量流和不同能源介质的行为和转换特征，进行钢铁厂能量流网络优化设计，配置了功能完善的能源中心，实现了系统节能和能源高效转换，钢铁厂的自发电率达到96%以上。

首钢京唐钢铁厂顶层设计中，还注重现代钢铁制造流程的功能拓展，通过减量化、再循环和再利用，铁素资源基本实现100%循环利用；通过系统节能和能量协同管控，能源转换效率得到大幅度提升，钢铁制造过程的余压、余热和余能，基本实现全量回收；通过实施清洁生产、节能减排、循环经济和低碳绿色，工艺过程实现超低排放和污水近"零排放"，烧结、球团、焦化、转炉等工艺装备还消纳处理了大量来自钢铁厂内部及社会的废弃物。

一期工程建设了 4 座 70 孔 7.63m 大容积焦炉，匹配了 2 套 260t/h 高温-高压干熄焦装置（CDQ），吨焦发电量设计为 100kW·h/t，弱黏结煤配加比例达到 35%以上，利用焦炉煤气制氢实现高值转化。5500m³ 高炉采用全烧高炉煤气的高风温技术[22]、煤气全干法除尘技术[23]、配置 36.5MW 高效 TRT 余压发电装置，回收炉顶煤气余压进行发电，达到 45kW·h/t 以上，高炉设计燃料比为 496kg/t，渣量设计值为 280kg/t。转炉煤气采用干法除尘技术高效回收蒸汽和煤气，降低新水消耗[24]。全量回收冶金煤气经过燃气-蒸汽联合发电（CCPP），提高发电效能约为 8%，再回收汽轮机排出的低品质、低能阶蒸汽，用作低温多效海水淡化装置的热源，利用钢铁厂自产余热，供应 5.0 万+3.5 万吨/天海水淡化装置，节约了大量淡水资源，钢铁厂水资源基本全部依靠海水淡化；每年利用余能余热发电量达到 3.4 亿千瓦时。

6.3 工程动态精准设计

首钢京唐钢铁厂的动态精准设计，在概念设计和顶层设计的基础上，注重建立时间-空间的协调关系，注重流程网络的构建与优化。在工程动态精准设计过程中，首先建构出合理的物质流流程网络及其静态物理结构，以此为基础，再研究和构建能量流网络和信息流网络。

以图论为工具构建制造流程的静态网络，采用排队论和动态运行甘特图等先进方法描述"界面"过程的运行动力学，制定了钢铁制造流程动态运行规则（设定运行程序），采用数学模型和计算软件仿真等手段，研究工序界面过程动力学及其比较优化（如铁水运输的优化调度问题）。通过物质流、能量流和信息流的协同优化，构建自感知、自学习、自适应、自决策、自执行的数字化信息系统，协调解决好物质流网络、能量流网络和信息流网络之间的"三网"链结、融合问题，构建起基于物理系统过程耗散优化的 CPS 系统，进而实现钢铁制造全流程实际运行的智能化。

以炼钢工程动态精准设计为例，通过铁水预处理、转炉冶炼、二次精炼和连铸等单元工序的工艺过程基础理论研究和技术研究，从而确定了基于铁水"三脱预处理"（脱硫、脱硅、脱磷）+转炉脱碳+二次精炼+高效连铸的洁净钢生产工艺路线和流程。应用冶金过程基础科学知识，通过对铁水中硫、硅、磷等元素脱除的冶金反应热力学分析，研究得出高温条件下有利于铁水脱硫反应的进行，而在相对"较低温度"高氧势的条件下则有利于氧化脱硅和氧化脱磷反应的进行。通过冶金过程动力学的仿真研究和传输过程的动态仿真，研究了 KR 脱硫装置在强搅拌的条件下，更有利于改善传质过程、促进铁水脱硫反应的进行；针对单吹颗粒镁和 KR 工艺的冶金反应工程学研究，采用冶金反应器动态仿真研究结果，选择了效率更高、处理能力大、处理时间短的 KR 脱硫工艺。同时根据脱除硅、磷、碳等元素的冶金机理和过程仿真研究，确定了采用"双跨"布置的 2 座 300t 脱硅/脱磷转炉和 3 座 300t 脱碳转炉，将转炉冶炼过程的冶金反应进行解析优化，把常规转炉的脱硫、脱硅、脱磷、脱碳等复合冶金功能，解析为由不同的工序装置分别完成，进而提高了冶金反应的速率和生产效率，有效降低了生产成本。例如脱硅/脱磷转炉的冶炼周期约为 20~25min，脱碳转炉的冶炼周期约为 30~35min，2 座前置的脱硅/脱磷转炉与 3 座脱碳转炉耦合匹配，在连续生产过程中，KR 脱硫、前置转炉脱硅/脱磷预处理、转炉脱碳、

RH 精炼和连铸的工序节奏匹配"时间步长"为 35min 左右，消除了相互等待、间歇过长的问题，有效解决了转炉和连铸之间的"炉机匹配"问题，从而构建出高效率、快节奏、低成本的洁净钢生产流程（见图 4、表 2 和图 5）。

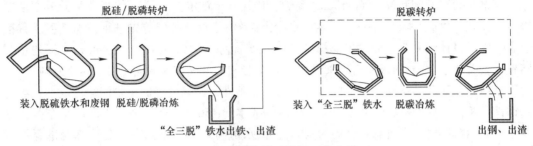

图 4　首钢京唐转炉炼钢工艺流程

表 2　首钢京唐炼钢厂主要工序配置

单元工序装备	数量
铁水 KR 脱硫装置/套	4
300t 铁水脱硅-脱磷转炉/座	2
300t 顶底复吹脱碳转炉/座	3
CAS 精炼装置/套	2
双工位 LF 钢包精炼炉/座	1
双工位 RH 真空处理装置/台	2
2150mm 双流板坯连铸机/台	2
1650mm 双流板坯连铸机/台	2

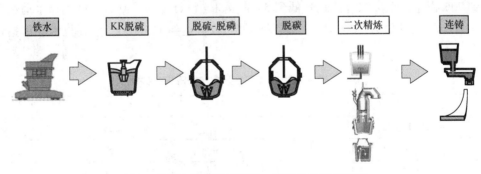

图 5　首钢京唐钢铁厂炼钢—连铸生产工艺流程

6.4　生产运行实绩

首钢京唐 1 号高炉于 2009 年 5 月 21 日建成投产，一期工程于 2010 年 6 月 26 日全面竣工、投产。沿海靠港建设 2 座高炉、1 个炼钢厂、2 条热轧生产线和 4 条冷轧生产线，产能为 870 万~920 万吨/年，形成了全部生产板带材的新一代可循环钢铁制造工艺流程。2019 年 4 月，首钢京唐二期一步工程顺利投产，在一期工程的基础上，增建了 2 条带式焙烧机球团生产线、2座焦炉、3号高炉、第二炼钢厂、4300mm中厚板轧机、ESP连铸-连轧生产线及其配套的高强镀锌线和能源公辅设施等，目前形成了约1350万吨/年的生产规模。

首钢京唐工程创新设计了5500m³高炉，构建基于铁水"三脱"的高效率低成本洁净钢生产流程，建成了5.0万+3.5万吨/天低温多效海水淡化装置等先进工艺与装备。首钢京唐钢铁厂设计中，为充分发挥靠海临港的区域优势，钢铁厂采用"一线形"工艺总图布置，设置2个30万吨级原料码头可承接大型船舶，成品码头可停靠1万~5万吨船舶，为原燃料运输和产品运输创造了极为有利的运输条件，大幅度降低了物流运输成本，与内陆钢铁厂相比，仅原燃料运输就可以节约炼铁成本约150~200元/吨。生产流程布置紧凑有序，原燃料运输皮带机总长度约为42km，铁水由高炉"一罐到底"到达炼钢运距仅有841~1250m（平均为900m），全厂占地面积为0.9m²/t，达到了流程紧凑、集约高效、布局合理的设计目标[25]。

首钢京唐钢铁厂热轧至冷轧之间，设计采用新型钢卷运输车，实现了短距离快捷运输，降低成本、节省占地。设计了能源储备、转化、输配的集中功能区，构建了全流程能量流网络和能源管控中心，实现了能源高效转换和系统节能。经过10年的生产运行和持续提升，高炉日产量达到12800t/d，利用系数稳定达到2.35t/(m³·d)以上，燃料比、焦比、煤比分别达到480kg/t、280kg/t、180kg/t，风温达到1250~1300℃，煤气利用率大于50%，TRT发电大于48kWh/t，渣量为215kg/t，工序能耗为385kgce/t。炼钢工序转炉冶炼周期为24~30min，转炉出钢温度降低到1647℃以下，碳氧积降低到0.0016%，连铸拉速为2.3~2.4m/min；炼钢过程[S]、[P]、[N]、[H]、[O]等有害元素总含量降低到45ppm的水平，与日本最先进的炼钢厂处于同一等级（40~45ppm），达到世界先进水平；2250mm宽带钢生产线采用多项先进技术[26]，热轧板坯热装率大于75%，冷热转换比达到70%以上。全厂综合能耗降低到579kgce/t（设计值为640kgce/t），新水消耗降低到2.41m³/t（设计值为3.84m³/t），转炉煤气回收量达到115m³/t。主要技术经济指标见表3。

表3 首钢京唐钢铁厂主要技术经济指标实绩

项目	数值
高炉利用系数/t·(m³·d)⁻¹	2.35
入炉焦比/kg·t⁻¹	280
煤比/kg·t⁻¹	180
燃料比/kg·t⁻¹	480
风温/℃	1245
渣量/kg·t⁻¹	215
转炉钢铁料消耗/kg·t⁻¹	1080
出钢温度/℃	1647
转炉出钢碳氧积/%	0.0016
低碳钢连铸最高拉速/m·min⁻¹	2.5

续表 3

项目	数值
热轧成材率/%	98.53
冷轧成材率/%	97.79
吨钢综合能耗/kgce·t^{-1}	579.36
吨钢综合电耗/kgce·t^{-1}	605.72
吨钢新水消耗/m^3·t^{-1}	2.41
转炉煤气回收/m^3·t^{-1}	114.76
实物劳动生产率/t·（人·月）$^{-1}$	966

7 结语

（1）钢铁是重要的结构材料和功能材料，钢铁产业是重要基础产业。冶金流程工程学是面向宏观领域的工程科学，是指导钢铁产业在面向低碳绿色可持续发展的工程思维、工程理念和工程方法论。

（2）钢铁制造流程的本构特征是开放的、不可逆的耗散结构。研究钢铁制造全流程的工程科学问题，必须以工程哲学的视野，从孤立系统和封闭系统迈入到开放体系的研究，从"三传一反"上升到"三流一态"。

（3）现代冶金学经已经形成了基础科学、技术科学和工程科学的知识体系。冶金流程工程学以系统论、协同学、牛顿力学、耗散结构理论为基础，建构了以钢铁冶金全流程和钢铁产业为研究对象的宏观层次的工程科学，学科发展具有重要的理论价值和实践意义。

（4）首钢京唐钢铁厂工程设计建设，在冶金流程工程学指导下，以全新的工程思维和理念，在概念设计、顶层设计和动态精准设计中，以建构物质流-能量流-信息流协同耦合的流程网络和运行结构为核心，注重信息物理系统（CPS）构建和优化。经过 10 年的运行实践，取得了显著的应用实绩。

参考文献

[1] 殷瑞钰. 冶金流程工程学（第 2 版）[M]. 北京：冶金工业出版社，2009：116-135.

[2] 殷瑞钰. 冶金流程集成理论与方法 [M]. 北京：冶金工业出版社，2013：120-123.

[3] 殷瑞钰. 冶金流程工程学 [M]. 北京：冶金工业出版社，2004：379-386.

[4] 殷瑞钰，李伯聪，栾恩杰，等. 工程知识论 [M]. 北京：高等教育出版社，2020：291-304.

[5] 张春霞，殷瑞钰，秦松，等. 循环经济社会中的中国钢厂 [J]. 钢铁，2011，46（7）：1-6.

[6] 张春霞，王海风，张寿荣，等. 中国钢铁工业绿色发展工程科技战略及对策 [J]. 钢铁，2015，50（10）：1-7.

[7] 殷瑞钰. 钢铁制造流程的本质、功能与钢厂未来发展模式 [J]. 中国科学 E 辑：技术科学，2008，38（9）：1365-1377.

[8] 殷瑞钰. 过程工程与制造流程 [J]. 钢铁，2014，49（7）：15-22.

[9] 徐匡迪. 中国钢铁工业的发展和技术创新 [J]. 钢铁，2008，43（2）：1-13.

[10] 殷瑞钰. 工程科学与冶金学 [J]. 工程研究-跨学科视野中的工程, 2020, 12 (5): 435-443.

[11] 鞭岩, 森山昭. 冶金反应工程学 [M]. 蔡志鹏, 谢裕生, 译. 北京: 科学出版社, 1981.

[12] 殷瑞钰, 张福明, 张寿荣, 等. 钢铁冶金工程知识研究与展望 [J]. 工程研究, 2019, 11 (5): 438-452.

[13] 殷瑞钰. "流"、流程网络与耗散结构——关于流程制造型制造流程物理系统的认识 [J]. 中国科学: 技术科学, 2018, 48 (2): 136-142.

[14] 殷瑞钰. 从开放系统、耗散结构到钢厂的能量流网络化集成 [J]. 中国冶金, 2010, 20 (8): 1-14.

[15] 殷瑞钰, 汪应洛, 李伯聪. 工程方法论 [M]. 北京: 高等教育出版社, 2007: 123-146.

[16] 张福明, 颉建新. 冶金工程设计的发展现状及展望 [J]. 钢铁, 2014, 49 (7): 41-48.

[17] 殷瑞钰. 关于智能化钢厂的讨论——从物理系统一侧出发讨论钢厂智能化 [J]. 钢铁, 2017, 52 (6): 1-12.

[18] 殷瑞钰, 张寿荣, 张福明, 等. 现代钢铁冶金工程设计方法研究 [J]. 工程研究, 2016, 8 (5): 502-510.

[19] 张福明, 崔幸超, 张德国, 等. 首钢京唐炼钢厂新一代工艺流程与应用实践 [J]. 炼钢, 2012, 28 (2): 1-6.

[20] 张福明, 钱世崇, 殷瑞钰. 钢铁厂流程结构优化与高炉大型化 [J]. 钢铁, 2012, 47 (7): 1-9.

[21] 张福明, 钱世崇, 张建, 等. 首钢京唐5500m^3高炉采用的新技术 [J]. 钢铁, 2011, 46 (2): 12-17.

[22] 张福明, 梅丛华, 银光宇, 等. 首钢京唐5500m^3高炉BSK顶燃式热风炉设计研究 [J]. 中国冶金, 2012, 22 (3): 27-32.

[23] 张福明. 大型高炉煤气干式布袋除尘技术研究 [J]. 炼铁, 2011, 30 (1): 1-5.

[24] 张福明, 张德国, 张凌义, 等. 大型转炉煤气干法除尘技术研究与应用 [J]. 钢铁, 2013, 48 (2): 1-9.

[25] 尚国普, 向春涛, 范明浩. 首钢京唐钢铁厂总图运输系统的创新及应用 [J]. 中国冶金, 2012, 22 (8): 1-6.

[26] 张福明, 颉建新. 首钢京唐2250mm热轧生产线采用的先进技术 [J]. 轧钢, 2012, 29 (1): 45-49.

Typical Application of Metallurgical Process Engineering

Zhang Fuming Xie Jianxin

Abstract: Iron & steel is an indispensable basic material for economic and social development, and iron & steel industry is an important basic industry of national economy. The physical essence and main characteristics of iron & steel manufacturing process is analyzed, the great significance of the function expansion for modern iron & steel manufacturing process in terms of metallurgical process engineering is expounded. The knowledge system of iron & steel metallurgical engineering and its developing evolution process are expounded, the formation and development course of basic science, technological science and engineering science knowledge system are discussed. The facts that metallurgical process engineering is macro metallurgy, and it is an engineering science facing the whole manufacturing process and industrial level of iron & steel metallurgy are described. The design system and method innovation of modern iron & steel metallurgical engineering under the guidance of metallurgical process engineering are investigated and analyzed. The philosophy, connotation and method for concept design, top level design and dynamic precision design connotation and method are expounded, and the engineering design innovation and its application achievements of Shougang Jingtang Iron & Steel Plant are introduced.

Keywords: iron & steel metallurgy; process engineering; engineering science; engineering design; dissipative structure

Innovation and Application on Pelletizing Technology of Large Travelling Grate Induration Machine

Zhang Fuming Wang Qusheng Han Zhiguo

Abstract: This paper belongs to pelletizing field, specialize in China's first large scale travelling grate machine pelletizing plant. The pelletizing process and technical equipment of 504m^2 travelling grate machine at Shougang Jingtang steel plant are introduced in this paper. In order to improve the pellet quality and performance, meet the requirement of two 5500m^3 huge blast furnaces stable operation, adopt reasonable burden composition, therefore a large scale travelling grate machine pelletizing plant with annual output 4.0 million tons is configured at Shougang Jingtang. In this project, a number of technological developments and innovations are implemented, such as the pelletizing process design, general layout, equipment development, energy saving and emission control, automatic operation system, etc. A series of advanced technologies and large scale equipment are developed and applied, such as the large drying rotary kiln, the balling disc with diameter 7.5 meter, the shuttle type green pellet distributor, the large travelling grate machine and the seal facility, the high efficiency burner and so on. After commissioning, the comprehensive operation system is established based on charging mixture investigation, balling, green pellet distribution, and baking temperature control. As a result, the operational parameters and pellet quality have achieved advanced technical level; huge operational benefit and magnificent environmental benefit have been obtained.

Keywords: pelletizing; travelling grate induration machine; raw material; balling disc; energy saving

1 Introduction

Shougang Jingtang Steel Works is the "Eleventh Five-Year" national key project in China, which is constructed as a new generation recycling steel plant according to the concept of circular economy[1]. In order to achieve stable production of huge blast furnace, optimize burden composition, improve the level of concentrate burden, the largest scale and most advanced 504m^2 travelling grate induration machine for pellet production is researched and applied in China. A "high efficiency, low cost, energy saving, emission reducing" high quality pellet

---

本文作者：张福明，王渠生，韩志国。原刊于 AISTech-Iron and Steel Technology Conference Proceedings, 2015: 402-412。

production platform is configured, to meet the requirements of the two large 5500m³ blast furnaces operated in high efficiency and stable production.

By means of technical theory research, analysis of process flow and function, engineering optimization of numerical simulation, industrial experimental research, development of key equipment in the field of metallurgical process, this paper, focused on pelletizing production process and technical characteristics of large travelling grate induration machine, systematically studies process optimization, nationalization of key equipment, precise control system of large travelling grate machine pelletizing technology, and provides an important technical guarantee for the successful application of this technology. A series of critical equipment are independently developed including heavy duty pallet, etc. A quantity of process technical equipment are firstly developed and applied, such as drying kiln, large balling disc, and so on.

2 Selection of Pelletizing Process

2.1 Determination of Reasonable Burden Composition

Shougang Jingtang Steel Works is the national major engineering project that aimed at construction of 21st century advanced steel works, in accordance with requirement of "new function, new technology and new process" of new generation steel works, with characteristics of resource-saving, environment-friendly, rapid development of circular economy, as well as independent innovation[2].

In order to realize stable production, optimization of burden composition, improvement of beneficiated burden level of the two Shougang Jingtang 5500m³ BFs, a high-quality and high-performance raw material production system characterized by "high efficiency, low cost, energy-saving and environment friendly" is established to meet the burden material requirements for the two 5500m³ huge BFs in stable production process of high efficiency and low consumption[3].

BF burden mainly consists of three categories, namely sinter, pellet and lump ore. Pellet, compared with sinter, has such advantages as high ferrous content, good reducibility, even particle size, high cold strength, low energy consumption, little fume emission, etc.

After researches on large-scale BF burden composition at home and abroad, the burden composition of Shougang Jingtang 5500m³ BF was determined to be 65% sinter, 25% pellet and 10% lump ore, and the production scale of pelletizing plant was determined to be 4.00Mt/a. The "high performance, high efficiency, low cost" pelletizing production technology is being developed and applied to satisfy pellet demand of Shougang Jingtang huge BFs for stable, energy-saving, highly efficient and long life production. It is key supporting technology for stable production of huge BFs[4].

2.2 Comparative Study of Agglomeration Process

The technology of beneficiated burden is important support to realize "consumption minimization" of iron-making industry. Due to causes of resource endowment and technology inheritance, China, for a long time, has been using sinter as main raw material of BF[5]. Sinter has always occupied an important position in BF burden as its low price. While due to deterioration of high quality ore fine resource and influence of environment and resource availability, it has become the trend to increase quantity of pellet in BF burden[6].

Compared with sintering process, the prominent technical advantages of pelletizing process are represented in following aspects:

(1) Low system air leakage, high circulating utilization of hot air, low energy consumption of production procedure with only 1/3 of energy consumption in comparison with that of conventional sintering process.

(2) High pellet grade, less BF slag, low coke rate and fuel ratio.

(3) Less pollutant and fume emission with environment protection effects, only 1/10 of collected dust in comparison with that of conventional sintering process.

Technologies of large scale pelletizing production mainly include grate-kiln technology and large travelling-grate machine technology (Fig. 1 and Fig. 2). Comparison of technical feature between these two technologies is referred in Table 1.

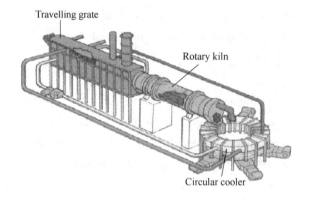

Fig. 1　Pelletizing production technology of grate-kiln

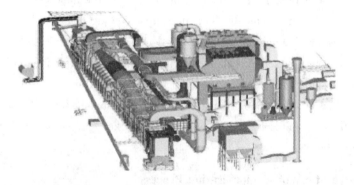

Fig. 2　Pelletizing production technology of travelling grate machine

Table 1 Technical feature comparison of pelletizing production technology

Item	Travelling grate machine	Rotary kiln	Evaluation
Capacity	Higher	High	Max. production scale of travelling grate machine is 7.5Mt/a and max. production scale of rotary kiln is 6.0Mt/a.
Quantity of main machines	1	3	The characteristics of travelling grate machine are intensive process flow, less heat dissipation, good circulation utilization of waste heat, high utilization efficiency of heat, less occupied area, low heat consumption and less dust emission.
Adaptability to raw material	Better, suitable for all raw materials	Worse, mainly or magnetite	Travelling grate machine has several burners arranged in preheating zone and induration zone. The process could be adjusted flexibly, which is suitable for material hard to be indurated. Less operation personnel and low running cost.
Pellet induration status	Static	Rolling	Pellets from rotary kiln need to be transported, thus there is a high requirement for dry ball strength. Powder and ring are easily produced. Pellet in travelling grate machine is indurated still with less added bentonite and higher pellet grade.
Production operation and control	Convenient	More complex	Rotary kiln has a long process flow, low heat utilization efficiency, slow thermal response speed and complicated process control. Travelling grate machine has high automatic control level with easy process control.

Production scale of Shougang Jingtang pelletizing plant is 4.00Mt/a. Technology of large travelling grate induration machine with an area of 504m^2 is adopted. Process flow of pelletizing production fully absorbs advanced technologies at local and abroad, pursuing easy and high efficient flow, compact and reasonable process. Main process flow includes such procedures as pre-proportioning, drying, roller pressing, preparation of flux and fuel, proportioning, balling, induration, and product classification. The process flow sheet is referred in Fig. 3.

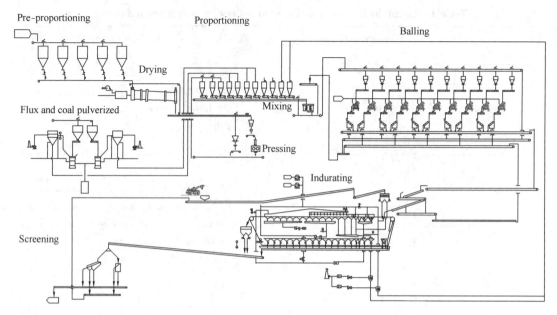

Fig. 3 Process flow diagram of Shougang Jingtang pelletizing plant

3 Innovation and Research of Engineering

3.1 Optimization of Process Flow

The process engineering, based on metallurgical process engineering theory and construction idea of Shougang Jingtang Steel Works, targets at optimization of process, function, structure and efficiency, takes dynamic precision design as guide principle, emphasizes top level design of process system, coordinative and high-efficient operating of material flow, energy flow and information flow and forms dynamic precision operation system of new-generation pelletizing plant with "flow", "flow network" and "operation procedure" as basic elements[7]. Precision design of raw material preparation sequence is emphasized and many regulation modes are available for precise proportioning of various high-quality pellets; function analysis and high-efficient, intensive configuration of main sequences, such as balling, preheating, induration, cooling, process fan are prominent, which lay foundations for green production with high efficiency, low energy consumption, high quality, and cleanness.

Plenty of researches have been conducted as to process flow, function integration, intensiveness and high efficiency, compact layout in the design. Close linkage has been realized for sequences as pre-proportioning, drying, rolling, preparation of flux and fuel, proportioning, balling, induration, cooling and products classification to shorten logistics transportation distance and reduce transfers of materials. United arrangement of buildings with the same or similar functions could decrease land area with continuous and compact layout. There is no transfer station in the whole plant, which is rare in new-built large-scale pelletizing plants at home and abroad. Its land occupation index per ton of ore is 0.016m²/t, ranking first in China. Layout of Shougang Jingtang Pelletizing Plant is referred in Fig. 4.

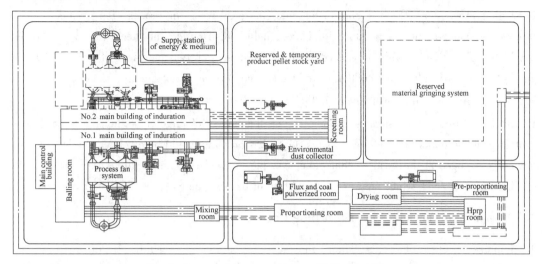

Fig. 4 Layout of Shougang Jingtang Pelletizing Plant

3.2 Optimization of Process Technology

The international and advanced technologies of large travelling grate machine are adopted, with total length of indurating machine 126m, width of pallet 4m, area of pallet 504m², material bed thickness 400mm, annual production capacity 4.00Mt/a.

3.2.1 Improvement of utilization ratio of hot air

Large travelling grate machine integrates seven process zones together, and completes drying, preheating, induration, homogenizing, and cooling process of green ballson pallets through circulating running of pallets. There are 2 sections of drying zone: up-draft drying (UDD) zone and down-draft drying (DDD) zone. Hot air from cooling zone Ⅱ is used for drying of green balls. Hot air from cooling zone Ⅰ and exterior heat source is used for preheating zone and induration zone. Homogenizing zone uses mainly self dissipated heat and hot air from cooling zone Ⅰ, rather than exterior heat source. Reasonable waste heat recycling system realizes utilization of heat in the most effective way. Effective and short process flow reduces radiating area the most with eminent decrease of heat consumption and improvement of energy recycling ratio.

3.2.2 Reasonable utilization of secondary energy source of iron and steel plant

This project, unlike foreign common practise that adopts heavy oil or natural gas as heat source, takes coke oven gas as exterior heat source for the indurating machine. 32 groups of burners are arranged at the fixed machine hood, which can realize flexible adjustment of process temperature through change of flow rate. In this way, requirements of different raw materials are satisfied the most. Meanwhile, the by-product gas is utilized most effectively, especially those gas with high calorific value so as to achieve high efficient conversion of energy[8].

3.2.3 Flexible adjustment and control of the system

Temperature and pressure of each process zone are balanced through process fan system in induration process. VVVF fans and many auto-control valves, in process fan system, realize flexible adjustment of temperature and reasonable distribution of heat in every process zone. Through reliable and advanced balling system, entry intensity of green ball is guaranteed. As a result, dust content is reduced, and service life of fans is extended.

3.2.4 Availability improvement of pallet

Hearth layer circulating system is adopted in the technology of large travelling grate machine to protect pallet so that service life of heat resistant parts on pallet could be extended. Pallet changing device is arranged at the head so that pallet can be changed on line rapidly within 5minutes for off-line maintenance, which improves availability of pallet effectively.

3.3 Application of Solid Fuel Burdening Process

To improve quality of pellet, fuel and flux preparation procedure is included in the design. Procedures of solid fuel burdening and dolomite burdening are arranged to adjust fuel proportioning and produce fluxed pellet based on different raw materials. The procedure of fuel burdening can increase porosity and reducibility of pellet, which used in BF production could improve productivity and decrease coke ratio. This pellet with high porosity and reducibility can reduce fuel consumption, temperature of grate bed of the indurating machine, and power consumption of the fans. Fig. 5 shows the energy consumption condition of the technology with fuel burdening.

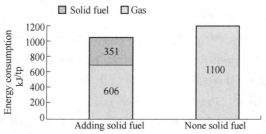

Fig. 5   Energy consumption condition of the technology with fuel burdening

From the aspect of process engineering, it is the first time to locate the fuel & flux preparation system near the proportioning room, the fuel and flux preparation collectors are arranged at top of bins in the proportioning room. In this way, powder of fuel and flux can be conveyed to bins in the proportioning room directly, which optimizes the process flow and simplifies equipment configuration.

New technology of hot fume self-circulating utilization in the flux and fuel preparation system is adopted to reduce system heat consumption and equipment specification of hot air stove, with less equipment investment. Oxygen content at inlet of the grinding machine is limited below 8% to guarantee safety of pulverized coal preparation system and reduce waste gas emission for the sake of environment protection.

3.4 Development of Pellet Shuttle Distributing Technology

Prepared green balls are conveyed stably and evenly to the indurating machine to avoid crushing of green balls in the conveying process. It is very important to guarantee strength of green balls. It is a key process to reduce transfer times of green balls and decrease falling height of them. A new type of distributing belt conveyor dedicated for pellet is developed, which will collect materials under balling disc and distribute green ball to the wide belt conveyor through single-direction shuttle distributor. An advanced process with distributing belt conveyor + wide belt conveyor + double-deck roller screen is applied. The distributing belt conveyor integrates the shuttle distributor and the collecting belt conveyor under the bailing disc into one belt conveyor. Its falling height can be minimized by means of control of diameter and height of the distributor head drum; Single stroke material distributing of green balls is realized through travelling back and forth of the distributor. In this way, number of transfer times and falling height of the green balls are decreased effectively, thus yield of green ball size is improved. The distributed material is even without piling on both sides. And conformity of green ball layer on the large travelling grate machine is guaranteed with good permeability; meanwhile, height of the building and occupied land area are decreased as well. The shuttle distributing technology effectively reduces number of transfer times and falling height of green balls, so that even thickness of material layer is guaranteed. The shuttle distributing technology is referred in Fig. 6 and Fig. 7.

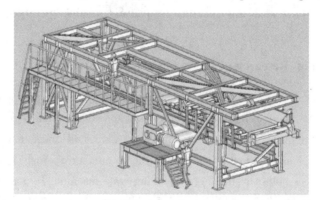

Fig. 6　3-D dynamic simulation engineering structure of the distributor

Fig. 7　Visual picture of distributor

3.5 Development and Application of Large-scale Fine Ore Drying Kiln

To satisfy normal production of the large travelling grate machine, the technology of large-scale fine ore drying kiln is developed with joint efforts from equipment manufacturer. A home-made super-large $\phi 5m \times 22m$ fine ore drying kiln is designed and manufactured to control material moisture. Procedure of material drying and roller pressing are considered on the basis of the general process configuration of the pelletizing production to guarantee stable moisture content of concentrate as main material of pellet, even particle size and specific surface area. So ball ability of raw material could be improved, while strength and quality of green balls are improved as well[9].

Maximum size $\phi 5m \times 22m$ fine ore drying kiln is developed the first time in China. Based on features of dry fine ore, a short and thick type rotary drying kiln with 4.4 length/diameter ratio is designed specially for this equipment, which can both strengthen drying effects, and reduce flow rate at discharging outlet. Consequently load of the dust catcher is reduced and dust emission is guaranteed up to the standard.

The hydraulic motor is firstly used as drive device of the dryer to have continuously variable transmission. Moisture of fine ore can be adjusted flexibly as per production requirement. Also, automatic heat-compensating system of the tail exhaust hood is designed for the first time to make sure gas temperature at the tail is above the dew point, so as to prevent low-temperature fume from producing dew. Above measures guarantee stable and reliable operation of the whole drying system.

3.6 Development and Research of Large Scale Balling System

The most important purpose of stable and smooth operation of pellet production is to guarantee quality of green ball. Capacity of balling disc with equal specification has a great effect on material balling. This project requires a variety of raw materials, and must have sufficient balling capacity. In case of poor balling quality, it causes a large number of crushed green ball as return, which directly influences output of the large travelling grate machine. To guarantee balling effect, the balling disc is firstly configured with a 7.5m diameter. Large scale balling disc which is matched with the large travelling grate machine. In comparison with conventional $\phi 6m$ balling disc, capacity of an individual balling disc is improved greatly with characteristics of increased diameter of balling disc, increased frequency of green ball rolling, improvement of balling effect, as well as decrease of balling disc quantity, less occupied land of the building and low project cost. By means of adjustment of balling disc angle and rotating speed, balling requirement of different materials can be satisfied.

A new balling disc with maximum diameter $\phi 7.5m$ has been firstly developed domestically with the manufacturer (Fig. 8 and Fig. 9). By this, diameter of the well proven balling disc in China is only 6m. Increase of balling disc diameter can intensify balling ability, improve balling effect and increase green pellet strength. Meanwhile, large-scale balling process can decrease balling disc quantity and reduce cost for equipment and civil works. In the design, angle of the balling disc can be adjusted, and VVVF is applied to adjust speed of the balling disc so as to satisfy change requirement of moisture and material. In addition, the brand new fixed scraper and balling disc support structure is specially considered and applied in order to meet demand of large-scale equipment. Research and development of this technology realizes domestic independent design and manufacture of the large scale balling disc, and has already popularized and applied in construction of pelletizing project.

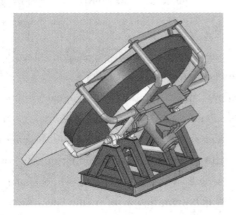

Fig. 8 3-D design structure of balling disc   Fig. 9 Site picture of balling disc

3.7 Integrated Application of Large Scale Technical Equipment

For well match with the 504m² large scale travelling grate machine, the φ1700mm diameter roller press, the 600m² ESP and the heat resisting fan with φ3.6m impeller diameter are firstly applied, in which, the 600m² ESP is the largest area and single unit electrostatic precipitator at home; all main process fans are localized to solve importation problems from the foreign companies of the previous large scale fan for pelletizing process. HV VVVF is applied in all process fans with characteristics of convenient control and regulation as well as effective decrease of power consumption.

3.7.1 Research of pallet kinematic mechanics

Pallet is the most critical equipment in the core of large travelling grate machine, and total weight of pallet is 17% of the complete indurating machine. Since it's hard to have direct measurement, detection of pallet is carried out only via the modelling established and program specially programmed by the known conditions under temperature and stress distribution change in working condition so as to calculate and develop distribution condition of temperature and stress in the complete cycle. On this basis, ANSYS software is used to have finite element stress analysis in order to analyse failure mode of pallet. In accordance with thermal condition of pallet, 3-D design and thermal stress finite element analysis are used to determine structure and grate bar reasonably (Fig. 10 and Fig. 11).

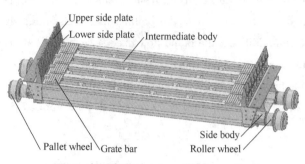 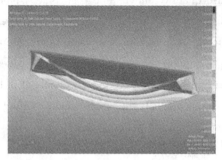

Fig. 10 3-D structural design drawing of pallet   Fig. 11 Finite element analysis diagram of pallet

Calculation and analysis show that pallet takes alternating mechanic stress and pulsating heat stress in one operational cycle, among them, thermal stress dominant factor; especially when it enters the high temperature zone, temperature rises rapidly with great temperature gradient. According to the calculation result, economical and reasonable material should be selected with satisfactory of strength requirement, and special requirements should be considered in aspects of chemical composition, organization status, casting and machining quality, and so on for the material in order to guarantee thermal strength of the main beam, while heat transfer study, design and development of new grate bar and optimization of grate bar structure are conducted to reduce direct heat transfer from grate bar to pallet beam and decrease hot load of the pallet beam. Optimization of grate bar structure can have reasonable guidance of wind flow, reduce ventilation resistance of the grate bed effectively and decrease blocking problem. Fig. 12 is pallet assembly and application picture.

Fig. 12 Pallet assembly and application situation

3.7.2 Innovation and application of self-lubricating roller sleeve technology

Learning from well-proven self-lubricating technology at sintering pallet in recent years, self-lubricating bearing structure is used for pallet roller bearing with characteristics of low manufacture cost, short manufacture cycle and long service life to reduce maintenance cost and improve operability of the indurating machine effectively.

Solid lubricant with molybdenum disulphide and tungsten disulphide is inlaid to the self-lubricating bearing, and the lubricant particle attached to the metal surface during friction process, which has lubrication action. It is not necessary to have grease during application, maintenance free, long service life, and equipment maintenance cost can be lowered availably.

3.7.3 Development of sealing technology of wind-box end and partition

In production practice, for further improvement of sealing performance, intensive study of working condition for wind-box end and partition sealing is carried out, and new wind-box end and partition sealing device of the large travelling grate machine is optimized to improve sealing effect and reduce air leakage of the system efficiently.

On the basis of production practice, sealing effect between hood of the large travelling grate machine and pallet is analysed with application of specific improvement measures, a new sealing technology between the machine hood and pallet is developed, and the sealing effect is further improved.

3.8 Research and Development of Energy Conservation and Environmental Protection Technology

Energy-saving and environmental protection is a significant symbol of modern metallurgical plant. Shougang Jingtang Pelletizing Plant adheres to the advanced concept in energy conservation and emission reduction as well as green and environment protection with application of pelletizing production process by the large travelling grate induration machine. Many advanced technologies are applied in aspects of energy conservation and emission reduction[10], clean production, recycle economy, etc. which have outstanding technological superiority on energy conservation and environmental protection. Main performances as follow:

(1) The large travelling grate machine completes all the process of drying, preheating, indurating and cooling in one sealing hood, while the complete large travelling grate machine is arranged in one enclosed building for effective reduction of dust leakage.

(2) High temperature hot air in the hood can be reused directly with characteristics of short distance pipeline, large return air pipe area and high heat utilization efficiency. The grate hood adopts multilayer heat resisting material and thermal insulation material with characteristics of effective reduction of surface heat dissipation and heat consumption. HV VVVF is applied to all the process fans with low power consumption.

(3) Thick phase pneumatic dust conveying system is applied to convey all collected dust to the material proportioning room for sufficient recovery and resource utilization. Fully closed conveying equipment and pipeline are used to avoid secondary dust generation in comparison with the conventional dust conveying process with great improvement of work condition and plant environment.

(4) The high efficient lifting blade, which is specially researched and manufactured, is applied in the first developed ore fine drying kiln, with application of high temperature cyclone

wind distribution device, so that heat exchange efficiency is improved, drying result is strengthened and heat consumption is reduced.

During design and research, the energy flow network and operation system are established based on high efficient conversion of energy source so as to materialize energy conservation. On principle of "reduction, recycling and reuse" energy consumption, the energy input-output model is established to realize optimization of energy consuming. Major technical innovations include:

(1) Coke oven gas produced from the iron and steel plant is used as fuel of pellet induration to optimize fuel structure, instead of high calorific value natural gas, to make the energy source structure of the iron and steel plant optimized, and efficiency of coke oven gas can be utilized in high efficient production.

(2) Burner applicable to coke oven gas is researched and manufactured with characteristics of reasonable design structure so as to realize regulation and control of low coefficient of excess air, guarantee complete combustion of fuel and sufficient flame temperature, reduce a quantity of production and emission of CO_2 and NO_x, so as to have high efficient clean combustion and decrease pollutant emission.

(3) Erection position and quantity of burner are considered reasonably in order to meet pellet indurating process, realize precise control of combustion and temperature field, reduce energy consumption and lower pollutant emission.

(4) The high efficient "airflow heat recovery and utilization system" is designed and developed, which can high efficiently use the heat energy at various steps produced from indurating process of the large travelling grate machine to realize reasonable matching and efficient utilization of energy at different procedures of pellet drying, preheating, indurating, etc.

(5) The technology of carbon-burdening into pellet is researched and developed, which take carbon dust and coke breeze as energy sources of the iron and steel plant, so that solid wastes are recycled and utilized comprehensively and efficiently, while pellet indurating quality is improved and energy consumption at various working procedures is lowered.

4 Application

Since Shougang Jingtang Pelletizing Plant was put into production in August, 2010, the pelletizing production process with large travelling grate machine has been studied and explored. The key technologies of stable production, high quality and high performance for pelletizing process have been developed under condition of different materials. The integrated control management technology system of multiple working procedures are established based on material proportioning research, balling, distribution and indurating control, etc.

(1) The perfect automatic test and control system are designed to make information of the overall process of the entire technology monitored online.

(2) The optimal mathematical model is built, as well as the close-loop control for material

distribution, close-loop control for temperature, close-loop control for wind system, balance control of hearth layer, and gas safety system are developed to have optimal operation of pelletizing production.

(3) Remote control and unattended operation for equipment are foreseen, so that labour productivity is improved greatly.

Optimization of production process is very important to ensure stable and smooth production. Innovations in the following aspects are mainly carried out:

(1) Centralized control and regulation by computer is applied in production process. The industrial TV monitor and expert control system management are used in major process systems. Such as fan system, materialthickness control system, hearth layer circulating system, burner system, and so on, with advantages of high automation level, stable production and low gas consumption.

(2) A series of optimization and innovations were carried out after it was put into production, for instance, feeding mode from malarial bin, feeding belt conveyor for balling disc, bottom lining of the balling disc, etc. And a crusher for green ball is provided. So that balling performance of the returns is improved and product pellet yield is further increased.

(3) The software for equipment management cycle is developed to have scheduled maintenance, change and service for wearing parts. Maintenance schedule can be adjusted in time as per the service experience. Off-line maintenance should be conducted to the qualified equipment as far as possible, so as to solve any problem before it occurs for guarantee of efficient operability.

(4) Job-site manning is adjusted according to production experience for raising labour productivity in order to achieve 98% or above of the annual availability.

After 4 years' stable operation of the plant since it was put into operation in August, 2010, it has completely reached the designed level: yearly availability is above 8%, pellet total ferrous content is close to 66%, compression strength of pellets gets to 3000N/pellet, energy consumption in the process is 17.11kgce/t and screening index stands on 0.3%. Main techno-economic indices reached to the international advanced level with remarkable environmental benefit and social benefit.

This project is integrated with advanced process and technologies at home and abroad with basis of independent design and integration. On basis of more than 10 years' development and research by Shougang, the international advanced process technologies and equipments are integrated, the process flow and key unit technology are optimized, so as to ensure reasonable process flow, complete process configuration, compact plant layout and high efficient production operation. Raw material preparation stage and precise regulation of indurating heat is especially emphasized in design and development process, so as to make it suitable for different kinds of raw material and create advantages to produce multi-variety and high-quality pellets. It has been proved through practices of production that this technology can produce acid and alkali pellets suitable for BFs in a high-efficiency, low energy-consumption, low-pollution way and solve technical difficulties thoroughly from process aspect, such as relatively low heat-resistant temperature of travelling grate and easy ringing of rotary kiln.

After put into operation, 4.00Mt/a pelletizing plant of Shougang Jingtang steel works mainly treats high-sulfur magnetite from Peru with relatively high alkali metal content. Due to reliable technical performance, good adaptability of the process, flexible control and regulation of large travelling grate machine, it has fitted with many kinds of raw materials, thus stable and fluent production running and advanced technical index are realized. Many kinds of raw materials are adopted to produce high-quality pellets stably. Needs for concentrate burden for high-efficiency and low consumption production of two Jingtang 5500m^3 huge BFs are fully satisfied. Those important indexes as energy consumption of pellet production, pellet quality, and equipment availability all reach advanced international level. The index comparison with advanced pelletizing line home and abroad is referred in Table 2.

Table 2 Main operational Parameters of pellet production

Item	Shougang Jingtang (TGIM)	Ansteel Pelletizing Plant (TGIM)	WISCO Pelletizing Plant (Rotary Kiln)	Shougang Mining Co. (Rotary Kiln)	Brazil CVRD (TGIM)	Iran GMI (TGIM)
Design capacity/Mt·a^{-1}	4.0	2.1	5.0	2.0	7.5	5.0
Availability/%	98.32	93.34	95.86	90.84	96.44	82.20
Total Fe/%	65.95	57.8	63.26	65.2	65.5	66.78
Compression strength, N/pellet	3000	2427	2584	2782	3300	2700
Screening index/%	0.35	2.82	2.31	0.73	2.47	3.35
Energy consumption of procedures/kgce·t^{-1}	17.11	35.12	25.51	17.55	30.78	19.10

5 Conclusion

Focused on design development and research of pelletizing production line, research of technical equipment as well as study of production and application of the first more than 4.00Mt/a large travelling grate machine, the high efficiency and low consumption pelletizing production technology by the large travelling grate machine in China are developed and come into being, with main technical innovations as follows:

The first large travelling grate machine with morethan 4.00Mt/a output for pelletizing production has been independently designed, integrated and manufactured in China, with innovations in aspects of functional optimization, process optimization, design optimization, efficiency optimization, etc., possession of independent intellectual property rights, as well as realization of high efficient and low consumption production.

A series of large scale technical equipment with independent intellectual property rights are developed, manufactured and applied, such as shuttle distribution equipment, pallet sealing device, large scale drying kiln, large scale balling disc, etc. Overall localization of large scale technical equipments are achieved, while the independent design innovation and equipment manufacture innovation are realized.

The pelletizing production process technique by large travelling grate machine is researched and explored. The key technologies of stable production, high quality and high performance for pelletizing process have been developed under condition of different materials. And the integrated control management technology system of multiple working procedures are established based on material proportioning research, pellet balling and indurating control, etc.

After 4 years' stable operation of the plant since Shougang Jingtang Pelletizing Plant was put into operation in August, 2010, it has completely reached the designed level, with yearly availability above 98%, pellet grade 65.95%, compression strength of pellets ≥ 3000N per pellet, and energy consumption in the process 17.11kgce/t. It reached to the international advanced level in comparison with similar technologies, with remarkable environmental benefit and social benefit.

Acknowledgements

The authors would like to thank our colleagues of Ms. Li Min, Mr. Huang Wenbin, Mr. Li Xiang and Mr. Zhang Weihua, etc. who participate the technical research and design works on this engineering.

References

[1] Zhang Fuming, Qian Shichong. Design, construction and application of 5, 500m^3 blast furnace at Shougang Jingtang [C]. AISTech 2014 Proceedings of the Iron & Steel Technology Conference, Indianapolis, Ind., USA, 2014: 753-765.

[2] Zhang Fuming, Cao Chaozhen, Meng Xianglong. Philosiphy and application on circulating economy for contemporary steel plant [J]. Advanced materials Research, 2013, 813: 196-200.

[3] Zhang Fuming, Cao Chaozhen, Meng Xianglong. Technological development orientation on ironmaking of contemporary blast furnace [J]. Advanced Materials Research, 2013, 813: 192-195.

[4] Zhang Shourong. Development and problems of ironmaking industry after entering the 21st century in China [J]. Ironmaking, 2012, 31 (1): 1-6.

[5] Ye Kuangwu. Current status and expectations of pelletizing production in China [J]. Sintering & Pelletizing, 2013, 28 (1): 1-7.

[6] Yang Jialong, Tan Suiqin, Wang Zhaoping, et al. Increase pellet proportion to optimize burden composition of BF [J]. Iron & Steel, 2005, 40 (10): 13-17.

[7] Yin Ruiyu. Theory and Method of Metallurgical Process Integration [M]. Beijing: Metallurgical Industry Press, 2013: 97-98.

[8] Englund D J, Davis R A. CFD Model of a straight-grate furnace for iron oxide pellet induration [J].

Minerals & Metallurgical Processing, 2014, 31 (4): 200-208.
[9] Umadevi T, Kumar Prachethan P, Kumar P, et al. Investigation of factors affecting pellet strength in straight grate induration machine [J]. Ironmaking and Steelmaking, 2008, 35 (5): 321-326.
[10] Li Guowei, Xia Leige, Qing Gele, et al. Study on thermal system and raw material plan of travelling grate induration machine at jingtang [J]. Sintering & Pelletizing, 2001, 36 (2): 20-24.

Research and Optimization of BF-BOF Interface Technology

Zhang Fuming Xie Jianxin Yan Zhanhui Zhang Lingyi

Abstract: The iron and steel manufacturing process is formed by reasonable matching and coupling through several sections. Ferrous mass flow, energy flow and information flow run dynamically in the whole steel manufacturing process. The interface technology of ironmaking and steelmaking is consisted by the unit processes of hot metal transportation and pretreatment between blast furnace and converter. In this paper, the traditional torpedo tank is cancelled and the hot metal is transported directly from blast furnace to steelmaking desulphurization process with the technology of multi-function direct transportation of hot metal ladle, and the "one ladle to the end" of hot metal transportation is realized. The hot metal direct transportation by ladle technique was adopted in Shougang Jingtang's two sets of $5500m^3$ blast furnace. Through the ferrous mass flow and process operational network dynamic ordering, the interface technology is coordinated and the flow network is rationalized.

Keywords: blast furnace; converter; interface technology; hot metal transportation; process optimization

1 Introduction

Iron and steel production process is a complete system composed of several different unit processes through orderly combination. The goal of iron and steel manufacturing is to achieve multi-objective collaborative optimization. It is necessary not only to adopt the advanced and reliable technological process and equipment, but also to optimize the layout of the process and simplify the technological links in order to achieve the goal of high efficiency, high quality, low consumption and low emission in iron and steel production. The interface technology between different processes is the key essential factor to optimize the process flow and layout of the general plan. The "interface technology" in the iron and steel production process refers to the connection-matching, coordination-buffer technology and corresponding equipment between the ironmaking, steelmaking, continuous casting, hot rolling and cold rolling processes etc. For a long time, blast furnace-converter process is concerned, different production processes of different steel

---

本文作者：张福明，颉建新，闫占辉，张凌义。原刊于 Proceedings of 2018 China Symposium on Sustainable Steelmaking Technology (CSST 2018). Tianjin, The Chinese Society for Metals (CSM), 2018: 114-117。

products and different interface technical forms between ironmaking and steelmaking are also different. For a long time, torpedo tankers have been applied widely in the transportation of hot metal in large steel plants. Torpedo tanker is a kind of mature hot metal transportation device, which has been developed for many years and has the advantages of stable operation, good heat insulation performance and long service life. However, in the transportation process of hot metal from blast furnace to converter, there are many technological links and large drop of hot metal temperature, which is far from the goal of high efficiency, fast rhythm, low cost and low energy consumption pursued by modern iron and steel metallurgical process.

The technology of "one ladle to the end" is applied in the interface of ironmaking and steelmaking, which means that the traditional torpedo tank is abolished, the hot metal is carried and transported directly by the hot metal ladle of steelmaking, the hot metal accepting, transporting, buffering, storing, pretreating, and charging to the converter are integrated into multifunctional ladle. By adopting the hot metal direct transportation technology of "one ladle to the end", the hot metal transfer station of steelmaking workshop is cancelled, the operation of hot metal pouring can be reduced, the process flow is shortened, the layout of general plan is optimized, and the temperature of hot metal can be increased. Therefore, the loss of hot metal can be reduced efficiently, the energy consumption can be reduced remarkably, and the smoke dust pollution can be reduced remarkably, the distinct economic and environmental benefits can be achieved. Therefore, the research and application on the technology of "one ladle to the end" hot metal direct transportation has a great significance for the new steel plant process optimization, and has a profound influence on the development of iron and steel industry.

In order to dispel the non-capital function of Beijing and conduct the 2008 Beijing Olympic Games successfully, it was decided that Shougang should move away from Beijing and build a new steel base in Caofeidian industrial district, Tangshan city, Hebei Province, named as Shougang Jingtang Iron and Steel Plant. The project is designed and built independently according to the concept of a new generation recyclable iron and steel process, and has a large steel project with advanced international level in the 21st century. It is also a large-scale modern steel factory which produces thin strip. The annual crude steel design production capacity of Jingtang plant is 8.7-9.2 million tons. Two blast furnaces, one steelmaking plant, two hot rolling lines and four cold rolling lines are configured as static physical frame of clean steel production process. Two 5500m^3 blast furnaces, "2+3" 300 tons converters of steelmaking, high efficiency and low cost clean steel production platform is designed and built, and the hot metal direct transportation technology is adopted in ironmaking and steelmaking process interface.

2 Process Research

2.1 Analysis of Technical Advantage

Compared with the traditional hot metal torpedo tanker, the transportation mode of "one ladle to the end" has the following technical advantages:

Reducing the temperature drop of hot metal, reducing the energy consumption, increasing the output by using 300 tons hot metal ladle to transport the hot metal directly, and reducing the hot metal from torpedo tank to ladle pouring operation. In particular, the temperature drop of hot metal from blast furnace to converter will be greatly reduced under the condition of ladle covered with cap during whole process. The higher hot metal temperature, the more conducive to KR desulphurization treatment and the slag skimming operation. It is beneficial to improve the efficiency of hot metal desulphurization and the smelting of desilicication and dephophorization converter to reduce the iron consumption and increase the output.

Shortening the process flow and simplifying the process of transporting hot metal with "one ladle to the end" process can cancel the hot metal transfer station, thus reducing the operational procedure of hot metal pouring to the ladle, simplifying the production process and speeding up the production rhythm. Also make steelmaking workshop layout more compact and reasonable.

Shortening the hoisting time of hot metal can reduce the hoisting height of hot metal hoisting by about 10 m because of canceling the transfer station, which can make the feeding span of 450 tons casting crane hoisting height. The average running speed of the crane main hook is 8m/min. The hot metal hoisting procedure can save 2.5minutes, which is helpful to reduce the occupation of the crane, speed up the production rhythm and improve the production efficiency.

When the traditional torpedo tanker is used to reduce the environmental pollution, the total exhuast volume of smoke and dust caused by hot metal transfer station is 960000m^3/h. If the technology of "one ladle to the end" is used to transport hot metal, it can avoid the smoke dust pollution caused by hot metal pouring, which is beneficial to clean production and environmental protection.

2.2 Study on Hot Metal Temperature Variation During Transportation

The temperature drop test of 260t torpedo tank and 200t hot metal ladle was carried out in Shougang No.2 steelmaking plant. The hot metal quantity in torpedo tank was 246t, and the average temperature drop of hot metal was 0.21℃/min. Fig. 1 shows the hot metal temperature drop test result of 260t torpedo tank.

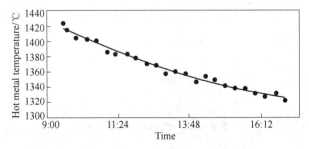

Fig. 1 Hot metal temperature drop test result of 260t torpedo tank

The hot metal temperature drop test of 210t ladle was carried out twice. For the first time,

200kg heat preservation agent was added to the surface of hot metal for the first time. The service life of the ladle was 882 heats, the amount of hot metal was 190t, and the average temperature dropped was 0.14℃/min. The test result is shown in Fig. 2. After adding 200kg heat insulating agent on the surface of hot metal for the second time, the cap of the hot metal ladle was simulated. Every time 20minutes was added with the heat insulating agent 20kg, the service life of the hot metal ladle was 215 heats, the average temperature dropped was 0.148℃/min, and the charging quantity of hot metal was 190t. The temperature change of hot metal ladle after simulated covering is shown in Fig. 3.

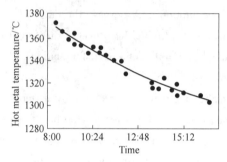

Fig. 2 Hot metal temperature drop test result of 210t ladle

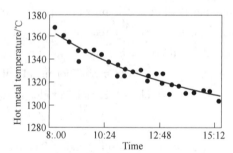

Fig. 3 Hot metal temperature drop test result of 210t ladle covered with cap simulation

The temperature measurement of 210t hot metal ladle without cover was carried out in Qian'an steel plant. After 74minutes at ambient temperature of 0℃, the temperature of empty ladle was reduced from 925℃ to 600℃. According to statistics, the cooling rate of hot metal ladle is 50-60℃/min, the cooling rate is 10-16.5℃/min at 900℃ and 5℃/min at 800-900℃. When the empty ladle was below 800℃, the cooling rate is 2.5-5℃/min.

3 Engineering Design and Innovation

3.1 Optimization of Layout and Transportation Scheme

The overall planning concept of the Jingtang steel plant is to take "flow" as the core, the principle of time and space is adopted effectively, the concept of "compact interface connection, rapid logistics transportation and dynamic-orderly production" is implemented in the whole process. In order to reduce the temperature drop of hot metal in the transportation process, the layout of ironmaking and steelmaking workshop is arranged in series, and the railway from blast furnace to steelmaking workshop by straight line, through the research and analysis of the layout characteristics of the general plan of the large advanced iron and steel plants at home and abroad, in order to reduce the temperature drop of the hot metal during the transportation process. In order to ensure the hot metal transportation to enter the steelmaking workshop quickly in time, the maximum conveying distance is 1300m, according to the 7km/h transporting speed, the maximum conveying time is 11.14minutes. The transportation time is shortened effectively, the temperature drop of hot metal is reduced, and good economic benefit will be obtained.

3.2 Development of Hot Metal Weighing Technology

There are four tapholes are configured for each blast furnace, and there are two sets of hot metal liquid level meter on the railway under each taphole. The liquid level meter has the weighing function and indicating the hot metal liquid level function. During the tapping process, the hot metal weight and height of ladle liquid level are displayed on the large screen in front of the cast house, which is convenient for the operator of cast house to control the amount of hot metal in the ladle. In accordance with the needs of steelmaking and smelting varieties, the hot metal content of each ladle is informed by computer network system or telephone to the blast furnace operator daily. The blast furnace operator strictly controls the amount of hot metal in the ladle according to the requirements. The quantity of hot metal in each ladle must meet the needs of steelmaking, and the weighing error shall not exceed 2 tons plus-minus. After each ladle is filled under the cast house, the ladle will be transported to steelmaking workshop, and the weight data is measured in the middle of 800 tons railway weighing meter, which is the basis data for ironmaking and steelmaking production. Fig. 4 shows the tapping process in Jingtang No. 1 blast furnace.

Fig. 4　Tapping process in Jingtang No. 1 blast furnace by ladle

3.3 Application

In the engineering design, the straight line distance between blast furnace and steelmaking plant is shortened to 900m, and the temperature drop of hot metal is reduced by 30-50℃, and the average temperature of hot metal entering KR desulfurization station reaches 1387.5℃. Due to the steelmaking plant hot metal pouring station is cancelled, the investment of the equipment and workshop can be reduced more than 40 million Yuan. The annual emission of smoke and dust can be reduced by about 4700 tons, and the annual production cost will be reduced by more than 60 million Yuan.

4 Conclusions

(1) In the process of ironmaking and steelmaking, the multifunctional hot metal ladle is applied to carry the hot metal, which is transported directly from the blast furnace to the steelmaking process. Multifunctional hot metal ladle integrates blast furnace hot metal acceptance, transportation, storage (buffer), hot metal insulation, hot metal pretreatment, converter charging, accurate measurement, accurate positioning and fast turnover.

(2) Compared with the torpedo tank mode, the advantages of the direct transportation process of multifunctional hot metal ladle are as follows: short transportation time and no need to go through the "pouring ladle" process of hot metal, which shortens the process flow and improves the production efficiency; The loss of hot metal and the loss of hot metal heat are reduced; energy consumption and smoke pollution are reduced; project construction investment is reduced and the number of molten iron tanks is reduced; the hot metal temperature is high and the activity coefficient of molten iron [S] is high, so high temperature desulphurization is realized. It is helpful to improve the desulphurization efficiency of hot metal pretreatment process.

(3) The application of hot metal directly conveying of Jingtang's 5500m^3 blast furnace has been successfully achieved, and the temperature of hot metal is increased by 30-50℃, and the successful application results have been achieved.

References

[1] Yin Ruiyu. Theory and Method of Metallurgical Process Flow Integration [M]. Beijing: Metallurgical Industry Press, 2016: 284-287.
[2] Zhang Fuming, Cui Xingchao, Zhang Deguo, et al. New generation process flow and application practice of Shougang Jingtang steel making plant [J]. Steelmaking, 2012, 28 (2): 1-6.

钢铁冶金工程技术创新

现代大型高炉关键技术的研究与创新

摘　要：无料钟炉顶、高风温和煤气干式除尘是现代大型高炉的关键技术。自主设计研制的大型高炉无料钟炉顶设备，通过三维设计优化、仿真模拟和试验测试，在设备可靠性、使用寿命等方面达到国际先进水平；自主开发的高炉煤气全干式布袋脉冲除尘工艺，在系统优化设计、煤气温度控制、除尘灰气力输送等方面取得技术突破；集成创新的高效长寿高风温技术，通过开发应用助燃空气高温预热技术、提高热风炉传热效率、热风炉系统结构优化等措施，在使用高炉煤气燃烧的条件下，风温达到1250℃。

关键词：高炉；炼铁；无料钟炉顶；热风炉；煤气干式布袋除尘

1　引言

近10年来，我国高炉大型化、高效化、现代化、长寿化、清洁化发展进程加快，炼铁工艺技术装备水平迅速提升，在系统总体设计、核心工艺技术、装备制造技术、数字化控制技术、生产技术及工程集成能力等方面取得重大进展，大型高炉关键技术自主创新取得显著成效。在大型高炉无料钟炉顶设备、煤气干式布袋除尘、高效长寿高风温等单元技术自主开发、系统技术集成方面形成突破，成为支撑现代大型高炉技术发展的关键技术装备。

2　大型高炉无料钟炉顶设备技术创新

采用无料钟炉顶装料设备是现代化高炉的重要技术特征。首钢自主设计研制的无料钟炉顶设备经历了20多年的创新发展历程，结合大型高炉生产技术的进步，在已有技术的基础上不断优化创新，攻克了大型高炉无料钟炉顶布料装置、齿轮箱冷却、设备工作可靠性及设备使用寿命等关键性技术难题，成为我国自主设计制造全部实现国产化并具有核心竞争力的关键技术装备。

2.1　开发新型齿轮箱冷却技术

布料溜槽传动齿轮箱是驱动并控制布料溜槽进行旋转和倾动，实现布料功能的核心关键设备。1998年，首钢4号高炉（2100m³）采用了首钢开发研制的水冷齿轮箱，该齿轮箱采用水冷盘管式间接冷却结构。生产实践中发现此种冷却结构冷却范围较小，水冷盘管阻力损失较大，对冷却水水质要求高，容易出现管道结垢现象，冷却效率随之降低。

针对上述技术缺陷，研究开发了新型齿轮箱冷却结构，将间接冷却改为直接冷却方

本文作者：张福明。原刊于《钢铁》，2009，44（4）：1-5。

式,同时将冷却范围扩大,提高了冷却效率。冷却水量可以提高到 25t/h 以上且不会发生冷却水溢漏,由于冷却效率提高,实际生产操作中冷却水量仅为 8t/h,在炉顶温度达到 800℃ 的极端情况下齿轮箱仍能正常工作。

研究开发了齿轮箱开路工业新水冷却系统,从根本上确保了冷却水水质,同时降低了冷却水温度,为齿轮箱冷却系统正常工作提供了可靠保障。冷却系统采用"U"形水封技术,与国外闭路循环冷却技术相比系统简化、流程紧凑,设备运行更加可靠,调节控制灵活,设备维护量少,节省投资。

2.2 研究优化炉料分布控制技术

采用现代化计算模型,自主开发了无料钟炉顶布料料流轨迹计算软件和仿真分析模型,修正了炉顶调节阀料流控制曲线和布料溜槽布料控制曲线,通过无料钟炉顶设备模拟布料试验进行验证,满足了高炉生产各种布料方式的要求。料流轨迹计算模型如图 1 所示,布料仿真分析模型如图 2 所示。

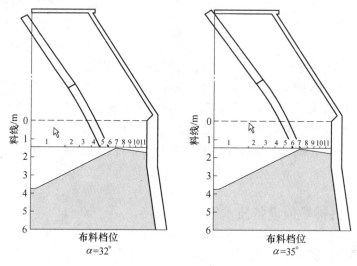

图 1 料流轨迹计算模型

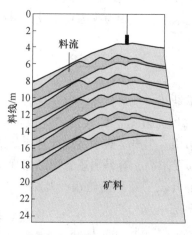

图 2 布料仿真分析模型

2.3 提高设备质量

采用先进的三维设计方式和有限元应力分析计算，对无料钟炉顶设备结构进行了设计优化，对炉顶设备结构、控制系统、耐磨材料等进行了系统研究和创新。开发了布料溜槽悬挂锁紧装置，彻底解决了无料钟炉顶布料溜槽脱落的问题，提高了高炉作业率；布料流槽倾角的控制精度达到了±0.25°，实现了角度的定量微调，保证了多环布料的灵活、准确和均匀，达到引进设备的控制水平；改进了布料溜槽、换向溜槽的结构及其衬板材质，采用镶嵌硬质合金衬板及独特的制造工艺，使用寿命达到产铁量 300 万吨以上，布料溜槽的使用寿命由 12 个月提高到 20 个月，换向溜槽的使用寿命由 3 个月提高到 24 个月[1]。

目前我国大型高炉无料钟炉顶设备已经全面实现国产化，并在 2650m³ 级高炉上得到成功应用，国产无料钟炉顶设备与引进设备的主要技术性能见表 1。

表 1 无料钟炉顶设备主要技术性能

设备性能	迁钢 1 号高炉	国内 A 高炉	国内 B 高炉	国内 C 高炉
高炉容积/m³	2650	2545	2580	3200
炉顶压力/MPa	0.25	0.25	0.2（最大 0.25）	0.25
料罐布置形式	并联	串联	串联	并联
料罐容积/m³	55	55	55	70
炉顶温度/℃	150~250，最高 800，持续时间 30min	150~250，最高 600，持续时间 30min	200，最高 500，持续时间 30min	200，最高 500，持续时间 30min
设备耐压能力/MPa	0.25±10%	0.25±10%	0.25±10%	0.25±10%
最大装料批数/批·h^{-1}	12	9.75	9	9
料流调节阀精度/(°)	位置传感器 0.1 定位精度±0.3	位置传感器 0.1 定位精度±0.3	位置传感器 0.1 定位精度±0.3	位置传感器 0.1 定位精度±0.3
料流调节阀下料速度/m³·s^{-1}	0.8，焦炭最大尺寸 80mm	0.7，焦炭最大尺寸 80mm	0.7，焦炭最大尺寸 80mm	0.7，焦炭最大尺寸 80mm
布料溜槽长度/m	3.5	4	4	4
布料溜槽转速/r·min^{-1}	10	8	8	8
布料溜槽倾动范围/(°)	5~70	2~53	2~53	2~53
布料溜槽使用寿命	300 万吨铁			

3 大型高炉煤气全干式布袋除尘技术

高炉煤气干式布袋除尘技术是 21 世纪高炉实现节能减排、清洁生产的重要技术创新，可以显著降低炼铁生产过程的新水消耗、减少环境污染，已成为现代高炉炼铁技术的发展方向。

高炉煤气干式布袋除尘技术已有 30 多年的发展历程。2007 年 1 月，我国自主开发的高炉煤气全干式低压脉冲布袋除尘技术在迁钢 2 号高炉（2650m³）获得成功，完全取消了备用的煤气湿式除尘系统。研究开发了煤气温度控制、除尘灰浓相气力输送、管道系统防腐等核心技术，使我国大型高炉煤气全干式布袋除尘技术达到国际先进水平。

3.1 优化集成工艺流程和系统配置

通过研究分析国内外高炉煤气干式布袋除尘技术应用实践,对加压煤气反吹除尘技术和低压脉冲除尘技术进行了系统的对比研究,设计开发了高炉煤气全干式低压脉冲布袋除尘工艺。迁钢2号高炉采用14个直径为4600mm的除尘箱体,箱体为双列布置方式,两列箱体中间设置荒煤气和净煤气管道,煤气管道按等流速原理设计,使进入各箱体的煤气量分配均匀,整个系统工艺布置紧凑、流程短捷顺畅、设备检修维护便利。

采用低滤速设计理念,确保系统运行安全可靠。每个箱体设滤袋250条,滤袋规格$\phi 160mm \times 7000mm$,单箱过滤面积$880m^2$,总过滤面积$12320m^2$。设计中加大了滤袋的直径和长度,高径比降低,滤袋结构尺寸更加合理;扩大了箱体直径,使除尘单元的处理能力提高,减少了箱体数量、建设投资和占地面积。表2是迁钢2号高炉煤气全干式布袋除尘工艺技术参数。

表2 迁钢2号高炉煤气全干式布袋除尘工艺技术参数

项 目	参 数
高炉容积$/m^3$	2650
煤气量$/m^3 \cdot h^{-1}$	50×10^4(最大)
炉顶压力$/MPa$	0.25
煤气温度$/℃$	100~220
箱体数量$/$个	14
箱体直径$/mm$	4600
滤袋条数$/$个	250
滤袋规格$/mm$	$\phi 160 \times 7000$
总过滤面积$/m^2$	12320
单箱过滤面积$/m^2$	880
标况滤速$/m \cdot min^{-1}$	0.68
工况滤速$/m \cdot min^{-1}$	0.31
净煤气含尘量$/mg \cdot m^{-3}$	≤5

3.2 煤气温度控制技术

煤气温度控制是布袋除尘技术的关键要素,正常状态下,煤气温度应控制在100~220℃,煤气温度过高、过低都会影响系统的正常运行。采用煤气干式布袋除尘技术,高炉操作要更加重视炉顶温度的调节控制。炉顶温度升高时采取炉顶雾化喷水降温措施,同时在高炉荒煤气管道上设置热管换热器,用水作为冷却介质,将高温煤气的热量通过热管传递,使水汽化吸收煤气热量,可以有效地解决煤气高温控制的技术难题。

煤气低温控制要采取提高入炉原燃料质量、加强炉体冷却设备的监控、合理控制炉顶温度、荒煤气管道保温等技术措施,在高炉开炉、复风时要采取有效措施,降低煤气中的含水量,使煤气温度控制在露点以上。目前正在研究开发煤气低温状态的高效快速换热技术,从而使煤气高温、低温都能够得到有效的控制。

3.3 煤气含尘量在线监测技术

煤气含尘量在线监测装置是监控高炉煤气布袋除尘系统运行的重要检测设备,研究开发的高炉煤气含尘量在线监测系统采用电荷感应原理。在流动粉体中,颗粒与颗粒、颗粒与管壁、颗粒与布袋之间因摩擦、碰撞产生静电荷,形成静电场,其静电场的变化即可反映粉尘含量的变化。煤气含尘量在线监测系统通过测量静电荷的变化,来判断布袋除尘系统运行是否正常。当布袋破损时,管道中气、固两相流粉尘含量增加,同时静电荷量强度增大,插入箱体输出管道中的传感器可以及时检测到电荷量值并输出到变送器,实现煤气含尘量的自动监测[2]。

迁钢2号高炉煤气含尘量在线监测装置的传感器表面采用特殊涂敷材料,避免了由于高炉煤气中含水导致传感器表面黏结灰尘,从而提高了该装置的检测精度和稳定性,解决大型高炉煤气干式除尘系统含尘量在线监测的技术难题。

3.4 煤气管道系统防腐技术

生产实践发现,采用高炉煤气干式布袋除尘技术,净煤气中的氯离子含量显著升高。这主要是由于高炉原燃料中的卤化物,在高炉冶炼过程中形成气态的HCl,当煤气温度达到露点时,气态HCl与凝结水结合,形成盐酸。通过检验分析煤气凝结水发现,其pH值低于7,有时甚至达到2~3,呈强酸性,对煤气管道和波纹补偿器具有强腐蚀性,造成煤气管道系统异常腐蚀。

为了防止煤气管道系统的异常腐蚀,对煤气管道波纹补偿器的材质、结构进行了技术攻关,材质由316L改进为800系列,提高了防腐性能;煤气管道内壁采用防腐涂料处理;在净煤气管道上设置了喷洒碱液或喷水系统等技术措施,经过一年的生产实践取得了显著效果。

3.5 除尘灰浓相气力输送技术

自主研究开发除尘灰浓相输送技术,利用氮气或净化后的煤气作为载气输送除尘灰,解决了传统的机械输灰工艺一系列技术缺陷,优化了工艺流程,降低了能源消耗,减少了二次污染,攻克了输灰管道磨损等技术难题。

4 高效长寿高风温技术

4.1 研究开发助燃空气高温预热技术

采用全烧高炉煤气实现高风温是世界性的技术难题,日本、欧洲及我国宝钢的高炉风温达到1250℃,均掺烧了部分高热值煤气。由于我国钢铁企业高热值煤气匮乏,大多数热风炉只能使用低热值的高炉煤气,为了实现高风温,开发了助燃空气高温预热技术[3]。

助燃空气高温预热工艺的技术原理是:设置两座助燃空气高温预热炉,通过燃烧低热值的高炉煤气将预热炉加热后,再用来预热热风炉使用的助燃空气。预热炉燃烧温度在1000℃以上,助燃空气可以被预热到600℃以上,同时利用热风炉烟气余热预热高炉煤气到200℃。由于提高了助燃空气、煤气的物理热,使热风炉拱顶温度也相应提高,从而可以有效地提高送风温度。助燃空气高温预热工艺流程图如图3所示。

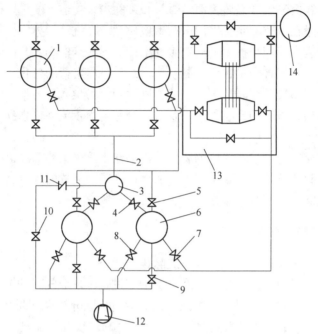

图3 助燃空气高温预热工艺流程图
1—热风炉;2—温度检测点;3—混合室;4—热助燃空气切断阀;5—烟气切断阀;
6—助燃空气高温预热炉;7—煤气切断阀;8—空气切断阀;9—冷助燃空气切断阀;
10—助燃空气旁通阀;11—助燃空气调节阀;12—助燃风机;13—煤气预热器;14—烟囱

本项技术的主要技术特点是:采用顶燃式热风炉作为助燃空气高温预热炉,充分发挥了顶燃式热风炉的技术优势;采用助燃空气高温预热技术与利用热风炉烟气余热预热高炉煤气技术相结合的助燃空气、煤气双预热工艺,有效提高了热风炉系统的综合热效率;助燃空气高温预热工艺技术成熟可靠,系统运行安全稳定,温度控制精准,是实现高风温的有效技术措施。

4.2 集成创新热风炉高风温技术

4.2.1 设计开发高效格子砖

为提高热风炉传热效率,加大格子砖的加热面积是提高热风炉换热能力的重要技术措施。通过优化格子砖结构,缩小格子砖孔径,加大单位体积格子砖的加热面积,提高格子砖的传热效率和热工性能。在迁钢2号高炉内燃式热风炉上采用直径为30mm、加热面积为47.08m²/m³的高效格子砖,格子砖加热面积提高24%,蓄热室高度比同类型热风炉降低约6m,同时减小了热风炉直径,降低了热风炉整体高度,节约工程投资约10%。

4.2.2 热风炉系统工艺设计优化

针对目前国内外热风炉系统的设计和技术缺陷,进行了设计优化和创新:热风炉系统采用紧凑型工艺布局,热风炉靠近高炉布置,缩短热风总管长度,降低热风总管的热损失,为提高风温创造了有利条件,降低了工程投资;在设计中对热风炉的高温、高压管道进行了系统的设计优化,通过管系受力计算,合理设置管道支架、波纹补偿器和拉杆,实现管道系统低应力设计,满足了高风温的使用要求;为防止晶间应力腐蚀的发生,热风炉高温区炉壳内壁采用环氧涂层和喷涂防酸涂料,热风炉壳采用细晶粒耐龟裂钢板,炉壳采用低应力设计,减少或消除炉壳焊接应力;改进热风总管的砌筑结构,水平管路与垂直管路连接处采用各自独立的砖衬结构,使两者之间的热膨胀不互相干涉,解决了孔口砖衬膨胀不均造成管道窜风的技术难题,满足了送风温度高于1250℃的使用要求。

迁钢2号高炉投产以来操作稳定顺行,在全烧高炉煤气的条件下,年平均风温达到1220℃,月平均热风温度达到1239℃。

5 结论

(1) 我国大型高炉工艺技术装备自主创新水平不断提升,大型高炉关键技术的研究开发与应用已经取得技术突破。

(2) 自行设计研制的大型高炉无料钟炉顶设备在设备可靠性、使用寿命、布料控制技术等方面达到国际先进水平,全面实现设备国产化。

(3) 自主开发的高炉煤气全干式布袋脉冲除尘工艺,在系统优化设计、煤气温度控制、除尘灰气力输送、管道防腐等方面取得技术突破,在生产实践中获得成功。

(4) 自主集成创新的高效长寿高风温技术,通过开发应用助燃空气高温预热技术、提高热风炉传热效率、热风炉系统结构优化等措施,在使用高炉煤气燃烧的条件下,风温已达到1250℃以上。

参考文献

[1] 毛庆武,张福明,姚轼,钱世崇,倪苹. 迁钢2号高炉工艺优化与技术创新 [C] //中国金属学会. 2008年全国炼铁生产会议暨炼铁年会文集,北京:中国金属学会,2008:90.

[2] 任绍峰,张福明,周为民,马维理. 大型高炉煤气含尘量在线监测系统的设计与实现 [J]. 冶金自动化,2008,32(S1):1.

[3] 毛庆武,张福明,张建,倪苹,梅丛华. 首钢高炉高风温技术进步 [C] //中国金属学会. 2007中国钢铁年会论文集,北京:冶金工业出版社,2007:3-20.

Study and Innovation of the Key Technologies on Modern Large Blast Furnace

Zhang Fuming

Abstract: The key technologies at modern large blast frunace are bell-less top, high blast temperature and dry bag filter dedusting of BF gas. Bell-less top equipment of large blast furnace designed and developed on our own has reached advanced level in the world in terms of equipment reliability and service life by 3D design optimum, simulation analysis and pilot trail; fully-dry impulse bag filter dedusting technology of BF gas, which is also developed by our own, has gained technical breakthroughs in terms of optimized system design, gas temperature control, pneumatic conveying of dedusting fines; the integrated innovative high-efficiency long-life high-temperature hot blast stove technology, through applying high-temperature preheating technology of combustion air, improving heat transfer efficiency of hot blast stove and optimizing structure of the hot blast stove system, enables the blast temperature to reach 1250℃ with BF gas as fuel.

Keywords: blast furnace; ironmaking; bell-less top; hot blast stove; dry bag filter dedusting

现代高炉高风温关键技术问题的认识与研究

摘　要：提高风温可以有效降低高炉燃料消耗，促进高炉生产稳定顺行，是绿色低碳炼铁技术的重要发展方向之一。研究了热风炉热量传输过程和传热特性，通过传热学机理的研究解析，阐述了热风炉加热面积与风温之间的关系，提出了提高热流通量以改善热风炉传热的观点。研究了热风炉理论燃烧温度、拱顶温度和风温之间的关系，介绍了利用低热值高炉煤气和回收热风炉烟气余热，通过耦合预热和能量梯级利用的技术方法，实现高风温的技术创新及实践。提出了实现热风炉智能化操作的技术要素，论述了合理控制拱顶温度和抑制 NO_x 大量生成的工艺方法，以及有效预防热风炉炉壳晶间应力腐蚀的技术措施。指出实现低热值煤气的高效利用和高值转化，提高风温、降低燃料比和 CO_2 排放，是未来高炉炼铁的关键共性技术。

关键词：炼铁；高炉；高风温；热风炉；拱顶温度；余热回收；能量转换

1　引言

第一次工业革命以后，热风炉开始被用于加热高炉鼓风。经过约 200 年的发展演进，高风温已成为现代高炉重要的技术特征。提高风温有利于高炉生产稳定顺行和高效低耗，热风炉工序属于典型的耗散结构体系，是开放的、动态变化的不可逆工艺过程和能量转换过程，通过物质、能量和信息连续的输入/输出，从而维持系统的稳定运行。现代热风炉的技术理念是提高热风温度和能量转换效率，实现清洁低碳燃烧，延长热风炉寿命，有效降低燃料消耗和污染物排放，实现多目标的集成优化[1]。2019 年我国主要钢铁企业高炉的平均风温为 1140 ℃，与国际先进水平相差约 100 ℃。本文针对高风温关键技术若干问题的解析探讨，提出了低碳清洁高风温技术的发展趋势和路径。

2　热风炉的热量传输

2.1　热风炉的传热原理

蓄热式热风炉的工作原理是燃烧与传热过程轮流交替进行。热风炉燃烧过程产生高温烟气，首先通过高温烟气将热量传递给热风炉蓄热室的格子砖，使格子砖蓄积足够的热量并达到设定的温度；再经过送风过程将格子砖的热量传递给高炉鼓风，将鼓风温度加热到 1000 ℃ 以上。实际上高炉热风炉是以格子砖为载体，经过吸热-蓄热-放热的交替过程，实现热量的储备与传输。图 1 是热风炉蓄热室加热-放热过程温度场数值模拟计算结果。

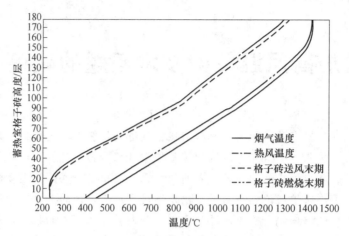

图 1 热风炉蓄热室格子砖传热过程温度场数学解析

20世纪90年代以前，我国高风温技术相对落后。用于热风炉燃烧的高热值燃料供给不足，导致长期以来我国高炉风温长期徘徊在1100~1150℃之间，是我国高炉炼铁与国际先进水平差距最大的技术指标[2]，图2为21世纪以来我国重点钢铁企业高炉炼铁主要技术指标的变化趋势。

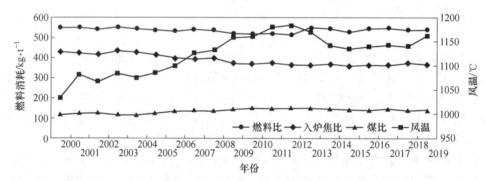

图 2 我国重点钢铁企业高炉主要技术指标变化趋势（2000~2019年）

基于换热器热交换理论，针对蓄热式热风炉蓄热室传热过程的理论研究，公式（1）给出了热风炉蓄热室吸收或释放的热量可以表述为：

$$Q_c = F_c \cdot k_c \cdot \Delta T_c = F_c \cdot q_c \tag{1}$$

式中 Q_c——热风炉吸收或释放的热量，kJ；

F_c——热风炉加热面积，m²；

k_c——高温烟气或冷风和格子砖之间的综合换热系数，kJ/(m²·h·K)；

ΔT_c——气体与格子砖之间的平均温差，K；

q_c——热流通量，kJ/(m²·h)。

公式（1）表明，热风炉的热量传输，与加热面积和热流通量相关，取决于热风炉加热面积、综合换热系数和换热温差。因此提高加热面积、换热系数和换热温差，有助于强化热风炉热量传输，进而提高风温。现代热风炉设计应以提高传热效率为关注点，注重提高热流通量强化热风炉传热过程，而不再单纯依靠增加热风炉加热面积来提高风温。

2.2 热风炉的加热面积

20世纪90年代,为了提高风温在高热值燃料匮乏的条件下,企图通过增加热风炉加热面积的方法提高热风炉换热能力。热风炉加热面积从 $80m^2/m^3$ 增加到 $100m^2/m^3$ 以上,高炉鼓风的加热面积也增加到 $50m^2/(m^3·min)$ 以上,但是高炉风温却始终停留在1050~1150℃之间。其根本原因在于,热风炉仅燃烧低热值高炉煤气,在助燃空气和煤气没有预热的条件下,热风炉理论燃烧温度和拱顶温度偏低,导致热风炉换热温差低、热流通量低,从根本上就不具备实现高风温的物质基础和技术条件。

图3统计了国内外部分高炉容积与热风炉加热面积的关系,从图3中可以看出 1000~6000m^3 高炉的热风炉加热面积,基本维持在 $70~90m^2/m^3$ 的区间,个别甚至超过了 $100m^2/m^3$。理论研究和生产实践表明,提高热风炉加热面积主要是通过增加格子砖使用量实现的,其结果是可以缩小拱顶温度与风温的温差。但是如果拱顶温度不能达到预期水平,仅靠提高热风炉加热面积这种单一的技术手段,还是无法从根本上解决高风温的技术难题。

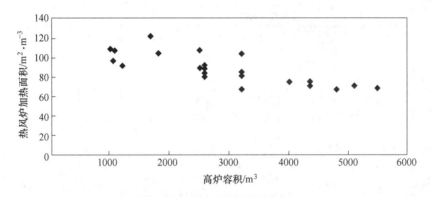

图3 高炉容积与热风炉加热面积的关系

从图3中可以看出,单位高炉容积的热风炉加热面积与高炉容积之间,并不是线性相关关系,1000~6000m^3 高炉其热风炉加热面积波动范围很大,且离散度很高。进入21世纪以来,我国有一批 4000m^3 以上高炉建成投产,到2019年底已经有25座容积大于 4000m^3 的大型高炉运行。从图3中可以看出,近些年新建的 4000m^3 和 5000m^3 级高炉的热风炉加热面积基本维持在 $60~80m^2/m^3$ 的范围内,单位高炉容积热风炉加热面积呈现减小趋势,表明新建大型热风炉已经不再沿承单纯提高加热面积的技术理念而提高风温。

表1列出了近年来我国新建不同容积级别高炉热风炉蓄热室的主要热工设计参数;表2为近年来我国已投产运行的 8 座 5000m^3 超大型高炉热风炉主要技术装备及参数;图4统计了我国部分高炉不同热风炉加热面积条件下的高炉风温 (2019年)。

表1 不同容积高炉热风炉的加热面积

高炉容积/m^3	1200	1800	2500	3200	4000	5500
入炉风量/$Nm^3·min^{-1}$	2800	4450	5200	6300	8000	9300

续表1

热风炉加热面积/m²	105324	184677	268602	306304	458243	383540
单位高炉容积加热面积/m²·m⁻³	90.80	103.75	107.4	127.6	114.6	69.73
单位鼓风加热面积/m²·(Nm³·min)⁻¹	37.62	41.50	51.7	48.6	57.3	41.24
热风炉格子砖质量/t	2702.85	4821.66	6051	6940	9417	10202
单位高炉容积格子砖质量/t·m⁻³	2.33	2.71	2.4	2.2	2.5	1.85
单位鼓风格子砖质量/t·(Nm³·min)⁻¹	0.97	1.08	1.16	1.1	1.18	1.10

表2 我国5000m³高炉热风炉主要技术装备及参数

钢铁厂	宝武湛江		沙钢	山钢日照		首钢京唐		
高炉炉号	1	2	4	1	2	1	2	3
高炉容积/m³	5050	5050	5800	5100	5100	5500	5500	5500
炉缸直径/m	14.5	14.5	15.3	14.6	14.6	15.5	15.5	15.2
设计风量/m³·min⁻¹	8000	8000	8300（最大9500）	8000	8000	9300	9300	9300
送风压力/MPa	0.55	0.55	0.52	0.50	0.50	0.54	0.54	0.54
热风炉结构	顶燃式	顶燃式	DME外燃式	顶燃式	顶燃式	顶燃式	顶燃式	顶燃式
热风炉座数	4	4	3	4	4	4	4	4
预热工艺	热管换热器+预热炉	热管换热器+预热炉	整体式换热器	板式换热器+预热炉	板式换热器+预热炉	热管换热器+预热炉	热管换热器+预热炉	板式换热器+预热炉
热风炉燃料	全高炉煤气	全高炉煤气	高炉煤气+转炉煤气	全高炉煤气	全高炉煤气	全高炉煤气	全高炉煤气	全高炉煤气
设计风温/℃	1300	1300	1250（最高1310）	1260（最高1300）	1260（最高1300）	1300	1300	1300
实际风温/℃（2019年）	1251	1240	1217	1248	1232	1240	1230	1250
投产时间/年-月	2015-9	2016-7	2009-10	2017-12	2019-4	2009-5	2010-6	2019-4

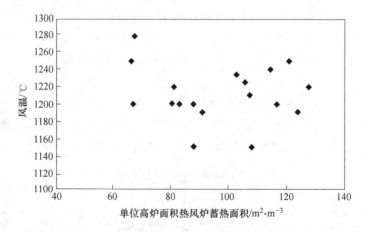

图4 高炉风温与热风炉加热面积的关系（2019年）

从图4中可以清晰看出，高炉风温与热风炉加热面积之间相关性不强，不同加热面积的热风炉风温都可以达到1200℃甚至更高，与此同时，还能发现相对较小的加热面积的热风炉风温反而达到1250℃以上。由此，可以说明高炉风温是一个综合技术指标，是多种要素非线性耦合作用的结果，并不是意味着提高热风炉加热面积就能够提高风温。

3 提高热风炉拱顶温度

3.1 提高理论燃烧温度

理论上提高热风炉传热的热流通量，就必须要提高综合换热系数，增加高温烟气与格子砖的换热温差、鼓风与格子砖的换热温差。在工程实践中，通过采用高效格子砖能够有效提高换热系数；提高热风炉拱顶温度和烟道温度能够增加烟气和鼓风与格子砖之间的温差；加快热风炉循环节奏、优化热风炉燃烧和送风操作等可以有效增加热风炉热流通量[3]。

热风炉输入的热量主要来源于煤气的化学热，以及煤气和助燃空气所携带的物理热（显热）。前者取决于煤气的品质和发热值，而后者则取决于煤气和助燃空气的预热温度。工程中一般假设化学热和物理热全部用于加热燃烧产物，在没有其他热损失的情况下，燃烧产物可能达到的温度称之为热风炉的理论燃烧温度。热风炉理论燃烧温度的计算公式表达为：

$$T_f = \frac{Q_h + c_g T_g + c_a T_a l}{c_w V_w} \tag{2}$$

式中 T_f——理论燃烧温度，K；

Q_h——煤气热值，kJ/m³；

c_g——煤气比热容，kJ/(m³·K)；

T_g——煤气温度，K；

c_a——助燃空气比热容，kJ/(m³·K)；

T_a——助燃空气温度，K；

l——燃烧1m³煤气所消耗的助燃空气体积，m³/m³；

c_w——烟气的比热容，kJ/(m³·K)；

V_w——燃烧1m³煤气所产生的烟气体积，m³/m³。

式（2）表明，通过提高助燃空气和煤气的物理热，可以提高热风炉的理论燃烧温度；而提高助燃空气和煤气物理热的有效措施则是提高二者的预热温度，图5显示了煤气和助燃空气预热条件下所获得的理论燃烧温度。

由式（2）可以看出，提高热风炉理论燃烧温度，除了提高助燃空气和煤气的预热温度，从而提高其输入的物理热之外，还可以通过提高燃烧器效能，降低燃烧空气过剩系数，降低助燃空气消耗量，进而减少燃烧产物体积等技术措施来提高理论燃烧温度。

热风炉生产运行过程，是一个动态变化、开放的、不可逆的耗散结构体系，理论燃烧温度是通过燃料燃烧和传热计算而得到的，是体系在假定的理想状态下可能达到的燃烧温度。而实际热风炉生产过程中，能够直接在线检测的则是热风炉的拱顶温度。拱顶温度是

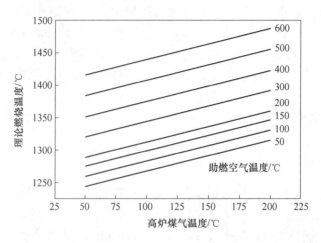

图5 理论燃烧温度与煤气和助燃空气预热温度的关系

通过设置在热风炉拱顶部位的热电偶或红外测温装置,实际检测出的热风炉拱顶或格子砖表面的温度,热风炉拱顶温度是热风炉内的最高温度,拱顶温度表征了热风炉所具备的温度水平和换热能力。工程设计中,热风炉拱顶温度与理论燃烧温度存在如下关系:

$$T_d = k \cdot T_f \quad (3)$$

式中　T_d——拱顶温度,K;

　　　k——温度系数,一般为 0.94~0.98;

　　　T_f——理论燃烧温度,K。

3.2 预热煤气和助燃空气

在以不同热值的高炉煤气为燃料、助燃空气过剩系数为1.1时,热风炉的理论燃烧温度与煤气和助燃空气预热温度的关系如图6所示。从图6中可以看出,煤气热值越低,达到相同的理论燃烧温度所需要的煤气和助燃空气预热温度就越高,燃烧热值为 3250kJ/Nm³ 的高炉煤气时,理论燃烧温度要达到1400℃需要将煤气和助燃空气均预热到220℃。

为了有效回收热风炉烟气余热、预热煤气和助燃空气,2000年以后我国普遍采用低温热管换热器,受限于热管的承压能力(约为4.0MPa),低温热管的工作温度一般都限制在250℃以下,因此煤气和助燃空气的预热温度,仅能达到180~200℃。热管在恶劣环境中工作,还受到积灰、腐蚀、析氢等综合因素的影响,低温热管换热器的热效率会随着运行时间的延续而不断衰减、直至失效,同时换热器的整体寿命也普遍较短,一般使用寿命为3~5年。

实践表明,提高热风炉烟气温度,能够有效提高热风温度,但是其代价是造成热风炉热效率下降,烟气带走了大量热能。热风炉烟气温度在燃烧周期内是从低到高动态变化的,燃烧末期热风炉最高烟气温度一般处于400~450℃的温度范围内,燃烧期平均烟气温度约为280~320℃,回收热风炉烟气余热,可将高炉煤气和空气预热到150~200℃,大约可以提高热风炉理论燃烧温度150℃。但是这项技术措施的缺陷在于:烟气带走了过多的热量,导致热风炉热效率下降;热风炉炉箅子的高温蠕变承受温度一般为500℃,不可能

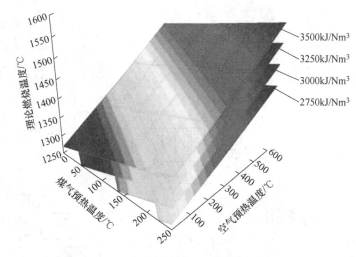

图 6　热风炉理论燃烧温度和双预热温度的关系

无限制提高烟道温度；通过烟气低温余热回收，再进行煤气和助燃空气预热，能源转换出现衰减，能源转换效率不高；常规热管换热器由于自身结构特点，热管管壁最高温度一般为 250℃ 左右，难于实现更高的预热温度。

为了有效提高热风炉拱顶温度，在燃烧低热值煤气条件下获得高风温，国内外都开展了许多研究和试验工作。德国迪林根钢铁厂在 20 世纪 80 年代进行了技术探索，在 5 号高炉上采用附加燃烧炉的组合式预热器（见图 7），将高炉煤气温度预热到 250℃、助燃空气温度预热到 500℃，在燃烧热值为 3000kJ/m³ 高炉煤气时，风温达到 1285℃[4]。由于引进技术适用性等多种原因，该工艺并没有在我国得到应用推广。20 世纪 90 年代，我国也开发应用了热风炉自身预热技术，利用送风后热风炉蓄热室的余热，再对助燃空气进行预热，使助燃空气温度达到 200℃ 以上。鞍钢等企业研发出前置燃烧炉的工艺方案，配置专用的燃烧炉将高温烟气（1000℃）与热风炉烟气（200℃）混合，通过提高并稳定烟气温度，配置管式换热器+热管换热器对煤气和助燃空气进行双预热[5]。令人遗憾的是，这些我国自主研发的热风炉预热工艺技术，由于装备可靠性、耐久性、运行稳定性、经济性等诸多问题，没能得到长期的实践和推广，最终被其他预热工艺和板式换热器所取代。

首钢京唐钢铁厂是进入新世纪以后，我国基于新一代可循环钢铁制造流程的理念，自主设计建造的第一个靠海临港的千万吨级现代化钢铁基地。在我国首座 5000m³ 巨型高炉和热风炉设计过程中，为了实现新一代钢铁厂高效能源转换和能量流网络结构优化，实现低热值煤气的高效利用，热风炉采用全高炉煤气燃烧，回收热风炉烟气余热预热煤气和助燃空气，研发应用助燃空气两级高效预热工艺，实现助燃空气低温预热与高温预热耦合工艺流程，提高煤气和助燃空气温度及输入的物理热，以实现 1300℃ 高风温[6~10]。该流程采用换热器回收热风炉烟气余热，将煤气和助燃空气温度均预热到约 200℃，一部分助燃空气再通过 2 座小型顶燃式高温预热炉，将其温度预热到 1000℃，与另一部分未经过高温预热炉的助燃空气进行混合后，温度达到 500~650℃，在全烧热值为 3000kJ/m³ 的高炉煤气工况条件下，热风炉拱顶温度能够达到 1400℃ 甚至更高，图 8 为首钢京唐 5500m³ 高炉热风炉煤气和助燃空气高效预热工艺流程。

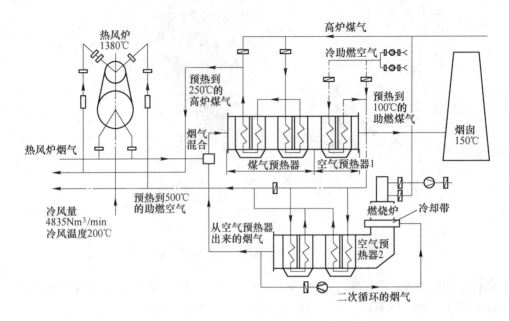

图 7 迪林根厂 5 号高炉热风炉双预热流程

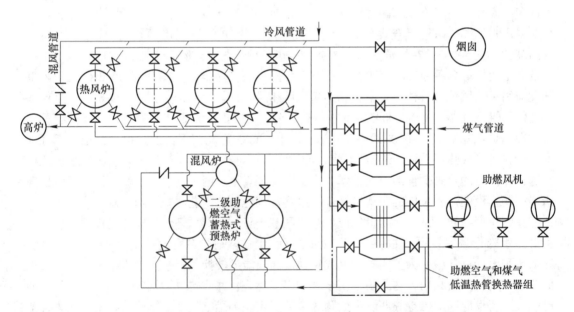

图 8 首钢京唐 5500m³ 高炉热风炉煤气和助燃空气双预热工艺流程

首钢京唐 5500m³ 高炉热风炉高效预热工艺流程稳定运行 11 年以上，在全烧高炉煤气条件下，风温长期稳定在 1250℃，最高月平均风温达到了 1300℃，煤气和助燃空气高效预热工艺与顶燃式热风炉相匹配，是我国近 10 年来高炉高风温技术发展的一个重要技术特征，为解决燃烧低热值高炉煤气条件下，实现 1250℃ 高风温关键技术难题，开创了一个新的技术途径和发展模式，取得了显著的应用成效和经济效益[11]。

4 优化热风炉操作

4.1 拱顶温度的控制

如前所述，随着技术进步和对高风温技术的深入研究，人们越来越清晰地认识到，提高风温是一项涉及专业门类广、跨学科领域和需要多工序协同的复杂工程问题。这个复杂的工程问题，不仅关系到高炉自身的生产运行，还关乎到整个钢铁制造流程能源流的动态调控。从某种程度上，提高风温已成为高炉炼铁工序的关键共性技术，对于高炉炼铁低碳绿色、节能减排和高效长寿，具有重要的现实意义和深远的战略意义。进而言之，高风温技术包括清洁高效燃烧技术、高效传热技术、高温-高压热风安全稳定输送技术，以及高温热风的应用技术等，还应当包括降低燃料消耗减少 CO_2 排放，减少热风炉烟气 SO_x 和 NO_x 的生成与排放，实现热风炉全流程智能化运行等命题。

提高热风炉拱顶温度是实现高风温的一个基本前提和必要条件。理论研究和生产实践表明，当拱顶温度超过 1420℃ 以上时，NO_x 会大量生成、体积分数急剧增高。这不仅造成严重的环境污染，还会导致热风炉炉壳出现晶界应力腐蚀，严重影响热风炉寿命和安全，成为当前限制热风炉提高风温的最大技术障碍之一。对此必须给予足够的重视，并着力解决限制性关键技术难题。从当前技术发展现状分析，将热风炉拱顶温度控制在 1380℃ 以下是比较安全的温度区间[12]，风温达到 1250℃ 以上热风炉的拱顶温度也不宜超过 1400℃，努力将拱顶温度与风温的差值控制在 100℃ 以下，提高热风炉能量转换效率和热能有效利用，最大限度降低热量耗散。根据热力型 NO_x 生成机理[13]，实测表明在热风炉拱顶温度低于 1400℃ 时，热风炉烟气中 NO_x 的质量分数一般可以控制在 100ppm 以内，甚至低于 70ppm，在这种工况下，可以有效抑制晶界应力腐蚀问题，同时减少污染物排放。

随着顶燃式热风炉技术的推广应用[14~17]，配置助燃空气高温预热技术，在全烧高炉煤气条件下，我国部分先进大型高炉风温水平已达到国际一流水平，特别是京唐、湛江和日照等 8 座 5000m^3 以上巨型高炉，其中 7 座高炉采用了顶燃式热风炉配置助燃空气高温预热技术，风温均达到了 1250℃ 甚至更高。在拱顶温度已不再是高风温的关键技术难题时，必须要将拱顶温度控制在合理的范围内，而不是无限度地提高拱顶温度。当拱顶温度超过 1420℃ 以上时，会造成 NO_x 急剧生成，不仅对生态环境造成严重污染，还极易造成热风炉炉壳的晶界应力腐蚀，造成高温区炉壳在焊缝和残余应力集中的部位，沿着晶界出现腐蚀和裂纹，使炉壳强度下降和失效，最终导致炉壳出现开裂、发红、窜风等破损，威胁热风炉安全生产和使用寿命[18]。

4.2 晶间应力腐蚀的预防

在热风炉操作过程中，建议将热风炉拱顶温度控制在 1380±20℃ 的合理范围内，进一步缩小拱顶温度与风温的差值，将其控制在 100℃ 以内，采用高效格子砖、涂装高辐射材料、提高鼓风流场分布均匀性等多种技术措施，有效提高热风炉的能量利用和转换效率，更加充分地利用好高温热量。图 9 为首钢京唐 1 号高炉 1 号热风炉工作周期内的温度变化过程。

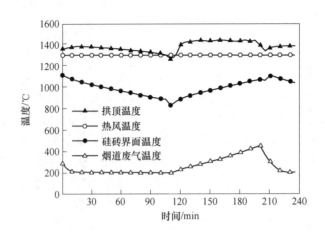

图 9 首钢京唐 1 号高炉 1 号热风炉温度变化曲线

在热风炉设计建造过程中，为了有效预防热风炉炉壳晶间应力腐蚀，应当采取以下措施：(1) 提高热风炉炉壳材质，优先采用细晶粒高强度、高性能的钢板，严格控制钢板材料缺陷；(2) 根据炉壳热-力加载条件下的有限元仿真分析结果，对强度薄弱的区域采取"补强"措施，减少炉壳应力集中；(3) 拱顶和高温区炉壳的结构设计应采用曲面平滑过渡，减少尖角连接结构，使其力学结构合理、受力均匀；(4) 炉壳钢板采用预装切割、保护性拼装焊接，冬季施工中必须对焊接钢板和焊条进行加热，降低焊接应力和变形，焊接后要对炉壳表面的焊缝打磨平滑；(5) 炉壳焊后应进行退火热处理，以消除焊接残余应力；(6) 炉壳内表面涂装抗酸腐蚀的耐酸涂料，隔断 NO_x 等酸性介质与炉壳内壁的接触；(7) 有条件的可在炉壳内表面敷设抗酸腐蚀的不锈钢钢箔，或采用不锈钢复合钢板，抵抗腐蚀性气体在冷凝结露后积聚所产生的破坏；(8) 采用炉壳外保温等措施，控制炉壳表面温度达到露点以上（一般炉壳温度控制在 100℃ 以上），防止等酸性物质的凝聚、沉积。总之，热风炉炉壳晶间应力腐蚀的预防是一项综合的技术，需要从系统全局出发，必须采取综合措施才能取得有效结果。

4.3 热风炉操作的优化

毋庸置疑，在当前技术条件下，高效清洁高风温技术是高炉炼铁实现低碳绿色的重要基础和支撑，进一步提高风温是现代高炉炼铁的关键共性技术[19]。未来高风温技术，不再追求单一的风温指标，而是以高效、低碳、清洁、低排放、绿色化、智能化多目标为关注点，实现多目标的集成优化。从热风炉生产操作的层次分析，应当做好以下几个方面：(1) 在约束的条件下，合理控制拱顶温度，努力将拱顶温度控制在 1380±20℃ 的范围内，缩小拱顶温度与风温的差值；(2) 采用数字化、信息化和智能化控制模型或方法[20,21]，实现热风炉系统的自感知、自适应、自决策和自执行；进而实现热风炉燃烧-换炉-送风全流程智能化精准调控和闭环运行；(3) 构建动态输入-输出的耗散结构结构体系，注重全过程物质流、能量流和信息流的高效耦合运行，例如关注拱顶温度与烟道温度的协同耦合，注重烟气成分的动态变化，建立数字映射模型，前置反馈到煤气和助燃空气的流量的输入与调节;(4)减少冷风的混入量,提高风温水平;(5)通过大数据挖掘和系统自学习,建立起热风炉强化燃烧、强化送风的动态甘特图,提高热风炉工作效率和转换周期,通过缩短送风周期进而提高风温;(6)4座热风炉发挥好交错并联送风的技术优势,使4座热风炉运行动态-有序、协同-连续、高效-节能;(7)合理调控热风炉燃烧的空燃比,降低空气过剩系数,降低能源消耗和CO_2、CO 的排放;(8)加强热风炉炉体和管道的监测与维护,密切关注炉壳、管道表面温度、应力的变化,杜绝安全事故、延长使用寿命。

5 结论

(1)高风温是现代高炉重要的技术特征之一,是现代高炉实现低碳绿色冶炼的重要技术支撑,也是未来高炉炼铁发展的关键共性技术。提高风温必须从传统的技术理念中摆脱出来,建立起耗散结构优化的能量高效转换的工程思维和认识。

(2)全燃烧低热值高炉煤气实现高风温是未来高炉炼铁技术的发展方向。顶燃式热风炉由于具有多工况适应性,与煤气预热和助燃空气高温预热工艺耦合匹配,是我国在 5000m^3 巨型高炉上实现的重大技术创新,全部燃烧低热值煤气时风温稳定达到1250℃以上,取得了显著的应用成效和实绩,对产业技术进步产生了重要的推动作用。

(3)在当前技术条件下,控制拱顶温度在合理范围内,采取综合措施有效抑制热风炉晶间应力腐蚀,缩小拱顶温度与风温的差值,采用智能化改进热风炉操作等是实现热风炉高风温和高效长寿的重要技术措施。

(4)高风温技术是系统集成技术,包括热风炉燃烧、气体运动、传热、热风输送和热风使用等多个单元技术组合,是 21 世纪中叶高炉炼铁技术进步的重要方向之一。提高风温的前提条件是必须具备足够高的拱顶温度和蓄热量,通过提高热风炉拱顶温度、优化热风炉燃烧、改善热风炉操作,降低热风炉自身能源消耗,提高能源转换效率,从而降低高炉炼铁工艺的 CO_2、CO 和 NO_x 排放。

参考文献

[1] 张福明. 高炉高风温技术发展与创新 [J]. 炼铁, 2013, 32 (6): 1-5.
[2] 张福明. 面向未来的低碳绿色高炉炼铁技术发展方向 [J]. 炼铁, 2013, 32 (6): 1-5.
[3] 吴启常, 沈明, 陈秀娟. 高风温热风炉设计理念的调整及相关问题讨论 [J]. 炼铁, 2008, 27 (4): 11-15.
[4] 金岩, 陈巍, 刘权兴. 高炉煤气和助燃空气 "双预热" 提高高炉风温的研究 [J]. 1998, 33 (6): 54-56.
[5] 华建社, 李静. 高炉热风炉的预热技术研究及其发展 [J]. 钢铁研究, 2008, 36 (5): 48-51.
[6] 张福明. 高风温低燃料比高炉冶炼工艺技术的发展前景 [J]. 中国冶金, 2013, 23 (2): 1-7.
[7] 张福明. 当代高炉炼铁技术若干问题的认识 [J]. 炼铁, 2012, 31 (5): 1-6.
[8] 张福明, 梅丛华, 银光宇, 等. 首钢京唐 5500m^3 高炉 BSK 顶燃式热风炉设计研究 [J]. 中国冶金, 2012, 22 (3): 27-32.

[9] 魏红旗. 首钢京唐1号高炉1300℃风温应用实践[J]. 炼铁, 2010, 29 (4): 7-10.
[10] 庄辉, 刘长江. 京唐1号高炉低燃料比冶炼技术[J]. 中国冶金, 2017, 27 (10): 49-53.
[11] 刘宏强, 张福明, 刘思雨, 等. 首钢京唐钢铁公司绿色低碳钢铁生产流程解析[J]. 钢铁, 2016, (12): 80-88.
[12] 吴启常, 吕宇来. 高风温长寿热风炉设计的一些问题[J]. 炼铁, 2006, 25 (3): 23-27.
[13] 张福明, 胡祖瑞, 程树森. 顶燃式热风炉高温低氧燃烧技术[J]. 钢铁, 2012, 47 (8): 74-80.
[14] 曹伟, 陈世坤, 王包明, 等. 宁钢1号高炉热风炉大修设计特点及生产实践[J]. 中国冶金, 2018, 28 (12): 45-47.
[15] 杨丽. 4800m^3高炉热风炉设计及运行研究[J]. 工业加热, 2014, 43 (3): 37-39.
[16] 廖建锋, 文辉正, 邹忠平, 等. 宝钢湛江5050m^3高炉工艺技术特点[J]. 炼铁, 2017, 36 (3): 19-23.
[17] 刘华平, 况维良, 刘红军, 等. 5100m^3高炉热风炉的设计[J]. 工业炉, 2019, 41 (1): 42-44.
[18] 武建龙, 陈辉, 孙健, 等. 大型高炉热风炉工作强度等若干问题的讨论[C]//中国金属学会. 第十届中国钢铁年会暨第六届宝钢学术年会论文集. 上海: 中国金属学会, 2015: 1-6.
[19] 张福明. 我国高炉炼铁技术装备发展成就与展望[J]. 钢铁, 2019, 54 (11): 1-8.
[20] 郝聚显, 赵贤聪, 韩玉召, 等. 热风炉煤气消耗量中期预测模型[J]. 中国冶金, 2018, 28 (2): 17-22.
[21] 杜玮, 沈海彬, 罗晓. 基于生产实践对热风炉高风温技术的几点思考[C]//中国金属学会. 2012年全国炼铁生产技术会议暨炼铁学术年会会议论文集(上). 无锡: 中国金属学会, 2012: 679-685.

Research and Cognition on Key Technical Problems of High Blast Temperature for Modern Blast Furnace

Zhang Fuming Yin Guangyu Li Xin

Abstract: Increasing the blast temperature can effectively reduce the fuel consumption of blast furnace (BF) and promote the BF stable and smooth production, which is one of the important developing directions of green-low carbon ironmaking technology. The heat transformation process and heat transfer characteristics of hot blast stove are researched. The relationship between heating surface and blast temperature of hot blast stove is expounded through the study and analysis of heat transfer mechanism, and the view point of increasing heat flux to improve heat transfer in hot blast stove is put forward. The relationships between theoretical combustion temperature, dome temperature and blast temperature of hot blast stove are studied. The technical innovation and practice of high blast temperature are introduced by applying low calorific value blast furnace gas and recovering flue gas waste heat of hot blast stove by coupling preheating and energy cascade utilization. Technical elements to realize intelligent operation of hot blast stove are put forward. The technological methods of controlling dome temperature and suppressing large amount of NO_x are discussed. High efficiency utilization and high value conversion of low calorific value gas, increasing blast temperature, reducing fuel ratio and CO_2 emission are the key-common technologies of blast furnace ironmaking in the future.

Keywords: ironmaking; blast furnace; high blast temperature; hot blast stove; dome temperature; waste heat recovery; energy transformation

首钢京唐 BSK 顶燃式热风炉的燃烧技术

摘 要：为开发 5500m³ 高炉 BSK 顶燃式热风炉技术，对顶燃式热风炉的燃烧机制和燃烧特性进行了研究。采用 CFD 数学仿真模拟研究了 BSK 顶燃式热风炉环形陶瓷燃烧器的燃烧机制，解析了顶燃式热风炉燃烧室内气体的混合、流动以及燃烧过程，计算分析了顶燃式热风炉燃烧过程的速度场、温度场以及浓度场分布。通过对实体热风炉的冷态测试，验证了 CFD 数学仿真计算的结果。研究结果表明，BSK 顶燃式热风炉采用旋流扩散燃烧技术使燃烧过程速度场、温度场和浓度场分布均匀对称，并可以有效控制火焰长度和火焰形状，使煤气在拱顶空间内充分燃烧。速度场、温度场和浓度场的分布与煤气和助燃空气的初始分布有直接关系。通过燃烧器喷嘴结构优化设计可以显著提高空气与煤气混合的均匀性，改善燃烧室内浓度、温度分布以及火焰形状。

关键词：顶燃式热风炉；燃烧器；燃烧机理；优化设计

1 引言

首钢京唐 5500m³ 高炉是我国第一次设计建设的 5000m³ 以上特大型高炉。设计中对改造型内燃式、外燃式和顶燃式 3 种结构形式的热风炉技术进行了全面深入的研究分析，在首钢顶燃式热风炉技术和卡鲁金顶燃式热风炉技术的基础上，综合两种技术的优势，设计开发了 BSK（Beijing Shougang Kalugin）型顶燃式热风炉技术，将顶燃式热风炉技术首次应用在 5000m³ 级特大型高炉上，在燃烧单一高炉煤气的条件下，风温可以达到 1300℃[1,2]。

2 顶燃式热风炉技术特征和燃烧特性

顶燃式热风炉是 20 世纪 70 年代为了克服传统内燃式和外燃式热风炉的技术缺陷而发展起来的一种新型结构的热风炉。其结构特点是取消了独立的燃烧室，将燃烧器直接置于热风炉的拱顶，利用拱顶空间作为燃烧室[1]，其结构设计充分吸收了内燃式、外燃式热风炉的技术优点。同内燃式、外燃式热风炉相比，顶燃式热风炉还具有以下优点：（1）取消了独立的燃烧室和隔墙，扩大了蓄热室容积，在相同的容量条件下，蓄热面积增加 25%～30%；（2）热风炉本体采用轴向对称结构，结构稳定性增强；（3）将大功率燃烧器安装在热风炉拱顶，使高温热量集中在拱顶区间，热损失减少，有利于提高拱顶温度；（4）顶燃式热风炉的工作过程是一种典型的逆向强化换热过程，提高了热风炉换热效率；（5）热风炉耐火材料砌体稳定性提高，热风炉寿命延长；（6）工艺布置紧凑，占地面积

本文作者：张福明，胡祖瑞，程树森，李欣。原刊于《钢铁》，2012，47（5）：75-81。

小，节省钢材和耐火材料，节约工程建设投资。

顶燃式热风炉将环形陶瓷燃烧器置于热风炉拱顶，利用热风炉拱顶空间作为燃烧室。环形陶瓷燃烧器由煤气环道、助燃空气环道、煤气喷嘴、助燃空气喷嘴组成。煤气和助燃空气以一定切向角度由喷嘴喷出，混合气体在陶瓷燃烧器预混室内旋流燃烧，使煤气和助燃空气混合均匀。由陶瓷燃烧器喉口到热风炉拱顶扩张段，横截面积突然增大，使得该区域出现较大的回流，该回流区的产生是顶燃式热风炉的关键技术所在。回流的存在一方面能够起到稳定火焰的作用，使得新补充的煤气和空气始终能与高温气体接触，连续燃烧；另一方面扩张段中心的回流切断了火焰的中心发展，使火焰沿着拱顶锥角空间扩张，从而缩短了火焰长度，使火焰前锋不会对蓄热室格子砖产生冲击。生产实践表明，顶燃式热风炉拱顶空间内煤气能实现完全燃烧，而且在所有工作制度下都不会出现燃烧脉动现象。由于燃烧火焰不直接接触耐火材料砌体，砌体不会出现局部过热现象，因此旋流扩散燃烧是顶燃式热风炉燃烧最主要的技术特征，与内燃式、外燃式热风炉燃烧具有显著的区别。

顶燃式热风炉环形陶瓷燃烧器布置在蓄热室顶部，与蓄热室在同一中心轴线上，可使烟气均匀进入格子砖。经测定，其不均匀度为±(3%~5%)，有效提高了热风炉蓄热室的利用率，并使得温度场在热风炉内部均匀分布，减少了温差热应力，有利于提高热风炉的寿命。

3 首钢京唐 5500m³ 高炉 BSK 顶燃式热风炉燃烧数学仿真计算

热风炉炉内处于高温高压状态，燃烧器结构复杂，很难用实际测量的方法获悉内部的气体速度、温度以及浓度分布状况[4]。目前国内外对热风炉炉内燃烧过程的研究普遍采用数学仿真的计算方法。数学仿真计算能建立与实际物体完全相同的物理模型，提供更为接近实际的计算数据；而且能够较容易地通过改变模型结构和操作条件对实际过程进行深入研究。随着现代计算机技术的快速发展，计算机仿真计算结果的精准度得到大幅度提高，因此在许多技术领域都得到了非常广泛的应用。

3.1 物理模型

本文研究的对象是首钢京唐 5500m³ 高炉所配置的 BSK 型顶燃式热风炉。针对 BSK 顶燃式热风炉建立了计算的物理模型。模型由煤气管道、空气管道、煤气环道、空气环道、半球型拱顶、喷嘴、燃烧室和蓄热室等部分组成。图1是 BSK 顶燃式热风炉物理模型的示意图。

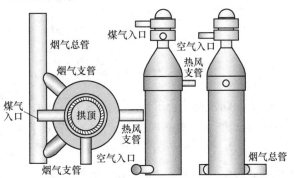

图 1 BSK 顶燃式热风炉示意图

3.2 控制方程

热风炉燃烧室中的燃烧系统属于湍流非预混燃烧系统。模拟湍流扩散燃烧需要将混合过程和脉动的影响有机地统一，目前采用混合分数 k-ε-g 模型[5,6]模拟非预混燃烧现象的结果与实际吻合较好。k-ε-g 模型所包含的要点有：（1）用 k-ε 模型模拟湍流流动；（2）采用简单化学反应系统（SCRS）和快速反应描述燃烧过程化学反应；（3）建立以 g（$g \equiv \overline{f'^2}$）为因变量的控制方程；（4）假设 f 的概率分布函数的形式，由求解微分方程得到的 $\bar{f}$ 和 g，确定 f 的概率分布函数 $P(f)$，把 f 写成 $\bar{f}$、g 和 $P(f)$ 的显函数；（5）基于快速反应假设，根据 f 求出燃料和氧化剂的质量分数的瞬时值，进而用概率分布函数得到燃料和氧化剂质量分数的平均值，这就避开了对燃料的平均质量分数控制方程的直接模化和求解；（6）燃烧过程滞止焓 $\tilde{h}$ 的控制微分方程；（7）燃烧过程反应物和产物的浓度控制微分方程。

k-ε-g 模型所包含的控制方程有以下 5 种。

3.2.1 连续性方程

$$\frac{\partial \rho}{\partial t} + \nabla \cdot (\rho U) = 0 \tag{1}$$

3.2.2 动量方程

$$\frac{\partial \rho U}{\partial t} + \nabla \cdot (\rho U \times U) - \nabla \cdot (\mu_{\text{eff}} \nabla U)$$

$$= -p' + \nabla \cdot [\mu_{\text{eff}} (\nabla U)^T] + B \tag{2}$$

式中，ρ 和 U 是主流的密度和平均速度；μ_{eff} 是有效黏度，定义为 $\mu_{\text{eff}} = \mu + \mu_t$，$\mu_t$ 为湍流黏度；B 为体积力。

3.2.3 k-ε 湍流双方程

$$\frac{\partial \rho k}{\partial t} + \nabla \cdot (\rho U k) - \nabla \cdot \left[\left(\mu + \frac{\mu_t}{\sigma_k}\right) \nabla k\right] = P + G - \rho \varepsilon \tag{3}$$

$$\frac{\partial \rho \varepsilon}{\partial t} + \nabla \cdot (\rho U \varepsilon) - \nabla \cdot \left[\left(\mu + \frac{\mu_t}{\sigma_\varepsilon}\right) \nabla \varepsilon\right]$$

$$= C_1 \frac{\varepsilon}{k} (P + C_3 \max(G, 0)) - C_2 \frac{\varepsilon^2}{k} \tag{4}$$

式中，k 为湍流动能；ε 为湍动耗散率；P 为剪切生成项；G 为体积生成项；σ_k 和 σ_ε 分别为湍流动能 k 和湍动耗散率 ε 对应的普朗特数，计算中取值为 $\sigma_k = 1.0$，$\sigma_\varepsilon = 1.3$；C_1、C_2、C_3 均为经验常数，$C_1 = 1.44$，$C_2 = 1.92$，$C_3 = 0.09$。

3.2.4 平均（时间平均）混合分数方程

$$\frac{\partial}{\partial t}(\rho \bar{f}) + \nabla \cdot (\rho \bar{v} \bar{f}) = \nabla \cdot \left(\frac{\mu_t}{\sigma_t} \nabla \bar{f}\right) + S_m + S_u \tag{5}$$

式中，S_m 和 S_u 为反应源项，S_m 仅指由液体燃料滴或反应颗粒传入气相引起的质量混合分数的变化，S_u 为自定义源项。由于本文只研究气体燃烧，因此式（5）中 S_m 和 S_u 均为 0。

3.2.5 平均混合分数均方差 $g = \overline{f'^2}$ 的守恒方程

$$\frac{\partial}{\partial t}(\rho \overline{f'^2}) + \nabla \cdot (\rho \bar{v} \overline{f'^2})$$
$$= \nabla \cdot \left(\frac{\mu_t}{\sigma_t} \nabla \overline{f'^2}\right) + C_g \mu_t (\nabla^2 \bar{f}) - C_d \rho \frac{\varepsilon}{k} \overline{f'^2} + S_u \quad (6)$$

混合分数 f 是来自燃料流的质量分数，可根据原子质量分数写为：

$$f = \frac{Z_i - Z_{i,ox}}{Z_{i,fu} - Z_{i,ox}} \quad (7)$$

式中，Z_i 为元素 i 的质量分数；下标 ox 表示氧化剂入口处的值；fu 表示燃料入口处的值。

3.3 边界条件

首钢京唐 5500m³ 高炉 BSK 顶燃式热风炉采用了助燃空气和煤气双预热技术，回收热风炉烟气余热来预热煤气和助燃空气，同时设置 2 座小型顶燃式预热炉来预热助燃空气，使助燃空气温度达到 550℃ 以上。在采用纯高炉煤气燃烧的条件下，热风炉风温可以达到 1300℃。高炉煤气成分见表 1，热风炉的设计参数见表 2。

表 1 高炉煤气成分（体积分数） （%）

CO	CO_2	N_2	CH_4	H_2
20.87	23.2	52.02	0.05	3.86

表 2 BSK 顶燃式热风炉设计参数

项 目	数 值
煤气入口流量/m³·h⁻¹	190000
煤气入口温度/℃	168
煤气入口压力/kPa	4.34
理论需要空气量/m³·h⁻¹	112860
空气入口流量/m³·h⁻¹	140534
空气入口温度/℃	566
空气入口压力/kPa	10.7
拱顶温度设定值/℃	1390

根据设计参数给出数学模型所对应的边界条件如下：煤气和助燃空气入口均为质量流量边界，助燃空气入口的质量流量为 50.36kg/s，湍流强度为 10%，入口温度 839K，平均混合分数和平均混合分数均方差都为 0；煤气入口的质量流量为 70.72kg/s，湍流强度为 10%，入口温度为 441K，平均混合分数为 1，平均混合分数均方差为 0。出口为压力出口边界，设定压强为大气压即表压为 0，设定回流湍流强度为 10%，平均混合分数和平均混合分数均方差均为 0。对于壁面边界，壁面为非滑移边界，传热采用第 2 类边界条件且壁

面绝热。流体密度采用 PDF 混合物模型给出，比热容采用混合定律给出，辐射模型采用 P1 模型。

3.4 计算结果与分析

3.4.1 速度分布

图 2 给出了燃烧室空间 $Y=0$ 和 $X=0$ 截面的速度分布云图。从图 2 中可以看出，整个空间内速度基本沿中心呈对称分布。随着高度的下降，整个预燃室煤气和空气的混合在径向方向逐渐均匀，从煤气喷嘴平面到空气喷嘴平面燃烧室的中心低速区逐渐减小，到达预燃室底部时，低速区已经基本消失。

在喉口区域，由于气体运动空间大幅度压缩，靠近炉墙的气体速度迅速升高，中心位置的速度大小基本不变。靠

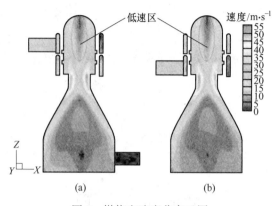

图 2 燃烧室速度分布云图
(a) $Y=0$；(b) $X=0$

近炉墙处气体速度很大，而且旋流非常强烈，对炉墙造成了强烈的冲刷，因此喉口处的炉墙在设计时需考虑到这一因素。

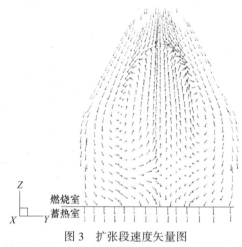

图 3 扩张段速度矢量图

另外，从图 2 中还能看到，当气体通过喉口以下的扩张段时，气体速度迅速减小，而且依然保持中心速度低、边缘速度高的分布规律，而且扩张段中心存在一个较大的低速区。图 3 是扩张段至燃烧室底部的速度矢量图。从图 3 中明显看出，由于从喉口到扩张段横截面积突然增大，使得该区域出现了一个较大的回流区，这个回流的产生是 BSK 顶燃式热风炉的关键技术所在。一方面回流能够起到稳定火焰的作用，使得新补充的煤气和空气始终能与高温气体接触；另一方面扩张段中心的回流切断了火焰的中心发展，

而使火焰沿着拱顶锥角空间扩张，从而缩短了火焰长度，使火焰不会对蓄热室格子砖产生冲击。当气体通过扩张段到达燃烧室底部时，竖直向下的速度基本占主导，这有利于高温烟气的热量利用。燃烧室底部中心与边缘的速度差异缩小，均匀性较好。

3.4.2 浓度分布

由图 4 可以看出，预燃室上部（包括半球拱顶内）均被煤气充满，CO 浓度和煤气入口浓度一致。在预燃室下部，助燃空气从 2 层空气喷嘴喷出，夹裹煤气，并依靠旋流作用切割煤气燃烧。由流场计算结果与分析可知，此处气流速度中等，混合强度一般，因此预燃室下部中心区域的 CO 浓度依然很高。

在喉口区域，由于靠近炉墙处的速度大于中心区域，即外围空气的速度远大于中心煤气速度，因此在喉口处空气以较大速度切割中心煤气，发生强烈混合和燃烧，从图中也能看到在喉口处 CO 浓度急剧下降。在扩张段，混合空间的增大使得煤气和助燃空气进一步混合、燃烧，扩张段中心的回流也切断了煤气的中心流动，使煤气向边缘流动，更好地与空气混合。图 4（a）中 $Y=0$ 平面是扩张段底部，CO 浓度分布略微不对称，右侧 CO 浓度要稍高于左侧。主要原因是上部各喷嘴流量不均匀，造成扩张段 CO 浓度出现偏差，但总体来讲浓度分布较为合理，煤气燃烧充分。

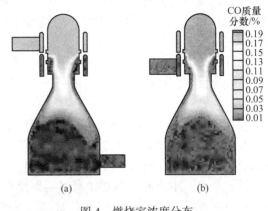

图 4　燃烧室浓度分布
（a）$Y=0$；（b）$X=0$

3.4.3　温度分布与火焰形状

燃烧室的温度分布与浓度分布密切相关。在图 4 中 CO 浓度较高的区域（如图 4 中预燃室上部包括半球拱顶和预燃室下部的中心区域），由于 CO 尚未燃烧完全，因此该区域温度较低。

喉口区域由于煤气和助燃空气发生了较为强烈的混合和燃烧，因此从喉口处温度便迅速升高，甚至中心区域的煤气温度也从原来的 215℃ 逐渐被加热到 600℃。

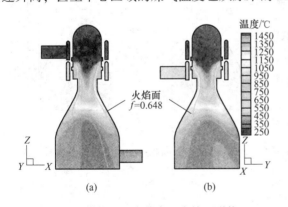

图 5　燃烧室温度分布和火焰面形状
（a）$Y=0$；（b）$X=0$

在扩张段向下区域，更多的煤气发生燃烧反应，拱顶温度迅速升高，在扩张段底部温度基本趋于稳定。图 5 是燃烧室底部的温度分布云图。由于燃烧充分，燃烧室底部均处于 1350℃ 以上的高温状态，另外 CO 浓度存在轻微的不均匀分布，使得左右两侧温度不是完全均匀。通过对底部截面温度进行面积分计算，求得底部截面的平均温度为 1419.7℃，整体温度分布状况良好。

根据燃烧 k-ε-g 模型，需引入混合分数的概念对温度场和浓度分布进行求解。顶燃式热风炉的燃烧是非预混燃烧，非预混燃烧可以看作简单化学系统，温度与混合分数呈线性关系。计算得出：当平均混合分数 $f<0.648$ 时，温度随着平均混合分数的增大线性递增；当 $f>0.648$ 时，温度随着平均混合分数的增大线性递减；仅当 $f=0.648$ 时，温度达到最大值 1447℃。$f=0.648$ 对应的区域即为火焰面，图 5 示出火焰面轮廓线。

火焰面的形状与炉内煤气和助燃空气的混合、燃烧、温度场、浓度分布都是相吻合的。在预燃室的下部，火焰面反映出煤气与助燃空气的交界面。由于该处产生的少量高温烟气将周围的助燃空气和煤气加热了近 200℃，因此此处温度大约为 1250℃。在扩张段底部，沿火焰面高度方向从上向下，气体温度呈一定梯度逐渐上升；靠近热风出口附近，CO 浓度高于燃烧室另一侧，因此火焰向热风出口侧发生轻微偏斜。在火焰面以下区域温度基本保持稳定。

4 燃烧器结构设计优化

燃烧室底部煤气浓度和烟气温度分布不均匀主要是环形陶瓷燃烧器空气喷嘴流量不均匀导致的。根据相关文献对类似问题的研究[7]可知，调节空气喷嘴的截面积能够调节空气的流量。在调整个别喷嘴的情况下，气体质量流量会随着喷嘴截面积的增大而增加。

图 6（a）是优化前空气第 1 层喷嘴平面 CO 浓度分布。根据其浓度分布情况分析，需要减小第 3 象限助燃空气的流量，增大第 2 象限助燃空气的流量。各喷嘴尺寸在原设计时完全一样，宽度均为 160mm，现将第 3 象限顺时针方向的第 2~4 个喷嘴宽度由 160mm 减小到 140mm，将第 2 象限顺时针方向的第 3~5 个喷嘴宽度由 160mm 增大到 180mm。优化后的结果如图 6（b）所示。整个截面 4 个象限空气和煤气基本呈对称分布，气体混合的均匀性得到大幅度提高[7]。

图 7 是优化后燃烧室温度分布和火焰面形状。对比图 5 优化之前的温度场和火焰面形状，不仅火焰偏斜的现象得到改善，而且扩张段以下区域的温度分布更加均匀。

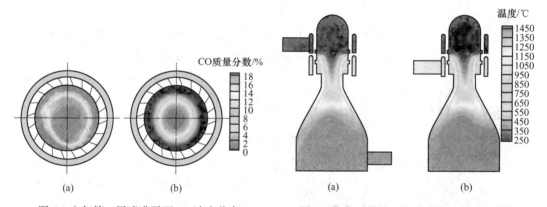

图 6 空气第 1 层喷嘴平面 CO 浓度分布
（a）优化前；（b）优化后

图 7 优化后燃烧室温度分布和火焰面形状

5 实际冷态测试验证

在首钢京唐 2 号高炉投产前，对 BSK 顶燃式热风炉进行了冷态工业测试。选取 2 号热风炉作为测试炉，测试的主要内容包括燃烧器煤气、空气喷嘴的流速以及蓄热室格子砖上表面的速度分布。测量煤气喷嘴流速时，将助燃空气引入 2 号热风炉的煤气管道，用助燃空气替代煤气进行实测。

5.1 喷嘴流速

图 8 是环形陶瓷燃烧器空气上下环喷嘴与煤气上下环喷嘴流速的实际测量值与仿真计

算值的比较。可以看出，空气喷嘴流速的实测值与计算值的分布规律大致相同，速度的绝对值略有差别，可能与测速仪在喷嘴截面处的测量角度以及气体的湍流有关。煤气喷嘴测速的实测值与计算值非常接近，除个别喷嘴外，二者在速度绝对值上都能较好吻合[8]。

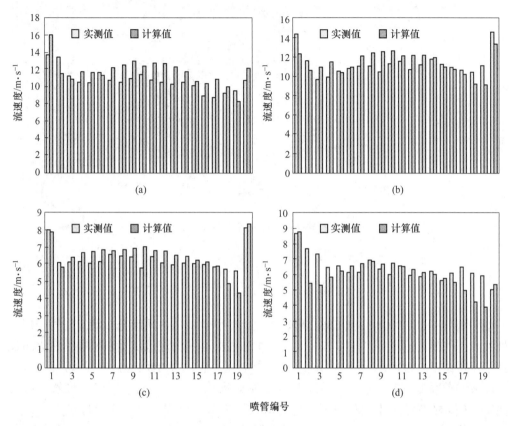

图 8 喷嘴的流速分布
(a) 空气上环；(b) 空气下环；(c) 煤气上环；(d) 煤气下环

5.2 格子砖上表面速度分布

格子砖上表面共取测点 337 个，其中：中心 1 点；圆周以 15°为夹角，取 24 个方向，每个方向取 14 个测点，共 336 个测点。

格子砖上表面的测点分布与实测结果如图 9 和图 10 所示。比较两图可以看出，一方面热风炉中部的速度比壁面处的速度小；另一方面，中部的速度也呈区域性分布，区域 S3~N1 与区域 W~S2 相比，速度稍大，区域 N2~N5 为过渡区。区域 S3~N1 的速度普遍大于 2m/s，

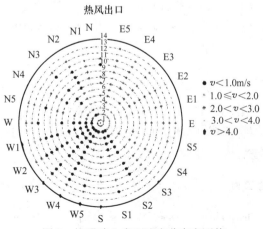

图 9 格子砖上表面速度分布实测值

区域 W~S2 的速度普遍小于 1.5m/s，区域 N2~N5 的速度介于二者之间。虽然具体到每个点的速度值有所不同，但整体的速度分布趋势是一致的。

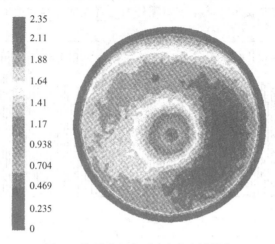

图 10　格子砖上表面速度分布计算值

6　结论

（1）首钢京唐 5500m³ 高炉 BSK 顶燃式热风炉燃烧室内的速度分布状况较为合理。喉口处的高速气流加速了空气与煤气的混合。扩张段中心的回流区域对稳焰和缩短火焰长度起到了重要作用。蓄热室表面边缘速度略高于中心，整齐均匀性良好。

（2）空气从喷嘴喷出后以较大速度切割中心煤气，进行混合、燃烧。虽然热风出口一侧 CO 浓度要略高于另一侧，但总体 CO 浓度分布较为合理，煤气燃烧充分。

（3）燃烧室底部均处于 1350℃ 以上的高温状态，CO 浓度出现轻微的不均匀性，使得左右两侧温度不是完全均匀，通过对底部截面温度进行面积分计算，求得底部截面的平均温度为 1419.7℃，整体温度分布状况良好。火焰面向热风出口轻微偏斜，基本达到短焰扩散燃烧的目的。

（4）数值仿真计算的结果与实际测量值吻合良好，具有较高的精确度。

（5）通过燃烧器喷嘴结构优化可以大幅度提高空气与煤气混合的均匀性，改善燃烧室内浓度、温度分布以及火焰面形状。

致谢

感谢共同参加热风炉冷态测试工作的毛庆武、钱世崇、银光宇、倪萍、戴建华、李林、梅丛华等的辛勤工作；感谢首钢京唐公司王涛、张卫东、周雁、韩向东、苏殿昌的大力支持。

参考文献

[1] Zhang Fuming, Qian Shichong, Zhang Jian, et al. Design of 5500m³ blast furnace at Shougang Jingtang [J]. Journal of Iron and Steel Research International, 2009, 16 (S2)：1029.

[2] 张福明,钱世崇,张建,等. 首钢京唐5500m³高炉采用的新技术[J]. 钢铁, 2011, 46 (2): 12.
[3] 张福明. 我国大型顶燃式热风炉技术进步[J]. 炼铁, 2002, 21 (5): 5.
[4] 项钟庸,郭庆第. 蓄热式热风炉[M]. 北京:冶金工业出版社, 1988.
[5] Li Baokua. Metallurgical Application of Advanced Fluid Dynamics [M]. Beijing: Metallurgical Industry Press, 2004.
[6] 胡祖瑞,程树森,张福明. 霍戈文内燃式热风炉传输现象研究[J]. 北京科技大学学报, 2010, 32 (8): 1053.
[7] 吴狄峰,程树森,赵宏博,等. 关于高炉风口面积调节方法的探讨[J]. 中国冶金, 2007 (12): 55.
[8] 张福明. 长寿高效热风炉的传输理论与设计研究[D]. 北京:北京科技大学, 2010.

Combustion Technology of BSK Dome Combustion Hot Blast Stove at Shougang Jingtang

Zhang Fuming   Hu Zurui   Cheng Shusen   Li Xin

Abstract: In order to develop the technology of the BSK dome combustion hot blast stove equipped in 5500m³ BF, the combustion mechanism and characteristic of dome combustion hot blast stove was investigated. The combustion mechanism of annular ceramic burner of BSK dome combustion hot blast stove was simulated through CFD mathematics simulation technology. The mixing, flow and combustion process of air and gas in dome combustion hot blast stove was analyzed, and the distribution of velocity field, temperature field and concentration field in the stove was also calculated. The calculated results by CFD mathematics simulation technology was verified through cold test on actual hot blast stove. The result shows that BSK dome combustion hot blast stove uses swirling diffusion combustion technology to make the distribution of velocity field, temperature field and concentration field uniformly and symmetrically, control the flame length and flame shape efficiently, as well as make the gas have sufficient in the dome. There is direct relationship of the distribution of velocity field, temperature field and concentration field with the initial distribution of gas and combustion air. Optimal design on structure of the burner nozzle is carried out so as to increase uniformity of air and gas mixing remarkably, and improve distribution of concentration and temperature as well as flame shape in the combustion chamber.
Keywords: dome combustion hot blast stove; burner; combustion mechanism; optimal design

顶燃式热风炉高温低氧燃烧技术

摘　要：NO_x 是制约热风炉实现高风温长寿的主要技术障碍。为有效抑制和降低热风炉燃烧过程生成的 NO_x，研究分析了 NO_x 的生成机制，运用热力型 NO_x 生成模型计算了热风炉燃烧过程 NO_x 生成速率和生成量，开发设计了基于高温低氧燃烧技术（HTAC）的新型顶燃式热风炉，采用 CFD 仿真模型对比研究了常规热风炉和高温低氧热风炉的燃烧过程和特性。计算得出两种热风炉的温度场分布和火焰形状、浓度场分布以及 NO_x 的浓度分布。研究结果表明，高温低氧热风炉的温度场分布均匀，在相同拱顶温度下，NO_x 生成量仅为 0.008%，比常规热风炉降低约 76%。高温低氧热风炉可以获得更高的风温并可以有效减少 NO_x 排放，实现热风炉高效长寿和节能减排。

关键词：顶燃式热风炉；高温低氧燃烧；高风温；低 NO_x 含量

1　引言

随着炼铁工业的技术发展，提高风温已成为现代高炉的重要技术特征。在高炉炼铁工艺中采用热风炉加热鼓风已有近 200 年历史，最早经过热风炉加热后的鼓风温度只有 149℃。随着技术的不断进步，目前高炉风温已达 1250～1300℃，提高风温可以大幅度降低高炉燃料消耗，节约焦炭，提高喷煤量，促进高炉生产稳定顺行，还可以充分利用低热值高炉煤气，提高能源利用效率，减少煤气放散和 CO_2 排放，节约能源，保护环境。因此，高风温是现代高炉实现强化冶炼、高效低耗、节能减排的重要技术措施。现代热风炉要求达到 1250℃ 以上的高风温，使用寿命要大于 30 年，同时要降低 CO_2、NO_x 等污染物的排放，实现热风炉长寿、高效、高风温和低排放。

高炉热风炉按结构形式分为内燃式、外燃式和顶燃式。顶燃式热风炉是 20 世纪 70 年代针对内燃式和外燃式热风炉的技术缺陷而创新发展的一种新型热风炉结构。顶燃式热风炉的特点是利用热风炉的拱顶空间作为燃烧室，取消了热风炉内部或外部独立设置的燃烧室。1978 年，首钢 2 号高炉（1327m³）率先采用了顶燃式热风炉，这是世界上第 1 座大型顶燃式热风炉实现工业化应用[1]。这种热风炉具有结构对称、温度区间分布合理、占地面积小、工程投资低等优点。但传统的顶燃式热风炉受燃烧空间的影响，容易造成拱顶局部高温，使燃烧室温度变化剧烈且温度分布不均匀，降低热风炉的传热效果和使用寿命。现有 3 种结构形式的热风炉均为常规热风炉，无论采用何种结构形式的燃烧器，其燃烧原理和特性并无本质差别。研究表明，热风炉拱顶温度达到 1400℃ 以上时，NO_x 大量生成，燃烧产物中的 NO_x 含量急剧升高，燃烧产物中的水蒸气在温度降低到露点以下时冷凝成液态水，NO_x 与冷凝水结合形成酸性腐蚀性介质，对热风炉炉壳钢板产生晶间应力

本文作者：张福明，胡祖瑞，程树森。原刊于《钢铁》，2012，47（8）：74-80。

腐蚀。因此现有的常规热风炉一般将拱顶温度控制在1420℃以下，旨在降低NO_x的含量，从而抑制炉壳晶间应力腐蚀，但由此却限制了风温的进一步提高。因此设计开发出一种改变常规热风炉燃烧过程，进一步提高风温，同时降低CO_2、NO_x排放的高风温高效长寿热风炉，已成为克服上述技术缺陷的必要条件。

2 热风炉燃烧过程NO_x的形成机制

燃料燃烧过程中生成的氮的氧化物总称为NO_x。NO_x主要包括N_2O、NO、NO_2、N_2O_3、NO_3和N_2O_4、N_2O_5等，燃烧生成的NO_x主要是NO和少量的NO_2。NO_x对人体、动物和植物都具有极大的危害，还会导致光化学雾、酸雨和臭氧损耗，对自然生态环境产生破坏作用，因此工业生产和燃料燃烧中要尽量减少NO_x的排放。

NO_x在燃烧过程的生成量受燃烧方式、空气混合比、燃烧温度等燃烧条件的影响很大。NO_x按其起源和生成途径可以分为热力型NO_x、快速型NO_x和燃料型NO_x。热力型NO_x是通过氧化燃烧空气中的N_2形成的；快速型NO_x是通过在火焰前锋面的快速反应形成的；燃料型NO_x是通过氧化燃料中的N形成的。高炉热风炉在燃烧高炉煤气条件下，由于高炉煤气中含氮化合物很少，因而极少生成燃料型NO_x，主要以生成热力型NO_x为主。

2.1 热力型NO_x的形成机制

热力型NO_x的形成是由一组高度依赖于温度的化学反应决定的，这也被称为广义的捷尔道维奇（Zeldovich）生成机制，该理论认为，在O_2-N_2-NO系统中，由氮分子形成的热力型NO_x的主要反应如下：

$$O + N_2 \underset{}{\overset{K_1}{\rightleftharpoons}} N + NO \quad (1)$$

$$O_2 + N \underset{}{\overset{K_2}{\rightleftharpoons}} O + NO \quad (2)$$

$$N + OH \underset{}{\overset{K_3}{\rightleftharpoons}} H + NO \quad (3)$$

反应（1）和反应（2）被称为捷尔道维奇生成机制。当燃烧过程有水蒸气时，燃烧产物中存在OH，此时NO也可以按反应（3）生成，因此被称为广义的捷尔道维奇生成机制。热力型NO_x生成的特点是生成反应比燃烧反应慢，主要是在火焰前锋的高温区间内生成NO_x。

大量研究结果表明[2~4]，NO的生成是在燃烧带之后靠近最高温度区间的燃烧产物中进行的。目前的研究结果也认为在燃烧带中有NO的生成反应进行。NO的浓度与燃烧产物的温度有关，而且无论燃烧反应结束还是正在进行，生成NO浓度最高的区域均处于温度最高的区间。研究还发现，NO的生成并不是瞬间完成的，燃烧产物在燃烧室停留时间越长，烟气中的NO浓度就越高，因此增加气流速度可以使NO浓度降低。总之，NO_x的生成主要与火焰的最高温度、N_2和O_2的浓度以及气体在高温区的停留时间等因素有关。

2.2 热力型NO_x的反应速率

热力型NO_x中NO的质量分数在95%左右，仅在局部有少量NO被氧化成NO_2。在热风炉燃烧的条件下，NO 生成反应尚未达到化学平衡，反应基本上服从阿累尼乌斯定律。Zeldovich 通过试验及推导确认 NO_x 的生成速率可表示为：

$$\frac{d\phi([NO_x])}{dt} = 3 \times 10^{14} \phi([N_2]) \phi([O_2])^{0.5} \times \exp[-542000/(RT)] \quad (4)$$

式中，$\phi([NO])$、$\phi([N_2])$、$\phi([O_2])$ 分别为 NO、N_2、O_2 的体积浓度，mol/cm^3；T 为反应温度，K；t 为时间，s；R 为通用气体常数，$J/(mol \cdot K)$。

由式（4）可以看出，NO 的产生量随着烟气在高温区内的停留时间延长而增加。氧浓度也直接影响 NO 的生成量。氧浓度越高，NO 的生成量越多；温度的升高也将提高 NO 的生成量。研究表明，当热风炉温度高于 1400℃ 时，NO 的生成量随火焰温度的升高急剧增加，此时温度对 NO 的生成具有决定性影响。

由反应式（1）~式（3）可推导 NO_x 的生成净速率，如式（5）所示。式（5）相对于式（4）考虑了中间产物的反应过程以及逆反应对 NO_x 浓度的影响，因此对于计算 NO_x 的生成量更为准确。

$$\frac{d\phi([NO_x])}{dt} = k_1 \phi([O]) \phi([N_2]) + k_2 \phi([N]) \phi([O_2]) + k_3 \phi([N]) \phi([OH]) - k_{-1} \phi([NO]) \phi(N) - k_{-2} \phi([NO]) \phi([O]) - k_{-3} \phi([NO]) \phi([H]) \quad (5)$$

式中，$\phi([NO])$、$\phi([O])$、$\phi([N_2])$、$\phi([N])$、$\phi([O_2])$、$\phi([OH])$ 分别是 NO、O、N_2、N、O_2、OH 的体积浓度，mol/m^3；k_1、k_2、k_3 为正反应的速率常数，$m^3/(mol \cdot s)$；k_{-1}、k_{-2}、k_{-3} 为相应的逆反应的速率常数，$m^3/(mol \cdot s)$。

反应（1）~反应（3）的速率常数已经在大量的试验研究中测得。这些研究数据已经过 Hanson 和 Salimian 等人的精确评估。在热力型 NO_x 生成模型中，式（5）中的速率系数分别为 $k_1 = 1.8 \times 10^8 e^{-38370/T}$，$k_{-1} = 3.8 \times 10^7 e^{-425/T}$，$k_2 = 1.8 \times 10^4 e^{-4680/T}$，$k_{-2} = 3.8 \times 10^3 e^{-20820/T}$，$k_3 = 7.1 \times 10^7 e^{-450/T}$，$k_{-3} = 1.7 \times 10^8 e^{-24560/T}$。

2.3 抑制热力型 NO_x 生成的措施

抑制热力型 NO_x 的燃烧技术包括低氧燃烧法、分段燃烧法和烟气再循环法等。这些方法的基本原理都为偏离化学当量燃烧法，即在局部的燃烧区域内使化学当量比不在燃烧反应化学当量比范围，从而抑制 NO_x 的生成。

特恩斯[5]用 NO_x 模型计算了稀释燃烧空气对降低 NO_x 生成速率的影响。在碳氢化合物的燃烧温度达到 1995℃ 时，计算得出 NO_x 生成速率为：

$$d\phi([NO_x])/dt = 19750 \times 10^{-6}/s$$

在采用 N_2 稀释燃烧用空气后，NO_x 的生成速率降低到：

$$d\phi([NO_x])/dt = 388 \times 10^{-6}/s$$

采用 N_2 稀释空气后，NO_x 生成量降低了近 50 倍。因此用惰性气体或不可燃气体稀释燃烧用空气中的氧浓度，可以抑制 NO_x 的生成，大幅度降低 NO_x 的浓度。这一研究结果表明，在低氧环境下燃烧可以有效抑制 NO_x 的生成，也是高温低氧热风炉开发研究的理论基础。

3 高温低氧热风炉的设计开发

3.1 高温空气燃烧技术（HTAC）

高温空气燃烧技术（HTAC）[6~8]是20世纪90年代开发成功的一项燃料燃烧领域中的新技术。HTAC包括两项基本技术措施：一是最大限度回收或称极限回收燃烧产物显热；二是燃料在低氧气氛下燃烧。燃料在高温和低氧空气中燃烧，燃烧过程和体系内的热工条件与常规的燃烧过程（空气为常温或低于600℃，氧的体积分数量不小于21%）具有显著的差异。这项技术为当今以燃烧为基础的能源转换带来变革性的发展，具有高效烟气余热回收和高预热空气温度、低NO_x排放等多重优越性，被认为是21世纪核心工业技术之一。

目前高温低氧燃烧技术已开始在轧钢加热炉上逐渐采用，但从未在高炉热风炉上得到应用。基于以上高温低氧燃烧理论，将高温低氧燃烧技术运用于高炉热风炉燃烧过程，使燃烧产生的烟气与高温预热后的助燃空气混合，可降低氧气浓度，实现热风炉高温低氧燃烧。

3.2 高温低氧顶燃式热风炉结构开发

高温空气燃烧技术的基本原理是使煤气在高温低氧气氛中燃烧。目前采用的助燃空气高温预热技术，已经能将助燃空气温度预热到800℃以上；通过采用煤气分级燃烧和高速气流卷吸燃烧产物，稀释反应区氧浓度，获得氧浓度（体积分数）低于15%的低氧气氛。煤气在这种高温低氧气氛中形成与传统燃烧过程完全不同的热力学条件，在与低氧气体作延缓状燃烧下释放热能，消除了传统燃烧过程中出现的局部高温高氧区。

热风炉高温低氧燃烧方式一方面使燃烧室内的温度整体升高且分布更加均匀，使煤气消耗显著降低，相应减少了CO_2等温室气体的排放；另一方面还有效抑制了热力型NO_x的生成。热力型NO_x的生成速度主要与燃烧过程中的火焰最高温度及氮、氧的浓度有关，其中温度是影响热力型NO_x的主要因素。在高温空气燃烧条件下，尽管热风炉内平均温度升高，但由于消除了传统燃烧的局部高温区；同时在热风炉内高温烟气与助燃空气旋流混合，降低了气氛中氮、氧的浓度；另外，在热风炉内气流速度高、燃烧速度快，因此NO_x排放浓度大幅度降低。

图1是顶燃式高温低氧热风炉的基本结构[9]。置于拱顶燃烧室的高温低氧燃烧器设有4层或以上的环状煤气、空气环道，每层环道上有一定数量的喷口。煤气和空气经喷口喷出，进入燃烧室内进行燃烧。各层喷口由上至下依次为：第1层为煤气喷口，第2层为空气喷口，第3层为空气喷口，第4层为煤气喷口。由于煤气、空气入口位置对煤气、空气喷口气流分配的均匀性影响较大，因此各煤气、空气喷口尺寸、间距根据煤气、空气入口管的数量和位置呈渐变分布或对称分布。

第1层煤气喷口喷出的煤气与第2层空气喷口喷出的空气在旋流扩散的条件下混合后燃烧，导致高温烟气向燃烧室下部流动；由第3层空气喷口喷出的空气与燃烧室内向下流动的高温烟气混合后，其温度可达到800~1000℃，氧浓度（体积分数）低于15%，形成

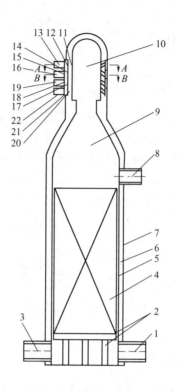

图1 高温低氧顶燃式热风炉的基本结构

1—冷风入口；2—炉箅子及支柱；3—烟气出口；4—格子砖；5—蓄热室；6—炉衬；7—炉壳；8—热风出口；9—燃烧室；10—高温低氧燃烧器；11—第1层煤气喷口；12—第1层煤气环道；13—第1层煤气入口；14—第1层空气喷口；15—第1层空气环道；16—第1层空气入口；17—第2层空气喷口；18—第2层空气环道；19—第2层空气入口；20—第2层煤气喷口；21—第2层煤气环道；22—第2层煤气入口

高温低氧的助燃空气，在燃烧室内向下旋转流动；由第4层煤气喷口喷出的煤气在燃烧室内高温低氧的气氛中燃烧，燃烧过程成为扩散控制反应，不再存在传统燃烧过程中出现的局部高温高氮区域，NO_x 的生成受到抑制。同时低氧状态下燃烧的火焰体积增大，在整个燃烧室内形成温度分布均匀的高温强辐射黑体，传热效率显著提高，NO_x 排放量大幅度降低，还可节约25%的燃料消耗，相应可减少 CO_2 排放。

4 高温低氧热风炉的燃烧特性

为研究高温低氧顶燃式热风炉的燃烧特性，建立了常规顶燃式热风炉与高温低氧顶燃式热风炉的物理模型以及湍流燃烧的数学模型。通过 CFD 仿真计算解析研究了2种热风炉燃烧室内的温度分布、浓度分布以及 NO_x 的生成量。

4.1 温度场与火焰形状

图2、图3是常规热风炉和高温低氧热风炉在理论燃烧温度均为1510℃时燃烧室内温度场和火焰形状的对比情况。其中图2（a）和图3（a）为常规顶燃式热风炉，图2（b）和图3（b）为高温低氧顶燃式热风炉。图2为 $X=0$ 和 $Y=0$ 两个中心截面的温度场和火

焰形状的对比情况。图中高温低氧热风炉燃烧室喉口段以下区域均处于1450℃以上的高温。与常规热风炉相比，燃烧效率的提高使得相同位置温度更高，而且几乎没有局部的高温区。图中实线代表火焰形状。对比发现，高温低氧热风炉的火焰形状更短，所涉及的燃烧室空间更大，在几乎整个燃烧室内形成弥散性火焰，使得温度分布均匀。从图3中也清晰地看到，燃烧室中下部相同截面位置，高温低氧热风炉的温度要高于常规热风炉，而且温度分布更加均匀[10]。

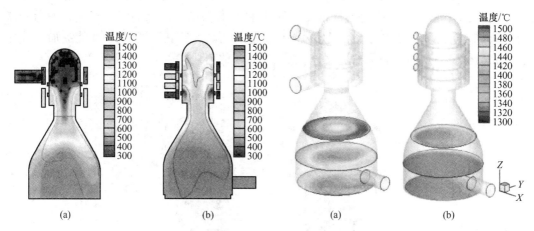

图2 2种热风炉中心截面温度分布和火焰形状的对比
(a) 常规顶燃式；(b) 高温低氧顶燃式

图3 2种热风炉燃烧室下部温度分布对比
(a) 常规顶燃式；(b) 高温低氧顶燃式

热风炉燃烧室底部（即蓄热室格子砖）上表面温度分布的均匀性对于热风炉而言非常重要。温度均匀分布的烟气能提高蓄热室格子砖的传热效率和延长格子砖寿命。图4是2种热风炉燃烧室底部径向温度比较。显而易见，高温低氧燃烧器最高温度与最低温度的差值减小，而且温度分布的均匀性得到明显提高。经计算，常规热风炉和高温低氧热风炉的温度均匀度分别为98.28%和99.56%。

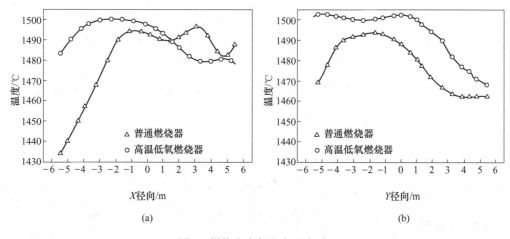

图4 燃烧室底部径向温度对比
(a) X径向；(b) Y径向

4.2 浓度分布

图 5 给出两种热风炉中心截面 CO 的浓度分布。对比发现，高温低氧热风炉燃烧室下部 CO 浓度明显降低，表明燃烧反应进行得更加充分。常规顶燃式热风炉在球顶空间内形成大片死区，大量煤气充斥在球顶空间内，该部分煤气不仅不能燃烧，而且在热风炉的换炉过程中，为防止送风期发生爆炸，还需要消耗大量氮气进行吹扫。采用高温低氧燃烧器以后，球顶空间的死区得到了有效利用，CO 浓度明显降低，换炉过程所消耗的氮气也相应大幅度减少。图 6 给出燃烧室底部 CO 和 O_2 的质量分数。显然高温低氧热风炉燃烧室底部 O_2 和 CO 的质量分数均比常规热风炉低，这表明高温低氧环境下，CO 能更加充分地与 O_2 混合燃烧，从而提高了燃烧效率，降低了 CO 的消耗。

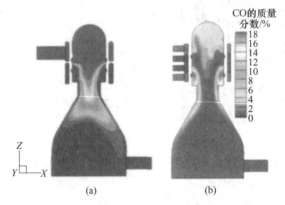

图 5 两种热风炉 CO 浓度分布的对比
(a) 常规顶燃式；(b) 高温低氧顶燃式

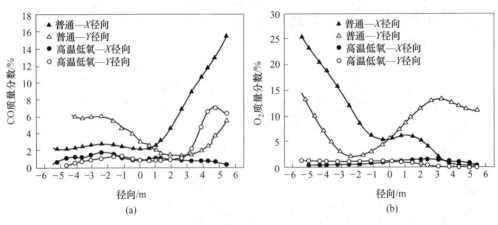

图 6 燃烧室底部 CO 和 O_2 浓度对比
(a) CO；(b) O_2

4.3 NO_x 生成量

图 7 是两种热风炉中心截面 NO_x 的浓度分布。图 8 是燃烧室底部径向的 NO_x 浓度。

从两图的对比中均可以看出,采用高温低氧燃烧器时,热风炉内最高 NO_x 体积分数从 0.033%左右降低到 0.008%左右,降低约76%,证明高温低氧燃烧技术显著抑制了 NO_x 在高温条件下的急剧生成。这在很大程度上可以减少燃烧期 NO_x 的排放。NO_x 浓度的降低可以减少 NO_x 在炉壳处与冷凝水结合形成酸性水溶液,从而有效抑制热风炉炉壳出现晶间应力腐蚀,延长热风炉的使用寿命。这也充分证实,高温低氧燃烧可以在控制 NO_x 生成的前提下使热风炉获得更高的拱顶温度,为进一步提高风温创造了条件。

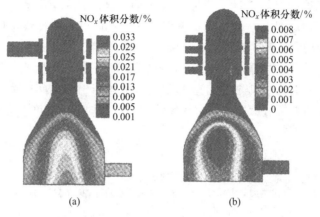

图7 2种热风炉 NO_x 浓度分布的对比

(a) 常规顶燃式;(b) 高温低氧顶燃式

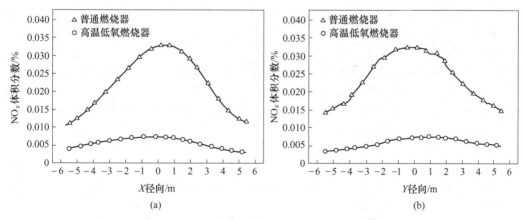

图8 燃烧室底部 NO_x 浓度对比

(a) X 径向;(b) Y 径向

5 结论

(1) 在高炉热风炉高温燃烧过程中,NO_x 的生成服从广义的捷尔道维奇生成机制,以生成热力型 NO_x 为主。当热风炉拱顶温度达到1400℃以上时,大量生成的 NO_x 与冷凝水结合后形成腐蚀性介质,造成热风炉炉壳晶间应力腐蚀,限制热风温度的提高,并缩短热风炉使用寿命。

(2) 基于高温空气燃烧技术（HTAC）设计开发的高温低氧顶燃式热风炉，采用助燃空气高温预热技术，可以将助燃空气温度预热到800℃以上。通过采用煤气分级燃烧和高速气流卷吸燃烧产物的技术措施，稀释了反应区氧浓度，获得氧体积分数低于15%的低氧气氛，从而创造高温低氧燃烧环境，实现高温空气燃烧技术在高炉热风炉中的应用。

(3) 通过对常规顶燃式热风炉与高温低氧顶燃式热风炉的仿真计算研究，分析了高温低氧顶燃式热风炉的燃烧特性。研究结果表明：高温低氧热风炉在高温低氧环境下，CO能够更充分地与O_2混合燃烧，提高燃烧效率，获得更高的燃烧温度且温度分布更加均匀合理。在实现相同风温的条件下，可减少CO的消耗，相应地减少CO_2排放。

(4) 高温低氧热风炉能显著抑制NO_x在高温条件下的急剧生成，很大程度上有效降低热风炉在燃烧期内NO_x的排放，有效抑制热风炉炉壳出现晶间应力腐蚀，延长热风炉的使用寿命。而且高温低氧燃烧可以使热风炉获得更高的拱顶温度，为进一步提高风温创造了有利条件。

参考文献

[1] 张福明. 我国大型顶燃式热风炉技术进步 [J]. 炼铁, 2002, 21 (5): 5.

[2] Zhang Xiaohui, Sun Rui, Sun Shaozeng, et al. Effects of stereo-staged combustion technique on NO_x emmision charactisctics [J]. Chinese Journal of Mechanical Engineering, 2009, 45 (2): 199.

[3] Xie Chongming. NO_x formation mechanism in the process of combustion and its control technology [J]. Guangzhou Chemical Industry, 2009, 37 (3): 161.

[4] Xia Xiaoxial, Wang Zhiqi, Xu Shunsheng. Numerical simulation on influence factors of NO_x emissions for pulverized coal boiler [J]. Journal of Central South University (Science and Technology), 2010, 41 (5): 2046.

[5] Sobisiak A. Performance characteristic of the novel low NO_x CGRI burner for use with high air preheat [J]. Combustion and Flame, 1998 (115): 93.

[6] Flamme M. Low NO_x combustion technologies for high temperature applications [J]. Energy Conversion and Management, 2001 (42): 1919.

[7] Guo Hongsheng. Numerical study of NO_x emission in high temperature air combustion [J]. JSME International Journal Series B: 1998, 41 (2): 134.

[8] Choi G, Katsuki M. Advanced low NO_x combustion using highly preheated air [J]. Energy Conversion and Management, 2001 (42): 639.

[9] 张福明, 程树森, 胡祖瑞, 等. 高温低氧顶燃式热风炉, 中国: ZL201020102450.9 [P]. 2010-11-17.

[10] 张福明. 长寿高效热风炉的传输理论与设计研究 [D]. 北京: 北京科技大学, 2010: 141.

High Temperature Air Combustion of Dome Combustion Hot Blast Stove

Zhang Fuming Hu Zurui Cheng Shusen

Abstract: NO_x is the major technical barrier to increase hot blast temperature and prolong campaign life of hot blast stove at present. In order to restrain the amount of NO_x formation during combustion process in the hot blast stove, the generation mechanism of NO_x production was investigated, and NO_x generation rate and amount in hot blast stove was calculated by means of thermodynamic generation model. A new dome combustion stove was developed based on high temperature air combustion (HTAC) technology. A comparison on the combustion process and characteristic of conventional hot blast stove and HTAC hot blast stove was performed by application of CFD simulation model. Temperature and concentration field distribution, flame shape and NO_x concentration distribution of two kinds of stove was calculated. The result shows quite symmetrical HTAC stove temperature field distribution. Under the same dome temperature, NO_x generation is 80ppm only, reduced by approx 76% in comparison with conventional stove. HTAC hot blast stove can get higher temperature and decrease NO_x emission efficiently, as well as realize long campaign life of hot blast stove and energy-saving and emission reduction.

Keywords: dome combustion hot blast stove; high temperature air combustion; high hot blast temperature; low NO_x content

热风炉送风期格子砖温度分布计算

摘　要：为优化热风炉设计，建立了热风炉蓄热室内传热过程的数学模型，制作了应用程序软件，探讨了不同的格子砖物性参数，如导热系数、比热容和密度等对格子砖温度场分布、气体温度分布以及格子砖蓄热量和放热量的影响。研究结果表明：设计热风炉时要综合考虑各种影响因素对气体和格砖温度分布的影响；格砖的温度梯度沿着径向在逐渐减小，设计热风炉时应选择合适的格砖厚度；空气和格砖内壁的温差在蓄热室不同材质格砖的交界面上存在波动；增大格砖导热系数和单纯增加格砖的密度和比热都会导致风温更低也更不稳定，但是只要延长燃烧时间就能充分发挥蓄热室的蓄热容和稳定送风能力；格砖热物性的影响程度随格砖增大而减小。

关键词：热风炉；蓄热室；格子砖；高风温

1　引言

近年来，我国高炉炼铁技术处于高速发展阶段，但是在热风温这项技术指标上却一直停滞不前。国际先进钢铁企业的热风温度在1300℃左右，而我国大部分钢铁企业热风温度在1100℃以下，这严重制约了我国高炉炼铁技术进步的前进步伐。提高热风温度是提高高炉产量、降低焦比和能源消耗的有效措施。一般每提高100℃的风温可以降低焦比15~20kg/t 铁，增加喷煤量约30kg/t 铁，所以提高风温既可以降低焦比，又可以增加喷煤量，从而获得良好的经济效益。

早在20世纪30年代，Hausen 就提出了描述热风炉蓄热室格砖与流过格孔的气体间的换热模型，并先后采用特征函数法、热极法对模型进行了解析求解。从60年代以后，BuTerfield[1~7]等人开始采用数值积分的方法对该问题进行了数值求解。为了全面研究热风炉，先后对热风炉蓄热室底部空腔内冷风流场进行了三维数值模拟和比较分析[8]、对热风炉鼓风室冷风气流分布进行了分析[9]、对热风炉拱顶空间内烟气分布进行了数值模拟计算[10]。

根据传热学和流体力学建立气体和格子砖温度分布的数学模型，设计出了专门用于热风炉蓄热室格子砖温度分布计算和设计的应用程序，通过改变格子砖物性参数，如导热系数、比热容和密度等，模拟计算出不同的温度分布，探讨不同的格子砖物性参数对格子砖温度场分布、气体温度分布和格子砖蓄热量的影响，从而为热风炉蓄热室格子砖的设计提供可靠的依据。

本文作者：郭敏雷，程树森，张福明，全强。原刊于《钢铁》，2008，43（6）：15-21。

2 物理模型和数学模型

2.1 物理模型

格子砖的物理模型如图 1 所示。该模型是将蓄热室看作由很多气体通道组成[6,7],气体通道内半径 (r_i) 为格孔水力学当量直径的一半,管道的外半径 (r_o) 则由式 (1) 来求得:

$$r_o = \sqrt{\frac{m_n}{\pi \rho_n N_c L_n} + \frac{D_h^2}{4}} \tag{1}$$

式中,m_n 为蓄热室每一段格子砖的重量,kg;ρ_n 为蓄热室相应格子砖的密度,kg/m³;N_c 为气体通道的个数;L_n 为蓄热室每一段格子砖的高度,m;D_h 为格子砖孔的水力学当量直径,m。

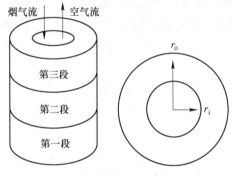

图 1 格子砖的几何模型

2.2 基本假设

(1) 忽略换炉所用时间,即上个周期结束后马上进入下个周期;
(2) 在燃烧期和送风期两个时期,气体入口温度保持稳定;
(3) 烟气和冷风的质量流量是气体的温度和压强的函数;
(4) 格子砖的热物性仅是温度的函数,气体的热物性则是温度和压力的函数;
(5) 气体速度在同一高度上分布均匀,是入口气体质量、气体密度、气体通道数目和水力学直径的函数;
(6) 蓄热室中同一高度上气流分布均匀,气体管道传热过程相同,各气体管道之间绝热。

2.3 数学模型

在模型[7]假设的基础上,格子砖采用二维非稳态传热方程,与气体在格孔中的换热进行耦合求解,具体如下:

(1) 气体传热方程:

$$\rho_g C_{p,g}\left(\frac{\partial T_g}{\partial t} + v_g \frac{\partial T_g}{\partial z}\right) + v_g \frac{\partial P}{\partial z} = \frac{4h}{D_h}(T_w - T_g) \quad (2)$$

(2) 格子砖内部二维非稳态传热控制方程：

$$\rho_s C_{p,s} \frac{\partial T_s}{\partial t} - \frac{1}{r}\frac{\partial}{\partial r}\left(rk\frac{\partial T_s}{\partial r}\right) - k_s \frac{\partial^2 T_s}{\partial z^2} = 0 \quad (3)$$

$$v_g = \frac{4\dot{m}_{in}}{\pi \rho_g N_c D_h^2} \quad (4)$$

式中，ρ_g，ρ_s 为气体和格子砖的密度，kg/m³；$C_{p,g}$，$C_{p,s}$ 为气体和格子砖的比热容，J/(kg·K)；v_g 为气体的流速，m/s；k_s 为格子砖导热系数，W/(m·K)；h 为对流换热系数，W/(m²·K)；$\dot{m}_{in}$ 为气体入口质量，kg；P 为气体的分压，kPa；N_c 为气体通道个数；D_h 为格子砖格孔的水力学直径，m。

整个过程中的热量传递共有三种方式：导热、对流和辐射。格子砖内部只有导热作用，烟气和格子砖有对流换热和辐射换热两种形式，燃烧期复合换热系数为对流换热系数与辐射换热系数的和，送风期只有对流换热系数，忽略辐射换热系数[8~10]。

对流换热系数由式 (5)、式 (6) 来进行计算：

烟气流动为湍流状态时：

$$\alpha_c = 0.86 C w^{0.8} D_h^{-0.333} T^{0.25} \quad (5)$$

烟气流动为层流状态时：

$$\alpha_c = (1.115 + 0.244 w D_h^{-0.6}) C T^{0.25} \quad (6)$$

式中，w 为气流的流速，m/s；T 为气体的绝对温度，K；D_h 为格孔的直径，m；C 为格孔表面特征系数，对于光滑的格孔，$C=1$，为了接近实际，C 取值范围 1.16~1.88。

辐射换热系数由式 (7)~式(12) 进行计算：

$$\alpha_r = C_b \varepsilon_s \frac{A\left(\frac{T_w}{100}\right)^4 - \varepsilon\left(\frac{T_g}{100}\right)^4}{T_w - T_g} \quad (7)$$

$$\varepsilon = \varepsilon(H_2O) + \varepsilon(CO_2) \quad (8)$$

$$A = A(H_2O) + A(CO_2) \quad (9)$$

$$A(H_2O) = \varepsilon(H_2O)\left(\frac{T_g}{T_w}\right)^{0.45} \quad (10)$$

$$A(CO_2) = \varepsilon(CO_2)\left(\frac{T_g}{T_w}\right)^{0.65} \quad (11)$$

$$l = 3.6 V/F \quad (12)$$

式中，$C_b = 5.67$，为绝对黑体的辐射系数，W/(m·s)；ε_s 为固体壁面的发射率；ε 为气体的发射率；A 为气体的吸收率；l 为气体辐射平均射线行程，m；V 为气体所充满的空间体积，m³；F 为围绕气体的容器表面积，m²；T_g 为烟气温度，K；T_w 为壁面温度，K。

边界条件：气体在燃烧期和送风期入口温度为定值。气体管道外表面和上下表面绝热，并且格子砖各段之间的交界面传热量相同，见式 (13)：

$$\left.\frac{dT_s^5}{dz}\right|_{z=0} = \left.\frac{dT_s^1}{dz}\right|_{z=L_1} = \left.\frac{dT_s}{dr}\right|_{r=r_0} = 0 \tag{13}$$

对格砖内壁上，边界为对流换热边界，公式描述见式（14）：

$$\left.\frac{dT_s}{dr}\right|_{r=r_i} = \frac{2h}{k_s(r_o^2/r_i^2 - 1)}(T_g - T_s|_{r=r_i}) \tag{14}$$

初始条件：气体在燃烧期入口温度为1350℃，在送风期入口温度为150℃。格子砖初始温度先设定为200℃，类似开炉计算，然后把上一周期结束的温度值作为下一周期格子砖的初始值进行计算，如此反复到两次送风期结束时温度大致相当，则认为热风炉已经正常运行，取得的值作为计算结果。

3 计算条件

本文初始数据采用国内某钢厂设计的热风炉来进行计算，其热工特性见表1。

表 1 热风炉主要热工参数

热工参数	特 性 值
蓄热室高/m	31.3
蓄热室面积/m²	36.3
蓄热室段数	3
各段长/m	14.9, 7.1, 9.3
格孔形状	圆孔
燃烧时间/min	45
送风时间/min	30
高炉风量/m³·min⁻¹	87180
烟气流量/m³·min⁻¹	94940
冷风入口温度/℃	150
蓄热室热风温度/℃	1350

程序中计算燃烧期辐射换热系数需要用到烟气的成分。计算中采用的烟气成分（体积分数）为 CO_2 占 24.5%，H_2O 占 7.3%，N_2 占 67.6%，O_2 占 0.6%。

蓄热室各段所用格子砖的材质参数见表2[11]。格子砖选用高效19孔砖，其水力学直径为 0.03m，活面积为 0.398m²/m²，填充系数为 0.602m³/m³，受热面积为 53.07m²/m³，当量厚度为 0.023m。

表 2 蓄热室格子砖材质参数

段号	材质	密度/kg·m⁻³	导热系数/W·(m·K)⁻¹	比热容/J·(kg·K)⁻¹
三	硅砖	1900	0.93+0.0007T	794.0+0.251T
二	低蠕变高铝砖	2500	1.51-0.00019T	836.8+0.234T
一	高密度黏土砖	2070	0.84+0.00052T	836.8+0.263T

4 计算结果及讨论

应用上述数学模型,自主开发了热风炉蓄热室格子砖温度分布模拟软件,分别计算了 12 种不同的情况,具体见表 3。

表 3 不同计算参数列表

项目	ρ_s/kg·m^{-3}	C_p/J·(kg·K)$^{-1}$	λ_s/W·(m·K)$^{-1}$	燃烧期 t/min
1	ρ_s	C_p	λ_s	45
2	ρ_s+100	同 1	同 1	同 1
3	ρ_s+200	同 1	同 1	同 1
4	ρ_s+500	同 1	同 1	同 1
5	ρ_s+500	同 1	同 1	75
6	同 1	C_p+100	同 1	同 1
7	同 1	C_p+200	同 1	同 1
8	同 1	C_p+500	同 1	同 1
9	同 1	C_p+500	同 1	75
10	同 1	同 1	$2\lambda_s$	同 1
11	同 1	同 1	$3\lambda_s$	同 1
12	同 1	同 1	$5\lambda_s$	同 1

注:因为蓄热室共分三段,表中对格子砖的物性参数的改变是对三层格子砖的物性参数进行同样的改变。

为了研究格子砖的物性参数的变化对格子砖温度分布、空气温度和格子砖蓄热量的影响规律,对于气体选取了出口处空气温度点(见 A 点)、蓄热室中间高度处空气温度点(见 B 点)两个点,对于格子砖是在格子砖模型的内壁、管道中部和管道外表面上分别取了蓄热室上表面温度点(分别为 D、G、J 点)、蓄热室中间高度处的温度点(分别为 E、H、K 点)和蓄热室下表面温度点(分别为 F、I、L 点)共 9 个点来进行研究,具体示意图如图 2 所示。

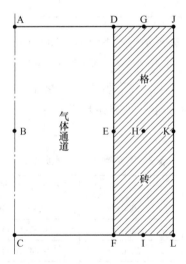

图 2 选取的温度点分布示意图

图 3 是不同导热系数下空气出口温度和其同高度上格砖上 3 个点随时间变化的曲线。

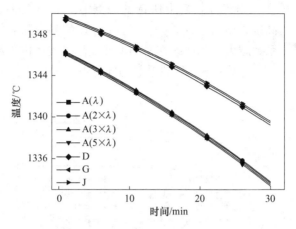

图 3 出口处空气温度和格砖温度随时间变化

图 3 中 A 表示出口处空气温度点，D、G、J 分别代表空气出口处格砖内壁、格砖中部和格砖外表面上的 3 个温度点。从图 3 可以看出，空气出口处空气和格砖的温度差在增大，这是因为随着热量的放出，格砖与冷风之间换热减弱，空气温度降低速率较大，因此出口处两者的温差将越来越大，从图中还可看出，导热系数越大，空气温度降低速率越大，这是因为导热系数大时，热量散失速度大，所以空气出口温度降低也越快。

从图 3 还可以看出格砖内壁与格砖中部的温差都要大于格砖中部与格砖外表面的温差，这说明格砖的温度梯度沿着径向在逐渐减小，如果设计的格砖当量厚度太大的话，格砖外部对燃烧期的蓄热和送风期的换热贡献都很小，造成浪费。

图 4 是不同导热系数时送风温度的温降时间的变化规律。从图 4 可以看出导热系数对送风温度的温降影响不大，但总体来说仍是导热系数大时温降也略大些。

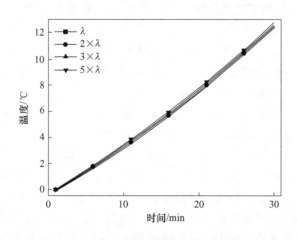

图 4 送风温度的温降随时间的变化

导热系数作为反应材料导热性能的主要指标，最直接的影响就是格砖传导传热的速度，导热系数大，可以迅速地将热量由中心传至表面，充分发挥其放热能力，使得蓄热体的中心温度与表面温度的差别很小。

图 5 上部的曲线是空气出口处格砖上所选的三点 D、G、J 的温度随时间变化曲线, 下部曲线是蓄热室中间高度处格砖上所选三点 E、H、K 的温度随时间变化曲线。从图 5 可以很明显地看出导热系数越大, 则曲线在径向上变化越平缓, 内外温度差别越小, 而且最大温差也不超过 1℃, 因此可在增加导热系数的同时适当增加格砖的厚度, 以提高其强度和燃烧期的蓄热量。

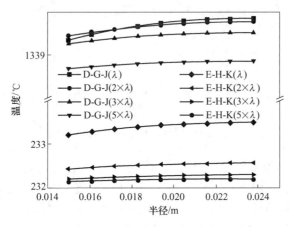

图 5 格砖的温度随径向的变化

从图中可以得出, 导热系数越大, 格砖的温度越低, 这再次说明增大格砖的导热系数能够加快传热速度。

从图 5 中还可以看出, 改变导热系数对它们温度的影响越来越小, 因此导热系数不是越大越好, 还要综合考虑其他因素和成本的影响。

格砖的密度和比热都是影响气体和格砖温度分布的重要参数, 因为在传热控制方程中, ρ_s 和 C_p 总是以乘积形式出现, 因此将这两个参数对比进行研究。

图 6~图 11 分别比较了不同的密度和比热条件下空气出口、中部温度和送风温度的温降以及蓄热量随时间的变化、空气和格砖内壁温度随高度的变化和格砖中间层温度随径向的变化。从总体来看, 密度和比热对空气和格砖温度分布以及蓄热量的影响是非常相似的, 只不过比热的影响程度比密度要稍微大一些。

图 6 中 A 代表出口处空气温度即送风温度, B 代表蓄热室中间高度处空气的温度。图中字母 t 代表燃烧期时间, 未标注时间 t 的燃烧期时间都是 45min, 下同。从图 6 中可以看出相同时间下中间高度处空气温度都是密度或比热越大温度越低, 但是出口处空气的温度变化率越大, 而蓄热室中间高度处空气的温度变化率却越小。这是因为密度和比热容越大, 在燃烧期内格砖蓄热能力越大, 燃烧期末格砖的温度越低, 由于该热风炉设计太高, 蓄热室中下部格砖的温度一直很低, 因此其处的格砖和空气的温度随格砖的密度和比热增大变化更加平缓, 但总体的送风温度受到燃烧期加热的格砖温度低的影响, 送风温度降低的速率也要更大一些。但是只要增大格砖密度和比热容的同时延长燃烧时间 (如从 45min 增至 75min), 那么由图中可以明确地看出: 无论蓄热室中间高度处还是出口处空气的温度都大大升高, 而且温度降落非常平缓, 送风温度非常稳定。

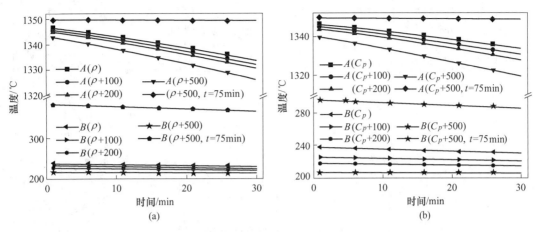

图 6 空气温度随时间的变化
(a) 密度；(b) 比热容

图 7 给出了不同条件下送风温度的温降随时间的变化规律。由图中可以看出，如果只是单纯增加格砖的密度或比热容，那么送风温度的降低速率同样会增大，导致格砖的蓄热能力在燃烧期没有得到充分地发挥，风温更低也更不稳定，但是只要增加格砖的密度或比热的同时延长燃烧时间，那么温降将变得非常小而且非常平缓，有利于稳定送风。

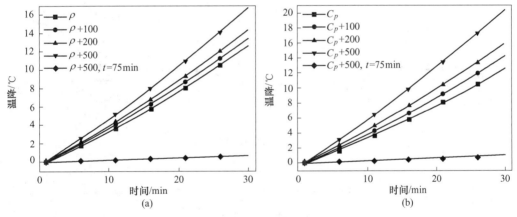

图 7 送风温度的温降随时间的变化
(a) 密度；(b) 比热容

从图 8 和图 9 中可以看出空气和格砖在高度上的温度分布是十分相似的，都是在相同的高度上密度或比热越大温度越低，而比热影响的程度要更大一些。单纯改变格砖的密度和比热容对空气和格砖的影响比较小，温度变化不大，而在增大格砖这两个热物性参数的同时延长燃烧时间则能大大提高空气和格砖整体的温度，达到良好的效果，但延长燃烧时间也要综合考虑不同层的格砖所设计的温度承受范围和操作制度等因素的影响。

图 10 除了能得出同样是密度或比热容越大格砖温度越低的规律外，而且还能看出密度和比热的增幅相同时，温度的降幅在减小，这说明随密度和比热容的增大，它们对温度分布的影响在减小，而且密度或比热越大温度曲线越趋于水平，这说明密度和比热容越

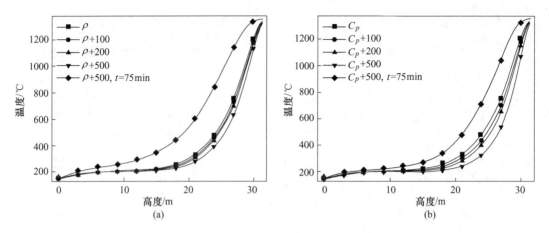

图 8 空气温度随高度的变化
(a) 密度；(b) 比热容

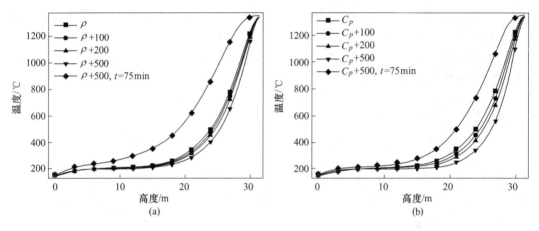

图 9 格砖内壁温度随高度的变化
(a) 密度；(b) 比热容

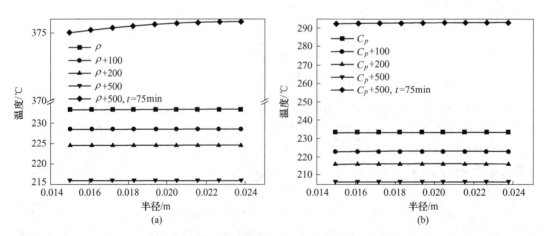

图 10 蓄热室中间高度处格砖温度随径向的变化
(a) 密度；(b) 比热容

大,格砖内外温差减小,即密度和比热一定程度上影响其本身的温度传播速度。同样,在增大密度和比热容的同时延长燃烧时间可以大幅度提高格砖的温度,并且由所得数据可知格砖内外的温差也同样增大了,这说明格砖的热量进一步向格砖内壁传输,有利于增强与空气的对流。

由图11可以清晰地看出,密度和比热容越大,格砖的放热量越小,这主要是因为单纯增大密度和比热容时格砖在燃烧期温度相对要低很多,但是只要同时再增加燃烧期时间就能使燃烧期的蓄热量和送风期的放热量都大大提高。

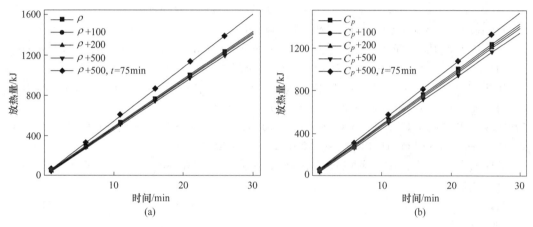

图11 格砖的放热量随时间变化
（a）密度；（b）比热容

5 结论

（1）设计热风炉时要综合考虑各种影响因素对气体和格砖温度分布的影响,才能使热风炉利用效率更高。

（2）格砖的温度梯度沿着径向在逐渐减小,如果设计的格砖当量厚度太大的话,格砖外部对燃烧期的蓄热和送风期的换热贡献都很小,造成浪费。

（3）热风炉蓄热室格子砖在送风过程中的温度差依然主要集中在高度方向上和随时间的变化上,而在径向上温差仍不大。

（4）蓄热体导热系数越大,蓄热体表面和内部的温度差别就越小,可以迅速地将热量由中心传至表面,如果条件允许可在增加导热系数的同时适当增加格砖的厚度,以提高其强度和燃烧期内的蓄热量,增大蓄热能力,导热系数越大,格砖的温度越低,送风温度温降越大,不过影响的程度都比较小。

（5）密度和比热容对气体和格砖的温度分布影响十分相似,并且都会使得蓄热体内外在燃烧期内达到饱和温度的时间向后延迟,进而使得送风期内气体和格砖的温度降低,放热量也低,送风温度的温降速率越大,风温更低也更不稳定,但是只要延长燃烧时间就能充分发挥蓄热室的蓄热和稳定送风能力,另外密度和比热容还会一定程度地影响格子砖的温度传播速度,它们的影响力也随其增大而减小。

参考文献

[1] Schofield J, Butterfield P, Young P A. Hot blast stoves [J]. Journal of the Iron and Steel Institute, 1961, 199 (39): 229-240.

[2] Willmott A J. Simulation of a thermal regenerator under conditions of variable mass flow [J]. International Journal of Heat and Mass Transfer, 1968, 11: 1105-1116.

[3] Willmott A J. The regenerative heat exchanger computer representation [J]. International Journal of Heat and Mass Transfer, 1969, 12: 997-1014.

[4] Kwakernaak H, Strijbos R C W, Tijssen P. Optimal operation of thermal regenerators [J]. IEEE Transactions on Automatic Control, 1969, 14: 728-731.

[5] Willmott A J, Hinchcliffe C. The effect of gas heat storage upon the performance of the thermal regenerator [J]. Int. J. Heat Mass Transfer, 1976, 19 (8): 821-826.

[6] Muske K R, Howse J W, Hansen G A, et al. Hot blast stove process model and model-based controller [J]. Journal of Iron and Steel Institute, 1999, 76 (6): 56-62.

[7] Muske K R, Howse J W, Hansen G A. Model-based control of A thermal regenerator part1: Dynamic model [J]. Computers and Chemical Engineer, 2000, 24 (11): 2519-2531.

[8] 胡日君, 程树森. 考贝式热风炉蓄热室底部空腔冷风分布的数值模拟 [J]. 过程工程学报, 2005, 5 (5): 490-494.

[9] 赵民革, 胡日君, 程树森, 等. 鼓风管对热风炉鼓风室气流分布的影响 [J]. 钢铁, 2005, 12 (40): 17-20.

[10] 胡日君, 程树森. 考贝式热风炉拱顶空间烟气分布的数值模拟 [J]. 北京科技大学学报, 2006, 4 (28): 338-342.

[11] 成兰伯. 高炉炼铁工艺及计算 [M]. 北京: 冶金工业出版社, 1991, 12: 511.

Calculation of Temperature Distribution in Checkers During "On Blast" Cycle in Hot Blast Stove

Guo Minlei Cheng Shusen Zhang Fuming Quan Qiang

Abstract: In order to optimaze the design of hot blast stove, a model of heat transfer in chamber of hot blast stove was developed and the software based on the model was established. The effect of the checker thermal characteristics such as thermal conductivity, heat capacity, density etc on the temperature of blast and checkers, the heat accumulation and heat release of checkers was determined. During "on blast", the temperature gradient in checkers is reduced along the radius, so a proper thickness of the checker should be chosen; increasing both the thermal conductivity and just only increasing the density or heat capacity makes the blast temperature lower and more unstable, however as long as the combustion time is long enough, the capability of regenerator and to accumulate heat and supply stable blast will be made full use. The effect of the checker thermal charateristics becomes smaller with the increase of "on blast" time.

Keywords: hot blast stove; regenerator; checker; hot blast temperature

顶燃式与内燃式热风炉燃烧过程物理量均匀性的定量比较

摘　要：本文利用CFD仿真计算软件，对霍戈文内燃式热风炉和首钢京唐BSK顶燃式热风炉燃烧室的流场、温度场进行了计算，对二者蓄热室上表面速度、温度分布以及热量的均匀性进行了定量比较，结果表明：BSK顶燃式热风炉流场的均匀性要远大于霍戈文内燃式，温度场方面BSK顶燃式热风炉和霍戈文内燃式热风炉的均匀性都很高，前者略大于后者，热量均匀性基本由速度决定，BSK顶燃式要优于霍戈文内燃式。
关键词：热风炉；燃烧过程；均匀性；定量比较

1　引言

20世纪50年代，我国高炉主要采用传统的内燃式热风炉。这种热风炉存在着诸多技术缺陷，且随着风温的提高而暴露得更加明显。为克服传统内燃式热风炉的技术缺陷，20世纪70年代，荷兰霍戈文公司（现为康利斯公司）对传统的内燃式热风炉进行优化和改进，开发了改造型内燃式热风炉，在欧美等地区得到应用并获得成功。

20世纪70年代末，首钢自行设计开发的顶燃式热风炉在首钢2号高炉（1780m^3）上得到成功应用，开创了大型高炉采用顶燃式热风炉的技术先例[1]。90年代俄罗斯卡鲁金（Kalugin）顶燃式热风炉（小拱顶结构）投入运行，进一步推动了顶燃式热风炉的大型化进程。顶燃式热风炉由于具有结构稳定性好、气流分布均匀、布置紧凑、占地面积小、投资省、热效率高、寿命长等优势，已在国内外100多座高炉上应用。北京首钢国际工程技术有限公司与卡鲁金公司合作，首次在首钢京唐5500m^3超大型高炉上成功应用了BSK顶燃式热风炉技术。

本文选取了内燃式和顶燃式热风炉作为研究对象，对热风炉内的流动、传热和燃烧现象运用数学仿真的方法进行了分析对比。其中内燃式热风炉选取的是原首钢2号高炉采用的霍戈文内燃式热风炉，顶燃式热风炉选取的是首钢京唐5500m^3高炉BSK顶燃式热风炉。

2　物理模型和数学模型

2.1　物理模型

图1是霍戈文内燃式热风炉的示意图，采用的是眼睛形燃烧室，其蓄热室上表面为月亮形，主要的结构尺寸有：热风炉的总高度为41.6m，燃烧室断面积9.7m^2，蓄热室断面

本文作者：张福明，胡祖瑞，程树森，毛庆武，钱世崇。原刊于《2010年全国炼铁生产技术会议暨炼铁年会论文集（上）》，2010：688-692。

积35.8m²。图2是BSK顶燃式热风炉的示意图，从喉口以下均为圆形截面。燃烧室的直径和高度分别为10.244m、6.362m，蓄热室高度为21.6m。

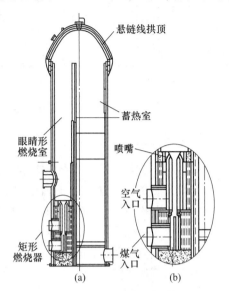

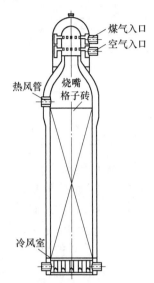

图1 霍戈文内燃式热风炉示意图　　　图2 BSK顶燃式热风炉示意图

其中内燃式采用的是眼睛形燃烧室，其蓄热室上表面为月亮形，而顶燃式热风炉喉口以下均为圆形截面。

2.2 数学模型

两种热风炉在燃烧方式上都可看作是非预混燃烧，因此在数学模型的选择上是类似的。实际中的燃烧过程是湍流和化学反应相互作用的结果，解决这一过程目前最常用的数学模型是 k-ε-g 模型。综合来说该模型将混合过程的控制作用和脉动的影响有机地统一，主要包括五个要点[2]，在这里不做详细叙述。

2.3 边界条件

霍戈文内燃式热风炉燃烧所用高炉煤气成分见表1。采用助燃空气和煤气预热技术，其煤气和空气管道入口处两种气体的相关参数见表2。BSK顶燃式热风炉燃烧所用高炉煤气成分见表3，其煤气和空气管道入口处两种气体的相关参数见表4。

表1 高炉煤气成分

组分	CO	CO_2	N_2	CH_4	H_2	H_2O
体积分数/%	20.33	16.63	50.46	0.44	1.16	10.98

表2 助燃空气和煤气相关参数

项目	流量/Nm³·h⁻¹	质量流量/kg·s⁻¹	预热温度/℃
空气	56190	20.14	600
煤气	73840	26.66	200

表3 高炉煤气成分

组分	H_2	CH_4	CO	N_2	CO_2	H_2O
体积分数/%	2.95	0.37	20.24	50.28	21.16	5

表4 助燃空气和煤气相关参数

项目	流量/$Nm^3 \cdot h^{-1}$	质量流量/$kg \cdot s^{-1}$	预热温度/℃
空气	118205	42.357	570
煤气	182952	68.10	215

表1~表4中的数据均为设计值，实际生产中的数据与其略有偏差。根据表中的数据给出数学模型所对应的边界条件，煤气和空气入口均为质量流量边界，湍流强度为10%。空气的平均混合分数和平均混合分数均方差为0；煤气的平均混合分数为1，平均混合分数均方差为0。出口设为压力出口边界，压强为一个大气压即表压为0，设定回流湍流强度为10%，平均混合分数和平均混合分数均方差均为0。壁面设置为非滑移边界，传热采用第二类边界条件且壁面绝热。流体密度采用PDF（概率密度函数）混合物模型给出，比热容采用混合定律给出。辐射模型为Discrete Ordinates模型[3]。

3 拱顶速度均匀性的比较

拱顶出口，即蓄热室上表面烟气分布均匀程度是热风炉的一个非常重要的指标，直接影响热风炉的热效率和热风温度。烟气在燃烧室出口横截面上的分布均匀程度用M_p[4]值表示。气体分布越均匀，M_p值越大，在理想均匀的条件下，M_p值为100%。

$$M_p = \frac{\overline{v} - \sum_{i=1}^{n}|v_i - \overline{v}|/n}{\overline{v}} \times 100\% \quad (1)$$

式中，$\overline{v}$为蓄热室上表面平均速度，m/s；v_i为取样点的速度，m/s；n为取样点的数量。

3.1 内燃式热风炉蓄热室上表面烟气速度分布均匀度

根据式（1）对蓄热室上表面选择了取样点，图3是内燃式热风炉蓄热室上表面速度云图以及取样点分布图。由于截面上的速度基本沿中心轴线上下对称，因此只在中心轴线的上部选择了8条取样线，编号分别从A~H，每条取样线上从下往上均匀选取了5个取样点。表5是各取样点的速度大小，单位为m/s。

根据表5中的数据计算出蓄热室上表面的平均速度为$\overline{v}$=8.41m/s。

由式（1）计算出速度分布的均匀度为：

图3 内燃式热风炉速度取样点示意图

$$M_{\mathrm{p}} = \frac{\overline{v} - \sum_{i=1}^{n} |v_i - \overline{v}|/n}{\overline{v}} \times 100\% = 48.98\%$$

表 5　内燃式热风炉各取样点的速度　　　　　　　　　　　　　　（m/s）

A	B	C	D	E	F	G	H
3.52	2.09	3.72	6.93	9.94	11.90	13.00	13.65
3.33	4.03	4.23	7.50	10.27	12.20	14.35	13.77
2.88	3.61	4.52	8.27	10.89	12.89	13.47	13.81
4.09	0.85	3.67	5.97	11.39	13.61	13.85	13.92
3.89	2.02	3.76	3.48	10.68	13.78	14.15	13.62

3.2　顶燃式热风炉蓄热室上表面烟气速度分布均匀度

图 4 是顶燃式热风炉蓄热室上表面的速度云图以及取样点分布图，截面各方向的速度分布基本都不对称，因此将 X 正向水平轴线每旋转 45°选取一条取样线，总共 8 条取样线，编号也是从 A~H，在每条取样线上从中心向边缘均匀选取 5 个取样点。表 6 是各取样点的速度大小，单位为 m/s。

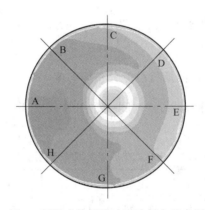

图 4　顶燃式热风炉速度取样点示意图

表 6　顶燃式热风炉各取样点的速度　　　　　　　　　　　　　　（m/s）

A	B	C	D	E	F	G	H
6.66	6.32	5.84	5.57	5.84	6.28	6.65	6.83
8.63	9.18	7.00	6.98	6.76	7.91	8.28	8.6
9.42	9.21	8.75	8.65	8.59	8.64	8.92	9.26
9.44	8.01	8.87	8.63	8.49	8.48	8.83	9.29
9.48	9.04	8.46	8.18	8.13	8.43	8.92	9.35

根据表 6 中的数据计算出蓄热室上表面的平均速度为 $\overline{v} = 8.12 \mathrm{m/s}$。

由式（1）计算出速度分布的均匀度为：

$$M_\mathrm{p} = \frac{\overline{v} - \sum_{i=1}^{n}|v_i - \overline{v}|/n}{\overline{v}} \times 100\% = 88.35\%$$

根据计算结果，内燃式蓄热室上表面的速度均匀度为 $M_\mathrm{p} = 48.98\%$，顶燃式的均匀度为 $M_\mathrm{p} = 88.35\%$，可以明显比较出 BSK 型顶燃式热风炉在蓄热室上表面的速度分布要优于霍戈文内燃式。

4 拱顶温度均匀性的比较

蓄热室上表面温度分布的均匀程度也是热风炉的一个非常重要的指标。高温烟气均匀地进入蓄热室格子砖，能使格子砖被高效利用，有利于提高格子砖热效率和延长格子砖寿命。为研究其均匀性，参照 M_p 值的定义，特定义参数 M_t 值，其定义式与上述 M_p 形式相同，只是将平均速度和实际速度转变为平均温度和实际温度。

$$M_\mathrm{t} = \frac{\overline{T} - \sum_{i=1}^{n}|T_i - \overline{T}|/n}{\overline{T}} \times 100\% \qquad (2)$$

式中，$\overline{T}$ 为蓄热室上表面平均温度，℃；T_i 为取样点的温度，℃；n 为取样点的数量。

4.1 内燃式热风炉蓄热室上表面烟气温度分布均匀度

烟气温度的取样点与速度取样点相同，如图 5 所示，表 7 是各取样点的温度。

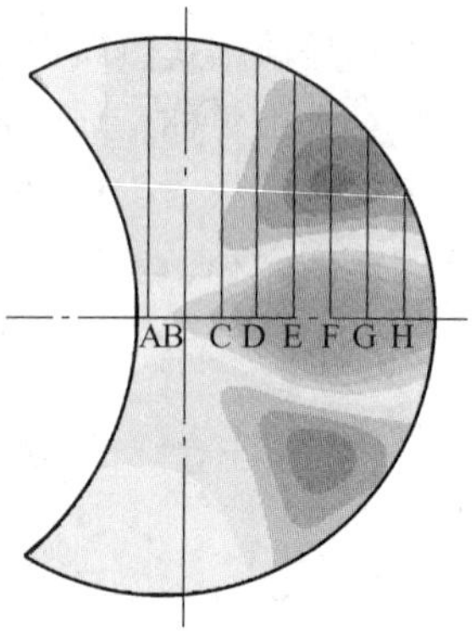

图 5 内燃式热风炉温度取样点示意图

表 7 内燃式热风炉各取样点的温度 (℃)

A	B	C	D	E	F	G	H
1432.11	1424.07	1405.77	1379.15	1371.45	1382.54	1391.18	1392.71
1443.04	1446.11	1447.88	1426.23	1407.33	1404.44	1406.48	1402.95
1443.50	1444.83	1451.59	1467.13	1465.68	1448.64	1436.85	1421.56
1441.49	1442.25	1445.53	1462.51	1480.12	1479.60	1465.69	1441.30
1440.67	1441.34	1443.22	1453.18	1464.23	1374.19	1474.88	1458.13

根据表 7 中的数据计算出蓄热室上表面的平均温度为 $\overline{T} = 1433.8$ ℃。

由式 (2) 计算出温度分布的均匀度为：

$$M_\mathrm{t} = \frac{\overline{T} - \sum_{i=1}^{n}|T_i - \overline{T}|/n}{\overline{T}} \times 100\% = 98.31\%$$

4.2 顶燃式热风炉蓄热室上表面烟气温度分布均匀度

顶燃式热风炉燃烧室出口截面烟气温度的取样点与其速度取样点相同,如图 6 所示,表 8 是各取样点的温度。

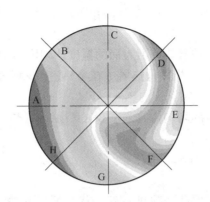

图 6 顶燃式热风炉温度取样点示意图

表 8 顶燃式热风炉各取样点的温度 (℃)

A	B	C	D	E	F	G	H
1455.53	1453.22	1450.01	1453.53	1459.43	1467.35	1469.71	1465.12
1438.63	1438.88	1439.33	1448.05	1461.40	1477.08	1476.61	1460.11
1421.93	1425.90	1433.71	1451.64	1473.97	1480.35	1472.37	1440.22
1407.16	1418.11	1437.68	1469.28	1472.77	1475.72	1454.90	1420.18
1400.34	1416.39	1447.06	1484.16	1446.79	1483.93	1437.72	1403.79

根据表 8 中的数据计算出蓄热室上表面的平均温度为 $\overline{T} = 1449.7$℃。

由式 (2) 计算出温度分布的均匀度为:

$$M_t = \frac{\overline{T} - \sum_{i=1}^{n} |T_i - \overline{T}|/n}{\overline{T}} \times 100\% = 98.71\%$$

5 进入格子砖的热量均匀性的比较

$$Q = C_p \cdot \rho A v \cdot T \tag{3}$$

式 (3) 是进入格子砖单个格孔的热量计算表达式。式中,Q 是热量,J;C_p 是烟气比热容,J/(mol·K);ρ 是密度,kg/m³;A 是单个格子砖格孔面积,m²;v 为气体速度,m/s;T 是烟气温度,K。

由于两种热风炉温度均匀度都较高,因此 C_p、ρ、T 均非常相近,进入格子砖单个格孔的热量 Q 以及其热量均匀性可看作只由速度 v 决定,因此顶燃式热风炉在蓄热室上表面的热量均匀性要优于内燃式热风炉。

6 结论

(1) 霍戈文内燃式热风炉蓄热室上表面的速度均匀度为 $M_p = 48.98\%$，BSK 顶燃式热风炉的均匀度为 $M_p = 88.35\%$，可以明显比较出 BSK 顶燃式热风炉在蓄热室上表面的速度分布要优于霍戈文内燃式。

(2) 霍戈文内燃式热风炉蓄热室上表面的温度均匀度为 $M_t = 98.31\%$，BSK 顶燃式的均匀度为 $M_t = 98.71\%$，可以看出两种热风炉在蓄热室上表面的温度均匀性都比较高，但 BSK 型顶燃式热风炉仍略优于霍戈文内燃式热风炉。

(3) 从进入格子砖热量的均匀性来看，BSK 型顶燃式热风炉都要优于霍戈文内燃式热风炉。

致谢

本研究受到国家科技支撑计划项目"新一代可循环钢铁流程工艺技术——长寿集约型冶金煤气干法除尘技术的开发"(2006BAE03A10) 的支持。

参考文献

[1] 张福明. 大型顶燃式热风炉的技术进步 [J]. 炼铁, 2002, 21 (5): 5-10.
[2] 王应时, 范维澄, 周力行, 等. 燃烧过程数值计算 [M]. 北京: 科学出版社, 1986 (1): 72.
[3] 郭敏雷. 顶燃式热风炉传热及气体燃烧的数值模拟 [D]. 北京: 北京科技大学, 2008.
[4] 陈冠军, 胡雄光, 钱凯, 等. 新型顶燃式热风炉燃烧技术的研究 [J]. 钢铁, 2009 (1): 79.

Quantitative Comparison on Uniformity of Combustion Physical Quantity in Top and Inner Combustion Hot Blast Stove

Zhang Fuming　Hu Zurui　Cheng Shusen　Mao Qingwu　Qian Shichong

Abstract: In this paper, CFD software is used to calculate the flow field and temperature field of the Hoogoven inner combustion hot blast stove and Shougang BSK top combustion hot blast stove. The quantitative comparison is carried out on the surface velocity, temperature distribution and heat of regenerator between them. The result shows the flow field uniformity of the BSK is better than the Hoogoven one. Although they both have high uniformity on temperature field, the former one is higher than the latter one, the heat uniformity is decided by the velocity, the former is also better than the latter.

Keywords: hot blast stove; combustion; uniformity; quantitative comparison

首钢京唐 5500m³ 高炉 BSK 顶燃式热风炉燃烧器分项冷态测试研究

摘　要：首钢京唐钢铁厂 5500m³ 高炉集中采用了一系列当今国际先进的综合技术，首次在 5500m³ 超大型高炉上成功应用 BSK 新型顶燃式热风炉。为充分掌握热风炉内流场分布状况，检验评价燃烧器气体分配均匀程度，北京首钢国际工程技术有限公司与首钢京唐钢铁公司炼铁作业部联合对京唐 5500m³ 高炉配置的 BSK 新型顶燃式热风炉进行了冷态测试。本文总结了单测空气喷口、单测煤气喷口的测试工作，通过对测试数据进行分析研究，整理出空气、煤气喷口的速度分布规律，经统计计算，证明燃烧器的整体喷口速度分布是比较均匀的，达到了热风炉燃烧器的性能要求。

关键词：大型高炉；热风炉；燃烧器；测试；流场

1 引言

由北京首钢国际工程技术有限公司设计的首钢京唐 5500m³ 超大型高炉，集中采用了精料、炉料分布控制、高风温、富氧大喷煤、长寿、环保等一系列当今国际先进的综合技术，实现了高炉的"大型化、高效化、长寿化、清洁化"。该项目集中采用了 60 余项自主研发设计的新技术、新工艺，实现了特大型高炉的工艺技术装备创新。首次在 5500m³ 超大型高炉上成功应用自主设计研制并全面实现国产化的并罐式无料钟炉顶设备；首次在 5500m³ 超大型高炉上成功应用高炉煤气全干法除尘技术；首次在 5500m³ 超大型高炉上成功应用 BSK 顶燃式热风炉技术等。

顶燃式热风炉又称无燃烧室热风炉，将燃烧器直接安装于热风炉的顶部。此前，4000m³ 以上的高炉配置的热风炉一般为外燃式或内燃式，顶燃式热风炉从未应用于 4000m³ 以上的高炉。随着顶燃式热风炉技术日渐成熟，它在经济和技术上的优势也逐渐体现出来。与内燃式、外燃式热风炉相比，顶燃式热风炉的主要优点是：热风炉炉壳结构对称，稳定性强；拱顶砌体与大墙隔开，拱顶直径小，结构热稳定性好；高温烟气流场分布均匀，蓄热室的利用率高；占地面积小，节约钢结构和耐火材料；布置形式灵活多样，紧凑合理等[1,2]。经综合研究论证和多方案对比，最终决定为首钢京唐 5500m³ 超大型高炉配置 4 座 BSK 新型顶燃式热风炉。

热风炉燃烧器作为热风炉的核心装置，是实现合理组织煤气燃烧的关键部件，它的技术性能、工作状况直接影响热风温度和热效率。顶燃式热风炉的最大特点是燃烧器置于热风炉顶部，空气和煤气自顶部进入热风炉，经燃烧器燃烧后产生的高温烟气向下进入蓄热

本文作者：张福明，李欣，毛庆武，钱世崇，银光宇，倪苹。原刊于《2010 年全国炼铁生产技术会议暨炼铁学术年会文集（上）》，2010：654-658。

室，为其提供热量。燃烧器的气体分配状况决定了燃料燃烧是否完全和烟气流场分布是否均匀，因此燃烧器结构的设计合理性非常重要。

2 测试方法与目的

目前，对燃烧器的性能研究主要有三种方法，包括实物测量法、模型实验法和数值模拟计算法。实物测量法就是对热风炉燃烧器进行冷态测试和热态实测[3]。这种方法首先要求现场具备测量条件，包括测试对象已安装调试到位，准备好满足测试要求的测试设备，同时测试不能影响正常生产。虽然实物测量法测量难度大，但得到的数据最直观准确，能够直接反映出燃烧器的工作状态。

顶燃式热风炉燃烧器的空气、煤气分配状况是燃料燃烧是否充分和高温烟气流场是否均匀有序的重要影响因素。对顶燃式热风炉燃烧器进行热态实测难度极大，尤其是测量速度分布，要求测量仪表耐1400℃以上的高温，并且不能干扰正常生产。而冷态测试要求的介质温度低，测试人员可直接在炉内进行测试，得到的数据真实反映出各喷口的气流分配状况。另外冷态测试可在投产之前进行，不会影响正常生产。

通过测试冷态工况下空气喷口和煤气喷口单独工作时各喷口的流速分布，描绘出空气上、下环，煤气上、下环的喷口速度变化图，同时引进统计学中的变异系数，以研究分析每环喷口速度分布与平均值的偏差程度，检验评价燃烧器气体分配均匀程度。

3 测试准备

冷态测试主要测试热风炉燃烧器煤气、空气喷口的气体流速，分析评价气流速度的均匀性。选用叶轮式风速仪，测速范围为 0~45m/s。还准备了高空安全带、安全绳、防尘面罩等劳保用具，对讲机等通信用具及数码相机等记录用具。

测试前，热风炉内已完成砌筑，施工用支架已拆除，因此需搭建测试用临时支架，并在空气、煤气喷口处设置临时平台，供测试使用。在首钢京唐钢铁公司炼铁作业部的大力协助下，测试现场的各项准备工作有序进行，完成了热风炉助燃风机和相关阀门等设备的安装调试，管道内部清理及封堵人孔等，确保了测试工作的顺利进行。

为准确得出空气、煤气每环喷口的速度分布，测试前对每个喷口编号，将测试结果与设计图对照，即可得出各喷口的速度分布规律。图1为首钢京唐BSK顶燃式热风炉燃烧器结构图，图2为燃烧器煤气、空气喷口编号图。

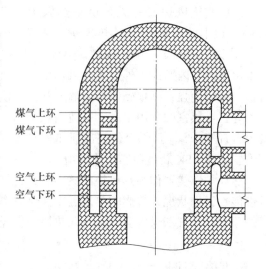

图1 BSK顶燃式热风炉燃烧器示意图

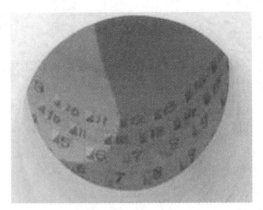

图 2 喷口编号示意图

4 空气喷口单工作状态冷态测试

首先测试空气喷口单工作状态下流速分布，介质为常温空气，开启 1 台助燃风机，空气流量约为 54000m³/h，压力 3200Pa，温度约 20℃，全部经空气喷口进入热风炉，利用叶轮式风速仪，对每个喷口的不同位置进行测试，取其平均值。空气上、下环的喷口速度分布如图 3 和图 4 所示。

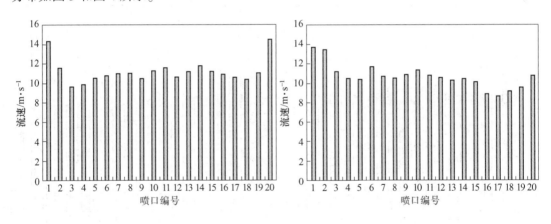

图 3 空气上环喷口速度分布　　　　图 4 空气下环喷口速度分布

空气上环的最高速度出现在正对空气支管上方的两个喷口 1 和 20，空气支管对面的喷口 10 和 11 速度也较高，但整体而言，喷口 2~19 的速度分布较为均匀，相差不大。

空气下环的最高速度出现在喷口 1、2 处，在喷口位置和旋转方向两个因素的综合影响下，这两个喷口的气流阻力最小。喷口 3~15 的速度基本维持在 10~12m/s 之间，分布比较均匀。受喷口位置和旋转方向的影响，喷口 16~19 的速度相对较小，但也在 9m/s 左右。喷口 20 的旋转方向虽与气体流向相反，但因靠近空气支管，速度也在 11m/s 左右。

空气上环基本是以空气支管中心线为界，呈对称分布；空气下环则呈非对称分布，但整体的速度分布都较为均匀。

引入统计学中的变异系数，计算各喷口流速与平均速度之间的偏差值，定义如下：

$$\mu = \frac{1}{\bar{v}}\sqrt{\frac{1}{n}\sum_{i=1}^{n}(v_i - \bar{v})^2}$$

式中 μ——速度的变异系数,数值越大,与平均速度的偏差越大,说明各喷口速度相差大,分布均匀性越差;

$\bar{v}$——喷口平均速度,m/s;

v_i——喷口 i 的速度值,m/s;

n——喷口数目。

将空气上、下环各喷口速度代入,计算得出:

空气上环,$\mu = 0.105$;

空气下环,$\mu = 0.114$。

空气上、下环的喷口速度变异系数都较小,说明虽然都有 1~2 个喷口速度偏高,但整体分布还是比较均匀的。

5 煤气喷口单工作状态冷态测试

测试煤气喷口单工作状态下流速分布,介质为常温空气,利用相邻热风炉将空气从空气管道接入煤气管道。同样开启一台助燃风机,空气流量为 54000m³/h,压力为 3200Pa,温度约 20℃,全部经煤气喷口进入热风炉。利用叶轮式风速仪,对每个喷口的不同位置进行测试,取其平均值。煤气上、下环的喷口速度分布如图 5 和图 6 所示。

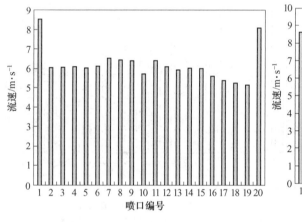
图 5 煤气上环喷口速度分布

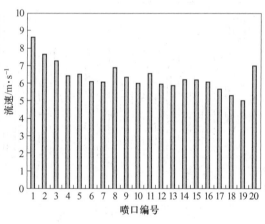
图 6 煤气下环喷口速度分布

虽然测试风量与单测空气喷口的风量相同,但气体管道路径长,泄漏和阻力损失大,因此煤气喷口速度整体偏低。

煤气上环的最高速度出现在正对煤气支管的两个喷口 1 和 20,大于 8m/s;喷口 2~19 的速度基本稳定在 5.5~6.5m/s 之间,分布比较均匀。

煤气下环的喷口 1 速度最大,它最靠近煤气支管,且气流阻力小。喷口 2、3 逐渐远离煤气支管,速度逐渐降低。喷口 4~20 的速度基本稳定在 5.5~6.5m/s 之间,分布比较均匀。

同样引入统计学中的变异系数,计算各喷口流速与平均速度之间的偏差值,将煤气上、下环各喷口速度代入,计算得出:

煤气上环，$\mu = 0.127$；
煤气下环，$\mu = 0.123$。

煤气上、下环的喷口速度变异系数都较小，说明虽然都有 2~3 个喷口速度偏高，但整体分布还是比较均匀的。

6 结论

(1) 对首钢京唐钢铁公司 5500m³ 超大型高炉配套 BSK 顶燃式热风炉燃烧器进行了空气、煤气喷口单工作状态冷态测试，取得了完整的燃烧器喷口速度分布数据；

(2) 空气上环以空气支管中心线为界，呈对称分布；空气下环呈非对称分布，除个别喷口外，整体速度分布均匀；

(3) 煤气上环以煤气支管中心线为界，基本呈对称分布；煤气下环呈非对称分布，除个别喷口外，整体速度分布均匀；

(4) 经计算，空气上、下环，煤气上、下环的喷口速度分布变异系数小，说明各喷口流速与平均速度相比偏差小，该燃烧器的整体喷口速度分布是比较均匀的，达到了热风炉燃烧器的性能要求。

致谢

参加此项工作的还有韩向东、梅丛华、戴建华、曹源、李林等，衷心感谢对本次测试工作给予大力支持的首钢京唐钢铁公司领导王毅、王涛、张卫东、周雁、宋静林等同志，感谢首钢京唐钢铁公司炼铁部热风炉技师苏殿昌和其他操作人员的大力配合。

参考文献

[1] 黄晋，林起礽. 首钢大型顶燃式热风炉设计 [J]. 首钢科技，1992 (2)：189.
[2] 张福明，毛庆武，等. 我国大型顶燃式热风炉技术进步 [J]. 设计通讯，2002，2：20-27.
[3] 项钟庸，王筱留，等. 高炉设计——炼铁工艺设计理论与实践 [M]. 北京：冶金工业出版社，2007：485-487.

Separate Cold-State Testing and Research of BSK Top Combustion Hot Air Stove Burner of Shougang Jingtang 5500m³ Blast Furnace

Zhang Fuming Li Xin Mao Qingwu Qian Shichong
Yin Guangyu Ni Ping

Abstract: A series of internationally advanced technologies have been adopted in the 5500m³ blast furnace of Shougang Jingtang Iron and Steel Corporation, including the successful installation of the new model BSK top combustion hot air stove in the ultra-large 5500m³ blast furnace ever for the first time. In order to acquire a thorough understanding of the flow field distribution in the hot air stove and verify and evaluate the distribution evenness of gases in the burner, Beijing Shougang International Engineering Technology Co, Ltd, together with Ironmaking Department of Shougang Jingtang Iron and Steel Corporation, have conducted a cold-state testing to the BSK top combustion hot air stove of the 5500m³ blast furnace of Jingtang. In the paper a summary of the single testing of the air port and that of the gas port is made, and the testing data are analyzed and studied to arrive at the velocity distribution pattern of the air port and the gas port. The statistics and calculation results show that the overall port velocity distribution of the burner is fairly even and reaches the performance requirements of hot air stove burner.

Keywords: large blast furnace; hot air stove; burner; test; flow field

高效长寿热风炉格子砖传热研究

摘　要：本文结合首钢京唐5500m³高炉特大型顶燃式热风炉，论述了顶燃式热风炉的技术优势和蓄热室传热的技术特征，研究了顶燃式热风炉传热理论，重点解析了格子砖的传热过程。通过建立蓄热室二维传热计算数学模型，对不同砖型的情况进行了计算，结果表明：减小格孔直径能提高蓄热室的热效率，合理的硅砖层厚度能延长送风时间，增加送风温度可调节范围，保护硅砖安全高效的运行。

关键词：顶燃式热风炉；格子砖；传热理论；硅砖层厚度

1　引言

近年来，随着顶燃式热风炉技术日臻完善，它在经济和技术上的优势逐渐体现出来。与内燃式、外燃式热风炉相比，顶燃式热风炉所具备的优势得到了广泛的认同。首钢、莱钢、济钢、天钢、唐钢等相继在1200~3200m³高炉上推广采用顶燃式热风炉技术，京唐钢铁厂将BSK顶燃式热风炉推广到5500m³的特大型高炉上，取得了顶燃式热风炉应用的新突破。

在燃烧期和送风期，顶燃式热风炉蓄热室的传热过程相对内燃式和外燃式热风炉都有着很明显的技术优势：（1）由于取消了燃烧室和隔墙，扩大了蓄热室容积，在相同占地条件下，蓄热面积增加了25%~30%[1]；（2）蓄热室布置在燃烧室的正下方，可保证烟气均匀进入格子砖，据测定其不均匀度为±3%~5%，大大提高了热风炉蓄热室的利用率，并使得温度沿拱顶、格子砖、内衬和炉壳均匀分布，减少了温差热应力，提高了热风炉的寿命。

首钢京唐5500m³特大型高炉配备的顶燃式热风炉采用结构独特的19孔薄壁小孔格子砖[2]。小孔格子砖的孔径为30mm，有柱型孔道和锥型孔道两种。柱型孔道的比表面积为48m²/m³，锥型孔道格子砖的比表面积为48.7m²/m³，均大于七孔格子砖的比表面积（38.06m²/m³）。因此，在蓄热室面积不变的前提下，采用19孔格子砖比采用七孔格子砖可增加蓄热面积26%左右。另外，由于砌筑后格子砖通道为宝塔形，可降低气体流动阻力，提高格子砖的传热效率。

本文利用蓄热室传热计算模型，通过编程计算探讨不同格子砖砖型和不同硅砖层厚度对蓄热室传热过程的影响。

本文作者：张福明，胡祖瑞，程树森，毛庆武，钱世崇。原刊于《2010年全国炼铁生产技术会议暨炼铁学术年会文集（上）》，2010，693-697。

2 蓄热室传热计算模型

2.1 物理模型

首钢京唐 5500m³ BSK 顶燃式热风炉采用 19 孔圆孔格子砖,每块砖为正六边形。表 1 是首钢京唐 5500m³ BSK 顶燃式热风炉蓄热室设计结构参数,其蓄热室格子砖总共分成 3 段,采用 3 种不同材质,每段格子砖的分布状况和材质参数见表 2。

表 1 首钢京唐 5500m³ BSK 顶燃式热风炉蓄热室设计参数

热风炉座数/座	蓄热室高度/m	热风炉蓄热室内径/m	格子砖加热面积/m²	活面积
4	21.6	10.9	95885	0.36

表 2 格子砖分布状况与物性参数

层数	高度/m	材质	C_p/J·(kg·K)⁻¹	密度/t·m⁻³	导热系数/W·(m·K)⁻¹
第一层	0.6	低蠕变黏土质	$836.8+0.234T$	2.1	$0.84+0.00052T$
第二层	10.56	高密度黏土质	$836.8+0.234T$	2.15	$0.84+0.00052T$
第三层	10.44	硅质	$794.0+0.251T$	1.8	$0.93+0.0007T$

对热风炉蓄热室的研究有多种模型,本文采用的方法是将截面形状复杂的格子砖通道简化为具有一定厚度的管道,将整个蓄热室视为由有限根管道组成的空间[3,4],以单个格子砖通道作为研究对象,建立模型,如图 1(a)所示。气体通道的直径为格孔的直径,格子砖的厚度可由整个蓄热室的活面积(又称有效通道截面积)计算得出。

由 $\dfrac{\pi d^2}{4} = \dfrac{\pi D^2}{4} \times s$,求得:$D = 0.05$ m

式中,d 为格子砖的内直径;D 为格子砖的外直径;s 为活面积。

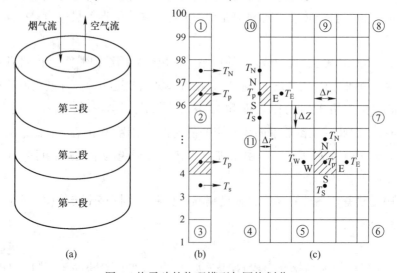

图 1 格子砖的物理模型与网格划分

2.2 数学模型

由于实际热风炉蓄热室中气体流动和传热的复杂性，数值模拟中常常需要通过一些假设来简化模型，本文的计算将采用如下基本假设[5]：忽略换炉所用时间，即上个周期结束后马上进入下个周期；蓄热室中同一高度上气流分布均匀，气体管道传热过程相同，各气体管道之间绝热；气体的密度是温度和压力的函数，格子砖的热物性仅是温度的函数；气体速度和温度在同一高度上分布均匀；在燃烧期和送风期两个时期，气体入口温度和气体流量保持稳定。在模型假设的基础上，气体的温度场计算采用一维非稳态对流传热方程[3]，格子砖采用二维非稳态传热方程[3]，对气体与格子砖的温度进行耦合求解，具体如下：

气体一维非稳态对流传热控制方程：

$$\rho_g C_{p,g}\left(\frac{\partial T_g}{\partial t}+v_g\frac{\partial T_g}{\partial z}\right)+v_g\frac{\partial P}{\partial z}=\frac{4\alpha}{d}(T_w-T_g)$$

格子砖内部二维非稳态传热控制方程：

$$\rho_s C_{p,s}\frac{\partial T_s}{\partial t}-\frac{1}{r}\frac{\partial}{\partial r}\left(r\lambda\frac{\partial T_s}{\partial r}\right)-\frac{\partial}{\partial z}\left(\lambda\frac{\partial T}{\partial z}\right)=0$$

式中，ρ_g、ρ_s 为气体和格子砖的密度；$C_{p,g}$、$C_{p,s}$ 为气体和格子砖的比热容；v_g 为气体的流速；P 为气体压力；λ 为格子砖导热系数；α 为对流换热系数。

按图 1(b)、(c) 所示分别对气体和格子砖进行网格划分，然后采用整体离散、整体求解的耦合方法进行计算[6]，整个过程编写成一套计算程序。

3 不同砖型的传热特性比较

3.1 计算方案

为了比较不同格子砖砖型传热性能的优劣，选择了几种孔径的格子砖，其主要热工参数见表3。应用所开发的程序，分别将上述四种格子砖的参数值输入程序，计算得到了每一种工况的温度场和蓄热量。

表3 不同类型格子砖的热工参数

项目	水力学直径/m	活面积	填充系数/$m^3 \cdot m^{-3}$	受热面积/$m^2 \cdot m^{-3}$	当量厚度/m
五孔砖	0.0583	0.434	0.566	29.78	0.038
七孔砖	0.043	0.409	0.591	38.05	0.031
高效19孔砖	0.03	0.398	0.602	53.07	0.023
高效37孔砖	0.02	0.354	0.646	70.80	0.018

3.2 计算结果及讨论

图2为燃烧期的烟气出口处的温度随时间的变化曲线。图3为送风期的空气出口处的温度随时间的变化曲线。从图2和图3可以看出，燃烧期烟气出口处的温度随时间呈增大

趋势，送风期的空气出口处的温度随时间呈减小趋势，而且格子砖孔数越多，出口处烟气的温度越低，鼓风的温度越高。图 4 为送风期出口处鼓风的温降随时间的变化曲线。从图中可以看出，孔数越多的砖型，鼓风的温降越小，变化也越平稳。

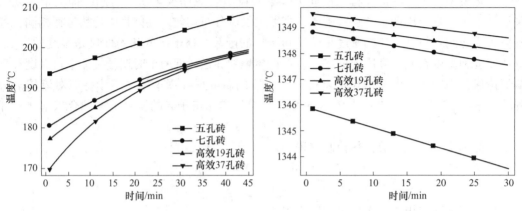

图 2　出口处烟气的温度随时间的变化　　　　图 3　出口处空气的温度随时间的变化

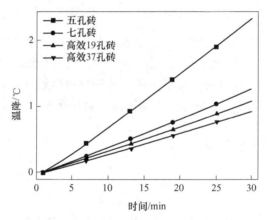

图 4　送风期鼓风的温降随时间的变化

从四种不同砖型格子砖的对比可以看出，在蓄热室结构尺寸、切换周期、热风流量、燃料流量均相同的条件下，孔数越多的格子砖由于孔径减小，换热面积增加、换热系数增大、热交换量增加，在燃烧期烟气温度能够更多地传递给蓄热体，使得排烟温度更低，而送风期的热风温度则更高，热风的温降也更平缓，风温更稳定。因此，在一定程度上，减小格孔能够提高蓄热室的热效率，但是减小格孔会导致气体通过格孔通道的阻力增大，气体压力损失增大，耗费更多的鼓风动能，并且小格孔还比较容易被堵塞，导致蓄热室利用率下降，因此对于砖型的选择并不是孔数越多越有利，要综合考虑蓄热室热效率、蓄热室利用率和使用寿命等各种因素的影响。

4　硅砖层厚度优化研究

蓄热室总高度为 21.6m，整个高度方向上温差很大。因此将格子砖分为 3 段，每段采用不同材质的格子砖，具体参数见表2。其中硅砖中石英晶体占95%以上，石英晶体转变中热膨胀率的变化如图5所示，当硅砖中石英晶体的温度位于870~1470℃区间时，硅砖的热膨胀率稳定在5.4%，而当温度低于870℃时，硅砖的热膨胀率迅速减小为0.5%。因此如果硅砖的工作温度长期在870℃上下波动，则会导致硅砖的破损、渣化，由于硅砖在蓄热室最上部，渣化的硅砖还会堵塞格子砖的格孔，对整个蓄热室都造成严重的破坏，因此保证硅砖一直在其允许的工况范围内工作，对于保证格子砖的长寿和高效是非常重要的。

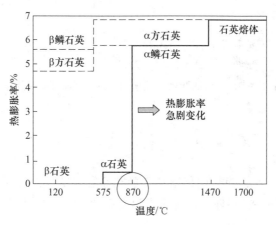

图5 石英晶体转变中的热膨胀率

4.1 计算方案

送风时，把最底部的硅砖温度恰好高于870℃时的热风温度定义为热风临界温度，设为T_0。热风临界温度为该工况下能接受的最高热风温度，如果此时继续送风，最底部的硅砖温度将低于870℃，对硅砖产生危险，因此热风临界温度越高，意味着热风炉允许的温度变化范围越窄，送风也就时间越短。因此尽可能地减小热风临界温度有利于提高延长送风时间，减少换炉次数，同时也能延长格子砖的寿命。采用不同拱顶温度加热热风炉，使燃烧期结束时烟气出口温度为450℃，然后计算送风过程中，当分别将硅砖层的厚度减少1m、2m，相应的将高密度格子砖的厚度增加1m、2m，相当于在整体蓄热室高度不变的情况下，将硅砖底部的标高提升了1m、2m。此时的热风临界温度设为T_1、T_2。本论文对三种不同硅砖层厚度下的热风临界温度T_0、T_1、T_2进行了计算。

4.2 计算结果及分析

图6给出了不同硅砖层厚度下热风临界温度随拱顶温度的变化，图中热风临界温度和拱顶温度基本呈线性规律变化，相同工况下，硅砖层厚度越高，热风临界温度越低。从图7能明显看到硅砖层厚度的增加使得热风可变温度范围大幅度提高，因此将硅砖层厚度减少2m，对于延长送风时间，增加送风温度可调节范围，保护硅砖安全高效地运行都大有益处。

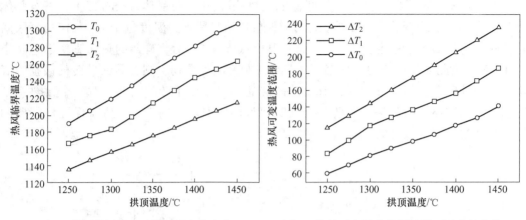

图 6 热风临界温度随拱顶温度的变化　　图 7 热风可变温度范围随拱顶温度的变化

5 结论

（1）格子砖孔径减小，换热面积增加、换热系数增大、热交换量增加，在燃烧期烟气温度能够更多地传递给蓄热体，使得排烟温度更低。而且送风期的热风温度会更高，热风的温降也更平缓，风温更稳定。

（2）对于砖型的选择并不是孔数越多越有利，要综合考虑蓄热室热效率、蓄热室利用率和使用寿命等各种因素的影响。

（3）通过对热风临界温度的计算发现，对于首钢京唐 $5500m^3$ 高炉所配置的顶燃式热风炉，当硅砖层厚度设计为 8.5m 左右时能使热风可变温度的范围大幅度提高，这对于适当延长送风时间，增加送风温度可调节范围，保护硅砖安全高效的运行都十分有益。

致谢

本研究受到国家科技支撑计划项目"新一代可循环钢铁流程工艺技术——长寿集约型冶金煤气干法除尘技术的开发"（2006BAE03A10）的支持。

参考文献

[1] 张福明. 大型顶燃式热风炉的技术进步 [J]. 炼铁，2002，21（5）：5-10.
[2] 郭敏雷. 顶燃式热风炉传热及气体燃烧的数值模拟 [D]. 北京：北京科技大学，2008.
[3] Muske K R, Howse J W, Hausen G A, et al. Model-based control of a thermal regenerator. Part 1：Dynamic model [J]. Computers & Chemical Engineer，2000，24：2519-2531.
[4] MUSKE K R, HOWSE J W, HANSEN G A, et al. Hot blast stove process model and model-based controller [J]. Journal of Iron and Steel Institute，1999，76（6）：56-62.
[5] 郭敏雷，程树森，张福明，等. 热风炉格子砖传热计算模拟软件开发及应用 [C]//中国钢铁年会论文集，成都，2008：822.
[6] 陶文铨. 数值传热学 [M]. 西安：西安交通大学出版社，2001（2）：485.

Heat Transfer Research on Checkers in High Efficiency and Long Campaign Life Blast Stove

Zhang Fuming   Hu Zurui   Cheng Shusen   Mao Qingwu   Qian Shichong

Abstract: The technology advantage and heat transfer characteristics in the regenerator are introduced in this paper with the oversized top combustion hot blast stove equipped in Shougang Jingtang 5500m^3 BF. Heat transfer theory is discussed here and analyses the process in the checkers. Built 2-dimension model for heat transfer and calculated under different brick shape. The result shows the diameter decrease can help improve the heat transfer efficiency. The reasonable thickness of the silica brick can prolong the gas time and enlarge the temperature period of gas to make sure the smooth status of silica brick.

Keywords: hot blast stove; checkers; heat transfer theory; height of silica brick

霍戈文内燃式热风炉传输现象的研究

摘　要：利用CFD仿真对首钢1780m³霍戈文内燃式热风炉进行了数值模拟研究，主要研究了空气和煤气混合前在矩形燃烧器中的流动，混合气体在燃烧室的燃烧、含量分布、温度分布、火焰形状以及拱顶的速度分布和温度分布。结果表明：空气喷嘴出口截面和煤气出口截面的流场都存在一定程度的不均匀性；沿燃烧室宽度方向，火焰高度变化剧烈；拱顶出口截面残余极少量的一氧化碳，蓄热室表面烟气速度分布不均，最高温差也很大。

关键词：霍戈文内燃式热风炉；流动；传热；燃烧；数值模拟

1　引言

热风炉是高炉炼铁生产中的重要设备，在炼铁过程中，提高热风炉送风温度，是降低焦比、增加生铁产量以及降低生铁成本的有效措施，也是提高高炉喷吹燃料的重要条件。因此提高热风温度是目前热风炉研究的焦点问题。提高风温的主要措施包括两点：一是提高理论燃烧温度，通过降低空气过剩系数，对空气、煤气进行预热可以达到理想的理论燃烧温度；二是降低理论燃烧温度与风温的差值，这需要在热风炉的结构上进行优化，提高蓄热室的蓄热面积的同时，使燃烧室出口截面获得均匀的速度分布和温度分布。另外，热风炉炉内的最高温度在1400℃以上，因此合理的火焰形状和温度分布是热风炉长寿的关键因素。

郭敏雷、程树森等对燃烧期[1]和送风期[2]格子砖温度分布进行了计算研究，但对燃烧器、燃烧室以及拱顶部分的研究却没有涉及。胡日君、程树森对考贝式热风炉拱顶空间烟气分布进行过数值模拟[3]，计算主要建立在拱顶入口截面速度分布均匀的假设之上，与实际情况有一定的偏差。本文中的流场计算从空气和煤气入口开始，与实际更加接近，计算更加准确，加上霍戈文式热风炉独特的悬链线式拱顶，都是对拱顶烟气分布研究的补充和完善。张胤、贺友多等分别对栅格式和套筒式陶瓷燃烧器的燃烧过程进行了模拟研究[4,5]，对矩形陶瓷燃烧器的计算没有涉及。而且主要研究的位置局限于燃烧室，对空气和煤气混合前在各自通道内的流动状况、出口截面的速度分布情况未作分析。

本文针对首钢1780m³高炉所配置的霍戈文内燃式热风炉，用计算流体力学的方法，采用稳态计算，对其燃烧期中后期达到稳定后的速度分布、温度分布以及火焰形状进行了数值模拟研究，并对提高风温和延长热风炉寿命给出了建议。

本文作者：胡祖瑞，程树森，张福明。原刊于《北京科技大学学报》，2010，32（8）：1053-1059。

2 物理模型

内燃式热风炉是目前国内使用最多的热风炉，首钢1780m³高炉所配置的霍戈文内燃式热风炉在传统的内燃式热风炉的基础上进行了改进。其中特有的结构包括：（1）矩形陶瓷燃烧器与眼睛形火井；（2）全脱开悬链线形拱顶与关节砖的设计；（3）板块式结构与分层自立式结构；（4）自密闭锁砖结构等。该热风炉主要的结构尺寸有：热风炉的总高度为41.6m，燃烧室断面面积为9.7m²，蓄热室断面面积为35.8m²，煤气和空气管道的直径分别为1.7m、1.8m；煤气管道底部有一尺寸为4000mm×3060mm×150mm的导流板。空气喷嘴的出口截面为336mm×100mm的矩形，与竖直方向的夹角为37.275°；左右两侧每侧有喷嘴22个，共44个，每个喷嘴之间间隔100mm；眼睛形燃烧室的圆弧半径为4.2m，高度为23.507m；拱顶底面半径为4.621m。

3 数学模型及边界条件

3.1 数学模型

首钢1780m³高炉所配置的霍戈文内燃式热风炉，助燃空气和煤气分别从两个入口进入热风炉中混合燃烧，属于非预混燃烧，解决这一过程目前最常用的数学模型是 $k\text{-}\varepsilon\text{-}g$ 模型。综合来说该模型将混合过程的控制作用和脉动的影响有机地统一，主要包括5个要点[6]，这里不作详细叙述。

3.2 边界条件

该热风炉对应的高炉设计风量为4200m³/min，燃烧所用高炉煤气成分见表1。

该热风炉采用助燃空气和煤气预热技术，其煤气和空气管道入口处2种气体的相关参数见表2。

表1 高炉煤气成分

组　分	体积分数/%	密度/kg·m⁻³	质量分数/%
CO	20.33	1.25	19.46
CO_2	16.63	1.97	25.09
N_2	50.46	1.2507	48.33
CH_4	0.44	0.714	0.24
H_2	1.16	0.0899	0.08
H_2O	10.98	0.81	6.80

表2 空气和煤气相关参数

气体	流量/m³·h⁻¹	质量流量/kg·s⁻¹	预热温度/℃
空气	56190	20.14	600
煤气	73840	26.66	200

表1和表2中的数据均为设计值，实际生产中的数据与其略有偏差。根据表2中的数据给出数学模型所对应的边界条件，煤气和空气入口均为质量流量边界，空气入口的质量流量为20.14kg/s，湍流强度为10%，温度600℃，平均混合质量分数和平均混合质量分数均方差都为0；煤气入口的质量流量为26.66kg/s，湍流强度为10%，温度200℃，平均混合质量分数为1，平均混合质量分数均方差为0。出口设为压力出口边界，压强为一个大气压即表压为0，设定回流湍流强度为10%，平均混合质量分数和平均混合质量分数均方差均为0。壁面设置为非滑移边界，传热采用第二类边界条件且壁面绝热。流体密度采用PDF（概率密度函数）混合物模型给出，比热容采用混合定律给出。辐射模型为Discrete Ordinates 模型[7]。

4 计算结果及分析

应用CFD仿真计算软件，对该热风炉燃烧室内气体的湍流扩散燃烧过程进行了深入而且系统的研究，得到了煤气和助燃空气混合前，在各自通道内的速度分布情况，以及燃烧室和拱顶烟气的速度场、温度场、含量分布和火焰的形状，对热风炉的设计以及改进提供了重要的依据。

4.1 煤气在矩形燃烧器中的流动状况

图1为矩形燃烧器的示意图，煤气通道的中心位置有一块长4m、宽0.15m、高3m的挡墙。挡墙上端是一小段收缩段，煤气通道的面积由3.3m²收缩到1.98m²。煤气从下部入口进入燃烧器，图2分别给出了煤气入口中心截面，空气入口中心截面和空气喷嘴底部截面的速度云图。从图中可以看出，煤气的速度基本沿入口中心线呈对称分布，随着高度的增加，横截面的速度分布总体趋于均匀化。图3和图4分别是煤气出口截面Y方向和X向上速度变化曲线图。从图3可以看出，Y方向上壁面附近的速度明显高于中间的速度。

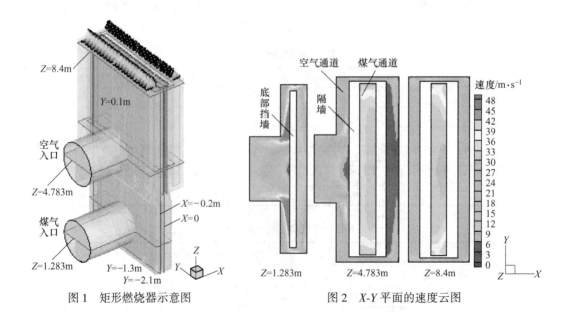

图1 矩形燃烧器示意图　　　　图2 X-Y平面的速度云图

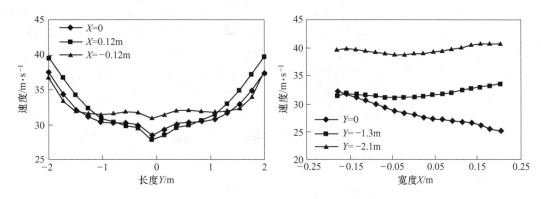

图 3　煤气出口截面 Y 方向的速度变化　　　图 4　煤气出口截面 X 方向的速度变化

其原因可从图 5 分析得出，煤气遇到挡墙后向各个方向放射状运动，挡墙两个角部的速度要明显高于中间位置的速度，这样的情形一直保持到煤气与空气混合，因此会出现图 5 所示的速度分布。从图 4 中可以看到靠近中心位置时（$Y=0$），入口一侧的速度要明显高于入口对侧的速度；而靠近壁面处（$Y=-2.1m$），入口一侧和入口对侧的速度差异则很小。

图 6 将 $Y=0.1m$ 和 $Y=-1.3m$ 平面收缩段附近的速度矢量图进行了对比：在 $Y=0.1m$ 平面，入口管道和挡墙的拐角处速度很大，随之通过收缩段后，靠近左端壁面的速度要远大于右端壁面的速度；而随着离中心的增加，对于 $Y=-1.3m$ 平面，收缩段上部左右两端壁面的速度差异则明显缩小。因此挡墙的结构，收缩段角度和高度以及煤气通道的高度存在一定的不合理性，导致了在煤气出口截面存在着一定程度的不均匀性。

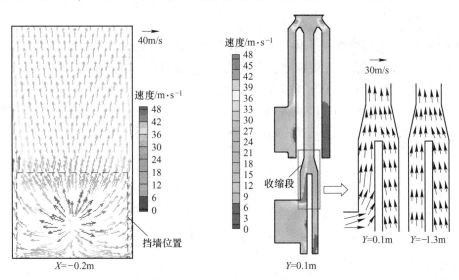

图 5　$X=-0.2m$ 平面的速度矢量图　　　图 6　X-Z 平面流动状况

4.2　空气在矩形燃烧器中的流动状况

空气从上部入口进入燃烧器后沿"回"形通道运动。图 2 中右侧的 2 张图说明整个空气通道内的速度基本沿入口中心线呈对称分布。在入口平面 $Z=4.783m$ 处，入口一侧的速度要远大于入口对侧的速度,随着高度的增加,在空气喷嘴底部,两侧的速度已基本达到均匀。因此空气通道的高度直接决定了各截面两侧速度的均匀性,也就决定了喷嘴出口截面速度的均匀性。图7给出了入口一侧和对侧各喷嘴的量纲1流量柱状图,其中量纲1流量是各喷嘴的流量与所有喷嘴平均流量的比值。实际每侧有22个喷嘴,由于流场沿X-Z平面对称,因此只给出了一半喷嘴的结果,编号从边缘到中心依次增大。结果显示:空气入口一侧的喷嘴流量全部大于平均流量,边缘附近的流量大于中心流量;而入口对侧的喷嘴除中心一个喷嘴等于平均流量外,其余全小于平均流量,而且靠近壁面的2个喷嘴流量远低于平均流量。由此看来,空气通道的高度还需要略微增大,使得每个空气喷嘴的流量尽可能均匀。

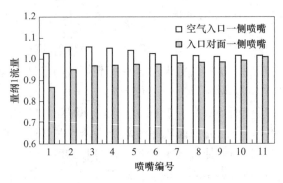

图7 各喷嘴的量纲1流量

4.3 燃烧室气体流动状况

图8给出了眼睛形燃烧室$Y=0$平面的速度矢量图。图8(a)是燃烧室底部的流场矢量图,从图中可以看到在靠近壁面位置形成了一个回旋区,这是由燃烧器的流体流动特性造成的,回旋区的存在有利于燃烧过程煤气与空气的混合,稳定燃烧以及烟气平稳的向上运动。图8(b)是燃烧室出口附近的流场矢量图,当烟气平稳上升至燃烧室出口附近时,气体分布基本达到均匀,中间的速度略高于两边,此部分结果与张胤、贺友多等的计算结果十分一致[4,8]。

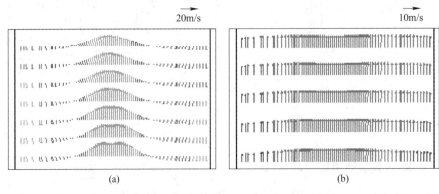

图8 燃烧室$Y=0$平面的速度矢量图
(a)燃烧室底部;(b)燃烧室出口附近

4.4 拱顶速度分布

气体经过燃烧室后达到拱顶,通过拱顶中心 X-Z 截面的速度云图和流线图如图 9(a)所示,云图标示了速度的大小,流线图标示了速度的方向。拱顶下部形成了一个半径约为 1.2m 的涡流,使得该处气体的速度很小,而靠近壁面处气体速度则相对较大。图 9(b)给出了烟气在拱顶出口截面的分布状况:速度分布基本沿 X 轴对称分布,但由于涡流的存在,右半部分烟气的速度要明显大于左半部分的烟气速度。这种速度的不均匀性是由拱顶的结构特点所决定的,虽然悬链线拱顶相对圆球形拱顶能减小这种不均匀性[3,9],但想要完全消除还需在结构上进行改进。

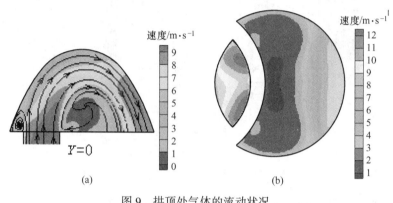

图 9 拱顶处气体的流动状况
(a) $Y=0$;(b) 拱顶出口横截面

4.5 燃烧室温度分布和火焰形状分析

图 10 为截止至拱顶底部出口,X-Z 平面和 Y-Z 平面的温度分布云图和火焰形状。由当量混合分数[5]可以得到如图中蓝色曲线所示的平均火焰面形状[10]。煤气和助燃空气在眼睛形燃烧室混合并发生剧烈燃烧,放出大量热量。如图 10(a)所示,$X=0$ 平面内温度分布和火焰形状基本关于 $Y=0$ 平面对称,火焰面呈现中间低,两边高的状态,这与燃烧室气体流动状态密切相关,图 8 中眼睛形燃烧室中心位置(即 $Y=0$ 平面)形成的回流,使得通过中心区域总的煤气量要小于边缘,因此中心火焰短于两侧。对于温度分布情况,由于火焰内部的主要成分是煤气,没有得到充分燃烧,因此温度较低;而在火焰表面附近,煤气充分燃烧,达到最高温度 1440℃,这与该热风炉所设计达到的峰温相吻合。另外需要说明的是,由于眼睛形燃烧室在划分六面体网格时出现不对称的状况,从而导致计算结果不是完全关于 $Y=0$ 对称,因此在不规则图形网格划分方面需要做细致的优化。图 10(b)、(c)、(d)选取了最短火焰和最长火焰位置的 X-Z 平面温度分布云图和火焰形状。各平面火焰宽度基本一致,与煤气通道出口的宽度很接近;最短的火焰长度为 8m 左右,最长的火焰长度为 28m 左右,而且已经伸入到了拱顶部分。对于拱顶的温度分布,$Y=0$ 平面处,温度分布与速度分布情况刚好相反,在涡流位置,由于与周围发生较少热交换导致温度热量积累,温度比靠近壁面处高出很多;相对 $Y=0$ 平面,$Y=-1.5$m 和 $Y=1.5$m 平面的平均温度显然要高出很多。

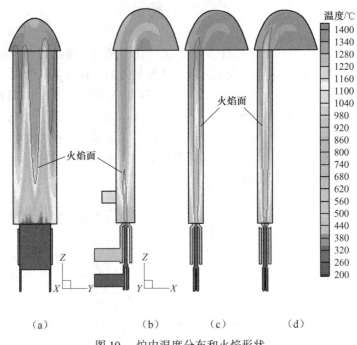

图 10　炉内温度分布和火焰形状

(a) $X=0$；(b) $Y=0$；(c) $Y=-1.5m$；(d) $Y=1.5m$

图 11 左侧给出了眼睛形燃烧室横截面上，温度随高度变化的云图。随着高度的增加，温度在整个横截面上变得均匀。图 11 右边给出了拱顶出口截面的温度分布，温度分布沿 X 轴对称分布；根据温度的大小能将整个出口截面分成 5 块区域，温度最低的区域集中在 X 轴附近，靠近壁面处；温度最高的区域集中在温度最低区域的上下两侧；剩下的区域温度较为均匀，而且介于最高温度区域和最低温度区域之间。另外整个截面的最大温差在 250℃ 左右，如此大的温差对提高热风温度和蓄热室的热效率都会产生不利的影响。因此很有必要对燃烧室和拱顶的结构进行改进。

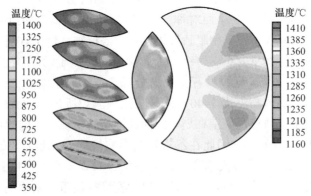

图 11　燃烧室和拱顶出口截面温度分布

4.6　燃烧室和拱顶含量分布

图 12 左侧给出了燃烧室高度方向 CO 的质量分数随高度变化的云图。在燃烧室底部，

煤气通道出口对应的位置CO的质量分数很高；随着高度的增加，CO含量逐渐减小，而且CO含量高的区域主要集中在眼睛形燃烧室的角部，这与燃烧室的温度分布和火焰形状也能很好地对应上；同时可以看到在燃烧室出口处仍然有CO剩余。但随着煤气和助燃空气在拱顶左侧的继续反应，会看到如图12右侧所示的拱顶CO含量分布云图，在拱顶右侧以及拱顶出口附近CO的质量分数已基本接近于零，避免了CO进入格子砖燃烧。

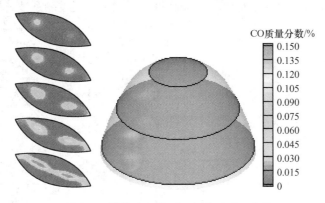

图12　燃烧室和拱顶一氧化碳质量分数

5　计算结果的验证

对首钢霍戈文内燃式热风炉2008年8月11日的生产数据进行了采样计算，具体工况参数见表3。实际生产中对拱顶处的平均温度和烟道中CO的残余量保持实时监测，通过计算拱顶截面所有结点温度和CO含量的平均值得到相应的平均值，由于蓄热室中CO含量不发生变化，因此拱顶截面CO的平均含量与烟道中一致。图13和图14将5种工况下拱顶平均温度、CO残余质量分数的监测值和计算值进行了比较，从图中可以看出通过数值模拟计算出的结果与实际测量值非常接近，最大误差在1%以内。因此能证明采用的数学模型比较准确，计算结果对指导生产有很大的实际意义，对下一步内燃式热风炉的改进工作研究奠定了基础。

表3　空气和煤气相关参数

工况	空气流量 /m³·h⁻¹	煤气流量 /m³·h⁻¹	空气温度/℃	煤气温度/℃	空气过剩系数 α
1	43574	64918	507	180	1.2120
2	41584	63356	498	180	1.1856
3	42472	65759	500	180	1.1670
4	41810	66012	501	180	1.1440
5	46248	73934	500	180	1.1300

表3中的空气过剩系数定义为实际助燃空气流量与理论所需量的比值，即

$$\alpha = V_1/V_0$$

其中，单位时间煤气完全燃烧所需要空气量的理论值计算如下：

$$V_0 = V_{gas} \times (V_{CO} \times 0.5 + V_{CH_4} \times 2 + V_{H_2} \times 0.5)/21\%$$

从图 13 和图 14 中不难发现，随着空气过剩系数的减小，拱顶温度升高，CO 残余量增大。当空气与煤气恰好完全燃烧时，烟气能达到的温度最高，而当空气过量之后，烟气温度则会随着空气量的增大而减小。由于燃烧器的结构复杂，需要有一定量的空气过剩来保证煤气的充分燃烧，同时为了尽可能的提高烟气温度，空气过剩系数一般为 1.1~1.3。

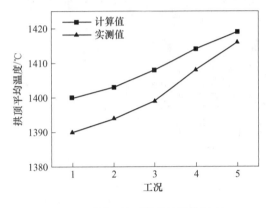

图 13　不同工况下拱顶的平均温度

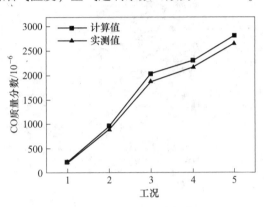

图 14　不同工况下一氧化碳残余质量分数

6　结论

（1）煤气通道出口处的速度分布情况为：Y 方向上中间的速度明显低于壁面附近的速度；X 方向上，靠近中心位置处入口一侧的速度要明显大于入口对侧的速度，而越靠近壁面，入口侧和入口对侧的速度差异则越小。挡墙结构，收缩段高度和角度以及煤气通道的高度存在一定的不合理性。

（2）空气通道各喷嘴的流量分布情况为：空气入口一侧的喷嘴流量全部大于平均流量，边缘附近的流量大于中心流量；而入口对侧的喷嘴除中心一个喷嘴等于平均流量外，其余全小于平均流量，而且靠近壁面的两个喷嘴流量远低于平均流量。因此空气通道的高度还需要略微增大，使得每个空气喷嘴的流量尽可能均匀。

（3）燃烧室的流场状况为：燃烧室底部靠近壁面位置形成了一个回旋区，有利于燃烧过程煤气与空气的混合和稳定燃烧。燃烧室出口附近，中间的速度略高于两边，基本达到均匀。

（4）拱顶的流场状况为：在拱顶空间一半高度靠近壁面的位置形成了两个对称的涡流，距离燃烧室中心约 2m 的区域形成了半径约为 1.2m 的涡流。在拱顶出口截面，从靠近右侧壁面到隔墙，烟气速度迅速减小。整个截面的速度分布很不均匀，因此拱顶的结构还有待改进。

（5）煤气和助燃空气燃烧较为充分，峰温的计算值与设计值基本吻合，燃烧室出口截面基本没有 CO 剩余，但温度分布较为不均匀，火焰长度在纵切面上波动很大，局部火焰长度太长，对拱顶耐火材料造成较大危害。其主要原因是由于煤气和空气通道出口处的速度分布不够合理。

参考文献

[1] 郭敏雷,程树森,张福明,等.热风炉燃烧期格子砖温度分布计算[C]//中国钢铁年会论文集.成都,2007:89.

[2] 郭敏雷,程树森,张福明,等.热风炉送风期格子砖温度分布计算[J].钢铁,2008,43(6):15.

[3] 胡日君,程树森.考贝式热风炉拱顶空间烟气分布的数值模拟[J].北京科技大学学报,2006,27(4):71.

[4] 张胤,贺友多,黄晓煜,等.新型栅格式陶瓷燃烧器燃烧过程数学模型研究[J].包头钢铁学院学报,2001,20(2):101.

[5] 张胤,贺友多,李士琦,等.预热对陶瓷燃烧器燃烧过程的影响[J].燃烧科学与技术,2001,7(3):267.

[6] 王应时,范维澄,周力行,等.燃烧过程数值计算[M].北京:科学出版社,1986:72.

[7] 陈冠军,胡雄光,钱凯,等.新型顶燃式热风炉燃烧技术的研究[J].钢铁,2009,44(1):79.

[8] 陈义胜,贺友多,贺真,等.热风炉眼睛形燃烧室内流动特性模拟研究[J],包头钢铁学院学报,2006,25(6):102.

[9] 唐兴智.高温长寿型内燃式热风炉的应用[J].工业加热,2007,36(5):47.

[10] 胡祖瑞,程树森,郭喜斌,等.宣钢旋流顶燃式热风炉传输现象的研究[C]//第13届冶金反应工程学会议论文集.包头,2009:243.

Transport Phenomenon of Hoogovens Internal Hot Blast Stove

Hu Zurui Cheng Shusen Zhang Fuming

Abstract: Simulation research was carried out on the 1780m^3 Hoogovens internal hot blast stove in Shougang using the CFD method. The main research contents mainly include the flow status of air and gas in the rectangular burners before mixing, combustion status of mixture gas, concentration and temperature distribution in the combustion chamber, flame shape, and the velocity and temperature distributions in the dome. The result shows that there is heterogeneity in some degree about flow filed in air nozzle outlet and gas outlet. Along the width direction of the combustion chamber, the flame height changes greatly. A little carbon monoxide remains in outlet of dome. On the surface of checker chamber, the velocity distribution of smoke is not uniform, and the maximum temperature difference is also very large.

Keywords: hot blast stove; flow; heat transfer; combustion; numerical simulation

热风炉格子砖活面积优化设计

摘　要：为了优化热风炉格子砖的活面积设计，通过二维传热数学模型对热风炉操作过程中格子砖传热现象进行了模拟，研究了混风比例、活面积等参数对格子砖传热性能的影响。结果表明：对于确定工作状态下的热风炉，存在一个使送风期风温保持平稳的最少总混风比例，相应平均风温为该工作状态下热风炉最高送风温度。对于单一孔径格子砖，随着活面积的增大，气-固界面换热加强；但蓄热体容积随之减小，高温区向下扩展，送风期温降加快，所需最少总混风比例增大，送风温度先增大后减小。对于不同孔径格子砖，随着格孔直径的减小，最高送风温度逐渐升高，最少总混风比例逐渐减小，最优活面积逐渐减小，对应当量厚度相对稳定。

关键词：热风炉；格子砖；活面积；混风操作；风温；操作周期；格子砖温度

1　引言

　　提高高炉热风炉的热风温度是改善高炉下部热制度，提高高炉产量，降低高炉焦比，降低炼铁能耗的有效措施[1,2]。很多研究者[3~9]通过数值模拟的方法计算热风炉蓄热室内温度场分布，研究了格子砖的物性参数对格子砖蓄热量、格子砖放热量、格子砖温度场分布和气体温度分布的影响。文献[10~12]中通过对格子砖与气体间换热过程的研究，提出了减小格子砖孔径、增大格子砖活面积（单位面积的格子砖的有效通道面积，m^2/m^2）以提高风温的建议。需要注意的是，对于一定孔径的格子砖，活面积的增大虽然可以增大格子砖换热面积，提高格子砖换热性能；但同时也伴随着蓄热体体积的减小，降低格子砖蓄热性能。

　　为了优化格子砖活面积设计，在本工作中，建立了蓄热室格子砖二维温度分布数学模型，采用元体平衡法计算蓄热室格子砖和气体温度场分布，通过分析不同混风曲线对蓄热室传热的影响，提出了一定状态下热风炉所能获得的最高送风温度及所需的最少混风比例，并结合不同状态下热风炉最高送风温度及最少混风比例对格子砖结构优化进行了分析。

2　物理模型和数学模型

2.1　物理模型

　　在计算中，将蓄热室看作由众多相同格孔组成的气体通道，格孔内径即为气体通道内

本文作者：颜坤，程树森，张福明。原刊于《耐火材料》，2017，51（5）：374-378。

径，通道的外径由式（1）求得：

$$d_e = \frac{d_i}{\sqrt{f}} \tag{1}$$

式中，d_e 为格孔外径，m；d_i 为格孔内径，m；f 为格子砖活面积。

2.2 基本假设

（1）不考虑换炉阶段蓄热室内热损失；
（2）在不同操作阶段，气体入口温度保持稳定；
（3）煤气、助燃空气质量流量保持稳定；
（4）混风流量和冷风流量随时间变化，且总量保持稳定；
（5）气体与格子砖的物性参数随温度变化；
（6）格孔内气体流速和温度在水平方向分布均匀。

2.3 数学模型

气体与格子砖在格孔中的换热方程如下[13]：
气体换热方程：

$$\rho_g c_{p,g} \left(\frac{\partial T}{\partial t} + v_g \frac{\partial T_g}{\partial z} \right) + v_g \frac{\partial p}{\partial z} = \frac{4\alpha}{d_i}(T_w - T_g) \tag{2}$$

固体换热方程：

$$\rho_s c_{p,s} \frac{\partial T_s}{\partial t} - \frac{1}{r} \frac{\partial}{\partial r}\left(r\lambda \frac{\partial T_s}{\partial r}\right) - \lambda \frac{\partial^2 T_s}{\partial z^2} = 0 \tag{3}$$

式中，t 为时间，s；z，r 为柱坐标系坐标，m；ρ_g 为气体密度，kg/m³；$c_{p,g}$ 为气体比热容，J/(kg·℃)；v_g 为气体速度，m/s；p 为气体压力，Pa；α 为综合换热系数，W/(m²·℃)；T_w 为格子砖壁体温度，℃；T_g 为气体温度，℃；T_s 为固体温度，℃；ρ_s 为格子砖密度，kg/m³；$c_{p,s}$ 为固体比热容，J/(kg·℃)；λ 为格子砖热导率，W/(m·℃)。

蓄热室内的热量传递考虑考虑格子砖内部导热以及烟气和格子砖之间对流换热和辐射换热[14]。其中，在燃烧期，烟气和格子砖之间考虑辐射和对流换热；而在送风期，气体主要成分为 N_2 和 O_2，不考虑辐射换热。对流换热系数计算公式如下：

湍流状态：

$$\alpha_c = 0.74 v_0^{0.8} d_i^{-0.333} T^{0.25} \tag{4}$$

层流状态：

$$\alpha_c = (0.96 + 0.21 v_0 d_i^{-0.6}) T^{0.25} \tag{5}$$

式中，α_c 为对流换热系数，W/(m²·℃)；v_0 为格子砖孔道内气体流速，m/s；d_i 为格孔内径，m；T 为气体的绝对温度，K。

辐射换热系数计算公式如下：

$$\alpha_r = \alpha_{r,CO_2} + \alpha_{r,H_2O} \tag{6}$$

$$\alpha_{r,CO_2} = \frac{B\left[q_{CO_2 \cdot T_y} - \left(\frac{T_y}{T_z}\right)^{0.65} q_{CO_2 \cdot T_z}\right]}{T_y - T_z} \tag{7}$$

$$\alpha_{r,H_2O} = \frac{B[q_{H_2O \cdot T_y} - q_{H_2O \cdot T_z}]}{T_y - T_z} \tag{8}$$

式中，α_r 为辐射换热系数，$W/(m^2 \cdot ℃)$；α_{r,CO_2}，α_{r,H_2O} 分别为二氧化碳和水的辐射换热系数；B 为耐火砖黑度，计算中取 0.75；T_y 和 T_z 分别为烟气温度和砖壁温度，K；$q_{CO_2 \cdot T_i}$ 和 $q_{H_2O \cdot T_i}$ 分别为二氧化碳和水在烟气温度为 T_i 时的辐射热，W/m^2。

假设混风流量和冷风流量随时间呈线性分布，则混风流量和冷风流量计算公式如下：

$$V_{total} = V_c + V_{mix}, \quad V_{mix} = V_{initial} + a \cdot t \tag{9}$$

式中，V_{total} 为高炉风量，m^3/h；V_c 为通过热风炉的冷风流量，m^3/h；V_{mix} 为混风流量，m^3/h；$V_{initial}$ 为最初混风流量，m^3/h；t 为送风周期中 t 时刻，min；a 为混风流量随时间变化的比例系数，$m^3/(h \cdot min)$。

混风后风温根据热风及混入冷风的流量和温度计算，计算方法如下：

$$T_{blast} = \frac{c_c T_c V_c + c_{mix} T_{mix} V_{mix}}{c_c V_c + c_{mix} V_{mix}} \tag{10}$$

式中，T_{blast} 为进入高炉的热风温度，℃；T_c 为混风前热风温度，℃；T_{mix} 为混风温度，℃；c_c 和 c_{mix} 分别为热风和混风的比热容，$J/(kg \cdot ℃)$。

2.4 计算条件

本研究中基于某钢厂热风炉参数对不同格子砖传热性能进行分析：热风炉蓄热室高度为 21m，高径比为 2.55，采用"两烧一送"操作制度，送风期及燃烧期时间分别为 45min 和 78min，煤气流量、助燃空气流量及高炉风量分别为 $60000m^3/h$、$40000m^3/h$ 和 $240000m^3/h$，煤气、助燃空气、冷风的预热温度分别为 190℃、400℃和 200℃。

3 结果与分析

3.1 最少混风比例与最高送风温度

在热风炉实际操作中，需要根据生产过程中实时参数调整热风炉操作参数，以获得较高的风温并保持相对稳定。调整混风曲线，改变送风期内不同时刻进入热风炉或混风室的风量，降低初期风温，提高后期风温，是保持热风风温在送风期内相对稳定的重要手段。对于不同的热风炉操作状态，所采用的混风曲线也不相同。因此，为准确比较不同格子砖的优劣，必须先确定混风曲线对风温的影响。

在无混风状态下，孔径 20mm 的格子砖在送风期内热风炉平均风温、最高风温（送风开始时刻风温）及最低风温（送风结束时刻风温）与活面积变化之间的关系如图 1 所示。可以看出：随着活面积的增大，平均风温先呈逐渐增大的趋势，但在活面积超过 0.6 时转为逐渐减小；最高风温逐渐增大，而最低风温逐渐减小，这导致风温温差逐渐增大。为获得能稳定供应高炉的热风，所需混入的冷风量也会随之增大。

在送风期内总混风比例保持不变的情况下，平均风温（混风后送风期内平均热风温度）与末期混风前风温（送风期最末时刻混风前的热风温度）随初始混风比例（送风开始时刻的混风比例）的变化如图 2 所示。可以看出：在保证热风炉总混风比例不变的前

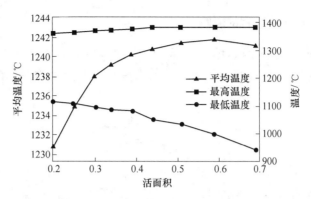

图 1 活面积对热风炉平均风温、最高风温和最低风温的影响

提下,平均风温及送风末期混风前风温变化较小。这说明在总混风比例不变时,进入热风炉蓄热室的冷风总风比例不变,在整个送风期内,冷风从蓄热室内带走的总热量相对稳定,同时,在送风末期,蓄热室内残留的热量相对稳定,因而平均风温及送风末期混风前风温相对稳定。因此,保持总混风比例不变,可以通过调整混风曲线使风温达到供应高炉的要求,最终送风温度接近相同总混风比例时的平均风温。但是,在总混风比例一定时,若送风末期混风前风温低于该混风比例时平均风温,无法通过调整混风曲线使风温达到供应高炉的要求。因此,为保证风温达到供应高炉要求,送风末期混风前风温不能低于送风期内平均风温。

孔径 20mm 的格子砖在活面积为 0.39 时,平均风温及末期风温与总混风比例的关系如图 3 所示。

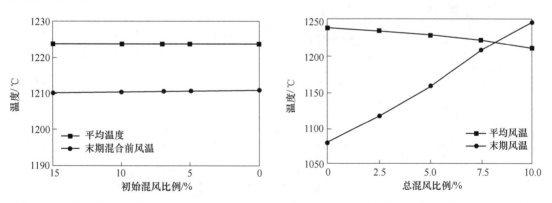

图 2 混风曲线对平均风温与末期混风前风温的影响　图 3 总混风比例对平均风温与末期风温的影响

从图 3 可以看出:随总混风比例的增加,进入蓄热室内的风量减小,带出总热量减少,混风后平均风温降低,同时残留在蓄热室内的热量增多,送风末期风温提高。在总混风比例约为 8% 时,送风期平均风温与送风末期风温相等。在总混风比例低于 8% 时,送风期平均风温高于送风末期风温,不满足送风条件,只有当总混风比例高于 8% 时,送风末期风温高于送风期平均风温,方能保证稳定送风。综上所述,对于不同工作状态下热风炉蓄热室,均存在一个最少总混风比例使风温满足送风条件,对应总混风比例下平均风温为该状态下最高送风温度。

3.2 活面积对格子砖传热过程的影响

孔径 20mm 格子砖的活面积对最高送风温度及最少总混风比例的影响如图 4 所示。可以看出：最高送风温度随活面积的增大呈先增大后减小的变化趋势，在活面积约为 0.44 时最大；最少混风比例基本上呈线性增大。

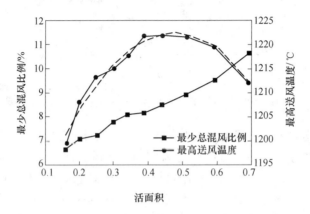

图 4 活面积对最高送风温度和最少总混风比例的影响

活面积对不同高度位置格孔在燃烧末期及送风末期的表面温度的影响如图 5 所示。其中，0m 代表蓄热室底部，21.0m 代表蓄热室顶部。可以看出：燃烧末期、送风末期蓄热室顶部、底部表面温度随活面积增大变化很小；在活面积较小时，燃烧末期、送风末期蓄热室中部表面温度相对稳定；当活面积超过 0.39 时，蓄热室中部燃烧末期表面温度逐渐升高，送风末期表面温度几乎保持不变。

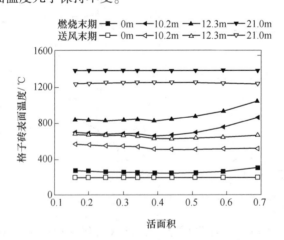

图 5 活面积对不同高度位置格孔在燃烧末期及送风末期的表面温度的影响

格子砖活面积对燃烧末期及送风末期格孔表面平均温度及格孔表面平均温差（燃烧末期格孔表面平均温度与送风末期格孔表面平均温度的差值）的影响如图 6 所示。可以看出：随着活面积的增大，燃烧末期格孔表面平均温度先缓慢降低，当活面积超过 0.39 时转为逐渐升高；送风末期格孔表面平均温度先缓慢降低，当活面积超过 0.39 时相对稳定，基本上保持在590℃左右。随着活面积的增大，格孔表面平均温差先呈缓慢增大的趋势，但当活面积增大到0.39时则呈加速增大的趋势。

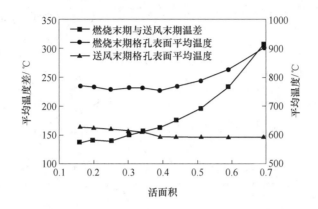

图6 活面积对格孔表面平均温度的影响

综上所述，随着活面积的增大，格子砖比换热面积逐渐增大，总换热量逐渐增大，当总蓄热体体积足够时，热量更多地向耐火材料中心传递，格子砖表面温度不发生显著变化，而送风温度逐渐升高；但随着活面积的继续增大，蓄热体容积进一步减小，换热面积继续增大，当蓄热体容积过小时，热量储存空间不足，蓄热室上部格子砖不能充分地吸收热量，富余热量将由较低位置格子砖吸收储存，格子砖高温区域向下延伸，操作周期中格孔表面平均温差快速升高，当转入送风状态后，由于格子砖表面温度更高，格子砖与冷风间温差更大，格子砖与冷风间热量交换加强，若保持混风曲线不变，送风初期由冷风带走的热量将增大，格子砖表面温度降低速度加快，后期格子砖与冷风间换热减弱，整个送风期内送风温度差值增大，为保持送风期内风温均匀，所需最少总混风比例增大，减少进入热风炉的冷风流量，整体送风温度降低。

3.3 不同孔径格子砖最优活面积的选择

为进一步研究活面积对格子砖传热性能的影响，计算了格孔直径分别为23mm、28mm、30mm和33mm格子砖的。不同格孔直径格子砖均表现出热风温度随着活面积的提高先增大、后减小的趋势，存在使热风温度达到最大值的临界活面积，且最少混风比例随活面积增大而升高。

最高送风温度及对应的最少混风比例、活面积和当量厚度随格孔直径的变化如图7所示。可以看出：随着格孔直径的减小，最高送风温度逐渐升高，而对应最少混风比例逐渐降低。格孔直径由20mm增大到33mm，最优活面积由0.44增大0.58左右，而所对应的当量厚度基本保持10mm左右，相对稳定。这是因为随着格子砖孔径的减小，相同活面积格子砖的比换热面积增大，格孔表面与气体间换热系数升高，格子砖在热风炉工作过程中总交换热量升高，送风温度升高。如前文所说，当蓄热室换热量过大时，格子砖表面高温区域向下扩展。因此，格孔直径越小，需要更多的蓄热体以保证热量更多地向耐材中心传递，最高送风温度所对应的活面积减小，而当量厚度保持相对稳定。

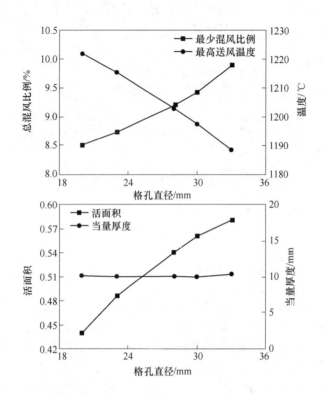

图7　不同孔径格子砖最高送风温度及对应的最少混风比例、活面积和当量厚度

4　结论

(1) 在总混风比例保持一定时，不同混风曲线对整个送风期内平均送风温度无明显影响。

(2) 随着总混风比例的增大，整个送风期内平均送风温度逐渐减小，而末期混风前风温逐渐增大。对于确定工作状态下的热风炉蓄热室，存在一个使送风期风温保持平稳的最少总混风比例；最少总混风比例应保证平均风温与末期混风前风温相等，该总混风比例下平均风温为对应工作状态热风炉最高送风温度。

(3) 对于单一孔径格子砖，随着活面积的增大，气-固界面换热加强，风温升高；但活面积过高时，蓄热空间过小，高温区向下扩展，送风期温降加快，所需最少总混风比例增大，风温降低；存在使送风温度最高的特定活面积。

(4) 对于不同孔径格子砖，随着格孔直径的减小，最高送风温度逐渐升高，最少总混风比例逐渐减小，最优活面积逐渐减小，最优活面积所对应当量厚度相对稳定。

参考文献

[1] 刘全兴. 我国高炉热风炉的新技术应用回顾与展望 [J]. 炼铁, 2007, 26 (2): 56-60.

[2] 钱世崇, 张福明, 李欣, 等. 大型高炉热风炉技术的比较分析 [J]. 钢铁, 2011, 46 (10): 1-7.

[3] 陈川,程树森,郭喜斌. 热风炉内格子砖气体的传热参数[J]. 过程工程学报, 2012, 12 (5): 765-769.

[4] 郭敏雷,程树森,张福明,等. 热风炉送风期格子砖温度分布计算[J]. 钢铁, 2008, 43 (6): 16-21.

[5] 陈冠军,张建良,马金芳,等. 高炉高风温的试验研究[J]. 钢铁研究学报, 2011, 23 (5): 15-19.

[6] 周惠敏,张浩,苍大强,等. 高辐射覆层对热风炉传热过程影响的数值模拟[J]. 钢铁研究学报, 2011, 23 (3): 6-10.

[7] 陈川,程树森,郭喜斌. 高炉风温影响因素研究[J]. 钢铁, 2013, 48 (4): 12-17.

[8] Kimura Y, Takatani K, Otsu N. Three-dimensional mathe-matical modeling and designing of hot stove [J]. ISIJ Int, 2010, 50 (7): 1040-1047.

[9] Zhong L C, Liu Q X, Wang W Z, et al. Computer simulation of heat transfer in regenerative chambers of self-preheating hot blast stoves [J]. ISIJ Int, 2004, 44 (5): 795-800.

[10] 吴启常,沈明,陈秀娟. 高风温热风炉设计理念的调整及相关问题讨论[J]. 炼铁, 2008, 27 (4): 11-15.

[11] 姜凤山,全强,李富朝,等. 新型高效蜂窝格子砖的研制与应用[J]. 炼铁, 2006, 25 (5): 31-34.

[12] 张红哲,李富朝,孙庚辰,等. 高风温长寿热风炉用高效小孔径格子砖的设计及选材[J]. 耐火材料, 2015, 49 (5): 376-380.

[13] 杨世铭,陶文铨. 传热学[M]. 4版. 北京: 高等教育出版社, 2009: 115-119.

[14] 《炼铁设计参考资料》编写组. 炼铁设计参考资料[M]. 北京: 冶金工业出版社, 1975: 446-448.

Optimization Design of Relative Open Area of Hot Blast Stove Checker Brick

Yan Kun Cheng Shusen Zhang Fuming

Abstract: A two-dimensional heat transfer mathematical model was developed to simulate the heat transfer in checker bricks in hot blast stove operation to optimize the design for relative open area. The influences of blast mixing ratio and relative open area on the heat transfer of checker bricks were studied. The results show that: for a hot blast stove working in a certain operation condition, there shall be a minimum blast mixing ratio which can keep the hot blast temperature stable; the corresponding average blast temperature is the highest under this operation condition; for checker bricks with a constant hole diameter, with the increase of relative open area, the heat transfer between air and brick surface enhances, but the total volume of regenerator decreases, as a result, the high temperature region of checker chamber downwards expands, the temperature decreases acceleratedly in blast period, requiring a higher minimum blast mixing ratio, the blast temperature increases firstly and then decreases; for the checker bricks with different hole diameters, with the hole diameters decreasing, the highest blast temperature increases, the minimum blast mixing ratio decreases, the optimum relative open area decreases and the corresponding equivalent thickness keeps almost stable.

Keywords: hot blast stove; checker brick; relative open area; mixed air operation; blast temperature; opera-tion period; checker brick temperature

首钢高炉出铁场设备的开发与应用

1 引言

"八五"期间,首钢4座高炉相继进行了现代化新技术扩容大修改造,高炉总容积由原来的 4139m³ 扩大到 9934m³。通过高炉新技术大修改造,首钢高炉实现了大型化、高效化、长寿化、机械化和自动化,高炉整体技术装备达到国内先进水平。

高炉出铁场设备的装备水平是机械化的重要技术特征。高炉炉前操作直接影响高炉的正常生产,因此炉前操作在高炉生产中占有十分重要的地位。炉前是高炉生产中高温多尘、环境恶劣、劳动强度大、操作繁重的岗位。提高炉前机械化程度,降低工人劳动强度,改善工作条件是现代化大型高炉设计和建设的一个主要技术要素。

2 首钢高炉出铁场设备的应用现状

目前,首钢有5座高炉生产,1~4号高炉相继在1991~1994年间进行了现代化新技术大修改造。设计中为提高高炉炉前机械化水平,自行设计开发了矮式液压泥炮、全液压开口机、气—液复合开口机、电动铁水摆动流槽、环行桥式起重机、液压吊盖机等多项专利设备,为实现高炉炉前的机械化、装备现代化、清洁化奠定了基础。首钢高炉出铁场设备配置见表1。

表1 首钢高炉出铁场设备配置

项 目	1号高炉	2号高炉	3号高炉	4号高炉
高炉有效容积/m³	2536	1780	2536	2100
投产时间	1994年8月	2002年5月	1993年6月	1992年5月
铁口数目/个	3	2	3	2
出铁场结构形式	圆形出铁场	圆形出铁场	圆形出铁场	圆形出铁场
液压泥炮	SGXP-400型 液压工作压力32MPa, 推力4000kN	SG-280型 液压工作压力25MPa, 推力2800kN	SGXP-400型 液压工作压力32MPa, 推力4000kN	SGXP-400型 液压工作压力32MPa, 推力4000kN
开口机	SGK型全液压开口机	气—液复合式开口机	SGK型全液压开口机	SGK型全液压开口机
吊盖机	悬臂式液压吊盖机	封闭式固定除尘罩	悬臂式液压吊盖机	悬臂式液压吊盖机
环行桥式起重机	L_k = 21000mm Q = 30t/5t	L_k = 18600mm Q = 30t/5t	L_k = 21000mm Q = 30t/5t	L_k = 21000mm Q = 30t/5t

---

本文作者:张福明,苏维,张建,王建涛。原刊于《设计通讯》,2003(2):15-21。

续表 1

项 目	1号高炉	2号高炉	3号高炉	4号高炉
铁水摆动流槽	3台电动铁水摆动流槽	2台电动铁水摆动流槽	3台电动铁水摆动流槽	2台电动铁水摆动流槽

3 首钢高炉出铁场设备的技术特点

3.1 矮式液压泥炮

首钢高炉实现大型化以后，高炉产量增加，炉顶压力均按 0.2MPa 设计，传统的电动泥炮已经难以满足高炉高效化生产的要求，首钢自行开发研制了 SGXP 型矮式液压泥炮。该设备具有如下特点：

(1) 泥炮具有较大的打泥推力，可以迅速地将致密的炮泥压入铁口，特别适用于高压的大中型高炉，对于无水炮泥的适用性很强。

(2) 泥炮结构简单，体积紧凑、矮小，无论在工作位置还是停放位置，均能处于风口平台以下，不影响风口平台的完整性，能适应炉前拥挤的现场环境。

(3) 在炉前粉尘大、铁水喷溅大、温度高的恶劣环境下，该泥炮具有很高的可靠性，极低的故障率，维护工作量小，大大提高了炉前的生产效率。

(4) 采用独特的斜基础设计结构，使压炮、锁炮等动作简化，降低了设备的复杂性。

(5) 操作简单、便捷，既可以实现手动，也可实现远程遥控操作。

(6) 液压系统采用了独特的回路及先进的阀件，实现了设备平稳起停，无压、无溢流、低功率的空载循环。采用手动变量泵可保证设备运动速度快，在任何负荷下都不会掉速。

(7) 采用国内航空工业引进、消化的先进技术，实现了设备的全部国产化。

(8) 作为易损件的炮嘴，使用了首钢设计院的专利技术，寿命长，损耗低。图 1 是 SGXP 型矮式液压泥炮设备总图，表 2 是首钢 1、3、4 号高炉采用的 SGXP-400 型矮式液压泥炮的技术性能。

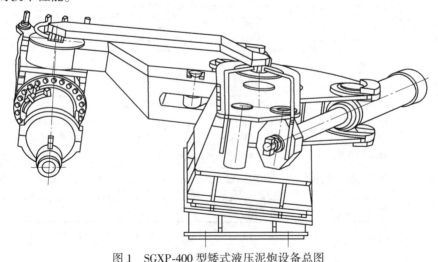

图 1 SGXP-400 型矮式液压泥炮设备总图

表 2　SGXP-400 液压泥炮性能参数

活塞总推力	4000kN	压炮力	360kN
泥缸有效容积	0.25m^3	回转角度	160°
堵泥单位压力	16.9MPa	回转半径	2500mm
系统油压	32MPa	堵泥时间	57s
炮嘴内径	ϕ150mm	转炮时间	19s
泥缸直径	ϕ58mm	炮身倾角	15°~18°
打泥油缸直径	ϕ450mm	转炮油缸直径	ϕ280mm

3.2　全液压开口机

首钢 1、3、4 号高炉均采用了首钢自行设计研制的 SGK 型全液压开口机。这种开口机是新一代多功能开口机，其结构紧凑、体积小、工作可靠，可在 300~3000m^3 的高炉上配套使用。

SGK 型全液压开口机采用矮式刚性结构，可以放置在风口平台之下，与矮式液压泥炮在铁口两侧呈对称布置。其特点是能力强、功率大、效率高、直线钻孔，打开铁口时间一般不超过 2min，全部操作通过三个手柄完成，操作安全、可靠、简便、容易掌握。该设备采用航空工业先进技术，开发了刚性、半刚性凿岩机输油管路系统取代了传统的橡胶软管输油管路系统，大大提高了开口机的耐高温性能，确保高炉的安全生产。首钢十几年的使用经验表明，液压开口机比气动开口机具有开铁口能力高，凿岩机寿命长，日常维护量小等明显优势。

SGK 型全液压开口机分为开口机设备本体和液压驱动装置两大部分。设备本体由斜基础、回转机构、钻进机构三大部分及回转油缸、液压马达、液压凿岩机、三四路铰、钎杆、钻头等组成。图 2 是 SGK 型全液压开口机设备总图。

SGK 型全液压开口机液压驱动装置由驱动泵组、油箱、阀台、操作控制台等组成。开口机通过液压操纵控制系统实行远距离人工操纵。开口机工作时，开口机回转机构的回转油缸驱动大臂回转，是开口机钻进机构由停放位置回转到工作位置，钎杆准确对准高炉出铁口，然后由液压马达通过链传动机构驱动钻进小车向前走行，推动液压凿岩机及钎杆按调定的轨迹冲击式钻通出铁口。铁口钻通后，钻进机构的钻进小车迅速退回，钎杆则迅速推出铁口，回转机构迅速回转，使钻进机构迅速退离主沟，回到开口机停放位置。表 3 是首钢高炉上采用的 SGK 型全液压开口机的主要技术性能。

2002 年首钢 2 号高炉技术改造中，在全液压开口机的基础上进行了改进和创新，开发研制成功了气-液复合开口机，目前已投入生产半年，使用情况基本正常，但使用的国产气动凿岩机的可靠性和使用寿命有待提高。

3.3　环行桥式起重机

环行桥式起重机是高炉圆形出铁场的专用吊装设备，可承担高炉炉前全部吊运工作，包括铁水摆动流槽、渣铁沟盖、炉前设备备件、耐火材料等。

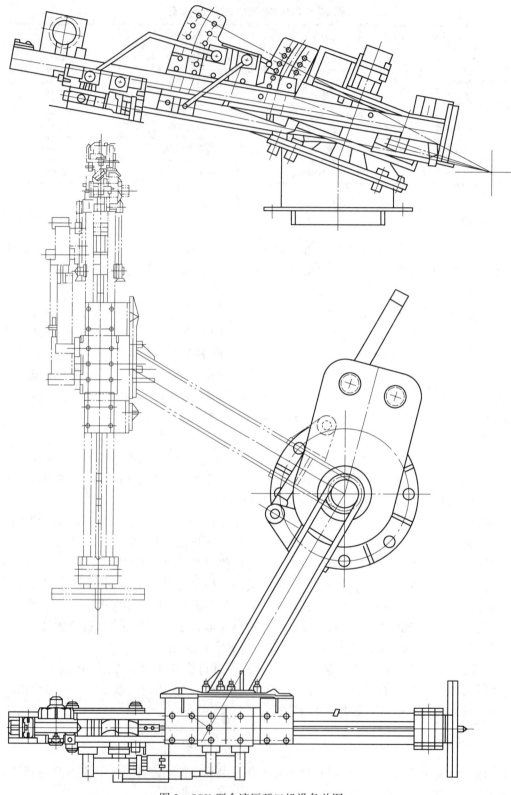

图 2 SGK 型全液压开口机设备总图

表3　SGK全液压开口机主要性能参数

回转角度			0°~110°		
回转时间			5s（退）/10s（进）		
钻头压紧力			<19000N		
钻杆送进速度			<9m/min		
钻杆退回速度			30~50m/min		
钻杆工作角度			6°~18°		
最大钻孔深度			3000mm		
凿岩机行程			3300mn		
回转油缸	型号	Y-HG$_1$-G140/100×1400LZ$_2$(315)-HL$_1$OT$_2$	油马达	型号	NHM6-600B
	油压	16MPa		额定压力	20MPa
	行程	1500mm		转速范围	4~500r/min
	缸径	160/110mm		流量	1.6~206L/min
凿岩机	型号	HYD300		旋转扭矩	100~300N/m
	冲击力	300J		流量	40~60L/min
	冲击频率	34~67Hz		旋转油压	4.5~15MPa
	冲击油压	19MPa		旋转流量	87L/min

环行桥式起重机围绕高炉在两条同心的环形轨道上行走。两条环行轨道分别安装在与高炉同心的内外环梁上，内环梁支撑在炉体框架分离柱上，外环梁支撑在出铁场厂房柱上。起重机沿环行轨道行走时，内外环的角速度必须相同，亦即内外环走行轮的线速度必须严格按比例，以保证起重机在环行轨道上行走时不啃轨。

大车采用分别驱动方式，每个驱动机构各有一台电动机，分别布置在靠近外环和内环走行轮处。为将驱动设备简化成一线布置，采用针轮摆线减速机，同时将齿型浮动轴改为十字轴万向联轴器，允许电动机轴同走行轮轴有一定交角。

两台电动机转速相同，控制了稳定运行。由此确定内外不同轮径和传动比值，即实现机械、电气双同步。一台电动机故障时大车仍能继续运行。在2002年首钢2号高炉技术改造中，为更精确地控制内外环大车轮的线速度差，大车轮采用交流变频电动机驱动，使大车启动时更平稳，行走时不跑偏不啃轨。

环行桥式起重机主要由桥架、运行机构、小车和电控设备等组成。用于首钢3号高炉的环行桥式起重机的主要技术性能见表4，环行桥式起重机的设备总图如图3所示。

3.4　电动铁水摆动流槽

对于大、中型高炉，出铁量增加，铁水罐增多，单线铁路长度增加，使铁沟延长，因此，采用铁水摆动流槽来缩短铁沟长度，减少出铁场面积，改善操作条件，减轻劳动强度，而且对于圆形出铁场铁水摆动流槽已成为不可缺少的设备。

电动铁水摆动流槽安装在出铁场铁水沟下边，将铁水分配到出铁场平台下边的铁水罐车中，和传统的铁沟、流嘴直接进入各铁水罐车的布置相比，摆动流槽具有以下优点：

表4 首钢3号高炉环行桥式起重机主要技术性能

项　目	参　数
起升能力	主钩30t，副钩5t
电源电压	380V
接电持续率	40%
起升高度	主钩18m，副钩20m
起升速度	主钩7.85m/min，副钩20m/min
运行速度	小车42.8m/min 大车外环约55m/min，内环约22m/min
大车走行轮直径	外环800mm，内环764mm
最大轮压	260kN
起重机质量	约40t

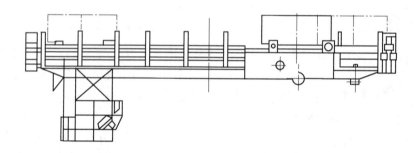

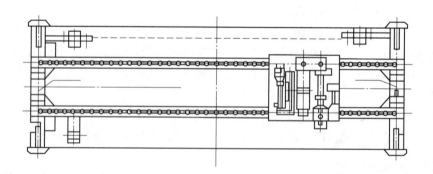

图3　环行桥式起重机设备总图

（1）可以大大缩短铁沟的长度，简化了出铁场的布置。
（2）减少了在高温、粉尘条件下转换铁沟闸板的作业。
（3）减轻了修补铁沟的作业，减少了耐火材料的消耗量。

（4）提高了炉前铁水的运输能力，使高炉车间和铁路布置更为简单特别适用于圆形出铁场的高炉。

（5）简化了铁水罐的除尘设施，改善了出铁场除尘效果。

电动铁水摆动流槽由流槽、支撑机构及摆动机构组成，流槽可以采用铸件或钢板焊接而成，内衬用浇注料浇注而成。传动装置用电机驱动减速机，通过扇形齿轮带动固定摆槽的曲拐轴和槽体一起转动。摆槽的摆动角度大小由主令控制器进行控制，当电机有故障时，可以用备用手动系统驱动槽体进行摆动以保证安全生产。首钢高炉用电动摆动流槽的主要技术性能见表5，设备总图如图4所示。

表5 首钢高炉电动铁水摆动流槽技术性能

规格	长度 4600mm	每次通铁量	5万~6万吨	制动器	TJ2-220-TH
	宽度 1900mm			主令控制器	LK4-054（1:1）
电动机	型号	YZ160-6-TH		流槽摆角	±25°
	功率	11kW		流槽摆动速度	0.3637r/min
	转速	953r/min		流槽摆动周期	29.32s
	负载持续率40%			减速机 ZD10-6-Ⅱ	速比3.55
总速比	电动速比 2619.8	开式齿轮		速比 5.475	适用高炉容积 1000~4000m³
	手动速比 7178.4	差动联节		手动速比 10.714	
蜗轮减速	速比 435			电动速比 1.1029	总质量 19.230t

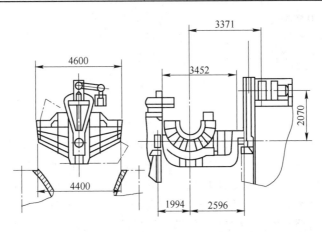

图4 首钢高炉电动铁水摆动流槽设备总图

3.5 液压吊盖机

主沟盖吊盖机是炉前主沟除尘沟盖的配套设备，用于主沟第一节除尘水冷沟盖的吊装平移。在铁口打开后，此设备将主沟除尘沟盖盖在主沟铁口前，捕集出铁过程中产生的烟尘，烟尘经除尘管道抽风吸走。在泥炮堵铁口前，再将除尘沟盖吊起移开。

由首钢开发研制的主沟盖吊盖机为全液压悬臂式，驱动为液压，作为动力源的液压站远离环境恶劣的炉前，寿命长，易于维护。设备结构简单紧凑，坚固耐用，操作方便。无论采用手动或电磁阀操作，均可保证除尘沟盖到位准确。液压吊盖机为悬臂结构，臂长约8m，其旋转、提升均为液压驱动，液压吊盖机固定在炉体框架柱上，围绕其设备本体的固定转轴旋转。整个设备安装在高炉大立柱上，不影响炉前布置和操作，由于不再需要其他安装支架等，故设备总质量较轻。

主沟盖吊盖机是由钢丝绳与主沟除尘沟盖联结，主沟盖内通水冷却，水管也经由吊盖机连通。

4 出铁场设备的应用实践

20世纪90年代以来，首钢在高炉前设备方面进行了大胆的创新和积极的技术探索，经过10余年的生产实践证实，首钢自行开发研制的炉前设备完全可以满足大型现代化高炉的高效化生产，设备运行稳定可靠，为提高炉前作业机械化、降低工人劳动强度、实现清洁化生产创造了有利条件，创造了显著的经济效益和社会效益。

矮式液压泥炮、全液压开口机、环行桥式起重机等炉前设备获得了国家专利，通过了冶金部技术成果鉴定，获得了国家发明奖和冶金部、北京市的科技进步奖。

除在首钢大型高炉上应用以外，还在唐钢2560m^3高炉，本钢2000m^3高炉，太钢1350m^3高炉，邯钢1260m^3、2000m^3高炉，莱钢750m^3高炉，承德钢厂350m^3高炉等国内知名企业的高炉上成功应用，而且在印度那柯钢铁公司600m^3高炉、巴塞尔钢铁公司650m^3高炉、苏佳纳钢铁公司350m^3高炉、津巴布韦钢铁公司1500m^3高炉等多座高炉上都取得了良好的应用效果。

近年来，在提高液压系统工作的可靠性，进一步降低设备故障率，延长设备易损部件的使用寿命等方面进行了许多有益的创新和优化完善。

5 结语

随着现代高炉大型化、高效化、长寿化、清洁化的技术进步，高炉出铁场设备的机械化水平应不断提高，以适应高炉生产的要求。首钢自行开发研制的出铁场设备，经过生产实践的长期检验，设备运行可靠，适应性强，故障率低，寿命长，是新一代的高炉炉前设备。随着高炉炼铁技术进步，首钢将进一步开发研制更新的炉前设备，不断改进完善现有技术，为推动首钢乃至我国高炉技术的持续发展做出更大的贡献。

CFD Study on Flow Characteristics of BOF Gas in Evaporating Cooler

Zhang Fuming   Cheng Shusen   Zhou Hong   Zhao Jingjing

Abstract: Basic oxygen furnace (BOF) gas dry dedusting technology is the main orientation of steelmaking industry. It has technical superiority in high efficiency energy conversion, energy saving and emission control and clean environmental protection fields, and it can decrease water consumption greatly, have high efficiency for steam and gas recovery, and reduce environmental pollution, and it is a key technology for realization of high-efficiency energy conversion in the contemporary steelmaking. The evaporating cooler is a key facility of temperature control for the BOF gas; it plays an important role in dedusting process. The CFD numerical simulation on flow characteristics of BOF gas in evaporating cooler has been accomplished. The gas velocity distribution, temperature distribution and dust distribution in the evaporating cooler are simulated by CFD. According to the result of research, a number of optimizations are applied in engineering, and have been proven by the outstanding performance in operating practice.

Keywords: computational fluid dynamics; BOF gas cleaning; evaporating cooler; optimum design

1 Introduction

BOF gas dry dedusting technology has advantages of high dedusting efficiency, obvious water saving, low energy consumption and operating cost, long service life and less maintenance. Particularly, it has significant advantages in aspects of reduction of fresh water consumption and energy consumption to realize sewage "zero emission". And ferrous dust after briquetting can be directly fed to converter so as to have wastes converting into resources for recycling. BOF gas after collection by the fume hood is led to evaporating cooling flue, and it is heat exchanged to recover high temperature gas heat before the evaporating cooler to make converter gas temperature cool down to 1000 degree centigrade below. Then it is fed into the evaporating cooler for cooling process to have secondary temperature lowering and primary dedusting of BOF gas. After that it comes to the dry electrostatic precipitator (abbreviated as ESP) for secondary dedusting process, the converter gas after ESP is boosted by the axial blower. Qualified gas after secondary temperature lowering by the gas cooler is sent to the BOF gas holder as secondary energy recovery.

2 Functions of the Evaporating Cooler

BOF gas has characteristics of high temperature, more dust and high CO content. Approximate 1600 degree centigrade high temperature BOF gas is cooled to 900-1000 degree centigrade via the flue of evaporating cooler, and enters into the evaporating cooler. Function of the evaporating cooler is to have once more temperature lowering and primary dust removal to gas. Many atomization nozzles are provided at the top of the evaporating cooler in well distributed way. Gas in the evaporating cooler is sufficient heat exchanged with countercurrent atomization solution drops, and gas temperature is lowered from 900 degree centigrade to 200 degree centigrade around. The gas after temperature lowering enters into the ESP for secondary dust removal.

Precise control of cooler water flow and arrangement mode of the atomization nozzles are the technical keys for high efficiency operation of the evaporating cooler. Cooling water flow depends on gas flow and temperature. If the cooling water flow is insufficient, gas cooling efficiency is not good, and gas temperature is high. These can affect ESP dedusting result and equipment service life seriously, and probability of gas explosion relief increases. If cooling water is too much, fume is aggregated and sludge is formed, so that work efficiency of the evaporating cooler is impacted. Arrangement mode of the atomizing nozzles can have corresponding adjustment based on flow characteristics of BOF gas inside the evaporating cooler. In addition, since gas generation and temperature changes during converter blowing process, reliable control model must be established to have precise adjustment on cooling water volume.

In order to realize precise control on gas temperature and cooling water volume, and reduce explosion relief caused by too high converter gas temperature. CFD simulating study is applied to analysis flow characteristics of BOF gas inside the evaporating cooler, and establish mathematical model on gas flow, water drop atomization and heat transfer inside the evaporating cooler so as to have the optimal evaporator design parameters and arrangement parameters of water spray devices.

3 Gas Flow Distribution in the Evaporating Cooler

The gas flow distribution in the evaporating cooler is shown in Fig. 1. The picture shows that velocity of gas in the evaporating cooling flue is high, and the flue shape is irregular. At X-Z section of the evaporating cooler, the high velocity zone of gas mainly concentrated upon external arc side part of the flue elbow, and velocity of gas flow along the internal arc side along the flue is less. After gas enters into the evaporating cooler, the velocity decreases, most gas flows along the left side of the evaporating cooler. And a small low temperature zone appears at the internal side of the evaporating cooler with velocity 6m/s below. At Y-Z section of the evaporating cooler, due to action of the bottom elbow of the evaporating cooler, the high temperature zone of gas at this section has a slight right offset. And at the left side of the evaporating cooler, also there is a small low velocity zone existed. Gas flows mainly along the right side of the evaporating cooler. At the outlet elbow at the bottom of the evaporating cooler, gas velocity increase somewhat, and still exist unequal phenomenon on velocity distribution, and the gas velocity at the internal arc side at the pipeline bend. Fig. 1 shows gas velocity distribution at the cross-section of different heights of the evaporative cooling flue and the evaporating cooler. Fig. 1 shows clearly that the gas velocity distribution at the cross section of the pipeline is uneven.

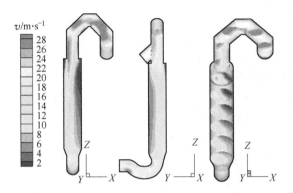

Fig. 1 Velocity distribution of different section in the evaporating cooler

Inside the evaporative cooling flue, the high temperature zone of gas at the cross-section of the conical pipe mainly concentrates upon the external arc side part of the elbow pipe. After gas enters into the evaporating cooler, gas temperature has somewhat decrease. But there still is serious impact one gas flow inside the evaporating cooler due to irregular cooling flue. The high velocity zone of gas mainly focuses on one side of the evaporating cooler, namely the external arc side direction of the flue. Therefore the gas velocity at another side of the evaporating cooler is low from beginning to end and extents to the bottom of the evaporating cooler. With gas flow to the bottom of the evaporating cooler, the low velocity zone becomes small, and the gas velocity distribution at the cross-section of the pipeline becomes even, and remains around 12m/s.

4 Temperature Distribution in the Evaporating Cooler

The distribution of temperature field in the evaporating cooler is shown in Fig. 2. Gas temperature at the model inlet is set to 1500 degree centigrade . Because of cooling of the evaporative cooling flue, temperature of gas near the pipeline wall is low, and that at the pipeline centre is high. After gas enters into the evaporating cooler, the temperature is decreased to approx. 1000 degree centigrade . In the evaporative cooling flue due to high velocity of gas at the external arc side part of the pipeline, and the high temperature gas changes rapidly so that the high temperature zone at the cross-section of the pipeline has offset to the external side part of the flue. After gas enters into the evaporating cooler, centre of the gas high temperature zone has right offset to the evaporating cooler, and temperature at the left side of the evaporating cooler is low accordingly. This is due to the impact of inverse flow in the evaporating cooler. Due to existing of inverse flow at right side of the evaporating cooler, the high temperature gas has strong heat transfer to the right side of the evaporating cooler, and these results in corresponding high of gas temperature in this zone. With gas flow at bottom of the evaporating cooler, gas temperature high slight decrease, and temperature distribution at the cross-section of the evaporating cooler becomes equal gradually. Temperature difference at the bottom of the evaporating cooler remains in the range 100 degree centigrade.

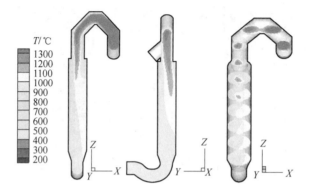

Fig. 2　Temperature distribution at different sections in the evaporating cooler

Calculation and analysis by means of velocity field and temperature field in the evaporating cooler shows that under condition of no water spray, the gas temperature in the evaporating cooler is high, with average temperature is approximate 900 degree centigrade. Irregularity of the evaporative cooling flue impacts more to the gas flow in the evaporating cooler. After gas enters into the evaporating cooler, gas moves downwards along the negative X of the evaporating cooler (namely the external arc direction of the flue), and the velocity at the positive X direction of the evaporating cooler is low. In here a big zone of inverse flow zone is produced so that the effective utilization volume of the evaporating cooler decreases greatly. With gas flow to the bottom of the evaporating cooler, gas velocity distribution at the cross-section of the evaporating cooler becomes more and more equal.

In actual production, reasonable design should be considered based on flow characteristics of BOF gas in the evaporating cooler. In order to improve uniformity of gas flow, the gas deflector plate is provided at front of the nozzles in the evaporating cooler. As per flow characteristics of gas at different places, different gas deflector plates should be considered to achieve the target of uniform distribution of velocity and temperature.

5　Optimization of Design of Evaporating Cooler

Mathematical model on fluid flow and temperature distribution in evaporating cooler is established, and CFD is used to numerical simulation. And the calculation result is used for primary design optimization for cooling device of the evaporating cooler. Calculation and analysis on gas flow field and temperature field in the evaporating cooler is carried out, study on equal distribution of gas and water vapour in the evaporating cooler is implemented, and design optimization on installation position, section distribution of water spray devices and gas velocity is executed. Mathematical simulation is carried out to dust movement in the evaporating cooler, and numerical value is used to calculate motion curve and settlement position of dust in the evaporating cooler, with deposit position of dust is presented accordingly.

The study shows there is important relevance between the arrangement mode of the converter evaporative cooling flue and the evaporating cooler. Arrangement of the evaporative cooling flue has great influences on gas flow status in the flue. When flue arrangement has less deflection, uneven phenomenon of gas flow in the flue can be decreased greatly, and this is beneficial to have even gas flow of the flue to enter into the gas evaporating cooler. On this account, there is facilitation of the optimization system if the gas flow deflection device is foreseen at the gas inlet of the evaporating cooler. Gas deflection plate is provided in front of the nozzles of the evaporating cooler to have efficient adjustment on gas flow distribution status, improvement of temperature field distribution of gas in the evaporating cooler. At the same initiate condition, the gas deflection plate is added to reduce outlet temperature of the evaporating cooler by 5-8 degree centigrade. Fig. 3 shows the velocity and temperature distribution status at different sections before and after the deflection device provided for the evaporating cooler; Fig. 4 shows dust distribution status at different sections before and after the deflection device provided for the evaporating cooler.

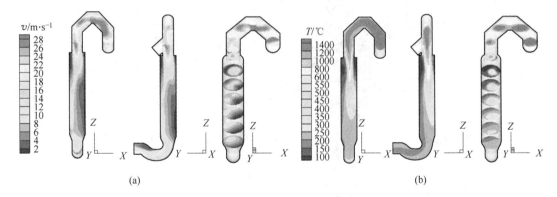

Fig. 3 Velocity and temperature distributions at different sections
with additional deflection device in the evaporating cooler
(a) Velocity distribution; (b) Temperature distribution

Fig. 5 shows the gas velocity distribution after additional deflection device is installed at different height in evaporating cooler; Fig. 6 shows the gas temperature distribution after additional deflection device is installed at different height in evaporating cooler. As a result, the CFD investigations proved the BOF gas velocity and temperature distributions are improved and optimized. The functions of BOF gas temperature control and primary dust removal of evaporating cooler are promoted. CFD numerical simulation is applied to study the gas flow field and distribution of temperature field in evaporating cooler. By taking such measures as reasonable

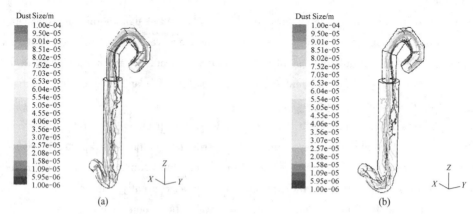

Fig. 4 Dust distribution at different sections in the evaporating cooler
(a) Without deflection device; (b) With additional deflection device

arrangement of evaporative cooling flue, gas guiding device and optimized spray nozzle arrangement etc., make the distribution of gas velocity and temperature in evaporating cooler even. The efficiency of evaporating cooler ensures stable and reliable operation of BOF gas dry dedusting system.

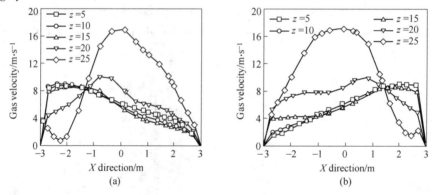

Fig. 5 Gas velocity distribution at different height in evaporating cooler
(a) Veloctity distrbution on X direction; (b) Veloctity distribution on X direction

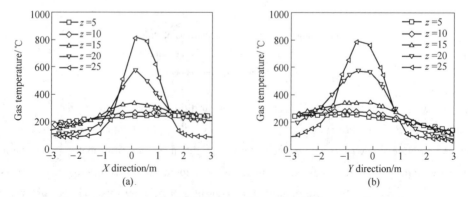

Fig. 6 Gas temperature distribution at different height in evaporating cooler
(a) Temperature distribution on X direction; (b) Temperature distribution on Y direction

6 Conclusions

The BOF gas flow in evaporating cooler is influenced sufficiently by evaporation cooling flue arrangement. After gas enters evaporating cooler, primarily along the outside of the tower with a larger velocity moves downward, and produces a bigger circumfluence in the inside of the evaporating cooler, greatly reduces the effective utilizable volume of evaporating cooler.

After optimization in evaporating coolerwith additional deflection device, the features of gas flow in evaporating cooler are improved significantly. The scope of low flow velocity zone is reduced significantly, gas velocity distribution on the cross-section is become more evenly, but the gas temperature is increased in evaporating cooler. Optimization of the baffle makes the droplets become more evenly distributed in the evaporating cooler; the number of drops near the walls is reduced significantly, and the movement distance is reduced significantly in the evaporating cooler.

The assembled baffle influences the dust movement in the evaporating cooler significantly. Before the baffle is not installed, dust move primarily along the lateral wall, and larger amount dust stick on the lateral wall, easily lead evaporating cooler lateral wall nodulation. After the baffle is installed, the dust removal ratio in the evaporating cooler is increased, larger amount dust at the bottom of evaporating cooler elbow pipe is collected, and the gas dust removal efficiency of 79.84%.

References

[1] Zhang Fuming, Cui Xingchao, Zhang Deguo. Construction of high-efficiency and low-cost clean steel production system in Shougang Jingtang [J]. Journal of Iron and Steel Research International. 2011, 18 (2): 42-51.
[2] Yin Ruiyu. Metallurgical Process Engineering [M]. Metallurgical Industrial Press, 2005: 325-328.

Green Iron and Steel Manufacturing Process of Shougang Jingtang Plant

Zhang Fuming Xie Jianxin

Abstract: Shougang Jingtang Iron & Steel Plant is the project which is built based on the concept of new generation recycling iron & steel process, with characteristics of "three functions", namely iron & steel product manufacture, high efficient conversion of energy, and disposal of wastes by reutilization. In engineering design, on the strength of theory of metallurgical process engineering, the advanced iron & steel manufacturing process of static process structure has been established by construction of two blast furnaces, one steel making plant and two hot rolling production lines, the advanced interface technology on ironmaking & steelmaking process has been developed, and the complete process of energy flow networking framing technology has been created. Energy consumption is reduced, with sufficient recovery of gas, heat and surplus energy during iron and steel manufacturing process, as well as emission reduction of dust and pollutant so as to achieve green iron and steel manufacturing and recycling economy, and energy utilization efficiency has an outstanding improvement. This paper introduces the green iron and steel manufacturing process of Shougang Jingtang Iron & Steel Plant, and establishment of its energy flow network.

Keywords: green iron and steel; engineering design; iron & steel manufacturing process; energy saving; circular economy

1 Introduction

For dispersal of non-capital functions in Beijing, and successful Beijing Olympics holding, it was decided that Shougang would be moved away from Beijing, a new iron and steel base, named Shougang Jingtang Iron & Steel Plant would be built in Caifeidian district, Tangshan, Hebei province. This project was built based on concept of new generation recycling iron and steel process, a large scale iron and steel project with the world advanced level in the 21st century. And it is the large scale modern iron and steel plant for all strip production built adjacent the sea and near the port. Shougang Jingtang Iron & Steel Plant has "three functions", namely: they pay attention to quality and high efficiency of iron and steel product manufacturing, while

---

本文作者：张福明，颉建新。原刊于 TMS 2017 Conference Proceedings (Energy Materials 2017). The Minerals, Metals & Materials Society, San Diego, 2017：17-29。

they focuses on high efficient and clean energy conversion, and disposal of a large quantity of social wastes by reutilization.

Design scale of Shougang Jingtang Iron & Steel is 8.70-9.20Mt/a crude steel. The construction was started on March 12, 2007, No. 1 blast furnace was blown in on May 21, 2009, and it was completely built and put into production smoothly on June 26, 2010. The plant was established with high efficiency and clean steel production process flow in physical frame by construction of two blast furnaces, one steel making plant, two sets of hot rolling mill and three sets of cold rolling mill. 5500m^3 blast furnace (the hearth diameter is 15.6m, hot metal daily output is 12,500t/d) was designed and built independently, with a batch of typical advanced process and equipment such as high efficiency and low cost clean steel production platform, 50,000t/d seawater desalination, and so on. It has important leading and demonstration effect for design and construction of the new generation and recycling iron and steel plant in the 21st century [1].

2 Process Flow Design and Structure Optimization

2.1 Analysis of Iron and Steel Manufacturing Process

2.1.1 Physical essential and feature of iron and steel manufacturing process

Physical essential of iron and steel manufacturing process is flow/rheologic process of material, energy and information in different time-space scale. That means to say, mass flow, under drive of energy flow, runs in dynamic-orderly mode along with specified "flow network" so as to achieve multiobjective optimization as per the configured "program". The optimized objectives include high quality product, low cost, high efficient and smooth production, high efficient energy utilization, low energy consumption, less emission, friendly environment, etc. Evolution and flow is the core of iron and steel flow operation.

Iron and steel manufacturing process is tasks in series of working procedures of various units, and production process cooperated and integrated with different working procedures. Normally, output of previous working procedures is input of subsequent working procedures, which are in mutual links and mutual buffer-matching. Iron and steel manufacturing process possesses complexity and integrality, the complexity presents two characteristics, namely, "complex variety" and "hierarchical structure".

2.1.2 Characteristic element of dynamic operation of iron and steel manufacturing process

Characteristic element of dynamic operation of iron and steel manufacturing process consists of "flow", "flow network" and "operational procedure", in which, "flow" means the main body of dynamic variation of the iron and steel manufacturing process operation, "flow network" (i.e., diagram structured by "joint" and "connector") means "flow" operation carrier and time-space boundary, and "operation program" means reaction of operation characteristics in information mode.

2.1.3 Operation characteristics of iron and steel manufacturing process

Therefore, it can be inferred that physical essential of iron and steel manufacturing process operation is the working procedure with relative units of class opening, far from equilibrium, irreversible, and different structures and functions, which is a process system established by means of no linear interaction and nesting structure. In this process system, ferrous mass flow (including iron ore, scrap, hot metal, molten steel, slab, steel product, etc.), under drive and action of energy flow (including coal, coke, electricity, steam, etc.), runs in accordance with specific "sequence" (including function sequence, time sequence, space sequence, time-space sequence and information flow regulating program) in specially designed complicated network structure (e.g. plane arrangement drawing of production workshop, general layout drawing etc.). The operation process of this kind process flow includes optimized integration to realize the operating elements and multi-target optimization of operating results.

Follow the guidance of integrated concept of overall dynamic-orderly, collaborative-continuous operation of iron and steel manufacturing process, the core ideology of iron and steel metallurgical engineering is that: analysis and optimization of working procedure function collection (including function collection of unit working procedure) are considered, as well as collaborative optimization of relationship collection between working procedures and reconstruction optimization of working procedures in the whole process, on the basis of capacity matching of dynamic operation of the upper stream and down stream working procedures,

2.2 Optimization of Elements

Selection and optimization of elements include these of technical elements, optimization of technical elements and collaborative optimization of basic elements in economy. In engineering design of Shougang Jingtang Iron & Steel Plant, selection and optimization of technical elements mainly include:

In selection of product rolling mill, Proposal 1 is provided with one set of hot strip mill and one set of plate mill, with production capacity approx. 7.0 Mt/a; Proposal 2 is provided with 2 sets of hot strip mill with production scale approx. 9.0 Mt/a. Different rolling mill configuration directly influences scale, process and equipment of the steel making plant, also influences structure and dynamic operation efficiency of the steel making plant, and then influences quantity, volume and layout of blast furnace. After systematic comparison and selection, as well as careful research and decision-making, Proposal 2 was finally adopted, namely 2 sets of hot strip mill were used, the iron and steel plant produces complete thin strip products, so that a professional production base has been built for strip products.

In design of process flow of the steel making plant, and in selection of production process of complete strip, intensive theoretical analysis and comparison study were carried out to the conventional process flow and the "high efficient, low cost and clean steel production process flow", which consists of "three-all-removal" pretreatment of total hot metal-steel making-secondary refining-high casting and constant casting CCM. After scientific verification and research finally, the high efficient, low cost and clean steel production process flow with "three-all-removal" pretreatment of total hot metal has been accepted [2].

In design of iron making process, as to quantity of blast furnace, in-depth research, analysis and demonstration were conducted, and the design concept with blast furnace large-sizing was established under the premise of process structure optimization of the iron and steel plant. Delicate comparative study and scientific verification were carried out whether two 5500m³ blast furnaces or three 4000m³ blast furnaces are built [3]. It is determined by study that the two 5500m³ blast furnaces were provided, and configured with two 550m² sintering machines, one 504m² straight grate for pellet production, and four 70-chambers 7.63m high coke ovens for material and fuel supply to the blast furnaces, so as to achieve "iron making system" process structure optimization with blast furnace as the center and optimal matching of process equipment. Furthermore, the process flow can be optimized, and project investment can be decreased, which are benefit to improve operation efficiency of ferrite material flow and carbon energy flow.

In study of interface technique between iron making plant-steel making plant, the multi-function hot metal ladle technology is finally chosen after a mass of investigation, survey and test, i. e. the multi-function hot metal ladle type direct transport process "one ladle through" is applied to decrease project investment, reduce hot metal temperature lowering and environment pollution, as well as improve efficiency of hot metal desulphurization pre-treatment [4].

In design of energy flow network structure, on the basis of process behaviour and conversion feature of energy flow and different energy medium operation, a perfect energy supply system and an energy conversion network system are designed. And the energy control center is designed and established based on real time monitor, on-line despatching, process control and centralized management [5]. Intensive study and systematical optimization are carried out to effective conversion of energy and optimal configuration of energy structure, with sufficient recovery and reuse of secondary energy from iron and steel manufacturing process, taking full advantage of waste heat and waste energy from the iron and steel plant for power generation with self power supply ratio from the iron and steel plant more than 96%, and "zero emission" of various associated gas during iron and steel metallurgical process can be achieved.

In design of recycling economy, green iron and steel manufacturing, as well as energy saving and emission control, the advanced large scale technological equipment is applied to improve production efficiency and energy utilization efficiency, and reduce energy consumption. The four 70-chambers 7.63m high coke ovens are provided with two sets of 260t/h high temperature and high pressure coke dry quenching device with power generation 105kW · h/t; The 5500m³ blast furnace uses complete low calorific value BFG as fuel, with application of the effective air and gas preheating technology and high blast temperature top-combustion hot blast stoves to have BF blast more than 1250 degree centigrade, max. monthly average blast temperature 1300 degree centigrade [6,7]; the full-dry dedusting process and TRT technology are adopted for BFG in innovative way, configured with the 36.5MW installed capacity Top Gas Recovery Turbine Unit (TRT), and dust in gas can be 2-4mg/m³, power generation 54kW · h/t; the dry dedusting technology is adopted to BOF gas[8], with gas recovery 110m³/t, steam recovery 97kg/t, and dust in gas can be 3-5mg/m³. Gas associated to metallurgical process of the iron and steel plant is recovered completely so as to be used for combustion and heating in the iron and steel manufacturing process. Surplus gas is used to power generation unit, and the "exhaust steam" after power generation is applied as heat source of the 50,000t/d low temperature multi-effect seawater desalination device.

2.3 Optimized Structure

A flow network with simple and smooth, systematic and intensive, high efficient operation characteristics has been established after optimized selection of a series of working procedure/equipment element, in particular, the physical frame "2-1-2" high efficiency iron and steel manufacturing process structure with two blast furnaces + one steel making plant and two sets of hot continuous casting and rolling mill is built, taking this as a core to build the dynamic operation structure of iron and steel manufacturing process with dynamic-orderly and collaborative-continuous features.

Main technological equipment of Shougang Jingtang Iron & Steel Plant includes: four 7.63m high coke ovens and two sets of 260t/h CDQ device, two 550m² sintering machines, one 504m² straight grate pellet production line, two 5500m³ blast furnaces; four sets of KR hot metal desulphurisation pre-treatment device and two 300t dephosphorization and desiliconization converters, three 300t decarburization converters, one LF, two CAS refining devices, two RH vacuum refining devices, two 2150mm 2-strand slab continuous casting machines, two 1650mm 2-strand slab continuous casting machines; one set of 2250mm hot continuous casting and rolling mill production line, one set of 1580mm hot continuous casting and rolling mill production line, each of 2230mm, 1700mm and 1420mm pickling and cold rolling combined unit each as well as the matched production lines of the continuous annealing train, box annealing, continuous hot dip galvanizing, cross cutting, etc. The iron and steel manufacturing process flow of Shougang Jingtang Iron & Steel Plant is shown in Fig. 1.

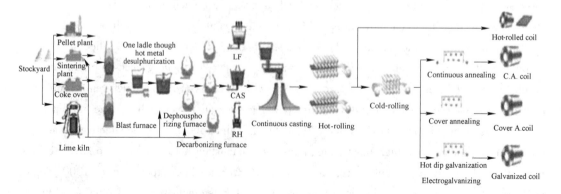

Fig. 1 Iron and steel manufacturing process flow of Shougang Jingtang Steel Plant

In design of the iron and steel plant, the compact, high efficient, smooth, aesthetic general layout has been realized to the maximum. High efficiency and collaboration is achieved for material flow, energy flow and information flow. Material transportation between working procedures with compact, intensive, efficient and fast features is realized. The raw material yard and product warehouse is closely arranged to wharf, realizing receiving, unloading and transferring with shortest distance for raw material and product; transportation distance from BF to the steel making desulphurizing station is only 900m; zero distance interface between CCM and hot rolling is realized; 1580mm hot rolling product warehouse is arranged closely to 1700mm cold rolling raw material warehouse, realizing compact layout; the occupied land area per ton of steel in the iron and steel plant is 0.9m^2, which reached international advanced level.

2.4 Function Expansion and Efficiency Optimization

At the same time of optimized selection and process structure optimization of working procedure/equipment element, function expansion and efficiency optimization must be drawn attention, i.e. the single function of conventional iron and steel plant is expanded to "three functions", and the functional connotation is more innovative. For instance, iron and steel manufacturing functions can be integrated to high efficient, low cost and clean steel production system; the energy conversion function should form the whole network and entire process high efficient conversion system with collaborative and efficient characteristics one the basis of omni-process energy flow network structure and feature of input/output dynamic operation optimization, so as to have scientific, reasonable and efficient recovery and reuse for further energy saving and emission control; in solid waste disposal aspect with consideration of recycling and circular economy, a chain of circular economy with iron and steel plant as the core is established in order to expand to the eco-industrial park and achieve integrative development of many industries [9].

3 Establishment and Optimization of Flow Network

3.1 Establishment of Flow Network

Setting flow network concept must guarantee simple, compact, intensive and smooth process arrangement drawing, general layout and others in the iron and steel plant. Taking this as static frame to make "flow" behaviour run in regular, dynamic, orderly, collaborative and consecutive, so as to achieve "minimum" dissipation during operation process. In engineering design of the iron and steel plant, flow network firstly reflected in material flow network, while study and establishment of energy flow network and information flow network should be highlighted. Fig. 2 and Fig. 3 are network flow diagram of material flow and energy flow operation in Shougang Jingtang Iron & Steel Plant respectively.

3.2 Energy Flow Network of Iron and Steel Manufacturing Process

Modern iron and steel metallurgical engineering design takes energy as an important factor

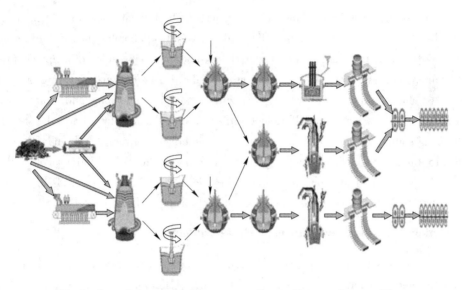

Fig. 2 Operation network of ferrous mass flow in Shougang Jingtang Plant

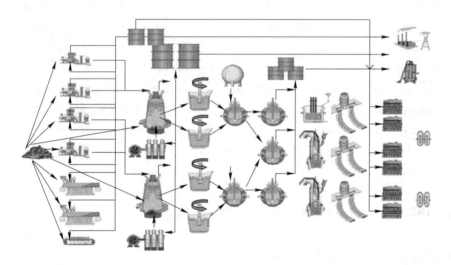

Fig. 3 Operation network of energy flow in Shougang Jingtang Plant

throughout the whole process. With consideration of correlation causality and dynamic cooperative with material flow, it is necessary to have study at level of energy flow behaviour and energy flow network. Concept of elements such as "flow", "flow network", "operational procedure" and so on must be established in energy research and engineering design, to study input/output behaviour of energy flow in open, non equilibrium and irreversible process, that is to say, some dynamic and isolated cross points should be used to calculate dynamic operation of energy flow in the flow network. And time-space-information concept of energy flow operation concerned to iron and steel manufacturing is included, but it should not be limited to concept of material-energy balance calculation. In process design and modification progress of iron and steel manufacturing, the material flow conversion progress as well as design of its "program" and "material flow network" should be paid attention, meanwhile, engineering design of the energy flow, energy conversion "program" and "energy flow network" should be attached in importance.

There are primary energy (mainly outsourced coal and others) and secondary energy (e. g, coke, electricity, oxygen, various gases, waste heat, surplus energy, etc.) in iron and steel manufacturing process. These form different starting nodes (for instance, material stock yard, blast furnace, coke oven, converter, etc.) respectively. Energy and medium from these starting "nodes" are conveyed to the terminal nodes (for instance, various terminal consumers, steam power plant, steam station, power generation station, etc.) of energy conversion via the connecting approaches (namely connectors) of conveying line, pipeline, etc. During conveying and conversion of energy flow, the necessary and effective intermediate surge tank (buffering system) should be provided, e. g, gas holder, boiler, pipeline, etc. so as to achieve buffering, collaboration and stability in aspects of time, space and energy stage and so on between starting nodes and terminal nodes of energy, so that energy flow network of iron and steel manufacturing process is built.

3.3 Development and Application of Interface Technology

The so called "interface technology" means connection-matching, collaboration-buffering technology and the relative device (equipment) between the main working procedures, such as iron making, steel making, CCM, hot rolling mill, etc. in iron and steel manufacturing process. "Interface technology" not only includes relative process and devices, but also a series of engineering technology of time-space reasonable configuration, device quantity (capacity) matching, etc. for arrangement plan and others.

The interface technology mainly shows in aspects of connection, matching, coordination, stability, and so on of basic parameters of production process material flow (including flowrate, composition, organization, shape, etc.), production process energy flow (including primary energy, secondary energy, energy consuming terminal, etc.), production process temperature, production process time, space position, etc.

The interface technology is collaborative optimization technology of relationship between working procedures developed on the basis of process design and innovation of function optimization of unit working procedure, optimization of operation procedure, optimization of flow network, etc. including relationship collaboration-optimization of adjacent working procedures, or relationship collaboration-optimization between multiple working procedures. The interface technology in form can be split to material-time/space interface technology, interface technology of material nature conversion, interface technology of energy/temperature conversion, etc.

One of the important ideas of dynamic and precise design method is to pay attention to optimization of device proper of various working procedures, especially in connection and matching relationship between devices of working procedures, as well as development and application of the interface technology. For example, the multi-function hot metal ladle direct transport technology is applied between working procedure interface of the iron making plant and steel making plant so that the interface connection between the upstream and downstream working procedures. Advanced design tools such as dynamic "Gantt Chart" and so on are applied to have careful design in advance to working procedure devices of the iron and steel manufacturing process and its dynamic operation, to make the actual operation achieve dynamic, orderly, collaborative and continuous one.

4 Operation Practice

Shougang Jingtang Iron & Steel Plant has built the efficient process structure with two blast furnaces + one steel making plant + two sets of hot tandem rolling mill by means of analysis, optimization and systematic integration of the working procedure-function-structure-efficiency, based on design and development of extra-large 5500m^3 BF bell-less top equipment, high blast temperature dome combustion hot blast stove, full dry dedusting technology in extra-large scale BF, hot metal "three-all-removal" pre-treatment, clean steel production platform, and application of the direct hot metal transport mode by multi-function hot metal ladle etc. The omni-process with characteristics of precise matching and dynamic coordination has been realized.

In comparison with iron & steel plants at home and abroad, obvious improvement is shown in operating efficiency of the system. By means of construction of production dispatching centre and energy management and control centre of the whole plant, the system with the whole plant flow-program-network were established. Through research and development of the direct hot metal transport technology with multi-function hot metal ladle "One ladle for the whole process" and the standard track gauge of iron making-steel making interface, the distance between BF and steel making is cut down to 900m hot metal temperature lowering is decreased by 30-50 degree centigrade, and average temperature of hot metal before the KR desulphurizing station can reach to 1370 degree centigrade. Because of cancellation of the hot metal reladling station and the hot metal reladling bay in the steel making building, area of the steel making plant is reduced by approx. 1150m^2, more than 40,000,000RMB investment is saved due to removal of reladling equipment and area reduction of the building, while fume and dust emission can be reduced by approx. 4700t each year, post manning can be decreased by approx. 108, and more than 60,000,000 RMB production operation cost can be lowered each year. The low-temperature multi-effect seawater desalination process with application of low temperature and low pressure turbonator was independently researched and developed so as to achieve "trigeneration of heat-water-electricity", and thermal efficiency of the system was improved to 82.23%. Switchover on three kinds of working condition can be carried out by the system, and it makes full use of different quality and abundant steam resources to reduce steam bleeding. The four sets 12,500t/d seawater desalination equipment with application of low-temperature and multiple-effect distillation technology were designed and applied. 17,500,000t/a fresh water resource can be replaced, and approx. 10Mt/a brine can be supplied to alkali plants in the society, which can reduce direct drainage of brine and realize reuse of resources. Where two sets of topping generator unit with installed capacity 25MW each and two sets of 12,500t/d low-temperature and multiple-effect distillation equipment can generate 350,000,000kW · h per annum. By application of the above mentioned advanced process technologies, enormous economic benefit can be created each year, with obvious social benefit and environmental benefit.

The large scale converter "2+3" clean steel production platform was built, which was well matched with the efficient CCM system. In comparison with conventional combined blowing process, the new process for clean steel production with characteristics of hot metal "three-all-removal" pre-treatment technique built by means of improvement of elementary reaction efficiency could increase by double of steel product cleanness, and total content of impurity elements of [S], [P], [H], [N] and [O] can be decreased to less than $70×10^{-6}$, consumption of lime for steel making is reduced to 25kg/t, slag produced from steel making process is lowered to 50kg/t. In comparison to the conventional top-bottom combined blowing converter production process, production cost advantage is obvious.

The dynamic and precise design as well as dynamic process optimization technology was studied and applied for the new generation iron and steel plant. By establishment of theoretical framework, physical model and simulation model of the dynamic-orderly operation, intensive analysis and research were carried out to various working procedures (systems) of iron and steel plant such as material and fuel consumption, production capacity matching, determination of various process parameters, consumption of energy and dynamics, arrangement of energy facilities, connection between different working procedures, and so on. Accurate calculation and optimized configuration were conducted in aspects of temperature, material composition and ingredient, operation rhythm, energy input/output, etc. By application of dynamic and precise design theory, the production operation system of Shougang Jingtang Iron & Steel Plant with dynamic, orderly, continuous, compact, precise and harmonious characteristics was established.

The practice shows that, hot metal tapped from blast furnace was entered to the steel making plant for desulphurization, desiliconization-dephosphorization, decarburization; secondary refining, CCM; and then hot rolled product by hot rolling working procedure. Dynamic and precise optimization was carried out by means of overall process operation rhythm, temperature regulating, composition control and working organization to control the whole production time within 400min.

As per the complete process energy conversion system which was established on basis of circular economy, Shougang Jingtang Iron & Steel Plant has achieved high efficiency energy conversion of waste heat and residual pressure and constituted the industrial chain of circular economy, giving play to remarkable economic and social benefits. The industry chain of comprehensive utilization of seawater cored on seawater desalination has been taken initial shape. The thermal type low-temperature multi-effect seawater desalination process was firstly applied to achieve high efficient energy conversion and utilization to make use of waste heat and residual energy from the iron and steel plant. Fresh water production capacity is 17,500,000t/a by means of the seawater desalination process, and power generation 340,000,000kW · h/a, which became a model of circular economy development of modern iron and steel plant. The industry chain is made up with seawater desalination and downstream salt making sector, and brine produced from seawater desalination is supplied to the alkali making enterprises adjacent to the iron and steel plant. Full recovery of residual energy from iron and steel manufacturing process for achievement of high efficient energy conversion. Two sets of 300MW power generation unit take mixed combustion of BFG, annual gas recovery is equivalent to approx. 370,000t standard coal, and CO_2 emission is reduced by 1,160,000t per annum. The residual heat resource recovered from the production process, besides self use of the steel plant, can be supplied to the peripheral enterprises. At present, hot water has been delivering to the heating system of the enterprises around the steel plant. Resourceful utilization of solid wastes "Zero emission" has been achieved by means of technical integration of high efficient recovery, resource regeneration and industrialization. Furthermore, deep processing is used to promote value of recycling product of solid wastes. 100% cyclic utilization comes true to various solid wastes, such as granulated BF slag, steel slag, coal ash, collected dust, scale from rolling mill, and so on.

Meanwhile, clean production and environmental protection have been vigorously strengthened, as well as development of low carbon green iron and steel manufacturing and circular economy. Investment of environmental protection of Shougang Jingtang Iron & Steel Plant is approx. 7,596,000,000RMB, occupying 11.21% of the total project investment. Among them, there are 128 offgas treatment facilities and 8 sets of wastewater treatment facility. The on-line environmental monitoring system was built to strengthen monitor of pollution source. So far, 14 on-line systems for fume monitoring were built and put into operation, including the captive power plant, desulphurization station of sintering process, dedusting system of blast furnace casthouses, dedusting system of secondary dedusting of the steel making process, etc. of the iron and steel plant.

5 Conclusion

The project of Shougang Jingtang Iron & Steel Plant was built based on concept of new generation recycling iron and steel process, a large scale iron and steel project with the world advanced level in the 21st century. And it is the large scale modern iron and steel plant for all strip production built adjacent the sea and near the port. It is the major demonstration project in aspects of structure adjustment, technology upgrading and transformation development of China's iron and steel industry in the 21st century. It has great significance to optimization of structure and achievement of sustainable development of China's iron and steel industry.

The advanced philosophy and method for iron and steel metallurgical engineering design were used in construction of Shougang Jingtang Iron & Steel Plant project, which is based on physical nature of iron and steel manufacturing process dynamic operation and deep understanding of development trend of iron and steel industry, focused on links of concept design, top-down design, dynamic and precise design, etc. of the iron and steel plant, so that the new generation green recycling iron & steel manufacturing process was independently designed and built. Shougang Jingtang Iron & Steel Plant possesses the "three functions", namely, iron and steel product manufacturing, high efficient energy conversion and disposal of solid wastes by reutilization, which is an important direction and approach for achievement of green transformation development of China's iron and steel industry in the 21st century.

Since Shougang Jingtang Iron & Steel Plant was put into production for 6 years, the production operation is stable and smooth, with advanced technical indices, the major technical indices have been achieved or exceeded the design level. The plant has been the demonstration project of new generation and green recycling iron & steel manufacturing process, leading the development direction of China's iron and steel industry, and it has far-reaching and great strategic significance to improve competitiveness and sustainable development of China iron and steel industry.

References

[1] Zhang Fuming, Cui Xingchao, Zhang Deguo, et al. New generation process flow and application practice of Shougang Jingtang Steel Making Plant [J]. Steelmaking, 2012, 28 (2): 1-6.

[2] Yin Ruiyu. Study on high efficient and low cost clean steel platform [J]. China Metallurgy, 2010, 20 (10): 1-10.

[3] Zhang Fuming, Qian Shichong, Zhang Jian, et al. New technologies applied in Shougang Jingtang 5500m^3 blast furnace [J]. Iron & Steel, 2011, 46 (2): 2-17.

[4] Qiu Jian, Tian Naiyuan. Shougang ironmaking-research on model of ironmaking/steelmaking interface [J]. Iron & Steel, 2004, 39 (4): 74-78.

[5] Gu Liyun. Theory and practice of the control and management system of energy in Shougang Jingtang united iron & steel Co., Ltd [J]. Metallurgical Automation, 2011, 35 (3): 24-28.

[6] Zhang Fuming, Hu Zurui, Cheng Shusen. Combustion technology of BSK dome combustion hot blast stove at Shougang Jingtang [J]. Iron and Steel, 2012, 47 (5): 75-81.

[7] Zhang Fuming, Mei Conghua, Yin Guangyu, et al. Design and research of BSK dome combustion hot blast stove of Shougang Jingtang 5500m^3 blast furnace [J]. China Metallurgy, 2012, 22 (3): 27-32.

[8] Zhang Fuming. Study on large scale blast furnace gas dry bag dedusting technology [J]. Ironmaking, 2011, 30 (1): 1-5.

[9] Yin Ruiyu. Theory and Method of Metallurgical Process Flow Integration [M]. Beijing: Metallurgical Industry Press, 2016: 284-287.

Innovation and Application on High Blast Temperature Technology of Shougang Blast Furnace

Zhang Fuming

Abstract: This paper sets forth the important significance of high blast temperature technology, introduces process equipment of Shougang hot blast stove and development status of high blast temperature technology, and analyses the technical concept and key technology for achievement of high blast temperature. This paper also presents the process flow of high blast temperature achieved by development of combustion air high temperature preheating technology under combustion of low calorific value blast furnace gas. Various methods of numerical simulation, test research and so on are applied for design and development of dome combustion hot blast stove technology. Design of hot blast stove regenerator chamber is optimized and high efficient checker brick is applied to improve heat transfer efficiency. Design concept of "no-overheating and low stress" is applied as basis for achievment of optimal design of high temperature hot blast pipeline and stable transfer of high temperature hot blast. Effective measures are taken to prevent intergranular stress corrosion of hot blast stove and prolonged service life of hot blast stove. Operation of blast furnace and hot blast stove is optimized to make high blast stove technology have better application effectiveness.

Keywords: blast furnace; ironmaking; high blast temperature; dome combustion hot blast stove; hot blast pipe

1 Introduction

Since stepping on 21st century, blast furnace (BF) ironmaking process is faced with many major development issues on restriction again by aspects of natural resources shortage, undersupply of energy resources, ecological environment protection, etc.[1] In the present fierce market competition environment at domestic and abroad, and in the face of severe situation and challenge, with consideration of sustainable development of BF ironmaking process in 21st century, it is a must to have remarkable breakthroughs on aspects of high efficiency and low consumption, energy saving and emission control, circular economy, low carbon metallurgy, clean and environment protection, etc. It is necessary to have further increase of blast temperature and reduction of fuel ratio in order to improve vitality and competitiveness of BF ironmaking technology. BF ironmaking production system with characteristics of high efficiency, low

---

本文作者：张福明。原刊于 Proceedings of the Iron & Steel Technology Conference, AISTech, 2014: 435-447。

consumption, low cost and low emission is established with effort in order to realize sustainable development of BF ironmaking process, and actualize low carbon smelting and circular economy.[2]

High blast temperature is one of the important technical features of modern BF ironmaking. Improvement of blast temperature is key and common technology of development of circulating economy, realization of low carbon smelting, energy saving and emission control, and sustainable development at present in iron and steel industry. High blast temperature has extreme important significance to improvement of integrative BF technical level, reduction of CO_2 emission, and leading the industry technology progress.

2 Development Status

For a long time, average blast temperature of BF in key steel plants in China has been wandering at 1,000-1,150℃, which is maximum difference of technical parameters in comparison with overseas advanced level. With innovation development of high blast temperature technology in China in recent years, average BF blast temperature presents rising trend. In 2012, the average blast temperature in major steel enterprises in China reached to 1,183℃, 4℃ improved in comparison with that in 2011. More than 50 BFs nationwide has annual average blast temperature more than 1,200℃. Annual average blast temperature of large scale BFs of Shougang Qiangang and Baosteel with volume 4,000m³ has achieved more than 1,250℃ blast temperature, which own preferable leading role. Fig. 1 shows variation tendency of main operational parameters of BFs of China's key steel enterprises.

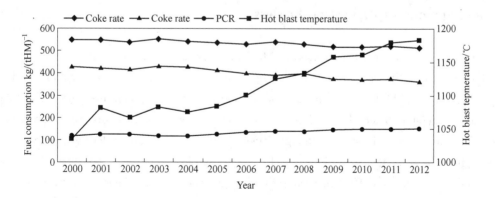

Fig. 1 Variation tendency of main operational parameters of BFs of China's key steel enterprises

Since 2004, Shougang has built new steel production bases in Qinhuangdao, Qian'an and Caofeidian in Hebei province one after another, and the original steel plant in Beijing area shut down completely in 2010. Production scale, product structure and technical equipment in the three steel plants, namely Shougang Shouqin, Qiangang and Jingtang are different. Although there are quite differences on BF capacity and number, as well as process and technical equipment, there are consistencies in aspects of development concept, research and application of the high blast temperature technology. Table 1 lists main technical configuration of BF hot blast stove in Shougang's new steel plants.

Table 1 Main technical equipment of BF hot blast stoves in Shougang's three steel plants

Items	Shouqin		Qiangang			Jingtang	
BF No.	1	2	1*	2	3	1	2
Effective volume/m^3	1,200	1,780	2,650	2,650	4,000	5,500	5,500
Hearth diameter/m	8.1	9.7	11.5	11.5	13.5	15.5	15.5
Structure of hot blast stove	Dome combustion	Dome combustion	Internal combustion	Internal combustion	Internal combustion	Dome combustion	Dome combustion
No. of hot blast stoves	3	3	3	3	4	4	4
Preheating process	Heat pipe exchanger + preheating stove	Heat pipe exchanger + preheating stove	Heat pipe exchanger	Heat pipe exchanger + preheating stove	Heat pipe exchanger + preheating stove	Heat pipe exchanger + preheating stove	Heat pipe exchanger + preheating stove
Fuel for hot blast stove	Full BFG	Full BFG	BFG + COG	Full BFG	Full BFG	Full BFG	Full BFG
Design maximum blast temperature/℃	1,250	1,250	1,250	1,250	1,280	1,300	1,300
Commissioning (y-m)	2004-6	2005-1	2004-10	2005-1	2010-1	2009-5	2010-6

* Note: Technical revamping was carried out for the hot blast stoves of this BF, and No. 4 hot blast stove (dome combustion) was newly built with application of full BFG process.

3 Development on Preheating Process

With technical progress of BF iron making, fuel ratio for large scale BF has been decreased to less than 520kg/t, BF gas utilization ratio improved by more than 45%, and calorific value of BF gas less than 3,000kJ/m^3. For realization of energy resources optimization and high efficient utilization of the steel plant, high calorific value coke oven gas and BOF gas are mainly used for production procedure of steel making, steel rolling, etc. High calorific value gas is undersupply. Therefore low calorific value BFgas has to be used for combustion of hot blast stove. Under the condition of no high calorific value gas enrichment, theoretic flame temperature and dome temperature are not high, so that it is hard to have more than 1,200℃ high blast temperature. This is one of the main reasons which restrict improvement of BF high blast temperature in China at present.

During design of Shougang Jingtang 5,500m^3 BFs, in order to reach 1,300℃ blast temperature under condition of single BF gas combustion for the hot blast stoves, the integrative technical measures of improvement of theoretic flame temperature, dome temperature and hot blast temperature of the hot blast stoves have been studied systematically.[3] At the basis of high temperature preheating technology with combustion air, two level preheating technology with high efficiency with long service life type gas and combustion air has been designed and developed.[4]

The technical principle of this process is: the separate type heat pipe exchanger is applied to recover fume waste heat of the hot blast stoves, gas and combustion air are preheated, temperature of the preheated gas and combustion air can reach to 200℃ approximately, this process is called as first stage double preheating. Two small hot blast stoves for preheating of combustion air are provided to improve combustion air temperature to more than 520℃. Both gas and combustion air after preheating by combustion air preheating stove should pass through first stage preheating process by the heat pipe exchange, and dome temperature of the preheating stove can reach to 1,300℃. Two preheating stoves work alternatively to heating some combustion air which is used for combustion process of the hot blast stoves. The combustion air after preheating process by the preheating stoves can have temperature 1,200℃. Then it is mixed with the combustion air after first stage preheating, so that the temperature of combustion air after mixing can controlled in 520-600℃. This process is called as second stage preheating. This process is one kind of self-circulating preheating one which improves physical heat of combustion air and gas obviously. Under condition of combustion with single BF gas dome temperature of the hot blast stoves can reach to 1,420℃ and even more so as to improve blasting temperature efficiently, and the overall thermal efficiency of the hot blast stove system improves obviously. This technology is applied in Jintgang's two huge BFs, and monthly average blast temperature can be 1,300℃.[5] Table 2 shows the main technical parameters of heat pipe exchanger. Table 3 indicates main technical parameters of the preheating stove for combustion air; and Fig. 2 shows the preheating process flow of gas and combustion air for hot blast stoves of Jingtang's 5,500m³ BF.

Table 2 Main technical parameters of heat pipe exchanger

Items	Fume	Gas	Air
Gas flow Nm³/h	700,000	505,349	319,000
Inlet temperature/℃	350	45	20
Outlet temperature/℃	165	215	200
Flow resistance/Pa	≤500	≤550	≤600
Maximum steam temperature inside the heat pipe/℃		260	248
Heat exchange area/m²	6, 811+14, 224 (high temperature side)		
Heat recovery/kW	21, 461+35, 373=56, 833		

Table 3 Main technical parameters of preheating stove for combustion air

Items	Data
Quantity of preheating stove for combustion air	2
Working cycle/min	90
Blasting time/min	45
Dome temperature/℃	1,315

Continued Table 3

Items	Data
Hot air outlet temperature/℃	1,200
Inlet air temperature/℃	190
Temperature of combustion air/℃	190
Gas temperature/℃	190
Average temperature of fume/℃	340
Maximum high temperature of fume/℃	450
Combustion air to be heated/Nm³·h⁻¹	129,149
Gas consumption volume/Nm³·h⁻¹	99,083
Combustion air consumption volume/Nm³·h⁻¹	68,888

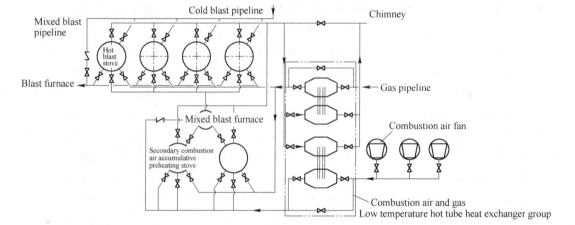

Fig. 2　Preheating process flow of gas and air for hot blast stove of Jingtang's 5,500m³ BF

4　Study on Process

As far back as 1979, dome combustion hot blast stove is applied to Shougang's No. 2 BF (1,327m³). In 2004, Shougang at basis of the existing technology of dome combustion hot blast stove, introduced Russian Kalugin dome combustion stove technology with circular ceramic burner, and the new type dome combustion hot blast stove was applied at Shouqin's No. 1 and No. 2 BF in succession. Better performance of application is achieved after the put into production of these 2 BFs with monthly average blast temperature 1,250℃. During design of Shougang Jingtang's 5,500m³ BF, Shougang dome combustion stove technology is combined with Kalugin dome combustion stove technology with integrative advantage of the two technologies, so that the dome combustion stove technology for huge BF is designed and developed. It is the first one worldwide to use the dome combustion stove technology at huge BF with volume more than 5,000m³. Main features of this new type dome combustion stove as follows:

(1) The circular ceramic burners are provided at the dome of the stove so as to have a better

adaptability to the working conditions. It is possible to have operation at various working condition with gas and combustion air with characteristics of strong combustion power, high combustion efficiency and longer service life. Circular ceramic burner adopts swirling diffusion combustion technology to ensure sufficient mixing and combustion of combustion air and gas so as to improve theoretic flame temperature and dome temperature.

(2) Dome space is used as combustion chamber, and independent combustion chamber structure is cancelled to strengthen thermolability of stove proper structure. Ceramic burner is located at dome position, and high temperature fume distributes equally under rational flow status so as to efficiently improve uniformity and heat transfer efficiency of high temperature fume at surface of checker bricks in the regenerator chamber.

(3) Regenerator chamber adopts high efficient checker bricks to reduce the diameter of checker hole in order to increase heating surface of checker bricks and improve heat transfer efficiency of hot blast stove.

(4) Waste heat from fume of the hot blast stoves is reused to preheat gas and combustion air, and the combustion air is preheated to 520℃ above again by the preheating stoves. Under condition of combustion with single BFG, dome temperature of the hot blast stoves can reach to 1,420℃ and the blast temperature can be 1,300℃.

(5) Design concept of "no-overheating and low stress" is used for hot blast stove high temperature and high pressure pipeline system. Pipeline system and refractory structure design are optimized to have stable transfer of 1,300℃ high temperature blast.

Shougang Jingtang's 5,500m³ BF is configured with 4 dome combustion stoves with design blast temperature 1,300℃, and dome temperature 1,420℃.[6] Silica brick is applied in high temperature zone of the hot blast stoves with design service life more than 25 years. Single BF gas is used as single fuel for the hot blast stoves, and the recovery devices for fume waste heat from the hot blast stoves are used for heating of gas and combustion air. Two small dome combustion stoves are applied for heating combustion air to make combustion air temperature at more than 520℃. High temperature valves for hot blast stoves are applied for demineralized water closed circulating cooling. And automatic control is considered for heating, blasting and changeover of the hot blast stoves. During normal operation of the 4 hot blast stoves, the staggered parallel operation mode with "two-stove burning and two-stove blasting" is applied. Under combustion of single BF gas, the blast temperature can reach to 1,300℃. Table 4 shows the main technical features of dome combustion stoves of Jingtang's No. 1 BF.

Table 4 Main technical parameters of dome combustion stove of Jingtang's No. 1 BF

Items	Data
No. of hot blast stoves set	4
Height of hot blast stove/m	49.22
Diameter of hot blast stoves/m	12.50
Total heating area of checker brick of one hot blast stove/m²	95,885

Continued Table 4

Items	Data
Heating area of checker brick of unit volume/$m^2 \cdot m^{-3}$	48
Pore diameter of checker brick/mm	30
Height of checker brick/m	21.48
Sectional area of the regenerator chamber/m^2	93.21
Hot blast temperature/℃	1,300
Dome temperature/℃	1,420 (maximum)
Design temperature of dome refractory/℃	1550
Fume temperature/℃	Maximum 450, normal average 368
Preheating temperature of combustion air/℃	520-600
Gas preheating temperature/℃	215
Cold blast temperature/℃	235
Cold blast volume/$Nm^3 \cdot min^{-1}$	9,300
Cold blast pressure/kPa	540
Blasting time/min	60
Burning time/min	48
Changeover time/min	12

5 Research on Transfer Theory

5.1 Computus Fluid Dynamic Simulating Research

In view of principles of mass conservation, momentum conservation and energy conservation of hydromechanics, initial condition and boundary condition are set up. CFD mathematical model is used for numerical calculation to heating process of the dome combustion stoves. Transfer theory of heating process of the dome combustion stove is studied. The speed field, concentrate field and temperature field distribution in the combustion chamber during heating process of the dome combustion stove are analysed. As the research results, the design structure of hot blast stove and burner are optimized to have more uniformity of high temperature distribution in the combustion chamber of hot blast stove.[7] Fig. 3 to Fig. 6 illustrate the simulating calculation results of velocity field distribution, flow field distribution, CO concentrate field distribution and temperature field distribution respectively.

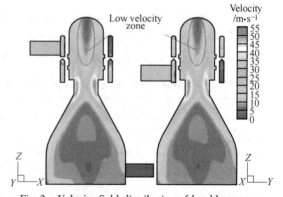

Fig. 3 Velocity field distribution of hot blast stove

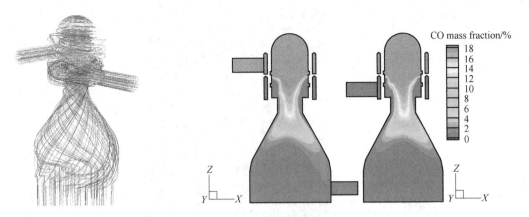

Fig. 4 Flow field distribution of hot blast stove   Fig. 5 CO concentration field distribution of hot blast stove

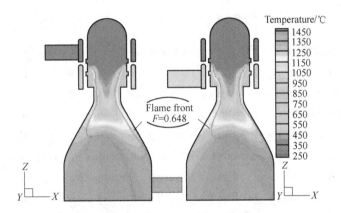

Fig. 6 Temperature field distribution of hot blast stove

5.2 Experimental Research of Physical Model

To verify correctness of numerical calculation results, physical model experiment device for dome combustion hot blast stove is established in accordance with design parameters of hot blast stove, initial condition and boundary condition of numerical calculation, as well as principle of similitude and geometric dimensioning 1∶10 proportion. Fig. 7 shows the photo of physical experimental model.

Simulation test of physical model is used to measure parameters of flow velocity, pressure, temperature and so on of test device of physical model, and comparison with CFD numerical calculation result is carried out so as to verify correctness of numerical calculation results. Gas flow velocity at different nozzles of annular burners is measured to test uniformity of gas flow distribution. Structure type and angle of combustion air and gas nozzle is adjusted to make equal gas flow velocity distribution so as to optimize design of real hot blast stove ceramic burner.[8]

Through physical model test study, the following targets can be achieved:

(1) The velocity, pressure and temperature etc. of physical model test facility can be

Fig. 7 Photo of the physical experimental model of hot blast stove

measured, and compared with CFD numerical calculation result in order to verify the correctness of numerical calculation results.

(2) According to similarity principle, normal temperature air is used to simulate BF gas, heated air is used to simulate combustion air, the flow field of physical model is put into second simulation zone by adjusting the flow rate of combustion air and gas, Euler Dimensionless Number is basically unchanged, actual pressure loss of hot blast stove is measured according to measured pressure loss in order to compare and verify the flow field numerical calculation result.

(3) Through measuring the gas flow rate at different injection position of circular burner, the evenness of gas distribution is verified.

(4) By adjusting the structure type and angle of combustion air and gas nozzle, the gas flow rate is distributed evenly, actual hot blast stove ceramic burner design is optimized. Through physical model simulation test, numerical calculation results are verified.

5.3 Investigation of Hot Condition Simulating Trial

For further study of transfer theory of combustion, gas flow and heat transfer process of dome combustion type hot blast stove, 2 hot blast stoves for hot condition trial based on prototypical of dome combustion hot blast stove of Jingtang's 5,500m^3 BF are established at basis of cold simulating model (Fig. 8). The "one stove burning and one stove blasting" working mode is applied to simulate real working process of the hot blast stove. The inner diameter of the 2 dome combustion type experimental stoves is in proportion 1 : 10 with real hot blast stove, with total height 11.3m. No.1 hot blast experimental stove adopts 37 hole checker bricks with 20mm hole diameter, and No.2 hot blast experimental stove adopts 19 hole checker bricks with 30mm hole diameter.

The 2 hot blast stoves for hot condition trial are provided with 289 temperature meters and corresponding gas flow meters for many times of hot condition test of hot blast stove. On-line

Fig. 8 Photo of the hot condition trial facility of hot blast stove

detection is used to test temperature variety of checker brick in the regenerator chamber at combustion and blasting statuses of hot blast stove, and with combination of pressure detection and fume composition detection toanalyse temperature field distribution status and combustion status in the experimental hot blast stoves. Hot condition test of hot blast stove directly verifies heating process of dome combustion stove and numerical calculation result of heat transfer theory so as to provide theoretic and experimental basis for optimal design of the hot blast stove and configure checker brick reasonably.

6 Research and Design on Engineering

6.1 Process Layout of Hot Blast Stove and 3D CAD Simulating Design

The dome combustion hot blast stove can adopt one line type or rectangular type process arrangement. The layout of hot blast stoves depends on the BF general layout in order to reduce occupied area, reduce pipe length, reduce hot blast stove building steel structure and reduce project investment. In the design, hot blast stove process arrangement is studied, and new type dome combustion hot blast stove rectangular process arrangement is designed and developed considering existing Shougang's rectangular layout. The new hot blast stove rectangular process layout has compact process layout, smooth and short flow process, reduces the hot blast main length significantly. 4 stoves are arranged in non-symmetrical rectangular type, hot blast vertical pipe is arranged outside the rectangular area of 4 stoves. Independent hot blast stove frame is arranged to support pipe and valves. One bridge crane can fulfill all the equipments maintenance requirement in hot blast stove area. Fig. 9 shows the Jingtang's No. 1 BF dome combustion hot blast stove rectangular process layout.

As dome combustion hot blast stove burner is arranged at dome position, the installation

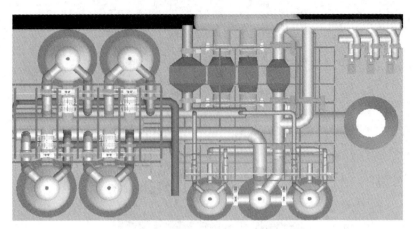

Fig. 9 Process layout of hot blast stove for Jingtang's No. 1 BF

positions of all pipes and valves are relatively higher. Influenced by the hot blast stove heat expansion, the pipe system stress and heat expansion shifting is more complicated, and the requirement on pipeline system design and stress calculation is more accurate. Especially the pipeline system design in hot blast stove area, reasonable pipe stress, easy installation, operation and maintenance shall be considered generally. In order to improve design accuracy and arrange pipe reasonably, three dimensional (3D) computer aid optimized design is adopted, compact arrangement and reasonable flow of hot blast stove is achieved. Fig. 10 shows the 3D CAD design drawing of Jingtang's No. 1 BF dome combustion hot blast stove.

Fig. 10 3D simulating design of dome combustion hot blast stove for Jingtang's No. 1 BF

6.2 Circular Ceramic Burner Structure Optimization

Dome is the burning space of dome combustion hot blast stove. In order to fully utilize the dome space, completely burn the gas, and distribute evenly the high temperature fume in regenerative chamber, cyclone pervasion combustion technology is adopted. The nozzles of circular ceramic burner gas and combustion air are arranged long tangent to periphery. Two rings of gas nozzles are arranged at the burner top with downward inclination, two rings of combustion air nozzles are arranged at burner bottom with upward inclination, the injected flow can be mixed inside pre-mixing chamber with certain velocity and revolve downwards, intensify the pervasion mixing of gas and combustion air, in order to achieve complete combustion of gas. In order to achieve the even distribution of gas, reasonable circular ceramic burner pre-mixing chamber and cone dome geometry structure is designed. Through fume flow shrinking, expansion, cyclone and return flow in dome space, the gas can be burned completely and high temperature fume can be distributed evenly. Fig. 11 shows the circular ceramic burning design structure.

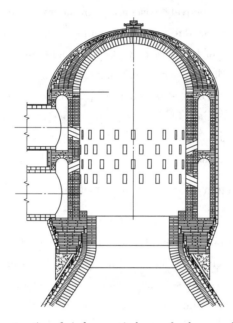

Fig. 11　Design construction of circle ceramic burner for dome combustion hot blast stove

New dome combustion hot blast stove circular ceramic burner is completely suitable with the combustion condition of preheating combustion air to 600℃, and preheating gas to 200℃. The burner adopts cyclone pervasion burning type, has big regulation range of combustion air and gas flow rate, complete and even mixing. Before high temperature fume enters regenerative chamber checker brick, the combustion is complete, which can restrain effectively the section high temperature zone in dome and reduce NO_x emission. After burning, the high temperature fume flow field is distributed evenly and properly. The results of CFD numerical calculation and physical model simulation test verify that flow field inside hot blast stove can fulfill the design requirement.

6.3　Extension of Service Life of Hot Blast Stove

Intercrystalline stress corrosion exists widely in high blast temperature hot blast stove. The main reason is that when hot blast stove dome temperature reaches above 1,400℃, the formation NO_x of increases quickly and its concentration reaches 350×10^{-6} or above. NO_x combines with the condensed water at hot blast stove inner wall, and forms corrosive acid. The corrosive acid penetrates deeply, expands and cracks along the crystal lattice from shell position with stress. The pulse pressure and tiredness stress during hot blast stove operation enhances corrosion and cracking process. The hot blast stove high temperature area shell intercrystalline stress cession becomes the restriction condition of influencing hot blast stove campaign life and improving blast temperature.

In dome combustion hot blast stove design, the maximum design temperature of hot blast stove dome is controlled within 1,450℃, and normal operating temperature is controlled within 1,420℃ in order to ensure safe temperature operation of hot blast stove, control NO_x formation, prevent intercrystalline stress corrosion effectively. Also low alloy microlite corrosion proof steel plate is adopted, welding stress generated during shell manufacturing is eliminated, corrosion proof painting is applied at inner wall of hot blast stove high temperature area shell, one layer of acid proof gunning material is used at corrosion proof painting in order to increase the intercrystalline stress corrosion proof ability of hot blast stove shell. Integrative protection measures such as thermal insulation and so on for the stove shell in high temperature zone of hot blast stove are considered to prevent generation of intercrystalline stress corrosion to the hot blast stove shell so as to prolong the service life of the hot blast stove and keep synchronization with the BF campaign life.

6.4 Hot Blast Stove Refractory Material Configuration

In order to fulfill the 1,300℃ high blast temperature of dome combustion hot blast stove and prolong its campaign life, hot blast refractory material and bricklaying structure is designed and optimized. Low stress design concept is applied and effective measures to eliminate or reduce refractory material system thermal stress, mechanical stress, phase change stress and pressure stress is adopted. According to working conditions of different hot blast stove zone, different refractory materials are used to achieve the optimized selection of hot blast stove refractory material. In hot blast stove high temperature zone, silicon bricks with high temperature proof and good creep performance are used. In middle temperature zone, andalusite bricks and low creep clay bricks are used. In low temperature zone, high density clay bricks are used. The optimized configuration of refractory material adapts with the working characters of dome combustion hot blast stove, improves the economic rationality of refractory material and reduces the project investment.

Low creepmullite bricks and insulation bricks composite bricklaying structure is adopted for hot blast main, hot blast branch and hot blast bustle. Two layers of indefinite form refractory material are used at internal surface of pipe shell. One layer of ceramic fiber felt is filled between pipe upper bricks and gunning material. The expansion joints at hot blast pipe bricklaying are filled with ceramic fiber material which can resist 1,420℃ high temperature and absorb the heat expansion of refractory material lining stably. In order to prolong the service life of hot blast main and branch, one layer of acid proof painting to prevent intercrystalline stress corrosion is applied at internal wall of hot blast pipe shell.

6.5 Hot Blast Pipeline Optimization

The hot blast branch, main and bustle of hot blast stove is high temperature and high pressure pipe, is the most complicated system of hot blast stove pipeline system. Hot blast pipe is the importantfactor to achieve high blast temperature, is the technical guarantee of stable transferring high temperature blasting. In the design, through hot blast stove piping system stress calculation analyse, the expansion displacement caused by operating temperature, environmental temperature and operating pressure is considered, low stress pipe design philosophy is adopted, hot blast pipe is optimized in order to fulfill the 1,300℃ high temperature blasting requirement. Different structure bellow compensator and tie-rod are arranged reasonably. The bellow compensator position of hot blast branch is arranged between hot blast valve and hot blast main. At the end of hot blast main, pressure equalizing type bellow compensator is arranged to optimize the pipe support and provide technical assurance for hot blast pipe stable operation.

Reasonable lining insulation structure is designed by heat transfer calculation of hot blast pipeline so as to reduce blast temperature loss and decrease pipeline wall temperature. Lining of working layer of hot blast pipeline adopts andalusite brick with excellent creep resistance performance, and the insulation layer adopts low bulk density high alumina brick and fireclay brick with excellent insulation performance. Reasonable refractory expansion joint and labyrinth seal structure are designed along length direction of hot blast pipeline to intake thermal expansion of refractory. The expansion joint is filled with ceramic fibre. Brick lining at top of the hot blast stove pipeline adopts rider brick structure to improve stability of bricklaying structure.

6.6 Design of Orifices of Hot Blast Stove

Working condition of various orifices of hot blast stove is extremely terrible, which restricts service life of hot blast stove and improvement of blast temperature. The hot blast outlet is composed of individual ring type composite brick. Between composite bricks, double tongue and grooves are used for strengthening. In the upper part of the composite brick there is half-ring type special arch bridge brick so as to relieve the pressure stress of the upper wall lining onto the composite brick. Composite bricks are applied for bricklaying of places like hot blast outlets of hot blast stove, junction at hot blast pipelines, etc.

7 Application

Optimization of hot blast stove operation is an important measure to have high blast temperature. Jingtang's No. 1 and No. 2 BF, and Qiangang's No. 3 BF adopt the staggered parallel operation mode with "two burning and two blasting" to reduce cold air to be mixed and decrease blast temperature fluctuation. Proportion of combustion air and gas is adjusted reasonably and combustion of hot blast stove is strengthened. Stoichiometric air coefficient is controlled in

1.05-1.15. Combustion air volume of hot blast stove for Jingtang's No. 1 and No. 2 BF should be kept at 140,000Nm^3/h, and BF gas is controlled within 170,000-200,000Nm^3/h. Blasting period is shortened and changeover number is increased appropriately. Changeover times per shift is increased from 7-8 times to 8-9 times. These kinds of technical measures are efficient support to have 1,280-1,300℃ high blast temperature. In order to have the BF to accept high blast temperature, a series of techniques for BF operation for satisfactory of high temperature production are studied and developed, mainly including:

(1) Application of concentrate burden technology improvement of sinter quality, reduction of chargeable powder, improvement of acceptance ability of high blast temperature by BF, optimization of burden structure, improvement of integrative grade and decrease of slag volume.

(2) Improvement of high top pressure and control of the bosh gas volume index of BF. Jingtang's No. 1 and No. 2 BF top pressure are controlled at 270kPa, and that of Qiangang's No. 3 BF is controlled at 250kPa.

(3) Control of reasonable theoretic flame temperature when the blast temperature of Qiangang's No. 2 and No. 3 BF reaches to 1280℃, measures of pulverized coal injection, oxygen enrichment and so on are taken, with theoretic combustion temperature 2,090-2,150℃, oxygen enrichment ratio 3.52%-3.81%, and coal ratio 173kg/t.[9,10] After the blast temperature of Jingtang's No. 1 BF reaches to 1,300℃, measures like improvement of ore and coke ratio, coal ratio and adjustment of blast humidity, etc. are taken with theoretic combustion temperature controlled within 2,100±50℃.

(4) Strengthening burden distribution control and realization of reasonable distribution of gas when the blast temperature reaches to 1,250℃ and plus, measures of increase of ore batch, improvement of ore-coke burden ratio, etc. are taken to open central gas flow and stabilize gas flow at edges, with gas utilization ratio to 50%-52%.

(5) Improvement of blast temperature stability and reduction of blast temperature fluctuation BF operation can stabilize ore-coke burden ratio based on fixation of blast temperature and control of theoretic combustion temperature to control theoretic combustion temperature fluctuation within 50℃.

The optimal operation mode under high temperature condition has been explored from long term production, practice and research. The blast temperature of large sized BFs of Jingtang and Qiangang are kept within 1,250±50℃ in long term stability status. Fuel ratio is decreased and BF high blast temperature has been achieved significant application effectiveness. Fig. 12 and Fig. 13 are the actual achievement of high blast temperature for one year after commissioning of Jingtang's No. 1 BF. Table 5 shows the main operational parameters of Jingtang's No. 1 and No. 2 BF in 2012 and 2013. Table 6 shows main operational parameters of Qiangang's No. 2 and No. 3 BF in 2012. Fig. 14 shows calculation result of temperature distribution in regenerator chamber of dome combustion hot blast stove for Jingtang's No. 1 BF, Fig. 15 shows the actual temperature variation curve of hot blast stove Jingtang's No. 1 BF. Fig. 15 indicates when blast temperature reaches to 1,300℃, the temperature variation of hot blast stove is stable, which demonstrates configuration rationality of checker brick in the regenerator chamber of hot blast stove.

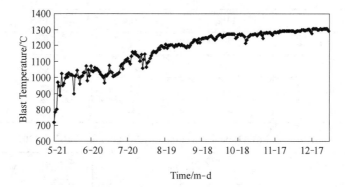

Fig. 12 Actual achievement of blast temperature of Jingtang's No. 1 BF after commissioning in 2009

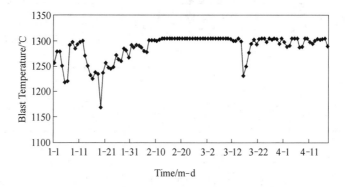

Fig. 13 Actual achievement of blast temperature of Jingtang's No. 1 BF from January to May, 2010

Table 5 Main operational parameters of Jingtang's BFs in 2012 and 2013

Items	No. 1 BF		No. 2 BF	
Year	2012	2013	2012	2013
Daily production/tHM · d^{-1}	11,995	12,552	12,313	12,104
Productivity/tHM · $(m^3 \cdot d)^{-1}$	2.21	2.26	2.24	2.20
Blast temperature/℃	1,223	1,235	1,234	1,214
Blast volume/$Nm^3 \cdot min^{-1}$	7,934	8,032	8,268	7,987
Blast pressure/kPa	433	448	447	445
Top pressure/kPa	228	254	260	254
Oxygen enrichment ratio/%	3.75	5.46	3.53	4.86
Total Fe in burden/%	58.94	59.2	58.95	58.7
CO utilization ratio/%	50.82	49.30	50.59	48.30

Continued Table 5

Items	No. 1 BF		No. 2 BF	
Year	2012	2013	2012	2013
Coke rate/kg · tHM^{-1}	309	308	304	317
PCR/kg · tHM^{-1}	148	154	157	153
Coke nut/kg · tHM^{-1}	28	32	29	30
Fuel ratio/kg · tHM^{-1}	485	494	490	500

Table 6 Main technical parameters of Qiangang's BFs in 2012

Items	No. 2 BF	No. 3 BF
Daily production/tHM · d^{-1}	6,090	9,173
Productivity/tHM · (m^3 · d)$^{-1}$	2.29	2.30
Blast temperature/℃	1,233	1,245
Blast volume/Nm^3 · min^{-1}	4,536	6,257
Blast pressure/kPa	345	398
Oxygen enrichment ratio/%	2.50	4.84
Coke rate/kg · tHM^{-1}	310	305
PCR/kg · tHM^{-1}	146	158
Coke nut/kg · tHM^{-1}	42	42
Fuel ratio/kg · tHM^{-1}	498	505

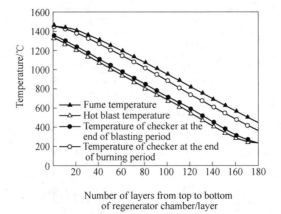

Fig. 14 Temperature distribution of regenerator chamber

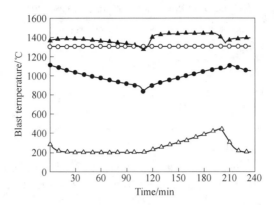

Fig. 15 Temperature variation curve of hot blast stove of Shougang's No. 1 BF
▲—Dome temperature; ○—Hot blast temperature; ●—Silica brick interface temperature; △—Temperature of fume gas

8 Conclusions

At beginning of the new century, modern BF ironmaking technology is faced with restrictions of resources shortage, undersupply of energy resources and ecological environment. For actualization of sustainable development of BF ironmaking process, it is a must to adopt technologies of high blast temperature, high oxygen enrichment and large scale pulverized coal injection to reduce fuel consumption and production cost. This is the only way for development of modern BF ironmaking industry.

High blast temperature is integrative technology and important technical support for reduction of fuel ratio and improvement of pulverized coal injection quantity. At present condition, low calorific value BFG is used to have more than 1,250℃ blast temperature, and it is the optimal technical measure for realization of high efficient utilization and high efficient energy resources conversion of low quality energy resources. For the key technology to systematically solve issues, such as acquisition of high blast temperature, as well as stable transfer and high efficient utilization of high temperature hot blast so on, high efficient and long service life hot blast stove is applied with high efficient checker brick for optimization of hot blast stove operation, guarantee of stable transfer of high temperature hot blast, extension of service life of hot blast stove so as to have high efficient utilization of high temperature hot blast.

Shougang has achieved good performance in aspects of study of high blast temperature technology, development and application of high efficient and long service life type dome combustion hot blast stove, application of high blast temperature technology, and technical innovation on high blast temperature came true.

Acknowledgements

The author would like to thank Prof. Cheng Shusen, Mr. Hu Zurui, Mr. Li Qin, Mr. Mei

Conghua, Mr. Yin Guangyu, etc. for their support and assistance; and many thanks to the experts from Jingtang and Qiangang for their support and assistance of operational information.

References

[1] Zhang Shourong. Development and problems of ironmaking industry after entering the 21st century in China [J]. Ironmaking, 2012, 31 (1): 1-6.

[2] Zhang Fuming. Cognition of some technical issues on contemporary blast furnace ironmaking [J]. Ironmaking, 2012, 31 (5): 1-6.

[3] Zhang Fuming, Qian Shichong, Zhang Jian. Design of 5500m^3 blast furnace at Shougang Jingtang [J]. Journal of Iron and Steel Research International, 2009, 16 (52): 1029-1033.

[4] Zhang Fuming, Mei Conghua, Yin Guangyu. Design study on BSK dome combustion hot blast stove of Shougang Jingtang 5500m^3 blast furnace [J]. China Metallurgy, 2012, 22 (3): 27-32.

[5] Wei Hongqi. Application and practices of 1300℃ air temperature in Shougang Jingtang united iron and steel Co., Ltd [J]. Ironmaking, 2010, 29 (4): 7-10.

[6] Zhang Fuming. Developing prospects on high temperature and low fuel ratio technologies for blast furnace ironmaking [J]. China Metallurgy, 2013, 23 (2): 1-7.

[7] Zhang Fuming, Hu Zurui, Cheng Shusen. Ombustion technology of BSK dome combustion hot blast stove at Shougang Jingtang [J]. Iron and Steel 2012, 47 (5): 75-81.

[8] Zhang Fuming, Mao Qingwu, Mei Conghua. Dome combustion hot blast for huge blast furnace [J]. Journal of Iron and Steel Research International, 2012, 19 (9): 1-7.

[9] Chen Guanjun, Ma Zejun, Zheng Jingxian. Trial research on 1280℃ high blast temperature of No. 2 blast furnace at Qiangang [J]. Ironmaking, 2010, 29 (4): 23-26.

[10] Ma Jinfang, Wan Lei, Jia Guoli. Technology of high blast air temperature for No. 2 BF in Shougang Qian'an iron and steel Co., Ltd. [J]. Iron and Steel, 2010, 46 (6): 26-31.

Construction and Practice on Energy Flow Network of New Generation Recyclable Iron and Steel Manufacturing Process

Zhang Fuming

Abstract: Shougang Jingtang iron and steel plant is a new generation recyclable iron and steel plant designed according to the concept and principle of circular economy. The steel plant is provided with the comprehensive functions of high quality steel product manufacture, high efficiency energy conversion and waste disposal. In order to realize the cooperation and high efficiency of iron and steel manufacturing process, a full process energy flow network with carbon flow as the core is designed and constructed to realize energy high efficiency conversion and low carbon green manufacturing. Since Jingtang steel plant was put into production, the efficiency of energy conversion has been continuously improved, and remarkable results have been obtained in high efficiency energy utilization and high value conversion. The emission of CO_2 and pollutants has been greatly reduced, and cleaning production and low-carbon metallurgy have been realized.

Keywords: iron and steel; energy saving; energy flow network; emission control; circular economy

1 Introduction

In order to solve the non-capital functions of Beijing at the beginning of the 21st century, Shougang decided to move out of Beijing region and build a new steel base, Jingtang Iron and Steel Plant, in Caofeidian industrial district, Tangshan city, Hebei province. The project is designed and built independently according to the concept of a new generation of recyclable iron and steel processes, and has a large steel project with advanced international technical merit in the 21st century. It is also a large scale modern steel plant which produces thin strip and sheet. Shougang Jingtang Iron and Steel Plant has "three functions"[1], it not only pays attention to high quality and high efficiency iron and steel products manufacturing function, but also pays attention to high efficiency, clean energy conversion function, handles large amount of social waste and realizes the function of reusing resources.

The design annual output of Shougang Jingtang Iron and Steel Plant is 8.7-9.2 million tons. The

---

本文作者：张福明。原刊于 TMS 2019 Conference Proceedings (10th International Symposium on High-Temperature Metallurgical Processing). The Minerals, Metals & Materials Society, San Antonio, 2019: 269-278。

engineering erection began on March 12, 2007, No. 1 blast furnace blew in on May 21, 2009, No. 2 blast furnace was put into production on June 26, 2010, and the complete iron and steel manufacturing process put into production smoothly. The high efficiency and clean steel production process with 2 blast furnaces, 1 steelmaking plant, 2 hot rolling lines and 4 cold rolling lines are integrated and constructed. 5500m^3 blast furnace, high efficiency and low cost clean steel production platform, 50000 tons per day seawater desalination facility and other representative advanced technological equipment have been designed and constructed. It plays an important role in guiding and demonstrating the design and construction of a new generation of recyclable steel plant in the 21st century.

2 Design Philosophy and Main Objectives

2.1 Engineering Concept

Shougang Jingtang Iron and Steel Plant is a modern large scale iron and steel project with international advanced level in the 21st century, the first real coastal port construction and the whole production of thin strip and sheet, according to the concept of the new generation of recyclable steel process and the independent design and construction of China. The engineering design concept and goal are: guided by the theory of metallurgical process engineering[2], taking the construction of the new generation recyclable steel plant "Three Functions" as the core design philosophy, according to the "high starting point, high standard and high requirement" put forward by the central government, achieve the "First Class" project construction objectives of product, management, environment and benefit. In engineering design, independent integrated innovation has been carried forward and Shougang Jingtang has become a clean steel manufacturing base, a circular economy demonstration base and an independent innovation demonstration base for the production of thin strip with high quality and high efficiency.

In engineering design, on the basis of deep understanding of the physical essence of the dynamic operation of steel plant and the development trend of iron and steel industry in the new century, the concept of a new generation of recyclable iron and steel technological process is followed. Based on the theory and method of modern iron and steel metallurgical engineering design, the conceptual design, top level design, dynamic precision design and three-dimensional simulation design of Shougang Jingtang Iron and Steel Plant project are innovated and practiced, a new generation of recyclable steel manufacturing process is designed and constructed, and become a new generation of recyclable steel manufacturing process demonstration project[3].

2.2 Engineering Objectives

The main goals of Shougang Jingtang project are according to the construction concept of the new generation recyclable iron and steel manufacturing process, research and development of a new producing process of clean steel with high efficiency and low cost, metallurgical gas dry dust removal and the high temperature with high pressure dry coke quenching power generation. In order to realize the effective utilization of low quality waste heat and social waste resources, the integrated technology is applied to Shougang Jingtang as a systematic technological integration, and strive to build the 21st century world-class advanced iron and steel plant.

The new generation recyclableiron and steel manufacturing process is not a representation or transformation of existing processes or facilities, but a new iron and steel manufacturing process based on the dynamic integration of substance flow, energy flow and information flow around the core concepts of high efficiency, low cost and recyclable process. To realize the three functions of product manufacture, energy conversion and disposal of social waste, and integrate construction in Shougang Jingtang.

(1) High efficiency and low cost clean steel manufacturing platform.

(2) High efficiency energy conversion and energy saving steel plant.

(3) Clean steel plant with recycling resources and energy.

The dynamic and precise concept of engineering designis established in the top level design of Shougang Jingtang. Around the dynamic running rule of "Flow" in the whole iron and steel manufacture process, the basic concept of process optimization is dynamic-orderly and synergetic-continuous. In the top level design, the principles of overall, hierarchical, dynamic, relevance and environmental adaptability are emphasized, including process or facility element selection, process structure optimization, function extension, as well as efficiency excellence.

2.2.1 Factor selection

In the capital, land, resource, market, environment, labor force and technology and so on, the reasonable matching of the basic elements of industry and economy is realized, and the technology integration system is built, and the development mode of interaction, cooperation, integration and evolution is realized.

2.2.2 Structure optimization

Advanced and large scale technical equipment are adopted and configured in Shougang Jingtang to promote the operational efficiency, product quality and economic benefit. Two blast furnaces, one steelmaking plant, two hot rolling plants and the corresponding four cold rolling and coating production line are constructed by adopting the most advanced technological process in the world and the "2-1-2-4" type manufacturing process structure. The annual production capacity of crude steel is between 8.7-9.2 million tons. The iron and steel manufacturing process flow of Shougang Jingtang Iron & Steel Plant is shown in Fig. 1.

2.2.3 Function extension

The "Three Functions" concept of new generation steel plant was carried out during the engineering design. The multi-functions of modern steel plant are explored from single function of steel products manufacturing to three functions and circular economic zone.

2.2.4 Efficiency excellence

The coordination relationship between time and space is established, the construction of process

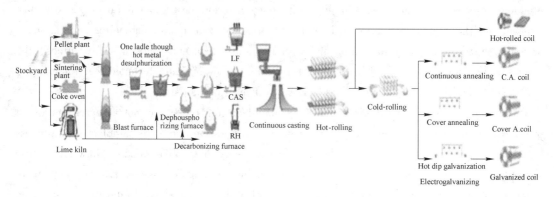

Fig. 1 Iron and steel manufacturing process flow of Shougang Jingtang Steel Plant

network and the relationship between processes are emphasized, and the new interface technologies between different operational unit are developed and applied. The integration and innovation of top level design are emphasized, pay attention to the stability, reliability and efficiency of the whole process.

3 Construction of Full Process Energy Flow Network

3.1 Energy Flow Network Design

In the design of energy flow network structure, according to the behavior and conversion characteristics of energy flow and different energy medium operating process, a perfect energy supply system and energy conversion network system are designed and built based on real-time monitoring and online scheduling. Process control, centrally managed by Energy Control Center[4]. The efficient conversion of energy and the optimal allocation of energy structure are studied in depth and the system is optimized. The secondary energy of the steel manufacturing process is fully recovered, and the waste heat of the steel plant is fully utilized to generate electricity. The power generation rate of iron and steel plant is more than 94%, and all kinds of associated gas in iron and steel metallurgical process are "zero emission".

All kinds of energy media are connected with each manufacturing process through pipeline, and each production process acts as the node on the energy network to become the pivot of energy medium conversion, transmission and storage, so as to realize the energy network management of the whole plant.

In the engineering design of circular economy, green manufacturing, energy saving and emission reducing, a lots of advanced large scale technology and equipment are adopted to improve production efficiency and energy use efficiency. The 2 sets of 260t/h high temperature and high pressure dry quenching (CDQ) device are configured to four series 70 batteries 7.63m coke oven, and the generating electricity up to 105kW · h/t. The low calorific value blast furnace gas (BFG) is applied for 5500m^3 blast furnace hot blast stove combustion, The high efficiency preheating technology of gas and combustion air is applied, the high air temperature dome combustion hot blast stoves are configured, so that the blast temperature reaches above 1250℃, and the highest monthly average blast temperature reaches 1300℃[5]. The blast furnace gas cleaning system is innovated by full dry de-dusting technology and high efficiency top gas recovery turbine (TRT) coupling technology[6]. The top gas residual pressure turbine generator set with installed capacity of 36.5MW, and the power generation is up to 50kW · h/tHM, the dust content in purified gas is 2-4mg/m^3. The dry dust removal technology is adopted in BOF gas de-dusting system[7], the BOF gas recovery reaches 96m^3/t and steam recovery reaches 115kg/t, dust content in purified gas is less than 10mg/m^3. All kinds of the associated gas in the metallurgical process of the iron and steel plant are recovered completely, used for the combustion and heating of the iron and steel manufacturing process, the surplus gas is used for generating electricity, and the "spent steam" after generating electricity as the heat source for the 50000t/d low temperature and multi-efficiency seawater desalination facility, is used for desalination of seawater.

Fig. 2 shows the network flow diagram of energy flow operation in Shougang Jingtang Steel Plant.

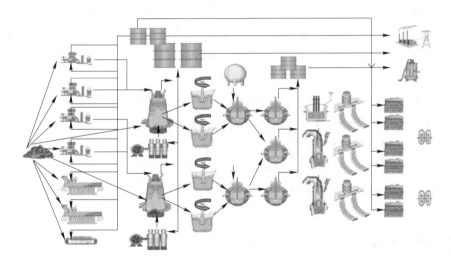

Fig. 2 Operation network of energy flow in Shougang Jingtang Plant

3.2 Technical Feature

(1) Energy management networking. Through the GIS (Geography Information System) simulation system to realize the accurate space and time positioning in the control of various energy media, and the network management information.

(2) Real-time on-line monitoring, scheduling and management are carried out for any energy medium through space and time systems, and operational decision making is modeled.

(3) The control and forecast model of energy network operation is developed to realize the intelligent management of energy network. Includes process model, cost model and decision model.

3.3 Operating Practice

(1) The systematic energy saving is achieved. The energy saving of the whole process system can save 91,600 tons standard coal per year.

(2) The gas release rate is reduced. BOF gas is "zero release", coke oven gas emission rate is 0.043% and blast furnace gas is less than 1.96%.

(3) The pressure fluctuation of gas pipeline network is reduced. The range of pressure fluctuation of blast furnace gas is reduced to ±0.3kPa.

(4) The steam pipe network runs stably. The fluctuation range of steam pressure is less than 0.1MPa, and the steam "zero release" is realized.

4 Energy Saving and Energy Recycling Practice

4.1 Technical Concept

In the design and production, the energy supply structure is optimized, the high price energy is replaced by the low price energy. The blast oxygen enrichment and pulverized coal injection technology are adopted to reduce the coke consumption, and the secondary energy (such as coke oven gas, blast furnace gas and converter gas) in the iron and steel production process could be fully recovered to reduce the costs of energy purchase and outsourced energy. All kinds of residual heat generated by various production processes (including sintering, hot blast stove, converter, heating furnace, etc.) are recovered and reutilized. The clean production is implemented, reduce primary energy consumption from the source and reduce emissions of various pollutants. It has been proved that energy saving and consumption reducing are the most effective and economic measures for environmental protection because of the reduction of all kinds of emissions from the iron and steel production source.

4.2 Technical Route and Basic Principles

(1) According to the 3R principle of circular economy of "Reduce, Reuse and Recycle", the emphasis of energy saving is on optimizing production technology. Through the active use of energy saving technology and technology in the design, the production process and product structure optimization.

(2) The new energy saving technologies are applied and popularized, such as dry coke quenching and metallurgical gas dry dust removal, to promote the reuse and recycling of resources and energy.

As a result of ironmaking system accounts for about 70% of the total energy consumption of iron and steel production, ironmaking process is the focus of energy saving. The overall implementation of the concentrate policy in the design mainly includes controlling the fluctuation range of grade of ferrous content of raw materials within ±0.3% and ore grade reaching more than 60%. Increasing coke strength, reducing coke ash and sulfur content. Giving play to the advantages of comprehensive synthesis raw material yard, stabilizing blast furnace raw material composition. Adopting the high blast oxygen enrichment, high blast temperature and high pulverized coal injection technologies to reduce the coke consumption. The TRT device power generation for recovery of top gas residual pressure and the waste heat of hot blast stove flue gas are configured to realize the high efficiency energy recovery and conversion. The reasonable burden composition of high basicity sinter ore with acid pellet (or lump ore) is adopted to ensure the blast furnace stable and smooth operation, to realize the long campaign life and to promote the hot metal quality.

(3) The secondary energy and residual pressure, waste heat resources are fully recycled, and the energy structure of enterprises is optimized. The recovery and utilization of blast furnace gas, coke oven gas and converter gas are improved. The comprehensive utilization of all kinds of energy medium on the basis of energy balance is emphasized.

4.3 High efficiency Energy Conversion and Highlight Energy Saving Technology

(1) Advanced large-scale modern technologies and equipment for energy saving and emission reducing are applied.

The main manufacturing process is configured advanced and large sized facilities, such as $5500m^3$ blast furnace, $500m^2$ sintering machine, $504m^2$ travelling grate induration machine, 7.63m coke oven, 300t converter, secondary refining, high efficiency slab continuous casting machine, 2250mm hot continuous rolling, 1580mm hot continuous rolling, 2230mm cold continuous rolling, 1700mm cold continuous rolling, 1420mm cold continuous rolling, hot galvanizing and electro tinning, coating production line, etc. These equipment are modern large scale production facilities with high production efficiency and low energy consumption.

1) The large scale and high efficiency equipment is integrated and applied.

2) The $5500m^3$ blast furnace is adopted, compare with the $4000m^3$ blast furnace, the capacity is increased 32% and the fuel ratio is reduced approximate 4.8%.

3) The $500m^2$ sintering machine compare with $360m^2$ sintering machine, the production capacity is increased by 62%, the energy consumption is reduced by 8%, the consumption of fuel and electricity is reduced by 10%.

4) The 7.63m coke oven configured with 260t/h dry coke quenching facility, compare with the 6.0m coke oven configured with 140t/h CDQ, the production efficiency is improved 50%, energy consumption is reduced 6.0%, and the CDQ power generation efficiency is improved 12%.

(2) Advanced energy saving and emission reducing technologies are adopted and innovated.

1) The 2 sets of 260t/h CDQ facility are equipped with 2 sets 30MW high temperature and high pressure steam generator. The power output reached 447million kWh and per ton coke was 113.5kW · h in 2016.

2) Dry gas dust removal technology is developed in 5500m^3 blast furnace. The dust content of purified gas reaches 2-3.5mg/m^3. Compared with wet dust removal technology, the water saving is 4000t/d and the electricity saving is 36000kW · h/d. The power generation is increased by more than 45%. The energy saving 7.24kg standard coal per ton hot metal, and annual emission reduction of CO_2 approximate 169100 tons.

3) 300t converter gas dry dust removal technology, the clean gas dust content is less than 10mg/m^3, compared with converter gas wet dust removal, save electricity and water about 1/3 respectively, reducing construction land about 1/2, energy saving reach 4.5kg standard coal per ton steel, and reduce annual CO_2 emissions approximate 107600 tons.

4.4 Practice and effect of Energy-saving Technology Innovation

(1) 5500m^3 super large blast furnace technology is adopted.

The dome combustion hot blast furnace technology is developed and adopted in 5500m^3 super large blast furnace in the world for the first time. The dome combustion hot blast stove can promote the heat transfer efficiency and reduce the fuel consumption. The hot blast temperature can achieve more than 1250℃ under the condition of only burning blast furnace gas. The comprehensive utilization ratio of energy is increased by 20% compare with other large blast furnace.

The 11000m^3/min blast volume (Standard Temperature and Pressure) axial intake blower is adopted, which is the largest in the world and energy saving and high efficiency at present, and a large capacity dehumidification system is provided. Compared with the traditional blast furnace blower, the power consumption is reduced greatly, the power can be saved by 8 million kWh per year, and it is beneficial to the stability of the blower operation in summer and to increase the stability of blast furnace smelting.

(2) Clean steel manufacturing platform is constructed.

The hot metal pretreatment of desulphurization, dephosphorization, desilication is developed and adopted firstly in China, and the 4 sets KR device for desulphurization, 2 sets of 300t converter for dephosphorization and desilication 3 sets of converter for steelmaking. The process route of "full three removal" is configured for clean steel manufacturing in Shougang Jingtang. A small amount slag steelmaking and total molten steel refining are adopted to create a clean steel production platform. The lime consumption is reduced to 30kg/t steel when the decarburized slag is returned to the dephosphorization converter.

In order to realize slag splashing and converter protection under the condition of less slag, the technology of triple slag circulation has created a solid foundation for the reduction and reuse of steel slag. By using ladle capping technology, the tapping temperature of converter can be reduced by 9℃, the consumption of coke oven gas is reduced by 6.5m^3/t, and the stable operating ratio is more than 99.97%. The total contents of five harmful elements (sulfur, phosphorus, nitrogen, hydrogen, oxygen) in steelmaking can reach the international advanced level of less than 45ppm.

(3) BF-BOF interface technology is developed and applied.

The hot metal transportation direct from blast furnace to steelmaking plant by open ladle technology is adopted. The temperature of hot metal to the desulfurizing station is 1390℃, compared with the torpedo tank, the temperature drop of hot metal transportation is reduced by 50℃, and the dust produced in the process of pouring the ladle is reduced by 4700 tons per year; the lifting height of the crane is reduced, and the electricity consumption is saved by 11.4 million kW·h/year. After put into production, the hot metal ladle cover measure is applied, and the drop of hot metal temperature is further reduced.

(4) Large scale oxygen production technology is configured.

The VAROX process is adopted in 75000m^3/h oxygen generator, it can adjust the load in extremely large range variable, the oxygen output can be adjusted in 36250-98750m^3/h in working condition to require the iron and steel enterprise discontinuous oxygen demand, achieves the oxygen zero emission. It can save the electricity 20 million kW·h and save energy 2458 tons standard coal per year.

(5) Energy management center is constructed to strengthen energy management.

The whole plant energy management center system (EMS) has been built into an intelligent control system which integrates process monitoring, energy management and energy scheduling. The system plays a significant role in reducing gas emission, improving environmental quality, reducing energy consumption, improving labor productivity and energy management level.

5 Conclusions

Shougang Jingtang Iron and Steel Plant design and construction are based on a new generation of recyclable iron and steel manufacturing process. It has three functions: product manufacturing, energy conversion and waste disposal. Through the optimization of the design, the large scale technical equipment and energy saving technology are adopted to effectively construct the energy flow network, to realize the high efficiency energy conversion and utilization. The energy consumption is greatly reduced, and the effect of energy saving and emission reducing is remarkable. The significant achievement and performance have been achieved.

References

[1] Yin Ruiyu. Theory and Method of Metallurgical Process Flow Integration [M]. Beijing: Metallurgical Industry Press, 2016: 284-287.

[2] Yin Ruiyu. Metallurgical Process Engineering [M]. Beijing: Metallurgical Industry Press, 2004: 379-386.

[3] Zhang Fuming, Xie Jianxin. Green iron and steel manufacturing process of Shougang Jingtang Plant [J]. TMS Energy Materials 2017 Conference Proceedings, 2017: 17-29.

[4] Gu Liyun. Theory and practice of the control and management system of energy in Shougang Jingtang united iron & steel Co., Ltd [J]. Metallurgical Automation, 2011, 35 (3): 24-28.

[5] Zhang Fuming, Mei Conghua, Yin Guangyu, et al. Design and research of BSK dome combustion hot blast

stove of Shougang Jingtang 5500m³ blast furnace [J]. China Metallurgy, 2012, 22 (3): 27-32.
[6] Zhang Fuming. Study on large scale blast furnace gas dry bag dedusting technology [J]. Ironmaking, 2011, 30 (1): 1-5.
[7] Zhang Fuming, Zhang Deguo, Zhang Lingyi, et al. Research and application on large BOF gas dry dedusting technology [J]. Iron and Steel, 2013, 48 (2): 1-9.

Analysis of Pulverized Coal and Natural Gas Injection in 5500m³ Blast Furnace in Shougang Jingtang

Meng Xianglong　Zhang Fuming　Wang Weiqiao　Li Lin　Dai Jianhua

Keywords: blast furnace; pulverized coal injection; natural gas injection; carbon emission

1 Introduction

The blast furnace (BF) ironmaking process is one of the most energy consumption in the metallurgical industries which represents about 70% of the energy input [1]. As the main reducing agent, the carbon is finally released to the environments as carbon dioxide to cause global warming [2]. Maximum of reducing energy consumption, improving energy efficiency and decreasing carbon dioxide release should be paid attention in BF process. Injecting fuels into BF through tuyere could replace some coke, decrease the hot metal cost, regulate hearth condition and reduce environment pollution. Many BFs in the world have adopted various injection fuels like coal, dead oil, tar, natural gas (NG), coke oven gas, dedusting ash and other reducing fuels at present [3-7].

In china, as a kind of high calorific value fuel, NG hasn't been injected into BF except for an experiment in a 620m³ BF of Chong Steel in late 1960s which was ceased for resource shortage, after then, researches on NG injection progressed slowly. Recently, abundant NG is exploited in scope of Shougang Jingtang, the quantity provided for BF injection is about $24 \times 10^4 m^3/d$ at present and may reach $36 \times 10^4 m^3/d$ or even more later. In order to maximize the utilization of resources, decrease the hot metal cost and improve the economic benefit, Shougang has carried out a series of research on NG injection for Jingtang No.2 BF.

2 Running Conditions Description of Jingtang No.2 BF

Pulverized coal injection (PCI) technology for BF was applied earlier in china. As a typical steelmaking enterprise, Shougang explored the PCI technology from 1960s in last century, and has established a complete PCI process by decades of development. All the BFs in Shougang are adopting PCI technology by now [8].

Shougang Jingtang Company has two 5500m³ BFs, which were commissioned in 2009.9 and 2010.6 respectively. In 2014, combining with oxygen enrichment blast technique, the pulverized coal injection rate (*PCR*) of No.2 BF was about 150kg/t, theoretical flame temperature (*TFT*) was above 2300℃, coke rate was less than 350kg/tHM. Operational characteristics of No.2 BF of Shougang Jingtang are presented in Table 1.

Table 1 Operational characteristics of No. 2 BF of Shougang Jingtang

Indexes	Jan.	Feb.	Mar.	Apr.	May.	Average
Productivity/t·d^{-1}	13041	12821	12948	12647	12178	12727
Coke rate/kg·tHM^{-1}	340.00	336.63	333.55	345.64	365.80	344.32
Coal injection rate/kg·tHM^{-1}	150.70	155.67	157.41	150.23	140.60	150.92
Slag/kg·tHM^{-1}	312	313	316	300	306	309
Blast parameters						
Blast consumption/m³·tHM^{-1}	924	932	922	927	978	937
Enriched O$_2$/%	4.99	4.84	4.99	4.69	4.44	4.79
Blast moisture/g·m^{-3}	2.59	2.70	3.60	4.20	8.90	4.39
Blast temperature/℃	1214	1217	1223	1217	1181	1210
Top gas analysis						
φCO_2/%	25.4	24.5	24.1	24.5	22.9	24.3
φCO/%	24.6	25.7	25.3	25.3	25.3	25.2
φH_2/%	2.6	2.5	2.5	2.8	2.6	2.6
Hot metal analysis						
Si/%	0.240	0.23	0.240	0.288	0.233	0.246
Mn/%	0.16	0.19	0.18	0.17	0.16	0.172
P/%	0.109	0.111	0.111	0.099	0.100	0.106
S/%	0.056	0.055	0.055	0.049	0.056	0.054

3 Key Parameters and Calculation Steps

The main component of NG is methane (CH_4), as an injected fuel, it could accelerate ore reduction, decrease coke consumption and reduce CO_2 emission. NG injection into BF is a method to achieve low carbon consumption and high efficiency operation [9-11]. The calorific value of NG prepared for Jingtang No.2 BF is higher than common in China, and the compositions and characteristics of NG are presented in Table 2.

Table 2 the compositions and characteristics of NG prepared for Jingtang No. 2 BF

Indexes	Value	Indexes	Value
Methane (CH_4)/mol %	74.91	Hexane (C_6H_{14})/mol %	0.24
Ethane (C_2H_6)/mol %	10.11	Heptane (C_7H_{16})/mol %	0.09
Propane (C_3H_8)/mol %	4.47	CO_2/mol %	6.80
Isobutene (C_4H_{10})/mol %	0.77	N_2/mol %	0.06
Normal butane (C_4H_{10})/mol %	1.47	O_2/mol %	0.00
Neopentane (C_5H_{12})/mol %	0.05	H_2S/mg·m^{-3}	79.08
Isopentane (C_5H_{12})/mol %	0.52	Density (20℃, 0.1013/MPa)/kg·m^{-3}	0.9303
Normal pentane (C_5H_{12})/mol %	0.51	Calorific value (20℃, 0.1013MPa)/MJ·m^{-3}	43.30

NG injectionon BF has been applied in many countries on the earth. In the former USSR, the first successful experiment of raw NG injection was realized in 1957 on No. 4 BF at the Petrovsky plant in Ukraine. Two years after the trial, the injection of NG together with hot blast enriched to 23%-24% oxygen was executed. Until 1977, 111 out of 136 BFs in the USSR used NG injection, with 86 of these using oxygen injection as well. In North America, regular use of NG did not begin until 1958 at the Colorado Fuel and Iron Company in Pueblo, and by 1970s more than 70 BFs were using NG injection [12,13]. In Japan, from 2004 to 2008, NG injection on 5000m^3 BF was realized with NG injection rate (NGR) was 20-50kg/tHM, the BF productivity reached to 2.56t/(m^3·d) together with increasing oxygen enrichment rate and controlling TFT.

3.1 Calculation of TFT

Theoretical flame temperature (TFT) in the combustion zone is the temperature that results from a combustion process in the raceway that occurs with any heat transfer or exchanges of kinetic or potential energy. TFT is higher than the actual temperature in the raceway because all heat losses are ignored in calculation, but it is still an integrated indicator of raceway thermal conditions.

Professor A. N. Ramm developed a TFT formula based on the law of conservation of energyin the raceway in 1958 [14]. In industry, simplified calculations of TFT have developed based on directly measured parameters of hot blast and other raw material such as the equation created by N. Dunaev and T. Kukhtin in 1977 [12].

Based on the law of conservation of heat, THT can be estimated from equation (1) by 1ton hot metal smelting.

$$TFT = \frac{Q_{in} - Q_{out}}{V_g \cdot C_g} \quad (1)$$

Where   TFT——Theoretical flame temperature, ℃;

Q_{in} ——Heat input in the raceway, kJ/tHM;

Q_{out} ——Heat output in the raceway, kJ/tHM;

V_g ——Gas volume in the raceway, m³/tHM;

C_g ——Average heat capacity of gas in the raceway, kJ/(m³·℃).

The heat input Q_{in} and heat output Q_{out} can be written as show in equation (2) and (3) with coal and NG injection.

$$Q_{in} = Q_c + Q_b + Q_{coal} + Q_{coke} + Q_{NG} \tag{2}$$

$$Q_{out} = Q_{res.\,W.\,coal} + Q_{res.\,W.\,b} + Q_{res.\,coal} + Q_{b.\,NG} \tag{3}$$

Where Q_c ——Calorific effect of carbon combustion to CO in the raceway, kJ/tHM;

Q_b ——Hot blast enthalpy, kJ/tHM;

Q_{coal} ——Enthalpy of injected coal in the raceway, kJ/tHM;

Q_{coke} ——Enthalpy of coke coming to the raceway, kJ/tHM;

Q_{NG} ——Enthalpy of injected NG in the raceway, kJ/tHM;

$Q_{res.\,W.\,b}$ ——Calorific effect by moisture thermal destruction of blast in the raceway, kJ/tHM;

$Q_{res.\,W.\,coal}$ ——Calorific effect by moisture thermal destruction of coal in the raceway, kJ/tHM;

$Q_{res.\,coal}$ ——Calorific effect by coal thermal destruction in the raceway, kJ/tHM;

$Q_{b.\,NG}$ ——Calorific effect of NG combustion to CO and H₂ in the raceway, kJ/tHM.

The final equation for *TFT* estimation can be written as

$$TFT = \frac{Q_c + Q_b + Q_{coal} + Q_{coke} + Q_{NG} - Q_{res.\,W.\,coal} - Q_{res.\,W.\,b} - Q_{res.\,coal} - Q_{b.\,NG}}{V_g \cdot C_g} \tag{4}$$

3.2 Degree of Direct Reduction of Wustite r_d

Degree of direct reduction, which is an indicator of direct reduction grade in BF, has two expressions. The r_d indicates the proportion of directly reduced Fe to total reduced Fe. The r_c indicates the proportion of oxygen through reaction FeO+C=Fe+CO to the total oxygen input with iron oxides. The r_d, which could reflect the principles of reduction process in BF, and be unaffected by the degree of ore oxidation, are more concise and practical, and used widely in China.

The r_d is determined from the equation (5) based on 1ton hot metal smelting.

$$r_d = 1 - r_{i(CO)} - r_{i(H_2)} \tag{5}$$

Where r_d ——Degree of direct reduction;

$r_{i(CO)}, r_{i(H_2)}$ ——Degree of indirect reduction by CO and H₂, respectively.

The $r_{i(CO)}$ can be calculated by equation (6).

$$r_{i(CO)} = \frac{56 C'_{CO_2}}{12 Fe.r} \tag{6}$$

Where Fe.r ——Iron content in hot metal, kg/tHM;

C'_{CO_2} ——Carbon content in CO₂ generated by FeO indirect reduction, kg/tHM.

The C'_{CO_2} can be calculated by equation (7).

$$C'_{CO_2} = \frac{12}{22.4}V_g \varphi_{CO_2} - 12\left(\frac{W(Fe_2O_3)}{160} + \frac{W(MnO_2)}{87} + \frac{W(CO_2)}{44}\right) \quad (7)$$

Where V_g ——Top gas volume per hot metal, m³/tHM;

φ_{CO_2} ——CO_2 volume fraction in top gas, m³/m³;

$W(CO_2)$ ——CO_2 mass inputted into BF, kg/tHM;

$W(Fe_2O_3)$, $W(MnO_2)$ ——Fe_2O_3 or MnO_2 mass in burden per hot metal, respectively, kg/tHM.

The $r_{i(H_2)}$ can be calculated by equation (8).

$$r_{i(H_2)} = \frac{56 \times [\sum H_2 - V_g(2\varphi_{H_2} + 4\varphi_{CH_4})/22.4]}{2Fe \cdot r} \quad (8)$$

Where φ_{H_2}, φ_{CH_4} ——H_2 or CH_4 volume fraction of top gas, m³/m³;

$\sum H_2$ ——total hydrogen mass inputted into BF per ton hot metal, kg/tHM.

The $\sum H_2$ can be calculated by equation (9).

$$\sum H_2 = \frac{2}{22.4}V_b\varphi' + K(H_2)_K + PCR\left[(H_2)_M + \frac{2}{18}(H_2O)_M\right] + NGR\left[(H_2)_N + \frac{2}{18}(H_2O)_N\right] \quad (9)$$

Where V_b ——Volume of the blast, m³/tHM;

φ' ——Moisture fraction of the blast, kg/m³;

K ——Coke consumption per ton hot metal, kg/tHM;

PCR ——Pulverized coal injection rate, kg/tHM;

NGR ——Natural gas injection rate, kg/tHM;

$(H_2)_M$, $(H_2)_K$, $(H_2)_N$ ——H_2 fraction in pulverized coal, coke or NG, respectively, kg/kg;

$(H_2O)_M$, $(H_2O)_N$ ——H_2O fraction in pulverized coal or NG, respectively, kg/kg.

The r_d formula developed by Professor A. N. Ramm which was based on theoretical calculation and actual operation data of BF is shown in equation (10).

$$r_d = \frac{r_d^0 \times 10^{-s\lambda}(0.648 + 0.01t_B^{0.5})}{0.96 + 4\varphi} \quad (10)$$

Where r_d^0 ——Degree of direct reduction when using coke smelting without any fuels injection;

t_B ——Blast temperature, ℃;

φ ——Moisture fraction of the blast, m³/m³;

s ——Reducing fuels injection rate, kg/kgHM;

λ ——Coefficient of injected fuels.

The λ can be calculated by equation (11).

$$\lambda = 0.2\overline{C} + 0.9\overline{H} \quad (11)$$

Where $\overline{C}$, $\overline{H}$ ——C or H mass fraction of injected fuels, respectively, kg/kg.

The r_d^0 of Jingtang No. 2 BF can be estimated as 0.5966 by equation (10) and operation date in the first five months in 2014.

3.3 Raceway Parameters

M. hatano developed a raceway shape mathematical model by analyzing single particle in 1976 in

Japan [15]. The model could estimate the depth, width and height of raceway by analyzing the force of blast penetrating power, coke gravity and reaction of raceway wall on single particle. Zhang guoli updated the constant of the model shown in equation (12)-equation (15) by experiment and actual operation data in 2006 in China [16].

$$D_r = 0.409 \times PF^{0.693} \times D_t \tag{12}$$

$$W_r = 2.631 \times \left(\frac{D_r}{D_t}\right) 0.331 \times D_t \tag{13}$$

$$\frac{4H_r^2 + D_r^2}{H_r \times D_t} = 8.780 \times \left(\frac{D_r}{D_t}\right)^{0.721} \tag{14}$$

$$V_r = 0.53 \times D_r \times W_r \times H_r \tag{15}$$

Where D_r, W_r, H_r——Depth, width or height of raceway, respectively, m;

V_r——Volume of raceway, m^3;

D_t——Diameter of tuyere, m;

PF——Penetration factor.

The PF can be calculated by equation (11).

$$PF = \frac{\rho_0}{\rho_s \times D_p} \left(\frac{q_V}{S_t}\right)^2 \frac{TFT + 273}{p_b \times 298} \tag{16}$$

Where ρ_0——Density of bosh gas, kg/m^3;

ρ_s——True density of coke, kg/m^3;

q_V——Flow of raceway, m^3/s;

p_b——Blast pressure, kPa;

D_p——Grain size of coke in raceway, mm;

S_t——Sum of the tuyere areas, m^2.

The raceway sharp parameters of Jingtang No. 2 BF calculated by equation (12)-equation (16) are that the depth D_r is 1.728m, the width W_r is 0.846m, the height H_r is 1.313m, and the volume V_r is 1.02m^3。

3.4 Degree of Utilization of H$_2$ and CO

In Cherepovets metallurgical plant in Russia, much of the researches in the influence of NGR from 0 to 100m^3/tHM on reduction process and top gas temperature has examined that H$_2$ content will increase from 1% to 6% with an additional increase of NGR. But the degree of utilization of H$_2$ (η_{H_2}) will still increase more although the content of H$_2$ increases much. Furthermore, NG injection will bring about the decrease of degree of direct reduction and the CO content in top gas, which results in the increase of the degree of utilization of CO (η_{CO}). The value of η_{H_2}/η_{CO} rise up from 0.5 to 1.0 with the NGR increase from 0 to 50-100kg/tHM. The linear relation between η_{H_2}/η_{CO} and NGR shown in equation (18) and equation (19) is adopted for simplifying.

$$\frac{\eta_{H_2}}{\eta_{CO}} = 0.5 + 0.01 NGR \qquad (NGR<50) \tag{17}$$

$$\frac{\eta_{H_2}}{\eta_{CO}} = 1 \quad (NGR \geqslant 50) \tag{18}$$

Where η_{H_2} ——Degree of utilization of H_2;

η_{CO} ——Degree of utilization of CO.

3.5 Calculation Steps

Calculation of BF process is based on the law of material and heat balance. Material balance calculation could indicate the relations between parameters as blast volume, oxygen enrichment rate, blast moisture, ore consumption, flux consumption and injected fuel consumption. The het balance calculation could indicate the heat consumption conditions to analyze the smelting process for improving the energy utilization and fuel consumption. Calculating as the steps shown in Fig. 1 could investigate the effects of changing injected fuels and operation parameters on smelting process by material and raceway heat balance calculation based on the current operation data of Jingtang No. 2 BF.

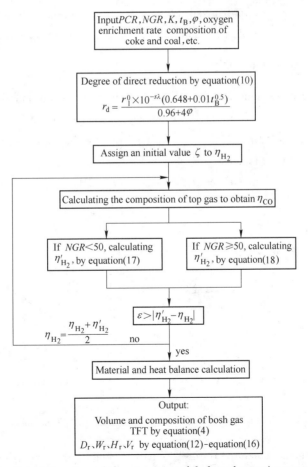

Fig. 1 Calculation steps of changing injected fuels and operation parameters

As known conditions in material balance calculation, r_d and η_{H_2} must be certain firstly. The calculating roadmap start from estimation of r_d by equation (10) after determining the parameters of *PCR*, *NGR*, coke consumption, blast temperature, blast moisture, oxygen enrichment rate and component of coke, pulverized coal and NG. Further, assign the η_{H_2} an initial value ζ, calculate the composition of top gas by balance calculation to obtain η_{CO}, then the η_{H_2} can be calculated by equation (17) or equation (18) marked η'_{H_2}. Adjusting η_{H_2} by interpolation and loop computing until the difference between η_{H_2} and η'_{H_2} is less than ε. After that, TFT, raceway shape and volume and component of bosh gas could be calculated by material balance and heat balance calculation again. Rationality of changing injected fuels and parameters may be discussed by contrasting the calculations with the current value.

4 Results and Discussion

BF operation is complex and multi-variable and analyzing the effects of changing certain parameters on smelting process while fixing the others is acceptable. The analysis is based on fixing the component of ore, pulverized coal, NG, flux, coke and hot metal, moreover, keeping coke consumption and blast moisture consistent with current operating conditions.

4.1 Injection of Natural Gas Only

Injection of NG changes the composition and volume of bosh gas as presented in Fig. 2. NG partial combustion through reactions (19) and (20) generates amount of CO and H_2, which results in bosh gas increase by about $55 m^3/tHM$ per $10 kg/tHM$ of *NGR* increase. Volume percentage of N_2 and H_2 decrease while CO's increase.

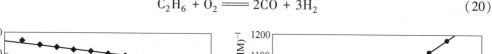

$$CH_4 + 0.5O_2 = CO + 2H_2 \quad (19)$$
$$C_2H_6 + O_2 = 2CO + 3H_2 \quad (20)$$

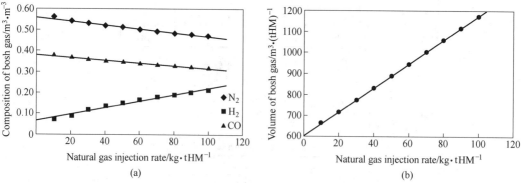

Fig. 2 Effect of *NGR* on composition of bosh gas (a) and effect of *NGR* on volume of bosh gas (b)

The value of heat energy generated during NG partial combustion is negative number because of the thermal destruction of hydrocarbon, which results in quantity of heat loss. As shown in Fig. 3, based on current operating conditions of Jingtang No. 2 BF, *TFT* decreases rapidly by 90 ℃ per

10kg/tHM of NGR increase. Bosh gas volume increases by NG combustion, whose effect on PF in equation (16) is greater than TFT's. The result is the volume of raceway increase by about 0.0334m³ per 10kg/tHM of NGR increase.

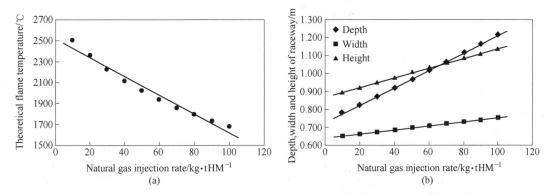

Fig. 3 Effect of NGR on TFT (a) and effect of NGR on raceway parameters (b)

Oxygen enrichment rate and blast temperature are significant parameters for BF operation too. As shown in Fig. 4, oxygen demand for combustion in raceway is relative fixed, so increase of oxygen enrichment rate lead to blast volume reducing which results in bosh gas and raceway volume decrease by 18m³/tHM and 0.0175m³ respectively per 1% oxygen enrichment rate increase, and TFT increase by 26℃. Furthermore, effects of oxygen enrichment rate increase on TFT increase and raceway volume decrease are greater in lower NGR. TFT increase too by 7℃ per 10℃ blast temperature increase which has little effect on volume of raceway and bosh gas.

Effects of PCR on TFT, raceway parameters and bosh gas volume could be estimated by the same method in Fig. 1. The results show that, TFT decrease by 18.5℃, raceway volume increase by 0.058m³, and bosh gas volume increase by 40m³/tHM per 10kg/tHM of PCR increase.

4.2 Co-injection of Pulverized Coal and Natural Gas

A standard should be set to judge the rationality of PCR and NGR when co-injection is adopted. Most of heat for smelting in BF comes from fuel combustion in raceway and high temperature blast. Actual temperature of raceway is so difficult to be measured that TFT is usually regarded as the main indicator of raceway thermal conditions. TFT, which is affected by blast temperature, moisture, types of injected fuel, fuel injection rate and oxygen enrichment rate etc., has a great influence on heat transfer, reduction, slag formation regimes, desulfurization and temperature and chemical composition of slag and hot metal. Rationality of changing operation parameters may be judged by analyzing deviation of TFT [17].

The effect of NG and pulverized coal on BF process is different. Fig. 5 shows that NG injection influences the TFT and bosh gas volume more considerably than PCI. The difference of effect of NG and pulverized coal is more obvious on TFT. Changing the injected fuel of Jingtang No. 2 BF from PCI only to co-injection of pulverized coal and NG may be achieved without altering blast

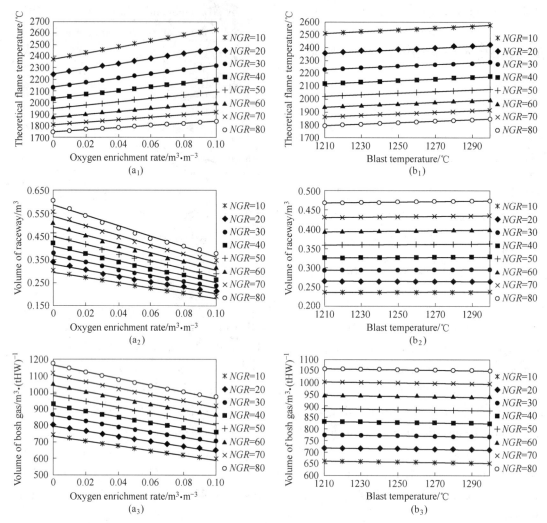

Fig. 4 (a_1) Effect of oxygen enrichment rate on *TFT*, (a_2) Effect of oxygen enrichment rate on volume of raceway, (a_3) Effect of oxygen enrichment rate on volume of bosh gas, (b_1) Effect of blast temperature on *TFT*, (b_2) Effect of blast temperature on volume of raceway and (b_3) Effect of blast temperature on volume of bosh gas

temperature by the following methods to prevent decrease of *TFT*: (1) Maintain *PCR* and increase *NGR*, meanwhile increase oxygen enrichment rate. (2) Increase *NGR*, meanwhile significantly decrease *PCR*. (3) Increase *NGR*, meanwhile decrease *PCR* slightly and increase oxygen enrichment rate. (4) Increase *NGR* and *PCR*, meanwhile increase oxygen enrichment rate.

Altering blast temperature may result in the risk of damage to the hot blast mains, refractories and valves, so the decision was made to change the parameters of *PCR*, *NGR* and oxygen enrichment rate without blast temperature which is kept around 1210℃ as the operation value at present. Based on the actual value of *TFT* and bosh gas volume of Jingtang No. 2 BF, the *TFT*

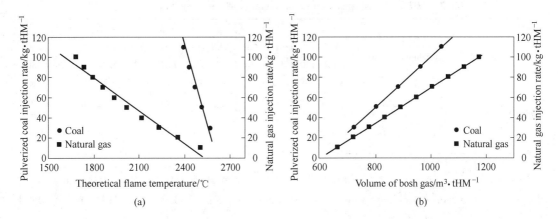

Fig. 5 Effect of *PCR* and *NGR* on *TFT* (a) and effect of *PCR* and *NGR* on volume of bosh gas (b)

and bosh gas volume after altering the operation parameters could be determined by equation (18) and equation (19).

$$a_1(PCR - PCR') + b_1(NGR - NGR') + c_1(\varepsilon - \varepsilon') < \Delta TFT \quad (18)$$

$$a_2(PCR - PCR') + b_2(NGR - NGR') + c_2(\varepsilon - \varepsilon') < \Delta V_g \quad (19)$$

Where PCR', PCR ——Pulverized coal injection rate at present or after altering, respectively, kg/tHM;

NGR', NGR ——Natural gas injection rate at present or after altering, respectively, kg/tHM;

ε', ε ——Oxygen enrichment rate at present or after altering, respectively, m^3/m^3;

a_1, b_1, c_1 ——Impact factor of *PCR*, *NGR*, oxygen enrichment rate on *TFT*, ℃/kg, respectively;

a_2, b_2, c_2 ——Impact factor of *PCR*, *NGR*, oxygen enrichment rate on bosh gas volume, m^3/kg, respectively;

ΔTFT ——Deviation of *TFT*, ℃;

ΔV_g ——Deviation of bosh gas volume, m^3.

Acceptable deviations of *TFT* and bosh gas volume for each BF are different. The minimum *TFT* for blast furnaces at Severstal Company in Russia SSC can be recommended to 1850-1900℃, which is 1880-1930℃ in Egyptian Iron and Steel Company considering the chemical composition of Egyptian iron ore and consequently the properties of produced slag[18]. In ArcelorMittal Burns Harbor Iron department in USA, *TFT* fluctuated from 1980℃ to 2123℃ in 2013 [12]. The scopes of *TFT* and bosh gas volume of Jingtang No. 2 BF in 2014 were 2143-2376℃ and 1045-1310m^3/tHM respectively, which may be used as the acceptable deviations for lack of experience.

The maximum NG provided for Jingtang No. 2 BF injection is about 30kg/tHM at present, for that reason, the *TFT* and bosh gas volume of several variants of BF operation were calculated in which the *NGR* was not more than 30kg/tHM (Table 3).

Table 3 TFT and bosh gas volume of variants of Jingtang No. 2 BF operation

Parameters	Actual operation conditions of Jingtang No. 2 BF at present	Case A: $NGR=10$, $PCR=150$, $\varepsilon=0.0479$.	Case B: $NGR=20$, $PCR=150$, $\varepsilon=0.0479$.	Case C: $NGR=30$, $PCR=150$, $\varepsilon=0.0479$.	Case D: $NGR=30$, $PCR=150$, $\varepsilon=0.06$.	Case E: $NGR=30$, $PCR=100$, $\varepsilon=0.0479$.	Case F: $NGR=30$, $PCR=80$, $\varepsilon=0.06$.
TFT/℃	2345.44	2268.22	2196.72	2131.07	2151.52	2152.58	2184.47
Bosh gas volume /m³·tHM⁻¹	1198.18	1251.16	1308.07	1364.95	1331.94	1168.19	1063.28

In case A, case B and case C, the NGR increases from 10kg/tHM to 30kg/tHM while fixing the PCR and oxygen enrichment rate, the TFT decrease from 2268.22℃ to 2131.07℃, which is out of the acceptable deviation, for compensating the decrease of TFT, the oxygen enrichment rate is raised to 0.06 and PCR is reduced to 100kg/tHM compared to case C to form case D and case E respectively. The bosh gas volume of case D increases too much which may cause flooding in BF, so the PCR is reduced to 80kg/tHM comparing case D to form case F. Case E and case F is better than the others. To be clear, it is just calculated for some cases and optimal operation condition is not limited in these 6 cases.

Compositions of top gas of case E and case F are calculated in Table 4. It shows that, contrasting to the actual operation data, the top gas volume decreases by about 100m³/tHM and 200m³/tHM, CO_2 emission decreases by 30m³/tHM and 32m³/tHM, CO emission decreases by 42m³/tHM and 70m³/tHM, carbon emission decreases by 26kg/tHM and 59kg/tHM, respectively in case E and case F. The altering of injection fuels has a positive effect on energy conservation and emission reduction, which is more remarkable in case F than that in case E.

Table 4 Compositions of top gas of case E and case F

Cases	Items	CH_4	H_2	CO_2	CO	N_2	Total
Case E	Volume/m³·tHM⁻¹	0.49	62.85	310.61	303.33	629.86	1307.14
	Volume percentage/%	0.04	4.81	23.76	23.21	48.19	100.00
Case F	Volume/m³·tHM⁻¹	0.48	58.35	308.96	275.93	560.03	1203.76
	Volume percentage/%	0.04	4.85	25.67	22.92	46.52	100.00
Actual value at present	Volume/m³·tHM⁻¹	0.53	38.32	340.66	345.18	680.45	1405.13
	Volume percentage/%	0.04	2.73	24.24	24.57	48.43	100.00

5 Further Researches

BF operation is complex and multi variable, and all the parameters are interrelated and interact on each other. Further researches based on theory and experiment will be on for achieving co-injection of pulverized coal and NG in Jingtang No. 2 BF including: (1) Acceptable scopes of TFT and bosh gas volume on Jingtang No. 2 BF. (2) Co-firing features of mixing pulverized coal and NG in raceway. (3) Effects of other operation parameters as blast temperature, moisture, top pressure and compositions of burden, etc. on BF process. (4) Effects of excess oxygen and soot on TFT and bosh gas volume. Excess oxygen is necessary since complete mixing of the injected fuels with the blow is impossible to maintain, and soot is generated by molecules of NG which are not participating in partial combustion. (5) Distribution and utilization of gas inside BF. (6) Analysis of economic benefit on altering injected fuels.

6 Conclusions

Key parameters as TFT, degree of direct reduction and degree of utilization of H_2 and CO were discussed, a mathematical calculation steps was introduced by which the effects of injected fuels, blast temperature, blast moisture and oxygen enrichment rate on BF process could be analyzed. Research on injected fuels of Jingtang No. 2 BF altering from pulverized coal injection only to co-injection of pulverized coal and NG shows that:

(1) NG injection has significant effect on TFT, raceway shape and compositions and volume of bosh gas, which will lead to a decreased activity of hearth and operation troubles. It is necessary to adjust blast temperature, blast moisture and oxygen enrichment rate to maintain TFT and hearth activity.

(2) Increase of both PCR and NGR lead to decrease of TFT and increase of bosh gas volume, for maintaining which PCR should be reduced simultaneously with NGR increasing.

(3) NG injection influences the TFT and bosh gas volume more considerably than PCI, so other operation parameters like oxygen enrichment rate should be adjusted combined with PCR altering.

(4) Oxycarbide emission decrease more obviously under co-injection of pulverized coal and NG than PCI only, which has a positive effect on energy conservation and emission reduction.

Acknowledgements

The author would like to thank Mr. Mao Qingwu, Mr. Fan Zhengyun, Mr. Zheng Pengchao, Ms. Liuran, etc. for their support and assistance and experts from Jingtang for their support of operation information.

References

[1] Jose Adilson de Castro, Hiroshi Nogami, Jun-ichiro Yagi. Numerical investigation of simultaneous injection of pulverized coal and natural gas with oxygen enrichment to the blast furnace [J]. ISIJ International, 2002, 42 (11): 1203-1211.

[2] Hiroshi Nogami, Jun-ichiro Yagi, Shin-ya Kitamura, Peter Richard Austin. Analysis on material and energy balances of ironmaking systems on blast furnace operations with metallic charging, top gas recycling and natural gas injection [J]. ISIJ International, 2006, 46 (12): 1759-1766.

[3] Zhang Liguo, Wang Zaiyi. Study and application of dust injection into BF [J]. China Metallurgy, 2012,

32 (3): 40-48.

[4] Chen Yongxing, Wang Guangwei, Zhang Jianliang, et al. Theoretical analysis of injecting coke oven gas with oxygen enriched into blast furnace [J]. Iron and Steel, 2012, 47 (2): 12-16.

[5] Bi Xuegong, Rao Changrun. Development status of simultaneous injection of agricultural and forestry residues and its perspective [J]. Henan Metallrgy, 2012, 20 (3): 1-5.

[6] Guo Tonglai, Liu Zhenggen. Numerical simulation of blast furnace raceway with coke oven gas injection [J]. Journal of Northeastern University (Science and Technology), 2012, 33 (7): 987-991.

[7] Gao Jianjun, Guo Peimin. Numerical simulation of injection of coke oven gas with oxygen enrichment to the blast furnace [J]. Iron Steel Vanadium Titanium, 2010, 31 (3): 1-5.

[8] Meng Xianglong, Zhang Fuming, Li Lin et al. Pulverized coal injection technology in large-sized blast furnace of Shougang [C]. AISTech 2014 Proceedings, 2014: 805-813.

[9] Halim K S A. Theoretical approach to change blast furnace regime with natural gas injection [J]. Journal Of Iron and Steel Research, International, 2013, 20 (9): 40-46.

[10] Chu Mansheng, GUO Xianzhen, Shen Fengman. Numerical analysis on blast furnace operation with reducing gases injection [J]. China Metallurgy, 2007, 17 (6): 34-39.

[11] Halim K S A. Effective utilization of using natural gas injection in the production of pig iron [J]. Materials Letters, 2007, 61 (14-15): 3281-3286.

[12] Michal J. Wojewodka, James P. Keith, Stephen D. Horvath, et al. Natural gas injection maximization on C and D blast furnace at Arcelor Mittal burns harbor [C]. AISTech 2014 Proceedings, 2014: 767-779.

[13] Babich A, Yaroshevskii S, Formoso A, et al. Co-injection of noncoking coal and natural gas in blast furnace [J]. ISIJ International, 1999, 39 (3): 229-238.

[14] Ramm A N. Modern blast furnace process [C]. metallurgiya, Moscow, Russian, 1980: 194, 200, 221.

[15] Hatano M, Fukuda M, Takeuchi M. An experimental study of the formation of raceway using a cold model [J]. ISIJ International, 1976, 62 (1): 25-32.

[16] Zhang Liguo, Liu Dejun, Zhang Lei, et al. Study on influence regularity of BF tuyere diameter and coke particle size on it [J]. Angang Technology, 2006 (1): 7-10.

[17] Zhang Jianliang, Qiu Jiayong, Guo Hongwei, et al. Research of theoretical flame temperature in blast furnace tuyere [J]. Iron and Steel, 2012, 47 (7): 10-14.

[18] Halim K S A, Andronov V N, Nasr M I. Blast furnace operation with natural gas injection andminimum theoretical flame temperature [J]. Ironmaking and steelmaking, 2009, 36 (1): 12-18.

Development and Application on Large Annular Shaft Kiln at Shougang

Zhou Hong Zhang Fuming Zhang Tao

Abstract: The current status of technical development of annular shaft kiln for limestone calcination production is described in this paper. Based on the existing technological accumulation, a new type annular shaft kiln with capacity of 600t/d has been developed independently by Beijing Shougang International Engineering Technology Co. Ltd. A number of key technologies of the new type large annular shaft kiln are researched and integrated, include thermal process system determination, lining refractory structure optimization, heat-exchanger configuration etc. The major advanced characteristics of new type annular shaft kiln are energy saving, higher efficiency, lower environmental pollution, lower emission, lower engineering investment, and lower operation cost. The excellent performances and application results in production have been achieved at Shougang Qiangang steel plant.

Keywords: annular shaft kiln; lining structure; active lime; energy saving

1 Introduction

Annular shaft kiln (abbreviated as ASK) process for limestone baking is invented by Germany's Karl BeceKenbach in 1960's. Because the kiln body is consists of two circular steel sleeves from the inside and outside, hence is named "Beckenbach ASK". The two types of ASK with capacity of 300t/d and 500t/d are introduced from Germany respectively. The 500t/d ASK technology is digested and absorbed completely in China, and the operation performances are stable and smooth[1]. This type ASK can be developed and designed completely in China, and a number of domestic steel enterprises can master the large and medium sized ASK. In 2007, WISCO introduced 600t/d ASK from Italy's Fercalx company[2], this ASK investment not only higher, but were put in quite a long time cannot be stable in commissioning, put into operation after the 3rd year annual production to 200,000 tons. Up to now, China has been the lack of localization of 600t/d above the capacity level of the ASK technology.

Beijing Shougang International Engineering Technology Co., Ltd. (BSIET) based on the Shougang's No. 1500t/d ASK which is the first introduced facility in China, develop and integrate the ASK technologies independently. A number of technical innovations are achieved for the new type large ASK with daily capacity 600t/d. Automatic control system, process flow and refractory materials design have been innovated and optimized. More than 10 sets of 300t/d and 500t/d ASK have been designed and built, and the sufficient engineering experiences have been accumulated from practice, therefore, the achievements provide a strong technological guarantee for develop large capacity ASK.

BSIET through the key technical challenges for comprehensive research, development, evaluation, and design, especially on middle and lower inner sleeve, ASK integral lining structure, thermal system design, research and develop 600t/d ASK inner structure design independently, construct the first fully realized localization of 600t/d ASK in China.

The thermal adjusting system design for 600t/d ASK is one of the technical difficulties. In according to the reference of 500t/d ASK thermal system to calculate 300t/d ASK thermal model, with large differences in the practical application of 300t/d ASK thermal system, the different parameters need to be adjusted in different proportions. In accordance with thermal experience 500t/d ASK to compute 600t/d ASK thermal system, needs to be adjusted and modified, a lot of investigations need to carry out in research and study deeply.

The second technical difficulty is expanding annular sleeve inside and outside diameters. Kiln body structure based on capacity requirements need to be expanded, and lime kiln calcination principle decided to ASK in a vertical direction changes of body length size should not be. Therefore, ASK capacity increase will have to adopt through the expansion of sleeve inside and outside diameters. Which not only inner annular sleeve designs put forward higher requirements, but also new demands on ASK lining structure design. No other than solve these problems of the key technology, can guarantee the ASK thermal system meet the requirements in limestone calcination, ensure the cooling effect of inner annular sleeve, ensure the stability and long service life of lining structure.

2 Design of Thermal Process System

By comparing the actual production parameters of the 300t/d and 500t/d ASK, amendments to the theoretical calculation results of the 600t/d ASK. according to the ASK calcination space requirements, the numbers of upper and lower burner, and the single burner of upper and lower combustion capacity are determined for 600t/d ASK. According to the ASK combustion chamber structure characteristics and calcination requirements, as well as the fuel gas flow velocity through the annular space of burner, determine the heat requirement of 600t/d ASK, and calculate the available fuel gas calorific value range. According to the main factors of raw materials and fuel gas supply condition, make a calculation to determine the mass balance and heat balance. The main factors include limestone, fuel gas, cooling air for lime, cooling air for inner sleeve, driving air, and waste flue gas etc. According to the designing calculation result, the cooling air fan, driving air fan, waste flue gas fan etc. will be determined and selected. As well as the corresponding fuel gas supply facility and relative pipeline system design will be carried out respectively.

3 Design of Lining Structure

According to production requirements, the annular space of ASK is determined by calculation of bricking-work inner and outer diameters, effective height, vertical height, diameter and height of preheating zone expanded band etc. parameters[3]. Under the precondition of arch bridge spans is constant, 600t/d ASK lining structure design ensure the same structure, ASK shell and inner sleeve while are expanded in diameter, curvature of the annular space is reduced.

In order to suit the variance of 600t/d ASK proper curvature, the arch-foot shape of joint between arch-foot and shell, arch-foot and inner sleeve are determined by particular calculation. Wedge-shaped bricks are designed accurately by means of three-dimensional design to ensure the stability and service life of lining structure for 600t/d ASK.

The cooling beam of shell and supporting ring beam of brickwork are applied in the ASK design. On the inner sleeve and ASK shell, through installed on the shell with a pillar of support ring and connecting bracket reinforcement. So that the expansion of refractory brickwork can be controlled, and the support stress can be reduced; on the one hand it can be reduced for the maintenance workload of the refractory lining. The cooling beam of ASK shell is applied in the design, aim to reduce the heat conduction between lining wall and shell steel structure of ASK, thus avoid the higher temperature fluctuation of refractory lining during the ASK shutdown because of heat supply significant reduction. It is proved this optimization of structure design play a protective role of refractory lining in the actual application.

The magnesia-alumina spinel bricks with double layers design structure under the bottom of arch bridge is adopted in the ASK design. This structure can promote the stress distribution uniformly on the arch bridge, avoid the risks of bricks deformation and dislocation because of material flow attack and impact, reinforce the refractory structure stability, and prolong the lining service life. In the ASK design, the upper and lower straight segments of combustion chamber have been adjusted and optimized. The optimized combustion chamber can improve the flame position in the combustion chamber effectively, and promote the smooth and reasonable connection between the gas channel out of the kiln body and injection nozzle.

Rational determination of ring-shaped space, reduce the Coanda effect in the design. On the basis of arch bridge span to meet technological and structural requirements, reduce the annular space of curvature. While the hot air penetrates the entire section of raw material burden, reduces the influence of wall effect further; therefore, the limestone calcination is improved more uniformly. Fig. 1 shows the three dimensions CAD simulating result of inner sleeve structure.

Fig. 1 Three dimensions CAD simulating result of inner sleeve structure

4 Design of Inner Sleeve and Heat-exchanger

The inner sleeve structure dimensions depend on space requirement of limestone calcinations. The labyrinth partitions within inner sleeve interlayer are reasonable arranged, so that the cooling air flows within the interlayer is simple, and all parts of the cooling air to reach inside the interlayer, and as possible as to reduce the cooling gas pressure loss.

Design of heat-exchanger according to the capacity relationship between waste flue gas and heat tube quantity, by determining the length of heat exchanger tube bundles in overall height, relative position, calculate the number of bundles of 600t/d ASK. As a result, the flow volume of driving air and waste flue gas will be determined. The technical parameters of heat-exchanger of 600t/d ASK are shown in Table 1.

Table 1 Main technical parameters of heat-exchanger of 600t/d ASK

Item	Standard flow volume /Nm$^3 \cdot$ h^{-1}	Temperature /℃	Pressure /kPa	Actual flow volume /m$^3 \cdot$ h^{-1}
Waste flue gas into heat-exchanger	16155	750	-3	56400
Waste flue gas out off heat-exchanger	16155	350	-3	34350
Air into heat-exchanger	9500	20	30	9500
Air out off heat-exchanger	9500	450	30	23442

5 Application in Production

Shougang Qiangang's 600t/d ASK started up on December 16, 2009. After put into production 2 weeks of continuous and stable operation, the design capacity is reached with daily output 600t/d design capacity for 24 days continual, and the maximum daily output upgrade to 660t/d, various economic-technical specifications have been reached or exceeded the design level. Up to now, the 600t/d ASK has been continuous and stable operate 32 months, various technical parameters have reached the design requirements, some of them are higher than the design level. The actual photo of Shougang Qiangang's 600t/d ASK is shown in Fig. 2.

A number of advanced operation parameters are achieved in production. Such as the actual heat consumption of lime calcination is 3980kJ/kg, less than the design value of heat consumption approximate 117kJ; top exhaust gas temperature is 130-180 degree centigrade in the design, but the actual top exhaust gas temperature is controlled during production in 80-130 degree centigrade; the discharging lime temperature is 80-150 degree centigrade, but the actual production process discharging lime temperature is controlled 40-80 degree centigrade, under the environment of low temperature in winter, lime temperature is controlled at 20 degree centigrade or less. It is proven by more than 2 years production, the engineering achieve energy saving and

Fig. 2 Actual photo of Shougang Qiangang's 600t/d ASK

environment friendly, the energy consumption and water consumption, and other technical-economic parameters reach the advanced level in China. Table 2 shows the comparisons of major technical-economic parameters of Shougang Qiangang 600t/d ASK.

Table 2 Major technical-economic parameters of 600t/d ASK

Item	Design data	Actual data in 2010
Limestone/t · t^{-1}	1.8	1.7
Heat consumption/kJ · kg^{-1}	4096	3980
Power consumption/kW · h · t^{-1}	28.0	27.0
Activity of lime/mL 4N-HCl	≥350	380
CO_2 remains/%	≤2.0	1.03
Qualified ratio/%	>95	99
+Water consumption/m^3 · t^{-1}	0.39	0.30
Power consumption/kW · h · t^{-1}	28	27
Compressed air/Nm^3 · t^{-1}	50	43.49
BOF gas/Nm^3 · t^{-1}	576	553
Synthesis energy consuption/kgce · t^{-1}	165	139.95

ASK compared to other type kiln, the most remarkable feature is the recovery and utilization of waste heat technology. The high-temperature exhaust heat can be recovered through the heat exchanger to preheat driving air, circulating air temperature is increased, thereby reducing consumption of BOF gas in combustion chamber, achieve the goal of energy saving and consumption reduction, refer to Fig. 3. In the design of 600t/d ASK, by optimizing and improving the operation of heat exchanger, making ASK waste heat utilization more efficient, energy-saving and consumption-reducing effect is even more significant, making it significantly reduce CO_2 emission. Major optimization and improvement projects are: firstly, increase the heat exchanger tube bundles, making the heat exchanging area increases of driving air and high temperature waste flue gas to absorb the heat sufficiently. Secondly, change the calcination process, namely, reduce exhaust emission rationally, increase the circulating air flow volume, reduce discharging lime temperature in order to reduce the overall heat emission of the ASK.

ASK operation under negative pressure characteristics and good performances of dust collection design capability can reduce dust concentration below $30mg/Nm^3$, practice shows that the 600t/d ASK dust emission concentration is less than $10mg/Nm^3$.

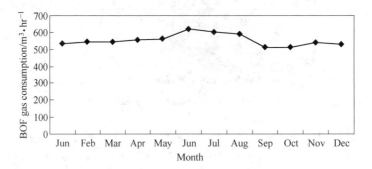

Fig. 3　BOF gas consumption in 2010

6　Conclusions

600t/d ASK of Shougang Qiangang has been researched and developed successfully, it filled the blank in domestic ASK of 600t/d in China, and its design level has reached the international advanced level. As a result, meet the lime production materials and energy balance requirement, guarantee the steel plant BOF, LF, and KR desulphurization process of active lime supply.

A number of advanced technologies are developed and applied in the new type 600t/d ASK. New shaped inner sleeve lining structure, heat exchanger configuration, thermal process determination etc. have been applied in the large ASK engineering design.

The new large ASK operation stable and smooth with a good permeability, thermal system easy to adjust. The lime quality is stable and the activity is higher than normal ASK. The effects of energy-saving and consumption-reducing are significant compared to the actual production and design.

References

[1] Su Tiansen. Iron and steel science and technology progress promote the metallurgical lime development [J]. Lime. 2005, 76: 1.

[2] Zhu Quanyi, Huang Youguo. Application of production on 600t/d annular shaft kiln at WISCO [J]. Lime, 2009, 93: 2.

[3] Zhang Deguo, Wang Xin, Zhou Hong. Improvement on lining of annular shaft kiln [C] //Refractory. 2007 Supplement, 5th International Conference of Refractory Proceedings, 2007, 41: 115.